Ayurveda
At The Turning Point

The Editors

Dr Chandra Kant Katiyar, born on September 27, 1954, is MD in Ayurveda and Ph.D. Pharmacology/ Rasa Shastra from Institute of Medical Sciences, Banaras Hindu University. For the last more than 25 years Dr Katiyar has been at the helm of affairs at India's best Research and Development-driven companies like Ranbaxy as Director, Herbal Drug Research and Dabur India Ltd as Head and Vice President of Dabur Research and Development Centre.Currently Dr Katiyar is CEO Health Care (Technical) of Emami Ltd.

Dr Katiyar has authoured a few books including "Modern Ayurveda: Milestones Beyond the Classical Age" published by CRC Press, USA with Mr C. P. Khare, and more than 10 book chapters, about 75 research papers and has about 20 patents to his credit. He is member of several committees of Government of India including ASU Drugs Technical Advisory Board, Ayurvedic Pharmacopoeia Committee, Herbal Committee for Indian Pharmacopoeia and South East Asia Expert Panel on Dietary Supplements for United States Pharmacopoeia.

C. P. Khare was born on March 10, 1932, and has been a herbalist for more than 60 years. His profile is among "All-Star Profiles" of Linkedin. His works have been cited in more than 1000 research papers, as recorded by Research Gate till April 28, 2017.

Reference works by C. P. Khare:

- Indian Herbal Medicine: Application of Research Findings, Third edition, Vishv Books, New Delhi, 2010.
- Indian Herbal Remedies/Encyclopedia of Indian Medicinal Plants,Two editions, Springer, 2004.
- Indian Medicinal Plants: An Illustrated Dictionary, Springer, 2007.
- Dictionary of Synonyms: Indian medicinal Plants, IK International Publishing House, New Delhi, 2011.
- The Modern Ayurveda: Milestones beyond the Classical Age (With C.K. Katiyar) CRC Press, 2012.
- AyurvedicPharmacopoeial Plant Drugs: Expanded therapeutics, CRC Press, 2016.

Ayurveda
At The Turning Point

Editors

Chandra Kant Katiyar

C. P. Khare

2018

Front Cover:
Crocus sativus Linn. flower and chemical structure of picrocrocin.

Kruger Brentt Publishers UK. LTD.
Company Number 9728962

Regd. Office: 68 St Margarets Road, Edgware, Middlesex HA8 9UU

Library of Congress Cataloging-in-Publication Data

Ayurveda : at the turning point / editors, Chandra Kant Katiyar, C.P. Khare.
pages cm
Contributed articles.
Includes index.
ISBN 978-1-78715-005-8 (HB)

1. Medicine, Ayurvedic--India. 2. Materia medica, Vegetable--India. I. Katiyar, Chandra Kant, editor. II. Khare, C. P., 1932- editor.

R605.A98 2017
DDC 615.5380954 23

For information on all our publications visit our website at http://krugerbrentt.com/

Foreword

At present, health and wellness, as opposed to curative healthcare, has emerged as one of the most difficult societal challenges, not just for India but for the entire world. Ayurvedic history gives us insight into the development of one of the Indian Medical systems, which can be traced back to several centuries BC. It also takes into account, the temporary decline of Ayurveda and the present renaissance of interest in it globally.

Ayurveda in the twenty-first century needs emergence of new ideas, adventures and liberation, in order to play its requisite role in the newly emerging medical pluralism. We can no longer live on the glory of the past. The critical look of Ayurveda must be regained to build progressive future in healthcare. New scientific evidence is genuinely important for quality, safety and efficacy of Ayurvedic drugs. We need more efforts on development of appropriate research method than aimlessly borrowing outdated conventional methods which may lead to distortion of Ayurveda with no benefits to either side. It is unlikely that Ayurveda can grow to the expectations of its stakeholders, in isolation. A myopic view may probably result in intensification of marginalisation, enhanced insecurity and self-pride based complacency in the sector.

China has very boldly integrated Traditional Chinese Medicine (TCM) and modern medicine. India also may need to discover its own model of integration. Given our strength in molecular biology, phytochemistry, biomedicine and Ayurveda, efforts should be directed to develop a national programme for transdisciplinary research to achieve global scientific leadership in complimentary and integrative healthcare. Fortunately, the Government of India is highly supportive to Ayurveda as evidenced by the creation of independent Ministry of AYUSH. Government-blessed think-tank and mission consisting of extraordinary strategists and visionaries from research, education, clinical practice, regulations, institution building and industry are necessary. There is a need for integrating Ayurveda and modern medicine in teaching and research, such an integration can enhance the development of Ayurveda and modern medicine.

In the light of the above, this book, "Ayurveda At The Turning Point," is a befitting publication for the benefit of all stakeholders.

Sukhdev Swami Handa

Former Director,
CSIR-Indian Institute of Integrative Medicine, Jammu
Chairman, Scientific Body, Pharmacopoeia Commission of Indian Medicine,
Ministry of AYUSH, Government of India
Chairman, United States Pharmacopoiea, Herbal Medicine Compendium, South Asia

Preface

Ayurveda has seen tremendous growth in popularity in postindependent India. However, during the early phase for three to four decades, this growth was happening in an unplanned manner. It got certain directions in 1994 when the Government of India created a new department of AYUSH under the Ministry of Health and Family Welfare. Regulated and planned promotion and growth of the AYUSH system in India and abroad was the mandate of the Department of AYUSH. To provide quality parameters Ayurvedic Pharmacopoeial Committee was formed and laid down the standards of raw materials to be used in manufacture of Ayurvedic medicines. The first committee was headed by Col. Sir R. N. Chopra. Keeping in view the tiny nature of the majority of Ayurvedic Industry, a conscious effort was made to keep the quality parameters to bare minimum but capable enough to identify right material. Developing Ayurvedic Pharmacopeia in India has been a herculean task and it has not been able to include all the official raw materials even now despite publishing seven volumes. Evolution of parameters for quality is perceptible from Vol 1 to Vol 8 but these could not be updated even now.

Ministry of AYUSH is putting maximum efforts on Classical (Generic) Ayurvedic products. However, market dynamics demonstrates that out of almost Rs 10,000 crore turnover of Ayurvedic drugs, Classical Medicines hardly contribute approximately Rs 1000 crore. The Ministry needs to review its priorities to have a balanced approach.

The Editors of this book earlier brought out a book titled "Modern Ayurveda: Milestone beyond the Classical Age" which was published in 2012 by CRC Press in the United States of America. This book reviewed the last classical phase of Ayurveda during the sixteenth century, and tried to present a blueprint for further development of Ayurveda during the post classical period.

The current book "Ayurveda At The Turning Point" is a natural extension of the same thought process but providing greater focus on contemporary research.

This book provides insights on revival of Ayurveda in India besides providing historical state of Ayurveda through the chapters written by Ms Uma Ganesan, Mr N. Gangadharan as well Mr C. P. Khare. Since the book predominately deals with the research in Ayurveda it has several chapters summarising the same from leading public research laboratories like Council for Scientific and Industrial Research, Indian Council of Medical Research and Central Council for Research in Ayurveda Sciences.

The book has dedicated a special chapter providing abstracts of researches published on Ayurvedic medicinal plants in India for the last about 15 years from Journals like Journal of Research in Ayurveda and Siddha (JRAS) and Indian Journal of Pharmacology, covering both ingredients as well as the formulations.

The book also has the chapters dealing with dynamics of nomenclature of classical Ayurvedic texts besides divergent uses of Ayurvedic plant drugs in North and South India. There are two chapters dedicated to standardisation of medicinal plants through analytical markers. An effort has been made to provide updated information in these chapters. The book further provides a chapter on bioactive phytoconstituents to supplement the scientific input. The detailed review of pre-clinical and clinical research of more than 30 most important medicinal plants have been provided in the book to help the researchers across academic institutions, research organisations and industries. Our earlier book "Modern Ayurveda: Milestones beyond Classical Age" had a chapter on Herb-Drug Interactions. The current book has one chapter on Herb-Drug Interaction Studies with the difference that this chapter deals with the interaction with the Ayurvedic medicinal plants or formulations and most of the studies described in this chapter have been conducted in India.

Ayurvedic Drug Development would remain incomplete without application of modern pharmaceutical principles. Hence, the book has a special chapter on "Challenges on Application on Pharmaceutics in development of Ayurvedic medicines," based on practical experiences from formulation development researchers.

We hope, this book would serve a valuable reference to the institutions, practitioners, researchers and industry.

Chandra Kant Katiyar

C. P. Khare

The Project

"Ayurveda At The Turning Point." was a project which was aimed at analyzing emerging trends during three centuries and could not be completed for two years, then Dr C. K. Katiyar took it over and completed it in another one year. He proved himself a rare blend of Ayurvedic wisdom and modern scientific insight, successfully went deep into all the layers of modern period of Ayurveda and broke a new path which I never expected from those who are still beckoning and worshiping the past.

Dr Katiyar did MD in Ayurveda, but is a different brand from those whom I was trying to contact during last four decades for speeding up modernization process of Indian classical medicines. I still remember three incidents which were the trigging factors in founding the Society for New Age Herbals and bringing open-minded Ayurvedic scholars like Dr Katiyar and pharmacologists, botanists and researchers of CSIR and ICMR on one platform.

In 1948, when I used to prepare classical Ayurvedic medicines in my village, I decided to prune the long list of herbs. To my surprise, the modified compound gave the same result. I could take it up again in 1999 with active participation of Dr V.K. Agarwal, Head of Pharmacology in The Wealth of India project, and Prof. Anwar Ahmad, the then Head of Unani Medicine, Delhi University. When I tried to seek the cooperation of the then Director of Central Council for Research in Ayurveda and Siddha (CCRAS), he shouted at me: who are you to touch our age-old formulations. The book was published in 2000 and was reviewed by the Herb Society, London. That proved the gateway for my next books which were published from Germany and the United States and UK.

After a few years, when I met Dr G. S. Lavekar in CCRAS, he appreciated my effort, but added that Ayurvedic medicine is not the herbal medicine (indicated towards the products of Himalaya Drug Company). Those were the days when I used to go to Ayurvedic institutes off and on to discuss various issues with Ayurvedic and Unani members. When I was sitting with an Ayurvedic scholar, a patient came in holding one X-ray. The Ayurvedic scholars tried to see the X-ray against tubelight, and wrote the prescription. He prescribed Himalaya Drug Company's Gerifort. I left the place thinking about the comments of Dr Lavekar and the future of Chyavanprash, Brahma Rasayan and…I missed one step while coming down. (After retirement, Dr Lavekar is promoting research-based herbal medicines, and is on Research Gate.)

Now back to CCRAS (now upgraded as Central Council for Research in Ayurvedic Sciences). I was sitting with the Deputy Director who was a pharmacologist. He was preparing a presentation on scientific achievements of the council. He said: you know how disgusting it is for a scientist to work with a *vaidya*! Only a *vaidya* can become the director of the Council and a Deputy Director will always remain a Deputy Director just because he is a scientist.

Among this flock, Dr Katiyar crossed the vermilion line, and filtered Ayurvedic wisdom into a scientific regime to bridge the gap between a *vaidya* and a scientist while heading herbal drug research in Dabur, Ranbaxy and Zandu.

Though this project was conceived and initiated by me, four decade's experience of Dr Katiyar is its life and blood. I am sure, "Ayurveda At The Turning Point" will change the course of a number of earlier concepts and misconcepts and prove a blueprint for much-awaited new curriculum of Ayurveda. It is bound to leave a mark, wherever Ayurveda is being taught and practised.

C. P. Khare

Founder President,
Society for New Age Herbals

Acknowledgement

The idea of writing this book came in 2012 after publishing our first book "Modern Ayurveda: Milestone beyond Classical Age" and overwhelming response it received from researchers, academicians and the universities in the United States and Europe. We started planning and preparing an outline for this book as a natural corollary.

We have also kept in view some of the published research papers and tried to provide them bigger exposure after obtaining proper permission. We are thankful to Dr Uma Ganesan and RK Mission Trust, Chennai, for permitting us to publish the article of late Dr N. Gangadharan.

We are thankful to Dr (Mrs) Neeraj Tandon, Sr. Deputy Director General, Indian Council of Medical Research, New Delhi and Dr A.K.S. Rawat from CSIR-National Botanical Research Institute, Lucknow, Dr Vikram Andrew Naharwar, the Chief Scientific Officer at Integrated Clinical Research Sciences, and to all other scholars, for their valued contributions.

One of the chapters titled "Ayurvedic Pharmacopeia of India: Ready Reckoner" was written when the editor (Chandra Kant Katiyar) was Head at Dabur Research and Development Centre at Sahidabad, Ghaziabad, Uttar Pradesh. The editors are thankful to Dr J.L.N. Shastri, Dr Arun Gupta and his team for compiling the information in this chapter.

The authors are thankful to Mr Gautam Gangopadhyay, Senior Executive Assistant at Emami R and D Centre for extending secretarial assistance in preparing the abstracts from Journal of Research in Ayurveda and Siddha.

Special thanks are due to Ms Arundhati Gupta, Executive Assistant to CEO Healthcare (Technical) at Emami Ltd, Kolkata, for extending secretarial support for several chapters of the book, beside making the manuscript presentable for the publication.

We express our gratitude to the members of Society for New Age Herbals for their continued encouragement and support for this project.

Chandra Kant Katiyar

C. P. Khare

The Contributors

A.K.S. Rawat

Dr A.K.S. Rawat is Scientist and Head of the Department of Pharmacognosy and Ethnopharmacology, CSIR-National Botanical Research Institute, Lucknow. He is reputed scientist in the field and has more than 50 publications in reputed scientific journals. Dr Rawat has been conferred with many awards and medals including Zandu Award for best Scientist in Medicinal Plants Research, endowed through Society of Ethnopharmacology.

Amitabha Dey

Dr Amitabha Dey is a Pharmacologist and completed his Ph.D. in Pharmacology from IIT (BHU), Varanasi and having in-depth knowledge on Ayurveda Natural Medicine, bioassay procedures and drug's mechanism of action. Dr Dey is presently working with Emami Limited, Kolkata. He has five years of Research experience in research and development on Ayurvedic medicinal products. Dr Dey has expertise in the field of preclinical research including *in vitro* enzymatic, protein and cell line bioassays, scientific input in research-related work for new drug development and scientific writing"

Arun Gupta

Dr Arun Gupta is a Medical Doctor holding degree of MD in Pharmacology. He has more than 20 years of Industrial Research experience and is heading department of Medical Services and Clinical Research at Dabur Research and Development Centre, Sahidabad, Ghaziabad. Currently he is the President of Society for New Age Herbals, a Delhi-based non-profit organization dedicated to academic and research-related aspects of herbals in India.

Avinash Narwaria

Master of Pharmacy from Banaras Hindu University with 21 years of experience in Research and Development of diverse products in the field of Pharmaceutical, Herbal Healthcare and Personal Care Products at Dabur India Ltd. Paras Pharmaceutical and Ranbaxy Research Laboratory. Currently working as Senior General Manager-Technical with Emami Limited at Kolkata, leading Formulation Development work of Healthcare-OTC and Ethical products.

Bibhuti Nath Bhatt

Dr Bibhuti Nath Bhatt (BAMS and MBA) is an Ayurvedic physician, having 15 years of Industrial experience in research and development on Ayurvedic medicinal products and in the field of clinical research. Presently working with Emami Ltd.

Deepa Gandhi

Dr Deepa Gandhi is MD in Ayurveda (Pharmacology and Pharmacognosy of medicinal herbs *i.e.* Dravyaguna Vigyan) and an expert Ayurveda Consultant. Working with Emami –Zandu and practising Ayurveda for more than 16 years in Mumbai, India. She has vast experience of herbal medicines and their applied uses. She has presented articles and given lectures at various places for awareness about Ayurveda in common man.

J.L.N. Sastry

Dr J.L.N. Sastry is MD in Ayurveda and has authoured multiple books on Dravya-Guna. He is having more than 20 years industrial experience and is Head of Dabur Research and Development Centre at Sahidabad, Ghaziabad, Uttar Pradesh.

Neeraj Tandon

Dr Neeraj Tandon is Ph.D. in Botany, and is having more than 25 years of experience as scientist in the field of medicinal plants at Indian Council of Medical Research, New Delhi. She is instrumental in publication of multiple volumes of "Review of Indian Medicinal Plants", "Quality Standards of Indian Medicinal Plants", "Safety Review" and "Phytochemical Research Standard" and "Perspective of Indian Medicinal Plants in the Area of Diabetes", "Liver and Filaria". Currently she is Head, Medicinal Plants Division, Publication, Information Technology, and Senior Deputy Director General of Indian Council of Medical Research.

N. Gangadharan

N. Gangadharan was a student of Ramkrishna Mission Vivekananda College, Mylapore, Chennai. For post-graduate studies he joined the Department of Sanskrit, University of Madras (Chennai). After earning his Ph.D. he joined the Sanskrit department of the University of Madras as Research Assistant in the New Catalogus Catalogorum Section and then became lecturer and subsequently retired as Professor of Sanskrit from the same department. He was closely associated with Kuppuswami Sastri Research Institute, and was a regular participant in the All India Oriental Conferences. Dr Gangadharan passed away on 3rd March 2002. (Input from Dr K. Srinivasan, Principal, Ramkrishna Mission Vivekananda College, Mylapore, Chennai).

Rahul Singh

Ph.D. in Organic Chemistry from Allahabad University having more than 22 years of professional experience in the field of Analytical development and Quality control of Ayurvedic/Herbal/Natural products. Having expertise in isolation and characterization of pure phytochemical compounds from Medicinal plants. Currently working as Head – Corporate Quality Assurance (HCD and Foods) and R and D (Analytical – HCD and Foods) – Emami Limited. Has more than 30 Research publications and 12 patents. Has workd as Head, Analytical Division of Dabur R&D Centre and Ranbaxy Reserach Labs earlier.

Subrata Pandit

Dr Subrata Pandit completed Ph.D. and post-doctoral research from School of Natural Product Studies, Jadavpur University. He made significant contributions in scientific community towards development of herbal medicinal products since several years. He had 22 international research articles, reviews and book chapters' publication in high-impact journals on Ayurvedic plants. Dr Pandit has vast industrial experience on preclinical studies for claim substantiation by *in vitro* bioassay and *in vivo* pharmacological activity evaluation. He extensively works on establishment of safety profile of Ayurvedic herbs through CYP 450 enzyme inhibition assay. He has made significant contribution to pharmacokinetic interaction on several Ayurvedic plants through drug metabolizing enzyme mediated herb-drug interaction studies. Dr Pandit also works extensively on chemo profiling, biomarker analysis of Ayurvedic plants for quality evaluation."

Satyajyoti Kanjilal

Dr Satyajyoti Kanjilal is Bachelors in Ayurvedic Medicine and Surgery with postgraduation in Clinical Research and Regulatory affairs, Psychology and Management. He has more than 15 years of experience in the field of Clinical research, Medical Affairs and Patient care in prestigious organization of All India Institute of Medical Sciences and Dabur India Ltd; and currently heading Medical and Clinical research group at Emami Ltd, Kolkata. He has authoured more than 10 publications in reputed journals and has presented his works in several national scientific forums.

Uma Ganesan

Uma Ganesan is a historian of modern South Asia, with teaching interests and experience in the histories of South Asia, the British Empire, and modern China and research interest in gender and women's history of modern South Asia, colonialism and nationalism. She received her Ph.D. in modern South Asian history from the University of Cincinnati in 2011. She was Visiting Assistant Professor of History at Franklin and Marshall College, Pennsylvania during 2013-2014. She has presented her work at various scholarly fora including the Midwest Conference for Asian Affairs (2010), the Madison South Asia Conference (2010 and 2013), the Association for Asian Studies Conference (2013 and 2014), the International Convention of Asia Scholars (2013), and the Lieden University Conference on South Asia (2013).

Vikram Naharwar

Dr Vikram Naharwar is a third-generation herbalist. For 23 years he has been the Director of India's first private-sector R&D laboratory, Amsar, which has been recognized by the Department of Scientific Industrial Research of India. He participated in the Conference on Medicinal and Aromatic Plants, Vienna, 1997, organized by the United Nations Industrial Development Organization (UNIDO). Dr Naharwar is the Chief Scientific Officer at Integrated Clinical Research Sciences, an Indian CRO that has conducted clinical trials for global pharmaceutical companies such as Pfizer, Aventis, GSK, and Panacea Biotech. In 2001, he was nominated by the Central Drug Laboratory, Government of India to the sub-committee of the Official Indian Pharmacopoeia. He is a member of the review committee of the Indian Herbal Pharmacopoeia. Dr Naharwar has worked with FONDEF, an initiative of the Chilean government, in a joint research programme to investigate botanical ingredients as antiviral agents.

Contents

1

Ayurveda at the Cross Roads

Chandra Kant Katiyar

In 1964, Composite Drug Research Scheme (CDRS), conceived for the first time by an eminent Ayurveda physician Vd. C. Dwarkanath, the then Adviser, Ministry of Health, Govt. of India, was started at the Indian Council of Medical Plants Research under which, it brought together scientists of different disciplines such as Pharmacognosy (Botany, Taxonomy, Morphology and Histology) Chemistry (Natural Products), Pharmacology and Clinical Medicine (both Ayurvedic and the Modern systems) on the same platform to study promising Ayurvedic plants in an integrated and coordinated manner. Nine main research circuits were established in different geographical locations of the country (depending on the best expertise available), each circuit consisting of 5 units, *viz.*, a Pharmacognosy unit (for collection, identification and authentication of the plant describing morphological and microscopic features), Chemistry unit (responsible for Phytochemical studies;) Pharmacology unit (to study the pharmacological activity of the extract and active principles isolated and characterized with special reference to the therapeutic claims made in Ayurveda), and the Clinical unit (involving both the Ayurvedic and clinician of modern medicine) for clinical validation of the Ayurvedic plant drugs already studied by the botanist, the chemist and the pharmacologist according to the therapeutic regimen recommended by the traditional Ayurvedic system. CDRS made notable success, within a short span of five and a half years in pharmacognostic, pharmacological and phytochemical studies, however, the same cannot be said of the clinical trials. Out of the 58 medicinal plants selected for study, 11 had reached an advanced stage of Pharmaconostic, Chemical and Pharmacological investigations. These Ayurvedic plants included *Asparagus racemosus, Clerodendron serratum, Commiphore wightii, Curcuma longa, Nardostachys jatamansi, Picrorhiza kurroa, Piper longum, Pluchea lanceolata, Saraca indica, Vanda roxburghii.* The CDRS thus marked an important milestone in Indian research on Ayurvedic medicinal plants, with a well-conceived integrated multidisciplinary approach, with central coordination. Subsequently, with the branching of the CCRIMH into several independent Councils (*viz.*, Central Council of Research of Ayurveda and Siddha (CCRAS); Central Council for Research in Unani Medicine (CCRUM); Central Council for Research in Homeopathy (CCRH); and Central Council for Research in Yoga and Naturopathy (CCRYN), all these Councils undertook research on various aspects of traditional medicine, while the Indian Council of Medical Research (ICMR) maintained a low profile in this field for nearly a decade and a half.

World Health Organization (WHO) gave the slogan "Health for All by 2000" in the eighties. WHO recognised that since the target of "Health for all by 2000" cannot be achieved without using traditional medicines, it decided to give focus on developing the policies on the same.

Taking leads from this highly ambitious objective, WHO signatory countries started giving impetus to their respective traditional systems of medicines and health practices. In India this impetus was visible by sudden spurt of policy initiatives. These policy initiatives led to interdisciplinary collaborative projects between modern science and Ayurveda streams.

In 1985-1986, major national effort was initiated by ICMR. This time, the approach was totally disease-oriented (in contrast to the drug-oriented approach of CDRS). Following a brainstorming session in 1984 convened by Dr G. V. Satyavati, former Director General, a list of 30 "refractory diseases" was prepared jointly by experts in modern medicine and Indian Systems of Medicine(ISM). Ayurveda and Unani experts identified 26 plant-based drugs to be used for six conditions selected (mainly single drugs except where the ISM specialists felt formulations would work better) of the 30 refractory diseases listed, only 6 were selected (anal fistula, diabetes mellitus, bronchial asthma, filariasis, viral hepatitis and urolithiasis) for multidisciplinary, multicentric research, keeping in view the budget allocation and availability of experts and infrastructure. The Council adopted a centrally coordinated integrated multidisciplinary Task Force strategy with a new disease-oriented strategy involving advanced Chemical, Pharmacological, Toxicological and Clinical Research on selected traditional remedies, combined with simultaneous quality control and standardisation studies on each drug. The clinical trials in various thrust areas were conducted at various medical institutes/hospitals in different parts of the country following uniform protocols and proformae developed by group of experts, in respective areas.

Long-term projects between Indian Council of Medical Research (ICMR), Ayurveda Cell in the Ministry of Health and Family Welfare, Council for Scientific and Industrial Research (CSIR) and Department of Science and Technology (DST) were conceived and launched. Several academic institutions also exhibited their interest and started researches on Indian Medicinal Plants. This trend was visible specially in the pharmacy and medical colleges, though these researches were limited to pre-clinical studies only.

Lack of proper execution of the policies led to the failure to meet the WHO target of "Health for all by 2000." Not only in the year 2000, even in 2016 it remained unachieved and there is no sign that this dream may be fulfilled even by 2020. This happened in spite of the fact that the focus on Ayurveda by the Governmentt of India has significantly increased. The Ayurvedic Cell in the Ministry of Health and Family Welfare was upgraded to the Department of AYUSH in 1994, headed by a Secretary, and later, in 2014. upgraded to the Ministry of AYUSH (Ayurveda, Yoga, Unani, Siddha, Homoeopathy, Naturopathy and Sowa Rigpa), headed by the Minister of State with independent charge.

In the meantime, much water has flown under the epitome of Ayurveda in terms of education, experimental and clinical research, and high-value commercialisation of *Panchkarma*, *Yoga*, Art of Living, and an all-out effort by the Government for globalisation.

One of my friends in a lighter vein says that Ayurveda has become so popular during the last few decades that it has become Maharshi Ayurveda, Massage Ayurveda, Kerala Ayurveda, Keraliya Ayurveda, Sri Sri Ayurveda, Panchakarma Ayurveda, Skin Ayurveda, Cosmetic Ayurveda, *etc.* but somewhere original Ayurveda seems to have been lost.

In this background, "*Ayurveda at the Cross Roads*" is aimed at initiating a discussions on the following broad topics:

- Purists vs Progressives
- North vs South
- Consumerism
- Infrastructure
- Industry
- Regulations
- Research
- Education
- Governance
- Globalisation
- Services
- Pharmacopoeia
- Quality standards
- Good Manufacturing Practice (GMP)
- Pharmaceutical dosage forms

Purists vs Progressives

Both, Purists and Progressives, have their own justifications. The issues between Purists and Progressives are visible at various levels. All India Ayurvedic Congress, about 100 year-old organisation, belongs to traditional vaidyas, while NIMA (National Integrated Medical Association) takes pride in representing ayurvedic physicians with integrated course (Bachelor of Ayurveda with Medicine and Surgery). Not only this, but quite recently in the field of education, there were two schools of thoughts on Ayurveda, Jamnagar School of Ayurveda and BHU (Banaras Hindu University) School of Ayurveda. While Jamnagar School of Ayurveda was perceived to be pursuing original basic and pure Ayurveda, BHU School of Ayurveda was perceived to be promoting integrated research and evidence-based Ayurveda.

The same divergent views were visible even in the industry where companies like Zandu and Dhoot Papeswar were perceived to be following traditional methods of manufacturing, on the other hand companies like Dabur and Himalaya Drug Company were perceived to be following modern mechanised automated methods of manufacturing.

This divergence of Purists and Progressives percolated education, research, clinical practice as well as in the Industry. Now, the void between these two approaches are slowly being reduced post the year 2000 and the proponents of both are coming together. However, the differences are so deep that in certain aspects Ayurveda would still remain at the crossroads.

North vs South

Ayurveda has traditionally been riddled by factionalism, be at education or practice. As if this was not enough, geographical locations added a new dimension to it. This led to North and South Divide in Ayurveda. This trend is perceptible not only in the Ayurvedic practices but also in the brand name of the products, for example Kwath in North India is the Kasayam in South India. Similarly, Taila becomes Thailam and Panchakarma gets limited to Poorvakarma like Abhyanga and added with Shirodhara in South India for convenience and better commercial propagation.

A whole chapter titled "Divergent uses of Classical Ayurvedic Plant Drugs in North and South India" has been written by Mr C P Khare, my associate of this book, to illustrate and expand the matter further.

However, as discussed above, even this Divide is now getting bridged in Ayurvedic practices of South India. Panchakarma, Shirodhara are getting popular in North India also. Similarly, Ayurvedic products from North India are getting popular in South India. These regional compulsions are understandable in terms of education. A few Ayurvedic colleges in South India are providing courses on Massage or Panchakarma and the people trained in such skills, especially nurses from Kerala, are actually in great demand. However, we do not know whether such courses have the approval of the Ministry of AYUSH or the appropriate universities.

This North vs South Divide actually has a historical background. Charaka and *Sushruta Samhita* are perceived to be Lexicons from North India and *Ashtang Hriday* and *Asthang Sangrah* by Vagbhatt belong to South India. The texts are more or less common. However, later on certain Ayurveda books were written in local languages, created regional practices and formulation akin to classical Ayurvedic products in South India. North India lacked in this, barring *Ayurveda Sar Sangraha* published by Baidyanath Bhawan.

Consumerism

Consumerism during the last two decades has played a major role in the growth of Ayurvedic products market. This growth has come from the perception that drugs and products of daily use contain harmful chemicals and are not safe for long time use. Consumers also believe that Ayurvedic products are safe since they do not contain "harmful chemicals". This belief has led to the growth of Ayurvedic products in the categories like medicine, cosmetics, nutraceuticals as well as food and beverages. We would not go into the discussion whether consumers' belief is right or wrong, but it is a fact that emergence of consumerism is a major factor in the growth of the products licensed as Ayurvedic medicines. This is what has led to expansion of the categories of Ayurvedic Products in over-the-counter (OTC) space. The competition amongst the companies is so intense that the consumers are taking Natural, Healthy and Ayurveda as synonyms. Companies are misusing this trust on tradition of Ayurveda and coming out with the products like Ayurvedic toothpaste, Ayurvedic shampoo, *etc.* where actual Ayurvedic ingredients are present in miniscule quantities which cannot deliver the benefits claimed on the labels. Probably there is a need of regulatory as well as consumer activism to remove the anomalies and protect the interest of the gullible consumers. Desire to remain healthy, desire to look beautiful, desire to maintain young skin, desire to change the complexion of the skin from dark to light, desire to reduce the weight without exercise and diet, desire to have long and strong hair without improving nutrition, desire to enhance memory without being attentive to studies, these are some of the common desires which the companies take benefit of. The companies make tall claims to target the consumers and since consumers have faith and trust on the long tradition of Ayurveda they tend to believe the claims on the products. There is a need to create awareness about products with tall claims besides creating awareness among consumers to follow right practices in lifestyle to achieve their desires.

In Ayurveda there is a word called "Pragyaparadh" which means that a person knows that a particular practice is bad for health but still gets engaged in it. Such consumers expect that consuming one or two pills a day will take care of the ills of such bad habits. This needs change.

Infrastructure

Ministry of AYUSH, Government of India has provided details of infrastructure for Ayurveda in India. As per its data there are 2833 Ayurveda hospitals registering a growth of 1.3 per cent per annum from 2012 to 2015. Number of dispensaries is 15555. Human resource indicators cover the details of AYUSH practitioners, among them 402079 were Ayurveda practitioners in 2015.

A considerable increase in Ayurveda Colleges/teaching institutions has been observed during 1993-2015. Maximum 279 Under Graduate Colleges with admission capacities for 15117 students belonged to Ayurveda. In 2007, there were 240 Ayurveda Colleges whereas it was 260 and 279 in 2011 and 2015 respectively. Maximum 112 Post Graduate colleges with their admission capacities for 3029 students belong to Ayurveda system. System-wise average admission capacity of PG in Ayurveda colleges in India during 2007-2015 were 37, 40 and 47 in 2007, 2011 and in 2015 respectively.

There were 7995 manufacturing units for Ayurvedic medicines in India in 2015.

Infrastructure has, therefore, grown, however, quality might be a moot point.

Industry

As per the Govt. of India statistics, there are more than 9000 Ayurvedic industries registered in India. Though there is no confirmed figure of the turnover of Ayurvedic medicines in India it is probably within the range of Rs.7,500 crores to Rs.10,000 crores in the organised sector. This figure does not include the value of medicines prepared and dispensed by the individual ayurvedic practitioners on their own and it also does not include the value of the medicines prepared by the State Govt. pharmacies for the distribution to the patients at primary or secondary health centres free of cost. 80 per cent turnover of Ayurvedic medicines comes from only 20 to 25 Ayurvedic industries. It does not include the revenue of Hospitality services like Panchakarma which itself has a reasonable size of business and earns foreign currency in Kerala. While the Pharma industry has grown from approximately Rs.25,000 crore to more than Rs.1,25,000 crore from 2002 to 2015, the Ayurvedic industry could grow from Rs.3500 crore to only approximately Rs.7,500 crores in the corresponding period. This shows dismal performance of the Ayurvedic industry despite the interest of the consumers in Ayurveda and its products.

A recent Government of India survey showed that 90 per cent of the population in India prefers allopathy as first line of treatment and not Ayurveda. Several stake-holders of Ayurveda have responded to this survey in defensive manner. Some of them said that the question asked were not well-structured and some of them said that essential deep dive was lacking. However, the fact is that Ayurvedic industry could not contribute as per expectations. It is tuned to highly protectionist approach by the Government.

Since more that 90 per cent of the Ayurvedic industry falls under small sector, they always tend to oppose any new and progressive regulation by the Government. of India. Be it GMP, changes in requirement for new product licensing including need of safety and efficacy studies or be it more stringent quality parameters or the move to control the advertisements on the Ayurvedic products. The industry always cries foul in the name of additional cost involvement though the regulations may actually be in the best interest of the industry itself.

The fraternity of Ayurvedic industry should do some introspection on whether they are happy and content with its growth. Unfortunately, some regional players have entered the market of Ayurvedic medicine as fly by night operators. And they have started launching the products exploiting sexual misconception of gullible consumers. A look at the advertisements in Saturday and Sunday editions of local newspapers tells its own story of how these regional players are tarnishing not only the name of Ayurveda but also continuing their uninterrupted disservice to the consumers. Unfortunately, in the name of protecting the interest of small players, industry associations also try to protect these unscrupulous players to some extent.

Appreciating the growing interest of consumers in Ayurvedic medicines, hardcore pharmaceutical companies like Lupin, Ranbaxy, Nicholas Piramal also made their foray into this segment and launched several Ayurvedic products. Scientific inputs were expected from Research and Development (R and D) wings of such pharmaceutical companies but unfortunately, these companies also tried to make booty out of loose regulations and they did not show the same seriousness to research on Ayurvedic products which under the same roof was shown to allopathic drug research.

It is surprising to know that even some of the multinational companies have changed the licence of otherwise Allopathic OTC products outside India to Ayurveda medicine in India to take benefit of lack of stringent regulations including no-sale license requirement for Ayurvedic medicines. These companies do so to expand their distribution reach so that they may be able to sell their products even at the *Kirana* shops. Though legally there is nothing wrong in it, however, the regulators need to do introspection on the impact of what they are promoting.

The industry is engaged in various ventures, like manufacturing products, running hospitals, clinical research organizations (CROs), providing health-related services. While manufacturing sector has grown to a particular level, services like Panchakarma, leech therapy, *agni karma* remain comparatively under-explored despite having huge growth potential. One sector, which started as an allied sector, but has done pretty well in this segment, is extract manufacturing. Some of the extract manufacturers are concentrating on export and are able to establish global leadership in the products like Boswellia, Ashwagandha, Curcumin, Garcinia, Triphala and Amla. These companies are already earning forex for the country. Some of Indian companies are dreaming big to bring Indian herbs like Ashwagandha on par with Ginseng in international markets. Ayurveda has several herbs having such potential. For example, Amla has a potential of being positioned as a super fruit from India. Industry may focus in future in this direction.

Regulations

The Drugs and Cosmetics Act of India was amended to include chapter IV(A) containing regulatory provisions for control of Ayurveda, Siddha and Unani (ASU) drugs in 1964. At that time the act had limited provisions having basic legal requirements, like definition of ASU Drug, patent or proprietary drugs, adulterated drugs, misbranded drugs, labelling requirement, *etc.* and did not include quality requirements in a reasonable manner.

Regulations took more than 30 years to include GMP requirements, quality control requirements besides requirements of clinical trial and toxicity studies for licensing of new Patent or Proprietary ASU Drugs.

Inclusion of GMP and stringent regulatory measures faced a lot of opposition from the industry associations claiming that small manufacturers had limited turnover and could not afford to set up quality-control labs or meet the requirements of GMP facilities. Bowing under pressure the Central Government diluted the provisions of originally proposed GMP and introduced the diluted version called Schedule T GMP. To comply with the demand of industry, the Govenent of India also approved more than 50 testing laboratories where Ayurvedic industries could send samples for testing raw material and finished products.

A group of senior passionate persons from the industry had strategic discussions on how to expand the usage of Ayurvedic ingredients as well as products. The group also discussed the reasons why some of the Indian R and D-based pharmaceutical companies entered and exited from business of Ayurvedic/herbal medicines. They also discussed why combination of vitamins are permitted with non-Ayurvedic herbs while vitamins and Ayurvedic herb combinations are not permitted.

Dr Nitya Anand, former Chairman, Scientific Committee of Indian Pharmacopeia Commission, Prof. S. S. Handa (Chairman, Pharmacopeia Commission of Indian Medicines), Dr C. K. Katiyar (Director, Herbal Drug Research, Ranbaxy Laboratory, Gurgaon) and Dr D. B. Narayana (Director, Regulatory Affairs, Asia, Hindustan Unilever R and D, Bangalore) discussed the status of Indian Herbal Drug Industry. It was deliberated as to why pharmaceutical industry decided to recede after taking a few steps forward. Lack of interest of BAMS doctors in prescribing generic Ayurvedic products, lack of Intellectual Property Rights (IPR) provisions for traditional medicine-based products, as well as judgment of Hon'ble Supreme Court in the matter of Ashwin Patel being misrepresented by Indian Medical Association for banning prescription of Ayurvedic medicines by allopathic doctors, were a few of the reasons which created impediment for the growth of herbal drug industry. After prolonged deliberations following strategic intiatives were decided to be pursued with vigour:

a) To introduce data protection for ASU Drugs as alterate to patents. The provision of data protection was passed by the ASU Drug Technical Advisory Board but was pending for publication by the Ministry of AYUSH even till end of 2016.
b) Creating and introducing Phytopharmaceuticals as a new category of drugs (Allopathy) under rule 122E of the Drug and Cosmetics Act.
c) To pursue FSSAI for permitting the use of Ayurvedic medicinal plants in Health supplements.

The group met the then Drugs Controller General of India, Dr Surendra Singh, and an Expert Committee was formed under the Chairmanship of Dr Nitya Anand to suggest recommendations on Phytopharmaceuticals.

The committee submitted its recommendations, and a draft notification was published by the Ministry of Health and Family Welfare, Government of India. Ministry of AYUSH opposed this notification vehemently, however, after lot of pursuasion and changing the definition of Phytopharmaceuticals the Gazette notification introducing Phytopharmaceutical as a new Drug under Allopathic medicines was published on November 24, 2015. It has been clearly differentiated from AYUSH Drugs. While domain of AYUSH drugs is up to crude water or alcohol or hydro-alcoholic extract, Phytopharmaceutical scope includes fractions of the crude extract with isolation, characterisation and identification of minimum four compounds in the same drug.

With the introduction of Phytopharmaceutical as a new Drug, India has become the first country in the world to have regulations of developing Drug from plants on par with new chemical entities. Hopefully, it would receive attention of R and D-based pharmaceutical companies as a rescue to develop new drugs at faster pace with lesser investment and thereby improving the productivity of their R and D efforts.

Another regulatory development has happened in the field of foods for which FSSAI has now approved the use of the herbal ingredients of Ayurveda in food products, such as health supplements, dietary supplements, food for special medical purposes, *etc.* Earlier, the herbs having origin in Ayurveda were not permitted to be combined with vitamins. After this Gazette Notification dated November 24, 2016 from the Ministry of Health and Family Welfare, such combinations are now permitted. It is expected that very shortly markets would be flooded with such products, giving boost to the herbal sector in India.

With this, India has become the first country in the Word to have such a progressive regulation. A few countries have already started following our initiative for the promotion of herbal sector.

Introduction of Phytopharmaceutical and permission of using Indian medicinal plant ingredients in the health food, dietary supplement and other categories of FSSAI has opened up a huge opportunity of developing new products.

Hopefully, these regulatory initiative of promoting the business would be fruitful and the sector would see growth by leaps and bounds in near future.

Researches

With the increasing awareness, interest of consumers on Ayurveda and its products has also attracted the attention of researchers, educationist as well as regulators, besides policy makers. Very often need for research on Ayurveda and its products has been highlighted by various stakeholders. They believe that only traditional or faith-based claims would not do and Ayurvedic products should be subjected to rigorous scientific studies and evaluation. With this objective, Government of India had formed a central agency to conduct and coordinate research on Ayurveda. This Agency is called Central Council for Research in Ayurv eda and Siddha (CCRAS). In 2015-16 this Council was bifurcated into two separate councils for Ayurveda and Siddha each. Currently researches on Ayurvedic products or ingredients are being conducted at Ayurvedic Institutions like CCRAS; Banaras Hindu University; Gujrat Ayurveda University, Jamnagar; National Institute of Ayurveda, Jaipur; Ayurveda University, Jodhpur; newly established All India Institute of Ayurveda, New Delhi, and few other institutions. Simultaniously, some of the medical colleges like All India Institute of Medical Sciences, New Delhi, Institute of Medial Sciences, Banaras Hindu University, Varanasi, are also conducting and publishing research on Ayurvedic products.

In addition, pharmacy colleges, specially their departments of Pharmacognosy, Phytochemisty as well as Pharmacology, are currently conducting researches and publishing papers, but mostly on single medicinal plants.

Public research Institutions like Council of Scientific and Industrial Research (CSIR), Indian Council of Medical Research (ICMR), Department of Science and Technology (DST) as well as Department of Bio Technology (DBT) besides Defence Research and Development Organisation (DRDO) are either involved in direct research or are extending financial support to research on medicinal plants and their products.

Ayurvedic industries are also involved in researches, but they are limited to the products and areas of their interest. Most of the times Ayurvedic industry does not publish their researches unless it serves their purpose of product promotion.

It is to be noted that though so many institutions are involved in the research, India lacks in publication of research papers on Ayurvedic products, treatments or ingredients in reputed high impact journals like *Science, Nature, Lancet, BMJ* (British Medical Journal) and *JAMA* (Journal of American Medical Association) *etc.* It raises a question on quality of researches on Ayurvedic medicines conducted in India vs that on TCM-based products in China which frequently get published in Journals like *Science.*

We need to do introspection and review and revise our research policy on Ayurveda and its products in India in a comprehensive manner. We also need to have one agency to coordinate research on Ayurveda from all across the institutions to prevent duplication of efforts.

Education

India had 279 Graduate and 112 Post Graduate Colleges of Ayurveda in 2015 as per data by the Ministry of AYUSH.

The graduation-level course is mostly integrated with the subjects of modern medicines also. Internationally, Ayurveda colleges and Ayurveda universities have come up across the continents including Europe and America recently.

However, great limitation is being felt for the faculty. Ayurvedic teaching institutions must enrich their teaching faculty. Otherwise on one side the system and practice would continue to grow but the quality of teachers would see the downward trend. It is to be noted that, though in the past few decades number of Ayurvedic manufacturers grew tremendously, education system lacks in creation of manpower to commensurate level. Joint effort of creating interdisciplinary manpower is also required, besides creating experts in Ayurveda through our education system.

Governance

The Govt. of India came out with the AYUSH policy for the first time in 2002 which paved the way and direction to move ahead in certain areas. The current political dispensation in India for the first time created an independent Ministry of AYUSH in Central Government in 2014. Before that in 1994 it was elevated to the level of a Department in the Ministry of Health and Family Welfare and there used to be only one position of Advisor in Ayurveda. Since Ministry of AYUSH is very new, it would take time for it to start delivering results and hopefully Government's focus would continue on promotion of Ayurveda the way it is done for yoga. AYUSH is the acronym of Ayurveda, Unani, Siddha, Homeopathy, Yoga, Naturopathy and recent addition is Sowa Rigpa. AYUSH is the main hub, while major responsibility of creating awareness about Ayurveda in India lies with the State Governments, since health is the subject of the State government.

Effort of Globalisation

Ministry of AYUSH, Government of India have been very aggressive in taking steps for globalisation of Ayurveda both in terms of its products, services as well as education and research. Declaration of 21st June as International Yoga Day is result of such aggressive approach.

Few years ago, Government of India took a unique initiative of creating Ayurveda Chairs at the universities of various countries throughout the world with a view to attract researchers, educationists besides the practitioners. It would help in creation of awareness of the system in these countries to a great extent. Government need to enhance dialogue with the drug regulatory authorities to achieve the target of recognition of Ayurveda for clinical practice. Recognition of Ayurvedic medicines would be a natural corollary.

Services

There are 1,46,036 health sub centres in 6,38,588 villages, 22370 Primary Health Centres, 26107 AYUSH dispensaries in 6345 block,/referral centres, teaching hospitals, Govt. hospitals and dispensaries in 29 states and Union Territories as per AYUSH report (Ministry of AYUSH).

In the recent decades there has been great spurt in number of Ayurvedic spas even in the star Hotels and Resorts in India. Emergence of Lifestyle Disorders and mere fact that they are preventable has created special space for Ayurveda. Service sector has taken note of it and is catching up fast not only with the consumers but also investors.

Pharmacopoeia

India is one of the few countries having its own Pharmacopoeia for Ayurvedic, Unani and Siddha medicines independently.

Ayurvedic Pharmacopoeia of India (API) has been published in two parts, part one deals with the raw materials and part two deals with the finished products. Part 1 of Ayurvedic Pharmacopoeia of India has eight volumes while part two has two volumes.

In addition to API, Indian Pharmacopoeia has also published Monographs of crude herbal raw material as well as extracts. Appreciating the increasing popularity of Indian Medicinal Plants United States Pharmacopoeia (USP) has also now included Monographs of few Indian Medicinal Plants like Ashwagandha.

In terms of quality standards, API has followed minimum quality standards sufficient enough to prevent adulteration. This strategy has probably been followed since majority of close to 9000 manufacturers of Ayurvedic medicines may not be able to afford full-fledged quality-control labs, therefore, the Pharmacopoeia has included basic yet important parameters only. Quality standard in Indian Pharmacopeia are more stringent than API. For example, analytical marker-based standardization is mandatory as per Indian Pharmacopoeia. USP has gone further to include minimum two analytical markers for quality standardization. API is evolving and volume eight dealing with the extracts has also now made it mandatory to use analytical markers for quality standardization.

Part two of the API is dedicated to Classical Ayurvedic formulations and their quality standards. However, since most of the commercial manufacturers are marketing pre-dominantly proprietary medicines, they need to develop their own in-house quality standards.

It is pertinent to share our views regarding inclusion of Ayurvedic parameters like *Rasa, Guna, Veerya, Vipak* in the API. It is to be appreciated that Pharmacopoeia is a book of standards and each parameter should be supported with the Standard Test Procedures (STPs) to avoid test protocol-induced variability. API does not provide STPs of such Ayurvedic parameters. APC may consider either to introduce their STPs or to declare them "non-mandatory."

API, IP and USP have different levels of quality standards. However, concerns are now-a-days being raised on safety as well. Therefore, it is pertinent that these Pharmacopoeia should consider including negative markers also to strengthen the safety parameters in future.

Quality Standards

As per WHO quality, safety and efficacy are the hallmarks of traditional medicines. Major raw materials for Ayurvedic medicines come from nature which leads to lot of variability in terms of chemical composition leading to requirement of laying down the limit of parameters within which raw material has to fit before it may be called as appropriate quality raw material. In case of herbal raw material adulteration, contaminants and innate chemotaxonomical variability may play the role for inconsistency in product.

Lately, the boundary of safety and quality has been narrowed, heavy metal, aflatoxins, pesticide residues as well as microbial testing have now become very important parameters of quality which otherwise may impact on the safety of the product. Therefore, standardization in certain cases is also now becoming synonymous with safety standards.

GMP

The Drugs and Cosmetics Act of India introduced schedule T GMP as mandatory condition to obtain manufacture for sale license of Ayurvedic products more than a decade ago.

Schedule T is based on basic principles of GMP instead of recommending very high standards. It is surprising to know that about 1000 Ayurvedic industries had to close down their operations because they could not meet even this basic requirement of GMP. It is to be noted that if Ayurvedic medicines are to be exported, even the countries like Nepal and Sri Lanka are demanding WHO GMP which is much above the Sch 'T' GMP.

GMP is the culture which the companies need to inculcate to move from people-driven to process-driven.

Pharmaceutical Dosage Forms

Ayurveda boasts of having as much as 126 dosage forms including food formats. However, the major dosage forms remain juice (*swaras*), decoction (*kwath*), hot infusion (*phant*), powder (*churna*) and cold infusion (*him* or *sheet*). These are the five basic dosage forms which were further advanced to *Avaleha, Vati Gutika, Asav Arishta, Anjan, Bhasma, Kshar etc.*

Now-a-days developments of modern pharmaceutical sciences are applied to further evolve Ayurvedic dosage forms into more convenient and palatable forms like tablets, pills, jelly, syrup, granules, capsules (hard and soft gelatin), *etc.*

Application of New Drug Delivery Systems (NDDS) should be considered for the benefit of contemporary consumers. For example, development of sustained release tablets, chewable tablets, soft-gelatin capsules, nano particles-based products ensuring faster absorption may be required to keep the dosage forms of Ayurvedic medicines adaptable, convenient, up to date and hi-tech.

Extraction technologies have undergone a sea of change from water to solvent extract. The journey has moved further to super critical extraction.

Areas of manufacturing, extraction, quality control, prevention of counterfeiting, *etc.* require technology adoption.

Some of the research opportunity areas for Ayurvedic product development include, development of natural preservative, natural foaming agent, taste-masking technology, cost-effective process of decontamination, modern fermentation technology in manufacture of *Asav Arishtas* and computerized furnaces for manufacture of *Bhasmas.*

These are top-of-mind recall topics when we discuss Ayurveda at the cross-roads. Lot has been done, but lot still needs to be done.

2

The Ayurvedic Revival Movement in India, 1885-1947*

Uma Ganesan

Ayurvedic revival movement of the late nineteenth and early twentieth century is historically an important period of colonial India.

On the one hand, nationalism provided the space or fertile ground for revival of Ayurveda. On the other the demand for swaraj or home rule, which increased in frequency and intensity during 1920s and 1930s, also entailed the projection of a modern, scientific, and progressive image of India in order to justify and legitimate that very demand. Therefore, efforts to revive Ayurveda during this period placed importance on establishing its scientific and progressive credentials. An analysis of the revivalist discourse can help to unravel the important connections among nationalism, identity, modernity, science, and medicine during these crucial decades of the twentieth century.

By the end of the nineteenth century, Ayurveda along with other indigenous systems of medicine such as Unani (Graeco-Arabic medicine)[1] and Siddha (South Indian Tamil traditional medicine)[2] were profoundly influenced by their encounters with Western medicine. The closing decades of the nineteenth century witnessed a proliferation of books on Ayurveda in English, Sanskrit, and vernacular languages as the proponents of Ayurveda "tried to transform the hitherto relatively inaccessible knowledge into social knowledge as well as a shared system of knowledge among the practitioners".[3] This outpouring of Ayurvedic publications roughly coincided with the rise of nationalism[4] in general and that of Hindu revivalist nationalism[5] in particular.[5] It was also a response to the colonial government's decision in 1835 to suspend the teaching of Ayurveda in the Calcutta Medical College.[6]

While of crucial significance to this early period, Ayurvedic publications, addressing the theme of Ayurvedic knowledge and history, continued to sustain the movement through the first half of the twentieth century. Also, Ayurvedic practitioners organized themselves as a professional interest group through the founding of the All India Ayurvedic Congress (A.I.A.C.) in 1907. The annual conferences organized by the A.I.A.C. provided another key forum for revival efforts. The proceedings of these conferences along with books, tracts, and journals provide useful source material for the historian interested in understanding how the proponents of Ayurveda positioned themselves and their art within the larger nationalist discourse that envisioned a modern, scientific, progressive future for India. Thus, the focus in this chapter is on how a male elite, doctors and *vaidyas* (Ayurvedic practitioners) in particular, envisioned the medical past, present, and future of India.

Three themes were central to the Ayurvedic revivalist discourse contained in the literature on Ayurveda and the conference proceedings of the A.I.A.C.: British Orientalism, the synthesis of medical systems, and institutionalization of Ayurveda. These themes were

-----

* The text of this chapter first appeared in Studies on Asia, series IV, 2010, when Uma Ganesan was at the University of Cincinnati, OH.

common to multiple reform and revival efforts in colonial India. The latter two themes must be seen as specific manifestations of the larger theme of Indians' attempts to preserve tradition while at the same time appealing to modernity and its corollaries of science, reason, and progress. Hence, as in the other movements, the interplay of these themes in the Ayurvedic revival movement generated the kinds of complexities, contradictions, paradoxes, and inconsistencies inherent to any project that attempted to articulate its needs, aspirations, and goals within the terms of the very structure (the Western medical system) it was responding to or reacting against. The story of Ayurvedic revival brings into focus the tension between structure and agency in a colonial setting: How far did the structure--the western colonial system--constrain the choices of these reform and revival movements?

British Orientalism

British Orientalism refers to a set of ideas and practices inaugurated under Warren Hastings, Governor-General of Bengal from 1773 to 1785 that sought to know and understand the languages and culture of India as a key step towards good governance. Deploying eighteenth century Enlightenment ideals, British Orientalists such as William Jones and Henry Thomas Colebrooke translated ancient texts, posited a Vedic golden age of Aryan Hindus, identified Sanskrit as the fount of Indian civilization, and in the process stimulated a vigorous intellectual and cultural 'renaissance' among the Bengali elite.[7]

This generally positive attitude to ancient Indian culture coexisted alongside the denigration of contemporary conditions which served to justify the British presence in India. In response to this, early reformers such as Ram Mohan Roy (1772-1833) sought to selectively reform and recast Hindu socio-religious practices while drawing upon the Orientalist discourse of early glory and achievement--a trend that would characterize several reform and revival movements in colonial India.[8]

Aryans, antiquity, Vedic civilization--ideas that were at the heart of British Orientalism--were invoked in the service of Ayurveda. In an early (1895) and widely acknowledged contribution to the Ayurvedic revival movement, Bhagvat Sinh Jee, the Maharaja of Gondal (a princely state in western India), Fellow of the Royal College of Physicians of Edinburgh, and later Vice-President of the Indian Medical Association, provided a brief sketch of Hindu achievements in astronomy, mathematics, chemistry, music, religion, philosophy, architecture, lexicography, grammar, and the art of war during a Vedic golden age and argued that "all this unmistakably proves that the Aryans were the most enlightened race in the dawn of history....When the state of civilization was so perfect, and when all sorts of useful sciences were regularly studied, there should be no wonder if the science of medicine too received its share of attention. This science forms part of the Vedas, and is called "Ayur Veda" or the "Science of Life."[9]

Thus, placing Ayurveda in the context of achievements in other sciences served to emphasize its normalcy. Similarly, in 1919, Nagendra Nath Sen Gupta, an Ayurvedic practitioner from Bengal and a prolific author of books on Ayurveda asserted that "it would be no exaggeration to say that of all nations of the earth, the Hindus first turned their attention to the study of disease and the means of its alleviation. The Vedas are undoubtedly the most ancient written records in the world. The *Ayurveda* or Science of Life is believed to have formed a part of the Vedas, vis., those that go by the name of the *Atharvas*."[10]

Indian elites responded to the colonial 'gaze' by articulating an indigenous version of a 'glorious' past of Hindu culture. As Michael Dodson argues, British Orientalism proceeded from being a hand-maiden of Britain's fledgling Indian empire to being linked with the emergence of early forms of Indian national identity and anticolonial cultural movements.[11] Ayurveda was one of the Sanskritic Vedic traditions that Indian nationalists, particularly Hindus, drew upon. Thus, for example, D. Chowry Muthu, an Associate of King's College, London and a consulting physician for the Thambaram Sanatorium in Madras which he had established in 1928, deployed the Orientalist discourse that emphasized a vibrant and strong Hindu-Aryan past:

"The history of Hindu medicine....takes us back to the very cradle of Aryan civilization....Whether it be in the domain of art or science, in poetry or philosophy, in religion or mythology, in commerce or manufacture, ancient India excelled in almost every department of human activity or enterprise for many, many centuries."[12]

Written in 1930, nearly a century after the British abandoned the Orientalist discourse, Muthu's account demonstrates the continued power of that discourse in lending credence to identity construction and national consciousness among elites in colonial India. Since Ayurvedic proponents argued that Hindu civilization and Hindu medicine could not be separated, this discourse had particular relevance for Ayurvedic revival efforts.

As the medical history of the Indo-Aryans forms an inseparable part of the history of their civilization, the proof of antiquity of their medicine can be found in the antiquity of their civilization....no nation on earth can vie with the Hindus in the antiquity of their civilization and religion....the history gathered from the recent rendering of the Vedic hymns...take[s] one back to immemorial antiquity.[13]

Ayurveda was the progenitor of medicine whose "glowing embers had lighted the torch of Arabian medicine, and through it the fire of European medicine."[14] Shiv Sharma, youngest president of the All India Ayurvedic Congress in 1938 elaborated: "The evidence adduced in favour of the indebtedness of the Greeks and Arabs to the Hindus in the science of medicine may also serve to prove that the system of Ayurveda stands unrivalled in its antiquity. The presence of the Ayurvedic principles in the hymns of the Vedas, the earliest records of human intellect, establishes for good the high antiquity and the originality of Hindu Medicine."[15] For Sharma "it was only too natural for the ancient Hindus to bring forth a highly developed system of medicine consistent with their other scientific achievements."

Placing Ayurveda in the context of overall early Hindu achievement was important to revivalists' attempts to explain contemporary degeneracy in terms of a general decline of Hindu civilization of which Ayurveda was but a part. Therefore, the contemporary condition of Ayurveda was a manifestation of the larger condition of Indian society, and not inherent deficiencies as a medical science.

The Society for the Resuscitation of Indian Literature in Calcutta, a prolific publisher of books on Hindu philosophy, theology, and literature--many of them translations of classics, also published one on Ayurveda in 1899, further emphasizing the notion that medicine was an integral part of classic Hindu civilization. This was also a reflection of the growing Hindu revivalist nationalism of the late nineteenth century. This book in keeping with the tendency to stress the antiquity of Ayurveda as an important first step in making a case for its revival argued that "for the first conception of the medical science the whole world is indebted to the Rishis of India"[17] and pointed out that "[like various other departments of science and literature, the Hindu medical system claims to be the first of its kind in the world and has lent much towards the advancement of the medical systems of other countries."[18] As David Arnold argues, in invoking the past Indians were not merely setting the historical record straight but were also "shaping contemporary identities and aspirations."[19] Thus, the apparent paradox of the "modern nation's return in the archaic" was resolved as Hindu intellectuals posited "religion as the embodiment of eternal and universal laws....forg[ing] difference into unity, multiplicity into singularity.... homogenous, whole and pure."[20]

Like the early reform movements that British Orientalism spawned, the Ayurvedic revival movement also reinterpreted and defended Hindu civilization in the light of modern European scientific thought. M. M. Gananath Sen, an Ayurvedic practitioner from Bengal who founded both a college for the study of Ayurveda and a pharmaceutical concern for manufacturing Ayurvedic medicine, pointed out that "the charge that Ayurveda is not a progressive system is not so much a charge against the science itself as against ourselves"[21] for "when the greater part of the world was submerged in the abyss of ignorance, it is the Indian sages who first understood the necessity of dissection of the human body in the education of Physicians and Surgeons."[22] Therefore, science was part of Hindu civilization and Ayurveda was not at odds with modernity: "If Ayurveda is not scientific, it is not worth pleading for."[23] In a similar vein, Jivaram Kalidas Shastri, Ayurvedic practitioner and royal physician of Gondal, posited that "To the Aryans, their Healing Art is as old as the Vedas which they regard as of divine revelation. Even those who speak of their human origin do not fail to recognize their antiquity in the remote past....The might[y] sages of old...shaped Ayurveda as a systematized science."[24]

The reference to science and antiquity in the same breath challenged modernity's sole claim to science and also pointed to the contradiction implicit in the British dismissal of Ayurveda. The contemporary condition of Ayurveda thus could not be used to judge its past glory. If science was the yardstick by which medical knowledge and practice was measured, then Ayurveda had the first and most ancient claim to it. This analysis of the revivalist discourse of Ayurveda reveals certain key shared features with other reform and revival movements that adopted the British Orientalist discourse of early achievement and subsequent decline of Hindu civilization, central to which was an appeal to an orginary, 'pure', Vedic/ Sanskritic tradition.

The Brahmo Samaj, founded in 1832 in Bengal, was an early example of this effort as it posited a monotheistic, philosophical version of Hinduism devoid of rituals and social practices such as caste and *sati* which were seen as later accretions. Similarly, the Arya Samaj, founded in 1875 in western India, advocated a return to the Vedas as the fount of Hindu civilization and rejection of all later accretions to the Vedas as degenerate.[25] Also, much like the Ayurvedic revival movement which attempted to cleanse Ayurveda of degeneracies and separate it from what were seen as "ignorant and superstitious folk practices"[26] the Brahmo Samajists and Arya Samajists privileged Sanskritic textual Brahminic traditions to the detriment of social practices and customs. Therefore, claiming legitimacy for Hinduism and Ayurveda involved recovering ancient texts to posit early glory and achievement. This again had its antecedents in British efforts to outlaw *sati* when they sought the help of learned Brahmins to interpret the scriptures for them, thus legitimating written texts as the sole criteria for judging the validity of Hindu customs. [27]

Thus, Indians countered the British discourse of "people with no history" by constructing the history of Hindu medicine and civilization central to which was the study of ancient books. Thus, these reform efforts functioned within an upper-caste, upper-class idiom that was Hindu and male.[28]

Synthesis of Medical Systems

As Ayurveda was pushed to a defensive position from which it could not extricate itself completely, its revival, occurring as it did under the glare of western medicine, depended on continuously answering its critics as well as justifying itself in terms set by its critics. G. Srinivasa Murti, Director of the School of Indian Medicine established in Madras in 1924, pointed out that the "violent denunciations" of Ayurveda cannot be relegated to an earlier period for "the Ayurvedic movement is even now being treated" to them "in some 'scientific' quarters."[29] Examples of this included Mr Pilcher's reference to the "preposterous Ayurvedic and Younani Systems of Medicine"[30] and Surgeon General (of Bombay) Hooton's characterization of Ayurveda as being based on "erroneous theories" that "cannot bear comparison with the modern system of Medicine founded on recent advances in Science."[31]

Stinging reactions to these views were evidence of the continuing need felt by the proponents of Ayurveda to defend it in terms of its antiquity and scientific basis. Thus, M. R. Samey, a passionate defender of Ayurveda and a harsh critic of British rule, challenging the Surgeon's denunciation of Ayurveda retorted, " The *worthy* in his 'Service' Throne felt uneasy at the rearing of his 'Enemy's head' and chooses to mollycoddle it in his 'Blue Book'. This philippic against [the]Indigenous System of Medicine is naively introduced into the Report by way of precious vituperation to be standardized as a classic monograph and Public Document"[32] and "What then remains of modern Medicine if our heritage is discarded?.who has given the first system of Medicine to the world?--The Hindus--Charaka and Sushrata. Egypt, Greece and Rome built their medical systems on these 'World Classics of Medicine' and England only painted it red and claimed it as her own.[45]

K. P. Sankara Pillai, writing in 1933, declared, "The freedom of our dear motherland remains in the resurrection of our ancient civilization and in the utter abolition of modern inventions with their incommodious paraphernalia. Our Rishis have practically shown us that man is completely self-sufficient. Then why not work for the realization of this, rather than follow the delusive meteor of occidental 'unrest'." [33]

Such a pungently reactionist discourse for the most part coexisted with the recognition of the need for accommodation and cooperation. Murti offered a proposal for cooperation: "In so far as the one common ideal of all Systems of Medicine is the preservation of health and prevention or cure of ill-health, there can really be but One System of Medicine, of which the many existing 'systems' are but parts, each part being more appropriately looked upon as a special 'school' of thought rather than as an independent System of Medicine. Consistently with this view, one would like to see that the future practitioners of India, no matter whatever denomination they belong to – Ayurveda, Unani, Siddha or European Medicine - as so schooled and trained so as to bring to bear on the problems of Health and Ill-health, not only the expert knowledge of their own systems but, as far as practicable, the best that is in other systems also....It is the true ideal of Ayurveda and must not be lost sight of on [any] account."[34]

Several motivations, compulsions, and understandings underpinned the move toward cooperation between the two systems of medicine in the noble cause of health and well-being. The recognition that Western medicine was here to stay due to its patronage by the colonial government, its appeal as a modern, rational, progressive, research-oriented system, and its value and usefulness in several areas of health and disease together with practical exigencies such as the fact that vaids and hakims remained the main source of medical succour to the millions in rural India prompted many of the Ayurvedic proponents to argue for a synthesis. An anonymous article in the JAHSM in 1927 defended provincial governments' support of Ayurveda in the following manner: "It was an outcome of the economic and business point of view of those provincial governments who have given State-aid to existing Ayurvedic institutions incorporating in their teaching what is lost or unknown to Ayurveda from modern medicine, so that the best of both systems are taught to the students."[35]

One of the tools lost to Ayurveda, both due to Buddhist abhorrence of it and popular prejudice against it, was surgery.[36] Therefore, its inclusion in the curriculum of Ayurvedic colleges was one way to enable Ayurveda to progress and keep up with Western medicine. Thus, for example C. G. Mahadeva argued: "There cannot be watertight compartments between the two systems of medicine. Both aim at alleviating human suffering....What is really good in one must be assimilated by the other....Surgeons...should be invited to take charge of surgical wards of hospitals attached to these [Ayurvedic] institutions. These surgeons should have strong ambitions to see that Ayurveda should be a progressive science and that latest surgery must be taught to Ayurvedic practitioners and students as it is taught in the sister institutions teaching Western Medicine."[37]

Some proponents engaged in a reasoned analysis of the pros and cons of the different systems of medicine to underscore the imperfections contained in all of them so that synthesis could be understood not as a sign of capitulation but of strength: "The real system of medicine should be one, unifying all these different schools and thereby forming a united system in which all the good things of all the schools should be given place, so that this united system may be more perfect than the disjointed individual sciences."[38]

These proponents were motivated by a desire to end the acrimony between vaids and doctors: "Our object is to put a stop to the habit of throwing stones at each other. It is also our object to request our brother Allopaths to study Ayurveda as it is, and incorporate what is good in Ayurveda into their system. We are also trying to remodel Ayurveda in the line of other sciences, purging it of all its defects, if any, and incorporating everything good, found in other allied sciences, into it, even at the sacrifice of the fetish, *prestige*."[39]

M.M. Gananatha Sen, too, while not conceding much to western medicine, acknowledged the importance of cooperation in the common cause of alleviating disease and suffering: "An open-hearted and liberal cooperation of both should be a source of great help to the profession as a whole and to the sufferers entrusted to our care."[40]

The Ayurvedic revival movement, like other reform and revival movements, was not a simple, linear isolated process of reviving a pristine, pre-colonial indigenous system but a complex one of emphasizing tradition while at the same time attending to the changed and changing conditions under colonialism. As Deepak Kumar argues, condemnation and appreciation of Western medicine coexisted in a complex relationship.[41]

An anonymous article in the June 1928 issue of the JAHSM argued:

"Medical Education in India should be so devised that it should take into account not only the present-day medical education but also medical knowledge of the past....While Ayurveda cannot move on in [an] old groove, Allopathy should not be accepted in toto for India. While we should absorb the pathology of the "seed of disease" from Allopathy, we must give the "pathology of the soil" in disease to modern medicine. The two angles are at present different but should be harmonized.[42]

Charles Leslie[43] contends that the rapid progress of Western medicine in India by the beginning of the twentieth century made any claim to a pure practice of Ayurvedic or Unani medicine not only anachronistic but also self-deceptive. Therefore, he argues, the revivalists had no choice but to contend with and adopt the theories and instruments of Western medicine. Thus, the bitter polemics and the acrimonious debate notwithstanding, ayurvedic revivalists - in keeping with the changed and changing conditions of colonialism and the vision of a modern, progressive future for India - argued for cooperation and synthesis.

Institutionalization of Ayurveda

A central criticism against the Ayurvedic revival movement levelled for example by Sir John Megaw, Director General of the Indian Medical Service, was that it was motivated by political considerations of nationalism and patriotism rather than by medical merit. Some revivalists held that 'medical patriotism' so to speak was both appropriate and normal. An example is a 1933 article by M. R. Samey who responded as follows:

"The false pretext of patriotism, which the gallant General makes 'out of Bounds' for medicine, has really been the prop and pillar of modern medicine and 'British Medicine' has been made what it is by the British Medical Council and British Medical Association whose very plinth and foundation is the false pretext of 'Patriotism'.... Why should Ayurveda alone and Aryan Medicine, of all, be denied the fostering and tender devotion of that noble sentiment 'Patriotism'." [44]

Here, on the one hand, patriotism was justified as a legitimate motivation behind the revival movement; on the other British denunciation of the revivalists was seen as a manifestation of British imperialist tendencies confronted as they were in the 1930s with a rising tide of nationalist sentiment. Therefore, their dismissal of the revivalist movement as the work of misguided nationalists was seen as equally politically motivated. A. Raman made the same argument later that year in an even angrier tone: "The Colonel's [Col. R.T.Baird, a retired officer of the Indian Medical Service] view that the public money spent in fostering the Ayurvedic and Unani systems in India is a waste shamelessly betray the imperialistic tendency of a Britisher in the I.M.S. who cannot change his angle of vision even after spending the best part of his life on the Indian soil, as a leopard cannot change its spot[s]."[45] Economic reasons along with nationalistic ones were put forward for attracting government support. M. K. Mukherjee argued that the less expensive Sanskrit education including ayurvedic education would provide a livelihood to those who could not afford English education reducing the unemployment rampant among Indian youth while at the same time making them "standard-bearers of our own national pride and individuality, which would resist the cultural onslaught of outsiders like veritable ramparts."[46] In response, Western doctors accused Ayurvedic proponents of being cocky and arrogant, refusing to admit imperfections and to learn from modern medicine:

"It is not that we refuse to lend a helping hand to our brothers who belong to the ancient systems, far from it, but can those who admit their imperfection offer to teach those who claim to be perfect? Let the practitioners of the systems place their cards on the table as we have done, let them admit that existing knowledge is far from perfect…let them join us in the search for truth and knowledge, and in the application of that knowledge and we shall receive them with open arms. Under existing conditions fellowship with them in medical practice is inconceivable."[47]

Also, under such conditions granting them equality with practitioners of modern medicine through medical registration acts would be a mistake, for the limited resources at the disposal of the state should be directed to modern medicine in keeping with India's goal of "admission to the commonwealth of advanced and civilized peoples"[48] This goal would certainly not be served by "a deliberate return to obsolete systems."[49] Thus, they argued that medical anti-imperialism would be self-defeating for Indians since it would oust India from the comity of 'civilized' nations. State aid to indigenous medicine was seen as capitulation to the "persistent clamour of Indian politicians."[50]

Some among the British, recognizing the connection between nationalist/religious sentiment and demands for state support and recognition of Ayurveda, advocated a middle ground that would enable indigenous practitioners to continue their services until such time as they could be completely replaced by allopathic practitioners. But in the long run "if the Madras Government has the interests of the Indian people genuinely at heart, it will expend its energies in planting modern science in the country…instead of endeavouring to stimulate the belated indigenous systems into renewed activity."[51]

According to this view, the choice was between progress and European science on the one hand and Indian backwardness and a "metaphysical rut"[52] on the other, since contemporary Ayurvedic knowledge and practice revealed features familiar to Western medicine from an earlier phase in its own progress. Here we see a familiar deployment of the modernist discourse that saw history in linear terms - as moving from the primitive to the developed, and the Orientalist discourse that saw Indian culture as metaphysical rather than scientific, and as passive and conservative rather than active and progressive. Thus, a value judgement was placed on Ayurveda, as it was trapped in an earlier phase of the development of medicine. Thus, from this standpoint 'tradition' and 'modernity' became value judgements, and Ayurvedic proponents' use of these categories to validate Ayurveda became self-defeating in some ways.

A good example of the official British stance on Ayurveda was a 1927 report the Surgeon General of Bombay who referred to "erroneous theories" and unscientific principles of Ayurveda to argue against government support to the indigenous systems of medicine: "To revert to any of the ancient systems of medicine would be a very retrograde step. They are not founded on scientific principles, and are entirely out-classed by modern scientific medicine."[53] Similarly, W.D. Sutherland of the Indian Medical Service had opined in 1919 that overdependence on and obedience to authority that refused to consider anything that might challenge that authority was Ayurveda's undoing.[54] Another critic held that Ayurveda with its "obsolete absurdities" could not legitimately lay claim to public money.[55]

Such views were not the sole province of the British. State aid to Ayurveda was a contentious issue that pitted not only the British against Indians but also Indians against Indians. Ralph Crozier characterizes the division among Indians as one between traditionalists and modernists or between "those who thought traditional medicine could be modernized and those who did not."[56]

K.C.K.E. Raja of the Madras Public Health Services asserted: "The methods of patient research into the causation, treatment and control of diseases pursued by Western medicine must remain the sheetanchor for man in his fight against ill-health…the Indian systems represent a stage of arrested development, whilst Western medicine is dynamic with all the life of a growing organism….even legitimate pride in the nation's past should not be permitted to keep India from the broad

road of orderly progress along which other nations of the world are marching towards a slow but increasing mastery over man's bodily ills."[57]

For the traditionalists, the recognition that the survival of Ayurveda was dependent on adopting some elements from Western medicine was also accompanied by the fear that Ayurveda would be swallowed up by western medicine if care was not taken to let it stand on its own as well. They advocated a more cautious approach to synthesis and cooperation. For example, an article in the September 1927 issue of the JAHSM while praising the Punjab Government's decision to initiate research on Ayurvedic therapeutics and dietetics argued that it should be conducted on modern lines in "Ayurvedic institutions where the very spirit of Ayurveda…is readily grasped and absorbed by research workers and such a procedure will be productive of better results, not in institutions attached to modern medical schools and colleges where the atmosphere is so un-Ayurvedic."[58]

Members of the Madras Ayurveda Sabha found that the Indian School of Medicine paid mere lip service to Ayurvedic study as "the value of the little knowledge of Ayurveda which the pupils got was lost like a drop of water in the sea of foreign studies and foreign methods of treatment with which the students were compelled to become familiar."[59] They urged the Government of Madras to institute reforms that would make it possible for interested students to study Ayurveda for its own sake and further be able to disseminate the knowledge thus gained "in a style attractive even to the modern world without outside interference."[60] Ayurvedic spirit should also inform its practice so that "Ayurvedic hospitals are not to be like officialised organisations, where grudging words, grim eyes, dry performance of duty, mercenary considerations, and alarmed-looking patients are the glaring features."[61] The author argued that Ayurveda could be modernized; implicit in this was the recognition that modernization was desirable and inescapable, and that it could occur meaningfully only when Ayurvedic research was isolated from Western medicine, for Western medicine was not only very hostile to Ayurveda but also very powerful.

A. Lakshmi Pathi, Ayurvedic practitioner from Madras, dispelled the fears that "the glamour of the Western system of Medicine would supplant the Ayurvedic and the Unani Systems instead of supplementing them"[62] in two ways: first by arguing that independence would usher in a "great period of renaissance"[63] when the "true value of the Indian Sciences"[64] would triumph over the glamour of Western medicine; second by arguing that any contribution that Western medicine may make to Ayurveda during their association would "be absorbed imperceptibly and may not probably be recognisable after some time."[65]

Medical registration was a key site of acrimony among Indians as it increasingly became the legitimate way for medical practitioners to gain recognition and status. According to Roger Jeffrey, it was a key political issue in the inter-war period.[66] The denial of registration to practitioners of indigenous systems of medicine by the Madras Medical Registration Act of 1914 was seen as gross discrimination: "[I]ndigenous practitioners who served about eighty percent of the people of the land were being treated as untouchables of the profession by the allopathic practitioners who considered themselves as 'seraphi illuminati'."[67] However, registration of medical practitioners also became a key site for contention among Ayurvedists, attesting to Panikkar's contention that "[t] he quest to revitalize indigenous medicine reflected a multi-pronged struggle for cultural hegemony, not only between the coloniser and the colonized, but also between classes within the colonised society."[68]

Graduates of the Madras Ayurvedic College, a private institution established in 1909 were pitted against graduates of the Government Indian Medical School established in 1924 over the issue of registration as 'A' class or 'B' class practitioner. Criticizing the attitude of L.I.Ms (Licentiate in Indian Medicine, a diploma granted by the Government Indian Medical School), N. Kesavacharlu, Secretary of the Madras Ayurvedic College Graduates' Association implored:

"It is beneath the dignity of an Indian medical practitioner to thus scandalize others of the same fold having known that some A.M.A.Cs [Associates of Madras Ayurveda College] are professors and demonstrators in their School….Let the L.I.Ms understand that 'A' class is not the prerogative of the Indian Medical School. It shall be open to all deserving, be they of private institutions or of Governments."[69]

Moreover, 'B' class was unacceptable since it put them on a par with hereditary practitioners of Indian medicine who had no institutional training but who were also granted 'B' class.[70]

This struggle reveals another division over and above the traditionalist-modernist one identified by Crozier. While graduates of the Government Indian Medical School fitted Crozier's definition of traditionalists for the School's goal was to produce future medical practitioners with the knowledge and training that will enable them "to bring to bear on the problems of Health and ill-health not only the expert knowledge of their own systems but, as far as practicable, the best that is in other systems also,"[71] the graduates of the Madras Ayurvedic College can be classified as proponents of 'pure' Ayurveda as they were products of an institution that had come into existence "at a time when Ayurveda was in oblivion" to demonstrate to the world the "scientific nature of the system".

Thus, purists, syncreticists, and modernists vied with each other for a share in India's medical future.

In the long run, the syncreticists with their middle-of-the-road approach seem to have influenced government policy on indigenous medicine towards the closing years of British rule and during the inaugural period of independent India.[73] As Crozier argues, the profound cultural crisis engendered by Western technical and material superiority made medicine, like religion and women's status, an important site of constructing cultural and national identity--an endeavour that had to take into account ideas about science, progress, and modernity.[73] Thus, argument for synthesis between the two systems that would preserve the core of the traditional system triumphed over total rejection of either system, for purists envisaged a complete turning back of the clock by returning to a pure Ayurveda and modernists saw no role for Ayurveda in India's triumphal march towards science, progress, and modernity. Just as the various reform movements, in response to the colonial 'gaze', adopted and adapted western idioms of rationalism, progress, modernity, privileged textual material over customary practices, and attempted to institutionalize traditions through law and the state,[74] the Ayurvedic revival movement laid claim to history and science, privileged Sanskritic textual material in keeping with British orientalism, and fought for state recognition and support.

Ayurveda, in spite of its imbrication in a Hindu cultural nationalist vision, found a place in the post-independence policies of the Government of India due to its appeal to science, modernity, and progress. What Poonam Bala argues in the case of the colonial period—that it was a medical "oligopoly"[75]—is also true of independent India, revealing a success story of sorts for the Ayurvedic revival movement. But, as Padma Srinivasan and Samshad Khan argue, this has to be weighed against the Structural Adjustments Programmes (SAP) of globalization that increases India's dependence on international organizations dominated by Western governments such as the World Bank for support in the area of healthcare.[76] Also, as Jeffrey argues, many of the issues that animated the medical political landscape during the colonial period continue to define medical policy-making now such as 'pure' vs. integrated training, registration of indigenous practitioners, and drug control,[77] revealing a continuity and not a once-for-all shift in policy in favour of Ayurveda after independence. However, the syncretist model that found a place in the immediate post-independence medical policy is poised for discursive stability and longevity.

References

1. Guy N. A. Attewell. *Refiguring Unani Tibb: Plural Healing in Late Colonial India.* Hyderabad, India: Orient Longman, 2007.; Seema Alavi. *Islam and Healing: Loss and Recovery of an Indo-Muslim Medical Tradition, 1600-1900.* New York: Palgrave Macmillan, 2008.
2. Richard S. Weiss. *Recipes for Immortality:Medicine, Religion, and Community in South India.* New York: Oxford University Press, 2009.
3. K.N. Panikkar. *Culture, Ideology, Hegemony: Intellectuals and Social Consciousness in Colonial India.* London: Anthem Press, 2002, 165.
4. The Indian National Congress, the organization that spearheaded the nationalist movement, was founded in 1885 in Bombay.
5. For an account of Hindu revivalist nationalism, see Tanika Sarkar, *Hindu Wife, Hindu Nation: Community, Religion, and Cultural Nationalism* (Bloomington, IN.: Indiana University Press, 2001).
6. Panikkar, *Culture, Ideology, Hegemony*, 150-151.
7. David Kopf. *British Orientalism and the Bengal Renaissance: The Dynamics of Indian Modernization, 1773-1835.* Berkeley: University of California Press, 1969.; Tony Ballantyne. *Orientalism and Race: Aryanism in the British Empire.* New York: Palgrave, 2002.; Thomas R. Trautmann. *Aryans and British India.* Berkeley: University of California Press, 1997.; Michael S. Dodson. *Orientalism, Empire, and National Culture: India, 1770-1880.* New York: Palgrave Macmillan, 2007.
8. See: Kenneth Jones. *Arya Dharm: Hindu Consciousness in 19th Century Punjab.* Berkeley: University of California Press, 1976.; David Lelyveld. *Aligarh's First Generation: Muslim Solidarity in British India.* Princeton, N.J.: Princeton University Press, 1977.; Barbara Metcalf, *Islamic Revival in British India: Deoband, 1860-1900.* Princeton, N.J.: Princeton University Press, 1982.; Harjot Oberoi. *The Construction of Religious Boundaries: Culture, Identity, and Diversity in the Sikh Tradition.* Chicago, IL.: University of Chicago Press, 1994.
9. Bhagvat Sinh Jee. *A Short History of Aryan Medical Science.* Gondal: Shree Bhagvat Sinh Jee Electric Printing Press, 1927., First Published in 1895, 22-23.
10. Nagendra Nath Sen Gupta. *The Ayurvedic System of Medicine or an Exposition, in English, of Hindu Medicine as Occurring in Charaka, Susruta, Bagbhata, and other Authoritative Sanskrit Works, Ancient and Modern, Vol. 1.* New Delhi: Logos Press, 1984. First Published in 1919.

11. See "Introduction: Histories of Empire, Histories of Knowledge," in Dodson, *Orientalism, Empire, and National Culture,* 1-17.
12. D. Chowry Muthu. *A Short Account of the Antiquity of Hindu Medicine and Civilization.* London: Bailliere, Tindall and Cox, 1930, 5-6.
13. D. Chowry Muthu. *A Short Account of the Antiquity of Hindu Medicine and Civilization.* London: Bailliere, Tindall and Cox, 1930, 8.
14. Ibid, 40.
15. Shiv Sharma. *The System of Ayurveda.* Delhi: Neeraj Publishing House, 1929, 90.
16. Shiv Sharma. *The System of Ayurveda.* Delhi: Neeraj Publishing House, 1929, 68.
17. The Society for the Resuscitation of Indian Literature, *Ayurveda or the Hindu System of Medical Science.* Calcutta: H. C. Dass, 1899, 12.
18. Ibid, 1.
19. David Arnold. "A Time for Science: Past and Present in the Reconstruction of Hindu Science, 1860-1920" in *Invoking the Past: The Uses of History in South Asia, ed. Daud Ali.* New Delhi: Oxford University Press, 1999, 157.
20. Gyan Prakash. "The Modern Nation's Return in the Archaic," *Critical Inquiry*, vol. 23, no.3, Spring 1997, 539.
21. M.M. Gananath Sen, *Hindu Medicine: An Address on Ayurveda Delivered at the Foundation Ceremony of Benares Hindu University in 1916.* Calcutta: Kalpataru Palace, 1937, 23.
22. Ibid, 11.
23. Sen. "The Scientific Basis of Ayurveda: An Address Delivered before the South Indian Medical Union, Madras (1923)", in *Lectures of M.M. Gananatha Sen Saraswati.* Varanasi: Chowkhamba Sanskrity Series Office, 2002, 2.
24. Jivaram Kalidas Shastri. *Presidential Address: 31st All India Ayurved Congress, Lahore, December 1942.* Gondal, 2nd Edition, 12.
25. See Jones, *Arya Dharm.*
26. David Arnold. *Science, Technology and Medicine in Colonial India.* Cambridge University Press, 2000, 179.
27. See Lata Mani. "Contentious Traditions: The Debate on Sati in Colonial India," in *Recasting Women: Essays in Indian Colonial History, ed.* Kumkum Sangari and Sudesh Vaid. New Jersey: Rutgers University Press, 1999, 88-126.
28. For other articulations of the ayurvedic past, present and future, see Kavita Sivaramakrishnan, "The Use of the Past in a Public Campaign: Ayurvedic Prachar in the Writings of Bhai Mohan Singh Vaid," in *Invoking the Past: The Uses of History in South Asia*, ed. Daud Ali. New Delhi: Oxford University Press, 1999, 178-191 and Charu Gupta, "Procreation and Pleasure: Writings of a Woman Ayurvedic Practitioner in Colonial North India," *Studies in History*, vol 21, no.1, February 2005, 17-44.
29. G. Srinivasa Murti. "Our Aims and Ideals", *The Journal of Ayurveda or the Hindu System of Medicine (JAHSM)*, vol. 1, no. 1, July 1924, 10.
30. "Mr Pilcher and Ayurveda", *JAHSM*, vol. 4, no. 3, September, 1927.
31. M. R. Samey, "Hooton Hoots out Ayurveda", *JAHSM*, vol. 4, no. 4, October, 1927, 121.
32. M. R. Samey, "Hooton Hoots out Ayurveda", *JAHSM*, vol. 4, no. 4, October, 1927, 122.
33. K.P. Sankara Pillai. "Ayurveda and some Western Medical Sciences," *JAHSM*, vol. 10, no. 4, October, 1933, 221.
34. Murti, "Our Aims and Ideals," *JAHSM*, vol. 1, no. 1, July 1924, 10.
35. "Pilcher", *JAHSM*, 4 (3), September, 1927.
36. Girindranath Mukhopadhyaya. *Ancient Indian Surgery: Surgical Instruments of the Hindus with a Comparative Study of the Surgical Instruments of the Greek, Roman, Arab, and the Modern European Surgeons, vol. 1.* New Delhi: Cosmo Publications, 1994.
37. C.G. Mahadeva. "Necessity of Introducing Western Surgery in Ayurveda," *JAHSM*, 6 (7), January 1930, 241-242.
38. "The Science of Medicine," *JAHSM*, 8 (5), November, 1930, 164.
39. "Science versus Empiricism," *JAHSM*, 9 (3), September 1932, 84.
40. Sen. *Address at Benares Hindu University*, 25.
41. Deepak Kumar. "Unequal Contenders, Uneven Ground: Medical Encounters in British India, 1820-1920," in *Western Medicine as Contested Knowledge*, ed. Andrew Cunningham and Bridie Andrews. New York: Manchester University Press, 1997.
42. "Scientific vs. Practical Medicine," *JAHSM*, 4 (12), June, 1928, 444.
43. Charles Leslie. "The Professionalization of Ayurvedic and Unani Medicine," *Transactions of the New York Academy of Sciences*, Ser. II, 30 (4) (February, 1968), 559-572.
44. M.R.Samey. "Medicine, Past, Present and Future or Ayurvedic Giants of Yesterday, Allopahic Generals of Today, and Heirs Presumptive of Medicine To-morrow," *JAHSM*, 10 (4), October, 1933, 126-127.
45. A. Raman, "I.M.S. and Ayurveda," *JAHSM*, 10 (5), 162.
46. M.K. Mukherjee. "Nation's Health," *JAHSM*, 11 (10), April, 1935, 384.
47. "The Ancient Systems of Medicine," *Indian Medical Gazette*, vol. 60, December 1925, 587-588.
48. Ibid, 587.
49. Ibid.

50. "Pilcher. "*JAHSM,* 4 (3), September 1927, 82.
51. "Indigenous Systems of Medicine in India: Ayurveda, Siddha, Unani," *The British Medical Journal*, September 15, 1923, 479.
52. Ibid, 480.
53. "Based on False Theories: Ayurvedic System, Surgeon General's Criticism," *JAHSM*, 4 (4), October 1927, 160.
54. W.D.Sutherland. "Charaka Samhita," *The Indian Medical Gazette*, 54, February 1919, 89.
55. "Indigenous Drugs," *The Indian Medical Gazette*, 54, March 1919, 101.
56. Ralph Crozier. "Medicine, Modernization, and Cultural Crisis in China and India," *Comparative Studies in Society and History,* 12 (3), July, 1970, 283.
57. K.C.K.E. Raja. "A Plea for a Forward Public-Health Policy in India," *The Indian Medical Gazette*, 72, July 1937, 434.
58. "Punjab Government and Indigenous Medicine," *JAHSM*, 4 (3), September 1927, 84.
59. "Medical New and Notes: Teaching of Indian Medicine in Madras," *Journal of the Indian Medical Association,* 1 (3), November, 1931, 111.
60. Ibid.
61. M.K. Mukherjee. "Whither Ayurveda!" *JAHSM,* 10 (3), September 1933, 83.
62. A. Lakshmi Pathi. *The Future of Ayurveda: Opening Address at the Round Table Conference for the Synthesis of the Medical Systems of India* (New Delhi, May 1947), 19.
63. A. Lakshmi Pathi. *The Future of Ayurveda: Opening Address at the Round Table Conference for the Synthesis of the Medical Systems of India* (New Delhi, May 1947), 20.
64. Ibid.
65. Ibid, 21.
66. Roger Jeffrey. "Recognizing India's Doctors: Institutionalization of Medical Dependency, 1918-39," *Modern Asian Studies*, 13 (2), 1979, 301-326.
67. Medical News and Notes: Recognition of Indian Systems of Medicine," *Journal of the Indian Medical Association*, 1 (6), February, 1932, 243.
68. Panikkar, *Culture, Hegemony, Ideology*, 175.
69. N. Kesavacharlu. "Medical Registration: The Sad Plight of Non-L.I.Ms," *JAHSM,* 9, May, 1935, 402-403.
70. N. Kesavacharlu. "Medical registration: Appeal to A.M.A.Cs," *JHASM,* 9 (10), April 1935, 361-362.
71. G. Srinivasa Murti. "The Government Indian Medical School, Madras," in *Handbook of The Indian Science Congress, Twenty-Seventh Session.* Madras: G.S.Press, 1940, 86.
72. Kesavacharlu. "Appeal to A.M.A.C.'s," *JAHSM.*
73. Crozier. *Medicine, Modernization, and Cultural Crisis.*
74. Kopf. *British Orientalism*; Jones. *Arya Dharm*; Lelyveld. *Aligarh's First Generation*; Metcalf. *Islamic Revival in British India*; Oberoi. *The Construction of Religious Boundaries.*
75. Poonam Bala. "Indian and Western Medicine: Rival Traditions in British India," in *Colonialism and Psychiatry,* ed. Dinesh Bhugra and Roland Little. Delhi: Oxford University Press, 2001.
76. Padma Srinivasan. "National Health Policy for Traditional Medicine in India," *World Health Forum*, vol. 16, 1995, 190-193; Samshad Khan. "Systems of Medicine and Nationalist Discourse in India: Towards 'New Horizons' in Medical Anthropology and History," *Social Science and Medicine*, 62, 2006, 2786-2797.
77. Roger Jeffrey. "Policies Towards Indigenous Healers in Independent India," *Social Science and Medicine*, 16, 1982, 1835-1841.

3

The State of Ayurveda in the Eighteenth and Nineteenth Centuries*

N. Gangadharan

Introduction

Ayurveda flourished well under the patronage of the Hindu rulers up to the advent of the Mohammedans in the 11th century. Then there was a succession of foreign rulers belonging to one dynasty or other until the advent of the British rule. During the Mohammedan rule royal patronage was extended to the Unani system of medicine. There were many physicians belonging to this system and many treatises were written by eminent men of those days relating to this system.[1] There also appeared Persian translations and adaptations of standard treatises on Ayurveda. After the death of Aurangazeb in 1707 A.D., the Mughal empire dwindled in size. Then the British were able to establish themselves firmly after overcoming the rival Western forces. The British rulers showed no interest in the preservation and propagation of the indigenous systems of medicine either Unani or Ayurveda. Hence, both these systems had suffered some setback.

Inspite of this, Ayurveda continued to be popular with the natives and treatises were being written. A fillip to these compositions was of course due to the patronage extended by the surviving Hindu stately kingdoms. That is why probably we find that even today Kerala, Gujarat and West Bengal have got flourishing Ayurvedic institutions. In the following pages we shall first make a survey of the medical works written during this period, assess the impact of the Western thought and then notice the contribution of Western Indologists in the recognition of this system.

Works Belonging to the Eighteenth Century

Under the patronage of Anandaraya-makhin, minister of King Sahaji (1684-1710) and Serfoji (1710-1728) of Tanjore, Vedakavi composed an allegorical play in Sanskrit in 7 acts called *Jivanandanam*[2] cleverly combining the *Advaita Vedanta* and the fundamentals of Ayurvedic science[3]. It deals with the ultimate bliss of the soul which results from the duel between the two rival forces, *viz.* disease on one hand and the human body on the other. The metaphor of the king and his enemies has been brought in here. Finally, Lord Siva manifests to the king and imparts the wisdom of *yoga*, the true knowledge of the essence of God and the self.

Raghunathapandita, a resident of Chamavati, modern Choul in Kolaba district of Bombay, composed his *Vaidya-vilasa* and *Cikitsamanjari*[4] in the beginning of the 18th century. These two works are handy guides of physicians.

We have then the work of Madhava Upadhyaya composed about the same period. Although he hailed

-----

* This chapter first appeared in Indian Journal of History of Science, 1982, 17(1), 154-163.

from Saurashtra he spent a greater part of his life at Kasi. He was an exponent of alchemy (*rasasastra*). His *Ayurvedaprakasa* deals with all the *rasa-samskaras* such as purification, incineration and use of all minerals and metals. He also asserts about the efficacy of the methods suggested as they have been personally tried.

Then we have the *Rajavallabha,* a lexicon in 6 chapters composed about the same period, describing good habits, properties of articles of food and drink, *etc.* This is more a handbook on personal hygiene than a *materia medica.*

Keladi Basavaraja composed his *Sivatattvaratnakara*[5], an elaborate encyclopaedic work in 1709 A.D. It is divided into 9 *kallolas,* each *kallola* being sub-divided into *angas*; four kinds of treatment; qualities of a physician; the *dosas* in the body; the time taken for digesting various kinds of foods; six kinds of tastes; their nature and effects, diagnosis, parts of the body and the things to be examined; various kinds of pulsebeats; how pulse works in different diseases; pulsebeats in various living beings; the cause of windiness; the cause of biliousness; kinds of fever and their effects; the things which reduce wind, bile and phlegm; treatment and drugs which produce various effects in the body; seasons suitable for using various kinds of medicines; weights, measures and doses; the qualities of food stuffs and herds and the preparations of medicine; how to test whether the preparations are satisfactory or not; doses and the duration of potency of various classes of medicines; mercury (*rasa*) and treatment by mercurial preparations; mica, pyrites (*maksika*), cowrie (*varati*), blue vitriol, lapis-lazuli, realgar, red chalk (*gairika*), yellow-orpiment, red lead, arsenic, antimony and mercury; their place or origin, nature, qualities, colour, uses, *etc.*; purification of mercury (*rasa*); its uses and effects when combined with other things; various methods of conversion of base metals into silver and gold; chemical laboratory and the proper arrangement of articles in it; serpents, their varieties and nature, life-period and changes at different stages; how to determine the kind of serpent which has bitten a person; the place affected by biting; the question of survival of the person bitten; poisonous bites by rats and spiders, and treatment by drugs.

We now enter a period where the contact of foreigners from Europe has been great. Even during the period of the Ayurveda writer Bhavamisra, *i.e.* in the 16th century, foreigners, mainly the Portuguese had come to India. Along with them came the syphilis, which was then referred to as *phirangiroga* because it was unknown to early writers on Ayurveda. Although Bhavamisra had referred to it in his *Bhavaprakasa* and has mentioned *Tob-chini* (China root) as its remedy, only the anonymous *Yogaratnakara* composed prior to 1746 A.D. has more references to the name and use of *Copa-cini*-such as *Copa-ciniprakasa, Copacini-curna, etc.* under treatment for *upadamsa.* But the *phirangaoganidana* of *Bhavaprakasa* does not find a place here.

This interesting work refers to many more foreign materials such as *Birija* and *Kabab* which are Unani terms[6]. This is the first Ayurvedic treatise which refers to tobacco and its uses, although tobacco had been introduced in India in the 15th-16th centuries. Tobacco leaves were kept folded under the aching tooth for relief from pain.

Although references were found in these earlier works to *Copacini,* only in the 19th century, the *Copaciniprakasa* was compiled under the patronage of the famous Ranjit Singh[7] (1780-1839). The New Catalogus Catalogorum records[8] the availability of 2 manuscripts of this work. Perhaps the *Cocaniprakasa*[9] ascribed to Madhusudana Sarman Gosvamin available in Alwar Library is same as this work. There is another work on this subject, namely, the *Covacinisevanavidhi.*[10]

Mir Jafar, the Nawab of Bengal (1757-1762) had a personal physician by name Ramasena Kavindramani. He composed commentaries on the *Rasendrasarasamgraha* of *Gopalakrsna*[11] and the *Rasendracintamani* of Ramacandra Guha.

The following works on medicine were composed during the 18th century. Govindadasa's *Bhaisajyarantnavali*[12] gives a collection of recipes. Herein we find references to new diseases like *vrkkaroga* (kidney diseases), and *mastiskaroga* (brain diseases). This has a Sanskrit commentary by Narendrnatha Misra of Lahore. *Rajavallabhiyadravyaguna*[13] of Narayanadasa was composed in 1760 A.D. *Prayogamrta*, the most extensive work on therapy was composed by his pupil Vaidyacintamani. Dhanapati's *Divyarasendrasara* and *Narayana's Vidyamrta* also belonged to this peiod.[14]

Works Belonging to the Nineteenth Century

The first half of the 19th century is quite important in the history of Indian medicine. The Mahratta king o Tanjore, Raja Serfoji (1798-1832), had deep interest in Indian medicine. Equipping himself with a full knowledge of the Western medicine and the native Ayurveda, he tested all indigenous recipes by actual administration to patients in a hospital where both English and Indian doctors worked in harmony. He selected some four thousand prescriptions as the most efficacious ones and had them written in Tamil verse form, classified into 18 volumes according to the diseases they relate to. These volumes appeared under the general title *Sarabhendra Vaidya murai*[15]. In one of the volumes relating to skin diseases, we find the method of treating cancer successfully.

About 25 titles of works composed in the 18th and 19th centuries are known from the historical books on Ayurveda[16]. A few works which have some special theme in them may be mentioned. *Paradakalpadruma* dealing with the use of mercury was composed by Ananta in the year 1792 A.D. *Vaidyakasarasamgraha* was composed in 1734 A.D. by Srikantha Sambhu. The *Vaidyavinoda*[17] of Sankarabhatta was composed in 1705 A.D. under the patronage of King Ramsingh of Jaipur. Godbole's *Nighanturatnakara*[18] and Sridattaram Chaube's *Brhannighnturatnakara* mention pineapple, tobacco and the examination of urine, while the latter has also given equivalents of the names of medicines in other Indian languages and English.[19]

Some Great Exponents of Ayurveda in the 19th Century

Some of the exponents of Ayurveda born in the 19th century made great name and established flourishing institutions of Ayurveda in the early part of the 20th century.

Kaviraj Ganga Prasad Sen[20], a great physician of Bengal was the editor of the magazine '*Ayurveda Sanjibani*' in Bengali. He had distinguished men as his patients. He manufactured medicines and exported them. He was the upholder of the dignity of the Kavirajas. He was honoured by Queen Victoria in 1877 with a Service Medal. Mahamahopadhyaya Bejoyratha Sen was his disciple. He revised and edited *Astangahrdaya* of Vagbhata. Kaviraj Hari Mohan Dasgupta and Kaviraj Kalish Chandra Sen were also his disciples. Kaviraj Amritalal Gupta, author of *Ayurvedasiksa* was the nephew and disciple of Kaviraj Kalish Chandra Sen. Mahamahopahdyaya Kaviraj Gananath Sen, M.A., L.M.S. Saraswati was he son and disciple of Biswanathe Kalpadruma of Banaras. Kaviraj Gananath Sen compiled *Pratyaksasarira*, *Siddhantanidana* and *Ayurveda paricaya* (in Bengali). He established the Kalpataru Ayurvedic works-a manufacturing concern. Kaviraj Jyotish Chandra Saraswati of Bengal wrote a commentary on *Susrutasamhita* which is now lost. Kaviraj Gananath Sen was the founder of "Viswanayh ayurveda Mahavidyalaya of Caclutta.' Kaviraj Jamini Bhusan Roy, M.A., M.B. was the founder of the 'Astanga Ayurveda Vidyalaya' of Calcutta.

Kaviraj Gangadhar Roy[21] (1798-1865 A.D.) often called as Gangadhar Kaviraj was the son of Bhavani Prasad Roy of Bengal. He was a great scholar, physician and unique teacher of Ayurveda who dedicated himself to the resuscitation of Ayurveda. He is a profile writer. Among his forty works, about a dozen works were on Ayurveda, among which the following may be mentioned as important. He commented on the medical chapters of the Agnipurana and on the Carakasamhita. He also composed the Pathyapathya[22], Bhaskarodaya[23] on pathology and Vaidyatattvaviniscaya[24].

He had brilliant galaxy of direct disciples (who had spread the science of Ayurvedic treatment throughout India) of whom the following may be mentioned together with others.

His most senior direct disciple was Kaviraj Gayanath Sen of Purulia village in the Birbhum district of Bengal. He was a great scholar and successful physician. His son Kaviraj Sitanath Sen was also a great physician of Ayurveda. Mahamahopadhyaya Dwarakanath Sen was another great direct disciple of Kaviraj Gangadhar Roy and a famous Ayurvedic physician. Kaviraj Jogindranath Sen, M.A., son of Dwarakanath Sen was a great scholar. He wrote a commentary on the *Carakasamhita* called *Carakopaskara*. He was honoured with the title '*Baidyaratna*' by the then Government for his scholarship. Kaviraj Parash Nath Sen of Banaras was another direct disciple of Kaviraj Gangadhar Roy. Kaviraj Rajendranath Sen of Calcutta, Kaviraj Jadunath Bhattacharya of Pubna, Kaviraj Gobinda Chandra Roy of Murshidabad and Kaviraj Sricharan Sen are all direct disciples of Kaviraj Gangadhar Roy. Kaviraj Haran Chandra Chakraborti, another disciple of Gangadhar Roy, was a great scholar, physician and surgeon of Ayurveda. He wrote a commentary *Susrutarthasandipan* on *Susrutasamhita*.

Haran Chandra Chakravorti, a student of the preceding scholar belonged to late 19th and early 20th century. He performed different types of operations as per Susruta. His commentary on Susruta, although the latest, was acclaimed on account of its practical approach.

Zandu Vittalji Bhat, born at Kathiawad in 1831, served as a physician to Jamsaheb of Navnagar. He started his *rasasala* at Jamnagar in 1865 which was later shifted to Baroda. This is now famous as the Zandu Pharmaceuticals.

H.H.Bhagvat Sinhji, the Maharaja at Gondal, born in 1865, studied medicine at Edinburgh University. He had a great ambition to expound the medical heritage of India to the minds of Eastern and Western scholars. His doctoral thesis, "A short History of Aryan Medical Sciences" secured him Fellowship of the Royal College of Physicians, Edinburgh. This is a very good reference book.

We may make a passing reference to Umeshchandra Datta, the Chief Librarian of the Government Sanskrit College, Calcutta, who compiled the first Ayurvedic Dictionary- *Vaidyasabdsasibdhu* under the inspiration of the then Vice-Chancellor of the Calcutta University.

Shankardaji Shastry Pade, born in 1867 in a village near Poona, scholar in Sanskrit and Ayurveda published ayurvedic magazines. He was able to persuade the Maharaja of Baroda to start an Ayurvedic College.

P. S. Warrier of Kerala, born in 1869, learnt Ayurveda and gained a working knowledge of the Western medicine. He founded the Aryavaidasala at Kottakkal in 1902. He wrote many works relating to Ayurveda.

Jeevaram Kalidas Shastry of Gujarat, was proficient in Ayurveda, especially in *rasasastra, mantravidya, yoga etc.* He lived at Bombay for some time and finally established himself as an ayurvedic physician at Gondal. He became the royal physician of the King of Gondal. He founded the *Rasasla ausadhalaya* in 1910. He had a good collection of manuscripts of texts on Ayurveda. He translated many Sanskrit works of Ayurveda into Gujarathi.

H.H. Kerala Varma (1864-1944), the Elayaraja of Cochin was well versed in Ayurveda. He translated two works on Toxicology into Malayalam. He compiled the *Visacikista* on the basis of Charaka, Susruta and Vagbhata.

The later part of the 19th century witnessed the effects of Indian writers to absorb material from the Western medical theories and to translate the Sanskrit medical works into the regional languages. An early example for the former is the attempt of Dr Bhaskar Govind Ghanekar in his *Susrutasamhita.*[25] to cite one example: He has used *svarnalavana* in the place of gold chloride.

Like the *Hindu Chemistry* of Acharya Prafullachandra Roy, Sir B.N.Seal compiled *Positive Sciences of the Ancient Hindus* containing themes, matters, chemistry *etc.* of Ayurveda. All those referred to above were born in the 19th century but shed lustre to both 19th and 20th centuries except Kaviraj Gangadhar Roy.

Western Studies on Ayurveda

We may now make a brief survey of the contributions of the Western indologists and physicians to the cause of Indian medicine.[26]

It was Sir William Jones who wrote the first article *"On the cure of elephantiasis and other disorders of the blood"*, being a translation from the Sanskrit original in 1785 and "*The design of a treatise on the plants of India*" in 1789. Richard Miller wrote his "*Disquisitions in the History of Medicine*" devoting a section to the conditions of healing in Hindustan in1811.

In 1823, H.H.Wilson wrote an article entitled "*On the Medical and Surgical Sciences of the Hindus*".

Martin Honigberger, born in Transylvania in 1795, was a court physician of Ranjit Singh in 1829. In his autobiography "Thirty five years in the East" published from London in 1852, we find that he has compiled a dictionary of medicinal plants, which he described in details, giving their names in various languages. Similarly, a German physician, Johannes Gerhard Konig (1728-85) from Kurtland in Baltic combined medical and botanical research.[27]

With the German physicians living and working in India during the 19th century interested in Indian medicine, it received a powerful stimulus. One of the first to study healing practices in India in general and their homoeopathic aspect in particular, was Rudolf Roth (1821-95). He paid his attention to *Madanavinoda*, a work dealing with the flora, fauna and medicinal remedies and to the earlier collection, *Carakasamhita*, of which he published a selection.[29] Similar to Miller's work is the *History of Medicine, Surgery and Anatomy* of William Hamilton published in 1831.

After Roth, Ernst Haas[29] (1835-82) made a study of the secrets of Indian medicinal treasures. In 1837, J.F. Royle wrote an article "*The Antiquity and Independent Origin of Hindu Medicine.*"

Renouard gave a brief account about the theory and practice in his "*Medicine of the Oriental Indians*" in 1836, translated by C.G. Cornegys in 1856. Then followed a comprehensive treatise called "*A Commentary on the Hindu System of Medicine*" by T.A Wise in 1845. Robley Dunglison wrote his "*History of medicine from the earliest ages to the commencement of the 19th century*" in 1872. Hermann Bass wrote his "*Outlines of the History of Medicine and the Medical Profession*". It was translated by H.E. Handerson in 1889. This book makes an analysis of the Indian medical thought and praises the standard of perfection achieved by this system.[30] Edward Berdose wrote on the "Origin and Growth of the healing Art" in 1893. It was followed by Edward Theodore Withingdon's "*A Popular History of the healing Art*" in 1894.

Discovery of much Importance

The 19th century was important in one respect that search for manuscript of Sanskrit texts were made and catalogues of collections of manuscripts in the possession of institutions and private individuals were prepared. One of the benefits of this continuous search was the discovery in the year 1890, of the *Navanitaka* (forming a part of the manuscripts discovered by Bower) in a Buddhistic *Stupa* in Kashgar (China) which gives valuable information about the early existence of Indian medicine. It is supposed to be the cream of all other earlier texts. The date of this is fixed in the 4th century.[31]

The Practical Aspect

We have an interesting account 32 of the Banian Hospital at Surat instituted to take care of injured and old animals that existed in the second half of the 18th century.

Plastic Surgery

Susruta's technique of making the new nose was being practised in India till recently. Description of this operation was published in the Gentleman's Magazine in 1794 (Brown, J.B. and McDowell, F. (1965), (Plastic Surgery of the Nose. Harles, C. Thomas, U.S.A., p.5), and after that a leading surgeon of London, Joseph Constantine Carpue published his results in 1816, which led to the worldwide publicity for the technique. Later, many modifications of the technique were adopted and a new science of plastic surgery took birth.[33]

From another source[34] we learn that the Indian medical men seem to have made considerable use of surgical techniques in different parts of India. According to Colonel Kyd "in Chinigrey (in which they are considered by us the least advanced) they often succeed, in removing ulcers and cutaneous irruptions of the worst kind, which have baffled the skill of our surgeons, by the process of inducing inflammation and by means directly opposite to ours, and which they have probably long been in possession of".

This is corroborated by the letters[35] of Dr Helenus Scott written to Sir Joseph Banks, President of Royal Society, London on January 19, 1792 on the prevalence of plastic surgery in Western India: "They practice with great success the operation of depressing the chrystalline lens when become 'opake' (opaque) and from time immemorial they have cut for the stone at the same place which they now do in Europe."

Inoculation

From some accounts[36] we find that "Inoculation against the smallpox seems to have been universal, if not throughout, in large parts of northern and southern India, till it was banned in Calcutta and other places under the Bengal Presidency (and perhaps elsewhere) from around 1802-03. The most detailed account of the practice of inoculation against the smallpox in India is by J.Z. Holwell written by him (in 1767 A.D.) for the College of Physicians in London.

Earlier Ro. Coult wrote to Dr Oliver Coult in "*An Account of the Diseases of Bengal*' (dated February 10, 1731) about the operation of inoculation of the smallpox as performed in Bengal.[37] It was called '*Tikha*' by the natives. The method is as follows:

"They take a little of the pus (from the mature pox) and dip in it the point of a pretty large sharp needle. Several punctures are made with this in the hollow under the deltoid (delloid?) muscle or sometimes in the forehead. They are then covered with a little paste made of boiled rice. This commonly features and comes to a small supporation and if not the operation has no effect and the person is still liable to have the smallpox. But if the punctures suppurate and no fever or eruption ensues then they are no longer subject to the injection."[38]

Teaching of Ayurveda

Although teaching and practice of Ayurveda continued to be encouraged by the native rulers and individuals proficient in the theory imparted to the younger generations it was only in the 1827 that regular arrangements were made to teach Ayurveda at Sanskrit College at Calcutta by the British rulers.

Conclusion

The above account which is only representative but not exhaustive shows that the ancient system of Ayurveda was somehow struggling for survival inspite of the cultural, economic and social impact of alien cultures. The main reason for its survival is, of course, the dormant undercurrent of the culture and civilization of the Indian people.

Bibilography

Atridev Vidyalankar, *Ayurved ka brhat itihas*, Lucknow, 1960.

Vaidya Bhagwan Dash, *Fundamentals of Ayurvedic Medicine*, Delhi, 1978.

Dharampal, *Indian Science and Technology in the Eighteenth Century*, Delhi, 1971.

Forbes, *Oriental Memoirs*, Vol.I, London, 1834.

O.P. Jaggi, *Scientists of Ancient India*, Delhi, 1966

O.P. Jaggi, *Medicine in Medieval India*, Delhi, 1973.

O.P. Jaggi, *Indian System of Medicine*, Delhi, 1973.

Julius Jolly, *Indian Medicine*, English translation from German, C.G. Kashikar, Poona, 1951.

Walter Leifer, *India and the Germans*, Bombay, 1977 (2nd edn.).

New Catalogus Catalogorum, University of Madras.

K.R. Srikanthamurthy, *Luminaries of Indian Medicine*, Mysore, 1968.

D.V. Subha Reddy, *Western Epitomes of Indian Medicine*, Hyderabad, 1966.

References

1. Jaggi O.P., *Medicine in Medieval India*, pp. 209-22.
2. Ptd. *Kavyamala* 27.1891, 1933 (2nd edn.); *Adyar Library series* 59, 1947.
3. A detailed account may be had from O.P. Jaggi, *Scientists of Ancient India*, pp. 115-20.
4. On this and the following three works see K.R. Srikanthamurthy, *Luminaries of Indian Medicine*, pp. 82-83.
5. See O. P. Jaggi, *Indian system of Medicine*, p. 43; Dikshit, *IJHS* 4.1 and 2.p. 11 (1970) *IJHM*. 5.2.37.
6. See O.P. Jaggi, *Indian system of Medicine*, p. 44; Atridev Vidyalankar, *Ayurved ka brhat itihas*, pp. 310-13.
7. Jolly, *Indian Medicine*, pp.1-2, 3.
8. Vol.VII. p. 86b.
9. See New Catalogus Catalogorum, Vol. VII, p.85b.
10. See *ibid.*, p. 86b.
11. Ptd. Calcutta, 1915. See *New Catalogus Catalogorum*, Vol. VI. p. 135b. See also P.C.Ray, *History of Hindu Chemistry* Vol. II.pp. lxxi-lxxii.
12. Ptd. Calcutta, 1893; in Malayalam script, Trivandrum, 1935. See K.R. Srikanthamurthy, *loc. cit.*, p. 83.
13. Ptd.Calcutta, 1868.
14. On these works see Atridev Vidyalankar, *Aryuveda* ka brhat itihas, p. 322.
15. Published by the Tanjore Saraswati Mahal Library, *Tanjore since 1952*. It may be pointed out that the efficacy of the recipes were reported by the doctors who had access to these volumes after their publication.
16. See Atridev Vidyalankar, *loc. cit.*, pp. 596-98.
17. Ptd. Bombay, 1913.
18. Ptd. Bombay.
18. See Atridev Vidyalankar, *loc. cit.*, pp.602-03.
20. On this and the succeeding writers see. K.R. Srikanthamurthy, *loc. cit.*, pp. 84 ff.
21. See *New Catalogus Catalogorum*, Vol. V. p. 202 a-b.
22. Ptd. Berhampore, 1869.
23. Ptd. Calcutta, 1909.
24. Ptd. See 10, Ptd. Bks., 1938, p. 862.
25. See Atrived Vidyalankar, loc. cit., p. 600.
26. For a brief epitome of these works see D. V.Subha Reddy, *Western epitomes of Indian medicine*, Hyderabad, 1966 and O.P. Jaggi, *Indian System of Medicine*, p.232.
27. See Walter Leifer, *India and the Germans*, pp.185 ff.
28. On this writer see Walter Leifer, *ibid.*, pp.185 ff.
29. See *ibid.*
30. See also O.P. Jaggi, *Indian System of Medicine*, p. 232.
31. See Jolly, *Indian Medicine* (English translation), p. 22.
32. Forbes, *Oriental Memoirs*, Vol. I. pp. 156-57.
33. O. P. Jaggi, *Indian System of Medicine*, p. 175.
34. See Dharampal, *Indian Science and Technology in the Eighteenth century*, Intro. p. XLIII.
35. See Dharampal, loc. cit., p. 268.
36. *Ibid.* intro. p. XLIV.
37. *Ibid.*p.141.
38. It is observed that this is found even today practised by the *Malis* in Bihar. See Vaidya Bhagwan Dash, *Fundamentals of Ayurvedic Medicine*, intro. p. xi.

4

Rationalisation of Medicine in India Based on Excerpts from Presidential Address by Col Sir Ram Nath Chopra during 35th Indian Science Congress, January, 1948, Patna

Condensed by C.P. Khare

Excerpts from Presidential address during 35th Indian Science Congress, January, 1948, Patna:

Ayurveda, which is stated to have been written about 2,000 years B. C., formed the foundation stone of medicine in India. Later, *one side believed that as the Ayurveda was an inspired Science, it was not possible to improve on it further by the wisdom of man.* On the other hand, the touching of dead body was regarded as polluting and therefore sinful and the dissection of bodies was given up, the result was that *studies of such basic subjects as anatomy and physiology were altogether neglected.* The effect of ignoring these basic studies, particularly on surgery can well be imagined.

Decline of the System

With the decline of Buddhism deterioration set in all round in knowledge, teaching and practice of medicine. Not only was information regarding many drugs lost, but recognition and identification of a *large number became impossible.* This was a great loss to Ayurvedic medicine, and whilst there, undoubtedly, is reason to be proud of its glorious past, it is not possible to view its present condition without a sense of apprehension.

It has been my good fortune to come contact with eminent scholars and practitioners of Indigenous Medicine, some of whom have also studied the Western medicine. They are equally concerned with the decline and sterility of the system but consider that as there is much in common between the old and new ideas concerning the etiology, pathogenesis and treatment of disease, the restoration and development of the Indigenous system would not be difficult. The old theories of causation of disease, according to them, can be justified in the light of recent advances in scientific medicine. Some even claim that there are indications that the old Hindu medicine had knowledge of the bacterial origin of disease, and that even the role of viruses was not unknown. They point out that in their system more attention has rightly been given to the state of the soil, that is the body, than to the seed of disease, that is, the microorganisms which become engrafted on it. In recognizing the importance of this soil factor, which has now been appreciated by Modern medicine, Indian nedicine was undoubtedly centuries ahead of the Western.

The Need for Investigation

The eminent scholars and practitioners of Indigenous Medicine, however realize fully the need for investigation and research, if their system of medicine is to be brought into line with the present-day requirements. With the lapse of time and changes in environments some diseases have become modified and perhaps their etiology and pathology altered. While some of the old diseases have disappeared new ones have come in. The great advance recently made in treatment too cannot be neglected. Unless fresh knowledge is acquired, their practice would be at a grave disadvantage. Unfortunately, very little has been done so far in this direction.

Many among the orthodox Vaidyas still believe in the divine origin of Ayurveda and resent the introduction of any innovations. They have resisted all attempts at the inclusion of new ideas and even of considering the possibility of any improvements in the practice of their ancient system.

The position with regard to Tibbi (Unani) medicine, both as regards its theory and practice, is no better than that of the Ayurvedic medicine. What has been said about the present-day status of Ayurvedic medicine applies with equal force to the literature and practice of Tibbi medicine.

Prejudices should be Discarded

That there is much in the Indigenous Medicine, especially in its material medica, which can contribute to the well-being of the people of the country, is beyond any doubt. Eminent Western scholars and the medical men have borne testimony to their efficacy and usefulness. With such a background as has been described above, Indigenous Medicine, it is believed, cannot play as effective a part nor take its proper place in the present day medical relief in this country, as its exponents would claim, and the people would rightly expect them to do. They will first have to put their house in order and become cognisant of the present-day environments and their requirements. The practitioners will have to be properly trained, and unauthorised practice will have to be rigidly eliminated. *The true spirit of research and discovery will have to be inculcated and irrationalism excluded from diagnosis and treatment of disease. The discoveries which have proved effective beyond doubt in the treatment of disease must be accepted and incorporated and all inherent prejudices discarded to achieve the one sublime object of the alleviation of human suffering.*

Indigenous Medicine not Obsolete

Indigenous Medicine is being practised by a vast number of practitioners. Unfortunately, only a small portion of these have received any proper systematic training in educational institutions set up for the purpose. Of the rest a few may have acquired some knowledge through connection with families which have either practised this art for generations or through having sold medicines. By far the largest number, however, have medicines and sell these to the ignorant and credulous masses.

Many of them are wandering pedlars who carry their stock-in-trade of medicines with them from place to place. Their nostrums have the virtue of cheapness no doubt, but they cost the people very dear indeed. There is evidence to prove that they are responsible for producing much misery and suffering. But it is this very class which at the present time attends to the needs of the major portion of the population, particularly in the rural areas which are so inadequately supplied by properly trained practitioners of any system.

If the existing state of affairs is considered without prejudice, there would appear to be a good deal of justification in favour of Indigenous Medicine. It should be given a fair opportunity to overhaul itself, to discard what is useless, and to bring up-to-date in the light of modern discoveries, what is intrinsically efficient and useful in it. It is also to be remembered that the people who still use Indigenous Medicine do not all belong to the ignorant and uneducated class.

A portion of the high intelligentsia in the country, who can think for themselves, believe in its efficacy, and it is reasonable to suppose that they must be getting some benefit out of it to think in this way.

One is, therefore, forced to conclude that *Western Medicine has not attempted to understand the* Indigenous systems, and has been carried away by the inherent prejudices of the foreigners who have *hiterto controlled the destinies of medical relief in this country.* While one can understand the European medical practitioner not understanding the value of Indigenous Medicine, *it is not possible* to endorse the views of the Indian practitioners of standing, when they assert that no notice should be *taken of it (Indigenous Medicine), as it is archaic and obsolete, and therefore more or less useless.* This attitude shows a lack of appreciation of the fundamentals and practice of their own system, and makes it essential to reorient their ideas with regard to the extension of medical relief.

Rationalization Process

I have always held that some of the distinctions drawn between the various systems of medicine practised in this country and which have given rise to the prejudices in the minds of the advocates of one system against even the good points of another, are unreasonable and unscientific. The universal and cosmopolitan nature of medicine does, of course, vary according to environment and with the advance of knowledge necessary adjustments have to be made. The only solution for rationalization of medicine is the evolution of a countrywide extension of a system, which can be regarded in the words of the Bhore Committee (1946), "neither as Eastern nor Western but in a corpus of scientific knowledge and practice belonging to the whole world to which every country has made its contribution."

Regarding this question I feel that a thoroughgoing synthesis may, at present, result in the almost complete submergence of the Indigenous into the Western System. For, the Western System is based on the surer foundations of Biological and Physical Sciences, and has all the recent facilities for diagnostics, cure and prevention at its command. Moreover, any real synthesis will take years to work out.

Western Medicine, which at present dominates the country and is the system recognized by the State, should discard its narrow outlook of contempt for anything which is not its own. The Indigenous Medicine, on the other hand, should discard its inherent prejudices and bring itself up-to-date by incorporating from other systems all that is of value. The practitioners of indigenous medicine should understand that in these days no claims of esoteric knowledge can be entertained, nor origin, antiquity and fancied utility urged as justifications.

In connection with the rationalization of medical practice in India two important questions suggest themselves. Firstly, can the practice of medicine be so regulated by the exponents of Modern and of the Indigenous Systems that the fullest possible use is made of the facilities available for diagnosis, treatment and prevention of disease? I have already alluded to this. The second question is, can a synthesis of Indigenous and Modern Systems of Medicine be attempted so as to promote the utilization of the knowledge from all available sources for the interpretation of health and disease and for diagnostic, curative and preventive purposes.

Partial Synthesis Needed

I believe that both extension and acceleration are possible through what may be called a partial synthesis of the two systems, in the elementary stages of our teaching. The present course of study in the Indigenous Medicine should be suitably curtailed on one side and enlarged on the other. On the side of curtailment, I suggest the teaching of subjects like Anatomy, Physiology and Pathology should be reduced to some extent. Their inclusion, on the present scale in the curriculum of the Indigenous Medicine leads to considerable confusion in the minds of the students. In passing it may be mentioned that the reduction in studies will go parallel to the reduction in the studies of students of Western Medicine, as proposed by the Bhore Committee. On the side of enlargement, I suggest that in addition to the basic principle of Ayurveda and Unani the students should be taught the basic principles of Western Medicine. They should also be given training in preventive health measures.

The next effect of this suggestion will be twofold. It will shorten considerably the period of study and thus lead to the training of a much larger number of qualified practitioners. And, secondly while giving the students a sufficient background of scientific knowledge with regard to the diagnosis, treatment and prevention of disease, it at the same time will make them conscious of their own limitations and of the necessity to appeal to higher practice in difficult cases. There are about a hundred thousand practitioners of Indigenous Medicine in India, many of whom could be quickly fitted for this purpose after suitable training.

Those who are likely to object to this curtailment and partial synthesis, should bear in mind that nearly 80 per cent of ailments are of a minor nature and can be dealt with by simple medical and surgical measures and require no advanced knowledge of the theory of diagnosis and treatment.

Moreover, the suggested training will enable the practitioner to become an integral part of the health services, and thus the administrative difficulties now being experienced in some provinces through a dual system of medical relief will be avoided. The services of such practitioners will be of particular value in the rural areas, which are now almost beyond the reach of Modern Medicine.

Rural medical relief will be considerably facilitated if some further steps are taken to standardize medical practice by prescribing uniform scales of drugs and medical appliances for institutions, their production in bulk and distribution under the auspices of the State. If all practitioners are properly registered and practice by non-registered practitioners prohibited, a reasonable standard of competence could be secured by prescribing and enforcing the necessary rules regarding an expeditious system of training and examination in respect of their qualifying diplomas.

While this partial and workable synthesis is taken in hand, I would earnestly suggest that careful research be made by the exponents of the Indigenous Medicine so as to link their system with the Modern Medicine. There is nothing derogatory in this. The work can only be taken up by the learned Vaidyas and Hakims and is bound to take time; and yet, *if this is not done their* systems are bound to become entirely obsolete.

The process of rationalization of their material medica should be comparatively easy and has already been taken up by a number of workers outside these systems.

A reference has, however, to be made here to the unreasonable attitude adopted by some of the indigenous practitioners towards the workers outside their own fold. They consider that investigation of their drugs by methods of chemical analysis and biological testing developed by Science, serves little purpose. It is opined that there is something mysterious in the action of "whole drugs" which cannot be investigated or elucidated by such tests. It is possible that there may be some such factors.

The discovery of antibiotics and hormones in plants to which no importance was previously attached may lend support to these views. But these should be explained and the mystery cleared by efforts of the exponents in the light of present knowledge. If they do this a complete synthesis will not be a remote possibility. If they fail, the world outside cannot be blamed if it refuses to believe their theories. The present-day world cannot accept any fantastic views whatever be their origin and however strong their following. *The result will be their complete extinction in the course of time.*

The outside workers should not be depressed by the hostile attitude and should go ahead. Every little contribution adds to our knowledge and may help materially towards the alleviation of human suffering.

Systematic Research Needed

Research must be carried out in a systematic manner. The materia medica of Indigenous Medicine consists predominantly of substances derived from the vegetable kingdom and practically all the plants used, grow in India. In the investigation of these plants the greatest difficulty I encountered in the beginning was that many plants mentioned in the literature baffled and defied recognition and identification. The descriptions in the old texts were in such vague and general terms that it was often impossible to be certain whether the specimens obtained were of the drug described. The identification of drugs is naturally not possible until prominent characteristics of each plant are establised.

There is no doubt, a large herbarium in the Botanical Gardens at Sibpur, Calcutta, but the specimens are so mixed up that to look up for a specimen is like hunting for a needle in a haystack.

In order, therefore to facilitate the work, on indigenous drugs, some years back, I decided to build up a reference herbarium containing authenticated specimens of all medicinal plants growing in the country. The collection of such a herbarium was slow at first, but was speeded up when grants for the purpose were sanctioned by the Indian Research Fund Association and the Imperial Council of Agricultural Research. A well-equipped botanical unit was established for making collections of plants from all parts of India and for their proper preservation and identification. By extensive investigations and collections in the field and by laborious studies in all the existing local herbaria in different parts of the country, about 10,000 specimens of nearly 2,000 of the common species of medicinal plants were collected. Several sheets of each species were prepared, and to ensure perpetuity and enhance and extend their utility to scientific workers, three more or less, complete sets of specimens were housed at the Forest Research Institute, Dehra Dun, the School of Tropical Medicine, Calcutta and at the Drug Research Laboratory, Kashmir.

A preliminary given in the old literature, could not enable the botanist to identify the plants and parts of plants which even in themselves do not invariably present the same characteristics and even the learned exponents of Indigenous Medicine cannot with certainty indicate which is the authentic specimen mentioned in the old texts. The distribution of many plants as described in the literature of Indigenous Medicine in the latter half of the 19th century is often vague and inaccuracies which have crept in have passed from one book to another. As a result, considerable confusion has arisen in the literature of Indigenous Medicine. Again, *many drugs are frequently sold under different names, and entirely different drugs often under the same name.* Very careful and detailed enquiries had, therefore, to be made before a plant could be taken up for investigation. In the work of identification help was at first obtained from the works of Western writers of the 19th century such as Jones, Ainslie, Roxburg, Wallich, Dymock, Watt and others who had carried out laborious studies to classify these plants. This also did not solve all the practical difficulties that arose.

The great handicap was that there has not been in this country (India) a proper organisation corresponding to the Bureaus of Plant Industry in advanced countries which collect and keep the information concerning plants up-to-date and encourage investigation and research. The Botanical Survey of India (Economic Products Section) and Forest Research Institute (Minor Products Section) do some scattered work in this connection, but the whole work must be unified and concentrated so that full collaboration with allied organizations can be established.

For collecting and supplying all information regarding plants, a Bureau of Plant Industry on the lines of that existing in the United States of America and in the U. S. S. R (Russia), should be established under the Ministry of Agriculture. The Bureau in

America can serve as a model. It carries on its activities under the Department of Agriculture in collaboration with agencies such as Bureau of Plant Quarantine, Federal Crop Insurance Corporation, Federal Surplus Commodities Corporation, Forest Services, Office of Foreign Agricultural Relations, Agricultural Marketing Service, Food and Drug Administration, *etc.* Any organization planning to stimulate the cultivation and development of medicinal plants in this country must collaborate with scientific workers in allied branches for the solution of interrelated problems. The functions which such a Bureau could usefully perform, are multifarious and should be worked out according to the requirements of the country. It could have information with regard to new species which can be successfully introduced and commercially developed and about markets in India and abroad. It should have knowledge regarding the quality and quantity of drugs growing in a state of nature and which are and can be successfully cultivated. Substitutes of pharmacopoeial drugs which might serve the same therapeutic purposes could be investigated under their auspices and brought into use. The Indian drug trade has seriously suffered because the quality of drugs has not been maintained and adulteration has been rife. The Bureau could exercise quality control and regulate drug trade by establishing drug emporia which could act as a central clearing house for authentic drugs.

It should be realized, that cultivation of medicinal plants is no novice's work, but needs specialised knowledge and guidance of a scientific organization. The soil, the season of planting, the gathering time, hybridization, plant diseases, *etc.*, are some of the important factors which call for expert attention in connection with the active principle of plants. The collaborative efforts of plant and culturists, pharmacognosists, pharmacists, pharmacologists, entomologists and chemists are essential.

Rationalisation of medicine is not possible till the drugs in use are standardized and controlled.

This applies with equal force to the drugs used in Indigenous Medicine.

It has been shown that a large number of drugs and pharmaceutical preparations marketed in this country, vary a good deal in regard to the potency claimed for them. While a certain amount of determination in active principles takes place through climatic factors and effects of storage, it has also to be admitted that often open and wilful adulteration of many remedies is being practised.

Investigation on an extensive scale showed that the position was even worse than what had been believed.

Need for a National Pharmacopoeia

The protagonists of Indigenous Medicine should realize that unless standards are established for drugs they use, either by their own methods or by the generally accepted chemical and biological assay, the efficacy of their drugs cannot be guaranteed. Establishment of standards for all drugs and their inclusion in the Indian Pharmacopeia of the future is absolutely essential before such drugs can play an effective role in a rational system of treatment.

A National Pharmacopoeia is primarily meant to meet the claims and to satisfy the needs of a particular group of physicians at a particular time. The object of a pharmacopoeia is, in the words of founders of the United States Pharmacopoeia, "to select from among substances which have medicinal power, those the utility of which is most fully established and best understood, and to form from these, preparations and compositions in which their powers may be exerted to the greatest advantage."

The modern pharmacopoeia is a book of standards, its fundamental objects are "to provide standards for drugs and medicines of therapeutic usefulness or pharmaceutic necessity, sufficiently used in medical practice; to lay down tests for the identity, quality and purity, to ensure as far as possible, uniformity in physical properties and active constituents." In other words, usage—rational usage and scientific usage are bases of judgement. Such criteria are no less applicable to Indigenous Medicine as to Western or any other system of medicine wherever practised.

The Drugs Enquiry Committee (1930) considered the question of compilation of an Indian Pharmacopoeia and came to the conclusion that there were cogent scientific reasons in favour of it. The methods of therapy vary in different countries. The raw materials from which medicinal preparations are made do not possess the same qualities, and may not be available so readily in one part of the country as in another. The effect of climatic conditions on the pharmaceutical processes has to be studied. Racial variations in dosage also have to be considered. For these reasons the pharmacopoeia of one country is not always applicable to another country. It is essential, therefore, that each country should evolve a pharmacopoeia best suited to its own peculiar climatic and racial factors. It should include the therapeutically active substances of known composition, of definite action, of well-established therapeutic use of known toxicity, and with necessary standards for determing safe maximum dosages. In the case of the drugs in use, it is essential that requisite standards should be established for strength and purity of the materials which are to be used in treatment.

Such a Pharmacopoeia is essential for rationalization of medicine in this country and will act as a bulwark against the present tendency towards irrational practice.

Sir Colonel Ram Nath Chopra (August 17, 1882-June 13, 1973)

He is known as Father of Indian Pharmacology; Doyen of Science and Medicine. On his 101th birth anniversary a fifty paise stamp was released.

After schooling at Jammu and Srinagar, College studies at Government College, Lahore, Chopra joined Downing College, Cambridge in 1903. He was awarded the Sc.D. degree of Cambridge University for his contributions to the science of medicine. The Royal College of Physicians of London elected him as a Fellow.

In 1921, Col. Chopra joined as First Professor of Pharmacology in Calcutta School of Tropical Medicine, became Director of the institution in 1934, also chaired the Pharmacology at Calcutta Medical College. He developed the Pharmacology laboratory equal to those in UK. Served the school for 20 years till 1941. After retirement in the same year the Government of Kashmir appointed him as Director of Medical Services and then as Director of Drug Research Laboratory where he served the lab till 1960.

He was first to establish a research centre which covered studies on Indigenous drugs, their chemical composition, *in vitro* and *in vivo* tests for the active principles, biochemical and biophysical changes in mammalian organism; and drug analysis. The Department of Pharmacology at Calcutta School of Tropical Medicines stood as landmark for researches, covering clinical evaluation of drugs, tropical medicine, therapeutics, experimental pharmacology, toxicology, drug standardization and biological assays, diagnostic services.

With his continuous research various indigenous drugs like ispaghula, kurchi, rauwolfia, psoralea, cobra venom, *etc.* were proved to have pharmacologically active principles and got place in Indian Pharmacopoeial List 1946 and Pharmacopoeia of India 1955. During his chairmanship of Drug Enquiry Committee in 1930-31 pharmacy profession had taken birth. With his recommendations Prof. Mahadev Lal Schroff got inspired and started Pharmacy course first time in India in 1932 at Banaras Hindu University.

Col. Chopra will be remembered for:

- Drugs Act 1940 which was later changed as Drugs and Cosmetics Act in 1962. Ayurvedic (including Siddha) and Unani drugs also came under its coverage in 1964.
- Indian Pharmacopoeial List 1946.
- The Pharmacy Act,1948.
- The Indian Pharmacopoeia first edition, 1955.
- Chairman of First Ayurvedic Pharmacopoeia Committee (1962-1972).

Important reference works by Col. R. N. Chopra:

Indigenous Drugs of India, Glossary of Indian Medicinal Plants, Poisonous Plants of India, Drug Addiction, with special reference to India, A Handbook of Medical Treatments with Prescriptions, The Shaping of Indian Science.

References

1. Chopra, R. N., Ghosh, B. N., Field for Research in India, Indigenous Drugs; *Ind. Med. Gazette,* Vol. 50, p. 99, 1923.
2. Report of Drugs Enquiry Committee (1930-31), Government of India Press, New Delhi, 1931.
3. Chopra R. N.; Indigenous Drugs of India, The Art Press, Calcutta, 1933.
4. Chopra, R. N.; Indigenous Drugs Inquiry; A review of the work, Indian Research Fund Association, New Delhi, 1939.
5. Mukerji, B.; Planning for India's Foreign Trade in Vegetable Drugs, *Indian Pharmacist,* 1945, Vol. I, No. 1. p. 31.
6. Report of the Health Survey and Development Committee, Government of India Press, New Delhi, 1946.
7. Chopra, R. N.; Indian Medicinal and Poisonous Plants; A brief review (1921-46) Indian Science Congress and Council of Scientific and Industrial Research, 1947.
8. Report of the Panel on Fine Chemicals, Drugs and Pharmaceuticals, Government of India Press, Simla, 1947.
9. Chopra R. N.; Badhwar, R. L., and Ghosh, S; *Poisonous Plants of India,* in two volumes. Government of India Press, Calcutta (in press).
10. Mukerji, B.; Scheme for Central Institute for Drug Research-Unpublished.

Complete text: http://www.sciencecongress.nic.in/html/pdf/e-book/April_May_2008.pdf

5

The Nomenclature Dynamics of Classical Ayurvedic Texts

C.P. Khare

After the 16th century, light at the end of the tunnel was seen during 18th century. A hazy view of a possible path let Ayurveda to explore possibilities for its revival in a totally changed world with a strong scientific approach during the British Period. Ayurvedic fraternity was desperate to equate classical herbs with botanical nomenclature which was assigned by Portuguese and botanists of the East India Company. It became an issue of survival. We remained unaware of the intricate pattern of this exercise.

How Traditional Indian Botany was Moulded

For the first time botanical names were introduced in India by Garcia de Orta (*Colloquios*, printed at Goa in 1563), C. d'Acosta (*Tractado*, Burgos, 1578), and Rhede van Drakenstein (*Hortus Malabaricus*, Amsterdam, 1682). *Hortus Malabaricus* was used by Linnaeus while proposing nomenclature of Indian plants.

Volume 5 of the thirteen-volumes of Sir William Jones (1746–94), published in 1807, contains Jones' researches into Indian botany, including the *Botanical Observations on Select Indian Plants*. A few samples of his observations will illustrate the experimental process of converting Sanskrit names into *possible* botanical names:

Muchukanda: "This plant differs a little from the *Pentopetes* of Linnaeus."

Bilva: "I had imagined that it belonged to the same class as *Durio*."

Sringataka: "It seems to be floating *Trapa* of Linnaeus."

The first half of the 19th century saw a number of European plant collectors, particularly superintendents of the Royal Botanic Garden, Calcutta. Francis Buchnan (1762-1829), Nathaniel Wallich (1815-1835), John Mc-Clelland (1805-1855), Hugh Falconer (1808-1865), William Griffith (1810-1846) and Thomas Thomson (1817-1878) collected plants extensively in Northern and Eastern India, Nepal, Burma and the Himalayas. Wallich collected 7,693 plant specimens which were incorporated in his work *Plantae Asiaticae Rariories* (1829-32). He prepared a catalogue of more than 20,000 specimens and published *Tentamen Flora Nepalensis Illustratae* (1824-26). Wallich was responsible for packing many of the specimens that came through the gardens on the way to England.

William Graffith, in a short period of 13 years, collected 9,000 plant species which formed basic material of *Flora Indica* (1855). Joint collection of Sir J.D. Hooker and Thomas Thomson contained 1,50,000 specimens of 9,500 species. *The Flora of British India* (7 volumes) was published between 1872 and 1897.

Earlier In 1801, Sir William Jones wrote a memoir entitled *Botanical observations on Selected Medicinal Plants*. This was followed by John Fleming's *Catalogue of Indian Medicinal Plants and Drugs* in 1810, Ainslie's *Materia Medica of Hindoostan* in 1813.

Most of the North Indian botanical names were introduced by William Roxburgh. William Roxburgh was a natural historian of East India Company in 1789. He moved to Calcutta (Kolkata) to be the superintendent of the Botanic Garden, the present Indian Botanical Garden at Sheopur, Howrah, and Kolkata. Working here, he was instrumental in introducing many plant and species to India and simultaneously sent many species to Kew, London.

In 1820 at the Mission Press in Serampore, William Carey posthumously edited and published vol.1 of Dr William Roxburgh's *Flora Indica or Descriptions of Indian Plants.* In 1824, Carey edited and published vol.2 of *Flora Indica*, including extensive remarks and contributions by Dr Nathaniel Wallich. *Flora Indica* contained descriptions of a large number of wild plants, as well as, a large collection of drawings of 2,533 species.

Roxburgh's investigations formed the basis of Sir J.D. Hooker's *Flora of British India* (1875-1897), which earned him the honour of the "Father of Indian Botany."

Sir J. D. Hooker's *Flora of British India* became the standard of scientific names and synonyms. The botanic diagnosis of the plants from *Ranunculaceae* to *Acanthaceae* has been entirely derived from *The Flora of British India.* Bentham and Hooker's *Genera Plantarum*; DeCandolle's *Prodromus*; Roxburgh's *Flora Indica;* The Linnaean Society's publications; Brandis' *Forest Flora* were the main source of scientific names.

As a contribution to Indian Pharmacopoeia, Moodeen Sheriff equated botanical (Linnaeus) names with synonyms of 12 languages, Sanskrit (in a few cases), Arabic, Persian, Hindustani, Dukhni, Tamil, Telugu, Malyalim, Canarese, Bengali, Maharatti, Guzratti and Burmese. Ashwagandha, Bala, Brahmi, Paashanabheda, Raasana, Shatavari and a host of important Ayurvedic plant drugs did not feature among 285 Sanskrit names. *Flacourtia cataphracta* Roxb. was equated with Taalisha, also with Barahmi, Tejpaat and Zarnab. This work was published as *Supplement to the Pharmacopoeia of India* in 1869.

Other important works are U.C. Dutt's translation of *Sanskrit Materia Medica*, Fluckiger and Hambury's *Pharmacographia*, Dymock's *Vegetable Materia Medica of Western India. Pharmacographia Indica* was published in 1890-93 under the joint editorship of Dymock, Warden and Hooper. *Dictionary of the Economic Products of India* (published in 1889-1904) by George Watt and Edgar Thursion was the most exhaustive compilation of commercial plants in India. Further work was taken up by the Wealth of India series under the auspices of the Council of Scientific and Industrial Research and from 1949 to 2009, 23 volumes were produced, making the series a monumental work on Indian botany.[1,2]

While concluding the summary of work done during early nineteenth century, we would like to point out that it was used as a conduit between Calcutta and Kew. Kew, as the centre of a network of colonial botanical gardens, trained up specialist botanists and sent them into different parts of the empire. Main aim was to assist the expansion of the British Empire.

On academic platform the contribution of British botanists was not free from controversies. For example, *by the time Flora Indica was published, 185 species and their descriptions had already appeared in books by other British botanists in London. They had assigned different systemic names than those used by Roxburgh, thus causing confusion about species in his catalogue.*[3]

Ayurvedic Nomenclature Dynamics

Now, let us try to understand, how Ayurvedic nomenclature dynamics was different from Linnaeus system.

M.V. Vishwanathan and associates provided an indepth study (the only study available in print) on this sensitive issue in *Indian Journal of Traditional Knowledge.*[4]

"The nomenclature of Ayurveda was not a binomial system as adopted by modern botany. In Ayurveda, there are many names for a single entity and a single name is used to denote many plants. So, it is essential to understand the way Ayurvedic nomenclature works. The total number of names pertaining to medicinal sources may be approximately 20,000 to 25,000. A particular plant will have a group of synonyms which may range from one to approximately fifty. Each name focuses on a very specific aspect of the plant. So, these names give a good picture of the various aspects of the plant including morphology, habit, habitat, qualities, biological actions, therapeutic uses and so on.

"This naming system was primarily designed to help a physician to select a plant for medicinal purposes. It was not designed to establish the taxonomical identity of a plant. The nomenclature of Ayurveda is, therefore, a therapeutic nomenclature based on a polynomial system of naming. In the literature the names are categorized as *Svarupa bodhaka* (revealing the form) or *Guna bodhaka* name (revealing the quality). *Guna bodhaka* names are names pertaining to qualities, actions, specific action in relation to therapeutic conditions, *etc.* The different types of names highlight different aspects of the drug sources."[4]

Now, we will quote a detailed example.

"*Tinospora cordifolia* is equated with *Guduci* along with several other Sanskrit names.

Names describing external characters (*Svarupa bodhaka*) are *Amrtavalli*—weak-stemmed plant, *Cakralaksana*—wheel like appearance on cross section, *Mandali*—circular shape, *Kundali*—stem gets entangled with twiner, *Nagakumari*—twining nature comparable to a snake, *Tantrika*—spreading nature, *Madhuparni*—honey-like leaf juice, *Chadmika*—thick foliage, *Syama*—smoky hue of the stem,*Vatsadani*—eaten by calves, *Dhara*—longitudinal groves on the stem, *Visalya*—no thorns, appendages, *Chinnaruha*—capacity of cut stems to regenerate fast, *Abdikahvaya*—growing near a reservoir of water.

The terms describing qualities (*Guna bodhaka*) are *Amrta*—an elixir, *Guduci*—that which protects, *Vayastha*—rejuvenative nature, *Jvarari*—anti-pyretic property, *Soumya*—benevolent in action."[4]

Though the Sanskrit literature provided elaborate description of herbs, confusion in the identification of their botanical sources was the result of poor understanding of the Sanskrit literature and misinterpretation by various commentators. This confusion has become confounded by the existence of several names for one drug and several drugs having one common name. A high percentage of plants used in Ayurvedic practice have some controversy or the other, due to lack of coordination between taxonomists lacking Ayurvedic knowledge and Sanskrit scholars having no academic knowledge of botany. It is not easy to correlate morphological characters with clinical applications and textual descriptions. A drug can also become controversial when the information available on the species is limited. Poor understanding of the nomenclature intricacies, misinterpretations, poor deciphering of the classical texts, poor field identification skills, dependence on semiliterate herb collectors and traders, wide chronological gaps between different classical texts, all lead to wrong identity. Due to similarity in the morphological characters two different species may be known by the same name in the vernacular languages, resulting in wrong identity. For example, in Tamil, both *Cressa cretica* Linn. and *Drosera indica* Linn. are known by the same name *Azhukanni* because of the similar feature of presence of dew like substance on the leaves. Diverse dialects and accent, lead to multiple identities. For example, *Matala* in Tamil refers to *Punica granatum* Linn., whereas in Kannada it pertains to *Citrus medica*. *Aralimara* is the Kannada name for *Ficus religiosa* Linn., but in few specific locations of Karnataka *Ficus* is known as *Ragi mara*. Ragi is the popular name for the cereal *Eleusine coracana* Gaertn. All these issues may mislead the botanist, researcher and scientists.[4]

A few examples of misinterpretations from current standard works.[5]

Putikaranja is equated with *Caesalpinia crista* Linn. in Ayurvedic Pharmacopoeia of India Vol. IV. In Vol. V, *Caesalpinia bonduc* (L.) Roxb. is equated with Lata karanja, while *Caesalpinia crista* Linn. is also equated with Latakaranja (Ayurvedic Formulary of India (AFI), Part I, second revised edition, page 332) and *Holoptelea integrifolia* Planch is equated Putikaranja, Putika, Putigandha, and Chirbilva (pages 310, 323, 335).

Three plant species are being used as Karanja or similar because flowers impart colour to water. *Pongamia prinnata* Pierre, a tree, is equated with Karanja, Karanjaka, Naktamal and Udakirya; *Holoptelea integrifolia* (Roxb.) Planch., also a tree, is equated with Chirbilva, Putika, Putikaranja, Putikaranja and Prakirya; *Caesalpinia bonduc*, a shrub, has been identified as Latakaranja, Kantaki karanja and Karanji.

Tamlaki is equated with *Phyllanthus fraternus* Webst. syn. *Phyllanthus niruri* Hook. f. non Linn., *P. niruri* sensu Hook. f. while Ayurvedic Formulary of India, Part I wrongly equated it with *Phyllanthus niruri* Linn. (page 327), which is an American species.

In Glossary of Indian Medicinal Plants (CSIR) *A. farnesiana* is equated with Arimedah, Gandh Babool and Guya-babool. The Wealth of India (CSIR) also equated it with Guya-babool and Gandh Babool. With the result, Ayurvedic literature treated *A. farnesiana* as Arimeda. The extract of *A. farnesiana* was found devoid of antibacterial and antifungal activity. Arimeda is correctly equated with *A. leucophloea* by scientists of Indian National Science Academy.

Ajmoda of Ayurvedic medicine is equated with *Trachyspermum roxburghianum* (DC.) Craib, syn. *Carum roxburghianum* Kurz. Central Council for Research in Ayurveda and Siddha accepted Vilaayati Ajmod (*Apium graveolens* Linn.) as its substitute in 1990, and as a drug source in 1996. In 2003, Ayurvedic Formulary of India equated Ajmod and Ajmoda with *Apium graveolens*.

In Ayurvedic Phamacopoeia of India, Virala and Tinduka are synonyms. Ayurvedic Formulary of India equated Tinduka with *Diospyros embryopteris* Pers. syn. *D. malabarica* (Desr.) Kostel (page 327); Virala with *D. tomentosa* Roxb. syn. *D. exsculpta* Buch.-Hem. (page 329). *D. embryopteris* and *D. tomentosa* are not synonyms.

Jayanti is equated with *Sesbania sesban* in Ayurvedic Formulary of India, Part I, page 314, while Jayanti root is quoted as Agnimantha on page 284. Agnimantha is equated with *Premna obtusifolia* R.Br.[5]

Taxus baccata Linn. is European Yew. Himalayan Yew is *T. wallichiana* Zucc. syn. *T. baccata* Linn. subsp. *wallichiana* (Zucc.) Pilgoe; T. baccata Hook. f. non Linn. In Ayurvedic Formulary of India (Part I, page 327) *Abies webbiana* Lindl. is equated with Talisha. *A. pindrow* Spach. and *Taxus baccata* Linn. have been recognized as its substitute. In South Indian compounds, *T. baccata* is used as Talispatra.[6]

In Ayurvedic Pharmacopoeia of India, *Ferula narthex* is equated with Hingu though its valatile oil is reported to be sulphur-free. Charaka used dried fruits of Hingu (Hinguka) in a gruel as a blood purifier and purgative (Charaka Samhita, Su. 2, 29)27 Hingu and Hingupatri in a medicinal ghee for insanity. Hinguparni (Ci 9, 63), Hingupatrika (Ci 9, 72), Hinguvatika (Ci 15, 108) of Charaka Samhita still remain unidentified.

A typical example is that of the source of the Ayurvedic drug Aswagandha, which is equated with *Withania somnifera* Dunal. In practice, the cultivated drug is used which is quite different in its morphology, odour, *etc.* from its wild source.[4] The Wealth of India, Vol X, 1976, also pointed out that "it is difficult to say whether the source of drug (Ashwagandha) was *Withania somnifera*."[6]

As Ashwagandha is an important plant drug of Ayurveda, we will discuss various issues related to it in detail.

According to "Ayurvedic Pharmacopoeia of India" (Part I, Vol.), Ashwagandha (equated with *Withania somnifera* Dunal) was used for *Klaibya* (male impotence), *Kshaya* (phthisis), *Daurbalya* (weakness), *Vaataroga* (diseases of the nervous system/neurological diseases) and *Shoth* (inflammation, oedema). These attributes do not match with Indian Ashwagandha's use in practice.[6] KN Kaul was first to point out that cultivated Ashwagandha of commerce, used in Ayurvedic compounds, is different from the wild variety not only in therapeutic properties but also in morphological characters. Kaul named the cultivated variety as *Withania ashwagandha.*[6] Research by Bilal Ahmad Mir, Sushma Koul, Amarjit Singh Soodan (Reproductive biology of *Withania ashwagandha* sp. *Novo*) support the delineation of *Withania ashwagandha* from *Withania somnifera* based on morphology, chemical profiling, crossability features, AFLP fingerprinting and ITS analysis (Mir *et al.*, 2010 and Kumar *et al.*, 2011). Earlier, Kaul (1957) and Atal and Schwarting (1962) suggested that *Withania somnifera* is an amalgamation of two or more taxa. Kaul (1957) proposed a separate species status, *W. ashwagandha* to the cultivated forms in the *Withania somnifera* germplasm. Subsequently, Sastry *et al.* (1960) and Dhalla *et al.* (1961) conducted anatomical, pharmacognostic and chemical studies on *Withania ashwagandha* and confirmed its distinct nature from its ally *Withania somnifera*. Negi *et al.* (2006) also confirmed the distinctness of cultivated types on the basis of AFLP and SAMPL studies. Three chemotypes of *Withania somnifera* have been identified. Which one was Ashwagandha of the classical age, is yet to be identified.[7,8] As of now, Ashwagandha of the twentieth century in a well-researched drug of Modern Ayurveda.[5] The problem is that classical reputation of Ashwagandha is being exploited by chemotypes which do not belong to India.

This analysis shows how polynomial nomenclature and lack of understanding of nomenclature dynamics of Ayurveda are one of the major issues which resulted into a number of controversial botanical names introduced by Linnaeus system.

Lifeline of Linnaean System

Now, let us try to understand Linnaean Classification System.[9]

Carolus Linnaeus (1707-1778) has been commonly described as "Father of Modern Taxonomy" for his contributions. His main contribution was to bring order to the mixed-up array of literature and system of classification which the 17th century botanists had confronted.

Linnaean Classification System has seven different levels. From smallest to largest, the levels are species, genus, family, order, class, phylum, and kingdom. Each of the ranking levels is called a taxon. A genus is a group of similar species, A family is a group of similar genera. An order is a group of similar families. A class is a group of similar orders. A phylum is a group of similar classes. A kingdom is a group of similar phyla.

The Linnaeus System classifies organisms based on overall similarities and differences to one another. Organisms in the same kingdom may have many differences, but they still have common traits with one another.

In the 1730s, Carolus Linnaeus developed a naming system, called binomial nomenclature. In binomial nomenclature, each species is assigned a two-part scientific name:

The first name refers to the genus. The second name refers to its species, which is chosen by the author. Name must be written italic, first name must begin with capital letter, and the second name must be written in small letters. It must be followed by author's name.

This is the nomenculature which we use today is essentially his. By means of the binomial nomenculator, the plants could be defined which had not been able to define before. He classified the plants due to their similarities and differences.

Before Linnaeus, scientists could use different names for the same kind of species. This nomenculature caused lots of disorder and confusion. The Linnaeus' nomenculator system brought a standization and easily understandible form of description.

Two works of Carl Linnaeus are regarded as the starting points of modern botanical and zoological taxonomy: the global flora Species *Plantarum*, published in 1753, and the tenth edition of *Systema Naturae* in 1758 including global fauna. Linnaeus introduced in these books a binary form of species names called "trivial names" for both plants and animals.

A Word of Caution

While concluding, we express our view that Ayurvedic scholars and taxonomists need to work together but should not have adopted Linnaeus system in a hurry to salvage a classical legacy. They should take into account Ayurvedic herbs' morphological characters, while correlating it with clinical applications and textual descriptions. The nomenclature of Ayurveda is a therapeutic nomenclature, based on a polynomial system of naming. Unless we understand this specific feature of classical texts, in North and South and in different regions of India, different species, even from different families, will be used in practice for the same Ayurvedic pharmacopoeial name.

References

1. Indigenous Drugs of India, Col. Sir R.N. Chopra *et al.*, 1933, 1958.
2. Development of Life Sciences in India in Eighteenth-Nineteenteeth Century, B.L. Jain, *Indian Journal of History of Science*, 17(1), 114-115.
3. Circulation of knowledge between Britain, India and China: The Early Modern World to the Twentieth Century, Ed. Bernard Lightman *et al.*, BRILL,2013; chapter: Between Calcutta and Kew, the divergent circulation of *Hortis Bengalensis* and *Flora Indica*, by Khyati Nagar.
4. A brief introduction to Ayurvedic system of medicine and some of its problems, MV Viswanathan *et al.*, *Indian Journal of Traditional Knowledge*, Vol 2(2)-April 2003-pp 159-169

   http://www.niscair.res.in/sciencecommunication/ResearchJournals/rejour/ijtk/Fulltextsearch/2003/April per cent 202003/IJTK-Vol per cent 202(2)-April per cent 202003-pp per cent 20159-169.htm
5. Ayurvedic Pharmacopoeial Plant Drugs: Expanded Therapeutics, C.P. Khare, CRC Press, 2016.
6. Indian Medicinal Plants: An Illustrated Dictionary, C.P. Khare, Springer, 2006.

   Indian Herbal Remedies, C.P. Khare, Springer, 2004.

   Irrational use of *Withania somnifera*, C.P. Khare, unpublished.
7. Bioorg Med Chem, Xu Ym *et al.*, 17 (6), 2009; 2210-2214. (a) Reproductive biology of *Withania ashwagandha* sp. *novo*, Bilal Ahmad Mir *et al.*
8. The delineation of *Withania ashwagandha* from *Withania somnifera*, based on morphology, chemical profiling, Mir *et al.*, 2010; Kumar *et al.*, 2011. Research articles, University of Pretoria (9495).
9. Bases on "Natural Products-the secondary metabolities," a powerpoint presentation by Prof. Dr Nezhun Gorden, Yýldýz Technical University, Faculty of Science and Arts, Department of Molecular Biology and Genetics, Istanbul/Turkey.

6

Divergent Uses of Classical Ayurvedic Plant Drugs in North and South India

C.P. Khare

Sanskrit names of Classical Ayurvedic plant drugs have been equated with botanical names by Ayurvedic scholars who were Ayurvedic practitioners or teachers in Ayurvedic colleges, but had scarce knowledge of plant taxonomy and lacked sound knowledge of modern botany and pharmacognosy and pharmacology (many field investigators never studied even English). Many of these Sanskrit scholars could not understand complexities of botanical identifications. They remained busy in *shaashtraarth* (academic discussion based on Sanskrit texts), sorting out multiple Sanskrit synonyms and multiple pharmacological actions ascribed to one plant drug in classical texts. Ayurvedic literature is overflowing with synonyms and homonyms. It was not an easy task for taxonomists to wade through it.

This has resulted into a number of anomalies. A classical compound manufactured in South India and North India might be available under the same classical drug name but with a few different herbs even of different families. A few examples will elucidate this situation. Amitaarishta of South India contains *Mollugu cerviana* instead of Fumaria parviflora. Chandanaasava of South India and North India contains five herbs of different botanical identities. Mahaamaasha Taila of North India and South India are totally different. South Indian product contains goat's meat as the main drug, while in North Indian product contains *Phaseolus mungo* as Maasha.

In South India, Ayurvedic plants have been equated with botanical names by plant taxonomists, S. Usman Ali, V.V, Sivarajan and Indira Balachandran. S. Usman Ali was the botanist of Captain Srinivasa Murthy Research Institute, Chennai, when he correlated regional and classical names with botanical names for the English version of Vaidya Yoga Ratnavali which was originally compiled by Vaidya Mulugu Ramalingayya Guru in Telugu. Arya Vaidyasala, Kottakkal, brought out a compendium of 500 species of Indian medicinal plants in1993. In 1994, V.V. Sivarajan and Indira Balachandran brought out another reference work on plant sources of Ayuvedic drugs in South India, especially Kerala. V.V. Sivarajan is a Professor of Botany at University of Calicut and Indira Balachandran is a research officer in Arya Vaidyasala herbal garden, Kottakkal. These works are based on extensive field surveys.

In North India, Thakur Balwant Singh produced remarkable results from his field and literature studies. Vaidya Daljit Singh, Dr V.G. Desai, Vaidya Bapalal G. Shah, Acharya Priyavrat Sharma, Vaidya Maya Ram Uniyal, Prof. K.C. Chunekar, Dr S.K. Sharma, Dr Asima Chatterjee and Satyesh Chandra Pakrashi made an allout effort to sort out controversies.

With this prologue, we present this study based on standard reference works, including Ayurvedic Pharmacopoeia of India (API), and Ayurvedic Formulary of India (AFI).

Achyranthes aspera Linn. (Fam. Amaranthaceae). Apaamaarga (API)

The white variety of Apaamaarga (Gaur-dand Apaamaarga) was recommended in Ashtaangahridaya (7th century AD) for promoting fertility, also for bearing a male child. *A. aspera* has been wrongly equated with White Apaamaarga of classical texts. Its stem bark exhibited 100 per cent abortifacient activity experimentally. In South India *Achyranthes prostrate* syn. *Cyathula prostrata*[2], known as Kshudra Apaamaarga (Cheria Kadaladi), is equated with Rakta Apamaarga. Both the species exhibit contraceptive properties[2] and indicate that classical Apaamaarga, which was used for promoting fertility, was some other species.

Aconitum heterophyllum Wall. ex. Royle (Fam. Ranunculaceae). Ativishaa (API)

In South India *Cryptocoryne spiralis* Fisch., corms known as Nattu Athividayum, and *C. retrospiralis* (Araceae) are used as Ativisha.[1,4] The only other non poisionous species is *A. palmatum* D. Don. (Prativishaa).[4]

Adiantum lunulatum Burm. (Fam. Polypodiaceae). Hamspadi (API)

In Kerala *Desmodium triflorum* (L.) DC. is used as Hamsapadi.[2,4,10]

In Tamil Nadu *Coldenia procumbens* L. is the source of Hamsapadi.[1]

Vitis pedeta Vahl (Vitiaceae) is also reported as being used as Hamspadi.[4]

Aerva lanata (Linn.) Juss. (Fam. Amaranthaceae). Pattura, Bhadra, Gorakshganja (API)

Aerva lanata is a substitute of Paashaanabheda (AFI). In Kerala *Rotula aquatica* Lour. and *Homonoia reparia* Lour. are used as Paashaanabheda.[2] *Aerva lanata* is used as Bhadra.[4] In AFI, Bhadraa is a synonym of Sthulailaa (*Amomum subulatum* Roxb.).

Chakrapani equated Pattura with Shaalincha which has been identified as *Alternanthera sissillis* (L.) R. Br. This has been accepted by Aya Vaidyasala, Kottakkal, Kerala.[4,3]

Albizzia lebbeck Benth. (Fam. Fabaceae). Shrisha (API)

Stem bark of *Albizia marginata* Merr. is used as Shirisha (Arya Vaidyasala, Kottakkal)[3] in Kerala and south India.[4] *Albizia odoratissima* Benth. is used as a Shrish or Krishna Shirisha.[4]

Alhagi pseudalhagi (Bieb). Desv. (Fam. Fabaceae). Yavaasaka (API)

In Kerala *Tragia involucrata* Linn. is being used as Yavasaa and Dhanvayaasa,[2]also as Duraalabha.[10]

Alpinia calcarata Rosc. (Fam. Zingiberaceae). Granthimuula, Shveta kulanjana (API)

In South India, rhizomes of *A. galanga* (Peroratta) and *A. calcarata* (Aratta) are used as Raasnaa.[2]

Alpinia galanga is a substitute of *Pluchea Lanceolata* Oliver and Hiem (Raasnaa of AFI).

Alternanthera sessilis (Linn.) R. Br. Matsyaakshi (API)

In different parts of Kerala *Alternanthera sessilis*, *Ginus oppositifolius* and *Portulaca oleracea* are used as Lonikaa.[2]

Amomum subulatum Roxb. (Fam. Zingiberaceae). Sthulailaa (API)

In Kerala, fruits of Pucedanum grande C.B. Clarke are also used as Sthulailaa.[4]

Angelica archangelica Linn. (Fam. Apiaceae). Chandaa, Laghu Choraka (API)

In Kerala *Costus specious* (Koenig) Smith is considered the exclusive source of Chandaa.[2]

Kampferia galanga Linn. is used as Choraka.[4] In AFI, Choraka is equated with *Angelica glauca* Edgw.

Anisomeles malabarica (L.) R. Br. ex Sims (Fam. Lamiaceae). Sprkkaa (API)

In practice, in Kerala, *Adenosma indiana* (Lour.) Merr. (Fam. Scrophulariaceae) is mainly used as Sprkkaa.[2]

AFI accepted *Anisomeles malabarica* as the main source and *Schizachyrum exile* Stapf. and *Delphinium zalil* Aitch and Hemsl as its substitutes (AFI, Part I, Page 325).

Anogeissus latifolia Wall. (Fam. Combretaceae). Dhava (API)

Though *A. latifolia* is the well-identified source of Dhava, the Indian Medical Practitioners' Cooperative Stores Ltd. (IMPCOPS), Chennai, is using *Syzygium hemisphericum* (Walp.) Alston, syn. *Eugenia hemispherica* Wight (Myrtacea) as Dhava since 1968.[1] In Tamil Nadu, White Rose Apple Wood, *Syzygium hemisphericum*, known as Vennavalmaram, is used as Dhava, while *A. latifolia* is in abundance in Madurai, Tirunelvali and Tamil Nadu, also in Andhra Pradesh, Karnataka and Kerala.[5(b)] Heartwoods of both (*A. latifolia* and

S. *hemisphericum*) were used for preparing cart wheels, which was the identification source of Dhava during classical period.

Argyreia nervosa (Burm.f.) Boj. Syn. _A. speciosa_ Sweet. (Fam. Convolvulaceae). Bastaantri (API), Vrdhadaaru (AFI)

Roots of *Ipomoea pescaprae* (L.) Sw., syn. *I. biloba* Chois. And *I. petaloidea* Chois. are common substitutes in South and North-west India respective.[9] *Argyreria nervosa* is found to be used as Murvaa in some parts of Kerala.[2]

Barleria prionitis Linn. (Fam. Acanthaceae). Sahachara (API)

In Kerala, other Acanthaceae spp. are used as Sahachara: *Nilgirianthus ciliatus* (Nees) Bremek., *Ecbolium viride* (Forsk.) Alston and *Justicia betonica* Linn.[2] In northerb part of Malabar, practitioners prefer *Calacanthus grandiflorus* (Dalz.) Radlk (Acanthaceae), as Shveta Sahachara, reportedly due to similarity of roots.[2] In Kerala, *Echoliums linneanum* Kurz. is used as Nila Sahachara and *Justicia betonica* Linn. as Shveta Sahachara.[1] (*Barleria prionitis* bears yellow, *B. cristata* Linn. white-reddish and *B. strigosa* Willd. blue flowers.)[10]

Benincasa hispida (Thunb.) Cogn. (Fam. Cucurbitaceae). Kushmaanda (API)

A distinct variety, locally known as 'Vaidya kumbhalam', is grown in parts of Malabar (Kerala) for use in Kushmaanda Lehyam.[5(b)] Kushmaandi is a separate plant equated with *Cucurbita pepo* Linn.[4]

Berberis aristata DC. (Fam. Berberidaceae). Daaruharidraa (API)

Identified substitutes are *B. asiatica* Roxb. ex DC. and *B. lyceum* Royle (AFI), also *B. chitria* Lindl., *B. ulicina* Hook and *B. vulgaris* Linn.[10]

The stem bark of *Coscinium fenestratum* Colebr. is used as a substitute in Kerala and Tamil Nadu.[9,4] Considered better than Berberis.[1] Known as Pita Chandana (AFI) and Ceylon Calumba or False Calumba.[1]

Bergenia ciliata (Haw.) Sternb. syn. _Bergenia ligulata_ (Wall.) Engl. (Fam. Saxifragaceae). Paashaanabheda (API)

Commonly accepted in the northern market as Paashaanabheda.

Aerva lanata Juss ex Schult is a validahriverated substitute (AFI). It is used in Tamil Nadu.[1] *Rotula aquatica* Lour. (Boraginaceae) and *Homonoia riparia* Lour., syn. *Adelia neriifolia* Roth. (Euphorbiaceae) are used in Kerala.[2] *Rotula aquatica* Lour. Is used in Karnataka[10] and is preferred in South.[3,4] According to IMPCOPS, Unani compounds contain Gentiana kurroo royle.[1]

Boerhaavia diffusa Linn. (Fam. Nyctaginaceae). Rakta Punarnavaa (API)

B. diffusa and *B. verticillata* Poir Shveta Punarnavaa (AFI) are used in Kerala.[3] *Trianthema portulacastrum* Linn., syn. *T. monogyna* (Fam. Aizoaceae), Shveta Punarnavaa, is sometimes used as a substitute in the South.[1,2] *Trianthema portulacastrum*, a rainy season annual, is equated with Varshaabhu and red and white forms of *Boerhaavia diffusa* with red and white Punarnavaa.[2,4]

Boerhaavia diffusa is also used as Kathilla (AFI).

Callicarpa macrophylla Vahl. (Fam. Verbenaceae). Priyangu (API)

Central Council for Research in Ayurveda and Siddha, in its monograph on Priyangu, covered both, *Calicarpa macrophylla* and *Aglaia roxburghiana* (1990).[7] Substitute of *Calicarpa macrophylla* is *Prunus mahaleb* Linn. (AFI). *Aglaia roxburghiana* is used as Priyangu in Uttar Pradesh.[10]

In Kerala, dried male flowers of *Myristica fragrans* Houtt. are sold as Priyangu.[3] It is reported that flower buds of *M. malabarica* Lam., *Orchrocarpus longifolius* Benth. and Hook. are sold in Chennai (Tamil Nadu) while fruits of *Zanthoxylum budrunga* Wall. in Kerala as Priyangu.[4] (Priyangu cereal is equated with *Setaria glauca* Beauv.).[10]

Cardiospermum halicacabum Linn. (Fam. Sapindaceae). Karnasphotaa (API)

In North India, *C. halicacabum* is used as Karnasphotaa, in West Bengal as Jyotishmati.[6]

Jyotishmati is a different and well identified plant equated with *Celastrus panniculatus* Willd. (AFI).

In South India, *Cardiospermum halicacabum* is used as Indravalli.[2] (According to taxonomists of South India, there is no mention of Indravalli in classical Ayurvedic texts.)[2] Indravalli and Hrivera are synonyms in AFI, equated with *Coleus vettiveroides* K.C. Jacob. Indravalli has been identified as Baloon Vine, Mudakkaruttan (Tamil) and Uzhingna (Malyalam).[1]

Careya arborea Roxb. (Fam. Lecythidaceae). Kumbhikah (API)

Kumbhi and Pezhu have been identified with *Careya arborea* and *Pistia stratiotes* Linn. (AFI, Part 1, page 318, 332.) *Pistia stratiotes* is Water Lettuce, while *Careya arborea* is a tree. Flowers of *C. arborea* are used in an aphrodisiac.[5(b)] Kumbhika is equated with Jala Kumbhi (*Pistia stratiotes* L.); Kumbha and Nikumbha with *Operculina turpethum* Linn. and Kumbhayoni with *Sesbania grandiflora* (L.) Pers.[4,6]

Cassia angustifolia Vahl (Fam. Leguminosae). Svarnapatri (API)

C. angustifolia Vahl and *C. acutifolia* Delite are considered two distinct species in a number of pharmacopoeias as Tinnevelly senna and Alexandian senna. However, both are considered to be synonyms of the single species *Cassia senna* Linn. (WHO, Kew bulletin, 1958).

Cissampelos pareira Linn. (Fam. Menispermaceae). Paathaa (API)

Cyclea peltata Diel is used as Paathaa in all formulations in Kerala. *Cissampelos pareira* is used in North India, but not in Ayurvedic practice in Kerala.[2] *Cyclea arnotii* Miers and *C. burmani* Miers are used as Paathaa substitutes in the East and South India. *Chondodendron tomentosum* Ruiz et Par. and *Stephania glabra* Miers are also used as Paathaa.[10]

Clerodendrum phlomidis Linn. (Fam. Verbenaceae). Agnimantha (API)

In AFI, Agnimantha is equated with *Premna intgrefolia* Linn., *Clerodendrum phlomidis* Linn. f. and *Premna mucronata* Roxb. as substitute drugs.

Agnimantha of eastern and central parts of India contains root and root bark of *Premna obstusifolia* R. Br. and drug source of northern and western region is *P. latifolia* Roxb.[9] Roots of *P. serratifolia* Linn. constitute the drug used in South India, especially in Kerala.[2,9]

Clitoria ternatea Linn. (Fam. Fabaceae). Aparaajitaa (API)

The market samples of Shankhapushpi from South India, consisted largely of *C. ternatea*, whereas from other regions contained dried herb of *Canscora desussata* Roem. and Schult., *Convolvulus microphyllus* Sieb. ex Spreng. (syn. *C. pluricaulis* Choisy), *Evolvulus alsinoides* L. and *Lavendula bipinnata* Kuntze (Syn. *L. burmanii Benth.*)[5(c)]

Kerala physicians do not discriminate between Aparaajitaa and Shankhapushpi, and use *Clitoria ternatea* in place of both. (Vishnukraantaa, a blue-flowered variety of Shankhapushpi, is treated as a synonym of Aparaajita in Bhavaprakasha, 16th century.)[2]

Coleus forskohlii Briq. syn. _C. barbatus_ Benth. (Fam. Lamiaceae). Gandira (API)

In Kerala, *Cayratia carnosa* Gagnep. is reported as being used as Gandira.[4]

Convolvulus pluricaulis Choisy (Fam. Convolvulaceae). Shankhapushpi (API)

Convolvulus pluricaulis is used as Shankhapushpi in North India;[2] *Clitoria ternatea* Linn. (Papilionaceae) and *Canscora decussata* Schult. in South India.[1,2,4] *Canscora decussata* and *Lavendula bipinnata* O. Ktze. are used as Shankhapushpi in Bengal;[4,6] *Evolvulus alsinoides* (Convolvulaceae) is treated as Vishnukrantaa, Vishnukrandi, Vishnugandhi of Siddha medicine.[7]

Coscinium fenestratum (Gaertn.) Colebr. (Fam. Menispermaceae). Kaaliyaka (API)

Stem bark of C. fenestratum is used as a substitute of Daaruharidraa in Tamil Nadu and Kerala,[1,9] also as an Indian substitute of true Columba (Jateorhiza palmate Miers).[4] Contains up to 3.5 per cent berberine, considered better than *Berberis*.[1] It has since long been accepted as a substitute for Berberis aristate DC.[6]

Cryptolepsis buchanani Roem. and Schult. (Fam. Axclepiadaceae). Krishna Saarivaa (API)

Ichnocarpus frutescens R. Br.[3,4] and *Decalepis hamiltonii* Wight and Arn. are used as Krishna Saarivaa in South India.[2,4] The roots of *Decalepis hamiltonii* form a better substitute for Saarivaa than *Ichnocarpus frutescens* and *Cryptolepsis buchanani*.[2]

Curculigo orchioides Gaertn. (Fam. Amaryllidaceae). Taalamuli (API)

In Kerala, in practice, *C. orchioides* is used for both, black and white variety of Mushali (known as Nilappana in Malayalam).[2] (Black variety of Mushali is equated with *Curculigo orchioides,* white variety with *Asparagus adscendens* Roxb., *Chlorophytum arundiancaeum* Baker and *Chlorophytum tuberosum* Baker.)[4,10]

Curcuma zedoaria Rosc. (Fam. Zingiberaceae). Karchuura (API)

The source of Karchuura in Kerala, atleast since the 17th century, has been *Kaempfera galanga* Linn.[2]

Dalbergia sissoo Roxb. (Fam. Fabaceae). Shimshapaa (API)

In Kerala, heartwood of *Xyla xylocarpa* Roxb. Taub. is wrongly used as Shimshapaa.[3,4]

Desmodium gangeticum DC. (Fam. Fabaceae). Shaaliparni (API)

Kerala physicians, by and large, accepted *Psuedarthria viscida* (L.) W. and A. as the source plant of Shaalaparni.[2]

Euphorbia dracunculoides Lam. (Fam. Euphorbiaceae). Saptalaa (API)

AFI equated Saptalaa and Saatalaa with *Euphorbia pilosa* Linn. and *E. dracunculoides* with Shankhini (Part 1, second edn, page 324), while on page 334, both *Euphorbia dracunculoides and Euphorbia pilosa* are equated with Saptalaa. *Clitoria ternatea* Linn. is used as Shankhapushpi and Shankhini both in Kerala.[3,4] In Kerala and Bengal, *Acacia concinna* (Willd.) DC. (Soapnut Acacia) is used as Saptalaa.[3,4]

Fagonia cretica Linn. (Fam. Zygophyllaceae). Dhanvayaasah (API)

In Kerala, *Tragia involucrata* Linn. is used for both Dhanvayaasa and Yavaasah.[2]

Ficus arnottiana Miq. (Fam. Moraceae). Nandi (API)

Nandi-ervatam of Kerala is considered a synonym of Nandivrksha. Nandivrksha is equated with *Cedrela toona* Roxb., also with *Ficus arnottiana* in AFI. Some scholars equated it with *F. retusa*. But in Kerala, *Tabernaemontana divaricata* (L.) Roem. and Schutt. is used as the drug source.[2]

Ficus lacor Buch.-Ham. (Fam. Moraceae). Plaksha (API)

Botanical source of Plaksha in Kerala is *F. microcarpa* Linn.[2] *F. retusa* sensu Hook. f. is also used.[3,4] In Tamil Nadu, *F. lacor* is used.[1]

Flacourtia indica Merr. syn. F. ramontchi L'Herit. (Fam. Flacourtiaceae). Sruvavrksha, Vikankata (API)

In Kerala, one plant drug Aghori, not mentioned in any of the Ayurvedic texts, is equated with *Flacourtia indica*.[2] *F. ramontchi* L'Herit. Is known as Governor's Plum.[5(a)]

Fumaria parviflora Lam. (Fam. Fumaraceae). Parpata (API)

Used in North India as Shahtaraa. In Kerala, *Hedyotis brachypoda* DC. (Rubiaceae), *H. corymbosa* (L.) Lam. (Rubiaceae) and *H. diffusa* Willd. (Rubiaceae) are generally accepted as Parpata.[2] Also in Maharashtra.[4] IMPCOPS, Chennai, is using *Mollugo cerviana* Ser. (Molluginaceae) as Parpataka.[1] *Justicia procumbens* Linn. (Acanthaceae) is used in Gujarat. *Polycarpea corymbosa* Lam. (Caryophyllaceae), *Glossocardia bosvallia* DC. (Compositae/Asteraceae and *Rungia repens* Nees (Acanthaceae) are also used as Parpata.[4]

Garcinia pedunculata Roxb. (Fam. Guttiferae). Amlavetasa

The fruits of *G. pedunculata* are used in Bengal and Assam as Amlavetas; almost everywhere else dried leaf stalks of *Rheum emodi* Wall. Ex Meissn. are is use.[6]

In Kerala, the plants used as Amlavetas are *Cissus repens* Lamk., *C. vitiginiea* Linn., *Ampelocissus latifolia* (Roxb.) Planch. and *Cayratia trifolia* (L.) Domin (all Vitaceae).[2] Fruits of *Garcinia indica* Choisy. are used Amlavetasa (Malabar tamarind) in Tamil Nadu.[1]

Gymnema sylvestre R. Br. (Fam. Asclepiadaceae). Meshashrngi (API)

Meshashrngi is equated with *G. sylvestre*. Kerala physicians use different plants as the source of Vishaanikaa (Meshashringa or Medaashringa): *Vallaris solanacea* (Roth) Kunize, *Cryptolepis buchanani* Roem. and Schult., *Aristolochia bracteolata* Ham., and *Dolichandrone falcata* (DC.) Seemann.[2]

Meshashrngi is a climber while Meshashrnga is a tree. Kerala physicians do not make a distinction between Meshashrngi and Meshashranga. They do not use *Gymnema sylvestre* as the source of the drug Meshashrngi. Meshashranga is equated with *Dolichandrone falcata* (Bignoniaceae), known as Medaashringi in Maharashtra and Gujarat, but *Dolichandrone falcate* is seldom found used in Kerala.

National Academy of Ayurveda equated Meshashrngi with *Dolichandrone falcate* while there a question mark against its equation with *Gymnema sylvestre*.[10]

Hedychium spicatum Ham. ex Smith (Fam. Zingiberaceae). Shati (API)

H. spicatum (a Himalayan plant) is not available in South India. *Curcuma zedoaria* Rosc. is used as a substitute.[1,9]

Hemidesmus indicus (L.) R. Br. (Fam. Asclepiadeceae). Shveta Saariva (API)

In Kerala, Tamil Nadu and Karnataka, aromatic roots of *Decalepis hamiltonii* Wight and Arn. are sold as Saarivaa.[2] *Cryptolepis buchanani* Roem and Schult. is equated with Krishna Saarivaa (AFI, Part 1, page 333), while *Ichnocarpus frutescens* R. Br. has also been validated as Krishna Saarivaa (AFI, Part 1, page 324).[1,2] *Ichnocarpus frutescens* is sold as Krishna Saarivaa in Bengal and Kerala.[3] *Decalepis hamiltonii* roots form a better substitute for Saarivaa than *Ichnocarpus frutescens* and *Cryptolepis buchanani.*[2] South Indian taxonomists, V.V. Sivarajan and Indira Balachandran identified *Hemidesmus indicus* as Krishna Saarivaa as the roots are blackish in colour and *Decalepis hamiltonii* as Shveta Saarivaa, the roots being brownish white in colour.[2]

Ipomoea digitata Linn. (Fam. Convolvulaceae). Kshiravidaari (API)

Kerala physicians accept *Ipomoea mauritiana* Jacq. as the source of Kshiravidari.[2]

Juniperus communis Linn. (Fam. Cupressaceae). Hapushaa (API)

Most of the authors equate Hapusha with *J. communis*, but in South India Sphaeranthus indicus Linn. is used as the drug source.[1,2] Kerala physician consider Hapushaa and Mundi to be synonymous.[2]

Leptadenia reticulata W. and A. (Fam. Asclepiadaceae). Jivanti (API)

Roots of Holostemma adakodien Schult. are used as Jivanti in South India, especially in Kerala.[2,4] *Desmotrichum fimbriatum* Blume is known as Jivanti in Hindi, Bengali and Marathi.[5(a)] It is sold as Swarnjivanti in Bengal.[4] Jivanti is a synonym of Guduchi, Abhayaa, Medaa, and Vrkshaadani. When the selected drug is a vegetable, then *Leptadenia reticulate* is the correct validation.

Madhuca indica J.F. Gmel. (Fam. Sapotaceae). Madhuuka (API)

M. longifolia (Koenig) Macb. syn. *Bassia longifolia* Koenig is equated with South Indian Mahua, Mowra Butter Tree.[4] It has been identified as Jala-mahuaa (Jala-madhuuka) and Madhuulaka.[10]

Marsdenia tenacissima Wight. and Arn. (Fam. Asclepiadaceae). Muurvaa (API)

In kerala, *Marsdenia volubilis* T. Cooke is used as Muurvaa and *M. tenacissima* as Morata, considering it as a variety.[4] But *Chonemorpha fragrans* (Moon) Alston is the accepted source of Muurvaa.[2,3,4] *Argyreia nervosa* (Burm. F.) Boj., Vrdhadaaruka, is also found to be used as Muurvaa in some parts of Kerala.[2] In Tamil Nadu, *Sanseveria roxburghiana* Schult. is used.[1]

Martynia annua Linn. (Fam. Martyniaceae). Kaakanaasikaa (API)

Seeds of *Anamirta paniculata* W and A. are used as Kaakanaasika in Tamil Nadu.[1] *Trichosanthes tricuspidata Lour.*, syn. *T. palmata* Roxb., is used as Kaakanaasaa.[3,4] National Academy of Ayurveda equated Kaakanaasikaa with *Pentatropis microphylla* W. and A. while there a question mark against its equation with *Martynia annua.*[10] One scholar of Ayuveda (Vaidya Bapalal) equated Trikantaka, a synonym of Gokshura, with *Martynia annua.*[4]

Merremia tridentata (L.) Hall. f. (Fam. Convolvulaceae). Matsyapatrikaa (API), Prasaarini keraliya (Non-classical Sanskritized name in API)

Prasaarini is officially equated with *Paederia foetida* Linn. (Fam. Rubiaceae), but in Tamil Nadu and Kerala, *M. tridentata* ssp. *tridentata* and subsp. *hastata* are the source of Prasaarani or Prasaarini.[1,2,10] South Indian plant drug is actually Tala Nili-Prasaarini.[5]

Mesua ferrea Linn. (Fam. Guttiferae). Naagakeshara (API)

In Tamil Nadu and adjacent states, tender fruits of *Cinnamomum wightii* or fruits of *Dillenia pentagyna* are used as Naagakeshara.[1] In South India, flower buds of *Mammea longifolia* Planch. and Triana, syn. *Ochrocarpus longifolius* Benth. and Hook. f. (Cluciaceae) are sold as Naagakeshara.[4,9] *Calophyllum inophyllum* Linn. is Punnaaga of South India.[10]

Mollugo cerviana Seringe (Fam. Aizoaceae). Grishmachatraka, Ushnasundara (API)

Mollugo oppositifolia Linn. is used as Parpata in Kerala.[10] *Hedyotis corymbosa* (Linn.) Lam., *H. brachypoda* (DC.) Sivar.et al. and *H. diffusa* Willd. (Rubiaceae) are also accepted as a source of Parpata.[2] *Hedyotis herbacea* Linn. is also reported as Parpata.[10] *M. cerviana* Ser. is the source of Parpata in Tamil Nadu.[1]

Nigella sativa Linn. (Fam. Ranunculaceae). Upakunchikaa (API)

In AFI, Ajaaji is equated with *Cuminum cyminum* Linn. (Shveta Jiraka) but in Kerala it is equated with Upakunchikaa (*Nigella sativa*, Kalaunji of North India).[4] *Carum carvi* Linn. is Krishna Jiraka[10] of the North while in Tamil Nadu Krishna Jiraka is equated with *Nigella sativa.*[1]

Ougeinia oojeinensis (Roxb.) Hochr. (Fam. Fabaceae). Tinishah (API)

In Kerala, *Melastoma malabathricum* Linn. and *Osbeckia aspera* (L.) Blume (Melastomataceae) are the main sources of Tinisha.[2]

Paederia foetida Linn. (Fam. Rubiaceae). Prasaarini (API)

In North India, *P. foetida* is used as Prasaarini. In Kerala and Karnataka, the source of the drug is *Merremia tridentata* (L.) Hall. f. subsp. *tridentata* and subsp. *hastata.*[2,3,10]

Pavonia odorata Willd. (Fam. Malvaceae). Gandhashiphaa, Picchila lomashah (API) (Nonclassical nomenclature based on English synonym, Fragrant Sticky Mallow.)

Pavonia odorata is equated with Hrivera and Baalaka.[10] In Kerala and Tamil Nadu, the main source of Baalaka and Hrivera was never *P. odorata*. *Coleus vettiveroides* Jacob and *C. zeylanicus* (Benth.) Cramer were used.[2] *Coleus vettiveroides* is cultivated in Kerala.[1]

Phyla nodiflora (Linn.) Greene (Fam. Verbenaceae). Jalapippali (API)

National Academy of Ayurveda equated *Phyla nodiflora* with Siddha Poduthalai. (In Formulary of Siddha Medicine, published by IMPCOPS, Poduthalai has been identified as Potuttali in Tamil, Bokkena in Telugu, Kattu Thippli and in Malyalam and Bhui okra in Hindi. API, Vol. V, quoted Malyalam synonym as Nirthippali. Thus Malyalam name has been linked with an unidentified Ayurvedic drug Jalapippali and Jalakarnaa.

Indian National Science Academy scientists equated Jalapippali with *Ranunculus aculeata* Pers. (Poison Buttercup). The Wealth of India equated Jalapippli with *Commelina salicifolia* (syn. *C. longifolia*) as Jalapippli based on its folk name, Jalapipari.[5] (*Commelina* spp. are known as Dayflower weed.) National Academy of Ayurveda identified it as a variety of Kanchata.[10] (Kanchata Ayurvedic synonyms: Toyapippali, Shakulaadani, Laangali and Shaaradi. Panikanchira is a folk synonym in Bengal.) Valid botanical identity of Jalapippali of Charaka Samhita is yet to be identified. Two varieties of Ayurvedic plant drug Gandira have been mentioned, *sthalaja* and *jalaja*. Vaidya Bapalal has equated *sthalaja* variety as *Coleus barbatus*, and *jalaja* variety as *Amaranthus mangostanus*. Jalaja Gandira and Jalapippali have been suggested as synonyms. (Priyavrt Sharma, Dravyaguna Vigyana, V, page 113).

Piper retrofractum Vahl. syn. *P. chaba* Hunter non Blume. (Fam. Piperaceae). Chavya (API)

In Kerala, roots of many wild spp., including *Piper brachystachym* Wall. Syn. *Chavika shpaerostachya* Mig., are used as Chavya.[3,4]

Pluchea lanceolata Oliver and Hiern. (Fam. Asteraceae). Raasnaa (API)

P. lanceolata is the official Raasnaa, substitute plant drug is *Alpinia galanga* Willd. which is used in South India (AFI, Part I, page 323). *Dodonaea viscosa* Linn. is used in Andhra Pradesh. *Polygonum grabrum* Willd. is sold since decades in Varanasi market as Raasanaa.[4] *Vanda roxburghii* R. Br. is used in Eastern India.[10] Bengal. *Heliotropium strigosum* Willd. is sold in Bihar.[4] Blepharispermum subsessile DC. Is used in Baster region of Madhya Pradesh.[10]

Plumbago indica Linn. syn. *P. rosea* Linn. (Fam. Plumbaginaceae). Rakta chitraka (API)

Plumbago zeylanica Linn. (white-flowered variety) is quoted as Chitraka in AFI (pages 310, 337). It is used in North India. *P. indica*, the red-flowered variety, is used in South India and Bengal.[2] Black (blue)-flowered variety, *P. auriculata* Lam. (=*P. capensis* Thumb.) is not known to be used as a source of Chitraka.[2] In South India, Rakta chiraka is considered to be therapeutically more active.[5(a,c)] The roots as well as the root bark of *P. indica* form an important indigenous drug, but less commonly used than those of *P. zeylanica*.[5(a,c)] (Plumbagin is present in both, about 0.9 per cent).[5(a)]

Portulaca oleracea Linn. (Fam. Portulacaceae). Kozuppaa (API, AFI) Brhat lonikaa (RAV)

Valid Ayurvedic name of Kozuppa is Loni and Lonikaa.[6] Kozuppaa is the bigger variety. At least three different plants are currently used as the source of Kozuppaa in different parts of Kerala: *Alternanthera sessilis* Linn., *Glinus oppositifolius* (Linn.) A. DC, and *Portulaca oleracea*.[2] *Alternanthera sessilis* (Linn.) R.Br. ex DC.(Matysaakshi) has been the source of Kozuppaa since long in Kerala.[2] *Portulaca quadrifida* Linn., the smaller variety of Lonikaa, is used as Pasalai in Siddha medicine of Tamil Nadu.[10]

Pueraria tuberosa DC. (Fam. Fabaceae). Vidaarikanda (API)

Ayurvedic texts mention two varieties of Vidaarikanda. Vidari is equated with *Pueraria tuberosa* and Kshira vidaari with *Ipomoea digitata* Linn. syn. *I.paniculata* R.Br. In Kerala, *Ipomoea digitata* is used for both.[4] Kerala physicians also accept *Ipomoea mauritiana* Jacq. as the source of Kshira vidaari. Tubers of *Adena homdala* (Gaertn.) de Wilde are also sold in the name of Vidaari. (The tuber is an unauthorized adulterant and is reported to be poisonous.) [2] In tamil Nadu *Ipomoea digitata* is preferred.[1] In Bihar, Bengal and Odisha, stem tubers of *Ipomoea paniculata* are more frequently used. Stem tubers of *Trichosanthes cordata* Roxb. are also sometimes sold as Vidaarikanda.[9]

Rauwolfia serpentina (Linn.) Benth.ex Kurz (Fam. Apocynaceae). Sarpagandhaa (API)

Roots of *Rauvolfia densiflora* Benth. and *R. micrantha* Hook. f. are sometimes found mixed with plant material sold as Sarpagandhaa in Kerala and Western India.[9] *R. micrantha* is known as Malabar

Rauvofolia. Found in Kerala up to an altitude of 300m. Roots of *Rauvolfia tetraphylla* Linn. syn. *R. canescens* Linn., cultivated in various parts of India, is employed as a substitute when *R. serpentina* root is not available. (In folk medicine, *Rauvolfia tetraphylla* is known as Badaa chaanda, *R. serpentina* as Chhotaa chaanda). Among major adulterants of *Rauwolfia serpentina* are thin roots of *Tabernaemontana divaricata* (L.) R. Br. (Apocynaceae).

Saccharum bengalense Retz. syn. *S. munja* Roxb. (Fam. Poaceae). Shara (API)

In Kerala, *Saccharum arundinaceum* Retz. is used as Shara. Arundo donax Linn. is also found used in many places as the drug source.[2] Trna panchamuula (the roots of five grasses) of Charaka Samhita contains Shara, while in South India, the group contains Kusha (*Imperata cylindrical* Beave.), Ikshu (*Saccharum officinarum* Linn.), Shaali (*Oryza sativa* Linn.), Darbha (*Desmostachya bipinnata* Staph.) and Kaandekshu (*Saccharum spontaneum* Linn.).[1]

Salix alba Linn. (Fam. Salicaceae). Shveta Vetas (API)

In Kerala, *Homonoia riparia* Lour. (Euphorbiaceae) is used as Vetasa or Jala-vetasa.[3,4]

In Dhanvantari Nighantu (13th century) text, Vetas-dwya (the Two Vetasa) have been mentioned. Vetasa is now equated with *Salix caprea* Linn.[4] and Jala vetasa with *Salix tertasperma* Roxb.[4] *Salix alba* belongs to North-western Himalayas. It still remains unexplained why Vetasa was included in the "Five Latex-bearing Trees" (Pancha-kshiri Vrksha) as a substitute.[4,6]

Saraca asoca (Rosc.) Dc. Willd. (Fam. Leguminosae). Ashoka (API)

Polyalthia longifolia (Sonn.) Thwaites (Annonaceae), introduced from Sri Lanka, is wrongly called Ashoka.[2] Its bark is the most common adulterant of Ayurvedic Ashoka bark.[9]

Saussurea lappa C.B. Clarke (Fam. Compositae). Kushtha (API)

Kustha is used as a substitute of Pushkar-muula (*Inula racemosa* Hook. f.)[4] Accepted source of Pushkar-muula in Kerala is *Psilanthus travancorensis* Leroy syn. *Coffea travancorenis* Wt. and Arn.[2] Kuth, commonly known as costus in trade, has no connection with the botanical genus Costus.[5(a)] *Costus speciosus* (Koenig) Smith is used as the source of Chandaa of Ayurvedic medicine in Kerala.

Scindapsus officinalis Schoott. (Fam. Araceae). Gajapippali (API)

In Kerala, sliced and dried inflorescence of *Balanophora indica* Wall. and pieces of stem (not fruits) of *Scindapus officinalis* are sold as Gajapippali.[4]

Sida rhombifolia Linn. (Fam. Malvaceae). Mahaa balaa (API)

While *Sida cordifolia* Linn. is widely used source of Balaa in northern parts of India, Kerala physicians have adopted *Sida rhombifolia* ssp. *retusa* for Balaa. *Sida acuta* Burm. F. is also widely used as adulterant in Kerala.[2] (Medicinal oils of Balaa are prepared mostly in Kerala and Tamil Nadu.)

Solanum nigrum Linn. (Fam. Solanaceae). Kaakamaachi (API)

Solanum nigrum auct. non. Linn. syn. *S. americanum Mill.*, is equated with Kaakamaachi in South India.[2] In Kerala, Kaakamaachi and Karintakkaali were treated as one drug. Karintakkaali has been identified as *Geophilla repens* (Linn.) I. M. Johnson, syn. *G. reniformis* D. Don (Fam. Rubiaceae).[4] However, these two are different plant drugs.

Sphaeranthus indicus Linn. (Fam. Asteraceae). Munditikaa (API)

In Kerala, *Sphaeranthus indicus* is equated with Hapushaa (*Juniperus communis* Linn.); red and white varieties of Hapushaa with *S. indicus* and *S. africanus* Linn. respectively.[2,3] This is not acceptable to other schools of Ayurveda.[4] (In the Wealth of India, Vol X, page 4, Hapushaa and Shveta Hapushaa are included among synonyms of *S. indicus* and *S. africans* Linn. respectively.)[5(a)]

Stereospermum chelonoides (L. f.) DC. Paatlai (API vol., IV), Paatalaa (API vol. III)

S. tetragonum DC. syn. *S. personatum* is used in Kerala as Paatalaa.[4]

Strychnos nux-vomica Linn. (Fam. Fabaceae). Vishamushti (API)

Strychnos nux-vomica is a tree, while *S. colubrina* Linn. is a climber of the Deccan peninsula, found from Konkan to Cochin. Its roots, seeds bark and wood contain strychnine and brucine. Also used as Nux-vomica.[2, 5(a)]

Symplocos racemosa Roxb. (Fam. Symplocaceae). Lodhra (API)

In Kerala, *S. cochinchinensis* (Lour.) S. Moore (= *Symplocos spicata* Roxb.) is used as Lodhra.[2] In different regions, *Symplocos crataegoides* Buch,-Ham. ex D. Don, *Symplocos reticulate* Grah ex C.B. Clarke *and Symplocos sumuntia* Buch.-Ham. ex D. Don are used as Lodhra.[10]

Synantherias sylvatica Schott Gen. Aocja syn. *Amorphophallus sylvaticus* (Roxb.) Kunth. (Fam. Araceae). Aranya suurana (API)

Suurana is equated with *Amorphophalus paeoniifolius* var. *campanulatus* (Decne) Sivad.; Aranya Suurana with *Amorphophanus paeoniifolius* var. *paeoniifolius*, syn. *Arum paenoiifolium* Dennst. (now scarce). Only var. *campanulatus* is used in Kerala.[2]

Tecomella undulata (Sm.) Seem. syn. *Tecoma undulata* G. Don (Fam. Bignoniaceae). Rohitaka (API)

In South India, *Aglaia polystachya* Wall. is used as Rohitaka.[1] In Gujarat, *Maba nigrescens* Dalz. and Gibbs. Is used as Rohitaka (Rohido).[10] *Aphanamixis polystachya* (Wall.) Parker syn *Amoora rohitaka* W. and A. is also an accepted substitutes of Rohitaka (AFI).

Teramnus labialis Spreng. (Fam. Fabaceae). Maashparni (API)

In Kerala, *Vigna radiata* var. *sublobata*, *Vigna dalzelliana*, *Vigna mungo*, *Vigna umbellata* and *Rhyncosia nummularia* are used as Maasaparni.[2]

Tragia involucrata Linn. (Fam. Euphorbiaceaee). Vrshchikaali (API)

In Kerala, *Tragia involucrata* is used as Duraalabhaa,5 and *Heliotropium indicum* as Vrshchikaali.[2]

Uraria picta Desv. (Fam. Fabaceae). Prshniparni (API)

In Kerala, *Desmodium gangeticum* DC. is used as Prshniparni and *Uraria picta* as Shaaliparni. However, *Pseudarthria viscida* (L.) W and A of the same family is preferred in Kerala as well as in Tamil Nadu, as Shaaliparni.[1,2]

Valeriana wallichii DC. (Fam. Valerianaceae). Tagara (API)

In Kerala, *Limanthemum cristatum* Griseb. (Gentianaceae) is used in certain parts as Tagara.[4] In Tamil Nadu, *Nymphoides macrospermum* Vasudevan (Menanthaceae) is sold in the market as Tagara.[1] But *V. wallichii* DC. is preferred.[4]

Vallaris solanacea Kuntze syn. *V. heynei* Spreng. (Fam. Apocynaceae). Aasphotaa (API)

In Kerala, *V. solanacea*, *Cryptolepis buchnani* Roem. and Schult, *Aristolochia bracteolata* Lam. are used in different places as Vishaanikaa.[2] *Pergularia extensa* N. E. Br. Is also used as Vishaanikaa.[10] It is known as Utthaamani in Tamil Nadu and Uttamaarani in North India. Used in Siddha medicine of Tamil Nadu.[10]

Bhavamisra (16th century) equated Aasphotaa with two drugs—Aparaajita (*Clitoria ternatea* Linn.) and Saarivaa (*Hemedismus indicus* R.Br.). AFI, Part I recognized *Hemesdesmus indicus* as a substitute drug of *Vallaris solanacea* (page 308). However, modern commentators have identified Aasphotaa as *V. solanacea*, in preference to Saarivaa.[4] *Vallaris glabra* Kuntze is also known as Aasphota.[10]

Vigna trilobata (L.) Verdc. syn. *Phaseolus trilobus* Ait. (Fam. Fabaceae). Mudagaparni (API)

In Kerala, *Centrosema pubescens* and *Vigna* spp., *V. pilosa*, *V. angularis*, *V. umbellata*, *V. vexillata* and *V. adenantha* are used as the source of Mudgaparni.[5] In Tamil Nadu, *Vigna trilobata* is used.[1]

Vitex negundo Linn. (Fam. Verbenaceae). Nirgundi seed=Renukaa (API)

In Tamil Nadu, *Piper aurantiacum* Wall ex. DC. fruits are used as Harenukaa.[1] Harenukaa and Renukaa are equated with *Vitex agnus-castus* Linn., and Nirgundi with Vitex negundo Linn. (AFI, Part I, pages 313 and 321.) *Piper aurantiacum* and *V. negundo* are different drugs.[5(a)]

Zanthoxylum armatum DC. (Fam. Rutaceae). Tejovati (API)

Z. rhetsa syn. *Z. limonella* (Dennst.) Alston; *Z. budrunga* Wall. ex. DC. are found in Kerala, Karnataka and Bangla Desh. Used mostly in South India. Known as Atitejani, Su-tejasi, while *Z. alatum* Roxb. (syn. of *Zanthoxylum armatum*) is Tejovati of Ayurvedic medicine.(AFI, Vol. 1, page 327.)

Abbreviations

API: The Ayurvedic Pharmacopoeia of India, Part I, Vol. I (1989), II (1993), III (2001), IV (2004), V (2006), VI (2008); Government of India, Ministry of Health and Family Welfare, New Delhi.

AFI: The Ayurvedic Formulary of India, Part I, Second Revised English Edn., 2003; Part II, First English Edn., 2000, AYUSH, Government of India, Ministry of Health and Family Welfare, New Delhi.

RAV: Rashtriya Ayurveda Vidyapeeth (An auponomous organization under Ministry of Health and Family Welfare, Government of India).

References

1. C.P. Khare, Ayurvedic Pharmacopoeial Plant Drugs: Expanded Therapeutics, CRC Press, 2015.
2. Vaidya Yoga Ratnavali (Formulary of Ayurvedic Medicines), The Indian Medical Practitioner's Coop. Pharmacy and Stores Ltd. (IMPCOPS), Chennai-600041; 2000.
3. Sivarajan V.V. and Indira Balachandran, Ayurvedic Drugs and their Plant Sources, Oxford and IBH Publishing Co. Pvt. Ltd., New Delhi, 1994.
4. Indian Medicinal Plants: A Compendium of 500 species, Arya Vaidyasala, Kottakkal, 1993-97,Orient Longman.
5. Sharma S.K., Chunekar K.C. and Hota N.P.,Plants of Bhava Prakash, Eds,1999, National Academi of Ayurveda (RAV), New Delhi.
6. The Wealth of India, Vol. II to XI(a), Revised Vol. 1 to 3(b), First Supplement Series, Vol. 1 to 5(c), Second Supplement Series, Vol. 1 to 3(c); 1949-2009; National Institute of Science Communication and Information Resources (Council of Scientific and Industrial Research), New Delhi-110012.
7. Thakur Balwant Singh and Chunekar K.C,Glossary of Vegetable Drugs in Brhittryi (Plants of Charaka Samhita, Sushruta Samhita and Ashtangahridaya), Chaukhamba Amarbharati Prakashan, Varanasi, 1999.
8. Malhotra S.C., Pharmacological Investigation of Certain Medicine Plants and Compound Formulations used in Ayurveda and Siddha, Central Council for Research in Ayurveda and Siddha (CCRAS), 1990.
9. Y.K. Sarin, Illustrated Manuel of Herbal Drugs used in Ayurveda, CSIR and ICMR, 1996.
10. Sharma S.K., Chunekar K.C., Medicinal Plants Used In Ayurveda, Rashtriya Ayurveda Vidyapeeth (National Academy of Ayurveda), New Delhi, 1998.

7

Indian Initiatives in the Revival of Traditional Systems of Medicine: An Overview and Recent Leads in Ayurveda*

Chandra Kant Katiyar

India has a rich cultural heritage and Ayurveda, the ancient medical wisdom, represents one aspect that enjoys the same prestige today as in the past. Ayurveda, the 'Science of Life', not only encompasses the preventive and curative aspects of diseases but also provides a unique approach of health promotion, leading to a healthy, active and long lifespan.

In its journey through the 20th Century, which has witnessed epoch-making discoveries in science leading to inventions and a whole gamut of technological advances, Ayurveda had to face several hurdles and challenges, and both cultural and physical onslaughts from within and outside the country. But it has successfully withstood all these due to its inherent strength, based on its own philosophy, science and ethical values. More so, the popularity of Ayurveda has not only withstood the test of times but has actually crossed the trans-national and cultural boundaries, and is being incorporated as a mainstream medical field besides attaining a global status.

The journey from the Vedic era to the Genomic era has been engrossing because of the concerted efforts and contributions of the leading Indian scientific institutions, Indian scientists from varied disciplines, Ayurvedic practitioners, visionary leaders and above all the faith and trust of the common man in Ayurveda, its principles, products and the belief that it can do no harm but only good.

Over the last six decades, several developments have taken place in the fields of infrastructure, education, research, regulatory controls, commerce and governance and globalization which have profoundly helped Ayurveda in its rejuvenation (re-awakening).

Pt. Madan Mohan Malviya, a visionary and academician of high order started a unique integrated course of Ayurveda and Modern Medicine, AMS at Banaras Hindu University (BHU) in 1925. After few decades this became a centre of excellence for postgraduate education in Ayurveda in the 60s. Banaras Hindu University has contributed a lot to development of Ayurveda. Later, Gujarat Ayurveda University, Jamnagar also helped in carrying forward the mantle of Ayurveda. While BHU School concentrated more on science and research oriented education, the Jamnagar School of Ayurveda decided to focus more on puristic Ayurveda. Besides academic institutions, private organizations like Arya Vaidyasala, Dabur and Zandu also made significant contributions in popularizing Ayurvedic treatments like Pancha karma as well as Ayurvedic products. Dabur's major contribution is into converting

---

* This article is being reproduced from "Science in India: Achievement and Aspirations : 75 years of the Academy" edited by Prof H Y Mohan Ram and P N Tandon and published by Indian National Science Academy, New Delhi in 2010, with due permission from Indian National Science Academy.

an age-old formulation of Chyawanprash into popular consumer health product among the Indian masses. Therefore, directly or indirectly both public and private institutions contributed to the growth of indigenous system of medicine in India over the decades.

Various government bodies such as the Indian Council of Medical Research (ICMR); Department of Indian Systems of Medicine and Homoeopathy (ISM and H), presently renamed Department of Ayurveda, Yoga and Naturopathy, Unani, Siddha and Homoeopathy (AYUSH); Central Council for Research in Indian Medicine and Homoeopathy (CCRIMH), now segregated into four different councils, *viz.* CCRAS (Ayurveda and Siddha), CCRUM (Unani Medicine), CCRH (Homoeopathy) and CCRYN (Yoga and Naturopathy); Council of Scientific and Industrial Research (CSIR); Department of Science and Technology (DST); Department of Biotechnology (DBT) and other affiliated advanced centres are actively involved in the research and development of Indian Systems of Medicine.

Infrastructure

The emphasis of the Department of AYUSH is on implementing the schemes which address the identified thrust areas such as upgradation of educational standards, quality control and standardization of drugs, improving the availability of raw material, research and development, and awareness generation about the efficacy of the systems in domestic and international spheres. The system-wise infrastructure pertaining to Indian Systems of Medicine as on April 1, 2007 is presented in Table 7.1.

Statutory regulatory control pertaining to education and practice is taken care of by the Central Council of Indian Medicine (CCIM) which was established through an Act of Parliament in 1970 while that pertaining to drug regulation is governed by the Drugs and Cosmetics Act of 1940 and rules thereunder of 1945 amended from time to time. A few colleges and universities have also started short-term courses on Ayurveda in Australia and Europe.

Following is the history of reorganization of various administrative bodies dealing with indigenous systems of traditional medicine.

The Department of AYUSH also finances CCRAS, CCRYN, CCRUM and CCRH. CCRAS came into existence after the bifurcation of the erstwhile Central Council for Research in Indian Medicine and Homoeopathy in the year 1978. CCRAS is an apex body in India for the coordination, development and promotion of research on scientific lines in fundamental and applied aspects of Ayurveda and Siddha systems of medicine. It also promotes and assists institutions of research for the study of diseases, their prevention and cure, especially with emphasis on covering the rural population of the country. CCRAS has been executing its research programmes through 38 constituent institutes/centres spread across India. However, the research carried out is limited and often with unknown scientific output.

TABLE 7.1: System-wise AYUSH Infrastructure Available in India (as on 01.04.2007)

Sl.No.	*Facility*	*Ayurveda*	*Unani*	*Siddha*	*Yoga*	*Naturopathy*	*Homoeopathy*	*Amchi*	*Total*
1.	Hospitals	2398	268	281	8	18	230	1	3204
2.	Beds	42963	4489	2401	135	722	10851	22	61583
3.	Dispensaries	13914	1010	464	71	56	5836	86	21437
4.	Regd. practitioners	453661	46558	6381		888	217850		725338
	(a) Institutionally qualified	324242	23982	2926		839	154240		506229
	(b) Non-institutionally qualified	129419	22576	3455		49	63610		219109
5.	AYUSH colleges (UG and PG)	242	40	8		10	185		485
	(a) Admission capacity	12216	1817	460		385	14509		29387
6.	Colleges (UG)	240	39	7		10	183		479
	(a) Admission capacity	11225	1750	350		385	13425		27135
7.	Colleges (PG)	62	7	3			33		105
	(a) Admission capacity	991	67	110			1084		2252
8.	Exclusive PG colleges	2	1	1			2		6
	(a) Admission capacity	40	28	30			99		197
9.	Manufacturing units	7621	321	325			628		8895

UG: Undergraduate; PG: Postgraduate.

CCRYN was established in 1978 for providing better opportunity for all-round development of Yoga and Naturopathy, independently according to their own doctrines and fundamental principles. This Council also undertakes education, training, research and other programmes in Yoga and Naturopathy and is also involved in initiating, aiding, developing and coordinating scientific research in fundamental and applied aspects of Yoga and Naturopathy.

CCRUM was established by the Ministry of Health and Family Welfare, Government of India, as an autonomous organization in 1979 to initiate, aid, develop and to coordinate scientific research in Unani system of medicine. The Council is engaged in the multifaceted research activities in the field of Unani medicine. The Council's research programme comprises clinical research, drug standardization, survey and cultivation of medicinal plants and literary research. These activities are being carried out through a network of 22 institutes/units functioning in different parts of the country.

CCRH is fully funded by the Government of India and is engaged in research in Homoeopathy. The Council functions through a network of 40 institutes/units located in different parts of the country. These institutes/units are engaged in research in various aspects of Homoeopathy such as clinical research; drug proving research (Homoeopathic pathogenetic trial); clinical verification research; drug standardization; and survey, collection and cultivation of medicinal plants.

National Institutions pertaining to individual system of traditional medicine have also been set up. These include:

- National Institute of Ayurveda, Jaipur was established in 1976 by the Government of India as an apex institute of Ayurveda in the country to develop high standards of teaching, training and research in all aspects of Ayurvedic system of medicine with a scientific approach. The Institute is engaged in teaching, clinical evaluation, training and research at under-graduate, post-graduate and Ph.D. levels. It also provides guidance for external Ph.D. scholars in Ayurveda by affiliation with the Rajasthan Ayurved University.
- National Institute of Naturopathy, Pune was set up in 1986. This institute has a Governing Body headed by the Union Minister for Health and Family Welfare as its President.
- National Institute of Unani Medicine, Bangalore was started in 1984 as a centre of excellence to develop and propagate Unani system of medicine. It is a joint venture of the Government of India and the State Government of Karnataka. It is affiliated with the Rajiv Gandhi University of Health Science, Bengaluru.
- National Institute of Siddha, Chennai was founded in 2005 and is an autonomous organization under the control of Department of AYUSH. The Institute conducts post-graduate education for students of Siddha system, provides medical care, conducts research, and develops, promotes and propagates Siddha system of medicine.
- National Institute of Homoeopathy, Kolkata was established in 1975 as an autonomous organization under the Ministry of Health and Family Welfare, Government of India. The Institute offers degree courses in Homoeopathy since 1987 and post-graduate courses since 1998-99. It was functioning under the University of Calcutta up to 2003-04. From 2004-05 onwards, it has been affiliated to the West Bengal University of Health Sciences. The Institute also conducts regular orientation/training courses for teachers and physicians.

A National Medicinal Plants Board (NMPB) was set up under the Department of AYUSH through a Government resolution in 2000. The Board is responsible for coordination with Ministries/Departments/Organizations/State and UT Governments for sustainable development of medicinal plants in general and specifically for drawing up policies and strategies for conservation, cost-effective cultivation, proper harvesting, processing, research and development, and marketing of raw material in order to protect, sustain and develop the medicinal plants sector.

Research

Sir Col. Ram Nath Chopra, also known as the father of Indian pharmacology, propagated the integration of Indian Systems of Medicine to take care of the health of Indian population in the first health policy document prepared after India won Independence. He was also the founder Director of the Indian Drug Laboratory and later of the Regional Research Laboratory, Jammu and started pharmacological research on Indian medicinal plants.

An Advisory Committee on indigenous drugs constituted in 1963 for the unique Composite Drug Research Scheme (CDRS) by the ICMR brought together, for the first time, experts in the Ayurvedic system of medicine, modern medicine and scientists (botanists and phytochemists) for selecting and screening of Indian medicinal plants for biological activity on the basis of their therapeutic claims. During 1964-70, ICMR, through the Ministry of Health and in collaboration with the then CCAR and CSIR, conceived, designed and technically implemented this scheme. From the Ayurvedic fraternity, Dr C. Dwarakanath was instrumental in carrying forward this scheme and subsequently the first group of 58 medicinal plants was subjected to investigation for pharmacognostic, phytochemical and pharmacological aspects and some of these reached an advanced stage of investigation. CDRS was the very first attempt at a multidisciplinary, integrated, coordinated research on medicinal plants. Under this scheme, Saptachakra (*Salacia macrosperma and Salacia prinoides*) showed promising results for diabetes mellitus. In 1970, CDRS was transferred to the newly constituted CCRIMH.

In 1983-84, ICMR initiated another project to review certain time-honoured traditional therapies, aimed at validation of Ayurvedic products and practices through product standardization and clinical validation. For this, the concept of reverse pharmacology was applied and goal oriented projects were formulated on traditional remedies for anal fistula, diabetes mellitus, viral hepatitis, bronchial asthma, urolithiasis, filariasis, Kala-azar and wound healing. Among these, '*Kshaara Sutra*', a medicated thread used for anal fistula was found to be safe, ambulatory and cost-effective alternative to surgery. Further- more, pharmacopoeial standards pertaining to '*Kshaara Sutra*' were also delineated, depicting proper quality control during production of the same. It was Prof P.J. Deshpande of Banaras Hindu University, Varanasi, who had done extensive research and had provided a scientific and standard method for the preparation of '*Kshaara Sutra*'. '*Kshaara Sutra*' is a special surgical procedure using medicated thread to treat fistula-in-ano. '*Kshaara Sutra*' has been standardized by ICMR and includes dipping of standardized size linen thread in latex of *Euphorbia neriifolia* with turmeric powder, dried and rolled in ash of *Achyranthes aspera*. Presently, '*Kshaara Sutra*' therapy is a very viable alternative even in refractory and relapsed cases of anal fistula. Another breakthrough was achieved with the plant Kutaki (*Picrorhiza kurroa*) as source of a hepatoprotective drug at Central Drug Research Laboratory (CDRI), Lucknow. In diabetes, the plant Vijaysara (*Pterocarpus marsupium*) gave consistently promising results. To promote research on traditional remedies, two advanced research centres — one for drug standardization at the Department of Pharmacognosy, University Institute of Pharmaceutical Sciences, Punjab University, Chandigarh, later shifted to Regional Research Laboratory (presently Indian Institute of Integrative Medicine), Jammu, and one for clinical pharmacology at KEM Hospital, Mumbai, were set up. Recently, two advanced centres — one for pharmacokinetics, bioavailability and herb-drug interaction studies at BYL Nair Hospital, Mumbai and one for standardization and quality control of selected herbal remedies/natural products at the National Institute of Pharmaceutical Education and Research (NIPER), Chandigarh, have been organized.

Indian systems of medicine have also contributed in the field of therapeutics and certain therapeutic regimen and therapeutic modalities have resulted in enormous utility in chronic degenerative disorders, neuro-degenerative disorders and auto-immune disorders. The modalities which have been used are Panch-karma, Shiro-dhara, Jalauka (Leech therapy) and Ashtanga Yoga. Another significant contribution of Ayurveda is the unique way of classifying human population based on individual constitution or 'Prakriti'. Ayurveda identifies principles of motion (Vata), metabolism (Pittha) and structure (Kapha) as discrete phenotypic groupings, elements of which may be found in all people, but which predominate in sufficiently differing degrees in individuals to form a three-fold body typology. This concept of 'Prakriti' based on 'Tridosha Theory' allows for individually suited treatment and lifestyle recommendations. This concept has recently been validated through genomic studies and has been published in the Journal of Translational Medicine (2008).

Efficacy of Ayurveda and Siddha drug formulations have been proven for the treatment of various diseases such as bronchial asthma, epilepsy, malaria and peptic ulcer. Some of the formulations developed and researched by CCRAS that have been clinically validated are AYUSH-64 for malaria, AYUSH-56 for epilepsy, AYUSH-82 for diabetes mellitus and 777 oil for psoriasis. Ayurveda has also been streamlined with Reproductive and Child Healthcare (RCH), programme and various Ayurvedic regimens have been included and propagated through various national campaigns. A few areas related to RCH, namely antenatal care, complications of pregnancy, postnatal care, care of the new born, infantile and childhood diseases and gynaecological disorders, have been identified for intervention. Five states, namely Himachal Pradesh, Rajasthan, Maharashtra, Karnataka and Tamil Nadu (Tamil Nadu for Siddha intervention and other States for Ayurveda) have been selected for this project on the basis of availability of Indian systems of medicine infrastructure. CCRAS has taken up the development and standardization activity of 16 drugs each of Ayurveda and Siddha systems to be used in the RCH programme.

Pippalyadi Yoga, a formulation for oral contraception, has been extensively studied by the Council in fertile female volunteers in the last two decades. Presently, this has been taken up by the Department of Family Welfare, Ministry of Health and Family Welfare, to evaluate its antifertility potential for inclusion in the National Population Control Programme. Phase-II multicentric clinical trials on '*Pippalyadi Yoga*' are being conducted at AIIMS, New Delhi; PGI, Chandigarh; JIPMER, Pondicherry and KEM Hospital, Mumbai. A water soluble fraction of neem seed (*Azadirachta indica*) containing sodium nimbinate had shown spermicidal activity in human sperms in an *in vitro* study. The efficacy of *Neem Oil* as a spermicidal agent has been taken up by the Council in fertile female volunteers at Central Research Institute in New Delhi.

CCRUM has a clinical research programme which is aimed at a critical appraisal of the theory of pathogenesis, symptomatology, clinical methods of diagnosis and prognosis, principles, lines and methods of treatment enunciated in the classical texts of Unani system of medicine. The diseases on which clinical trials have been undertaken include vitiligo, eczema, psoriasis, chronic urticaria, infective hepatitis, urolithiasis, duodenal ulcer, chronic diarrhoea, infantile diarrhoea, helminthiasis, malaria, amoebic dysentery, kala-azar, filariasis, diabetes mellitus, essential hypertension, obesity, rheumatoid arthritis, sinusitis, bronchial asthma, gingivitis, dental plaque, pyorrhea, menstrual disorders, leucorrhoea, hyperlipidemia and chronic stable angina. The Council has developed potential drugs for the treatment of some common diseases having national priority such as malaria, filariasis, infective hepatitis and infantile diarrhoea.

CCRH has developed a plan and protocol based on Double Blind Technique in Drug Proving. Proving of a drug substance is a process unique to Homoeopathy. Unlike conventional medicine where animal experimentation forms the basis of evaluation of drug pathogenesis, homoeopathic medicines are proved on healthy human volunteers, including controls, from both sexes. The entire process takes about 12-24 months and has to be repeated more than once at different places and in different settings. The Council has undertaken drug proving programme on a priority. The main objective of the Council is to find out the proving symptoms of indigenous and partially proved homoeopathic drugs on healthy human volunteers. The Council has completed proving the efficacy of 76 drugs, out of which 35 are indigenous. The data on 62 drugs has been published by the Council so far and it is planned to report data on eight drugs shortly. Proving data of 6 drugs is under compilation. Apart from the above, clinical verification of 11 plants, *viz. Achyranthes aspera, Aegle marmelos, Boerhavia diffusa, Caesalpinia bonducella, Carica papaya, Embelia ribes, Centella asiatica, Asteracantha longifolia, Nyctanthes arbor- tristis, Saraca indica* and *Terminalia chebula* has been done by the Council.

Regulations

In an effort to globalize the system and its products, the Department of AYUSH has strictly focussed its attention on standardization and quality control of drugs. Further, displaying on the label of the container or package of Ayurveda, Siddha and Unani preparations, the true list of ingredients (official and botanical names) used in the manufacture of the preparation, together with the quantity of each of the ingredients incorporated therein, has been made mandatory. Good Manufacturing Practices (GMP) have been notified under 'Schedule T' of the Drugs and Cosmetics Rules, 1945 and testing for heavy metals, *viz.* mercury, arsenic, lead and cadmium, in all purely herbal Ayurvedic, Siddha and Unani drugs has been made mandatory for export purposes with effect from January 1, 2006. All these measures have been introduced to give greater impetus to consumer awareness, consumer and doctor benefit, acceptance in the globalized markets and to ensure safety which is of utmost concern while using Ayurveda, Siddha or Unani medicines.

The Ayurvedic Pharmacopoeia Committee (APC) was constituted in 1962 with the aim of preparing the Ayurvedic Pharmacopoeia of India on single and compound drugs and to prescribe the working standards for compound Ayurvedic formulations including tests for identity, purity and quality so as to ensure uniformity of the finished formulations. So far, six volumes of Ayurvedic Pharmacopoeia of India covering 519 plants, have been published, each in the form of a monograph dealing with single drugs. Table 7.2 provides the break-up of the plants covered in individual volumes of Ayurvedic Pharmacopoeia of India.

TABLE 7.2: Main Content of Ayurvedic Pharmacopoeia of India.

Volume No.	*Year of Publication*	*No. of Plants Covered*
I	1990	78
II	1999	80
III	2001	100
IV	2004	68
V	2006	92
VI	2008	101

The Council has also published 'Ayurvedic Pharmacopoeia of India' (Formulations; Part–2) covering 50 formulations. To bring uniformity among the manufacturers and to follow the same formula of ingredients in the same proportion, two parts of Ayurvedic Formulary of India (covering 635 formulations) have been published in Hindi and English separately and a third part covering 500 formulations is under preparation.

Some of the other projects undertaken by APC are:

- Development of standard operating procedures.
- Development of pharmacopoeial standards.
- Assigning of shelf life to formulations.
- Chemo-profiling and bio-efficacy evaluation of Ayurvedic herbal drugs and formulations.
- Effect of treatment of herbs by gamma radiation for the prevention of microbial growth on drying or storage.
- Development of standard operating procedures and pharmacopoeial standard for extracts of Ayurvedic, Siddha and Unani (ASU) medicinal plants.
- Standardization of genuine/authentic samples of metals and minerals used as raw material for production of Ayurvedic drugs.
- Estimation of heavy metals, microbial load and pesticide residues in single drugs of plant origin.
- Publication of an Extra Ayurvedic Pharmacopoeia of India (Namatah/Anuyukta Dravyas). There are certain single plant drugs which are being used in traditional practices of health care but do not find mention in the 56 authoritative textbooks of Ayurveda as mentioned in the First Schedule of the Drugs and Cosmetics Act, 1945. These plant drugs have been selected for inclusion in the Extra Ayurvedic Pharmacopoeia of India.

On the recommendation of APC a notification for protection of ASU drugs and formulations from microbial contamination by gamma radiation has been issued by the Government of India and a dosage of 5-10 Gy has been recommended.

Government of India Initiatives

National

In the X five year plan, proposals have been put forward for validation of traditional knowledge pertaining to Shakhotak (*Streblus asper*) for wucherarian and bancroftian filariasis, Varuna (*Crataeva magna*) for benign hypertrophy of prostate and a compound formulation for cancer.

CSIR has contributed significantly in providing a visionary leadership to the cause of Indian Systems of Medicine. Several laboratories of CSIR have been conducting research on Indian medicinal plants and traditional medicines under a coordinated New Millennium Indian Technology Leadership Initiative (NMITLI) which is a public-private partnership effort within the R and D domain in the country. It looks beyond today's technology and thus seeks to build, capture and retain for India a leadership position by synergizing the best competencies of publicly-funded R and D institutions, academia and private industry. The strategy adopted for NMITLI is to obtain an inverse risk-investment profile *i.e.* low investment – high risk technology areas with investments increasing as developments take place and the projects move up on the innovation curve with reduction in risks. Subsequently, with reference to Indian Systems of Medicine, the NMITLI project was aimed at developing herbal preparations on the concept of reverse pharmacology for global positioning for degenerative disorders, diabetes mellitus type II (NIDDM), osteoarthritis and rheumatoid arthritis, and common hepatic disorders with emphasis on hepatocellular protection. Various CSIR laboratories were involved in developing standardized herb-based Ayurvedic products and multicentric clinical trials along with safety studies have been conducted to ascertain the efficacy and safety of the formulations. Other NMITLI projects that are underway are development of an oral herbal formulation for the treatment of psoriasis and pharmacological and genomic investigations on Ashwagandha (*Withania somnifera*).

The Golden Triangle Partnership (GTP) concept emerged in a National Workshop on Ayurveda Research organized at Chitrakoot in May 2003 when it was decided to set up an integrated technology mission for the development of Ayurveda and traditional medical knowledge based on synchronized working of modern medicine, traditional medicine and modern science with special budgetary support. Subsequently in July 2004, the Department of AYUSH, CSIR and ICMR decided to work together under a tripartite agreement to achieve safe, effective and standardized classical Ayurvedic products for the identified disease conditions and to develop new Ayurvedic and herbal products effective in disease conditions of national/global importance. It was also decided to utilize appropriate technologies to develop single, poly-herbal and herbo-mineral products and to develop products which have IPR potentials. The individual roles have been defined — the Department of AYUSH would be giving the technical guidance regarding formulations to be used, CSIR will carry out the standardization and pre-clinical studies and ICMR would be conducting the clinical trials. At the time of inception of GTP Scheme, the number of identified disease conditions were 12 which have now been increased to 28 in the revised scheme. The proposed disease conditions are attention deficit hyperactive disorder (ADHD) in children, anxiety neurosis, oligospermia, osteoporosis, rheumatoid arthritis, osteoarthritis, immunomodulation for HIV/AIDS, menopausal manifestations, premenstrual tension, allergic bronchial asthma, male infertility, female infertility, hypertension, dyslipidaemia, stress-induced chronic insomnia, psoriasis, irritable bowel syndrome, senile macular degeneration, retinopathy, malaria, urolithiasis, benign prostrate hypertrophy, early chronic renal failure, filariasis, leishmaniasis, diabetes mellitus, obesity and certain identified cancer conditions. In addition to this, standardization, safety and toxicity studies of eight commonly used Rasa Yogas (herbo-mineral/metallic preparations) are being identified for standardization.

The Department of Science and Technology initiated the Drugs and Pharmaceuticals Research Programme (DPRP) in 1994-95 for promoting industry-institutional collaboration in the drugs and pharmaceuticals sector. This programme aims at enhancing capabilities of institutions and the Indian drugs and pharmaceuticals industry towards development of new drugs in all systems of medicine. Emphasis has been laid on Indian Systems of Medicine and some of the areas that have been identified are: development of herbal drugs as adaptogens/immunomodulators, process validation and biological evaluation of Asava and Arishtas with special reference to inoculum bearing herbs, bio-efficacy and analytical evaluation of herbal active molecules, development of standardized single plant formulations for commonly encountered diseases associated with high morbidity and mortality, *viz.* diarrhoea, pancreatitis, gastritis and ischaemic heart disease, and development of standardized metallic and herbo-mineral formulations based on toxicological, pharmacological and process chemistry investigations.

Pharmaceutical Export Promotion Council (Pharmexcil) with an aim to prepare a road map for AYUSH industry has formed an exclusive cell for export promotion of AYUSH products. Subsequent to this, Pharmexcil has constituted a National Committee in the field of Ayurvedic medicines to guide the industry to march ahead with the basic objective of promoting exports to developed countries.

The Department of AYUSH is actively pursuing the proposal for establishing an All India Institute of Ayurveda in New Delhi, which would be an apex Ayurveda institute for postgraduate education, research and healthcare.

Global Arena

An Indo-US forum has been created for exchange of ideas and proposals for exploring the opportunities of collaborative projects. As a follow-up to this and also as a success measure to this forum, Mayo Clinic has expressed interest in doing research on Ayurvedic products. The Government of India is also promoting Ayurveda through Embassies, particularly in Europe, and has organized exhibitions and promotive lectures on Ayurveda. CSIR has gone ahead by setting up a Translational Collaborative Research Programme with the University of Mississippi. Collaboration of Indian Pharmacopoeia Commission and Pharmexcil is being set up with the United States Pharmacopoeia Committee for preparation of quality standards on some Indian medicinal plants. Similar collaborative alliances with the British Pharmacopoeia Commission are also underway for preparing monographs on Indian medicinal plants for inclusion in the British Pharmacopoeia.

Some Publications on Indian Medicinal Plants

It is important to state that several Indian medicinal plants were listed in the Indian Pharmacopoeia of 1966, but were gradually omitted due to lack of assays to ensure reproducibility. Now with the involvement of various private R and D institutions, the quality parameters and standardization assays have been developed for around 20 Indian medicinal plants and these have been incorporated in the Addendum to the Indian Pharmacopoeia since 2005. Some more exhaustive monographs on Indian medicinal plants that encompass various aspects such as Ayurvedic description, macroscopic and microscopic features, geographical distribution, phytochemistry, pharmacology, toxicology and clinical studies have been prepared by a number of institutions. These include: Reviews on Indian Medicinal Plants (16 Volumes) by ICMR, covering under alphabets A to M. It is an ongoing activity to prepare further monographs on Indian medicinal plants under alphabets N to Z. Earlier, ICMR had published Medicinal Plants of India Vols 1 and 2 in 1976 and 1987 respectively and CCRAS had published a database on Research on Medicinal Plants of India (5 Volumes). ICMR has also brought out 14 volumes of Quality Standards of Indian Medicinal Plants whose standards are set as per guidelines of the World

Health Organization wherein conventional and modern scientific approaches have been followed and their standards have been developed at various established and reputed laboratories in the country. ICMR has plans to develop medicinal plants monographs on diseases of public health importance, *viz.* filariasis, malaria, kala-azar, liver disorders, diabetes mellitus, inflammation and immunomodulation, and at the backdrop of this, a book on liver disorders titled 'Perspectives of Indian Medicinal Plants in the Management of Liver Disorders, Lymphatic Filariasis and Diabetes Mellitus' has been published in 2008, 2012 and 2014.

Excellent Leads

Ayurveda, though being considered to be an experiential science, has evolved through the realms of metaphysics and has withstood the tests and rigours of the 21st century. Though ancient classical scriptures had depicted Ayurveda as a complete system of medicine having a wholistic approach, until around three decades ago it was basically known to be a rural man's medicine and was even considered in a disparaging manner. Of late, a good deal of research has been done to prove the concepts, therapeutic regimens, therapies and other modalities pertaining to Ayurveda and a good deal of support has been provided by the Government of India. Scientists, academicians and researchers from allied disciplines have started to work independently and in collaboration to seek more knowledge from the Ayurvedic medicinal plants and concepts in a strategic manner. As a result, certain excellent leads have emerged which include Guggulu for hypercholesterolemia, *Boswellia* for inflammatory disorders, Arjuna for cardioprotection, turmeric for wound healing and antioxidant and anticancer properties, Kutaki for hepatoprotection, *Kshaara-Sutra* for ano-rectal disorders, and Panch-karma for neurodegenerative disorders. Interest has been generated not only in India but also at the global level and certain universities and institutes in India as well as abroad have started relevant research activities. It can be said that the revival of a glorified age-old scientific wisdom *i.e.* Ayurveda has taken place but has still to go a long way to be treaded so that the leads that are available today can be utilized by the populations across geographies for the betterment of the health of the humans.

8

Indian Council of Medical Research Initiatives in the Area of Medicinal Plants Research

Neeraj Tandon

The active role played by Indian Council of Medical Research (ICMR) in research on indigenous drugs from its very inception marks the beginning of a new era. Significant support was given by the Council to the pioneering researches of the late Sir Ram Nath Chopra (known today fondly as the Father of Indian Pharmacology) in the formative years of his career in India. The Council had been supporting various research programs in the area of medicinal plants and Traditional Medicine since 1929.

In 1964, the Indian Council of Medical Research tried to give a new orientation to research on Indian medicinal plants by initiating the Composite Drug Research Scheme under, which, for the first time, it bought together scientists of different disciplines such as Botany (including Pharmacognosy), Chemistry, Pharmacology and Clinical Medicine (including both Ayurvedic and the Modern systems) on the same platform to study medicinal plants of promise in an integrated and coordinated manner.

Nine main research circuits were established in different geographical locations of the country (depending on the expertise available), each circuit consisting of 5 units, *viz.* a Pharmacognosy unit (in which the botanist identified the plant and described the characteristic morphological features), Chemistry unit (wherein the phytochemist analysed the plant and isolated the active principle), Pharmacology unit (in which the pharmacologist studied the pharmacological activity of the active principle isolated from the plant by the chemist, with special reference to the therapeutic claims made), and the Clinical unit (in which the Ayurvedic team treated patients of particular disease with the drug already studied by the botanist, the chemist and the pharmacologist according to the therapeutic regimen recommended by the traditional Ayurvedic system, while the modern medical team made their own diagnosis and kept a record of the day-to-day progress and the ultimate outcome of the treatment).

This was a unique experiment undertaken not only to study the therapeutic efficacy of single drug regimens as practiced by the Ayurvedic physicians from centuries past, but also to establish scientifically the botanical, chemical and pharmacological basis for the use of a particular plant in specific clinical conditions. During the course of 5 years that this unique scheme was in operation under the Council, it was found that nearly 18 plants, out of a batch of 58 selected for study, had reached an advanced stage of investigation. Under this scheme, Saptachakra (*Salacia macrosperma* and *Salacia prinoides*) showed promising results for diabetes mellitus.

Subsequently, with the branching of the CCRIMH into several independent Councils (*viz.* Central Council of Research of Ayurveda and Siddha, CCRAS: Central Council for Research in Unani Medicine, CCRUM: Central Council for Research in Homoeopathy, CCRH; and Central Council for Research in Yoga and Naturopathy, CCRYN), all these Councils undertook research on various aspects of traditional medicine, while the ICMR maintained a low profile in this field for nearly a decade and a half. The Council, however, did not completely lose its basic interest in the scientific evaluation of the time-honoured ancient systems of medicine and continued to support open-ended research in this area.

In the wake of the worldwide interest in the Traditional system of medicine in recent years, the ICMR was persuaded by the scientific community to revive its research efforts in this area. On the recommendations of the Scientific Advisory Board of the Council in 1982, a Scientific Advisory Group on Traditional Medicine Research was constituted, comprising outstanding experts in Ayurveda, Siddha, Unani (including representatives of CCRAS and CCRUM), on one hand, and experts in different branches of modern science (including pharmacologists, chemists, botanists and clinicians), on the other. This Advisory Group met for the first time in November 1983 and identified a number of priority areas in Traditional medicine for intensive research by the ICMR in the coming years.

A new approach that was disease oriented (in contrast to the conventional drug based approach) for a comprehensive scientific evaluation/appraisal of therapeutic claims of few reputed and time honored traditional remedies against certain refractory diseases was adopted. Based on careful planning through the Council's Scientific Advisory Group on Traditional Medicine Research and various Task Forces and Study Groups, ICMR identified 8 thrust areas for indepth research, where the Traditional system could offer an alternate therapy. These 8 areas were: (i) anal fistula, (ii) diabetes mellitus (iii) viral hepatitis, (iv) filariasis, (v) urolithiasis, (vi) bronchial asthma, (vii) kala azar and (viii) wound healing. Of these, eventually research could be undertaken only on the first six thrust areas.

In these efforts, the Council adopted a centrally coordinated integrated multidisciplinary Task Force Strategy involving advanced Chemical, Pharmacological, Toxicological and Clinical Research on selected traditional remedies, combined with simultaneous quality control and standardization studies on each drug. The clinical trials in various thrust areas were conducted at different Medical Institutes/Hospitals in different parts of the country following a uniform protocol and proformae developed by a group of experts, in the respective area. The guiding force behind the formulation, programming and implementation of the programme was Dr G.V. Satyavati, Sr. Deputy Director General of ICMR.

The objectives of each trial were clearly identified and in most cases these involved proving the efficacy and safety of the traditional drugs vis.a-vis available modern drugs wherever possible. As most of the clinical trials in this field were multicentric and double blind, efforts had to be made to involve not only experts in traditional as well as the Allopathic systems of medicine but also experts in pharmaceutical sciences. In addition, in this excerise not only experts from various disciplines and systems of medicine, but also different research agencies working on medicinal plants were involved.

A centrally coordinated task force strategy was formulated for multidisciplinary, integrated studies (through clinical as well as experimental projects) on nearly 26 carefully selected indigenous drugs. Most of the clinical trials were multicentric and, wherever possible, double blind in nature. These were closely monitored by the ICMR Hqrs and a specially set up Central Biostatistical Monitoring Unit (CBMU) located at IRMS, Madras. Experimental studies (including indepth chemical, pharmacological and toxicological investigations) were undertaken mainly at the ICMR Centre for Advanced Pharmacological Research on Traditional Remedies set up at the CSIR-CDRI, Lucknow.

The other important notable activity initiated was the Human Resource Development to generate trained manpower to strengthen indigenous capabilities/ facilities through the Training programmes and Workshops at national level in the areas of Quality Control and Standardisation of Traditional remedies, and biostatistical techniques/methodologies in the conduct of clinical trials of herbal drugs. These initiatives of the Council were first such efforts by any national agency in the area of Traditional Medicine Research in this country at national level.

The pragmatic novel approach/strategy adopted yielded in this short span of time, rich dividends leading to several encouraging leads and fruitful results.

The most outstanding achievements in the area of traditional medicine research has been the authentication of the efficacy of *Kshaara sootra* (Ayurvedic medicted thread described in *Sushruta Samhita*, 600 B. C.) in the management of anal fistula, as compared to surgery. Randomised multicentric trials in over 500 patients at 6 centres showed a success rate of 92 per cent in the *Kshaara sootra* treated group as compared to 89 per cent in patients subjected to surgery, though the healing time was longer with the hread, as compared to surgery, the recurrence rate of fistula was less (4 per cent) in *Kshaara sootra* group than in surgery group (11 per cent). *Kshaara sootra* technique was a safe, acceptable, cost effective and ambulatory alternative to surgery for the management of anal fistula. Furthermore, pharmacopoeial standards pertaining to '*Kshaara sutra*' were also delineated, depicting proper quality control during production of the same. It was Prof P.J. Deshpande of Banaras Hindu University, Varanasi, who had done extensive research and had provided a scientific method for the preparation of '*Kshaara sutra*'. '*Kshaara sutra*' has been standardized byICMR and includes dipping of standardized size linen thread in the milky latex of Snuhi (*Euphorbia neriifolia*), a thorny succulent plant, with turmeric powder, dried and rolled in the ash of Apamarga(*Achyranthes aspera*). Presently, '*Kshaara sutra*'therapy is a very viable alternative even in refractory and relapsed cases of anal fistula. Later on, with a view to transfer the *Kshaara sootra* manufacturing technology to the pharmaceutical industry, physio-chemical standards were worked out in respect of its individual ingredients as well as the finished product (*Kshaara sootra*); shelf life was also studied. Monographs and a Dossier covering various aspects were prepared; a model for mechanised coating of the thread with the ingredients (used in the preparation of *Kshaara sootra*) was fabricated at the ICMR Centre for Advanced Research at Regional Research Laboratory, Jammu and a patent was obtained for the machine for the production of a coated threads.

Diabetes mellitus was the other area, where multicentric flexible dose, double blind randomised clinical trial of an Ayurvedic drug progressed at four centres. Results were very encouraging with respect to *Pterocarpus marsupium* and four patents were obtained which included the process for isolation of Pteroside form *Pterocarpus marsupium.*

In the area of Filariasis, Shakotak (*Streblus asper*) revealed encouraging micro- and macrofilaricidal activity in experimental studies. Interesting leads were also obtained in the area of leishmaniasis, adaptogens and wound healing, in advanced experimental studies.

The Centre for Advanced Pharmacological Research on Traditional Remedies, set up at CDRI, Lucknow continued, systematic botanical, phytochemical, pharmacological and toxicological studies on all the plant drugs undergoing clinical trials (and also a few others not yet taken up for clinical studies). This centre provided very vital and useful feedback through results of the chemical, pharmacological and toxicological studies to all the concerned clinical projects. Along with the Drug Manufacturing and Quality Control Unit set up at Punjab University, Chandigarh, this Advanced Centre played a crucial role in assisting not only monitoring the toxicity/side effects of the drugs but also in testing and maintaining the quality and ensuring standardization of each batch of the drugs prepared and supplied for clinical trials.

In the area of viral hepatitis on the iridoid glycosides isolated from root/rhizomes of *Picrorhiza kurroa* (designated as *Picroliv*) revealed highly encouraging hepatoprotective action against a variety of hepatotoxic agents in animal models. The plant also revealed potent choleretic and anticholestatic activities, as well as immune-stimulant effect, apart from binding with HBs and HBe antigens as well as HBV-DNA. Picroliv was found to be twice as potent as silymarin (another well known plant product) and safe in subacute toxicity, terratogenic and mutagenic studies. Experimental studies on another reputed Ayurvedic drug, *Phyllanthus amarus* also revealed marked hepatoprotective activity in the *Plasmodium berghii* model in Mastomys and significant anti HBsAg activity *in vitro* and *in vivo.*

Subsequent to the monumental publications of R. N. Chopra (1956, 1958) and K. M. Nadkarni (1954), there had been no concerted efforts to collect and publish the results of scientific studies carried out on medicinal plants. Data being scattered widely in various journals relating to different disciplines like Botany, Chemistry, Pharmacology, Pharmacy, Medicine, *etc.* had always been a handicap to research workers interested in medicinal plants. With a view to meeting the long-felt need and demand of researches in this field, the ICMR under the leadership of Dr G. V. Satyavati, the then Sr. Deputy Director General and later Director General, the ICMR published a compilation of data on scientific work conducted in India on medicinal plants in two volumes of Monograph, covering a total of about 900 plant species (Alphabet A to P as per their botanical names). There have been several other publications in the last few decades. The coverage of plants in these and other publications was selective or focused primarily on aspects like agronomy, botany, cultivation and chemistry. There was, however, less focus on pharmacological, clinical, toxicological and drug development aspects. In view of the global resurgence of interest and multifaceted use of medicinal plants, the programmes needed to be looked with different perspective lending support to IPR protection, selling priorities in R and D, promote planned and coordinated research, credence to plant-based drugs, global acceptance, boost trade, exploitation of untapped plants and development of plant drugs for better health care.

In this changed scenario, the Council, particularly in need of the country's rich biodiversity, revived its efforts with greater thrust. A functionally independent medicinal plants unit was established in 1999 to give an impetus with the focused programs. The thrust areas included establishment of knowledge base/information resource and development of databases on various aspects related to medicinal plants/plants based drugs; Quality Standards of important medicinal plants; Integration of leads from ancient knowledge/wisdom/concepts, modern system of medicine and evidence generated through scientific studies; Retrieval, analysis and dissemination of information related to medicinal plants, plants based drugs through traditional and digital means; Human resource development/strengthening of existing infrastructure and facilities, training programs, workshops, interagency programs *etc.*

From the year 2000, ICMR consolidated the Indian research efforts in the area of medicinal plants and presented a series of Review Monographs with particular focus on their medicinal potential. The present series of Review monographs is an effort by ICMR in this direction. Thirteen volumes of the Reviews on Indian Medicinal Plants (with botanical names A-K) covering information on 3679 plant species and carrying 56964 original references have been published. The work is in progress with the remaining plants to complete the series till Z alphabet.

Another important programme on Standardisation of raw materials which has been one of the major impediments in wider acceptance of herbal drugs was initiated by the Council in the year 2000. In an effort to address this issue monographs on Quality Standards of important medicinal plants used by the industry involving several reputed research institutes in the country were prepared. Special emphasis was laid on chromatographic finger printing of the extracts and assay using phytochemical references standards as one of the parameters of identity, purity and quality under this programme. The endeavour yielded very fruitful results evidenced by the publication of 13 volumes of Quality Standards of Indian Medicinal Plants containing 449 plants. The work continues to progress on remaining potential plants required by the industry.

Another programme on isolation of Phytochemical Reference Standards (PRS), a key factor in standardization was initiated in 2006 from selected medicinal plants. The procedure of isolation was optimised and characterized both on the basis of chromatography and spectroscopy for the benefit of these interested in standardizing drugs. Information on first thirty marker compounds was presented in 2010 as the first volume on 'Phytochemical Reference Standards of Selected Indian Medicinal Plants'. It was greatly appreciated and well received in India and abroad by all those actively involved in the field. Encouraged by the response, efforts to characterize the PRS were continued with much more vigor which resulted in the preparation of next 60 monographs. These monographs are presented in the second and third volumes which were brought out in 2012 and 2014. The work is continuing on other important PRS.

A program on preparation of Disease based monographs by integrating the data from ancient knowledge, modern system of medicine and evidence generated through scientific studies on medicinal plant on diseases of public health importance initiated in the year 2000. The envisaged monographs incorporate information on the diseases (including etiopathogenesis) of public health importance and the plant drugs as given in the ancient texts of indigenous systems and Allopathic system of medicine as well as the multidisciplinary research data generated through various scientific studies on such plant drugs with focus on pharmacological toxicological, clinical, phytochemical and pharmacognostic studies with complete references on the work cited.

Under this endeavour of the Council, published two monographs on "Perspectives of Indian medicinal plants in the management of Liver disorders in 2008, Lymphatic Filariasis in 2012 and Diabetes Mellitus in 2014. These specialized monographs provide an insight of the three diseases from the point of the view of Ayurveda, Unani and Allopathic systems of medicine, rationale of use of plant drugs of traditional Indian systems of medicine leading eventually to concepts of therapy and therapeutic agents of the three diseases.

Under the Inter-Agency Programme, as recommended by an expert group comprising experts as well as representatives of various scientific agencies (CSIR, ICAR, CCRAS, ICMR *etc.*), the preparation of an 'Illustrated Manual of commonly used Indian plant Drugs' was entrusted to the RRL (CSIR) Jammu. This manual included useful scientific information of relevance to the lay public on 150 common plant drugs identified by a Sub-Group of experts, apart from 50 plant drugs derived from the Ayurvedic Formulary, as finalized by the Ayurvedic Pharmacopia Committee and published in 1996.

C new initiative to bring out the compendium of Safety of Medicinal Plants with the objective to review and document the safety-related scientific information on important medicinal plants which are commonly used by the industry either as single herb or as part of the polyherbal formulation is presently under preparation. Under this programme information related to botanical name, common name of the plant (Ayurveda, Hindi and English), family name, parts used in the traditional medicine (*i.e.* Ayurveda), historical use of the plant, scientific studies related to therapeutic uses (both preclinical and clinical), safety information derived from acute, subacute and chronic toxicity studies, safety information from pharmacological studies, information on genotoxicity, reproductive studies in animals, safety information derived from clinical studies (*i.e.* adverse drug reactions), case reports on toxic effects in human and herb drug-interaction both in animals and humans have been reviewed and are being compiled.

New initiatives have also been taken by the Division on pursuing research from leads from already generated in adhoc projects funded by ICMR and its institutes in the last 5 years in the area of Medicinal Plants in a translational mode so as to reach the level of product for the benefit of health of the people. To formulate goal oriented, time bound research projects on Investigations on Plant products for diseases of public health importance with translational approach, the Task Force strategy has been proposed for designing plant product oriented experimental studies, preclinical investigations and multicentric clinical trials. At present two areas *viz.*, stress induced sleep disorders and diabetes and its complications are being worked at.

Indian Council of Medical Research has earned a unique place for itself due to its publications on Medicinal Plants.

9

Contribution of Council of Scientific and Industrial Research in Research and Development of Ayurvedic Plant Drugs

A.K.S. Rawat

Council of Scientific and Industrial Research (CSIR), Department of Science and technology, Government of India, established in 1942, provide a leading role in developing India's science and technology capability in most of the areas of science for the betterment of mankind. In order to know how important is the plants as a source of pharmacologically active substances for the human health, nutraceutical and cosmaceuticals, a numbers of CSIR labs have been working to obtain novel lead molecules from medicinal plants and to develop them as modern drugs. Apart from this, recently a number of multiinstitutional programmes have successfully completed and initiated to developed effective and standardized plant-based formulation and developed a range of Ayurvedic formulations for healthcare through their biological laboratories *viz.*CSIR-Central Drug Research Institute (CDRI), Central Institute of Medicinal and Aromatic Plants (CIMAP), Indian Institute of Intregrative Medicine (IIIM), CSIR-National Botanical Research Institute (NBRI), CSIR-Indian Institute of Toxicology Research (IITR), CSIR-Institute of Himalayan Bioresource Technology (IHBT), CSIR-Indian Institute of Chemical Technology (IICT), CSIR-Indian Institute of Chemical Biology (IICB), CSIR-Central Food Technology Research Institute (CFTRI) and North East Institute of Science and Technology (NEIST), Jorhat (formerly known as RRL Jorhat). Further, in developing quality monographs of single raw drugs/polyherbal formulations for the preparation of Ayurvedic, Unani and Siddha Pharmacopoeias of India, identification of adulteration and substitute of raw drugs of classical Ayurvedic formulation, TKDL, Golden Triangle Project (GTP) and cultivation practices, conservation and releasing of new high-yielding varieties of plants mentioned in Ayurveda are some other areas where especially NBRI, CIMAP, IHBT, IIM, IITR, IICT, NEIST are involved and promoting scientific validation of medicinal plants used as Ayurvedic medicines.

Traditional Knowledge Digital Library (TKDL) is a collaborative project between Council of Scientific and Industrial Research (CSIR), Ministry of Science and Technology and Department of AYUSH, Ministry of Health and Family Welfare, and is being implemented at CSIR. An interdisciplinary team of Traditional Medicine (Ayurveda, Unani, Siddha and Yoga) experts, patent examiners, information technology experts, scientists and technical officers are involved in creation of TKDL for Indian Systems of Medicine.

The project TKDL involves documentation of the traditional knowledge available in public domain in the form of existing literature related to Ayurveda, Unani, Siddha and Yoga, in digitized format in five international languages, English, German, French, Japanese and Spanish. Traditional Knowledge Resource Classification (TKRC), an innovative structured classification system for the purpose of systematic arrangement, dissemination and retrieval has been evolved for about 25,000 subgroups against few subgroups that was available in earlier version of the International Patent Classification (IPC), related to medicinal plants, minerals, animal resources, effects and diseases, methods of preparations, mode of administration.

TKDL provides information on traditional knowledge existing in the country in languages and format understandable by patent examiners at International Patent Offices (IPOs), so as to prevent the grant of wrong patents. TKDL thus, acts as a bridge between the traditional knowledge information existing in local languages and the patent examiners at IPOs. TKDL database contains more than 2.90 lakh formulations from the texts of traditional medicine systems of India which are Ayurveda, Unani and Siddha.

Phytochemical studies

A study of bisbenzyl isoquinoline alkaloid, hayatin, from *Cissampelos pareira* (Zakhm-e-hayaat) was one of the major achievements during the first decade of these investigations.

Studies on *Rivea cuneata* provided fractions, which lowered blood sugar in experimental animals.

A method for estimation of reserpine from *Rauvolfia serpentina* by alkaline hydrolysis and paper chromatography and Ultraviolet estimation was established. Investigation on *Rauvolfia canescens* yielded sarpagine and a new alkaloid raunescine. Subsequently, the metabolism of reserpine in rats was studied and its metabolites, reserpic acid and reserpic acid methylester identified. The distribution and excretion of the drug was also studies in rats identified as epireserpine.

Nardostachys jatamansi yielded large crystals, about one inch long, of a sesquiterpine designated as jatamansic acid. One of these was cut in half with a hacksaw and the structural work was carried out on half a crystal. On the basis of Infrared, Ultraviolet and Nuclear Magnetic Resonance, used for the first time for investigations in India, jatamansic acid was assigned a structure. In this structure, the isopropyl and methyl groups were erroneously placed due to the incorrect structure assigned to an azulene isolated on selenium dehydrogenation.

Effectiveness of psoralen, a furanocoumarin from *Psorolea corylifolia*, in the treatment of leucoderma, prompted investigations on its mechanism of action and safety evaluation.

So far, more than 4500 samples of terrestrial plants and 360 samples of marine flora and fauna have been screened. The biological activity data of 3789 plant samples have so far been published whereas in case of marine organisms biological screening data of 372 samples has been published. Out of these 128 samples belonged to CDRI and the rest were supplied by the other laboratories participating in the Department of Defence (DOD) project, This broad screening has led to the identification of anticancer, cardiovascular, Central nervous system (CNS) depressant, diuretic, spasmolytic, anti-inflammatory, antifertility, antimicrobial, antituberculosis, hypoglycaemic, antiviral, antileishmanial, immunomodulatory, hypolipidaemic, hepatoprotective, wound-healing, antiallergic nootropic and adaptogenic activities in 738 plant samples and 42 marine organisms. Bioassay-linked chemical investigations on the identified active plants and marine organisms with desired order of activities were undertaken which have resulted in the identification of some active constituents whose structures have been established.

Pharmacognostic studies

The pharmacognostic studies on the following 23 Indian Medicinal plants with respect to their identification and evaluation were carried out both in whole plant and its powdered state between in 1956 to 1963 at Department of Botany CDRI, Lucknow. The plants are: *Coculus villousus* DC., *Melaleuca leucodendron* L., *Ricinus communis* L., *Argyreia speciosa* Sweet, *Salmalia malabarica* DC., *Coccinia indica* Wt. and Arn., *Erythrina varigata L.* var. *orientalis* (L.) Merrill, *Vitex negundo* L., *Ficus glomerata* Roxb., *Curculigo orchioides* Gaertn., *Xanthium strumarium* L., *Ervatamia coronaria* Stapf., *Onosma echioides* L., *Agrimonia eupatorium* L., *Clerodendrum serratum* (L.) Moon, *Vallaris solanacea* O. Ktze., *Justicia gendarussa* Burm. f., *Astercantha longifolia* Nees, *Piper longum* L. (Bisht, 1963); *Streblus asper* Lour., *Carissa carandas* and *Carissa spinarum* L.

In addition to above studies the classification of drugs from pharmacognostic point of view was also established. The cultivation of *Rauvolfia serpentina* Benth. ex Kurz, in small scale under local conditions was undertaken in order to increase the alkaloids and the effect of growth hormones was also observed on some drug plants in respect to their growth, morphology and physiology. The organoleptic studies of about 20 Indian pharmacopoeial drugs were also carried out for their characterization,

After the lapse of about 20 years, at department of Botany CDRI, Lucknow, pharmacognostic study was carried out again on some biologically active plants like *Coleus forskohli* Briq., *Ferula jaeschkeana* Vatke and on 'Ratanjot' (*Arnebia* spp.) and related *Boraginaceous* spp.

Quality Control and Authentication of Herbal Drugs

The CSIR-National Botanical Research Institute, Lucknow has developed quality control parameters of more than 250 raw single Ayrvedic drugs and standardized about 30 polyherbal classical formulations in the last three decades. A number of important raw drugs evaluated. *'Aakashvalli', 'Apamarg', 'Arjun', 'Amra-haridra', 'Ativisha', 'Bala', 'Banafshan', 'Bhuiamla', 'Brahmi', 'Chiraita' 'Daruharidra', 'Deodar', 'Dugdhika', 'Dugdhpheni', 'Gurmar', 'Gokhru' 'Guruchi', 'Hansraj', 'Jivak', ' Jivanti', 'Jangali Kuth', Kalmegh, 'Kuth', 'Kalimusli', 'Kapikacchu', 'Meda', 'Mahameda', 'Mustak', 'Pittapapra', 'Pippali', 'Rakta Punarnava', 'Raasna', 'Resha-Khatmi', 'Renuka', 'Rishbhak', 'Shweta Punarnava', 'Satawari', 'Salampanja', 'Safed musli', 'Senna', 'Talishpatra', 'Tukhm-e-Khatmi', 'Tulsi', 'Vacha', 'Vidarikand'* were standardized pharmacognostically. Pharmacognostic evaluation of *Berberis aristata, Acorus calamus, Arnebia nobilis, Berberis asiatica* root, *Gymnema sylvestre, Leucas aspera, Pygmaeopremna herbacea, Curcuma haritha, Cassia angustifolia* seed, *Piper longum* fruits, *Lycopodium clavatum* stem, *Berberis lyceum, Abrus precatorius, Cordia macleodii, Curcuma zedoaria,* and *Berberis umbellata* has already been published in various national and international journals.

Identification of substitutes for traded drug 'Chirayata' (*Swertia* species) using pharmacognostical parameters were also carried out. In addition, the identification markers of 'Dashmoola' and its possible adulterants/substitutes were developed.

'Dashmoola' consists of ten root drugs, classically divided into two groups the 'Brihat Panchmool' (*Aegle marmelos, Premna integrifolia, Stereospermum suaveolens, Gmelina arborea* and *Oroxylum indicum*) and the 'Laghu Panchmool' (*Tribulus terrestris, Desmodium gangeticum, Solanum indicum, Solanum xanthocarpum* and *Uraria picta*).

The author and his associates, in 2012, carried out pharmacognostical evaluation *viz.* botanical study, physicochemical parameters and High performance thin layer chromatography (HPTLC) analysis and antioxidant studies to make a comparison among stem barks of four *Ficus* species *viz. F. carica, F. religiosa, F. glomerata* and *F. retusa*. Determinations of various physicochemical constants were carried out according to the methods provided in Ayurvedic Pharmacopoeia of India (API). The macro and microscopical character of these *Ficus* species also shows moderate variation. Tannin content was found to be maximum in *F. religiosa* (6.36 per cent) and minimum in *F. carica* (0.14 per cent). This result was supported by microscopical studies which showed the presence of numerous dark brown cell contents in case of *F. religiosa*. Total phenolic content was also found to be maximum in *F. religiosa* (20.57 per cent) and minimum in *F. carica* (3.04 per cent). Sugar content was found to be maximum in *F. religiosa* (2.04 per cent) and minimum in *Ficus glomerata* (1.36 per cent). Starch content was found to be maximum in *F. carica* (6.44 per cent) and minimum in *F. glomerata* (1.49 per cent).

HPTLC analysis showed the presence of β-sitosterol and lupeol in the ethanolic extract of bark of all the four *Ficus* species, The R_f values of β-sitosterol and lupeol was found to be 0.46 and 0.62 respectively using toluene: ethyl acetate (80: 20 v/v) which is clearly visualized in HPTLC chromatogram and in densitometric chromatogram. Concentration of β-sitosterol was found maximum in *F. carica* (0.131 per cent) and minimum in *F. glomerata* (0.041 per cent) and conc. of lupeol was found to be maximum in *F. retusa* (0.069 per cent) and minimum in *F. religiosa* (0.020 per cent).

Search for Biological Activities of Active Principles

Asparagus racemosus, Curculigo orchioides, Withania somnifera were investigated for adaptogenic activity besides *Strychnos nux-vomica.*

Colenol, a diterpene isolated from *Coleus forskohlii* has been shown to activate adenylate cyclase.

Ethanolic extracts of the seeds of the plant *Nyctanthes arbo-rtristis* has shown promising antiallergic activity.

Solanum xanthocarpum is a traditional remedy for the treatment of asthma. The crude extract and alkaloidal fraction of *S. xanthocarpum* were evaluated for their ability to prevent mast cell degranulation, to antagonize Protocatechuc acid (PCA) in rats and mice and protective effect in autacoid induced anaphylactic test in rats. On chromatography the alkaloidal fraction yielded a pure crystalline alkaloid, solanine which had 84 per cent anti PCA activity at 30 mg/kg, p.o. in rats.

During general biological screening, a potent antiallergic activity was observed in the ethanolic extract of the wood of *Cedrus deodara.*

The plant *Andrographis paniculata* is used in folklore medicines in India for multiple clinical conditions including allergic manifestations. The diterpenes were further evaluated for antiallergic activity.

Stem bark of *Albizzia lebbeck* is an antiasthamtic drug of Ayurvedic medicine. The immunomodulatory effect of *Albizzia lebbeck* was evaluated by studying humoral and cell-mediated responses.

Ailanthus malabarica stem bark is used in dyspepsia and its resin is considered of value in the treatment of dysentery and bronchitis.

An ethanolic extractive of leaves and stems of the plant *Annona Squamosa* exhibited anticancer activity. From the alkaloidal fraction of *Cocculus pendulus*, showing anticancer as well as hypotensive activities, cocsulinin, a new anticancer active principle was isolated and characterized along with pendulin, cocsolin and cocsuline. The chloroform insoluble fraction of the plant *Ipomoea leari* furnished a new anticancer constituent, ipolearoside.

Central nervous system depressant activity was confirmed in *Centella asiatica*, an Ayurvedic remedy for memory improvement. Two triterpenoid saponins, asiaticoside and brahmoside were isolated from the plant. Jatamansic acid, the active constituent of the plant *Nordostachys jatamansi* was characterized as a sesquiterpenoid acid. D-glucoside has been found to be responsible for the activity in the plant *Colocosia fornicate. Cocculus hirsutus* yielded known alkaloida-trilobine, codaurine.

An iridoid glucoside mixture *kutkin* isolated from *Picrorhiza kurrooa* showed diuretic activity.

Some traditional drugs have been evaluated for hepatoprotective activity against variety of hepatotoxicants *viz.* paracetamol and galactosamine induced hepatic damage in rats and mastomys. Choleretic and anticholestatic effect against paracetamol-induced cholestasis in conscious rats were also evaluated. *Ricinus communis* leaves, due to N-demethylricinin, exhibited the dose dependent activity by restoring the altered levels of several enzymatic and non-enzymatic parmeters in the serum and liver of rat. *Phyllanthus amarus* and *Phyllanthus maderaspatensis* showed sigifiant hepatoprotective activity against galactosamine and paracetamol induced hepatotoxicity.

The ethanolic extract of the plant *Swertia chirata* was found to possess significant hypoglycaemic effect. A new xanthone-chiratol and two known xanthones, swerchirin and 7-O-methyl swertianin have been isolated from the active fraction of the plant. Two new triterpenoid glycosides trichonin and santholin were obtained from another hypoglycaemic plant, *Trichosanthes palmate.* Both the compounds possess significant hypoglycaemic tactivity. *Aegle marmelos*, initially exhibited moderate hypoglycaemic activity which, however, could not be confirmed in later stages.

Cissampelos pareira was one of the earliest plants investigated for neuromuscular-blocking activity.Its roots are used in heart troubles, asthma, dysentery and intestinal tuberculosis. Water soluble fraction of the total alkaloids possesses neuro-muscular-blocking activity. It yielded a bisbenzyl isoquinoline alkaloid, hayatin.

The plant *Achyranthus aspera* is reputed to be a purgative, useful in piles, boils, skin eruption and colic. The crude extractive of its seeds showed high order of activity. Two saponins A and B were isolated from this plant. Saponin A was found to be more active.

Apigenin-8-C-(2"-O-xylosyl) glucopyranoside was characterized as an active principle (active at 50 mg/ml) from *Amorphophyllus companulatus.*

The oxytocic activity was localized in a new flavonoid arjunalone from *Terminalia arjuna.*

Five sesquiterpene alcohols of himachalane series *viz.* himachalol, allohimachalol, centdarol, isocentdarol and himadarol (unidentified) were obtained from the wood of *Cedrus deodara*, while himadarol was an isomer of either of the himachalols, centdarol.

The coumarins have shown a good degree of spasmolytic activity in *Bonninghausenia albiflora* and *Heracleum thomsonii.* Bergapten has been identified as active principle in *B.albiflora*, while angelicin has been found to be active principle of *H. thomsonii.*

The active chloroform soluble fraction of *Cymbidium giganteum* yielded a demethoxy bisbenzyl compound, gigantol.

The steam non-volatile fraction of the leaves of *Aegle marmelos* yielded aegleamide, which showed a moderate spasmolytic activity. *Stephania glabra* was found to be a good source of protoberberine alkaloids, corynoidine, jatrorhizine, magnoflorine, N-methyl corydolmine and cyclanoline chloride but all the alkaloids lacked spasmolytic activity. Some other plants viz *Rheum webbianum, Saussurea albescens, Verbena hybrid, Thermopsis barbata, Axyris amaranthoides* showed spasmolytic activity in one or other fraction, but neither any active principle nor any novel compounds could be isolated from these plants.

Based on the traditional use the plant, *Calotropis procera* was selected for evaluation of wound-healing activity at CDRI, Lucknow. Topical application of 1.0 per cent sterile solution of the latex of *C. procera* twice daily for sevebn days significantly augumented the healing process by markedly increasing collagen, promoting epithelisation and angiogenesis, leading to reduction in wound area. *Centella asiatica* is used in the indigenous system of medicine as a tonic in skin diseases and leprosy. Asiaticoside showed promising wound-healing activity whereas madecassoside was found to be inactive. Asiaticoside exhibits significant wound-healing activity in normal as well as delayed healing models and is the main active constituent of *Centella asiatica.*

Euphorbia neriifolia, one of the constituents of *Kshaarasootra,* which is used in the traditional system of Indian medicine to heal anal-fistula, was evaluated for its possible wound-healing activity in experimental animals. The chloroform and water soluble fractions were evaporated to dryness. The fraction facilitated the healing Deoxyribonucleic acid (DNA) content, epithelization and angiogenesis.

Among several drugs used in Indian medicine, the oleoresin of *Commiphora mukul* (gum guggulu) appears to be the most promising lipid-lowering agent. The ethyl acetate, soluble and insoluble fractions were obtained. The activity was found only in the ethyl acetate soluble fraction designated as guggulipid while the insoluble fraction had no activity, instead it showed hepatotoxicity.

Various species of Coleus have been described in Ayurveda as remedies for the treatment of heart diseases, spasmolytic, painful micturition and convulsions. The hypotensive activity of the plant *Coleus forskohlii* was localized in a diterpene, coleonol.

The herb Brahmi as a nervine tonic has been validated at CDRI, Lucknow, by its pharmacological evaluation on animals. The ethanolic extract of the herb was found to improve the performance of rats in several test models of memory and learning. Initially, the chemical investigations were focused on bacoside A, since it was the major product and appeared to be a single compound. Bacoside A mainly comprised two sets of saponins. One set was derived from pseudojujbogenin, which on acid hydrolysis furnished four triterpenoid transformation products, bacogenin A1. From the non-polar fraction of the ethanolic extract a diglycoside, bacoside A1 was also isolated. bacoside B was found to be a triglycoside having the same aglycone, *viz.* pseudojujubogenin, as bacoside A_2.

A standardized herbal preparation containing 50 per cent bacosides was released in 1996 as a memory enhancer and is being marketed under the name "Memory Plus." An Indian patent has been filed for this preparation.

The scientific study of *Bacopa monnieri* started at Central Drug Research Institute (CDRI), Lucknow. Chemical investigations showed that Bacosides A and B were the main constituents of the plant, of which Bacoside A is predominant. Bacoside A was further shown to have different moieties as sapogenins. On acid hydrolysis, bacosides yield a mixture of aglycones, bacogenin A_1, A_2, A_3, among which the major component was ebelin lactone. Bacogenin A_5, a rearranged sapogenin was also isolated and identified. Two new triterpenoid saponins, Bacoside A_3 and A_2, were also isolated and identified from *B. monnieri*. Seasonal variation studies in the chemical constituents were also conducted.

Later, CDRI, Lucknow, developed a unique single plant-based natural memory enhancer formulation - BESEB ("Bacosides Enriched Standardized Extract of Bacopa") from the herb *Bacopa monniera.* The process for making enrichment of the active constituents (Bacosides A and B) in BESEB has been patented by CDRI (CSIR). BESEB was successfully commercialized.

The root and rhizome of the plant *Picrorhiza* forms an integral ingredient of several Indian herbal preparations used in the treatment of liver disorders. A bitter crystalline product, designated as kutkin was isolated from the plant. It was later identified as a mixture of two iridoid glycosides, picroside-I and kutkoside in the ratio of 1:2. Kutkin was found to possess significant protective activity against hepatic damage induced at CSIR-NBRI.

Picrorhiza contain picroside-I and picroside-II, which are known bioactive metabolites. In our study a simple highly precise method has been established for the simultaneous determination of picrosides (picroside-I and picroside-II) in two different *Picrorhiza* species *viz. P. kurroa* and *P. scrophulariiflora.*

Picroliv possessed marked immunostimulant activity in doses of 5 and 10mg/kg X 7 days as evidenced by Plaque forming cell (PFC) assay, Haemagglutinin (HA) titre, macrophage migratin index, Delayed-type hypersensitivity (DTH) response, macrophage activation and mitogenic response of lumphocytes. A nonspecific immunostimulatory response was also observed against *Leishmania donovani* infection in mastomys at the dose of 10mg/kg X 7 days of picroliv.

Turmeric is widely used in India in the form of poultice as an anti-inflammatory agent. Anti-inflammatory activity of curcumin, the major active constituent of turmeric, has been tested against carrageenin induced oedema in mice and rat against formalin induced oedema in mice. Apart from efficacy in acute and chronic inflammation curcumin had a low ulcerogenic index.

Streblus asper was found to possess significant *in vitro* and *in vivo* macrofilariacidal activity against *Litomosoides carinii* and *Brugia malayi* in rodents.

Central Drug Research Institute, Lucknow developed a drug from *Commiphora mukul* for the treatment of hyperlepidemia and is marketed in India as well as other countries.

Products Developed by Central Drug Research Institute

Development of new leads from plant source mentioned in classical text of Ayurveda resulted in developing number of modern drugs at CDRI since its inception. Herbal medicines have been developed and successfully commercialized are given in Table 9.1.

CSIR-NBRI has developed several Intellectual Property Right (IPR) covered novel scientifically validated and standardized herbal products/formulations based on Ayurvedic as well as traditional knowledge, such as plant-based colour for lipstick, herbal health drink, antiulcer, anticough syrups, ointment for cuts, burn and wound.

Identification of Five Elite Chemotypes

***Acorus calamus*:** Accession (NBA-10) was identified with max. volatile oil content- (9.5 per cent) from North-West Himalayan zone at 1097 meter altitude, with average temperature 23°C to 2°C and rainfall 13 to 428 mm, high altitude clay soil.

Chemotype-1: α asarone rich-NBA-3 (16.82 per cent) (North-West Himalayan).

Chemotype-2: β asarone rich-NBA-2 (92.60 per cent) (North-West Himalayan).

***Tribulus terrestris*:** Three chemotypes were identified from Arid zone (sandy-loamy soil with average temperature 42°C to 9°C and low rainfall 1 to 310 mm) of India.

Chemotype-1: Prototribestin rich-(0.6350 per cent) NBT-06 of Arid zone.

Chemotype-2: Tribulosin rich- (1.076 per cent) NBT-06 of Arid zone.

Chemotype-3: Rutin rich- (0.5429 per cent) NBT-04 of Arid zone.

Two alternative species of *Tribulus terrestris* have also been identified from the arid zone of country with common secondary metabolites (protodioscin, prototribestein and trubulosin).

***Centella asiatica* elite chemotype(s):** Under this project, total of 72 accessions have been collected and studied from different phyto-geographical zones of the country. Among all the collected 72 accessions, two morphotypes (CA-09 and 10) has been identified. Qualitative phytochemical screening and quantitative estimation of secondary metabolites using conventional methods of HPTLC on these 72 accessions for their bioactive chemical constituents showed elite accessions CA-15 and CA-45 for Asiaticoside, CA-45 and CA-51 for Madecassoside and CA-07 for Asiatic acid. Among all the collected 72 accessions, three elite accessions have been identified.

TABLE 9.1: Herbal Medicines have been Developed and Successfully Commercialized

Sl.No.	*Product/Trade Name*	*Source*	*Use*	*Licensee*
1.	Memory Sure, Keen Mind, Memo Plus Gold	BESEB: Bacosides Enriched Standardized Extract of *Bacopa monniera* (Brahmi)	Memory improvement	Lumen Marketing Co., Chennai
2.	Gugulip	Gum of *Commiphora mukul* (Guggul)	Hypolipidemic	Cipla Ltd., Mumbai Nicholas Piramal Ind. Ltd., Mumbai
3.	Dilex C	Seed husk of *Plantago ovate* (Isabgol)	Cervical dilatation (MPT)	Unichem Labs., Mumbai
4.	Consap cream	Nuts of *Sapindus mukorossi* (Reetha)	Spermicidal	HLL Life Care Ltd., Thiruvananthapuram
5.	Picroliv	Standardized fraction: *Picrorhiza kurrooa* (Kutki)	Hepatoprotective	DIL, Thane
6.	CDR134D123 and F194	Marine Plant, *Xylocarpus granatum* (Mangrove)	Antidiabetic	TVC Sky Shop Ltd., Mumbai
7.	Herbal medicament	Rhizomes of *Curcuma longa* (Turmeric)	Prevention of Stroke (ischemic or hemorrahagic)	Themis Medicare Ltd., Mumbai
8.	(Dietary supplement)	Standardized fraction (F147) stem bark of *Butea monosperma*	Nutraceutical and Dietary Supplement for optimum bone health.	Natural Remedies, Bangaluru

New Plant Varieties Developed by IIIM (Formerly named RRL) Jammu

Cymbopogons **(Gandhatrina):** KALAM—A new citral rich (78-83 per cent) variety for drought prone areas. Released by His Excellency Dr A. P. J. Abdul Kalam on 26th June, 2003. The variety is suitable for cultivation both under rainfed and irrigated conditions. Tawi Rosa—A new geraniol rich variety of Cymbopogon released on 26th September, 2003. The main feature of this strain is that it can withstand moisture stress level of 15 per cent and contains total alcohol in the range of 80-85 per cent, calculated as geraniol 70-75 per cent and geranyl acetate 10-15 per cent. Apart from the main chemical constituents, it contains 3-5 per cent ocimene which is used in high grade perfumery. Jamrosa—*C. khasianus* and *C. nardus* var. *confertiflorus* x *C. jawarancusa*, both the varieties can be successfully cultivated in sub-tropical and tropical regions of India. The varieties yield up to 0.4 per cent oil with 60-65 per cent geraniol and 20-25 per cent geranyl acetate.

***Ocimums*:** Developed three varieties rich in eugenol, linalool and methyl cinnamate.

***Mentha longifolia*:** *Mentha longifolia* cultivar developed. Gives about 0.5 per cent oil containing about 65 per cent carvone. This variety has been transferred to South India Mints and Aromatic Products, Trivulveli (T.N.) where it is growing over an area of about 200 acres.

***Withania somnifera*:** WSR—An improved strain WSR is developed and released for commercial cultivation by IIIM, Jammu. The dry biomass yield of this newly developed variety shows 200 per cent increase in yield over the conventionally grown varieties.

Dioscorea **(Chupri)composite (S2-58):** The strain developed by IIIM, Jammu grows vigorously and contains 3.6 per cent diosgenin.

Apium graveolens **(RRL-85-1):** IIIM has developed this variety of *Apium graveolens*. The herb essential oil contains phellandrene as principal constituant and up to 30 per cent carvone.

Piper nigrum Linn., ***Piper longum*** Linn. and ***Zingiber officinale*** Rosc.(composite drug, Ayurvedic Trikatu): Contains active constituent piperine, the bio enhancer. Through a study on curcumin and its polyherbal combination with trikatu, the traditional use of trikatu has been validated. Moreover, developing the polymeric delivery system for this polyherbal combination apprehend the bioefficiacy and bioavailability of curcumin with sustained action, rescue bioenhancers from degradation. Thus, bioenhancer concepts will ultimately prove beneficial by reducing the dose of the drug, and ensure safety and stability of the formulation.

References

Rastogi, S. and D. K. Kulshreshtha. 1999. "Bacoside A_2—A Triterpenoid Saponin from *Bacopa monniera*," *Indian J Chem.* 38B(1-3): 353-356.

Rastogi, S.; R. Pal and D. K. Kulshreshtha. 1994. "Bacoside A_3—A Triterpenoid Saponin from *Bacopa monniera*," *Phytochemistry*, 36: 133-137.

Satyavati, G. V. 1987. "Gum Guggul: the Success Story of an Ancient Insight Leading to an Modern Discovery," *Indian Council of Medicinal Research Bulletin*. 17(1): 1-5.

Satyavati, G. V.; C. Dwarkanath and S. Tripathi. 1969. "Experimental Studies on the Hypocholesterolemic Effect of *Commiphora mukul*," *Indian J. Med. Res.*, 57(10): 1950-1962.

Singh, H. K.; B. N. Dhawan. 1982. "Effect of *Bacopa monniera* (Brahmi) Extract on Avoidance Responses in Rats" *J Ethanopharmacol.*, 5: 205-214.

Singh, H. K.; R. P. Rastogi; R. C. Srimal and B. N.Dhawan. 1988. "Effect of Bacosides A and B on Avoidance Responses in Rats," *Phytother Res.*, 2: 70-75.

Verma, S. K. and A. Bordia. 1988. "Effect of *Commiphora mukul* (gum guggul) in patients with hyperlipidaemia with special reference to HDL cholesterol," *Indian J. Med. Res.*, 87: 356-360.

10

Selected Researches on Ayurvedic Medicinal Plants Published in Indian Journals

C.P. Khare, Chandra Kant Katiyar, Satyajyoti Kanjilal, Deepa Gandhi, B.N. Bhatta and Amitabha Dey

According to World Health Organization (WHO), majority of the population in developing countries depends primarily on herbal medicines for basic healthcare. The traditional systems of medicine or Complementary and Alternative Medicine are widely used to treat present-day health and medical problems. Ayurveda, the Indian system of traditional medicine practised today has its roots in the Vedic-age thinking. Ayurveda follows its own distinctive idea and methodologies to address issues of human ailments.

Since the last few decades important changes have occurred in Ayurvedic practice, research and manufacturing of Ayurvedic formulations. Research in Ayurveda that began to search for new entities from Ayurvedic herbs and formulations based on pharmacological claims and chemical moieties has come a long way with recent advances in biomedicine and technology. It is understood that deviation from the basic principles and concepts had a negative impact and therefore incorporation at the basic level of development of hypothesis of the research is felt necessary.

Much of Ayurvedic research is being carried out on single herb extracts, poly-herbal formulations, combined treatments and disease-specific therapies. However, the path remains uncertain in terms of standardization of products along with safety and efficacy for universal acceptance. Resurgence of interest in herbal/Ayurvedic medicines in the past few decades has attracted the attention of mainstream medical researchers. Some of the enthusiastic researchers started treating herbal drugs on par with Allopathic medicines for the purpose of conducting clinical trials. This led to variability of results and criticism that Ayurvedic principles were ignored during the studies.

When a new chemical entity is subjected to clinical trials objective is to define the dose, duration, indication and side-effect profile of the same. However, when a traditional herbal medicine is subjected to clinical trials in most of the cases their dose is known, duration of use is known, indications are known but side-effect profile is not known. Also, the validation of efficacy is additional objective in such cases. Therefore, clinical trials on traditional medicine-based products need significantly different approach. There is an urgent need to understand Ayurvedic principles of treatment and search for inclusive solutions to design clinical studies.

Provided below are the abstracts from Ayurvedic research articles published in Indian Journal of Pharmacology and in Journal of Research in Ayurveda and Siddha over few decades with the objective of bringing forward efforts on validating the efficacy of Ayurvedic Medicines. This chapter is designed by selecting abstracts on diferent medicinal plants in single or in combination to give an insight of the enormous researches that have been conducted in the field so far. In addition to these two journals there are a lot more researches available in public domain and many researches on-going in various sophisticated laboratories. This chapter is limited to the selected researches only. The tabulated disease-wise summaries of Ayurvedic research are provided in Table 10.1.

TABLE 10.1: Disease-wise Summaries of Ayurvedic Research

Sl.No.	*Disease/Illness*	*Ayurvedic Formulations/Herbs Studied*	*Number of Studies*
1.	Diabetes and Hyperlipidemia	*Trigonella foenum graecum, Boerhavia diffusa, OB-200G, Cocculus hirsutus, Cassia kleinii, Piper betle, Ficus hispida, Coccinia indica, Curcumin, Satureja hortensis, Asparagus racemosus, HK-07, Ceiba pentandra, SH-01D, Persea americana, Rutin, Phyllanthus amarus, Acacia catechu, Pongamia pinnata, Dolichos biflorus, 7-hydroxycoumarin, Acorus calamus, Morus alba, Ichnocarpus frutescens, Salvadora oleoides, Trichosanthes cucumerina, Parthenium hysterophorus, Triticum aestivum, Michelia champaca, Cassia glauca, Bauhinia variegate, Barleria prionitis, Cassia occidentalis, Phyllanthus reticulatus, Flacourtia jangomas, Pterocarpus marsupium, Amrita-Pippali-Nimba, Palasha Pushpadi Churna, Chanderprabha Vati, Trivang Bhasma with Vijayasara Kwatha, Nishamalaki, M-93, Ptero-carpus marsupium, Coccinia Indica, Ayush-82 and Shudha Shilajittu, Saussurea lappa, Coscintium fenestratum, Ficus bengalensis, Triphala, Coccinia indica, Emblica officinalis, Commiphora mukul, Terminalia chebula.*	59
2.	Inflammation, Pain and Arthritis	*Dalbergia sissoo, Azadirachta indica, Madhuka longifolia, Curcuma amada, Hemigraphis colorata, Achyranthes bidentate, Zingiber zerumbet, Solanum melongena, Martynia diandra, Strobilanthus callosus, Strobilanthus ixiocephala, Carum copticum, Leucas aspera, Spilanthes acmella, Centella asiatica, Sapindus trifoliatus, Curcuma longa, Vernonia arborea, Elephantopus scaber, Boswellia serrata, Aloe vera, Vitex negundo, chandraprabha vati and Maha yogaraja guggulu, Kalanchoe crenata, Eugenol, Zingiber officinale, Sida cordifolia, Silybum marianum, Momordica cymbalaria, Cocos nucifera, Argyreia speciose, Bauhinia purpurea, Holoptelea integrifolia, Ficus bengalensis, Kaurenic acid, Vitex leucoxylon, Swertia chirata, Heliotropium indicum, Leucas aspera, Calotropis procera, Withania somnifera, Alangium lamarkii, Strobilanthes heyneanus, Ocimum sanctum, Picrorhiza kurroa.*	46
3.	Heart Diseases	ABANA[*], *Ocimum basilicum, Viscum album, Allium sativum, Apocynin, Ascorbic acid, Mammea africana, Acorus calamus, Commiphore mukul, Centella asiatica Inula racemose, Hypericum perforatum, Terminalia Arjuna,* Brahmyadi Ghana Vati, *Balanites roxburghii.*	15
4.	Gastric Diseases	*Gingko biloba, Shankha bhasma, Synclisia scabrida, Hippophae rhamnoides, Cassia nigricans, Zingiber officinale, Tephrosia purpurea, Camellia sinensis, Amoora rohituka,* Sodium curcuminate, *Pterocarpus marsupium, Cyperus rotundus, Cardiospermum halicacabum, Euphorbia prostrata, Stachytarpheta jamaicensis, Polyalthia longifolia, Benincasa hispida, Trilepisium madagascariense, Dalbergia sissoo, Asparagus racemosus, Berberis aristata, Bombax malabaricum, Gossypium herbaceum, Holarrhena anti-dysenterica, Myristica fragrance and Punica granatum, Emblica officinalis, Glycyrrhiza glabra.*	29
5.	Anxiety and Depression	*Myristica fragrans,* Glycyrrhizin, *Morus alba, Ocimum sanctum, Camellia sinensis, Sphaeranthus indicus,* NR-ANX-C, *Kielmeyera coriacea, Kielmeyera coriacea,* Trans-01, *Bacopa monnieri, Convolvulus pluricaulis, Celastrus paniculatus, Acorus calamus, Mucuna pruriens, Withania somnifera, Vitex negundo.*	19
6.	Cancer	*Grewia tiliaefolia, Ehrlich ascites, Lantana camara, Emilia sonchifolia, Ipomoea aquatic, Ouratea, Luxemburgia, Tephrosia purpurea, Cassia occidentalis, Trigonella foenum-graecum, Ixora javanica*	9
7.	Stress Disorders	*Garcinia cambogia, Syzygium aromaticum, Vitex negundo, Pfaffia glomerata, Globularia alypum, Solanum pseudocapsicum, Trianthema portulacastrum, Taraxacum officinale, Acacia arabica, Piper nigrum, Embelia ribes, Morus alba, Phyllanthus amarus, Eugenia caryophyllus,* Brahma Rasayana, *Ocimum sanctum, Salvia verbenaca, Paronychia argentea, gossypin and nevadensin, Withania Somnifera, Convolvulus pluricaulis, Phyllanthus emblica.*	29
8.	Liver Disease	*Andrographis paniculata, HD-03, Ginkgo biloba, Mamordica subangulata, Naragamia alata, Curculigo orchioides, Azadirachta indica, Himoliv, Panchagavya ghrita, Glycyrrhiza glabra, Adhatoda vasica, Pterocarpus marsupium, Vitis vinifera, Leucas aspera Tylophora indica, Allium sativum* and *Piper longum, Terminalia glaucescens, Cleome viscosa, Eugenia jambolana, Scoparia dulcis, Plantago major, Achyranthes aspera, Cissus quadrangularis, Hygrophila auriculata, Silymarin, Pterocarpus santalinus, Gallus domesticus, Withania somnifera, Tinospora cordifolia, Picrorhiza kurroa, Kutkin.*	37

Contd...

TABLE 10.1–*Contd...*

Sl.No.	*Disease/Illness*	*Ayurvedic Formulations/Herbs Studied*	*Number of Studies*
9.	Immune Diseases	*Tinospora cordifolia, Yi Shen Juan Bi, Pluchea lanceolate, Hemidesmus indicus, Ficus benghalensis, Heracleum nepalense, DLH-721A and DLH-721B, Trichopus zeylanicus, Immu-21, RV08, DLH-3041, Ashtamangal Ghrita, Haridradi ghrita, Rubia cordifolia, Selaginella, Justicia gendarussa, Plumbago indica, Aloe vera,* and *Aegle marmelos, Sphatika, Ocimum sanctum.*	19
10.	Renal Diseases	*Mimusops elengi,* NR-AG-I and NR-AG-II, *Lawsonia Innermis, Plectranthus amboinicus, Kalanchoe pinnata, Boerhavia diffusa, Tribulus terrestris.*	7
11.	Infectious Diseases	Canova, *Artemisia indica, Alangium salviifolium, Pallavicinia lyellii, Amorphophallus campanulatus, Gracilaria changii, Evolvulus nummularius, Microglossa angolensis, Gymnema sylvestre, Eclipta prostrata, Lawsonia inermis, Saraca indica, Syzygium cumini, Terminalia belerica, Allium sativum, and Datura stramonium, Chloranthus erectus, Ruta graveolens, Psidium guajava, Salvadora persica, Centella asiatica, Eucalyptus globulus, Achromanes difformis, Cleome rutidosperma, Cymbopogon citratus, Piper umbellatum, Mellotus appositofolius, Mangifera indicus and Annona muricata, Quercus infectoria, Elephantopus scaber, Croton zambesicus, Calotropis procera, Azadirachta indica, Coriandrum Sativum, Stachytarpheta cayennensis, Anthocleista djalonensis.*	26
12.	Reproductive Health	*Vanda tessellate,* Menotab (M-3119), *Aegle marmelos, Thespesia populnea, Piper guineense, Labisia pumila, Plumbago rosea, Areca catechu, Mondia whitei, Momordica cymbalaria.*	10
13.	Neuroprotection or CNS Disease	*Rhazya stricta, Glycyrrhiza glabra, Myristica fragrans, Tinospora cordifolia, Bramhi Ghrita, Benincasa hispida, Bacopa monneri, Evolvulus alsinoids, Acorus calamus, Saussurea lappa, Sapindus trifoliatus, Hypericum perforatum, Ocimum sanctum, NR-ANX-C, Withania somnifera, Embelia ribes, Morus alba,* Stevioside, *Prunus amygdalus, Centella asiatica,* Dashmool.	28
14.	Toxicity Studies	Navbal Rasayan, *Aloe vera, Hippophae rhamnoides,* trolox and quercetin, *Cleistanthus collinus, Boswellia dalzielii, Nepeta hindostana.*	7
15.	Miscellaneous	*Synclisia scabrida, Vitis vinifera, Trichopus zeylanicus, Benincasa hispida, Tinospora cordifolia, Pistia stratiotes, Cuminum cyminum, Chonemorpha macrophylla, Moringa oleifera, Spilanthes acmella, Asteracantha longifolia, Arishta* and *Asava, Acanthophyllum squarrosum, Zingiber officinale, Curcuma longa, Gjridhrasi, Guggulu and Pippali, Trichosanthes dioica, Ashokarista, Boerhaavia diffusa, Pterocarpus santalinus, Withanaia somnifera, Solanum xanthocarpur, Mucunaa pruriens.*	43

DIABETES AND HYPERLIPIDEMIA

Hypolipidemic Effect of Fenugreek: A Clinical Study

M. Prasanna; Indian Journal of Pharmacology 2000; 32: 34-36.

Objective: To investigate the hypolipidemic effect of fenugreek in hyper-cholesterolaemic patients.

Methods: Fenugreek (*Trigonella foenum graecum*) seeds (FG) were powdered and extracted with hexane to remove its lipid content and alcohol to remove the saponins. This powder was used for the study. The patients were divided into 3 groups of 6 each as follows: Group I received placebo 50 g (rice powder and Bengal gram powder in equal measures); Group II -placebo 25 g + FG 25 g and Group III –FG 50 g. Patients were directed to take each 50 g pack orally before lunch and dinner every day for 20 days. Blood samples were collected after overnight fasting on 0, 10th and 20th days during test period and estimated for lipid profile.

Results: There were no significant changes in lipid profile of group I patients. In groups II and III serum cholesterol, triglycerides and VLDL levels were significantly decreased when compared to group I.

Conclusion: FG powder given orally before food at 25 and 50 g twice a day may have hypolipidemic effect in hypercholesterolaemic patients.

Hypoglycaemic Effect of the Aqueous Extract of *Boerhavia diffusa* Leaves

M.A. Chude, O.E. Orisakwe, O.J. Afonne, K.S. Gamaniel, O.H. Vongtau, E. Obi; Indian Journal of Pharmacology 2001; 33: 215-216.

The present study aims at investigating the effects of *B. diffusa* aqueous leaf extract on the blood sugar level of rats with a view to elucidating the rationale behind its use in the management of diabetes by herbalists. The baseline plasma glucose levels were determined prior to administration. The alloxan-induced diabetic rats were divided into four groups of six male rats each. The test groups received 100, 200 and 400 mg/kg aqueous extract, while the control group received appropriate volumes of water orally respectively. At 0, 2, 4, 6, 8 and 24 hours, 0.5 ml of blood from the tail vein of the rats was dropped on the reagent pad of the one touch strip. The strip was inserted into a one-touch brand meter and the reading noted. The aqueous extract was found to contain flavonoids, glycosides, tannins, saponins and proteins. The extract showed non-dose dependent hypoglycemic activity. The peak activity of the extract was observed at 6th post-drug administration. While the 400 mg/kg dose caused a maximum per cent reduction of 21.56 in glucose level at 6h, the 100 and 200 mg/kg doses of the extract showed more hypoglycemic effects which were significant at $p < 0.05$ from the initial value, with percentage decreases of 38.07 per cent and 51.95 per cent respectively. The aqueous leaf extract of *B. diffusa* produced nondose-related decreases in blood glucose level in alloxan-induced diabetic rats. Thus, the hypoglycemic effect produced by the extract of *B. diffusa* leaves may be due to the glycosides, flavonoids, tannins and saponins present in the extract.

Antiobesity Effect of a Polyherbal Formulation, OB-200G in Female Rats Fed on Cafeteria and Atherogenic Diets

Gurpreet Kaur, S.K. Kulkarni; Indian Journal of Pharmacology 2000; 32: 294-299.

Objective: To study the antiobesity effect of OB-200G, a polyherbal formulation in female Wistar rats fed on cafeteria and atherogenic diets.

Methods: Female rats were fed cafeteria diet (highly palatable, energy-rich animal diet that includes a variety of human snack foods) and atherogenic diet for 40 days. OB-200G was administered in a dose of 400 mg/kg, p.o., twice a day to the drug treatment groups. The effect of OB-200G on following parameters was recorded - body weight, rectal temperature, locomotor activity and various biochemical parameters like serum glucose, total cholesterol and triglyceride levels.

Results: There was a significant ($p < 0.05$) reduction in body weight, increase in body temperature, locomotor activity and serum glucose levels after treatment with OB-200G in cafeteria diet and atherogenic diet fed rats. Treatment with OB-200G also significantly ($p < 0.05$) decreased total cholesterol in rats fed with atherogenic diet.

Conclusion: OB-200G, a polyherbal formulation exhibited antiobesity effect in cafeteria and atherogenic diet fed rats

Antihyperglycemic Activity of Aqueous Extract of Leaves of *Cocculus hirsutus* (L.) Diels in Alloxan-induced Diabetic Mice

Badole S, Patel N, Bodhankar S, Jain B, Bhardwaj S.; Indian Journal of Pharmacology. 2006; 38: 49-53.

Objective: To evaluate the antihyperglycemic activity of aqueous extract of leaves of *Cocculus hirsutus* (L.) Diels in alloxan-induced diabetic mice.

Materials and Methods: Alloxan-induced (70 mg/kg, i.v.) diabetic mice were given aqueous leaf extract (250, 500, and 1000 mg/kg, p.o., $n = 6$) of *C. hirsutus* or vehicle (distilled water, 10 ml/kg, p.o.) or standard drug glyburide (10 mg/kg, p.o.) for 28 days. Blood samples were withdrawn by retro-orbital puncture and were analyzed for serum glucose on 0th, 7th, 14th, 21st, and 28th days by glucose oxidase/peroxidase method. In oral glucose tolerance test, glucose (2.5 g/kg, p.o.) was administered to nondiabetic control, glyburide (10 mg/kg, p.o.), and aqueous extract of *C. hirsutus* (1000 mg/kg, p.o.) treated mice. The serum glucose level was analyzed at 0, 30, 60, and 120 min after drug administration.

Results: The aqueous leaf extract of *C. hirsutus* (250, 500, and 1000 mg/kg, p.o.) showed significant ($P < 0.01$) reduction of serum glucose level in alloxan-induced diabetic mice at 28th day. In oral glucose tolerance test, aqueous extract of *C. hirsutus* increased the glucose tolerance.

Conclusion: It is concluded that *C. hirsutus* has significant antihyperglycemic activity as it lowers serum glucose level in diabetic mice and significantly increases glucose tolerance.

Anti-hyperglycaemic Activity of *Cassia kleinii* Leaf Extract in Glucose Fed Normal Rats and Alloxan-Induced Diabetic Rats

Babu V, Gangadevi T, Subramoniam A.; Indian Journal of Pharmacology 2002; 34: 409-415.

Objective: To study the effect of *Cassia kleinii* on serum glucose levels in both normal and diabetic rats.

Methods: The extracts of dried root (suspension in water) and leaf (water, alcohol and n-hexane) were screened for their effects on serum glucose levels in glucose overloaded rats. The most active extract (alcohol extract of leaf) was tested for antidiabetes activity in alloxan-induced diabetic rats and for hypoglycaemic activity in normal fasted rats.

Results: The plant leaf as well as its alcohol extract (but not the other extracts) exhibited concentration dependent antihyperglycaemic effect in glucose loaded rats. However, the extract did not show hypoglycaemic effect in fasted normal rats. In alloxan-induced diabetic rats the extract (200 mg/kg) showed remarkable efficacy.

Conclusion: The study reveals for the first time the antihyperglycaemic activity of *Cassia kleinii* leaf (alcohol extract) in both glucose fed hyperglycaemic and alloxan-induced diabetic rats. The extract seems promising for the development of a phytomedicine for diabetes mellitus.

Antidiabetic Activity of Ethanol Extract of *Cassia kleinii* Leaf in Streptozotocin-Induced Diabetic Rats and Isolation of an Active Fraction and Toxicity Evaluation of the Extract

Babu V, Gangadevi T, Subramoniam A.; Indian Journal of Pharmacology 2003; 35: 290-296.

Objective: (i) To study the efficacy of *Cassia kleinii* leaf (ethanol extract) in streptozotocin diabetic rats. (ii) To isolate the active fraction from the ethanol extract. (iii) To evaluate acute and short-term general toxicity in male mice.

Methods: (i) The alcohol extract of the plant leaf was tested for its efficacy in streptozotocin-induced diabetic rats. (ii) The extract was evaluated for its acute and short-term general toxicity in male mice. (iii) The aqueous suspension of ethanol extract was successively extracted with petroleum ether, chloroform, ethyl acetate and butanol. Each fraction was tested for antihyperglycemic activity using glucose tolerance test in rats. The active chloroform and ethyl acetate fractions were subjected to qualitative chemical analysis.

Results: In streptozotocin-diabetic rats the alcohol extract (200 mg/kg) showed significant antidiabetic property as judged from body weight, serum glucose, lipids, cholesterol and urea, and liver glycogen levels. However, the extract did not significantly influence the levels of serum insulin in both diabetic and normoglycemic rats. The alcohol extract was devoid of any conspicuous acute and short-term general toxicity in mice. The anti-hyperglycemic activity was found predominantly in the chloroform fraction of alcohol extract, which contained terpenoids, coumarins and saponins.

Conclusion: The active fraction of *Cassia kleinii* leaf extract is very promising to develop standardized phytomedicine for diabetes mellitus.

Modulation of Oxidative Stress Parameters by Treatment with *Piper betle* Leaf in Streptozotocin Induced Diabetic Rats

Santhakumari P, Prakasam A, Pugalendi KV.; Indian Journal of Pharmacology 2003; 35: 373-378.

Objectives: To study the effect of oral administration of Piper betle leaf powder suspension on lipid peroxidation and antioxidants in streptozotocin (STZ) diabetic rats.

Methods: Male Wistar strain rats were orally administered the leaf suspension of P. betle, (75 and 150 mg/kg body weight) for 30 days. Plasma and erythrocytes were separated out and the liver and kidney were homogenized in ice-cold buffer and the assays of thiobarbituric acid reactive substances (TBARS), hydroperoxides, glutathione (GSH), superoxide dismutase (SOD EC 1.11.1.1), catalase (CAT EC 1.11.1.6) and glutathione peroxidase (GPx EC 1.11.1.9) were performed in the supernatant obtained from liver and kidney of control and STZ diabetic rats. Plasma TBARS, hydroperoxides, ascorbic acid and alphatocopherol were measured.

Results: Oral administration of P. betle (75 and 150 mg/kg body weight) for 30 days resulted in a significant reduction in plasma thiobarbituric acid reactive substances (TBARS), hydroperoxides, alpha-tocopherol and significant improvement in glutathione, superoxide dismutase, catalase and glutathione peroxidase in the liver and kidney of STZ diabetic rats when compared with untreated diabetic rats. The antioxidant effect of P. betle at 75 mg/kg for 30 days was found to be comparable to glibenclamide in diabetic rats.

Conclusion: The leaf suspension of P. betle 75 mg/kg body weight showed significant antioxidant effects in STZ diabetic rats.

Hypoglycemic Activity of *Ficus hispida* (Bark) in Normal and Diabetic Albino Rats

Ghosh R, Sharatchandra Kh., Rita S, Thokchom IS.; Indian Journal of Pharmacology, 2004; 36: 222-225.

Objective: To find out the hypoglycemic activity of *Ficus hispida* Linn. (bark) in normal and diabetic albino rats and to evaluate its probable mechanism of hypoglycemic activity if any.

Material and Methods: Albino rats were divided into groups (n=6) receiving different treatments consisting of vehicle, water-soluble portion of the ethanol extract of Ficus hispida bark (FH) (1.25 g/kg) and standard antidiabetic drugs, glibenclamide (0.5 mg/kg) and 0.24 units of insulin (0.62 ml of 0.40 units/ml). Blood glucose was estimated by the glucose oxidase method in both normal and alloxan-induced diabetic rats before and 2 h after the administration of drugs. To find out the probable mechanism of action of FH as a hypoglycemic agent, i) the glycogen content of the liver, skeletal muscle and cardiac muscle, and ii) glucose uptake by isolated rat hemi-diaphragm were estimated.

Results: FH showed significant reduction of blood glucose level both in the normal ($P<0.01$) and diabetic ($P<0.001$) rats. However, the reduction in the blood glucose level was less than that of the standard drug, glibenclamide. FH also increased the uptake of glucose by rat hemi-diaphragm significantly ($P<0.001$). There was a significant increase in the glycogen content of the liver ($P<0.05$), skeletal muscle ($P<0.01$) and cardiac muscle (P<0.001). The amount of glycogen present in the cardiac muscle was more than the glycogen present in the skeletal muscle and liver.

Conclusion: FH has significant hypoglycemic activity. Increased glycogenesis and enhanced peripheral uptake of glucose are the probable mechanisms involved in its hypoglycemic activity.

The Hypoglycemic Activity of *Coccinia indica* Wight and Arn. and its Influence on Certain Biochemical Parameters

Dhanabal SP, Koate CK, Ramanathan M, Elango K, Suresh B.; Indian Journal of Pharmacology, 2004; 36: 244-250.

Coccinia indica Wight and Arn. (Cucurbitaceae), which is grown abundantly in India have been widely used in the traditional treatment of diabetes mellitus. In the present study it was planned to test a few fractions and delineate the most active fraction of *Coccinia indica*. The alcoholic extract of *Coccinia indica* was found to be more active in reducing blood glucose level; this extract was subjected to further fractionation and evaluation on antidiabetic activity and biochemical parameters. Among the fractions tested, only the toluene sub-fraction was found to be effective in reducing blood sugar level. The toluene fraction prevented the elevation of lipid profile significantly (P<0.001) in comparison to control diabetic rats, prevented the elevation of AST (P<0.01) and ALT (P<0.02) levels but not that of ALP. Toluene fraction could not attenuate decrease in the plasma protein content. The restoration of AST and ALT to their normal levels by the toluene fraction may also indicate the revival of insulin secretion to near normal levels. The results of the present study indicate that the toluene fraction was the only active fraction. The active principles in this fraction were found to be triterpenes which may be responsible for the antidiabetic activity and correction of the altered metabolic functions. The mechanism of action of these principle(s) may be due to their cell restorative properties against alloxan-induced damage.

Effect of Curcumin on Triton WR 1339 Induced Hypercholesterolemia in Mice

Majithiya JB, Parmar AN, Balaraman R.; Indian Journal of Pharmacology, 2004; 36: 381-384.

Curcumin (diferuloylmethane), a major component of turmeric, is a yellow pigment obtained from rhizomes of Curcuma longa, is commonly used in Indian cuisine as a spice and food-colouring agent. The study was done to see the effect of curcumin on hyperlipidemia induced by triton WR 1339 (Tyloxapol: a nonionic detergent, oxyethylated tertiary octyl phenol formaldehyde polymer). Hyperlipidemia was induced by single intravenous injection of 200 mg/kg of triton WR 1339 in normal saline. Control animals were injected with normal saline. The animals were divided into different groups in different dose levels of curcumin. Treatment with curcumin (100 mg/kg) caused 6.2 per cent and 5.0 per cent reduction in total cholesterol and triglycerides respectively. Treatment with (200 and 400 mg/kg) of curcumin caused a dose dependent change in total cholesterol and triglycerides. Control mice treated with curcumin had no significant change in total cholesterol and triglycerides. The hypolipidemic effect of curcumin administration could be due to an increased catabolism of cholesterol into bile acids.

Influence of Flavonoids Isolated from *Satureja hortensis* L. on Hypercholesterolemic Rabbits

Mchedlishvili D, Kuchukashvili Z, Tabatadze T, Davitaia G.; Indian Journal of Pharmacology 2005; 37: 259-60.

The effect of flavonoids from *S. hortensis* L. on serum cholesterol of rabbits was investigated in the present study. The experiments were carried out for 8 weeks in 20 male, albino, New Zealand rabbits weighing between 2.1 and 2.5 kg. Rabbits were housed individually in stainless steel cages in an air-conditioned room (23 ± 1 oC) under a 12 h light/dark cycle and were acclimated for 2 weeks. Study divided into 4 groups and studied for anticholestimic activity. The results are expressed as mean±SD. Inter-group comparisons among the 4 groups were determined by one-way ANOVA followed by Scheffe's 'F' test. All analyses were performed using the GraphPad Prism software version 4.01; P<0.05 was considered statistically significant. Administration of flavonoids isolated from *S. hortensis* L. along with cholesterol in rabbits has resulted in a significant attenuation of rise in serum cholesterol value after 8 weeks, when compared to the cholesterol alone group. This suggests the cholesterol-lowering effect of fraction F in a situation of rising serum cholesterol. Though the hypolipidemic properties of flavonoids are well known such an effect of flavonoids from *S. hortensis* L. has not yet been shown.

Hypolipidemic and Antioxidant Activities of *Asparagus racemosus* in Hypercholesteremic Rats

Visavadiya NP, RL Narasimhacharya AV.; Indian Journal of Pharmacology 2005; 37: 376-80.

To study the efficacy of *Asparagus racemosus* in reducing the cholesterol levels and as an antioxidant in hypercholesteremic rats.

Materials and Methods: Hypercholesteremia was induced in normal rats by including 0.75 g per cent cholesterol and 1.5 g per cent bile salt in normal diet and were used for the experiments. Dried root powder of Asparagus racemosus was administered as feed supplement at 5 g per cent and 10 g per cent dose levels to the hypercholesteremic rats. Plasma and liver lipid profiles, hepatic HMG-CoA reductase, bile acid, malondialdehyde, ascorbic acid, catalase and SOD, fecal bile acid, cholesterol and neutral sterols were estimated using standard methods.

Results: Feed supplementation with 5 g per cent and 10 g per cent *Asparagus racemosus* resulted in a significant decline in plasma and hepatic lipid profiles. The feed supplementation increased the HMG-CoA reductase activity and bile acid production in both groups (5 and 10 g per cent supplemented groups) with concomitant increase in fecal bile acid and fecal cholesterol excretion. The activities of catalase, SOD and ascorbic acid content increased significantly in both the experimental groups (5 and 10 g per cent supplemented groups). On the other hand, the concentration of malondialdehyde in these groups (5 and 10 g per cent supplemented groups) decreased significantly, indicating decreased lipid peroxidation.

Conclusion: The present study demonstrates that addition of *Asparagus racemosus* root powder at 5 g per cent and 10 g per cent level as feed supplement reduces the plasma and hepatic lipid (cholesterol) levels and also decreases lipid peroxidation.

Antihistaminic and Antianaphylactic Activity of HK-07, a Herbal Formulation

Gopumadhavan S, Md Rafiq, Venkataranganna MV, Mitra SK.; Indian Journal of Pharmacology 2005; 37: 300-303.

Objective: To study the antianaphylactic, antihistaminic and mast cell stabilization activity of HK-07 in experimental animals.

Materials and Methods: HK-07 is a polyherbal formulation containing extracts of various plant constituents. The compound HK-07 was evaluated using Wistar rats and Duncan Hartley guinea pigs. The antianaphylactic activity was investigated in rats using the active anaphylaxis model. The effect on mast cell stabilization was performed by ex vivo challenge of antigen in sensitized rat intestinal mesenteries. Antihistaminic activity was studied in guinea pigs using histamine-induced bronchospasm where preconvulsive dyspnea was used as an end point following exposure to histamine aerosol. Dose response studies of HK-07 were conducted at 125, 250, and 500 mg/kg, p.o. in anaphylactic shock-induced bronchospasm in rats. The optimal dose level was used for the remaining experimental models.

Results: Treatment with HK-07 at 125, 250, and 500 mg/kg, p.o. showed significant reduction in signs and severity of symptoms ($P < 0.05$), onset ($P < 0.001$) and mortality rate ($P < 0.05$) following anaphylactic shock-induced bronchospasm. HK-07 also significantly reduced the serum IgE levels ($P < 0.001$) in animals compared to untreated controls. Treatment of sensitized animals with HK-07 at 500 mg/kg, p.o. for 2 weeks resulted in a significant reduction in the number of disrupted mast cells ($P < 0.001$) when challenged with an antigen (horse serum). HK-07 significantly prolonged the latent period of convulsion ($P < 0.008$) as compared to control following exposure of guinea pigs to histamine aerosol.

Conclusion: The findings from various studies reveal that the antihistaminic and antianaphylactic activity of HK-07 may be due to the mast cell stabilizing potential, suppression of IgE, and inhibition of release of inflammatory mediators.

Hypoglycaemic Effect of Methylene Chloride/ Methanol Root Extract of *Ceiba pentandra* in Normal and Diabetic Rats

Djomeni Dzeufiet P D, Tedong L, Asongalem E A, Dimo T, Sokeng S D, Kamtchouing P.; Indian Journal of Pharmacology. 2006; 38: 194-7.

Objective: The current study examined the effects of the methylene chloride/methanol extract of root bark of *Ceiba pentandra* (L.) in normal and streptozotocin-induced diabetic rats.

Materials and Methods: Diabetes was induced by intravenous streptozotocin (55 mg/kg) in adult male albino Wistar rats. Single and multiple dose studies were carried out. Blood glucose levels were determined after oral administration of graded doses of *C. pentandra* (40, 75, 150 and 300 mg/kg) in fasting normal and diabetic groups for the single dose study; and before and at the end of day 3 of the treatment period for the multiple dose study.

Results: In both the groups, the extract (40 and 75 mg/kg) significantly reduced the blood glucose 5 hours after administration, in a consistent and time-dependent manner. *C. pentandra* at the lower dose (40 mg/kg) produced 40 per cent and 48.9 per cent lowering of blood-glucose in normal and diabetic rats, respectively compared to the initial values. In the multiple dose studies, the diabetic rats were treated orally by gavage, twice a day for three days. On day 3, *C. pentandra* (40 and 75 mg/kg) significantly decreased blood and urine glucose, compared to initial values. With 40 and 75 mg/kg of drug, the 14 h fasting blood glucose concentration was reduced by 59.8 per cent and 42.8 per cent with corresponding reductions of urine glucose levels by 95.7 per cent and 63.6 per cent, respectively.

Conclusion: These results indicate that *C. pentandra* possesses a hypoglycaemic effect. The plant extract is capable of ameliorating hyperglycaemia in streptozotocin-induced diabetic rats and is a potential source for isolation of new orally active agent(s) for diabetes mellitus.

Prevention of Dexamethasone- and Fructose-Induced Insulin Resistance in Rats by SH-01D, a Herbal Preparation

Md. Shalam, Harish M S, Farhana S A.; Indian Journal of Pharmacology. 2006; 38(6): 419-22.

Objective: To investigate the preventive effect of SH-01D, a herbomineral preparation, on the development of insulin resistance induced by dexamethasone and fructose, in rats.

Materials and Methods: Two models of insulin resistance were used (dexamethasone 10 mg/kg, s.c. once daily and fructose 10 per cent w/v, p.o., *ad libitum*) in rats for a period of 10 and 20 days, respectively. Two doses of SH-01D (30 mg and 60 mg/kg, p.o.) were used. At the end of the experimental period, serum biochemical parameters like insulin, glucose, triglycerides, LDL, HDL and cholesterol were studied. Liver and muscle glycogen were estimated in the fructose model after sacrificing the animals.

Results: In both the models, SH-01D at 60 mg/kg showed significant effect. Fructose feeding increased serum biochemical parameters and decreased liver and skeletal muscle glycogen levels. Dexamethasone caused an increase in serum glucose, triglyceride levels and a decrease in body weight. In fructose-fed rats, SH-01D at 60 mg/kg significantly prevented (a) the increase in serum biochemical parameters and (b) the decrease in glycogen levels. In the dexamethasone model, SH-01D prevented the rise in serum glucose and triglycerides and improved the body weight.

Conclusion: The present study indicates that SH-01D may be useful in the management of insulin resistance.

Hypoglycemic Activity of Aqueous Leaf Extract of *Persea americana* Mill

Antia BS, Okokon JE, Okon PA.; Indian Journal of Pharmacology 2005; 37: 325-6.

A study was conducted in albino Wistar rats to find out hypoglycemic activity of aqueous leaf extract of *Persea americana* Mill. The phytochemical screening of the aqueous leaf extract of *Persea americana* revealed that the extract contained various pharmacologically active compounds such as saponins, tannins, phlobatannins, flavonoids, alkaloids, and polysaccharides. The observation confirms the use of this plant in ethnomedical practice for diabetes management and demonstrated that long-term treatment for 7 days was more effective than single dose acute treatment.

Partial Protective Effect of Rutin on Multiple Low Dose Streptozotocin-Induced Diabetes in Mice

Srinivasan K, Kaul C L, Ramarao P.; Indian Journal of Pharmacology 2005; 37: 327-8.

A study was conducted in Type 1 diabetes induced in Swiss albino mice. Rutin (200 mg/kg BW) was used for 28 days. The plasma was analyzed for glucose concentration by the glucose oxidase-peroxidase (GOD-POD) method, using commercially available spectrophotometric kit. Results showed that the mice treated with MLDS demonstrated a significant ($P<0.05$) and progressive rise in PGL attaining peak at twenty-first day after STZ injection as compared to vehicle-treated normal control mice. Also, rutin significantly but not completely prevented the development of hyperglycemia in MLDS mice as shown by significant ($P<0.05$) reduction in PGL at day 21 and 28. However, rutin per se did not significantly alter the PGL in normal mice ruling out its hypoglycemic activity

In vitro Study on α-amylase Inhibitory Activity of an Indian Medicinal Plant, *Phyllanthus amarus*

Tamil IG, Dineshkumar B, Nandhakumar M, Senthilkumar M, Mitra A.; Indian Journal of Pharmacology. 2010; 42: 280-2.

Objective: The objective of this study was to evaluate the α-amylase inhibitory activity of different extracts of *Phyllanthus amarus* against porcine pancreatic amylase *in vitro.*

Materials and Methods: The plant extracts were prepared sequentially with ethanol, chloroform, and hexane. Each extract was evaporated using rotary evaporator, under reduced pressure. Different concentrations (10, 20, 40, 60, 80, and 100 μg/mL) of each extract were made by using dimethyl sulfoxide (DMSO) and subjected to α-amylase inhibitory assay using starch azure as a substrate. The absorbance was read at 595 nm using spectrophotometer. Using this method, the percentage of α-amylase inhibitory activity and IC(50) values of each extract was calculated.

Results: The chloroform extract failed to inhibit α-amylase activity. However, the ethanol and hexane extracts of *P. amarus* exhibited appreciable α-amylase inhibitory activity with an IC50 values 36.05 ± 4.01 μg/mL and 48.92 ± 3.43 μg/mL, respectively, when compared with acarbose (IC_{50} value 83.33 ± 0.34 μg/mL).

Conclusion: This study supports the ayurvedic concept that ethanol and hexane extracts of *P. amarus* exhibit considerable α-amylase inhibitory activities. Further, this study supports its usage in ethnomedicines for management of diabetes.

Anti-pyretic, Antidiarrhoeal, Hypoglycaemic and Hepatoprotective Activities of Ethyl Acetate Extract of *Acacia catechu* Willd. in Albino Rats

Ray D, Kh. Sharatchandra, Thokchom I S.; Indian Journal of Pharmacology. 2006; 38(6): 408-13.

Objective: To evaluate the anti-pyretic, antidiarrhoeal, hypoglycaemic and hepatoprotective effects of the ethyl acetate extract of *Acacia catechu* in experimental animal models.

Materials and Methods: Ethyl acetate extract of *Acacia catechu* was evaluated for anti-pyretic activity in yeast-induced pyrexia and for antidiarrhoeal activity in castor oil-induced diarrhoea in albino rats. Hypoglycaemic activity was studied in both normal and alloxan (120 mg/kg, s.c.) induced diabetic albino rats. The hepatoprotective potential of *Acacia catechu* was evaluated by CCl_4 induced hepatotoxicity in albino rats.

Results: Single administration of the ethyl acetate extract of *Acacia catechu* at doses of 250 and 500 mg/kg, p.o. showed significant anti-pyretic activity ($P < 0.01$) in albino rats. *Acacia catechu* at a dose of 250 mg/kg, p.o. (single dose) has been found to possess highly significant antidiarrhoeal property ($P < 0.001$) in respect of latent period of onset of diarrhoea, average number of stool passed and purging index. Significant reduction of blood glucose level was observed in nondiabetic albino rats following single dose treatment with the test drug at a dose of 500 mg/kg, p.o. ($P < 0.01$). Significant reduction of blood glucose level was also evident in diabetic rats at doses of 250 and 500 mg/kg ($P < 0.001$). Highly significant hepatoprotective activity was also observed when the extract of *Acacia catechu* (250 mg/kg) was administered prophylactically for seven days ($P < 0.001$).

Conclusion: The present study shows that ethyl acetate extract of *Acacia catechu* (cutch/katha) has significant anti-pyretic, antidiarrhoeal, hypoglycaemic and hepatoprotective properties.

Effect of *Pongamia pinnata* Flowers on Blood Glucose and Oxidative Stress in Alloxan Induced Diabetic Rats

Punitha R, Vasudevan K, Manoharan S.; Indian Journal of Pharmacology. 2006; 38(1): 62-3.

Diabetes mellitus is a major metabolic syndrome characterized by derangement in carbohydrate metabolism associated with defect in insulin secretion or action. Alloxan is widely used to induce diabetes mellitus in experimental animals, owing to its ability to destroy the β-cells of pancreas possibly by generating excess reactive oxygen species. Free radical-mediated biomembrane lipid peroxidation has been implicated in the pathogenesis of many pathological conditions including diabetes mellitus and its complications. Overproductions of lipid peroxidation by-products and insufficient antioxidant potential have been reported in both experimental and human diabetes mellitus.

Medicinal plants and their bioactive constituents are used for the treatment of diabetes mellitus throughout the world, especially in countries where access to the conventional treatment of diabetes mellitus is inadequate. A few side effects associated with the use of insulin (hypoglycemia) and oral hypoglycemic agents prompted us to search new bioactive principles from antidiabetic plants used in traditional medicine. Although several medicinal plants have gained importance for the treatment of diabetes mellitus, many remain to be scientifically investigated. *Pongamia pinnata* (Linn.) Pierre is a medium-sized glabrous tree popularly known as *Karanja* in Hindi, Indian beech in English, and *Pongam* in Tamil. *P. pinnata* is an important medicinal plant chiefly found in tidal forests of India and has been largely used in the traditional Indian system of medicine (Ayurveda) for bronchitis, whooping cough, rheumatic arthritis, and diabetes. Despite its prominence in ayurvedic medicine, there is a dearth of scientific data on its antihyperglycemic and antilipidperoxidative effects in diabetes mellitus. Thus, the present study was focused on the antihyperglycemic and antilipidperoxidative activities of aqueous extract of *P. pinnata* flowers (PpFAet) in alloxan-induced diabetic rats.

Albino Wistar male rats (7-8 week old, weighing 150-200 g) were used in the present study and housed in the Central Animal House with 12 h light/dark cycle. Standard pellet (Mysore Snack Feed Ltd, Mysore, India) was used as a basal diet during the experiment. The control and experimental animals were provided food and water *ad libitum*. The Institutional Animal Ethics Committee of the Annamalai University, Annamalai Nagar, India, approved the experimental design. A total of 30 rats were divided into 5 groups of 6 each and treated as follows.

- Group I-control (2 ml distilled water, orally).
- Group II-diabetic control (alloxan, 150 mg/kg, i.p).
- Group III-diabetic + PpFAet (300 mg/kg, orally).
- Group IV-diabetic+ glibenclamide (600 μg/kg, orally).
- Group V-PpFAet + distilled water (300 mg/kg, orally).

The diabetic condition was assessed by determining the blood glucose concentration 3 and 5 days after alloxan treatment. The rats with blood glucose level above 260 mg/dl and urinary sugar (+++) were selected for study (Groups II-IV). The dose (300 mg/kg, orally) was standardized after a pilot study with different doses of the PpFAet extract to assess the antihyperglycemic and antilipidperoxidative effects in alloxan-induced diabetic rats. After the experimental period, all animals were sacrificed by cervical dislocation and biochemical studies conducted in blood, plasma, and liver samples using colorimetric methods. Plasma insulin was assayed by ELISA method using Boehinger Mannheim Gmbh kit. The data are expressed as mean±SD. Statistical comparisons were performed by one-way anova followed by Dunnett's test. The results were considered statistically significant if $P < 0.05$.

A significant decrease in the level of blood glucose and glucose-6-phosphatase activity and a significant increase in the plasma insulin level and hexokinase activity were noted at the end of the experimental period in diabetic rats treated with PpFAet (300 mg/kg, orally). PpFAet also significantly reduced the thiobarbituric acid-reactive substances (TBARS) level and enhanced the antioxidants status in induced diabetic rats after 45 days of treatment. PpFAet showed antihyperglycemic and antilipidperoxidative effects in a manner similar to that of the reference drug glibenclamide in alloxan-induced diabetic rats.

Diabetes mellitus is a chronic metabolic disorder characterized by hyperglycemia associated with several other factors including dyslipidemia, which are involved in the development of micro- and macrovascular complications. In the present study, the orally administered PpFAet (300 mg/orally) to diabetic rats showed a significant antihyperglycemic activity as well as significantly increased the plasma insulin level. The extract also normalized the activities of glucose-6-phosphatase and hexokinase in diabetic rats. Any defect in insulin and glucagon secretion brings about changes in glucose homeostasis and activities of key enzymes of glucose metabolism. Decreased hexokinase activity and increased glucose-6-phosphatase activity in diabetic rats is probably owing to insulin deficiency. Enhanced hexokinase activity and decreased glucose-6-phosphatase activity in PpFAet-treated diabetic rats, therefore, suggest its stimulatory effects on insulin secretion from remnant pancreatic β-cells, which in turn promotes greater utilization of blood glucose by the liver, muscle, and adipose tissues of diabetic rats.

Lipid peroxide-mediated tissue damage has been demonstrated in diabetes mellitus. Several studies indicated that the alloxan-induced diabetes mellitus involves the degeneration of islet β-cells by accumulation of cytotoxic free radicals such as superoxides. The chronic hyperglycemia in diabetes mellitus enhances the production of ROS owing to glucose auto-oxidation, protein glycation, and glyco-oxidation, which in turn cause severe tissue damage.

Several studies have shown a decrease in nonenzymatic antioxidants (vitamin C and glutathione) and lowered activities of enzymatic antioxidants (superoxide dismutase, catalase, and glutathione peroxidase) in the plasma of alloxan-induced diabetic rats. Our results lend credibility to these observations. The observed increase in enzymatic antioxidants activities and decline in lipid peroxide concentration in PpFAet-treated rats suggest its potent antilipid-peroxidative and antioxidant effects in diabetic rats. PpFAet not only reduced the blood glucose level and lipid peroxides, but also enhanced the status of antioxidants to an extent similar to that of the reference drug glibenclamide in alloxan-induced diabetic rats. Further studies are warranted to isolate and characterize the antidiabetic principle from the PpFAet.

Antioxidant Potential of Methanolic Extract of *Dolichos biflorus* Linn in High-fat diet Fed Rabbits

Muthu A K, Sethupathy S, Manavalan R, Karar P K.; Indian Journal of Pharmacology. 2006; 38(2): 131-2.

Dolichos biflorus Linn (Fabaceae), is commonly known as Kollu in Tamil and horse gram in English. It has been reported to lower lipids in rats. A high-fat diet induces oxidative stress in the cells by producing reactive oxygen species. Therefore, in this study, the influence of the *Dolichos biflorus* extract on high-fat diet (HFD) induced oxidative stress in rabbits, has been investigated.

Whole plants of *D. biflorus* were collected from Sankaran Koil, Tirunelveli district of Tamilnadu, India. Taxonomic identification was made by the Botanical

Survey of Medicinal Plant Unit, Siddha, Government of India, Palayamkottai, Tamilnadu. Four month-old whole plants were dried in the shade, segregated, and pulverized by a mechanical grinder and passed through a 40 mesh sieve. The powder was extracted by methanol in Soxhlet apparatus by continuous hot percolation method. After filtration through Whatmann filter paper No 40, the filtrate was vacuum dried at 35 to 40°C. The extracts were stored in screw cap vials at 4°C until further use. The extractive value of the methanolic extract was 8.13 per cent w/w. The methanolic extract of *D. biflorus* was subjected to preliminary phytochemical screening to find out the presence of active principles. The extracts were suspended in 2 per cent tween 80.

New Zealand white rabbits, weighing 900-1050 g were procured from the central animal house, Rajah Muthiah Medical College, Annamalai University. The animals were kept in cages, 2 per cage, with 12:12 h light/dark cycle at 25 ± 2°C. The animals were maintained on their respective diets and water *ad libitum*. Animal ethical committee's clearance was obtained for the study.

Rabbits were divided into the following five groups, with six rabbits in each group:

- Group I: (Control): Standard chow diet.
- Group II: High-fat diet (HFD)
- Group III: High-fat diet plus methanolic extract of *D. biflorus* (dose I -200 mg/kg body weight)
- Group IV: High-fat diet plus methanolic extract of *D. biflorus* (dose II -400 mg/kg body weight)
- Group V: High-fat diet plus standard drug atorvastatin (1.2 mg/kg body weight).

The compositions of the two diets were as follows:

Wheat flour 22.5 per cent, roasted Bengal gram powder 60 per cent, skimmed milk powder 5 per cent, casein 4 per cent, refined oil 4 per cent, salt mixture with starch 4 per cent, and vitamin and choline mixture 0.5 per cent.

High-Fat Diet

Wheatflour 20.5 per cent, roasted Bengal gram 52.6 per cent, skimmed milk powder 5 per cent, casein 4 per cent, refined oil 4 per cent, coconut oil 9 per cent, salt mixture with starch 4 per cent, vitamin and choline mixture 0.5 per cent, and cholesterol 0.4 per cent.

Rabbits in groups III and IV were orally fed the methanolic extracts of *D. biflorus* dose I and dose II, respectively, and rabbits in group V were fed standard drug atorvastatin. The *D. biflorus* extracts and atorvastatin were suspended in 2 per cent tween 80 separately and fed to the respective groups of rabbits by oral incubation. At the end of 11 weeks, all the animals were sacrificed by cervical decapitation after overnight fasting.

Portions of the tissues from liver, heart, and aorta were blotted, weighed, and homogenized with methanol (3 volumes). The lipid extract obtained by the method of Folch *et al.*, was used for the estimation of thiobarbituric acid reactive substances (TBARS). Another portion of the tissues was homogenized with phosphate buffer saline and used for the estimation of reduced glutathione (GSH), catalase (CAT), and superoxide dismutase (SOD).

Results were expressed as mean ± SE of 6 rabbits in each group. One-way analysis of variance (ANOVA) with Scheffe's multiple comparisons test, were used to determine the statistical significance. $P<0.05$ was considered significant.

The average body weight was found to have increased in high-fat diet fed rabbits compared with those in control group. After administration of two doses of *D. biflorus* extract it was found to have decreased. TBARS had significantly increased and GSH had significantly decreased in liver, heart, and aorta of rabbits fed HFD compared with those in control group I. Administration of *D. biflorus* extract significantly lowered the level of TBARS and enhanced the level of GSH. Higher dose of the plant extract was found to be more effective and showed comparable results with standard drug atorvastatin on these two parameters. Activities of antioxidant enzymes, that is, SOD and CAT in different groups are given in. These two enzymes showed a marked reduction in activity in liver, heart, and aorta of rabbits in group II which had been fed HFD. Supplementation of *D. biflorus* extract with HFD significantly improved the activities of SOD and CAT in the above tissues of rabbits in groups III and IV as compared with group II. Preliminary phytochemical study shows that methanolic extract of the experimental plant contains phytoconstituents, such as, alkaloids, steroids, flavonoids, and isoflavone.

Elevated levels of TBARS in liver, heart, and aorta in group II rabbits are a clear manifestation of excessive formation of free radical and activation of lipid peroxidation. The significant decline in the level of TBARS, in rabbits administered with *D. biflorus* (Groups III and IV), unveils the antioxidant potential of the *D. biflorus* extract. Both doses of the *D. biflorus*extract (groups III and IV) help to restore the GSH levels near normal. The higher dose has more effect, which was comparable to atorvastatin. Increase in GSH concentration in animals treated with *D. biflorus* extract may be due to increased activity of the enzyme, glutathione reductase, which catalyses the conversion of oxidized glutathione to reduced glutathione in liver. It may also be due to enhanced synthesis/transport of GSH.

Restoration of the activities of SOD and CAT to near normal, observed in tissues *of* rabbits supplemented with *D. biflorous* may be due to the removal of toxic intermediates by the plant extract in HFD fed rabbits. The components of the plant extract may also directly activate the antioxidant enzymes. It is concluded that administration of *D. biflorus* manifests a protective action against HFD induced oxidative stress in different tissues in rabbits. However, further studies are needed to isolate the active principles, elucidate their structures, and determine their pharmacological activities.

Impact of Umbelliferone (7-hydroxycoumarin) on Hepatic Marker Enzymes in Streptozotocin Diabetic Rats

Ramesh B, Pugalendi K V.; Indian Journal of Pharmacology 2006; 38(3): 209-10.

Diabetes mellitus is by far the most common of endocrine disorders and a major threat to health care, worldwide. The increase of free radical mediated-toxicity is well documented in streptozotocin (STZ)-diabetic rats. The liver is the main effector organ for maintaining plasma glucose levels within narrow limits. Hyperglycemia can generate a redox imbalance inside the cells, especially in the liver. A model antidiabetic drug should possess both hypoglycemic and antioxidant properties, without any adverse effects. Plant drugs are frequently considered to be less toxic than synthetic ones. Plant derived phenolic coumarins might play a role as dietary antioxidants because of their presence in the human diet, especially in fruits and vegetables. Umbelliferone (UMB, 7-hydroxycoumarin), a benzopyrone in nature, is a derivative of coumarin. Our previous studies have shown that UMB had both antihyperglycemic and antioxidant properties in diabetic rats. The objective of the present study is to analyze the effect of UMB on serum hepatic marker enzymes, total protein, liver weight and glycogen content in STZ-diabetic rats. The structure of UMB is depicted below. Male Wistar albino rats (weight 180-200 g) were procured from the Central Animal House, Department of Experimental Medicine, Rajah Muthiah Medical College and Hospital, Annamalai University. The study was carried out in accordance with Indian National Law on Animal Care and Use and was approved by the Ethical Committee of Rajah Muthiah Medical College and Hospital (Reg. No: 160/1999/CPCSEA*), Annamalai University, Annamalainagar, Tamil Nadu, India.

Streptozotocin was purchased from Sigma-Aldrich, St. Louis, USA. UMB was procured from Carl Roth GmbH and Co, Germany.

After an overnight fast, the rats were injected with a single dose of STZ (40 mg/kg, b.w.) intraperitoneally (i.p.). The animals with blood glucose above 235 mg/dL were considered to be diabetic and used for the experiment.

The animals were randomly divided into 5 groups of six animals each, as given below. Dimethyl sulphoxide (DMSO) was used as a vehicle.

- Group I: Normal control received 10 per cent DMSO (i.p.) only.
- Group II: Normal + UMB (30 mg/kg/b.w., i.p.) in 10 per cent DMSO.
- Group III: Diabetic control (10 per cent DMSO (i.p.)).
- Group IV: Diabetic + UMB (30 mg/kg/b.wt., i.p.) in 10 per cent DMSO.
- Group V: Diabetic + glibenclamide (600 μg/kg/b.w., i.p.) in 10 per cent DMSO.

After 45 days of treatment, the 12 h-fasted animals were sacrificed by decapitation. Blood was collected for the estimation of hepatic marker enzymes and total proteins. The liver was collected for the determination of weight and glycogen content. The activities of serum aspartate aminotransferase (AST), alanine aminotransferase (ALT) and alkaline phosphatase (ALP) and the level of total proteins were estimated by using commercially available kits (Boehringer Mannheim, Mannheim, Germany). The activity of gamma glutamyl transferase (GGT) was measured by the method of Rosalki and Rau. The level of liver glycogen was estimated by the method of Morales *et al.*

The results are expressed as mean ± SD (n=6 rats/group). Data were analysed by one-way analysis of variance (ANOVA), followed by Duncan's multiple range test (DMRT). The statistical significance was set at $P < 0.05$. A marked decrease in liver weight was observed in diabetic rats, which may be due to an increased glycogen breakdown and gluconeogenesis with protein degradation. Protein synthesis decreases in the absence of insulin, partially because the transport of amino acids into muscle is diminished (amino acids serve as gluconeogenic substrate). Thus, insulin deficient persons are in negative nitrogen balance. Treatment with UMB elevated liver weight, glycogen content and plasma proteins, which may be due to increased plasma insulin level. Insulin favours glycogenesis and generally has an anabolic effect on protein metabolism, in that, it stimulates protein synthesis and retards protein degradation.

Enzymes directly associated with the conversion of amino acids to keto acids are ALT and AST. ALT and AST activities are used as the indicators of hepatocyte damage. Diabetic rats have increased activities of these enzymes, which may be due to hepatic damage and deficiency of insulin. Insulin suppresses the genes encoding gluconeogenic enzymes. ALT is a gluconeogenic enzyme and it is possible that ALT is an indicator of impaired insulin signaling. Treatment with UMB decreased the activities of these enzymes, by its insulin secretory and antioxidant properties.

ALP is present in all tissues of the body, especially in the cell membrane and the levels are high in the liver, kidney, bone and placenta. New enzyme synthesis mainly occurs in the hepatocytes adjacent to the biliary canaliculi. In our study, the activity of ALP increased in diabetic rats as compared with normal control rats, which could be due to an increased release from the hepatocytes damaged by diabetes-induced oxidative insult. In UMB treated rats, the activity of ALP also reversed to near normal level.

GGT reflects the biliary tract function and may act to transport amino acids and peptides into the cells, in the form of gamma glutamyl peptides. It has reported that high level of hepatic enzyme, GGT, is associated with later development of diabetes. In our study, the activity of GGT increased, which may be associated with diabetes. Treatment with UMB reversed the activity of GGT, which reflects the strong protective effect of UMB. Thus, our results have shown that treatment with UMB reversed hepatic marker enzymes, total proteins, liver weight and glycogen content to near normalcy. It reflects that UMB has a protective effect against liver cell damage in STZ-diabetic rats.

Efficacy Study of the Bioactive Fraction (F-3) of *Acorus calamus* in Hyperlipidemia

D'Souza T, Mengi SA, Hassarajani S, Chattopadhayay S.; Indian Journal of Pharmacology. 2007; 39: 196-200.

Objective: To investigate the effect of the bioactive F-3 fraction from the rhizomes of *Acorus calamus* in experimentally induced hyperlipidemic rats.

Materials and Methods: Doses of 10, 20 and 40 mg/kg of the bioactive fraction were evaluated for its effect on the lipid profile and fibrinogen levels in diet-induced hyperlipidemia. Additionally, apoprotein A1 and apoprotein B levels were estimated using immune-turbidimetric assays. Furthermore, the bioactive F-3 fraction was investigated for its mechanism of action by estimating HMG-CoA reductase activity and fecal cholesterol levels. Besides evaluating the free radical-scavenging activity using the Diphenyl picryl hydrazyl (DPPH) method, the high performance thin layer chromatography (HPTLC) fingerprint of the bioactive fraction was also developed.

Results: At doses of 20 and 40 mg/kg, the bioactive fraction significantly ($P < 0.05$) decreased the total cholesterol (TC) and low-density lipoprotein (LDL) levels. The bioactive F-3 fraction also attenuated the raised plasma fibrinogen levels. Fecal cholesterol excretion was significantly ($P < 0.05$) enhanced by the F-3 fraction while 3-hydroxy-3-methyl-glutaryl-CoA reductase (HMG-CoA reductase) activity was depressed. Furthermore, the F-3 fraction also possessed an appreciable free radical scavenging activity.

Conclusion: The results of the present study revealed that the bioactive F-3 fraction demonstrated its cholesterol-reducing effect by increasing fecal cholesterol excretion and decreasing cholesterol biosynthesis in the liver. Additionally, the effects on fibrinogen levels and free radicals indicate that the bioactive F-3 fraction could have a potentially beneficial effect in atherosclerosis associated with hyperlipidemia.

Evaluation of Hypoglycemic Effect of *Morus alba* in an Animal Model

Mohammadi J, Naik PR.; Indian Journal of Pharmacololy. 2008; 40: 15-8.

Objective: The objective of the present investigation was to evaluate the therapeutic efficacy of mulberry leaves in an animal model of diabetes.

Materials and Methods: Animals were treated with mulberry leaf extract 400 mg and 600 mg/kg body weight for 35 days. Blood glucose, glycosylated hemoglobin, triglyceride, LDL, VLDL, HDL, blood urea, cholesterol, number of β cells, and diameter of the islets of Langerhans were measured at the beginning and at the end of the experiment.

Results: Blood glucose level and other parameters (except HDL) were elevated in the diabetic group, but were brought to control group level in the diabetic group treated with 600 mg/kg body weight of mulberry leaf extract. The diameter of the islets and the number of β cells were reduced in the diabetic group; both parameters were brought to control group level after treatment with mulberry leaf extract.

Conclusion: Mulberry leaf extract, at a dose of 600 mg/kg body weight, has therapeutic effects in diabetes-induced Wistar rats and can restore the diminished β cell numbers.

Antidiabetic Activity of Aqueous Root Extract of *Ichnocarpus frutescens* in Streptozotocin-Nicotinamide Induced Type-II Diabetes in Rats

Barik R, Jain S, Qwatra D, Joshi A, Tripathi GS, Goyal R.; Indian Journal of Pharmacology. 2008; 40: 19-22.

Objective: To evaluate the antidiabetic activity of aqueous extract of roots of *Ichnocarpus frutescens* in streptozotocin-nicotinamide induced type-II diabetes in rats.

Materials and Methods: Streptozotocin-nicotinamide induced type-II diabetic rats (n = 6) were administered aqueous root extract (250 and 500 mg/kg, p.o.) of *Ichnocarpus frutescens* or vehicle (gum acacia solution) or standard drug glibenclamide (0.25 mg/kg) for 15 days. Blood samples were collected by retro-orbital puncture and were analyzed for serum glucose on days 0, 5, 10, and 15 by using glucose oxidase-peroxidase reactive strips and a glucometer. For oral glucose tolerance test, glucose (2 g/kg, p.o.) was administered to nondiabetic control rats and the rats treated with glibenclamide (10 mg/kg, p.o.) and aqueous root extract of *Ichnocarpus frutescens*. The serum glucose levels were analyzed at 0, 30, 60, and 120 min after drug administration. The effect of the extract on the body weight of the diabetic rats was also observed.

Results: The aqueous root extract of *Ichnocarpus frutescens* (250 and 500 mg/kg, p.o.) induced significant reduction (P < 0.05) of fasting blood glucose levels in streptozotocin-nicotinamide induced type-II diabetic rats on the 10(th) and 15(th) days. In the oral glucose tolerance test, the extract increased the glucose tolerance. It also brought about an increase in the body weight of diabetic rats.

Conclusion: It is concluded that *Ichnocarpus frutescens* has significant antidiabetic activity as it lowers the fasting blood sugar level in diabetic rats and increases the glucose tolerance.

Hypoglycemic and Hypolipidemic Activity of Ethanolic Extract of *Salvadora oleoides* in Normal and Alloxan-Induced Diabetic Rats

Yadav JP, Saini S, Kalia AN, Dangi AS. Indian Journal of Pharmacology. 2008; 40: 23-7.

Objective: To find out the hypoglycemic and hypolipidemic activity of an ethanolic extract of the aerial part of *Salvadora oleoides* Decne in euglycemic and alloxan-induced diabetic albino rats.

Materials and Methods: Diabetes was induced in albino rats by administration of alloxan monohydrate (120 mg/kg, i.p.). Normal as well as diabetic albino rats were divided into groups (n = 6) receiving different treatments: vehicle (control), ethanolic extract (1 g and 2 g/kg b.w), and standard antidiabetic drug tolbutamide (0.5 g/kg b.w.). Blood samples were collected by cardiac puncture and were analyzed for blood glucose and lipid profile on days 0, 7, 14, and 21.

Results: The ethanolic extract of *S. oleoides* produced significant reduction (P < 0.001) in blood glucose and also had beneficial effects (P < 0.001) on the lipid profile in euglycemic as well as alloxan-induced diabetic rats at the end of the treatment period (21(st) day). However, the reduction in the blood glucose and improvement in lipid profile was less than that achieved with the standard drug tolbutamide.

Conclusion: We concluded that an ethanolic extract of *S. oleoides* is effective in controlling blood glucose levels and improves lipid profile in euglycemic as well as diabetic rats.

Mechanisms Responsible for the Vascular Effect of Aqueous *Trigonella foenum-graecum* Leaf Extract in Diabetic Rats

Mahdavi MR, Roghani M, Baluchnejadmojarad T.; Indian Journal of Pharmacology. 2008; 40: 59-63.

Background and *Objective:* Since a beneficial vascular effect of aqueous leaf extract of *Trigonella foenum-graecum* (TFG) has previously been reported, this study was conducted to evaluate the underlying mechanisms, including the role of nitric oxide (NO) and cyclooxygenase pathways, in diabetic rats.

Materials and Methods: Male Wistar rats were divided into control, extract-treated control, diabetic, and extract-treated diabetic groups. Diabetes was induced by a single i.p. injection of streptozotocin (STZ; 60 mg/kg). Treatment groups received TFG extract (200 mg/kg; ip.) every other day for 1 month. Contractile reactivity of the thoracic aorta to KCl and noradrenaline (NA) and relaxation response to acetylcholine (ACh) were determined. For determination of the participation of NO and prostaglandins in the relaxation response to ACh, aortic rings were incubated for 30 min before the experiment with N-nitro-l-arginine methyl ester (L-NAME) and/or indomethacin (INDO).

Results: The diabetic state significantly increased the maximum contractile response to KCl and NA (P < 0.01-0.005) and reduced the maximum relaxation due to ACh (P < 0.01) as compared to controls and treatment with TFG extract in the diabetic group significantly improved these changes relative to the untreated diabetic group

(P < 0.05). With L-NAME pretreatment, no significant difference between diabetic and extract-treated diabetic groups was found out. On the other hand, there was a significant difference between these two groups following INDO pretreatment (P < 0.05).

Conclusion: Intraperitoneal administration of aqueous leaf extract of TFG for one month could improve some functional indices of the vascular system in the diabetic state and endothelium-derived prostaglandins are essential in this respect.

Trichosanthes Cucumerina Linn. Improves Glucose Tolerance and Tissue Glycogen in Non-Insulin Dependent *Diabetes mellitus* Induced Rats

Kirana H, Srinivasan BP.; Indian Journal of Pharmacology. 2008; 40: 103-6.

Objective: To study the effect of *Trichosanthes cucumerina* Linn. on non-insulin dependent diabetes mellitus induced rats.

Materials and Methods: Non-Insulin Dependent Diabetes Mellitus (NIDDM) was induced by administering streptozotocin (90 mg/kg, i.p.) in neonatal rat model. NIDDM animals were treated with aqueous extract of *Trichosanthes cucumerina* (100 mg/kg/day) orally for six weeks. Parameters such as fasting blood glucose, Oral Glucose Tolerance Test (OGTT) and tissue glycogen content were evaluated.

Results: Aqueous extract of *Trichosanthes cucumerina* significantly (P<0.01) decreased the elevated blood glucose of NIDDM induced rats. OGTT of NIDDM animals showed glucose intolerance. Blood glucose of diabetic animals reached peak at 45 min and remains high even after 2h. In case of *Trichosanthes cucumerina* treated group, the blood glucose reached peak level at 30 min, followed by decrease in glucose level up to 2h. The drug has significantly (P<0.01) reduced the postprandial blood glucose of diabetic animals. Glycogen content of insulin dependent tissues such as liver and skeletal muscle was found to be improved by 62 per cent and 58.8 per cent respectively with *Trichosanthes cucumerina* as compared to NIDDM control.

Conclusion: Studies revealed that, *Trichosanthes cucumerina* possess antidiabetic activity. The drug improved the oral glucose tolerance of NIDDM subjects. Increase in tissue glycogen content indicates the effect of the drug on the uptake of glucose by the peripheral tissues to reduce insulin resistance of NIDDM.

Hypoglycemic Effect of Aqueous Extract of *Parthenium hysterophorus* L. in Normal and Alloxan Induced Diabetic Rats

Patel VS, Chitra V, Prasanna PL, Krishnaraju V.; Indian Journal of Pharmacology. 2008; 40: 183-5.

Objectives: To study the effects of *Parthenium hysterophorus* L. flower on serum glucose level in normal and alloxan induced diabetic rats.

Materials and Methods: Albino rats were divided into six groups of six animals each, three groups of normal animals receiving different treatments consisting of vehicle, aqueous extract of *Parthenium hysterophorus* L. flower (100 mg/kg) and the standard antidiabetic drug, glibenclamide (0.5 mg/kg). The same treatment was given to the other three groups comprising alloxan-induced diabetic animals. Fasting blood glucose level was estimated using the glucose oxidase method in normal and alloxan induced diabetic rats, before and 2 h after the administration of drugs.

Results: Parthenium hysterophorus L. showed significant reduction in blood glucose level in the diabetic (P<0.01) rats. However, the reduction in blood glucose level with aqueous extract was less than with the standard drug glibenclamide. The extract showed less hypoglycemic effect in fasted normal rats, (P<0.05).

Conclusion: The study reveals that the active fraction of *Parthenium hysterophorus* L. flower extract is very promising for developing standardized phytomedicine for diabetes mellitus.

Effect of Fresh *Triticum aestivum* Grass Juice on Lipid Profile of Normal Rats

Kothari S, Jain AK, Mehta SC, Tonpay SD.; Indian Journal of Pharmacology. 2008; 40: 235-6.

Objective: To study the hypolipidemic activity of fresh grass juice of *Triticum aestivum* in normal rats.

Materials and Methods: Freshly prepared *Triticum aestivum* grass juice was administered to normal rats at the dose of 5 ml/kg and 10 ml/kg orally once daily for 21 days. Blood samples were collected after 24 hours of last administration and used for estimation of lipid profile. Fresh grass juice was also subjected to preliminary phytochemical screening.

Results: Fresh grass juice administration produced dose related significant (P < 0.05) reduction in total chloesterol, triglycerides, low density lipoprotein-cholesterol and very low density lipoprotein-cholesterol levels in normal rats as compared to control. Preliminary phytochemical screening revealed presence of alkaloids, tannins, saponins and sterols in *Triticum aestivum* grass.

Conclusion: The results of the present study Indicate hypolipidemic activity of fresh *Triticum aestivum* grass juice.

Antidiabetic Activity of Flower Buds of *Michelia champaca* Linn.

Jarald EE, Joshi SB, Jain DC.; Indian Journal of Pharmacology. 2008; 40: 256-60.

Objective: To identify the antihyperglycemic activity of various extracts, petroleum ether (60-80°), chloroform, acetone, ethanol, aqueous and crude aqueous, of the flower buds of *Michelia champaca*, and to identify the antidiabetic activity of active antihyperglycemic extract.

Materials and Methods: Plant extracts were tested for antihyperglycemic activity in glucose overloaded hyperglycemic rats. The effective antihyperglycemic extract was tested for its hypoglycemic activity at two-dose levels, 200 and 400 mg/kg respectively. To confirm its utility in the higher model, the effective extract of *M. champaca* was subjected to antidiabetic study in alloxan induced diabetic model at two dose levels, 200 and 400 mg/kg respectively. The biochemical parameters, glucose, urea, creatinine, serum cholesterol, serum triglyceride, high density lipoprotein, low density lipoprotein, hemoglobin and glycosylated hemoglobin were also assessed in the experimental animals.

Results: The ethanolic extract of *M. champaca* exhibited significant antihyperglycemic activity but did not produce hypoglycemia in fasted normal rats. Apart from this extract, the crude aqueous and petroleum ether extracts were found active only at the end of the first hour. Treatment of diabetic rats with ethanolic extract of this plant restored the elevated biochemical parameters significantly ($P<0.05$) ($P<0.01$) and the activity was found dose dependent.

Conclusion: This study supports the traditional claim and the ethanolic extract of this plant could be added in traditional preparations for the ailment of various diabetes-associated complications.

Effect of an Isolated Active Compound (Cg-1) of *Cassia glauca* Leaf on Blood Glucose, Lipid Profile, and Atherogenic Index in Diabetic Rats

Mazumder PM, Farswan M, Parcha V.; Indian Journal of Pharmacology. 2009; 41: 182-6.

Objectives: The objective of present study was to evaluate the effect of active principle (Cg-1) from *Cassia glauca* leaf on serum glucose and lipid profile in normal and diabetic rats.

Materials and Methods: Diabetes was induced by streptozotocin in neonates. Oral administration of petroleum ether, chloroform, acetone, and methanol of *C. glauca* leaf (100 mg/kg, p.o.) for 21 days caused a decrease in fasting blood glucose (FBG) in diabetic rats. Among all the extracts, acetone extract was found to lower the FBG level significantly in diabetic rats. Glibenclamide was used as standard antidiabetic drug (5 mg/kg, p.o). Acetone extract was subjected to column chromatography that led to isolation of an active principle, which was given trivial name Cg-1. Cg-1 (50 mg/kg, p.o.) was studied for its hypoglycemic and hypolipidemic potential. The unpaired t-test and analysis of variance (ANOVA) followed by post hoc test was used for statistical analysis.

Results: Cg-1 caused a significant reduction in FBG level. It also caused reduction in cholesterol, triglycerides, and LDL levels and improvement in the atherogenic index and HDL level in diabetic rats.

Conclusion: Improvement in the FBG and the atherogenic index by Cg-1 indicates that Cg-1 has cardioprotective potential along with antidiabetic activity and provides a scientific rationale for the use as an antidiabetic agent.

In vitro Antioxidant and Antihyperlipidemic Activities of *Bauhinia variegata* Linn.

Rajani GP, Ashok P. Indian Journal of Pharmacology. 2009; 41: 227-32.

Objectives: To evaluate the ethanolic and aqueous extracts of *Bauhinia variegata* Linn. for *in vitro* antioxidant and antihyperlipidemic activity.

Materials and Methods: Ethanolic and aqueous extracts of the stem bark and root of *B. variegata* Linn. were prepared and assessed for *in vitro* antioxidant activity by various methods namely total reducing power, scavenging of various free radicals such as 1,2-diphenyl-2-picrylhydrazyl (DPPH), super oxide, nitric oxide, and hydrogen peroxide. The percentage scavenging of various free radicals were compared with standard antioxidants such as ascorbic acid and butylated hydroxyl anisole (BHA). The extracts were also evaluated for antihyperlipidemic activity in Triton WR-1339 (iso-octyl polyoxyethylene phenol)-induced hyperlipidemic albino rats by estimating serum triglyceride, very low density lipids (VLDL), cholesterol, low-density lipids (LDL), and high-density lipid (HDL) levels.

Result: Significant antioxidant activity was observed in all the methods, ($P < 0.01$) for reducing power and (P

< 0.001) for scavenging DPPH, super oxide, nitric oxide, and hydrogen peroxide radicals. The extracts showed significant reduction ($P < 0.01$) in cholesterol at 6 and 24 h and ($P < 0.05$) at 48 h. There was significant reduction ($P < 0.01$) in triglyceride level at 6, 24, and 48 h. The VLDL level was also significantly ($P < 0.05$) reduced from 24 h and maximum reduction ($P < 0.01$) was seen at 48 h. There was significant increase ($P < 0.01$) in HDL at 6, 24, and 48 h.

Conclusion: From the results, it is evident that alcoholic and aqueous extracts of *B. variegata* Linn. can effectively decrease plasma cholesterol, triglyceride, LDL, and VLDL and increase plasma HDL levels. In addition, the alcoholic and aqueous extracts have shown significant antioxidant activity. By the virtue of its antioxidant activity, *B. variegata* Linn. may show antihyperlipidemic activity.

Protective Effect of *Cassia glauca* Linn. on the Serum Glucose and Hepatic Enzymes Level in Streptozotocin Induced NIDDM in rats

Farswan M, Mazumder PM, Percha V. Indian Journal of Pharmacology. 2009; 41: 19-22.

Objective: The objective of the present study was to investigate the hypoglycemic and hepatoprotective effect of *Cassia glauca* leaf extracts on normal and non-insulin dependent diabetes mellitus (NIDDM) in rats. The study was further carried out to investigate the effect of different fractions of the active extract of *Cassia glauca*, on normal and NIDDM rats, and the effect of active fraction on the blood glucose and hepatic enzymes level.

Methods: Diabetes was induced by streptozotocin (STZ) at a dose of 90mg/kg, i.p. in neonates. Different extracts of *Cassia glauca* (100mg/kg, p.o.) were administered to the diabetic rats. Acetone extract was found to lower the serum glucose level significantly in diabetic rats. Further, the acetone extract was subjected to column chromatography and four fractions were obtained on the basis of TLC. All the four fractions (100mg/kg, p.o.) were administered to the diabetic rats. Fraction 1 (F1) caused the maximum reduction in the blood glucose level. The results of the test were compared with the standard antidiabetic drug glibenclamide (5mg/kg, p.o.).

Results: Fraction 1 of acetone extract caused a significant reduction in the levels of hepatic enzyme Aspartate transaminase (AST), alanine transaminase (ALT), creatine kinase (CK), and lactate dehydrogenase (LDH) in STZ-induced diabetic rats.

Conclusion: Improvement in the blood sugar level and normalization of liver functions by *Cassia glauca* indicates that the plant has hepatoprotective potential, along with antidiabetic activity, and it provides a scientific rationale for the use of *Cassia glauca* as an antidiabetic agent.

A Study of the Antidiabetic Activity of *Barleria prionitis* Linn.

Dheer R, Bhatnagar P. Indian Journal of Pharmacology. 2010; 42: 70-3.

Objectives: To study the antidiabetic activity of *Barleria prionitis* Linn. in normal and alloxan-induced diabetic rats.

Materials and Methods: Alcoholic extract of leaf and root of *B. prionitis* was tested for their antidiabetic activity. Albino rats were divided into six groups of six animals each. In three groups, diabetes was induced using alloxan monohydrate (150 mg/kg b.w., i.p.) and all the rats were given different treatments consisting of vehicle, alcoholic extract of leaves, and alcoholic extract roots of *B. prionitis* Linn. (200 mg/kg) for 14 days. The same treatment was given to the other three groups, comprising non-diabetic (normal) animals. Blood glucose level, glycosylated hemoglobin, liver glycogen, serum insulin, and body weight were estimated in normal and alloxan-induced diabetic rats, before and 2 weeks after administration of drugs.

Results: Animals treated with the alcoholic extract of leaves of *B. prionitis* Linn showed a significant decrease in blood glucose level ($P<0.01$) and glycosylated hemoglobin ($P<0.01$). A significant increase was observed in serum insulin level ($P<0.01$) and liver glycogen level ($P<0.05$), whereas the decrease in the body weight was arrested by administration of leaf extract to the animals. The alcoholic extract of roots showed a moderate but non-significant antidiabetic activity in experimental animals.

Conclusion: The study reveals that the alcoholic leaf extract of *B. prionitis* could be added in the list of herbal preparations beneficial in diabetes mellitus.

Antidiabetic Activity of *Cassia occidentalis* (Linn.) in normal and alloxan-induced diabetic rats.

Verma L, Khatri A, Kaushik B, Patil UK, Pawar RS. Indian Journal of Pharmacology. 2010; 42: 224-8.

Objective: To evaluate the hypoglycemic activity of various extracts, petroleum ether, chloroform and aqueous extract of *Cassia occidentalis* in normal and alloxan-induced diabetic rats.

Materials and Methods: Petroleum ether, chloroform and aqueous extract of whole plant of *Cassia occidentalis* were orally tested at the dose of 200 mg/kg for hypoglycemic effect in normal and alloxan-induced diabetic rats. In addition, changes in body weight, serum cholesterol, triglyceride and total protein levels, assessed in the ethanol extract-treated diabetic rats, were compared with diabetic control and normal animals. Histopathological observations during 21 days treatment were also evaluated.

Results: Aqueous extract of *C. occidentalis* produced a significant reduction in fasting blood glucose levels in the normal and alloxan-induced diabetic rats. Apart from aqueous extract, petroleum ether extract showed activity from day 14 and chloroform extract showed activity from 7 days. Significant differences were observed in serum lipid profiles (cholesterol and triglyceride), serum protein, and changes in body weight by aqueous extract treated-diabetic animals, when compared with the diabetic control and normal animals. Concurrent histopathological studies of the pancreas of these animals showed comparable regeneration by extract which were earlier necrosed by alloxan.

Conclusion: Aqueous extract of *C. occidentalis* exhibited significant antihyperglycemic activity in normal and alloxan-induced diabetic rats. They also showed improvement in parameters like body weight and serum lipid profiles as well as histopathological studies showed regeneration of β-cells of pancreas and so might be of value in diabetes treatment.

Effect of Saturated Fatty Acid-Rich Dietary Vegetable Oils on Lipid Profile, Antioxidant Enzymes and Glucose Tolerance in Diabetic Rats

Kochikuzhyil BM, Devi K, Fattepur SR.; Indian Journal of Pharmacology. 2010; 42: 142-5.

Objective: To study the effect of saturated fatty acid (SFA)-rich dietary vegetable oils on the lipid profile, endogenous antioxidant enzymes and glucose tolerance in type 2 diabetic rats.

Materials and Methods: Type 2 diabetes was induced by administering streptozotocin (90 mg/kg, i.p.) in neonatal rats. Twenty-eight-day-old normal (N) and diabetic (D) male Wistar rats were fed for 45 days with a fat-enriched special diet (10 per cent) prepared with coconut oil (CO) - lauric acid-rich SFA, palm oil (PO) - palmitic acid-rich SFA and groundnut oil (GNO) - control (N and D). Lipid profile, endogenous antioxidant enzymes and oral glucose tolerance tests were monitored.

Results: D rats fed with CO (D + CO) exhibited a significant decrease in the total cholesterol and non-high-density lipoprotein cholesterol. Besides, they also showed a trend toward improving antioxidant enzymes and glucose tolerance as compared to the D + GNO group, whereas D + PO treatment aggravated the dyslipidemic condition while causing a significant decrease in the superoxide dismutase levels when compared to N rats fed with GNO (N + GNO). D + PO treatment also impaired the glucose tolerance when compared to N + GNO and D + GNO.

Conclusion: The type of FA in the dietary oil determines its deleterious or beneficial effects. Lauric acid present in CO may protect against diabetes-induced dyslipidemia.

Effects of *Phyllanthus reticulatus* on Lipid Profile and Oxidative Stress in Hypercholesterolemic Albino Rats

Maruthappan V, Shree KS.; Indian Journal of Pharmacology. 2010; 42: 388-91.

Objective: This study was designed to investigate the effect of *Phyllanthus reticulatus* on lipid profile and oxidative stress in hypercholesterolemic albino rats.

Materials and Methods: Hypercholesterolemia was induced in albino rats by administration of atherogenic diet for 2 weeks. Experimental rats were divided into different groups: normal, hypercholesterolemic control and *P. reticulatus* treated (250 and 500 mg/kg body weight doses for 45 days). After the treatment period of 45(th) day triglyceride, VLDL-cholesterol, HDL-cholesterol, total cholesterol (TC), LDL-cholesterol and oxidative stress (protein carbonyl) were assayed and compared with hypercholesterolemic control.

Results: The aqueous extract of *P. reticulatus* (250 mg and 500 mg/kg) produced significant reduction ($P < 0.05$) in triglyceride, VLDL-cholesterol, total cholesterol (TC), LDL-cholesterol and oxidative stress (protein carbonyl) while increased HDL-cholesterol in atherogenic diet-induced hypercholesterolemic rats at the end of the treatment period (45 days). However, the reduction in the above parameters was comparable with hypercholesterolemic control. Thus, aqueous extract of *P. reticulatus* is effective in controlling TC, lipid profile and oxidative stress in hypercholesterolemic animals.

Conclusion: The results suggest the aqueous extract of *P. reticulatus* can be utilized for prevention of atherosclerosis in hypercholesterolemic patients.

Evaluation of Antidiabetic Potential of Leaves and Stem of *Flacourtia jangomas* in Streptozotocin-Induced Diabetic Rats

Singh AK, Singh J.; Indian Journal of Pharmacology. 2010; 42: 301-5.

Objectives: To study the efficacy of combination of *Flacourtia jangomas* leaf and stem (1:1) methanolic extract (MEFJ) in streptozotocin (STZ)-induced diabetic rats and to investigate the qualitative phytochemical present in the extract. The study also aims to evaluate acute and short-term general toxicity of the extract in rats.

Material and Methods: MEFJ of leaves and stem was subjected to preliminary qualitative phytochemical investigations by using standard procedures. The extract (400 mg/kg p.o.) was screened for antidiabetic activity in STZ-induced diabetic rats (30 mg/kg, i.p.). Acute oral toxicity study for the test extract of the plant was carried out using OECD/OCED guideline 425.

Results: Phytochemical analysis of MEFJ of leaves and stem revealed the presence of flavonoids, saponins, carbohydrates, steroids, tannins, and phenolic compounds. In acute toxicity study, no toxic symptoms were observed for MEFJ up to dose 2000 mg/kg. Oral administration of MEFJ for 21 days exhibited highly significant ($P < 0.01$) hypoglycemic activity and also correction of altered biochemical parameters, namely cholesterol and triglycerides significantly ($P < 0.05$). Urine analysis on 1(st) day showed the presence of glucose and traces of ketone in the entire group except normal control group. However, on 21(st) day glucose and ketone traces were absent in MEFJ- and glibenclamide-treated groups while they were present in diabetic control. The data were analyzed using analysis of variance followed by Dunnett's test.

Conclusion: The observations confirm that methanolic extract of the leaf and stem of the plant has antidiabetic activity and is also involved in correction of altered biological parameters. It also warrants further investigation to isolate and identify the hypoglycemic principles in this plant so as to elucidate their mode of action.

The Study of Aqueous Extract of *Pterocarpus marsupium* Roxb. on Cytokine TNF-α in Type 2 Diabetic Rats

Halagappa K, Girish HN, Srinivasan BP.; Indian Journal of Pharmacology. 2010; 42: 392-6.

Objective: This study was designed to investigate the effect of aqueous extract of *Pterocarpus marsupium* Roxb. on elevated inflammatory cytokine, tumor necrosis factor (TNF)-α in type 2 diabetic rats.

Materials and Methods: Type 2 diabetes was induced by administering streptozotocin (90 mg/kg, i.p.) in a neonatal rat model. Aqueous extract of *P. marsupium* at a dose of 100 and 200 mg/kg was given orally to desired group of animals for a period of 4 weeks. After 4 weeks of drug treatment, parameters such as fasting blood glucose, postprandial blood glucose, and TNF-α in serum were analyzed.

Results: Aqueous extract of *P. marsupium* at both doses, *i.e.*, 100 and 200 mg/kg, decreased the fasting and postprandial blood glucose in type 2 diabetic rats. The 200 mg/kg had more pronounced effect on postprandial hyperglycemia. The drug also improved the body weight of diabetic animals. Cytokine TNF-α was found to be elevated in untreated diabetic rats due to chronic systemic inflammation. The aqueous extract at both doses significantly ($P < 0.001$) decreased the elevated TNF-α level in type 2 diabetic rats.

Conclusion: Modulation of cytokine TNF-α by the rasayana drug *P. marsupium* is related with its potential anti-diabetic activity.

Clinical Evaluation of the Effect of Amrita-Pippali-Nimba Yoga in Diabetes Mellitus with Special Reference to the Role of *Agni* and *Ojas*

P. S. Mehra, R. H. Singh; J Res Ayur Siddha. 2001; 22: 183-197.

Diabetes mellitus is recognized as a global health hazard, which seeks great attention from the present day practitioners and researchers. Many of the NIDDM patients in recent time are observed to be transferring into insulin requirement group. Though there are several anti-diabetic drugs available in the market, drug resistance and adverse effect are posing major problems. Therefore, the search for a suitable anti-diabetic drug from ayurvedic resources continues. The ayurvedic physicians since antiquity have the knowledge of Prameha - Madhumeha, a disease synonymous with diabetes mellitus of the Modern medicine.

Apart from various aetiological factors responsible for the causation of Prameha the impaired Agni *i.e.*, biofire system of the body has been conceived as one of the important events by Vagbhatta in the pathogenesis of this disease, resulting in loss of Apara Ojas. Hence the patients of Prameha are believed to possess impaired state of Agni and reduced state of Ojas *i.e.*, Ojah Ksaya.

The Present study is conducted with the aim to promote Ojas and to maintain the Agni in equilibrium. Amrita (*Tinspora ccordifolia* (wild.) Miers), Pippali (*Piper longum* Linn.) and Nimba (*Azadirachta indica* A. Juss.) are well known for their Rasayana and antidiabetic activity. The clinical trial with these drugs in the present study showed notable symptomatic relief, reduction in blood sugar level and promotion of Agni and Ojas grade scores in cases of diabetes mellitus.

Clinical Evaluation of Palasha Pushpadi Churna in the Management of Madhumera Roga (Diabetes Mellitus)

J Res Ayur Siddha. 2000; 21: 1-10.

Thirty patients of Madhumeha (non-insulin dependent diabetes mellitus) a clinical trial was conducted Palasha Pushpadi (churna in the dose of 5 g trice a day with leukwarm water for a period of two months. Significant improvement in he clinical symptoms was found in the study. Fasting blood sugar level was decreased by 30.69 per cent and post-prandial blood sugar level was decreased by 25.15 per cent. A decreased of 72.37 per cent was also noted in the urinary sugar level. No side effect or toxic effect was reported during the trial.

Clinical Evaluation of Single and Herbo-Mineral Compound Drugs in the Management of Madhumeha

Naresh Kumar, Anil Kumar, M. L. Sharma; J Res Ayur Siddha. 1999; 20: 1-9.

In the present study three different drug therapies were put to trial in 111 ambulatory NIDDM cases. Group A-Ayush-82 tabs., group B-Chanderprabha Vati, Trivang Bhasma with Vijayasara Kwatha and group C-Methika Curna, effects of all these therapies were found statistically highly significant (P<0.001) in the reduction of FBS and PPBS along with remarkable improvement in the presenting signs and symptoms of the disease. the drug combination of group-A was found comparatively most effective in the correction of maximum of the presenting symptoms along with it is having better anti-cholesterolemic effect whereas the single drug of group-C was found comparatively equally effective in lowering the FBS, PPBS levels and in polyurea. No case under the study in any of the groups remained unchanged along with the blood sugar level (FBS and PPBS) in any of the case did no fall below the normal range during or after the course of the therapy.

Nishaamalaki in Madhumeha (NIDDM): A Clinical Study

G. C. Nanda, K. K. Chopra, D. P. Sahu, M. M. Padhi; J Res Ayur Siddha. 1998, XIX (1-2), 34-40.

The paper presents observations and results on the clinical trial of Nishamalaki in reducing blood sugar and urine sugar levels in the management of Madhumeha (NIDDM) patients. The observations indicate that the disease is more commonly affecting the patients between the age range of 40-60 years and more in female patients. The effects of the drug in the clinical trials have been significantly positive in reducing the blood sugar and urine sugar levels. The drug was also found to be effective in controlling the clinical sign and symptoms like frequency of micturition *etc.* Efforts should be made to explain the mechanism of action of the drug on pharmacological and both modern as well as Ayurvedic pharmacodynamic principles. The scope for further work on Madhumeha has also been suggested.

A Clinical Trial of M-93 Compound in the Management of Madhumeha (Diabetes Mellitus)

N. Kumar, A. Kumar; J Res Ayur Siddha. 1995; 16: 102-107.

Keeping in view the complications and intricate aetiopathogenesis of the disease, a herbal folklore combination was tried in 30 cases of non-insulin dependant cases of Diabetes mellitus. The analysis of the patients was made on the basis of classically mentioned symptomatology of the disease along with the required laboratory investigations. In the cases under study there was 33.33 per cent complete, 30 per cent marked relief, 26.67 per cent moderate relief and 10 per cent mild relief after the course of treatment. The effect of the drug combination on the cardinal symptoms of the disease is appreciable especially in the complaint of polyuria which impoved remarkably in almost all the cases just after two to five days of treatment.

Vijaysar, *Pterocarpus marsupium* in the Treatment of Madhumeha (Diabetes Mellitus): A Clinical Trial

S. Rajasekharan, S. N. Tuli; J Res Ayur Siddha. 1976; 11.

Capsule Vijaysar is prepared from *Pterocarpus marsupium* (Beejak), a common tree growing in central and southernparts of India and Sree Lanka. Beejak, Pethsar, Pethasalak, Bendookapuspa, Priyak, Sarjak, Acana *etc.* are the Sanskrit synonyms of the Pterocarpus marsupium Roxb. Bhavamisra (16th century AD_ has escribed it as a useful medicine for Prameha. Wood of the *Pterocarpus marsupium* is used as astringent, tonic, for external application in inflammation and headache *etc.* (Chopra *et al.*, 1956). Sharma (1959) has mentioned in his text-book "Dravyagunavigyana" about the Karma (action) of *Pterocarpus marsupium* especially Abhyantara Panchana Samsthen Sthambhaniya, Mutra Vaha Samsthan Mutra Samgrahaniya and Mehaha. Trivedi (1971) and Shah *et al.* (1972) have carried out some chemical nvestigations of the heartwood of *Pterocarpus marsupium* Roxb. Shah *et al.* (1972) have isolated some chemical investigations of the heartwood of *Pterocarpus marsupium* Roxb. Shah *et al.* (1972) have isolaed some flavanoi compounds. From pharmacological investigation of one of these compounds, named Bianol, they found that it has no anthelmintic activity on earth worms, no significant effect on blood pressure and respiration of dog. The study of the hypoglycemic effect of this substance according to them, is under process.

Efficacy of *Coccinia indica* W. and A. in Diabetes Mellitus

S. M. Kamble, G. S. Jyotishi, P. L. Kamalakar, S. M. Vaidya; J Res Ayur Siddha. 1996; 17: 77-84.

Diabetic patients become susceptible to complications like hyperlipidemia and various types of bacterial infections. Many attempts have been made to manage diabetes by use of synthetic drugs,but all have met with limited success due to their side effects.

Therefore, the concept of use of plant drugs have become the potent and safe alternatives. In view of this, trial of indigenous preparation from *Coccinia indica* W. and A. leaves has been taken up on 30 non insulin dependent (NIDDM) diabetic patients. The drugs was found to exert protective influence against hyperlipidemia along with thecontrol of hyperglycemia.

An Effective Ayurvedic Hypoglycemic Formulation

V. N. Pandey, S. S. Rajagopalan, D. P. Chowdhury; J Res Ayur Siddha. 1995; 16: 1-14.

In 80 Non-Insuline Dependent Diabetes Mellitus (NIDDM) cases, an Ayurvedic formulation code named Ayush-82 and Shudha Shilajittu were orally administered for a period of 24 weeks. Fasting and post prandial blood sugar were estimated at 6th weekly intervals. There was statistically significant reduction in both fasting and post prandial blood sugar in both males and females. On physician's rating, there was "good response" in 74 per cent of the cases.

Clinical Assessment of the Effects of Sandana (Sandal) Podi – in the Treatment of Diabetes Mellitus (Neerazhiv)

Ravi Shankar, R. K. Singhal; J Res Ayur Siddha. 1994; 15: 89-97.

Sandana Podi-a Siddha drug was clinically tried on the ambulatory patients of non-insulin dependent Diabetes mellitus (NIDDM) attending the diabetic linic at Safdarjang Hospital, New Delhi. During preliminary screening, those cases in which reduction in calories intake controlled the disease effective and those cases with serious complications like Ketoacidosis, Nephropathy, Neuropathy or Retinopathy etc. were excluded. The cases taken up for study were mild and moderate type. Twenty patients were treated with Sandanapodi after recording detailed history and through clinical examination. 300mg. of drug filled in an empty gelatin capsule was administered two times a day for forty five days. The total calories intake for 24 hours was decided at he rate of 25 calories per kilogram of ideal body weight. Seventy six percent of the patients belonged to the age group of 40-60 years. Twenty four per cent were familiar incidence cases. The study showed mean fall of 49.8 mg at fasting and mean fall of 110.45mg after two hours of food intake (post prandial) with 45 days of treatment with this drug, which was found to be statistically significant (p,0.10) both at fasting and post prandial level. No side or toxic effects of the drug were seen during the course of study.

Medoroga and Medodosha (Obesity and Lipid Disorders)

Bharati, Adarsh Kumar, Singhal R. K; J Res Ayur Siddha.

In ancient Ayurvedic literature, aetiology of most of the diseases is described in terms of Ahara and Vihara. Besides this general description, there are diseases in which special emphasis is laid on Ahara (*i.e.* Food) and Vihara (*i.e.* life style). Medoroga and Prameha are few among these. If we go through Sushruta Sutra Sthana Chapter 15 (Doshadhatu Mala Kshaya Vriddhi Vijnaniya). Ahara rasa is described to be the main cause of obesity.

Hypoglycaemic Activity of *Coscinium Fenestratum* (Gaertn.) Colber

Mahapatra B.; J RES AYUR SIDDHA. 1997; 18; 89-96.

Decoction of the stems of *Coscinium fenestratum* (Gaertn.) Colebr. was fed to fasting rabbits in order to detect its potential hypoglycemic activity. Coscinium in 4g/kg body weight showed the maximum potency of 73.75 percent as percentage of tolbutamide at the 5thhour of feeding and this result was statisically significant whereas Coscinium in 2g/kg and 8g/kg did not show any significant result. In G.T.T studies in rabbits coscinium in 4g/kg and also in 8g/kg body weight showed a pattern of activity which is comparable to that of tolbutamide 250mg/kg body weight. On allozan induced diabetic rabbits Coscinium in 2g/kg, 4g/kg and 8g/kg body weight exhibited a significant hypoglycemic activity. This significant result was at 5th hour for 2g/kg dose, both at 3rd and 5th hour for 4g/kg does and at 1st hour for 8g/kg body weight dose while for tolbutamde it was not significant statistically.

Study of Antidiabetic Effects of Alcoholic Extract of *Ficus bengalensis* (Linn.) on Alloxan Diabetic Albino Rats

Neera Singh, S. D. Tyani, S. C. Agarwal; J RES AYUR SIDDHA. 8: 56-62.

Alcoholic extract was prepared from the stem bark of *Ficus bengalensis* (Linn.) and was studied for its long-term feeding effects on blood sugar, serum cholesterol, urea and total protein levels of albino rats, which were made diabetic through i.v. injection of alloxan monohydrate. Different doses (25, 50 and 75 mg/day/100g body wt. of the rate) over different durations showed highly significant hypoglycemic potential. The most interesting features of the study have been-1. The blood sugar once lowered remained unaltered when the treatment was discontinued and 2. The blood sugar at no stage fell down below normal level. 3. This extract is also able to bring down the level of serum cholesterol and blood urea.

Hypoglycaemic and Toxicity Studies of Triphala: A Siddha Drug

D. Ghosh, R. Uma, P. Thejomoorthy, G. Veluchamy; J RES AYUR SIDDHA. 11: 78-89.

The drug Triphala which is a combination of equal parts of three myrobalans namely chebulic, beleric and embelic has been studied for hypoglycemic activity, acture and sub-acture toxicities. The drug produced significant hypoglycaemia only in the dose of 90 mg./kg. orally which was approximately 45 per cent of tolbutamide potency. No acute toxicity was observed even with the maximum dose of 10 g/kg orally in both the species of rats and mice. There was no adverse effect found in the body growth, feed and water intake, rather the body growth rate and feed intake were observed to be significantly enhanced. The relative weights of the vital organs did not register any significant difference except the weight of the heart was increased in the dose of 500 mg/kg. Haematological parameters also did not indicate any adverse effects. The biochemical parameters mostly indicted the changes favourable towards the body growth rather than any adverse effect. A further detailed study is suggested.

Hypoglycaemic Activity of *Coccinia indica* W and A

Pillai NR, Ghosh D, Uma R, Kumar AA.; J Res Ayur Siddha.

In the present study, both the juice and aqueous extract of various parts of *Coccinia indica* W and A and fruits of Coccinia india Var. palmata W and A were studied systematically for their hypoglyceamic activity in rabbits. The juice and decoction of leaves and stem of Coccinia indica W and A and decoction of the fruits in 20ml/kg. dose level showed significant hypoglycemic response in fasting rabbits. But the fruits of Coccinia Indica Var. palmata found to be more potent than that of Coccinia indica W and A. The root of the plant did not show any activity.

The Relationship of the Post-absorptive State to the Hypoglycemic Action Studies on *Ficus benghalensis*

Shroti DS.; Ind. J. Med. Res. 48, 2, March 1960.

Various indigenous plants are reputed to have anti-diabetic properties and many of them have been screened from time to time for their hypoglycemic action, as reviewed by Lewis (1949) and Mukherjee (1957). Most of these attempts have yielded unsatisfactory results and the possibility of securing safe and cheap oral anti-diabetic drugs from plant sources has remained distant. This laboratory also has screened some plant during the last fou or five years (Aiman, 1954, 1944, 1956, 1957). The present study was undertaken as a part of the screening programme at the ICMR Drug Research Unit. the two plants studied by us have been previously investigated by Gujral *et al.* (1954) whoc have reported negative results in alloxan diabetic rabbits. In view of the repeatedly negative results obtained in such studies and of some observations previously reported by us (Kulkarni, Shrotri and Aiman, 1958) on the importance of the period of fasting in rabbits, we found it necessary to make certain alterations on the usual methods and have studied the effects of theses on the results obtained.

Experimental Study of Hypolipidaemic activity of Kustha (*Saussurea lappa* C.S. Clarke)

O. P. Upadhyay, J. K. Ojha, H. S. Bajpal; J RES AYUR SIDDHA. 1993, XV(1-2), 52-63.

It is well established that raised serum lipids level are associated with an increased risk of subsequent development of Ischaemic Heart Disease. Any drug acting as hypolipidaemic could be made useful in several concerning ailments. In this communications, preliminary attempts were made and the drug *Saussurea lappa*, has been selected on the basis of its indication in Medoroga because of its Lekhana-karma.

27 rabbits of either sex (wt. 2.5 to 3.0 kg) were selected for present study. The dietary hyperlipidaemia was produced by cholesterol feeding for 4 weeks. The drug responses were evaluated on biochemical parameters and histopathological study of heart and liver tissues. the aqueous extract of S. lappa was administered orally and

effect was compared with a known hypolipidaemic agent (Atromid S.R)

The hypolipidaemic effect of S. lappa on dietary hyperlipidaemia were observed in this experiment. The drug reduced serum cholesterol to 258.79±17.19 from 347.83±19.69 in group A. 219.14±11.46 deom 360.10±19.07 in group B and 166.14±11.46 from 327.18±9.58 mg per cent in group C. Similarly serum triglycerides value were also reduced to 179.36±7.73 from 207.79±5.14, 95.44±4.66 from 201.86±7.70 and 99.86±3.28 from 203.80±8.59 mg per cent respectively in group A, B nd C effect of the drug when statistically evaluated, the reduction in serum cholesterol and triglycerides were noted significantly ($p<0.01$). While comparing group B and C the reduction was highly significant ($p<0.001$). These positive results confirm the claims of our ancient scientist, Charaka and Vagbhatta.

Effect of Amla (*Emblica officianalis*) on the Development of Atherosclerosis on Hypercholesterolemic Rabbits

Jasjit K. Mand, G. L. Soni, P. P. Gupta, Rattan Singh; J RES AYUR SIDDHA, April-June, 1991.

Role of Amla (*Emblica officinalis*) in prevention of development of atherosclerosis in hypercholesterolemic rabbits has been studied by feeding Amla to the hypercholesterolemic rabbits for 12 weeks. Feeding of Amla showed a two-prong effect, its feeding increased the lipid mobilization and catabolism and retarded the deposition of lipids in the extahepatic tissues. Feeding of Amla initially raised the plasma lipids and cholesterol level but by the end of 12 weeks their levels were reduced significantly below the levels in the control group. Lipid levels of liver were also significantly lowered. Though lipid levels of aorta increased during this period but the increase was much less in Amla fed animals as compared to the control group. The degree of atherosclerosis at the end of 12 weeks of Amla feeding was lower as compared to that in the control group.

Hypocholesterolaemic Action of Three Guggulu (*Commiphora mukul*) Preparations: A Comparative Study

R. B. Nair, R. P. Pilai, B. K. R. Pilai, S. Vijayaraj, C. P. R. Nair; J RES AYUR SIDDHA. XV (3-4), 155-166.

In this paper, a comparative study of the hypocholesterolaemic effect of three Guggulu preparations, *viz.* Kaisora Guggulu (K.G), Yogaraja Guggulu (Y.G) and Nava Guggulu (N.G) on albino rats is given. The minimum effective dose of the Guggulu preparation is 1g/kg. b.w. of rats bringing about significant reduction in serum and liver cholesterol levels. Among the three preparations, Kaisora Guggulu seems to have greater hypocholesterolaemic and anti-obese action compared with the other two drug preparations.

Effect of Bala Haritaki on Hyperchloesterolaemia

Rajeev Sood, A. K. Sharma; J RES AYUR SIDDHA. 2000; XXI (1-2), 11-18

Bala Haritaki (*Terminalia chebula* Retz.) was given to 15 subjects in the dose of 2gm. Twice a day in powder form for 6 weeks with the lukewarm water. The results obtained showed statistically significant reduction in the level of serum cholesterol, serum triglycerides, total lipids, low density lipids (LDL), very low density lipids (VLDL).

There was statistically significant increase in the level of high density lipids (HDL). However, there was no change in the body weight and adipose tissue thickness at the level of(i) triceps, (ii)nape of neck and (iii) abdomen just below umbilicus after therapy.

INFLAMMATION, PAIN AND ARTHRITIS

Analgesic and Anti-pyretic Activities of *Dalbergia sissoo* Leaves

S. W. Hajare, Suresh Chandra, S. K. Tandan, J. Sarma, J. Lal, A. G. Telang; Indian Journal of Pharmacology 2000; 32: 357-360.

Objective: To evaluate the analgesic and anti-pyretic activities of alcoholic extract of *Dalbergia sissoo* leaves.

Methods: The peripheral analgesic activity of *Dalbergia sissoo* leaves (SLE; 100, 300 and 1000 mg/kg) was studied using acetic acid-induced writhing in mice and by Randall-Selitto assay. The central analgesic activity of SLE was studied using hot-plate method and tail-clip test in mice. The anti-pyretic activity of SLE was studied in Brewer's yeast-induced pyrexia in rats.

Results: SLE significantly decreased the writhing movements in mice in acetic acid-induced writhing test. SLE (1000 mg/kg) significantly increased the pain threshold capacity in rats in Randall-Selitto assay and the reaction time in hot-plate test but not in tail-clip test. It also showed significant anti-pyretic activity in Brewer's yeast-induced pyrexia in rats throughout the observation period of 6 h.

Conclusion: SLE may have analgesic and anti-pyretic activities.

Antinociceptive Activity of *Azadirachta indica* (neem) in Rats

P. Khosla, Bhanwra Sangeeta, J. Singh, R.K. Srivastava; Indian Journal of Pharmacology 2000; 32: 372-374.

Objective: To study the antinociceptive effect of *A.indica* in rats.

Methods: Analgesia was evaluated using tail flick reaction time to thermal stimulus and glacial acetic acid (GAA) induced writhing. *A. indica* leaf extract (500mg/kg) and *A. indica* seed oil (2ml/kg) were given orally by an intragastric tube. Naloxone was given to study the mechanism of antinociceptive effect.

Results: Tail-flick reaction time was significantly increased in rats with both leaf extract and seed oil. Naloxone pretreatment partially reversed the antinociceptive effect of leaf extract and seed oil. GAA induced writhing was reduced with both leaf extract and seed oil. Pretreatment with naloxone partially reversed the inhibitory effect of leaf extract and seed oil on GAA induced writhing. Leaf extract was more potent than seed oil.

Conclusion: The results indicate that both the preparations of *A. indica* (leaf extract and seed oil) possess antinociceptive activity in rats, leaf extract being more potent.

Analgesic Effect of Aqueous and Alcoholic Extracts of *Madhuka longifolia* (koeing)

Dinesh Chandra; Indian Journal of Pharmacology 2001; 33: 108-111.

Objective: To screen the analgesic effect of aqueous and alcoholic extract of *Madhuka longifolia* and elucidate its probable mechanism of action.

Method: The analgesic effect was screened through tail flick, hot plate and chemical writhing methods. The probable mechanism of action through opioid receptors was elucidated by i.m. administration of naloxone - specific antagonists 30 min before the last dose of aqueous or alcoholic extract of *M. longifolia.*

Result: Graded doses of both aqueous and alcoholic extract of *M. longifolia* (4.0 to 64.0 mg/kg, i.m. X 3 days) produced dose dependent analgesic effect in all the three nociceptive methods carried out either in rats or mice. The analgesic effect exhibited by both the extracts was not antagonized by naloxone in rats only.

Conclusion: The analgesic effect exhibited by both aqueous and alcoholic extracts does not mediate through opioid receptors.

Anti-inflammatory Activity of *Curcuma amada* Roxb. in Albino Rats

A.M. Mujumdar, D.G. Naik, C.N. Dandge, H.M. Puntambekar; Indian Journal of Pharmacology 2000; 32: 375-377.

Objective: To study the anti-inflammatory activity of *Curcuma amada* rhizome extract in albino rats.

Methods: Rhizomes of *Curcuma amada* were extracted and subjected to spectroscopic studies. The extract was screened for anti-inflammatory activity in albino rats using acute carrageenan paw oedema and chronic granuloma pouch model.

Results: The extract showed presence of chemical compounds with hydroxyl, ester, carbonyl and olefin funtionalities and exhibited dose dependant anti-inflammatory activity in acute and chronic models.

Conclusion: The extract of *Curcuma amada* rhizomes showed anti-inflammatory activity in acute and chronic administration in albino rats.

Effect of *Hemigraphis colorata* (blume) H. G. Hallier Leaf on Wound Healing and Inflammation in Mice

A. Subramoniam, D. A. Evans, S. Rajasekharan, G. Sreekandan Nair; Indian Journal of Pharmacology 2001; 33: 283-285.

Objective: To study the wound healing and anti-inflammatory properties of *Hemigraphis colorata* (leaf).

Methods: The wound healing property of *H. colorata* leaf paste (topical application) or suspension (p.o., 1g/kg) was studied, using excision wound which was inflicted by cutting away 500 mm2 of the skin on the anterio-dorsal side of mouse. The anti- inflammatory activity of the leaf was evaluated using carrageenan induced paw oedema model in mice.

Results: H. colorata leaf paste when applied on the wound promoted wound healing in mice but oral administration was ineffective. The wound contraction and epithelialisation was faster in the leaf paste applied on mice. The leaf suspension or paste was devoid of anti-inflammatory activity.

Conclusion: H. colorata leaf paste promotes excision wound healing in mice.

Effect of Alcoholic Extract of *Achyranthes bidentata* Blume on Acute and Sub-acute Inflammation

T. Vetrichelvan, M. Jegadeesan; Indian Journal of Pharmacology 2002; 34: 115-118.

Objective: To study the anti-inflammatory activity of alcoholic extract of *Achyranthes bidentata* on carrageenan-induced hind paw oedema and cotton pellet granuloma models in Swiss male rats.

Methods: The hind paw oedema was produced by subplantar injection of carrageenan and the paw volume was measured plethysmographically at 0, 1, 2, 3, 4 and 5 h. In sub-acute model, cotton pellet granuloma was produced by implantation of 50±1 mg sterile cotton in axilla under ether anaesthesia. The animals were fed with ethanolic extract at various dose levels (125, 250, 375 and 500 mg/kg). Diclofenac sodium was used as a standard drug.

Results: The alcoholic extract (375 and 500 mg/kg) showed maximum inhibition of oedema by 63.52 per cent and 79.73 per cent at the end of 3 h in acute model of inflammation, respectively. Using a chronic test, the granuloma pouch in rats, the extract exhibited a 50.76 per cent and 57.49 per cent reduction in granuloma weight.

Conclusion: Achyranthes bidentata possesses anti-inflammatory effects in both acute and sub acute inflammation.

Anti-Inflammatory Property of Ethanol and Water Extracts of *Zingiber zerumbet*

Somchit MN, Nur Shukriyah MH.; Indian Journal of Pharmacology 2003; 35: 181-182.

Zingiber zerumbet (L) Sm. known as lempoyang, wild ginger belongs to Zingiberaceae family, which is a widely cultivated plant in village gardens throughout the tropics for its medicinal properties.The objective of this present study was to evaluate the anti-inflammatory activity of *Zingiber zerumbet* ethanol and aqueous extracts in rats. Male Sprague Dawley rats (180 to 200 g) were obtained from Institute of Medical Research, Kuala Lumpur, Malaysia. Aqueous and ethanol extracts from the rhizomes were given upto 500 mg/kg, i.p. in rats. Prostaglandin E2 (PGE2) at 100 ng/ml was injected (0.01 ml/rat) in the left hind paw to induce oedema. Anti-inflammatory effects were observed against an acute (PGE2-induced paw edema) model of inflammation when rats were pre-treated with 50 and 100 mg/kg water extracts of *Zingiber zerumbet*. The extracts were devoid of any toxicity upto 500 mg/kg in rats. The anti-inflammatory effect of water extract was similar to the reference NSAID mefenamic acid where the percentage anti-inflammatory effects were 46.8 per cent and 51.7 per cent respectively. The anti-inflammatory activity of extracts appears to be significant in early phases of the inflammatory process, similar to the reference NSAID.

Anti-pyretic and Analgesic Effect of Leaves of *Solanum melongena* Linn. in Rodents

Mutalik S, Paridhavi K, Mallikarjuna Rao C, Udupa N; Indian Journal of Pharmacology 2003; 35: 312-315.

Objective: To investigate the anti-pyretic and analgesic activity of dry residue of leaf juice of *Solanum melongena.*

Methods: The preliminary phytochemical screening of the dry residue was carried out by chemical tests, spectrophotometric and thin layer chromatographic methods. Acute toxicity study was performed in mice after administration of the dry residue orally in graded doses (0.5-4 g/kg body weight). Anti-pyretic activity of dry residue of *S. melongena* (100, 250 and 500 mg/kg doses) was carried out on yeast induced pyrexia in rats. Analgesic activity of dry residue was evaluated in mice using the acetic acid induced writhing test at 100, 250 and 500 mg/kg doses.

Results: The preliminary phytochemical screening of the dry residue showed the presence of flavonoids, alkaloids, tannins and steroids. In acute toxicity study, no mortality was observed at a dose as high as 4 g/kg. The dry residue of fresh juice produced significant anti-pyretic effect in a dose dependent manner and an appreciable anti-pyretic effect was noticed at 500 mg/kg dose. A dose dependent analgesic activity was observed with *S. melongena* and significant effect was observed at 500 mg/kg dose.

Conclusion: The present study demonstrates the potential anti-pyretic and analgesic effect of *S. melongena*, further supporting the claims by traditional medicine practitioners.

Antinociceptive Activity of *Martynia diandra* Glox.

Chatpalliwar VA, Joharapurkar AA, Wanjari MM, Chakraborty RR, Kharkar VT.; Indian Journal of Pharmacology 2003; 35: 320-321.

Martynia diandra Glox. (vaghnakhi or vinchoo) is one of the herb that is employed in ancient Indian medicine against inflammation. The antinociceptive activity of partially purified ethanolic extract of the aerial parts of the plant against acetic acid and formalin-induced nociception was investigated. Animals were pretreated orally with the extract of *Martynia diandra* one hour prior to formalin challenge. Pain response was observed as paw licking from 0 to 40 min after the challenge. Animals were pre-treated orally with the extract of *Martynia diandra* one hour before the intraperitoneal injection of acetic acid to induce writhing. Control animals received same amount of saline. Animals were placed in separate boxes and the number of abdominal constrictions was cumulatively counted over a period of 20 min. Antinociceptive activity was expressed as the reduction in the number of constrictions between control and pretreated animals. In the groups, pretreated with increasing concentration of *Martynia diandra*, there was a marked, dose-related inhibition of both the phases of formalin-induced pain. The oral treatment of animals with the partially purified extract of Martynia diandra produced a dose-dependent and significant inhibition of acetic acid induced abdominal constrictions. The present investigation indicates that the partially purified ethanolic extract of *Martynia diandra* produces a pronounced, dose-dependent antinociception, in chemical models of nociception

Anti-inflammatory and Antiarthritic Activities of Lupeol and 19α-h Lupeol Isolated from *Strobilanthus callosus* and *Strobilanthus ixiocephala* Roots

Agarwal RB, Rangari VD.; Indian Journal of Pharmacology 2003; 35: 384-387.

Objective: To study the anti-inflammatory and antiarthritic activities of lupeol and 19 a-H- lupeol isolated from the roots of *Strobilanthus callosus* and *Strobilanthus ixiocephala* respectively.

Methods: The anti-inflammatory activity was evaluated using carrageenan induced rat paw oedema model for acute inflammation and cotton pellet granuloma model for chronic inflammation. Antiarthritic activity was carried out using Freund's adjuvant-induced arthritis model. Prednisolone was used as a standard drug.

Results: The lupeol in the doses of 200, 400 and 800 mg/kg produced a dose dependent inhibition *i.e.* 24 per cent, 40 per cent and 72 per cent where as 19 α -H- lupeol showed 21 per cent, 47 per cent and 62 per cent inhibition after 24 h in acute model of inflammation. In chronic model of granuloma pouch in rats, lupeol exhibited 33 per cent and 19 α -Hlupeol, 38 per cent reduction in granuloma weight. In arthritis model, lupeol exhibited 29 per cent and 19 a-H- lupeol 33 per cent inhibition after 21 days respectively.

Conclusion: Both lupeol and 19 α -H- lupeol isolated from *Strobilanthus callosus* and *Strobilanthus ixiocephala* exhibit significant anti-inflammatory and antiarthritic activities respectively.

Anti-inflammatory Potential of the Seeds of *Carum copticum* Linn.

Thangam C, Dhananjayan R.; Indian Journal of Pharmacology 2003; 35: 388-391.

Objective: To investigate the anti-inflammatory principles of the total alcoholic extract (TAE) and total aqueous extract (TAQ) of the seeds of *Carum copticum*. Linn. (Umbelliferae).

Methods: Anti-inflammatory potential was evaluated using acute rat model (carrageenan induced rat paw oedema) and a sub acute rat model (cotton pellet induced granuloma). Aspirin (ASA) (150 mg/kg) and anti-inflammatory drug phenyl butazone (PBZ) (150 mg/kg) were used as standard positive controls.

Results: TAE and TAQ in 100 mg/kg doses exhibited significant ($P<0.001$) anti-inflammatory activity in both the animal models. In carragenan induced rat paw oedema, ASA and PBZ showed an inhibition of 45.23 per cent and 43.83 per cent respectively, while TAE and TAQ extracts showed an inhibition of 38.32 per cent and 41.11 per cent. In cotton pellet induced granuloma studies also TAE and TAQ produced 38.05 per cent and 43.87 per cent inhibition of the pellets weight respectively whereas ASA and PBS produced 44.69 per cent and 42.04 per cent inhibition. The weights of the adrenal glands were found to be significantly increased in TAE and TAQ treated animals (25.53 per cent and 32.2 per cent) where as ASA and PBS showed an increase of 18.86 per cent and 10.00 per cent respectively.

Conclusion: TAE and TAQ extracts from the seeds of *Carum copticum* Linn. exhibit significant anti-inflammatory potential.

Anti-Inflammatory Activity of different Fractions of *Leucas aspera* Spreng

Goudgaon NM, Basavaraj NR, Vijayalaxmi A.; Indian Journal of Pharmacology 2003; 35: 397-398

L. aspera Spreng. (F- Labiatae), is an annual herb found throughout India as a weed in cultivated fields, wastelands and roadsides. The juice of the leaves is used as local application for psoriasis, chronic skin eruptions and chronic rheumatism. In the present investigation, the crude extract was fractionated and the anti-inflammatory activity of crude, alkaloid and non-alkaliod fractions of *L. aspera* was studied. Anti-inflammatory activity was studied by formalin induced rat hind paw oedema, measured by plethysmograph (mercury displacement method). Wistar strain rats of either sex weighing between

150-200 g were divided into five groups of six animals each. The first group served as the control and received the vehicle *i.e.* Tween-80, second group of animals were administered with standard drug phenylbutazone, 200 mg/kg body weight, (subcutaneous).The third, fourth and fifth groups of animals were treated with crude extract, alkaloid and non-alkaloid fractions of *L. aspera* at a dose of 200 mg/kg body weight, orally. The volume of paw oedema was measured in control, standard and treated groups accordingly 1, 2, 3 and 4 h after formalin injection. The percent inhibition of oedema was calculated. Phenylbutazone showed highest anti-inflammatory activity followed by alkaloid fraction and crude extract. To conclude, the alkaloid fraction of the crude ethanolic extract of *L. aspera* is accountable for the anti-inflammatory activity.

Preliminary Studies on Anti-inflammatory and Analgesic Activities of *Spilanthes acmella* in Experimental Animal Models

Chakraborty A, Devi RKB, Rita S, Sharatchandra Kh., Singh Th I.; Indian Journal of Pharmacology, 2004; 36: 148-150.

Objective: To evaluate the anti-inflammatory and analgesic activities of the aqueous extract of *Spilanthes acmella* (SPA) in experimental animal models.

Material and Methods: SPA was evaluated for anti-inflammatory action by carrageenan-induced rat paw edema. The analgesic activity was tested by acetic acid-induced writhing response in albino mice and tail flick method in albino rats.

Results: The aqueous extract of SPA in doses of 100, 200 and 400 mg/kg showed 52.6, 54.4 and 56.1 per cent inhibition of paw edema respectively at the end of three hours and the percentage of protection from writhing was 46.9, 51.0 and 65.6 respectively. In the tail flick model, the aqueous extract of SPA in the above doses increased the pain threshold significantly after 30 min, 1, 2 and 4 h of administration. SPA showed dose-dependent action in all the experimental models.

Conclusion: The present study indicates that SPA has significant anti-inflammatory and analgesic properties.

Antinociceptive and Anti-inflammatory Effects of *Centella asiatica*

Somchit MN, Sulaiman MR, Zuraini A, Samsuddin L, Somchit N, Israf DA, Moin S.; Indian Journal of Pharmacology, 2004; 36: 377-380.

Objective: To evaluate the effects of *Centella asiatica* (CA) upon pain (antinociception) and inflammation in rodent models.

Material and Methods: The antinociceptive activity of the water extract of CA (10, 30, 100 and 300 mg/kg) was studied, using acetic acid-induced writhing and hot-plate method in mice. The anti-inflammatory activity of CA was studied in rats by prostaglandin E2-induced paw edema.

Results: Water extract of CA revealed significant antinociceptive activity with both the models. The activity was statistically similar to aspirin but less potent than morphine. The CA extract also revealed significant anti-inflammatory activity. This effect was statistically similar to the non-steroidal anti-inflammatory drug, mefenamic acid.

Conclusion: These results suggest that the water extract of CA possesses antinociceptive and anti-inflammatory activities.

Study of the Antinociceptive Effect of Neem Leaf Extract and its Interaction with Morphine in Mice

Patel J P, Hemavathi K G, Bhatt JD.; Indian Journal of Pharmacology 2005; 37: 37-8.

The objectives of the present work were to study the antinociceptive effects of neem leaf extract and its interaction with morphine andto delineate the probable site of action of neem leaf extract, using an opioid antagonist, naloxone.

The results of the present study revealed the antinociceptive effect of NLE in the pain model of the tail-flick test due to thermal stimulation. Neem leaves have been reported to relieve pain by opioidergic as well as other mechanisms. These results could thus have a potential clinical implication. Thus patients can benefit from relief of pain, using either morphine or NLE alone or their combination with lesser adverse effects.

Pharmacological Investigations of *Sapindus trifoliatus* in Various *In vitro* and *In vivo* Models of Inflammation

Arulmozhi D K, Veeranjaneyulu A, Bodhankar S L, Arora S K.; Indian Journal of Pharmacology 2005; 37: 96-102.

To investigate the effect of lyophilized aqueous extract of pericarps of Sapindus trifoliatus (ST) in various *in vitro* and *in vivo* inflammatory models. ST was studied for its *in vitro* inhibitory activity against 5-lipoxygenase (5-LO), cyclo-oxygenase (COX), leukotriene B4 (LTB4) and nitric oxide synthase (NOS). At doses 20 and 100 mg/kg, i.p. ST was evaluated in acute pedal inflammation induced by carrageenan, histamine, serotonin and zymosan in rats and mice. Further, the effect of topical application of the extract

(1 mg and 5 mg) on ear inflammation induced by various inflammatory agents like -O-tetradecanoyl-phorbol 13-acetate (TPA) or capsaicin or arachidonic oxazolone or dinitrofluorobenzene (DNFB) was also investigated. Results showed that *in vitro* evaluation of the extract revealed its inhibitory activity against the major inflammatory mediators 5-LO, COX, LTB4 and NOS. The extract significantly inhibited the pedal inflammation produced by carrageenan, histamine, serotonin and zymosan. Further, topical application of ST significantly inhibited the ear inflammation induced by acute and multiple applications of TPA and acute application of capsaicin or arachidonic acid. However, the extract failed to inhibit ear inflammation induced by oxazolone or DNFB. It was concluded that ST has anti-inflammatory activity possibly mediated through 5-LO and COX pathways.

Curcumin: A Natural Anti-inflammatory Agent

Kohli K, Ali J, Ansari M J, Raheman Z.; Indian Journal of Pharmacology 2005; 37: 141-7

Extensive scientific research on curcumin, a natural compound present in the rhizomes of plant *Curcuma longa* Linn., demonstrated its anti-inflammatory action. Curcumin was found to inhibit arachidonic acid metabolism, cyclooxygenase, lipoxygenase, cytokines (Interleukins and tumour necrosis factor) Nuclear factor-kB and release of steroidal hormones. Curcumin was reported to stabilize lysosomal membrane and cause uncoupling of oxidative phosphorylation besides having strong oxygen radical scavenging activity, which was responsible for its anti-inflammatory property. In various animal studies, a dose range of 100-200 mg/kg body weight exhibited good anti-inflammatory activity and seemed to have negligible adverse effect on human systems. Oral LD_{50} in mice was found to be more than 2.0 g/kg body weight.

Evaluation of Wound-Healing Potency of *Vernonia arborea* Hk.

Manjunatha B K, Vidya S M, Rashmi K V, Mankani K L, Shilpa H J, Singh SJ.; Indian Journal of Pharmacology 2005; 37: 223-6.

To investigate the comparative wound-healing potency of aqueous and methanol leaf extracts of *Vernonia arborea* Hk. Excision, incision and dead space wound models were used to evaluate the wound-healing activity of *Vernonia arborea* Hk., on Swiss Wistar strain rats of either sex. Results showed that Aqueous and methanol leaf extracts promoted the wound-healing activity significantly in all the wound models studied. High rate of wound contraction, decrease in the period for epithelialisation, high skin breaking strength and granulation strength, increase in dry granulation tissue weight, elevated hydroxyproline content and increased collagenation in histopathological section were observed in animals treated with methanol leaf extract and aqueous leaf extract when compared to the control group of animals. So concluded that methanol and aqueous leaf extracts of *Vernonia arborea* Hk. promote wound-healing activity. Methanol extract possesses better wound-healing property than the aqueous extract.

Wound-Healing Activity of the Leaf Extracts and Deoxyelephantopin Isolated from *Elephantopus scaber* Linn.

S D J Singh, V Krishna, KL Mankani, BK Manjunatha, SM Vidya, YN Manohara; Indian Journal of Pharmacology. 2005; 37: 238-242

Objective: To evaluate the wound healing activity of the leaf extracts and deoxyelephantopin isolated from *Elephantopus scaber* Linn.

Materials and Methods: The effect of aqueous ethanol extracts and the isolated compound deoxyelephantopin from *E. scaber* Linn. (Asteraceae) was evaluated on excision, incision, and dead space wound models in rats. The wound-healing activity was assessed by the rate of wound contraction, period of epithelialization, skin-breaking strength, weight of the granulation tissue, and collagen content. Histological study of the granulation tissue was carried out to know the extent of collagen formation in the wound tissue.

Results: The ethanol extract and the isolated constituent deoxyelephantopin of *E. scaber* promoted wound-healing activity in all the three wound models. Significant ($P <0.01$) increase in the rate of wound contraction on day 16 (98.8 per cent, $P <0.01$), skin-breaking strength (412 g, $P <0.01$), and weight of the granulation tissue on day 10 (74 mg/100 g, $P <0.01$) were observed with deoxyelephantopin-treated animals. In ethanol extract-treated animals, the rate of wound contraction on day 16, skin-breaking strength, and weight of the granulation tissue on day 10 ($P <0.01$) were 92.4 per cent, 380 g, and 61.67 mg/100 g, respectively. Histological studies of the granulation tissue also evidenced the healing process by the presence of a lesser number of chronic inflammatory cells, lesser edema, and increased collagenation than the control.

Conclusion: The wound-healing activity was more significant in deoxyelephantopin-treated animals.

Analgesic Activity of acetyl-11-keto-beta-boswellic acid, a 5-lipoxygenase-enzyme Inhibitor

Bishnoi M, Patil C S, Kumar A, Kulkarni S K.; Indian Journal of Pharmacology 2005; 37: 255-6.

Acetyl-11-keto-beta-boswellic acid (AKBA) is one of the four major pentacyclic triterpenic acids present in the acidic extract of the *Boswellia serrata* gum resin that is used for a variety of inflammatory disorders, such as rheumatoid arthritis, osteoarthritis, and cervical spondylosis. AKBA is a novel, highly specific inhibitor of 5-lipoxygenase, the key enzyme for leukotriene biosynthesis. It inhibits 5-LOX either directly interacting with the enzyme itself, or interacting with 5-lipoxygenase activating proteins (FLAP). In the present study, there was a dose-dependent increase in the analgesic activity of AKBA in acetic acid-induced writhing. In case of tail flick test there was no difference in the analgesic effect of 100 and 200 mg/kg dose of AKBA. AKBA showed antinociceptive activity as early as 30 min which was increased up to 60 min and after that the effect declined. The effect of AKBA at 200 mg/kg, p.o. was more pronounced in tail-flick test rather than acetic-acid-induced writhing as compared to nimesulide (2 mg/kg, p.o.). AKBA showed lesser duration of action. In conclusion, further evaluation and toxicity test are needed to prove that AKBA possesses antinociceptive property.

Effect of *Aloe vera* on Nitric Oxide Production by Macrophages during Inflammation

Sarkar D, Dutta A, Das M, Sarkar K, Mandal C, Chatterjee M.; Indian Journal of Pharmacology 2005; 37: 371-5.

To demonstrate the mechanism of action mediating the acute and chronic anti-inflammatory activity of leafy exudate of *Aloe vera* (AVL) in animal models of inflammation.

Materials and Methods: The acute anti-inflammatory activity of AVL was evaluated using carrageenan and dextran as phlogistic agents while its chronic anti-inflammatory effect was investigated in a complete Freund's adjuvant-induced model of arthritis. The degree of inflammation in all models was measured plethysmographically. The effect of AVL on nitric oxide production in mouse peritoneal macrophages was measured by the Griess reagent.

Results: AVL (25 mg/kg) significantly reduced carrageenan and dextran-induced pedal edema in rats by 61.9 per cent and 61.7 per cent, respectively. In the Freund's adjuvant-induced model of chronic inflammation, AVL showed chronic anti-inflammatory activity but failed to decrease the arthritic index indicating the absence of anti-arthritic activity. AVL (10 μg/ml) caused a decrease in NO production in macrophages without causing toxicity.

Conclusion: AVL possesses acute and chronic anti-inflammatory activity, which is partly mediated by reduced production of NO, which in turn prevents the release of inflammatory mediators.

Histomorphological Changes Induced by *Vitex negundo* in Albino Rats

Tandon V, Gupta RK.; Indian Journal of Pharmacology, 2004; 36: 175-180

Vitex negundo Linn. (VN) has been investigated extensively for its anti-inflammatory and analgesic activities. The present study was undertaken to evaluate the histomorphological changes produced in various organs and to ascertain the type of cyclooxygenase inhibition produced. The acute toxicity study was carried out by administering VN leaf extract orally in graded doses (1-10 g/kg, body weight) to seven groups of animals each consisting of six animals. LD_{50} of the extract was determined by graphical method and the histomorphological changes in vital organs were studied. The present study indicated that oral LD50 dose of VN leaf extract is 7.58 g/kg, b.wt of rats. The stomach showed no histomorphological changes in any of the doses of the extract. VN which is known to act by prostaglandin inhibition, may be expected to cause gastric damage but on the contrary it produced no histomorphological changes in the stomach even in toxic doses. This may be due to a selective COX-2 inhibition that might be responsible for the NSAID-like activity. Dose-dependent histomorphological changes were observed in the specimens of the heart, liver and lung. From the histomorphological examination it seems that the major toxic assault of VN was on the heart.

Anti-inflammatory Activity of Two Ayurvedic Formulations Containing Guggul

Bagul M S, Srinivasa H, Kanaki N S, Rajani M.; Indian Journal of Pharmacology 2005; 37: 399-400

Study was conducted to evaluated the anti-inflammatory activity of two important polyherbal Ayurvedic formulations *viz.*, chandraprabha vati (CPV) and Maha yogaraja guggulu (MYG) in rat paw edema model.

CPV and MYG showed dose-dependent anti-inflammatory activity with a maximum of 45 per cent and 49 per cent in paw edema, respectively, at a dose of

500 mg/kg. Prepared samples showed significantly better activity as compared to the commercial samples (P<0.01).

Anti-inflammatory Activity of Leaf Extracts of *Kalanchoe crenata* Andr.

Dimo T, Fotio AL, Nguelefack T B, Asongalem E A, Kamtchouing P.; Indian Journal of Pharmacology 2006; 38(2): 115-9.

Objective: To evaluate the acute and chronic anti-inflammatory properties of leaf extracts of *Kalanchoe crenata* in rats.

Material and Methods: The methylene chloride/ methanol extract of *K. crenata* was extracted by using hexane, methylene chloride, ethyl acetate, and n-butanol. The anti-inflammatory profile of these extracts was investigated on the basis of paw edema induced by carrageenan. The n-butanol fraction (most potent) was further assessed through acute inflammatory models induced by histamine, serotonin, and formalin. The chronic anti-inflammatory and the ulcerogenic activities of the n-butanol fraction were also examined.

Results: The oral administration of n-butanol fraction (600 mg/kg) caused a maximum inhibition of about 45 per cent in paw edema induced by carrageenan. The n-butanol fraction also exhibited acute anti-inflammatory activity on paw edema induced by histamine (47.51 per cent), serotonin (54.71 per cent), and formalin-(40.00 per cent). In the chronic inflammation model, this extract showed maximum inhibition of 61.26 per cent on the ninth day of treatment. The ulcerogenic assessment showed that ulcer indices after oral treatment with n-butanol fraction were zero and 0.4±0.2, for the 300 and 600 mg/kg doses, respectively.

Conclusion: On the basis of these findings, it may be inferred that *K. crenata* is an anti-inflammatory and antiarthritic agent that blocks histamine and serotonin pathways. The results are in agreement with the traditional use of the plant in inflammatory conditions.

Effect of Eugenol on Animal Models of Nociception

Kurian R, Arulmozhi D K, Veeranjaneyulu A, Bodhankar S L.; Indian Journal of Pharmacology 2006; 38: 341-5.

Objective: To investigate the antinociceptive potential of eugenol on different pain models in mice.

Materials and Methods : Eugenol was evaluated (1-100 mg/kg, i.p.) in various experimentally induced pain models like, formalin induced hyperalgesia, acetic acid induced abdominal constrictions, and thermal pain experiment using Eddy's hot plate.

Results: Eugenol significantly inhibited acetic acid induced abdominal constrictions, with the maximal effect (92.73 per cent inhibition) at 100 mg/kg. In formalin induced paw licking pain model, eugenol exhibited more pronounced antinociceptive effect in the inflammatory phase than the neurogenic phase (maximal effect was 70.33 per cent and 42.22 per cent, respectively, at 100 mg/kg, i.p). A mild reduction in the pain response latency at 100 mg/kg, i.p. dose of eugenol was observed in the hotplate thermal pain studies in mice. In the rotarod motor coordination experiment eugenol reduced the endurance time at the dose of 100 mg/kg, i.p.

Conclusion: The data suggest that eugenol exerts antinociceptive activity in different experimental models of pain in mice.

Anti-inflammatory and Antinociceptive Activities of *Zingiber officinale* Roscoe Essential Oil in Experimental Animal Models

Vendruscolo A, Takaki I, Bersani-Amado L E, Dantas J A, Bersani-Amado C A, Cuman R K.; Indian Journal of Pharmacology. 2006; 38: 58-9.

Ginger, *Zingiber officinale* Roscoe (Zingiberaceae), in folk medicine has been used against pain, inflammation, arthritis, urinary infections, and gastrointestinal disorders. The oil of ginger is a mixture of constituents, consisting of monoterpenes (phellandrene, camphene, cineole, citral, and borneol) and sesquiterpenes (zingiberene, zingiberol, zingiberenol, β-bisabolene, sesquiphellandrene, and others). Aldehydes and alcohols are also present.

Gingerol and its analogs found in rhizome extracts are responsible for many pharmacological activities. Few works have reported the properties of ginger essential oil (GEO). However, several types of terpene compounds are known to present anti-inflammatory and antinoceptive activities. The aim of the present study was to evaluate the anti-inflammatory and analgesic effects of GEO administered orally in rodents. Groups of 10 male Swiss mice (25-30 g) and male Wistar rats (190-230 g) were used for evaluation of the antinoceptive and anti-inflammatory effects, respectively. All animals were housed in groups of five and maintained in standardized conditions (12/12 h light/dark cycle, 25°C) with free access to water and food. The protocol for these experiments was approved and was in accordance with the guidelines of the Brazilian Committee of Animal Experimentation.

Fresh rhizomes of *Z. officinale* were collected from the herbarium of the State University of Maringα, identified, and authenticated. GEO was obtained from

250 g of rhizomes by conventional steam distillation using Clevenger apparatus during 3 h. The oil obtained was kept refrigerated and protected form direct light.

Pleurisy was induced in anesthetized mice by intraperitoneal (i.p.) injection of carrageenan (200 μg/cavity). Four hours later, the rats were sacrificed and the exudate was collected to determine the total volume and leukocyte number. Exudates smears were prepared, air-dried, and fixed with Rosenfeld stain for leukocyte differential count. The parameters studied were leukocyte migration and fluid leakage. GEO (100, 200, and 500 mg/kg, p.o.) and indomethacin (5 mg/kg, p.o.) were administered 30 min before the test.

The antinociceptive activity of the GEO was assessed using the writhing test. Acetic acid solution (10 ml/kg, 0.6 per cent) was i.p. injected and abdominal muscles constriction together with stretching of the hind limbs was counted over a period of 20 min, starting immediately after acetic acid injection. GEO (50, 100, and 200 mg/kg, p.o.) and indomethacin (5 mg/kg, p.o.) were administered 30 min before the acid injection. Antinociceptive activity was expressed as the percentage of inhibition of writhings compared with control animals.

The hot-plate test was performed to measure response latencies. The hot plate was maintained at 55.0 ± 1°C. The time taken (s) to cause a discomfort reaction (licking paws or jumping) was recorded as the response latency 0, 15, 30, 60, and 90 min after administration of GEO (100 and 200 mg/kg, p.o.), meperidine (50 mg/kg, i.p.) or saline solution 0.9 per cent (control group). A latency period of 25 s was defined as complete analgesia and the experiment was stopped if it exceeded the latency period in order to avoid injury.

Data are reported as mean ± SEM. Statistical differences in all groups were determined using one-way ANOVA. P values <0.05 were considered significant.

In the pleurisy test, indomethacin and GEO 200 and 500 mg/kg reduced significantly the exudate volume ($P<0.05$ and $P < 0.001$) without promoting alteration of total leukocyte migration. Data suggest that GEO does not have influence on cells' recruitment, different to that observed for others essential oils. The anti-inflammatory activities of compounds obtained from GEO have been reported by other investigations using ginger extract. These anti-inflammatory actions could be owing to the inhibition of prostaglandin release, and hence ginger may act in a way similar to other nonsteroidal anti-inflammatory drugs which interfere with prostaglandin biosynthesis. Gingerol has been reported to have anti-inflammatory actions, which include suppression of both cyclooxygenase and lipooxygenase metabolites of arachidonic acid. Furthermore, constituents of essential oils obtained from many other plants have been proposed to have anti-inflammatory activity.

Essential oils' constituents such as (-)-linalool antagonize different pain responses elicited by exposure to a chemical stimulus such as acetic-induced, by a thermal stimulus or by a tissue injury produced by formalin injection. It has showed in a report that analgesic and anti-pyretic properties from ginger extracts in a range of laboratory animals. In the present experiments, GEO (50, 100, and 200 mg/kg, p.o.) and indomethacin significantly suppressed the acetic acid-induced writhing response in a dose-dependent manner. Maximum inhibition of GEO was observed at the dose of 200 mg/kg.

In the hot-plate test, the time course of the antinociceptive reaction produced by saline or GEO (100 and 200 mg/kg) administration did not result in significant prolongation of the response latency as observed for meperidine group animals (data not shown).

GEO was found to contain monoterpenes and sesquiterpenes as principal compounds, suggesting that the anti-inflammatory and analgesic effects could be correlated to these essential oil constituents. Further studies are needed to reveal the mechanisms of action for these activities of GEO.

Analgesic and Anti-inflammatory Activities of *Sida cordifolia* Linn.

Sutradhar R K, Matior Rahman A, Ahmad M U, Datta B K, Bachar S C, Saha A. Indian Journal of Pharmacology 2006; 38: 207-8.

Sida cordifolia Linn is a herb belonging to the family *Malvaceae*. The water extract of the whole plant is used in the treatment of rheumatism. Earlier, phytochemical studies of its roots have shown the presence of ephedrine, vasicinol, vasicinone and N-methyl tryptophan.The objective of the current study is to evaluate the analgesic and anti-inflammatory activities of different extracts of *Sida cordifolia* Linn. (SIC).

The aerial parts of SIC were collected from the south-eastern region of Bangladesh. The air-dried powder of the plant (5.5 kg) was successively extracted with chloroform (3x72 h), methanol (3x72 h) and 80 per cent ethanol (3x72 h). Chloroform and methanol extracts were evaporated to dryness under reduced pressure at 40°C to yield extracts A and B, respectively. The 80 per cent ethanol extract C was concentrated to one-third of its volume and was partitioned with hexane, dichloromethane, ethyl acetate and butanol. Evaporation of the hexane, dichloromethane, ethyl acetate and butanol extracts, under reduced pressure at 40°C, yielded the dry extracts D, E, F and G respectively. After acid base treatment, the methanol extract B afforded the basic extract H and the neutral extract I.

Long Evans rats (150-200 g) and Swiss albino mice (25-30 g) of either sex were collected from the International Centre for Diarrhoeal Diseases and Research, Bangladesh (ICDDR, B). The animals were kept in polyvinyl cages under controlled room temperature (25±2°C) for 7 days and supplied with ICDDR, B formulated food pellets and water *ad libitum*.

No adverse effect or mortality was detected in the Swiss albino mice up to 4 g/kg, p.o., for any of the extract of SIC during the 24 h observation period.

The pre-screened Swiss albino mice employed for the acetic acid induced writhing test were divided into groups. The inhibition of the writhing reflex in mice by the plant extracts (*p.o.* at a dose of 100 and 200 mg/kg, body weight) were compared against the standard analgesic, aminopyrine 50 mg/kg, p.o. The analgesic activity was assessed by calculating the number of writhing reflexes for 10 min, occurring immediately after 0.1 ml/10 g of intraperitoneal acetic acid (0.7 per cent).

In carrageenan induced rat paw edema the rats were divided into groups. Acute inflammation was produced by subplantar injection of 0.1 ml of 1 per cent suspension of carrageenan with 2 per cent gum acacia in normal saline, in the right hind paw of the rats, one hour after oral administration of the drugs. The paw volume was measured plethysmometrically (Ugo Basile, Italy) at 1, 2, 3, 4 and 24 h after the carrageenan injection. The plant extracts were given orally (100 and 200 mg/kg body weight) in suspension form. Phenylbutazone suspended in 2 per cent gum acacia at a dose of 100 mg/kg, p.o., was used as the standard anti-inflammatory drug.

The results were analyzed for statistical significance using one-way ANOVA followed by Dunnett's test. $P < 0.05$ was considered significant.

From the experimental data, it is found that the extracts A, B, D, E, F, H and I in doses of 100 and 200 mg/kg body weight showed significant inhibition of writhing reflexes *i.e.* (58.86, 66.53 per cent), (45.56, 52.81 per cent), (48.78, 55.64 per cent), (26.20, 56.44 per cent), (26.61, 52.43 per cent), (43.55, 56.06 per cent) and (41.13, 54.85 per cent), respectively with the statistical significance of ($P < 0.01$). Among the SIC, the maximum and minimum analgesic activity was exhibited by chloroform extract A and butanol extract G respectively.

Results show that the extracts A, B, F, G and H exhibited sufficient inhibition of paw edema of 33.61, 32.97, 34.46, 39.35 and 40.85 per cent, respectively at the end of the fourth hour. The activities of different SIC extracts were comparable to the standard drug, phenylbutazone. In this experiment, the lower dose 100 mg/kg did not show any significant anti-inflammatory activity (data not given).

The exact mechanism(s) of the analgesic and anti-inflammatory activities of the extracts is/are yet to be elucidated.

Anti-inflammatory Activity of Leaf and Leaf Callus of *Silybum marianum* (L.) Gaertn. in albino rats.

Balian S, Ahmad S, Zafar R.; Indian Journal of Pharmacology 2006; 38: 213-4.

Silybum marianum (L.) Gaertn. is an important medicinal plant of family *Compositae*, commonly known as Milk-thistle or St. Mary's thistle. The plant and its extracts are reported to possess hepatoprotective, antioxidant, anticancer, anti-inflammatory and antidiabetic properties. It contains flavonolignan Silymarin, which is an important bioactive principle having anticancer, anti-inflammatory, antioxidant and immunomodulatory effects. However, till date, no anti-inflammatory activity has been carried out on tissue cultures developed from *S. marianum*. Therefore, it was thought worthwhile, to determine the anti-inflammatory activity of plant cultures developed *in vitro*. In the present investigation, the leaf callus of plant has been successfully developed and maintained for six months. The methanolic extract of dried leaf callus was examined for anti-inflammatory activity, using carrageenan and formalin- induced rat paw oedema models, which was also compared with that of leaf extract. Leaves were collected from the plants grown in the herbal garden of Hamdard University, New Delhi, identified by Department of Botany and the voucher specimen was kept in the herbarium of the University.

The immature leaves were washed with water and detergent, followed by rinsing with double distilled water, to remove the detergent. The cleaned leaves were then transferred aseptically to Mercuric chloride solution (0.5 per cent w/v) and stirred for five minutes. Then these were removed and washed six times with double distilled water, for complete removal of chemical sterilent. The sterilized leaves were then transferred in culture tubes (Borosil glass works Ltd.) containing Murashige and Skoog (MS) medium (Sigma chemicals), supplemented with various growth hormones like indole acetic acid (IAA), Indole butyric acid (IBA), 2,4-dichlorophenoxyacetic acid (2,4-D), naphthalene acetic acid (NAA), 6-benzyl adenine (6-BA) and kinetin (Sigma chemicals) in different concentrations and kept in a BOD incubator (25±2°C temprature, 16 and 8 h light/dark cycle). Out of several hormonal combinations tried, MS medium supplemented with IAA + IBA + 2, 4-D and kinetin (1ppm each), showed best results for initiation of creamy soft and friable leaf callus, within 12-16 days. The leaf calli initiated on the above medium, were further developed and maintained for six months on the same medium. Dried and powdered leaves (35 g) and callus (20 g), were separately extracted with methanol in soxhlet apparatus (Borosil glass works Ltd.), for four hrs. The methanolic extracts were evaporated to dryness under vacuum evaporator (Scientific system, New Delhi) and the residue obtained (4.00 g and 1.85 g respectively, for leaf and leaf callus), was triturated with gum acacia in distilled water (1:1) and administered to adult female Wistar albino rats by oral route (100 mg/kg, body wt).

Forty-eight female Wistar albino rats weighing 150-200 g were used. The rats were housed in colony cages in an animal house, at an ambient temperature of 25±2°C, with 12 h light/dark cycle. The rats were allowed standard laboratory feed and water *ad libitum*.

Preliminary phytochemical screening of methanol extracts of leaf and leaf callus were carried out for the detection of phytoconstituents, using standard chemical tests. Alkaloids, amino acids, flavonoids, carbohydrates, phenolics, steroids and tannins, were detected in both the extracts. HPTLC fingerprints of methanolic extracts were established using CAMAG HPTLC and chloroform: acetone: formic acid (9:2:1) as solvent system, which showed presence of 10 spots (Rf-value: 0.03, 0.07, 0.10, 0.19, 0.42, 0.51, 0.59, 0.66, 0.74 and 0.79) and 8 spots (Rf-value: 0.03, 0.07, 0.19, 0.42, 0.52, 0.59, 0.67 and 0.75) respectively at 254 nm wavelength.

In carrageenan- induced paw odema model, groups of rats were orally administered with the leaf extract (100 mg/kg, bw), leaf callus extract (100 mg/kg, b.w.), Aspirin (150 mg/kg) or saline, 1 h before administration of an intradermal injection of carrageenan (0.1 ml of a 1 per cent in 0.9 per cent saline), into the plantar surface of the right hind paw. The doses of extracts were chosen, based on those used in an earlier study. The paw volume up to a fixed mark at the level of lateral malleolus was measured by recording the volume displacement by digital plethysmometer (UGO-BASILE-7140 Barcelona), just before and three hours after the injection of carrageenan. The average percent increase in paw volume of each group was calculated and compared with that of the control (saline) and aspirin groups.

In formalin-induced paw odema model, the same procedure was carried out, except that 0.05 ml of 1 per cent formalin was injected, instead of carrageenan.

The data were analyzed using one-way analysis of variance (ANOVA), followed by Dunnett's test. $P < 0.01$ was considered as statistically significant. The data are expressed as mean ± SEM. The results are shown in.

The leaf and leaf callus of *Silybum marianum* (L.) Gaertn. inhibited the formation of paw oedema to significant levels in rats treated either with carrageenan or formalin. At a dose of 100 mg/kg orally, the leaf extract produced 74 per cent inhibition, while leaf callus produced 93.9 per cent inhibition in case of the carrageenan-induced oedema ($P < 0.01$) and there was 85.61 per cent inhibition in leaf extract and 91.27 per cent inhibition in leaf callus extract, in formalin-induced oedema ($P < 0.01$). The per cent inhibition showed by leaf callus extracts (100 mg/kg) was found to be more than that of reference standard *i.e.*, aspirin (93.9 per cent *Vs* 78.79 per cent inhibition in carrageenan-induced rat paw oedema,and 91.27 per cent *Vs* 86.86 per cent inhibition in formalin-induced rat paw oedema).

The *in vitro* culture- generated callus extract showed maximum inhibition in rat paw oedema, which is due to presence of higher amount of secondary metabolites, as compared to natural plant leaf. Our results strongly suggest that the methanolic extract of leaf and leaf callus of *Silybum marianum* possesses a potent anti-inflammatory activity that could inhibit the acute inflammation in rat paw, induced either by carrageenan or formalin.

Open, Randomized, Controlled Clinical Trial of *Boswellia serrata* Extract as Compared to Valdecoxib in Osteoarthritis of Knee

Sontakke S, Thawani V, Pimpalkhute S, Kabra P, Babhulkar S, Hingorani L.; Indian Journal of Pharmacology. 2007; 39: 27-29.

Objective: To compare the efficacy, safety and tolerability of *Boswellia serrata* extract (BSE) in osteoarthritis (OA) knee with valdecoxib, a selective COX-2 inhibitor.

Materials and Methods: In a randomized, prospective, open-label, comparative study the efficacy, safety and tolerability of BSE was compared with valdecoxib in 66 patients of OA of knee for six months. The patients were assessed by WOMAC scale at baseline and thereafter at monthly interval till 1 month after drug discontinuation. Antero-posterior radiographs of affected knee joint were taken at baseline and after 6 months.

Results: In BSE group the pain, stiffness, difficulty in performing daily activities showed statistically significant improvement with two months of therapy which even lasted till one month after stopping the intervention. In valdecoxib group the statistically significant improvement in all parameters was reported after one month of therapy but the effect persisted only as long as drug therapy continued. Three patients from BSE group and two from valdecoxib group complained of acidity. One patient from BSE group complained of diarrhoea and abdominal cramps.

Conclusion: BSE showed a slower onset of action but the effect persisted even after stopping therapy while the action of valdecoxib became evident faster but waned rapidly after stopping the treatment.

Antiimplantation Activity of the Ethanolic Root Extract of *Momordica cymbalaria* Fenzl in Rats

Koneri R, Saraswati CD, Balaraman R, Ajeesha EA.; Indian Journal of Pharmacology. 2007; 39: 90 -96.

Objective : To evaluate the antiimplantation activity of the ethanolic root extract of *Momordica cymbalaria* Fenzl.

Materials and Methods: The acute oral toxicity study was performed according to the OPPTS guidelines. The ethanolic root extract was investigated for antiimplantation, estrogenic and progestrogenic activities at doses of 250 and 500 mg/kg body weight. Antiimplantation activity was studied on successive stages of embryogenesis. Estrogenic studies were carried out by examining uterine weight, histoarchitecture of uterus, vaginal cornification, uterine content of glucose, cholesterol and alkaline phosphatase levels in immature rats. Progestrogenic activity assay was performed by pregnancy maintenance in rats and Clauberg's test (endometrial proliferation assay) in immature rabbits.

Results : Both doses of the ethanolic root extract exhibited highly significant ($P < 0.001$) antiimplantation activity. However, an investigation of the estrogenic activity did not show any increase in uterine weight or vaginal cornification. The histoarchitecture (uterotrophic changes) such as thickness of endometrium and height of endometrial epithelium was unaltered in treated rats. There were no increases in the uterine content of glucose, cholesterol or alkaline phosphatase levels when compared with the control group. Pregnancy was not maintained in the pregnancy maintenance test for progestrogenic activity. Uterine proliferation was not seen in Clauberg's test (endometrial proliferation assay) for progestrogenic activity in immature rabbits.

Conclusion : The ethanolic root extract of *Momordica cymbalaria* Fenzl exhibited antiimplantation activity but this is not due to estrogenic or progestrogenic activities.

Burn Wound Healing Property of *Cocos nucifera*: An Appraisal

Srivastava P, Durgaprasad S.; Indian Journal of Pharmacology. 2008; 40: 144-6.

Objectives: The study was undertaken to evaluate the burn wound healing property of oil of *Cocos nucifera* and to compare the effect of the combination of oil of *Cocos nucifera* and silver sulphadiazine with silver sulphadiazine alone.

Materials and Methods: Partial thickness burn wounds were inflicted upon four groups of six rats each. Group I was assigned as control, Group II received the standard silver sulphadiazine. Group III was given pure oil of *Cocos nucifera*, and Group IV received the combination of the oil and the standard. The parameters observed were epithelialization period and percentage of wound contraction.

Results: It was noted that there was significant improvement in burn wound contraction in the group treated with the combination of *Cocos nucifera* and silver sulphadiazine. The period of epithelialization also decreased significantly in groups III and IV.

Conclusion: It is concluded that oil of *Cocos nucifera* is an effective burn wound-healing agent.

Analgesic and Anti-inflammatory Activity of *Argyreia speciosa* Root

Bachhav RS, Gulecha VS, Upasani CD.; Indian Journal of Pharmacology. 2009; 41: 158-61.

Objective: To study analgesic and anti-inflammatory activities of a methanolic extract (ME) of *Argyreia speciosa* (AS) root powder.

Materials and Methods: The study was carried out using male albino mice (20-25 g) and male wistar rats (100-150gm). The ME was prepared using soxhlet extraction process. The effect of ME of *A. speciosa* was investigated for analgesic activity using acetic acid-induced abdominal constriction, tail immersion method and hot plate method. The anti-inflammatory activity of ME of AS roots was studied using carrageenan-induced rat paw edema.

Result: The ME of *A. speciosa* root was used in pain and inflammation models. The analgesic activity of AS at the dose of (30,100, and 300 mg/kg p.o) showed significant ($P<0.01$) decrease in acetic acid-induced writhing, whereas ME of *A. speciosa* at the dose of (100, 300 mg/kg p.o) showed significant ($P<0.01$) increase in latency to tail flick in tail immersion method and elevated mean basal reaction time in hot plate method. The ME of the *A. speciosa* at doses (30, 100, and 300mg/kg) showed significant (P < 0.01) inhibition of carrageenan induced hind paw edema in rats.

Conclusion: The ME of *A. speciosa* showed significant analgesic and anti-inflammatory activity in mice and rat.

Screening of *Bauhinia purpurea* Linn. for Analgesic and Anti-inflammatory Activities

C.S. Shreedhara, V.P. Vaidya, H.M. Vagdevi, K.P. Latha1, K.S. Muralikrishna, A.M. Krupanidhi.; Indian Journal of Pharmacology. 2009; 41: 75-79.

Objectives: Ethanol extract of the stem of *Bauhinia purpurea* Linn. was subjected to analgesic and anti-inflammatory activities in animal models.

Materials and Methods: Albino Wistar rats and mice were the experimental animals respectively. Different CNS depressant paradigms like analgesic activity (determined by Eddy's hot-plate method and acetic acid writhing method) and anti-inflammatory activity determined by carrageenan induced paw edema using plethysmometer in albino rats) were carried out, following the intra-peritoneal administration of ethanol extract of *Bauhinia purpurea* Linn. (BP) at the dose level of 50 mg/kg and 100 mg/kg.

Results: The analgesic and anti-inflammatory activities of ethanol extracts of BP were significant (P < 0.001). The maximum analgesic effect was observed at 120 min at the dose of 100 mg/kg (i.p.) and was comparable to that of standard analgin (150 mg/kg) and the percentage of edema inhibition effect was 46.4 per cent and 77 per cent for 50 mg/kg and 100 mg/kg (i.p) respectively. Anti-inflammatory activity was compared with standard Diclofenac sodium (5 mg/kg).

Conclusion: Ethanol extract of *Bauhinia purpurea* has shown significant analgesic and anti-inflammatory activities at the dose of 100 mg/kg and was comparable with corresponding standard drugs. The activity was attributed to the presence of phytoconstituents in the tested extract.

Studies on Anti-inflammatory Effect of Aqueous Extract of Leaves of *Holoptelea integrifolia*, Planch. in Rats

Sharma S, Lakshmi K.S., Patidar A, Chaudhary A, Dhaker S.; Indian Journal of Pharmacology. 2009; 41: 87-88.

Objectives: The purpose of the present study was to investigate the anti-inflammatory properties of aqueous extract of the leaves of *H. integrifolia*, Planch.

Materials and Methods: The hind paw edema was produced in rats by subplanter injection of carageenan. The aqueous extract of *H. integrifolia*, Planch. (AHI) at dose (250 and 500 mg/kg p.o) was given to observe per cent inhibition of paw edema which were comparable with indomethacin (10 mg/kg p.o) used as a reference drug.

Results: The extract administered orally at doses of 250 and 500 mglkg p.o produced a significant (P < 0.05) dose dependent inhibition of edema formation

Conclusions: A significant percentage inhibition of paw edema by the aqueous extract of leaves of *H. integrifolia*, Planch. and its almost nearby same per cent inhibition with indomethacin suggest its usefulness as an anti-inflammatory agent.

Effects of Ethanol Extract of *Ficus bengalensis* (Bark) on Inflammatory Bowel Disease

Patel MA, Patel PK, Patel MB.; Indian Journal of Pharmacology. 2010; 42: 214-8.

Objective: The present study was designed to evaluate the effects of ethanol extract of *Ficus bengalensis* Linn. bark (AEFB) on inflammatory bowel disease (IBD).

Materials and Methods: Effects of AEFB were studied on 2,4,6-trinitrobenzenesulfonic acid (TNBS, 0.25 ml 120 mg/ml in 50 per cent ethanol intrarectally, on first day only) induced IBD in rats. The effects of co-administration of prednisolone (2 mg/kg) and AEFB (250, 500 mg/kg) for 21 days were evaluated. Animals sacrificed at end of the experiment and various histopathological parameters like colon mucosal damage index (CMDI) and disease activity index (DAI) were assessed. In the colon homogenate malondialdehyde (MDA), myeloperoxidase (MPO), superoxide dismutase (SOD), and nitric oxide (NO) levels and in mesentery per cent mast cell protection was also measured.

Results: Rats treated with only TNBS showed more score of CMDI and DAI, higher MDA, NO, MPO, and lower SOD activity as compared to the control group. Treatment with AEFB significantly declined both indices scores and decreased the MPO, MDA, NO, and increased the SOD activity. AEFB also increased the per cent mast cell protection compared to alone TNBS-treated animals.

Conclusion: In our study, we found that AEFB has a significant protective effect in the IBD in rats that is comparable to that of prednisolone and may be because of the presence of flavonoids, terpenoids, and phenolic compounds.

Kaurenic Acid: An *In vivo* Experimental Study of its Anti-inflammatory and Anti-pyretic Effects

Sosa-Sequera MC, Suárez O, Daló NL.; Indian Journal of Pharmacology. 2010; 42: 293-6.

Objective: This study was designed to investigate the anti-inflammatory and anti-pyretic effects of kaurenic acid (KA), a tetracyclic diterpenoid carboxylic acid, using *in vivo* experimental animal models.

Material and Methods: The anti-inflammatory activity of KA was evaluated in rats, using egg albumin-induced paw edema (acute test) and Freund's complete adjuvant-induced paw edema (subacute test), whereas the anti-pyretic effect was studied in rabbits by peptone-induced pyresis. Acute and subacute toxicity of KA were analyzed in NMRI mice.

Results: KA showed anti-inflammatory and anti-pyretic properties, and the effect caused was significantly dose-related (P<0.001) in both cases. The mean lethal doses of KA were 439.2 and 344.6 mg/kg for acute and subacute toxicity, respectively.

Conclusion: On the basis of these findings, it may be inferred that KA has an anti-inflammatory and anti-pyretic potential.

Screening of Anti-inflammatory and Anti-pyretic Activity of *Vitex leucoxylon* Linn.

Shukla P, Shukla P, Mishra SB, Gopalakrishna B.; Indian Journal of Pharmacology. 2010; 42: 409-11.

Objective: This study was designed to evaluate the anti-inflammatory and anti-pyretic activity of ethyl acetate extract of *Vitex leucoxylon* Linn. in various animal experimental models.

Materials and Methods: Ethyl acetate extract of *V. leucoxylon* Linn. evaluated for anti-inflammatory activity in carrageenan, mediator-induced rat paw edema, and cotton pellet-induced granuloma model. The anti-pyretic activity was evaluated by yeast-induced pyrexia model.

Results: Single administration of the ethyl acetate extract of *V. leucoxylon* Linn. at dose of 500 mg/kg p.o. showed significant (P < 0.001) inhibition of rat paw edema. The ethyl acetate extract showed significant anti-pyretic activity in brewer yeast-induced pyrexia in rats throughout the observation period of 4 h.

Conclusion: This study shows that ethyl acetate extract of *V. leucoxylon* Linn. has significant anti-inflammatory and anti-pyretic activity.

Assessment of the Anti-inflammatory Effects of *Swertia chirata* in Acute and Chronic Experimental Models in Male Albino Rats

Shivaji Banerjee, Tapas Kumar Sur, Suvra Mandal, Prabhash Chandra Das, Sridhar Sikdar; Indian Journal of Pharmacology 2000; 32: 21-24

Objectives: To study the anti-inflammatory effect of xanthone derivative (1,5-dihydroxy-3,8-dimethoxy xanthone) of *Swertia chirata* (SC-I) in acute, sub-acute and chronic experimental models in male albino rats.

Methods: Aerial parts of *Swertia chirata* were extracted with organic solvent and purified by chromatographic procedure. SC-I was studied in carrageenin-induced hind paw oedema in rats and the paw volume was measured plethysmometrically at 0 and 3 h after injection. The compound was subjected to turpentine oil-induced granuloma pouch in rats. The pouch was opened on day 7 under anaesthesia and the exudate collected by a syringe was measured. The drug was also investigated in formalin induced oedema models in rats. Degree of inflammation was measured plethysmometrically on day 1 and 7 and compared with control and standard drug, diclofenac. All the drugs were administered orally.

Results: The higher dose of SC-I significantly reduced carrageenin-induced pedal oedema (57 per cent) and formalin-induced pedal oedema in rats (58 per cent). SC-I also decreased exudate volume (35 per cent) in turpentine oil-induced granuloma formation in comparison to control.

Conclusion: 1,5-dihydroxy-3,8-dimethoxy xanthone of *Swertia chirata* showed significant anti-inflammatory action in acute, sub-acute and chronic experimental models in rats.

Anti-inflammatory Activity of *Heliotropium indicum* Linn. and *Leucas aspera* Spreng in Albino Rats

K. Srinivas, M.E.B. Rao, S.S. Rao; Indian Journal of Pharmacology 2000; 32: 37-

Objective: To study the anti-inflammatory effect of *Heliotropium indicum*, and *Leucas aspera* on carrageenan induced hind paw oedema and cotton pellet granuloma in rats.

Methods: Hind paw oedema was produced by subplantar injection of carrageenin and paw volume was measured plethysmometrically at '0' and '3' hours intervals after injection. Cotton pellet granuloma was produced by implantation of 50 + 1 mg sterile cotton in each axilla under ether anaesthesia. The animals were treated with *H. indicum* and *L. aspera* and the standard drugs *viz.*, acetylsalicylic acid and phenylbutazone.

Results: H. indicum and *L. aspera* produced significant anti-inflammatory effect in both acute and subacute models of inflammation. In acute inflammation, *L. aspera* was more effective than acetylsalicylic acid. However in subacute inflammation, these two drugs were found to be less effective than phenylbutazone.

Conclusion: H. indicum and *L. aspera* possess anti-inflammatory effects in both acute and subacute inflammation.

Anti-pyretic Effect of Latex of *Calotropis procera*

Soneera Dewan, Suresh Kumar, Vijay L. Kumar; Indian Journal of Pharmacology 2000; 32: 252.

As the latex of *Calotropis procera* possessing anti-inflammatory and analgesic properties may also exhibit anti-pyretic effect we have carried out this study to test the anti-pyretic effect of latex of *C. procera* in the rat model. The latex was collected from the twigs of *C. procera* growing in the wild and was air-dried under shade. The dry latex (DL) was triturated with gum acacia in water (1:1), filtered and used. Fever was induced in male albino rats weighting 150 g. Freeze-dried Baker's yeast was administered as 20 per cent suspension in 0.9 per cent saline (1gm/kg, s.c.) in the nape. Four hours after administration of yeast, either dose of DL (250 or 500 mg/kg), aspirin (200 mg/kg) or saline were administered orally in 1 ml volume. Body temperature (°F) was measured at 0, 3, 4 and 6 hours through rectal route using a digital thermometer. The basal temperature at 0 h was mean of three consecutive readings. Administration of yeast produced an increase in rectal temperature from 97.32 + 0.19°F which reached to its maximum in 4 h (100.02 + 0.27°F). There was no further rise in temperature at 6 h in the control group and the mean temperature remained at 99.74 + 0.15°F. Administration of DL-250 mg/kg and 500 mg/kg at 4 h produced a significant ($P <0.05$) decline in rectal temperature to 98.50 + 0.29°F and 98.45 + 0.60°F respectively. The anti-pyretic effect was compared with that of aspirin, which was found to be more potent and brought down the temperature to 96.9 + 0.38 °F ($P <0.001$). This study along with our earlier findings, on anti-inflammatory and analgesic effect of DL, suggests that DL has actions similar to aspirin.

Clinical Evaluation of Suranjan Shallaki Yoga in the Management of Amavata (Rheumatoid Arthritis)

L. K. Sharma, G. V. Venkateshwarlu, T. Maheswar, Kiran V. Kale; J Res Ayur Siddha. 2004; XXV (3-4): 33-46.

A clinical study with a combination of Suranjan (*Colchicum luteum*) churna (500 mg. TDS) and Shallaki (*Boswelia serrata*) tablet (500mg. TDS) was carried out in O.P.D and I.P.D of Regional Research Institute (Ayurveda), Nagpur as per the C.C.R.A.S protocol during 1999 to 2004. A total number of 91 cases of Amavata (R.A) have been studied for a period of 6 weeks. Significant clinical improvement was noted as 17 (18.68 per cent) cases had Good Response, 29(31.87 per cent) cases shown Fair response, 21(23.07 per cent) cases shown poor response, 09(9.89 per cent) cases not responded to the treatment and rest are dropouts. The result of the study shows that the herbal formulation can be used as an effective remedy for Amavata (Rheumatoid Arthritis).

Clinical Evaluation of Ashvgandha in the Management of Amavata

T. Bikshapathi, Krishna Kumari; J Res Ayur Siddha. 1999; 20: 46.

Rheumatoid Arthritis is a auto immune disorder. The auto immune disorders are protean group acquired disease in which genetic factors appears to play a role. They have in common widespread immunologic and inflammatory alternations of connective tissue. In ayurveda the disease Amavata bears a nearer to-relation with rheumatoid arthritis. Ashvagandha (*Withania somnifera* Linn. Dunal) which is widely used drug in Indian system of Medicine was selected to assess the therapeutic efficacy in cases of Amavata. It is Tikta, Katu and Madhura in Rasa, Ushna Virya and alleviates Vata and Kapha Dosha. Ashvagandha also possesses analgesic, anti-inflammatory and immunomodulatory properties. This paper deals with aetio-pathogenesis, clinical observations and results in 77 patients of Amavata (Rheumatoid arthritis) in details.

Anti-inflammatory Activity of the Leaf of *Alangium lamarkii* Thwenum

Sangameswaran B, Balakrishnan B, Arul B Jayakar B.; J Res Ayur Siddha. 2004; 25: 79-83.

The effect of ethanolic extract of leaf of *Alangium lamarkii* was investigated in rat to evaluate the anti-inflammatory activity. Carrageenin induced rat paw edema and cotton pellet granuloma methods were employed to test anti-inflammatory activity. The result indicated that the ethanolic extract (250gm/kg) produced significant (P<0.001) anti-inflammatory activity when compared to control.

Analgesic, Anti-inflammatory and Immuno Suppressant Effect of *Strobilanthes heyneanus* Nees. Stem

Ravishankar B, Nair RB,. Sasikala CK; J Res Ayur Siddha. 8: 53-63.

Ninety per cent Ethanol (ETE) and aqueous extracts (AQE) of *S. Heyneanus* Nees. Stem were studied for possible analgesic, anti-inflammatory and immunosuppressant effects using various experimental models. Neither of the extracts could raise the threshold of rail-flick response. However, both produced significant decrease in number of acetic acid induced writhing in mice.

Both ETE and AQE suppressed carrageen in hind paw oedema and cotton pellet granuloma formation in rats. In mice treated with extracts marked suppression of antibody formation against sheep red blood cells was noted. The study indicates that ETE and AQE contain active principles having marked anti-inflammatory and immunosuppressant effects, which may be responsible for the reported clinical efficacy of the medicinal preparation of the plant material in inflammatory conditions.

Anti-asthmatic and Anti-inflammatory Activity of *Ocimum sanctum* Linn.

Singh S, Agarwal SS.; J Res Ayur Siddha, Jul-Sept., 1991.

Antiasthmatic activity of 50 per cent ethanolic extracts of dried and fresh leaves, volatile and fixed oils of *Ocimum sanctum* (O.S) was evaluated against histamine and acetylcholine induced pre-convulsive dyspnoea (PCD) in guinea pigs. The 50 per cent ethanolic extract of fish leave, volatile oil (extracted from fresh leave of O.S) and fixed oil (extracted from the seeds of O.S) significantly protected the guinea pigs against histamine and acetylcholine induced PCD. These extracts also inhibited the hind paw oedema in rats against carrageenan, serotonin, histamine and PGE-e induced inflammation. However, the 50 per cent ethanolic extract of dried leaves did not protect the guinea pigs against histamine-induced PCD.

Clinical Evaluation of *Picrorhiza Kurroa* Royle, Exbenth (Kutki) in the Management of Chronic Obstructive Airway Disease

Bikshapathi T, Tripathi SN, Pandey BL.; J Res Ayur Siddha. 1996; 17: 126-148.

The trial was carried out using crude rhizome power, as monotherapy in high doses. The select patient population also was such that may be expected to be resilient enough to exhibit drug influence reasonably over an evaluation period of two weeks. The trial gives special emphasis on constitutional trials of patients as a coralory to the Tridosha concept of Ayurvedic therapy. Interesting leads were delivered by such an approach regarding the Pharmacology of *Picrorhiza kurroa* Royle, ex Benth. therapy of chronic obstructive airway diseases. The immunomodulatory, anti-allergic, psychoneuro endrocrine and broncho-secretary perspectives of therapeutic benefits to airway disorders of inflammatory and allergic origin were suggested in agreement to earlier indications from the experiment studies

CARDIOPROTECTIVE ACTIVITY

Protective Effect of ABANA®, a Poly-herbal Formulation, on Isoproterenol-Induced Myocardial Infarction in Rats

C. Sheela Sasikumar, C.S. Shyamala Devi; Indian Journal of Pharmacology 2000; 32: 198-201.

Objective: To find out the possible role of lipid peroxidation and glutathione in the pathogenesis of myocardial infarction and the protective role of Abana, a polyherbal drug.

Methods: The effect of Abana pretreatment (75 mg/100 g) for a period of 60 days on isoproterenol (20 mg/100 g s.c. twice at an interval of 24 hrs) induced lipid peroxidation was studied in rats. Marker enzymes levels such as creatine kinase, lactate dehydrogenase, alanine transaminase and aspartate transaminase were assessed in serum and heart homogenate. Glutathione content and lipid peroxide level were also estimated.

Results: In isoproterenol administered rats, a significant decrease was observed in the levels of marker enzymes in the heart with a corresponding increase in their levels in serum. Lipid peroxide level measured in terms of "TBA reactants" increased significantly in serum and heart. In rats pretreated with Abana, the alterations observed in the marker enzymes and lipid peroxide levels were minimum on isoproterenol administration, and the levels were retained at near normal values.

Conclusion: Abana pretreatment may offer protection in experimental myocardial infarction induced by isoproterenolol.

Cardiac Stimulant Activity of *Ocimum basilicum* Linn. Extracts

Muralidharan A, Dhananjayan R.; Indian Journal of Pharmacology, 2004; 36: 163-166.

Objective: To evaluate the cardiac effects of extracts derived from the aerial parts of *Ocimum basilicum* Linn.

Material and Methods: The aerial parts of *Ocimum basilicum* Linn. were extracted with 95 per cent ethanol and double distilled water. The extracts were screened for their effects on frog-heart *in situ* preparation. Enzyme studies such as Na^{+}/K^{+} ATPase, Ca^{2+} ATPase and Mg2+ ATPase were done on the heart tissue aspartate transaminase (AST), alanine transaminase (ALT), lactate dehydrogenase (LDH) and creatine phosphokinase (CPK) were estimated in the heart tissue and serum of albino rats after administering the extracts for 7days.

Results: The alcoholic extract produced significant positive ionotropic and negative chronotropic actions on frog heart. The positive ionotropic effect was selectively inhibited by nifedipine. A significant decrease in membrane Na^{+}/K^{+} ATPase, Mg^{2+} ATPase and an increase in Ca^{2+} ATPase pointed the basis for the cardiotonic effect. The aqueous extract produced positive chronotropic and positive ionotropic effects which were antagonized by propranolol indicating that these might have been mediated through ß-adrenergic receptors. Nifedipine also blocks the action of the aqueous extract.

Conclusion: The alcoholic extract exhibited a cardiotonic effect and the aqueous extract produced a ß-adrenergic effect.

Effect of Crude Aqueous Leaf Extract of *Viscum album* (Mistletoe) in Hypertensive Rats

Ofem OE, Eno AE, Imoru J, Nkanu E, Unoh F, Ibu JO.; Indian Journal of Pharmacology. 2007; 39: 15-19.

Objectives: To study the effect of the crude aqueous extract from *Viscum album* (mistletoe) leaves on arterial blood pressure (BP) and heart rate (HR) in albino Wistar rats under pentobarbitone anesthesia.

Materials and Methods: About 42 male rats (130-150 g) were randomly divided into three batches, as normotensives (NMT, n=18), renal artery-occluded hypertensives, (ROH, n=12), salt-induced hypertensives, (SIH, n=12), The normotensives were further divided into three groups, untreated (control), sham-operated and extract-treated subgroups (n=6 per subgroup) while the ROH and SIH groups were also divided into the treated and untreated subgroups of six rats each. The extract (150 mg/kg) was administered via the oral route, once daily for six weeks. Propranolol (0.5 mg/kg i.v.), atropine (1.5 mg/kg i.v.) and noradrenaline (1.0 mg/kg i.v.) were also administered to elucidate the probable mechanism of action of the extract.

Results: The results showed that the control MAP and HR in the normotensives were 97.50±3.20 mmHg and 440.00±12.60 beats/min, respectively. The crude extract produced a significant decrease in BP *i.e.*, 11.28, 23.98 and 18.80 per cent in the NMT, ROH and SIH treated subgroups. The depression produced by the extract on the corresponding HR was not significant in the normotensive, ROH or SIH subgroups. Propranolol blocked the action of the extract on BP. However, atropine did not prevent the extract-induced depression of BP. The extract blocked noradrenaline-induced increase in BP in the NMT.

Conclusion: Our data suggest that the mistletoe extract produces antihypertensive effect without alteration in HR, possibly involving sympathetic mechanism.

The Effect of *Allium sativum* on Ischemic Preconditioning and Ischemia Reperfusion Induced Cardiac Injury

Bhatti R, Singh K, Ishar MP, Singh J. Indian Journal of Pharmacology. 2008; 40: 261-5.

In the present study, the effect of garlic (*Allium sativum*) extract on ischemic preconditioning and ischemia-reperfusion induced cardiac injury has been studied. Hearts from adult albino rats of Wistar strain were isolated and immediately mounted on Langendorff's apparatus for retrograde perfusion. After 15 minutes of stabilization, the hearts were subjected to four episodes of 5 min ischemia, interspersed with 5 min reperfusion (to complete the protocol of ischemic preconditioning), 30 min global ischemia, followed by 120 min of reperfusion. In the control and treated groups, respective interventions were given instead of ischemic preconditioning. The magnitude of cardiac injury was quantified by measuring Lactate Dehydrogenase and creatine kinase concentration in the coronary effluent and myocardial infarct size by macroscopic volume method. Our study demonstrates that garlic extract exaggerates the cardio protection offered by ischemic preconditioning and per se treatment with garlic extract also protects the myocardium against ischemia reperfusion induced cardiac injury.

Apocynin Improves Endothelial Function and Prevents the Development of Hypertension in Fructose Fed Rat

Unger BS, Patil BM.; Indian Journal of Pharmacology. 2009; 41: 208-12.

Background and *Objectives:* Exaggerated production of superoxide and inactivation of nitric oxide have been implicated in pathogenesis of hypertension. NAD(P)H oxidase is one of the major source of reactive oxygen species in vasculature. In the present study, we aimed to determine the effect of chronic administration of

Apocynin an NAD(P)H oxidase inhibitor on endothelial function and hypertension in fructose-fed rat.

Materials and Methods: Endothelial function, vascular superoxide, and nitric oxide production/ bioavailability in aortas from fructose-fed rats and age-matched controls treated with or without apocynin were assessed using isometric tension studies in organ chambers. Systolic blood pressure was measured by the tail cuff method.

Results: In fructose-fed rats, acetylcholine-induced relaxation was impaired, vascular superoxide production was increased, and nitric oxide bioavailability was decreased along with an increase in systolic blood pressure compared to controls. Apocynin treatment prevented the increased generation of superoxide, decreased nitric oxide bioavailability, impaired acetylcholine-induced relaxation, and elevation of systolic blood pressure.

Conclusion: Chronic administration of apocynin improves the endothelial function by reducing oxidative stress, improving NO bioavailability, and prevents the development hypertension in fructose-fed rat.

Effect of Ascorbic Acid Supplementation on Nitric Oxide Metabolites and Systolic Blood Pressure in Rats Exposed to Lead

Mohammad A, Ali N, Reza B, Ali K.; Indian Journal of Pharmacology. 2010; 42: 78-81.

Background: Extended exposure to low levels of lead causes high blood pressure in human and laboratory animals. The mechanism is not completely recognized, but it is relatively implicated with generation of free radicals, oxidant agents such as ROS, and decrease of available nitric oxide (NO). In this study, we have demonstrated the effect of ascorbic acid as an antioxidant on nitric oxide metabolites and systolic blood pressure in rats exposed to low levels of lead.

Materials and Methods: The adult male Wistar rats weighing 200-250 g were divided into four groups: control, lead acetate (receiving 100 ppm lead acetate in drinking water), lead acetate plus ascorbic acid (receiving 100 ppm lead acetate and 1 g/l ascorbic acid in drinking water), and ascorbic acid (receiving 1 g/l ascorbic acid in drinking water) groups. The animals were anesthetized with ketamin/xylazine (50 and 7 mg/kg, respectively, ip) and systolic blood pressure was then measured from the tail of the animals by a sphygmomanometer. Nitric oxide levels in serum were measured indirectly by evaluation of its stable metabolites (total nitrite and nitrate (NOchi)).

Results: After 8 and 12 weeks, systolic blood pressure in the lead acetate group was significantly elevated compared to the control group. Ascorbic acid supplementation could prevent the systolic blood pressure rise in the lead acetate plus ascorbic acid group and there was no significant difference relative to the control group. The serum NOchi levels in lead acetate group significantly decreased in relation to the control group, but this reduction was not significantly different between the lead acetate plus ascorbic acid group and the control group.

Conclusion: Results of this study suggest that ascorbic acid as an antioxidant prevents the lead induced hypertension. This effect may be mediated by inhibition of NOchi oxidation and thereby increasing availability of NO.

Renal Effects of *Mammea africana* Sabine (Guttiferae) Stem Bark Methanol/Methylene Chloride Extract on L-NAME Hypertensive Rats

Nguelefack-Mbuyo EP, Dimo T, Nguelefack TB, Dongmo AB, Kamtchouing P, Kamanyi A.; Indian Journal of Pharmacology. 2010; 42: 208-13.

Objective: The present study aims at evaluating the effects of methanol/methylene chloride extract of the stem bark of *Mammea africana* on the renal function of L-NAME treated rats.

Material and Methods: Normotensive male Wistar rats were divided into five groups respectively treated with distilled water, L-NAME (40 mg/kg/day), L-NAME + L-arginine (100 mg/kg/day), L-NAME + captopril (20 mg/kg/day) or L-NAME + *M. africana* extract (200 mg/ kg/day) for 30 days. Systolic blood pressure was measured before and at the end of treatment. Body weight was measured at the end of each week. Urine was collected 6 and 24 h after the first administration and further on day 15 and 30 of treatment for creatinine, sodium and potassium quantification, while plasma was collected at the end of treatment for the creatinine assay. ANOVA two way followed by Bonferonni or one way followed by Tukey were used for statistical analysis.

Results: M. africana successfully prevented the rise in blood pressure and the acute natriuresis and diuresis induced by L-NAME. When given chronically, the extract produced a sustained antinatriuretic effect, a non-significant increase in urine excretion and reduced the glomerular hyperfiltration induced by L-NAME.

Conclusions: The above results suggest that the methanol/methylene chloride extract of the stem bark of *M. africana* may protect kidney against renal dysfunction and further demonstrate that its antihypertensive effect does not depend on a diuretic or natriuretic activity.

Controlled Clinical Trial of the Lekhaniya Drug Vacha (*Acorus calamus*) in Cases of Ischaemic Heart Diseases

Pratibha Mamgain, R. H. Singh; J Res Ayur Siddha. 1994; 15: 35-51.

Indigenous drug Vaca (*Acorus calamus*) is investigated for it's role in the management of Ischaemic Heart Diseases, in 45 patients suffering from I.H.D. Results obtained were compared with standard established drug Guggulu (*Commiphore mukul*) and placebo. 45 I.H.D. patients were grouped into three sets randomly. To first group Vaca (*Acorus calamus*) Vati was administered internally. Group 2nd received Gugguly (*Commiphora mukul*) vati and IIIrd group received a placebo capsule. Encouraging improvement was noted in 1st and IInd group. These results were found approximately similar.

Individuals at Risk Coronary Heart Disease (CHD) Prevention and Management by an Indigenous Compound

Aruna Agarwal, S. P. Dixit, G. P. Dubey; J RES AYUR SIDDHA. 2001; 22: 228-242.

A variety of risk factors have been suspected for causing the coronary heart disease. 406 cases of both sex groups with age range of 35 to 55years were selected from three distinct localities of Varanasi. Individuals' who reported single or more risk factors of CHD were isolated from the population of the particular areas.

After a detailed preliminary screening of the subjects, various physical, psychologicals and biochemical measurements were carried out. On the basis of initial findings, the cases who showed abnormal lipid profile with dominant psychological involvement were given the organic extract of *Inula racemosa* Hook.F (Pushkarmool), *Commiphora mukul* Hook. ex Stocks Guggulu). *Centella asiatica* Linn. (Mandukaparni) and *Hypericum perforatum* Linn. (Basant) in prescribed doses continuously for 6 months. Correction in the lipid profile including triglycerides, blood pressure and the psychological factors like anxiety and depression to a significant level following test drug treatment indicated the cardio-protective and therapeutic effects of the present formulations.

Hence, by modifying the coronary risk factors, the incidence of CHD can be minimized to a great extent as well as the test formulation may also be advocated in the prevention and management of CHD.

Management of Vyanabala Vaisamya (Essential Hypertension) with Indigenous Drugs: A Comparative Study

Bharathi K, Swamy RK; J Res Ayur Siddha. 2005; 26: 23-24.

Hypertension is becoming more prevalent in the society mainly due to increased incidence of invisible stresses of life. It is one of the several risk factors for the development of cardiovascular disease in the long run. Essential or primary hypertension is the arterial hypertension (high blood pressure) with no definable cause. In Ayurveda, it is not described as such, but, depending on the features involved, it appears to be comparable to the derangement of Vyana vata, hence the name "Vyana bala vaishamya". In this direction the Central Council for Research in Ayurveda and Siddha has been conducting clinical trials at various level with different formulae. In the present study, two formulae are taken up for trial. In the first group, a combination of Arjuna. Vaca, Brahmi, Jatamansi (AVBJ) and in the second group a combination of Chandra prabhavati, Sweta parpati and Punarnava mandura (CPS) have been put on clinical trials. In comparing the results, the second group appears to have a slight margin over the first group as far as the effectiveness is concerned.

Effect of *Terminalia arjuna* W and A on Regression of LVH in Hypertensives: A Clinical Study

B. Chandra Sekhara Rao, R. H. Singh; J Res Ayur Siddha. 2001; 22: 216-227.

Hypertensives presented with left ventricular hypertrophy appear to be associated with many abnormal cardiac events including cardiac arrest. The effect of Arjuna Kwatha along with atenolol on left ventricular hypertrophy in hypertensive individuals was studied by echocardiography. The study infers improvement on regression of left ventricular mass as indicated by change in left ventricular posterior wall thickness (LVPWT), inter ventricular septal thickness (IVST), left ventricular internal dimension (LVID). Decrease in LV mass with Arjuna Kwatha along with stenolol (treated group.) was seen after 3 months and maintained even after a period of 6 monhs. Whereas in those patients whoc were on atenolol (control group), the regression in LV mass seen was insignificant. The patients included under control group the trial was continued for 9 months, but the effect seen after 9 months were not significantly different from those at 3 months. Thus we conclude that *Terminalia arjuna* W. and.A. is able to directly cause regression of increased LV mass.

Management of Raktavata vis-à-vis Arterial Hypertension with Brahmyadi Ghana Vati

Sura Tarangini Rath, Radhakanta Mishra, Braja Kishore Das; J RES AYUR SIDDHA. 1999; XX (1-2), 29-45.

An open case control study comprising of 40 patients suffering from all varieties of essential hypertension having diastolic blood pressure from 96 to 155 mm. Hg. Has been carried out. Cases were dijvided in trial and control groups consisting of 20 patients each. Trial groups patients were treated with Brahmyadi Ghana Vati 500 mg. tablet twice daily and control group patients ere treated with atenolol 50 mg. tablet twice daily. All patients were advised to curtail their intake of salt and fact. The total effect obtained by Brahmyadi Ghana Vati was 30 per cent marked improvement, 30 per cent moderate improvement, 25 per cent slight improvement and 15 per cent no improvement. Analysis shows that the trial drug is significantly effective to reduce hypertension with P value less than 0.001. The total effect obtained by atenolol treatment was 80 per cent mark improvement and 20 per cent moderate improvement.

Vyanabala Vaishamya/Hypertension

Bharati, Adarsh Kumar, Singhal R. K.; J Res Ayur Siddha.

Vyanabala Vaishamya or Hypertension is not described as such in classical Ayurvedic literature. Some Ayurvedic scholars call it Rakta-gata-vata. According to a WHO Expert Committee on Hypertension control met in Geneva (1994), it is the commonest cardiovascular disorders, posing major public health challenge to society. It is one of the major risk factors for cardiovascular mortality, which accounts for 20-50 per cent of all deaths. The control of hypertension requires its diagnosis with pharmacological and non-pharmacological methods of treatment. WHO has been concerned with hypertension since 1950s.

A Clinical Trial on Vyana Bala Vaishamya (Hypertension) by Ayurvedic Drugs

Subal Kumar Maity, Anukul Chandra Kar, M. Mruthyumjay Rao, Arun Kumar Mishra; J RES AYUR SIDDHA. 2000; XXI (1-2), 1-10.

Vyana Bala Vatshamya hypertension is a common health problem occuring in about 2-15 per cent of population in India and many other countriejs. So many Ayurvedic physicians are treating this disease successfully. But to give rationals to the effectiveness of Ayurvedic drugs and to provide statistically viable scientific data on the subject, a blind trial had been conducted during the year 1992-1999 at R.R.I (Ay.). in Calcutta as per the direction of CCRAS, New Delhi. Tagaaradi and Ushiradi Churna was undertaken for clinical trial and it was found that Ushiradi chauna was more effective than Tagaradi churna.

Pharmacological Studies on Fruit Pulp of *Balanites roxburghii* Planch

Rao TS, Kumar S, Sharma JN, Gupta SK.; J Res Ayur Siddha. 7: 47-61.

The aqueous solution of ethanolic extract of fruit pulp (pericarp of the fruit) of *Balanites roxburghii* Planch. (BR-A) on intravenous administration produced a triphasic response on the blood pressure of anaesthetized dogs and cats. The triphasic response consisted of an initial brief fall followed by a brief rise and then a prolonged fall in blood pressure. BR-A increased the urinary bladder pressure of anaesthetized female dogs. It exhibited contractile effect on nictitating membrance of anaes thetized cats and frog rectus abdominis muscle, rabbit intestine and guinea-pig ileum *in vitro*. Purgative action of BR-A was seen only in lethal doses in rats and mice. LD_{50} of BR-A in mice was 5.85 g/kg (oral) and 0.33g/kg (ip). The involvement of nicotinic receptors of autonomic ganglia; and skeletal muscle in some of the observed pharmacological effects of BR-A have been discussed.

GASTROPROTECTIVE AND ANTIDIARRHOEAL ACTIVITY

Effect of *Gingko biloba* Extract on Ethanol-Induced Gastric Mucosal Lesions in Rats

Reshma Shetty, K. Vijay Kumar, M.U.R. Naidu, K.S. Ratnakar; Indian Journal of Pharmacology 2000; 32: 313-317.

Objective: To investigate the effect of *Gingko biloba* extract (Gbe) on ethanol induced gastric mucosal lesions in rats.

Methods: Male Wistar rats used in the study were divided into 3 groups and fasted for 24 hours; Group one received oral normal saline and served as control; group two received 1.0ml of 80 per cent ethanol orally and group three received Gbe (300 mg/kg) orally 1 hour before ethanol (80 per cent, 1.0 ml) administration.

Results: Ethanol treated rats showed marked gross mucosal lesions in the stomach and these were characterized by multiple red bands (patches). Histologically, ethanol treated rats showed ulcerated mucosa with marked mucosal haemorrhage and destruction of glandular elements. Pretreatment with Gbe showed protection against ethanol-induced gastric mucosal damage.

Conclusion: Our data suggests that supplementation of Gbe may be useful in preventing the ethanol induced gastric ulcers.

Anti-ulcer Effect of *Shankha bhasma* in Rats: A Preliminary Study

S. Pandit, T.K. Sur, U. Jana, D. Bhattacharyya, P. K. Debnath; Indian Journal of Pharmacology 2000; 32: 378-380.

Objective: To investigate the anti-peptic ulcer effect of *Shankha bhasma* (conch shell ash) in rats.

Methods: Gastric ulcers were induced in rats by indomethacin and cold restraint stress, and the effect of two different doses of Shankha bhasma was studied. The response of the bhasma on ulcer index, lipid peroxidation (thiobarbituric acid reacting substances TBARS) in gastric tissue and serum calcium was determined.

Results: Shankha bhasma caused significant reduction in ulcer index in both the indomethacin and cold restraint models. TBARS of stomach in indomethacin treated rat was also reduced by Shankha bhasma but serum calcium level was not altered.

Conclusion: Shankha bhasma induced dose dependent protection against experimental gastric ulcers.

Investigation of the Biochemical Evidence for the Antiulcerogenic Activity of *Synclisia scabrida*

Obi E, J.K. Emeh, O.E. Orisakwe, O.J. Afonne, N.A. Ilondu, P.U. Agbasi; Indian Journal of Pharmacology 2000; 32: 381-383.

Objective: The antiulcerogenic activity of flavonoid (A) and alkaloid (B) fractions of the S. scabrida and their effect on alkaline phosphatase activity were studied.

Methods: The ethanol extract of *S. scabrida* was subjected to a column and preparative thin layer chromatography to obtain partially pure fractions of A and B which was subjected to antiulcerogenic screening and effects on alkaline phosphatase activity using aspirin and 0.6N NaOH ulcer models.

Results: The results showed that fractions A and B significantly ($P<0.05$) reduced both the ulcer index and alkaline phosphatase activity when compared with aspirin or 0.6N NaOH only.

Conclusion: This study seems to implicate alkaline phosphatase as a biochemical evidence of the antiulceroragenic activity of *S. scabrida*.

The Effects of *Hippophae rhamnoides* L. Extract on Ethanol-Induced Gastric Lesion and Gastric Tissue Glutathione Level in Rats: A Comparative Study with Melatonin and Omeprazole

Halis Süleyman, Mehmet Emin Büyükokuroglu, Mehmet Koruk, Fatith Akçay, Ahmet Kiziltunç, Akçahan Gepdiremen; Indian Journal of Pharmacology 2001; 33: 77-81.

Objective: To investigate and compare the effects of a hexanoic extract obtained from fresh fruit of *Hippophae rhamnoides* L. (HRe-1) melatonin and omeprazole on ethanol-induced gastric ulcer and on the levels of gastric tissue glutathione (GSH).

Methods: Fifty albino Wistar male rats were used. Gastric lesion was produced by ethanol. GSH levels of gastric tissue were determined according to Griffith method.

Results: Mean number of ulcer foci was 12.3+0.8 in ethanol group, 3.1+0.5 in HRe-1 and 4.3+0.67 melatonin groups. Mean ulcer area was 5.4+0.86 mm^2 in HRe-1, 20.5+0.72 mm^2 in omeprazole, 7.0+0.93 mm^2 in melatonin and 29.3+1.32 mm^2 in ethanol groups ($p< 0.001$: ethanol group vs other groups). Gastric tissue GSH levels of HRe-1 and melatonin groups were fairly close to the normal values. Additionally, this level was significantly reduced in omeprazole and ethanol groups. While there was no difference in terms of mean ulcer area and number of ulcer foci, between melatonin and HRe-1 groups, gastric tissue GSH levels were found significantly higher in HRe-1 than in melatonin groups.

Conclusion: HRe-1 has some benefical effects, even more potent than melatonin, on gastric tissue GSH levels and on the prevention of ethanol-induced ulcer formation in rats.

Pharmacological Activity of the Methanolic Extract of *Cassia nigricans* Leaves

F.C. Chidume, K. Gamaniel, S. Amos, P. Akah, O. Obodozie, C. Wambebe; Indian Journal of Pharmacology 2001; 33: 350-356.

Objective: To evaluate the pharmacological activity of the methanolic extract of *C. nigricans* leaves on gastrointestinal smooth muscles, pain and inflammation using laboratory animals.

Methods: The investigations were carried out using the combined effects of cold stress and aspirininduced ulceration, and egg albumin induced hind paw oedema in rats, acetic acid-induced writhing and gastrointestinal transit in mice as well as effects on the isolated rabbit jejunum.

Results: The extract markedly protected rats against cold-stress and aspirin-induced gastric mucosal damage. The extract exhibited significant ($p < 0.05$) anti-inflammatory and anti-nociceptive activities in rats and mice respectively. Furthermore, the extract decreased the amplitude of contraction of the isolated rabbit jejunum and inhibited histamine-induced contractions, but did not affect ACh induced responses. The intraperitoneal LD50 values of the extract were 210+4.5 mg/kg in mice.

Conclusion: The extract shows good analgesic, anti-inflammatory effects and protected rats against gastric mucosal damage. The anti-ulcer activity might be via histaminergic receptor inhibition.

Anti-Ulcer Drugs from Indigenous Sources with Emphasis on *Musa sapientum*, *Tamra bhasma*, *Asparagus racemosus* and *Zingiber officinale*

R.K. Goel, K. Sairam; Indian Journal of Pharmacology 2002; 34: 100-110.

Sula, Parinamasula and Amlapitta are clinical entities recognized by ayurveda, akin to peptic ulcer and functional dyspepsia. Many indigenous drugs have been advocated in ayurveda for treatment of dyspepsia. Our laboratory has been engaged in screening of various indigenous herbal and metallic drugs for their potential use in peptic ulcer diseases, taking lead from Ayurveda and have reported anti-ulcer and ulcer healing properties of *Tectona grandis* (lapachol), *Rhamnus procumbens* (kaempferol), *Rhamnus triquerta* (emodin), *Withania somnifera* (acylsteryl glycoside), Shilajit (fulvic acid and carboxymethoxybiphenyl), *Datura fastuosa* (withafastuosin E), *Fluggea microcarpa* and *Aegle marmelos* (pyrano- and iso- coumarins) *etc.*, along with their mechanism of action. The present article includes the detailed exploration of ulcer protective and healing effects of unripe plantain banana, tambrabhasma and *Asparagus racemosus* on various models of experimental gastroduodenal ulceration and patients with peptic ulcer. Their effects on mucin secretion, mucosal cell shedding, cell proliferation, anti-oxidant activity, glycoproteins, and PG synthesis have been reported. Clinical trials of these drugs for evaluating their potential ulcer healing effects in peptic ulcer patients have been done. Their potential ulcer protective effects both, experimental and clinical seemed to be due to their predominant effects on various mucosal defensive factors rather than on the offensive acid-pepsin secretion. Thus, the above herbal/herbo-mineral drugs do have potential usefulness for treatment of peptic ulcer diseases.

Antiulcer Activity of *Tephrosia purpurea* in Rats

S.S. Deshpande, G.B. Shah, N.S. Parmar; Indian Journal of Pharmacology 2003; 35: 168-172.

Objective: To study the antiulcer activity of aqueous extract of roots of *Tephrosia purpurea* (AETP) using different models of gastric and duodenal ulceration in rats.

Methods: Antiulcer activity of AETP was studied in rats in which gastric ulcers were induced by oral administration of ethanol or 0.6 M HCl or indomethacin or by pyloric ligation and duodenal ulcers were induced by oral administration of cysteamine HCl. AETP was administered in the dose of 1 to 20 mg/kg orally 30 min prior to ulcer induction. The antiulcer activity was assessed by determining and comparing the ulcer index in the test drug group with that of the vehicle control group. Gastric total acid output and pepsin activity were estimated in the pylorus ligated rats. Omeprazole was used as a reference drug.

Results: The ulcer index in the AETP treated animals was found to be significantly less in all the models compared to vehicle control animals. This antiulcer property was more prominent in animals in whom ulcers were induced by HCl, indomethacin and pyloric ligation. Omeprazole (8 mg/kg) produced a significant gastric and duodenal ulcer protection when compared with the control group. The anti-ulcer activity of AETP was however, less than that of omeprazole.

Conclusion: Our results suggest that AETP possesses significant antiulcer property which could be either due to cytoprotective action of the drug or by strengthening of gastric and duodenal mucosa and thus enhancing mucosal defense.

Cytoprotection Mediated Antiulcer Effect of Tea Root Extract

Maity S, Chaudhuri T, Vedasiromoni JR, Ganguly DK.; Indian Journal of Pharmacology 2003; 35: 213-219.

Objective: To study the cytoprotective effect of tea root (*Camellia sinensis* var. *assamica*) extract using ethanol-induced rat gastric ulcer as an experimental model.

Methods: Tea root extract (TRE) was administered intraperitoneally to rats for 10 days at a dose of 10 mg/kg/day. Ulceration was induced in rats by administering 50 per cent ethanol intragastrically. On day 11, the stomach was examined for ulcer by the severity of hemorrhagic erosions in acid secreting glandular mucosa. Total acid and peptic activity were determined in gastric juice using hemoglobin as substrate. Reduce glutathione (GSH) and glutathione peroxidase (GPX) were also estimated from gastric mucosa.

Results: Pretreatment with TRE for 10 days significantly reduced the incidence and severity of gastric erosions induced by ethanol. TRE treatment also favorably altered changes in volume and peptic activity of gastric juice in ethanol-treated animals. Single administration of succimer (60 mg/kg, i.g.), the standard sulfhydryl containing anti-ulcer agent used as a reference drug, was also effective. Reduction of gastric erosions caused by TRE was reversed by 25 mg/kg, i.p. of N-omega-monomethyl-L-arginine methyl ester (L-NMMA). Furthermore, the levels of GSH and GPX were significantly decreased after treatment with ethanol, and this decrease was prevented by TRE pretreatment.

Conclusion: The study provides evidence for possible involvement of both glutathione and nitric oxide in the TRE-mediated cytoprotection against ethanol-induced ulceration.

Effect of the Crude Extracts of *Amoora rohituka* Stem Bark on Gastrointestinal Transit in Mice

Chowdhury R, Rashid RB.; Indian Journal of Pharmacology 2003; 35: 304-307.

Objective: To study the laxative effects of organic extracts of *Amoora rohituka* stem bark and their actions on gastrointestinal transit in mice.

Methods: The animals were fed with crude extract suspensions and examined for hourly laxation for 5 h with the withdrawal of food and water. In GI motility test, the test materials were given intra-peritoneally followed by administration of $BaSO_4$ milk after 15 min. The treated mice were sacrificed, the small intestine was removed and its length was measured from the pyloric sphincter to the ileocaecal junction.

Results: The petroleum ether, dichloromethane and methanol extracts of *A. rohituka* demonstrated good laxative potential at 400, 250 and 400 mg/kg respectively and the data obtained after 1 h of drug administration were statistically significant. The petroleum ether and methanol extracts also showed significant gastrointestinal hypermotility following barium sulphate milk in mice. The data showed dose dependency and were well correlated with the findings of laxative screening.

Conclusion: The crude extracts of *A. rohituka* have laxative principle(s) comparable to those of a stimulant laxative, sennoside B.

Protective Effect of a Polyherbal Drug, Ambrex in Ethanol Induced Gastric Mucosal Lesions in Experimental Rats

Narayan S, Devi RS, Jainu M, Sabitha KE, Shyamala Devi CS.; Indian J Pharmacology, 2004; 36: 34-37.

Objective: To investigate the protective effect of ambrex in ethanol-induced gastric mucosal lesions in rats.

Material and Methods: Ethanol-induced gastric mucosal lesions in male Wistar rats were used to evaluate gastric ulcer protective effect of Ambrex (40 mg/kg/day p.o. for 15 days). The response to Ambrex was assessed from ulcer index, cell proliferation, histopathological changes and alkaline phosphatase (ALP) activity.

Results: Ambrex pretreatment showed protection against ethanol-induced gastric mucosal damage, a significant reduction in the ulcer index and ALP activity, and an increase in the DNA content.

Conclusion: Ambrex offers protection against ethanol-induced gastric ulcers.

Influence of Sodium Curcuminate on Castor Oil-induced Diarrhoea in Rats

Gnanasekar N, Perianayagam JB.; Indian Journal of Pharmacology, 2004; 36: 175-180.

The present study was designed to investigate the mechanism that might account for the anti-inflammatory action of sodium curcuminate by castor oil induced diarrhoea in rats. The sodium salt of curcumin (Loba chemie) was prepared. Albino rats (Wistar) of either sex were divided into six groups of six animals each. Sodium curcuminate was administered (0.1, 0.2, 0.6 and 1 mg/kg) orally to the first four groups. The fifth group received the standard anti-inflammatory agent, indomethacin (10 mg/kg, i.p.) and the sixth group water (vehicle control). One hour after treatment, each animal received 1 ml of castor oil (CDH, Mumbai) orally by gavage and was then observed for defecation. The rats were observed over a 4 h period for the assessment of characteristic diarrhoea droppings in the transparent plastic dishes placed beneath the individual rat cages. Sodium curcuminate dose-dependently inhibited the occurrence of diarrhoea as compared to the vehicle-treated control rats. The highest dose of sodium curcuminate (1 mg/kg) significantly inhibited castor oil-induced diarrhoea with the percentage inhibition of 80.03, which was comparable to that of indomethacin (10 mg/kg). The delay of castor oil-induced diarrhoea and the inhibition of carrageenin induced inflammation by sodium curcuminate may be related to the inhibition of prostaglandin synthesis.

Effects of *Pterocarpus marsupium* on NIDDM-Induced Rat Gastric Ulceration and Mucosal Offensive and Defensive Factors

Joshi MC, Dorababu M, Prabha T, Kumar MM, Goel RK.; Indian Journal of Pharmacology, 2004; 36: 296-302.

Objective: To evaluate the vulnerability of gastric mucosa to ulceration in non-insulin-dependent diabetes mellitus (NIDDM) rats vis-à-vis the protective effects of the methanolic extract of *Pterocarpus marsupium* heartwood (PMS, an antidiabetic herbal plant).

Material and Methods: NIDDM was produced in 5-day-old rat pups by administering streptozotocin (70 mg/kg, i.p). The animals showing blood glucose level > 140 mg/dl after 12 weeks of STZ administration were considered as NIDDM positive rats. The effective hypoglycemic dose of PMS (750 mg/kg/day, p.o.) for 6 days was studied for its gastric ulcer (GU) protective effects against cold restraint stress (CRS), aspirin (ASP), ethanol (EtOH) and pylorus ligation (PL)-induced GU both in normal (NR) and NIDDM rats. To ascertain the mechanism of action, the effects of NIDDM and that of PMS treatment in NIDDM rats on mucosal offensive acid-pepsin, free-radicals (LPO,NO) and defensive mucin secretion, cell shedding, cell proliferation, glycoproteins and antioxidant enzymes (SOD and CAT) were studied.

Results: PMS (750 mg/kg) decreased the blood sugar level both in NR and NIDDM rats. NIDDM rats exhibited an increased propensity to GU, induced by CRS, ASP, EtOH and PL. Though, PMS did not protect the NR rats against GU induced by the above methods it reversed their increased propensity in NIDDM rats. NIDDM PL-rats showed an increase in acid-pepsin secretion, cell shedding and decrease in mucin secretion and mucosal glycoproteins with little effect on cell proliferation. PMS treatment in NIDDM rats reversed the acid-pepsin secretion, enhanced mucin and mucosal glycoproteins and decreased cell shedding without any effect on cell proliferation. NIDDMCRS rats showed a significant increase in LPO and NO and a decrease in SOD and CAT levels, which were, reversed by PMS treatment.

Conclusion: NIDDM increased the propensity to GU by affecting both offensive (increased) and defensive (decreased) mucosal factors. Though PMS, a hypoglycemic agent, did not show any protection against ulceration induced by CRS, ASP, EtOH and PL in normal rats, it protected the mucosa against the same in NIDDM rats by affecting the above mucosal offensive and defensive factors.

Study of the Action of *Cyperus rotundus* Root Decoction on the Adherence and Enterotoxin Production of Diarrhoeagenic *Escherichia coli*

P.G. Daswani, T.J. Birdi, N.H. Antia; Indian Journal of Pharmacology 2001; 33: 116-117.

The action of a decoction of *Cyperus rotundus* (Linn.) (Cyperaceae), Musta known to have antidiarrhoeal activity2 on adherence and enterotoxin production of 2 groups of E.coli *viz.* EPEC and ETEC was studied. C. rotundus decoction did not affect adherence. A significant inhibition in labile toxin production was noted at 24 hours at a 1:2 dilution and at 72 hours at 1:2 and 1:100 dilution. Stable toxin was inhibited at 1:10, 1:100 and 1:1000 dilutions, maximum inhibition seen at 1:1000. Interestingly, an inverse correlation was observed between the stable toxin production and the concentration of the decoction. Here divergent bioassays targeting the important stages in the pathogenesis of bacterial diarrhoea have been used. *C. rotundus* selectively affected bacterial toxin production without affecting adherence or killing. Hence its effectiveness would be most visible against ETEC strains.

Pharmacological Investigation of *Cardiospermum halicacabum* (Linn.) in different Animal Models of Diarrhoea

Rao N V, Prakash K C, Shanta Kumar S M.; Indian Journal of Pharmacology 2006; 38: 346-9.

Objective: To evaluate the antidiarrhoeal activity of whole plant extracts of *Cardiospermum halicacabum* (Linn.) in rats.

Materials and Methods: Petroleum ether (PeCH) and alcoholic (AlCH) extracts of whole plant of *Cardiospermum halicacabum* (Linn.) were prepared, with successive extraction in soxhlet apparatus and aqueous (AqCH) extract, by the maceration process. LD50 studies for all the three extracts were carried out up to the dose limit of 2000 mg/kg in albino mice. One-fifth of the maximum dose of LD_{50} of each extract was selected to study the antidiarrhoeal activity in different experimental models such as castor oil-induced diarrhoea, prostaglandin E2 (PGE2)-induced enteropooling and charcoal meal test in rats.

Results: Preliminary phytochemical studies revealed the presence of sterols, carbohydrates, tannins and triterpenes in the PeCH extract; sterols, saponins, carbohydrates, flavonoids and tannins in the AlCH extract; sterols, saponins, carbohydrates, flavonoids and tannins in the AqCH extract. No mortality was observed with any of the three extracts up to the maximum dose of 2000 mg/kg. Further, all the three extracts at 400 mg/kg, p.o. had significantly ($P < 0.01$) reduced the fecal output in castor oil-induced diarrhoea, intestinal secretions in PGE2 -induced enteropooling and peristaltic movement in charcoal meal test, indicating antidiarrhoeal activity.

Conclusion: The present study revealed the antidiarrhoeal activity of the extracts of *Cardiospermum halicacabum*, which may be due to the presence of phytochemical constituents such as sterols, tannins, flavonoids and triterpenes.

Activity of Aqueous Ethanol Extract of *Euphorbia prostrata* Ait on *Shigella dysenteriae* Type 1-induced Diarrhoea in Rats

Rene K, Hortense GK, Pascal W, Alexis MNJ, Vidal PE, Archange FTM, Christine FM.; Indian Journal of Pharmacology. 2007; 39: 240-244.

Aim: Euphorbia prostrata (Euphorbiaceae) is traditionally used in Cameroon for the treatment of many diseases, including diarrhoea. We investigated the acute toxicity and effect of the aqueous ethanol extract of the plant on gastrointestinal propulsion, *in vitro* bacterial growth and *in vivo* bacillary dysentery.

Materials and Methods: Diarrhoea was induced by oral administration of 12 x 10^8 Shigella dysenteriae type 1 (Sd1) cells. Diarrhoeic rats were treated for 5 days with 10, 20 or 40 mg/kg extract or 20 mg/kg norfloxacin. The faeces frequencies and the number of Sd1 were assessed and the death rate recorded.

Results: The aqueous ethanol extract of *E. prostrata* was not toxic. *In vitro*, the minimal inhibitory and minimal bactericidal concentrations of the extract were 3,500 and 12,000 µg/ml, respectively. *In vivo*, diarrhoea went along with increase in faeces frequency ($P < 0.01$ by the 3rd day), increase in the bacterial population to a maximum on the 2nd day after infection ($P < 0.01$). The death rate in diarrheic control group was 100 per cent by day 6. *E. prostrata* extracts (20 and 40 mg/kg), like norfloxacin, reduced the bacterial growth ($P < 0.01$), so that by the 6thday Sd1 density was <100 and no death was recorded. There was a significant ($P < 0.01$) reduction in faeces frequencies. The extract exhibited notable ($P < 0.01$) inhibition of intestinal propulsion.

Conclusion: The results suggest that *E. prostrata* possesses bactericidal and antidiarrhoeic properties and could be a therapeutic alternative for diarrhoeas of bacterial etiology.

Antidiarrhoeal and Antimicrobial Activities of *Stachytarpheta jamaicensis* Leaves

Sasidharan S, Yoga Latha L, Zuraini Z, Suryani S, Sangetha S, Shirley L.; Indian Journal of Pharmacology. 2007; 39: 245–248.

Objective: To evaluate the antidiarrhoeal and antimicrobial activity of the extract of *Stachytarpheta jamaicensis* leaves.

Materials and Methods: The methanolic extract of leaves of *S. jamaicensis* was prepared, with successive extraction in soxhlet apparatus with 300 ml of methanol for 24 h. The methanol extract of the leaves of *S. jamaicensis* (250 and 500 mg/kg) was studied for antidiarrhoeal activity using castor oil and magnesium sulphate-induced diarrhoea models in mice. The antimicrobial activity of the extract (10 mg/ml) was determined by disk diffusion method.

Results: At the doses of 250 and 500 mg/kg, the methanol extract showed significant antidiarrhoeal activity ($P < 0.05$). When tested for antibacterial activity, the methanol extract displayed moderate inhibitory activity against *Escherichia coli, Staphylococcus epidermis* and *Pseudomonas aeruginosa*, with an MIC value of 5.00 mg/ml.

Conclusion: On the basis of these findings, it can be assumed that *S. jamaicensis* leaves could be a potential source for novel 'lead' discovery for antidiarrhoeal drug development.

Protective Effect of Palm Vitamin E and α-tocopherol against Gastric Lesions-Induced by Water Immersion Restraint Stress in Sprague-Dawley Rats

Ibrahim IA, Yusof K, Ismail NM, Fahami NA. Indian Journal of Pharmacology. 2008; 40: 73-7.

Objective: Stress can lead to various changes in the gastrointestinal tract of rats. The present study was designed to compare the effect of palm vitamin E (PVE) and α-tocopherol (α-TF) supplementations on the gastric parameters important in maintaining gastric mucosal integrity in rats exposed to water immersion restraint stress (WRS). These parameters include gastric acidity, plasma gastrin level, gastric prostaglandin E(2) (PGE(2)), and gastric lesions.

Materials and Methods: Sixty male Sprague-Dawley rats (200-250 g) were divided into three equal groups: a control group, which received a normal rat diet (RC), and two treatment groups, receiving oral supplementation of either PVE or α-TF at 60 mg/kg body weight for 28 days. Each group was further divided into two groups:

the nonstress and stress groups. The stress groups were subjected to 3.5 h of WRS once at the end of the treatment period. Blood samples were then taken to measure the gastrin level, after which the rats were killed. Gastric juice was collected for measurement of gastric acidity and gastric tissue was taken for measurement of gastric mucosal lesions and PGE(2).

Results: Exposure to stress resulted in the production of gastric lesions. PVE and α-TF lowered the lesion indices as compared to the stress control group. Stress reduced gastric acidity but pretreatment with PVE and α-TF prevented this reduction. The gastrin levels in the stress group were lower as compared to that in the nonstress control. However, following treatment with PVE and α-TF, gastrin levels increased and approached the normal level. There was also a significant reduction in the gastric PGE(2) content with stress exposure, but this reduction was blocked with treatment with both PVE and α-TF.

Conclusion: In conclusion, WRS leads to a reduction in the gastric acidity, gastrin level, and gastric PGE(2) level and there is increased formation of gastric lesions. Supplementation with either PVE or α-TF reduces the formation of gastric lesions, possibly by blocking the changes in the gastric acidity, gastrin, and gastric PGE(2) induced by stress. No significant difference between PVE and α-TF was observed.

Evalution of Anti-ulcer Activity of *Polyalthia longifolia* (Sonn.) Thwaites in Experimental Animals

Malairajan P1, Gopalakrishnan G, Narasimhan S, Veni KJ. Indian Journal of Pharmacology. 2008; 40: 126-8.

Objective: To evaluate the anti-ulcer activity of ethanol extract of leaves of *Polyalthia longifolia* (Sonn.) Thwaites.

Materials and Methods: The ethanol extract of *Polyalthia longifolia* was investigated for its anti-ulcer activity against aspirin plus pylorous ligation induced gastric ulcer in rats, HCl -Ethanol-induced ulcer in mice and water immersion stress induced ulcer in rats at 300 mg/kg body weight p.o.

Results: A significant ($P < 0.01$, $P < 0.001$) anti-ulcer activity was observed in all the models. Pylorous ligation showed significant ($P< 0.01$) reduction in gastric volume, free acidity and ulcer index as compared to control. It also showed 89.71 per cent ulcer inhibition in HCl- Ethanol induced ulcer and 95.3 per cent ulcer protection index in stress induced ulcer.

Conclusion: This present study indicates that *P. longifolia* leaves extract have potential anti-ulcer activity in the three models tested.

Gastroprotective Effect of *Benincasa hispida* Fruit Extract

Rachchh MA, Jain SM.; Indian Journal of Pharmacology. 2008; 40: 271-5.

Objectives: The antiulcer activity of *Benincasa hispida* (Thunb.) Cogn. fruit was evaluated in rats against ethanol-induced gastric mucosal damage, pylorus ligated (PL) gastric ulcers, and cold restraint-stress (CRS)-induced gastric ulcer models.

Methods: Petroleum ether and methanol extracts were administrated orally at the dose of 300 mg/kg, and omeprazole (reference standard) at the dose of 20 mg/kg. Ulcer index was common parameter studied in all the models. Further, vascular permeability was evaluated in ethanol model, and effect on lipid peroxidation, *viz.* melondialdehyde (MDA) content, superoxide dismutase (SOD), and catalase (CAT) levels were studied in CRS model.

Results: Both the extracts produced significant reduction in ulcer index ($P < 0.05$) in all the models and the results were comparable with that of omeprazole-treated group. Further, significant reduction in vascular permeability ($P < 0.05$) was observed. In CRS model, MDA content was significantly reduced along with increase in CAT levels as compared to control group.

Conclusion: Petroleum ether and methanol extracts of *B. hispida* possess significant antiulcer as well as antioxidant property.

Role of Serotonergic Mechanism in Gastric Contractions Induced by Indian Red Scorpion (*Mesobuthus tamulus*) Venom

A.K. Tiwari, M.B. Mandal, S.B. Deshpande.; Indian Journal of Pharmacology. 2009; 41: 255-257.

Aim: Gastric dysfunctions are commonly seen after scorpion envenomation, and the underlying mechanisms are not clear. Therefore, the present study was undertaken to investigate the effect of Indian red scorpion (*Mesobuthus tamulus*, MBT) venom on gastric fundus muscle contraction and the underlying mechanisms involved.

Materials and Methods: In vitro isometric contraction was recorded from gastric fundus muscle strips on a chart recorder. The tissue was exposed to different concentrations of serotonin or crude MBT venom. The contractile responses to venom were expressed as the percentage of maximum contraction produced by serotonin at the beginning of each experiment. The contractile responses to 1.0 μg/ml of crude MBT venom were ascertained in the absence or presence of serotonin antagonist, methysergide.

Results: Serotonin produced concentration-dependent fundus contractions (0.004–4.0 μM), and maximum contractile response was observed at 4.0 μM of serotonin. Hence, the contractile response obtained at 4.0 μM of serotonin was taken for normalization. The crude MBT venom (0.1–1.0 μg/ml) produced a concentration-dependent increase in fundus contractions (as per cent of maximum fundus contraction produced by serotonin at 4.0 μM). The maximum response was observed at 1.0 μg/ml of crude venom and a further increase in the concentration, up to 3.0 μg/ml, did not increase the response. In a separate series of experiments, pre-treatment with methysergide (1.0 μM) significantly attenuated the contractile response elicited by the venom (1.0 μg/ml) ($P<0.05$) and blocked the serotonin (4.0 μM) response.

Conclusion: The results suggest that the crude MBT venom produces gastric fundus contractions by partially involving serotonin.

Antidiarrhoeal Activity of Extracts and Compound from *Trilepisium madagascariense* Stem Bark

Teke GN, Kuiate JR, Kueté V, Teponno RB, Tapondjou LA, Vilarem G.; Indian Journal of Pharmacology. 2010; 42: 157-63.

Objective: The present study was performed to evaluate the preventive and curative antidiarrhoeal effects of the methanol extract, fractions and compound from the stem bark of *Trilepisium madagascariense* in rats.

Materials and Methods: The methanol extract from the stem bark of *T. madagascariense*, its fractions (n-hexane, ethyl acetate, n-butanol and aqueous residue) and compound (obtained from further column chromatography of the ethyl acetate fraction) were evaluated for the antidiarrhoeal activity in rats. These test samples (at 100, 200 and 400 mg/kg for the extract and fractions and 2.5 mg/kg for compound) were assayed on the latent periods, purging indices and fecal frequencies in castor oil-induced diarrhoea. Gastrointestinal transit and castor oil-induced enteropooling assays were conducted. Shigella-induced diarrhoea was assayed. Blood chemistry and fecal Shigella load were examined.

Results: The fractionation of the ethyl acetate fraction from the methanol extract of *T. madagascariense* afforded a known compound [isoliquiritigenin (1)]. Compound 1 increased the latent period of diarrhoea induction (179.40 min) compared to the saline control (60.80 min). The purging indices, fecal frequencies and intestinal enteropooling decreased with an increase in the dose of test samples. The blood cell counts, sera creatinine and fecal Shigella load decreased significantly ($P \leq 0.05$) in the plant extract-treated rats compared to the saline control.

Conclusion: The results of our study, being reported for the first time, provide clear evidence that the methanol extract, fractions and isoliquiritigenin from *T. madagascariense* stem bark possess antidiarrhoeal activities.

Studies on *Dalbergia sissoo* (Roxb.) Leaves: Possible Mechanism(s) of Action in Infectious Diarrhoea

Brijesh S, Daswani P G, Tetali P, Antia N H, Birdi TJ.; Indian Journal of Pharmacology 2006; 38: 120-4.

Objective: Several medicinal plants have been evaluated for their antidiarrhoeal activity. Most studies evaluated their effect on intestinal motility and antimicrobial activity and, therefore, did not take into account the pathogenesis of infectious diarrhoea. Features of infectious diarrhoea like abdominal pain, cramps, inflammation, and passage of blood/mucus in the stools are the combined effect of one or more virulence factors of the infecting organism. The effect of medicinal plants on the microbial virulent features can serve as marker(s) for testing their efficacy. In this study, we evaluated the effect of a decoction of dried leaves of *Dalbergia sissoo* on aspects of pathogenicity, that is, colonisation to intestinal epithelial cells and production/action of enterotoxins. This was done to define its possible mechanism(s) of action in infectious diarrhoea.

Materials and Methods: Antibacterial, antiprotozoal, and antiviral activities of the plant decoction were checked by agar dilution method, tube dilution method, and neutral red uptake assay, respectively. Cholera toxin (CT) and *Escherichia coli* labile toxin (LT) were assayed by ganglioside monosialic acid receptor ELISA. Suckling mouse assay was used to assess *E. coli* stable toxin (ST). As a measure of colonisation, the effect against adherence of *E. coli* and invasion of *E. coli* and *Shigella flexneri* to HEp-2 cells were studied.

Results: The decoction had no antibacterial, antiprotozoal, and antiviral activity. It reduced the production and the binding of CT and bacterial adherence and invasion.

Conclusion: This study showed that *D. sissoo* is antidiarrhoeal as it affects bacterial virulence. However, it has no antimicrobial activity.

Comparative Study of Three Regimen Containing Satavari on Amlapitta (Acid Dyspepsia with or without Ulcer)

T. N. Pande, S. S. Rajagopalan; J RES AYUR SIDDHA. 1993; 15: 23-24.

By adopting Bowalekar's model, an index of Discomfort (I.D) is computed, the assessment being scientifically codified on the basis of symptoms and their severity, in 109 cases of Amlapitta (Acid dyspepsia with or without ulcer) treated with three regimens, all containing Satavari (*Asparagus racemosus*) as an essential constituent. All the three treatments are shown to be equally effective in mitigating the discomfort. Satavari alone having marginally better effect.

Recent Advances in the Management of Amlapitta-Parinama Sula (Non-ulcer dyspepsia and peptic ulcer disease)

K. P. Singh, R. H. Singh; J Res Ayur Siddha. 1985; 6: 132-147.

Amlapitta and Parinama Sula clinically akin to non-ulcer dyspepsia and peptic ulcer disease are vividly described in Ayurveda. Peptic ulcer is a chronic and recurrent disease. Besides high cost and potential toxicity, modern drugs provide a transient and temporary relief. Ayurveda advocates various herbal and herbomineral preparations for these ailments. Plant drugs particularly Amalaki, Satavari, Madhuyasthi and Bhringaraja have given substantial relief to the patients with peptic ulcer. Fruits like Narikela and Kadali both have been fund quite effective. A herbomineral preparation namely Sutasekhara Rasa and Tamra Bhasma appear useful in the management of peptic ulcer. Various fat preparations namely Dadimadya Ghrita, Tikta Ghrita, Indukanta Ghrita and Kusadi Ghrita and even 'Ksira Basti" also cause considerable relief in peptic ulcer patients. Contrary to toxicity of modern drug these Ayurvedic drugs are not safe but also possess additional beneficial effects on general body system.

A Comparison of different Drug Schedules under different Groups of Grahani Roga

Naresh Kumar, Anil Kumar.; J Res Ayur Siddha. 1997; 18: 79-88.

Agni occupies an important place in the maintenance of a healthful life and is undoubtedly the hallmark of the very existence of the living organism. Grahani as a structure is the seat of Pacaka Pitta. As Agni is the pivot of metabolism, any factor that impairs Agni surely and certainly affects Grahani.

The objective of the present study was to compare the efficacy of two different drug schedules in the treatment of Grahani Roga. Thirty cases each were studied under two different groups. The results of the study indicate that none of the cases under study was non responder to the treatment. However, the cases treated under group II were benefitted more as compared to the cases under group I.

Clinical evaluation of Sunthi (*Zingiber officinale*) in the treatment of Grahni Roga

G. C. Nanda, N. S. Tekari, Prem Kishore; J Res Ayur Siddha. 1993; 14: 34-44.

Grahani Roga (Malabsorption Syndrome), a common disease of the people of the area, has been taken for extensive studies. A clinical trial of commonly available drug Sunthi (*Zingiber officinale*) has been put to trial on a series of 111 patients. The observations suggest high incidence in your patients, relatively moderate duration of chronicity and the presence of most of cardinal features of the disease. The effect of the treatment has been obvious within a short period. The regulation of bowel habits, improvement in general health including anemia and body weight and, improvement in the gastrointestinal function has been noted. The observations highlight the importance of simple herbal treatment of the disease

In vitro Action of Selected Medicinal Plants against Microorganisms Involved in Human Gastrointestinal Infections

Pereira ML, Sirsat SM, Antarkar DS, Vaidya AB.; J Res Ayur Siddha. 14: 149-153.

The anti-bacterial activity of the drugs *Berberis aristata, Bombax malabaricum, Gossypium herbaceum,* Holarrhena anti-dysenterica, *Myristica* fragrance and *Punica granatum* has been tested *in vitro* against four species of Shiegella and *E. coli.* The effective of *Punica granatum* has been noted more among all.

Clinical Trial on Satavari (*Asoaragus racemosus* Wild.) in Duodenal Ulcer Disease

Singh KP, Singh RH.; J Res Ayur Siddha. 7: 91-100.

Clinical efficacy of Satavari (*Asparagus racemosus*) was evaluated in 32 patients with proved duodenal ulcer disease. Drop out rate was 18.75 per cent. Root power of Satavari (12.0 g/day in four divided doses) was given for an average duration of six weeks. Criteria of assessment were symptomatic relief, reduction in gastric acidity response (both AHT and FTM) and radiologic as well as endoscopic improvement. Satavari relieved most of the symptoms in majority of the patient promptly and persistently. Satavari did not exhibit antacid activity, foreover, Satavari does not appear to possess strong anti-secretory effect since it inhibited basal output by only 48 per cent, histamine induced maximum output by 38 per cent and alcohol induced secretion by only 32 per cent. Approximately three fourth patients had rediologic and endoscopic improvement. Probably, the ulcer healing effect of drug Satavari appears due to its direct healing effect probably via strengthening the mucosal resistance or cytoprotection.

Role of Amalaki (*Emblica officinalis* Linn.) Rasayan in Experimental Peptic Ulcer

Naheed Banu, V. Patel, J. P. N. Chnsouria, O. P. Malhotra, K. N. Udupa; J Res Ayur Siddha, Jan-Mar, 1982.

The enhanced levels of acctylcholine, histaminasc and decreased level of acctylcholinestcrase are recorded in the duodenal ulcer. These altered biochemical parameters are found reverted towards their control values after pre and post-treatment with an indigenous drug Amalaki rasayana. The study reveals that Amalaki (Officinalis emblica Linn.) has got both prophylactic and curative property and can be used for the treatment of peptic ulcer.

NEUROPROTECTIVE ACTIVITY

Concentration of Amino Acids in Brains of Mice Treated with the Traditional Medicinal Plant *Rhazya stricta* Secne

B.H. Ali, A.A. Al-Qarawi, H. M.Mousa, A.K. Bashir, M.O.M. Tanira, M. Patel, R. Bayoumi; Indian Journal of Pharmacology 2000; 32: 253-254.

Rhazya stricta Decne (family Apocyaceae) is a medicinal plant used traditionally in some Asian countries to treat diabetes, helminthiasis, inflammatory conditions and other diseases. In mice and rats, leaf extracts were found to cause sedation, analgesia, decreased motor activity, anti-depressant-like activity in the forced swimming test, complex effects on brain endogenous monoamine oxidase activity and centrally-mediated hypotension. The biochemical basis of most of these actions is not well known. Here, we measured, apparently for the first time, the concentration of various inhibitory and excitatory amino acids in different brain regions of mice treated acutely and subchronically with the plant extract. The plant leaves were collected from Al-Ain region, United Arab Emirates, and authenticated at The National Herbarium of The UAE University, where voucher specimen is deposited. A lyophilized leaf extract was prepared as described earlier, and given orally to OT mice (n= 7 per group)at doses of 2, 4 and 8 g/kg. The animals were killed 45 min thereafter. Other groups of mice were given the plant extract in the drinking water at concentrations of 0.25, 0.5 and 1.0 per cent for 21 days and killed 24 h after the last dose. The concentrations of amino acids (taurine, aspartic acid, glycine, lysine, gamma amino butyric acid, leucine, methionine, proline, threonine, glutamic acid, serine, glutamine, tryptamine, ornithine, arginine, histidine, phenylalanine, tyrosine, tryptophan) and urea were measured in four brain areas: cerebellum, cerebrum, pons and medulla and the remaining part of the brain. The amino acid concentrations were measured, using ion exchange chromatography on a Beckman 6300 automatic amino acid analyzer (Paolo Ato, USA). The method is based on the post column derivatization by ninhydrin colouring agent at 135°C. Derivitized amino acids were detected at 570 nm, whereas proline and hydroxyproline were detected at 440nm. Preliminary experiments have been conducted to validate the methods of estimation of the amino acids concentration in mice brains. The results of this work indicated that the acute and subchronic treatments did not significantly affect the concentration of any of the amino acids tested. Some of the amino acids measured here have important excitatory or inhibitory functions in the brain. Thus, it may be concluded that the previously reported central actions of *R. stricta* are probably not related to the cerebral concentrations of these amino acids.

Anti-cataleptic, Anti-anxiety and Anti-depressant Activity of Gold Preparations Used in Indian Systems of Medicine

Sonia Bajaj, S.B Vohora; Indian Journal of Pharmacology 2000; 32: 339-346.

Objectives: To study traditional gold preparations for anti-cataleptic, anti-anxiety and anti-depressant effects.

Methods: Swarna Bhasma used in Ayurveda, Kushta Tila Kalan used in Unani-Tibb and Auranofin used in modern medicine were subjected to videopath analyzer, vogel conflict/anxiometer, elevated plus maze, and social behavioural deficit tests for anxiolytic activity, behavioural despair and learned helplessness tests for anti-depressant activity, haloperidol-induced catalepsy tests for neuroleptic activity, and maximum tolerated dose, gross behavioural observations and hematological parameters for safety evaluation in rats and mice.

Results: The test drugs caused significant increase in punished drinking episodes in anxiometer and open arm entries and time in elevated plus maze and decrease in behavioural deficit. A decrease in immobility time in forced swimming test, normalization of shock-induced escape failures in learned helplessness test, and reduction of haloperidol-induced catalepsy scores were also noted in treated animals. The maximum tolerated doses were found to be more than 80 times the effective doses and no weight loss or untoward effects were observed on gross behaviour and hematological parameters.

Conclusions: Traditional gold preparations used in Ayurveda and Unani-Tibb exhibited anxiolytic, anti-depressant and anticataleptic actions with wide margin of safety.

Anticonvulsant Activity of Roots and Rhizomes of *Glycyrrhiza glabra*

Shirish D. Ambawade, Veena S. Kasture, Sanjay B. Kasture; Indian Journal of Pharmacology 2002; 34: 251-255.

Objective: To study the anticonvulsant activity of ethanolic extract of *Glycyrrhiza glabra* in albino rats and mice.

Methods: The anticonvulsant activity of ethanolic extract of roots and rhizomes of *Glycyrrhiza glabra* (10, 30, 100 and 500 mg/kg, i.p.) in mice was assessed using maximum electroshock seizure (MES) test and pentylenetetrazol (PTZ) using albino mice. The lithium-pilocarpine model of status epilepticus was also used to assess the anticonvulsant activity in rats.

Results: The ethanolic extract of G. glabra did not reduce the duration of tonic hindleg extension in the MES test even in the dose of 500 mg/kg. However, the extract significantly and dose-dependently delayed the onset of clonic convulsions induced by pentylenetetrazol. The dose of 100 mg/kg afforded protection to all animals. The extract also protected rats against seizures induced by lithium-pilocarpine.

Conclusion: The ethanolic extract of *G. glabra* inhibits PTZ and lithium-pilocarpine-induced convulsions but not MES-induced convulsions.

Anticonvulsant and Behavioural Actions of *Myristica fragrans* Seeds

Sonavane G.S., Palekar R.C., Kasture V.S., Kasture S.B.; Indian Journal of Pharmacology 2002; 34: 332-338.

Objective: To investigate anticonvulsant, cataleptic and sedative properties of n-hexane fraction of acetone insoluble part of petroleum ether extract of Myristica fragrans (MF) seeds.

Methods: The anticonvulsant activity of MF (10, 30, 100 mg/kg i.p.) was studied against seizures induced by maximum electroshock (MES), pentylenetetrazol (PTZ), picrotoxin, and lithium sulphate-pilocarpine nitrate (Li-Pilo). The effect on gross behaviour, motor coordination, haloperidol-induced catalepsy (using Bar test) and pentobarbitone-induced sleep was studied.

Results: MF inhibited seizures induced by MES, PTZ, and Li-Pilo. However, picrotoxin-induced seizures were not inhibited. The haloperidol-induced catalepsy was potentiated but motor coordination and pentobarbitone-induced sleep were not affected significantly.

Conclusion: MF has complex actions on the central nervous system. Although it exhibited anticonvulsant activity against MES, PTZ and lithium-pilocarpine, it failed to inhibit picrotoxin-induced seizures. MF reduced central dopaminergic activity but was without any effect on pentobarbitone-induced sleep.

Effect of *Tinospora cordifolia* on Learning and Memory in Normal and Memory-Deficit Rats

Agarwal A, Malini S., Bairy KL, Rao MS.; Indian Journal of Pharmacology 2002; 34: 339-349.

Objective: To study the effect of *Tinospora cordifolia* (Tc) on learning and memory in normal and cyclosporine induced memory deficit rats.

Methods: Alcoholic and aqueous extracts of the whole plant of *Tinospora cordifolia* was administered orally for 15 days in two groups of rats. Cyclosporine 15, 25 mg/kg, i.p. was administered on alternate days for 10 days. Combination of cyclosporine 25 mg/kg, i.p. for 10 days and Tc alcoholic 200 mg/kg and Tc aqueous 100 mg/kg were administered in two different groups of rats. At the end of treatment, learning and memory was assessed using Hebb William maze and passive avoidance task. The locomotor activity was assessed using open field chamber. The immune status was studied using DNCB skin sensitivity test. Histopathological examination of hippocampus was done.

Results: Both alcoholic and aqueous extracts of Tc produced a decrease in learning scores in Hebb William maze and retention memory indicating enhancement of learning and memory. However, cyclosporine at both the doses increased the learning scores in Hebb William maze and decrease in retention time in the passive avoidance task suggesting a memory deficit. The combination of cyclosporine and Tc produced a decrease in learning scores in Hebb William maze and increase latency in passive avoidance task compared to cyclosporine alone treated rats. The histopathological examination of hippocampus in cyclosporine treated rats showed neurodegenerative changes which were protected by the Tc.

Conclusion: Tc enhances cognition (learning and memory) in normal rats. Cyclosporine induced memory deficit was successfully overcome by Tc.

Effect of *Bramhi Ghrita*, an Polyherbal Formulation on Learning and Memory Paradigms in Experimental Animals

Achliya G, Barabde U, Wadodkar S, Dorle A.; Indian Journal of Pharmacology, 2004; 36: 159-162.

Objective: To investigate the neuropsychopharmacological effect of a polyherbal formulation Bramhi Ghrita (BG) on learning and memory processes in rats by elevated plus maze, and in mice by Morris water maze model.

Material and Methods: BG contains *Bacopa monneri* (Bramhi), Evolvulus alsinoids, *Acorus calamus, Saussurea lappa* and cow's ghee. Its effect (30, 50 and 100 mg/kg, p.o.) was tested on learning and memory processes. The activity of BG on memory acquisition and retention was studied using elevated plus maze model (EPM) in rats, and on spatial memory using Morris water maze model (MWM) in mice. The alcoholic extract of *Bacopa monneri* (40 mg/kg, p.o.) was also administered to one group of animals. The results were compared with the vehicle-treated group.

Results: Administration of Bramhi Ghrita (50 and 100 mg/kg, p.o.) showed significant reduction in transfer latency in EPM and escape latency in MWM as compared with the control group.

Conclusion: BG may act as a memory enhancer formulation and may also be useful as a supportive adjuvant in the treatment of impaired memory functions.

Possible Anorectic Effect of Methanol Extract of *Benincasa hispida* (Thunb). Cogn, fruit

Kumar A, Vimalavathini R.; Indian Journal of Pharmacology, 2004; 36: 348-350.

Objective: To investigate the anorectic effect of the methanol extract of *Benincasa hispida* (MEBH) in Swiss albino mice.

Material and Methods: Fasted mice were administered with various doses of MEBH (0.2-1 g/kg, i.p.), and the food intake was measured hourly for a period of 7 h. In another experiment, the percentage of gastric emptying at 4th h was determined after the administration of MEBH (0.2-1 g/kg, i.p.) in different set of mice which had free access to preweighed food for either 1, 2 or 4 h.

Results: MEBH significantly reduced the cumulative food intake over a 7 h period in a dose-dependent manner. The percentage reduction of cumulative food intake at 7th h for MEBH with 0.2, 0.6 and 1 g/kg was 27 per cent, 38 per cent and 54 per cent respectively. The 4 h gastric emptying was not significantly influenced by MEBH when compared to control.

Conclusion: The present study reveals for the first time a possible anorectic activity of *Benincasa hispida*, most probably mediated through the CNS without affecting the gastric emptying. However, further studies are required to find its potential as an antiobesity agent.

Evaluation of CNS Activity of *Bramhi Ghrita*

Achliya GS, Wadodkar SG, Dorle AK.; Indian Journal of Pharmacology 2005; 37: 33-6.

To eavaluate the CNS activity of Bramhi Ghrita, a polyherbal formulation containing *Bacopa monneri, Evolvulus alsinoids, Acorus calamus, Saussurea lappa* and cow's ghee. *Materials and Methods:* The effect of Bramhi Ghrita on motor coordination, behaviour, sleep, convulsions, locomotion and analgesia was evaluated in mice using standard procedures.

Results: The formulation exhibited reduced alertness, spontaneous locomotor activity and reactivity. It also antagonized the behavioural effects of d-amphetamine, potentiated the pentobarbitone-induced sleep and increased the pain threshold. Bramhi Ghrita protected mice from maximum electroshock and pentylene tetrazole-induced convulsions.

Investigations of *Sapindus trifoliatus* in Dopaminergic and Serotonergic Systems: Putative Antimigraine Mchanisms

Arulmozhi D K, Veeranjaneyulu A, Bodhankar S L, Arora S K.; Indian Journal of Pharmacology 2005; 37: 120-5.

To evaluate the potential dopaminergic and serotonergic receptor-mediated modulatory effect of the aqueous extract of *Sapindus trifoliatus* [(ST), (family: Sapindaceae)], a traditional phytomedicine used in the treatment of hemicrania (migraine), using animal models and receptor assays.

Materials and Methods: ST (at 20 and 100 mg/kg, i.p. doses) was evaluated for its effect on apomorphine-induced climbing behaviour, 5-hydroxytryptophan (l-5-HTP)-induced serotonin syndrome, and MK-801-induced hyperactivity in mice. The radio ligand-binding studies for various receptors and enzymes were carried out (outsourced) using standard procedures at 250 μg/ml concentration of ST.

Results: ST significantly inhibited the apomorphine-induced climbing behaviour, the l-5-HTP-induced serotonin syndrome and MK-801-induced hyperactivity in mice. In the receptor radioligand-binding studies,

ST exhibited affinity towards dopamine D2, 5-HT2A receptors.

Conclusion: The results of the behavioural studies in mice indicate that ST modulated D2 and 5-HT2A receptor-mediated paradigms. The radio ligand binding studies supported these observations, suggesting the possible involvement of dopaminergic and serotonergic mechanisms in the antimigraine activity of ST.

The Effect of *Hypericum perforatum* Extract against the Neurochemical and Behavioural Changes Induced by 1-methyl-4-phenyl-1,2,3,6-tetrahydropyridine (MPTP) in Mice

Mohanasundari M, Sethupathy S, Sabesan M.; Indian Journal of Pharmacology 2006; 38: 266-70.

Objective: Hypericum perforatum extract (HPE), known for its anti-depressant effect, has been explored in the present study for its protective role against MPTP induced neurotoxicity.

Materials and Methods: Mice were treated with 20 mg/kg of MPTP, four injections i.p., at 2 h intervals within 24 h. HPE was administered at different doses of 100, 200 and 300 mg/kg (p.o) in different groups once a day for seven days and the dose on the first day was given 30 min prior to first MPTP injection. Striatal dopamine (DA) and its metabolites, antioxidant status were analysed. The behavioural changes were studied using the rotarod test, hang test and narrow beam test.

Results: HPE significantly ($P < 0.05$) improved the behavioural activities, striatal neurotransmitter levels and striatal antioxidant status in a dose dependent manner and significantly ($P < 0.05$) reduced TBARS levels.

Conclusion: HPE possesses significant antioxidant activity and renders neuroprotection which was more pronounced at the dose of 300 mg/kg against MPTP induced neurotoxicity.

Effect of Ethanolic Leaf Extract of *Ocimum sanctum* on Haloperidol-Induced Catalepsy in Albino Mice

S Pemminati, V Nair, P Dorababu, HN Gopalakrishna, MRSM Pai; Indian Journal of Pharmacology. 2007; 39: 87-89.

Neuroleptic drugs used in the treatment of schizophrenia and other affective disorders are known to produce extrapyramidal side effects. Catalepsy induced by these drugs in animals has been used as a model for the extrapyramidal side effects associated with antipsychotic agents in human beings. In the present study, we have attempted to evaluate the protective effect of the ethanolic leaf extract of *Ocimum sanctum* (OS) on haloperidol (1.0 mg/kg, intraperitoneal administration)-induced catalepsy in mice by employing the standard bar test. Mice were allocated to seven groups, each group containing six animals. The effects of the test drug OS (at 1.75, 4.25 and 8.5 mg/kg doses) and the standard drugs, scopolamine (1.0 mg/kg) and ondansetron (0.5 and 1.0 mg/kg doses) were assessed after single and repeat dose administration for seven days, 30 minutes prior to the haloperidol. The results suggest that OS has a protective effect against haloperidol-induced catalepsy, which is comparable to the standard drugs used for the same purpose. Our study indicates that OS could be used to prevent drug-induced extrapyramidal side effects.

Evaluation of the Antioxidant Potential of NR-ANX-C (a polyhedral formulation) and its Individual Constituents in Reversing Haloperidol-induced Catalepsy in Mice

Albina Arjuman, Vinod Nair, HN Gopalakrishna, M Nandini.; Indian Journal of Pharmacology. 2007; 39: 151 -154.

Objective: To evaluate the possible role of the antioxidant activity of the polyherbal formulation, NR-ANX-C and its individual components in reversing haloperidol-induced catalepsy in Swiss albino mice.

Materials and Methods : Catalepsy was induced with haloperidol (1 mg/kg *i.p*) in 13 groups of male albino mice ($n = 6$/group). Three groups received NR-ANX-C (10, 25, 50 mg/kg), three groups received *Withania somnifera* (1.7, 4.25, 8.5 mg/kg), another three groups *Ocimum sanctum* (1.7, 4.25, 8.5 mg/kg), three other groups *Camellia sinensis* (3.4, 8.5, 17 mg/kg) and one group received the vehicle (1 per cent *Gum acacia*) orally, 30 minutes prior to haloperidol administration, for a duration of seven days. Animals were sacrificed on the seventh day and superoxide dismutase (SOD) activity was estimated in the brain.

Results: A significant ($P < 0.01$) reduction in the cataleptic scores was observed in all the drug-treated groups as compared to the control, with maximum reduction in the NR-ANX-C 25 mg/kg group. Similarly, a reduction in SOD activity was observed in the NR-ANX-C-, *O. sanctum*- and the *W. somnifera*-treated groups. An increase in SOD activity was observed in the *C. sinensis*-treated groups.

Conclusion: With the exception of *C. sinensis*, the antioxidant potential of NR-ANX-C and its individual constituents has contributed to the reduction in the oxidative stress and the catalepsy induced by haloperidol administration.

Protective Effect of Aqueous Extract of *Embelia ribes* Burm Fruits in Middle Cerebral Artery Occlusion-Induced Focal Cerebral Ischemia in Rats

Bhandari U, Ansari MN.; Indian Journal of Pharmacology. 2008; 40: 215-20.

Objective: The present study was carried out to evaluate the neuroprotective effect of the aqueous extract of *Embelia ribes*, in focal ischemic brain.

Materials and Methods: Adult male Wistar albino rats were fed with the aqueous extract of *Embelia ribes* (100 and 200 mg/kg, p.o.) for 30 days. After 30 days of feeding, all the animals were anaesthetized with chloral hydrate (400 mg/kg, i.p.). The right middle cerebral artery was occluded with a 4-0 suture for 2 h. The suture was removed after 2 h, to allow reperfusion injury. The animals were used for grip strength measurement, biochemical estimation in serum and brain tissue (hippocampus and frontal cortex) and cerebral infarct size measurement.

Results: In the ischemic group, a significant ($P < 0.01$) alteration in the markers of oxidative damage (thiobarbituric acid reactive substances (TBARS); reduced glutathione (GSH); glutathione peroxidase (GPx); glutathione reductase (GR); and, glutathione-S-transferase (GST)) was observed in the hippocampus and frontal cortex, as compared to sham operated rats. We observed that the animals treated with the aqueous extract of *Embelia ribes* had a significant ($P < 0.01$) increase in the poststroke grip-strength activity. Further, supplementation with aqueous extract of *Embelia ribes* reversed the levels/activities of the above mentioned biochemical parameters significantly ($P< 0.01$) and also resulted in decreased cerebral infarct area, as compared to the ischemic group.

Conclusion: The results of our study, for the first time, provide clear evidence that aqueous extract of *Embelia ribes* pretreatment ameliorates cerebral ischemia/reperfusion injury and enhances the antioxidant defense against middle cerebral artery occlusion-induced cerebral infarction in rats; it exhibits neuroprotective property.

Anti-dopaminergic Effect of the Methanolic Extract of *Morus alba* L. Leaves

Yadav AV, Nade VS.; Indian Journal of Pharmacology. 2008; 40: 221-6.

Objective: To evaluate the effect of methanolic extract of *Morus alba* L. leaves on dopaminergic function.

Materials and Methods: The effect of the methanolic extract of *Morus alba* L. leaves was evaluated on haloperidol and metoclopramide induced catalepsy, foot shock-induced aggression, amphetamine-induced stereotyped behaviour and phenobarbitone induced sleeping in mice. In each of these tests, the extract was administered in doses of 50, 100 and 200 mg/kg, i.p., 30 min before performing the test in mice. Further, the inhibitory effect of the extract on dopamine was studied using isolated rat vas deferens.

Results: The extract produced significant dose dependent potentiation of haloperidol (1 mg/kg, i.p.) and metoclopramide (20 mg/kg, i.p.) induced catalepsy in mice. The extract significantly reduced number of fights and increased latency to fights in foot shock-induced aggression; it also decreased amphetamine (1 mg/kg, i.p.) induced stereotyped behaviour in a dose dependent manner. The sleeping time induced by phenobarbitone (50 mg/kg, i.p.) too was prolonged. The extract inhibited contractions produced by dopamine on isolated rat vas deferens.

Conclusion: The results suggest that the methanolic extract of *Morus alba* L. possesses antidopaminergic activity. Further neurochemical investigation can explore the mechanism of action of the plant drug with respect to antidopaminergic functions and help to establish the plant as an antipsychotic agent.

Antiamnesic Effect of Stevioside in Scopolamine-Treated Rats

Sharma D, Puri M, Tiwary AK, Singh N, Jaggi AS.; Indian Journal of Pharmacology. 2010; 42: 164-7.

The present study was undertaken to explore the potential of stevioside in memory dysfunction of rats. Memory impairment was produced by scopolamine (0.5 mg/kg, i.p.) in animals. Morris water maze (MWM) test was employed to assess learning and memory. Brain acetylcholinestrase enzyme (AChE) activity was measured to assess the central cholinergic activity. The levels of brain thiobarbituric acid-reactive species (TBARS) and reduced glutathione (GSH) were estimated to assess the degree of oxidative stress. Scopolamine administration induced significant impairment of learning and memory in rats, as indicated by a marked decrease in MWM performance. Scopolamine administration also produced a significant enhancement of brain AChE activity and brain oxidative stress (increase in TBARS and decrease in GSH) levels. Pretreatment of stevioside (250 mg/kg dose orally) significantly reversed scopolamine-induced learning and memory deficits along with attenuation of scopolamine-induced rise in brain AChE activity and brain oxidative stress levels. It may be concluded that stevioside exerts a memory-preservative effect in cognitive deficits of rats possibly through its multiple actions.

Efficacy Study of *Prunus amygdalus* (Almond) Nuts in Scopolamine-Induced Amnesia in Rats

Kulkarni KS, Kasture SB, Mengi SA.; Indian Journal of Pharmacology. 2010; 42: 168-73.

Objective: Cognitive disorders such as amnesia, attention deficit and Alzheimer's disease are emerging nightmares in the field of medicine because no exact cure exists for them, as existing nootropic agents (piractam, tacrine, metrifonate) have several limitations. The present study was undertaken to investigate the effect of *Prunus amygdalus* (PA) nuts on cognitive functions, total cholesterol levels and cholinesterase (ChE) activity in scopolamine-induced amnesia in rats.

Materials and Methods: The paste of PA nuts was administered orally at three doses (150, 300 and 600 mg/kg) for 7 and 14 consecutive days to the respective groups of rats. Piracetam (200 mg/kg) was used as a standard nootropic agent. Learning and memory parameters were evaluated, using elevated plus maze (EPM), passive avoidance and motor activity paradigms. Brain ChE activity and serum biochemical parameters like total cholesterol, total triglycerides and glucose were evaluated.

Results: It was observed that PA at the above-mentioned doses after 7 and 14 days of administration in the respective groups significantly reversed scopolamine (1 mg/kg i.p.)-induced amnesia, as evidenced by a decrease in the transfer latency in the EPM task and step-down latency in the passive avoidance task. PA reduced the brain ChE activity in rats. PA also exhibited a remarkable cholesterol and triglyceride-lowering property and slight increase in glucose levels in the present study.

Conclusion: As diminished cholinergic transmission and increase in cholesterol levels appear to be responsible for the development of amyloid plaques and dementia in Alzheimer patients, PA may prove to be a useful memory-restorative agent. It would be worthwhile to explore the potential of this plant in the management of Alzheimer's disease.

The Antiepileptic Effect of *Centella asiatica* on the Activities of Na/K, Mg and Ca-ATPases in Rat Brain during Pentylenetetrazol-Induced Epilepsy

G V1, K SP, V L, Rajendra W.; Indian Journal of Pharmacology. 2010; 42: 82-6.

Background: To study the anticonvulsant effect of different extracts of *Centella asiatica* (CA) in male albino rats with reference to $Na^{(+)}/K^{(+)}$, $Mg^{(2+)}$ and $Ca^{(2+)}$-ATPase activities.

Materials and Methods: Male Wistar rats (150+/-25 g b.w.) were divided into seven groups of six each *i.e.* (a) control rats treated with saline, (b) pentylenetetrazol (PTZ)-induced epileptic group (60 mg/kg, i.p.), (c) epileptic group pretreated with n-hexane extract (n-HE), (d) epileptic group pretreated with chloroform extract (CE), (e) epileptic group pretreated with ethyl acetate extract (EAE), (f) epileptic group pretreated with n-butanol extract (n-BE), and (g) epileptic group pretreated with aqueous extract (AE).

Results: The activities of three ATPases were decreased in different regions of brain during PTZ-induced epilepsy and were increased in epileptic rats pretreated with different extracts of CA except AE.

Conclusion: The extracts of *C. asiatica*, except AE, possess anticonvulsant and neuroprotective activity and thus can be used for effective management in treatment of epileptic seizures.

Clinical Evaluation of Dashmool in Various Neurological Disorders

Tripathi K, Upadhyay L.; J Res Ayur Siddha. 1996; 17: 85-90.

Effect of Dashmool (Ten plants roots) on management of sensory and motor disorder pertaining to sympathetic and parasympathetic outflow amongst the patients presenting with primary neurological disorders have been investigated. Significant improvement in nerve conduction velocity was observed. The pattern of "H" reflex also improved in all these patients along with clinical response. Out of 50 patients of nutritional neuropathy 40 patients showed improvement in nerve conduction velocity. Similarly, lumbago sciatica syndrome also showed remarkable improvement.

Three Controlled Clinical Studies on Unmada (Schizophrenia)

Ventataram BS, Mukundan H, Devidas KV.; J Res Ayur Siddha.

For centuries, Ayurvedic practitioners have been using, among others, a number of medicinal herbs, severally or in combination in different internal and external forms of administation in various diseases including Manasa Evam Vatavikara - Psychiatric and Neurological disorders. In order to verify the clinical efficacy of some of these herbs/herbal formulations scientifically, the Ayurvedic Research Unit (ARU) was established by the CCRAS in the NIMHANS - DU, Bangalore, the country's premier institution dedicated to the study and management of these conditions. The unit carried out several pilot as well as controlled studies relating to Schizophrenia (Unmada), Anxiety Neurosis (Chittodwega), Psychogenic Headache (Vataja - shirashoola), Epilepsy (Apasmara), Encephalytis (Abhinyasajwara), Duchenne's Muscular Dystrophy (Mamsavaata), to emntion a few.

Effect of Medhya Rasayana Drug Mandukaparni on Cognitive Functions and Social Adaptability of Mentally Retarded Children

S.C. Agarwal, R. H. Singh; J Res Ayur Siddha. 1997; 18: 97-107.

Mental retardation is a major health hazard in children, prevalent throughout the world. Depending upon the severity of brain domage, the degree of mental retardation varies widely. Besides self-recovery and compensation there is no definite treatment for this condition. The Ayurvedic texts describe mental retardation by the term Jadatva which can be correlated with mental retardation including mental subnormality and related disabilities. Ayurveda describes a series of restorative remedies or Rasayanas specifically for promotion of mental health and cognitive functions. Such Rasayanas are called Medhya Rasayana.

Clinical Evaluation of Medhya Rasayana Effect of Mandukaparni (*Centella asiatica*) a Scientific Study

Sharma AK, Sharma CM, Sharma UK.; J Res Ayur Siddha. 2005; 26: 32-44.

The Comparative study of the administration of mandukaparni Syrup with placebo in 100 subjects of minor disturbance in cerebral higher functions revealed that the drug Mandukaparni (*Centella asiatica*) produced significant improvement in all the factors of Medhya Rasanyana *viz.* Dhee, Dhriti, Smriti, Prabha, Varna, Swara, Dehabala, Indriyabala, Arogya and cerebral higher functions *viz.* attention, memory, calculation, abstract thought, spatial appreciation and visual and body perception. No statistically significant changes/ improvement were observed in most of the parameters of assessment in the subjects treated with placebo (Glucose).

Effect of Medhya Rasayana drug Mandukaparni on cognitive functions and social adaptability of mentally retarded children

Agarwal SC., Singh RH.; J Res Ayur Siddha. 1997; 18: 97-107.

Mental retardation is a major health hazard in children, prevalent throughout the world. Depending upon the severity of brain damage, the degree of mental retardation varies widely. Besides self recovery and compensation there is no definite treatment for this condition. The Ayurvedic texts describe mental retardation by the term Jadatva which can be correlated with mental retardation including mental sub normality and related disabilities. Ayurveda describes a series of restorative remedies or Rasayanas specifically for promotion of mental health and cognitive functions. Such Rasayanas are called Medhya Rasayana.

A Clinical and Experimental Study on Medhya Effect on Aindri (*Bacopa monnieri* Linn.)

Yadava RK, Singh RH.; J Res Ayur Siddha. 1996; 17: 1-15.

In the present study a classical Medhya Rasayana drug Aindri, now identified as *Bacopa monnieri* L. has been selected for the evaluation of its Medhya property in terms of improvement in psychological parameters like anxiety, depression, memory span *etc.* and physiological parameters like pulse, blood pressure, rate of respiration as well as improvement in symptoms along with its effect on the learning behaviour in albino rats. The study has been conducted in two parts *viz.*, clinical and experimental. 36 individuals were selected for the clinical study and they were randomly divided into two groups. Group I consists of 18 apparently normal subject while group II contained 18 patients of Cittodvega vis-a-vis anxiety neurosis. Total extract of the drug Aindri (*B. monnieri* L.) was administered in both the groups in a dose of 1.5 g. Representing 7.5 g of dry crude drug daily for a period of four weeks. The trial treatment produced significant improvement in the level of anxiety and depression, mental fatigue rate (errors and performance) and memory span, along with the positive effects on systolic blood pressure and rate of respiration. The drug also caused significant improvement in symptoms of anxiety *e.g.*, anxiety nervousness, palpitation, headache, insomnia *etc.* In experimental study Aindri showed barbiturate hypnosis potentiation effect with improved learning behaviour in albino rats.

Study to Evaluate the Effect of a Micro (Suksma Medicine Derived from Brahmi (*Herpestris monierra*) on Students of Average Intelligence

Abhang R.; J Res Ayur Siddha. 1993; 14: 10-24.

A double-blind controlled study was carried out to evaluate the effect of a micro (suksma) medicine derived from Brahmi (*Herpestris monierra*) by 9 months treatment on 110 boy students in the age range of 10-13 years and having average I.Q (100). Various factors of intelligence were measured by standardized intelligence tests before and after treatment. Mean difference in post-testing and pre-testing I.Q scores in experimental group were statistically significant as compared to the control group on following I.Q tests : Memory (Direct) test, Arithmetic test and 4 subtests of Budhimapana test. The result is interesting since in previous studies Ayurvedic Medhya Rasayanas were NOT found effective in enhancing intelligence of average students. Long-term studies may prove still more fruitful.

A Clinical Study of Medhya Rasayana Therapy in the Management of Convulsive Disorders

Diwvedi KK, Singh RH.; J Res Ayur Siddha. 1992; 12: 97-106.

The Medhya Rasayana effect appears to consist of general brain tonic. These drugs are traditionally used in different types of mental diseases and other ailments by Ayurvedic physicians. So selection of Medhya Rasayana drugs for therapeutic evaluation in Apasmara was made, Mandukaparni, Sankhapuspi, Guduchi, Yastimadhu, Aswagandha, *etc.* are some of the Medhya drugs which are used in the form of different preparations. The clinical trial over a period of one and half year showed significant improvement in these patients in terms of frequency, duration and severity of seizures. There seems to be a substantial scope of using Medhya Rasayana rugs as safe and moderately effective remedy of indigenous origin for treatment of such patients. The results in relation to prakrit shows that beneficial effect of the treatment were more pronounced in Vataja prakrit.

A study to evaluate the effect of a micro (Suksma) medicine from a Medhya Rasayana on intelligence of mentally retarded children using psychological and biochemical parameters

Abhang R.; J Res Ayur Siddha. 1992; 12: 35-47.

A double-blind controlled study was carried out to evaluate the effect of a micro (suksma) medicine derived from Mandukaparni (Centella asiatica) by 9 months treatment on 30 mentally retarded children of 9-13 years and having I.Q 55-90. Their intelligence and performance was measured by Binet test and Senguin form board test respectively. Some biochemical parameters *viz.* choline esterase, serum proteins, eletrolytes and creatinine were analysed, using blood and urine tests. An analysis of mean differences in post-testing and pre-testing scores of psychological and biochemical parameters revealed that in both the psychological tests. I.Q.s decreased in control group while they increase in experimental group. These mean differences were statistically significant on both I.Q tests but not in respect of any of the bio-chemical parameters. These results interpret that: The medicine may be having its beneficial action (As indicated by I.Q. scores) on Buddhi through Sadhaka Pitta or Indriyaposana dravyas and not through dhatuposana dravyas mainly which were tested through biochemical parameters in the present study.

Comparative Biochemical Studies on the Effect of Four Medhya Rasayana Drugs Described by Charaka on some Cenetral Neurotransmitters in Normal and Stressed Rats

J Res Indian Med Yoga Homoeo. 1979; 14: 7-14.

All Rasayana drugs of Ayurveda are claimed to produce general restorative effects on the body and to improve the mental functions. These drugs are believed to induce immunity and to prevent the occurrence of disease and aaging. Medhya Rasayanas form a special class of Rasayana drugs which are supposed to influence more the mental functions than the body. Some such drugs have been studies in the recent past and have been found to possess varying degrees of anti-anxiety and adaptogenic effects (Singh and Mehta, 1977). The four Medhya Rasayanas described by Caraka are Sankhapushpi, Mandukaparni, Yastimadhu and Gudduchi. Among these, the drug Mandukaparni has been in great controversy with another drug, Brahmi. a pharmacognostic and comparative pharmacological study on Mandukaparni vs Brahmi has indicated that they are similar in action but Brahmi is a relatively more potent drug (Singh and Sinha 1978). Therefore, in the present comparative studies on four Medhya

Rasayanas we included Brahmi (*Bacopa monniera* Linn.) in place of Mandukaparni. These drugs are identified as Sankhapuspi, Mandukaparni, Brahmi, Yastimadhu and Gudduchi.

Some CNS Effects of *Hydrocotyle asiatica* Linn.

Agarwal SS.; J Res Ayur Siddha. 1981; 2.

The alcoholic extract of *Hydrocotyle asiatica* Linn. has been investigated for its neuro-pharmacological actions. The drug when given alone, showed varying degrees of sedation. It showed only 20 per cent mortality even at a dose of 10 g/kg orally. It significantly potentiated the sodium peentobarbitone hypnosis and reduced amphe=toxicity but did not show any protection against metrazol induced or electro-shock seizures. The drug produced hypothermia which was maximum after two hours. It also did not show any protection against trmorine.

ANTI-ANXIETY AND ANTI-DEPRESSANT ACTIVITY

Behavioural Actions of *Myristica fragrans* Seeds

Ganeshchandra Sonavane, Vikram Sarveiya, Veena Kasture, Sanjay B. Kasture; Indian Journal of Pharmacology 2001; 33: 417-424.

Objective: To study behavioural effects of the acetone soluble part of n-hexane extract of *Myristica fragrans* (MF).

Methods: MF (10-100 mg/kg, i.p.) was administered to a group of mice 30min before placing them individually on the elevated plus maze or open field-test apparatus or hole board apparatus. The effect of MF was observed on the duration of occupancy of mice in the open and closed arm of the elevated plus maze and also on the number of entries in both the closed and open arms. The effect of MF was observed on the number of rearing, and the number of squares traversed in the open field apparatus and the number of head poking in the hole board apparatus. The effect of MF was also observed on the pentobarbitone-induced sleep and haloperidol-induced catalepsy. Diazepam and ondansetron were used as reference anxiolytic agents. The analgesic activity of MF was assessed using Eddy's hot plate and acetic acid -induced writhing in mice.

Results: The MF reduced the occupancy in the open arm and increased the occupancy in closed arm. The number of entries was reduced in both the arms. The anxiolytic agents ondansetron, a 5-HT3 receptor antagonist, (0.5 mg/kg, i.p.) and diazepam (1 mg/kg, i.p.) reduced the time spent in the closed arm. These effects of diazepam and ondansetron on the time spent on the closed arm were reversed by MF. MF also reduced the number of rearing and the number of squares traversed in the open field test and number of head poking in the hole board apparatus. The MF increased the duration of pentobarbitone induced sleep and increased the severity of haloperidol-induced catalepsy. The extract reduced the acetic acid induced writhing and increased reaction time when placed on the hot plate. The extract exhibited a wide margin of safety and did not produce mortality in dose upto 3.0 g/kg, i.p.

Conclusion: MF has anxiogenic, sedative and analgesic activities. Though the drugs having anxiogenic activity cannot have any therapeutic use, they can be used as experimental tools to study the anxiolytic activity of other compounds. Further research should be aimed at isolating the active principle responsible for anxiogenic action and the mechanism by which it induces anxiety.

Evaluation of Anti-depressant-like Activity of Glycyrrhizin in Mice

Dhingra D, Sharma A.; Indian Journal of Pharmacology 2005; 37: 390-4.

Objective: To investigate the anti-depressant-like effect of glycyrrhizin (glycyrrhizic acid ammonium) in mice.

Materials and Methods: Glycyrrhizin (1.5, 3.0 and 6.0 mg/kg, i.p.) was administered once daily for seven successive days to separate groups of young male Swiss albino mice. The immobility periods of control and treated mice were recorded in forced swim test (FST) and tail suspension test (TST). Effect of sulpiride (50 mg/kg, i.p.; a selective D2 receptor antagonist), prazosin (62.5 μg/kg, i.p.; an a1-adrenoceptor antagonist) and p-chlorophenylalanine (100 mg/kg, i.p.; an inhibitor of serotonin synthesis) on anti-depressant-like effect of glycyrrhizin in TST was also studied. The anti-depressant-like effect of glycyrrhizin was compared to that of imipramine (15 mg/kg, i.p.) and fluoxetine (20 mg/kg, i.p.) administered for seven successive days.

Results: Glycyrrhizin produced significant anti-depressant-like effect at a dose of 3.0 mg/kg administered for seven successive days, as indicated by reduction in the immobility times of mice in both FST and TST. Glycyrrhizin did not show significant effect on locomotor activity of mice. The efficacy of glycyrrhizin was found to be comparable to that of imipramine and fluoxetine. Sulpiride and prazosin significantly attenuated the glycyrrhizin-induced anti-depressant-like effect in TST. On the other hand, p-chlorophenylalanine did not reverse anti-depressant-like effect of glycyrrhizin. This suggests that the anti-depressant-like effect of glycyrrhizin seems to be mediated by an increase in brain norepinephrine and dopamine, but not by an increase in serotonin.

Conclusion: The results of the present study indicate the involvement of adrenergic and dopaminergic systems in the anti-depressant-like effect of glycyrrhizin.

Effect of *Morus alba* L. (Mulberry) Leaves on Anxiety in Mice

Yadav AV, Kawale LA, Nade VS.; Indian Journal of Pharmacology. 2008; 40: 32-6.

Objective: The aim of the present work is to evaluate the anxiolytic effect of a methanolic extract of *Morus alba* L. leaves in mice.

Materials and Methods: The hole-board test, elevated plus-maze paradigm, open field test, and light/dark paradigm were used to assess the anxiolytic activity of the methanolic extract of *M. alba* L. *Morus alba* extract (50, 100, and 200 mg/kg, i.p.) and diazepam (1 mg/kg, i.p.) were administered 30 min before the tests.

Results: The results showed that the methanolic extract of *M. alba* significantly increased the number and duration of head poking in the hole-board test. In the elevated plus-maze, the extract significantly increased the exploration of the open arm in similar way to that of diazepam. At a dose of 200 mg/kg i.p. the extract significantly increased both the time spent in and the entries into the open arm by mice. Further, in the open field test, the extract significantly increased rearing, assisted rearing, and number of squares traversed, all of which are demonstrations of exploratory behaviour. In the light/dark paradigm, the extract produced significant increase in time spent in the lighted box as compared to vehicle. The spontaneous locomotor activity count, measured using an actophotometer, was significantly decreased in animals pretreated with *M. alba* extract, indicating a remarkable sedative effect of the plant.

Conclusion: The results of the present study suggest that a methanolic extract of *M. alba* leaves may possess an anxiolytic effect.

Effects of *Ocimum sanctum* and *Camellia sinensis* on Stress-Induced Anxiety and Depression in Male Albino *Rattus norvegicus*

Tabassum I, Siddiqui ZN, Rizvi SJ.; Indian Journal of Pharmacology. 2010; 42: 283-8.

Objective: The aim of this study was to study the ameliorative effects of *Ocimum sanctum* and *Camellia sinensis* on stress-induced anxiety and depression.

Materials and Methods: The study was carried out using male albino rats (200 ± 50 g). The effect of *O. sanctum* and *C. sinensis* was evaluated for anxiety and depression using elevated plus maze (EPM) test, open field test (OFT), forced swim test (FST), and tail suspension test (TST).

Results: Restraint stress (3 h/day for six consecutive days) induced a significant reduction in both the percentage number of entries and time spent in open arms in EPM, and these changes were reversed with post-treatment of aqueous extract of *O. sanctum* and *C. sinensis* (100 mg/kg for 6 days). Restraint stress-induced (a) increased latency and (b) decreased ambulation and rearing were also reversed by *O. sanctum* and *C. sinensis* in OFT. A significant increase in immobility period was observed in FST and TST after restraint stress. *O. sanctum* and *C. sinensis* significantly reduced the immobility times of rats in FST and TST.

Conclusion: O. sanctum and *C. sinensis* possess anxiolytic and anti-depressant activities.

Pharmacological Evaluation of the Extracts of *Sphaeranthus indicus* Flowers on Anxiolytic Activity in Mice

Ambavade S D, Mhetre N A, Tate V D, Bodhankar S L.; Indian Journal of Pharmacology 2006; 38: 254-9.

Objective: The objective of the study was to investigate the anxiolytic activity of petroleum ether, alcohol and water extracts, obtained from the flowers of *Sphaeranthus indicus* Linn. in mice.

Materials and Methods: Elevated plus maze (EPM), open field test (OFT) and foot-shock induced aggression (FSIA) were the screening tests used to assess the anxiolytic activity of the extracts on mice. Diazepam (1 mg/kg) served as the standard anxiolytic agent.

Results: The animals receiving extracts or diazepam (1 mg/kg) showed an increase in the time spent, percent entries and total entries in the open arm of the EPM; increased ambulation, activity at centre and total locomotion in the OFT; and decreased fighting bouts in the FSIA, suggesting anxiolytic activity. Petroleum ether extract (10 mg/kg), alcoholic extract (10 mg/kg) and water extract (30 mg/kg) resulted in prominent activity in the mice. Petroleum ether extract (10 mg/kg) resulted in more prominent anxiolytic activity in the EPM and OFT than ethanolic or water extracts, but was less than that produced by diazepam (1 mg/kg).

Conclusion: Petroleum ether extract of *S. indicus* flowers produces prominent anxiolytic activity in mice.

Anti-anxiety Activity of NR-ANX-C, a Polyherbal Preparation in Rats

Gopala Krishna H N, Sangha R B, Misra N, Pai M.; Indian Journal of Pharmacology 2006; 38: 330-5.

Objective: To study the anxiolytic-like activity of NR-ANX-C, a polyherbal product, in rats.

Materials and Methods: Inbred, male, Wistar albino rats weighing between 150 and 180 g were used. The standard anxiolytic, diazepam (0.5 mg and 1 mg/kg), and the test drug, NR-ANX-C powder (5, 10 and 20 mg/kg), were dissolved/suspended in 1 per cent gum acacia solution and administered orally. In acute study the vehicle and the drugs were given sixty minutes prior to the experiment, while in the chronic study they were given twice daily for 10 days with the last dose one hour prior to the experiments (elevated plus maze and light and dark box).

Results: Acute (10 and 20 mg/kg) as well as chronic administration (5,10 and 20 mg/kg) of NR-ANX-C increased the number of entries, the time spent, and the rears in open arms of elevated plus maze model. Similarly, in light/dark box paradigm, at higher doses the test drug increased the time spent (10 and 20 mg/kg) and the number of rears (20 mg/kg) and decreased the duration of immobility (20 mg/kg).On the other hand, chronic administration of all the doses (5, 10 and 20 mg/kg), of the test compound increased the time spent and the number of rears in bright chamber and decreased the duration of immobility. At lower doses (5 and 10 mg/kg), the test compound increased the number of entries into bright chamber. Locomotor activity in the open field test was not affected at all by the doses tested in acute study. On repeated administration, however, the test drug increased the locomotor activity. These changes are similar to those induced by the standard anxiolytic diazepam.

Conclusion: NR-ANX-C exhibited anxiolytic-like activity comparable to that of diazepam.

Behavioural Effects of *Kielmeyera coriacea* Extract in Rats

Martins J, Otobone F J, Sela V R, Obici S, Trombelli M A, Garcia Cortez D A, Audi E A.; Indian Journal of Pharmacology. 2006; 38: 427-8.

Kielmeyera coriacea Mart. is a tree that belongs to the family *Clusiacea*, popularly known in Brazil as "pau santo". *Hypericum perforatum*, a plant from the same family, is considered an effective alternative treatment for moderate depression. Earlier we have reported that chronic administration of the ethanolic stem extract of *Kielmeyera coriacea* in rats reduced immobility time in the forced-swimming test (FST) without altering locomotor activity in the open-field test (OFT), and this effect was mediated through serotoninergic mechanism. The ethanolic extract of *Kielmeyera coriacea* stem was purified by vacuum chromatography on silica gel eluted with hexane and dichloromethane (DcM) to yield a semi-pure DcM fraction with high degree of purity, and the effect of chronic administration of DcM on FST or OFT was investigated in the present study.

The plant collected from Mogi-Guaçu (SP, Brazil) in July 1999 was authenticated by an expert. A voucher specimen (#SP298-463) was deposited with the Herbarium of the State Botanical Institute, SP, Brazil. The extract and the semi-pure DcM fraction (patent application # 001342 with the National Patents Institute (INPI) on October 9, 2002.) were analysed by HPLC-UV and spectroscopy (NMR), to identify probable active substances such as: (1) kielcorin; (2) swertinin; (3) 1, 3, 7-trihydroxy-2-(3-methylbut-2-enyl)-xanthone; (4) 1, 3, 5-trihydroxy-2-(3-methylbut-2-enyl)-xanthone and (5) and (6) mixture of complex triterpenes.

Male *Wistar* rats (55 days old, 240-270 g) were housed in groups of four per cage and maintained on a 12:12 h light:dark cycle (lights on at 7:00 h) in controlled temperature (22±1°C), with food and water freely available, and were acclimated for 3 days before the start of the treatment. All experiments were carried out between 8:00 or 12:00 h. The animals were treated with different doses (4.0, 5.0 and 6.0 mg/kg) of the semi-pure DcM fraction, nortriptyline (both dissolved in saline containing 0.2 per cent Tween 80 used as a vehicle), or vehicle (control group). All treatments were given i.g. for 45 days. The drug doses and treatment period were based on the inferences drawn from pilot studies.

In the FST the animals were placed individually in an open cylindrical container (diameter 30 cm, height 60 cm, containing 45-50 cm of water at 25±1°C) for 15 min (pre-test), followed 24 h later by a 5-min test (on day 44). After 30 s for acclimatization, the test session was videotaped for subsequent measurement of the time of immobility by a trained observer. After 24 h, each animal was placed in the OFT (on day 45). During a 5-min period, the number of squares visited was recorded using Royce's validation criteria. The experimental procedures adopted were approved by the UEM Ethics Committee (# 084-02/COBEA), and follow the norms recommended as international guiding principles for Biomedical Research Involving Animals (CIMS), Geneva, 1985.

The data were analysed using one-way ANOVA followed by Dunnett's test. A value of $P < 0.05$ was considered statistically significant. The results are expressed as mean±SEM. The semi-pure DcM fraction did not induce a consistent and significant decrease in immobility time in the FST when administered by i.g. route in sub-acute (24, 12 and 1h before the test) or chronic periods (15 or 30 days) at doses of 2.0, 4.0, 6.0 or 8.0 mg/kg (results not shown). After 45 days of i.g. treatment, at dose of 6.0 mg/kg, but not at dose of 8.0 mg/kg, DcM fraction produced a significant antiimmobility effect in the FST, without altering the crossings number in the OFT. For this reason, the doses of 4.0, 5.0 and 6.0 mg/kg were selected for our study. We observed that the lower effective dose of DcM fraction to produce antiimmobility effect in the FST ($F_{(4,33)} = 5.074$, $P = 0.0027$) after 45 days treatment was 5.0 mg/kg. The crossings number in the OFT was not altered by different doses used ($F_{(4,33)} = 2.283$, $P = 0.0812$). The inactivity of DcM in dose of 8.0 mg/kg in FST could probably be due to a mixture of unspecific activity or due to active substances present in this fraction but detected in the higher dose used.

The effect of the semi-pure fraction in reducing immobility time in the FST at doses of 5.0 and 6.0 mg/kg is comparable to the *Kielmeyera coriacea* ethanolic extract at a dose of 60.0 mg/kg. These results suggest that the semi-pure DcM fraction possesses an anti-depressant-like drug profile, and contains the component or components responsible for antiimmobility effect detected with the *Kielmeyera coriacea* ethanolic extract.

In accordance with this suggestion, the analysis of the extract and the semi-pure DcM fraction by HPLC-UV revealed the presence of several xanthones (1-4). These xanthones may be responsible for the anti-depressant-like effect detected in the ethanolic extract and in the semi-pure DcM fraction of *Kielmeyera coriacea* in the FST.

In conclusion, the present study shows that the semi-pure DcM fraction is ac tive orally and suggests an anti-depressant-like drug profile. The xanthones may be responsible for the anti-depressant-like action detected in the semi-pure DcM fraction in this study. Further studies are in progress to identify the mechanisms underlying the pharmacological activity observed.

Role of 5-HT$_{1A}$ Receptors in Anti-depressant-like Effect of Dichloromethane Fraction of *Kielmeyera coriacea* in Rats Subjected to the Forced Swim Test

Otobone FJ, Sela VR, Obici S, Moreira LY, Cortez DAG, Audi EA,; Indian Journal of Pharmacology. 2007; 39: 75 -79.

Objective: We examined the involvement of 5-HT neurotransmission on the anti-depressant-like effect of the dichloromethane (DcM) fraction of an extract from *Kielmeyera coriacea* stems.

Materials and Methods: Male Wistar rats treated chronically (45 days, gavage) with the DcM fraction received an intradorsal raphe nucleus (DRN) microinjection of saline or 5-HT$_{1A}$ receptor ligands and were evaluated in the forced swimming test (FST) and in the open-field test (OFT).

Results: The DcM fraction (5.0 mg/kg) reduced immobility time in the FST without altering locomotion in the OFT. IntraDRN microinjection of the 5-HT$_{1A}$ receptor agonist, (+)-8-OH-DPAT (0.10; 0.20 or 0.33 μg) increased immobility time and reduced locomotion at the higher dose whereas the 5-HT1A antagonists, (-)-pindolol (0.10; 0.20 or 0.40 μg) or WAY100635 (0.11; 0.22 or 0.43 μg) did not produce any effect in the behavioural tests. IntraDRN (+)-8-OH-DPAT (0.20 or 0.33 μg) in rats treated with the DcM fraction (5.0 mg/kg) blocked the changes in the immobility time or in locomotion produced by each drug. Intra-DRN (-)-pindolol (0.10 μg) or WAY100635 (0.43 μg) in rats treated with a subactive dose of the DcM fraction (4.0 mg/kg) synergistically reduced immobility time in the FST.

Conclusion: The DcM fraction of *Kielmeyera coriacea* produced an anti-depressant-like effect in the FST and interacted with 5-HT$_{1A}$ receptor ligands. Activation of 5-HT$_{1A}$receptors into DRN by (+) 8-OH-DPAT produced detectable changes in the FST or in the OFT.

Pharmacological and Biochemical Evidence for the Anti-depressant Effect of the Herbal Preparation Trans-01

Shalam Md., Shantakumar SM, Narasu LM.; Indian Journal of Pharmacology. 2007; 39: 231-234.

In this study, Trans-01, a polyherbal formulation, was explored for its anti-depressant properties, using the forced swim test (FST), tail suspension test (TST) and forced swimming stress (FSS)-induced alterations in serum corticosterone levels. For this purpose, the effect different doses of Trans-01 (25, 50, 75 and 100 mg/kg; PO) were studied. Trans-01 was found to safe up to a dose of 5000 mg/kg since no mortality was observed within 48 h of administration. In TST, Trans-01 showed a dose-dependent decrease in immobility time, which is an indication of its anti-depressant effect; this finding was further reinforced in the FST, where a significant effect on immobility was witnessed. However, to explore the possible mechanism of action of Trans-01, the FSS was used to induce corticosterone levels; Trans-01 significantly attenuated the elevated corticosteroid levels. A locomotor activity test was carried out to ascertain whether the anti-depressant effect of Trans-01 included general body stimulation. These results indicate that Trans-01 can be a potential candidate for managing depression. However further studies are required to substantiate the same which are underway in our lab.

A Conceptual and Clinical Study on the Scope of Medhya Rasayana and Vajikarana Therapy in Manas Roga with Special Reference to the Anti-anxiety and Anti-depressant Activity of Certain drugs

Singh RH., Tripathi RK.; J Res Educ Indian Med. 1982; 1(1): 23-28.

The critical examination of the classical Ayurvedic literature indicates a substantial psychological element in the concept of Medhya Rasayana and Vajikarana Therapy of Ayurveda. The Medhya Rasayana effect appears to consist of general brain tonic effect of which, the anti-anxiety effect appears to be an important aspect. Similarly the Vajikara effect appears to refer more to a state of mood elevation than its aphrodisiac effect in the conventional sense. Hence, it is postulated that the Medhya Rasayan and Vajikarana drugs of Ayurveda should be explored as the potential source for developing anti-anxiety and anti-depressant drugs respectively. The clinical and experimental studies done on the anti-anxiety effect of several Medhya Rasayana drugs and on the anti-depressant effect of the Viijikarana drug, Kapikacchu supplement this hypothesis.

Studies on the Anti-anxiety Effect of the Medhya Rasayana Drug Brahmi (*Bacopa monnieri* wettst.) – Part. I.

Singh RH, Singh L.; J Res Ayur Siddha. 1980; 1: 133–148.

Brahmi (*Bacopa monnieri* wettst.) a popular drug described in Indian Medicine as brain tonic *i.e.*, Medhya Rasayana, has been clinically tried in 35 cases of anxiety neurosis. One month treatment with this drug provides significant relief in symptom besides a quantitative reduction in the level of anxiety, (aml) adjustment and disability leading to improved mental functions studies in terms of mental fatigue and immediate memory span. The treatment has also produced reduction in the level of urinary VMA and costicodis. Thus, this drug appears to be an anti-anxiety agent having adaptogenic effect.

Studies on Psychotropic Effect of the Medhya Rasayana Drug, Shankhpushpi (*Convolvulus pluricaulis*) Part. I. Clinical Studies

Singh RH, Mehta AK.; J Res Indian Med Yoga Homoeo 1977; 12: 18-25.

A large number of indigenous drugs have been described as Medhya Rasayana. The Medhya Rasayana drugs are considered to promote the mental functions in addition to their general rejuvenative effect. Sankhapuspi has been considered as the best Medhya Rasayana by Charaka. In view of the possibility that an optimum degree of tranquility of mind may result into an improved mental function and also in view of the previous studies reported about such drugs, it was considered desirable to study the psychotropic effect of this drug clinically as well as experimentally. In clinical studies the drug was tried in patients of anxiety neurosis because in such patients a marked psychoneurotic presentation enables one to demonstrate a more pronounced psychotropic effects of a test drug.

Anti-anxiety Effect of a Classical Ayurvedic Compound, Brahma Rsayana

Ajay Kumar Sharma; J RES AYUR SIDDHA. 2002; XXIII (3-4), 38-48.

Anxiety neurosis is one of the major psychosomatic disorders, commonly seen in clinical practise. The aetiopathogenesis and the clinical presentation of anxiety neurosis as known today, resembles the clinical condition described in Ayurveda as Chittodvega.

The demographic profile was worked out in 30 cases of anxiety neurosis (GAD) by retrospective study while constitutional, clinical, laboratory and therapeutic profile was studied in 15 cases of anxiety neurosis (GAD). The overall results of treatment suggested that Brahma Rasayana is more effective in enhancing the perceptual discrimation and psychomotor performance than the other two control drugs. The data on bio-chemical parameters indicate that Brahma Rasayana offers protection to the stress induced metabollic changes by decreased urinary output of 17-OHCS and VMA. Hence the ayurvedic compound drug Brahma Rasayana can be safely prescribed for fairly longer periods without the fear of physical or psychological dependence, in acute and chronic states of anxiety, in old age with no apprehensions about the age related hazards and in cases presenting both the somatic and psychological symptoms.

Anti-anxiety Effect of an Ayurvedic Compound Drug: A Cross-Over Trial

K. Kuppuranja, C. Seshadri, V. Rajagopalan, Kanchana Srinivasa, R. Sitaraman, Janaki Indurthi, S. Venkataraghavan; J RES AYUR SIDDHA. XIII (3-4),107-116.

The results of a double blind study with a sequential cross-over design, comparing the efficacy of an Ayurvedic preparation with the modern control, *viz.* Diazepam and Placebo in 12 patients of Generalized Anxiety Disorder, are presented in this paper. Findings of the psychological parameters show that the Ayurvedic drug is more effective in enhancing the preceptual discrimination and psycho-motor performance than the other two control drugs. The data on bio-chemical parameters indicate that the Ayurvedic drug offers protection to the stress induced metabolic changes by decreased urinary outputs of 17 OHCS and VMA.

A Clinical Study on Depressive Illness and its Ayurvedic Management

Sheelendra Kumar Gupta; J RES AYUR SIDDHA. 2002; XXIII (3-4), 82-93.

The depressive illness is one of the most common psychiatric disorders, with a trend of rising incidence in recent years throughout the world and is still a serious challenge to the medical profession. The Ayurvedic classics describe depressive illness in different contexts. One can visualize the description of major depression through the classic description of Kaphaja-Unmada. Other forms of depression seem to have been described in brief in terms of Cittavasada. There are a number of anti-depressant drugs available in Modern Medicine with many side effects. In this context it seems fairly essential to develop an ideal ayurvedic management for depressive illness, which may promise a treatment with minimal side effects. The present study represents a short-term time bound project and is based on the demographic and clinical study of 93 patients, besides a clinical tial of a test drug Yogabala Curna (a compound herbal formulation of Kapikacchu, Jyotismati and Vaca) in 65 cases of depressive illness. Yogabala appears to be a safe and effective drug for the treatment of mild to moderate depressive illness comparable to the efficacy of St. John's Wortan European herb with proven anti-depressant action.

Controlled Clinical Trial of Jyotishmati (*Celastrus paniculatus* wild.) in Cases of Depressive Illness

Shekhar Baranwal, Sanjay Gupta, R. H. Singh; J RES AYUR SIDDHA. 2001; XXII (1-2), 35-47.

Mental illness has become the most disturbing health problem in modern industrialized society. Depressive illness is the commonest mental disease to be faced in phychiatry as well as in general medical practice. Because depression is a major health hazard in today's society, attempts have always been made to develop treatment strategies for such patients. A number of chemical drugs have already been introduced for the treatment of depression mostly based on the bio-chemical basis of depression. Although these drugs are therapeutically quite effective but their action is associated with major side effects, including drug dependence and drug resistance. On the other hand a number of Ayurvedic drugs are also in vogue in clinical practice. Jyotishmati is a classical Medhya drug with Ushna and Tikshna property. It is Vata-Kapha-Shamaka. Jyotishmati has been described as Medhya and Smritiprada and is indicated in the treatment of various Manas rogas. With this background a clinical trial was conducted to explore the possible anti-depressant effect of Jyotishmati. The clinical trial of Jyotishmati in 60 cases of depressive illness over a period of six weeks treatment showed significant improvement in these patients in terms of severity of depression, memory and adjustment.

Clinical Study on an Indigenous Drug VACA (*Acorus calamus*) in the Treatment of Depressive Illness

A.K. Tripathi, R. H. Singh; J RES AYUR SIDDHA. 1995; XVI (1-2), 24-34.

Depression is by far the commonest disease to be tackled in psychiatric practice. It is important to point out that depression is not only a pure psychiatric problem but its a general medical practitioner often encounters with depressive illness with other general medical problems. In recent years a number of chemicalpharmacological remedies have been introduced for treatment of depressive illness. Unfortunately, these drugs have very low safety margin and their use is invariably associated with major side effects, including the problems of drug resistance and drug-dependence. Vaca is a classical Lekhana and Madhya drug with Usna and Tiksna property.

It is Vata Kapha Samaka. Hence, in the present study Vaca was clinically tried for its possible anti-depressant effects. The clinical trial of Vaca in fifty cases of depression over a period of six weeks treatment and one and half year follow up showed significant improvement in these patients in terms of the degree of severity of depression and better rehabilitation.

Depressive Illness: A Therapeutic Evaluation with Herbal Drugs

R. H. Singh, S. K. Nath, P. B. Behere; J RES AYUR SIDDHA. XI (1-2) 1-6.

Depressive illness, a global mental health problem at present, constantly thrusts to search suitable drugs for its management. In consequence many modern therapeutic agents have come up in the recent years, but have proved hazardous to the health in terms of addiction and other side effects in long-term use. Two herbal drugs namely Asvagandha (*W. somnifera*) and Kapikacchu (*M. prurients*) were clinically tried in 25 cases of Depressive illness for two months period with encouraging results showing notable symptomatic improvement, decrease in the degree of anxiety and depression. Though it is a preliminary work, but has left a good lead for future study.

Pharmacology of *Vitex negundo* Linn. (Nirgundi) Root

Ravishankar, R. Bhaskaran Nair, C. K. Sasikala; J RES AYUR SIDDHA. VII (1 and 2), 62-77.

Petroleum ether 60-80°C)PE), Chloroform (CHE), n-Butanol (BE), 90 per cent Ethanol (ETE) extracts and cold aqueous infusion (CAI) of the root were studied. PE and BE produced moderate CNS depression, PE prolonged pentobarbijtone sleep, CHE shortened Diazepam sleep, other extracts had no effect. BE and ETE produced marked anti - parkinsonian effect (oxatremorline test), CHE markedly suppressed carrageenin pedal oedema in rats. BE and CAI antagonised acetic acid writhing, ETE significantly raised threshold of tail-flick response. PE protected mice against pentylenectrazol convulsions. None of the extract possesses antipsychotic and anti-depressant effects. They also did not affect SMA and forced locomotor acvitity in mice, and failed to protect mice against electric shock and strychmine induced convulsions. Study reveals that *Vitex negundo* Linn. root extracts possess interesting pharmacological activities and suggests the necessity of isolation of active principles in them.

ANTICANCER AND CYTOTOXIC ACTIVITY

In vitro Cytotoxic Properties of *Grewia tiliaefolia* Bark and Lupeol

Badami S, Vijayan P, Mathew N, Chandrashekhar R, Godavarthi A, Dhanaraj SA, Suresh B.; Indian Journal of Pharmacology. 2003; 35: 250-251.

Several members of the species *Grewia* (Tiliaceae) are being used traditionally for the treatment of a large number of disease conditions. Among them Grewia villosa is reported to possess anticancer activity. *Grewia tiliaefolia* bark is being used in traditional medicine1, and so far no investigation has been carried out for its anticancer property. Three tri-terpenoids, *viz.*, betulin, friedelin and lupeol were isolated from its bark. The *in vitro* cytotoxic properties of its 50 per cent methanolic extract, its fractions and lupeol isolated from it were studied. The fractions isolated were investigated for cytotoxic properties in three cell lines, Vero (a normal African green monkey kidney cell line), HEp-2 (a human larynx epithelial carcinoma cell line; and B16F10 (Mouse melanoma cell line). CTC50 values for lupeol were found to be 196±11.6, 330±16.5 and 302±14.7 mg/ml respectively, for Vero, B16F10 and HEp-2 cell lines. The other samples gave CTC50 values ranging from 98-202 mg/ml in Vero, 460-975 mg/ml in B16F10 and 398-475 mg/ml in HEp-2 cell Lines. The results of cytotoxicity studies and the CTC50 values thus obtained, indicate that the tested samples are cytotoxic only when the cultures are exposed to very high concentrations. High concentrations of any compound, under normal conditions are cytotoxic to cell cultures. Hence, the results showed weak cytotoxic properties of the extracts of *Grewia tiliaefolia* and lupeol.

Antitumour Activity of *Indigofera aspalathoides* on *Ehrlich ascites* Carcinoma in Mice

Rajkapoor B, Jayakar B, Murugesh N.; Indian Journal of Pharmacology, 2004; 36: 38-40.

Objective: To evaluate the antitumor activity of the ethanol extract of Indigofera aspalathoides (EIA) in mice.

Material and Methods: The antitumor activity of EIA was evaluated against the *Ehrlich ascites* carcinoma (EAC) tumor model. The activity was assessed using survival time, peritoneal cell count, hematological studies, solid tumor mass and *in vitro* cytotoxicity.

Results: Oral administration of EIA increased the survival time and normal peritoneal cell count. Hematological parameters, protein and PCV, which were altered by tumour inoculation, were restored. Solid tumour mass was also significantly reduced. EIA was found to be cytotoxic in the *in vitro* model.

Conclusion: EIA possesses significant antitumor activity.

In vitro Cytotoxic Activity of *Lantana camara* Linn.

Raghu C, Ashok G., Dhanaraj SA, Suresh B, Vijayan P.; Indian Journal of Pharmacology, 2004; 36: 93-95.

Lantana camara Linn. is a large evergreen strong-smelling herb, native of tropical America, but now naturalized in many parts of India. The crude methanolic extract of different parts of this plant was studied for its *in vitro* cytotoxic potential. Investigation was taken up against four cancerous cell lines *viz.* HEp-2, B16F10, A-549 and DLA and a normal NRK-49F cell line using standard procedures. Of the five methanol extracts obtained from different parts of *Lantana camara*, the leaf extract exhibited comparatively more cytotoxic activity against all the five cell lines tested. The human lung carcinoma cell line, A-549 was found to be more susceptible with a CTC50 value of 48.1 – 58.5 mg/ml extract. The other four extracts showed less activity as indicated by the relatively high CTC50 values. In the short-term toxicity studies, the methanol extracts of the root with 191.5 ± 5.1 µg/ml and leaf with 219.5 ± 8.4 µg/ml, showed moderate activity against DLA cells after 3 h of exposure. The extracts of the stem, fruit and flowers of Lantana camara, showed less activity with CTC50 values, 268.7 ± 10.2, 492.7 ± 14.4 and > 1000 µg/ml respectively. The results obtained from the present study show that the extract of the leaf of *Lantana camara* is cytotoxic in nature and may possess antitumour activity.

Induction of Cell-Specific Apoptosis and Protection from Dalton's Lymphoma Challenge in Mice by an Active Fraction from *Emilia sonchifolia*

Shylesh BS, Nair SA, Subramoniam A.; Indian Journal of Pharmacology 2005; 37: 232-7.

Objective: To isolate an active anticancer fraction from *Emilia sonchifolia* and to determine the mechanism of its anticancer activity.

Materials and Methods: The anticancer principle was separated using thin layer chromatography (TLC) from the most active n-hexane extract and chemically analysed. The anticancer efficacy of n-hexane extract was determined in mice using Dalton's lymphoma ascitic (DLA) cells. Cytotoxicity of the extracts and isolates to macrophages, thymocytes and DLA cells was measured using Trypan blue exclusion method, MTT (3-[4,5-Dimethylthiazol-2-yl]-2,5-diphenyl tetrazolium bromide) assay, DNA ladder assay and DNA synthesis in culture. Short-term toxicity evaluation of the active fraction was also carried out in mice.

Results: The hexane extract was found to be most active and it showed *in vitro* cytotoxicity to DLA and thymocytes, but not to macrophages. In a concentration and time-dependent manner, it induced membrane blebbing, nuclear condensation, DNA ladder formation, and formation of apoptotic bodies which are characteristic to apoptotic cell death. The n-hexane fraction protected 50 per cent of mice challenged intraperitoneally with 106 DLA cells. This fraction did not exhibit conspicuous adverse toxic symptoms in mice. An active terpene fraction was separated from the n-hexane extract by TLC. This isolate induced apoptotic cell death in DLA cells at 0.8 µg per mL level.

Conclusion: An anticancer terpene fraction was isolated by TLC from *Emilia sonchifolia* that induced cell-specific apoptosis and appears to be a promising anticancer agent.

In vitro Cytotoxic Properties of *Ipomoea aquatica* Leaf

Prasad KN, Ashok G, Raghu C, Shivamurthy GR, Vijayan P, Aradhya SM.; Indian Journal of Pharmacology. 2005; 37: 397-8.

A study carried out to test anticancer activity of Ipomoea bahiensis. Researcher studied the *in vitro* cytotoxic properties of its crude methanolic extract (CME), its column fraction (CF) and purified bioactive compound *i.e.*, 7-O-β-D-glucopyranosyl-dihydroquercetin-3-O-a-D-glucopyranoside (DHQG) isolated from it. Each test sample (CME, CF and DHQG) were separately dissolved in 1 per cent dimethyl sulphoxide (DMSO).The CME, CF and DHQG were investigated for cytotoxic properties against normal and cancer cell lines. DHQG showed cytotoxicity towards cell cultures with CTC50 values of 387 mg/ml against normal Vero cell line, where as 156 and 394 mg/ml, against Hep-2 and A-549 cell lines respectively. The CME and CF gave CTC50 values ranging from 41-332 mg/ml in Vero, 46 - 114 mg/ml in Hep-2 and 44 - 230 mg/ml in A-549 cell lines The results of CTC50 values thus obtained indicate that DHQG showed cytotoxicity towards cancer cell lines tested.

Antitumour Activity of Biflavonoids from Ouratea and Luxemburgia on Human Cancer Cell Lines

de Souza Daniel JF, Cristina C, Fernandes A, Grivicich I, da Rocha AB, de Carvalho MG.; Indian Journal of Pharmacology. 2007; 39: 180 – 183.

The biflavones 7,7"-dimethyllanaraflavone (1), agathisflavone (2), and 7"-methylagathisflavone (3) isolated from the leaves of Ouratea hexasperma and luxenchalcone (4) isolated from the leaves and branches of *Luxemburgia octandra*, as well as a mixture of 7,7"-dimethyllanaraflavone and 7"-methylagathisflavone, were assayed against HT-29 colon adenocarcinoma, NCl-H460 non-small cell lung carcinoma, MCF-7 breast cancer cell, OVCAR-3 ovarian adenocarcinoma cells, and RXF-393 renal cell carcinoma. The results show significant activities, particularly for 7,7"-dimethyllanaraflavone (IC_{50} 0.77 ± 0.08, 2.42 ± 0.22, and 2.59 ± 0.32 mg/ml for NCl-H460, MCF-7, and OVCAR-3, respectively), and for 7"-methylagathisflavone (IC_{50} values of 4 mg/ml). Luxenchalcone revealed significant cytotoxicity on the five cell lines tested.

Anticarcinogenic and Anti-lipidperoxidative Effects of *Tephrosia purpurea* (Linn.) Pers. in 7, 12-dimethylbenz(a)anthracene (DMBA) Induced Hamster Buccal Pouch Carcinoma

Kavitha K, Manoharan S.; Indian Journal of Pharmacology 2006; 38(3): 185-9.

Objectives: To investigate the chemopreventive potential and antilipidperoxidative effects of ethanolic root extract of *Tephrosia purpurea* (Linn.) Pers. (TpEt) on 7,12-dimethylbenz(a)anthracene (DMBA)- induced hamster buccal pouch carcinoma.

Materials and Methods: Oral squamous cell carcinoma was developed in the buccal pouch of Syrian golden hamsters, by painting with 0.5 per cent DMBA in liquid paraffin, thrice a week, for 14 weeks. The tumor incidence, volume and burden were determined. Oral administration of TpEt at a dose of 300 mg/kg, b.w., to DMBA (on alternate days for 14 weeks)- painted animals significantly prevented the incidence, volume and burden of the tumour.

Results: TpEt showed potent antilipidperoxidative effect, as well as enhanced the antioxidant status in DMBA- painted animals.

Conclusion: TpEt has potent chemopreventive efficacy and significant antilipidperoxidative effect, in DMBA-induced oral carcinogenesis. Further studies are needed to isolate and characterize the bioactive principle.

Evaluation of *Cassia occidentalis* for *In vitro* Cytotoxicity against Human Cancer Cell Lines and Antibacterial Activity

Bhagat M, Saxena AK.; Indian Journal of Pharmacology. 2010 Aug; 42(4): 234-7.

Objective: To evaluate the *in vitro* cytotoxicity and antibacterial properties of *Cassia occidentalis* (whole plant) via alcoholic, hydro-alcoholic, and aqueous extracts against eight human cancer cell lines from six different tissues and four bacterial strains.

Material and Methods: In vitro cytotoxicity against the human cancer cells, cultured for 48h in presence of different concentrations *C. occidentalis* extracts and percentage of cell viability, was evaluated using the sulforhodamine-B (SRB) assay. The antibacterial activity was performed using the standard protocol against bacterial strains.

Results: It was observed that aqueous extract of *C. occidentalis* (whole plant) had more potential than hydro-alcoholic and alcoholic extracts against HCT-15, SW-620, PC-3, MCF-7, SiHa, and OVCAR-5 human cancer cell lines at 100, 30, and 10 μg/ml in a dose-dependent manner. The hydro-alcoholic extract showed potential against *Bacillus subtillis*.

Conclusion: The plant can be explored for the possible development of lead molecules for drug discovery.

Anti-cancerous Reagents from some Selected Indian Medicinal Plants I: Screening Studies against Sarcoma 180 ascites

Jain SC, Purohit M.; J Res Ayur Siddha. 8: 70-73.

Ten plant species belonging to different families have been tested against Sarcoma 180 ascites by Total Packed Cell Volume Method. *Trigonella foenum-graecum* Linn. showed some antitumor activity.

Potential Antitumour Activity of *Ixora javanica* Leaf Extract

Nair SC, Panikar B, Panikkar KR; J Res Ayur Siddha, April-June, 1991.

The extract of Ixora javanica leaves was found to prossess anti-cancer activity. 51 chromium release cyto-toxicity assay was used to quantify the extent of cellular membrane damage 72.42±1.3, 61.36±4.21 and 70.05±0.74 cytolysis was observed using Sarcoma - 180 (S-180) Ehrich ascites carcinoma (EAC) and Dalton"s lymphoma ascites (DLA) tumour cells (*in vitro*) in presence of 40, 100 and 60 microgram in one m.1

MEM of the drug. Inhibition in the growth of K 562 (myelogenious) leukemia cells in culture was noted.

ANTIOXIDANT AND ANTI-STRESS ACTIVITY

The Non-dulating Effect of *Garcinia cambogia* Extract on Ethanol Induced Peroxidative Damage in Rats

P. Mahendran, C.S. Shyamala Devi; Indian Journal of Pharmacology 2001; 33: 87-91.

Objective: To determine the modulating effect of *Garcinia cambogia* fruit extract on ethanol induced peroxidative damage in rats.

Method: Male albino rats weighing 125 to 150g were administered ethanol (7.11g per kg body weight/day) for 45 days. Ethanol administered rats were treated concomitantly with *Garcinia cambogia* fruit extract (1g/kg body weight/day) orally for 45 days. After the experimental period the antioxidant enzymes, LPO, conjugated diene in the liver tissue, serum AST, ALT and alkaline phosphatase and lipid levels in both serum and liver tissue were estimated.

Results: Co-treatment of the rats with *Garcinia cambogia* significantly inhibited the rise in lipid levels and also the peroxidative damage caused by ethanol, which is evident from the improved antioxidant status. The levels of serum AST, ALT and alkaline phosphatase were maintained at near normalcy in *Garcinia cambogia* treated rats.

Conclusion: The imbalance in lipid metabolism could be the reason for increase in lipid peroxidation. In our present study the treatment with *Garcinia cambogia* fruit extract resulted in reduction of both serum and liver lipid to near normalcy. This hypolipidemic property of *Garcinia cambogia* in turn reduces the peroxidative damage, enhanced by ethanol.

Effect of Vitamin E, Vitamin C and Spirulina on the Levels of Membrane Bound Enzymes and Lipids in some Organs of Rats Exposed to Lead

C.D. Upasani, R. Balaraman; Indian Journal of Pharmacology. 2001; 33: 185-191.

Objectives: To study the effect of lead alone and its combination with vitamin E, vitamin C and spirulina on the levels of membrane bound enzymes and lipids in some organs of rats.

Methods: Lead acetate (100 ppm) alone and its combinations with vitamin E, vitamin C or spirulina were fed to the rats for thirty days. Na^+-K^+-ATPase, Ca^{++}-ATPase, Mg^{++}-ATPase were estimated in liver and kidney of rats. Similarly, the tissue lipids (Cholesterol, Triglycerides and Phospholipids) were also measured in the liver, lung, heart and kidney of rats.

Results: Lead acetate significantly ($p<0.001$) inhibited the levels of membrane bound enzymes in the liver and kidney of rats. Further, there was a significant ($p<0.001$) increase in the levels of cholesterol, triglyceride and phospholipid in the liver, lung, heart and kidney of animals exposed to lead. Simultaneous administration of vitamin E (50 IU/kg), vitamin C (800 mg/kg) or spirulina (1500 mg/kg) along with lead restored the levels of membrane-bound enzymes as well as the lipids in the animal tissues to normal levels.

Conclusion: It is concluded that vitamin E, C or spirulina had a significant antioxidant activity thereby protecting the organs from the lead-induced toxicity. Lead vitamin E vitamin C spirulina ATPases.

Antioxidant Potential of the *Syzygium aromaticum* (Gaertn.) Linn. (Cloves) in Rats Fed with High-Fat Diet

Shyamala M.P., Venukumar M.R., Latha M.S.; Indian Journal of Pharmacology 2003; 35: 99-103.

Objective: To combat oxidative stress due to hyperlipidemia, administration of moderate quantity of cloves in diet has been suggested. The objective of the present study was to assess the antioxidant efficacy of cloves in rats fed with high-fat diet (HFD).

Methods: In hyperlipidemic rats, oxidative damage was studied by assessing parameters such as thiobarbituric acid reactive substances (TBARS), diene conjugate (CD), superoxide dismutase (SOD), catalase (CAT), glutathione peroxidase (GPx), reduced glutathione (GSH) and glutathione-S-transferase (GST) in liver and kidneys, and also aspartate amino transferase (AST), alanine amino transferase (ALT), urea, triglycerides, and phospholipids in serum/liver/kidneys. The effect of co-administration of cloves on the above parameters was further investigated.

Results: Lipid peroxidation as evidenced by an increment in the values of TBARS, CD, urea, lipid profiles, AST and ALT and also a distinct diminution of levels of GSH in hyperlipidemic rats was found to be nullified by co-administration of cloves as these parameters registered a tendency to retrieve towards near normalcy. Antioxidant enzymes such as SOD, CAT, GPX and GST too showed enhanced activities on co-administration of cloves.

Conclusions: These results substantiate the use of moderate quantity of cloves in diet as an antioxidant in offering protection against hyperlipidemia.

Effect of *Vitex negundo* on Oxidative Stress

Tandon V, Gupta R K.; Indian Journal of Pharmacology 2005; 37: 38-40.

Present study was undertaken to investigate the effect of VN leaf extract on oxidative stress in albino rats. Findings of the present study suggest that VN leaf extract can produce reduction of oxidative stress by reducing lipid peroxidation whereas it has failed to modulate endogenous antioxidant enzyme (SOD) activity. The leaves of VN are known to possess various antioxidant chemical constituents like flavonoids, vitamin C and carotene, which might possibly be responsible for the reduction of lipid peroxidation produced by it in the present study. The limitation of the present study was that no positive control was taken. This is the first report which has indicated that VN can produce reduction of oxidative stress mainly by reducing lipid peroxidation, which needs to be substantiated by a detailed study.

Free Radical Scavenging Activity of *Pfaffia glomerata* (Spreng.) Pederson (Amaranthaceae)

de Souza Daniel J F, Alves K Z, da Silva Jacques D, da Silva e Souza P V, de Carvalho M G, Freire R B, Ferreira D T, Freire M F.; Indian Journal of Pharmacology 2005; 37: 174-8.

To evaluate the free radical scavenging and cytotoxic activities of the butanolic (BuOH) extract, methanolic (MeOH) extract and 20-hydroxyecdysone extracted from the roots of *Pfaffia glomerata*. Fhe free radical scavenging activity of EC was higher than that present in the BuOH fraction. The MeOH extract showed a remarkable pro-oxidant activity. The EC-free radical reaction-inhibition was almost twice of that of the control α-Tocopherol (aT). The Trypan blue exclusion assay confirmed toxicity of the MeOH extract, whose lethality surpassed 80 per cent of the treated macrophages after 1 h of 0.01 mg exposure per 106 cells. The study shows the antioxidant effect of the Brazilian Ginseng. The scavenging effect was evidenced for EC as well the BuOH fraction. The MeOH extract showed cytotoxicity on mice peritoneal macrophages. Such toxicity is probably due to ginsenosides present in this latter fraction and warrants further toxicological evaluation of the Brazilian Ginseng roots.

In vitro Antioxidant Effect of *Globularia alypum* L. Hydromethanolic Extract

Khlifi S, Hachimi Y E, Khalil A, Es-Safi N, Abbouyi A E.; Indian Journal of Pharmacology 2005; 37: 227-31.

To investigate the *in vitro* antioxidant activity of the hydromethanolic extract of aerial parts (leaves and stems) of *Globularia alypum* L. toward linoleic acid emulsion and human low-density lipoproteins (LDL) peroxidation. Lipid peroxidation was carried out in the presence of *G. alypum* hydromethanolic extract (10 and 100 µg of extract/ml). $CuSO_4$ (10 µM) was used as the oxidation initiator. Conjugated dienes (CD) formation and oxygen consumption were assessed for monitoring the antioxidant properties of the plant extract. Butylated hydroxytoluene at 50 µg/ml was used as standard antioxidant. Quantification of total polyphenolic compounds was carried out according to the Folin-Ciocalteu method. Results showed that the hydromethanolic extract of *G. alypum* exhibited significant antioxidant effect. There was a significant inhibition of CD formation in copper ions-mediated linoleic acid emulsion as well as human LDL peroxidation. Analysis of the plant extract revealed a high amount of polyphenols, suggesting a possible role of these compounds in the antioxidant properties.

In vitro Antioxidant Properties of *Solanum pseudocapsicum* Leaf Extracts

Badami S, Prakash O, Dongre S H, Suresh B.; Indian Journal of Pharmacology 2005; 37: 251-2.

The phytochemical tests indicated the presence of alkaloids, glycosides, tannins, and flavonoids in the crude methanolic extract. Several such compounds were known to possess potent antioxidant activity. Some of these constituents have already been isolated from this plant. Hence, the observed antioxidant activity may be due to the presence of any of these constituents. The plant exhibited strong anticancer, hepatoprotective and several other activities. These properties may be due to its antioxidant activity. The crude methanolic extract merits further experiments *in vivo*.

Evaluation of the Antioxidant Activity of *Trianthema portulacastrum* L.

Kumar G, Banu G S, Pandian M R.; Indian Journal of Pharmacology 2005; 37: 331-3

This study was conducted on Male Wistar albino rats (N=24), treatment group was treated with ethanolic extracts of *T. portulacastrum* (100 and 200 mg/kg, p.o) in 50 per cent w/v sucrose for 10 days in single dose. The results show that pretreatment of rats with 100 mg or 200 mg/kg, p.o., of ethanolic extracts of *T. portulacastrum* prevented significantly the paracetamol- and thioacetamide- induced reduction of blood and liver glutathione, liver Na-K-ATPase level. TBARS of toxicants-treated animals were significantly higher than the control animals. Administration of the ethanolic extract markedly decreased the level of TBARS. The degree of protection was more with the higher dose of the extract. In conclusion, the hepatoprotective effect of *T. portulacastrum* alcohol extract against paracetamol- and thioacetamide-induced hepatotoxicity in rats appears to be related to the inhibition of lipid peroxidative processes and to the prevention of GSH depletion.

In vivo Antioxidant Activity of Hydroalcoholic Extract of *Taraxacum officinale* Roots in Rats

Sumanth M, Rana A C.; Indian Journal of Pharmacology 2006; 38(1): 54-5.

In the traditional system of medicine, there are a number of plants which are used in the treatment of liver disorders. Their extract, fractions, and active constituents exhibit marked hepatoprotective action, which has been related to their antioxidant properties. Traditionally, *Taraxacum officinale*, WEBER, family Compositae, commonly known as "Dandelion" has been used as a remedy for jaundice, other disorders of the liver and gall bladder, and to counteract water retention.

Free radicals are reactive molecules involved in many physiological processes and human diseases, such as cancer, ageing, arthritis, Parkinson's syndrome, ischemia and liver injury. The elevation of free radical levels seen during the liver damage is owing to enhanced production of free radicals and decreased scavenging potential of the cells. A variety of intrinsic antioxidants (reduced glutathione, superoxide dismutase (SOD), catalase and peroxidase) are present in the organism, which protect them from oxidative stress, thereby forming the first line of defence. The present study was undertaken to evaluate antioxidant activity of 70 per cent hydroalcoholic extract of roots of *T. officinale* (TO) in rats.

The plant TO was collected from the Gandhi Krishi Vigyan Kendra (GKVK), Bangalore, in the month of December 2002 and was authenticated by Dr Yoganarsimhan, Scientist, Regional Research Centre [Ay], Bangalore. The dry powder of roots of TO was extracted with 70 per cent alcohol in a soxhlet apparatus for 24 h at approximately 60°C. The extract was then concentrated by distilling the solvent below 60°C and dried in a dessicator. A suspension was prepared by using equal volumes of the extract and gum acacia for administration to rats using oral gague. The chemicals used for investigation of antioxidant activity were CCl_4 (Quality Fine Chemicals Ltd. India), Liv-52 (Himalaya, India), hydroxylamine hydrochloride (Sigma, India), nitro-blue tetrazoleum (NBT) (Sigma, India), hydrogen peroxide, EDTA, and Ellman's reagent (Sigma, India). All other chemicals obtained from local sources were of analytical grade.

Albino rats of Wistar strain, weighing 100-150 g, maintained on normal diet (Amrut Laboratory Animal Feeds, Bangalore) and water *ad libitum*, were divided into five groups of six animals each. Before starting the experiment, permission from the Institutional Animal Ethics Committee was obtained.

Group-I animals served as normal control, treated with distilled water. Group-II animals served as hepatotoxic control, treated with CCl_4 in a single dose of 1.5 ml/kg, i.p., to produce acute hepatotoxicity. Group III served as a standard group, and was administered Liv-52 in a dose of 56 mg/kg, p.o. Group-IV and -V animals were treated with daily doses of 50 and 100 mg/kg, p.o., respectively, of TO extract for 7 days. The animals of Groups III-V were given single dose of CCl_4, 1.5 ml/kg, i.p., 6 h after the last treatment. On day 8 the rats were sacrificed by carotid bleeding and liver was rapidly excised, rinsed in ice-cold saline, and a 10 per cent w/v homogenate was prepared using 0.15 *M* KCl, centrifuged at 800 g for 10 min at 4°C. The supernatant obtained was used for the estimation of catalase, peroxidase, and lipid peroxidation. Further, the homogenate was centrifuged at 1000 g for 20 min at 4°C and the supernatant was used for estimation of SOD and glutathione. Estimation of SOD was done by autoxidation of hydroxylamine at pH 10.2, which was accompanied by reduction of NBT, and the nitrite produced in the presence of EDTA was detected colorimetrically. One enzymatic unit of SOD is the amount in the form of proteins present in 100 µl of 10 per cent liver homogenate required to inhibit the reduction of 24 mM NBT by 50 per cent and is expressed as units per milligram of protein.

Catalase activity was estimated by determining the decomposition of H_2O_2 at 240 nm in an assay mixture containing phosphate buffer. One international unit of catalase utilized is that amount that catalyzes the decomposition of 1 m *M* H_2O_2/min/mg of protein at 37°C. Catalase activity was calculated using the millimolar extinction coefficient of 0.07 and expressed in terms of micromole per minute per milligram of protein.

Glutathione was estimated using Ellman's reagent (5,5¢-dithiobis-(2-nitrobenzoic acid) [DTNB]). The sulphydryl groups present in glutathione forms a colored complex with DTNB, which was measured colorimetrically at 412 nm. The amount of glutathione was determined using its molar extinction coefficient of 13600/m/cm and expressed in terms of µmol/mg of protein. Peroxidase estimation is based on periodide formation. Periodide can be spectrophotometrically determined at 353 nm, and this is directly proportional to the peroxidase concentration in the reaction mixture containing approximate amounts of H_2O_2 and enzyme. One unit of peroxidase activity is defined as the change in absorbance per minute and expressed in terms of units per milligram of protein.

Malondialdehyde (MDA), a secondary product of *lipid peroxidation*, reacts with thiobarbituric acid at pH 3.5. The red pigment produced was extracted in *n* -butanol-pyridine mixture, and estimated by measuring the absorbance at 532 nm. Results were subjected to one-way ANOVA. $P<0.05$ was considered significant. The *post hoc* analysis was carried out by Dunnet's multiple comparison tests.

As shown, CCl_4 treatment decreased SOD, catalase, glutathione, and peroxidase and increased lipid peroxidation. Pretreatment with 100 mg/kg (p.o.) of TO extract improved the SOD, catalase, glutathione, and peroxidase levels significantly and reduced lipid peroxidation. SOD is a ubiquitous cellular enzyme that dismutates superoxide radical to H_2O_2 and oxygen and is one of the chief cellular defence mechanisms. The H_2O_2 formed by SOD and other processes is scavenged by catalase that catalyzes the dismutation of H_2O_2 into water and molecular oxygen. Thus, the antioxidant enzyme catalase is responsible for detoxification of H_2O_2. Glutathione is a tripeptide of glycine, glutamic acid, and cysteine. Glutathione is an important naturally occurring antioxidant as it prevents the hydrogen of sulfhydryl group to be abstracted instead of methylene hydrogen of unsaturated lipids. Therefore, levels of glutathione are of critical importance in tissue injury caused by toxic substances. The antioxidant enzymes and glutathione form the first line of defence against free radical-induced damage, offer protection against free radicals, and thereby maintain low levels of lipid peroxide. Peroxidase is an enzyme that catalyzes the reduction of hydroperoxides, including hydrogen peroxides, and functions to protect the cell from peroxidative damage. As the TO extract, in the dose of 100 mg/kg, p.o., has improved the SOD, catalase, glutathione, and peroxidase levels significantly, which were comparable with Liv 52. We conclude that the hydroalcoholic extract from the root of TO possesses antioxidant activity, confirming the traditional use of the plant in treatment of liver disorders.

Antioxidant Activity of Ethyl Acetate Soluble Fraction of *Acacia arabica* Bark in Rats

Sundaram R, Mitra SK.; Indian Journal of Pharmacology. 2007; 39: 33-38.

Objective: To study the antioxidant activity of various extracts and fractions of *Acacia arabica* by *in vitro* and *in vivo* experimental models.

Materials and Methods: Various solvent extracts were prepared by Soxhlet extraction. Extract fractionations were done by solvent extraction and flash chromatographic separation. *In vitro* lipid peroxidation was carried out by tertiary butyl hydroperoxide -induced lipid peroxidation. The most active fractions were identified and standardized by thin layer chromatography (TLC). *In vivo* experiments on the most active fraction were carried out with 50, 100, and 150 mg/kg, p.o. doses, in carbon tetrachloride (CCl_4)-induced hepatotoxicity, in rats. Various biochemical parameters like serum aspartate aminotransferase (AST), serum alanine aminotransferase (ALT), superoxide dismutase (SOD), catalase, glutathione peroxidase (GSH-Px), glutathione (GSH), and lipid peroxidation were estimated.

Results: Flash chromatographic fractions 2-6 of ethyl acetate extract exhibited maximum activity with *in vitro* lipid peroxidation. *In vivo* evaluation of this active fraction (AA) in CCl_4-induced hepatotoxicity for 19 days at a dose of 150 mg/kg offered marked liver protection, which was evident by significant changes in lipid peroxidation, glutathione, superoxide dismutase and catalase ($P<0.01$). The treatment also showed significant changes in AST, ALT, and GSH-Px levels ($P<0.05$). At lower doses, the protection was not consistent.

Conclusion: The polyphenol-rich active fraction of *Acacia arabica* is a potent free radical scavenger and hepatoprotective and protects TBH-induced lipid peroxidation and CCl_4 -induced hepatic damage.

In vitro Antioxidant Activity of Pet Ether Extract of Black Pepper

Singh R, Singh N, Saini BS, Rao HS.; Indian Journal of Pharmacology. 2008; 40: 147-51.

Objective: To investigate the *in vitro* antioxidant activity of different fractions (R1, R2 and R3) obtained from pet ether extract of black pepper fruits (*Piper nigrum* Linn.)

Materials and Methods: The fractions R1, R2 and R3 were eluted from pet ether and ethyl acetate in the ratio of 6:4, 5:5 and 4:6, respectively. 1,1-Diphenyl-2-picryl-hydrazyl (DPPH) radical, superoxide anion radical, nitric oxide radical, and hydroxyl radical scavenging assays were carried out to evaluate the antioxidant potential of the extract.

Results: The free radical scavenging activity of the different fractions of pet ether extract of *P. nigrum* (PEPN) increased in a concentration dependent manner. The R3 and R2 fraction of PEPN in 500 microg/ml inhibited the peroxidation of a linoleic acid emulsion by 60.48+/-3.33 per cent and 58.89+/-2.51 per cent, respectively. In DPPH free radical scavenging assay, the activity of R3 and R2 were found to be almost similar. The R3 (100microg/ml) fraction of PEPN inhibited 55.68+/-4.48 per cent nitric oxide radicals generated from sodium nitroprusside, whereas curcumin in the same concentration inhibited 84.27+/-4.12 per cent. Moreover, PEPN scavenged the superoxide radical generated by the Xanthine/Xanthine oxidase system. The fraction R2 and R3 in the doses of 1000microg/ml inhibited 61.04+/-5.11 per cent and 63.56+/-4.17 per cent, respectively. The hydroxyl radical was generated by Fenton's reaction. The amounts of total phenolic compounds were determined and 56.98 microg pyrocatechol phenol equivalents were detected in one mg of R3.

Conclusions: *P. nigrum* could be considered as a potential source of natural antioxidant.

The Effect of Aqueous Extract of *Embelia ribes* Burm on Serum Homocysteine, Lipids and Oxidative Enzymes in Methionine Induced Hyperhomocysteinemia

Bhandari U, Ansari MN, Islam F, Tripathi CD. Indian Journal of Pharmacology. 2008; 40: 152-7.

Objective: The present study was designed to evaluate the effect of the aqueous extract of *Embelia ribes* Burm fruits on methionine-induced hyperhomocysteinemia, hyperlipidemia and oxidative stress in albino rats.

Materials and Methods: Adult male Wistar albino rats were fed with the aqueous extract of *Embelia ribes* (100 and 200 mg/kg, p.o.) for 30 days. Hyperhomocysteinemia was induced by methionine treatment (1 g/kg, p.o.) for 30 days and folic acid (100 mg/kg, p.o.) was used as a standard drug. The animals were evaluated for various biochemical parameters in serum and brain homogenates, followed by histopathological studies at the end of the study.

Results: Administration of methionine (1 g/kg, p.o.) for 30 days to vehicle control rats produced significant increase ($P < 0.01$) in homocysteine, lactate dehydrogenase (LDH), total cholesterol, triglycerides, low density lipoprotein (LDL-C), very low density lipoprotein (VLDL-C) levels in serum and lipid peroxides (LPO) levels in brain homogenates, with reduction in high density lipoprotein (HDL-C) levels in serum, and glutathione (GSH) content in brain homogenates, as compared to vehicle control rats. Administration of the aqueous extract of Embelia ribes (100 and 200 mg/kg, p.o.) for 30 days, to hyperhomocysteinemic rats, significantly ($P < 0.01$) decreased the levels of homocysteine, LDH, total cholesterol, triglycerides, LDL-C and VLDL-C and increased the HDL-C levels in serum. In addition, a significant ($P < 0.01$) decrease in LPO levels with increase in GSH content was observed in hyperhomocysteinemic rats treated with the aqueous extract of Embelia ribes. The results were comparable to those obtained with folic acid, a standard antihyperhomocysteinemic drug.

Conclusion: The present results provide clear evidence that the aqueous extract of *Embelia ribes* treatment enhances the antioxidant defence against methionine-induced hyperhomocysteinemia, hyperlipidemia and oxidative stress in brain.

Adaptogenic Effect of *Morus alba* on Chronic Footshock-Induced Stress in Rats

Nade VS, Kawale LA, Naik RA, Yadav AV.; Indian Journal of Pharmacology. 2009; 41: 246-51.

Objective: The objective of the present study was to evaluate the adaptogenic property of the ethyl acetate-soluble fraction of methanol extract of Morus alba roots against a rat model of chronic stress (CS).

Materials and Methods: Rats were exposed to stress procedure for 21 days. The stress procedure was mild, unpredictable footshock, administered for 1 h once daily for 21 days. Rats were administered with the ethyl acetate soluble fraction of methanol extract of M. alba roots (25, 50 and 100 mg/kg p.o) 1 h before footshock for 21 days and behavioural parameters were evaluated for cognitive dysfunction and depression, using elevated plus maze and despair swim test, respectively. On day 21, rats were sacrificed immediately after stress and blood was collected for biochemical estimation. The adrenal gland and spleen were dissected for organ weight and the stomach was dissected for ulcer score.

Results: CS significantly induced cognitive deficit, mental depression and hyperglycemia and increased blood corticosterone levels, gastric ulcerations and adrenal gland weight, but decreased the splenic weight. Pre-treatments with the ethyl acetate soluble fraction of methanol extract of *M. alba* roots (25, 50 and 100 mg/kg, p.o.) significantly attenuated the CS-induced perturbations. Diazepam (1 mg/kg, p.o.) was used as the standard antistress drug.

Conclusion: The results indicate that *M. alba* possesses significant adaptogenic activity, indicating its possible clinical utility as an antistress agent.

Antioxidant Potential of Aqueous Extract of *Phyllanthus amarus* in Rats

Karuna R, Reddy SS, Baskar R, Saralakumari D.; Indian Journal of Pharmacology.2009; 41: 64-7.

Objective: Increased levels of oxidative stress may be implicated in the etiology of many pathological conditions. Protective antioxidant action imparted by many plant extracts and plant products make them promising therapeutic drugs for free radical-induced pathologies. In this study we assessed the antioxidant potential of *Phyllanthus amarus* (Euphorbiaceae).

Materials and Methods: Experimental rats were divided into two groups: Control and *Phyllanthus amarus* (*P. amarus*) treated. Treated rats received *P. amarus* aqueous extract (PAAEt) at a dose of 200 mg/kg body wt/day for 8 weeks. After the treatment period of 8 weeks lipid peroxidation (LPO), vitamin C, uric acid and reduced glutathione (GSH) were estimated in plasma and antioxidant enzymes: Glutathione peroxidase (GPx), catalase (CAT) and superoxide dismutase (SOD) were also assayed. Genotoxicity of PAAEt was assessed by single cell gel electrophoresis (SCGE) of lymphocytes under both *in vitro* and *in vivo* conditions. The protective role of PAAEt against hydrogen peroxide (H_2O_2), streptozotocin (STZ) and nitric oxide generating system induced lymphocyte DNA damage was also assessed by SCGE.

Results: PAAEt treated rats showed a significant decrease in plasma LPO and a significant increase in plasma vitamin C, uric acid, GSH levels and GPx, CAT and SOD activities. SCGE experiment reveals that PAAEt was devoid of genotoxicity and had a significant protective effect against H_2O_2, STZ and nitric oxide (NO) induced lymphocyte DNA damage.

Conclusion: The results suggest the non-toxic nature of PAAEt and consumption of PAAEt can be linked to improved antioxidant status and reduction in the risk of oxidative stress.

Anti-Stress Activity of Hydro-Alcoholic Extract of *Eugenia caryophyllus* Buds (Clove)

Singh AK, Dhamanigi SS, Asad M.; Indian Journal of Pharmacology. 2009; 41: 28-31.

Objective: The present study was undertaken to evaluate the anti-stress effect of the hydro-alcoholic extract of clove.

Methodology: The anti-stress effect was evaluated on cold restraint induced gastric ulcers, sound stress induced biochemical changes and anoxic stress induced convulsions. Clove extract was administered orally at two different doses of 100 and 200 mg/kg. Zeetress, a known anti-stress formulation (14 mg/kg p.o) was used as the standard drug.

Results: Both the doses of clove extract showed good anti-stress effect in all the tested models. The clove extract reduced the development of cold restraint-induced gastric ulcers and prevented the biochemical changes induced by sound stress such as increase in plasma levels of aspartate aminotransferase, alanine aminotransferase, alkaline phosphatase, glucose, cholesterol and corticosterone. Clove extract was also effective in increasing the latency of anoxic stress-induced convulsions in mice.

Conclusion: The hydro-alcoholic extract of clove at doses of 100 and 200 mg/kg orally possesses good anti-stress activity.

Brahma Rasayana Enhances *in vivo* Antioxidant Status in Cold-stressed Chickens (*Gallus gallus domesticus*)

Ramnath V, Rekha PS.; Indian Journal of Pharmacology. 2009; 41: 115-9.

Objective: To evaluate the antioxidant status of chicken during cold stress and to investigate if there are any beneficial effects of Brahma Rasayana supplementation in cold stressed chicken.

Materials and Methods: Activities of enzymatic and levels of non-enzymatic antioxidants in blood/serum and liver tissue were evaluated in chicken exposed to cold (4±10°C and relative humidity of 40±5 per cent, for six consecutive hours daily, for 5 or 10 days). The antioxidant properties of Brahma Rasayana (BR) supplementation (2 g/kg daily, orally) during cold stress was also studied.

Results: There was a significant ($P < 0.05$) decrease in antioxidant enzyme in the blood, such as, superoxide dismutase (SOD), glutathione peroxidase (GPX), glutathione reductase (GR), and serum reduced glutathione (GSH) in cold stressed chicken. Serum and liver lipid peroxidation levels were significantly ($P < 0.05$) higher in cold stressed untreated chickens when compared to the treated and unstressed groups. There was also a significant ($P < 0.05$) increase in the antioxidant enzymes in the blood, such as, catalase (CAT) and SOD, in the liver CAT and SOD, and in GPX and GR in BR-treated cold stressed chicken, when compared to the untreated controls.

Conclusions: Results of the present study conclude that in chicken, BR supplementation during cold stress brings about enhanced actions of the enzymatic and non-enzymatic antioxidants, which nullify the undesired side effects of free radicals generated during cold stress.

Effect of *Ocimum sanctum*, Ascorbic Acid, and Verapamil on Macrophage Function and Oxidative Stress in Mice Exposed to Cocaine

Bhattacharya SK, Rathi N, Mahajan P, Tripathi AK, Paudel KR, Rauniar GP, Das BP.; Indian Journal of Pharmacology. 2009; 41: 134-9.

Objective: To investigate the effect of *Ocimum sanctum*, ascorbic acid, and verapamil on macrophage function and oxidative stress in experimental animals exposed to cocaine.

Materials and Methods: Mice were used in this study and were divided randomly into different groups of six animals each. They were either treated with intraperitoneal injection of saline or cocaine hydrochloride or an oral feeding of oil of *Ocimum sanctum*, ascorbic acid or verapamil, or both (ascorbic acid and verapamil), and were evaluated for a respiratory burst of macrophages, superoxide and nitric oxide (NO) production, estimation of TNF-alpha in the serum and supernatant of cultured macrophages, estimation of lipid peroxidation (malondialdehyde- MDA) in the serum, and superoxide dismutase activity in the erythrocytes.

Results: Unstimulated respiratory burst as well as superoxide production was enhanced on treatment with cocaine and all the three drugs were found to attenuate this enhancement. The bactericidal capacity of macrophages decreased significantly on chronic cocaine exposure, as it was associated with decreased respiratory burst and superoxide production. There was a significant decrease in NO production by macrophages on chronic cocaine exposure and all the test drugs were found to restore nitrite formation to a normal level. There was an increase in the malonylodialdehyde (MDA) level and decrease in the superoxide dismutase level on chronic cocaine exposure, and all the three drugs effectively decreased the MDA level and increased superoxide dismutase level. There was an increase in serum TNF-alpha on chronic cocaine exposure, which was decreased significantly by ascorbic acid and verapamil.

Conclusion: O. sanctum, ascorbic acid, and verapamil were equally effective in improving the macrophage function and reducing oxidative stress. These findings suggested that *O. sanctum*, ascorbic acid, and verapamil attenuated acute and chronic cocaine-mediated effects.

In vitro Antioxidant Properties of *Salvia verbenaca* L. Hydromethanolic Extract

Khlifi S, El Hachimi Y, Khalil A, Es-Safi N, Belahyan A, Tellal R, El Abbouyi A.; Indian Journal of Pharmacology 2006; 38: 276-80.

Objective: To investigate the *in vitro* antioxidant activity of the hydromethanolic extract of the aerial parts (leaves and stems) of *Salvia verbenaca* L. towards fatty acids (linoleic and linolenic acids) and human, low density lipoproteins (LDL) peroxidation.

Materials and Methods: Lipid peroxidation was carried out in the presence of the *S. verbenaca* L. hydromethanolic extract (10 and 100 µg of extract/ml). $CuSO_4$ (10 µM) was used as the oxidation initiator. Conjugated dienes (CD) formation, oxygen consumption and thiobarbituric acid reactive substances (TBARS) formation were assessed to monitor the antioxidant properties of the plant extract. Butylated hydroxytoluene (BHT) at 50 µg/ml was used as a standard antioxidant. The quantification of total polyphenolic compounds was carried out, according to the Folin-Ciocalteu method.

Results: The hydromethanolic extract of *S. verbenaca* showed a significant antioxidant effect at 100 µg/mL. A strong inhibition of oxygen consumption (92 per cent, $P < 0.001$) and CD formation of LDL peroxidation (92 per cent, $P < 0.001$) as well as TBARS formation of linolenic acid oxidation (93 per cent, $P < 0.001$) were observed. The quantitative analysis revealed that the extract used contained a high amount of phenolic compounds, suggesting a possible role of these products in the observed antioxidant properties.

Conclusion: S. verbenaca could be considered as a potential source of natural antioxidants.

Chlorpyrifos-Induced Oxidative Stress and Tissue Damage in the Liver, Kidney, Brain and Fetus in Pregnant Rats: The Protective Role of the Butanolic Extract of *Paronychia argentea* L.

Zam D a, Meraihi Z, Tebibel S, Benayssa W, Benayache F, Benayache S, Vlietinck AJ.; Indian Journal of Pharmacology. 2007; 39: 145-150.

Objective: Toxicity of pesticides is thought to be due to reactive oxygen species (ROS). Due to their antioxidant property, polyphenols in plant extracts may afford protection from pesticide toxicity. In the present study, we evaluated the protective effect of a butanolic extract of *Paronychia argentea* L. against toxicity caused by the organophosphorus pesticide, chloropyriphos ethyl (CE).

Materials and Methods: Pregnant albino Wistar rats were used. Pesticide and plant extract were administered daily by oral gavage from the 6th to the 15th day of gestation. Plasma and tissue malondialdehyde (MDA), blood reduced glutathione (GSH) and erythrocyte superoxide dismutase (SOD) activities were estimated. MDA levels were estimated in plasma and different organs (liver, kidney, brain, placenta and in the fetuses and their livers) as an indicator of lipid peroxydation (LPO).

Results: The data showed a significant increase in plasma and tissue LPO levels in animals treated with the pesticide while the effect was attenuated by the plant extract (CE-ex). Also, CE caused a significant decrease in antioxidant enzyme activity and this effect was partially reversed in groups treated with the plant extract. The pesticide-induced embryotoxicity and resulted in resorption, fetal death and a reduced implant number.

Conclusion: It can be concluded that CE can lead to an increase in LPO production in adult and fetal tissues, while treatment with the plant extract leads to protection against CE toxicity. The decrease in LPO levels and the increase in GSH and SOD enzyme activities after treatment with the plant extract revealed its antioxidant property.

Free Radical Scavenging Activity of Gossypin and Nevadensin: An *In vitro* Evaluation

Ganapaty S, Chandrashekhar VM, Chitme HR, Narsu ML.; Indian Journal of Pharmacology. 2007; 39: 281–283.

Objectives: The antioxidant potential of gossypin and nevadensin, two flavonoid compounds, were evaluated by *in vitro* methods.

Materials and Methods: Gossypin, nevadensin, and the reference standard, butylated hydroxyl toluene (BHT), were evaluated for DPPH (1, 1-diphenyl-2-picrylhydrazyl), nitric oxide, superoxide, and hydroxyl radical scavenging activity.

Results: Gossypin and BHT showed the potential for significant DPPH radical inhibition of up to 88.52 and 91.45 per cent at 100 µg/ml concentration. With a 100 µg/ml concentration of gossypin, the *in vitro* nitric oxide, superoxide, and hydroxyl radical scavenging activity was found to be 74.00, 74.22, and 67.15 per cent, respectively; and with 100 µg/ml of BHT the corresponding values were 82.24, 81.76, and 73.03 per cent of inhibition, respectively.

Conclusion: The study results showed that gossypin has significant antioxidant activity.

Anti-stress Activity of *Ocimum sanctum* Linn.

K.P. Bhargava, N. Singh; J RES AYUR SIDDHA. 1981; 443-451.

The plant *Ocimum sanctum* has been found to possess adaptogenic properties when tested against a battery of tests in mice and rats. The drug increases the physical endurance (increased survival time) of swimming mice, prevented stress-induced ulcers in rats, protected mice and rats against carbon tetrachloride induced hepatotoxicity, prevented milk induced leucocytosis in mice. The result of the study indicates that *Ocimum sanctum* induces a state of non-specific increased resistance against a variety of stress induced biological changes in animals.

Studies on Rasayana Therapy and Anti-stress Effect of Ashwagandha (*Withania somnifera*): A Scientific Study

Ajay Kumar Sharma and Rajesh Kumar; J RES AYUR SIDDHA. 2005; XXVI (3-4), 54-73.

For the study of Rasayana and anti-stress effect of Ashwagandha, 50 patients suffering from stress disorder were treated with Ashwagandha Churna for 3 months. The observations revealed statistically significant improvement in various factors indicating Rasayana Prabhava *viz.* Smriti, Medha, Arogya, Prabha, Varna, Swara, Dehabala, Indriyabala and stress after the therapy.

It can be concluded from the present study that the Ashwagandha (*Withania somnifera*) is a good Rasayana drug as it improves the mental faculties, physical strength, possesses potent anti-stress activity due to its psychotropic and tranquillizing effects. Ashwagandha is a potent immunomodulator drug.

Effect of Shankhapuspi on Experimental Stress

G. C. Prasad, R. G. Gupta, D. N. Srivastava, A. K. Tandon, R. S. Wahi, K. N. Udupa; J RES AYUR SIDDHA. 1994.

Apart from the basic studies on various types of stress some attention hitherto been given regarding the management of the such diseases. Ganong (1961) has observed that chronic administration of large doses of tranquilizer had failed to inhibit the response to the stress of immobilization. Bitz and Ganong (1963) have used chlorpromazine in psychic and in few physically produced stress and they found that it stimulates the ACTH secretion in the blood which further failed to block the adrenocortical response to physical or psychological stress of immobilization. Where as reserpine decreases the adrenocortical response to an emotional stress in monkey. While studying he endocrine response after different type of stress, we came across with anindigenous drug which is called by the name of Shankhapushpi. this drug is known to have some transqullizing as well as hypotensive effect. After extensive phyto-chemical analysis an active pinciple was isolated and its effect on stress produced by different method has been studied.

Clinical Evaluation of Rasayana Prabhava of Amalaki Rasayana

Ankad VV, Sharma AK.; J Res Ayur Siddha. 2002; 23: 22-30.

According to principles of Ayurveda Rasayana Therapy of Ayurveda produces fundamental changes in the body an mind and improves psychosomatic make-up of individuals. It prevents premature ageing, improves mental functions and promotes immunity of the body. In the present context Amalaki Rasayana has been used to evaluate its Rasayana Prabhava in patients of obesity and cachexia.

It was observed that the administration of Amalaki Rasayana in cachexic patients led to gain in body weight, where as its administration in obese patients led to reduction in body weight. It was observed that regular use of Amalaki Rasayana in human beings produced significant reduction in the level of S. Cholesterol. S. triglycerides, LDL and VLDL where as there was significant improvement in the level of the HDL. Macmatological investigations revealed improvement in Hbgm of level with decrease in ESR level.

It can be concluded that Amalaki Rasayana possesses potent Rasayana properties like Medhavarodhaka, Balavardhaka and Vayasthapaka Guna (Anabolic effects of Rasayana therapy). Amalaki Rasayana is a potent hypolipidaemic, cardiotonic, cardioprotective, rejuvenating and immunomodulator drug.

Rasayana Therapy of Ayurveda and Aging

Sharma AK.; J Res Ayur Siddha. 2002; 23: 49-60.

Rasayana therapy is an important division of Ashtanga Ayurveda. Rasayana therapy of Ayurveda is mainly used for maintaining the health of healthy individuals. The concept of Rasyana therapy is as old as Vedas. Unfortunately gradually it lost its independent identity in post independent era. Recently there is increased awareness among the masses about the therapeutic uses of Rasayana therapy especially regarding anti-aging and adaptogenic effect of several Rasayana drugs. In the present times, the sign of aging are seen much earlier and they progress rather rapidly reducing the life span considerably. This is called as premature aging which can be prevented by Rasayana therapy. Rasayana is that process by which all the body tissues are nourished. Consequently Rasayana help in regeneration, revival and revitalization of Dhatus. On the other hand, the Rasayana can be described as a particular measure by which one can gain immunity against diseases and aging. 30 apparently healthy individuals in the age group of 60-80 years were selected for the study and were subjected for clinical trial of Amalaki for 45 days. It may be concluded from the present trial that Amalaki (Emblica officinalis Garten) is a mild to moderately effective remedy for the management ofstress and old age related common problems which is possible probably due to its Prabahva, being a Rasayana drug. Thus Amalaki seems to be a useful remedy for use in aged persons to promote the quality of life and to reduce the rate of ageing. Amalaki seems to be anti-stress and adaptogenic in nature besides having possible immunomodulating effect.

Scientific Study – Rasayana (Rejuvenative) Effect of Amalaki

Dhananjay NK, Shandilya MK, Mishra DS.; J Res Ayur Siddha. 2002; 21: 19-26.

Ayurveda is unique system of medicine which have holistic approach to treat individuals. Ayurveda is divided into eight branches. Rasayana is one of them. Rasayana is a therapy mainly aiming for prevention from untimely ageing (Akalaja Jara) and diseases which deteriorate the life in old age. Properlu used Rasayana drugs, diet, regime gives wonderful results. Amalaki (*Emblica officinalis* Gaertn.) said to be best Vayasthapana drug. 10 healthy volunteers of age group 40-60 years old were treated with Amalaki juice. Rejuvenation effect was assessed on symptomatic, haematolocal, bio-chemical, physiological, functional leve. Amalaki offers best rejuvenative effect.

Studies on the Rasayana Effect of a Geriatric Formulation in Apparently Normal Aged Persons

Shetty RS, Sesshadri C, T. P. Sundaresen, V. Rajagopalan, Kanchana Srinivasan, R. Sitaraman, R. Revathy, K. Janaki, B. Rama Rao; J Res Ayur Siddha. 1997; 18: 108-117.

The Rasayana (healthful longevity) effect of a compound preparation consisting of 6 drugs in 38 healthy male volunteers is assessed in this paper. Extensive clinical, psychological, bio-chemical and anthropometric tests are selected as parameters of assessment. The treatment period is six months. Results indicate that this compound drug is a comprehensive and balanced drug formulation, capable of restoring the age related impaired functions, *viz.* joint pains, memory deficts, *etc.* It can be prescribed as a geriatric tonic, as a psycho-stimulant, and for enhancing the non-specific immunity.

Physiological Endocrine and Metabolic Studies on the Effect of Rasayan Therapy in Aged Persons

Varma MD, Singh RH, Udupa KN.; JRIM, 1973; 8(2).

Healthful longevity has been the cherished wish of human being Rasayan Tantra of Ayurveda claims to impart this effect. Chyavanprash, a popular Rasayan has been given trial in the aged people to evaluate its Rasayan effects. Its three months' regular use in 9 old volunteers in the age group of 60-70 years has shown promising results. Significant improvement has been observed in the physical, physiological and haematological status of the volunteers. There has also been marked improvement in the protein metabolism, adrenal and testicular functions, positive nitrogen balance and decreased connective tissue break down leading to promotion of general well-being.

Effect of Shankshpuspi on Experimental Stress

Prasad GC, Gupta RG, Srivastava DN, Tandon AK, Wahi RS, Udupa KN.; JRIM, 1974; 9 (2).

Apart from the basic studies on various types of stress some attention hitherto been given regarding the management of the diseases. Ganong (1961) has observed that chronic administration of large doses of tranquilizer had failed to inhibit the response to the stress of immobilization. Bitz and Ganong (1963) have used chlorpromazine in psychic and in few phyusically produced stress and they found that it stimulates the ACTH secrection in the blood which further failed to block the adrenocortical response to physical or psychological stress of immobilization. Whereas, reserpine decreases the adrenocortical response to an emotional stress in monkey. While studying he endocrine response after different type of stress, we came across with anindigenous drug which is called by the name of Shankhapushpi. This drug is known to have some transqullizing as well as hypotensive effect. After extensive phytochemical analysis an active pinciple was isolated and its effect on stress produced by different method has been studied.

HEPATOPROTECTIVE ACTIVITY

Hepatoprotective and Toxicological Evaluation of *Andrographis paniculata* on Severe Liver Damage

Neha Trivedi, U.M. Rawal; Indian Journal of Pharmacology 2000; 32: 288-293.

Objectives: To study the hepatoprotective effect of aqueous extract of *Andrographis paniculata* (AP) on hexachlorocyclohexane (BHC) induced severe liver damage in Swiss male mice.

Methods: The aqueous extract of Andrographis paniculata was given orally to the animals with liver damage induced by hexachlorocyclohexane (BHC). The hepatoprotective activity was monitored by estimating serum ALT and AST and other parameters like alkaline phosphatase, g-Glutamyl transpeptidase, glutathione and lipid peroxidase.

Results: Aqueous extract of *Andrographis paniculata* inhibited BHC induced liver toxicity in Swiss male mice as assessed by the biochemical values.

Conclusions: Aqueous extract of Andrographis paniculata has significant hepatoprotective activity.

Effect of HD-03 on Levels of Various Enzymes in Paracetamol-Induced Liver Damage in Rats

Venkatesha Udupa, Kala Suhas Kulkarni, Md. Rafiq, S. Gopumadhavan, M.V. Venkataranganna, S.K. Mitra; Indian Journal of Pharmacology 2000; 32: 361-364.

Objective: To review the evidence for a role of HD-03, a polyherbal formulation in regulating the sodium pumps in hepatic injury induced by paracetamol.

Methods: Alterations in sodium pump was induced by chronic administration of paracetamol at the dose of 500 and 1000 mg/kg, b. wt. for 28 days. Serum alanine aminotransferase (ALT), aspartate aminotransferase (AST), liver glutathione, glycogen and Na^+-K^+-ATPase activity estimation and histology of liver were studied in rats.

Results: Chronic administration of paracetamol for 4 weeks to rats produced dose dependent increase in ALT, AST and reduction in liver Na^+-K^+-ATPase activity, glycogen and glutathione levels, indicating the hepatocellular damage. Histological evaluation supported this change with evidence of swelling, hydropic degeneration and necrosis of the hepatocytes. These changes were reversed with simultaneous administration of paracetamol and HD-03 at 750 mg/kg, for 28 days.

Conclusion: Reversal of Na^+-K^+-ATPase, glycogen, glutathione levels and restricted hepatic damage in HD-03 treated animals confirms the hepatoprotective effect of HD-03. Thus, Na^+-K^+-ATPase may be considered as a marker to evaluate the hepatoprotective effects of various herbs.

Hepatoprotective Effects of *Ginkgo biloba* against Carbon Tetrachloride Induced Hepatic Injury in Rats

K. Ashok Shenoy, S. N. Somayaji, K. L. Bairy; Indian Journal of Pharmacology 2001; 33: 260-266.

Objectives: To assess the protective activity of *Ginkgo biloba* (GB) against CCl_4 induced hepatotoxicity in rats and probe into its mechanism of action.

Methods: Liver damage was induced in Wistar rats by administering (150-250 g) CCl_4 (0.5 ml/kg, i.p.)once daily for 7 days. GB (50 mg/kg, i.p.) was given for one week. Silymarin (200 mg/kg, p.o.) was givenas a reference drug. Levels of marker enzymes (AST, ALT, ALP) and total proteins (TP), albumin (Alb)were estimated in serum. A probe into the mechanism of action was attempted by estimating thiobarbituric acid reactive substances (TBARS) and glutathione (GSH) levels in liver homogenates in order to evaluate the degree of lipid peroxidation. Histopathological studies were also done to confirm the biochemical changes.

Results: The mean + SEM serum AST, ALT, ALP levels in control animals were 66.8 + 4.2, 31.1 + 2.0 and 445.3 + 23.1 IU/L respectively whereas in CCl4 treated rats, the level rose to 319.6 + 22.7, 192.8 +16.0 and 809.3 + 65.3 IU/L respectively. GB reduced the AST, ALT and ALP levels to 55.5 + 5.3, 36.5 +3.6 and 489.6 + 43.9 IU/L respectively. Silymarin reduced AST, ALT and ALP levels to 51.8 + 5.2, 30.8 +3.4 and 437.8 + 35.7 IU/L respectively. There was a significant decrease in serum TP and Alb levels after CCl_4, which was reversed by GB and silymarin. The tissue mean + SEM values of TBARS and GSH in control animals were 3.1 + 0.1 nmol of malondialdehyde/g of wet tissue and 1.9 + 0.1 mg/g of wet tissue respectively. In CCl_4 treated animals, the TBARS and GSH levels were 3.91 + 0.41 and 1.97 + 0.11respectively. GB reduced TBARS to 2.4 + 0.09 nmol of MDA/g of wet tissue and increased GSH level to 2.4 + 0.1 mg/g of wet tissue. Silymarin reduced TBARS to 2.1 + 0.2 nmol of MDA/g of wet tissue and increased GSH level to 2.5 + 0.17 mg/g of wet tissue.

Conclusion: GB has protected the liver from CCl_4 damage. Probable mechanism of action is by protection against oxidative damage produced by CCl_4.

Preliminary Studies on the Hepatoprotective Activity of *Mamordica subangulata* and *Naragamia alata*

V.V.Asha; Indian Journal of Pharmacology 2001; 33: 276-279.

Objective: To study the hepatoprotective and choleretic activities of *Mamordica subangulata and Naragamia alata.*

Methods: The hepatoprotective activity of *Mamordica subangulata* (leaf) and *Naragamia alata* (whole plant) suspension was studied using paracetamol overdose induced liver damage in rats. The effect of the plant suspensions on bile flow was studied in anaesthetised normal rats by surgical cannulation of bile duct with polyethylene tubing. The drug was given intraduodenally after 1 hour bile collection.

Results: Mamordica subangulata leaf suspension (500mg/kg,fresh weight;50 mg/kg, dry wt) protected rats from paracetamol induced liver damage as judged from serum marker enzyme activities. It also stimulated bile flow in normal rats. *Naragamia alata* was inactive in protecting rats from paracetamol -induced hepatotoxicity.

Conclusion: A suspension of *Mamordica subangulata* leaf (dry or fresh) can protect rats from paracetamol-induced hepatotoxicity.

Hepatoprotective Effect of the Methanolic Extract of *Curculigo orchioides* in CCl_4-Treated Male Rats

M.R. Venukumar, M.S. Latha; Indian Journal of Pharmacology 2002; 34: 269-275.

Objective: To evaluate the hepatoprotective effect of the methanolic extract of *Curculigo orchioides* rhizomes (MEC) in rats treated with carbon tetrachloride.

Methods: In hepatotoxic rats, liver damage was studied by assessing parameters such as aspartate aminotransferase (AST), alanine aminotransferase (ALT), alkaline phosphatase (ALP) and gamma glutamyl transpeptidase (GGT) in serum, and concentrations of total proteins, total lipids, phospholipids, triglycerides and cholesterol in both serum and liver. The effect of co-administration of MEC on the above parameters was further investigated. Histopathological study of the liver in experimental animals was also undertaken.

Results: Hepatic damage as evidenced by a rise in the levels of AST, ALT, ALP and GGT in serum, and also changes observed in other biochemical parameters in serum and liver showed a tendency to attain near normalcy in animals co-administered with MEC. The normal values for AST (IU/L), ALP (IU/L), protein (g/100 ml) and total lipids (mg/100 ml) in serum (*i.e.*, 21.24, 71.04, 6.72 and 136.54 respectively) were found to alter towards values 33.61, 128.11, 4.83 and 266.91 in hepatotoxic rats. These parameters attained near-normal values (*i.e.*, 23.82, 80.3, 6.22 and 152.24 for AST, ALP, protein and total lipids respectively) in MEC co-administered rats. Profound steatosis, ballooning degeneration and nodule formation observed in the hepatic architecture of CCl4 treated rats were found to acquire near-normalcy in drug co-administered rats, thus corroborating the biochemical observations.

Conclusion: The study substantiates the hepatoprotective potential of MEC.

Effect of Aqueous Extract of *Azadirachta indica* Leaves on Hepatotoxicity Induced by Antitubercular Drugs in Rats

B.P. Kale, M.A. Kothekar, H.P. Tayade, J.B. Jaju, M. Mateenuddin; Indian Journal of Pharmacology 2003; 35: 177-180.

Objective: To assess the hepatoprotective activity of *Azadirachta indica* (AI) aqueous leaf extract on antitubercular drugs-induced hepatotoxicity in albino rats.

Methods: Hepatotoxicity was induced in rats by combination of isoniazid, rifampicin and pyrazinamide given orally as suspension for 30 days. Treatment groups received AI aqueous leaf extract along with antitubercular drugs. In the second phase of study the effect of AI aqueous leaf extract on established hepatotoxicity was studied by giving the extract for 20 days after withdrawal of antitubercular drugs. Liver damage was assessed by biochemical and histological parameters.

Results: AI aqueous leaf extract significantly ($P<0.05$) prevented changes in the serum levels of bilirubin, protein, alanine aminotransferase, aspartate aminotransferase and alkaline phosphatase. Similarly it significantly prevented the histological changes as compared to the group receiving antitubercular drugs. It also significantly reversed the biochemical and histological changes.

Conclusion: AI aqueous leaf extract significantly prevents and reverses the hepatotoxic damage induced by antitubercular drugs in rats.

Prevention of Carbon Tetrachloride Induced Hepatotoxicity in Rats by Himoliv, a Polyherbal Formulation

Bhattacharyya D., Mukherjee R., Pandit S., Das N., Sur T.K.; Indian Journal of Pharmacology 2003; 35: 183-185.

Himoliv (HV) is a polyherbal ayurvedic product claimed to be useful in hepatitis, jaundice and biliary dysfunction. The aim of the present study was to evaluate the antioxidant effect of HV on carbon tetrachloride (CCl_4) induced hepatotoxicity. Adult male albino rats (150-175 g) of Wistar strain were used. A total of 30 animals were divided into 5 groups (n = 6 in each group). Group I served as vehicle control and group II as CCl_4-treated control which received normal saline 5 ml/kg, for 9 days. Group III and Group IV were pretreated with HV at the dose of 0.5 ml/kg and 1.0 ml/kg, orally respectively for 9 days, while group V was pretreated with silymarin at 25 mg/kg, p.o., for 9 days. Liver damage was induced in these rats with 1:1 (v/v) mixture of CCl_4 in olive oil. Pretreatment with the test drug HV in both doses as well as pretreatment with standard drug silymarin significantly ($p<0.01$) reduced these liver enzyme levels dose dependently, showing that HV has hepatoprotective action. Histological studies indicated that pretreatment with HV (0.5-1.0 ml/kg) protected the hepatocytes from damage induced by CCl_4, with mild fatty changes in the hepatic parenchymal cells, which corroborated the changes observed in the hepatic enzymes.

Hepatoprotective Activity of Panchagavya Ghrita against Carbon Tetrachloride Induced Hepatotoxicity in Rats

Achliya GS, Kotagale NR, Wadodkar SG, Dorle AK.; Indian Journal of Pharmacology 2003; 35: 308-311.

Objective: To investigate the hepatoprotective activity of Panchagavya Ghrita (PG) against CCl_4 induced hepatotoxicity.

Methods: The hepatoprotective activity of PG was tested against carbontetrachloride induced hepatotoxicity in albino rats. The degree of protection was determined by measuring levels of serum marker enzymes like serum glutamate oxaloacetate transaminase (SGOT) serum glutamate pyruvate transaminase (SGPT), alkaline phosphatase (ALP) and acid phosphatase (ACP). The histological studies were also carried out. Silymarin was used as the standard drug for comparison.

Results: Administration of Panchagavya Ghrita (150-300 mg/kg, p.o.) markedly prevented CCl_4 induced elevation of levels of serum GPT, GOT, ACP and ALP. The results are comparable to that of silymarin. A comparative histopathological study of liver exhibited almost normal architecture, as compared to control group.

Conclusion: Treatment with Panchagavya Ghrita significantly reduced the CCl_4 induced hepatotoxicity. A comparative histological study of liver from different groups further confirmed the hepatoprotective activity of Panchagavya Ghrita.

Protective Activity of *Glycyrrhiza glabra* Linn. on Carbon Tetrachloride-Induced Peroxidative Damage

Rajesh MG, Latha MS.; Indian Journal of Pharmacology, 2004; 36: 284-287.

Objective: To evaluate the potential efficacy of *Glycyrrhiza glabra* Linn. (Fabaceae) in protecting tissues from peroxidative damage in CCl_4-intoxicated rats.

Material and Methods: Peroxidative hepatic damage in rats was studied by assessing parameters such as thiobarbituric acid reactive substances (TBARS), conjugated dienes (CD), superoxidedismutase (SOD), catalase (CAT), glutathione-S-transferase (GST), glutathione peroxidase (GSH-Px) and glutathione (GSH) in liver and kidneys. The effect of co-administration of *G. glabra* on the above parameters and histopathological findings of the liver in experimental animals was studied.

Results: The increased lipid peroxide formation in the tissues of CCl_4-treated rats was significantly inhibited by G. glabra. The observed decreased antioxidant enzyme activities of SOD, CAT, GSHPx, GST, and antioxidant concentration of glutathione were nearly normalized by *G. glabra* treatment. Carbon tetrachloride-induced damage produces alteration in the antioxidant status of the tissues, which is manifested by abnormal histopathology. *G. glabra* restored all these changes.

Conclusion: Glycyrrhiza glabra is a potential antioxidant and attenuates the hepatotoxic effect of CCl_4.

Prevention of Carbon Tetrachloride Induced Hepatotoxicity in Rats by *Adhatoda vasica* Leaves

Pandit S, Sur TK, Jana U, Debnath PK, Sen S, Bhattacharyya D.; Indian Journal of Pharmacology, 2004; 36: 312-320.

The plant *Adhatoda vasica* Nees (AV) of the Acanthaceae family has been used for thousands of years in India. Extracts of the leaves of AV are extensively used in cough. the aim of the present study was to evaluate the antioxidant effect of AV in carbon tetrachloride (CCl_4)-induced hepatotoxicity in rats. It is well established that CCl_4 is metabolized in the liver to the highly reactive trichloromethyl radical and this free radical leads to auto-oxidation of the fatty acids present in the cytoplasmic membrane phospholipids and causes functional and morphological changes in the cell membrane. This is evidenced by an elevation of the serum marker enzymes namely SGOT, SGPT and ALP in CCl_4-treated rats.7-8 Pretreatment with the test drug AV in both doses as well as pretreatment with standard drug silymarin significantly ($P<0.01$) reduced these liver enzyme levels dose dependently, showing that AV has hepatoprotective action. Histopathological findings indicated that pretreatment with AV (100 and 200 mg/kg) offered protection to the hepatocytes from damage induced by CCl_4, with mild fatty changes in the hepatic parenchymal cells, which corroborated the changes observed in the hepatic enzymes.

Evaluation of Hepatoprotective Activity of Stem Bark of *Pterocarpus marsupium* Roxb.

Mankani K L, Krishna V, Manjunatha B K, Vidya S M, Jagadeesh Singh S D, Manohara Y N, Raheman AU, Avinash K R.; Indian Journal of Pharmacology 2005; 37: 165-8.

To evaluate the hepato-protective activity of *Pterocarpus marsupium* stem bark extracts against carbon tetrachloride (CCl_4)-induced hepatotoxicity. In methanol extract-treated animals, the toxic effect of CCl_4 was controlled significantly by restoration of the levels of serum bilirubin, protein and enzymes as compared to the normal and the standard drug silymarin-treated groups. Histology of the liver sections of the animals treated with the extracts showed the presence of normal hepatic cords, absence of necrosis and fatty infiltration, which further evidenced the hepatoprotective activity. Results concluded that methanol extract of the stem bark of *P. marsupium* possesses significant hepatoprotective activity.

Antihepatotoxic Effect of Grape Seed Oil in Rat

Uma Maheswari M, Rao P G.; Indian Journal of Pharmacology 2005; 37: 179-82.

Objectives: To study the effect of oral administration of grape seed oil (GSO) against carbontetrachloride (CCl_4)-induced hepatotoxicity in rats.

Methods: Liver damage was induced in male Wistar rats (150–250 g) by administering CCl_4 (0.5 ml/kg, i.p.) once per day for 7 days and the extent of damage was studied by assessing biochemical parameters such as alanine aminotransferase (ALT), aspartate aminotransferase (AST), and alkaline phosphatase (ALP) in serum and concentrations of malondialdehyde (MDA), hydroperoxides, glutathione (GSH), catalase (CAT), superoxide dismutase (SOD), and total protein (TP) in liver. The effect of co-administration of GSO (3.7 g/kg, orally) on the above parameters was further investigated and compared with a vitamin E (100 mg/kg, orally) treated group. Histopatholgical studies of the experimental animals were also done.

Results: Oral administration of GSO (3.7 g/kg, body weight orally) for 7 days resulted in a significant reduction in serum AST, ALT, and ALP levels and liver MDA and hydroperoxides and significant improvement in glutathione, SOD, CAT, and TP, when compared with CCl_4 damaged rats. The antioxidant effect of GSO at 3.7 g/kg for 7 days was found to be comparable with vitamin E (100 mg/kg, orally) in CCl_4-treated rats. Profound fatty degeneration, fibrosis, and necrosis observed in the hepatic architecture of CCl_4-treated rats were found to acquire near – normalcy in drug co-administered rats.

Conclusion: The GSO has protected the liver from CCl_4 damage. Probable mechanism of action may be due to the protection against oxidative damage produced by CCl_4.

Possible Mechanism of Hepatoprotective Activity of *Azadirachta indica* Leaf Extract against Paracetamol-Induced Hepatic Damage in Rats: Part III

Chattopadhyay R R, Bandyopadhyay M.; Indian Journal of Pharmacology 2005; 37: 184-5.

In the present study the effects of *A. indica* leaf extract on antioxidant enzymes have been investigated to elucidate the possible mechanism of its hepatoprotective activity. Administration of *A. indica* leaf extract significantly enhanced the hepatic level of glutathione dependent enzymes and superoxide dismutase and catalase activity suggesting that the hepatoprotective effect of the extract on paracetamol induced hepatoxicity may be due to its antioxidant activity. Thus it can be inferred that *A. indica* leaf extract may be a promising hepatoprotective agent and this activity may be due to its antioxidant activity. Further studies involving the extract and/or its chemical constituents are needed to pinpoint the findings. This report may serve as a prelude to this aspect.

Effect of *Leucas aspera* on Hepatotoxicity in Rats

Mangathayaru K, Grace X F, Bhavani M, Meignanam E, Rajasekhar Karna S L, Pradeep Kumar D.; Indian Journal of Pharmacology 2005; 37: 329-30.

Swiss albino mice were used for toxicity study, while the hepatoprotective study was carried out in adult male Wistar rats (150-200 g). For hepatoprotective study, a total of 30 rats were divided into five groups (n=6 in each group). Treatment group treated with methanolic extract of *Leucas aspera* (LA), (200 and 400 mg/kg, p.o.; respectively) for 5 days. Comparative histopathological study of the liver from different groups of rats corroborated the hepatoprotective efficacy of LA. Because hepatotoxic effect of CCl_4 is due to oxidative damage by free radical generation, antioxidant property is claimed to be one of the mechanisms of hepatoprotective drugs. Further flavonoids have been suggested to act as antioxidants by free radical scavenging. Thus the hepatoprotective activity of LA may be attributed to the presence of flavonoids, though it is to be confirmed.

Hepatoprotective Activity of Alcoholic and Aqueous Extracts of Leaves of *Tylophora indica* (Linn.) in Rats

Gujrati V, Patel N, Rao VN, Nandakumar K, Gouda TS, Shalam Md., Shanta Kumar SM.; Indian Journal of Pharmacology. 2007; 39: 43-47.

Objective: To investigate the hepatoprotective activity of alcoholic (ALLT) and aqueous (AQLT) extracts of leaves of *Tylophora indica* (asclepiadaceae) against ethanol-induced hepatotoxicity.

Materials and Methods: Leaf powder of *Tylophora indica* was successively extracted with alcohol and water. Preliminary phytochemical tests were done and the LD50 values for both extracts determined. The hepatoprotective activity of the ALLT and AQLT were assessed in ethanol-induced hepatotoxic rats.

Results: The ALLT showed presence of alkaloids, carbohydrates, steroids, saponins and triterpenes, while alkaloids, carbohydrates and saponins were present with AQLT. The ALLT did not produce any mortality even at 5000 mg/kg while LD50 of AQLT was found to be 3162 mg/kg. Ethanol produced significant changes in physical (increased liver weight and volume), biochemical (increase in serum alanine transaminase, aspartate transaminase, alkaline phosphatase, direct bilirubin, total bilirubin, cholesterol, triglycerides and decrease in total protein and albumin level), histological (damage to hepatocytes) and functional (thiopentone-induced sleeping time) liver parameters. Pretreatment with ALLT or AQLT extract significantly prevented the physical, biochemical, histological and functional changes induced by ethanol in the liver.

Conclusion: The present study indicates that ALLT and AQLT extracts possessed hepatoprotective activity. The alcoholic extract was found to exhibit greater hepatoprotective activity than the aqueous extract.

Paralytic Effect of Alcoholic Extract of *Allium sativum* and *Piper longum* on Liver Amphistome, *Gigantocotyle explanatum*

Singh TU, Kumar D, Tandan SK.; Indian Journal of Pharmacology. 2008; 40: 64-8.

Objective: To investigate the effects of alcoholic extract of *Allium sativum* and *Piper longum* on the muscular activity of a parasitic amphistome, *Gigantocotyle explanatum.*

Materials and Methods: Amphistomes were isometrically mounted to record the spontaneous muscular activity by using Chart 4 software program (Power Lab, AD Instruments, Australia) and to examine the effects of cumulative doses (100, 300, 1000, and 3000 μg/ml) of the plant extracts on the amplitude (g), frequency (per 10 min), and baseline tension (g) of the spontaneous muscular activity of the amphistome.

Results: Alcoholic extract of *A. sativum* produced significant reduction in the frequency and amplitude of contractile activity of the amphistome at 1000 and 3000 μg/ml bath concentrations. Complete paralysis of the amphistome was observed after 15 min of addition of 3000 μg/ml concentration. Alcoholic extract of *P. longum* also caused paralysis following 15-20 min exposure of the amphistome to 3000 μg/ml concentration. In both the cases the amphistomes did not recover from paralysis following 2-3 washes.

Conclusion: The observations demonstrate the paralytic effect of alcoholic extract of *A. sativum* and *P. longum* on *G. explanatum.*

Antioxidant Potential of the Methanol-Methylene Chloride Extract of *Terminalia glaucescens* Leaves on Mice Liver in Streptozotocin-induced Stress

Njomen GB, Kamgang R, Oyono JL, Njikam N. Indian Journal of Pharmacology. 2008; 40: 266-70.

Aim: The antioxidant effect of the methanol-methylene chloride extract of *Terminalia glaucescens* (Combretaceae) leaves was investigated in streptozotocin (STZ)-induced oxidative stress.

Methods: Oxidative stress was induced in mice by a daily dose of STZ (45 mg/kg body weight i.p.) for five days. From day one, before STZ injection, normal and diabetic-test mice received an oral dose of the extract (100 or 300 mg/kg b.w.) daily. Plasma metabolites, lipid peroxidation, and antioxidant enzymes in the liver were assessed and gain in body weight recorded.

Results: In normal mice the plant extract reduced food and water intake, blood glucose and LDL-C level and body weight gain, did not affect the lipid peroxidation in the liver, while the antioxidant enzyme activities seemed increased. Blood glucose was decreased ($P < 0.05$) in normal mice treated with 300 mg/kg extract. Diabetic mice pretreated with 100 mg/kg extract as diabetic control mice (DC) showed significant ($P < 0.001$) body weight loss, polyphagia and polydipsia, high plasma glucose level, decrease in the liver catalase, peroxidase, and superoxide dismutase activities, and increase in lipid peroxidation. The HDL-C level was lowered ($P < 0.05$) whereas LDL-C increased. In 300 mg/kg extract-pretreated diabetic mice the extract prevented body weight loss, increase of blood glucose level, lipid peroxidation in liver, food and water intake, and lowering of plasma HDL-C level and liver antioxidants; this extract prevented LDL-C level increase.

Conclusion: These results indicate that *T. glaucescens* protects against STZ-induced oxidative stress and could thus explain its traditional use for diabetes and obesity treatment or management.

Evaluation of Hepatoprotective Activity of *Cleome viscosa* Linn. Extract

Gupta NK, Dixit VK.; Indian Journal of Pharmacology. 2009; 41: 36-40.

Objectives: To evaluate the hepatoprotective activity of ethanolic extract of *Cleome viscosa* Linn. (Capparidaceae) against carbon tetrachloride (CCl_4) induced hepatotoxicity in experimental animal models.

Materials and Methods: Leaf powder of Cleome viscosa was extracted with ethanol. The hepatoprotective activity of the extract was assessed in CCl_4 induced hepatotoxicity in rats. Various biochemical parameters were estimated and histopathological studies were also performed on rat liver. The hepatoprotective activity was also supported by determining a functional parameter, *i.e.* thiopental-induced sleep of mice poisoned with CCl_4.

Results: The test material was found effective as hepatoprotective, through *in vivo* and histopathological studies. The extract was found to be effective in shortening the thiopental induced sleep in mice poisoned with CCl_4. The hepatoprotective effect of ethanolic extract was comparable to that of silymarin, a standard hepatoprotective agent.

Conclusion: The results of the present study show that ethanolic extract of *Cleome viscosa* has significant hepatoprotective activity.

Hepatoprotective Activity of *Eugenia jambolana* Lam. in Carbon Tetrachloride Treated Rats

Sisodia SS, Bhatnagar M.; Indian Journal of Pharmacology. 2009; 41: 23-7.

Objective: To estimate the hepatoprotective effects of the methanolic seed extract of *Eugenia jambolana* Lam. (Myrtaceae), in Wistar albino rats treated with carbon tetrachloride (CCl_4).

Materials and Methods: Liver damage in rats treated with CCl(4) (1ml/kg/Bw, administered subcutaneously, on alternate days for one week) was studied by assessing parameters such as serum glutamate oxaloacetate transaminase (SGOT), serum glutamate pyruvate transaminase (SGPT), alkaline phosphatase (ALP), acid phosphatase (ACP) and bilirubin (total and direct). The effect of co-administration of *Eugenia jambolana* Lam. (doses 100, 200 and 400 mg/kg p. o.) on the above parameters was investigated. These biochemical observations were supplemented by weight and histological examination of liver sections. Liv.52((R)) was used as positive control. Data were analyzed by one way ANOVA, followed by Scheff's/Dunnett's Test.

Results: Administration of *Eugenia jambolana* Lam. (doses 100, 200 and 400 mg/kg p. o.) significantly prevented carbon tetrachloride induced elevation of serum SGOT, SGPT, ALP, ACP and bilirubin (total and direct) level. Histological examination of the liver section revealed hepatic regeneration, after administration of various doses of *Eugenia jambolana* Lam. The results were comparable to that of Liv.52(R).

Conclusion: The study suggests preventive action of *Eugenia jambolana* Lam. in carbon tetrachloride induced liver toxicity. Hepatic cell regeneration process was dose-dependent.

Hepatoprotective Activity of Petroleum Ether, Diethyl Ether, and Methanol Extract of *Scoparia dulcis* L. against CCl_4-induced Acute Liver Injury in Mice

Praveen TK, Dharmaraj S, Bajaj J, Dhanabal SP, Manimaran S, Nanjan MJ, Razdan R.; Indian Journal of Pharmacology. 2009; 41: 110-4.

Objectives: The present study was aimed at assessing the hepatoprotective activity of 1:1:1 petroleum ether, diethyl ether, and methanol (PDM) extract of *Scoparia dulcis* L. against carbon tetrachloride-induced acute liver injury in mice.

Materials and Methods: The PDM extract (50, 200, and 800 mg/kg, p.o.) and standard, silymarin (100 mg/kg, p.o) were tested for their antihepatotoxic activity against CCl4-induced acute liver injury in mice. The hepatoprotective activity was evaluated by measuring aspartate aminotransferase, alanine aminotransferase, alkaline phosphatase, and total proteins in serum, glycogen, lipid peroxides, superoxide dismutase, and glutathione reductase levels in liver homogenate and by histopathological analysis of the liver tissue. In addition, the extract was also evaluated for its *in vitro* antioxidant activity using 1, 1-Diphenyl-2-picrylhydrazyl-scavenging assay.

Results: The extract at the dose of 800 mg/kg, p.o., significantly prevented CCl_4-induced changes in the serum and liver biochemistry ($P < 0.05$) and changes in liver histopathology. The above results are comparable to standard, silymarin (100 mg/kg, p.o.). In the *in vitro* 1, 1-diphenyl-2-picrylhydrazyl scavenging assay, the extract showed good free radical scavenging potential (IC 50 38.9 +/- 1.0 mug/ml).

Conclusions: The results of the study indicate that the PDM extract of *Scoparia dulcis* L. possesses potential hepatoprotective activity, which may be attributed to its free radical scavenging potential, due to the terpenoid constituents.

Hepatoprotective and Anti-inflammatory Activities of *Plantago major* L.

Türel I, Ozbek H, Erten R, Oner AC, Cengiz N, Yilmaz O. Indian Journal of Pharmacology. 2009; 41: 120-4.

Objective: The aim of this study was to investigate anti-inflammatory and hepatoprotective activities of *Plantago major* L. (PM).

Materials and Methods: Anti-inflammatory activity: Control and reference groups were administered isotonic saline solution (ISS) and indomethacin, respectively. *Plantago major* groups were injected PM in doses of 5 mg/kg (PM-I), 10 mg/kg (PM-II), 20 mg/kg (PM-III) and 25 mg/kg (PM-IV). Before and three hours after the injections, the volume of right hind-paw of rats was measured using a plethysmometer. HEPATOPROTECTIVE ACTIVITY: The hepatotoxicity was induced by carbon tetrachloride (CCl_4) administration. Control, CCl_4 and reference groups received isotonic saline solution, CCl_4 and silibinin, respectively. Plantago major groups received CCl_4 (0.8 ml/kg) and PM in doses of 10, 20 and 25 mg/kg, respectively for seven days. Blood samples and liver were collected on the 8th day after the animals were killed.

Results: Plantago major had an anti-inflammatory effect matching to that of control group at doses of 20 and 25 mg/kg. It was found that reduction in the inflammation was 90.01 per cent with indomethacin, 3.10 per cent with PM-I, 41.56 per cent with PM-II, 45.87 per cent with PM-III and 49.76 per cent with PM-IV. Median effective dose (ED50) value of PM was found to be 7.507 mg/kg. Plantago major (25 mg/kg) significantly reduced the serum alanine aminotransferase (ALT) and aspartate aminotransferase (AST) levels when compared to the CCl_4 group. The histopathological findings showed a significant difference between the PM (25 mg/kg) and CCl_4 groups.

Conclusion: The results showed that PM had a considerable anti-inflammatory and hepatoprotective activities.

Amelioration Effects against N-nitrosodiethylamine and CCl_4-Induced Hepatocarcinogenesis in Swiss Albino Rats by Whole Plant Extract of *Achyranthes aspera*

Kartik R, Rao ChV, Trivedi SP, Pushpangadan P, Reddy GD.; Indian Journal of Pharmacology. 2010; 42: 370-5.

Objective: The prevalence of oxidative stress may be implicated in the etiology of many pathological conditions. Protective antioxidant action imparted by many plant extracts and plant products make them a promising therapeutic drug for free-radical-induced pathologies. In this study, we assessed the antioxidant potential and suppressive effects of *Achyranthes aspera* by evaluating the hepatic diagnostic markers on chemical-induced hepatocarcinogenesis.

Materials and Methods: The *in vivo* model of hepatocarcinogenesis was studied in Swiss albino rats. Experimental rats were divided into five groups: control, positive control (NDEA and CCl_4, *A. aspera* treated (100, 200, and 400 mg/kg b.w.). At 20 weeks after the administration of NDEA and CCl_4, treated rats received *A. aspera* extract (AAE) at a dose of 100, 200, and 400 mg/kg once daily route. At the end of 24 weeks, the liver and relative liver weight and body weight were estimated. Lipid peroxidation (LPO), superoxide dismutase (SOD), catalase (CAT), glutathione peroxidase (GPx), glutathione-S-transferase (GST), and reduced glutathione (GSH) were assayed. The hepatic diagnostic markers namely serum glutamic oxaloacetic transminase (AST), serum glutamic pyruvate transminase (ALT), serum alkaline phosphatase (ALP), gamma glutamyl transpeptidase (GGT), and bilirubin (BL) were also assayed, and the histopathological studies were investigated in control, positive control, and experimental groups.

Results: The extract did not show acute toxicity and the per se effect of the extract showed decrease in LPO, demonstrating antioxidant potential and furthermore no change in the hepatic diagnosis markers was observed. Administration of AAE suppressed hepatic diagnostic and oxidative stress markers as revealed by decrease in NDEA and CCl_4 -induced elevated levels of SGPT, SGOT, SALP, GGT, bilirubin, and LPO. There was also a significant elevation in the levels of SOD, CAT, GPx, GST, and GSH as observed after AAE treatment. The liver and relative liver weight were decreased after treatment with AAE in comparison to positive control group. The architecture of hepatic tissue was normalized upon treatment with extract at different dose graded at 100, 200, and 400 mg/kg. b.w. in comparison to positive control group.

Conclusion: These results suggest that *A. aspera* significantly alleviate hepatic diagnostic and oxidative stress markers which signify its protective effect against NDEA and CCl_4-induced two-stage hepatocarcinogenesis.

Evaluation of Hepatoprotective Activity of *Cissus quadrangularis* Stem Extract against Isoniazid-induced Liver Damage in Rats

Viswanatha Swamy AH, Kulkarni RV, Thippeswamy AH, Koti BC, Gore A.; Indian Journal of Pharmacology. 2010; 42: 397-400.

Objective: The study was designed to investigate the hepatoprotective activity of methanol extract of *Cissus quadrangularis* (CQ) against isoniazid-induced hepatotoxicity in rats.

Materials and Methods: The successive petroleum ether (60-80°C) and methanol extracts of *C. quadrangularis* were used. Hepatic damage was induced in Wistar rats by administering isoniazid (54 mg/kg, p.o.) once daily for 30 days. Simultaneously, CQ (500 mg/kg p.o) was administered 1 h prior to the administration of isoniazid (54 mg/kg, p.o.) once daily for 30 days. Silymarin (50 mg/kg p.o) was used as a reference drug.

Results: Elevated levels of aspartate transaminase, alanine transaminase, alkaline posphatase, and bilirubin following isoniazid administration were significantly lowered due to pretreatment with CQ. Isoniazid administration significantly increased lipid peroxidation (LPO) and decreased antioxidant activities such as reduced glutathione, superoxide dismutase, and catalase. Pretreatment of rats with CQ significantly decreased LPO and increased the antioxidant activities.

Conclusion: The results of this study indicated that the hepatoprotective effect of CQ might be attributed to its antioxidant property.

In vitro and *in vivo* Hepatoprotective Effects of the Total Alkaloid Fraction of *Hygrophila auriculata* Leaves

Raj VP, Chandrasekhar RH, P V, S A D, Rao MC, Rao VJ, Nitesh K.; Indian Journal of Pharmacology. 2010; 42: 99-104.

Objective: To investigate the total alkaloid fraction of the methanol extract of leaves of *Hygrophila auriculata* for its hepatoprotective activity against CCl_4-induced toxicity in freshly isolated rat hepatocytes, HepG2 cells, and animal models.

Materials and Methods: Mature leaves of *H. auriculata* were collected, authenticated, and subjected to methanolic extraction followed by isolation of total alkaloid fraction. Freshly isolated rat hepatocytes were exposed to CCl_4 (1 per cent) along with/without various concentrations of the total alkaloid fraction (80-40 microg/ml). Protection of human liver-derived HepG2 cells against CCl_4-induced damage was determined by the MTT assay. Twenty-four healthy Wistar albino rats (150-200 g) of either sex were used for the *in vivo* investigations. Liver damage was induced by administration of 30 per cent CCl_4 suspended in olive oil (1 ml/kg body weight, i.p).

Results: The antihepatotoxic effect of the total alkaloid fraction was observed in freshly isolated rat hepatocytes at very low concentrations (80-40 microg/ml). A dose-dependent increase in the percentage viability was observed when CCl_4-exposed HepG2 cells were treated with different concentrations of the total alkaloid fraction. Its *in vivo* hepatoprotective effect at 80 mg/kg body weight was comparable with that of the standard Silymarin at 250 mg/kg body weight.

Conclusion: The total alkaloid fraction was able to normalize the biochemical levels which were altered due to CCl_4 intoxication.

Olanzapine-Induced Hepatopathy in Albino Rats: A Newer Model for Screening Putative Hepatoprotective Agents, Namely Silymarin

Sengupta P, Bagchi C, Sharma A, Majumdar G, Dutta C, Tripathi S.; Indian Journal of Pharmacology. 2010; 42: 376-9.

Backgrounds: This study was conducted to establish olanzapine-induced hepatopathy in Wistar albino rats as a newer model to screen putative hepatoprotective agents namely silymarin.

Materials and Methods: Albino rats were divided into three groups, namely vehicle control group (CG), olanzapine-treated group (OZ), and olanzapine plus silymarin (OZS) treated groups. Both the OZ and OZS groups were treated with the same dose of intraperitoneal olanzapine for 6 weeks and group OZS additionally received oral silymarin. Baseline and terminal hepatic enzymes (SGOT, SGPT, and ALP) were measured in all three groups.

Results: Histopathological examination of livers of both OZ and OZS groups showed degenerative changes, whereas those of control group showed normal architecture. Liver enzyme levels showed statistically significant rise in comparison to the control group as well as the respective base line values in both the test groups, but the differences in the rise of liver enzymes between the two test groups were not statistically significant.

Conclusion: Olanzapine-induced hepatopathy in rats can be used as a model for screening putative hepatoprotective agents and in our setting silymarin has failed to provide any hepatoprotection.

Hepatoprotective Activity of *Pterocarpus santalinus* L.f., an Endangered Medicinal Plant

Manjunatha BK.; Indian Journal of Pharmacology. 2006; 38: 25-28.

Objective: To evaluate the hepatoprotective activity of crude aqueous and ethanol stem bark extracts of *Pterocarpus santalinus* (Fabaceae) using CCl_4 induced hepatic damage in male Wistar albino rats.

Materials and Methods: The aqueous (45 mg/ml) and ethanol (30 mg/ml) extracts of stem bark in 1 per cent gum tragacanth was administered orally for 14 days and the hepatoprotective activity studied in CCl_4 induced hepatic damage model. The hepatoprotective activity was assessed using various biochemical parameters like serum bilirubin, protein, alanine transaminase, aspartate transaminase and alkaline phosphatase along with histopathological studies of liver tissue.

Results: There was a significant increase in serum levels of bilirubin, alanine transaminase, aspartate transaminase and alkaline phosphatase with a decrease in total protein level, in the CCl_4 treated animals, reflecting liver injury. In the aqueous and ethanol extracts treated animals there was a decrease in serum levels of the markers and significant increase in total protein, indicating the recovery of hepatic cells. Histological study of aqueous extract treated group exhibited moderate accumulation of fatty lobules and cellular necrosis where as ethanol extract treated animals revealed normal hepatic cords without any cellular necrosis and fatty infiltration.

Conclusion: The ethanol and aqueous stem bark extract of *P. santalinus* afforded significant protection against CCl4 induced hepatocellular injury.

Diclofenac-Induced Biochemical and Histopathological Changes in White Leghorn Birds (*Gallus domesticus*)

Jain T, Koley KM, Vadlamudi VP, Ghosh RC, Roy S, Tiwari S, Sahu U.; Indian Journal of Pharmacology. 2009; 41: 237-41.

Objective: To evaluate diclofenac-induced biochemical and histopathological changes in White Leghorn birds.

Materials and Methods: Six-week-old birds were equally divided into three groups of six birds each. Group I served as control and received vehicle orally. The birds of Group II and III were orally administered with a single low (2 mg/kg) and high dose (20 mg/kg) of diclofenac sodium, respectively, and were observed for 7 days. The acute toxicity was assessed by observing the clinical signs and symptoms, mortality, alterations in blood biochemistry, and necropsy findings.

Results: The birds of Group II showed only mild symptoms of diarrhoea. In Group III, 50 per cent of birds died in between 24 and 36 h post-treatment showing the symptoms of segregatory behaviour, lethargy, terminal anorexia, and severe bloody diarrhoea. The birds of Group II and the surviving birds of Group III showed a significantly ($P<0.05$) increased plasma uric acid, creatinine and plasma glutamic pyruvic transaminase (PGPT), and decreased total protein and albumin at 12 and 24 h post-treatment which returned to the normal levels at 36 h post-treatment. The dead birds of the high-dose group also showed similar pattern of biochemical changes at 12 and 24 h post-treatment and revealed extensive visceral gout with characteristic histopathological lesions in liver, kidney, heart, spleen, and intestine on post-mortem.

Conclusion: The results indicate that diclofenac sodium has hepatotoxic, nephrotoxic, and visceral gout inducing potentials in White Leghorn birds, especially at higher dose.

Evaluation of Hepatoprotective Effect of *Piper-longum* (Pippali) and *Withania somnifera* (Ashwagandha) in Hepatotoxicity Induced by Anti-tubercular Drugs

Chhajed S, Baghel MS, Ravishankar B, Singh G.; J Res Ayur Siddha. Jul-Sep, 1991.

Effect of concomitant administration of *Piper longum* (Pippali) and *Withania somnifera* (Ashvagandha), two well known rasayana drugs were evaluated against hepatotoxicity induced by rifampn and isoniazid administration in mice. Rifampin an isoniazid treatment induced marked histopathological changes which were significantly inhibited by treatment with *P. longum* and *W. somnifera*. No significant changes were noted in the biochemical and ponderal parameters studies. Results of the study indicate that rasayana drugs can be used as important adjuvants with antitubercular drug therapy.

Clinical Evaluation of Athimathuram in Viral Hepatitis: A Controlled Study

S. Rajalakshmi, G. Sivanandam, G. Veluchamy; J RES AYUR SIDDHA. 1997; XVIII (3-4), 141-146.

A clinical trial was carried out with placebo control on two identical groups of patients of viral hepatitis to evaluate the effect of Athimathuram (*Glycyrrhiza glabra* Linn.) against viral hepatitis. The clinical and biochemical clearance of the disease was accelerated in drug group when compared with its placebo control as evidenced by their laboratory parameters.

Role of *Tinospora cordifolia* (wild) Miers. (Guduchi) in the Treatment of Infective Hepatitis

S. Prakash and N. P. Rai; J RES AYUR SIDDHA. 1996; XVII, (1-2), 58-68.

In this series of clinical studies 20 patients of infective hepatitis were selected on the basis of clinical and biochemical findings. Four tablets (500mg. each) thrice in a day, orally with fresh water were given to the patients for 4 weeks, comparison between before and after treatment of these patients (N=20) were showed that Guduchi plays an important role in relieving the symptoms as well as normalization of altered liver function test. Maximum laboratory changes which were highly significant *i.e.*, serumbilirubin alkaline phosphatase, S.G.O.T and S.G.P.T ($P < 0.001$). The majority of cases *i.e.*, 15 cases (75 per cent) showed cured and 5 cases (25 per cent) improved. The mode of action of *Tinospora cordifolia* (Wild.) Miers. is under consideration and not any side or toxic effect was observed during the clinical study. It is an ideal, cheaper herbal preparation as per Ayurvedic principles, which is easily and frequently available in every part of our country.

Effect of *Tinospora cordifolia* (Wild) Miers (Amrita) on Kamala Roga (Jaundice)

Mahendra Prasad, N. P. Rai, K. Tripathi; J RES AYUR SIDDHA. 1996; XVII (1-2), 69-76.

Considering the chronic nature and complications of hepatitis and non-availability of its satisfactory treatment, it is necessary to look for drugs from indigenous system of medicine. Accordingly Amrita (*Tinospora cordifolia* (Wild). Miers. Has been taken up on 20 patients of Kamala Roga (Jaundice of viral origin). The drug provided good clinical relief as well as biochemical improvement.

Effect of Kadugurohini (*Picrorhiza kurroa* Royle) in the Treatment of Viral Hepatitis: A Double-Blind Study with Placebo Control

S. Rajalakshmi, G. Sivanandam, G. Veluchamy; J RES AYUR SIDDHA. 1997; XVIII (3-4), 141-146.

A double blind clinical trial with placebo control on two identical groups was conducted to evaluate the efficacy of Kadugurohini (*Picrorhiza kurroa* Royle ex Benth) on viral hepatitis. The clinical and biochemical clearance was 100 per cent and highly significant ($P<0.001$) in case of Kadugurohini group. But in placebo group only partial clearnce was observed and the bio-chemical clearance was not significant($P>0.05$).

A Comparative Study of Ayurvedic Drugs *Picorrhiza kurroa* (Kutaki) and *Berberis aristats* (Daru Haridra) in Acute Viral Hepatitis at Varanasi (India)

D. S. Singh, S. S. Gupta, S. A. Ansari, R. H. Singh; J RES AYUR SIDDHA, Oct-Dec., 1991.

A prospective double-blind controlled trial of two potent indigenous drugs, *B. aristata* (Daru Haridra) and *P. kurroa* (Kutaki) was conducted on 42 uncomplicated cases of acute viral hepatitis. It was observed that clinical and biochemical recovery was earlier and better in drugs treated cases as compared to controls. Further, observed therapeutic response was superior in patients who received Kutaki as compared to Daru Haridra.

Advances in Hepatology: A Review of Recent Researches in Ayurveda

Suresh Kumar, V. N. Pandey, Gurdeep Singh, K. P. Singh, G. N. Chaturvedi; J RES AYUR SIDDHA. 1991.

Liver diseases with their management are well detailed in Ayurvedic texts. Ayurvedic physicians have been successfully treating various liver diseases with Arogyawardhani, Kutaki, Kumari asava, Daruhaidra, Kalmegh and other indigenous single and compound drugs.

But reports on clinical and experimental trials to evaluate the role of these Ayurvedic drugs are not many. Recently certain clinical and experimental researches has been conducted in various research institutions. Kutaki has been reported with 95 per cent excellent results in a study on twenty patients of infective hepatitis. It is also reported to have increased billiary output, decrease specific gravity and viscosity of bile with normalization of biochemical constituents of bile in experimental dogs. Two other compounds drugs, Kutaki compound and Arogyawardhani have also showed very encouraging results when clinically evaluated for the treatment of hepatocellular jaundice, chronic hepatitis, viral hepatitis and certain other hepato billiary diseases. Daruharidra an Kumariashava have also exhibited remarkable results when evaluated for he treatment of hepatocellular jaundice. Recently Kalmegha was tried in twenty patients of infective hepatitis with 80 per cent cure rate. These advances are overall encouraging and certainly invite the attention of specialized therapists and researchers of this field.

A Preliminary Open Trial on Interferon Stimulator (SNMC) Derived from *Glycyrrhiza glabra* in the Treatment of Subacute Hepatic Failure

S. K. Acharya, S. Dasarathy, A. Tandon, Y. K. Joshi, B. N. Tandon; J RES AYUR SIDDHA. 1993; 69-74.

The efficacy of the interferon stimulator named Stronger New Minophagen - C (SNMC) derived form the plant *G. glabra* was studied at a dose of 40 or 100 ml. daily for 30 days followed by thrice weekly intravenously for 8 wk in 18 patients of subacute hepatic failure due to viral hepatitis. The survival rate amongst these patients was 72.2 per cent, as compared to the earlier reported rate of 31.3 per cent in 98 patients who received supportive therapy (P<0.01). Death in favour of the five patients was due to associated infections leading to hepatorenal failure and terminal coma. Further studies are necessary to standardize the dose and duration of therapy with SNMC in subacute hepatic failure.

Hepatoprotective Activity of Kutkin: The Iridoid Glycoside Mixture of *Picrorhiza kurroa*

Ansari RA, Aswal BS, Chander R, Dhawan BN, Garg NK, Kapoor NK, Kulshreshtha DK, Mehdi H, Mehrotra BN, Patnaik GK, et al., J Res Ayur Siddha. 1988, 401-404.

The alcoholic extract of the root and rhizome of *P. kurrooa* exhibited hepatoprotective activity in rat a mastomys. The active principle was identified as kuthin and the kutkin in-free fractions of the extract were found to be devoid of any activity. Kutkin showed significant hepatoprotective activity in hepatic damage induced by galactosamine (in rats) and Plasmodium berghei (in mastomys) as assessed by changes in several serum and liver biochemical parameters.

Controlled Eexperimental Study of Ghanasatwa

Kumar D, Dube CB, Srivastava PS; J Res Ayur Siddha. 1981; 1

In the present study 34 healthy adult rabbits were taken. Leaving 6 rabbits, rest were administered CCl_4, to damage the liver. Then they were divided in four groups: Group A was given neither CCl_4 not any drug. Group B was challenged with CCl_4 and was given not any frug. Group C was challenged with CCl_4, and treated with Liv-52 syrup. Group D was challenged with CCl_4 and treated with E.alba ghanasatwa, 90mg./kg BW. After follow up with laboratory and hisjtopathological investigations, the results are that (1) Liv-52 is capable to normalize the liver, challenged with CCl4 and (2) the *E. alba* Ghanasatwa too is also capable to normalize the same.

ACTIVITY ON IMMUNE FUNCTIONS

Immunomodulatory Effect of *Tinospora cordifolia* Extract in Human Immuno-deficiency Virus Positive Patients

Kalikar MV, Thawani VR, Varadpande UK, Sontakke SD, Singh RP, Khiyani RK.; Indian Journal of Pharmacology. 2008; 40: 107-10.

Objectives: To assess the safety and efficacy of TCE in human immuno-deficiency virus positive patients.

Materials and Methods: Efficacy of *Tinospora cordifolia* extract (TCE) in HIV positive patients was assessed in randomized double blind placebo controlled trial. 68 HIV positive participants were randomly assigned to two groups to receive either TCE or placebo for six months. After clinical examination TLC, DLC, ESR, platelet count, hemoglobin and CD4 count were done. The hematological investigations were repeated at bimonthly intervals and CD4 count was repeated at the end of the study. Patients were clinically reviewed at monthly intervals for compliance, refill and ADR monitoring. The drugs were decoded at the end of the trial.

Results: TCE treatment caused significant reduction in eosinophil count and hemoglobin percentage. 60 per cent patients receiving TCE and 20 per cent on placebo reported decrease in the incidence of various symptoms associated with disease. Some of the common complaints reported by patients on TCE were anorexia, nausea, vomiting and weakness.

Conclusion: Tinospora cordifolia extract, a plant derived immunostimulant, significantly affected the symptoms of HIV. This was validated by clinical evaluation. However not all of the objective parameters studied by us, back this up. *Tinospora cordifolia* could be used as an adjunct to HIV/AIDS management.

Immunomodulatory Activity of a Chinese Herbal Drug Yi Shen Juan Bi in Adjuvant Arthritis

Perera PK, Li Y, Peng C, Fang W, Han C.; Indian Journal of Pharmacology. 2010; 42: 65-9.

Objective: To investigate the immunomodulating mechanisms of a Chinese herbal medicine Yi Shen Juan Bi (YJB) in treatment of adjuvant arthritis (AA) in rats.

Materials and Methods: Levels of serum tumor necrosis factor alpha (TNF-alpha) and interleukin-1beta (IL-1beta) were measured by the Enzyme-Linked Immunosorbent Assay (ELISA). Expression of TNF-alpha mRNA and IL-1beta mRNA in synovial cells was measured with the semi-quantitative technique of reverse transcription-polymerase chain reaction (RT-PCR), while caspase-3 was examined by western blot analysis.

Results: The administration of YJB significantly decreased the production of serum TNF-alpha and IL-1beta. It also decreased significantly the TNF-alpha mRNA, IL-1beta mRNA, and caspase-3 expression in synoviocytes.

Conclusions: YJB produces the immunomodulatory effects by downregulating the over-activated cytokines, while it activates caspase-3, which is the key executioner of apoptosis in the immune system. This may be the one of the underlying mechanisms that explains how YJB treats the rheumatoid arthritis.

Immunosuppressive Properties of *Pluchea lanceolata* Leaves

Bhagwat DP, Kharya MD, Bani S, Kaul A, Kour K, Chauhan PS, Suri KA, Satti NK. Indian Journal of Pharmacology. 2010; 42: 21-6.

Objective: To investigate the immunosuppressive potential of *Pluchea lanceolata* 50 per cent ethanolic extract (PL) and its bioactive chloroform fraction (PLC).

Materials and Methods: Preliminary screening of the *Pluchea lanceolata* 50 per cent ethanolic extract (PL) was carried out with basic models of immunomodulation, such as, the humoral antibody response (hemagglutination antibody titers), cell-mediated immune response (delayed-type hypersensitivity), skin allograft rejection test, *in vitro* (C. albicans method), and *in vivo* phagocytosis (carbon clearance test). The extract was then fractionated with chloroform, n-butanol, and water to receive the respective fractions by partitioning. These fractions were employed for flow cytometry to study the T-cell specific immunosuppressive potential of these fractions.

Results: Oral administration of PL at doses of 50 to 800 mg/kg in mice, with sheep red blood cells (SRBC) as an antigen, inhibited both humoral and cell-mediated immune responses, as evidenced by the production of the circulating antibody titer and delayed-type hypersensitiviy reaction results, respectively, and the immune suppression was statistically significant ($P < 0.01$) in Balb/C mice. PL also decreased the process of phagocytosis both *in vitro* (31.23 per cent) and ex vivo (32.81 per cent) and delayed the graft rejection time (30.76 per cent). To study the T-cell-specific activities, chloroform, n-butanol, and water fractions from *P. lanceolata* were tested for T-cell specific immunosuppressive evaluation, wherein only the chloroform fraction (PLC) showed significant ($P < 0.01$) suppression of CD8+/CD4+ T-cell surface markers and intracellular Th1 (IL-2 and IFN-(Y)) cytokines at 25 - 200 mg/kg p.o. doses. PLC, however, did not show significant suppression of the Th2 (IL-4) cytokine.

Conclusion: The findings from the present investigation reveal that *P. lanceolata* causes immunosuppression by inhibiting Th1 cytokines.

Effect of *Hemidesmus indicus* (Anantmool) Extract on IgG Production and Adenosine Deaminase Activity of Human Lymphocytes *in vitro*

Kainthla R P, Kashyap R S, Deopujari J Y, Purohit H J, Taori G M, Daginawala H F.; Indian Journal of Pharmacology 2006; 38: 190-3.

Objective: To investigate the effect of *Hemidesmus indicus* extract on activities of human peripheral blood lymphocytes *in vitro*.

Materials and Methods: The total extract of the raw herb was obtained by methanol: isopropyl alcohol: acetone extraction and used at different concentrations. Human peripheral blood lymphocytes (PBLs) were isolated, stimulated to proliferate using phytohaemagglutinin (PHA) or lipopolysaccharide (LPS), with and without different concentration of herbal extracts. Adenosine deaminase (ADA) activity and immunoglobulin (IgG) secretion from cultured PBLs were studied with the herbal extracts and appropriate controls.

Results: Hemidesmus indicus extract stimulated the cell proliferation at 1 mg/ml concentration significantly, after 72 h in culture. Viability of extract-treated PBLs was also maintained after culture. The extract increased the IgG production from cultured PBLs, when used at 1 mg/ml concentration. It also increased the ADA activity of PBLs after 72 h in culture.

Conclusion: An immunomodulatory activity of *H. indicus,* related to IgG secretion and ADA activity, is revealed during the study. The herbal extract has shown to promote the release of IgG by lymphocytes and also the ADA activity after 72 h of culture.

Evaluation of the Immunomodulatory Activity of the Methanol Extract of *Ficus benghalensis* Roots in Rats

Gabhe S Y, Tatke P A, Khan T A.; Indian Journal of Pharmacology 2006; 38: 271-5.

Objective: To evaluate the immunomodulatory activity of the aerial roots of *Ficus benghalensis* (Family Moraceae).

Materials and Methods: Various extracts of the aerial roots of *Ficus benghalensis* were evaluated for potential immunomodulatory activity, using the *in vitro* polymorphonuclear leucocyte (human neutrophils) function test. The methanol extract was evaluated for immunomodulatory activity in *in vivo* studies, using rats as the animal model. The extracts were tested for hypersensitivity and hemagglutination reactions, using sheep red blood cells (SRBC) as the antigen. Distilled water served as a control in all the tests.

Results: The successive methanol and water extracts exhibited a significant increase in the percentage phagocytosis versus the control. In the *in vivo* studies, the successive methanol extract was found to exhibit a dose related increase in the hypersensitivity reaction, to the SRBC antigen, at concentrations of 100 and 200 mg/kg. It also resulted in a significant increase in the antibody titer value, to SRBC, at doses of 100 and 200 mg/kg in animal studies.

Conclusion: The successive methanol extract was found to stimulate cell-mediated and antibody mediated immune responses in rats. It also enhanced the phagocytic function of the human neutrophils, *in vitro.*

Stimulation of Immune Function Activity by the Alcoholic Root Extract of *Heracleum nepalense* D. Don.

Dash S, Nath L K, Bhise S, Kar P, Bhattacharya S.; Indian Journal of Pharmacology 2006; 38: 336-40.

Objective: To assess the immunostimulatory activity of *H. nepalense,* using different *in vitro* and *in vivo* experimental models.

Materials and Methods: The immunostimulatory potential of the test compound was investigated by *in vitro*, phagocytic index and lymphocyte viability tests, using interferon a-2b, a known immunostimulant drug, as the standard. Other tests such as carbon clearance, antibody titer and delayed type hypersensitivity were studied in mice, using levimasole as the standard.

Results: The dried root extract (1000 μg/ml) and isolated quercetin glycoside (50 μg/ml) significantly increased the *in vitro* phagocytic index and lymphocyte viability in all assays. They also showed a significant increase in antibody titer, carbon clearance and delayed type hypersensitivity in mice.

Conclusion: H. nepalense exhibited a dose-dependent immunostimulant effect, which could be attributed to the flavonoid content or due to the combination with other component(s).

Effect of DLH-721A and DLH-721B (Polyherbal Formulations) on Rat Mesenteric Mast Cell Degranulation

K. Padmalatha, B.V. Venkataraman, R. Roopa; Indian Journal of Pharmacology 2000; 32: 7-10.

Objective: To study the antianaphylactic effect of DLH-721A and DLH-721B (herbal formulation) on the rat mesenteric mast cells.

Methods: The study was carried out on the rat mesenteries sensitised with sheep serum to induce mast cell degranulation. Mesenteries pretreated with prednisolone, disodium cromoglycate, DLH-721A and DLH-721B were analysed for the mast cell degranulation during the anaphylactic reactions.

Results: Treatment with DLH-721A and DLH-721B showed beneficial effect on degranulation of actively and passively sensitised mesenteric mast cells. The protective effect was comparable with that of prednisolone and disodium cromoglycate.

Conclusion: DLH-721A and DLH-721B may have anti-anaphylactic activity on the rat-mesenteric mast cells.

Effect of *Trichopus zeylanicus* Gaertn (Active fraction) on Phagocytosis by Permacrophages and Humoral Immune Response in Mice

A. Subramoniam, D.A. Evans, S. Rajasekharan, P. Pushpangadan; Indian Journal of Pharmacology 2000; 32: 221-225.

Objectives: To evaluate whether *Trichopus zeylanicus* (active fraction) influences (1) phagocytosis by mice peritoneal macrophages (2) antibody dependent complement mediated cytotoxicity to Ehrlichs ascitic carcinoma (EAC) cells and (3) humoral antibody response in mice.

Methods: Phagocytosis of opsonized sheep RBC by peritoneal macrophages obtained from control or *T. zeylanicus* (active fraction) treated (10-40 mg/kg, daily, p.o., 5 days) mice was determined. The *in vitro* effect of the drug on macrophage phagocytosis was also studied. The effect of the drug on antibody-dependent and complement-mediated toxicity (ACC) to EAC cells was carried out, using antiserum obtained from drug treated (p.o., daily for 5 days) mice which were challenged with or without EAC cells. The effect of the drug on humoral immune response in mice was studied by measuring haemagglutination antibody titre and Jerne's plaque forming assay, using sheep RBC as antigens.

Results: The drug treatment to mice resulted in stimulation of phagocytosis by peritoneal macrophages; but *in vitro* treatment to macrophages did not influence their phagocytic efficacy. The drug treatment enhanced ACC, haemagglutinating antibody titre and the number of antibody-producing spleen cells in mice.

Conclusion: The drug stimulates macrophage phagocytosis in mice; but not under *in vitro* conditions. The drug also enhances ACC-mediated cancer cell-killing and humoral antibody response in mice.

Effect of Immu-21, a Herbal Formulation on Granulocytemacrophage Colony Stimulating Factors, Macrophage Maturation and Splenic Plaque Forming Cells in Experimental Animals

Chatterjee S.; Indian Journal of Pharmacology 2001; 33: 442-444.

Objective: To investigate the mechanism of immunostimulating action of Immu-21.

Methods: Swiss albino mice were treated with Immu-21 at various doses (25, 50, 100 mg/kg, orally) for 15 and 30 days. Granulocyte-Macrophage Colony Stimulating Factors (GM-CSF) was estimated in the serum of the experimental animals. Numbers of esterase positive cells in the bone marrow were counted. Plaque forming cells in spleen of the experimental mice were estimated following sheep RBC challenge.

Results: Immu-21 significantly enhanced the GM-CSF activity, number of esterase positive bone marrow cells and plaque forming cells of spleen in experimental animals. The immunostimulatory effect of Immu-21 is dose-dependent.

Conclusion: The present study suggests that Immu-21 modulates macrophage maturation and function. Stimulation of plaque-forming cell production from spleen by this product helps in stimulating humoral arm of immunity in experimental animals.

Immunostimulant Profile of a Polyherbal Formulation RV08

M. Rajendra Babu, R.V. Krishna Rao, A. Annapurna, D. Ravi Krishna Babu; Indian Journal of Pharmacology 2001; 33: 454-455.

The present work was done on RV08, a polyherbal formulation developed by one of the authors (RVKR). Polyherbal formulations have been developed with a view to counteract the immunodeficient disorders. RV08 contains *Aspargus recemosus, Mucuna pruriens, Withania somnifera, Bombax malbaricum, Sphaeranthus indicus, Butea frondosa, Clerodenrum serratum and Sida cordifolia.* RV08 was evaluated for its ability to potentiate both specific and non-specific host defence responses. Six Swiss albino mice of either sex weighing 25-35 g were acclimatised in the laboratory for 3 weeks. The formulation was administered orally at a dose of 50 mg/kg for a period of 20 days as a supension in 5 per cent gum acacia. RV08 causes significant rise in blood lymphocyte count with 5 days administration and the rise is maintained at more or less same level after 10 days and 20 days administration. Splenic lymphocyte count was increased significantly after 10 days administration of RV08 and further rise is observed with 20 days administration. Peritoneal macrophage count was increased significantly with 10 days administration of RV08 and the rise is progressed throughout the experimental period *i.e.*, with 20 days administration. Footpad thickness test indicated that RV08 fails to show significant increase in paw volume with SRBC challenge in drug-treated (0.08+0.009) compared to control animals (0.063+0.003). The rise in serum antibody titre in drug treated mice (4.33+0.212) was not found to be significant when compared to control (3.83+0.308). In the present study, RV08 significantly increased the peritoneal macrophage count. RV08 showed significant rise of blood lymphocyte and splenic lymphocyte count in experimental animals which strongly suggest the possible involvement of RV08 as first line of defence through immunomodulation of lymphoid cells. RV08 is able to potentiate non-specific host defence mechanisms in mice (50 mg/kg, orally) with 20 days administration. RV08 failed to show significant rise in circulatory antibody titre and foot pad volume when administered for 5 days (50 mg/kg, orally). So the formulation RV08 is need to be studied with chronic administration and after secondary immunisation to assess its potentiatory activity on specific immunological responses.

Antianaphylactic Effect of DLH-3041 (Polyherbal Formulation) on Rat Mesenteric Mast Cell Degranulation

Padmalatha K, Venkataraman BV, Roopa R.; Indian Journal of Pharmacology 2002; 34: 119-122.

Objective: To study the antianaphylactic effect of DLH-3041 (Polyherbal Formulation) on the rat mesenteric mast cells.

Methods: The study was carried out on the rat mesenteries sensitized with sheep serum to induce mast cell degranulation. Mesenteries pretreated with prednisolone, disodium cromoglycate, DLH-3041 were analyzed for the mast cell degranulation during the anaphylactic reactions.

Results: Treatment with DLH-3041 showed beneficial effect on degranulation of actively ($P<0.005$) and passively ($P<0.05$-0.005) sensitized mesenteric mast cells. The protective effect was comparable with that of prednisolone and disodium cromoglycate and was also observed after one week of withdrawal of the compound.

Conclusion: Antianaphylactic activity of DLH-3041 may be possibly due to the membrane stabilizing potential, suppression of antibody production and inhibition of antigen-induced histamine release.

Immunostimulant Activity of *Ashtamangal Ghrita* in Rats

Fulzele S.V., Bhurchandi P.M., Kanoje V.M., Joshi S.B., Dorle A.K.; Indian Journal of Pharmacology 2002; 34: 194-197.

Objective: To study the Immunostimulant effect of an Indian Ayurvedic Polyherbal formulation, *Ashtamangal ghrita* (AG) in healthy albino rats.

Methods: AG was administered orally at doses of 150 mg/kg/day and 300 mg/kg/day to healthy rats. The assessment of immunostimulant activity was carried out by testing the humoral (antibody titre) and cellular (foot pad swelling) immune responses to the antigenic challenges with sheep RBCs and by neutrophil adhesion test.

Results: Orally administered AG showed a significant increase of test parameters *viz.* neutrophil adhesion, haemagglutinating antibody titre (HAT) and delayed type hypersensitivity (DTH) response. In rats immunized with sheep RBC, AG enhanced the humoral antibody response to the antigen and significantly potentiated the cellular immunity by facilitating the footpad thickness response to sheep RBC in sensitized rats. With a dose of 300 mg/kg/day the values of HAT and DTH responses were 455.08 ± 0.75 and 31.0 ± 10.72 respectively, in comparison to the control group. These differences were statistically significant.

Conclusion: The study demonstrates the immunostimulant activity of *Ashtamangal ghrita* in rats.

Study the Immunomodulatory Activity of *Haridradi ghrita* in Rats

Fulzele S.V., Satturwar P.M., Joshi S.B., Dorle A.K.; Indian Journal of Pharmacology 2003; 35: 51-54.

Objective: To study the immunomodulatory effect of a polyherbal formulation, Haridradi Ghrita (HG) in rats.

Methods: Haridradi Ghrita was administered orally at doses of 50, 100, 200 and 300 mg/kg/day to healthy rats divided into five groups consisting of six animals each. The assessment of immunomodulatory activity was carried out by testing the humoral (antibody titre) and cellular (foot pad swelling) immune responses to the antigenic challenge by sheep RBCs and by neutrophil adhesion test.

Results: On oral administration HG showed a significant increase in neutrophil adhesion and delayed type hypersensitivity (DTH) response whereas the humoral response to sheep RBCs was unaffected. Thus HG significantly potentiated the cellular immunity by facilitating the footpad thickness response to sheep RBCs in sensitized rats. With a dose of 200 and 300 mg/kg/day the DTH response (mean+SD per cent increase in paw volume) was 10.52 ± 3.12 and 14.50 ± 2.38 respectively, in comparison to the corresponding value of 6.01 ± 1.85 for the untreated control group. These differences in DTH response were statistically significant ($P<0.05$).

Conclusion: The study demonstrates that HG shows preferential stimulation of the components of cell-mediated immunity and shows no effect on the humoral immunity.

In vivo Evaluation of Antioxidant Activity of Alcoholic Extract of *Rubia cordifolia* Linn. and its Influence on Ethanol-induced Immunosuppression

Joharapurkar AA., Zambad SP, Wanjari MM, Umathe SN.; Indian Journal of Pharmacology 2003; 35: 232-236.

Objective: To evaluate the *in vivo* antioxidant activity of alcoholic extract of the roots of *Rubia cordifolia* Linn. (RC) and to study its influence on ethanol-induced impairment of immune responses.

Methods: The ethanol-treated (2 g/kg, 20 per cent w/v, p.o., daily for four weeks) rats concurrently received either RC or a combination of vitamin E and C (each 100 mg/kg, p.o.) daily for the same period. The parameters like phagocytosis, total leukocyte count (TLC), humoral and cell-mediated immune responses, lipid peroxidation (LPO), reduced glutathione (GSH) content, superoxide dismutase (SOD) and catalase (CAT) activities were assessed.

Results: Chronic administration of ethanol decreased the humoral and cell-mediated immune response, phagocytosis, phagocytosis index, TLC, GSH, CAT and SOD activities and increased the LPO. These influences of ethanol were prevented by concurrent daily administration of RC and the effect was comparable with that of the combination of vitamin E and C.

Conclusion: The ethanol-induced immunosuppression is due to oxidative stress and *Rubia cordifolia* can prevent the same by virtue of its *in vivo* antioxidant property.

Preliminary Studies on the Immunomodulatory and Antioxidant Properties of *Selaginella* Species

Gayathri V, Asha V V, Subramoniam A.; Indian Journal of Pharmacology 2005; 37: 381-5.

Aim: To evaluate the immunomodulatory and antioxidant properties, if any, of *S. elaginella* involvens, *S. delicatula* and *S. wightii.*

Materials and Methods: Immunomodulatory activity of the whole plants (water suspension) was studied in mice immunized with sheep RBC. The plant extracts were tested for their effect on lipid peroxidation (*in vitro* and *in vivo* in mice), and for *in vitro* hydroxyl radical scavenging activity. The most promising extract (water extract of *S. involvens*) was evaluated for its short-term toxicity in mice.

Results: The dried suspension (500 mg/kg) of the three plants did not influence humoral antibody titre and the number of antibody secreting cells in the mouse spleen. However, the plant suspensions as well as the water extracts (and not the other extracts) of the plants remarkably increased the weight of thymus in adult mice, and not in suckling mice. This effect was very marked in the case of *S. involvens* compared to the other two species. Although the water extract of all the three plants showed varying degrees of antioxidant activity, the antilipid peroxidation activity of *S. involvens* (water extract) was remarkable [EC50: 2 μg/ml]. This extract did not exhibit any conspicuous toxicity in mice in general, short-term toxicity evaluation. At high dose, serum cholesterol level was significantly reduced.

Conclusion: Out of the three Selaginella species studied, the water extract of *S. involvens* has promising thymus growth stimulatory activity in adult mice and remarkable antilipid peroxidation property; these observations are of interest in view of tribal and folklore belief that this plant prolongs life span.

Immunosuppressive Effect of Medicinal Plants of Kolli Hills on Mitogen-Stimulated Proliferation of the Human Peripheral Blood Mononuclear Cells *In vitro*

Arokiyaraj S, Perinbam K, Agastian P, Balaraju K.; Indian Journal of Pharmacology. 2007; 39: 180–183.

Four medicinal plant species were collected from the Kolli hills of Tamil Nadu and were screened for their immunosuppressive effect. The plants were shade dried and extracted with methanol. The crude methanol extracts were tested for inhibition of lymphocyte proliferation via lymphocyte proliferation assay by ^{3}thymidine uptake. The test plants were *Justicia gendarussa, Plumbago indica, Aloe vera,* and *Aegle marmelos.* Among the plants tested *J. gendarussa* (100 μg/ml) showed the highest lymphocyte inhibition (84 per cent). Sequential extraction of *J. gendarussa* in various solvents (n-hexane, benzene, ethyl acetate, chloroform, acetone, ethanol, and water) confirmed that all of the above extracts at 50 μg/ml, aqueous extract inhibited lymphocyte proliferation. Further, 17 high performance liquid chromatography fractions were collected for the aqueous extract and fraction no. 15 showed maximum inhibition of lymphocyte proliferation. The present study indicates that these extracts should be investigated further for the possible presence of immunosuppressive components.

Clinical Study on Balya Effect of Vidanga Compound with Reference to Infant's Immunity

Haichandan BK.; J RES AYUR SIDDHA.

Vidanga compound an Ayurvedic recipe prepared in syrup base, comprises of 19 Balya drugs as described in

Ayurvedic tests and administered in the dose of 5 drops twice daily to 15 number of infants from the 11th day of birth for a period f six months. The aim of study is to assess its imuno-enhancive effect. In control group, 15 number of infants were given a particular conventional multi-vitamin drops. The effect of the medicines administered to both groups were assessed through alteration in immune - globulins (lgG, IgM, IgA) level, serum protein level, changes in anthropometric value and morbidity incidence (severity and frequency). The study reveals good response in 80 per cent cases of infants those were administered with Vidanga compound as compared to 60 per cent response in infants administered with multi vitamin drops. In treated group 6.67 per cent cases have shown poor response as compare to 20 per cent in control group. The findings of the study proved effectiveness of Vidanga compound in enhancing immune response in infants.

Effect of Sphatika in the Management of Tundikeri (Tonsillitis-Streptococcal Infection)

R. K. Yadava, R. H. Singh; J RES AYUR SIDDHA. 1996; XVII (3-4), 149-156.

A clinical study was conducted on 36 cases to evaluate the effect of Sphatika in the management of Tonsillitis. The study was based on the symptomatology of the disease and all the cases under study were subject to throat swab test for positive evidence of streptococcal infection.

Purified *Sphatika* was prescribed both for local application as well as for internal application, thrice daily for seven days. The Study reveals that Sphatika bneing Kasaya, Katu, Tikta, Usna, Ksariya and having antibiotic properties can be a very effective and cheap remedy in the management of tonsillitis.

Effect of *Ocimum sanctum* Linn. on Humoral Immune Responses

Mediratta PK, Dewan V, Bhattacharya SK, Gupta VS, Maiti PC, Sen P.; J Res Ayur Siddha. 1988; 384-386.

Ocimum sanctum Linn. Also known as the sacred Basil is extensively used in the Ayurvedic system of medicine for various ailments including allergic conditions. Recently the plant has been reported to evince significant antistress properties. It is known that some of the psycho-social situations and stressful conditions that modify the susceptibility of an indivildual to a variety of illnesses including infectious and allergic disorders also influence the immune processes. The beneficial effects of *O. sanctum* could, therefore, be due to its direct or indirect effect on the immune system. The present study was undertaken to investigate the effects of *O. sanctum* on immune responses with particular reference to humoral immune responses in the experimental animals.

NEPHROPROTECTIVE ACTIVITY

Antiurolithiatic and Antioxidant Activity of *Mimusops elengi* on Ethylene Glycol-induced Urolithiasis in Rats

Ashok P, Koti BC, Vishwanathswamy AH.; Indian Journal of Pharmacology. 2010; 42: 380-3.

Objective: To evaluate the potential of *Mimusops elengi* in the treatment of renal calculi.

Materials and Methods: Petroleum ether, chloroform, and alcohol extracts of *Mimusops elengi* bark were evaluated for antiurolithiatic and antioxidant activity in male albino Wistar rats. Ethylene glycol (0.75 per cent) in drinking water was fed to all the groups (Groups II-IX) except normal control (Group I) for 28 days to induce urolithiasis for curative (CR) and preventive (PR) regimen. Groups IV, V, and VI served as CR, and groups VII, VIII, and IX as PR were treated with different extracts of M. elengi bark. Groups I, II, and III served as normal control, positive control (hyperurolithiatic), and standard (cystone 750 mg/kg), respectively. Oxalate, calcium, and phosphate were monitored in the urine and kidney. Serum BUN, creatinine, and uric acid were also recorded. *In vivo* antioxidant parameters such as lipid peroxidation (MDA), glutathione (GSH), superoxide dismutase (SOD), and catalase (CAT) were also monitored.

Results: All the extracts of *M. elengi* were safe orally and exhibited no gross behavioural changes in the rats. In hypercalculi animals, the oxalate, calcium, and phosphate excretion grossly increased. However, the increased deposition of stone-forming constituents in the kidneys of calculogenic rats were significantly ($P < 0.001$) lowered by curative and preventive treatment with alcohol extract (AlE) of *M. elengi*. It was also observed that alcoholic extract of *M. elengi* produced significant ($P < 0.001$) decrease in MDA, and increased GSH, SOD, and CAT. These results confirm that AlE of *M. elengi* possess potent antiurolithiatic activity.

Conclusion: The results obtained suggest potential usefulness of the AlE of *M. elengi* bark as an antiurolithiatic agent.

Effect of NR-AG-I and NR-AG-II (Polyherbal formulations) on Diuretic Activity in Rat

D.S. Samiulla, M.S. Harish; Indian Journal of Pharmacology 2000; 32: 112-113.

Objective: To study the comparative effect of NR-AG-I and NR-AG-II (polyherbal formulations) for diuretic activity on healthy albino rats.

Methods: The study was carried out on normal rats using frusemide as a standard reference drug. Rats were treated with frusemide, NR-AG-I and NR-AG-II. Urine was collected and its volume was recorded. Urinary levels of sodium, potassium and chloride were estimated.

Results: Treatment with NR-AG-II produced diuresis. The urine output increased from 4.4 to 9.1 ml/24 hrs. The level of electrolytes in urine also increased. NR-AG-I did not show any diuretic activity.

Conclusion: NR-AG-II has good diuretic activity on rats in the above experimental model.

Inhibitory Concentrations of *Lawsonia innermis* Dry Powder for Urinary Pathogens

Bhuvaneswari K, Gnana Poongothai S, Kuruvilla A, B. Appala R.; Indian Journal of Pharmacology. 2002; 34: 260-263.

Objective: To study the possibility of *in vitro* antimicrobial activity of *Lawsonia Innermis* (LI) leaves.

Method: Varying doses of LI suspensions (both powdered dried leaves and suspensions of fresh leaves) were tested for their antimicrobial activity against the urinary pathogens isolated from the patients urinary sample namely *E. coli, Pr. mirabilis, K. pneumoniae, Ps. aeroginosa* and *Staph. aureus,* using both tube turbidity standards and disc-diffusion method and their effects were observed.

Results: LI dried leaves suspension - Antimicrobial activity - Broth dilution method: Gram (-): 55-85 mg/ml and for Gram (+): >95 mg/ml. Disc diffusion method: >95 mg/ml for both Gram (-) and Gram (+) organisms. LI fresh leaves suspension - Antimicrobial activity - Disc diffusion method: *E. coli*: 10 mg/disc and for S. aureus: 25 mg/disc.

Conclusion: LI leaves have definite antimicrobial activity against the common urinary pathogens and the leaf components which are responsible for this action have to be isolated.

Modulatory Effect of *Plectranthus amboinicus* Lour. on Ethylene Glycol-induced Nephrolithiasis in Rats

Alvin Jose M, Ibrahim, Janardhanan S.; Indian Journal of Pharmacology 2005; 37: 43-4.

Present study was undertaken to evaluate the antilithiotic activity of the concentrated fresh juice of the leaves of Plectranthus amboinicus Lour. Histopathological studies clearly revealed that the tissue samples from the control group (G1) shows tubules with single epithelial lining along the margin and were of normal size. In G2 (lithiotic control), all the tubules showed the presence of crystals, there was marked dilatation of the tubules and total degeneration of the epithelial lining with infiltration of inflammatory cells into the interstitial space. In G3 (test Group) the specimen showed characters similar to the control group. Urine analysis showed a significant elevation of calcium, oxalates and total proteins level in the lithiotic control group (G2), when compared to normal control. The test group (G3), showed a significant reduction in all the parameters almost comparable with normal control. The urine and histopathological results clearly revealed the antilithiotic activity of P. amboinicus, particularly of calcium oxalate origin. Further research is needed to explore the exact active principle(s) responsible for the antilithiotic activity and the mechanism of action.

Protective Effect of *Kalanchoe pinnata* Pers. (Crassulaceae) on Gentamicin-Induced Nephrotoxicity in Rats

Harlalka GV, Patil CR, Patil MR.; Indian Journal of Pharmacology. 2007; 39: 201–205.

Objective: The present study was undertaken to evaluate the aqueous extract of *K. pinnata* for its protective effects on gentamicin-induced nephrotoxicity in rats.

Materials and Methods: Nephrotoxicity was induced in Wistar rats by intraperitoneal administration of gentamicin 100 mg/kg/day for eight days. Effect of concurrent administration of *K. pinnata* leaf extract at a dose of 125 mg/kg/day given by intraperitoneal route was determined using serum and urinary creatinine and blood urea nitrogen as indicators of kidney damage. The study groups contained six rats in each group. As nephrotoxicity of gentamicin is known to involve induction of oxidative stress, *in vitro* antioxidant activity and free radical-scavenging activity of this extract were evaluated.

Result: It was observed that the aqueous extract of *K. pinnata* leaves significantly protects rat kidneys from gentamicin-induced histopathological changes. Gentamicin-induced glomerular congestion, peritubular and blood vessel congestion, epithelial desquamation, accumulation of inflammatory cells and necrosis of the kidney cells were found to be reduced in the group receiving the leaf extract of *K. pinnata* along with gentamicin. This extract also normalized the gentamicin-induced increases in urine and plasma creatinine, blood urea and blood urea nitrogen levels. *In vitro* studies revealed that the *K. pinnata* leaf extract possesses significant antioxidant as well as oxidative radical scavenging activities.

Conclusion: It is proposed that the nephroprotective effect of the aqueous extract of *K. pinnata* leaves in gentamicin-induced nephrotoxicity may involve its antioxidant and oxidative radical scavenging activities.

Recent Approach in Clinical and Experimental Evaluation of Diuretic Action of Punarnava (*B. diffusa*) with Special Reference to Nephritic Syndrome

Singh RP, Singh RG, Shukla KP, Pandey BL, Usha, Singh RH.; J Res Ayur Siddha, Jan-March, 1992.

Various Ayurvedic herbal diuretics have been used by physicians since ancient times. Over past 50 years, many clinical and experimental studies have been conducted on these herbs. These studies showed usefulness of herbal medicines in treatment of various renal disorders. Herbal drugs, Punarnava (*B. diffusa*), Shara (*E. cynosorids*) *etc.* have been observed to possess diuretic effect. Results are reported herein of our clinical, experimental and immunological studies on Punarnava. The observations reveal equivalent diuretic effect to Furosemide. Punarnava increases the serum protein level and reduces urinary protein excretion in patients of Nephrotic syndrome. Increase was noted, also in the level of immunoglobulins and lower immune complex after one month of medication in patients of Nephrotic syndrome. Clinically Punarnava proved to be useful and safe drug in patients of Nephrotic syndrome.

Experimental Evaluation of Diuretic Action of Herbal Drug (*Tribulus terrestris* Linn.) on Albino Rats

Singh RG, Singh RP, Usha, Shukla KP, Singh P.; J Res Ayur Siddha, Jan-March, 1991.

The study was designed to evaluate the diuretic action of *Tribulus terrestris* Linn. (Gokshura) on the albino rat which has been claimed a diuretic in ancient Indian literature. The standardization was done by thin-layer chromatography. Study included 3 groups, containing 6 animals in each group designated as A,B,A respectively, namely control to and 20 times of the human dose. Human dose was crude drug. 1 g/kg/ day of body weight/day in the form of decoction. The weight showed progressive increase in all the groups with sluggish behaviour especially in group "C" on 15th and 30th Day. The heart rate and respiratory rate did not change much in the Group "B" and "C". The urinary pH became alkaline in all the groups with marked increase in sodium and potassium specially in group "C". The light microscopy showed normal glomerulus with distal tubular necrosis. In 20 times done on 30th day, showed evidence of renal infarction which was of cogulative types. The liver showed evidence of venous congestion and focal hepatic degeneration.

ANTIMICROBIAL ACTIVITY

In vivo and *In vitro* Effects of the Canova Medicine on Experimental Infection with *Paracoccidioides brasiliensis* in Mice

Takahachi G, Maluf M, Svidzinski T, Dalalio M, Bersani-Amado C A, Cuman R.; Indian Journal of Pharmacology 2006; 38(5): 350-4.

Objective: To evaluate the *in vivo* and *in vitro* activity of Canova in experimental infection with *Paracoccidioides brasiliensis.*

Materials and Methods: Mice infected with *P. brasiliensis*were treated with Canova for 17 weeks. Follow-up measures included the determination of total antibodies, global and differential leukocyte counts. Further, nitric oxide production was determined by adding macrophage cultures to different concentrations of Canova in the presence or absence of *P. brasiliensis.*

Results: The data revealed the protective effect of Canova in *P. brasiliensis*- infected animals. A higher nitric oxide production was found in the Canova- treated cultures.

Conclusion: These data suggest that Canova activates the macrophages by a way that depends, at least in part, on nitric oxide.

Antipromastigote Activity of an Ethanolic Extract of Leaves of *Artemisia indica*

Ganguly S, Bandyopadhyay S, Bera A, Chatterjee M.; Indian Journal of Pharmacology 2006; 38: 64-5.

Leishmania is a digenetic protozoan parasite responsible for cutaneous, mucocutaneous, or visceral leishmaniasis infecting almost 12 million people worldwide, 350 million remaining at risk, and importantly, the burden of the visceral form is borne primarily by the Indian subcontinent. Sodium antimony gluconate (SAG) has been the first line of treatment for leishmaniasis, but in recent years, an alarming increase in nonresponsiveness almost to epidemic proportions in Bihar, India, has led to the development of several new antileishmanial drugs that include amphotericin B (fast gaining acceptability as the primary drug of choice), miltefosine, and paromomycin. Viewed against this backdrop, plant-derived products are an attractive option, and herein, we report the antileishmanial efficacy of an ethanolic extract of an indigenous medicinal plant *Artemisia indica. A. indica* has been used for general malaise and fevers of unknown origin, whereas artemisinins, the sesquiterpine lactones isolated from *A. annua* have been used to treat multidrug-resistant malaria, analogs of which have been reported to exhibit both antimalarial and antileishmanial activity.

The leaves of *A. indica* were collected from the Kumaon area near Mukteswar, Uttarakhand, India. The leaves were air-dried, crushed into powder, and extracted with 90 per cent ethanol. The solution obtained was filtered thrice; the filtrate was pooled and evaporated in a rotary evaporator. A stock solution (10 mg/ml in 20 per cent DMSO) was prepared and stored at 4°C until use. *Leishmania* promastigotes from seven strains, as indicated in were routinely cultured at 24°C in M-199 medium supplemented with 10 per cent foetal calf serum and gentamicin (200 µg/ml). To study the *in vitro* effect of an ethanolic extract of *A. indica* on *Leishmania* promastigotes, exponentially growing parasites were resuspended in 96-well tissue culture plates (2 x 105/200 µl/well). The plates were incubated at 24°C for 6 h followed by the addition of an ethanolic extract of *A. indica* (0-1.0 mg/ml) and incubated for an additional 48 h. At the end of 48-h incubation, the parasite viability was checked using 3-(4,5 dimethylthiazol-2-yl)-5-(3-carboxy-methoxyphenyl)-2-(4-sulfonyl)-2*H*-tetrazolium (MTS), inner salt, and phenazonium methosulphate (PMS). MTS (2.0 mg/ml) and PMS (0.92 mg/ml) in a ratio of 5:1 was added (20 µl per well) and the plates were incubated for 3 h at 37°C. The resultant absorbances were measured at 490 nm in an ELISA reader. Accordingly, the specific absorbance that represented formazan production was calculated by subtraction of background absorbance from total absorbance. The mean percent viability was calculated as follows:

Mean-specific absorbance of treated parasites x 100
Mean-specific absorbance of untreated parasites

Accordingly, the IC50 for each drug, *i.e.*, the concentration of drug that decreased the percent viability by 50 per cent was graphically extrapolated by plotting percent viability against the respective drug concentration. All the experiments and protocols described in the present letter were approved by the Institutional Animal Ethical Committee and are in accordance with guidelines of the Committee for the Purpose of Control and Supervision of Experiments on Animals. The viability of promastigotes has a proportional relationship with the formazan complex formed as the conversion of MTS to formazan by the mitochondrial dehydrogenases is only achievable by viable cells, in the presence of the electron coupler PMS. As evident from *A. indica* showed a pronounced leishmanicidal activity in all the *Leishmania* strains studied, the IC50 ranging from 0.21 to 0.58 mg/ml, indicating its effectiveness in all three forms of leishmaniasis. To put the obtained results into perspective, the IC50 values of two established antileishmanial drugs, amphotericin B and miltefosine, were determined in the *Leishmania* strains used in this study. The values for amphotericin B ranged from 36 to 61 nM, whereas those for miltefosine varied between 11.5 and 27.5 µM (mentioned in a personal communication). Further confirmatory studies will be undertaken in the amastigote form as also the active principles in *A. indica* contributing to the observed antileishmanial activity will be delineated. In this regard, it is worthwhile to isolate artemisinin, a sesquiterpene lactone, analogs of which have displayed anti-leishmanial activity. Such studies are going on.

A Study of the Antimicrobial Activity of *Alangium salviifolium*

Pandian M R, Banu G S, Kumar G.; Indian Journal of Pharmacology 2006; 38: 203-4.

A certain interest in medicinal plants has been shown, even though the emphasis persists in research of synthetic compounds. These substances are potentially toxic and are not free of side effects on the host. This has urged microbiologists all over the world for formulation of new antimicrobial agents and evaluation of the efficacy of natural plant products as the substitute for chemical antimicrobial agents.

Alangium salviifolium Linn. (Alangiaceae) is a small deciduous tree or shrub, which grows in the wild throughout the hotter parts of India. The major phytochemical constitutes of the plant are alangine A and B, alangicine, markindine, lamarckinine and emetine. The root of *Alangium salviifolium* has been used in the Indian system of medicine as an acrid, diuretic, astringent and antidote for several poisons. The fruits (mucosa) of the plant are useful in treating burning sensation and haemorrhages. However, no scientific evidence is available regarding its antimicrobial activity. An investigation of *Alangium salviifolium* as an antiinfective agent is the objective of our present study.

The root of the plant was collected during May 1993 from Namakkal Dt. A specimen was deposited in the Rapinat Herbarium, St. Joseph's College; Tiruchirapalli. The shadow-dried root was macerated overnight with solvents butanol and ethanol in a 1:5 drug:solvent ratio x 3. Exhaustive extraction with the solvent was carried out by the cold extraction procedure. The respective extracts thus obtained were evaporated to dryness and stored in amber-coloured storage vials at 4-5°C until they were used for the experiment.

Ten Gram postive and Gram negative ATCC (American Type Culture Collection) bacterialisolates, were used in the present study.[5]

The isolates are: *Bacillus cereus (11778), Bacillus pumilus (14884), Bacillus subtilis (6633), Bordetella bronchiseptica ca (4617), Micrococcus luteus (9341), Staphylococcus epidermidis (6538), Escherichia coli (10536), Klebsiella pneumoniae (10031), Pseudomonas aeruginosa (9027) and Enterococcus faecalis (8043).*

Agar dilution method with working concentration of 1, 2 and 4 mg/ml of butanol and ethanol extracts, were used for the study. Standard antibiotic ciprofloxacin (Cadila Pharmaceuticals, India) at 4 μg/ml concentration, was used as positive control.

Butanol extract of the plant showed growth inhibitory effect at 4 mg/ml concentrations in all the bacterial isolates tested, except *Klebsiella pneumonia,* where it showed 75 per cent inhibition. Lower concentration of the extract showed concentration-dependent inhibition effect. At 2 mg/ml, 50 per cent inhibition in all the cultures was seen, while at 1 mg/ml, it was completely ineffective, when compared with the positive control (ciprofloxacin) and control (nutrient medium without antibiotic or plant extract). Inhibitory effect of the ethanol extract with all the three concentrations was not found on any of the cultures used for the experiment, except *Micrococcus luteus,* where it showed 50 per cent inhibition at 2 mg/ml and complete inhibition at 4 mg/ml of the concentration. The results of the study confirm the antimicrobial potential of the butanol extract of *Alangium salviifolium*. However, further detailed studies are required.

In vivo Efficacy of an Antifungal Fraction from *Pallavicinia lyellii,* a Liverwort

Subhisha S, Subramoniam A.; Indian Journal of Pharmacology 2006; 38(3): 211-2.

Almost all the antifungal agents, currently in use, have toxic side effects and are relatively expensive. Therefore, there is an urgent need to determine the *in vivo* efficacy of the active fraction or the isolate from the liverwort. Therefore, in the present study, the efficacy of the active fraction against aspergillosis caused by *A. fumigatus* in immuno-compromised mice was studied. The active fraction was also subjected to short-term general toxicity evaluation in mice. Mice were reared in the animal house facility of the institute and were fed with standard rodent pellets (Lipton and Co. Bangalore) and water, *ad libitium*. They were maintained under standard laboratory conditions, temperature (25-28°C), humidity (50-80 per cent) and 12 h light/dark cycle. The animal house and breeding facility has been registered with CPCSEA (Committee for the Purpose of Control and Supervision of Experiments on Animals), Government of India and CPCSEA guidelines are followed (IAEC approval obtained). *Pallavicinia lyellii* was collected with their rhizoid on sunny days of June, from the forests near Palode, Thiruvananthapuram District, Kerala State. The plant was identified by a bryophyte taxonomist of TBGRI and a voucher specimen was deposited. The alcohol extract of the powder of *P. lyellii* was prepared and subjected to fractionation with n-hexane to obtain the active hexane fraction. This fraction was used for *in vivo* efficacy and short-term toxicity evaluation in mice. To study short-term toxicity, 4 groups of mice, each containing 6 male mice (20-25 g, body weight) were used. One group was kept as the control group and Groups 2, 3 and 4 received 100, 200 and 400 mg/kg of the active fraction, respectively. The drug was administered daily for 15 days (p.o.). The control group received 1 per cent Tween 80 in an identical manner. The behaviour of the animals was observed daily for one hour in the forenoon (10 to 11 a.m.) for 14 days. The behavioural parameters observed were convulsion, grooming, hyperactivity, and sedation, loss of the writhing reflex, heart rate and respiratory rate. Initial and final body weights, water and food intake and state of stools were observed. The animals were killed on the 15th day. Hematological and serum biochemical parameters (glutamate pyruvate transaminase [GPT], glutamate oxaloacetate transaminase [GOT], urea, glucose, cholesterol, triglyceride and protein) were determined following standard methods. Hemoglobin was measured using hemoglobinometer with comparison standards. Liver, spleen, kidneys and heart were dissected, weighed and observed for pathological and morphological changes. The peritoneal macrophages and total leucocytes were counted as described elsewhere. Mice are normally resistant to the oral or intra-peritoneal route of infection with *Aspergillus sp* and other fungi. This resistance can be reduced by cortisone treatment, which suppresses the immune function. Therefore, the mice were inoculated subcutaneously with a single dose of 5 mg of hydrocortisone (Samarth Pharma (P) Ltd, Mumbai) and intramuscularly with 30,000 units of long-acting penicillin 2 days before intraperitoneal *A. fumigatus* spore challenge. The animals were challenged with different quantities (0.01, 0.1, 0.5, 1 and 2 million) of viable spores to determine the minimum number of spores required for 100 per cent mortality. 0.1 million spores were found to be sufficient to kill the mice. To determine anti *Aspergillus fumigatus* activity, 54 male mice weighing 20-25 g were divided into 9 groups of 6 mice in each group. One group was kept as normal control, without any treatment. Seven groups were treated with hydrocortisone and penicillin 48 h after hydrocortisone treatment, 6 groups were challenged with 10^5spores per mouse and one group was kept as (hydrocortisone and penicillin) control without spore challenge. Three groups of spore-challenged mice received different doses (25, 50 and 100 mg/kg) of the active n-hexane fraction of alcohol extract; daily for 10 days, p.o. and the 4th and 5th spore-challenged groups received 50 and 100 mg/kg (daily for 10 days, p.o. ketoconazole (Nizral, Janssen-Cilag Pharmaceuticals, Mumbai), respectively; and the 6th spore- challenged group received the vehicle (1 per cent Tween 80) and served as control. (1 per cent Tween 80 was used as a vehicle for both the herbal drug and ketoconazole). One group of normal mice also received the active fraction (100 mg/kg). Mortality was observed daily for 30 days. In short-term limited toxicity evaluation, the active fraction administration for 15 days did not influence any of the parameters studied, in any of the doses used. The active fraction also did not result in any change in general behaviour, body weight, stool state organ weight (liver, kidneys and spleen), food and water intake, hemoglobin, leucocyte count, peritoneal macrophage count, serum protein, urea, GPT, GOT, alkaline phosphatase, total cholesterol, triglyceride and glucose (data not given). As shown in, the administration of 10^5 spores per mouse to hydrocortisone-treated mice resulted in 100 per cent mortality within 6 days. The active fraction at a dose of 100 mg/kg, p.o. daily, starting from the day of fungal challenge, protected all the fungal-challenged mice. The antifungal activity was dose dependent. Lower dose (50 mg/kg) protected 4 out of 6 mice, while the lowest dose tried (25 mg/kg) was ineffective. The antifungal activity of the active fraction was comparable to the standard drug, ketoconazole. The antifungal fraction from *P. lyellii* was found to be effective against aspergillosis-induced mortality in immuno-compromised mice. We have already shown the direct antifungal activity of the herbal drug against the fungus under *in vitro* conditions. Further, there is no report on the immuno-modulatory activity of such liverworts. Therefore, the observed protection from aspergillosis-induced mortality is likely to be due to the antifungal activity. This herbal drug appears to be non toxic whereas the antifungal drugs in current use, including ketoconazole, have reported toxic side effects after prolonged use. *P. lyellii* is distributed in India and the biomass can be easily obtained. Thus, *P. lyellii* is likely to be an attractive material for developing invaluable antifungal drugs.

Antibacterial, Antifungal and Cytotoxic Activities of Amblyone Isolated from *Amorphophallus campanulatus*

Khan A, Rahman M, Islam MS. Indian Journal of Pharmacology. 2008; 40: 41-4.

Objective: To assess the *in vitro* antibacterial, antifungal and cytotoxic activities of amblyone, a triterpenoid isolated from *Amorphophallus campanulatus* (Roxb).

Methods: Disc diffusion technique was used for *in vitro* antibacterial and antifungal screening. Cytotoxicity was determined against brine shrimp nauplii. In addition, minimum inhibitory concentration (MIC) was determined using serial dilution technique to determine the antibacterial potency.

Results: Large zones of inhibition were observed in disc diffusion antibacterial screening against four Gram-positive bacteria (*Bacillus subtilis, Bacillus megaterium, Staphylococcus aureus and Streptococcus pyogenes*) and six Gram-negative bacteria (*Escherichia coli, Shigella dysenteriae, Shigella sonnei, Shigella flexneri, Pseudomonas aeruginosa and Salmonella typhi*). The MIC values against these bacteria ranged from 8 to 64 μg/ml. In antifungal screening, the compound showed small zones of inhibition against *Aspergillus flavus, Aspergillus niger and Rhizopus aryzae. Candida albicans* was resistant against the compound. In the cytotoxicity determination, LC(50) of the compound against brine shrimp nauplii was 13.25 μg/ml.

Conclusions: These results suggest that the compound has good antibacterial activity against the tested bacteria, moderate cytotoxicity against brine shrimp nauplii and insignificant antifungal activity against the tested fungi.

Preliminary Isolation and *in vitro* Antiyeast Activity of Active Fraction from Crude Extract of *Gracilaria changii*

Sasidharan S, Darah I, Noordin MK. Indian Journal of Pharmacology. 2008; 40: 227-9.

Objective: To isolate the active fraction from crude extract of *Gracilaria changii* and to determine its *in vitro* antifungal activity.

Materials and Methods: The active fraction was isolated from the crude extract of *G. changii* by various purification procedures such as column chromatography, thin layer chromatography, bioauthograph, *etc.* The *in vitro* antifungal activity (*Candida albicans*) of the active fraction (1.00, 0.50, and 0.25 mg/ml) was studied by disc diffusion method and the effect of the active fraction on the morphology of yeast was done by scanning electron microscope (SEM) studies.

Results: An active fraction with remarkable antifungal activity was separated from the crude extract. The active fraction was effective as a fungicide against *C. albicans* and showed a dose-dependent antifungal activity. A Scanning Electron Microscope (SEM) study confirmed the fungicidal effect of *G. changii* active fraction on *C. albicans*, by changing the normal morphology of *C. albicans.*

Conclusion: From *G. changii* crude extract, an active fraction with remarkable *in vitro* antifungal activity has been isolated.

Antibacterial and Antioxidant Activity of Methanol Extract of *Evolvulus nummularius*

Pavithra PS, Sreevidya N, Verma RS.; Indian Journal of Pharmacology. 2009; 41: 233-6.

Objective: To evaluate the antibacterial and antioxidant activity of methanol extract of *Evolvulus nummularius* (L) L.

Materials and Methods: Disc diffusion and broth serial dilution tests were used to determine the antibacterial activity of the methanol extract against two Gram-positive bacterial strains (*Bacillus subtilus* NCIM 2718, *Staphylococcus aureus* ATCC 25923) and three Gram-negative bacterial strains (*Pseudomonas aeruginosa* ATCC 27853, *Klebsiella pneumoniae* ATCC 70063 and *Escherichia coli* ATCC 25922). The methanol extract was subjected to preliminary phytochemical analysis. Free radical-scavenging activity of the methanol extract at different concentrations was determined with 2, 2-diphenyl-1picrylhydrazyl (DPPH).

Results: The susceptible organisms to the methanol extract were *Escherichia coli* (MIC=12.50 mg/ml) and *Bacillus subtilus* (MIC=3.125 mg/ml) and the most resistant strains were *Staphylococcus aureus, Klebsiella pneumoniae* and *Pseudomonas aeruginosa.* The methanol extracts exhibited radical scavenging activity with IC50 of 350 mug/ml.

Conclusion: The results from the study show that methanol extract of *E. nummularius* has antibacterial activity. The antioxidant activity may be attributed to the presence of tannins, flavonoids and triterpenoids in the methanol extract. The antibacterial and antioxidant activity exhibited by the methanol extract can be corroborated to the usage of this plant in Indian folk medicine.

Antimicrobial Clerodane Diterpenoids from *Microglossa angolensis* Oliv. et Hiern.

Tamokou JD, Kuiate JR, Tene M, Tane P.; Indian Journal of Pharmacology. 2009; 41: 60-3.

Objective: To identify the antimicrobial components present in *Microglossa angolensis* following fractionation of the methylene chloride extract of the aerial part of this plant.

Materials and Methods: The plant was dried and extracted by percolation with methylene chloride. The dry extract was fractionated and purified by silica gel column chromatography. The isolated compounds were identified by comparison of their Nuclear Magnetic Resonance (NMR) spectral data with those reported in the literature. Antimicrobial activity was assayed by broth macro-dilution method.

Results: The crude extract of *M. angolensis* displayed significant antifungal and antibacterial activities (MIC = 312.50-1250mug/ml). 6beta-(2-methylbut-2(Z)-enoyl)-3alpha,4alpha,15,16-bis-epoxy-8beta,10betaH-ent-cleroda-13(16),14-dien-20,12-olide and spinasterol were the most active compounds (MIC = 1.56-100mug/ml) and the most sensitive microorganisms were Enterococcus faecalis and Candida tropicalis for bacteria and yeasts respectively.

Conclusion: The isolation of these active antibacterial and antifungal principles supports the use of *M. angolensis* in traditional medicine for the treatment of gastrointestinal disorders.

Leishmanicidal Activity of Saponins Isolated from the Leaves of *Eclipta prostrata* and *Gymnema sylvestre*

Khanna VG, Kannabiran K, Getti G.; Indian Journal of Pharmacology. 2009; 41: 32-5.

Objective: To evaluate the leishmanicidal activity of saponin, dasyscyphin C of *Eclipta prostrata* and sapogenin, gymnemagenol from *Gymnema sylvestre* leaves under *in vitro* conditions.

Materials and Methods: Dasyscyphin C/ Gymnemagenol were dissolved in phosphate buffered saline (PBS) and diluted with liquid medium to obtain concentrations ranging from 1000 to 15 mug/ml. The leishmanicidal activity against leishmanial parasites, *Leishmania major, Leishmania aethiopica* and *Leishmania tropica* promastigotes was studied by the MTS assay.

Result: The Dasyscyphin C isolated from *E. prostrata* showed good leishmanicidal activity at 1000mug/ml concentration, with the IC(50) value of 450mug/ml against *L. major* promastigote and the percentage of parasitic death was 73; whereas, gymnemagenol of *G. sylvestre* showed only 52 per cent parasitic death at 1000 mug/ml concentration. The other *Leishmania* species, *L. aethiopica* and *L. tropica* promastigotes, were less sensitive to the saponins of *E. prostrata* and *G. sylvestre.*

Conclusion: From this study, it can be concluded that the dasyscyphin C of *E. prostrata* has significant leishmanicidal activity against *L. major* promastigote.

Vibriocidal Activity of Certain Medicinal Plants Used in Indian Folklore Medicine by Tribals of Mahakoshal Region of Central India

Sharma A, Patel VK, Chaturvedi AN.; Indian Journal of Pharmacology. 2009; 41: 129-133.

Objectives: Screening of the medicinal plants and determination of minimum inhibitory concentration (MIC) against *Vibrio cholerae* and *Vibrio parahaemolyticus.*

Materials and Methods: A simple *in vitro* screening assay was employed for the standard strain of Vibrio cholerae, 12 isolates of Vibrio cholerae non-O1, and Vibrio parahaemolyticus. Aqueous and organic solvent extracts of different parts of the plants were investigated by using the disk diffusion method. Extracts from 16 medicinal plants were selected on account of the reported traditional uses for the treatment of cholera and gastrointestinal diseases, and they were assayed for vibriocidal activities.

Results: The different extracts differed significantly in their vibriocidal properties with respect to different solvents. The MIC values of the plant extracts against test bacteria were found to be in the range of 2.5-20 mg/ml.

Conclusions: The results indicated that *Lawsonia inermis, Saraca indica, Syzygium cumini, Terminalia belerica, Allium sativum, and Datura stramonium* served as broad-spectrum vibriocidal agents.

Anti-pyretic and Antibacterial Activity of *Chloranthus erectus* (Buch.-Ham.) Verdcourt Leaf Extract: A Popular Folk Medicine of Arunachal Pradesh

Tag H, Namsa ND, Mandal M, Kalita P, Das AK, Mandal SC.; Indian Journal of Pharmacology. 2010; 42: 273-6.

Objective: The main objective of this work was to study the anti-pyretic and antibacterial activity of *C. erectus* (Buch.-Ham.) Verdcourt leaf extract in an experimental albino rat model.

Materials and Methods: The methanol extract of *C. erectus* leaf (MECEL) was evaluated for its anti-pyretic potential on normal body temperature and Brewer's yeast-induced pyrexia in albino rat's model. While the antibacterial activity of MECEL against five Gram (-) and three Gram (+) bacterial strains and antimycotic activity was investigated against four fungi using agar disk diffusion and microdilution methods.

Results: Yeast suspension (10 mL/kg b.w.) elevated rectal temperature after 19 h of subcutaneous injection. Oral administration of MECEL at 100 and 200 mg/kg b.w. showed significant reduction of normal rectal body temperature and yeast-provoked elevated temperature (38.8 ± 0.2 and 37.6 ± 0.4, respectively, at 2-3 h) in a dose-dependent manner, and the effect was comparable to that of the standard anti-pyretic drug-paracetamol (150 mg/kg b.w.). MECEL at 2 mg/disk showed broad spectrum of growth inhibition activity against both groups of bacteria. However, MECEL was not effective against the yeast strains tested in this study.

Conclusion: This study revealed that the methanol extract of *C. erectus* exhibited significant anti-pyretic activity in the tested models and antibacterial activity as well, and may provide the scientific rationale for its popular use as anti-pyretic agent in Khamptis' folk medicines.

Ruta graveolens L. Toxicity in Vampirolepis Nana-Infected Mice

Freire RB, Borba HR, Coelho CD.; Indian Journal of Pharmacology. 2010; 42: 345-50.

Objective: To determine possible toxic effects of *Ruta graveolens* hydroalcoholic extract in gastrointestinal parasitic infection.

Materials and Methods: A total of 100 g plant leaves and seeds were powdered and extracted with 1500 mL alcohol/water and administered by gavage to Swiss albino mice infected with Vampirolepis nana. Anti-parasitic evaluation and toxicity assays were carried out in six groups of ten animals each. Treatments were scheduled with both the leaves and the seeds' extracts at doses of 2.5, 5, and 10 mg per gram body weight. Toxicity was comparatively analyzed to a vehicle control group (n = 10) and to a Praziquantel® treated. On the fifth day, all the individuals were killed by euthanasia and parasite scores were correlated, giving rise to a relative percentage of elimination to each treatment. Toxicity was achieved by hematology and by clinical chemistry determinations.

Results: The use of the *R. graveolens* hydroalcoholic extract to treat V. nana-infected mice resulted in a mild-to-moderate hepatoxicity associated to a poor anti-parasitic effect. The major proglottids elimination (E per cent) was achieved at the lowest crude extract concentration with a mild anti-parasitic efficacy from the highest dose; that did not cause a significant elimination of parasites. A decrease of circulating polymorphonuclear-neutrophils associated with a normochromic-normocytic anemia was detected as the extract dose was augmented. The blood aspartate-aminotransferase and alanine-aminotransferase tended be slightly augmented with 100 mg *R. graveolens* extract.

Conclusion: R. graveolens is an unsafe natural anti-parasitic medicine as its active constituents may be poorly extracted by the popular crude herb infusion. Although it presented a mild anti-parasitic effect in mice, symptoms of natural-products-induced liver disease confirmed that its self-medication should be avoided.

Anticestodal Efficacy of *Psidium guajava* against Experimental *Hymenolepis diminuta* Infection in Rats

Tangpu TV, Yadav AK.; Indian Journal of Pharmacology. 2006; 38: 29-32.

Objective: To investigate the anticestodal efficacy of *Psidium guajava* L. leaf extract.

Materials and Methods: Anticestodal efficacy was evaluated using experimental *Hymenolepis diminuta* infection in rats. The leaf extract was administered orally to different groups of experimentally infected *H. diminuta* infections in rats. The efficacy was adjudged in terms of parasite eggs/g (EPG) of faeces count before and after treatment, direct count of surviving worms remaining in small intestines after completion of treatment and by host clearance of parasite. In all the experiments, the effect of leaf extract was compared with a standard anticestodal drug, praziquantel (PZQ).

Results: The leaf extract showed reduction in parasite EPG of faeces count in a dose-dependent manner. It further showed comparatively low recovery of worms including scolices in the small intestine and host clearance of parasite in a dose dependent manner. In all the experimental models the anticestodal efficacy of leaf extract was significantly comparable with that of PZQ.

Conclusion: The leaf extract of *P. guajava* possesses anticestodal efficacy. Study supports its folk medicinal use in the treatment of intestinal-worm infections in northeastern part of India.

Identification and Quantification of some Potentially Antimicrobial Anionic Components in Miswak Extract

Ismail A. Darout, Alfred A. Christy, Nils Skaug, Per K. Egeberg; Indian Journal of Pharmacology 2000; 32: 11-14.

Objective: To identify and quantify some potential antimicrobial anionic components in Salvadora persica root and stem aqueous extracts.

Methods: Extraction of powdered root and stem samples was performed by soaking the powder in sterile de-ionised distilled water for 24 h at 4 oC. Each 100mg of the freeze-dried extract was reconstituted with 10 ml de-ionised distilled water and filtered through a 0.45 mm cellulose acetate filter. The anionic components of the filtered extracts were identified and quantified by capillary electrophoresis.

Results: The root and stem extracts contained chloride, sulphate, thiocyanate and nitrate in the following concentrations (w/w per cent) in stem and root extracts, respectively: 6.84 per cent and 4.64 per cent, 20.1 per cent and 19.85 per cent, 0.38 and 0.28 per cent, and 0.05 per cent and 0.05 per cent. Only the differences in chloride were statistically significant ($p < 0.05$).

Conclusion: S. persica contains potential antimicrobial anionic components and that capillary electrophoresis is a convenient method for their identification and quantification.

An *In vitro* Study of the Effect of *Centella asiatica* (Indian pennywort) on Enteric Pathogens

Mamtha B, Kavitha K, Srinivasan KK, Shivananda PG.; Indian Journal of Pharmacology, 2004; 36: 41-44.

Diarrhoea is a major public health problem in developing countries. The present study was undertaken to find the antibacterial activity of *Centella asiatica* against a battery of enteric pathogens. Punch well and agar dilution methods along with viable cell count were carried out. The working concentrations of the extracts were 100, 200, 300 and 400 mg/ml respectively. Broad spectrum activity of the herb was observed at different concentrations against a battery of enteric pathogens. But the inhibitory effect of the extract was best demonstrated at a concentration of 400 mg/ml of the agar. Of the two methods used, the punch-well method yielded better results than the agar-dilution method. Viable cell-count method was used to study whether the observed inhibition was bactericidal or bacteriostatic in action. In case of Vibrio cholerae 01, Shigella species and Staphylococcus aureus, the alcoholic extract was bactericidal within 2 h.

A Study of the Antimicrobial Activity of Oil of Eucalyptus

Trivedi NA, Hotchandani SC.; Indian J Pharmacol, 2004; 36: 93-95.

Oil of eucalyptus (containing 63 per cent of eucalyptol) was obtained from the local market and used for the study. Organisms were isolated from the pus samples of patients from the surgery department. The organisms were *Klebsiella* spp., *Proteus* spp., *Pseudomonas* spp., *E. coli,* and *S. aureus.* Only those organisms resistant to conventional antimicrobials (tobramycin, gentamicin, amikacin, ciprofloxacin, chloramphenicol and cefotaxime) were used for the study. The data show that, *E. coli* and *Klebsiella* spp. were sensitive to 5 μl; *S. aureus* to 25 μl while *Pseudomonas* and *Proteus* spp. required 50 ml of eucalyptus oil. With an increasing dose of oil of eucalyptus, the resulting diameter of the zone of inhibition increased for all the organisms. The results of the study revealed that oil of eucalyptus has antibacterial activity against Gram-positive as well as Gram-negative bacteria resistant to commonly used antimicrobial agents.

Antiplasmodial Activity of Seven Plants Used in African Folk Medicine

Bidla G, Titanji VPK, Joko B, El-Ghazali G, Bolad A, Berzins K.; Indian J. of Pharmacology, 2004; 36: 244-250.

Currently, there is a considerable increase in mortality caused by malaria due to the rapid spread of drug-resistant strains of *P. falciparum.* It is important, therefore, that new antimalarial drugs are developed to cope with the spread of Resistance. The plants, *Achromanes difformis, Cleome rutidosperma, Cymbopogon citratus, Piper umbellatum, Mellotus appositofolius, Mangifera indicus and Annona muricata* were collected from various parts of Cameroon and identified in the Limbe Botanic Garden. In addition to the extract, each test well contained 1 per cent parasitized blood Group O of 2 per cent hematocrit in malaria culture medium. Growth inhibition was assessed after 24 h of culture by fluorescent microscopic examination. While Piper umbellatum and Mellotus appositofolius extracts had moderate activity against the parasite with 40 μg/ml of each giving 70 per cent and 57 per cent inhibition, respectively, extracts from *Cymbopogon citratus, Mangifera indicus* and *Annona muricata* were found to possess greater effects on the growth with 20 μg/ml of each giving 57.9 per cent, 50.4 per cent and 67 per cent inhibition, respectively. *Achromanes difformis* and *Cleome rutidosperma* extracts showed the least antiplasmodial activity even with 40 μg/ml of each resulting in 32.4 per cent and 31.6 per cent inhibition, respectively.

The Potential of Aqueous and Acetone Extracts of Galls of *Quercus infectoria* as Antibacterial Agents

Basri DF, Fan S H.; Indian Journal of Pharmacology 2005; 37: 26-9.

Aim: To evaluate the antibacterial potential of aqueous and acetone extracts of galls of *Quercus infectoria* by determination of Minimum Inhibitory Concentration (MIC) and Minimum Bactericidal Concentration (MBC) values.

Materials and Methods: The extracts from the galls of *Q. infectoria* at 10 mg/ml were screened against three Gram-positive bacteria (*Staphylococcus aureus* ATCC 25923, *Staphylococcus epidermidis* and *Bacillus subtilis*) and three Gram-negative bacteria (*Escherichia coli* NCTC 12079 serotype O157:H7, *Salmonella typhimurium* NCTC 74 and *Pseudomonas aeruginosa* ATCC 27853). The MIC of the extracts were then determined using the twofold serial microdilution technique at a concentration ranging from 5 mg/ml to 0.0024 mg/ml. The MBC values were finally obtained from the MIC microtiter wells which showed no turbidity after 24 hrs of incubation by subculturing method.

Results: Out of the six bacterial species tested, *S. aureus* was the most susceptible. On the other hand, the extracts showed weak inhibitory effect against *S. epidermidis, B. subtilis, S. typhimurium* and *P. aeruginosa* while there was no inhibition zone observed for *E. coli* O157. The MIC values of the extracts ranged from 0.0781 mg/ml to 1.25 mg/ml whereas the MBC values ranged from 0.3125 mg/ml to 2.50 mg/ml. The MBC values of aqueous extract against *S. aureus* and *S. typhimurium* were higher than their MIC values. The MBC value of acetone extract against *S. aureus* was also higher than its MIC value. Interestingly, however, the MIC and MBC values of acetone extract against *S. typhimurium* were the same (1.25 mg/ml).

Conclusion: The aqueous and acetone extracts displayed similarities in their antimicrobial activity on the bacterial species and as such, the galls of *Quercus infectoria* are potentially good source of antimicrobial agents.

A Study of the Antimicrobial Activity of *Elephantopus scaber*

Avani K, Neeta S.; Indian Journal of Pharmacology 2005; 37: 126-7.

To find out safer microbicides and for preventing environmental degradation a study of the antimicrobial activity of *Elephantopus scaber* was carried out. Inhibitory effect of the petroleum ether extract with all the three concentrations was not found on any of the cultures used for the experiment except *Micrococcus luteus* it showed 50 per cent inhibition at 2 mg/ml and complete inhibition at 4 mg/ml of the concentration. The results of the study confirm the antimicrobial potential of the ethyl acetate extract of *E. scaber*.

Pharmacological Screening and Evaluation of Antiplasmodial Activity of *Croton zambesicus* against *Plasmodium berghei* Infection in Mice

Okokon JE, Ofodum KC, Ajibesin KK, Danladi B, Gamaniel KS.; Indian Journal of Pharmacology. 2005; 37: 243-246.

Objective: To evaluate the antiplasmodial activity of leaf extract of *Croton zambesicus* on chloroquine-sensitive *Plasmodium berghei* infection in mice and to confirm its traditional use as a malarial remedy in Africa.

Materials and Methods: The ethanolic leaf extract of Croton zambesicus (50-200 mg/kg) was screened for blood schizontocidal activity against chloroquine-sensitive Plasmodium berghei berghei infection in mice. The schizontocidal activity during early and established infections as well as the repository activity were investigated.

Results: The extract demonstrated a dose-dependent chemosuppression or schizontocidal effect during early and in established infections, and also had repository activity. The activity was lower than that of the standard drugs (chloroquine 5 mg/kg, pyrimethamine 1.2 mg/kg/day).

Conclusion: The leaf extract possesses considerable antiplasmodial activity, which can be exploited in malaria therapy.

Inhibitory Effect of Extracts of Latex of *Calotropis procera* against *Candida albicans*: A Preliminary Study

Sehgal R, Arya S, Kumar VL.; Indian Journal of Pharmacology 2005; 37: 334-5.

The latex collected from the aerial parts of the plant was dried (DL), soxhlated successively with petroleum ether (PE), methanol and water and dried under vacuum and dissolved in sterile water (methanol and aqueous extract) and 0.05 per cent Tween-20 (PE extract). The standard antifungal drugs with low, moderate and high efficacy against *C. albicans* namely, griseofulvin, clotrimazole and nystatin were dissolved in 1 per cent di-methylsulfoxide. Study demonstrated that the anticandidial activity of latex suggests that it might be effective against other fungal strains.

Anticandidal Activity of *Azadirachta indica*

Charmaine Lloyd A C, Menon T, Umamaheshwari K.; Indian Journal of Pharmacology 2005; 37: 386-9.

Aim: To study the antifungal activity of 10 different extracts of seed kernels of *Azadirachta indica* A. Juss (Meliaceae) on *Candida* sps. isolated from immunocompromised patients.

Materials and Methods: The extractants used were hexane, methanol, chloroform, water, petroleum ether, dichloromethane, acetone and absolute alcohol. The products of a successive extraction procedure involving hexane, chloroform and methanol were also tested for anticandidal activity. The minimum inhibitory concentration was tested by broth dilution method at concentrations ranging from 1 to 0.0625 mg/ml.

Results: The ethanol extract of commercial neem seed oil, ethanol extract of neem seed kernels and the hexane extract showed best results. All strains were resistant to methanol: chloroform: water extracts and chloroform extracts of the successive extraction procedure.

Conclusion: The hexane and alcoholic extracts of neem seed seem to be promising anticandidal agents.

In vitro Antifungal Activity of the Essential Oil of *Coriandrum sativum* Linn.

Garg SC, Siddiqui N.; J Res Ayur Siddha, Jul-Sep, 1992.

The essential oil derived from the seeds of *Coriandrum sativum* L. has been studied *in vitro* for its antifungal activity against 18 fungal organisms using "filter paper disc. Agar diffusion technique". The oil has shown moderate-to-excellent activity against the test fungi. The strong antilfungal activity against *Alternaria alternata, Curularia lunata, Pestalotia psidi, Phytophihora parasitica, Trichodema viride* and *Collectotrichum capsici* can be exploited after detailed in vivi studies. The use of essential oils as fungicides may reduce the pollution as these are biodegradable.

ANTIMALARIAL ACTIVITY

In vivo Antimalarial Activity of Ethanolic Leaf Extract of *Stachytarpheta cayennensis*

Okokon JE, Ettebong E, Antia BS. Indian Journal of Pharmacology. 2008; 40: 111-3.

Objective: To evaluate the *in vivo* antiplasmodial activity of the ethanol leaf extract of *Stachytarpheta cayennensis* in the treatment of various ailment in Niger Delta region of Nigeria, in Plasmodium berghei-infected mice.

Materials and Methods: The ethanolic leaf extract of *Stachytarpheta cayennensis* (90-270 mg/kg/day) was screened for blood schizonticidal activity against chloroquine sensitive Plasmodium berghei berghei in mice. The schizonticidal effect during early and established infections was investigated.

Results: Stachytarpheta cayennensis (90-270 mg/kg/day) exhibited significant (P< 0.05) blood schizonticidal activity both in 4-day early infection test and in established infection with a considerable mean survival time comparable to that of the standard drug, chloroquine, 5 mg/kg/day.

Conclusion: The leaf extract possesses significant (P< 0.05) antiplasmodial activity which confirms its use in folkloric medicine in the treatment of malaria.

Evaluation of the *In vivo* Antimalarial Activity of Ethanolic Leaf and Stembark Extracts of *Anthocleista djalonensis*

Bassey AS, Okokon JE, Etim EI, Umoh FU, Bassey E.; Indian Journal of Pharmacology. 2009; 41: 258-61.

Objective: To evaluate the *in vivo* antimalarial activities of ethanolic leaf and stembark extracts of *Anthocleista djalonensis* used traditionally as malarial remedy in Southern Nigeria in mice infected with *Plasmodium berghei berghei.*

Methods: The ethanolic extracts of the *A. djalonensis* leaf (1000 - 3000 mg/kg/day) and stembark (220 - 660 mg/kg/day) were screened for blood schizonticidal activity against chloroquine-sensitive *P. berghei* in mice. The schizonticidal effect during early and established infections was investigated.

Results: The A. djalonensis leaf extract (1000 - 3000 mg/kg/day) exhibited a significant antiplasmodial activity both in the 4-day early infection test and in the established infection with a considerable mean survival time, which was incomparable to that of the standard drug, chloroquine (5 mg/kg/day). The stembark extract (220 - 660 mg/kg/day) also demonstrated a promising blood schizontocidal activity in early and established infections.

Conclusion: These plant extracts possess considerable antiplasmodial activities, which justify their use in ethnomedicine and can be exploited in malaria therapy.

REPRODUCTIVE HEALTH

Aphrodisiac Activity of *Vanda tessellata* (roxb.) Hook. Ex don Extract in Male Mice

P.K. Suresh Kumar, A. Subramoniam, P. Pushpangadan; Indian Journal of Pharmacology 2000; 32: 300-304.

Objective: To study the effect of *V. tessellata* on the sexual behaviour of male mice and general toxicity, if any, in mice.

Methods: An aqueous suspension (2 g/kg, wet wt.) or extract (water or alcohol, 200 mg/kg) of root, flower or leaf of *V. tessellata* was administered (p.o.) to male mice and 1 hr, after administration their mounting behaviour was observed. The most active extract (alcohol extract of flower) was administered (50 or 200 mg/kg, p.o.) to different groups of male mice and their mounting behaviour, mating performance and reproductive performance were determined. The general short-term toxicity of the alcohol extract in male mice was also determined.

Results: The flower and, to some extent, the root, but not the leaf of V. tessellata was found to stimulate the mounting behaviour of male mice. This activity was found in the alcohol extract of the flower. This extract (50 or 200 mg/kg) also increased mating performance in the mice. The pups fathered by the extract treated mice were found to be normal with an increasing trend in the male/female ratio of these pups. The alcohol extract was devoid of any conspicuous general toxicity.

Conclusion: The alcohol extract of *V. tessellata* flower stimulates the sexual behaviour of male mice.

Uterotrophic Effect of Menotab (M-3119): A Preclinical Ctudy

S. Gopumadhavan, M.V. Venkataranganna, Mohamed Rafiq, S.J. Seshadri, S.K. Mitra; Indian Journal of Pharmacology 2002; 34: 237-243.

Objective: To study the uterotrophic effect and toxicity of Menotab (M-3119) in rats.

Methods: Uterotrophic effect of menotab (M-3119) was studied at doses of 250 and 500 mg/kg b. wt., p.o., in the ovariectomized rats. Menotab (M-3119) was administered as an aqueous suspension for 21 days. The parameters studied were uterine weight, uterine glycogen, serum estrogen, serum progesterone and histopathology of uterus. For acute toxicity study rats were administered with menotab (M-3119) at a limit test dose of 5000 mg/kg b. wt., p.o. In subchronic toxicity study, rats received menotab (M-3119) at doses of 1000 or 3000 mg/kg b. wt., p.o. for 90 days. Daily cage side observation, weekly body weight, feed intake and behavioural changes were recorded. Hematological, biochemical and histopathological changes of target organs were evaluated at term.

Results: Treatment with menotab (M-3119) in ovariectomized rats showed a dose-dependent increase in uterine weight and glycogen levels. No significant difference was observed in hormone levels. Histopathological evaluation of uterus showed changes characteristic of atrophy of uterus in ovariectomized rats, which were ameliorated in rats treated with menotab (M-3119). Acute and subchronic toxicity studies revealed no adverse effects.

Conclusion: The evidence of uterotrophic activity, as indicated by uterine weight, uterine glycogen and characteristic histological changes without altering hormone levels following menotab (M-3119) treatment, suggests that it could possibly act directly on the estrogen receptors without enhancing the endogenous hormone levels. Acute and subchronic toxicity studies reflected non-toxic nature of menotab (M-3119). The findings indicate the potential and safe use of menotab (M-3119) in the treatment of post-menopausal symptoms.

Effect of *Aegle marmelos* Leaf on Rat Sperm Motility: An *in vitro* Study

Sur T.K., Pandit S., Pramanik T., Bhattacharyya D.; Indian Journal of Pharmacology 2002; 34: 276-277.

Aegle marmelos Corr. (Beng. Bael), is a popular plant and cosmopolitan in distribution. *A. marmelos* have rich medicinal properties. Earlier studies in laboratory have been shown that ethanolic extract of A. marmelos leaf possess anti-spermatogenic activity in rats. The present investigation has been carried out to find the activity of *A. marmelos* leaf on rat sperm motility through *in vitro* study. Fresh leaves of *A. marmelos* were used. Different dilutions (1, 2.5, 5 and 10 per cent) were prepared in buffer saline (pH 7.4) before the experiment. Six healthy adult male albino Wistar rats (180+5g) were used in this study. At least 80 per cent initial normal sperm motility and 2.5x107/cc sperm count was considered for the selection of samples. Wet drop technique was applied to study the motility of spermatozoa. The motility of sperm was observed at various time intervals up to 150 seconds. In control, 10 ml of buffer saline was used instead of plant extract. Initial motility of control sperm was 82.3 per cent, which remained almost static (80.5 per cent) up to 150 seconds as observed in the present study. But, sperm motility appears to decrease with the time and significantly so with the increasing concentration of *A. marmelos*. To conclude, Increases in concentration of water extract of *A. marmelos* decrease the complete immobility time of sperms. Therefore, it possesses antimotility action on spermatozoa in rats.

Preliminary Study on the Antiimplantation Activity of Compounds from the Extracts of Seeds of *Thespesia populnea*

Ghosh K, Bhattacharya TK.; Indian Journal of Pharmacology, 2004; 36: 288-291.

Objective: To evaluate the preliminary antiimplantation activity of isolated pure principles from successive extracts of petroleum-ether (PE) and ethyl acetate (EAc) and subsequent crude alcoholic extract of seeds of *T. populnea* in female albino rats.

Material and Methods: Graded doses of the active principles and the crude alcoholic extract (in 1 per cent gum acacia suspension) were tested for possible antiimplantation activity in Sprague-Dawley female rats of normal estrus cycle after overnight cohabitation with males of proven fertility. The day when spermatozoa were detected in vaginal smear was treated as 1st day of pregnancy. The compounds were administered to female rats from the 1st day to the 7th day of pregnancy. On the 10th day, the rats were laparotomized under light anesthesia and the numbers of implantation sites and corpora lutea were noted.

Results: Chromatographic pure principle from PE extract showed significant antiimplantation activity (60 per cent) at the dose of 110 mg/kg, b.w while that from EAc extract showed 48.6 per cent effect at the same dose. In contrast, the final alcoholic extract showed no such significant action.

Conclusion: The active principles from PE and EAc extracts showed significant anti-implantation activity and they were found to be a mixture of two groups of long-chain fatty acids from GLC.

Effect of the Aqueous Extract of Dry Fruits of *Piper guineense* on the Reproductive Function of Adult Male Rats

Mbongue F G, Kamtchouing P, Essame O J, Yewah P M, Dimo T, Lontsi D.; Indian Journal of Pharmacology 2005; 37: 30-2.

In many developing countries, traditional medicines are widely utilized in the treatment of various ailments on an empirical basis. A variety of plants have been used for the treatment of ulcer, hypertension, diabetes and male reproductive function. *Litsea chinensis* and *Ochis maculata* are used for their aphrodisiac activity. The plant of Striga orobanchioides has antiandrogenic and antispermatogenic effects. Leaves of *Hibiscus macranthus* and *Basella alba* have androgenic activity. Thus medicinal plants are used for the treatment of disorders linked to male infertility. Some of the factors responsible for this infertility are linked with hormonal secretion, erectile impotence, disorders of ejaculation, and toxic effects on the testes and accessory sex organs. Leaves of *Piper guineense* are used for respiratory infections and for female infertility while its fruits are used as an aphrodisiac. However, previous studies in our laboratory have shown that the aqueous extract of *Piper guineense* fruits at 122.5 mg/kg have been shown to stimulate sexual behaviour of mature male rats by decreasing mount and intromission latencies and by increasing mounting, anogenital sniffing and penile erection index. Since Piper guineense has an impact on penile erection and copulatory behaviour which are controlled by androgens, the present study was undertaken to evaluate the effects of the dry fruits of *Piper guineense* on some male reproductive parameters such as the secretory activities of the testis and some accessory sexual organs which are also controlled by androgens.

Evaluation of the Female Reproductive Toxicity of the Aqueous Extract of *Labisia pumila* var. Alata in Rats

Ezumi M, Amrah S S, Suhaimi A, Mohsin S.; Indian Journal of Pharmacology 2006; 38: 355-6.

Labisia pumila var. alata (LPA) or Kacip Fatimah (KF), as it is popularly known in Malaysia, is a very popular herb among the local women. Traditionally, the water decoction of the root or the whole plant of KF is consumed by Malay women for induction and facilitation of labour. Currently, many commercial products, containing this herb, have emerged in the Malaysian market, claiming to enhance vitality and libido. However, there is no scientific data on their quality, safety and efficacy to substantiate such claims. Studies supported by the Government of Malaysia, conducted at various universities and institutes, are in various stages of progress, such as the extract preparation-standardisation (undergoing patenting), authentication and evaluation of safety and efficacy. Reports have shown that LPA displayed a non-significant response to *in vitro* estrogen activity and had appreciable amount of iron. In addition, LPA root and leaves were found to contain two novel benzoquinoid compounds 1, 2 as major components. More information regarding this herb is expected to be available in the near future.

The objectives of the present study are to evaluate the female reproductive toxicity and potential effect of KF in inducing labour in rats. A standardised aqueous extract of LPA, at doses of 2-800 mg/kg/day, was administered, to determine the safety and efficacy of this herb. The general acute and sub-acute (28 days) toxicity studies of the same extract in rats had already been performed by a team at the Herbal Medicine Research Centre of the Institute for Medical Research, Kuala Lumpur, Malaysia. The results of the study revealed that the estimated LD_{50} of the extract is more than 5 g/kg, body weight and the extract produced no significant adverse effects (personal communication). A chronic toxicity study is ongoing and at present, no deleterious effects have been observed in rats. A Phase II clinical trial, conducted in postmenopausal women, by a research team at the School of Medical Sciences, Universiti Sains, Malaysia, concluded that the therapeutic dose of the extract is 2.5 mg/kg/day (personal communication).

Forty eight female Sprague Dawley rats were used in the Segment I (female reproductive toxicity) study. Rats with a regular estrous cycle (4-6 days) were given vehicle (distilled water) as control or LPA at 2, 20, 200, 400 or 800 mg/kg daily, by gavage, 10 days prior to mating, during mating (a maximum period of 10 days), throughout gestation and lactation periods of 7 days. Dams were permitted to deliver their litters, naturally. At birth (Day 1), the pups were individually counted, weighed, examined for external malformation and sexed. Dams and foetuses were sacrificed on Day 7, postpartum. The parameters measured are presented in. Data were analysed using SPSS version 11.0. Data on maternal body weight, throughout the study, were analysed using General Linear Model Repeated Measures. Mean days of estrous cycle, length of pregnancy (days), pregnancy index, number of pups, both at birth and on Day 7 lactation (litter size), pups' body weight during lactation and number of implantation sites per litter were analysed, using the one-way ANOVA, followed by the Scheffe test, if differences were found. Additionally, the Kruskal-Wallis test (non-parametric), followed by the Mann Whitney test (when appropriate), and were used to assess the live birth index, viability index and percentage of post-implantation death. The level of significance was set at 5 per cent. The parametric data were expressed as mean±SEM or ratio, while the non-parametric data were expressed as median (Interquartile range).

Results indicated that the LPA extracts did not alter the general health or estrous cycle of rats. All studied animals proceeded towards successful mating and pregnancies. The mean duration of pregnancy (in days) was shortened to 21 days in animals which received the herbal extract at a dose of 20 mg/kg/day and above. However, it was not statistically significant. All pregnant rats delivered normally with no evidence of prematurity or abortion, suggesting that the extract is not an abortifacient or causes prostaglandin-like activity when consumed orally. None of the rats exhibited a significant amount of foetal resorption, indicating that the herb was non-toxic to the foetuses. Statistically, no test agent-related changes in the maternal body weight, number of implantations, litter size and pup body weights were observed. No significant difference in pup sex ratio, live birth index, pup viability index and percentage of post-implantation death, was noted in this study.

The present findings indicated that the water-based extracts of LPA do not pose any significant reproductive toxicity or complication in pregnancy, delivery and early pup growth in rats. The no observable adverse effect level (NOAEL) of the extract in this study is 800 mg/kg/day. A closer observation of the duration of the pregnancy (in hours) and parturition time were not evaluated in this study to support the traditional claim of this herb.

Antiovulatory and Estrogenic Activity of *Plumbago rosea* Leaves in Female Albino Rats

Sheeja E, Joshi SB, Jain DC.; Indian Journal of Pharmacology. 2009; 41: 273-7.

Objective: To evaluate the effect of petroleum ether (60-80 degrees), chloroform, acetone, ethanol and aqueous extracts of *Plumbago rosea* leaves on the estrous cycle and to identify the estrogenic activity of active acetone and ethanol extracts in female albino rats.

Methods: plant extracts were tested for their effect on the estrous cycle at two dose levels: 200 and 400 mg/kg, respectively. The effective acetone and ethanol extracts were further studied on estrogenic activity in rats. Histological studies of the uterus were carried out to confirm their estrogenic activity.

Results: The acetone and ethanol extracts were most effective in interrupting the normal estrous cycle of the rats ($P<0.05$, <0.01, <0.001). These later exhibited prolonged diestrous stage of the estrous cycle with consequent temporary inhibition of ovulation. The antiovulatory activity was reversible on discontinuation of treatment. Both the extracts showed significant estrogenic and antiestrogenic activity.

Conclusion: The acetone and ethanolic extracts of *P. rosea* leaves have an antifertility activity.

Antiovulatory and Abortifacient Effects of *Areca catechu* (Betel nut) in Female Rats

Shrestha J, Shanbhag T, Shenoy S, Amuthan A, Prabhu K, Sharma S, Banerjee S, Kafle S.; Indian Journal of Pharmacology. 2010; 42: 306-11.

Objectives: To study the antiovulatory and abortifacient effects of ethanolic extract of *Areca catechu* in female rats.

Materials and Methods: For antiovulatory effect, ethanolic extract of A. catechu at 100 and 300 mg/kg doses was administered orally for 15 days. Vaginal smears were examined daily microscopically for estrus cycle. Rats were sacrificed on 16th day. Ovarian weight, cholesterol estimation, and histopathological studies were done. Abortifacient activity was studied in rats at 100 and 300 mg/kg doses administered orally from 6th to 15th day of pregnancy. Rats were laparotomised on 19th day. The number of implantation sites and live fetuses were observed in both horns of the uterus.

Results: The extract of *A. catechu* showed a significant decrease in the duration of estrus at 100 mg/kg ($P = 0.015$) and 300 mg/kg doses ($P = 0.002$) as compared with control. Metestrus phase was also significantly reduced at 100 mg/kg ($P = 0.024$) and 300 mg/kg doses ($P = 0.002$). There was a significant increase in proestrus ($P < 0.001$) phase. However, diestrus phase was unchanged. Histopathological study of the ovaries showed mainly primordial, primary, and secondary follicles in the test groups as compared to control. There was also a significant ($P = 0.002$) decrease in ovarian weight and a significant ($P = 0.021$) increase in ovarian cholesterol level at 100 mg/kg dose. In the study to evaluate abortifacient effect, the mean percentage of abortion with 100 and 300 mg/kg doses were 75.5 per cent and 72.22 per cent, respectively, which was significantly ($P = 0.008$ and $P = 0.006$, respectively) increased when compared with control.

Conclusion: The ethanolic extract of *A. catechu* at doses of 100 and 300 mg/kg has antiovulatory and abortifacient effects.

Effects of *Mondia whitei* Extracts on the Contractile Responses of Isolated Rat vas Deferens to Potassium Chloride and Adrenaline

Watcho P, Fotsing D, Zelefack F, Nguelefack T B, Kamtchouing P, Tsamo E, Kamanyi A.; Indian Journal of Pharmacology. 2006; 38: 33-37.

Objective: To investigate the effects of the methylene chloride:methanol (CH_2Cl_2:MeOH, 1:1) extract of the dried roots of *Mondia whitei* Linn and its hexane and methanol fractions on potassium chloride (KCl) and adrenaline (Adr)-induced contractions of rat vas deferens.

Materials and Methods: Isolated strips of normal adult rat vas deferens were mounted in a Ugo Basile single-organ bath containing Krebs solution. Cumulative concentration-response curves of KCl (1-7 x 10-2 *M*) and adrenaline (1.21-8.45 x 10-7 *M*) were established in the absence and presence of *M. whitei* (50-400 µg/ml). In separate experiments, after obtaining a stable plateau of contractions with KCl (60 m *M*), *M. whitei* samples (50-400 µg/ml) were added cumulatively to relax the preparation. In KCl (60 m *M*), containing depolarizing medium, cumulative concentration-contraction curve to $CaCl_2$ (2-14 x 10-2 *M*) was elicited in the absence and presence of the hexane fraction of *M. whitei* (50-400 μg/ml).

Results: All the *M. whitei* samples produced rightward shift of the concentration-response curves to KCl and Adr. At high concentration of the plant extracts (400 μg/ml), a decrease of the maximal response to the contractile agents was observed compared with that obtained with the control. All the three extracts produced concentration-dependent relaxation of the plateau of contraction induced by KCl and the hexane fraction appeared to be the more potent. In calcium-free physiological salt solution, the hexane fraction of *M. whitei* produced rightward shift to the concentration-response curve to $CaCl_2$ and completely abolished the contractile effect of calcium at high concentration (400 μg/ml).

Conclusion: It is concluded that *M. whitei* extracts antagonized the contractile responses to KCl and Adr in isolated rat vas deferens, which could be due to the blockade of voltage-operated calcium channels.

Antiovulatory and Abortifacient Potential of the Ethanolic Extract of Roots of *Momordica cymbalaria* Fenzl in rats.

Koneri R, Balaraman R, Saraswati C D.; Indian Journal of Pharmacology 2006; 38: 111-4.

Objective: To study the antiovulatory and abortifacient activity of the ethanolic extract of roots of *Momordica cymbalaria* Fenzl.

Materials and Methods: Female Wistar albino rats (150 to 200 g) with at least three regular estrous cycles were administered ethanolic extracts of roots of *Momordica cymbalaria* Fenzl. at two doses 250 and 500 mg/kg orally for 15 days. Control group received vehicle (tween 80 1 per cent, p.o. daily). Animals were sacrificed on 16th day. One ovary was subjected to histopathological studies and the other for biochemical studies. Abortifacient study was done in another set of three groups of animals. The extracts at doses of 250 and 500 mg/kg were administered orally through gastric gavage from the day 6 to day15 of pregnancy (the period of organogenesis). The animals were laparotomised under light ether anesthesia and semi- sterile conditions on day 19th of pregnancy. Both horns of the uterus were observed for the number of implantation sites, resorptions, dead and alive foetus.

Results: Highly significant (P<.001) decrease in the duration of estrous cycle and metaestrous phase and increase in proestrous phase was seen, but diestrous phase was unchanged in both 250 and 500 mg treated group when compared to untreated group. Significant decrease in the ovarian weight and a highly significant increase in serum cholesterol with 250 mg/kg dose were seen. Histology of ovary showed an increase in preovulatory and atretic follicles. Ethanolic extract showed a dose dependent abortifacient effect in pregnant rats during organogenesis period. At 250 mg/kg ethanolic extract did not show any abortifacient activity but reduced the number of viable foetus and resorptions with no change in the foetal weight when compared with control group. At 500 mg/kg ethanolic extract showed highly significant (P< 0.001) abortifacient activity.

Conclusion: The ethanolic extract at both doses (250 and 500 mg/kg) showed antiovulatory activity. It is abortifacient at 500 mg/kg but not at 250 mg/kg.

TOXICOLOGICAL STUDIES

Toxicological and Pharmacological Study of Navbal Rasayan: A Metal Based Formulation

Dinesh Chandra, A.K. Mandal; Indian Journal of Pharmacology 2000; 32: 369-371.

Objectives: Toxicological and pharmacological studies of "Navbal Rasayan" a metal based Ayurvedic formulation used for the treatment of multiple sclerosis, have been carried out.

Methods: Acute and chronic toxicities studies were conducted in rats with oral graded doses (0.37 to 3.0 g/kg) of Navbal Rasayan (NR). *In vitro* study was carried out on isolated guinea pig ileum obtained from control or animal pretreated with oral NR, 1.5 g/kg for 3 days. Dose responses with acetylcholine, histamine and 5-HT were obtained. The analgesic activity of oral NR 1.5 g/kg for 3 days was evaluated in albino mice against acetic acid induced writhings. The hypnotic activity was measured with pentobarbital after oral pretreatment with 1.5 and 3.0 g/kg NR for 3 days. The anti-convulsant activity was observed in rats against i.p. pentylenetetrazol induced seizures.

Results: Oral administration of graded doses of NR (upto 3.0 g/kg) did not produce any acute or chronic toxicities in rats. Oral pretreatment of guinea pigs with NR (1.5 g/kg for 3 days) increased 36.43 time the histamine dose while the agonistic effect of acetylcholine and 5-hydroxytryptamine was completely attenuated. Further, NR neither exhibited any analgesic, sedative effect or anticonvulsant effect in rodents.

Conclusion: Acute and chronic toxicity studies with NR in animals do not show any toxic effect/untowand effect during. But it has a powerful non-specific inhibitory effect on guinea pig ileum.

Protective Effect of *Aloe vera* L. gel against Sulphur Mustard-Induced Aystemic Toxicity and Skin Lesions

Anshoo G, Singh S, Kulkarni A S, Pant S C, Vijayaraghavan R.; Indian Journal of Pharmacology 2005; 37: 103-10.

Sulphur mustard (SM), chemically 2,2'-dichloro diethyl sulphide, is an incapacitating and extremely toxic chemical warfare agent, and causes serious blisters on contact with human skin. SM forms sulphonium ion in the body that alkylates DNA and several other macromolecules, and induces oxidative stress. The aim of this study was to evaluate the protective effect of *Aloe vera* L. gel against SM-induced systemic toxicity and skin lesions.

Materials and Methods: Aloe vera gel was given (250, 500 and 1000 mg/kg) orally to mice as three doses, one immediately after SM administration by percutaneous route, and the other two doses on the next two days. Protective index was calculated with and without Aloe vera gel treatment. *Aloe vera* gel was also given orally as three doses with 3 LD50 SM and the animals were sacrificed for biochemical and histological evaluation, 7 days after SM administration. In another set of experiment *Aloe vera* gel was liberally applied on the SM administered skin site and the animals were sacrificed after 14 days to detect its protective effect on the skin lesions induced by SM.

Results: The protection given by *Aloe vera* gel was marginal. 1000 mg/kg dose of *Aloe vera* gel gave a protection index of 2.8. SM significantly decreased reduced glutathione (GSH), oxidised glutathione (GSSG) and WBC count, and significantly increased malondialdehyde (MDA) level, RBC count and Hb concentration. *Aloe vera* gel offered protection only in the increase of MDA level by SM. Severe damage was observed in the histology of liver, spleen and skin following SM administration, and 1000 mg/kg of *Aloe vera* gel partially protected the lesions. However, topical application of *Aloe vera* gel showed better protection of the skin lesions induced by SM.

Conclusion: The study shows that percutaneous administration of SM induces oxidative stress and oral administration of *Aloe vera* gel could only partially protect it. Topical application of *Aloe vera* gel may be beneficial for protecting the skin lesions induced by SM.

Comparative Evaluation of some Flavonoids and Tocopherol Acetate against the Systemic Toxicity Induced by Sulphur Mustard

Vijayaraghavan R, Gautam A, Sharma M, Satish HT, Pant SC, Ganesan K.; Indian Journal of Pharmacology. 2008; 40: 114-20.

Objective: To evaluate the protective value of quercetin, gossypin, *Hippophae rhamnoides* (HR) flavone and tocopherol acetate against the systemic toxicity of percutaneously administered sulphur mustard (SM) in mice.

Materials and Methods: Quercetin, gossypin, HR flavone or tocopherol acetate (200 mg/kg, i.p.) were administered just before percutaneous administration of SM and protection against the SM lethality was evaluated. In another experiment quercetin, gossypin, HR flavone or tocopherol acetate were administered against 2 LD(50) SM. The animals were sacrificed seven days post SM administration and various biochemical parameters were estimated.

Results: The protection against the lethality of SM was very good with the flavonoids (quercetin = 4.7 folds; gossypin = 6.7 folds and HR flavone = 5.6 folds), compared to no protection with tocopherol acetate (0.7 fold). SM (2 LD(50)) showed decrease in reduced and oxidised glutathione (GSH and GSSG) levels, and an increase in malondialdehyde level (MDA). Oxidative stress enzymes like glutathione peroxidase, glutathione reductase and superoxide dismutase were significantly decreased. The total antioxidant status was also significantly decreased. Additionally, there was a significant increase in red blood corpuscles and hemoglobin content. All the flavonoids significantly protected the GSH, GSSG and MDA, and also the hematological variables. Tocopherol acetate failed to offer any protection in those parameters. Gossypin protected glutathione peroxidase, while HR flavone protected both glutathione reductase and glutathione peroxidase significantly. The decrease in body weight induced by SM and the histological lesions in liver and spleen were also significantly protected by the flavonoids but not by tocopherol acetate.

Conclusion: The present study supports that SM induces oxidative stress and flavonoids are promising cytoprotectants against this toxic effect.

Effect of Trolox and Quercetin on Sulfur Mustard-Induced Cytotoxicity in Human Peripheral Blood Lymphocytes

Bhattacharya R, Tulsawani R K, Vijayaraghavan R.; Indian Journal of Pharmacology. 2006; 38: 38-42.

Objective: To evaluate the protective activity of antioxidants, *viz.* trolox and quercetin, against sulfur mustard (SM)-induced cytotoxicity.

Materials and Methods: Cytotoxicity of various concentrations (20-640 μM) of SM, in the presence or absence of 10 μM trolox or quercetin (-0.5, 0, or +0.5 h) was determined in human peripheral blood lymphocytes after 6-h exposure. Cell viability was measured by Trypan blue dye exclusion (TBDE). Further, a cytotoxic concentration of SM (80 μM) was challenged by the two antidotes (-0.5 h) and cell viability was measured by TBDE and leakage of intracellular lactate dehydrogenase (LDH). Mitochondrial integrity and peroxide levels were measured by 3-4,5-dimethyl thiazol- *Z* -yl)-2,5-diphenyltetrazolium bromide and 2',7'-dichlorofluoroscin diacetate assay, respectively. Morphological changes of cells exposed to 320 μM SM (with or without antidotes) were also visualized under light microscope.

Results: On the basis of TBDE, SM caused cell death of approximately 50 per cent at 80 μM and 100 per cent at 640 μM, respectively. Pretreatment of trolox conferred significant protection compared with quercetin. Also, pretreatment of trolox significantly reduced cell death and LDH leakage caused by 80 μM SM but did not prevent the loss of mitochondrial integrity. Trolox significantly reduced the levels of peroxides generated by SM. The better protection offered by trolox was evidenced in cell morphology studies too.

Conclusion: Pretreatment (-0.5 h) of trolox afforded significant protection against SM-induced cytotoxicity in human lymphocytes. The protection was related to the antioxidant property of trolox, a water soluble analogue of α-tocopherol.

Cleistanthus collinus Induces Type I Distal Renal Tubular Acidosis and Type II Respiratory Failure in Rats

Maneksh D, Sidharthan A, Kettimuthu K, Kanthakumar P, Lourthuraj AA, Ramachandran A, Subramani S.; Indian Journal of Pharmacology. 2010; 42: 178-84.

Background and Purpose: A water decoction of the poisonous shrub *Cleistanthus collinus* is used for suicidal purposes. The mortality rate is 28 per cent. The clinical profile includes distal renal tubular acidosis (DRTA) and respiratory failure. The mechanism of toxicity is unclear.

Objectives: To demonstrate features of *C. collinus* toxicity in a rat model and to identify its mechanism(s) of action.

Materials and Methods: Rats were anesthetized and the carotid artery was cannulated. Electrocardiogram and respiratory movements were recorded. Either aqueous extract of *C. collinus* or control solution was administered intraperitoneally. Serial measurements of blood gases, electrolytes and urinary pH were made. Isolated brush border and basolateral membranes from rat kidney were incubated with *C. collinus* extract and reduction in ATPase activity was assessed. Venous blood samples from human volunteers and rats were incubated with an acetone extract of *C. collinus* and plasma potassium was estimated as an assay for sodium-potassium pump activity.

Results: The mortality was 100 per cent in tests and 17 per cent in controls. Terminal event in test animals was respiratory arrest. Controls had metabolic acidosis, respiratory compensation acidic urine and hyperkalemia. Test animals showed respiratory acidosis, alkaline urine and low blood potassium as compared to controls. *C. collinus* extract inhibited ATPase activity in rat kidney. Plasma K^+ did not increase in human blood incubated with *C. collinus* extract.

Conclusions and Implications: Active principles of *C. collinus* inhibit proton pumps in the renal brush border, resulting in type I DRTA in rats. There is no inhibition of sodium-potassium pump activity. Test animals develop respiratory acidosis, and the immediate cause of death is respiratory arrest.

Toxicological Studies of Aqueous Stem Bark Extract of *Boswellia dalzielii* in Albino Rats

Etuk E U, Agaie B M, Onyeyili P A, Ottah C U.; Indian Journal of Pharmacology. 2006; 38: 359-60.

Boswellia dalzielii (Frankincense) is a tree plant. It is abundantly found in north-western Nigeria, where the Hausa speaking people refer to it as *Hano* or *Harrabi.* This plant is very popular among the locals as a potent source of ethnomedicine. The extract from its leaves is used for the treatment of diarrhoea in poultry. The root decoction of *B. dalzielii* and *Daniella oliveri* is used for wound healing. The fresh bark is eaten to induce vomiting and relieve symptoms of giddiness and palpitations. The root decoction of the plant boiled along with *Hibiscus sabdariffa* is used for the treatment of syphilis. The fragrant gum resin from the plant is used locally for fumigation of clothes and houses and as a deodorant. Despite the widespread uses of this plant in treating a plethora of human and animal diseases in this environment, neither its phytochemistry nor toxicological profile has been studied and reported to our knowledge. The present effort is, therefore, aimed at investigating the chemical constituent and the toxicity profile of the stem bark extract of the plant following oral administration in rats.

The plant material was collected from Zuru in Kebbi state of Nigeria in September 2004. A Botanist in the Biological Sciences Department of Usmanu Danfodiyo University, Sokoto confirmed the identity of the plant and a voucher specimen was deposited at the department's Herbarium. The stem bark was carefully separated, air dried to a constant weight and pulverised to a dry powder using a wooden mortar. About 200 g of the powder was macerated in 1500 ml of distilled water for 24 h. The liquid filtrate was concentrated in vacuum at 40°C and the percentage yield of the extract was found to be 6.8 per cent (w/w). The extract was chemically tested for the presence of different chemical constituents using standard methods.

The up and down procedure was adopted to evaluate the acute toxicity of the extract after oral administration in the rats. Five adult, female, non-pregnant rats were randomly selected for this experiment. The animals were marked and housed individually in cages. They were fasted overnight, but allowed free access to water before the administration of a freshly prepared extract orally at a single dose of 3000 mg. kg^{-1}. Each rat was sequentially dosed with the extract and observed for 48 h for signs of acute toxicity or instant death. This was followed by repeated dose (subchronic) study for 28 days.

Twenty-four rats were randomly selected, weighed and divided into four groups of six rats each. The animals in group 1 (control) received 2 ml of normal saline daily, while those in groups II, III and IV received 900, 1800 and 2700 mg.kg^{-1} of the extract, respectively, through the oral route for 28 days. The extract was administered at the same time daily and the doses were adjusted as necessary to reflect the changes in the weights of the animals. The animals were fed and watered adequately during the treatment period. They were weighed weekly and constantly observed for signs of morbidity and mortality. At the end of the study period (28 days), the animals were anaesthesised with chloroform and blood samples immediately collected by cardiac puncture for haematological and biochemical analysis. Necropsy of all the animals was carried out and selected organs like the heart, lungs, liver and kidneys were removed and preserved. The organs were physically examined and weighed and samples were collected for histopathological examinations.

All the data collected during this study were expressed as the mean±SEM. One-way analysis of variance (ANOVA) with subsequent Dunnett's test was used to detect further differences between groups. P <0.05 was considered significant.

The aqueous stem bark extract of *Boswellia dalzeilii* contained various pharmacologically active compounds, such as, alkaloids, tannins, saponins and anthraquinones. The oral administration of 3000 mg.kg^{-1} of the extract did not produce any sign of acute toxicity or instant death in any of the five rats tested during the observation period. This suggests that the median lethal dose (LD_{50}) of the extract using the revised up and down procedure is greater than 3000 mg.kg^{-1}. The fact that the LD_{50} of aqueous extract of *Boswellia dalzeilii* is above 3000 mg.kg $^{-1}$ is an indication that the extract could be considered relatively safe, especially when given orally. Absorption may not be complete due to inherent factors limiting absorption from the gastrointestinal tract.

In the subchronic toxicity studies, the extract did not produce any lethality among the tested animals when varying doses of 900, 1800 and 2700 mg.kg^{-1}sub, were administered orally, daily for 28 days. The rats treated with the highest dose of the extract (2700 mg.kg^{-1}sub), however, recorded significant (P > 0.05) reductions in the percentage weight gain as compared with those in the control group. In addition, the organs' weight (kidney, liver, heart and lungs) of the animals treated with 2700 mg.kg^{-1} of the extract showed significant decease when compared with that of the control and the other treatment groups. The haematological analysis provided the results presented in. There were no significant (P >0.01) changes in the packed cell volume (PCV), white blood cell count (WBC) and red blood cell count (RBC) between the groups of rats treated with 900 and 1800 mg.kg^{-1} of the extract and the control group. There were significant (P <0.01) reductions, however, in the packed cell volume and red blood cell count (P <0.05) of the rats treated with the highest dose of the extract (2700 mg.kg^{-1}). The extract treatment did not change the values of serum electrotypes (sodium and potassium) in the treated animals when compared with the control group. There was a significant increase (P <0.01) in the serum urea level of the rats treated with 2700 mg.kg^{-1} of the extract when compared with the control and the other treated groups. The other biochemical indices like hepatic enzymes and serum proteins remained unchanged. Histopathological examinations of the tissue samples taken from the kidneys, heart, lungs and liver of both the rats treated with the extract and the untreated rats revealed no pathological lesions.

The above findings suggest that prolonged oral administration of very high doses of the aqueous stem bark extract of *Boswellia dalzielii* may be associated with increased risk of toxicity.

Some Pharmacological and Toxicological Aspects of *Nepeta hindostana*

R.B., T. Khanna, K. Kheterpal, M. Imran and D.K. Balani.; JREIM. 1991; 10: 27-30.

Nepeta hindostana is an erect ascending herb of family Iabiatae and popularly known as "Badranjboys". The alcoholic extract of *N. Hindostana* has been reported to be a cardiac stimulant on both normal and hypodynamic heart of frog and rabbit. The alcoholic extract provided significant protection to rats with isoproterenol induced myocardial necrosis. In the present study aqu. extract of *N. hindostana* (Dose 50mg./kg.) given intravenously reduced the blood pressure in dogs by 26.87 per cent. Effect of *N. hindostana* (aqu. extract) on CNS parameters viz behaviour, grooming, spontaneous motor activity (SMR), rectal temperature, righting reflex *etc.* were studied following Irwin"s chart. It reduced the SMR, rearing and grooming while, righting reflex and rectal temperature were not affected in rats in doses upto 1 g/kg. It increase the barbiturate sleeping time in mice (from 4.14 min. to 93.66 min.) thus experimentally supplementing the empirical finding to its sedative action. Toxicity studies are necessary to ensure that the drug (herb) is optimally safe and effective within the dose recommended for the human. Toxicological studies revealed its complete safety on single administration. In doses varying from 125mg/kg. p.o to 4g./kg. p.o in mice and rats, observed for 1 week. Thus encouraging for chronic toxicity studies in higher species.

OTHER ACTIVITIES

Some Pharmacological Properties of *Synclisia scabrida* III

Afonne OJ, Orisakwe OE, Obi E, Orish C, Akumka DD.; Indian Journal of Pharmacology. 2000; 32: 239-241.

Objective: To screen for the pharmacological effects of *S. scabrida* on behaviour, temperature and blood coagulation.

Methods: Aqueous leaf extract of *S. scabrida* (55-440 mg/kg, p.o.) was given to albino mice to test for the central effects, while the aqueous and ethanol extracts mixed with fresh normal human blood were used for the blood coagulation study.

Results: The aqueous extract produced slight motor activity but gave rise to transient hyperthermia, while prolonging the prothrombin time of human blood.

Conclusion: Aqueous extract of *S. scabrida* produces transient hyperthermia, but no motor activity. It also possesses some anticoagulant properties.

Effect of Grapefruit Juice and Cimetidine on the Concentration of Chloroquine in Plasma of Chickens

Ali BH, Al-Qarawi A, Mousa HM.; Indian Journal of Pharmacology. 2001; 33: 289-290.

In the present study chloroquine concentration in chickens treated with the drug with and without grapefruit juice were measured and for comparative purpose, with the drug metabolising enzyme inhibitor cimetidine. Fifteen adult male chickens (*Gallus domseticus*) weighing 900-1020 g were obtained from the King Saud University Farm in Buraydah and used in this study. The birds were randomly divided into three equal groups designated 1, 2 and 3. Group 1: The birds were given distilled water (4 ml/kg) orally using a crop tube. After one hour, chloroquine was administered orally at a dose of 100 mg/kg. Group 2: Chickens were treated as above, except that the distilled water was replaced with freshly squeezed grapefruit juice (4 ml/kg). Group 3: The birds were treated as in group 1, except that distilled water was replaced with cimetidine at an intramuscular dose of 150 mg/kg. Blood was collected from the wing vein using heparinized syringes 0.5, 1, 2, and 4 h post chloroquine treatment and immediately centrifuged at 900 g for 10 min. The drug concentration was measured fluorometrically. In the grapefruit juice-pretreated birds, the concentration of chloroquine in plasma was increased by about 7, 19, 23 and 37 per cent, 0.5, 1, 2 and 4 h after its administration, respectively. In birds pretreated with cimetidine the concentration of chloroquine was increased by about 9, 12, 25 and 21 per cent at 0.5, 1, 2 and 4 h respectively. At 2 and 4 h the increase was significant ($P < 0.05$), in groups 2 and 3 when compared to control. The present results show that grapefruit juice increased plasma concentration levels by 7-37 per cent. Relatively slight increases (9-21 per cent) in chloroquine levels were, however, observed in birds pretreated with cimetidine. Though the observed interaction here resulted in 37 per cent increase in chloroquine concentration, no overt signs of toxicity in any treated bird were seen. Clearly, it is worthwhile to study the interaction in human volunteers taking chloroquine and grapefruit juice.

Effect of *Trichopus zeylanicus* Leaf Extract on the Energy Metabolism in Mice during Exercise and at Rest

D. A. Evans, A. Subramoniam, S. Rajasekharan, P. Pushpangadan; Indian Journal of Pharmacology 2002; 34: 32-37.

Objective: To investigate the effect of the anti-fatigue agent, Trichopus zeylanicus leaf (alcohol extract) on energy metabolism in mice during exercise and at rest.

Methods: Trichopus zeylanicus leaf (alcohol extract) was orally administered to male adult mice. One hour after the extract administration (25, 50, 100 and 200 mg/kg), plasma glucose level was determined or subjected to swimming performance test. The effects of an optimum dose (100 mg/kg) of the extract on plasma glucose, free fatty acids (FFA), pyruvic acid (PA) and lactic acid (LA) levels were determined at rest (1 hr after drug administration) and after swimming exercise for 45 and 90 min.

Results: The alcohol extract of *Trichopus zeylanicus* leaf (100 mg/kg) decreased plasma glucose levels (1 hr after the administration) and increased the swimming performance of mice which was maximum at 100 mg/kg. At a dose of 100 mg/kg, the extract decreased plasma glucose levels and increased the levels of FFA without significant changes in the levels of PA and LA in the resting mice. In contrast, after exercise for 90 min, glucose level was found to be higher whereas the levels of FFA, LA and PA were found to decrease compared to control level.

Conclusion: T. zeylanicus leaf (alcohol extract) influences fuel metabolism in mice at rest as well as during exercise. It stimulates utilization of fatty acids during exercise.

Effect of Methanolic Extract of *Benincasa hispida* against Histamine and Acetylcholine induced Bronchospasm in Guinea Pigs

D. Anil Kumar, P. Ramu; Indian Journal of Pharmacology 2002; 34: 365-366.

Benincasa hispida commonly known as ash gourd belonging to Cucurbitaceae family, is employed as a main ingredient in Kusmanda lehyam in Ayurvedic system of medicine. The lehyam is used as a rejuvenative agent and in nervous disorders. The study was planned to investigate the methanol extract of *Benincasa hispida* (MEBH) fruit for possible antihistaminic or anticholinergic activity. Adult guinea pigs of either sex (400-600 g) obtained from King Institute, Chennai were used for the experiment. The methanol extract of *Benincasa hispida* (MEBH) was used for the pharmacological studies by dissolving each time with distilled water. Experimental bronchospasm was induced by exposing the animals to histamine acid phosphate 0.25 per cent or acetylcholine chloride. The mean increase in exposition time against histamine challenge was significantly ($p<0.001$) increased with the increasing doses of methanol extract of *Benincasa hispida* (MEBH). Methanol extract of *Benincasa hispida* (MEBH) showed excellent protection in guinea pigs against the histamine-induced bronchospasm even at a very low dose, 50 mg/kg, p.o. However, even at a higher dose level 400 mg/kg, MEBH did not offer any significant protection against acetylcholine challenge. Therefore, it can be deduced that MEBH is unlikely to have antimuscarinic action. Thus the results suggest that the protective effect against bronchospsam induced by histamine aerosol may be mediated by antihistaminic activity (H_1 receptor antagonism).

Ginger as an Antiemetic in Nausea and Vomiting Induced by Chemotherapy: A Randomized, Cross-Over, Double Blind Study

S. Sontakke, V. Thawani, M.S. Naik; Indian Journal of Pharmacology 2003; 35: 32-36.

Objective: To study the antiemetic effect of ginger root on nausea and vomiting induced by Cyclophosphamide.

Methods: A randomized, prospective, cross-over, double-blind study was carried out in patients receiving cyclophosphamide in combination with other chemotherapeutic agents. Patients with atleast two episodes of vomiting in the previous cycle were included. The patients were randomly assigned to receive one of the three antiemetics: ginger, metoclopramide or ondansetron in the first cycle. They were admitted in the ward for 24 h and observed for the incidence of nausea and vomiting and adverse effects if any, were recorded. Patients were crossed over to receive the other antiemetic treatments during the two successive cycles of chemotherapy.

Results: Complete control of nausea was achieved in 62 per cent of patients on ginger, 58 per cent with metoclopramide and 86 per cent with ondansetron. Complete control of vomiting was achieved in 68 per cent of patients on ginger, 64 per cent with metoclopramide and 86 per cent with ondansetron. No adverse effects attributable to ginger were recorded.

Conclusion: Powdered ginger root in the dose used was found to be effective in reducing nausea and vomiting induced by low dose cyclophosphamide in combination with drugs causing mild emesis. The antiemetic efficacy of ginger was found to be equal to that of metoclopramide but ondansetron was found to be superior than the other two.

Chemistry and Medicinal Properties of *Tinospora cordifolia* (Guduchi)

Singh SS, Pandey SC, Srivastava S, Gupta VS, Patro B, Ghosh AC.; Indian Journal of Pharmacology 2003; 35: 83-91.

Tinospora cordifolia (Guduchi) is a widely used shrub in folk and ayurvedic systems of medicine. This review presents a detailed survey of the literature on chemistry and medicinal properties of *Tinospora cordifolia*. The chemical constituents reported from this shrub belong to different classes such as alkaloids, diterpenoid lactones, glycosides, steroids, sesquiterpenoid, phenolics, aliphatic compounds and polysaccharides. The notable medicinal properties reported are anti-diabetic, anti-periodic, anti-spasmodic, anti-inflammatory, anti-arthritic, anti-oxidant, anti-allergic, anti-stress, anti-leprotic, anti-malarial, hepatoprotective, immunomodulatory and anti-neoplastic activities.

Antidermatophytic Activity of *Pistia stratiotes*

Premkumar V G, Shyamsundar D.; Indian Journal of Pharmacology 2005; 37: 127-8.

Extract of P. stratiotes leaves has been prepared. P. stratiotes methanolic extract was found to be the most active against the dermatophytes *T. rubrum*, *T. mentagrophytes* and *E. floccosum* with MIC and MFC values of 250 μg/ml, while against *M. gypseum* and *M. nanum*, the values were 125 μg/ml. The values are same for all the 15 replicate experiments. The results show that the trichophyton and epidermophyton species are more resistant to the extract and were inhibited at a higher dosage compared to the microsporum species. The results of the present work indicate that *P. stratiotes* leaves possess antifungal properties, which explains the use of this plant in folk medicine for the treatment of various diseases whose symptoms might involve fungal infections.

Emollient and Antipruritic Effect of Itch Cream in Dermatological Disorders: A Randomized Controlled Trial

Chatterjee S, Datta R N, Bhattacharyya D, Bandopadhyay S K.; Indian Journal of Pharmacology 2005; 37: 253-4.

The present study was a prospective, unicentric, open-label, randomized, controlled study conducted at the dermatology OPD of SSKM Hospital, Kolkata with prior approval from the Institutional Ethics Committee The objectives were to assess the efficacy and safety of Itch cream as an emollient and topical antipruritic for symptomatic relief in various xerotic and pruritic dermatological disorders like atopic dermatitis, senile pruritus, and ichthyosis. Inclusion criteria: (a) children (2-12 years) of either sex with atopic dermatitis, ichthyosis vulgaris, and impetigo. (b) Adults (>12-70 years) of both sex with senile pruritus, ichthyosis vulgaris, or other xerotic diseases. The results suggest that Itch cream has antipruritic and emollient effects in patients of atopic dermatitis, ichthyosis vulgaris, and other xerotic diseases of mild severity. Its efficacy as an emollient was comparable to the nonherbal comparator. The symptomatic benefits achieved could be attributed to the herbal ingredients of the product that have been quoted to possess efficacy in different dermatological conditions in traditional medicinal literature. However, there is no evidence to support that this formulation has any curative potential in the treatment of the above diseases and it can only be used as an add-on therapy to the existent treatment modalities

Relaxant Effect of *Cuminum cyminum* on Guinea Pig Tracheal Chains and its Possible Mechanism(s)

Boskabady M H, Kiani S, Azizi H.; Indian Journal of Pharmacology 2005; 37: 111-5.

To examine the relaxant effects of the macerated and aqueous extracts of *Cuminum cyminum* on the tracheal chains of guinea pig.

Materials and Methods: The relaxant effects of cumulative concentrations of macerated and aqueous extracts (0.25, 0.5, 0.75 and 1.0 g per cent) in comparison with saline and theophylline (0.25, 0.5, 0.75, and 1.0 mM) were examined on pre-contracted tracheal chains of guinea pigs under different conditions.

Results: In Group 1 experiments (contracted by KCl) only the last two concentrations of theophylline and the highest concentration of macerated extract showed significant relaxant effect compared to that of saline ($P<0.001$ and $P<0.05$ for theophylline and macerated extract respectively). The effects of the last two concentrations of theophylline in this group were significantly greater than those of the macerated and aqueous extracts ($P<0.001$). However, in Group 2 experiments (contracted by methacholine) both the extracts and theophylline showed concentration-dependent relaxant effect compared to that of saline ($P<0.05$ to $P<0.001$). The effects of the two last concentrations of both extracts were significantly lower than those of theophylline in Group 2 experiments ($P<0.05$ to $P<0.001$). In Group 3 (non-incubated, contracted by methacholine) the extracts of *Cuminum cyminum* did not show any relaxant effect of tracheal chains. The relaxant effects of macerated and aqueous extracts in Groups 1 and 3 were significantly lower than those of Group 2 ($P<0.05$ to $P<0.001$). However, the effects of different concentrations of theophylline obtained in Group 1 and 2 were not significantly different. There was a significant correlation between the effects and concentrations of theophylline in Groups 1 and 2, macerated extract in Groups 2 and 3 and aqueous extract in Group 1 ($P<0.05$ to $P<0.001$).

Conclusion: These results show a potent relaxant effect of *Cuminum cyminum* on guinea pig tracheal chains which may be due to a stimulatory effect of the plant on β-adrenoceptors and/or an inhibitory effect on histamine H1 receptors.

Skeletal Muscle Relaxant Effect of *Chonemorpha macrophylla* in Experimental Animals

Roy RK, Ray N M, Das A K.; Indian Journal of Pharmacology 2005; 37: 116-9.

Objective: To study the skeletal muscle relaxant property of *Chonemorpha macrophylla* (CM).

Materials and Method: The skeletal muscle relaxant effect of the alcoholic extract of CM was studied on isolated frog rectus abdominis muscle, isolated rat phrenic nerve diaphragm muscle preparation and in intact young chicks. The parameter studied in the isolated muscle or isolated nerve muscle preparations was the extent of inhibition of acetylcholine or electrically-induced contraction of skeletal muscles. In intact chicks, the drug was administered i.v. in wing veins and the onset, duration and nature of paralysis were recorded. In all the experiments, the effect of the drug was compared with that of gallamine and succinylcholine.

Results: The alcoholic extract of CM reduced the acetylcholine-induced contraction of isolated frog rectus abdominis and electrically stimulated contractions of rat phrenic nerve diaphragm in a dose-dependent manner. In unanaesthetized chicks, it produced spastic type of paralysis with extension of the neck and limbs. The effects were similar to the effects of succinylcholine but different from those of gallamine.

Conclusion: The alcoholic extract of CM possesses skeletal muscle relaxant property. It produces depolarizing type of muscle paralysis similar to that produced by succinylcholine.

Antiasthmatic Activity of *Moringa oleifera* Lam: A Clinical Study

Agrawal B, Mehta A. Indian Journal of Pharmacology. 2008; 40: 28-31.

The present study was carried out to investigate the efficacy and safety of seed kernels of *Moringa oleifera* in the treatment of bronchial asthma. Twenty patients of either sex with mild-to-moderate asthma were given finely powdered dried seed kernels in dose of 3 g for 3 weeks. The clinical efficacy with respect to symptoms and respiratory functions were assessed using a spirometer prior to and at the end of the treatment. Hematological parameters were not changed markedly by treatment with *M. oleifera*. However, the majority of patients showed a significant increase in hemoglobin (Hb) values and Erythrocyte sedimentation rate (ESR) was significantly reduced. Significant improvement was also observed in symptom score and severity of asthmatic attacks. Treatment with the drug for 3 weeks produced significant improvement in forced vital capacity, forced expiratory volume in one second, and peak expiratory flow rate values by 32.97 ± 6.03 per cent, 30.05 ± 8.12 per cent, and 32.09 ± 11.75 per cent, respectively, in asthmatic subjects. Improvement was also observed in per cent predicted values. None of the patients showed any adverse effects with *M. oleifera*. The results of the present study suggest the usefulness of *M. oleifera* seed kernel in patients of bronchial asthma.

Preliminary Studies on Local Anesthetic and Anti-pyretic Activities of *Spilanthes acmella* Murr. in Experimental Animal Models

Chakraborty A, Devi BR, Sanjebam R, Khumbong S, Thokchom IS.; Indian Journal of Pharmacology. 2010 Oct; 42(5): 277-9.

Objective: Spilanthes acmella Murr. (Family: Compositae) is a herb that grows throughout the tropics. It is used in the treatment of rheumatism, fever, sore throat, and hemorrhoids. A tincture of the flowers is used to relieve toothache. The leaves and flowers produce numbness of the tongue when eaten as salad. The present study was undertaken to evaluate the local anesthetic and anti-pyretic activities of S. acmella in experimental animal models.

Materials and Methods: Aqueous extract of *S. acmella* Murr. (SAM) was tested for local anesthetic action by (i) intracutaneous wheal in guinea pigs and (ii) plexus anesthesia in frogs. In both the models, 2 per cent xylocaine was used as the standard drug. The anti-pyretic activity was determined by yeast-induced pyrexia in rats. Aspirin 300 mg/kg was used as the standard drug.

Results: The test drug in concentrations of 10 per cent and 20 per cent produced 70.36 per cent and 87.02 per cent anesthesia respectively by the intracutaneous wheal compared to 97.22 per cent anesthetic effect produced by 2 per cent xylocaine ($P<0.001$). The mean onset of anesthesia with the test drug was 5.33 ± 0.57 min compared to 2.75 ± 0.31 min ($P<0.001$) for the standard drug in the plexus anesthesia model. In the anti-pyretic model, ASA in doses of 100, 200, and 400 mg produced dose-dependent reduction in mean temperature at various hours of observation.

Conclusion: The present study shows that SAM has significant local anaesthetic and anti-pyretic activities.

Effect of *Asteracantha longifolia* on Haematological Parameters in Rats

Pawar R S, Jain A P, Kashaw S, Singhai A.; Indian Journal of Pharmacology. 2006; 38: 285-6.

Anaemia is a common nutritional disorder, mainly caused by iron deficiency. *Asteracantha longifolia* Nees (Family Acanthaceae) is a source of the ayurvedic drug, 'Kokilaaksha,' and the Unani drug, Talimakhana. *Asteracantha longifolia* is useful in treating diseases of the blood. The aim of the study was to investigate petroleum ether extract plant leaves of *Asteracantha longifolia* on haematological parameters in anaemic rats. The study was performed with due permission from the Institutional Animal Ethical Committee (Registration No. 397/01/ab/CPCSEA). Male albino rats (90-110 g) of either sex, maintained under standard conditions, were used. The fresh leaves of *Asteracantha longifolia* were authenticated at the Department of Botany, Dr H.S. Gour Vishwavidyalaya, Sagar (MP). The powdered leaves (600 g) were extracted in the Soxhlet apparatus for 18 h with petroleum ether (60-80°C). One gram of petroleum ether extract was mixed with 10 ml of groundnut oil.

The LD_{50} study was performed using albino rats (90-110 g) of either sex. The test animals were divided into six groups of six rats in each group. The drugs were administered i.p., according to body weight. Groups I, II, III, IV, V and VI received petroleum ether extract in doses of 250, 500, 750, 1000, 1250 and 1500 mg/kg, respectively. Groups I, II, III were found to be healthy after seven days. In Group IV, one rat died after a day of extract administration while the remaining five died after three days. In Group VI, three albino rats died after 24 h, while two died after three days. The LD_{50} studies revealed that albino rats tolerated a considerably high dose of petroleum ether extract (1000 mg/kg, body weight, i.p.), without any manifestations.

A haematological study was performed using Swiss albino rats (90-110 g) of either sex. The test animals were divided into five groups of six rats in each group. Group I was kept as the control group (administered vehicle only), Group II was the cyclophosphamide (CP) control group (3 mg/kg, body weight, i.p.). Group III was given petroleum ether extract alone (500 mg/kg, body weight, i.p.), Group IV was given CP (3 mg/kg) and petroleum extract at dose of 250 mg/kg, body weight, i.p. and Group V was treated with CP (3 mg/kg) and 500 mg/kg, body weight, i.p. of petroleum ether extract

Haematological parameters were evaluated in the anaemic animal model. The animals were maintained in standard conditions (temperature, relative humidity, light/day cycles) and given a normal diet and water *ad libitum*. Anaemia was induced by cyclophosphamide (3 mg/kg, body weight) given i.p. for seven days. On the seventh day, blood samples were collected from the retro-orbital plexus vein of the rats' eyes in vials containing EDTA as the anticoagulant. These samples were evaluated for haematological parameters (erythrocyte count, leukocyte count, haemoglobin count and haematocrit value) using haematology cell counter (ERMA, Japan), repeatedly (five times) to check the reproducibility of results. A significant lowering in blood parameters was observed. After seven days, cyclophosphamide was withdrawn from all groups and the groups were treated with petroleum ether extract of *A. longifolia* once a day, at the doses of 250 and 500 mg/kg body weight continuously up to 15 days

Blood was collected on the 22nd day and evaluated for haematological parameters. All values are expressed as mean ± SEM. The data were statistically analysed using the one-way ANOVA, followed by the Dunnett's test. The cyclophosphamide (CP) control group (Group II) showed significant (P >0.05) decrease in blood parameters as compared to the vehicle control group (Group I), without drug treatment. The comparison of Group II with Groups III, IV and V exhibited a significant (P <0.05) increase in haematological parameters after seven days. Cyclophosphamide-induced anaemia in Group II was not restored to normal counts even after the discontinuation of the drug after seven days. Upon continuation of petroleum ether extract at 250 and 500 mg/kg, body weight, respectively, for the next 15 days there was a significant (P <0.05) improvement in haematological parameters in Groups III, IV and V as compared to Group II.

In rats, CP has a bone marrow suppressive effect and induces aplastic anaemia. CP treatment (3 mg/kg, body weight, i.p.) resulted in the significant lowering of haematological parameters on the seventh day. All the haematological indices were restored to almost normal counts after continuous administration of the extract. These investigations validate the use of the leaves of this plant, in the Indian system of medicine, for haematopoietic activity. In the present work, we have not studied the toxic effects caused by the extract. We aimed at investigating the haematopoietic activity of the plant extract and the results of the present study led to the conclusion that *A. longifolia* has great potential as a haematopoietic agent.

Quantitative Parameters of different Brands of Asava and Arishta Used in Ayurvedic Medicine: An Assessment

Weerasooriya W, Liyanage JA, Pandya SS.; Indian Journal of Pharmacology. 2006; 38: 365.

Arishta and *Asava* have been used as medicines for over 3000 years to treat various disorders and are also taken as appetisers and stimulants. Due to their medicinal value, sweet taste, and easy availability people are prone to consume higher doses of these drugs for longer periods. The manufacture and sale of *Arishta* and *Asava* occupies an important place in the ayurvedic pharmaceutical industry. The preparation and sale of 34 varieties of *Arishta* and 25 varieties of *Asava* has been legalised and listed in the official Ayurveda pharmacopoeia of Sri Lanka.

Information on the quantitative parameters of *Asava* and *Arishta* to guarantee the quality and the safety of the product to the consumer is less. Therefore, establishing quality and standard parameters like alcohol level, pH, acid value and other constituents of these preparations are highly significant.

The objective of this study was to determine the level of alcohol, acidity and pH in commercially available *Ashvagandarishta* and *Aravindasava* to establish a routine procedure for standardisation of these Ayurvedic preparations.

Twenty brands of commercially available *Ashvagandarishta* and *Aravindasavaya* stored in sealed bottles under room temperature were randomly collected (island-wide). They were tested for percentage of total alcohol (v/v), ethanol, pH, and acid value. Acid value indicates the total acids present in the product. Acids are produced during preparation (especially in the fermentation process) and storage (oxidation of alcohols) and are responsible for the sour taste of those preparations. Electric ebulliometer (Model no 01121E, Laboratories Dujardin Sallerone co., France) was used to determine the total alcohol content. Boiling points of the samples were tested and the alcohol percentages (v/v) were calculated using the scale disk of the ebulliometer. For the estimation of ethanol (v/v), 100 ml of each *Ashvagandharishta* and *Aravindasava* was distilled at 78.5°C. The evaporate was condensed into a flask, and the volume was measured. A bench type pH meter with microprocessor (Hanna, Italy serial no 249752) was used for the pH measurement after calibration with buffer solutions of pH 4 and 7 (Hanna). The pH values of the commercially available brands of *Ashvagandharishta* and *Aravindasavaya* were measured at the time of opening the bottle and seven days and 14 days after opening the bottle. During the 14 days, the drug bottles kept at room temperature were shaken well manually to mimic the normal practice during consumption and opened two to three times per day. Each brand of *Arishta/Asava* (10 ml) was diluted with distilled water (20 ml) and titrated with 0.1M sodium hydroxide. The acid concentration was calculated using phenolphthalein as the indicator.

Alcohol percentage (v/v) in each of the tested brands of *Ashvagandharishta* was different. The maximum was 13.13 per cent and the minimum was 7.27 per cent, with a mean value of 9.96±1.41 per cent. The maximum and the minimum values of the tested brands of *Aravindasava* were 13.00 per cent and 7.7 per cent, with a mean of 9.95±1.44 per cent. The results highlighted that the levels of alcohol in *Ashvagandharista* and *Aravindasava* were lower than those in fortified wines and distilled spirits. These contain 18 to 21 per cent and 40 to 50 per cent alcohol, respectively. In the commercially available *Ashvagandharishta*, the mean isolated ethanol was 6.55±0.87 per cent, which is given by the minimum of 5.17 per cent and the maximum of 8.23 per cent. For *Aravindasava*, these values were 7.85±1.31 per cent, 5.67 per cent and 11.03 per cent, respectively. Changing of the pH with time after the bottles were opened and acid values of commercially available *Ashvagandarishta* and *Aravindasava* are given in. Measured pH and acid values indicate that both *Ashvagandarishta* and *Aravindasava* have weak acidic properties. Therefore it can be concluded that the recorded levels of alcohol, acidity, and pH in commercially available *Ashvagandarishta* and *Aravindasava* could be used to establish and formulate procedures for standardization and quality control of these ayurvedic preparations.

Isolation, Characterization and Study of Enhancing Effects on Nasal Absorption of Insulin in Rat of the Total Saponin from *Acanthophyllum squarrosum*

Sajadi Tabassi SA, Hosseinzadeh H, Ramezani M, Moghimipour E, Mohajeri SA.; Indian Journal of Pharmacology. 2007; 39: 226–230.

Objective: Isolation of the total saponins from *Acanthophyllum squarrosum* Boiss. and investigation of its surface activity, haemolytic effects on human erythrocytes, as well as enhancing potentials on intranasal insulin absorption in rat as compared to two other enhancers, *i.e.*, Quillaja total saponin (QTS) and sodium cholate (SC).

Materials and Methods: The decrease in blood glucose levels in five fasting rats following nasal administration of regular insulin solutions in the presence or absence of enhancers was determined by glucometric strips and used as an indication of insulin absorption.

Results: The results showed that Acanthophyllum total saponin (ATS) decreased surface tension of water to about 50 dyne/cm and caused complete haemolysis of human RBCs at a concentration of 250 μg/ml. Following the instillation of solutions containing insulin and different absorption enhancers into the right nostril of rats, the percentage decrease in initial blood glucose was as follows: 72.46 per cent (±2.39 per cent) for ATS, 63.22 per cent (±11.06 per cent) for QTS and 60.06 per cent (±14.93 per cent) for SC. Percentage lowering of initial blood glucose concentrations against time showed that ATS exerts a stronger effect than the two other enhancers, although the difference was not statistically significant ($P > 0.05$).

Conclusion: ATS has a considerable absorption enhancing effect and can possibly be used to increase insulin bioavailability via the nasal route. However, the potential toxic effects of this saponin on nasal mucosa should be further evaluated.

Clinical Evaluation of Sunthi (*Zingiber officinale*) in the Treatment of Grahni Roga

Nanda GC, Tekari NS, Kishore P.; J Res Ayur Siddha. 14; No. 1-2, 34-44.

Grahani Roga (Malabsorption Syndrome), a common disease of the people of the area, has been taken for extensive studies. A clinical trial of commonly available drug Sunthi (*Zingiber officinale*) has been put to trial on a series of 111 patients. The observations suggest high incidence in your patients, relatively moderate duration of chronicity and the presence of most of cardinal features of the disease. The effect of the treatment has been obvious within a short period. The regulation of bowel habits, improvement in general health including anaemia and body wieght and, improvlement in the Gastro intestinal function has been noted. The observations highlight the improtance of simple herbal treatment of the disease.

A Clinical Trial of Volatile Oil of *Curcuma longa* Linn. (Haridra) in Cases of Bronchial Asthma (*Tamaka Swasa*)

Jain JP, Naqvi SMA, Sharma KD.; J Res Ayur Siddha. Vol 11, No. 1-2: 20-30.

Seven cases of bronchial asthma (Tamak Swasa) and one case of tropical eosinophilia were observed with the treatment with volatile oil of *Curcuma longa*, given by deep intramuscular injection. The response was good in 2 cases of bronchial asthma, fair in another 2 cases, while other 3 cases did not show any response. The case of tropical eosinophilia did not show any response at all. The trial was discontinued as side effects like excruciating pain (injection abscess in I case) and kidney irritation or damage as suggested by pus cells, casts, epithelial cells and R.B.C.S in the urine, was noticed in nearly all the cases. Two cases of bronchial asthma were treated with the volatile oil of *Curcuma longa* given as emulsion with milk. One case of bronchial asthma showed fair response while in the other response was poor.

A Clinical Trial on the Efficacy of Shirisa Twak Kwatha in the Management of Tamaka Swasa

Chandrakar A, Kale KV, Prasad RD, Mishra R, Audichya KC.; J Res Ayur Siddha. 2005; 26: 52-58.

A preliminary study was undertaken with Sirisha twak kwath at Maharao Shekhaji Central Research Institute (Ayurveda), Jaipur as a part of CCRAS project to evaluate its efficacy in Tamaka swasa as per the subjective parameters only. A total number of 13 patients were selected from OPD of Maharao Shekhaji Central Research Institute (Ayurveda), as per the inclusion criteria designed for this trial. Shirisa twak was selected for this study since it is having the Swasahara, Kasahara and having antispasmodic and antihistaminic effect as suggested by pharmacological study. This drug was administered in the dosage of 30ml. In Kwatha form three times daily. The patients who were taking inhalers during severe dyspnoea were allowed to take whenever needed. The effect of drug was found significant.

Preliminary Study of Anti-bacterial Effects of an Ayurvedic Recipe Rathakalka

Amarasinghe APG, Widanapathirana S.; J Res Ayur Siddha. 2005; 26: 1-5.

Rathakalka is a popular recipe prescribed routinely for infants by traditional physicians in Sri Lanka. This particular receipe is being used in infant at a dose of 250mg. Twice a day orally for two main purposes, one use of it as a preventive medicament to protect children from certain diseases during infancy. And he other ue of it as a curative measure in infant and young children against certain diseases specially skin diseases, upper respiratory tract infection and fever. One of the main causes for above conditions is bacterial infection. The objective of this experiment was to study the anti-bacterial activity of Rathakalka. 0.1 g of Rathakalka was dissolved in 5ml. of distilled water and filter sterilized, using a Hemmings filter. This sterile water soluble extract of Rathakalka was used in this study. Anti-bacterial assay was performed using cylinder plate method in Nutrient agar and incubated for 48 hours at room temperature. Bacterial species used were *Staphylococcus aureus. Pseudomonas aeruginosa, Escherichia coli* and *Listeria monocytogenes*, all obtain from the culture collection of the Department of Micro-biology, University of Kelaniya.

Ayurveda and Sukshma Medicines

Abhang R; J Res Ayur Siddha. Jan-June, 1985.

The study explores the principle of Sukshmikaran and the processes to prepare the Sukshma medicines as found in the Ayurveda. It is explained how these medicines act and why they are effective in Sukshma (minute) doses. Then follows an accunt of the various methods to prepare Sukshma medicines adopted by different researchers up to date. With the help of this knowledge, principle of Sukshmikaran is developed further the Sukshma medicines are prepared in two forms - (1) Using alcohol as vehicle; tinctures are prepared from about 100 herbal medicines for common ailments, children's complaints, female disorders, skin diseases *etc.* (2) Using milk-sugar as vehicle; triturations are developed from Bhasmas and Resaushadhis. These Sukshma medicines are used in clinical treatments. Sukshma medicines are effective and have many advantages over other preparation. They are, cost-effective, devoid of side-effects, suitable for delicate constitutions, compact in form and so on. These Sukshma medicines will be useful as simple drugs o achieve the goal of "Health for All" by 2000 A.D. This paper discusses the concept of Sukshamikaran. Sukshma medicines developed at different times in the past and their advantages as an integral part of Ayurveda.

A Clinical Study on Gjridhrasi (Sciatica) and its Management with Nirgundi

Sthiti Srujani Mishra, N. C. Dash, B. K. Das; J Res Ayur Siddha. 2003; 24: 42-50.

About 7 million people are temporarily out of work at any given time due to low back pain that often accompanies to leg pain (sciatica) and it is the most common cause of disability for persons under the age of 50. Searching out a remedy for sciatica still under process and administration of *Viten nirgundo* Linn. externally as well as internally discloses a remarkable efficacy established through clinical and statistical review which recorded as 70 per cent of maximum improvement, 20 per cent of moderate improvement and 10 per cent of mild improvement.

Pharmacological Studies on Dasamula Kvatha – Part II

Gupta RA, Singh BN, Singh RN.; J Res Ayur Siddha. 5: 38-50.

The study indicates that Dasamula Kvatha extract effectively produced like aspirin analgesic effect. The aspirin-like effect is further supported by the signijficant anti-pyretic effect and mild anti-inflammatory effect in rats against carageenin induced oedema. It supports the use of Dasamula Kvatha in various clinical conditions like pain, pleurodynia, backache, gout, pyrexia, sciatica, headache and sotha (inflammation and oedema) as recommended in our Ayurvedic texts.

Evaluation of the Efficacy of the Kaanchanaara Guggulu and Pippali Vardhmaana Rasayana in the Management of Hypothyroidism vis-à-vis Agnimandya

Sharma AK, Keswani P, Kankran K.; J Res Ayur Siddha. 2005; 26; 6-22.

Although no specific description of Thyroid gland and its disorders is available in Ayurvedic texts. But minute observations reveal that in the Thyroid hormone there is striking parallel functions of Kayagni in Dhatus. Hypo-secretion of Thyroid gland may lead to a clinical condition called Hypothyroidism which is manifested in the form of Mandagni at the systemic and cellular levels resulting in an increase of Dhatus. The study was conducted on 30 clinically diagnosed and confirmed patients of Hypothyroidism. 30 patients were randomly divided into 3 groups. 10 patients of I group were given tab. thyroxine Sodium *i.e.*, Allopathic Therapy which revealed 37.36 per cent of improvement. 10 patients of II group were given Tab. Kaanchanaara Guggulu and Pippali Vardhmaana Rasayana *i.e.* Ayurvedic Therapy which showed 68.67 per cent of improvement. 10 patients of III group were given Tab. Thyroxine Sodium. tab. Kaanchanaara Guggulu and Pippali Vardhmaana Rasayana simultaneously which revealed 79.73 per cent of improvement. The Clinical study on all the 30 registered patients of Hypothyroidism revealed that there was significant improvement in all the clinical symptoms but the percentage of improvement was mild in Allopathic therapy treated patients *i.e.* I group, moderate in Ayurvedic Therapy treated patients *i.e.* II Group and maximum in Mixed Therapy treated patients *i.e.*, III Group.

Effect of Ghrita Kumari in Alpartava

Bharathi K, Tewari PV.; J Res Ayur Siddha. 2002; 23: 68-75.

Alparatava is a common clinco-pathological entity encountered by gynecologists especially along with infertility (hypo-ovarian functions cases). As Alparatava is a cardinal symptom of Vataja Artava Dusti, indicates the main Dosha here is Vata. In allopathy it is correlated with scanty mensturation and can be seen in two conditions *i.e.*, hypomenorrhoea and oligomenorrhoea. The drug Ghirta Kumari is having the capacity to normalize Vata and also having specific Prabhava to increase the amount of Artava, so it has been selected a drug of choice. In the present work 40 cases studies to evaluate the efficacy of Ghrita Kumari as a single drug. 16 (40.00 per cent) cases completely cured and 14(35.00 per cent) cases showed improvement, 10(25.00 per cent) cases got relieved from Alparatava 0(00.00 per cent) of cases reported under no-change category.

A Study of Psychiatric Symptoms or Geriatric Patients and the Response to Ayurvedic Therapy

Diwvedi KK, Agarwal S, Singh RH.; J Res Ayur Siddha. 2001; 22: 198-207.

The increasing number of elderly people in the population has already been posing a number of problems in the western world and geriatrics has been emerging as a major medical specialty. In India too the population aging is becoming a more and more prominent issue. Hence there is a need of developing strategies for care of the aged. In this context the Rasayana therapy of Ayurveda has drawn the attention world over. The present study presents a psychiatric symptom profile of the aged people and the results of a short-term clinical trial of Rasayana remedies. Ashvagandha and Kapikacchu in aged persons in terms of their mental health and congnitive functions. Use of Ashvagandha Rasayana exhibits notable beneficial effects.

Preliminary Observations on Serum Biochemical Parameters of Albino Rabbits Fed on Seeds of *Trichosanthes dioica* (Roxb.)

Sharma G., Pant MC.; J Res Ayur Siddha.

Effects of feeding of T. dioica seeds at the level of 1 g per cent for eight weeks were studies in the normal, healthy, make albino rabbits. The seeds significantly lowered blood sugar, serum cholesterol, triglycerides and increased the level of phospholipids and HDL-cholesterol.

Treatment of Pradar with Ashokarista

Geeta, M. A. Amma, V. I. Nalini; J Res Ayur Siddha.

The cases selected were only Pradara Asrigdara-disfunctional uterine bleeding after ruling out the surgical conditions. The efficacy of Asokarista was studied in 22 cases of Pradara out of which Vatadhika 6. Pittadhika 5 and Kaphadhika 11. Even though the number of cases studied was less, the result was encouraging as the relief was noticed in Vatadhika dn Kaphadhika cases within 3 days after starting the treatment and 5-6 days in Pittadhika cases. Treatment was continued for 3 months. The result was encouraging as relief was noticed in Vatadhika and Kaphadhika cases, within 3 days after starting the medicine.

Clinical and Experimental Evaluation of endocrine response of Herbal Drugs

Tripathi SN, Tripathi CM.; J Res Ayur Siddha. 13: 174-178.

An attempt has been made to study the effect of selected Ayurvedic herbal drugs for their endocrinal response. A series of clinical and experimental studies have been taken up on bio-chemical histo-pathological and radioactive isotopes. Thyroid stimulating effect of Dhanayaka, Marica, anti-histaminate effect of Kantakari, are some of the important observations. The glucometabolism and hepatoprotective properties of Kalmegha are also significant.

Effects of *Boerhaavia diffusa* Linn. Extract on the Activities of Enzyme Systems

Goswami P, Sharma TC.; J Res Ayur Siddha. 1992, 13: No. 3-4, 48-55.

Varying quantities of ethanol-chloroform-dry ether crystalline extracts of the plant *Boerhaavia diffusa* Linn. were added into the *in vitro* system and enzyme activities of amylase and cholinesterase were estimated. Enhancement of the activities of amylase in the direct proportion to the amount upto 4.5mg in the system of the crystalline extract added was noted. Cholinesterase activity was noted to increase at lower amount but with the higher dose the activity was inhibited and finally no activity was noted. Significance of findings was discussed.

Pharmacological Studies on Raktachandana (*Pterocarpus santalinus* Linn.)

Verma RR, Vijayamma N.; J Res Ayur Siddha. 12: 190-199.

A decoction of the heartwood of *Pterocarpus santalinus* Linn. Was studied for its hypnosis potentiating, tranquillising, analgesic, anti-pyretic, anticonvulsant, local anaesthetic, anti-inflammatory and diuretic activities. In doses varying fom 1g to 8 g/kg (in terms of crude material) this decoction produced potentiation of pentobarbijtone induced hypnosis in mice., blocked conditions avoidance response in trained rats and showed anti-convulsant and anti-inflammatory activities.

Therapeutic Effect of Coconut Shell Extract in Dermatophytosis

Nair PKS, Pilai NGK, Kurup PB, Nair CPR.; J Res Ayur Siddha. 8: 46-52.

The effect of alcoholic extract of coconut shell, 2 per cent ointment with petroleum jelly as an external medication on thirty one cases of dermatophytosis is observed in this study. The decrease of symptoms, itching, burning sensation erythema, vesicles, oozing and hyperpigmentation are observed in 2nd week, 4th week and 6th week and compared with the initial stage. The drug showed significant antifungal activity.

Teratological Evaluation of Two Ayurvedic Medicines, Vi-musti Vati and Suddha Tankana in Rats

Sethi N, Nath D, Singh RK, Dayal R.; J Res Ayur Siddha. 8: 64-69.

Two commonly used Ayurvedic medicines, Vismusti Vati (antitumoral) and Suddha Tankana (anti-fertility) were studied in our laboratory for anti-fertility effect. We found 63.52 per cent and 33.33$ fertility - 175mg and 300mg./kg. aqueous solutions respectively were given orally from day 1 to day 7 of postmating period but with increasing postmating period, the antifertility activity gradually reduced and the born foetuses showed gross remarkable external morphological and skeletal defects.

Pharmacological Studies on Lodhrasava

Ravishankar B, Sasikala CK.; J Res Ayur Siddha. 7: 33-46.

Lodhrasava, a compound Ayurvedic preparation was evaluated in view of its observed sedative and tranquilising effects in patients. It has no sedative, anti-psychotic, anti-convulsant, anti-depressant and anti-parkinsonian effect. This study failed to corroborate clinical findings.

Pharmacological Studies on Dasamula Kvatha – Part II

Gupta RA, Singh BN, Singh RN.; J Res Ayur Siddha. 5: 38-50.

The Dasamula Kvatha extract produced C.N.S depressant effect. It reduced the spontaneous motor activity, potentiated the pentobarbitone hypnosis and antagonised the amphetamine induced hyperactivity, in mice. It also possesses a tranquillo-sedative activity like a major tranquillizer as it blocked the condition avoidance response in rats. It reduced the normal body temperature which further supports the tranquillization effects of the compound. Further study is needed to detect its usefulness in insanity descrbed in Ayurvedic texts.

Effect of Aswagandha (*Withanaia somnifera* dunal) on the Process of Ageing in Human Volunteers

Kuppurajan K, Rajagopalan SS, Ritaraman R, Rajagopalan V, Janaki K, Revathi R, Venkataraghavan S.; J Res Ayur Siddha. 1980; Vol. 1, No. 2.

A double blind clinical trial to study the effect of Aswagandha (*Withania somnifera* dunal) on the prevention of process of ageing in 101 male healthy adults in the age group of 50-59 years has been completed. The result indicate that the increase in haemoglobin, RBC, hair melanin and seated stature in the treated group is statistically significant in comparison to the placebo. The decrease in serum cholesterol is more and in nail calcium it is less in the treated group as compared to the placebo and this difference is statistically significant. The decrease in Erythrocyte Sedimentation Rate (ESR) is much higher in the treated group than in the placebo and this difference is statistically significant.

Unani Drugs in Ayurvedic Material Medica

Ali M.; J Res Ayur Siddha, Oct-Dec, 1992.

About one hundred Unani drugs have been added in the Ayurvedic materia medica. The Unani names of about seventy drugs have been adopted as such or with slight modification by the Ayurvedic physicians. There are some Unani drugs like *Acacia nilotica, Aloe barbaensis, Althaea officinalis, Anisomeles malabarica, Ariemisia siversiana, Asparagus racemosa, Atropa bellodonna, Berberis lycium, Carica papaya, Fumaria officinalis, Glycyrrhiza glabra, Haloxylon mulliforum Hibiscus rosa-sinensis, Ipomoea nil, Lodoicea maldivica, Nymphoea mouchali, Operalina turpethu, Pimpinella anisum, Polygonum aviculare, Pterocarpus marsupium, Smilax zeylanica, Storculia urans, Tamarix aphylla, Tamarix troupii* and *Vateria indica* whose Ayurvedic names differ from the Unani names. These names seem to be specified to some obscure regions. In addition to herbal drugs, some Greek-Arabic medicaments of animal and mineral orgins have been added to the Indian System of Medicine: Ayurveda.

Effect of an Herbal Compound: Thyrocap in the Patients of Simple Diffuse Goiter

Pandia RK, Gupta RC, Prasad GC.; J Res Ayur Siddha. Oct-Dec, 1992.

Thyrocap is a herbal preparation (consisting solid extract of *Bouhinia variegata*, *Commphora mukul*, *Glycyrrhiza glabra* an *Convalvulus plauricaulis* 100mg. Each) for the treatment of different types of goitre. It has been tried clinically on the patients of simple diffuse goitre. Treatment with thyrocap one capsule thrice a day for three months indicate that it is safe, effective and promising combination of herbs. The treatment showed marked symptomatic as well as biochemical improvement in patients of simple diffuse goiter.

Kanchna Guggulu: A Critical Review

Pandit RK, Kumar S, Sharma L.; J Res Ayur Siddha, Jul-Sep, 1992.

Kanchnar Guggulu is an ethical preparation advocated for the management of vaious glandular swellings like, galgand, gandmala, granthi and arbuda *etc.* The main Ayurvedic text of Bhaishjayakalpna *i.e.*, Bhava Prakash, Yoga Ratnakar, Vangsen Samhita, Sharangdhar Samhita and Bhaishajya Ratnavali has given almost similar description regarding its contents and indication. Most of the ingredients are Tridoshahara Kaphvataanashoka.

Physico-chemical Standardization of Habb-e-Mumsik

Ahmad J, Siddiqui ZS, Zaman A.; J Res Ayur Siddha, Jul-Sep, 1992.

Habb-e-Mumsik is zn important Unani compound preparation used for the treatment of excessive nocturnal emission. The standardization was carried out according to the parameters laid down by CCRUM and also includes organoleptic properties. The ash values, total fact, saponification and iodine values were performed. Successive extraction in petrol, chloroform and alcohol were carried out. Qualitative estimation of different secondary metabolities and parameters like resin, tanin, alkaloid, crude fiber and values of thin layer chromatography of different extractives have been reported in the present communication.

Therapeutic Efficacy of Punarnava (*Boerhaavia repanda* wild.) Root Powder

Singh SP.; J Res Ayur Siddha. 1991

Boerhaavia repanda willd. (Fam. Nyctagenaceae) is an indigenous plant and is popular as a home remedy for various ailments. The root power of this plant showed 100 per cent and 90 per cent therapeutic efficacy against leucorrhea and spermatorrhea respectively, fed at 500 mg dose twice/day for 25 days. The root powder also showed curative efficiency against helminth cases in children and adults at doses of 250 mg and 500mg/day for 5 days respectively. The efficacy was 90- per cent in children and 100 per cent in adults.

Drug Potential and Conservation Notion of Medicinal Plants of Western Himalaya

Pandey NK.; J Res Ayur Siddha, Jul-Sep, 1991.

Western Himalaya is endowed with phyto-geographical variations and procured about 65-70 per cent of herbal drug for welfare of our society. In this paper attempt is being made to enumerate 231 recognized and established medicinal plants. The environmental conservation of this zone is also discussed. An ecology development programme aimed at restoring ecosystem through management strategies *i.e.*, 'Plan to compartmentalization' the land scape so as to simultaneously maintain its high productivity. Some possible solutions for conservation of medicinal plants have been proposed and applied.

A Clinical Trial of Kantakari (*Solanum xanthocarpur*) in Case of Kasa Roga

Jain JP.; J Res Ayur Siddha.

Kantakati is effective in both Vataslesmik Kasa with slesmapradhan and vataslesmik Kasa with vatapradhan. There has been either good(complete) or fair (significant) response in 45 per cent of case of Vataslosmik Kasa with slesma pradhanata. Kantakari has a definite effect in diminishing the intensity of cough and dyspnoea and nearly 50-60 per cent cases have shown either good (complete) or fait (significant) relief. The amount of sputum also diminished by 50 per cent or more in about 50 per cent of cases. There was 50 per cent or more relief in difficulty in expectoration in only 20 per cent of cases. There was 50 per cent or more reduction in auscultatory physical signs in lungs in only 40 per cent of cases. The reduction in airway resistance as assessed by pulmonary functions has been found to be insignificant on statistical analysis. There has been either good) complete or fair (significant) response in more than 55 per cent of the cases of Vataslesmik Kasa Vatapradhan. There has been either complete of significant relief in cough and dyspnoes in about 56-62 per cent of cases. The amount of sputum diminished by 50 per cent or more in about 50 per cent of cases. There was 50 per cent or more relief in difficulty in expectoration in about 50 per cent of cases. There was 50 per cent or more reduction in auscultatory physical lung signs in more than 45 per cent of cases. The reduction in airway resistance as assessed by pulmonary function has been found to be insignificant on statistical analysis.

Some Pharmacological Actions of *Convolvulus pluricaulis* Chois: An Indigenous Herb

Sharma VN, Barar FSK, Khanna NK, Mahawar MM.; Ind. J. Med. Res. 1965, Vol. 53, No. 9, Sept., p. 871-76.

The indigenous plant, *Convolvulus pluricaulis* Chois, grows throughout the plains of India and is a hairy perennial herbbelonging to the natural order Convolvulacea. It flowers during the months of September and October. The flowers are white to light pink in shade. The common Indian name of the plant is Sankhapushpi (Sanskrit) and it also known as Pograng, Gorakhpinaw, Bephuli and Dodak. The Ayurvedic system of medcine advocates its use as a brain tonic in some forms of insanity and neurasthenia. Pharmacological investigations of this plant have been extended to the activity of the central nerous system. The commnucation includes the results of study undertaken with the extract of the entire plant.

Effect of Antiparkinson Drug HP-200 (*Mucunaa pruriens*) on the Central Monoaminergic Neuro Transmitters

Manyam BV, Dhanasekran M, Hare TA.; Drugs Aging, 2004; 21(11): 687-709.

HP-200, which contains *Mucuna prurients* endocarp, has been shown to be effective in the treatment of Parkinson"s disease. *Mucuna prurients* endocarp has also been shown to be more effective compared to synthetic levodopa in an animal model of Parkinson's disease. The present study was designed to elucidate the long-term effect of *Mucuna prurients* endocarp in HP-200 onmonoaminergic neurotransmitters and its metabolite in various regions of the rat brain. HP-200 at a dose of 2.5, 5.0 or 10.0 g/kg./day wa mixed with rat chow and fed daily ad lib to Sprage -Dawley rats (n=6 for each group) for 52 weeks. controls (n=6) received no drug. Random assignment was made for doses and control. The rats were sacrificed at the end of 52 weeks and the neurotransmitters were analyzed in the cortex, hippocampus, substantia nigra and striatum. Oral administration of *Mucuna prurients* endocarp in the form of HP-200 had a significant effect on dopamine content in the cortex with no significant effect on levodopa, norepinephrine or dopamine,serotonin, and their metabolities - HVA, DOPAC and 5-HIAA in the nigrostriatal tract. the failure of *Mucuna prurients* endocarp to significantly affect dopamine metabolism in the striatonigral tract along withits ability to improve Parkinsonian symptoms in the 6-hydroxydopamine animal model and humans may suggest that it antiparkinson effect may be due to components other than levodopa.

11

Problems of Pharmacopoeial Standardization of Ayurvedic Drugs

Chandra Kant Katiyar and Rahul Singh

Herbal Drugs: A Composite Review

Plants have played a significant role in maintaining human health and improving the quality of human life for thousands of years. From ancient times, plants are not only used for daily food, fuel, and shelter but also used for the prophylactic and/or therapeutic action, such as for the treatment of different diseases. Any species of plant useful in maintenance of health, detection and prevention or cure of disease is known as medicinal plant. Previously the mechanism of action of medicinal plants was not clear. Now scientists from different parts of the globe are actively engaged in the identification of chemical constituents of medicinal plants, their mechanism of action and toxicity, if any. If the active constituents present in the plants are established, cost effective nontoxic medicine can be generated which being of natural origin, would be more acceptable to teeming the human population. In recent years more people throughout world are turning to use medicinal plant products in healthcare system. Worldwide requirement of alternative medicine has resulted in growth of natural product markets and interest in traditional systems of medicine. Active phytoconstituents – bearing plants are known as plant/herbal drugs. In the last century, over 100 pharmaceutical products have been discovered based on the information obtained from the traditional healers.

The World Health Organization (WHO) estimated that nearly 80 per cent of the earth's inhabitants rely on traditional medicines for their primary healthcare needs, and most of this therapy involves the use of plant extracts or their active components. The widespread use of herbal remedies and healthcare preparations, as those described in ancient texts such as the Vedas and the Bible, and obtained from commonly used traditional herbs and medicinal plants, has been traced to the occurrence of natural products with medicinal properties.

Trade Situation

With the growing interest in plant-based medicine over the last 20 years, herbal medicine has been enjoying a renaissance throughout the world. The global market of herbal medicinal products was estimated at approximately US$ 60billion in the year 2000. Demand for herbal products has been growing at the rate of 7 per cent per year and is expected to reach US$ 5 trillion by 2050. More than 50 medicinal plants are traded extensively in the international market.

Export of herbal products and essential oils from India is more than Rs. 2 billion; 15 herbal drugs and essential oils are regularly exported from India. Within the country, the turnover of herbal drugs was estimated to be Rs. 2000 crore, which includes classical formulations of Ayurveda, Unani, Siddha, Homeo, proprietary medicines and over-the-counter products.

How to Handle the Situation

Despite the promise that plant-based medicine exhibited, the one major obstacle in using plant-based drugs has been the reproducibility of the activity.

In old times the traditional medicine used to be a personalized one, with the healers preparing the medicines on an individual basis, where the quality of the medicine and hence the safety and efficacy were taken care of completely. Large-scale production of herbal drugs has only started in the last 100 years or so. Now that the commercialization of the herbal medicine has happened, the onus of maintaining their quality falls to a large extent on the scientists and to a certain extent on the manufacturers. In many countries, the herbal market is poorly regulated. In this scenario, the assurance of safety, quality and efficacy of medicinal plants and herbal products has become an important issue.

The herbal raw material is prone to a lot of variation due to several factors, the important ones being the identity of the plants and seasonal variation (which has a bearing on the time of collection), the ecotypic, genotypic and chemotypic variations, drying and storage conditions and the presence of xenobiotics. The National Centre for Complementary and Alternative Medicine and the WHO stress the importance of the qualitative and quantitative methods for characterizing the samples, quantification of the biomarkers and/or chemical markers and the fingerprint profiles. It is indeed a challenging task to develop suitable standards for herbal drugs. The advancements in modern methods of analysis and the development of their application have made it possible to solve many of these problems. Extremely valuable are techniques like high-performance thin-layer chromatography (HPTLC), gas chromatography (GC), mass spectrometry (MS) high-performance liquid chromatography (HPLC), LC-MS, and GC-MS.

As mentioned above, development of standards for plant-based drugs is a challenging task and it needs innovative and creative approaches, different from the routine methods [Raskin *et.al;* 2002]. Starting from sourcing of the raw material, standardization, preparation of the extracts, to formulation of the extracts into suitable dosage form, the problems vary with each plant species and part of the plant that is being used. At each and every step, phytochemical profiles have to be generated and a multiple-marker-based standardization strategy needs to be adopted to minimize batch-to-batch variation and to maintain quality and ensure safety and efficacy.

Standardization of Herbal Drugs

As commercialization of the herbal medicine has happened, assurance of safety, quality and efficacy of medicinal plants and herbal products has become an important issue. Standardization as defined by American Herbal Product association: "Standardization refers to the body of information and control necessary to product material of reasonable consistency".

This achieved through minimizing the inherent variation of natural product composition through quality assurance practices applied to agricultural and manufacturing processes (Waldesch *et al.*, 2003). Methods of standardization should take into consideration all aspects that contribute to the quality of the herbal drugs, namely correct identity of the sample, organoleptic evaluation, pharmacognostic evaluation, volatile matter, quantitative evaluation (ash values, extractive values), phytochemical evaluation, test for the presence of xenobiotics, microbial load testing, toxicity testing, and biological activity. Of these, the phytochemical profile is of special significance since it has a direct effect on the activity of the herbal drugs. The fingerprint profiles serve as guideline to the phytochemical profile of the drug in ensuring the quality, while quantification of the marker compound/s would serve as an additional parameter in assessing the quality of the sample.

Plants contain several hundred constituents and most of them are present at very low concentrations. Apart from this, plant constituents vary considerably depending on several factors that impair the quality control of phytotherapeutic agents. (Jablonski, D., 2004) Quality control and standardization of herbal medicines involve several steps. However, the source and quality of raw materials play a pivotal role in guaranteeing the quality and stability of herbal preparations. Other factors such as the use of fresh plants, temperature, light exposure, water availability, nutrients, period and time of collection, method of collecting, drying, packing, storage and transportation of raw material, age and part of the plant collected, etc., can greatly affect the quality and consequently the therapeutic value of herbal medicines. Some plant constituents are heat labile and the plants containing them need to be dried at low temperatures. Also, other active principles are destroyed by enzymatic processes that continue for long periods of time after plant collection. This explains why frequently the composition of herbal-based drugs is quite variable.

Standardization of Herbal Drugs: Recent Practices

The WHO Guidelines for quality assessment of plant materials are not followed by most of the 8,500 licensed manufacturers of herbal formulations, since most of these are small-scale manufacturers with limited resources and knowledge. However, few medium and large-scale manufactures undertake some quality control testing of herbal materials. In most cases the traditional doctor attached to the manufacturing company approves the materials based on the organoleptic characters and their knowledge of plants. This is highly inadequate and steps must be taken by the manufacturers to undertake effective quality-control checks on the plant materials in order to manufacture herbal products with consistent clinical efficacy. In modern medicine, a great majority of drugs have been analyzed and their authenticity confirmed by chemical and instrumental analysis. Often the quality control and quality assurance of the drugs are confirmed by these tests and no biological screening is done. However, in the case of herbal drugs it is different since no standard pharmacopoeial methods are available for their identity, when they are in multi-herbal formulations and more so when the extracts of herbs have been used in the formulations. However, regular botanical identification and phytochemical testing shall be of immense help if carried out for both raw materials as well as for the formulations.

The recent advances which occurred in the processes of purification, isolation and structure elucidation of naturally occurring substances have made it possible to establish appropriate strategies for the analysis of quality and the process of standardization of herbal preparations in order to maintain as much as possible the homogeneity of the plant extract. Among others, thin-layer chromatography, gas chromatography, high performance liquid chromatography, mass spectrometry, infrared-spectrometry, ultraviolet/visible spectrometry, etc., used alone or in combination, can be successfully used for standardization and to control the quality of both the raw material and the finished herbal drugs. (Lazarowych, N.J. and P. Pekos, 1998).

Standardization of Herbal Formulation

Standardization of herbal formulation requires implementation of Good Manufacturing Practices (GMP) (WHO guideline, 1996). In addition, study of various parameters such as pharmaco-dynamics, pharmacokinetics, dosage, stability, self-life, toxicity evaluation, chemical profiling of the herbal formulations is considered essential (Mosihuzzaman *et al.*, 2008). Heavy metals, pesticides and aflatoxins contamination, implementation of Good Agricultural Practices (GAP) in herbal drug standardization are equally important. (Bauer, 1998).

Present Guidelines for the Standardization of Herbal Drugs

The guidelines set by WHO: Botanical characters, sensory evaluation, foreign organic matter, microscopic, histological, histochemical assessment, quantitative measurements. Physical and chemical identity, fingerprints chromatography, ash values, extractive values, moisture content, volatile oil and alkaloids tests, quantitative estimation of bioactive, estimation of biological activity, the values of bitterness, astringency hemolytic index, a factor swelling, foaming index, detail-toxicity, pesticides residues, heavy metals, microbial contamination as viable count total, pathogens such as *E. coli, Salmonella, P. aeruginosa*, *S. aureus*, *Enterobacteriaceae* (Shrikumar *et al.*, 2006).

Safety and efficacy assessment for any pharmaceutical must be taken into account for the quality of the proposed formulations. Minimum standards for acceptable quality are generally laid down in Pharmacopoeial monographs, which provide all the details of the acceptable substance and give the details of significant tests to determine its identity and purity.

Status of Marker-Based Standardization in Indian Pharmacopoeia

The Indian Pharmacopoeia includes Pharmacopoeial specifications with monographs for some medicinal plants (approx. 50) being most commonly used as therapeutic agents. The specifications include the name of the drug (along with its common name), its biological source (Latin name), the part of the plant under consideration, its description, macroscopic and microscopic study, identification, several physico-chemical quality control parameters and assay with respect to the phytochemical reference standards or botanical reference standards. The details of some herbs along with marker based specifications are given in the Table 11.1.

TABLE 11.1: Details of some Herbs Along with Marker-Based Specifications

Sl.No.	Name of the Herb	Scientific Name	Marker-Based Standardization	Instruments Used
1.	Acacia	*Acacia nilotica*	–	–
2.	Amalaki	*Emblica officinalis*	Gallic acid: NLT (Not Less Than) 1.0 per cent w/w	HPLC
3.	Arachis oil	*Arachis lypogaea*	–	–
4.	Amra	*Mangifera indica*	Mangiferin: 1.5 per cent w/w	HPLC
5.	Arjuna	*Terminalia arjuna*	Arjunetin: NLT 0.02 per cent w/w	HPLC
6.	Artemisia	*Artemisia annua*	Artemisinin: NLT 0.8 per cent w/w	HPLC
7.	Ashwagandha	*Withania somnifera*	Total withanolide A and Withaferin A: NLT 0.02 per cent	HPLC
8.	Belladonna Dry Extract	*Atropa belladonna*	Total alkaloids calculated as hyoscyamine: 0.95-1.05 per cent w/w	Titrimetry
9.	Belladonna leaf	*Atropa belladonna*	Total alkaloids calculated as hyoscyamine: NLT 0.3 per cent w/w	Titrimetry
10.	Bibhitaki	*Terminalia belerica*	Ellagic acid: NLT 0.3 per cent w/wGallic acid: 0.75 per cent w/w	HPLC
11.	Bhringaraj	*Eclipta alba*	Wedelactone: NLT 0.1 per cent w/w	HPLC
12.	Bhuiamla	*Phyllanthus amarus*	Total Phyllanthin and Hypophyllanthin: NLT 0.25 per cent w/w	HPLC
13.	Brahmi	*Bacopa monnieri*	Bacoside A: NLT 2.5 per cent w/w	HPLC
14.	Clove oil	*Syzygium aromaticum*	Phenolic substances, chiefly eugenol: 85.0-95.0 per cent w/w	GLC
15.	Coleus	*Coleus forskohlii*	Forskolin: NLT 0.4 per cent w/w	HPLC
16.	Eucalyptus oil	*Eucalyptus globules*	Cineole: NLT 60.0 per cent w/w	GLC
17.	Garcinia	*Garcinia cambogia*	Hydroxycitric acid and hydroxycitric acid lactone: NLT 12.0 per cent w/w	HPLC
18.	Gokhru	*Tribulus terestris*	Diosgenin: NLT 0.5 per cent w/w	HPLC
19.	Guar gum	*Cyamopsis tetragonolobus*	–	–
20.	Gudmar	*Gymnena sylvestre*	Gymnemic acid as gymnemagenin: NLT 1.0 per cent w/w	HPLC
21.	Guduchi	*Tinospora cordofolia*	Cordifolioside A: NLT 0.20 per cent w/w	HPLC
22.	Guggul Resin	*Commphora wightii*	Guggulsterone (Z and E):1.00 – 1.50 per cent w/w	HPLC
23.	Gugulipid	*Ethyl acetate* extractive of guggul resin	Guggulsterone (Z and E): 4.00 -6.00 per cent w/w	HPLC
24.	Haridra	*Curcuma longa*	Curcumin – NLT 1.50 per cent	HPLC
25.	Haritaki	*Terminalia chebula*	Chebulinic acid –5.00 – 12.50 per cent w/w	HPLC
26.	Isapgula Husk	*Plantago ovata*		
27.	Kalmegh	*Andrographis paniculata*	Andrographolide: NLT 1.00per cent w/w	HPLC
28.	Kunduru	*Boswellia serrata*	Total 11 Keto- β Boswellic acid and Acetyl 11 Keto - β Boswellic acid– NLT 1.00 per cent w/w	HPLC
29.	Kutki	*Piccorhiza kurroa*	Kutkin – NLT 5.00 per cent	HPLC
30.	Lasuna	*Allium sativum*	Alliin – NLT 0.20 per cent w/w	HPLC
31.	Malt Extract		Protien : NLT 4.00 per cent w/w	Titrimetric
32.	Mandukaparni	*Centella asiatica*	Asiaticoside : NLT 0.50per cent w/w	HPLC
33.	Manjistha	*Rubia cordifolia*	Rubiadin : NLT 0.02 per cent w/w	HPLC
34.	Maricha	*Piper nigrum*	Piperine : NLT 2.50 per cent w/w	HPLC
35.	Mentha Oil	*Mentha* sp.	Menthol : NLT 50.00 per cent w/w	GLC, Titrimetric

Contd...

Table 11.1–*Contd...*

Sl.No.	*Name of the Herb*	*Scientific Name*	*Marker-Based Standardization*	*Instruments Used*
36.	Opium	*Papaver somniferum*	Morphine : NLT 10.00 per cent w/w Codeine : NLT 2.00 per cent w/w	HPLC
37.	Opium powder		Morphine : 9.50 – 10.50 per cent w/w	HPLC
38.	Papain			
39.	Peppermint oil	*Mentha piperita*	Esters calculated as Menthyl acetate: 4.50 – 10.00 per cent w/wFree alcohol calculated as Menthol : NLT 44.00 per cent w/w Ketone calculated as Menthone: 15.00 - 32.00 per cent w/w	
40.	Pippali Large	*Piper longum*	Piperine : NLT 1.00 per cent w/w	HPLC
41.	Pippali small	*Piper longum*	Piperine : NLT 0.40 per cent w/w	HPLC
42.	Punarnava	*Boerhavia diffusa*	Boeravinone: NLT 0.005per cent w/w	
43.	Sarpagandha	*Rauwolfia serpentina*	Reserpine: NLT 0.15 per cent w/w	HPLC
44.	Senna pods	*Cassia senna*	Sennoside A and B :NLT 1.00 per cent w/w	HPLC
45.	Senna leaf	*Cassia angustifolia* *Cassia senna*	Sennoside A and B :NLT 1.00 per cent w/w	HPLC
46.	Shatavari	*Asperagus racemosus*	Shatavarin IV: NLT 0.10per cent w/w	HPTLC
47.	Shati	*Hedychium spicatum*	p-methoxy cinnamic acid methyl ester: NLT 0.80per cent w/w	HPLC
48.	Sunthi	*Zingiber officinale*	Total gingerols: NLT 0.80per cent w/w	HPLC
49.	Tulasi	*Oscimum sanctum*	Eugenol: NLT 0.40 per cent w/w	HPLC
50.	Vasaka	*Adhatoda vasica*	Vasicine : NLT 0.60 per cent w/w	HPLC
51.	Yasti	*Glycyrrhiza glabra*	Glycyrrhizinic acid: NLT 3.00 per cent w/w	HPLC

Types of Marker and its Standardization Using Chromatographic Techniques

Markers can be classified in different categories depending upon their availability and bioactivity in various plants. This has been summed up in the Table 11.2.

Table 11.2

Claim	*Specific to the Plant*	*Totally Responsible for Biological Activity*	*Partially Responsible for Biological Activity*	*No Link with Bio Activity*
Type I	✔	✔	–	–
Type II	✔	–	✔	–
Type III	✔	–	–	✔
Type IV	–	✔	–	–
Type V	–	–	✔	–

Type I: These are marker compounds found specifically in the plant and are totally responsible for its biological activity; *e.g.*: Sennoside A and B in *Cassia angustifolia* (Swarnapatra).

Type II: These are marker compounds found specifically in the plant and are partially responsible for its biological activity; *e.g.*: Withaferin A in *Withania somnifera* (Ashwagandha).

Type III: These are marker compounds found specifically in the plant and not at all responsible for its biological activity; *e.g.*: Vasicine in *Adhatoda vasica* (Vasaka).

Type IV: These are marker compounds found in many plant species (not specific) and are totally responsible for its biological activity; *e.g.*: Piperine in several Piper species (*e.g. Piper longum, Piper nigrum etc.*).

Type V: These are marker compounds found specifically in the plant and are partially responsible for its biological activity; *e.g.*: Quercetin can be found in various fruits and vegetables — particularly citrus fruits, apples, onions, parsley, sage, tea, and red wine — are the primary dietary sources of quercetin. Olive oil, grapes, dark cherries, and dark berries — such as blueberries, blackberries, and bilberries — are also high in flavonoids, including quercetin.

Problems of Pharmacopoeial Standardization of Ayurvedic Drugs

Ayurveda offers not only a safe treatment but also a holistic approach towards humanity. As the use of herbal products got increased in folds, so its standardization became impediment in its acceptance. The complex nature, volatility, stability, and inherent variability of drugs challenge to establish its quality control through modern analytical techniques. The different areas of problem incurring in Ayurvedic drugs are:

1. Identification of Stable Marker Compound

Selection of marker compound is crucial in authenticating and also in quantification of an herb for quality control purposes. But it is more important to identify a stable marker which will remain stable throughout the process. For example: Valtrate in *Valeriana wallichii* is its marker compound but it degrades at high temperature, so it becomes a problem for herbal drug preparation which undergoes through high temperature and also in its course of stability study.

2. Complex Matrix Formulation

Poly-herbal formulation or Ayurvedic preparation like avaleha, asava/arishta, constitute of large number herbs along with high percentage of sugar which makes difficult for an analyst to extract the desired compound for its quantification. This is because of the interference of the similar nature of other compounds which also get extracted. For example, estimation of Piperine in Chyavan-avaleha samples or Vitamins in asava/arishta samples, extraction get hindrance due to interference of complex nature of sugar moieties with target compounds.

3. Cost Related to Marker Based Analysis

Quantification related to marker compounds are always costly than other any types of analysis. This type of analysis involves the use of sophisticated instruments like HPLC, GC, LC-MS along with considerable amount of solvents both of which are highly cost bearing to the new growing companies or small-scale industries where monetary figure matters.

4. Presence of less Quantity of Markers

The marker compounds are present in very small amounts in any herbs. When any herbal formulation is prepared, using that herb the amount of marker compounds becomes even lesser so it becomes more tedious to estimate the marker compounds in herbal drugs/formulation to assess the quality.

Conclusion

With the global increase in the demand for plant-derived medicine as an alternative to synthetic medicine, there is a need to ensure the quality of the herbal drugs using modern analytical techniques, for therapeutic efficacy and safety.

Various methods of standardization and testing are needed immediately in the interest of both the manufacturer and the consumer. In the present business and industrial scenario, and considering the interest and faith that people have in herbal drugs, the need for the development of the modernized methods of standardization cannot be over emphasized in the interest of small and medium-size manufacturers.

References

Bauer R. Quality criteria and standardization of phytopharmaceuticals: Can acceptable drug standard can be achieved. *J Drug Inform* 1998; 32: 101-110.

Jablonski, D., 2004. Extinction: past and present. *Nature*, 427: 589.

Lazarowych, N.J. and P. Pekos. Use of fingerprinting and marker compounds for identification and standardization of botanical drugs: strategies for applying pharmaceutical HPLC analysis to herbal products. *Drug Information Journal* 1998; 32: 497-512.

Mosihuzzaman M and Choudhary MI. Protocols on safety, efficacy, standardization, and documentation of herbal medicine. *Pure Appl Chem* 2008; 80(10): 2195–2230.

Raskin I, Ribnicky DM, Komarnytsky S, *et al., Trends in Biotechnol* 2002; 20: 522.

Shrikumar S, Maheshwari U, Sughanti A and Ravi TK.WHO guidelines for standardization of herbal drugs. *Pharminfo. net* 2006; 2: 78-81.

Waldesch F.G., Konigswinter B.S. and Remagen H.B.,. Herbal medicinal products- Scientific and regulatory basis for development quality assurance and marketing authorization, published by medpharm stuttagart and CRC press, Washington DC, 2003 : 37-52.

12

Marker Standardization for Herbs in Commerce

Vikram Andrew Naharwar

Ayurveda has faced many challenges over the past century. The emergence of modern pharmaceuticals that are highly researched, proven, licensed, regulated and stringently monitored for quality based on regulatory norms established by Governments has given patients a reason to doubt the efficacy, safety and quality of Ayurvedic products that lack all these parameters. The meek effort by AYUSH, diluted by various compulsions and a lack of expertise, has failed to take Ayurveda to a level of a viable alternative therapy.

The WHO has established guidelines that attempt to lay down protocols for the quality control of herbal drugs, but also suggested in-house standards where regulatory standards do not exist. I wonder if the Governments around the world would be that generous with Pharmaceuticals companies to permit manufacturing and sale of drugs without legal parameters. The keepers of Ayurveda and the Ayurvedic industry have acted like the proverbial ostrich inside the heap of more than 2000 year old classical texts. The world is changing fast and consumers are demanding answers from companies that ask them to spend money on their products. The adage that all herbs, even of similar morphology, and Ayurvedic products are safe is the biggest misconception. The Industry has avoided the quality protocols that can identify and eliminate dangerous substances. Another critical problem which is difficult to tackle is of adulteration and spiking with synthetic molecules. Several Indian companies have already been caught spiking products meant for North America by the American authorities, but AYUSH has not taken any action against those who tarnished the reputation of AYURVEDA.

Modern Ayurveda versus Traditional Ayurveda: The Potential and the Risks

It is commonly assumed that the entire wisdom of Ayurveda was miraculously written much like the 10 commandments. Ayurveda has developed over thousands of years, each generation of Ayurvedic physicians rewriting earlier presumptions with updated information based on observation and clinical experience. This repository of clinical data is unsurpassed in human history. While trying to modernise Ayurveda for making it acceptable to the western mind, it should not lead to a loss of heritage and traditional wisdom. Modernisation does not mean transforming Ayurveda into a system alien to the sub continent.

The advent of modern technology should be used to prove traditional claims, modernize archaic production processes, establish quality control norms and provide standardization which is critical for functional claims since doses cannot vary and still provide the same response. Modern scientific technologies such as HTS provide leads to bioactive compounds and through this process one can rapidly identify active compounds, antibodies, or genes that modulate a particular biomolecular pathway. The results of these experiments provide starting points for understanding the interaction or role of a particular biochemical process in biology.

Purification of bio-active compounds hold a potential for drug discovery and synthesis however should not be included as part of Ayurveda that deals with a unique way of determining pharmacological process. Purification/synthesis which has proved successful in several cases in the past is an upstream effort of Ayurvedic development and standardization.

One of the major challenges facing Ayurveda is uniformity and consistency. Batch to batch variations are unacceptable when delivered in dose forms, yet a fraction of Ayurvedic companies have the ability to deliver batch to batch uniformity. Modern day Standardized Full Spectrum Extracts that are produced with great technological skills yet retaining all naturally occurring compounds allows Ayurvedic companies to deliver batch to batch consistency whilst retaining the essence of Ayurveda. HPTLC and IR/LC-MS are important tools in this effort.

Another area of concern is "convenient labelling", this is where companies deliberately mislead consumers and regulatory authorities by listing herbs not included in the product and using herbs not listed in the label. Lax regulatory and IP laws promote this activity which is widespread in the industry, fingerprinting of polyherbal formulations is an effective way of neutralizing this unethical practise and establishing strong IP laws will help innovative companies develop and declare the true formula.

Following measure that must be taken for standardization of herbal drugs (individual or poly herbal).

Botanical Identification

The first step is to correctly identify the botanical matter to be used. Ayurveda is based on multiple ingredients sourced from different geographies and climatic zones. Whilst it would be preferable to source herbs from their traditional habitat, huge commercial demands have led to unscrupulous collection and cultivation of different chemotypes outside natural habitats, raising questions about the chemical profile of such herbs.

To overcome this problem, there is comprehensive documentation available for the identification by microscopy and macroscopy of most botanical drugs used commercially. Sensory evaluation by trained Ayurvedic experts can also establish physical properties documented in Ayurvedic texts.

Evaluation of foreign matter is another major issue due to the lack of proper collection/handling facilities. Adulteration protocols establish the limits of permissible extraneous matter but dangerous adulterants are not possible to remove and become a hazard during processing of drugs.

Chemical Standardization

Modern techniques for establishing chemical integrity of medicinal plants is well developed. HPTLC profiles are the most effective way of identifying herbs for authenticity and adulteration. Even minute amounts of adulterants show up in HPTLC/LC-MS profiles, however an emerging threat is spiking with bio similar compounds, and therefore a technician needs to have significant experience in understanding the natural biochemical composition of each plant and the relative percentages of each compound vis-a-vis other naturally occurring molecules. LC-MS is a highly effective albeit expensive tool for QC.

UV Spectral Analysis

Simple UV-VIS data will establish batch to batch quality control of herbs/herbal drugs. Historic spectral data is available and can be used to establish industry wide norms.

Concentration Validation

When herbal extracts are used, they carry a declaration of ratio as a guideline for manufacturers to establish norms for doses. These ratios differ from manufacturer to manufacturer and is one of the biggest problems in establishing quality for Ayurvedic manufacturer. Instances of exaggerated claims of 200:1 for extracts of *Glycyrrhiza glabra* (indicating 200 kilos of herbal mass is equivalent to 1 kilo of extract) is common. Simple arithmetic shows the falsehood (price of herbs x 200 should be at least 40 per cent higher than cost of extract), chemical evaluation for known markers (for example glycyrrhizin that is present around 4 per cent in root) should be validated by the declared ratio (200 x 4=800 per cent which is impossible as 100 per cent makes it pure). This is one area of concern for an industry struggling to establish its credibility.

Extractive Data

To establish production norms for the recovery of all bioactive molecules identified by HPTLC profiles, extractive values are to be set to determine optimum extraction. Loss of compounds affect efficacy as well as safety and therefore establishing an industry wide average of three simple sigma limits is necessary. Some companies claim 200:1 raios to impress customers who end up using much less of the extract, but in reality these extracts are highly suspicious as they do not carry any validation of the actives.

Ash Values

Establishing ash values including acid insoluble ash is imperative. These are simple tests that can be carried out in all basic laboratories.

Heavy Metals

An area of major concern is the increasing amount of heavy metals in herbal drugs. Analyzing heavy metal contamination is necessary and even if ICP (Inductively Coupled Plasma) is not available, AAS (Atomic Absorption Spectrophotometers) should be the norm. The contamination by mercury, lead, arsenic and cadmium is a major threat to the health of consumers.

Microbial Analysis

Herbs and herbal drugs are prone to high exposure to microbial contamination. *E. coli*, *Salmonella*, *Staphylococcus aureus* are constant threat to the safety of herbal products as are yeast and mould. Total aerobic plate count is also necessary. Norms for microbial limits have been proposed, however these are far too liberal to allow Ayurveda to be considered a serious alternative therapy.

Pesticide Residues

It is necessary to establish norms for pesticide residues that are tested by GC. These are well established tests and can be conducted in any public laboratory.

Aflatoxins

Herbs are often contaminated with aflatoxins that pose a real threat to human safety. These can be tested by TLC using stock solutions.

Pre-clinical/Clinical Data

Most of the Ayurvedic companies claim cures, only a small number have data to support these claims, the rest rely on historic data that has no relevance to the processing undertaken by individual companies. Historic data can always be challenged if not supported by scientific studies of ingredients.

List of Herbs with Marker Standardization

Botanical Name/Part Used	*Chemical Marker(s)*
Abies webbiana leaf	Alkaloids >0.5 per cent
Abroma augusta root	Alkaloids 0.15 per cent Tannins 2.4 per cent-4 per cent
Abrus precatorius seed	Glycosides 15 per cent Alkaloids 1 per cent
Abutilon indicum seed	Mucilage 10 per cent
Acacia arabica bark	Tannins 40 per cent
Acacia catechu gum	Tannins 60 per cent Catechins 20 per cent
Acacia concinna pods	Saponins 10 per cent-20 per cent
Achillea millefolium seeds	Bitters 3 per cent
Achyranthes aspera plant	Saponins 3 per cent
Aconitum heterophyllum plant	Alkaloid 1 per cent-1.5 per cent
Aconitum spp. root	Alkaloid 2 per cent
Acorus calamus rhizome	Vol. Oil 1 per cent
Adhatoda vasica leaf	Alkaloids 0.5 per cent-2.5 per cent
Aegle marmelos leaf	Tannins 7.5 per cent
Aegle marmelos unripe fruit	Tannins 5 per cent Mucilage 10 per cent-15 per cent
Albizzia lebbeck bark	Tannins 15 per cent
Allium cepa bulbs	Quercetin 5 per cent
Allium sativum bulbs	Alliin 1.5 per cent-2.5 per cent by HPLC
Aloe vera leaf	Aloin 1.5 per cent Polysaccharides 50 per cent

Contd...

Contd...

Botanical Name/Part Used	*Chemical Marker(s)*
Alpinia galanga rhizome	Shogoal >4 per cent by HPLC
Alstonia scholaris bark	Alkaloid 0.3 per cent
Amomum aromaticum fruit	Volatile oil> 0.2 per cent
Anacyclus pyrethrum flowers	Alkaloid 0.5 per cent
Andrographis paniculata plant	Andrographolides >10 per cent
Anethum sowa seed	Vol.oil 3 per cent
Aphanamixis rohituka bark	Tannins 10 per cent
Apium graveolens seed	Vol. oil 5 per cent
Areca catechu nut	Tannins> 30 per cent
Argyreia speciosa root	Resin 4.5 per cent
Aristolochia indica root	Alkaloid> 0.15 per cent
Asparagus adscendens root	Saponin >10 per cent
Asparagus racemosus root	Saponins >15 per cent
Asphaltum	TLC for Benzopyrones
Asteracantha longifolia herb	Alkaloid 0.35 per cent
Atropa belladonna leaf	Alkaloid 0.95 per cent-1.6 per cent
Atropa belladonna root	Alkaloid 3 per cent
Azadirachta indica bark	Bitters 1 per cent
Azadirachta indica leaf	Bitters 2.5 per cent
Bacopa monnieri herb	Bacosides 15 per cent-50 per cent, Alkaloid 0.5 per cent
Bambusa arundinacea Manna	Silicates> 20-70 per cent
Barleria prionitis herb	Tannins 5 per cent, Alkaloid 0.5 per cent

Contd...

Contd...

Botanical Name/Part Used	Chemical Marker(s)
Bauhinia variegata bark	Tannins 25 per cent
Berberis aristata root	Berberine >8 per cent
Bergenia ligulata root	Tannins 10 per cent
Boerhaavia diffusa root	Alkaloid 0.01 per cent-0.08 per cent
Bombax malabaricum Bark/gum	Tannins> 2.5 per cent
Boswellia serrata gum	Boswellic acids >50 per cent
Butea frondosa flowers	Glycosides >8 per cent
Caesalpinia bonducella nut	Bonducin 2.5 per cent
Caesalpinia sappan wood	Tannins >10 per cent Saponin > 3 per cent
Calendula officinalis flower	Saponin > 15 per cent Flavones> 10 per cent
Calotropis gigantea root	Alkaloid 0.3 per cent
Camellia sinensis	Polyphenols >45 per cent by HPLC
Camellia sinensis leaf	Polyphenols> 45 per cent
Capparis spinosa root	Glycosides 15 per cent
Capsicum spp. fruit	Capsaicin >3 per cent
Carica papaya seed	Saponin >10 per cent Alkaloids >1 per cent
Casearia esculenta root	Tannins > 2 per cent
Cassia angustifolia leaf	Sennosides 15 per cent
Cassia auriculata seed	Tannins>2 per cent
Cassia angustifolia pods	Sennosides 2.5 per cent
Cassia fistula fruit	Oxymethyl-anthraquinones > 1 per cent
Cassia occidentalis fruit	Oxymethyl-anthraquinones 1.2 per cent
Cedrus deodara wood	Vol. Oil 1 per cent
Celastrus paniculatus seed	Alkaloid 0.1 per cent
Centella asiatica	Asiaticosides >10 per cent, Asiatic acid 2 per cent
Centratherum anthelminticum seed	Bitter >1.5 per cent
Cephaelis ipecacuanha root	Alkaloid 1 per cent
Cephalandra indica leaf	Resin 2.5 per cent
Chlorophytum arundinaceum root	Saponin >20 per cent Mucilage>10 per cent
Cicer arietinum seeds	Protein 15 per cent
Cichorium intybus seeds	Bitters >1 per cent
Cinchona ledgeriana bark	Alkaloids 4 per cent
Cinnamomum zeylanicum bark	Volatile Oil >1 per cent Flavones 7 per cent
Cinnamomum cassia leaf	Volatile oil> 1 per cent
Cissus quadrangularis stem	Ketosterones >2.5 per cent
Citrullus colocynthis fruit	Bitters 4.5 per cent

Contd...

Contd...

Botanical Name/Part Used	Chemical Marker(s)
Citrullus colocynthis root	Bitters 5 per cent
Citrus aurantium peel	Flavones >4 per cent
Citrus medica peel	Acidity as Citric acid 2 per cent
Clerodendrum phlomidis root	Alkaloids >0.2 per cent
Colchicum luteum corms	Alkaloid >0.5 per cent
Coleus forskohlii root	Forskolin >2.4 per cent by HPLC.
Commiphora mukul gum	Guggul Sterones 2.5 per cent to 7 per cent by HPLC
Commiphora myrrha gum	Vol. Oil 5 per cent
Convolvulus pluricolus herb	Bitters > 3 per cent
Coriandrum sativum seeds	Vol. Oil 1 per cent
Crataegus oxycantha fruit	Saponin > 5 per cent
Crataeva nurvala bark	Saponin >2.5 per cent
Crocus sativus stigma	Crocin
Cucumis sativus seeds	Mucilage 24 per cent
Cuminum cyminum fruit	Vol. Oil 1 per cent
Curcuma amada rhizome	Vol.Oil 10 per cent
Curcuma longa rhizome	Vol oil >10 per cent Curcumin >10 per cent, 95 per cent
Curcuma zedoaria Rhizome	Volatile oil> 1 per cent
Curculigo orchioides root	Saponins 20 per cent, Mucilages 30 per cent
Cynodon dactylon herb	Tannins 5 per cent Alkaloid +ive
Cyperus scariosus root	Alkaloid 0.15 per cent-0.5 per cent
DGL	Flavones 1 per cent Glycyrrhizin<3 per cent
Dashmool	Tannins 2 per cent
Datura stramonium leaf	Alkaloid 0.5 per cent-0.7 per cent
Daucus carota seed	Alkaloid 0.4 per cent, Flavones 5 per cent
Desmodium gangeticum herb	Tannins > 5 per cent
Dioscorea deltoids tuber	Flavones > 5 per cent
Dolichos biflorus seeds	Saponin 20 per cent
Eclipta alba plant	Nor-wedelolactone >3 per cent
Elettaria cardamomum fruit	Volatile oil> 1 per cent
Embelia ribes seed	Tannins 1 per cent, Embelin 1 per cent
Emblica officinalis fruit	Tannins >20 per cent Ellagic acid >5 per cent
Enicostemma littorale herb	Bitters 4 per cent
Ephedra vulgaris herb	Alkaloid 0.2 per cent-0.5 per cent
Eugenia jambolana seed	Saponin 4 per cent, Alkaloid 0.4 per cent

Contd...

Contd...

Botanical Name/Part Used	*Chemical Marker(s)*
Euphorbia hirta herb	Flavone >6 per cent
Evolvulus alsinoides plant	Bitters >2.5 per cent
Ferula foetida resin	Vol. Oil 1.5 per cent
Ficus carica fruit	Bitters > 1 per cent
Ficus religiosa bark	Tannins > 5 per cent
Ficus racemosa leaf	Tannins 5 per cent
Ficus racemosa bark	Tannins 10 per cent
Foeniculum vulgare fruit	Vol. Oil >1 per cent
Fumaria officinalis herb	Bitters >1 per cent
Garcinia cambogia rind	HCA 50 per cent by HPLC
Gloriosa superba root	Alkaloids> 0.2 per cent
Glycyrrhzia glabra root	Glycyrrhizin by Garratt method >15 per cent/24 per cent
Gmelina arborea leaf	Alkaloids> 0.15 per cent
Gossypium herbaceum root	Alkaloid 0.4 per cent, Flavone 1.5 per cent
Gymnema slyvestre leaf	Gymnemic acids 25 per cent-75 per cent
Hedychium spicatum rhizome	Volatile oil >1 per cent
Hemidesmus indicus root	Saponin 7.5 per cent
Hibiscus rosa sinensis flower	Hydroxycitric acid > 30 per cent
Holarrhena antidysenterica seed	Alkaloid 3 per cent
Holarrhena antidysenterica bark	Alkaloid >4 per cent
Hydrastis Indian root	Hydrastin 1 per cent
Hyoscyamus niger herb	Alkaloid 0.28 per cent-0.3 per cent
Hypericum perforatum herb	Hypericin 0.3 per cent
Hyssopus officinalis	Flavones >5 per cent
Indigofera tinctoria plant	Saponins 10 per cent
Inula racemosa root	Alantolactone 2.5 per cent by HPLC
Ipomoea digitata root	Saponin > 5 per cent Flavones> 15 per cent
Juglans regia bark	Tannins 12 per cent
Lavandula stoechus herb	Volatile oil> 1 per cent
Lawsonia alba leaf	Tannins >5 per cent
Leptadenia reticulata herb	Alkaloid 0.5 per cent
Litsea polyantha bark	Mucilage>15 per cent
Lobelia nicotianaefolia leaf	Alkaloid 0.5 per cent
Mangifera indica bark	Tannins 20 per cent Mangiferin 10 per cent by HPLC
Matricaria chamomilla Flower	Vol. Oil 1 per cent,
Mentha spicata	Flavones >2 per cent
Mesua ferrea fruit and Flower bud	Bitters>2.5 per cent
Mimosa pudica herb	Tannins > 3 per cent Mucilage >25 per cent

Contd...

Contd...

Botanical Name/Part Used	*Chemical Marker(s)*
Mimusops elegni bark	Tannins 10 per cent
Momordica charantia fruit	Bitters >2.5 per cent Charantin >0.4 per cent
Morinda citrifolia fruit	Morindin 15 per cent, Scopoletin 0.17 per cent
Moringa oleifera bark	Glycosides >5 per cent
Moringa oleifera leaf	Alkaloid 0.1 per cent
Mucuna pruriens seeds	L-Dopa by IP method >15 per cent, 10 per cent by HPLC
Murraya koenigi Leaf	Alkaloid> 0.15 per cent
Myrica nagi bark	Tannins 5 per cent
Myristica fragrans fruit	Vol. Oil 4 per cent
Nardostachys jatamansi root	Vol. Oil 0.1 per cent-0.5 per cent
Nelumbo nucifera seed	Saponin 30 per cent
Nigella sativa seed	Bitters 3 per cent Saponin 15 per cent
Nyctanthes arbortristis leaf	Alkaloid 0.15 per cent
Ocimum basilicum leaf	Tannins 5 per cent, Vol. Oil 1.5 per cent
Ocimum sanctum herb/leaf	Tannins >7 per cent, Ursolic acid >2 per cent
Oldenlandia corymbosa plant	Alkaloid 0.35 per cent
Onosma bracteatum Plant	Tannins >5 per cent
Onosma echioides root	Alkamine 1 per cent
Operculina turpethum root	Resin >10 per cent
Opuntia dillenii herb	Tannins 10 per cent
Orchis mascula tubers	Saponin 4 per cent, Mucilage 20 per cent
Oroxyllum indicum Bark	Alkaloid> 0.35 per cent
Passiflora incarnata leaf	Flavones>5 per cent
Pedalium murex fruit	Saponin >10 per cent
Peganum harmala	Alkaloids > 7 per cent
Phoenix dactylifera	Sugar >10 per cent
Phyllanthus niruri herb	Bitters 1.4 per cent-2 per cent
Picrorhiza kurrooa root	Kutkin 4 per cent-6 per cent, Bitters>6 per cent
Piper chaba fruit	Tannins >0.5 per cent
Piper cubeba fruit	Vol. Oil 8 per cent, Piperine 10 per cent
Piper longum fruit	Vol. Oil 10 per cent, Piperine 5 per cent
Piper nigrum fruit	Vol. Oil 7 per cent, Piperine >10 per cent
Pistacia integerrima galls	Tannins 20 per cent
Pluchea lanceolata leaf	Flavones > 5 per cent

Contd...

Contd...

Botanical Name/Part Used	Chemical Marker(s)
Plumbago indica root	Alkaloids 0.08 per cent-0.15 per cent
Podophyllum emodi rhizome	Resin >20 per cent
Polygala chinensis root	Saponin 5 per cent
Pongamia glabra seed	Fixed oil 10 per cent
Prunus serotina bark	Tannins 10 per cent
Psidium guajava leaf	Tannins >10 per cent
Psoralea corylifolia seed	Psoralen >5 per cent
Pterocarpus marsupium wood	Alkaloid 0.4 per cent, Pterostilbene 4 per cent-5 per cent Flavones 5 per cent
Pterocarpus santalinus bark	Santalin >3 per cent
Pueraria tuberosa root	Flavones 6 per cent
Punica granatum rind	Ellagic acid by HPLC >15 per cent
Putranjiva roxburghii berries	Alkaloid 0.5 per cent
Pyrus malus fruit	Mallic acid 4 per cent
Quercus infectoria galls	Tannins 40 per cent
Randia dumetorum fruit	Oleonolic acid >5 per cent
Rauwolfia serpentina root	Reserpine >8 per cent
Rheum emodi root	Oxymethyl anthraquinones > 5 per cent
Ricinus communis leaf	Alkaloid 2.5 per cent
Ricinus communis root	Alkaloid 0.5 per cent
Rubia cordifolia root	Tannins >2 per cent
Salacia reticulata root	Glycosides >15 per cent Flavones >5 per cent
Salvadora persica Shoots, leaf	Alkaloids>0.2 per cent Tannins >1.5 per cent Saponin >10 per cent
Santalum album wood	Vol. Oil 2 per cent
Sapindus trifoliatus fruit cortex	Saponin 20 per cent
Saraca indica bark	Tannins >2.5 per cent
Saussurea lappa root	Alkaloid >0.2 per cent
Semecarpus anacardium kernal	Anacardic acid 0.4 per cent
Sida cordifolia root	Alkaloid >0.1 per cent
Silacia raticulata Root bark	Flavones > 5 per cent Saponin > 20 per cent
Silybum marianum seeds	Flavones>50 per cent, Silymarin> 50 per cent
Smilax china root	Saponin >5 per cent
Solanum indicum Plant	Bitters >1.5 per cent
Solanum nigrum berries	Alkaloids 0.2 per cent Bitters >1 per cent
Solanum xanthocarpum plant	Solasodin >1 per cent
Spinacia oleracea leaves	Vit. K. by TLC

Contd...

Contd...

Botanical Name/Part Used	Chemical Marker(s)
Strychnos nux vomica seeds	Alkaloid 3 per cent-6 per cent
Swertia chirata plant	Bitters 4 per cent
Symplocos racemosa bark	Alkaloid 0.5 per cent
Tamarindus indica fruit	Tartaric acid >5 per cent
Tamarix gallica	Tannins >10 per cent
Taraxacum officinale herb	Bitters 2.5 per cent Alkaloid 0.5 per cent
Tecomella undulata bark	Tannins 10 per cent
Tephrosia purpurea plant	Rutin 2 per cent-6 per cent by HPLC
Terminalia arjuna bark	Tannins >25 per cent, Arjunic acid >1 per cent
Terminalia belerica fruit	Tannins >10 per cent
Terminalia chebula fruit	Tannins 20 per cent-40 per cent
Tinospora cordifolia root	Bitters >1.5 per cent
Trachyspermum ammi fruit	Vol. Oil 10 per cent
Tribulus terrestris fruit	Saponins 15 per cent-45 per cent
Trichosanthes dioica leaf	Saponins 10 per cent
Trigonella foenum-graecum seed	Saponins 15 per cent-40 per cent
Trikatu	Vol. Oil >1 per cent Piperine >2 per cent
Triphala	Tannins >25 per cent
Triticum sativum bran	Vit B1 =40 μg/gB2= 6.28 μg/g B5= 1130 μg/g
Triticum sativum germ	Vit. E 0.8 per cent by HPLC
Tylophora indica leaf	Alkaloid >0.1 per cent
Urginea indica bulbs	Glycosides 10 per cent
Urtica dioica plant	Citric acid> 3 per cent
Valeriana wallichii root	Valeric acid 0.8 per cent
Vetiveria zizanioides root	Volatile oil >1 per cent
Viburnum spp. bark	Tannins >1.4 per cent
Vinca rosea root/leaf	Alkaloids > 1.5 per cent
Viola odorata flowers	Saponins 2.5 per cent
Vitex negundo leaf	Alkaloids 0.15 per cent
Vitis vinifera fruit/seed	Polyphenols >20 per cent by HPLC
Wedelia calendulacea herb	Tannins 5 per cent, Saponins 2 per cent
Withania somnifera root	Withanolides >2.5 per cent
Woodfordia floribunda flowers	Tannins 7 per cent
Wrightia tinctoria seed	Alkaloids > 3 per cent
Zingiber officinale rhizome	Gingerols by HPLC >5 per cent
Zizyphus jujuba fruit	Citric acid > 5 per cent
Zizypus sativa/Vulgaris fruit	Tannins> 5 per cent

13

Bioactive Phytoconstituents as Natural Medicine

C.P. Khare

While investigating chemical constituents of herbs and screening their biological activities experimentally and clinically, a number of phytochemicals were found to exhibit possibly safer and better results than traditional herbal medicines. From polyherbal compounds containing different herbs of different families and different profiles, and from the new era of standardized herbs, in many cases threatened with contamination and adulterants, a number of bioactive phytoconstituents are emerging as a viable natural medicine all over the West.

5-HTP

5-HTP, derived from the seeds of African plant *Griffionia simplicifolia* is available in the USA as 50 and 100 mg capsules. 5-HTP (5-Hydroxytryptophan) is related to both L-tryptophan and serotonin.

In the body L-tryptophan is converted to 5-HTP, which can then be converted to serotonin. 5-HTP readily crosses the blood-brain barrier and increases central nervous system synthesis of serotonin. Serotonin plays a significant role in depression, anxiety and aggression.[2] A number of open and randomized controlled trials suggest that 5-HTP may be as effective as standard anti-depressants.[3] More than 20 5-HTP products are available in the USA, inspite of reports of contaminants from peak X family.[2]

Griffionia contains several indole derivatives including 5-Hydroxy-L-tryptophan, indole-3-acetylaspartic acid and 5-hydroxy indole-3-acetic acid (5-HIAA).[4] Till 2007, 40 studies have evaluated the clinical effects of 5-HTP on depression.

For depression, dosage may be 50 mg three times a day. For insomnia, the dosage is usually 100-300 mg before bedtime. In primary fibromyalgia syndrome (PFS), 100 mg three times daily has been used.[4]

Griffionia simplicifolia (Vahl ex DC.) Baill, Fam. Caesalpiniaceae/Leguminosae, in found in tropical Africa, most prevalently in Ghana. Ghana is exporting raw material mostly to Germany. The ripe seed contain as much as 20 per cent 5-HTP and lectins of interest in cancer and neurological research.[5]

Bromelain

Bromelain, a proteolytic enzyme found in the stem and fruit of the pineapple plant (*Ananas comosus*), is extracted from pineapple juice pulp. Its constituents are a mixture of basic glycoproteins similar to papain.[7]

Bromelain is used as a digestive aid and has been found to be an effective mucolytic agent, as well as an anti-inflammatory mediator. It reduces platelet aggregation, edema (from decreased vascular permeability), and other prostaglandin-related phenomena.[1] In various animal experiments (egg white-, carrageen-, dextran-, and yeast-induced edemas, traumatic edema, adrenalin-caused edema of the lungs), an edema-inhibiting effect was demonstrated with high doses of bromelain upon oral and intraperitoneal administration.[7] (German) Commission E approves the use of bromelain to quell surgical swelling, particularly nasal sinus swelling.[7] Bromelain may also therapeutically influence fibrinolysis, tumor growth, drug absorption, blood coagulation and the debridement of third-degree burns.[7] In Europe, a patented tape that contains bromelain is used for debriding escharotic skin.[7]

In rats, bromelain absorption rate is about 50 per cent. Human rates have not been documented. One study using 19 healthy human males suggests that small amounts of undegraded bromelain may pass through the gastrointestinal tract intact.[7]

Recommended dose: 250-750 mg tablet, twice daily, between meals.[1] The therapeutic dose for allergic rhinitis ranges from 400 to 500 mg, three times a day (bromelain with a potency of 1800-2000 milk-clotting units).[3] Bromelain tablet/capsule 500 mg (200 GDU), 500 mg (1000 GDU), 500 mg (2400MCU/gm), 500 mg (2400 GDU/gm), 375 mg (2400 GDU/3600 MCU) are available in the USA. For Osteoarthritis, a combination product containing bromelain 90 mg, trypsin 48 mg and rutin 100 mg is available as Phlogenzym or Wobenzyme PS. Recommended dose is 2 tablets 3 times daily. For acute knee pain, bromelain 200 mg to 400 mg daily for 30 days has been used.[2] Most authorities recommend bromelain on an empty stomach but no comparison data have evaluated its effect when taken with or between meals[3]

Confusion may arise with bromelain dosage because bromelain activity is defined in a number of ways, including rorer units, gelatin-dissolving units (GDU), milk-clotting units (MCU) and Federation Internationale Pharmaceutique standards.[5] The food Chemistry Codex (FCC) officially recognizes the use of milk clotting units (MCU). For most indications, the recommended MCU range is 1200 to 1800.[15]

Capsaicin

The major component of capsicum peppers is the glycoside capsaicinoid or capsaicin (N-vanillyl-8-methyl-6-[E]-nonemamide).[5] Naturally occurring capsaicin exists only in trans-stereoisomer form. However, the cis-isomer, known as civamide, is more potent and causes less irritation than trans-capsaicin. Civamide is currently an investigational drug for migraine, osteoarthritis and other pain-related conditions.[2]

Human trials have investigated capsaicin's use as a treatment for chronic post-herpetic neuralgia, its effects on normal skin and affected dermatomes in herpes zoster, the somesthetic and electrophysiologic effects topical capsaicin, its use in the treatment of painful diabetic neuropathy, the effect of local capsaicin treatment for chronic rhinopathy, its use in the management of surgical neuropathic pain in cancer patients, and the effect of topical capsaicin on substance P immunoreactivity.[7] A meta-analysis of trials of topical capsaicin for the treatment of diabetic neuropathy, osteoarthritis, post-herpetic neuralgia, and psoriasis has been published.[10] A systematic review in BMJ included 16 placebo-controlled studies (15 double-blind, one single-blind) involving a total of 1556 patients suffering from chronic pain due to either neuropathic conditions or musculoskeletal disorders.[11]

Capsaicin reduces the liver's production of certain kinds of p50, an enzyme which activate environmental contaminants such as carbon tetrachloride into having a carcinogenic effect. Capsaicin may reduce the carcinogenic effect of air pollution.[1]

For pain syndrome, including rheumatoid and osteoarthritis, neuropathy and fibromyalgia, creams containing 0.025 per cent to 0.075 per cent capsaicin are topically applied 3-4 times daily.[15] For cluster headache, 0.1 mL of a 10 mM capsaicin suspension, providing 300 mcg/day of capsaicin, applied to the ipsilateral nostril, has been used. For migraine headache, application of capsaicin 0.075 per cent to the nasal mucosa is recommended.[2]

Products labeled capsaicin sometimes include nonivamide which is an adulterant or pelargonic acid and vanillylamide, referred to as "synthetic capsaicin."

Curcumin

The major yellow pigment of Turmeric (*Curcuma longa*) has been identified as curcumin (diferuloylmethane), a phenolic antioxidant. Unlike most natural antioxidants that contain either beta-diketone or polyphenolic functional groups, curcumin possesses both active moieties. Its superior antioxidant activity has been attributed to this structural combination.[5]

In a randomized, double-blind, cross-over study, 18 patients with rheumatoid arthritis were treated daily for 2 weeks with 1200 mg of curcumin or 300 mg of phenylbutazone. In both phases of the study, significant improvement was observed in morning stiffness, walking time and joint swelling ($p<0.05$) compared to baseline values. In another randomized study, following surgery for inguinal hernia or hydrocele, 45 patients were treated daily for 5 days, beginning on the post-operative day with 1200 mg of curcumin or 300 mg of phenylbutazone or placebo. Improvements in typical post-operative symptoms were observed in 84 per cent of patients in curcumin group, 86 per cent in phenylbutazone group and 62 per cent in placebo group.[12]

Curcumin exhibits anti-inflammatory possibly by inhibiting cyclooxygenase-2 (COX-2), prostaglandins, and leukotrienes. But the bioavailability of curcumin is very low after oral administration[2] It is hydrophobic and cannot be given intravenously; disappears rapidly from tissues after intraperitoneal administration.[5] Oral administration of a single dose of 2 g of curcumin to 8 human volunteers led to undetectable or very low serum concentration. Concomitant administration of 20 mg of piperine significantly increased the serum concentrations.[8a]

Daily oral dose of curcumin 3600 mg have been advocated for use in clinical trials.[5] 400 mg t.i.d. is also used. (Turmeric products are mostly standardized to contain not less than 3 per cent curcumin. The usual dose of 1200 mg/day of curcumin is equal to about 40 g/day. A heaping teaspoonful of powdered turmeric is about 4 g)[6] Bromelain is sometimes recommended to enhance curcumin absorption[1], but there is no reliable evidence to support this.[2] Curcumin-bromelain combinations are best taken on an empty stomach 20 minutes before meals or between meals.[15] 2.5 mg Black Pepper extract has also been added in a 500 mg Curcumin Complex.[2] Providing curcumin in lipid base such as lecithin, fish oil, or essential fatty acids may also increase absorption, when taken with meals.[15]

Theracurmin is the latest variant of curcumin. It is claimed to be the most bioavailable form of curcumin capable of increasing blood levels of curcumin, significantly greater than other forms of curcumin. It is a colloidal mixture produced by reducing the particle size of curcumin by over 100 times and emulsified with a natural vegetable gum. Studies have shown that Theracurmin is not only better absorbed, but it also stays in the blood far longer than any other commercial form of curcumin.

To test the hypothesis that curcumin is a potent natural approach to not only decrease joint pain, but also slow the progression of the condition, Japanese researchers at Kyoto Medical Centre enrolled 50 patients over 40 years old with knee osteoarthritis confirmed by X-ray. The patients took either Theracurmin providing 180 mg/day of curcumin or a placebo daily for 8 weeks. Blood biochemistry analyses were performed before and after 8 weeks of each intervention to evaluate safety. The patients' knee symptoms were evaluated at 0, 2, 4, 6, and 8 weeks by the knee scoring system of the Japanese Orthopedic Association and also the Japanese Knee Osteoarthritis Measure, the knee pain visual analog scale (VAS), and the need for nonsteroidal anti-inflammatory drugs.

Results showed that knee pain scores were significantly lower in the Theracurmin group than in the placebo group in patients with moderate to severe symptoms. Theracurmin also lowered the use of celecoxib (Celebrex) much more significantly than placebo. While 60 per cent of the placebo group still relied on Celebrex for adequate pain relief at the 8-week mark, only 32 per cent of the Theracurmin group still needed the NSAID, and there was a definite strong trend for eventual discontinuation.[9]

Hesperidin

Hesperidin is primarily derived from citrus fruits and is known as a citrus bioflavonoid, closely related to quercetin, rutin and diosmin. Used for vascular conditions such as hemorrhoids, varicose veins and venous stasis.

Hesperidin inhibits phosphodiesterase and increases intracellular cyclic adenosine monophosphate (cAMP), which causes decreased production of inflammatory prostaglandins E2 and F2 and thromboxane B2. Hesperidin's analgesic effect seems to work through peripheral rather than central mechanisms.[2]

In one clinical study, 94 women who suffered from hot flashes were given a formula containing 900 mg of hesperidin, 300 mg SS of hesperidin methyl chalcone (another citrus flavonoid), and 1200 mg of vitamin C daily. At the end of one month, symptions of hot flashes were relieved in 53 per cent of the patients and reduced in 34 per cent.[13]

Hesperidin capsules contain hesperidine methyl chalcone 250 mg, bromelain 2400 GDU/g and Vitamin C 9 mg. For hemorrhoids, hesperidin 150 mg plus diosmin 1350 mg twice daily for 4 days, followed by hesperidin 100 mg and diosmin 900 mg twice daily for 3 days, has been used. For venous stasis ulcers, a combination of 100 mg hesperidin and 900 mg diosmin daily has been used up to 2 months. Diosmin and hesperidin, extracted from rutaceae species, have been used in clinical trials.

Lycopene

Lycopene is the most abundant carotenoid in tomatoes (*Lycopersicon esculentum*). Raw tomatoes contain 3.1 mg per 100 g. Mechanical treatment with heat helps release lycopene from the tomato matrix, improving its bioavailability. Tomato sauce and ketchup contain lycopene 33 to 66 mg per 100 g.[5] Heat processing of tomato paste, juice, ketchup induces the isomerization of lycopene from trans- to cis-configuration. The cis isomer has better bioavailability. In North America, 85 per cent dietary lycopene comes from tomato-derived products such as tomato juice or paste. One cup of tomato juice (240 ml) provides about 23 mg of lycopene.[2]

The majority of evidence supported the use of lycopene in cancer prevention. Lycopene levels have been shown to be inversely proportional to cancers of the prostate, cervix, pancreas, and stomach.[5] Lycopene has the most potent antioxidant activity of any common carotenoid. It might reduce cancer risk by scavenging free radicals and quenching singlet oxygen, which prevents oxidative damage to DNA. Preliminary evidence suggests that lycopene may decrease cholesterol synthesis and increase removal of LDL cholesterol from circulation. (Studies have found no correlation between lycopene intake and MI risk.)[2]

Lycopene absorption in humans is approximately 10 per cent to 30 per cent, with the remaining excreted. Lucopene administred as a pure compound has been studied in clinical trials at dosages of 13 to 75 mg/day.

Lycopene is mostly available in capsule and softgel form, with dosage guidelines ranging from 10 to 30 mg taken twice daily with meals. Lycopene is also incorporated in multivitamin and multimineral products.[5,2] Dosing recommendations of lycopene range from 5 to 10 mg daily, for hypercholesterolemia 60 mg daily. Many combinations products for support of the prostate contain lycopene with zinc and Saw palmetto.[4]

OPCs (Oligomeric Proanthocyanidins)

OPCs are usually derived from grape (*Vitis vinifera*) seeds. The highest concentration of proanthocyanidins is found in the skin or membrane of the grape seed. Total proanthocyanidin content consumed in 100 g of dry grape seed is approximately 3500 mg. The most active proanthocyanidins are those bound to other proanthocyanidins, mixture of proanthocyanidin dimers, trimers, tetramers and large molecules such as PCO.[5]

Grape seed extract has shown positive results in the treatment of peripheral venous insufficiency, varicose veins, capillary fragility, disorders of retina (in clinical studies)[13] also in diabetic retinopathy, edema, ocular stress and premenstrual syndrome.[4]

The majority of trials in humans have studied the use of grape seed extract as an antioxidant and for various cardiovascular disorders. Studies also suggest its use as a chemoprotective and cytoprotective agent.[2,5] OPCs inhibited the proteolytic enzymes collagenase, elastase, hyaluronidase, and beta-glucuronidase, which are involved in the breakdown of structural components of the vasculature and skin.[2]

Extracts of grape seed have been studied at doses of 50 to 900 mg/day for its antioxidant, cardiovascular and nutritional effects as well as venous insufficiency and ophthalmologic complaints in Europe.[5] However, grape seed polyphenols 1000 mg/day did not exhibit a significant effect on blood pressure in hypertensive patients.[2]

Preliminary evidence suggests grape seed proanthocyanidins may provide greater protection against reactive oxygen species, free radical-induced lipid peroxidation, and DNA damage than the combination of vitamin E, vitamin C, and beta-carotene or a combination of vitamin E and vitamin C.[2]

For chronic venous insufficiency grape seed extract proanthocyanidin doses of 150-300 mg/day have been used; for reducing ocular stress due to glare 200-300 mg/day[2] In the USA, OPC 30 mg capsules contain OPC 95 per cent extract from grape seed. OPC capsules are also available with Red Wine extract, Pine bark extract (Pycnogenol), Bilberry extract, citrus bioflavonoids extracts.[2] Grape seed extracts are high in tannins and may interfere with iron absorption.

Papain

The latex from unripe papaya (*Carica papaya*) is rich in two enzymes, papain and chymopapain. As the fruit ripens, papain and chymopapain dissipate and neither is present in the ripe fruit.[5] Purified papain enzyme is a thiol protease, obtaining enzymatic activity from the sulthydryl group of the cysteine residue. It is poorly active in extreme acidic or basic media, can be digested by pepsin, and is poorly absorbed when ingested orally. Thus it may have unpredictable activity when exposed to gastric contents. Papain hydrolyzes proteins, small peptidases, amides, and some esters.[6]

The use of oral papain tablets has been evaluated in several controlled clinical trials for post-traumatic inflammation. Papaya proteolytic enzyme tablets appeared to reduce postoperative edema, inflammation, or pain when compared to placebo.[6] But results of analgesic and anti-inflammatory effects are contradictory. The fibrinogenous effect has not been sufficiently proved.[4]

Clinical trials have used 1500 mg papain (2520 FIP units) per day to treat inflammation and swelling following trauma and surgery.[2] In a trial in India (Ind J Cancer, 1999,36), patients undergoing radiation therapy for head and neck cancer were given an oral proprietary product (Wobe-Mugos, 3 tablets three times for 7 weeks0, which contained 60 mg of papain along with bromelain, trypsin, chymotrypsin and pancreatin. Significantly less mucositis and dysphagia from radiation therapy was reported.[6]

Papaya enzyme chewing tablets contain 250 mg papaya powder, 150 mg of pineapple juice powder, and 10 mg of papain. One tablet is to be chewed up to three times daily, preferably after a meal. Papain powder is also available, each 1/8 tsp serving contains papain 362.5 mg.[2]

Pycnogenol

The name Pycnogenol is a trade mark of the British company Horphag Research Ltd. for a complex proprietary mixture of water-soluble proanthocyanidins derived from the bark of the European coastal pine, *Pinus pinaster*, which grows along the coast of southwest France in Gascogne.[5]

Pycnogenol is composed of 80 per cent to 85 per cent proanthocyanidins, the monomers catechin and taxifolin (5 per cent), and phenolic acids, including derivatives of benzoic and cinnamic acids (2 per cent to 4 per cent). Using a patented process,condensed tannins are removed. In some studies Pycnogenol-related compounds are designated procyanidiol oligomers (PCOs). The oligomers range from monomers to dodecamers. The phenolic acids are derivatives of benzoic and cinnamic acids. Tablets ans capsules must contain a minimum of 20 mg Pycnogenol. Maritime pine extract is monographed in the United States Pharmacopoeia.[5]

In chronic venous insufficiency procyanidins in Pycnogenol reduce capillary permeability. Pycnogenol seems to recycle ascorbyl radicals and tocopheryl radicals, helping to maintain vitamin C and E levels. Some research suggests that Pycnogenol might be useful in the prevention of cardiovascular disease. *In vitro*, Pycnogenol prevents oxidation of LDL cholesterol and protects DNA from damage by free radicals. Pycnogenol inhibits epinephrine-induced platelet aggregation, such as that seen in smokers, but Pycnogenol does not affect smoking related increases in blood pressure or heart rate. In non-smokers, it decreases serum thromboxane B2 and reduces systolic blood pressure. Pycnogenol might also increase production of nitric oxide from vascular endothelial cells which can lead to vasodilatation and possibly reduce the potential for atherogenesis and thrombus formation. For asthma, Pycnogenol is thought to be beneficial due to its anti-inflammatory and antioxidant effects. (Anti-inflammatory effects of Pycnogenol are controversial.)[2] Antioxidant properties of Pycnogenol have been well described in laboratory studies and are considered to be responsible for the majority of clinical effects of the extract. However, clinical studies evaluating changes in antioxidant status after Pycnogenol administration in humans have produced equivocal results.

Doses of 60 to 300 mg daily of Pycnogenol have been studied in clinical trials, most commonly at 150 mg/day.[5] For chronic venous insufficiency 100 mg thrice daily, for diabetic and other retinopathies 50 mg thrice daily, for coronary artery disease 150 mg thrice daily have been used.[2]

Pycnogenol tablets and capsules contain extract standardized to contain 65-75 per cent proanthocyanins, also available with citrus bioflavonoids (40 per cent hesperidin), vitamin C, selenium, grape seed extract, and Coenzyme Q-10.[2]

Quercetin

Buckwheat (*Fagopyrum esculentum*) tea has a high concentration of quercetin.[2] Also abundant in black tea, blue-green algae, broccoli, onions, red apple and red wine.[1]

Quercetin has antioxidant, anti-inflammatory, nitric oxide inhibitor, and tyrosine kinase inhibitor (leading to inhibition of the division and growth of T-cells and some cancer cells) activity.[2] It inhibits inflammatory processes, mainly by its action on membranes of specific immune cells. This membrane stabilization prevents mast cells and basophils from de granulating and releasing inflammatory mediators. Quercetin inhibited antigen-induced human basophil histamine release.[14] A Japanese study of mast cells from nasal mucosa of people with perennial allergic rhinitis found quercetin to be almost twice as effective as sodium cromoglycate in inhibiting antigen-stimulated histamine release.[3]

Quercetin reduces capillary fragility. Alters intestinal cell homeostasis of copper, iron and manganese. Might inhibit collagen-and ADP-induced platelet aggregation. Quercetin also demonstrated activity against retroviruses, Herpes simplex, polio, parainfluenza and respiratory syncytial viruses.[2] Inhibition of liver glutathione-S-transferase activity has been demonstrated. This enzyme plays an important part in the development of resistance to various cancer chemotherapy agents. In a phase 1 clinical trial, lymphocyte protein kinase phosphorylation was inhibited by quercetin in 9 of 11 cancer patients. Based on several studies, it appears that quercetin donates an electron to enzyme-bound alpha tocopherol, preventing LDL oxidation. Quercetin's antiviral activity may be related to its ability to bind viral protein coat and to interfere with viral nucleic acid synthesis (DNA).[4]

The oral absorption of quercetin is highly variably. Glycoside forms of quercetin are not absorbed from the gastrointestinal tract and must be hydrolysed, permitting absorption of the aglycone form of quercetin, 64 to 80 per cent.[2]

Recommended dose is 400-500 mg three times daily;[2] for allergic rhinitis from 250 to 600 mg 2 or 3 times/day, taken 5-10 minutes before a meal. Querticin synergizes well with bromelain.[3]

Capsules of quercetin from *Saphora japonica* (Japanese Pagoda tree) leaf and quercetin dehydrate from *Dimorphandra mollis* seed (found in South America), also with Vitamin C, bromelain (enzyme activity, 2400 GDU/g), Citrus Bioflavonoid Complex, and pycnogenol are available in the USA.

(Quercetin, rutin and hesperidin are essential bioflavonoids and accessory nutrients to form a Vitamin C Complex, previously known as Vitamin P factor. They function synergistically with Vitamin C in regard to maintaining healthy capillaries, to help form collagen in connective tissue, to help heal wounds, and to support a healthy immune system. Rutin and/or hesperidin, when low, frequently result in predictable, and even side-specific medical problems that include a greater risk for vascular degeneration, bruising/capillary fragility, nose bleeds, varicose veins, periodontal bleeding, hemorrhoids and aneurism.)

Rutin

The major sources of rutin include Buckwheat (*Fagopyrum esculentum*), Japanese Pagoda Tree (*Saphora japonica*) and *Eucalyptus macrohyncha*, also leaves of several species of eucalyptus, lime tree flowers, elder flowers, hawthorn leaves and flowers.[2]

Rutin is used as a vascular protectant for reducing capillary permeability and fragility, varicose veins, internal bleeding, hemorrhoids, inflammatory bowel disease, to prevent to strokes, and for prophylaxis of mucositis associated with cancer treatments. Rutin has also been demonstrated to lower intraocular pressure when used as an adjunct in patients who are unresponsive to glaucoma medication alone.[13] After oral administration, rutin is hydrolyzed in the gastrointestinal tract to release quercetin which is responsible for its biological activity.[2]

The most common form of quercetin is rutin, in which quercetin is bound to a glucose-rhamnose moiety.[2] Rutin-Quercetin tablets contain quercetin (from rutin) 250 mg and rutin (from *Fava D. anta*) 500 mg. Rutin 500 mg tablet contains rutin from *Saphora japonica* leaf. In combination with trypsin and bromelain, rutin is used for osteoarthritis. For relieving symptoms of edema associated with chronic venous insufficiency, 500 mg twice daily, has been used.

References

1. *Prescription for Herbal Healing*, Phyllis A. Balch, Avery, New York.
2. *Natural Medicines Comprehensive Database*, Therapeutic Research Faculty, Stockton, CA., 2007,20013.
3. *Complementary and Alternative Medicine Secrets*, Wendy Kohatsu, Hanley and Belfus, Philadelphia.
4. *PDR for Herbal Medicine*, Fourth Edition, Thomson, Montvale, NJ. Fourth Edition.
5. *The Review of Natural Products*, 7th Edition, Facts and Comparisons, Wolters Kluwer Health, Missouri. 7th Edition.
6. *Evidence-based Herbal Medicine*, Michael Rotblatt, Irwin Ziment, Hanley and Belfus, Philadelphia.
7. *Herbal Medicine*: Expanded Commission E Monograph, Mark Blumenthal, American Botanical Council, Austin, TX.
8. *ESCOP* (The European Scientific Cooperative on Phytotherapy) *Monograph*: The Scientific Foundation for Herbal Medicinal Products, Second edition, 2003; 8a Second edition Supplement, 2009.
9. Nakagawa Y, Mukai S, Yamada S, *et al.*, Short-term effects of highly-bioavailable curcumin for treating knee osteoarthritis: a randomized, double-blind, placebo-controlled prospective study. *J Orthop Sci.* 2014 Oct 13. (Epub ahead of print, courtesy Dr Michael Murray.)
10. The effectiveness of topically applied capsaicin, A meta-analysis, Zhang WY *et al., Eur J Clin Pharmacol* 46 96); 517-522.
11. Systematic review of topical capsaicin for the treatment of chronic pain, Mason L *et al., BMJ*, 2004, 328: 991-995. in
12. Cited in reference 8 in the monograph of *Curcuma longa*, and in reference 8a in the monograph of *Curcuma xanthorrhiza* (Javanese turmeric). Also cited in *WHO monographs on Selected Medicinal Plants*, Vol. 1, 1999.
13. *Encyclopedia of Natural Medicine*, Second edition, Michael Murray and Joseph Pizzorno, Prima Health, Rocklin, CA.
14. Quercetin: An inhibitor of antigen-induced human basophil histamine release, E. Middleton *et al., J Immunol*, 127 (1981): 546-550. Cited in Ref. 13.
15. *Healing Power of Herbs*, Michael T. Murray, Prima Health, Rocklin, CA.

14

Pharmacological and Clinical Researches on Frequently Used Ayurvedic Medicinal Plants

Chandra Kant Katiyar, Satyajyoti Kanjilal and Amitabha Dey

Introduction

Scientific studies have revived interest in herbal medicines. Published information available on clinical and preclinical research on 35 important frequently used Ayurvedic medicinal plants has been collected.

Databases such as Scopus, PubMed, MED-LINE, SciFinder, Google Scholar *etc.* have been searched to compile the research data on 35 Ayurvedic medicinal plants listed in table 14.1

TABLE 14.1: List of Medicinal Plants

Sl.No.	*Name of the Plant*	*Scientific Name*
1.	Amla	*Emblica officinalis* Linn.
2.	Arjuna	*Terminalia arjuna* (Roxb.) Wight and Arn.
3.	Ashoka	*Saraca asoca* (Roxb.) Willd.
4.	Ashwagandha	*Withania somnifera* (L.) Dunal
5.	Bala	*Sida cordifolia* Linn.
6.	Chirayata	*Boswellia serrata* Triana and Planch.
7.	Chitrak	*Plumbo zylenica* Linn.
8.	Draksha	*Vitis vinifera* Linn.
9.	Ginger	*Zingiber officinale* Roscoe
10.	Goksura	*Tribulus terrestris* Linn.
11.	Guduchi	*Tinospora cordifolia* (Thunb.) Miers
12.	Haritaki	*Terminalia chebula* Retz.
13.	Hadjod	*Cissus quadrangularis* Linn.
14.	Jatamansi	*Nardostachys jatamansi* (D.Don) DC.
15.	Kalmegh	*Andrographis paniculata* (Burm.f.) Wall. ex Nees
16.	Kantkari	*Solanum xanthocarpum* Schrad. and H. Wendl.
17.	Kutki	*Pichrohiza kurroa* Royle ex Benth.
18.	Lavang	*Syzygium aromaticum* (L.) Merrill and Perry
19.	Licorice	*Glycyrrhiza glabra* Linn.
20.	Madhuka	*Madhuca indica* J. F. Gmel
21.	Methi	*Trigonella foenum graecum* Linn.
22.	Musali	*Chlorophytum borivilianum* Linn.
23.	Musta	*Cyperus rotundus* Linn.
24.	Neem	*Azadirahta indica* A.Juss.
25.	Pippali	*Piper longum* Linn.
26.	Punarnava	*Boerhaavia diffusa* L. nom. cons.
27.	Rasana	*Pluchea lanceolate* (DC.) Oliv. and Hiern
28.	Salacia	*Salacia oblonga* Wall., *Salacia reticulata* Wight
29.	Sarpagandha	*Rauwolfia serpentina* (L.) Benth. ex Kurz
30.	Satavari	*Asparagus racemosus* Willd.
31.	Tagar	*Valeriana wallichii*
32.	Turmeric	*Curcuma longa* Linn.
33.	Tvaka	*Cinnamomum zeylanicum* Breyne
34.	Vacha	*Acorus calamus* Linn.
35.	Vasa	*Adhatoda vasika* Nees

TABLE 14.1

Sl.No.	Title	Type of Study	Study Design	Form and Dose	Summarized Efficacy	Conclusion	Reference
					AMLA (*Emblica officinalis* Linn.)		
					Anti-arthritic activity		
1.	Induction of apoptosis of human primary osteoclasts treated with extracts from the medicinal plant *Emblica officinalis*	*In vivo*	Potential in treatment of rheumatoid arthritis and osteoporosis by activating programmed cell death of human primary osteoclasts.	Extracts of *Emblica officinalis*	The effects of extracts from *Emblica officinalis* on differentiation and survival of human primary Ocs cultures obtained from peripheral blood were determined by tartrate-acid resistant acid phosphatase (TRAP)- positivity and colorimetric MTT assay. Extracts of Emblica officinalis were able to induce programmed cell death of mature OCs, without altering, at the concentrations employed in our study, the process of osteoclastogenesis. *Emblica officinalis* increased the expression levels of Fas, a critical member of the apoptotic pathway. Gel shift experiments demonstrated that *Emblica officinalis* extracts act by interfering with NF-kB activity, a transcription factor involved in osteoclast biology. The data obtained demonstrate that *Emblica officinalis* extracts selectively compete with the binding of transcription factor NF-kB to its specific target DNA sequences. This effect might explain the observed effects of *Emblica officinalis* on the expression levels of interleukin-6, a NF-kB specific target gene.	Induction of apoptosis of osteoclasts could be an important strategy both in interfering with rheumatoid arthritis complications of the bone skeleton leading to joint destruction, and preventing and reducing osteoporosis.	Letizia Penolazzi *et al.*, BMC Complementary and Alternative Medicine 2008, 8:59
					Antioxidant Activity		
2.	Antioxidant and hyaluronidase inhibitory activities of diverse phenolics in *Phyllanthus emblica*	*In vitro* and *in vivo*	Fifty-eight phenolic compounds isolated from *Phyllanthus emblica* were screened and compared for their *in vitro* and *in vivo* antioxidant properties, as well as hyaluronidase (HAase) inhibitory activities.	Phenolic compounds	Among them, 20 compounds showed to be promising antioxidants due to the stronger scavenging activity in both DPPH radical and Danio rerio reactive oxygen species assays, while nine compounds were potential HAase inhibitors with 100-fold stronger activities than that of the positive control, DSCG. The structure activity relationship was discussed.	Few compounds were having promising antioxidants and were potential HAase inhibitors	Xu M *et al.*, Nat Prod Res. 2016 Feb 12: 1-4.
3.	Chemical and antioxidant evaluation of Indian gooseberry (*Emblica officinalis* Gaertn., syn. *Phyllanthus emblica* L.) supplements	*In vitro*	RP-HPLC coupled with photodiode array detection. The iron(III) reduction and 1,1-diphenyl-2-picrylhydrazyl and superoxide anion radical scavenging and hydroxyl radical scavenging activity	*E. officinalis* fruit extracts	Total phenol, total flavonoid and total tannin assays were tested positive. The presence of predominantly (poly)phenolic analytes, *e.g.* ellagic and gallic acids and corilagin, was confirmed by RP-HPLC coupled with photodiode array detection.The extract were effective in the iron(III) reduction and 1,1-diphenyl-2-picrylhydrazyl and superoxide anion radical scavenging and hydroxyl radical scavenging activity.	The extracts demonstrated varying degrees of antioxidative efficacy	Poltanov EA *et al.*, Phytother Res. 2009 Sep; 23(9): 1309-15

Contd...

TABLE 14.2–*Contd....*

Sl.No.	*Title*	*Type of Study*	*Study Design*	*Form and Dose*	*Summarized Efficacy*	*Conclusion*	*Reference*
					Natural cognitive enhancer		
4.	Exploring the Effect of *Phyllanthus emblica* L. on Cognitive Performance, Brain Antioxidant Markers and Acetylcholinesterase Activity in Rats: Promising Natural Gift for the Mitigation of Alzheimer's Disease	*In vivo*	Extracts were given to Swiss albino male rats for 12 days and its effect on cognitive functions, brain antioxidant enzymes, and AChE activity determined. Learning and memory enhancing activity of EEPE fruit was examined by using passive avoidance test and rewarded alternation test. Antioxidant potentiality was evaluated by measuring the activity of antioxidant enzymes such as superoxide dismutase (SOD), catalase (CAT), glutathione peroxidase (GSH-Px), glutathione reductase, reduced glutathione (GSH), glutathione-S-transferase, and the contents of thiobarbituric acid reactive substances (TBARS) in entire brain tissue homogenates. AChE activity was determined using colorimetric method.	Ethanolic extracts of *Phyllanthus emblica* (EEPE) ripe (EEPEr) and EEPE unripe (EEPEu) fruits at 100 and 200 mg/kg b.w.	Administration of the highest dose (*i.e.*, 200 mg/kg b.w.) of EEPEr fruit significantly ($p < 0.01$) and both lowest and highest doses (*i.e.*, 100 and 200 mg/kg b.w.) of EEPEu fruit markedly ($p < 0.05$, $p < 0.001$) increased step-through latency in rats on 6th, 11th, and 12th day with respect to the control group. For aforementioned doses, the percentage of memory retention (MR) was considerably ($p < 0.05$, $p < 0.01$) increased in rats on 10th, 11th, and 12th days with respect to the control group. The highest dose (*i.e.*, 200 mg/kg b.w.) of EEPEr fruit suggestively ($p < 0.05$) and both lowest and highest doses (*i.e.*, 100 and 200 mg/kg b.w.) of EEPEu fruit suggestively ($p < 0.05$, $p < 0.01$, $p < 0.001$) increased the levels of SOD, CAT, GSH, GSH-Px and expressively ($p < 0.01$) decreased the TBARS level compared to the control group. Treatment with the highest dose (*i.e.*, 200 mg/kg b.w.) of EEPEr fruit significantly ($p < 0.05$) and both lowest and highest doses (*i.e.*, 100 and 200 mg/kg b.w.) of EEPEu fruit markedly ($p < 0.01$, $p < 0.001$) decreased the level of AChE activity compared to that of the control group.	EEPE fruit possesses an excellent source for natural cognitive enhancer which could be developed in the treatment of AD and other neurodegenerative diseases	Uddin MS *et al.*, Ann Neurosci. 2016, Oct; 23(4): 218-229.
					Anti-osteoporosis Activity		
5.	Preventive role of *Emblica officinalis* and *Cissus quadrangularis* on bone loss in osteoporosis	*In vivo*	Species Wistar Rat N= 40 Group 5 Groups Control Group (SHAM operated), OVX control	CQ 500 mg/kg/day and for EO 200 mg/kg/day 30 days	Treatment with EO and CQ significantly increased the serum ALP levels, while the serum TRAP and hydroxyproline levels were significantly restored towards normal level. Loss of bone mass and strength due to osteoporosis was significantly reduced with EO and CQ treatments. The results clearly depicts that the	EO and CQ can effectively reduce the bone loss and increase the bone strength, thus they can be used to	Srinivas Rao, *et al.*, Int J Pharm Pharm Sci, 2013; 5(4):

Contd...

TABLE 14.2–*Contd....*

Sl.No.	*Title*	*Type of Study*	*Study Design*	*Form and Dose*	*Summarized Efficacy*	*Conclusion*	*Reference*
			Group/OVX+ Ralixifen, OVX+ EO/OVX+ CQ/OVX+ CQ and EO. (for 6 weeks)		individual effects of EO and CQ are comparable with that of Raloxifene. Further when compared to effects of individual plant extracts, the combined effect of EO and CQ were equipotent.	treat the bone degenerative disorders	465-470
					Anti-inflammatory Activity		
6.	Anti-inflammatory and analgesic activities of the water extract from the fruit of *Phyllanthus emblica* Linn.	*In vivo*	Model Ethyl phenyl-propiolate (EPP)-induced and Arachidonic acid (AA)-induced ear edema, carrageenan-induced and cotton pellet-induced Species Sprague-Dawley rats and ICR Mice Dosage 150/300/600 mg/kg *P. emblica* extract orally Analgesic activity tested in mice using formalin test	Extract 150/300/ 600mg/kg	PEE has a significant analgesic activity in a dose-dependent manner on both the early and late phase of formalin test. The anti-inflammatory and analgesic mechanism of activity of the standardized Aqu. PEE is similar to NSAIDs rather than to steroidal drugs.	PEE has anti-inflammatory and analgesic activity.	Jaijoy K, *et al.*, International Journal of Applied Research in Natural Products. 2010; 3 (2): 28-35.
					Anti-hypoglycaemic Effect		
7.	Investigation into mechanism of action of anti-diabetic activity of *Emblica officinalis* on streptozotocin induced type I diabetic rat	*In vivo*	Streptozotocin induced type 2 diabetic rat	Fresh juice and hydro-alcoholic extract of fruits	Treatment with fresh juice and hydro-alcoholic extract significantly reduced elevated fasting glucose and AUC glucose levels in type I diabetic rats. It produced significant increase in serum insulin level and AUC insulin of diabetic rats compared to that of diabetic control.	Fresh juice and hydro-alcoholic extract of *E. officinalis* fruits possesses potential anti-diabetic activity in STZ induce type 1 diabetic rat.	PR Tirgar *et al.*, October – December 2010 RJPBCS 1(4) Page No. 672.
8.	Anti-diabetic activity of *Emblica officinalis* in animal models	*In vivo*	Streptozotocin induced type 2 diabetic rats	Aqueous seed extract suspended in distilled water at doses of 100, 200, 300, and 400 mg/kg Body weight	Treatment with *E. officinalis* aqueous seed extract reduces the blood glucose level and improves glucose tolerance in both normal and diabetic animals. the antihyperglycemic effect of the aqueous seeds extract might be due to the activation of existing β cells of the pancreas.	*E. officinalis* aqueous seed extract possesses potential anti-diabetic activity in STZ induce type 2 diabetic rats.	Mehta S, *et al.*, Pharmaceutical Biology, 2009; 47(11): 1050–1055.

Contd...

TABLE 14.2–*Contd....*

Sl.No.	*Title*	*Type of Study*	*Study Design*	*Form and Dose*	*Summarized Efficacy*	*Conclusion*	*Reference*
9.	Antidiabetic and antioxidant potential of *Emblica officinalis* Gaertn. Leaves extract in streptozotocin-induced type-2 diabetes mellitus (T2DM) rats	*In vivo*	Streptozotocin induced type 2 diabetic rat	Hydro-methanolic (20:80) extract of leaves of *Emblica officinalis* at a dose of 100, 200, 300 and 400 mg/kg b.w. daily for 45 days	Treatment with *E. officinalis* significant decrease in fasting blood glucose and increase insulin level as compared with the diabetic rats. Also it significantly reduced serum creatinine, serum urea, SGOT, SGPT and lipid profile. The treatment also resulted in a significant increase in reduced glutathione, glutathione peroxidase, superoxide dismutase, catalase, and decrease LPO level in the liver and kidney of diabetic rats.	*Emblica officials* treated group may effectively normalize the impaired antioxidant status in streptozotocin induced diabetes at dose dependent manner than the glibenclamide treated groups. The extract exerted rapid protective effects against lipid peroxidation by scavenging of free radicals and reducing the risk of diabetic complications.	Nain P, *et al.*, J Ethnopharmacol. 2012; 142(1): 65-71.
					Immunomodulatory activity		
10.	Immunomodulatory activity of Âmalaki Rasâyana: An experimental evaluation	*In vivo*	Cyclophosphamide induced immuno-suppression in rats	The test formulations (*Âmalaki Rasâyana*) was administered after addition of honey and ghee at dose of 450 mg/kg rat for 9 consecutive days.	*Âmalaki Rasâyana* possesses significant immunostimulant activity and moderate cytoprotective activity. AR21 was found to have better activity profile in terms of both immunostimulant as well as cytoprotective activity.	Among AR7 and AR21 formulations; AR21 was found to have better activity profile in terms of both immunostimulant as well as cytoprotective activity. Hence, it can be concluded that it is better to give 21 Bhâvanas to get desired pharmacological activity.	Rajani J, *et al.*, Anc Sci Life. 2012; 32(2): 93–98.
				ARJUNA (*Terminalia arjuna* (Roxb.) Wight and Arn.)			
					Cardio-protective Activity		
1.	Cardioprotective effect of the alcoholic extract of *Terminalia arjuna* bark in an *in vivo* model of myocardial ischemic reperfusion injury	*In vivo*	Species: Albino Wistar rats Model:- Isoproterenol induced myocardial injury. N=24. Group:- 4, Control and 3 groups with dosage of extract, 3.4/6.75/9.75 mg/kg 6 days/week	Alcoholic extract of *Terminalia arjuna* bark administered orally in three different doses, by	The 6.75 mg/kg arjuna treatment group (baseline) shows a significant increase in myocardial TBARS as well as endogenous antioxidants (GSH, SOD, and catalase), but not in the other treatment groups. In histological studies, all the groups, except the isoproterenol treated group, showed preserved myocardium.	Arjuna has antioxidant property and showes cardioprotective activity.	K. Karthikeyan, *et al.*, Life Sciences 003; 73: 2727–2739.

Contd...

TABLE 14.2–*Contd....*

Sl.No.	Title	Type of Study	Study Design	Form and Dose	Summarized Efficacy	Conclusion	Reference
			Duration:- 4 weeks. Parameter:- investigated for biochemical and histological studies. Superoxide dismutase level also find out	gastric gavage [3.4 mg/kg, 6.75 mg/kg and 9.75 mg/kg] 6 days/week for 4 weeks.			
2.	Catecholamine-induced myocardial fibrosis and oxidative stress is attenuated by *Terminalia arjuna*	*In vivo*	Species: Wistar rats Model:- Isoprenaline induced LVH. N=24. Group:- 5, control, positive control (Captopril), arjuna extract 65/125/250 mg/kg Duration:- 28 days Parameter:- investigated for artifibrotic and antioxidant effect	Aqueous extract of *T. arjuna* bark 63, 125 and 250 mg/kg given orally for 28 days	Isoprenaline caused fibrosis, increased oxidative stress and cardiac hypertrophy (increased heart weight: body weight ratio and cardiomyocyte diameter). The *T. arjuna* bark extract and captopril significantly prevented the isoprenaline-induced increase in oxidative stress and decline in endogenous anti-oxidant level. Both also prevented fibrosis but not the increase in heart weight: body weight ratio.	*T. arjuna* possess cardio-protective action by inhibiting angiotensin converting enzyme.	Santosh Kumar, *et al.* J Pharm Pharmacol, 2009; 61(11): 1529–1536.
3.	Cardioprotective effect of *Terminalia arjuna* on caffeine induced coronary heart disease	*In vivo*	Species : Wistar rats Model:- Isoprenaline induced LVH. N= 12 Group :- 3, control, two groups (B, C) treated with caffeine, and C group treated with 6.75 mg Arjun extract Duration:- 4 Weeks Parameter:- investigated for Serum lipid	Alcoholic extract of Bark of *Terminalia arjuna* (6.75mg/kg of body weight) for 4 weeks	When compared to the control rats, co-treatment of rats with caffeine and *Terminalia arjuna* resulted in an increase in HDL-cholesterol, decrease in serum total cholesterol, triglycerides, LDL cholesterol and VLDL cholesterol as compared to caffeine treated animals.	*T. arjuna* has protective effects against caffeine induced coronary heart disease and may have potential as a cardio-protective agent.	Asha and Taju *et al.*, IJPSR, 2012; 3(1): 150-153.
4.	Screening of cardio-protective effect of *Terminalia arjuna* Linn. bark in isoproterenol - Induced myocardial infarction in experi-mental animals	*In vivo*	Isoproterenol induced myocardial infarction in rats	Ethanol and aqueous extracts of *T. arjuna* bark 250 mg/kg through the oral route	The oral administration of T. arjuna bark extracts significantly restored the level of total cholesterol, triglyceride, LDL, HDL and Myocardial and serum LDH, CK, AST. The extract effect was compared with standard drug verapamil which also offered similar protection in biochemical and histopathological changes.	*Terminalia arjuna* bark possess significant cardioprotective activity.	V. Sivakumar and S. Rajesh Kumar; International Journal of Pharma Sciences and Research (IJPSR) Vol 5 No 06 Jun 2014; 262-268.

Contd...

TABLE 14.2–*Contd....*

Sl.No.	Title	Type of Study	Study Design	Form and Dose	Summarized Efficacy	Conclusion	Reference
5.	Mechanistic clues in the cardioprotective effect of *Terminalia arjuna* bark extract in isoproterenol-induced chronic heart failure in rats	*In vivo*	Isoproterenol (ISO)-induced chronic heart failure (CHF).	*T. arjuna* bark extract (500 mg/kg, p.o) for 15 days	Rats showed decline in maximal rate of rise and fall of left ventricular pressure, cardiac contractility index, cardiac output and rise in LV end-diastolic pressure. *T. arjuna* bark extract treatment significantly attenuated cardiac dysfunction and myocardial injury in CHF rats. Cardioprotective action of *T. arjuna* was comparable to fluvastatin, a synthetic drug.	*T. arjuna* has cardio-protective activity.	Parveen A1, *et al.* Cardiovasc Toxicol. 2011 Mar; 11(1): 48-57.
					Cholesterol Lowering Effect		
6.	Anti-Atherogenic Activity of Ethanolic Fraction of *Terminalia arjuna* Bark on Hypercholesterolemic Rabbits	*In vivo*	Species : Swiss albino mice and Newzeland rabbits N= 20. Group :- 5, control, HFD, HFD and Atorvastanin, rest group with extract 100/200 mg Arjun extract. Duration:- 72 days Parameter: investigated for Serum lipid para-meters and atherogenic index	Ethanolic fraction of *T. arjuna*; 100 and 200 mg/kg body weight	*T. arjuna* significantly decreases total cholesterol (TC), low density lipoprotein (LDL) and triglycerides (TG) levels and increases high density lipoprotein (HDL) and lessens athero-sclerotic lesion in aorta.	*T. arjuna* extract can effectively prevent the progress of atherosclerosis and thereby have cardioprotective and anti-atherogenic effects.	Saravanan Subramaniam *et al.*, Evidence-Based Complementary and Alternative Medicine; 2011; 2011: Article ID 487916, 8 pages.
					Platelet aggregation		
7.	Evaluation of *Terminalia arjuna* on cardiovascular parameters and platelet aggregation in patients with Type II diabetes mellitus	Clinical	Randomized, double-blind, parallel group placebo-controlled study in patients with type 2 diabetes mellitus	*Terminalia arjuna* 500mg capsule thrice daily for 8 weeks	There was a highly significant increase in mean cardiac output from 4.34±0.38 to 4.86±0.20 (Lt/min) along with reduction in mean systemic vascular resistance from 1729±93.52 to 1484± 115.5 (dyne.sec/cm5) in *Terminalia arjuna* group. Significant inhibition of platelet aggregation compared to placebo was observed. Treatment was well tolerated.	*Terminalia arjuna* by producing significant increase in mean cardiac output along with reduction in mean systemic vascular resis-tance, and also significant inhibition of platelet aggregation, can be a good therapeutic option for reducing cardio-vascular morbidity in diabetic subjects.	Usharani *et al.*, Research Journal of Life Sciences (May- 2013), Vol. 01, Issue 02, pp. 7-12.

Contd...

TABLE 14.2–*Contd....*

Sl.No.	Title	Type of Study	Study Design	Form and Dose	Summarized Efficacy	Conclusion	Reference
					Anti-hypertensive Activity		
8.	Possible mechanisms of hypotension produced 70 per cent alcoholic extract of *Terminalia arjuna* (L.) in anaesthetized dogs	*In vivo*	Species : Dog, thiopental sodium model N= 8 Group :- 7, each dog feeded with 5/6/8/10/15 mg Arjun extract and in 2 dogs 20mg/kg Parameter: investigated for hypotension	Alcoholic extract of *T. arjuna* 6 mg/kg, intravenous	*T. arjuna* produced dose-dependent hypotension. The hypotension produced by the extract was blocked by propranolol but not by atropine or mepyramine maleate.	*Terminalia arjuna* cause hypotension by virtue of active compound(s) possessing adrenergic β2-receptor agonist action and/or that act directly on the heart muscle.	Srinivas Nammi, *et al.* BMC Complementary and Alternative Medicine. 2003; 3.
9.	Effects of *Withania somnifera* (Ashwagandha) and *Terminalia arjuna* (Arjuna) on physical performance and cardio-respiratory endurance in healthy young adults	Clinical	Type:- Single-blind controlled randomized trial Sample Size 40 Subjects with cardiac problem Study Group :- 4 Control/ Ashwagandha/Arjuna/ Ashwagandha and Arjuna 500 mg/day Duration of Study 8 weeks Parameters kinematic measuring system used to measure velocity	500mg Capsules/day for 8 weeks	Treatment with *T. arjuna* demonstrated significant increase in maximum oxygen consumption capacity. The systolic blood pressure fell significantly from The average absolute power also increased significantly.	*T. arjuna* may prove useful to improve cardiovascular endurance and lowering systolic blood pressure. It is safe for young adults.	Jaspal Singh Sandhu, *et al.* Int J Ayurveda Res. 2010; 1(3): 144–149.
					Anti-hypertensive Activity		
10.	Efficacy of *Terminalia arjuna* in chronic stable angina: a double-blind, placebo-controlled, crossover study comparing *Terminalia arjuna* with isosorbide mononitrate	Clinical	Type:- Double-blind, randomized trial Sample Size 58 Subjects with chronic stable angina; Dosage: 500mg/kg Duration of Study 1 weeks Parameters Clinical, biochemical and tread-mill exercise were evaluated	*Terminalia arjuna* extract (500 mg 8 hourly) for one week each	Significant decrease in both the frequency of angina and need for isosorbide dinitrate was observed. Improvement in treadmill exercise test parameters compared to placebo was noted. Increase in total duration of exercise, decrease in maximal ST depression during the longest equivalent stages of submaximal exercise and time to recovery was noted. No significant untoward effects were reported.	*Terminalia arjuna* improves the clinical and treadmill exercise parameters and is beneficial in patients of angina.	Bharani A, *et al.* Indian Heart J. 2002; 54(2): 170-5.

Contd...

TABLE 14.2–*Contd....*

Sl.No.	*Title*	*Type of Study*	*Study Design*	*Form and Dose*	*Summarized Efficacy*	*Conclusion*	*Reference*
					Improves Coronary Flow		
11.	Experimental evaluation of *Terminalia arjuna* (Aqueous extract) on cardiovascular system in comparison to digoxin	*In vivo*	Heart rate and amplitude of frog's heart *in situ*; Heart rate and amplitude of hypodynamic frog's heart *in situ.*; Heart rate and amplitude of isolated perfused rabbit heart; Coronary flow of isolated perfused rabbit heart.	Aqueous extract *Terminalia arjuna* (100 μg, 200 μg, 400 μg, 800 μg)	*Terminalia arjuna* (Aq.E) increased the force of contraction of cardiac muscle in frog's heart *in situ*, hypodynamic frog's heart *in situ* and isolated perfused rabbit heart. It also increased the coronary flow at a 400 μg dose in isolated perfused rabbit heart along with dose dependent bradycardia.	*Terminalia arjuna* (Aq.E) produced cardiotonic effects along with increase in the coronary flow in experimental animals.	Dr Prem Verma, *et al.*, IOSR Journal of Dental and Medical Sciences (IOSR-JDMS); Volume 7(2) (May.- Jun. 2013), 48-51.
					ASHOKA (*Saraca asoca* (Roxb.) Willd.)		
					Estrogenic activity		
1.	Activity based evaluation of a traditional Ayurvedic medicinal plant: *Saraca asoca* (Roxb.) de Wilde flowers as estrogenic agents using ovariectomized rat model		Estrogenic potential using ovariectomized (OVX) female albino Wistar rat model.	Standardized ethanolic extract of *Saraca asoca* flowers (SAF) in doses of 100mg/kg, 200mg/kg and 400mg/kg body weight in distilled water as a vehicle, orally once a day for two weeks.	HPTLC revealed the presence of markers like quercetin, kaempferol, β-sitosterol and luteolin from the ethanolic extract of SAF. The content of the four markers was found to be 1.543mg/g, 0.924mg/g, 4.481mg/g and 2.349mg/g, respectively. SAF extract was found to be safe at an oral dose of 2000mg/kg body weight in rats. Among the three doses administered to ovariectomized rats, treatment with high dose was found to be more efficacious when compared with ovariectomized rats.	The findings of this study firmly support the estrogenic potency of ethanolic extract of SAF which may be by the reason of phyto-estrogens.	S. Gauri *et al.*, J Ethno-pharmacol. 2017 Jan 4; 195: 324-333.

Contd...

TABLE 14.2–*Contd....*

Sl.No.	Title	Type of Study	Study Design	Form and Dose	Summarized Efficacy	Conclusion	Reference
					Anti-oxidant activity		
2.	Lignan glycosides and flavonoids from *Saraca asoca* with antioxidant activity	*In vitro*	Five lignan glycosides, lyoniside, nudiposide, 5-methoxy-9-b-xylo pyranosyl-isolariciresinol, icariside E3, and schizandriside, and three flavonoids, epicatechin, Epiafzelechin, epicatechin and procyanidin B2, together with b-sitosterol glucoside, were isolated from a methyl alcohol (MeOH) extract of *Saraca asoca* dried bark. Their structures were determined by 1D and 2D nuclear magnetic resonance (NMR) and mass spectroscopic analysis. Antioxidant activities were evaluated by1,1-diphenyl-2-picryl-hydrazyl (DPPH) radical-scavenging assay.	Methyl alcohol (MeOH) extract of *Saraca asoca*	Based on the medicinal uses, the extract was tested for antioxidant activity using DPPH radical-scavenging assay that showed a potent effect with an IC50 value of 25 lg/ml. The lignan glycosides were obtained from S. asoca for the first time. The IC50 values of DPPH radical-scavenging assay for compounds and the positive control quercetin were 104, 85, 44, 75, 55, 50, 55, 40 and 30 lM, respectively.	All the isolated compounds exhibited moderate antioxidant activity that might be responsible together for the therapeutic efficacy of this herb.	Samir Kumar Sadhu *et al.*, J Nat Med DOI 0.1007/s 11418-007-0182-3.
					Hypolipidemic, hypoglycemic and antioxidant potential		
3.	Hypolipidemic, hypoglycemic and antioxidant potential of *Saraca asoca* ethanolic leaves extract in streptozotocin induced-experimental diabetes	*In vivo*	Experimental hyperglycemia was produced in rats by single dose of Streptozotocin (55 mg/kg). Assessment of blood glucose, TBARS and GSH was used as marker of hyperglycemia and oxidative stress respectively. Serum lipid profile (total cholesterol, HDL and triglyceride) were used as markers of dyslipidaemia.	*Saraca asoca* ethanolic leaves extract 100, 200 and 400 mg/kg	Treatment with *Saraca asoca* ethanolic extract normalizes the altered lipid profile and reduced the elevated glucose level in dose dependent manner. Further it attenuates the diabetes-induced renal oxidative stress.	Concurrent administrations of *Saraca asoca* ethanolic extract reducing the lipid alteration, decreasing the renal oxidative stress and certainly provide hypolipidemic, hypoglycemic and antioxidant effect.	Akash Jain *et al.*, Int J Pharm Pharm Sci, Vol 5, Suppl 1, 302-305.

Contd...

TABLE 14.2–*Contd....*

Sl.No.	*Title*	*Type of Study*	*Study Design*	*Form and Dose*	*Summarized Efficacy*	*Conclusion*	*Reference*
4.	Antioxidant, antiglycation and inhibitory potential of *Saraca asoca* flowers against the enzymes linked to type 2 Diabetes and LDL oxidation	*In vitro*	Antioxidant capacity of *Saraca asoca* flowers (SAF) was evaluated by estimating total antioxidant activity (TAA) and its protective effects against the oxidative stress induced by H_2O_2 on C_2C1_2 cells. Cytotoxicity by MTT assay and markers of oxidative stress: reduced glutathione (GSH), malondialdehyde (MDA) and reactive oxygenspecies (ROS) were measured. antiglycation and inhibitory potential against alpha glucosidase, alpha amylase and LDL oxidation were also measured.	*Saraca asoca* Flower (SAF) Extract	Pre-treatment of C_2C1_2 cells with SAFprevented the increased formation of MDA and depletionof GSH induced by H_2O_2. The increased ROS generation induced by H_2O_2 was also reduced by a pretreatment with SAF. Significant inhibitory potential against alpha glucosidase and alpha amylase enzymes revealed the therapeutic potential of SAF asan antihyperglycemic agent. SAF also demonstrated potent antiglycation property and inhibited LDL oxidation under *in vitro* conditions.	The overall results demonstratet hat SAF can be used as an ideal natural remedy for preventing oxidative stress and other complications associated with diabetes.	A. Prathapan *et al.,* European Review for Medical and Pharmacological Sciences, 2012; 16: 57-65.
					Anti-inflammatory activity		
5.	Anti-inflammatory flavanol glycosides from *Saraca asoca* bark	*In vivo*	Anti-TNF-a activity of compounds wastested on whole-blood assay and methanol extract was studied for carrageenan-induced rat paw oedema.	Methanol extract of *S. asoca* bark	Methanol extract of *S. asoca* bark after chromatographic separations yielded leucopelargonidin(1), leucocyanidin (2), 3,5-dimethoxy epicatechin (3), epicatechin (4), catechin (5), 3- deoxyepicatechin-3-O-b-D-glucopyranoside (6), lyoniside (7), 3- deoxycatechin-3-O-a-Lrhamnopyranoside(8), epigallocatechin (9) and gallocatechin (10). Anti TNF alpha activity - the methanol extract (100 mg/mL) as such showed an inhibition of 26 per cent, while pure 6 and 8 showed 26–34 per cent and 17–26 per cent activity, respectively, at 50, 100 and 200 mg/mL. methanol extract showed a maximum of 17 per cent inhibition while compounds 6 and 8 showed 52 per cent and 33 per cent, respectively, at 5 h in Paw odema model.	Ten isolated compounds from *S. asoca* bark have been characterised by using spectroscopic methods, and two of them (6 and 8) have been found to be anti-inflammatory, substantiating the traditional use of *S. asoca* bark in gynaecological problems of women.	F.Ahmad *et al.,* Natural Product Research, 2015 http://dx.doi.org/10.1080/14786419.2015.1023728.

Contd...

TABLE 14.2–*Contd....*

Sl.No.	Title	Type of Study	Study Design	Form and Dose	Summarized Efficacy	Conclusion	Reference
6.	Effect of *Saraca asoca* (Asoka) on estradiol-induced keratinizing metaplasia in rat uterus	*In vivo*	Estrogen-induced endometrial thickening of rat uterus. Endometrial thickening was induced by intraperitoneal injection of estradiol (20 μg/kg b.wt) to 8-day-old immature rats for alternate 5 days. Uterus endometrial thickening was analyzed histopathologically and serum estrogen level by radioimmunoassay (RIA). Cyclooxygenase (COX-2) expression in rat uterus was also estimated by Western blot. Anti-inflammatory activity of the extract was analyzed by formalin- and carrageenan-elicited paw edema models in mouse.	Methanolic extract (200 mg/kg b. wt) from *S. asoca* bark	Uterus endometrium proliferation and keratinized metaplasia with seven to eight stratified epithelial layers on day 16 was observed in rats administered with estradiol. Treatment with *S. asoca* reduced the thickening to two to four layers and the serum estrogen level diminished significantly to 82.9±12.87 pg/mL compared to rats administered with estrogen alone (111.2±10.68 pg/mL). A reduction of formalin- and carrageenan-induced paw edema in mouse by *S. asoca* extract was observed. Lower level of lipopolysaccharides (LPS)-induced COX-2 enzyme in rat uterus by the extract further confirms its anti-inflammatory activity.	Present study reveals the antiproliferative and antikeratinizing effects of *S. asoca* in uterus endometrium possibly through its anti-inflammatory properties.	A P Shahid *et al.*, J Basic Clin Physiol Pharmacol. 2015 Sep; 26(5): 509-15.
7.	Therapeutic effects of acetone extract of *Saraca asoca* seeds on rats with adjuvant-induced arthritis via attenuating inflammatory responses	*In vivo*	Adjuvant-induced arthritic rats; Regular treatment up to 21 days	Acetone extract of *Saraca asoca* seeds	The extract (at 300 and 500mg/kg doses) increases RBC and Hb, decreases WBC, ESR, and prostaglandin levels in blood, and restores body weight when compared with control (normal saline) and standard (Indomethacin) groups. Significant ($p < 0.05$) inhibitory effect was observed especially at higher dose on paw edema, ankle joint inflammation, and hydroxyproline and glucosamine concentrations in urine. Normal radiological images of joint and histopathological analysis of joint, liver, stomach, and kidney also confirmed its significant nontoxic, antiarthritic, and anti-inflammatory effect.	*Saraca asoca* has significant nontoxic, anti-arthritic, and anti-inflammatory pharmacological effect which is comparable to the standard drugsin a dose-dependant manner, possibly due to the presence of flavonoidic chemical compounds and the resultant lowering in the level of prostaglandin in the blood.	M.Gupta *et al.*, Hindawi Publishing Corporation ISRN Rheumatology Volume 2014, Article ID 959687, 12 pages.

Contd...

TABLE 14.2–*Contd....*

Sl.No.	*Title*	*Type of Study*	*Study Design*	*Form and Dose*	*Summarized Efficacy*	*Conclusion*	*Reference*
8.	Therapeutic effect of *Saraca asoca* (Roxb.) Wilde on lysosomal enzymes and collagen metabolism in adjuvant induced arthritis	*In vivo*	Adjuvant induced arthritis by assessing paw swelling, body weight, the levels of lysosomal enzymes, protein bound carbohydrates, serum cytokines, urinary collagen and histo-pathology of joints.	Methanol extract of *Saraca asoca*	*S. asoca* methanol extract at doses of 50, 100 and 200 mg/kg reduced the paw thickness and elevated the mean body weight of arthritic rats. The treatment of S. asoca showed a significant reduction in the levels of both plasma and liver lysosomal enzymes. The protein bound carbohydrates and urinary collagen contents were also decreased at a significant level by the treatment of *S. asoca* methanol extract. The histopathological study of the joints showed the anti-arthritic property of *S. asoca* which nearly normalized the histological architecture of the joints. Further, the anti-arthritic activity was established by measuring the levels of cytokines in both arthritic and treated rats. The treatment of *S. asoca* reduced the levels of pro-inflammatory cytokines.	*S. asoca* methanol extract was capable of amelio-rating the conditions of arthritis in adjuvant induced arthritic rats.	S. Saravanan *et al.*, Inflammo-pharmacology. 2011 Dec; 19(6): 317-25.
					Antipyretic effect		
9.	Pharmacognostical, phytochemical and pharmacological evalua-tion for the antipyretic effect of the seeds of *Saraca asoca* Roxb.	*In vivo*	*Saraca asoca* seed was studied for pharmaco-gnostical, phytochemical and other recommended methods for standardi-zations. Also, the acetone extract of the seeds was evaluated for acute toxicity study and anti-pyretic activity using Brewer's yeast induced pyrexia in Wistar rats at oral doses of 300 mg/kg and 500 mg/kg.	*Saraca asoca* seed acetone extract	After phytochemical screening, the acetone extract showed the presence of saponin, tannins and flavonoids which inhibit pyrexia. The therapeutic efficacy achieved at both the dose levels of the research drug and standard drug aspirin (100 mg/kg) showed significant (P<0.01) antipyretic activity when compared to the control group. The highly significant antipyreticeffect exhibited at the dose of 500 mg/kg was also found to be sustainable in nature.	The antipyretic effect of the acetone extract showed significant results in rats at the dose of 500 mg/kg after following the standard pharmaco-gnostical and phyto-chemical methods.	S.Sasmal *et al.*, Asian Pac J Trop Biomed 2012; 2(10): 782-786.
					Antimutagenic and genoprotective effects		
10.	Antimutagenic and genoprotective effects of *Saraca asoca* bark extract	*In vitro* and *in vivo*	SAE were evaluated for antimutagenic property in Salmonella strains (TA97a, TA98, TA100, and TA102), in the presence and absence of metabolic activation (S9). The SAE was also studied for antigenotoxic property against cyclophosphamide (CP) in Swiss albino male.	*Saraca asoca* bark extract	The study reveals antimutagenic property of the bark extract in Salmonella strains in the presence and absence of metabolic activation (S9). The study reports antigenotoxic property of the bark extract against CP *in vivo*. Thiobarbituric acid reactive species assay on the bark extract revealed antioxidant property. HPLC revealed the presence of two peaks corresponding to gallic acid and (-)-epicatechin, respectively.	The study clearly reveals the antimutagenic and antigenotoxic properties of SAE.	D. Nag *et al.*, Toxicol Ind Health. 2015 Aug; 31(8): 696-703.

Contd...

TABLE 14.2–*Contd....*

Sl.No.	Title	Type of Study	Study Design	Form and Dose	Summarized Efficacy	Conclusion	Reference
			mice *in vivo*. The extract was analyzed using high-performance liquid chromatography (HPLC).				
				ASHWAGANDHA (*Withania somnifera* (L.) Dunal)			
					Anti-stress activity		
1.	A biological active constituent of *Withania somnifera* (ashwagandha) with anti - stress activity	*In vivo*	Passive rat experimental model, where the animals are subjected to multiple stress of cold, hypoxia, restraint (C-H-R)	*W. somnifera* water suspension (360 mg/kg bw) and compound X (20mg/kg bw)	The effect on the fall and recovery of colonic temperature was noted. There was an increase of ≈38 per cent and ≈54 per cent in the time taken to attain T_{rec} 23°C by rats given a single dose of fresh aqueous suspension and biologically active constituent (Compound X) respectively, where as decrease in the recovery time to attain T_{rec} 37°C is ≈13 per cent and ≈33 per cent respectively, as compared to control group.	*Withania somnifera* can withstand the multiple stress of C-H-R	Parvinder Kaur *et al.*, Indian Journal of Clinical Biochemistry Year : 2001 Volume : 16 Issue : 2 Page : 195-198.
2.	Effect of Ashwagandha on lipid peroxidation in stress induced animal	*In vivo*	Effect on lipid peroxidation (LPO) in stress-induced animals. Elevation of LPO was observed in rabbits and mice after intravenous administration of 0.2 microg/kg of lipo-polysaccharide (LPS: from Klebsiella pneumoniae) and 100 microg/kg of peptidoglycan (PGN: from *Staphylococcus aureus*), respectively. The peak was reached immediately after PGN and 2-6 h after LPS administration.	Aqueous suspension of root extract of ashwagandha (100 mg/kg)	Oral administration of Ashwagandha prevented the rise in LPO in rabbits and mice.	This suggests the anti-stress activity.	Ayant N Dhuley, Journal of Ethnopharmacology Year : 1998 Volume : 60 Issue : 02 Page : 173-178.

Contd...

TABLE 14.2–*Contd....*

Sl.No.	*Title*	*Type of Study*	*Study Design*	*Form and Dose*	*Summarized Efficacy*	*Conclusion*	*Reference*
3.	Antistressor effect of *Withania somnifera*	*In vivo*	Cold water swimming stress test in Wistar strain albino rats		The results indicate that the drug treated animals show better stress tolerance.	*Withania somnifera* possesses potent anti-stress properties.	R Archana *et al.*, Journal of ethno-pharmacology, Vol. 64, Issue 1, 1 January 1998, Pages 91–93.
					Anxiolytic activity		
4.	A double blid, placebo control evaluation of the anxiolytic efficacy of an ethanolic extract of *Withania somnifera*	Clinical	Double-blind, placebo-controlled study	Ethanolic extract of Aswagandha		The drug was well-tolerated and has useful anxiolytic potential.	Chitranjan Andrade *et al.*, Indian Journal of Psychiatry Year : 2000 Volume : 42 Issue : 3 Page : 295-301.
5	Mechanisms of cardio-protective effect of *Withania somnifera* in experimentally induced myocardial infarction.	*In vivo*	Isoprenaline-induced myonecrosis in rats. Haemodynamic para-meters were recorded and the hearts were processed for histo-pathological and biochemical studies.	Hydro-alcoholic extract *Withania somnifera* (25, 50 and 100 mg/kg) at a dose of 100 mg/kg, orally for 4 weeks	*Withania somnifera* and vitamin E treatment significantly prevented myonecrosis as indicated by significant reduction in the infiltration of inflammatory cells, vacuolar changes as well as oedema as compared to the isoprenaline control group. A marked increase in the glutathione contents, inhibition of lipid peroxi-dation and preserved membrane integrity; and restoration in lactate dehydrogenase and creatinine phosphokinase enzymes was observed.	*Withania somnifera* exerts cardioprotective effect.	Mohanty *et al.*, Basic Clin Pharmacol Toxicol. 2004 Apr; 94(4): 184-90.
6.	A study on evalution of antidepressant effect of imipramine adjunct with Aswagandha and Bramhi	*In vivo*	Learned helplessness test (LHT) and swimming in Forced swimming test (FST) in animal model	Ashwagandh powder 50, 100,150 mg/kg	Ashwagandha showed highly significant ($p<0.01$) result on individual use. But combination of Bramhi and Ashwagandha in low doses with low dose of imipramine gave a highly significant result ($p<0.01$)	Ashwagandha had significant antidepressant action.	T Maity *et al.*, Nepal Med Coll J 2011; 13(4): 250-253.

Contd...

TABLE 14.2–*Contd....*

Sl.No.	Title	Type of Study	Study Design	Form and Dose	Summarized Efficacy	Conclusion	Reference
7.	Anxiolytic-antidepressant activity of *Withania somnifera* glycowith anolides: an experimental study	*In vivo*	Forced swim-induced 'behavioural despair' and 'Learned helplessness' tests.	20 and 50 mg/kg, Glycowith-anolides (WSG), isolated from WS roots, orally once daily for 5 days	WSG induced an anxiolytic effect, comparable to that produced by lorazepam, in the elevated plus-maze, social interaction and feeding latency in an unfamiliar environment, tests. WSG also exhibited an antidepressant effect, comparable with that induced by imipramine, in the forced swim-induced 'behavioural despair' and 'learned helplessness' tests.	*Withania somnifera* possesses potent Anxiolytic-anti-depressant activity.	S.K. Bhattacharya *et al.*, Phyto-medicine Vol. 7, Issue 6, December 2000, Pages 463–469.
8.	A prospective, rando-mized double-blind, placebo-controlled study of safety and efficacy of a high-concentration full-spectrum extract of *Ashwagandha* root in reducing stress and anxiety in adults	Clinical	Single center, prospective, double-blind, randomized, placebo-controlled trial; Measurement of serum cortisol, and assessing their scores on standard stress-assessment questionnaires.; Sample: 64 subjects with a history of chronic stress, age between 18 and 54 years	300 mg of high-concen-tration extract from the root of the *Ashwagandha;* Dosage: one capsule twice a day for a period of 60 days.	A significant reduction in scores on all the stress-assessment scales (Perceived stress scale) on Day 60, relative to the placebo group was noted. The serum cortisol levels were substantially reduced in the Ashwagandha group, relative to the placebo group. No serious adverse events were reported.	Ashwagandha safely and effectively improves an individual's resistance towards stress and thereby improves self-assessed quality of life.	Indian J Psychol Med. 2012 Jul-Sep; 34(3): 255–262.; K. Chandra-sekhar, Jyoti Kapoor and Sridhar Anishetty.
9.	Effects of *Withania somnifera* (Ashwagandha) and *Terminalia arjuna* (Arjuna) on physical performance and cardio-respiratory endurance in healthy young adults	Clinical	Randomized controlled, parallel group, single blinded study. Maximum velocity, average absolute and relative Power and balance - by Kinematic Measuring System (KMS) 20-second wobble board test (Kinematic); maxi-mum oxygen consumption (VO2 max) by Computer controlled Vista Turbo Trainer machine and blood pressure in humans were measured. Sample: Forty healthy individuals of either sex aged between 18 to 25 years years.	500mg Capsule, one capsule per day for 8 weeks	After treatment with *Withania somnifera*, maximum oxygen consumption increased significantly from 13.54±2.46 to 14.47±2.28 (*P*=0.005). Similarly, the maximum velocity increased from 5.37±0.75 to 5.53±0.70 (*P*=0.005), the average absolute power from 711.90±221.62 to 774.79±247.42 (*P*=0.002) and average relative power from 11.10±3.17 to 12.22±3.40 (*P*=0.007). However, there was no significant improvement in balance and blood pressure.	*Withania somnifera* may prove useful for generalized weakness and to improve speed and lower limb muscular strength and neuro-muscular co-ordination. It is safe for young adults when given for mentioned dosage and duration.	Int J Ayurveda Res. 2010 Jul-Sep; 1(3): 144–149; Jaspal Singh Sandhu, Biren Shah *et al.*

Contd...

TABLE 14.2–*Contd....*

Sl.No.	*Title*	*Type of Study*	*Study Design*	*Form and Dose*	*Summarized Efficacy*	*Conclusion*	*Reference*
10.	Effects of eight-week supplementation of *Ashwagandha* on cardio-respiratory endurance in elite Indian cyclists	Clinical	Double blind placebo controlled randomized; Treadmill test -to measure maximal aerobic capacity (VO_2 max), metabolic equivalent, respiratory exchange ratio (RER), and total time for the athlete to reach his exhaustion stage; Sample: Forty, Age 18- 27 years.	500 mg capsules of aqueous roots of *Ashwagandha* twice daily for eight weeks	There was significant improvement in the experimental group in VO2 max ($t = 5.356$; $P < 0.001$), METS ($t = 4.483$; $P < 0.001$), and time for exhaustion on treadmill ($t = 4.813$; $P < 0.001$) in comparison to the placebo group which did not show any change with respect to their baseline parameters.	Ashwgandha help in improvement in aerobic performance with regard to cardiorespiratory and cardiovascular endurance.	J Ayurveda Integr Med. 2012 Oct-Dec; 3(4): 209–214.; Shweta Shenoy, Udesh Chaskar, *et al.*
11.	Exploratory study to evaluate tolerability, safety, and activity of *Ashwagandha* (*Withania somnifera*) in healthy volunteers	Clinical	Prospective, open-labeled, variable doses in volunteers; Sample: Eighteen healthy volunteers, age:18-30 years.; Muscle activity was measured by hand grip strength, quadriceps strength, and back extensor force. Exercise tolerance was determined using cycle ergometry. Lean body weight and fat per cent were computed from skin fold thickness measurement. Adverse events were recorded, as volunteered by the subjects.	WS capsules (aqueous extract, 8:1) daily in two divided doses with increase in daily dosage every 10 days for 30 days (750 mg/ day × 10 days, 1000 mg/ day×10 days, 1250 mg/ day × 10 days)	In six subjects, improvement in quality of sleep was found. Reduction in total- and LDL- cholesterol and increase of strength in muscle activity was significant. Total body fat percentage showed a reduction trend. WS, in escalated dose, was tolerated well. The formulation appeared safe and strengthened muscle activity.	WS showed muscle strengthening, lipid lowering, and improved the quality of sleep. It is well tolerated.	Ashwinikumar A. Raut, Nirmala N. Rege *et al.*; J Ayurveda Integr Med. 2012 Jul-Sep; 3(3): 111–114.
					Anti-depressant activity		
12.	Evaluation of anti-depressant activity of aqueous extract of *Withania somnifera* [Aswagandha] roots in albino mice	*In vivo*	Forced swim test (FST) in albino mice.	Queous extract of *Withania somnifera* 30,40,50 mg/kg	Dose dependent decrease in immobility time in FST, maximum effect being observed with WS 50 mg/kg.	WS is effective in depressive disorders.	Dr P. Bharathi *et al.*, IOSR Volume 10, Issue 1 Ver. IV (Jan -Feb. 2015), pp. 27-29.

Contd...

TABLE 14.2–*Contd....*

Sl.No.	Title	Type of Study	Study Design	Form and Dose	Summarized Efficacy	Conclusion	Reference
					Psychomotor performance		
13.	Effect of standardized aqueous extract of *Withania somnifera* on tests of cognitive and psychomotor performance in healthy human participants	Clinical	Study Design - Double blind, multidose, placebo-controlled, crossover study. Sample size - 20 healthy volunteers Study Duration - 14 days	*Withania somnifera* extract 250 mg/Cap. Two capsules twice daily	Significant improvements were observed in reaction times with simple reaction, choice discrimination, digit symbol substitution, digit vigilance, and card sorting tests with *Withania somnifera* extract compared to placebo. However, no effect can be seen with the finger tapping test.	*Withania somnifera* extract can improve cognitive and psycho-motor performance.	Usharani Pingali *et al.*, Pharmacognosy Res. 2014 Jan-Mar; 6(1): 12–18.
					Anti-osteoporosis Activity (Preclinical)		
14.	The effects of dietary 1, 25-dihydroxycholecalciferol and hydro-alcoholic extract of *Withania somnifera* root on bone mineralisation, strength and histological characteristics in broiler chickens	*In vivo*	Species:- Broiler chickens; N=308; Group +ve (with adequate calcium); -ve (with 30 per cent reduced calcium) 3 groups with WS (0, 75, 150 mg/kg) each repeated for 5 times. 2 concentrations of 1, 25 (OH) 2 D3 (0 and 0.5 μg/kg diet) for 42 days	75-150 mg/kg for 21 days	On 21 days tibia calcium is high in WS supplemented. Ca is significantly high in WS and 2, 25 (OH) 2 D3. WS supplementation improved Ca and P retention, bone calcification and mechanical properties with no adverse effects on performance.	WS has calcium retntion and bone calcification property	M.T. Mirakzehia, *et al.*, British Poultry Science. 2013; 54, (6).
15.	*Withania somnifera* improves bone calcification in calcium-deficient ovariectomized rats.	*In vivo*	Species:-Sprague Dawley rats N=36 Group: 2 SHAM operated (n = 12) or ovariectomized (n = 12) and treated with WS/vehicle (65 mg/kg), orally for 16 weeks (n = 12). Urinary excretion of calcium (Ca) and phosphorus (P) and serum levels of Ca, P and alkaline phosphatase (ALP) were measured.	WS extract 65mg/kg for 16 weeks	Ovariectomized (OVX) rats showed a significant increase in serum ALP levels and urinary Ca and P excretion. Ash analysis showed a reduction in ash weight, percent ash, ash Ca, ash P and ash magnesium levels in the OVX group. WS treatment markedly prevented the calcium and P reduction changes in OVX rats.	WS extract is useful for treatment of osteoporosis.	Nagareddy PR, Lakshmana M, J Pharm Pharmacol. 2006; 58(4): 513-519.

Contd...

TABLE 14.2–*Contd....*

Sl.No.	*Title*	*Type of Study*	*Study Design*	*Form and Dose*	*Summarized Efficacy*	*Conclusion*	*Reference*
					BALA (*Sida cordifolia* Linn.)		
1.	Evaluation of Anti-inflammatory activities of *Sida cordifolia* Linn. in Albino rats	*In vivo*	Ethanolic extract of *Sida cordifolia* Linn for acute and sub-acute anti-inflammatory property in albino rats and compared with the referencedrug indo-methacin.	100 mg/kg and 200 mg/kg	Ethanolic extract of *Sida cordifolia* Linn showed statistically significant acute and sub- acute anti-inflammatory effects matching to that of control group at doses of 100 mg/kg and 200 mg/kg ($p<0.05$). It was found that percentage reduction in the pawoedema was 58.13 per cent with Indomethacin, 48.83 per cent and 53.48 per cent with 100 mg/kg and 200 mg/kg *Sida cordifolia* Linn respectively. Percentage reduction in the granuloma formation was 60.2 per cent with Indomethacin, 54.7 per cent and 56 per cent with 100 mg/kg and 200 mg/kg *Sida cordifolia* Linn respectively.	*Sida cordifolia* Linn could be a source of valuable anti-inflammatory drugs.	Praveen Panchakshari-math *et al.*, J. Chem. Pharm. Res., 2011, 3(6): 136-142.
2.	Analgesic and anti-inflammatory activities of Analgesic and anti-inflammatory activities of *Sida cordifolia* Linn.	*In vivo*	The objective of the current study is to evaluate the analgesic and anti-inflammatory activities of different extracts of *Sida cordifolia* Linn.	100 and 200 mg/kg	The inhibition of the writhing reflex in mice by theplant extracts (p.o. at a dose of 100 and 200 mg/kg, body weight) were compared against the standard analgesic, aminopyrine 50 mg/kg, p.o. The analgesic activity was assessed by calculating the number of writhing reflexes for 10 min, occurring immediately after 0.1 ml/ 10 g of intraperitoneal acetic acid (0.7 per cent). The results were analyzed for statistical significance using one-way ANOVA followed by Dunnett's test. $P<0.05$ was considered significant	Results show that the extracts exhibited sufficient inhibition of paw edema.	R.K. Sutradhar *et al.*, Indian J Pharmacol June 2006 Vol 38 Issue 3 207-08.
3.	Anti-inflammatory, analgesic activity and acute toxicity of *Sida cordifolia* L. (Malva-branca)	*In vivo*	The anti-inflammatory, analgesic effects and acute toxicity of an aqueous extract of *S. cordifolia* were evaluated in animal models.	400 mg/kg administered orally	The aqueous extract (AE) showed a significant inhibition of carrageenin-induced rat paw edema. The AE also increased the latency period for mice in the hot plate test, and inhibited the number of writhes produced by acetic acid	*Sida cardifolia* having Anti-inflammatory, analgesic activity properties and low acute toxicity in mice.	Franzotti EM, J Ethno-pharmacol. 2000 Sep; 72(1-2): 273-7.
4.	Analgesic, antiinflam-matory and hypogly-caemic activities of *Sida cordifolia*	*In vivo*	To assess the analgesic, antiinflammatory and hypoglycaemic activities of *Sida cordifolia.*	600 mg/kg orally	The ethyl acetate extract of root (SCR-E) showed comparable antiinflammatory activity with indomethacin and possessed significantly higher activity when compared with that of the methanol extract of the root part (SCR-M). The ethyl acetate extract of both root and aerial parts of *Sida cordifolia* (SCR-E and SCA-E) showed very good central and peripheral analgesic activities. The methanol extract of root (SCR-M) was found to possess significant hypoglycaemic activity.	*Sida cardifolia* possessed Analgesic, antiinflam-matory and hypogly-caemic activities.	Kanth VR *et al.*, Phytother Res. 1999 Feb; 13(1): 75-7.

Contd...

TABLE 14.2–*Contd....*

Sl.No.	Title	Type of Study	Study Design	Form and Dose	Summarized Efficacy	Conclusion	Reference
5.	Evaluation of the anti-oxidant activity of *Sida cordifolia*		To evaluate the anti-oxidant potential of ethanol extracts of *Sida cordifolia* leaf, stem, root, and whole plant compared with standard antioxidants such as BHA, α-tocopherol, and ascorbic acid.	1 mg	Ethanol extracts were found to be a good scavenger of DPPH radical in the order roots > stem > leaves > whole plant with values 76.62 per cent, 63.87 per cent, 58 per cent and 29 per cent at a dose of 1 mg, respectively All extracts of *Sida cordifolia.* (SC) have effective reducing power and free-radical scavenging activity.	*S. cordifolia* is a potential source of natural anti-oxidants.	K. Dhalwal, *et al.* Pharmaceutical Biology, 2005, Vol. 43, No. 9: Pages 754-761.
6.	Free radical scavenging capacity, anticandicidal effect of bioactive compounds from *Sida cordifolia* L., in combination with nystatin and clotrimazole and their effect on specific immune response in rats		To evaluate the possible beneficial antioxidant, anticandicidal and immunostimulatory potencies of the total alkaloid fractions from *Sida cordifolia* L. (Malvaceae)		The results obtained in this **antifungal activity** were interesting and indicated a synergistic effect between alkaloid compounds and the antifungal references such as Nystatin and Clotrimazole. **Antioxidant capacity** noticed that the reduction capacity of DPPH radicals obtained the best result comparatively to the others methods of free radical scavenging. Our results showed a low **immuno stimulatory** effect.	Study showed that alkaloid compounds in combination with antifungal references (Nystatin and Clotrimazole) exhibited antimicrobial effects against candida strains tested.	Maurice Ouédraogo *et al.*, Ouédraogo *et al.*, Annals of Clinical Microbiology and Antimicrobials 2012, 11: 33.
7.	Effect of the aqueous extract of *Sida cordifolia* on liver regeneration afterpartial hepatectomy	*In vivo*	Liver regeneration evaluated by immuno-histochemical staining for proliferating cell nuclear antigen (PCNA) using monoclonal primary anti-PCNA antibody (PC-10; DAKO A/S, Glostrup, Denmark) on formalin-fixed and paraffin-embedded liver tissues.	Aqueous extract of *Sida cordifolia* 100, 200 and 400 mg/kg bw.	All animals were randomly assigned to 4 groups, which consisted of 5 rats each. Control group animals were submitted to oral administration of distilled water at the time of surgery. Rats of Sida100, Sida200 and Sida400 groups underwent partial hepatectomy and oral administration of aqueous extract of *Sida cordifolia* were given. Sida100 and Sida200 groups disclosed higher liver regeneration indices than control group ($p<0.001$ and $p<0.05$, respectively). it stimulates liver regeneration after 67 per cent partial hepatectomy in rats.	*Sida cordifolia* is effective liver regeneration.	Renata Lemos Silva *et al.* Acta Cirúrgica Brasileira - Vol 21 (Suplemento 1) 2006 - 37.
8.	Antistress, adoptogenic activity of *Sida cordifolia* roots in mice	*In vivo*	Ethanol extract of roots of *Sida cordifolia* was evaluated for antistress, adaptogenic activity using cold restraint stress and swim endurance in mice.	40 per cent ethanolic roots extract of *Sida cordifolia* was dissolved in distilled water	Mice pretreated with ethanolic roots extract of *Sida cordifolia* show significant improvement in the swimming time. Cold stress typically increases total leukocyte count, eosinophils and basophils. The mode of action of adaptogens is basically associated with stress system. Adaptogen increase the capacity of stress to respond to the external signals of activating and deactivating mediators of stress response subsequently. The stress induced increase in total WBC count is decreased by roots extract	Mice pretreated with extract of *Sida cordifolia* showed significant improvement in the swim duration and reduced the elevated WBC, blood glucose and plasma cortisone.	Sumanth M. and Mustafa SS, Indian J Pharm Sci. 2009; 71(3): 323–324.

Contd...

TABLE 14.2–*Contd....*

Sl.No.	*Title*	*Type of Study*	*Study Design*	*Form and Dose*	*Summarized Efficacy*	*Conclusion*	*Reference*
				(10 mg/ml) for oral administration to animals. Administered dose of *Sida cordifolia* was 100 mg/kg orally, using oral gavage, for 7 days	of *Sida cordifolia*, indicating antistress, adaptogenic activity. *Sida cordifolia* reduced plasma cortisol level as well as blood glucose level, exhibiting antistress activity. Hence, it can be concluded that *Sida Cordifolia* roots possess antistress, and adaptogenic activity, hence can be categorized as plant adaptogen.		
9.	Evaluation of anti-arthritic activity of ethanolic extract of *Sida Cardifolia*	*In vivo*	50 per cent ethanolic extract of *Sida cardifolia* was tested on arthritic rats. Arthritis was induced by Complete Freund's Adjuvant method.	250 and 500mg/kg b.w doses of ethanolic extract of *Sida cardifolia* were selected, suspended in 1 per cent CMC solution and administered orally for 12 days	The high dose of ethanolic extract of *Sida cardifolia* exhibited a significant anti-arthritic activity by reducing serum biochemical parameters like ALP, SGOT, SGPT levels and reduced the haematological parameters like ESR and WBC and increases the RBC and Hb levels in FA induced arthritis models in rats. Radiological and histological studies revealed that near a normal structure of paw and knee joint respectively with the high dose of the extract in FA induced arthritis	*Sida cardifolia* extract showed potent anti-antioxidant and anti-inflammatory activity when compared with standard drug deprenyl.	Polireddy DM, International Journal of Scientific and Technology Research 2015; 4(11): 86-96.
10.	Cartilage protective effect of *Sida cordifolia* L. and *Piper longum* L. is through modulation of MMPs and TIMP	*In vivo*	Aqueous extracts of *S. cordifolia* roots and *P. longum* inflorescence were evaluated against DPPH radical scavenging activity, lipid peroxidation method, collagenase type II induced osteoarthritis model in rats.	Powders of herbs were administered orally at the dose of 270 mg/kg in the form of suspension prepared in water. Standard drug indomethacin was given at the dose of 3 mg/kg b. wt.	Results showed presence of significant amount of phenolics in *S. cordifolia* and *P. longum*, which might have been responsible for their anti-oxidant activity. Treatment with *S. cordifolia*, *P. longum* and indomethacin up-regulated SOD, CAT, GPx and PON-1 expression in synovium of osteoarthritic knee joint thus protecting tissue from deleterious effects of free radicals. *S. cordifolia* administration showed upregulation of TIMP with corresponding reduction in MMP-3 and -9, an ideal situation to restore MMP-TIMP balance. This activity of the extracts may be potentiated given to their ability to inhibit cyclooxygenase (COX) leading to the inhibition of prostaglandin (PGE) synthesis thereby preventing cartilage degeneration	Oral administration of *S. cordifolia* and *P. longum* aqueous suspension may have therapeutic effects in osteoarthritis whereas cartilage protection was also achieved through controlling MMPs/TIMP and anti-oxidant levels in the synovium.	Nirmal PS, *et al.*, International Journal of Advanced Research 2015; 3(11): 480-488.

Contd...

TABLE 14.2–*Contd....*

Sl.No.	Title	Type of Study	Study Design	Form and Dose	Summarized Efficacy	Conclusion	Reference
					CHIRAYATA (*Boswellia serrata* Triana and Planch.)		
					Anti-Arthritic activity		
1.	Clinical evaluation of *Boswellia serrata* (shallaki) resin in the management of sandhivata (osteoarthritis)	Clinical	Open, randomised, n-56 in two group.	500 mg capsule of Shallaki, 6 g per day (in three divided doses) with luke-warm water	In the present study, 56 patients fulfilling the diagnostic criteria of Sandhigata vata, divided into two groups. Patients of first group were administered with 500 mg capsule of Shallaki, 6 g per day (in three divided doses) with lukewarm water (n=29) and the second group) capsule Shallaki as above along with local application of Shallaki ointment on the affected joints (n=23).	After a course of therapy for 2 months, symptomatic improvement was observed in both the groups at various levels with promising results in the patients of first group.	PK Gupta *et al.,* AYU. 2011; 32(4): 478-482.
2.	A 32-week randomized, placebo-controlled clinical evaluation of ra-11, an ayurvedic drug, on osteoarthritis of the knees	Clinical	Randomized, placebo-controlled clinical, n- 358		A total of 358 patients with chronic knee pain were screened free-of-cost in "arthritis camps" in an Indian metropolis. Ninety patients with primary OA of the knees were found eligible to enroll into a randomized, double-blind, placebo-controlled, parallel efficacy, single-center, 32-week drug trial Concurrent analgesics/nonsteroidal antiinflammatory drugs and steroids in any form were not allowed. The WOMAC section on "physical function difficulty" was modified for Indian use and validated before the trial. Routine laboratory testing was primarily done to monitor drug safety. At baseline, the groups (active = 45, placebo = 45) were well matched for several measures. Efficacy: Compared with placebo, the mean reduction in pain VAS at week 16 (active = 2.7, placebo = 1.3) and week 32 (active = 2.8, placebo = 1.8) in the active group was significantly ($P < 0.05$, analysis of variance [ANOVA]) better. Similarly, the improvement in the WOMAC scores at week 16 and week 32 were also significantly superior ($P < 0.01$, ANOVA) in the active group. 2)	This controlled drug trial demonstrates the potential efficacy and safety of RA- 11 in the symptomatic treatment of OA knees over 32 weeks of therapy.	Chopra A, *et al.,* Journal of Clinical Rheumatology. 2004; 10(5): 236-245.
3.	*Boswellia serrata* extract attenuates inflammatory mediators and oxidative stress in collagen induced arthritis	*In vivo*	To evaluate the anti-oxidant and antiarthritic activity of *Boswellia serrata* gum resin extract (BSE) in collagen induced arthritis.	100 and 200 mg/kg body weight once daily for 21 days.	Arthritis was induced in male Wistar rats by collagen induced arthritis (CIA) method. BSE wasadministered at doses of 100 and 200 mg/kg body weight once daily for 21 days. The effects of treatment in the rats were assessed by biochemical (articular elastase, MPO, LPO, GSH, catalase, SOD and NO), inflammatory mediators (IL-1, IL-6, TNF-, IL-10, IFN- and PGE2), and histological studies in joints. BSE was effective in bringing significant changes on all the parameters (articular elastase, MPO, LPO, GSH, catalase, SOD and NO) studied. Oral administration of BSE resulted in significantly reduced levels of inflammatory mediators (IL-1, IL-6, TNF-, IFN- and PGE2), and increased level of IL-10. The protectiveeffects of BSE against RA were also evident from the decrease in arthritis scoring and bone histology.	The protective effect of *Boswellia serrata* extract on arthritis in rats might be mediated via the modulation of immune system.	Umar S, *et al.,* Phytomedicine. 2014; 21(6): 847-56.

Contd...

TABLE 14.2–*Contd....*

Sl.No.	*Title*	*Type of Study*	*Study Design*	*Form and Dose*	*Summarized Efficacy*	*Conclusion*	*Reference*
4.	Evaluation of anti-inflammatory activity of *Boswellia serrata* on carrageenan induced paw edema in albino Wistar rats	*In vivo*	To evaluate the anti-inflammatory activity of *B. serrata* against carrageenan induced paw edema.	50, 100, and 200 mg/kg/bw	Treatment of *B. serrata* inhibited the edema and decreases the cellular infiltrates probably by inhibiting the inflammatory mediators, and found to be greater at higher concentration *i.e.,* 200 mg/kg/bw as compared to standard drug (Indomethacin 10 mg/kg/bw).	*B. serrata* has high anti-inflammatory activity and supports its usage in traditional medicine as herbal anti-inflammatory medicine	Ismail SM *et al.,* Int J Res Med Sci. 2016; 4(7): 2980-2986.
5.	Anti-arthritic activity of *B. serrata* in Freund's Adjuvant induced arthritis in rats	*In vivo*	Methanolic extract of *Boswellia serrata* was evaluated for anti-arthritic action using Freund's complete adjuvant (FCA) induced arthritis in rats	100mg/kg and 200 mg/kg p.o for 21 days	Significant inhibition of paw edema volume was observed from day 5th to 21st in the groups treated with *B. serrata*. Significant increase in all the inflammatory markers was found in the rats treated by FCA. At dose of 100mg/kg and 200 mg/kg *B. serrata* display noticeable decrease in all inflammatory markers. A better activity of *Boswellia serrata* was found at 200 mg/kg dose against FCA induced arthritis in rats.	*Boswellia serrata* has significant anti-arthritic activity against FCA induced arthritis in rats.	Ziyaurrahman AR *et al.,* International Journal of Advances in Pharmacy Medicine and Bioallied Sciences. 2015; 3 (1): 29-31.
					Antidiabetic and Hypolipidemic Activity		
6.	Effect of *Boswellia serrata* supplementation on blood lipid, hepatic enzymes and fructosamine levels in type 2 diabetic patients	Clinical	To investigate the anti-diabetic, hypolipidemic and hepatoprotective effects of supplementation of *Boswellia serrata* in type 2 diabetic patients. Blood samples were taken at the beginning of the study and after 6 weeks. Blood levels of fructosamine, lipid profiles as well as hepatic enzyme in type 2 diabetic patients were measured.	*Boswellia serrata* gum resin in amount of 900 mg daily for 6 weeks were orally administered (as three 300 mg doses) in intervention group and the control group did not receive anything.	Treatment of diabetic patient with *Boswellia serrata* was caused to significant increase in blood HDL levels as well as a remarkable decrease in cholesterol, LDL, fructosamine SGPT and SGOT levels after 6 weeks.	*Boswellia serrata* supplementation can be beneficial in controlling blood lipid parameters in patients with type 2 diabetes.	Ahangarpour A *et al.,* J Diabetes Metab Disord. 2014; 13: 29.

Contd...

TABLE 14.2–*Contd....*

Sl.No.	Title	Type of Study	Study Design	Form and Dose	Summarized Efficacy	Conclusion	Reference
7.	The antioxidant capacity and anti-diabetic effect of *Boswellia serrata* Triana and planch aqueous extract in fertile female diabetic rats and the possible effects on reproduction and histological changes in the liver and kidneys	*In vivo*	The effects of *Boswellia serrata* aqueous extract on blood glucose and the complications of diabetes in the liver and kidneys and examined the impact of plant on reproduction in diabetic rats.	200, 400, and 600 mg/kg *Boswellia serrata* extract for 17 days	The Administration of *Boswellia serrata* in diabetic rats significantly decreased the level of blood glucose and HbA1c after 17th days. In the diabetic group treated with *Boswellia serrata*, separated necrosis of hepatocytes, anarchism of liver plates, and lymphocytic inflammation were improved. Diabetic complications were not seen and the severity of damage was reduced.	*Boswellia serrata* extract has the antidiabetic effects and can prevent the complications of diabetes in the kidneys and liver.	Azemi ME *et al.*, Jundishapur J Nat Pharm Prod. 2012 Autumn; 7(4): 168–175.
					Diuretic Activity		
8.	Evaluation of diuretic activity of *Boswellia serrata* leaf extracts in albino mice	*In vivo*	To evaluate the diuretic activity of petroleum ether, chloroform, ethanol and aqueous leaf extracts *of Boswellia serrata* in albino mice. The volume of urine and excretion of electrolytes in the urine was measured to assess the diuretic activity.	Petroleum ether extract [250 mg/kg p. o] chloroform extract [200 mg/kg p. o] ethanol extract [250 mg/kg p. o] aqueous extract [500 mg/kg p. o] Furosemide [10mg/kg p. o]	Petroleum ether extract showed excretion of Na^+ and K^+ while Chloroform extract showed excretion of Cl^- ions. Ethanolic group portrayed low excretion of both electrolytes and urine volume. In the petroleum ether extract and chloroform extract, BUN levels were significantly more when compared to other groups. Petroleum ether extract substantiated a near to equal value in comparison with the standard drug [furosemide].	The leaves of *Boswellia serrata* showed diuretic activity with no toxic effects on the kidneys.	Nayeem M and Quadri MFA, Int J Pharm Sci., 2015; 7(2): 502-505
					Immunomodulatory Activity		
9.	Immunomodulatory activity of boswellic acids of *Boswellia serrata* Roxb.	*In vivo*	Extract of gum resin of *B. serrata* containing 60 per cent acetyl 11-keto beta boswellic acid (AKBA) along with other constituents such as 11-keto beta-boswellic acid (KBA), acetyl beta-boswellic acid and beta-boswellic acid has been evaluated for antiana-	Three graded oral doses (20, 40 and 80 mg/kg, po) of *B. serrata* extract was used for this study.	The extract inhibited the passive paw anaphylaxis reaction in rats in dose dependant manner. A significant inhibition in the compound 48/80 induced degranulation of mast cells in dose-dependant manner was observed thus showing mast cell stabilizing activity.	The results suggest that *B. serrata* extract have promising antianaphylactic and mast cell stabilizing activity.	Pungle P *et al.*, Indian J Exp Biol. 2003; 41(12): 1450-2.

Contd...

TABLE 14.2–*Contd....*

Sl.No.	*Title*	*Type of Study*	*Study Design*	*Form and Dose*	*Summarized Efficacy*	*Conclusion*	*Reference*
			phylactic and mast cell stabilizing activity using passive paw anaphylaxis and compound 48/80 induced degranulation of mast cell methods in rats				
10.	Immunomodulatory activity of boswellic acids (Pentacyclic triterpene acids) from *Boswellia serrata*	*In vivo*	Boswellic acids, a mixture of pentacyclic triterpene acids (BA) obtained from *Boswellia serrata* Roxb., have been investigated for their effect on cell mediated and humoral components of the immune system and the immunotoxicological potential in mice	Three graded oral doses (25, 50 and 100 mg/kg, po) of *B. serrata* extract was used for this study.	A single oral administration of *B. serrata* extract inhibited the expression of the 24h delayed type hypersensitivity (DTH) reaction and primary humoral response to SRBC in mice. In a multiple oral dose schedule BA reduced the development of the 24h DTH reaction and complement fixing antibody titres and slightly enhanced the humoral antibody synthesis. In concentrations greater than 3.9μg/mL BA inhibited proliferative responsiveness of splenocytes to mitogens and alloantigen. Preincubation of macrophages with different concentrations of BA enhanced the phagocytic function of adherent macrophages. Prolonged oral administration of BA (21 days) increased the body weight, total leukocyte counts and humoral antibody titres in rats.	The results suggest that *B. serrata* extract have promising immunomodulatory activity.	Sharma ML, *et al.*, Phytotherapy Research. 1996; 10(2): 107–112.
					Antiulcer Activity		
11.	Evaluation of antiulcer activity of *Boswellia serrata* bark extracts using aspirin induced ulcer model in albino rats	*In vivo*	The effect of bark extracts of *Boswellia serrata* was evaluated in aspirin induced ulceration (200mg/kg) in albino rats. Antiulcer activity was evaluated by measuring ulcer index and percentage of ulcer healing.	Petroleum ether (250 mg/kg) and aqueous extracts (250mg/kg) of bark of *Boswellia serrata* was administered through oral route	Pretreatment with petroleum ether and aqueous extracts of bark of *Boswellia serrata* significantly inhibits ulceration induced by high dose of Aspirin. Histopathological findings also confirm the antiulcer activity of *Boswellia serrata* bark extracts in albino rats.	*Boswellia serrata* having a potential to inhibits gastric ulceration and could be a safe alternative to conventional antiulcer treatment.	Zeeyauddin K. *et al.*, J Med Allied Sci. 2011; 1(1): 14-20.
					CHITRAK (*Plumbo zylenica* Linn).		
1.	*Plumbago zeylanica* L. (chitrak) a gastro-intestinal flora normalizer	*In vitro*	To assess how Chitrak roots could act as digestive stimulant they were tested for the		It has been established that the feeding of Chitrak root stimulates the proliferation of coliform bacteria in mice.The data has been statistically analysed and the increase in coliform bacteria due to Chitrak feeding has been shown to be significant at 5 per cent	Result of the study indicates that the claims of the ancients that Chitrak is 'digestive	M.A. Iyengar1, G.S. Pendse *et al.*, Planta Med 1966;

Contd...

TABLE 14.2–*Contd....*

Sl.No.	*Title*	*Type of Study*	*Study Design*	*Form and Dose*	*Summarized Efficacy*	*Conclusion*	*Reference*
			digestive enzymes and vitamins.		level. This coliform proliferation action of Chitrak has been shown to be similar to that of 'Mexaform' a CIBA drug used to counteract the bacterial action of antibiotics.	stimulant' and an appetiser' is most 'probably due to its action as an Intestinal Flora Normalizer in the animal body.	14(3): 337-351.
					Immunosuppressive Properties		
2.	Immunosuppressive properties of aqueous extract of *Plumbago zeylanica* in Balb/c mice	*In vivo*	To identification of potential immuno-suppressive property by testing the aqueous extracts of the plant parts by inducing ovalbumin (OVA) specific IgG antibody responses in a murine system.		The aqueous root extract of *Plumbago zeylanica* (PZE) exhibited the significant suppression of OVA-specificIgG antibody response determined by enzyme-linked immunosorbent assay (ELISA). PZE also suppressed the anti-OVA antibody response in dose dependent manner. PZE is potent in exerting the suppressive effect on the down regulation of anti-OVA antibody and T cell responses and in all the three haplotypes of the mice.	*P. zeylanica* extract suppressed T cell responses in Balb/c mice.	Aparanji Poosarla *et al.* Journal of Medicinal Plants Research Vol. 4(20).
					Anti-inflammatory Activity		
3.	Bioassay-guided isolation of anti-inflammatory and antinociceptive compound from *Plumbago zeylanica* leaf	*In vivo*	To validate the anti-inflammatory and anti-nociceptive activities of various leaf extracts by using *in vivo* experi-mental models	200 and 400 mg/kg, p.o.	The acetone extract significantly ($p < 0.01$) reduced inflammation in the rats when compared to the control group. As for the analgesia effect, the acetone and petroleum ether extracts significantly ($p < 0.01$) decreased the pain stimulus.	The acetone extract of *P. zeylanica* having anti-inflammatory and analgesic activites.	Sheeja E *et al.,* Pharm Biol. 2010 Apr; 48(4): 381-7.
4.	Anti-inflammatory and cytotoxic effects of extract from *Plumbago zeylanica*	*In vivo*	The root of *P. zeylanica* extracted with methanol was used for determining the anti-inflammatory effects by using Carrageenin induced raw paw edema model.	Methanolic Roots extracts of *P. zeylanica* at doses 300 and 500 mg/kg was adminis-tered orally	The methanolic extracts at 300 and 500 mg/kg produced 31.03 and 60.3 per cent inhibition of acute inflammation, respectively, confirming that *P. zeylanica* roots are effective against acute inflammation. For the evaluation of cytotoxicity, the crude dichloromethane extract was subjected to silica gel column chro-matography and 120 fractions were collected. Their structures were elucidated with the help of spectroscopic techniques. High performance liquid chromatography (HPLC) was performed to determine the purity of gugultetrol-18-ferrulate in crude extract and the structure of betasitosterol and gugultetrol-18-ferrulate was identified using nuclear magnetic resonance spectroscopy analysis (1H and 13C NMR), Infra-red and mass spectroscopy. The lethal concentration (LC50) value was observed for crude extract, betasitosterol, gugultetrol-18-ferrulate and it was found to be 90, 75 and 65 ppm, respectively.	The use of Plumbago species as an effective anti-inflammatory agent and its cytotoxic effects have been ascertained and proved.	Arunachalam KD *et al.,* African Journal of Microbiology Research. 2012; 4(12): 1239-1245.

Contd...

TABLE 14.2–*Contd....*

Sl.No.	*Title*	*Type of Study*	*Study Design*	*Form and Dose*	*Summarized Efficacy*	*Conclusion*	*Reference*
5	Activity of *Plumbago zeylanica* Linn. root and *Holoptelea integrifolia* Roxb. bark pastes in acute and chronic paw inflammation in Wistar rat	*In vivo*	An acute anti-inflammatory effect was evaluated by using Carrageenan induced raw paw edema model and chronic inflammation was developed by using Complete Freund's Adjuvant (CFA).	Pastes of *Plumbago zeylanica* roots and *Holoptelea integrifolia* bark was applied topically once every day to the inflamed area of the paw	The formulations did not show any dermal toxicity and found to be safe. Both the pastes significantly suppressed, carrageenan-induced paw edema at 6th hour and *Holoptelea integrifolia* appears to be more effective than *Plumbago zeylanica*. Significant reduction was observed in paw volume, ankle joint circumference and animal body weight gained.	The tested formulations (*P. zeylanica* root and *H. integrifolia* bark pastes) showed significant antiinflammatory activity.	Kumar D *et al.*, J Ayurveda Integr Med. 2014; 5(1): 33–37.
					Antidiabetic Activity		
6.	Antidiabetic effect of plumbagin isolated from *Plumbago zeylanica* L. root and its effect on GLUT4 translocation in streptozotocin-induced diabetic rats.	*In vivo*	To evaluate the anti-diabetic effects of plumbagin isolated from *P. zeylanica* L. root and its effect on GLUT4 translocation in STZ-induced diabetic rats. An oral glucose tolerance test was performed on 21st day. The effect of plumbagin on body weight, blood glucose, plasma insulin, total protein, urea, creatinine, liver glycogen, plasma enzymes (SGOT, SGPT and ALP) and carbohydrate metabolism enzymes (glucose-6-phosphatase, fructose-1,6-bisphosphatase and hexokinase) were investigated. GLUT4 mRNA and protein expression in skeletal muscles were also studied.	Plumbagin (15 and 30 mg/kg b wt) was orally administered to STZ-induced diabetic rats for 28 days.	Plumbagin significantly reduced the blood glucose and significantly altered all other biochemical parameters to near normal. Further, plumbagin increased the activity of hexokinase and decreased the activities of glucose-6-phosphatase and fructose-1, 6-bisphosphatase significantly in treated diabetic rats. Enhanced GLUT4 mRNA and protein expression were observed in diabetic rats after treatment with plumbagin.	The results indicated that plumbagin enhanced GLUT4 translocation and contributed to glucose homeostasis. It could be further probed for use as a drug to treat diabetes.	Sunil C *et al.*, Food Chem Toxicol. 2012; 50(12): 4356-63.

Contd...

TABLE 14.2–*Contd....*

Sl.No.	*Title*	*Type of Study*	*Study Design*	*Form and Dose*	*Summarized Efficacy*	*Conclusion*	*Reference*
7.	Evaluation of anti-diabetic activity of leaf extract of *Plumbago zeylanica* in alloxan - induced diabetic rats	*In vivo*	To evaluate the anti-diabetic effects of *P. zeylanica* L. root in alloxan induced diabetic rats.	Methanolic roots extract of *P. zeylanicain* (100 and 200 mg/kg) was orally administered to alloxan induced diabetic rats for 21 days.	Treatment with Methanolic roots extract of *Plumbago zeylanicain* significantly reduced the blood glucose and significantly altered all other biochemical parameters to near normal.	*Plumbago zeylanica* have prominent antidiabetic effect in experimental diabetes and can therefore be used as an alternative remedy for the treatment of diabetes mellitus.	Dadi VK, Pharma Tutor, Art-1137.
					Antiulcer Activity		
8.	Anti-ulcer activity of *Plumbago zeylanica* Linn. root extract	*In vivo*	Evaluate the anti-ulcer activity of the aqueous extract of the roots of *Plumbago zeylanica* Linn. on aspirin and indomethacin-induced acute gastric ulceration in albino Wistar rats. The anti-ulcer activity of the aqueous root extract was assessed by determining and comparing the ulcer score, ulcer index and percentage protection of the extract with that of the negative and positive control groups. Omeprazole was used as standard drug.	The extract at doses of 25, 50 and 100 mg/kg	The extract produced statistically significant and dose-dependent inhibition of aspirin induced gastric mucosal damage. In the indomethacin induced ulcer, it was only at doses of 50 and 100 mg/kg respectively that the extract exhibited significant dose dependent inhibition of the gastric mucosal damage. Oral acute toxicity testing showed oral LD50 to be greater than 5000 mg/kg, indicative of the wide margin of safety of the root extract.	Aqueous extract of the roots of *Plumbago zeylanica* Linn. possesses anti-ulcer activity.	Dadul FK *et al.*, Journal of Natural Product and Plant Resources. 2012; 2(5): 563.
9.	Anti-helicobacter pylori and cytotoxic activity of detoxified root of *Plumbago auriculata, Plumbago indica* and *Plumbago zeylanica*	*In vitro*	Ethanol extract of *Plumbago auriculata, Plumbago indica* and *Plumbago zeylanica* are possible activity against *H. pylori* and cytotoxicity activity with MTT assay in HGE-17 cell lines.	50-250 µg/ml	These three plants ethanol extract have dose dependent cytotoxicity activity in HGE-17 cell lines. Zone of inhibition test of these Plumbaginales plants ethanol extract against *H. pylori* have significant activity. *Plumbago indica* (10 mg) have more activity compared to other two plants. Three Plumbaginales detoxified plants root have cytotoxicity in HGE-17 cell lines and anti-bacterial activity in *H. pylori.*	The three detoxified *P. indica, P. zeylanica* and *P. auriculata* can be considered as a source of compounds with anti *H. pylori* and cytotoxic activity.	Paul AS *et al.*, J Phyto-pharmacol 2013; 2 (3): 4-8.

Contd...

TABLE 14.2–*Contd....*

Sl.No.	Title	Type of Study	Study Design	Form and Dose	Summarized Efficacy	Conclusion	Reference
					Antihyperlipidemic Activity		
10.	Antihyperlipidemic effect of aqueous extract of *Plumbago zeylanica* roots in diet-induced hyperlipidemic rat	*In vivo*	This study examined the antihyperlipidemic effect of the aqueous extract of *Plumbago zeylanica* Linn. (Plumbaginaceae) roots in diet-induced hyper-lipidemic rats.	The aqueous extract at the dose of 20, 40, and 80mg/kg was administered orally	The aqueous extract was found to ameliorate the hyperlipidemic condition as evidenced by a reduction of cholesterol and triglyceride levels. The standards fenofibrate (20mg kg^{-1}) and atorvastatin (8mg/kg) were also found to exhibit significant cholesterol and triglyceride lowering effect. Further, the aqueous extract at all doses demonstrated a significant increase in fecal cholesterol excretion indicating a reduction in intestinal choles-terol absorption. Additionally, the activity of lipogenic enzymes like HMGCoA reductase in the liver remained significantly low on treatment of aqueous extract (80mg/kg) thus decreases the cholesterogenesis. The aqueous extract (20, 40 and 80mg kg^{-1}) also significantly reduced the total lipid content in the liver. Moreover, the aqueous extract demonstrated a potential anti-oxidant capacity in DPPH and TBARS *in vitro* antioxidant assay.	Thus the results suggest a beneficial role of aqueous extract of *Plumbago zeylanica* roots in ameliorating the hyperlipidemic condition leading to atheroscle-rosis.	Pendurkar SR and Mengi SA, Pharmaceutical Biology, 2009; 47(10): 1004–1010.
					DRAKSHA (*Vitis vinifera* Linn.)		
					Cardio protective activity		
1.	Protective effect of *Vitis vinifera* against myocardial ischemia induced by Isoproterenol in rats	*In vivo*	Isoproterenol induced myocardial ischemia. The activities of cardiac marker enzymes such as aspartate aminotrans-ferase (AST), alanine aminotransferase (ALT), lactate dehydrogenase (LDH), creatine phospho-kinase (CPK) were analyzed in heart and plasma. Cardiac protein as troponin T was estimated in serum, the levels of lipid peroxide products (thiobarbituric acid reactive substances) (TBARS), reduced glutathione (GSH) were analyzed in heart and plasma.	Seed 500 mg/kg body weight for 28 days	In ISO-treated group, shrinkage of cardiac markers in plasma and elevated lipid peroxidation where accompanied by decreased content of reduced glutathione in heart and plasma. The prior administration of Vitis vinifera significantly prevented the iso-proterenol-induced alterations and restored the cardiac markers.	Findings indicate the cardioprotective activities of *Vitis vinifera* during isoproterenol-induced myocardial ischemia.	S. Velavan *et al.* Pharmaco-logyonline 2008; 3: 958-967.

Contd...

TABLE 14.2–*Contd....*

Sl.No.	Title	Type of Study	Study Design	Form and Dose	Summarized Efficacy	Conclusion	Reference
2.	Reduction of myocardial ischemia reperfusion injury with regular consumption of grapes.	*In vivo*	Time-matched control experiments were performed by feeding the animals 45 microg/100 g of glucose plus 45 microg/ 100 g of fructose per day for 3 weeks. After 21 days, rats were killed and the hearts excised and perfused via working mode. Hearts were made ischemic for 30 min followed by 2 h of reperfusion.	Standardized Grape extract 100 mg/kg and at 200 mg/kg	Grapes provided significant cardioprotection as evidenced by improved post ischemic ventricular recovery (aortic flow, developed pressure, the maximum first derivative of the developed pressure) and reduced amount of myocardial infar ction. *In vitro* studies demonstrated that the SGE could directly scavenge superoxide and hydroxyl radicals that are formed in the ischemic reperfused myocardium.	Grape extract are resistant to myocardial ischemia reperfusion injury, suggesting a cardioprotective role of grapes.	Ann N Y Acad Sci. 2002 May; 957: 302-7.; Cui J, Cordis GA, *et al.*
					Antioxidant and Nootropic Activity		
3.	Antibacterial and anti-oxidant activities of grape (*Vitis vinifera*) seed extracts	*In vitro*	Evaluation of the anti-oxidant capacities of grape seed extracts by the formation of phos-phomolybdenum complex method and antibacterial activity by pour plate method		Gram-positive bacteria were completely inhibited at 850–1000 ppm, while Gram-negative bacteria were inhibited at 1250–1500 ppm concentration. Radical-scavenging activity of grape seed extracts of acetone:water:acetic acid (90:9.5:0.5) and methanol: water:acetic acid (90:9.5:0.5) were compared with BHA at 25 and 50 ppm concentrations by HPLC method using 1, 1-diphenyl-2-picrylhydrazyl (DPPH).	Anti-oxidant activity and antibacterial activity of grape seed extract were determine through formation of phospho-molybdenum complex method by pour plate method.	G.K. Jaya-prakasha, Tamil Selvi, K.K. Sakariah Antibacterial and antioxidant activities of grape (*Vitis vinifera*) seed extracts Food Research International Volume 36, Issue 2, 2003, Pages 117–122.
4.	Evaluation of cognitive enhancing activity of *Vitis vinifera* Linn. on albino rats	*In vivo*	Study was designed to evaluate the cognitive enhancing activity of *Vitis vinifera* Linn.	200mg/kg. B.W.	There was significant increase in step down latency and decrease in the transfers' latency and also decrease in acetylcholinesterase enzyme activity, which was as effective as that of standard drug.	It conclude that fruits of *Vitis vinifera* contain favonoid and hence possess significant cognitive test formu-lation act as antioxidants and might be helpful in stabilizing the cardiac membrane	G.K. Jaya-prakasha, Tamil Selvi, K.K. Sakariah Antibacterial and antioxidant activities of grape (*Vitis vinifera*) seed

Contd...

TABLE 14.2–*Contd....*

Sl.No.	*Title*	*Type of Study*	*Study Design*	*Form and Dose*	*Summarized Efficacy*	*Conclusion*	*Reference*
							extracts Food Research International Volume 36, Issue 2, 2003, Pages 117–122.
5.	Adaptogenic and noo-tropic activities of aqueous extract of *Vitis vinifera* (grape seed): an experimental study in rat model	*In vitro*	In the study the seed extract of *V. vinifera* was evaluated for antistress activity in normal and stress induced rats. Furthermore, the extract was studied for nootropic activity in rats and *in vitro* antioxidant potential to correlate its antistress activity.	Daily administration of *V. vinifera* at doses of 100, 200 and 300 mg/kg body	The cognition, as determined by the acquisition, retention and recovery in rats was observed to be dose dependent. The extract also produced significant inhibition of hydroxyl radicals in comparison to ascorbic acid in a dose dependent manner.	The study provides scientific support for the antistress (adaptogenic), antioxidant and noo-tropic activities of *V. vinifera* seed extract.	Satyanarayana Sreemantula, *et al.* BMC Complementary and Alternative Medicine 2005, 5:1 doi: 10.1186/1472-6882-5-1.
					Antidiabetic Activity		
6.	*In vivo* assessment of antidiabetic and anti-oxidant activities of grapevine leaves (*Vitis vinifera*) in diabetic rats	*In vivo*	Antihyperglycaemic effect of the of the aqueous extract from the leaves of *Vitis vinifera* L. were evaluated by using oral glucose tolerance test, Streptozotocin induced type 2 diabetes rat model and lipid peroxidation in liver, kidney and heart tissues. Blood glucose levels were measured according to the glucose oxidase method.	The dried aqueous leaves extract (250 and 500 mg/kg, p.o.) was administered for 15 days.	The dose-depended antidiabetic activity experiments revealed that the aqueous extract possessed a remarkable hypoglycaemic effect at 500 mg/kg dose in diabetic rats. the initial antidiabetic activity was observed on 5th day and continued to increase in all groups during the experimental period. The observed effect with aqueous extract was more pronounced (32.4 per cent) than that of reference drug, tolbutamide (28.0 per cent). *Vitis vinifera* aqueous extract was found more active in increasing body weight than tolbutamide when both were compared with the control group.	*Vitis vinifera* leaves showing potential anti-diabetic and antioxidant activities and could be a safe alternative medicine for controlling blood sugar level in diabetics.	Orhan N *et al.*, J Ethno-pharmacol 2006; 108: 280–286.
7.	Evaluation of antidiabetic activity of *Vitis vinifera* stem bark	*In vivo*	Antihyperglycaemic effect of the of the chloroform and ethanol extracts from the stem bark of *Vitis vinifera* L. were evaluated by using oral glucose tolerance	The dried extract (100 and 200 mg/kg, p.o.) was administered for 21 days.	Blood glucose levels, estimated in 16 h fasting diabetic rats (FBG) was reduced significantly upon treatment with the stem bark extracts. The levels of Total serum cholesterol (TC), Low density lipoprotein cholesterol (LDLc), Very Low density lipoprotein cholesterol (VLDLc) and Triglycerides (TG) were significantly decreased, whereas the High density lipoprotein cholesterol (HDL-c) was markedly increased in diabetic rats treated with	*V. vinifera* extracts exhibited antidiabetic activity in addition to antidiabetic and anti-hyperlipidemic effects the *V. vinifera* stem bark extracts also possess.	Ahmed M *et al.*, J Pharmacy Res 2012; 5(11): 5239-5242.

Contd...

TABLE 14.2–*Contd....*

Sl.No.	*Title*	*Type of Study*	*Study Design*	*Form and Dose*	*Summarized Efficacy*	*Conclusion*	*Reference*
			test and alloxan induced type 2 diabetes rat model. Experimental diabetes was induced in rats with alloxan (80 mg/kg, body weight).		*Vitis vinifera*. Pretreatment with plant extracts improved the SOD, catalase and peroxidase levels significantly and reduced lipid peroxidation comparable to standard drug treated group of animals. Chloroform and ethanol extracts could effectively normalize the enzyme activities; the values were high for ethanolic extract treated group. The extracts were effective in a dose dependent manner with highest activity at 200 mg/kg body weight concentration. Similar results were obtained in case of lipid peroxidation.	antioxidant potential that may be beneficial for correcting the hyperglycemia and preventing diabetic complications due to lipid peroxidation and free radicals.	
8.	Antidiabetic effect of grape seed extract (*Vitis vinifera*) in high fructose-fed rats	*In vivo*	To investigate the anti-hyperglycemic effects of grape seed extract in high fructose-fed rats.	Experimental animals receives, a high-fructose diet supplemented with 0.5, 1 and 2 per cent GSE for 8 weeks	The diabetic rats supplemented with 1 per cent GSE significantly decreased fasting plasma glucose level after 8 weeks of administration. Moreover, the diabetic rats supplemented with GSE at 1 and 2 per cent improved glucose intolerance when compared to the diabetic control group.	Grape seed extract reduced hyperglycemia in high fructose-induced diabetic rats. Thus, grape seed extract may be useful in the prevention for type 2 diabetes.	Suwannaphet W *et al.*, Thai J Pharmacol 2009; 31(1): 53.
					Gastric inflammatory diseases and Anti-inflammatory activity		
9.	The effect of *in vitro* gastrointestinal digestion on the anti-inflammatory activity of *Vitis vinifera* L. leaves.	*In vitro*	*In vitro* models of gastric and intestinal inflammation. The extract was characterized by a validated HPLC-DAD method, and tested on human epithelial gastric (AGS) and intestinal (Caco-2) cells with the aim to investigate the inhibitory effect on IL-8 secretion and promoter activity, before and after *in vitro* gastric or gastrointestinal digestion.	*Vitis vinifera* L. water extract (VVWE) from dried leaves	Water extract from red vine leaves inhibits TNFα-induced IL-8 secretion and expression in human gastric epithelial cells; the effect should be maintained, although to a lesser extent, after gastric digestion. In contrast, the effect after intestinal digestion is dramatically decreased since degradation of the active components in the gut does not allow the extract to efficiently counteract TNFα or IL-1β induced IL-8 expression and the NF-κB pathway. The main molecular target of VVWE at the gastric level includes TNFα-induced activation of NF-κB and occurs at concentrations easily reachable after plant food supplements consumption based on red vine leaf water extract as the ingredient.	Plant food supplements containing water extracts from *Vitis vinifera* L. leaves could be useful to inhibit/attenuate gastric inflammation inhibiting IL-8 secretion and expression through impairment of the NF-κB pathway.	Sangiovanni E *et al.*, Food Funct. 2015 Aug; 6(3): 2453-63.

Contd...

TABLE 14.2–*Contd....*

Sl.No.	Title	Type of Study	Study Design	Form and Dose	Summarized Efficacy	Conclusion	Reference
10.	Evaluation of the anti-inflammatory activity of raisins (*Vitis vinifera* L.) in human gastric epithelial cells: a comparative study.	*In vitro*	Anti-inflammatory activity of five raisin extracts focusing on Interleukin (IL)-8 and Nuclear Factor (NF)-κB pathway. Raisin extracts were characterized by High Performance Liquid Chromatography-Diode Array Detector (HPLC-DAD) analysis and screened for their ability to inhibit Tumor necrosis factor (TNF)α-induced IL-8 release and promoter activity in human gastric epithelial cells.	Five raisin extracts	Turkish variety significantly inhibited TNFα-induced IL-8 release, and the effect was due to the impairment of the corresponding promoter activity. Hydro-alcoholic extracts from fruits and seeds were individually tested on IL-8 and NF-κB pathway. Seed extract inhibited IL-8 and NF-κB pathway, showing higher potency with respect to the fruit. Although the main effect was due to the presence of seeds, the fruit showed significant activity as well.	Consumption of selected varieties of raisins could confer a beneficial effect against gastric inflammatory diseases.	Di Lorenzo C *et al.*, Int J Mol Sci. 2016 Jul; 17(7): 1156.
11.	Analgesic and anti-inflammatory activity of methanolic extract of *Vitis vinifera* leaves.	*In vivo*	Anti-inflammatory activity was studied by using carrageenan and histamine induced oedema right hind paw volume while the analgesic effect was evaluated using formalin-induced pain and tail flick nociception response.	Methanolic extract of *Vitis vinifera* leaves at 100, 200 and 400 mg/kg body weight were administered orally for anti-inflammatory and analgesic activities in rats and mice	Methanolic extract showed significant and dose dependent analgesic and anti-inflammatory activity. The potential analgesic and anti-inflammatory by methanolic extract was comparatively less than that of diclofenac (10 mg/kg p.o.).	Methanolic extract *V. vinifera* possess significant analgesic and anti-inflammatory activity.	Singh J. *et al.*, Pharmacologyonline 2009; 3: 496-504.
				GINGER (*Zingiber officinale* Roscoe)			
					Anti-Arthritic activity		
1.	Evaluation of the effect of gydroalcoholic extract of *Zingiber officinale* rhizomes in rat collagen-induced arthritis.	*In vivo*	The purpose of this study was to investigate the potential of hydroalcoholic extract of *Zingiber officinale* (extract) to	50 mg/kg/day intraperitoneally for 26 days	*Z. officinale* extract in doses higher than 50 mg/kg/day intraperitoneally starting from the dose of booster immunization and for 26 days can ameliorate the clinical scores, disease incidence, joint temperature and swelling, and cartilage destruction, together with reduction of serum levels of interleukin (IL)-1 β, IL-2, IL-6,	*Z. officinale* extract a good alternative to non-steroidal anti-inflammatory drugs for patients with rheumatoidarthritis.	Fouda AM, Berika MY. Basic Clin Pharmacol Toxicol.

Contd...

TABLE 14.2–*Contd....*

Sl.No.	Title	Type of Study	Study Design	Form and Dose	Summarized Efficacy	Conclusion	Reference
			ameliorate inflammatory process in rat collagen-induced arthritis.		tumour necrosis factor- α, and anti-CII antibodies. Moreover, *Z. officinale* extract at the dose of 200 mg/kg/day was superior to 2 mg/kg/day of indomethacin at most of the measured parameters.	Rheumatoid.	2009; 104(3): 262-71.
2.	Comparing the effects of ginger (*Zingiber officinale*) extract and ibuprofen on patients with osteo-arthritis	Clinical	To assess the effects of ginger extract as an alternative to NSAIDs and as a supplement drug in the symptomatic treatment of OA.	30 mg ginger extract	The improvement of symptoms (defined as reduction in the mean change) was superior in the ginger extract and ibuprofen groups than the placebo group. VAS scores and gelling or regressive pain after rising the scores were significantly higher in the PL group than both the GE and IBP groups, a month after the treatment ($P < 0.0001$). However, there was no significant difference in VAS and gelling pain scores between the ginger extract and the ibuprofen groups.	Ginger extract and ibuprofen were significantly more effective than the placebo in the symptomatic treatment of OA, while there was no significant difference between the ginger extract and ibuprofen groups in a test for multiple comparison.	Masoud Haghighi *et al.*, Archives of Iranian Medicine, 2005; 8(4): 267–271.
3.	Efficacy and tolerability of ginger (*Zingiber officinale*) in patients of osteoarthritis of knee	Clinical	To evaluate the safety and efficacy of ginger in management of OA.	Ginger 750 mg	Sixty patients of OA of knee were enrolled in randomized open label study and divided into three groups of 20 each. Group I received tab. Diclofenac 50 mg and cap. placebo, group II received cap. ginger 750 mg and cap. placebo and group III received cap. ginger 750 mg and tab. diclofenac 50 mg. The assessment of efficacy was done at every 2 weeks till 12 weeks, by using Western Ontario and McMaster Universities osteo-arthritis (WOMAC) index, Visual Analogue Scale (VAS) and the safety assessment was done by noting adverse events during the study. The analysis of WOMAC score and VAS score in all the three groups showed statistically significant improvement with time in all groups. On comparison among three groups, group III patients who received both ginger and diclofenac showed numerically superior improvement than the individual treatments.	Ginger powder has add-on effect on reducing the symptoms of OA of knee with acceptable safety profile.	Gill Paramdeep, *et al.* Indian J Physiol Pharmacol 2013; 57(2): 177–183.
					Anti-inflammatory Activity		
4.	Anti-inflammatory effects of *Zingiber officinale* in type 2 diabetic patients	Clinical	To evaluate the effects of ginger (*Zingiber officinale*) on pro-inflammatory cytokines (IL-6 and TNF-α) and the acute phase protein hs-CRP in DM2 patients as a randomized double-blind placebo controlled	All patients were randomly assigned to two groups of 32 subjects in each to receive either ginger or	Ginger supplementation significantly reduced the levels of TNF-α IL-6 and hs-CRP in ginger group in comparison to baseline. Moreover, the analysis of covariance showed that the group received ginger supplementation significantly lowered TNF- α and hs-CRP concentrations in comparison to control group.	Ginger supplementation in oral administration reduced inflammation in type 2 diabetic patients. So it may be a good remedy to diminish the risk of some chronic complications of diabetes.	Mahluji S *et al.*, Adv Pharm Bull. 2013; 3(2): 273–276.

Contd...

TABLE 14.2–*Contd....*

Sl.No.	*Title*	*Type of Study*	*Study Design*	*Form and Dose*	*Summarized Efficacy*	*Conclusion*	*Reference*
			trial. A total of 64 DM2 patients randomly were assigned to ginger or placebo groups and received 2 tablets/day of each for 2 months. The concentrations of IL-6, TNF-α and hs-CRP in blood samples were analyzed before and after the intervention.	placebo one tablet twice a day immediately after lunch and dinner for 8 weeks.			
5.	Evaluation of the anti-inflammatory effect of *Zingiber officinale* (Ginger) root in rats	*In vivo*	In this study, the anti-inflammatory activity of *Zingiber officinale* alone and in combination with indomethacin was studied using carrageenan-induced rat paw oedema.	Aqueous extract of *Zingiber officinale* (200 mg/kg or 400 mg/kg) was administered alone and in combination with indomethacin (25 mg/kg)	Indomethacin, ginger 200 mg/kg and ginger 400 mg/kg displayed values of 95 per cent, 89.5 per cent and 92.6 per cent inhibition of paw oedema respectively. The combinations of indomethacin with ginger 200 mg/kg and indomethacin with ginger 400 mg/kg displayed values of 95 per cent and 97.5 per cent inhibition of paw oedema respectively. These results indicate a similarity in the anti-inflammatory profile of ginger and indomethacin, and furthermore an enhanced anti-inflammatory profile when both are combined.	As ginger root showed significant anti-inflammatory activity in the model studied, it can be investigated further as a promising anti-inflammatory agent.	Zaman SU and Mirje MM, International Journal of Life Sciences Biotechnology and Pharma Research. 2014; 3(1): 292-298.
6.	Analgesic, antiinflammatory and hypoglycaemic effects of ethanol extract of *Zingiber officinale* (Roscoe) rhizomes (Zingiberaceae) in mice and rats	*In vivo*	The analgesic effect of ZOE was evaluated by 'hot-plate' and 'acetic acid' analgesic test methods in mice; while the antiinflammatory and hypoglycaemic effects of the plant extract were investigated in rats, using fresh egg albumin-induced pedal oedema, and streptozotocin (STZ)-induced diabetes mellitus models.	50, 100, 200, 400 and 800 mg/kg p.o.	The plant extract (ZOE, 50-800 mg/kg p.o.) significantly inhibited fresh egg albumin-induced acute inflammation, and caused dose-related, significant hypoglycaemia in diabetic rats. Treatment with extract produce significant analgesic effects against thermally and chemically induced nociceptive pain in mice.	*Zingiber officinale* rhizomes ethanol extract possesses analgesic, anti-inflammatory and hypoglycaemic properties; and thus lend pharmacological support to folkloric, ethnomedical uses of ginger in the treatment and/or management of painful, arthritic inflammatory conditions, as well as in the management and/or control of type 2 diabetes mellitus.	Ojewole JA, Phytother Res. 2006 Sep; 20(9): 764-72.

Contd...

TABLE 14.2–*Contd....*

Sl.No.	Title	Type of Study	Study Design	Form and Dose	Summarized Efficacy	Conclusion	Reference
7.	Anti-inflammatory effect of the hydralcoholic extract of *Zingiber officinale* rhizomes on rat paw and skin edema.	*In vivo*	Carrageenan induced and serotonin induced paw edema	186, 310 and 620 mg/kg, p.o.	The carrageenan-, compound 48/80- or serotonin-induced rat paw edema were inhibited significantly by the intraperitoneal administration of alcoholic ginger extract. Ginger extract was also effective in inhibiting 48/80-induced rat skin edema at doses of 0.6 and 1.8 mg/site. Rat skin edema induced by substance P or bradikinin was not affected by treatment with *Z. officinalle* extract. The intraperitoneal administration of ginger extract (186 mg/kg(-1) body wt.) 1 h prior to serotonin injections, reduced significantly the serotonin-induced rat skin edema.	Crude extract of *Zingiber officinale* was able to reduce rat paw and skin edema induced by carrageenan, 48/80 compound and serotonin. The antiedematogenic activity seems to be related, at least partially, to an antagonism of the serotonin receptor.	Penna SC, *et al.*, Phyto-medicine. 2003; 10(5): 381-5.
					Antiulcer Activity		
8.	Protective effects of ginger and marshmallow extracts on indomethacin-induced peptic ulcer in rats.	*In vivo*	To investigate the protective effects of two natural extracts, namely ginger and marshmallow extracts, on indomethacin-induced gastric ulcer in rats.	Treatment groups receiving famotidine (20 mg/kg), ginger (100 mg/kg), and marsh-mallow (100 mg/kg). Treatments were given orally on a daily basis for 14 days prior to a single intra-peritoneal administration of indometh-acin (20 mg/kg).	Indomethacin administration resulted in significant ulcerogenic effect evidenced by significant elevations in ulcer number, ulcer index, and blood superoxide dismutase activity accompanied by significant decreases in gastric mucosal nitric oxide and gluta-thione levels. In addition, elevations in gastric mucosal lipid peroxides and histamine content were observed. Alternatively, pretreatment with famotidine, ginger or marshmallow signifi-cantly corrected macroscopic and biochemical findings, supported microscopically by results of histopathological study.	These results demon-strate that administration of either ginger or marsh-mallow extract could protect against indo-methacin-induced peptic ulcer in rats presumably via their antioxidant properties and inhibition of hista-mine release.	Zaghlool SS, *et al.*, J Nat Sci Biol Med. 2015; 6(2): 421–428.
9.	Gastroprotective activity of ginger *Zingiber officinale* rosc., in albino rats.	*In vivo*	Gastric ulcers were produced by ulcerogenic agents including indo-methacin, aspirin and reserpine, beside hypo-thermic restraint stress and by pylorus ligated Shay rat technique.	500 mg/kg administered orally	The extract in the dose of 500 mg/kg orally exert highly signifi-cant cytoprotection against 80 per cent ethanol, 0.6M HC1, 0.2M NaOH and 25 per cent NaCl induced gastric lesions. The extract also prevented the occurrence of gastric ulcers induced by non-steroidal anti-inflammatory drugs (NSAIDs) and hypothermic restraint stress.	These observations suggest cytoprotective and anti-ulcerogenic effect of the ginger.	Al-Yahya MA *et al.*, Am J Chin Med. 1989; 17(1-2): 51-6.

Contd...

TABLE 14.2–*Contd....*

Sl.No.	*Title*	*Type of Study*	*Study Design*	*Form and Dose*	*Summarized Efficacy*	*Conclusion*	*Reference*
					Immunomodulatory activity		
10.	Immunomodulatory activity of *Zingiber officinale* Roscoe, *Salvia officinalis* L. and *Syzygium aromaticum* L. essential oils: evidence for humor- and cell-mediated responses.	*In vitro*	The immunomodulatory effect of ginger, *Zingiber officinale* (Zingiberaceae), sage, *Salvia officinalis* (Lamiaceae) and clove, *Syzygium aromaticum* (Myrtaceae), essential oils were evaluated by studying humor- and cell-mediated immune responses. Essential oils were administered to mice (once a day, orally, for a week) previously immunized with sheep red blood cells (SRBCs).		Clove essential oil increased the total white blood cell (WBC) count and enhanced the delayed-type hypersensitivity (DTH) response in mice. Moreover, it restored cellular and humoral immune responses in cyclophosphamide-immunosuppressed mice in a dose-dependent manner. Ginger essential oil recovered the humoral immune response in immunosuppressed mice. Contrary to the ginger essential oil response, sage essential oil did not show any immunomodulatory activity.	Our findings establish that the immunostimulatory activity found in mice treated with clove essential oil is due to improvement in humor- and cell-mediated immune response mechanisms.	Carrasco FR *et al.*, J Pharm Pharmacol. 2009; 61(7): 961-7.
				GOKSURA (*Tribulus terrestris* Linn)			
					Antiinflammatory and Antiarthritic Activity		
1.	Anti-inflammatory and antimicrobial activities of methanolic extract of *Tribulus terrestris* Linn. plant.	*In vitro*	The methanol extract of *Tribulus terrestris* plant was screened for anti-inflammatory and anti-microbial activity.		The methanolic extract showed a significant inhibition on the growth of Gram (+) and Gram (-) bacteria at concentrations of 200 8g/mL and 400 8g/mL, respectively. A dose dependent inhibition of rat paw volume by methanolic extract of *T. terrestris* in Carrageenan induced inflammation in rats was observed, which is comparable with standard drug, diclofenac sodium.	Methanolic extract of *Tribulus terrestris* plant might have antimicrobial and anti-inflammatory activities.	B. Baburao *et al.*, Int. J. Chem. Sci.: 7(3), 2009, 1867-1872.
2.	Anti-arthritic activity of *Tribulus terrestris* studied in Freund's Adjuvant induced arthritic rats.	*In vivo*	The anti-arthritic activity of *Tribulus terrestris* was evaluated using Frund's complete adjuvant (FCA) induced arthritis in rats.	The herbal extracts at dose 200mg/kg and 300 mg/kg p.o was administered for 21 days after the injection of FCA in the rats paws.	A significant inhibition of paw edema volume was observed from day 5th to 21st in the treated groups. The biochemical parameters like erythrocyte sedimentation rate (ESR), alkaline phosphatase (ALP), acid phospatase (ACP) and total WBC count was observed which are the major markers of arthritis. A significant increase in the level of all the markers were found in the arthritic rats where as in case of prednisolan and *Tribulus terrestris* treated groups a marked decrease in the level was observed. A better anti-arthritic activity was obtained at 300 mg/kg dose of Tribulus terrestris.	The methanolic extact of *Tribulus terrestris* has anti-arthritic activity.	Mishra NK, *et al.*, J Pharm Education Res 2013; 4(1): 41.

Contd...

TABLE 14.2–*Contd....*

Sl.No.	*Title*	*Type of Study*	*Study Design*	*Form and Dose*	*Summarized Efficacy*	*Conclusion*	*Reference*
3.	Anti-inflammatory activity of *Tribulus terrestris* in RAW264.7 Cells (54.2).	*In vitro*	To investigate the ethanol extract of *Tribulus terrestris* (ETT) inhibited expression of COX-2 and iNOS in lipopolysaccharide (LPS) - stimulated RAW264.7 cells.		ETT significantly decreased release of NO production in a dose-dependent manner in LPS-stimulated RAW264.7 cells. ETT also suppressed expression of proinflammtory cytokines, such as TNF-α and IL-4 in macrophage cell line.	ETT inhibits expression of mediators related to inflammation and expression of inflammatory cytokines, which has a beneficial effect on various inflammatory diseases.	Oh JS *et al.*, J Immunol. 2012; 188 (S1): 54.2.
					Hepatoprotective Activity		
4.	Hepatoprotective activity of *Tribulus terrestris* extract against acetaminophen-induced toxicity in a freshwater fish (*Oreochromis mossambicus*).	*In vitro*	To investigate the potential protective role of *Tribulus terrestris* in acetaminophen induced hepatotoxicity in *Oreochromis mossambicus.*		The plant extract (250 mg/kg) showed a remarkable hepatoprotective activity against acetaminophen-induced hepatotoxicity. The levels of all enzymes have significantly ($p<0.05$) increased in acetaminophen-treated fish tissues. The elevated levels of these enzymes were significantly controlled by the treatment of *T. terrestris* extract (250 kg/mg). Histopathological changes of liver, gill and muscle samples were compared with respective controls.	Study specifies the hepatoprotective and antioxidant properties of *T. terrestris* against acetaminophen-induced toxicity in freshwater fish, *O. mossambicus.*	Kavitha P *et al.* *In vitro* Cell Dev Biol Anim. 2011 Dec; 47(10): 698-706.
					Antiulcer Activity		
5.	Gastroprotective effects of fruits of *Tribulus terrestris* L. In pylorus-ligated wistar rat model.	*In vivo*	Gastroprotective (*i.e.* antiulcer and anti-secretory) potential of methanolic extract of TT fruits was evaluated in pylorus-ligated rat model of Wistar rat.	The methanolic extract of TT was tested orally at the doses of 150, 300 and 600 mg/kg	The methanolic extract at the doses of 300 and 600 mg/kg produced more significant inhibition when gastric ulcerations were induced by pylorus ligation respectively.	The methanol extract of the fruits of *Tribulus terrestris* L. possess gastroprotective *i.e.* antiulcer and anti-secretory effect.	Ansari JA *et al.*, Int J Pharm Sci Res. 2012; 4(1): 411-14.
					Antibacterial Activity		
6.	*In vitro* antibacterial activity of water and ethanol extract of *Tribulus terrestris* on the growth of *Pseudomonas aeruginosa* by disc diffusion test.	*In vitro*	Study was conducted to determine the antibacterial activity of *Tribulus terrestris* against *Pseudomonas aeruginosa*		Study was conducted to determine the antibacterial activity of *Tribulus terrestris* against *Pseudomonas aeruginosa.*	Ethanol extract of *Tribulus terrestris* showed maximum inhibition zone *i.e.* 27.33 mm as compared to aqueous extract. Aqueous extract of *Tribulus terrestris* showed inhibition zone from 13.66 to 25.66mm but ethanol extract showed 14.66 to 27.33 mm inhibition zone.	*Tribulus terrestris* is an effective medicinal plant which shows effective antibacterial activity against *Pseudomonas aeruginosa.*

Contd...

TABLE 14.2–*Contd....*

Sl.No.	*Title*	*Type of Study*	*Study Design*	*Form and Dose*	*Summarized Efficacy*	*Conclusion*	*Reference*
					Antidiabetic Activity		
7.	Hypoglycemic and hypo-lipidemic effects of alcoholic extract of *Tribulus alatus* in strepto-zotocin-induced diabetic rats: a comparative study with *T. terrestris* (Caltrop).	*In vivo*	Study was conducted to determine the anti-diabetic activity of alcoholic extract of *Tribulus alatus* and *T. terrestris* in strepto-zotocin-induced diabetic rats.	Both the extracts were tested orally at the dose 50 mg/kg for 21 days	The extracts of both *T. alatus* and *T. terrestris* significantly decrease fasting glucose level in diabetic rats. After 4 and 6 hr, *T. alatus* extract showed significant reduction in glucose level as compared to *T. terrestris*. After 3 weeks of treatment with *T. alatus* extract, glucose level was significantly decreased to the normal level. Both the extracts also caused a significant decrease in the levels of glycosylated hemoglobin, total cholesterol, triglycerides and LDL-cholesterol. The percent of reduction in rats treated with *T. alatus* extract was significantly higher than that of the rats treated with *T. terrestris*.	Alcoholic extract of *T. alatus* and *T. terrestris.* possesses hypoglycemic activity in type-1 model of diabetes.	El-Tantawy WH and Hassanin LA. Indian J Exp Biol. 2007; 45(9): 785-90.
8.	Anti-hyperglycaemic activity of *Tribulus terrestris* L aerial part extract in glucose-loaded normal rabbits.	*In vivo*	Investigate the anti-hyperglycaemic activity of methanol extract of *Tribulus terrestris* L. in glucose-loaded normal rabbits.	The metha-nolic extract of TT was given orally at the dose of 250 mg/kg	On comparing within groups, a single dose of the methanol extract of *Tribulus terrestris* L. lowered Fasting blood glucose (FBG) to levels comparable to that of glibenclamide, and reaching the initial level (0 h) at 2 h. FBG were significantly lowered at 2 and 3 h in both glibenclamide and extract groups as compared with their respective glucose levels at 0.5 h. Onthe other hand, on comparing between groups, both glibenclamide and methanol extract significantly lowered the rise in blood glucose at 1 h, 2 h and 3 h with respect to the hyperglycaemic control group.	The methanol extract of the aerial parts of *T. terrestris* possesses potential anti hypergly-caemic activity in glucose loaded normal rabbits.	El-Shaibany A *et al.*, Tropical Journal of Pharmaceutical Research. 2015; 14 (12): 2263-2268.
9.	α-glucosidase and aldose reductase inhibitory activity *in vitro* and antidiabetic activity *in vivo* of *Tribulus terrestris* L. (Dunal).	*In vivo*	Study was done for a comparative evaluation of α-glucosidase and aldose reductase inhibitory activity of *Tribulus terrestris* and *Curculigo orchioides*.	Diabetic rats were treated for 30 days with 100 mg/kg body weight of ethanolic extract of *T. terrestris* and *C. orchioides*	Both TT and CO were effective in inhibiting α glucosidase and aldose reductase. There was a reduction in fasting blood glucose level observed as well. Treatment with extracts could also ameliorate the loss of body weight as observed in case of hyper-glycemic control group animals.	The xtract of *T. terrestris* possesses potential anti-hyperglycaemic activity through α-glucosidase and aldose reductase inhibitory action.	Lamba *et al.*, Int J Pharm Pharm Sci, 2011; 3(3): 270-272.
10.	Efficacy of *Tribulus terrestris* extract and metformin on fertility indices and oxidative stress of testicular tissue in streptozotocin-induced diabetic male rats.	*In vivo*	To evaluating the effect of *Tribulus terrestris* on different parameters of oxidative stress and enzymatic/non-enzy-matic antioxidant as well as the number, viability and abnormalities of sperm in testis tissues of male rats after induction of diabetes.	Diabetic rats were treated with 10 mg/kg body weight of aque-ous extract of *T. terrestris.* The treatments were continued for 5 days/week for 60 days.	*T. terrestris* was noticed to reduce the oxidative stress levels, and restore antioxidant enzyme activity in testis tissues as well as to improve the lipid profile content in serum. Histological analysis showed that *T. terrestris* treatment decreased testis tubular damage, and restored it to normal morphology.	TT extract as compared to metformin has potential effect against spermatotoxic and testicular toxicity and it can improve redox state in diabetic male rats.	Tag HM *et al.*, Afr J Pharm Pharmacol. 2015; 9(48): 1088-98.

Contd...

TABLE 14.2–*Contd....*

Sl.No.	Title	Type of Study	Study Design	Form and Dose	Summarized Efficacy	Conclusion	Reference
					GUDUCHI (*Tinospora cordifolia* (Thunb.) Miers)		
					Nootropic Activity		
1.	Nootropic activity of saponins obtained from *Tinospora cordifolia* stem in scopolamine induced amnesia.	*In vivo*	Study was designed to evaluate the nootropic property of n-butanolic (TBF) fraction of ethanolic extract of *Tinospora cordifolia* stem which contain saponin.	Ethanolic extract of *Tinospora cordifolia* stem 100 mg/kg and 200 mg/kg.	Study showed significant increase in transfer latency by TBF on day -2 and Day-9. The TBF also increases step down latency, object recognition index significantly. Anti-cholinesterase activity of TBF was evaluated by estimation of acetylcholinesteras (AchE) concentration in mice brain after 7-days treatment with TBF.	The result showed decreased in AchE concentration indicating involvement of cholinergic system innootropic activity of TBF.	Hemant D. Une, International Journal of Pharma Research and Review, Feb 2014; 3(2):28-35.
					Immunomodulatory Activity		
2.	Immunomodulatory effect of *Tinospora cordifolia* extract in human immunodeficiency virus positive patients.	Clinical	To assess the safety and efficacy of *Tinospora cordifolia* extract (**TCE**) in human immunodeficiency virus positive patients.	TCE 300 mg per tablet three times a day for six months.	TCE treatment caused significant reduction in eosinophil count and hemoglobin percentage. 60 per cent patients receiving TCE and 20 per cent on placebo reported decrease in the incidence of various symptoms associated with disease. Some of the common complaints reported by patients on TCE were anorexia, nausea, vomiting and weakness.	*Tinospora cordifolia* extract, a plant derived immunostimulant, significantly affected the symptoms of HIV. *Tinospora cordifolia* could be used as an adjunct to HIV/AIDS management.	M.V. Kalikar J Pharmacol. Jun 2008; 40(3): 107–110.
3.	Immunomodulatory effects of *Tinospora cordifolia* (Guduchi) on macrophage activation.	*In vitro*	Study was designed to evaluate Immunomodulatory effects of *Tinospora cordifolia* on macrophage activation in J774A cells.		Enhanced secretion of lysozyme by macrophage cell line J774A on treatment with *Tinospora cordifolia* and lipopolysacharide was observed, suggesting activated state of macrophages. Enhanced lysozyme production was reported at different time intervals (24 hrs and 48 hrs). This led to check the effect of the drug on the functional activity of macrophage with respect to microbicidal properties by disk diffusion antibiotic sensitivity test. The enhanced inhibitory effects of *T. cordifolia* (direct effect) and *T. cordifolia* treated cell supernatant (indirect effect) on the bacteria (*E. coli*) indicates the susceptibility of bacteria.	The *T. cordifolia* to be used as immunomodulator for activation of macrophages.	More P and Pai K, Biology and Medicine, 2011; 3(2): 134-140.
4.	Study of Immunomodulatory activity of *Tinospora cordifolia* extract.	*In vitro*	The extracts of *T. cordifolia* for the stimulation of the immune defense system using an *in vitro* phagocytosis method. Human PMN cells were used in the assay along with Candida culture as an antigen.	*T. cordifolia* extract	The tested extracts were able to stimulate the PMN cells for phagocytosis of added Candida cells. The per cent phagocytosis of the test samples as compared to the control proves potent immunostimulatory activity of the *T. cordifolia* extract	The assay results substantiate the use of *T. cordifolia* as an immunomodulatory agent	Salkar K. *et al.*, International Journal of Advances in Pharmacy, Biology and Chemistry. 2014; 3(4): 880-83.

Contd...

TABLE 14.2–*Contd....*

Sl.No.	*Title*	*Type of Study*	*Study Design*	*Form and Dose*	*Summarized Efficacy*	*Conclusion*	*Reference*
5.	Immunomodulatory active compounds from *Tinospora cordifolia.*	*In vitro*	The immunomodulatory activity of different extracts, fractions and isolated compounds in relation to phagocytosis and reactive oxygen species production in human neutrophil cells have been investigated using the PMN phagocytic function studies, NBT, NO and chemiluminescence assay.		The results obtained indicate that ethyl acetate, water fractions and hot water extract exhibited significant immunomodulatory activity with an increase in percentage phagocyctosis. Chromatographic purification of these fraction led to the isolation of a mixture of two compounds 2, 3 isolated for the first time from natural source and five known compounds 1, 4-7 which were characterized as 11-hydroxymustakone (2), N-methyl-2-pyrrolidone (3), N-formylannonain (1), cordifolioside A (4), magnoflorine (5), tinocordiside (6), syringin (7) by nuclear magnetic resonance (NMR) and mass spectrometry (MS) and comparing the spectral data with reported one. Cordifolioside A and syringin have been reported to possess immunomodulatory activity. Other five compounds showed significant enhancement in phagocytic activity and increase in nitric oxide and reactive oxygen species generation at concentration 0.1-2.5 μg/ml.	Seven immunomodulatory active compounds belonging to different classes have been isolated and characterised indicating that the immunomodulatory activity of *Tinospora cordifolia* may be attributed to the synergistic effect of group of compounds.	Sharma U *et al.*, J Ethnopharmacol. 2012; 141(3): 918-26.
6.	Immunomodulatory activity of Guduchi Ghana (Aqueous extract of *Tinospora cordifolia* Miers).	*In vitro*	Assessment of the immunomodulatory activity of aqueous extract of *Tinospora cordifolia* was done by haemagglutination antibody titre method for humoral immunity and footpad swelling method for cell mediated immunity on wistar albino rats.	Aqueous extract of *Tinospora cordifolia*	Guduchi Ghana prepared by classically was found to possess significant immunostimulatory action on immune system but market sample of it exhibited significant immunosuppression effect in dose dependent manner when compare with control group at a dose of 50 mg/kg orally.	*Tinospora cordifolia* has potent immunomodulatory activity.	Umretia B *et al.*, NJIRM 2013; 4(3): 90-96.
7.	Pharmacological study of *Tinospora cordifolia* as an immunomodulator.	*In vivo*	To evaluate immunomodulatory activity *Tinospora cordifolia* alcoholic extract by Delayed Type Hypersentivity (DTH), bone marrow cellularity and the α-esterase cells and zinc sulphate turbidity test.	Orally administration of *T. cordifolia* alcoholic extract (100 mg/kg, p. o)	There was distinct increase in foot pad thickness after treatment with *T. cordifolia* alcoholic extracts which indicates immunomodulatory effects of T. cordifolia as compared to vehicle and cyclophosphamide treated groups. Also significant increase in the WBC counts and bone marrow cells significantly indicating stimulatory effect on haeomopoetic system. In zinc sulphate turbidity test *T. cordifolia* treated rats serum showed the more turbidity (cloudy) which indicate the increase in the immunoglobulin level as compared to vehicle, SRBC sensitized and cyclophosphamide treated group.	*Tinoposra cordifolia* shows potent immunomodulatory action.	Aher VD and Wahi A, Int J Curr Pharm Res, 2010; 2(4): 52-54.

Contd...

TABLE 14.2–*Contd....*

Sl.No.	Title	Type of Study	Study Design	Form and Dose	Summarized Efficacy	Conclusion	Reference
8.	Effect of aqueous extract of *Tinospora cordifolia* on functions of peritoneal macrophages isolated from CCl_4 intoxicated male albino mice.	*In vivo*	The present study focuses on the immunostimulant properties of *Tinospora cordifolia* extract that are exerted on circulating macrophages isolated from CCl_4 (0.5 ml/kg body weight) intoxicated male albino mice. Apart from damaging the liver system, carbon tetra-chloride also inhibits macrophage functions thus, creating an immuno-compromised state, as is evident from the present study. Such cell functions include cell morphology, adhesion property, phago-cytosis, enzyme release (myeloperoxidase or MPO), nitric oxide (NO) release, intracellular survival of ingested bacteria and DNA fragme-ntation in peritoneal macrophages isolated from these immunocompro-mised mice.	*T. cordifolia* extract was tested for acute toxicity at the given dose (150 mg/kg body weight) by lactate dehydro-genase (LDH) assay.	The number of morphologically altered macrophages was increased in mice exposed to CCl_4. Administration of CCl_4 (i.p.) also reduced the phagocytosis, cell adhesion, MPO release, NO release properties of circulating macrophages of mice. The DNA fragmentation of peritoneal macrophages was observed to be higher in CCl_4 intoxicated mice. The bacterial killing capacity of peritoneal macrophages was also adversely affected by CCl_4. However oral administration of aqueous fraction of *Tinospora cordifolia* stem parts at a dose of 40 mg/kg body weight (*in vivo*) in CCl_4 exposed mice ameliorated the effect of CCl_4, as the percentage of morphologically altered macrophages, phagocytosis activity, cell adhesion, MPO release, NO release, DNA fragmen-tation and intracellular killing capacity of CCl_4 intoxicated peritoneal macrophages came closer to those of the control group. No acute toxicity was identified in oral administration of the aqueous extract of *Tinospora cordifolia* at a dose of 150 mg/kg body weight.	Polar fractions of *Tinospora cordifolia* stem parts contain major bioactive compounds, which directly act on peritoneal macrophages and have been found to boost the non-specific host defenses of the immune system.	Sengupta M *et al.*, BMC Complement Altern Med. 2011; 11: 102.
					Hypoglycaemic activity		
9.	Hypoglycaemic and other related actions of *Tinospora cordifolia* roots in alloxan-induced diabetic rats.	*In vivo*	Alloxan diabetic rats	Aqueous *T. cordifolia* root extract	Oral administration of an aqueous *T. cordifolia* root extract (TCREt) to alloxan diabetic rats caused a significant reduction in blood glucose and brain lipids. The extract caused an increase in body weight, total haemoglobin and hepatic hexokinase. The root extract also lowers hepatic glucose-6-phosphatase and serum acid phosphatase, alkaline phosphatase, and lactate dehydrogenase in diabetic rats.	TCREt has hypoglycemic and hypolipidaemic effect.	Stanely P *et al.*, J Ethnophar-macol. 2000 Apr; 70(1):9-15.

Contd...

TABLE 14.2–*Contd....*

Sl.No.	Title	Type of Study	Study Design	Form and Dose	Summarized Efficacy	Conclusion	Reference
					Malarious splenomegaly		
10.	*Tinospora cordifolia* as an adjuvant drug in the treatment of hyper-reactive malarious splenomegaly – case reports.	Clinical case study	Treatment was given and patients were observed up to six months. Improvement was gauzed by measuring spleen enlargement, Hb, serum IgM and well-being in three cases of hyper-reactive malarious splenomegaly in District Hospital, Daltonganj town, Jharkhand, India. These cases were partial/slow responders to the conventional antimalarial drug chloroquine.	Aqueous extract of *T. cordifolia* (500 mg) was added to chloroquine (CQ) base (300 mg) weekly and CQ prophylaxis	Addition of extract of *T. cordifolia* for the first six weeks to chloroquine showed regression of spleen by 37–50 per cent after six weeks and 45–69 per cent after six months from the start of treatment. Likewise decrease in IgM and increase in Hb as well as wellbeing (Karnofsky performance scale) were observed.	The results of the present study paves a new sight in the treatment of hyper-reactive malarious splenomegaly, however, large-scale trial is required to confirm the beneficial effect of *T. cordifolia* extract in combination with chloroquine.	Ranjan Kumar Singh, J Vect Borne Dis 42, March 2005, pp. 36–38.
				HARITAKI (*Terminalia chebula* Retz.)			
					Hypolipidemic activity		
1.	Anti-hyperlipidemic activity of aqueos extract of *Terminalia chebula* and Gaumutra in rich cholesterol diet rats	*In vivo*	Hyperlipidemia was induced by giving high cholesterol diet for thirty days in standard rat chow diet. Rats on high cholesterol diet showed significant increase in serum and tissue cholesterol, LDL-C, VLDL-C, triglyceride, atherogenic index and decrease HDL-C levels.	Aqueous extract of *Terminalia chebula* (300mg/kg, p.o) for 30 days	Showed significant decrease in serum and tissue serum and tissue cholesterol, LDL-C, VLDL-C, triglyceride, atherogenic index and increase HDL-C levels. Histological study showed that *Terminalia chebula* caused decrease in aortic plaque and fatty liver formation but its combination with Gaumutra showed no significant effect in aorta and liver as compared to high cholesterol diet fed rats.	Haritaki has anti-hyperlipidemic activity.	Dipa A. Israni *et al.* Pharma Science Monitor. 2010; 1(1): 48-59.
2.	Hypolipidemic activity of haritaki (*Terminalia chebula*) in atherogenic diet induced hyperlipidemic rats.	*In vivo*		Suspension of Haritaki (*Terminalia chebula*) (1.05 and 2.10 mg/kg body weight) for 14 days	Haritaki at 1.05 and 2.10 mg/kg b.wt. concentrations showed statistically significant decreased in total cholesterol, triglyceride, total protein level and decrease in the atherogenic index.	Haritaki was found to possess hypolipidemic activity.	Maruthappan V, *et al.*, J Adv Pharm Technol Res. 2010; 1(2): 229-235.

Contd...

TABLE 14.2–*Contd....*

Sl.No.	Title	Type of Study	Study Design	Form and Dose	Summarized Efficacy	Conclusion	Reference
					Antidiabetic activity		
3.	Anti-diabetic activity of fruits of *Terminalia chebula* on streptozotocin induced diabetic rats.	*In vivo*	To evaluate the anti-diabetic potential of *Terminalia chebula* fruits on streptozotocin (STZ)-induced experimental diabetes in rats.	Oral administration of ethanolic extract of the fruits (200 mg/kg) for 30 days	Treatment with *Terminalia chebula* fruits significantly reduced the levels of blood glucose and glycosylated hemoglobin in diabetic rats. Determination of plasma insulin levels revealed the insulin stimulating action of the fruit extract. Also, the alterations observed in the activities of carbohydrate and glycogen metabolising enzymes were reverted back to near normal after 30 days of treatment with the extract. Electron microscopic studies showed significant morphological changes in the mitochondria and endoplasmic reticulum of pancreatic β cells of STZ-induced diabetic rats. Also, a decrease in the number of secretory granules of β-cells was observed in the STZ-induced diabetic rats and a these pathological abnormalities were normalized after treatment with *T. chebula* extract.	The ethanolic extract of *T. chebula* fruit has potential hypoglycemic action in STZ-induced diabetic rats and the effect was found to be more effective than glibenclamide.	Senthil Kumar GP *et al.*, Journal of Health Science, 2006; 52(3): 283–291.
4.	Antidiabetic and renoprotective effects of the chloroform extract of *Terminalia chebula* Retz. seeds in streptozotocin-induced diabetic rats.	*In vivo*	The blood glucose lowering activity of the chloroform extract was determined in streptozotocin-induced diabetic rats.	Chloroform extract of *T. chebula* seed powder was orally administered at the doses of 100, 200 and 300 mg/kg for 4 weeks	The chloroform extract of T. chebula seeds produced dose-dependent reduction in blood glucose of diabetic rats and comparable with that of standard drug, glibenclamide in short term study. It also produced significant reduction in blood glucose in long term study. Significant renoprotective activity is observed in *T. chebula* treated rats. The results indicate a prolonged action in reduction of blood glucose by *T. chebula* and is probably mediated through enhanced secretion of insulin from the β-cells of Langerhans or through extra pancreatic mechanism. The probable mechanism of potent renoprotective actions of *T. chebula* has to be evaluated.	The present studies indicated a significant antidiabetic and renoprotective effects with the chloroform extract of *Terminalia chebula* and support its traditional usage in the control of diabetes and its complications.	Rao NK and Nammi S, BMC Complement Altern Med. 2006; 6: 17.
5.	Evaluation of mechanism of anti-diabetic activity of *Terminalia chebula* on alloxan and adrenaline induced diabetic albino rats.	*In vivo*	To evaluate the anti-diabetic action of ethanolic pulp extract of a locally available plant, *Terminalia chebula* (AETC) on alloxan induced diabetic rats.	Ethanolic pulp extract of *Terminalia chebula* was administered orally at dose 100 mg/kg for 4 weeks	*Terminalia chebula* showed significant anti-hyperglycemic effect without hypoglycemic action in normal rats, and efficacy was lower than glibenclamide in alloxan model but higher in adrenaline induced model. The glycogen content of liver significantly increased in diabetes induced albino rat which may be due to insulin like action of ingredients present in *Terminalia chebula. T. chebula* also showed reduction in blood glucose level on adrenaline induced hyperglycemia resulting from inhibition of α2 receptor of pancreatic β-cells, thus promoting further insulin release.	The ethanolic pulp extract of *Terminalia chebula* fruit has significant anti-diabetic activity probably due to insulin like action of its constituents and promotion of insulin release.	Borgohain R *et al.*, Int J Pharm Bio Sci 2012; 3(3): 256-266.

Contd...

TABLE 14.2–*Contd....*

Sl.No.	*Title*	*Type of Study*	*Study Design*	*Form and Dose*	*Summarized Efficacy*	*Conclusion*	*Reference*
					Cardiotonic activity		
6.	Cardiotonic activity of *Terminalia chebula* bark on isolated frog's heart.	*Ex vivo*	Isolated rat heart perfussion technique.	Aqueous extract of *Terminalia chebula* bark	A significant increase in height of force of concentration.	Haritaki has cardiotonic activity.	Rabindra Babu P. *et al.* (2012) Int. J. Res. Farm. Sci., 3(1), 24-28.
					Antiulcer Activity		
7.	Antiulcerogenic activity of *Terminalia chebula* fruit in experimentally induced ulcer in rats.	*In vivo*	Aspirin, ethanol and cold restraint stress-induced ulcer methods in rats were used for the study. The effects of the extract on gastric secretions, pH, total and free acidity using pylorus ligated methods were also evaluated.	Animal were treated orally with doses of 200 and 500 mg/kg *T. chebula* hydro-alcoholic extract	Animals pretreated with doses of 200 and 500mg/kg hydro-alcoholic extract showed significant reduction in lesion index, total affected area and percentage of lesion in comparison with control group in the aspirin, ethanol and cold restraint stress-induced ulcer models. Similarly extracts increased mucus production in aspirin and ethanol-induced ulcer models. At doses of 200 and 500mg/kg of *T. chebula* extract showed antisecretory activity in pylorus ligated model, which lead to a reduction in the gastric juice volume, free acidity, total acidity, and significantly increased gastric pH.	These findings indicate that hydroalcoholic extract of the fruit *T. chebula* displays potential antiulcerogenic activity.	Sharma P *et al.*, Pharm Biol. 2011; 49(3): 262-8.
8.	Evaluation of anti-ulcer activity of methanolic extract of *Terminalia chebula* fruits in experimental rats.	*In vivo*	The anti-ulcer activity of methanolic extract of *T. chebula* fruits METC was investigated in pylorus ligation and ethanol induced ulcer models in wistar rats.	Animal were treated with 250,500 mg/kg p.o	Treatment with *T. chebula* fruit extract significant inhibition of the gastric lesions induced by Pylorus ligation induced ulcer and Ethanol induced gastric ulcer.The extract showed significant (P<0.01) reduction in gastric volume, free acidity and ulcer index as compared to control. This present study indicates that *T. chebula* fruit extract have potential anti-ulcer activity in the both models	Methanolic extract was found to possess anti-ulcerogenic as well as ulcer healing properties, which might be due to its antisecretory activity.	Raju D *et al.*, J Pharm Sci and Res. 2009; 1(3): 101-107.
					Antidiarrhoeal Activity		
9.	Assessment of the anti-diarrhoeal properties of the aqueous extract and its soluble fractions of Chebulae Fructus (*Terminalia chebula* fruits).	*In vivo*	The antidiarrhoeal effect of *Terminalia chebula* fruits was investigated by determining the wet dropping, intestinal transit in BALB/c mice and enteropooling in Wister rats. The anti-diarrhoeal fraction was determined by castor oil-induced diarrhoea and its main constituents were identified by HPLC-ESI-MS.	Animal were treated with 200, 400 and 800 mg/kg p.o	The extract at doses of 200, 400 and 800mg/kg reduced the diarrhoea by 9.1, 40.0 and 58.2 per cent and inhibited intestinal transit by 18.3, 24.1 and 35.7 per cent, respectively. Additionally, the CFAE (200, 400 and 800mg/kg) decreased the volume of enteropooling by 47.1, 58.8 and 64.7 per cent, respectively. Mice treated with castor oil presented morphological alterations in the small intestine and the liver. However, the lesions of mice treated with CFAE were alleviated. Moreover, the ethyl acetate fraction was the active fraction of CFAE, the fraction (41.7, 83.4 and 166.8 mg/kg) reduced the diarrhoea by 9.1, 38.2 and 54.5 per cent, respectively. The major components of the ethyl acetate fraction were tannins, including gallic acid, 3, 4, 6-tri-O-galloyl-β-d-Glc, corilagin and ellagic acid according to the HPLC-ESI-MS analysis.	The CFAE possessed antidiarrhoeal property and the ethyl acetate fraction was its main active fraction.	Sheng Z *et al.*, Pharm Biol. 2016 Sep; 54(9): 1847-56.

Contd...

TABLE 14.2–*Contd....*

Sl.No.	*Title*	*Type of Study*	*Study Design*	*Form and Dose*	*Summarized Efficacy*	*Conclusion*	*Reference*
					Antioxidant Activity		
10.	Antioxidant activity of polyphenolic extract of *Terminalia chebula* Retzius fruits.	*In vitro*	The polyphenolic extract of *T. chebula* fruits was evaluated for antioxidant activity by determining the reducing power, total antioxidant capacity, DPPH radical concentration, nitric oxide radical concentration and hydrogen peroxide scavenging activity. Moreover, the phytochemical characterisation of the extract was also measured by determining the total phenolic, flavonoid, tannin and ascorbic acid contents. Characterisation of the extract was also performed by HPLC profiling with the standard gallic acid.		DPPH radical concentration (IC50 14 μg/mL), nitric oxide radical concentration (IC50 30.51 μg/mL) and hydrogen peroxide scavenging activity (IC50 265.53 μg/mL). The present study demonstrated that the extract has significant reducing capacity and nitric oxide scavenging activity. It also scavenges hydrogen peroxide-induced radicals. The activity of the extract may be due to the total polyphenolic content. The antioxidant activity of the extract is significantly higher than the standard ascorbic acid, and its activity is concentration-dependent.	It is concluded that a polyphenolic-rich fraction of *T. chebula* fruits is a potential source of natural antioxidants.	Saha S and Verma RJ, Journal of Taibah University for Science. 2016; 10(6): 805-812.
				HADJOD (*Cissus quadrangularis* Linn.)			
					Anti-osteoporosis Activity (Preclinical)		
1.	Protective effect of *Cissus quadrangularis* Linn. on diabetes induced delayed fetal skeletal ossification.	*In vivo*	Model: Diabetic streptozotocin Species: Wistar rats (pregnant) Groups 3 Normal/ diabetic/diabetic and CQ To examine the pattern of skeletal ossification alizarin red S–alcian blue staining is done. Length of long bone also calculated.	CQ extract 500 mg/kg	CQ extract can influence the fetal ossification process and prevent diabetes-induced delay in the onset of the ossification during fetal life. It has fetal osteogenesis effect. CQ ext helps in increase in body weight.	CQ is effective in skeletal development in the offspring of diabetic rats and proves its osteoporotic activity.	Sirasanagandla SR, Ranganath Pai KS, Potu BK, Bhat KM. J Ayurveda Integr Med. 2014; 5(1): 25–32.

Contd...

TABLE 14.2–*Contd....*

Sl.No.	Title	Type of Study	Study Design	Form and Dose	Summarized Efficacy	Conclusion	Reference
2.	Antiosteoporotic effect of sequential extracts and freeze- dried juice of *Cissus quadrangularis* L. in ovariectomized mice .	*In vivo*	Species: ICR mice (Outbred mice) N=24 Divided into 8 groups: sham-operated, ovx-control, estradiol (E 2)-treated and five sequential extracts (hexane, dichloromethane, ethanol, and water) and freeze-dried. BMD of the femur and tibia, serum levels of osteocalcin and TRAP5b and histomorphological change of lumbar spine were determined.	CQ Dose equivalent to 5g of crude powder/kg/ day for 8 weeks.	The BMD of the femur and tibia in the hexane-treated group were elevated same as sham-operated group also lowest serum levels of osteocalcin and TRAP5b observed in this group, compared to the ovx-control and E 2 -treated groups, representing a decrease in the bone turnover rate.	Hexane extract of *Cissus quadrangularis* has antiosteoporostic property.	Thanika Pathom-wichaiwat, *et al.*, Asian Biomedicine 2012; 6(3): 377-384.
3.	Anti-osteoporotic activity of the petroleum ether extract of *Cissus quadrangularis* Linn. in ovariectomized wistar rats.	*In vivo*	Model: Ovariectomized Species: Wistar rats N= 30 Groups: 5, Normal control, SHAM group surgery but no treatment, Ovariectomized (OVX) ovariectomy and normal saline, Ovariectomized + raloxifene (OVX + RAL), Ovariectomy + *Cissus quadrangularis* (OVX + CQ) The right femur was used for biomechanical analysis, and the left femur for histo-morphometrical analysis.	CQ 500mg/kg for 90 days	CQ significantly increased the force required to break the femur and significantly increased the thickness of both cortical and trabecular bone. This action of CQ and Raloxifene groups has similar action. It enhances mineralization of bone.	CQ extract has anti-osteoporotic activity and safe in using elderly women also in bone related disorders.	Potu BK, Rao MS, Nampurath GK, Rao M, Soubhagya C, Nayak R, Chang Gung Med J. 2010; 33: 252-257.
4.	Anti-osteoporotic effect of ethanol extract of *Cissus quadrangularis* Linn. on ovariectomized rat.	*In vivo*	Model: Ovariectomized rat Groups: 5. 1st Control-Nonovari-ectomy; rest all ovari-ectomy; 2 control,	CQ extract 500 and 750 mg/kg	The biomechanical, biochemical and histopathological parameters showed that the ethanol extract had a definite anti-osteoporotic effect.	CQ has anti-osteo-porotic effect.	Shirwaikar A, Khan S, Malini S J Ethnophar-macol. 2003; 89(2-3): 245-50[i].

Contd...

TABLE 14.2–*Contd....*

Sl.No.	Title	Type of Study	Study Design	Form and Dose	Summarized Efficacy	Conclusion	Reference
			3- Raloxifen; 4- 5 CQ extract 500-750mg/kg Biomechanical/ biochemical and histopathological parameter observed.				
5.	Screening of various fractions of the ethanol extract of *Cissus quadrangularis* Linn. for their possible antiosteoporotic activity.	*In vivo*	Model: Ovarirectomized rats Species: Wistar rats N= 72 Groups: 12 Group SHAM operated control; Group 2 ovariectomized control; Group 3 Raloxifen, Group 4 to 12 were fed with extracts; Assessment by bio-chemical, biomechanical and histopathological examination.	Total n-hexane fraction (150 and 300 mg/kg/ day), unsaponified n-hexane fraction (100 and 200 mg/kg/ day), chloroform fraction (75 and 150 mg/ kg/day), ethyl acetate fraction (125 and 150 mg/kg/day) and total ethanol extract (750 mg/kg/ day).	All treated groups showed a moderate to significant reduction in (Tartrate Resistant acid Phosphate) TRAP (Best in Choroform one) and a moderate to significant elevation in Alkaline phosphatase. It showed effect in antiosteoclastic activity without affecting the osteoblastic function as reflected in the bone strength assessed by biomechanical parameters. Histopathological studies also showed the normalization of bone architecture.	CQ extract is used in prevention of management of osteoporosis.	S Khan, J Trivedi, A Shirwaikar Pharma Nutrition; 04/2011.
					Anti-osteoporosis Activity (Clinical)		
6.	Management of Colles fracture by Asthishrankhala a clinical study.	Randomized uni-central clinical study	**Type:** Randomized uni-central study **Sample Size** 30 Subjects selected with Colles fracture **Study Group**: 4 groups Osteoseal; Moringa; Harjor; placebo;	CQ powder 3-4 gm/BD per day for 6 weeks.	The group with both the external and internal use of CQ was best among individual use. *C. quadrangularis* is effective in management of Colles fracture and it is safe cost effective and free from any side effects.	CQ has anti osteoporotic property.	Arawatti S, Jain N, Nirmalakar U, International Research Journal of Pharmacy, 2012; 3 (10).

Contd...

TABLE 14.2–*Contd....*

Sl.No.	*Title*	*Type of Study*	*Study Design*	*Form and Dose*	*Summarized Efficacy*	*Conclusion*	*Reference*
			Parameters Testing tenderness and callus assessment **Duration of Study**: 4 week. Group A with Asthishrunkala ext. application (2 wk), Group B oral CQ powder with milk (6 wk), Group C with both oral and LA. Immobilization 4 wk.				
					Fracture Healing Activity (Preclinical)		
7.	Effect of *Cissus quadrangularis* in accelerating healing process of experimentally fractured radius-ulna of dog.	*In vivo*	Species: Mongrel Dog; N=8; Group =2 (CQ extract and another group with saline water); Group A Control, Group B; Model: Experimentally fractured radius-ulna by radiological, histological and biochemical parameters pretening to serum calcium level; Parameters: Histopathological and radiological investigations on 11th days was done.	CQ extract 50mg/kg	CQ treated animals revealed faster initiation of healing process than the control animals on radiological and histopathological examinations. The treated group also revealed a decrease in serum calcium level to a greater extent than the control group. Healing was almost complete on 21st day of fracture in the treated animals and remained incomplete in the control animals.	*C. quadrangularis* has a fracture healing property.	Deka DK, Lahon LC, Saikia J, Mukit A, Indian Journal of Pharmacology, 1994; 26(1): 44-45.
					Fracture Healing Activity (Clinical)		
8.	Clinical evaluation of *Cissus quadrangularis* and *Moringa oleifera* and osteoseal as osteogenic agents in mandibular fracture.	Clinical study	**Type:** Comparative study Sample Size 44 Subjects selected with mandibular fracture; Study Group: 4 groups Osteoseal; Moringa; Harjor; placebo Duration of Study: 6 week. Parametes Testing serum calcium, total and ionic, phosphorus, radiograph.	2 capsule twice a day	Harjod showed reduction in Pain, Swelling, Tenderness, Mobility however it was maximum in Osteoseal. Also there was increased in the serum calcium and phosphorus level. It has been observed that CQ increases rate of fracture healing and enhances entire remodeling process.	*C. quadrangularis* has fracture healing property.	Singh V, Singh N, Pal US, Dhasmana S, Mohammad S, Singh N, National J Maxillofac Surg. 2011; 2(2): 132–136.

Contd...

TABLE 14.2–*Contd....*

Sl.No.	Title	Type of Study	Study Design	Form and Dose	Summarized Efficacy	Conclusion	Reference
9.	Herbal remedies for mandibular fracture healing.	Clinical study	Type: Comparative study Sample Size 29 Subjects selected with mandibular fracture Study Group: 3 groups *Ocimum sanctum, Cissus quadrangularis* and Control Duration of Study: 4-8 week. Parameters Testing tensile strength and period of immobilization.	CQ powder 3.5 gm/8 hrly till clinical fracture healing	A significant increase in alkaline phosphatase and serum calcium was seen in CQ group. The period of immobilization was the lowest in the *OS* group followed by the *CQ* group. The tensile strength measured in terms of biting force was maximum in the *CQ* group.	*Cissus quadrangularis* has better fracture healing property.	Mohammad S, Pal US, Pradhan R, Singh N, National J Maxillofacal Surg. 2014; 5(1): 35-8.
10.	Osteogenic potential of *Cissus qudrangularis* assessed with osteo-pontin expression.	Fracture	Type: Comparative study Sample Size 60; Subjects selected with mandibular fracture Study Group: 2 groups *Cissus quadrangularis*/ Control Duration of Study: 8 week. Parameters Testing Clinical examination and radiography.	Each cap 300 CQ powder, 2 cap. twice a day	Significant reduction in pain, swelling, and mobility at fracture site. serum calcium level and alkaline phosphatase level increased in CQ group. Systemic use of CQ in rats caused complete restoration of normal composition of bone, after fracture in 4 weeks, while the controls required 6 weeks.	CQ accelerates fracture healing and also causes early remodeling of fracture callus.	Singh N, Singh V, Mehta G. National J Maxillofacal Surgery. 2013; 4(1): 52-6.
					JATAMANSI (*Nardostachys jatamansi* (D.Don) DC.)		
					Sleep quality		
1.	Comparative study of sedative and anxiolytic effects of herbal extracts of *Hypericum perforatum* with *Nardostachys jatamansi* in rats.	*In vivo*	Ketamine induced in Wister rats.	Intraperitoneally in dose of 100 mg/kg, 200 mg/kg and 400 mg/kg	Significant increase in sleep time induced by ketamine in the treatment groups with high and low doses of valerian extracts and the hypericum is significant at the 0.01 level.	Valerian exbitit significant sedative and anxiolytic effects.	Ali Rezaei, *et al.*, Zahedan J Res Med Sci 2014; 16(3): 40-43.
					Anti-Stress and Anti-anxiety activity		
2.	Stress modulating anti-oxidant effect of *Nardostachys jatamansi.*	*In vitro*	Cold restraint stress induced alterations in wister rat for free radicle scavenging activity.	*Nardostachys jatamansi* ethanolic extract (NJE) 200 and 500 mg/kg, orally	NJE showed antioxidant activity and significantly reversed the stress-induced elevation of lipid peroxidation (LPO) and nitric oxide (NO) levels and decrease in catalase activity in the brain. NJE also significantly altered stress-induced increase in adrenal and spleen weights and decrease in level of ascorbic acid in adrenal gland.	*Nardostachys jatamansi* possesses significant antistress activity.	Nazmun Lyle *et al.*, Indian Journal of Biochemistry and Biophysics 2009; 46:93-98.

Contd...

TABLE 14.2–*Contd....*

Sl.No.	*Title*	*Type of Study*	*Study Design*	*Form and Dose*	*Summarized Efficacy*	*Conclusion*	*Reference*
3.	Anxiolytic effects of *Nardostachys jatamansi* DC in mice.	*In vivo*	The effect on anxiety was assessed using Elevated plus maze and open field test. The behavioural testing was recorded using the Any maze software, Stoelting Co., USA.	Aqueous and 70 per cent ethanolic extracts at doses 125, 250 and 500 mg/kg body weight were administered through intraperitoneal route.	Pre-treatment of mice Aqueous and 70 per cent ethanolic extracts significantly alleviated anxiety by increasing the number of entries on the open arm, time spent on open arms, percent time spent on open arms and decreasing the time spent on closed arms in the elevated plus maze test. Moreover, the 70 per cent ethanolic extract also significantly increased the locomotor activity as observed in the open field test confirming the anxiolytic effects in mice.	The anxiolytic properties of the herb in mice suggest that the herb could serve as a new approach in the treatment of anxiety.	Razack S and Khanum F, Annals of Phytomedicine 2012; 1(2): 67-73.
					Anti-depressant activity		
4.	Comparative study of antidepressant activity of methanolic extract of *Nardostachys jatamansi* DC Rhizome on normal and sleep deprived mice.	*In vivo*	Forced swim test, Tail suspension test and locomotor activity in inbred male Swiss rat.	Methanlolic extract of jatamansi 200 and 400 mg/kg, p.o	MENJ (200 and 400 mg/kg, p.o) produced significant (P<0.001) antidepressant likeeffect in normal and sleep deprived mice in both TST and FST and their efficacies were found to be comparable to imipramine (10 mg/kg, p.o). It did not show any significant change in locomotor functions of mice as compared to normal control. However, it significantly (P<0.01) improved the locomotor activity in case of sleep deprivation which is comparable to normal control.	Jatamansi has anti-depressant activity.	Habibur Rahman *et al.*, Der Pharmacia Lettre, 2010; 2(5): 441-449.
5.	Inhibition of MAO and GABA: probable mechanisms for anti-depressant-like activity of *Nardostachys jatamansi* DC. in mice.	*In vivo*	To evaluate antidepressant activity of Ethanolic extract of *Nardostachys jatamansi* tail suspension and forced swim tests.	Ethanolic extract (100, 200 and 400 mg/kg, po) of *N. jatamansi* administered for 14 successive days to Swiss young albino mice	The efficacy of the extract was found to be comparable to imipramine (15 mg/kg, po) and sertraline (20 mg/kg, po). Ethanolic extract did not show any significant change on locomotor activity of mice as compared to control; hence it did not produce any motor effects. Further, the extract decreased the whole brain MAO-A and MAO-B activities as compared to control, thus increased the levels of monoamines. The anti-depressant effect of the extract was also significantly reversed by pretreatment of animals with baclofen (GABAB agonist); when tested in tail suspension test.	The results suggested that the antidepressant-like effect of the extract may also be due to interaction with GABAB receptors, resulting in decrease in the levels of GABA in mouse brain. Thus, the extract may have potential therapeutic value for the management of mental depression.	Dhingra D and Goyal PK, Indian J Exp Biol. 2008; 46 (4): 212-8.
					Nootropic activity		
6.	*Nardostacys jatamansi* DC protects from the loss of memory and cognition deficits in sleep.	*In vivo*	Sleep deprived (SD) amnesic mice via Elevated Plus Maze test, Step down Passive Shock	Methanolic extract of *Nardostacys jatamansi*	Behavioral changes were evaluated against normal control and negative control animals using Elevated plus maze, passive shock avoidance, Y-maze, Morris water-maze and object recognition tests. Methanolic extract of *Nardostacys jatamansi* treated groups	Methanolic extract of *Nardostacys jatamansi* exerts a protective effect against loss of memory	Habibur Rahman *et al.*, International Journal of

Contd...

TABLE 14.2–*Contd....*

Sl.No.	*Title*	*Type of Study*	*Study Design*	*Form and Dose*	*Summarized Efficacy*	*Conclusion*	*Reference*
	deprived alzheimer's disease (ad) mice model		avoidance test, Morris water maze task and Object recognition task.	200 and 400 mg/kg, orally for 14 days	showed a significant improvement in learning and cognition parameters in behavioral tests.	and cognitive deficits due to sleep deprivation.	Pharmaceutical Sciences Review and Research, 2010; 5(3): Article-029.
7.	Non-linear dose effect relationship in anxiolytic and nootropic activity of lithium carbonate and *Nardostachys jatamansi* in rats.	*In vivo*	Elevated plus maze test and object recognition test.	Acute administrations of lithium carbonate (10 mg/kg i. p.) NJE (100 and 200 mg/kg i. p.) given separately for 21 days	Acute administration of lithium carbonate and NJE significantly increased time spent in the open arm indicating reduced anxiety. Lithium carbonate significantly reduced anxiolytic activity of NJE. Rats receiving lithium carbonate (for 21 days) and NJE acutely on 21st day, showed significantly diminished anxiolytic activity of NJE. NJE used alone, exhibited nootropic activity in the elevated plus maze and object recognition test. In the elevated plus maze, acute administration of lithium carbonate (10 mg/kg) and NJE (100 mg/kg) showed significant improvement in nootropic activity and diminished nootropic response to NJE (200 mg/kg). Similar biphasic effect was observed when rats received lithium carbonate for 21 days and NJE on 21st day. In the object recognition test, acute administration of lithium carbonate and NJE improved nootropic activity but in rats receiving lithium for 21 days, NJE (100 mg/kg) improved and (200 mg/kg) reduced nootropic activity.	In conclusion, lithium carbonate and *N. jatamansi* extract, exhibited anxiolytic and nootropic activity when used alone; anxiolytic activity of NJE decreased when given with lithium carbonate; lower dose of NJE and lithium carbonate showed improved nootropic activity but higher dose of NJE decreased nootropic activity.	Kasture SB *et al.*, Orient Pharm Exp Med. 2014; 14(4): 357–362.
					Anti-inflammatory Activity		
8.	Evaluation of anti-inflammatory potential of *Nardostachys jatamansi* rhizome in experimental rodents.	*In vivo*	To evaluate the anti-inflammatory effect of *N. jatamansi* rhizome against acute (Carrageenin-induced hind paw oedema in rats and Autacoids-induced hind paw oedema in rats), subacute (Formaldehyde-induced hind paw volume) and chronic (Cotton pellet granuloma in rats) models of inflammation in experimental animals.	*N. jatamansi* rhizome extract (150 and 300 mg/kg, p.o.) and the reference drugs phenylbutazone (100 mg/kg, p.o.) and acetylsalicylic acid (300 mg/kg, p.o.) were used	In acute inflammation as produced by carrageenin 29.06 per cent and 55.81 per cent, by histamine 25.0 per cent and 39.28 per cent, by 5-hydroxytryptamine 21.37 per cent and 36.95 per cent and by prostaglandin E2- induced hind paw oedema 31.03 per cent and 44.82 per cent protection was observed. While in subacute anti-inflammatory models using formaldehyde-induced hind paw oedema (after 1.5 h) 13.88 per cent and 33.33 per cent and in chronic anti-inflammatory model using cotton pellet granuloma 7.4 per cent and 17.58 per cent protection from inflammation was observed. *N. jatamansi* rhizome extract also inhibited the inflammatory mediators (nitric oxide by 12.81 per cent and 38.41 per cent, by prostaglandin E2 12.58 per cent and 47.82 per cent while by TNF-α 13.51 per cent and 41.89 per cent) produced in the pouch.	The results of this study strongly indicate the protective effect of *N. jatamansi* rhizome extract against acute, subacute and chronic models of inflammation, which may be attributed to its anti-inflammatory potential.	Singh RK *et al.*, Journal of Coastal Life Medicine 2014; 2(1): 38-43.

Contd...

TABLE 14.2–*Contd....*

Sl.No.	*Title*	*Type of Study*	*Study Design*	*Form and Dose*	*Summarized Efficacy*	*Conclusion*	*Reference*
9.	Beneficial effects of fractions of *Nardostachys jatamansi* on lipopoly-saccharide-induced inflammatory response.	*In vivo*	To evaluate the anti-inflammatory effect of *N. jatamansi* roots against lipopolysaccharide (LPS) induced acute inflammatory model.	*N. jatamansi* roots extract (0.1μg/ml, 1μg/ml and 10μg/ml) was administered by intraperitoneal (i.p.) injection	NJ-6 fraction inhibited LPS-induced production of NO. The NJ-3, NJ-4, and NJ-6 fractions also inhibited the production of cytokines, such as IL-1β, IL-6, and TNF-α. However, NJ-1, NJ-3, NJ-4, and NJ-6 showed differential inhibitory mechanisms against LPS-induced inflammatory responses. NJ-1, NJ-3, and NJ-4 inhibited LPS-induced activation of c-jun NH2-terminal kinase (JNK) and p38 but did not affect activation of extracellular signal-regulated kinase (ERK) or NF-κB. On the other hand, NJ-6 inhibited activation of MAPKs and NF-κB. In addition, *in vivo* experiments revealed that administration of NJ-1, NJ-3, NJ-4, and NJ-6 reduced LPS-induced endotoxin shock, with NJ-6 especially showing a marked protective effect.	Among the 6 fractions isolated, NJ-1, NJ-3, NJ-4, and NJ-6 showed the inhibition of NO, and NJ-3, NJ-4, and NJ-6 showed the inhibition of cytokines. In addition, NJ-4 and NJ-6 exhibited dramatic prevention against LPS-endotoxin shock. These results suggest that NJ-4 and NJ-6 may be effective and beneficial candidates to ameliorate LPS-induced inflammation.	Bae GS *et al.*, Evid Based Complement Alternat Med. 2014; 2014: 837-835.
10	Antioxidant and diuretic activity of *Nardostachys jatamansi* Dc. roots.	*In vivo* and *in vitro*	*In vitro* antioxidant activity was measured by using hydrogen peroxide scavenging activity, reducing power assay, DPPH scavenging activity. The diuretic effect of the extracts was evaluated by measuring urine volume, electrolyte concentration, pH, density, conductivity, saluretic index, diuretic index and carbonic anhydrase inhibition ratio.	Ethyl acetate extract of *N. jatamansi* DC roots administered in rats at 100mg/kg, 200mg/kg and 500 mg/kg, orally and compared with furosemide (20mg/Kg, intraperito neally) as the standard	Ethyl acetate extract of *Nardostachys jatamansi* DC roots showed significant Anti-oxidant activity. Extract induced strong diuresis at higher doses and was not accompanied with a reduction in urinary K^+ levels. Further, there was no alkalization of urine. Collectively, these observations suggest that the extract is not acting as potassium-sparing diuretics. Extract showed strong diuretic index, saluretic index, natriuretic index and weak carbonic anhydrase inhibition ratio.	*Nardostachys jatamansi* DC roots extract exhibit anti-oxidant and diuretic activity and could be a potential medicine for renal or hypertensive syndromes.	Kendre DG *et al.*, American Journal of Pharmacy and Health Research. 2014; 2(9): 84-96.

Contd...

TABLE 14.2–*Contd....*

Sl.No.	*Title*	*Type of Study*	*Study Design*	*Form and Dose*	*Summarized Efficacy*	*Conclusion*	*Reference*
				KALMEGH (*Andrographis paniculata* (Burm.f.) Wall. ex Nees)			
					Anti-inflammatory Activity		
1.	Andrographolide prevents oxygen radical production by human neutrophils: Possible mechanism(s) involved in its anti-inflammatory effect.	*In vitro*	N-formyl-methionyl-leucyl-phenylalanine (fMLP)-induced adhesion and trans-migration of isolated peripheral human neutrophils.	Androgra-pholide 0.1-10 microM	Andrographolide pretreatment significantly decreased fMLP-induced up-expression of both CD11b and CD18. PMA triggered remarkable ROS production and adhesion, and were partially reversed by ANDRO. This indicated that a PKC-dependent mechanism might be interfered by ANDRO.	The prevention of ROS production through, modulation of PKC-dependent pathway could confer ANDRO the ability to down-regulate Mac-1 up-expression that is essential for neutrophil adhesion and trans-migration.	Shen, Y.C., *et al.*, British Journal of Pharmacology, 2002; 135(2): 399-406.
					Upper Respiratory Tract Infection		
2.	A randomized double blind placebo controlled clinical evaluation of extract of *Andrographis paniculata* (Kalm Cold) in patients with uncomplicated upper respiratory tract infection.	Clinical study	Randomized double blind study; Sample Size: 223; Nine self evaluated symptoms of cough, expectoration, nasal discharge, headache, fever, sore throat, earache, malaise/fatigue and sleep disturbance were scored.	Kalm Cold tablet 200 mg/day for 5 days	In both the treatments, mean scores of all symptoms showed a decreasing trend from day 1 to day 3 but from day 3 to day 5 most of the symptoms in placebo treated group either remained unchanged (cough, headache and earache) or got aggravated (sore throat and sleep disturbance) whereas in KalmCold treated group all symptoms showed a decreasing trend. In both placebo and KalmCold treated groups, there were only a few minor adverse effects with no significant difference in occurrence ($Z = 0.63$; $p > 0.05$). The comparison of overall efficacy of KalmCold over placebo was found to be significant ($p <$ or $= 0.05$) and it was 2.1 times (52.7 per cent) higher than placebo.	KalmCold is effective in reducing symptoms of upper respiratory tract infection.	Saxena RC, *et al.*, Phyto-medicine. 2010; 17(3-4): 178-185.
3.	*Andrographis paniculata* in the treatment of upper respiratory tract infections: a systematic review of safety and efficacy.	Clinical studies-Review	Seven double-blind, controlled trials (n = 896) were evaluated for efficacy. All types of study design were eligible for inclusion in the safety review.	Studies of single or combined preparations of oral *A. paniculata* were included. The daily dose of androgra-pholide ranged from 48 to	*A. paniculata* is superior to placebo in alleviating the subjective symptoms of uncomplicated upper respiratory tract infection. There is also preliminary evidence of a preventative effect. Adverse events were described as mild, infrequent and reversible.	*A. paniculata* may be a safe and efficacious treatment for the relief of symptoms of uncomp-licated upper respiratory tract infection.	Coon JT and Ernst E; Planta Med. 2004; 70(4): 293-298.

Contd...

TABLE 14.2–*Contd....*

Sl.No.	Title	Type of Study	Study Design	Form and Dose	Summarized Efficacy	Conclusion	Reference
				360 mg/day in the efficacy review and from 11 mg/day to 10 mg/kg per day for studies included in the safety review			
4.	*Andrographis paniculata* in the symptomatic treatment of uncomplicated upper respiratory tract infection: systematic review of randomized controlled trials.	RCT - Review	Systematic review of the literature and meta-analysis of randomized controlled trials.	-	A total of 433 patients reported in three trials were included in the statistical analysis. Andrographis paniculata in fixed combination was more effective than placebo. The difference in effects between *A. paniculata* and placebo was 10.85 points (95 per cent CI 10.36-11.34 points, P<0.0001) in favour of *A. paniculata.*	*A. paniculata* extract alone or in combination may be more effective than placebo and may be an appropriate alternative treatment of uncomplicated acute upper respiratory tract infection.	Poolsup N, *et al.*, J Clin Pharm Ther. 2004; 29(1): 37-45.
5.	Use of visual analogue scale measurements (VAS) to asses the effectiveness of standardized *Andrographis paniculata* extract SHA-10 in reducing the symptoms of common cold. A randomized double blind-placebo study.	Clinical Study	Randomized double blind study; Sample Size: 158 adult patients of both sexes; Evaluation was done at day 0, 2, and 4 of headache, tiredness, earache, sleeplessness, sore throat, nasal secretion, phlegm, frequency and intensity of cough.	*Andrographis paniculata* dried extract 1200 mg/day for 5 days	At day 4, a significant decrease in the intensity of all symptoms was observed for the Andrographis paniculata group. The higher Odds ratio values were for the following parameters: sore throat (OR = 3.59; 95 per cent CI 2.04-5.35), nasal secretion (OR = 3.27; 95 per cent CI 2.31-4.62) and earache (OR = 3.11; 95 per cent CI 2.01-4.80) for Andrographis paniculata treatment over placebo, respectively.	*Andrographis paniculata* had a high degree of effectiveness in reducing the prevalence and intensity of the symptoms in uncomplicated common cold beginning at day two of treatment. No adverse effects were observed or reported.	Caceres, D.D., Hancke *et al.*, Phytomedicine. 1999; 6(4): 217-223.
6.	Comparative controlled study of *Andrographis paniculata* fixed combination, Kan Jang and an Echinacea preparation as adjuvant, in the treatment of uncomplicated respiratory disease in children.	Clinical Study	Three-arm study; Sample Size: 130 children aged between 4 and 11 years; 3 groups - standard treatment. Adjuvant group with herbal tablets and adjuvant control group with *Echinacea purpurea.*	Kan Jang (Standardized *Andrographis paniculata* extract); Immunal (*Echinacea purpurea* extract), over a period of 10 days.	It was found that the adjuvant treatment with Kan Jang was significantly more effective than Echinacea purpurea, when started at an early stage of uncomplicated common colds. The effect was more in amount of nasal secretion g/day and nasal congestion. It also accelerated the recovery time, whereas *Echinacea purpurea* did not show the same efficacy.	Standardized *A. paniculata* extract has good results in common cold than standard treatment. There was no side effects or adverse reactions reported.	Spasov, A.A., *et al.*, Phytother. Res. 2004; 18(1): 47-53.

Contd...

TABLE 14.2–*Contd....*

Sl.No.	*Title*	*Type of Study*	*Study Design*	*Form and Dose*	*Summarized Efficacy*	*Conclusion*	*Reference*
7.	A double blind, placebo-controlled study of *Andrographis paniculata* fixed combination Kan Jang in the treatment of acute upper respiratory tract infections including sinusitis.	Clinical Study	A double blind, placebo-controlled, parallel-group clinical study; Sample Size 185; Temperature, headache, muscle aches, throat symptoms, cough, nasal symptoms, general malaise and eye symptoms were taken as outcome measures with given scores	Duration 5 days	The total score analysis showed a highly significant improvement in the Andrographis group versus the placebo. The individual symptoms of headache and nasal and throat symptoms together with general malaise showed the most significant improvement while cough and eye symptoms did not differ significantly between the groups. Temperature was moderately reduced in the Andrographis group.	*A. paniculata* has a positive effect in the treatment of acute upper respiratory tract infections and also relieves the inflammatory symptoms of sinusitis. The drug is well tolerated.	Gabrielian, E.S., *et al.*, Phyto-medicine. 2002; 9(7): 589-597.
8.	A randomized, contro-lled study of Kan Jang versus amantadine in the treatment of influenza in Volgograd.	Clinical Study	Two randomized, parallel-group clinical studies with Kan Jang and a control group were performed. The pilot study was performed on 540 patients with 71 Kan Jang-treated patients with the second phase conducted enrolling 66 patients.		The differences in the duration of sick leave and frequency of post-influenza complications indicate that the Kan Jang phyto-preparation not only contributes to quicker recovery, but also reduces the risk of post-influenza complications. Kan Jang was well tolerated by patients.	Andrographis extract is effective in reducing symtoms of influenza.	Kulichenko, L.L., *et al.* J.Herb. Pharma-cother. 2003; 3(1): 77-93.
9.	Controlled clinical study of standardized *Andrographis paniculata* extract in common cold - a pilot trial.	Clinical Study	Double blind controlled clinical study; Sample size: 50 patients		Both subjective symptoms as well as duration of the symptoms were significantly reduced.	Andrographis decreases the subjective symptoms of common cold as well as shortens the period of sick leave significantly.	Melchior J *et al.*, Phyto-medicine, 1997, 3: 315-318.
					Safety		
10.	Evaluation of the geno-toxic potential and acute oral toxicity of standar-dized extract of *Andrographis paniculata* (KalmCold™).	*In vitro* and *In vivo*	Genotoxicity was perfor-med through three different *in vitro* tests: Ames, chromosome aberration (CA), and micronucleus (MN).; Oral toxicity in female rats observed for 14 days.	Ames test was performed at different strengths and acute oral toxicity study at 5000 mg/kg of KalmCold	Results of Ames test confirmed that KalmCold did not induce mutations both in the presence and absence of S9 in *Salmonella typhimurium* mutant strains TA98 and TAMix. In CA and MN, KalmCold did not induce clastogenicity in CHO-K1 cells *in vitro*. No toxicity was observed in the acute toxicity study.	*A. paniculata* is geno-toxically safe and did not produce any treatment-related toxic effects in rats.	C.V. Chandra-sekaran, *et al.*, Food and Chemical Toxicology, 2009; 47(8): 1892–1902.

Contd...

TABLE 14.2–*Contd....*

Sl.No.	*Title*	*Type of Study*	*Study Design*	*Form and Dose*	*Summarized Efficacy*	*Conclusion*	*Reference*
					KANTKARI (*Solanum xanthocarpum* Schrad. and H. Wendl.)		
					Bronchial asthma		
1.	A pilot study on the clinical efficacy of *Solanum xanthocarpum* and *Solanum trilobatum* in Bronchial asthma.	Clinical	A pilot study was undertaken to investigate the clinical efficacy and safety of a single dose of the above herbs in mild to moderate bronchial asthma.	Oral administration of 300 mg powder of whole plant	Treatment with either *S. xanthocarpum* or *S. trilobatum* significantly improved the various parameters of pulmonary function in asthmatic subjects. However, the effect was less when compared to that of deriphylline or salbutamol.	Study confirmed the traditional claim for the usefulness of these herbs in bronchial asthma.	Govindan S, J. Ethnopharmacol. 1999; 66(2): 205-10.
					Anti-inflammatory Activity		
2.	Evaluation of anti-inflammatory activity of *Solanum xanthocarpum* Schrad and Wendl (Kantakāri) extract in laboratory animals.	*In vivo*	In continuation of search for potent natural anti-inflammatory agents, the present research work was planned to evaluate the anti-inflammatory activity of ethanol extract of *S. xanthocarpum* whole plant.	10, 30 and 100 mg/kg p.o for 7 days	Administration of 100 mg/kg p.o for 7 day reduced the granuloma formation in cotton pellet granuloma model.	Results support the traditional use of plant for anti-inflammatory activity	Shraddha K. More *et al.*, Anc Sci Life. 2013; 32(4): 222–226.
3.	Study of the synergistic anti-inflammatory activity of *Solanum xanthocarpum* Schrad and Wendl and *Cassia fistula* Linn.	*In vivo*	The water extract of dried fruits of *Solanum xanthocarpum* Schrad and Wendl and dried pulp of *Cassia fistula* Linn was prepared. The anti-inflammatory activity of these extracts was investigated using the carragenan-induced paw edema model in rats individually and in two different combinations.	Oral single doses of individual plant extracts ranged from 100 to 500 mg/kg and a combination of both plant extracts were administered	It was observed that extracts of dried fruits of *Solanum xanthocarpum* showed more anti-inflammatory activity than dried fruits of *Cassia fistula* Linn. Both the extracts showed maximum anti-inflammatory activity at 500 mg/kg dose. Among the different dose combinations of both the extracts, the 1:1 combination at the 500 mg/kg dose showed maximum percentage inhibition of 75 per cent, which was comparable with the positive control, diclofenac sodium, which showed 81 per cent inhibition.	As revealed by the isobolograms, both the combinations fell below the additivity line, which indicates synergistic interactions between *Solanum xanthocarpum* and *Cassia fistula* extracts. Interaction indices of both combinations were observed to be <1, which re-demonstrated the synergistic effects of the combination.	Anwikar S and Bhitre M, Int J Ayurveda Res. 2010; 1(3): 167–171.

Contd...

TABLE 14.2–*Contd....*

Sl.No.	Title	Type of Study	Study Design	Form and Dose	Summarized Efficacy	Conclusion	Reference
					Oxidative potential of cauda epididymal spermatozoa		
4.	Effect of *Solanum surattense* seed on the oxidative potential of cauda epididymal spermatozoa.	*In vivo*	To evaluate the effect of aqueous seed extract of *Solanum surattense* on the oxidative potential of cauda epididymal spermatozoa.	Oral dosage of 10 mg/kg b.w. for 15 days	The activity levels of the enzymes AST and ALT, which are considered to be the androgenicity in the sperm suspension, were depleted in the extract fed rats. The activity level of the enzyme ICDH, was reduced significantly in the treated group ($P<0.001$).	It can be concluded that the oral administration of the aqueous seed extract of *S. surattense* can deplete the oxidative stress of cauda epididymal spermatozoa in albino rats.	T Thirumalai, *et al.*, Asian Pac J Trop Biomed. Jan 2012; 2(1): 21–23.
					Antimicrobial Activity		
5.	Evaluation of antimicrobial potential of different extracts of *Solanum xanthocarpum* Schrad. and Wendl.	*In vitro*	Antimicrobial activity of the aqueous and organic solvent extracts of different parts (roots, stems, leaves and fruits) of *S. xanthocarpum* against Gram-positive and Gram negative bacteria and a fungus was evaluated. Plant extracts of *S. xanthocarpum* were prepared in distilled water and in organic solvents, *viz.* ethanol, benzene, acetone and methanol. Agar well diffusion technique was used to assess the antimicrobial activity of various extracts against Gram-positive (*Staphylococcus aureus, S. epidermidis*), Gram-negative (*Escherichia coli, Pseudomonas aeruginosa*) bacteria and the fungus *Aspergillus niger*. The diameter of zone of inhibition was taken as an indicator of antimicrobial effect.		Except aqueous extracts of different parts of *S. xanthocarpum*, extracts prepared in organic solvents showed antimicrobial activity against the test organisms. A strong inhibition of *P. aeruginosa* was caused by the ethanolic and methanolic extracts of *S. xanthocarpum*.	*S. xanthocarpum* could be considered as a potential source of natural antimicrobials.	Salar RK and Suchitra, African Journal of Microbiology Research 2009; 3(3): 97-100.

Contd...

TABLE 14.2–*Contd....*

Sl.No.	Title	Type of Study	Study Design	Form and Dose	Summarized Efficacy	Conclusion	Reference
6.	Anti-fungal activity of *Solanum xanthocarpum* (Kantkari) leaf extract.	*In vitro*	The distil water (mother) extract and hexanic extract of *S. xanthocarpum* were tested against *A. niger* and *C. albicans*. The well diffusion and growth inhibition in broth methods were used for antifungal activity.		The distil water extract of *S. xanthocarpum* was not effective against both the fungal species but hexanic extract was effective against *C. albicans*. The zones of inhibition against *C. albicans* were seen in hexanic extract. Maximum growth inhibition was found in 500 ug/ml and minimum growth inhibition found in 100 ug/ml concentration of hexanic extract of *S. xanthocarpum* against *C. albicans*.	Haxen extract of *Solanum xanthocarpum* leaf showed good anti-fungal activity against the *Candida albicans* in solid and broth media.	Gaherwal S. *et al.*, World Journal of Zoology 2014; 9 (2): 111-114.
7.	Antibacterial activity of *Solanum xanthocarpum* leaf extract.	*In vitro*	The antibacterial activity of plant *S. xanthocarpum* was evaluated against some selected human pathogenic microorganisms (*Escherichia coli, Yersinia pestis, Pseudomonas aeruginosa* and *Staphylococcus aureus*) following agar-well diffusion method. Two solvents methanol and acetone were used for extraction of different bioactive constituents from fresh leaves.	Different concentrations (30 per cent, 50 per cent, 70 per cent and 100 per cent) of plant extract was used	Methanolic as well as acetone leaf extracts of *S. xanthocarpum* were quite effective in inhibiting the growth of *Staphylococcus aureus* which is a serious human pathogen causing infections in wounds.	*S. xanthocarpum* could be considered as a potential source of natural antimicrobials.	Rana S, *et al.*, Int J Curr Microbiol App Sci. 2016; 5(4): 323-328.
					Antidiabetic Activity		
8.	Studies on hypoglycaemic activity of *Solanum xanthocarpum* Schrad. and Wendl. fruit extract in rats.	*In vivo*	Screening for the hypoglycaemic activity was assessed on normoglycaemic, alloxan treated hyperglycaemic and glucose loaded rats along with *in vitro* study on glucose utilization by isolated rat hemidiaphragm. The various haematological and biochemical parameters were also studied.	Test extract was administered at doses of 100 and 200 mg/kg orally for 21 days	The extract was found to possess significant hypoglycaemic activity when compared with the reference standard glibenclamide. The *in vitro* study on glucose utilization by isolated rat hemidiaphragm suggests that the aqueous extract may have direct insulin like activity which enhances the peripheral utilization of glucose and have extra pancreatic effect. The toxicity studies report safety usage of the plant extract.	The present study reveals the hypoglycaemic potential of the *S. xanthocarpum* fruits used by the tribes since time immemorial.	D.M. Kar *et al.*, J Ethnopharmacol. 2006; 108: 251–256.

Contd...

TABLE 14.2–*Contd....*

Sl.No.	*Title*	*Type of Study*	*Study Design*	*Form and Dose*	*Summarized Efficacy*	*Conclusion*	*Reference*
9.	Antihyperglycemic and antioxidant effects of *Solanum xanthocarpum* leaves (field grown and *in vitro* raised) extracts on alloxan induced diabetic rats.	*In vivo*	The antidiabetic activity of the crude methanol extracts of the field grown and *in vitro* raised leaves of *S. xanthocarpum* was tested against alloxan induced diabetic rats. The antidiabetic efficacy was validated through various biochemical parameters and the antioxidant effect was also determined.	Test extract was administered at different doses 100–200 mg/kg bw 25 days	The results revealed that the methanol extracts of both the leaves (field grown and *in vitro* raised) of *S. xanthocarpum* was efficient anti-hyperglycemic agents at a concentration of 200 mg/kg bw and possess potent antioxidant activity. However, *S. xanthocarpum* raised leaves exhibit higher efficacy than the field grown leaves in all tested concentrations. Proximal composition and mineral analysis of *S. xanthocarpum* revealed higher concentration of contents in *in vitro* rasied *S. xanthocarpum* than field grown *S. xanthocarpum.*	The leaves extracts of *S. xanthocarpum* can be a potential candidate in treating the hyperglycemic conditions and suits to be an agent to reduce oxidative stress.	K Poongothai *et al.*, Asian Pacific Journal of Tropical Medicine. 2011; 4(10): 778-785.
10.	Antidiabetic and anti-hyperlipidaemic effects of *Solanum xanthocarpum* total extract in alloxan induced diabetic rats.	*In vivo*	The antidiabetic and antihyperlipidaemic effects of aqueous-methanol (40:60) extract of *S. xanthocarpum* whole plant (Sx) were investigated in normoglycemic, glucose fed, and alloxan-induced diabetic rats	Sx, administered at doses of 200 mg/kg and 400 mg/kg for 7 days	In normoglycemic rats, the powdered extract of Sx resulted in reduction of blood glucose level. In glucose fed diabetic rats, the reduction of blood glucose level was also achieved dose dependently and significantly compared to the vehicle control group. The treatment showed significant dose-dependent percentage blood glucose lowering in the diabetic rats.	Sx plant extract has potent antidiabetic and antihyperlipidaemic activity and was found to improve the blood glucose and lipid levels of alloxan induced diabetic rats at the end of the treatment period.	Gupta AK, *et al.*, Pharmacologyonline 2009; 1: 484-497.
					KUTKI (*Pichrohiza kurroa* Royle ex Benth)		
					Antidiabetic activity		
1.	Antidiabetic activity of standardized extract of *Picrorhiza kurroain* rat model of NIDDM.	*In vivo*	*Diabetes mellitus* was induced with streptozotocin-nicotinamide. Fasting blood glucose levels, liver glycogen level, Oral glucose tolerance and lipid profiles were measured after two week treatment.	Aqueous extract of *Picrorhiza kurroa* PkE (100 and 200 mg/kg, p.o.) for 14 days	PkE treatment induced significant reduction ($p < 0.001$) inelevated fasting blood glucose level in streptozotocin nicotin amide induced type-2 diabetic rats. In oralglucose tolerance test, oral administration of PkE increased the glucose tolerance. PkE treatment also significantly ($p < 0.001$) reversed the weight loss associated with streptozotocin treatment.	Standardized extract of *Picrorhiza kurroa* possess significant antidiabetic activity in streptozotocin-nicotin amide induced type-2 diabetes mellitus in rats	G M Husain *et al.*, Drug Discov Ther. 009 Jun; 3(3): 88-92.
					Antioxidant Activity		
2.	Evaluation of antioxidant activity of *Picrorhiza kurroa* (Leaves) extracts.	*In vitro*	All the extracts and isolated compounds were evaluated for its antioxidant activity using two	*Picrorhiza kurroa* (Leaves) extracts	Butanol and ethyl acetate extract showed geater antioxidant activity as compared to ethanol extract. Compound 1 and ascorbic acid showed nearly similar antioxidant activity where as 2 showed no activity at standard concentration. The IC50	Antioxidant and radical scavenging activity of parent extract, fractions, and isolated compound	K. Kant *et al.*, Indian J Pharm Sci. 2013 May; 75(3): 324-9.

Contd...

TABLE 14.2–*Contd....*

Sl.No.	Title	Type of Study	Study Design	Form and Dose	Summarized Efficacy	Conclusion	Reference
			assays, 2,2-diphenyl-1-picrylhydrazyl radical and 2,2'-azino-bis(3-ethylbenzothiazoline-6-sulphonic acid) assay.		values for 2,2-diphenyl-1-picrylhydrazyl radical and 2,2'-azino-bis(3-ethylbenzothiazoline-6-sulphonic acid) assay for ascorbic acid, compound 1, ethanol extract and its different fractions (ethyl acetate and butanol) were found to be 0.81, 1.04, 67.48, 39.58, 37.12 and 2.59, 4.02, 48.36, 33.24, 29.48 μg, respectively.	of *P. kurroa* leaves indicate its role toward various oxidative stress related diseases, as a food supplement and source of natural antioxidants.	
					Immunostimulant properties		
3.	Protective effects of *Picrorhiza kurroa* on cyclophosphamide-induced immuno-suppression in mice.	*In vivo*	Cyclophasphamide-induced immuno-suppression. The mice were antigenically challenged with SRBC (0.5x109 cells/ml/100 g) on 10th day intrapari-toneally. Cellular immune response (Foot pad reaction test) the edema was induced by injecting SRBC (0.025x109 cells) in left paw, and 0.025 ml of saline was injected in right paw.	Ethanolic and aqueous extract of the rhizomes of *Picrorrhiza kurroa*	The plant extract showed protectiveeffects on humoral immunity. The change in percentage deduction in footpad volume was also found significant (P<0.001). Administration of extract remarkably ameliorated both cellular and humoral antibody response.	Extracts possessed promising immuno-stimulant properties. But, the alcoholic extract is more potent than aqueous extract in producing delayed type hypersensitivity response.	A. Hussain *et al.*, Pharma-cognosy Research, 2013; 5: 30-5.
					Hepato-protective activity		
4.	A study of standardized extracts of *Picrorhiza kurroa* Royle ex Benth in experimental non-alcoholic fatty liver disease.	*In vivo*	Male Wistar rats were challenged with 30 per cent high fat diet (HFD) –butter, for 2 weeks to induce NAFLD. There were three control groups (n = 6/group) – vehicle with a regular diet, vehicle with HFD, and HFD with silymarin.	Hydroalcoholic extract of *Picrorhiza kurroa* given in two doses *viz.*, 200 mg/kg and 400 mg/kg b.i.d., p.o. for a period of 4 weeks	Histopathology showed that the P. kurroa extract brought about a reversal of the fatty infiltration of the liver (mg/g) and a lowering of the quantity of hepatic lipids (mg/g) compared to that in the HFD control group (38.33 ± 5.35 for 200mg/kg; 29.44 ± 8.49 for 400 mg/kg of *P. kurroa* vs.130.07 ± 6.36mg/g of liver tissue in the HFD control group; P<0.001). Compared to the standard dose of the known hepatoprotective silymarin, *P. kurroa* reduced the lipid content (mg/g) of the liver more significantly at the dose of 400 mg/kg (57.71 ±12.45mg/kg vs. 29.44 ± 8.49 for the silymarin group vs. 400mg/kg of *P. kurroa*, P<0.001).	Intervention with standardized plant extracts of *P. kurroa* regressed several features of NAFLD like lipid content of the liver tissue, morphological regression of fatty infiltration, hypolipidemic activity, and reduction of cholestatis.	N. Sapna., J Ayurveda Integr Med. 2010 Jul-Sep; 1(3): 203–210.

Contd...

TABLE 14.2–*Contd....*

Sl.No.	Title	Type of Study	Study Design	Form and Dose	Summarized Efficacy	Conclusion	Reference
5.	Protective effect of Picroliv against hydrazine-induced hyperlipidemia and hepatic steatosis in rats.	*In vivo*	Hydrazine (Hz)-induced hyperlipidemia was evaluated in rats. Hz administration (50 mg/kg, i.p.) caused an increase in triglyceride (TG), cholesterol (CHO), free fatty acids (FFA), and total lipids (TL) in both the plasma and liver tissueof rats accompanied by a fall in phospholipids (PL) in the liver tissue 24 h.	Picroliv 50 mg/kg, p.o.)	The abnormality of the lipid levels was prevented by simultaneous treatment of Picroliv. The effect of increase in the mobility of TG and TL from adipose tissue, because of hepatic steatosis by non-hepatocellular factors (such as mobilization of depot fats) was also prevented by simultaneous treatment of PIC with Hz.	Picroliv prevents Hz-induced hyperlipidemia, hepatic steatosis, and mobilization of lipids from depot fats.	P. Viveka-nandan *et al.*, Drug and Chemical Toxicology, 2007; 30: 241–252.
6.	Effects of *Picrorrhiza rhizoma* water extracts on the subacute liver damages induced by carbon tetrachloride.	*In vivo*	Carbon tetrachloride (CCl_4)-induced sub-acute hepatic damage, induced by subcutaneous injection of CCl_4 (0.15 mL/kg of body weight) in pure olive oil (7.92 per cent, vol/vol) three times a week for 10 weeks.	*Picrorrhiza rhizoma* (PR) water extract 50, 100, or 200 mg/kg	Ten weeks of CCl(4) injections caused subacute hepatic damage, featuring significantly less body weight gain and hepatic protein contents and higher liver weight, serum AST and ALT levels, and hepatic malondialdehyde and hydroxyproline contents with subacute hepatic damage-related histopathology of the liver. However, the CCl(4)-induced toxic effects were dramatically and dose-dependently inhibited by PR extract treatment.	Oral administration of PR extracts significantly reduced CCl_4-induced subacute hepatic damage in rats, probably by exerting a protective effect against hepato-cellular necrosis via its free radical scavenging ability.	H.S. Lee *et al.*, J Med Food. 2007 Mar; 10(1): 110-7.
					Nephro-protective activity		
7.	Evaluation of therapeutic potential of *Picrorhiza kurroa* glycosidal extract against nimesulide nephrotoxicity: A pilot study.	*In vivo*	The mice were divided in to 4 groups. One group was given only PK while the other three groups were given nimesulide in a dosage of 750 mg/kg body weight for 3 days to induce nephrotoxicity and protective effect of Pk was noted in nime-sulide induced nephro-toxicity groups. Bio-chemical assessment by measuring serum urea and creatinine and histology was done.	*Picrorhiza kurroa* glycosidal extract 250 mg/kg and 500 mg/kg pk for 14 days	Out of 20 mice, 19 mice survived. Only 1 mouse of nimesulide group died. Mean serum urea of nimesulide group was 60 mg/dl and was decreased to 23 mg/dl and 25 mg/dl by two doses of Pk. Mean creatinine in group 2 was 0.55 mg/dl and was decreased to 0.21 and 0.19 mg/dl bytwo doses of Pk.	Nimesulide is a potential nephrotoxic drug and its toxic effects on kidney can be minimized by using glycosidal extract of *Picrorhiza kurroa.*	A. Siddiqi *et al.*, J Ayub Med Coll Abbottabad 2015; 27(2).

Contd...

TABLE 14.2–Contd....

Sl.No.	Title	Type of Study	Study Design	Form and Dose	Summarized Efficacy	Conclusion	Reference
					Hepato- and renal protective activity		
8.	Effect of picroliv on cadmium-induced hepatic and renal damage in the rat.	*In vivo*	Male rats exposed to CdCl2 (0.5 mg/kg, sc), 5 days/week for 18 weeks. The Cd altered oxidative stress indices, such as increased lipid peroxidation and membrane fluidity, reduced levels of non-protein sulphydryls (NPSHs), and Na+K+ ATPase activity in the liver and kidney.	Picroliv—a standardized extract of *Picrorhiza kurroa* at two doses (6 and 12 mg/kg, po) was given to the cadmium (Cd)-administered group for the last 4 weeks (*i.e.*, weeks 15-18)	The oxidative stress indices were found close to the control values by Picroliv treatment, suggesting its antioxidant potential. Picroliv lowered the derranged alkaline phosphatase (ALP), alanine aminotransferase (ALT), aspartate aminotransferase (AST), gamma-glutamyl transpeptidase (GGT) and lactate dehydrogenase (LDH) levels. Bile flow and biliary Cd also increased as a result of Picroliv's choleretic property. The Cd-induced serum urea and urinary excretion of proteins, calcium (Ca), Cd and enzymes, such as N-acetyl-beta-D-glucosaminidase (NAG) and LDH, were less marked on Picroliv treatment, indicating recovery from nephrotoxicity. Organ uptake of Cd and essential metals by Cd exposure was reduced on Picroliv treatment. Cd-induced hepatic metallothionein (MT) was lowered by Picroliv, whereas renal MT was unaltered. Cd-induced hepatic damage was also minimized. However, the renal morphological changes were marginally protected by Picroliv.	The study has provided clear evidence for the hepato- and renal protective efficacy of Picroliv against experimental Cd toxicity.	N. Yadav N and S. Khandelwal; Hum Exp Toxicol. 2006 Oct; 25(10): 581-91.
					Cardioprotective activity		
9.	Cardioprotective effect of root extract of *Picrorhiza kurroa* (Royle Ex Benth) against isoproterenol-induced cardiotoxicity in rats.	*In vivo*	Isoproterenol-induced cardiotoxicity in rats.	*Picrorhiza kurroa* root extract 200 mg/kg	Isoproterenol (ISP) administration resulted in hemodynamic and left ventricular dysfunction, oxidative stress, and lipid peroxidation. Such cardiac dysfunction was significantly prevented by *P. kurroa* root extract pre-treatment. Pre-treatment significantly attenuated the ISP-induced oxidative stress by restoring myocardial superoxide dismutase, catalase, and glutathione peroxidase enzymes except reduced glutathione content, controlled the rise in lipid peroxidation, thereby prevented leakage of myocyte creatine kinase-MB and lactate dehydrogenase enzymes.	*P. kurroa* root extract possesses significant cardioprotective effect, which may be attributed to its antioxidant, anti-peroxidative, and myocardial preservative properties.	M. Nandave *et al.*, Indian J Exp Biol. 2013 Sep; 51(9): 694-701.
					Anti-arthritic activity		
10.	*Picrorhiza kurroa* inhibits experimental arthritis through inhibition of pro-inflammatory cytokines, angiogenesis and MMPs.	*In vivo*	Formaldehyde and adjuvant-induced arthritis (AIA) in rat.	*Picrorhiza kurroa* rhizome extract	In AIA-induced arthritic rat, treatment with PKRE considerably decreased synovial expression of interleukin-1β (IL-1β), interleukin-6 (IL-6), tumor necrosis factor receptor-1 (TNF-R1) and vascular endothelial growth factor as compared with control. There was significant suppression of oxidative and inflammatory markers as there was decreased malonaldehyde, Nitric oxide, tumor necrosis factor alpha levels accompanied with increased	The results demonstrated the anti-arthritic activity of PKRE against experimental arthritis, and the underlying mechanism behind this efficacy might be mediated by	R. Kumar *et al.*, Phytother Res. 2016 Jan; 30(1): 112-9.

Contd...

TABLE 14.2–*Contd....*

Sl.No.	*Title*	*Type of Study*	*Study Design*	*Form and Dose*	*Summarized Efficacy*	*Conclusion*	*Reference*
					glutathione and superoxide dismutase, catalase activities. Additionally, PKRE significantly inhibited the expression of degrading enzymes, matrix metalloproteinases-3 and matrix metalloproteinases-9 in AIA-induced arthritic rat. Histopathology of paw tissue displayed decreased inflammatory cell infiltration as compared with control.	inhibition of inflammatory mediators and angiogenesis, improvement of the synovium redox status and decreased expression of matrix metalloproteinases.	
				LAVANG (*Syzygium aromaticum* (L.) Merrill and Perry)			
1.	The effects of *Syzygium aromaticum* (clove) on learning and memory in mice.	*In vivo*	In order to evaluate the beneficial effect of clove on learning and memory, an experimental study was conducted in normal male mice. In this study a Shuttle Box device used to evaluate the active avoidance learning and memory in mice.	Ethanol extract of clove (100, 200 and 400 mg/kg)	Data showed that during the learning procedure the mean numbers of free shock trials in the test groups were increased compared with the control group In the short- and long-term memory assessments the animals in the test groups received less shocks than the control group and the differences were significant in the case of the 100 and 200 mg/kg clove extract.	Findings indicate that acute administrations of ethanolic extracts of clove enhance the learning and memory recallability in mice in an inverse dose-dependent manner.	Mohammad Hossein Dashti-R, Abbas Morshedi, The effects of *Syzygium aromaticum* (clove) on learning and memory in mice/Asian J. of Traditional Medicines, 2009, 4 (4).
2.	Gastroprotective activity of essential oil of the *Syzygium aromaticum* and its major component eugenol in different animal models.	*In vivo*	The aims of this study were: to identify and quantify the main component of the essential oil, and to evaluate its antiulcer activity using different animal models.	*S. aromaticum* oil and eugenol were administered at doses of 50, 100, and 250 mg/kg, respectively	The results show that there was no significant effect on the volume of gastric juice and total acidity. However, the quantification of free gastric mucus showed that theclove oil and eugenol were capable of significantly enhancing mucus production.	The results of this study show that essential oil of *S. aromaticum*, as well as its main component (eugenol), possesses antiulcer activity.	Santin JR *et al.*, Naunyn Schmiedebergs Arch Pharmacol. 2011 Feb; 383(2): 149-58.
3.	Immunomodulatory activity of *Zingiber officinale* Roscoe, *Salvia officinalis* L. and *Syzygium aromaticum* L. essential oils.	*In vivo*	The immunomodulatory effect of ginger, *Zingiber officinale* (Zingiberaceae), sage, *Salvia officinalis* (Lamiaceae) and clove, *Syzygium aromaticum* (Myrtaceae), essential oils were evaluated by studying humor- and cell-mediated immune responses.		Clove essential oil increased the total white blood cell (WBC) count and enhanced the delayed-type hypersensitivity (DTH) response in mice. Moreover, it restored cellular and humoral immune responses in cyclophosphamide-immunosuppressed mice in a dose-dependent manner.	Findings establish that the immunostimulatory activity found in mice treated with clove essential oil is due to improvement in humor- and cell-mediated immune response mechanisms.	Fábio Ricardo Carrasco, Gustavo *et al.*, Journal of Pharmacy and Pharmacology, Volume 61, Issue 7, pages 961–967, July 2009.

Contd...

TABLE 14.2–*Contd....*

Sl.No.	Title	Type of Study	Study Design	Form and Dose	Summarized Efficacy	Conclusion	Reference
4.	Antioxidant potential of the *Syzygium aromaticum* (Gaertn.) Linn. (cloves) in rats fed with high fat diet	*In vivo*	to assess the antioxidant efficacy of cloves in rats fed with high fat diet (HFD)		Lipid peroxidation as evidenced by an increment in the values of TBARS, CD, urea, lipid profiles, AST and ALT in hyperlipidemic rats was found to benullified by co-administration of cloves antioxidant enzymes showed enhanced activities on co-administration of cloves.	Results substantiate the use of moderate quantity of cloves in diet as an antioxidant in offering protection against hyperlipidemia.	M.P. Shyamala *et al.*, Indian Journal of Pharmacology 2003; 35: 99-103
5.	Hypoglycemic effects of clove (*Syzygium aromaticum* flower buds) on genetically diabetic KK-Ay mice and identification of the active ingredients.	*In vivo*	Evaluation of antidiabetic activity of Clove EtOH extract by using genetically induced type 2 diabetic KK-A(y) mice	The mice were fed the clove EtOH extract at 0.5 g/100 g diet or pioglitazone at 0.02 g/100 g diet in the treated groups for 3 weeks	*In vitro* evaluation showed the extract had human peroxisome proliferator-activated receptor (PPAR)-γ ligand-binding activity in a GAL4-PPAR-γ chimera assay. Bioassay-guided fractionation of the EtOH extract resulted in the isolation of eight compounds, of which dehydrodieugenol (2) and dehydrodieugenol B (3) had potent PPAR-γ ligand-binding activities, whereas oleanolic acid (4), a major constituent in the EtOH extract, had moderate activity. Furthermore, 2 and 3 were shown to stimulate 3T3-L1 preadipocyte differentiation through PPAR-γ activation.	These results indicate that clove has potential as a functional food ingredient for the prevention of type 2 diabetes and that 2-4 mainly contribute to its hypoglycemic effects via PPAR-γ activation.	Kuroda M *et al.*, J Nat Med. 2012; 66(2): 394-9.
6.	Chemical composition and antidiabetic activity of essential oils obtained from two spices (*Syzygium aromaticum* and *Cuminum cyminum*)	*In vitro*	To evaluate antidiabetic potential of *Syzygium aromaticum* and *Cuminum cyminum* essential oils and their emulsions by alpha amylase inhibition assay.	Antidiabetic activity of *C. cyminum* and *S. aromaticum* was examined in dose dependent mode (1 to 100 µg/mL).	The maximum antidiabetic activity for *S. aromaticum* and *C. cyminum* essential oils was noted at the highest dose (100 µg/mL). Five emulsions (essential oil + surfactant [tween 80] + co-surfactant [ethanol] + water) of different concentrations for *S. aromaticum* (A1 to A5) and C. cyminum (B1 to B5) essential oils were formulated. Among different emulsions, A5 of *S. aromaticum* and B5 of C. cyminum essential oil exhibited a maximum antidiabetic activity with 95.30 and 83.09 per cent inhibition of α-amylase, respectively. Moreover, the analysis of essential oils showed that eugenol (18.7 per cent) and α-pinene (18.8 per cent) were the major components of *S. aromaticum* and *C. cyminum* essential oils, respectively.	These results indicate that clove has potential as a functional food ingredient for the prevention of type 2 diabetes.	Tahir HU *et al.*, International Journal of Food Properties. 2016; 19(10): 2156-2164.
7.	*In vitro* inhibition activity of polyphenol-rich extracts from *Syzygium aromaticum* (L.) Merr. and Perry (Clove) buds against carbohydrate hydrolyzing enzymes linked to type 2 diabetes and Fe^{2+}-induced lipid peroxidation in rat pancreas	*In vitro*	To investigate and compare the inhibitory properties of free and bound phenolic extracts of clove bud against carbohydrate hydrolyzing enzymes (alpha-amylase and alpha-glucosidase) and Fe2+-induced lipid peroxidation in rat pancreas *in vitro*.	Free and bound phenolic extracts of clove bud	The result revealed that both extracts inhibited alpha-amylase and alpha-glucosidase in a dose-dependent manner. However, the alpha-glucosidase inhibitory activity of the extracts were significantly (P<0.05) higher than their alpha-amylase inhibitory activity. The free phenolics (31.67 mg/g) and flavonoid (17.28 mg/g) contents were significantly (P<0.05) higher than bound phenolic (23.52 mg/g) and flavonoid (13.70 mg/g) contents. Both extracts also exhibited high antioxidant activities as typified by their high reducing power, 1,1 diphenyl-2- picrylhydrazyl (DPPH) and 2, 2-azinobis-3-ethylbenzo-thiazoline-6-sulfonate (ABTS) radical scavenging abilities, as well as inhibition of Fe2+-induced lipid peroxidation in rat pancreas *in vitro*.	This study reports a strong correlation between the phenolic content of clove bud and the enzyme inhibitory activities. Furthermore, the inhibition of key enzymes linked to type-2 diabetes (alpha-amylase and alpha-glucosidase) coupled with strong antioxidant	

Contd...

TABLE 14.2–*Contd....*

Sl.No.	Title	Type of Study	Study Design	Form and Dose	Summarized Efficacy	Conclusion	Reference
						properties of the tropical cloves phenolics could be part or possible mechanism through which the clove bud elicits its antidiabetic potentials.	Adefegha SA and Oboh G, Asian Pac J Trop Biomed. 2012; 2(10): 774–781.
8.	Clove extract inhibits tumor growth and promotes cell cycle arrest and apoptosis	*In vivo* and *In vitro*	Investigated the *in vitro* and *in vivo* antitumor effects and biological mechanisms of ethyl acetate extract of cloves (EAEC) and the potential bioactive components responsible for its anti-tumor activity. The effects of EAEC on cell growth, cell cycle distribution, and apoptosis were investigated using human cancer cell lines. The molecular changes associated with the effects of EAEC were analyzed by Western blot and (qRT)-PCR analysis. The *in vivo* effect of EAEC and its bioactive component was investi-gated using the HT-29 tumor xenograft model.	Ethyl acetate extract of cloves (EAEC) and bioactive components	Both EAEC and OA display cytotoxicity against several human cancer cell lines. Interestingly, EAEC was superior to OA and the chemotherapeutic agent 5-fluorouracil at suppressing growth of colon tumor xenografts. EAEC promoted G0/G1 cell cycle arrest and induced apoptosis in a dose-dependent manner. Treatment with EAEC and OA selectively increased protein expression of $p21^{WAF1/Cip1}$ and γ-H2AX and downregulated expression of cell cycle-regulated proteins. Moreover, many of these changes were at the mRNA level, suggesting transcriptional regulation by EAEC treatment.	The ethyl acetate extract of cloves displayed anti-tumor activity both *in vitro* and *in vivo*. Clove extract may represent a novel therapeutic herb for cancer treatment, and OA is one of the compo-nents responsible for part of its antitumor activity.	Liu H, *et al.*, Oncol Res. Author manuscript; available in PMC 2014 Aug 14. Published in final edited form as: Oncol Res. 2014; 21(5): 247–259.
9.	Safety and anti-ulcero-genic activity of a novel polyphenol-rich extract of clove buds (*Syzygium aromaticum* L).	*In vivo*	To derive their standar-dized polyphenol-rich extracts as a water soluble free flowing powder (Clovinol) suitable for functional food appli-cations, without the issues of its characteristic pungency and aroma.	Clovinol 25 and 50 mg/kg orally	Clovinol showed significant antioxidant and anti-inflammatory effects as measured by cellular antioxidant levels, and the ability to inhibit carrageenan-induced paw swelling in mice. Further investigations revealed its significant anti-ulcerogenic activity (>97 per cent inhibition of ethanol-induced stomach ulcers in Wistar rats when orally administered at 100 mg per kg b.w.) and up regulation of *in vivo* antioxidants such as superoxide dismutase (SOD), glutathione (GSH), and catalase (CAT). Clovinol also reduced the extent of lipid peroxidation among ulcer induced	The present contribution has demonstrated the safety and efficacy of clove buds derived poly-phenol-rich standardized extract (Clovinol) as an alcohol-induced stomach ulcer protective/curative agent. The potent anti-	Issac A *et al.*, Food Funct. 2015; 6(3): 842-52.

Contd...

TABLE 14.2–Contd....

Sl.No.	Title	Type of Study	Study Design	Form and Dose	Summarized Efficacy	Conclusion	Reference
			The extract was characterized by electrospray ionization-time of flight mass spectrometry (ESI-TOF-MS), and investigated for *in vivo* antioxidant, anti-inflammatory and anti-ulcerogenic activities. Anti-inflammatory activity: Carrageenan induced paw oedema model Ethanol induced ulcer model.		rats, indicating its usefulness in ameliorating oxidative stress and improving gastrointestinal health, especially upon chronic alcohol consumption.	ulcerogenic activity of Clovinol was found to be mainly due to its high antioxidant and anti-inflammatory effects in addition to its mucous producing effect.	
10.	Immunomodulatory effects of clove (*Syzygium aromaticum*) constituents on macrophages: *in vitro* evaluations of aqueous and ethanolic components.	*In vitro*	To investigate potential suppressive effects on mouse macrophages using doses (ranging from 0.001-1000μg/ml) of each material freshly prepared in the laboratory, cell survival and production of nitric oxide (NO), tumor necrosis factor (TNF)-α, interleukin (IL)-6, and IL-12 by the treated cells (that in all cases also had received LPS stimulation) were measured.	Clove (*Syzygium aromaticum*) ethanolic extracted essential oil (containing eugenol) or its water-soluble extract 0.001-1000 μg/ml	NO release by LPS-stimulated macrophages was generally significantly suppressed by either material; in contrast, low (*i.e.* 0.001-1 μg/ml) doses of either extract class appeared to enhance NO release by non-LPS (unstimulated)-treated macrophages. Among LPS-stimulated cells, TNFα release was also significantly affected by each extract; the ethanolic extract was suppressive at all doses tested, while the aqueous material was so up to 1μg/ml and then became stimulatory. In contrast, nearly every dose of either extract appeared to stimulate IL-6 release from the LPS-treated cells. Effects on IL-12 production were overall inconsistent; in general, the ethanolic extract tended to be stimulatory of production by the LPS-treated cells. The data for the aqueous material showed no discernable pattern of effect.	Clove extracts do not have a distinct cytotoxic activity, but do impart potential anti- and pro-oxidant effects in cells, depending on their concentrations and on the activation state of the macrophages themselves at the time of exposure to the extracts. The impact of the extracts on macrophage cytokine release also displays a pattern of dose-relatedness.	Dibazar SP *et al.*, J Immunotoxicol. 2015 Apr-Jun; 12(2):124-31.
					LICORICE (*Glycyrrhiza glabra* Linn)		
					Antiulcer Activity		
1.	Gastro protective and antioxidant potential of *Glycyrrhiza glabra* on	*In vivo*	Model: Aspirin induced gastric ulcer Species: Male albino mice.	*G. glabra* extract. 250mg-	*G. glabra* shows gastro protective potential with significant reduction in the ulcer score, acid output, and gastric volume. pH of gastric mucosa increases significantly at the dose of	*G. glabra* has gastro protective and anti-oxidant activity.	B. Aslam, *et al.*, Int J Curr Microbiol

Contd...

TABLE 14.2–*Contd....*

Sl.No.	*Title*	*Type of Study*	*Study Design*	*Form and Dose*	*Summarized Efficacy*	*Conclusion*	*Reference*
	experimentally induced gastric ulcers in albino mice.		N=36. Groups 6:- control, Aspirin; Omeprazole; GG. Glabra ext. 250, 500, and 750 mg/kg, with aspirin. Parameter: Total oxidant status (TOS), Total anti-oxidant capacity (TAC), Catalase (CAT), and Malondialdehyde (MDA), Ulcer score, gastric volume, gastric pH, and total acid output.	750mg/kg once a day for 7 days	750 mg/kg WRT aspirin treated group.Also significant increase in TAC and CAT activity and decreased in the levels of TOS and MDA which indicate reduction in gastric damage.		Applied Sci. 2015; 4(2): 451-460.
2.	Antiulcer properties of *Glycyrrhiza glabra* L. extract on experimental models of gastric ulcer in mice.	*In vivo*	Model: 1 HCl/ethanol and indomethacin induced gastric ulcer Species: Swiss mice Groups 6 each for 2 models Negative control, positive control, Saline and Omeprazole/ camitidine. Rest 4 Gr. ext. of Gly. Glabra (GG) 50, 100, 150 and 200 mg/kg. ulcer index calculated in all groups.	*G. glabra* extract 50 to 200mg/kg	*Glycyrrhiza glabra* L. extract has anti-ulcer capacity. It is dose dependent and 150-200 mg/kg doses of GG extract reduces ulcer better than omeprazole and cimetidine.	*G. Glabra* extract has anti ulcerogenic effect that could be associated with increase gastric mucosal defensive factors.	Ghader Jalilzadeh-Amin, *et al.*, Iranian Journal of Pharma-ceutical Research (Autumn 2015).
3.	Comparative effective-ness of *Glycyrrhiza glabra* vs. omeprazole and misoprostol for the treatment of aspirin-induced gastric ulcers.	*In vivo*	Model: Aspirin induced gastric ulcer Species: Wistar albino rats Group 2 Prophylaxis and "treatment" groups. subdivided into 4. Control and treated with licorice ext. and in combination with Ome-prazole and misoprostol. Ulcer score calculated.	Licorice extract 2.5 gm/kg for 4 weeks	Licorice extract has better effect in treatment of aspirin induced gastric ulcers. However Misoprostol found to be more effect than omeprazole and licorice in prevention of aspirin induced gastric ulcer.	*Glycyrrhiza glabra* has antiulcer property. Can be use as an inexpensive alternative to miso-prostol and omeprazole.	Mesut Sancar, *et al.*, African Journal of Pharmacy and Pharmacology 2010; 3:615-620.

Contd...

TABLE 14.2–*Contd....*

Sl.No.	Title	Type of Study	Study Design	Form and Dose	Summarized Efficacy	Conclusion	Reference
4.	Licorice: A possible anti-inflammatory and anti-ulcer drug	*In vivo*	Model: Indomethacin induced ulcer Speceies: Albino rats N=36; Group: 4 (N=6) Control, Famotidine (FT), Aqu. Licorice ext (100mg/kg), Combination (FT 0.57 mg/kg and 100mg/kg licorice), Parameter: Ulceration index.	Licorice extract 100mg/kg	Combination of FT and licorice extract has better antiulcer effect than individual treatment.	Addition of Licorice extract enhances anti-ulcer property of famotidine.	Adel M. Aly, *et al.*, AAPS Pharm Sci Tech. 2005; 6(1): E74–E82.
					Anti-*Helicobacter pylori* Activity		
5.	Effect of GutGard in the management of *Helicobacter pylori*: A randomized double blind placebo controlled study.	Clinical study	**Type:** Double Blind Placebo Controlled Study **Sample Size** 107 Subjects selected with positive *H. pylori* stool antigen test (HpSA) and ^{13}C-urea breath test (^{13}C-UBT) **Study Group**: Two groups Placebo (N=52) and with Gutgard (N=55) **Parametes Testing** for HpSA and 13CUBT (carbon13 urea breath test **Duration of Study**: 30th and 60th day.	1 capsule/ day (150 mg *G. glabra* extract) for 60 days	In GutGard treated group HpSA and ^{13}C-UBT test become negative in 56 per cent and in control only 4 per cent. That indicates GutGard is better in management of *H. pylori* infection.	GutGard is effective in the management of *H. pylori.*	Sreenivasulu Puram, *et al.*, Evidence-based Complementary Medicine. 2013; Article ID-263805.
6.	*In vitro* anti-*Helicobacter pylori* activity of a flavonoid rich extract of *Glycyrrhiza glabra* and its probable mechanisms of action	*In vitro*	Agar dilution and micro-broth dilution methods were used to determine the minimum inhibitory concentration of GutGard against *Helicobacter pylori*. Protein synthesis, DNA gyrase, dihydrofolate reductase assays and anti-adhesion assay in human gastric mucosal cell line were performed to understand the mechanisms of anti-*Helicobacter pylori* activity of Gut Gard	–	GutGard and glabridine exhibited anti-*Helicobacter pylori* activity in both agar dilution and microbroth dilution methods. Glabridin flavonide in gutgard gas has better activity than omeprazole and other derivatives of licorice. DGL and MAG[1] did not show activity even at 250 μg/ml concentration.	GutGard acts against *Helicobacter pylori* possibly by inhibiting protein synthesis, DNA gyrase and dihydrofolate reductase.	Asha MK, *et al.*, J Ethno-pharmacol. 2013; 145(2): 581-586.

Contd...

TABLE 14.2–*Contd....*

Sl.No.	*Title*	*Type of Study*	*Study Design*	*Form and Dose*	*Summarized Efficacy*	*Conclusion*	*Reference*
7.	Anti-*Helicobacter pylori* Properties of GutGard.	*In vivo*	Species: Mongolian gerbils mouse Samples size: N=30) and C57BL/6 (=30) Models infected with bacteria by intragastric inoculation (2×10^9 CFU/gerbil) 3 times for 5 days. Groups: 5 Groups Gr 1 control Gr2 *H. pylori* infection, Gr 3, 4, 5 orally 15, 30 and 60 mg/kg GutGard. Biopsy samples of the gastric mucosa assayed by bacterial identification via urease, catalase and ELISA assays as well as immune-histochemistry (IHC).	Gurgard 15-60 mg/kg 6 times/ week for 8 weeks	GutGard inhibited *H. pylori* colonization in gastric mucosa, however it is dose dependent.	GutGard may be useful as an agent to prevent *H. pylori* infection.	Jae Min Kim, *et al.*, Prev Nutr Food Sci. 2013; 18(2): 104-110.
					Antiulcer Activity		
8.	Anti-ulcer and anti-oxidant activity of GutGard.	*In vivo*	Wistar rat divided into 6 groups (N=6), Gr 1 Control (DM water 10mg/kg), Gr II Negative control (pyloric ligation ulcer/Indomethacin 40mg/kg BWT), Gr III Positive control Omeprezole 10mg/kg, Gr IV/V/VI treated with Gutgard 12.5/25/50 mg/kg per BWT, by oral route.	Gutgard 12.5/ 25/50 mg/kg per BWT, for 30 days	Gut Gard has protective effect against pylorus ligation, cold restraint stress and indomethacin induced ulcer. Gut Gard at the dose 12.5, 25, 50 mg/kg reduces volume of gastric juice, total acidity ulcer index and increase gastric juice pH so acts an anti-ulcer genic. Antioxidant parameters are reduced in stomach tissue damaged by indomethacin. But treatment of gut Gard has antiulcer genic effect that indicates gutgard stimulate prosta-glandin or similar to it substance.	Licorice will be useful in the treatment of gastric ulcer.	Moumita Mukherjee, and Natarajan Bhaskaran, Indian Journal of Experimental Biology 2010; 48(3): 269-74.
					Functional dyspepsia		
9.	An Extract of *Glycyrrhiza glabra* (GutGard) alleviates symptoms of functional dyspepsia: A randomized,	Clinical study	Type: Double blind placebo controlled study Sample size 50 study Group: Two groups Placebo (N=25) and with	1 Capsule (75mg) twice a day for 30 days	The GutGard group a showed a significant decrease ($P \leq .05$) in the Nepean dyspepsia index (NDI) on day 15 and 30, compared to placebo. GutGard was found to be safe and well-tolerated by all patients.	Gut Gard has efficacy in the management of functional dyspepsia.	Kadur Ramamurthy Raveendra, Jayachandra, *et al.*,

Contd...

TABLE 14.2–*Contd....*

Sl.No.	*Title*	*Type of Study*	*Study Design*	*Form and Dose*	*Summarized Efficacy*	*Conclusion*	*Reference*
	double- blind, placebo-controlled study.		Gutgard (N=25) investigated on the basis of Gastrointestinal subjective questionnaires and Nepean dyspepsia index Duration of study: 30 day.				Evidence-based Complementary and Alternative Medicine. 2012.
					DEGLYCYRRHIZINISED LIQUORICE (DGL)		
					Antiulcer Activity		
10.	Antiulcer activities of liquorice and its derivatives in experimental gastric lesion induced by ibuprofen in rats.	*In vivo*	Model Ibuprofen induced Species: Albino wistar rats Group: 7 groups Control Gr. Saline Rest 6 Groups treated with granules coated with licorice, DGL and its derivatives. Ulcer index measured.	Licorice extracts and DGL derivatives 60mg/kg.	Ibuprofen coated with liquorice, DGL or enoxolone reduced the number and size of ulcers, lowering the ulcer index from 1.86 to 1 and the incidence of ulceration also reduced from 100 to 59 per cent.	Licorice has antiulcer property.	Ahmad Reza Dehpour *et al.*, International Journal of Pharmaceutics 1995; 119(2): 133-138 ·
11.	Deglycyrrhizinised liquorice (DGL) and the renewal of rat stomach epithelium.	*In vivo*	Model: Pylorus ligated rats Species: Wistar rats N=24, Groups: 2 subdivided into 2 hald with DGL 0.85mg/kg and half control. 2nd group SHAM operated and out of taht half got DGL treatment. After abdominal wall suturing all rats were given with 1 microCi/g [^{3}H] thymidine.	DGL 0.85 mg/kg	Deglycyrrhizinised liquorice (DGL) increased the number of fundus glands in which labelled mucus secreting cells occur as well as the total number of labelled mucus cells per gland. DGL stimulates and/or accelerates the differentiation to glandular cells as well as mucus formation and secretion. The increase in proliferation observed in the epithelium of the forestomach might then be ascribed to an improvement of the environment in the stomach as a consequence of enhanced mucus secretion under the influence of DGL.	DGL influence the mucus secretion and accelerate proliferation of it in the forestomach so it helps to improve stomach environment.	Jan Van Marle, *et al.*, European Journal of Pharmacology 1981; 72(2–3): 219–225.

Contd...

TABLE 14.2–*Contd....*

Sl.No.	*Title*	*Type of Study*	*Study Design*	*Form and Dose*	*Summarized Efficacy*	*Conclusion*	*Reference*
					MADHUKA (*Madhuca indica* J.F.GMEL)		
					Anti-inflammatory and Analgesic Activity		
1.	Investigation of anti-inflammatory, analgesic and antipyretic properties of *Madhuca indica* GMEL	*In vivo*	Carrageenan induced oedema right hind paw volume for inflammation acetic acid-induced abdominal pains *i.e.,* nociception response for analgesic and the brewer's yeast-induced pyrexia model was used for antipyretic investigation.	Methanolic extract 50, 100 and 200 mg kg^{-1} body weight	The extract at all the doses used and the indomethacin significantly inhibited carrageenan induced inflammation in a manner that was not dose dependent. The extract reduced the acetic acid induced pain licking. While the 50 and 100 mg kg-1 body weight of the extract reduced the brewer's yeast provoked elevated body temperature in rats after 60 min that of 200 mg kg^{-1} body weight manifested from 30 min.	The results suggest a potential benefit of *M. indica* methanolic extract in treating conditions associated with inflammation, pain and fever.	Shekhawat N, Vijayvergia R, International Journal of Molecular Medicine and Advance Sciences. 2010; 6(2): 26-30.
					Antioxidant Activity		
2.	Evaluation of *in vitro* antioxidant activity of a triterepne isolated from *Madhuca longifolia* L. leaves	*In vitro*	By using reducing power assay, super oxide radical scavenging activity, Hydroxyl radical scavenging activity		Methanolic extract and derivative of Madhucic Acid, effectively scavenged free radicals at all different concentrations showed its potent antioxidant activity. Results were compared to standard antioxidant such as butylated hydroxyl anisole.	The result indicated that *Madhuca longifolia* L. leaves has significant natural radical scavengers.	Swamy *et al.,* Int J Pharm Pharm Sci, 5(4): 389-391.
					Antidiabetic Activity		
3.	Antihyperglycemic and antioxidant activity of ethanolic extract of *Madhuca longifolia* bark.	*In vivo* and *in vitro*	Glucose loaded and streptozotocin induced diabetic rats and free radical scavenging activity using 1, 1-diphenyl-2-picrylhydrazil (DPPH), reducing power assay and superoxide scavenging activity.	Ethanolic bark extract of *Madhuca longifolia* 100 and 200 mg/kg body weight for AHG effect.	The extract exhibited a dose dependent hypoglycemic activity in all three animal models as compared with the standard anti-diabetic agent glibenclamide. The hypoglycemia produced by the extract may be due to the increased glucose uptake at the tissue level and/or an increase in pancreatic β-cell function, or due to inhibition of intestinal glucose absorption and a good source of compounds with antioxidant properties.	The study indicated the ethanolic extract of *Madhuca longifolia* to be a potential anti-diabetic andantioxidant properties and the extract also exhibited significant free radical scavenging activity and superoxide scavenging activity.	Srirangam Prasharth *et al.,* International Journal of Pharmaceutical Sciences Review and Research, 2010; 5(3).
					Antiulcer Activity		
4.	Antiulcer activity of aqueous extract of leaves of *Madhuca indica* J. F. Gmel against naproxen induced gastric mucosal injury in rats	*In vivo*	Naproxen (30 mg/kg, p.o) induced gastric ulcer in rats was used to evaluate antiulcer activity. Ulcerated area was measured by Image J	Tested drug was administered at dose of 100, 200 and 400	After 4 week treatment period, desired aim was achieved using aqueous extract of plant of *M. indica* at the dose of 200 and 400 mg/kg, p.o. showed significant reduction in ulcerated area and ulcer index as compared to control group. Omeprazole (30 mg/kg, p.o.) was more effective in reducing ulcerated area after 30 days treatment period. In addition, SOD, GSH, NO	Aqueous extract of *M. indica* J.F. Gmel leaves is effective in gastric ulcer protection.	Mohod SM and Bodhankar SL, Journal of Acute Disease, 2013; 2(2): 127-33.

Contd...

TABLE 14.2–*Contd....*

Sl.No.	Title	Type of Study	Study Design	Form and Dose	Summarized Efficacy	Conclusion	Reference
			software. Various anti-oxidant parameter like SOD, GSH, MDA, MPO, NO and histamine were also determined.	mg/kg, p.o.	significantly increased; MDA, MPO content significantly lowered when compared with control group. Histamine content didn't show any significant change at all the three doses.		
5.	Anti-ulcer activity of petroleum ether extract of leaves of *Madhuca indica* J. F. Gmel against pylorus ligation and naproxen-induced gastric mucosal injury in rats.	*In vivo*	to study the antiulcer activity of Madhuca using the pylorus ligation and naproxen-induced gastric ulcer models in rats.	Tested drug was administered at dose 100, 200 and 400 mg/kg, p.o	In pylorus ligation, the extract provided significant ulcer protective effect as evinced through significant increase in gastric pH and mucin content of the stomach along with reduction in total acidity and pepsin activity. After 4 week treatment period, petroleum ether extract of plant of *Madhuca indica* showed significant reduction in ulcerated area and ulcer index as compared to control group in naproxen induced ulcer model. Moreover, ulcerated area was reduced significantly in all two models.	Petroleum ether extract of *Madhuca indica* leaves possesses antiulcer activity which can be attributed to its ability to increase the protective layer of mucin and decrease the damaging and or digestive effects of pepsin and acid.	Mohod SM and Bodhankar SL, Der Pharmacia Lettre, 2013, 5 (2): 205-211.
6.	Evaluation of antiulcer activity of ethanolic extract of *Madhuca longifolia* flowers in experimental rats.	*In vivo*	To investigate antiulcer activity of ethanolic extract of *Madhuca longifolia* flowers in pylorus ligated ulceration in the albino rats.	Tested drug was administered at doses of 100, 200, 300 mg/kg. p.o.	The ethanolic extract of *Madhuca longifolia* flowers produced significant inhibition of the gastric fluid volume, free acidity, and total acidity.	The antiulcer properties of the extract may be attributed to the presence of phytochemicals like flavonoids (quercetin), alkaloids and tannins present in the plant extract with various biological activities	Kalaivani M and Jegadeesan M, International Journal of Scientific and Research Publications, 2013; 3(6): 1-7.
7.	Anti-ulcer and anti-oxidant activity of leaves of *Madhuca indica* in rats.	*In vivo*	To evaluate the anti-ulcer and antioxidant activity of *M. indica* in rats using pylorus-ligation and on ethanol-induced gastric mucosal injury in rats.	Tested drug was administered at doses 75, 150 and 300 mg/kg, p.o.	The significant reduction in ulcer index in both the models along with an increase in the pH of the gastric fluid and mucin content of stomach, and the acid secretory parameters such as total acidity and volume of gastric fluid were also significantly reduced along with reduction in the pepsin activity in pylorus ligated rats proved the anti-ulcer activity of *M. indica.* The increase in the levels of superoxide dismutase, catalase and reduced glutathione and decrease in lipid peroxidation in both the models proved the antioxidant activity of *M. indica.*	*M. indica* possesses anti-ulcer activity, which can be attributed to its anti-oxidant mechanism of action	Chidrewar GU, *et al.*, Oriental Pharmacy and Experimental Medicine. 2010; 10(1): 13-20.
8.	Gastroprotective potential of Pentahydroxy flavone isolated from *Madhuca indica* J. F.	*In vivo*	To investigate the efficacy and possible mechanism of *Madhuca indica* J. F. Gmel. leaves methanolic	D3 (2.5, 5 and 10 mg/kg, p.o.) were admin-	Administration of 3,5,7,3',4'-Pentahydroxy flavone (5 and 10 mg/kg) significantly and dose-dependently inhibited ($P<0.01$ and $P<0.001$) acetic acid induced an alteration in the antioxidant enzyme. It also significantly and dose-dependently down-	MI-ALC possessed potent antiulcer activity due to the presence of 3,5,7,3',4'-Pentahydroxy	Mohod SM *et al.*, J Ethnopharmacol.

Contd...

TABLE 14.2–*Contd....*

Sl.No.	Title	Type of Study	Study Design	Form and Dose	Summarized Efficacy	Conclusion	Reference
	Gmel. leaves against acetic acid-induced ulcer in rats: The role of oxido-inflammatory and prostaglandins markers.		extract (MI-ALC) and its isolated chloroform fraction (D3) against experimental induced gastric ulcers.D3 was isolated from MI-ALC, well characterized (HPTLC, FT-IR, (1)H-NMR and LC-MS). At the end of treatment, rats were sacrificed to collect the stomach sample for evaluation of antioxidant (SOD, GSH, and MDA) enzyme, oxido-inflammatory (TNF-α, IL-1, iNOs) and prostaglandins (COX-II) markers by using RT-PCR.	istered for the period of 14 days.	regulated gastric oxido-inflammatory and prostaglandins markers. Histopathological aberration induced in the stomach also attenuated by 3,5,7,3',4'-Pentahydroxy flavone treatment.	flavone via its oxido-inflammatory and prostaglandins modulatory potential.	2016; 182: 150-9.
					Antimicrobial Activity		
9.	Antimicrobial activity of alcoholic extract of leaves and flowers of *Madhuca longifolia.*	*In vitro*	The alcoholic extract of leaves and flowers of *Madhuca longifolia* were screened for antimicrobial activities against *Staphylococcus aureus, Bacillus subtilis, Escherichia coli, Pseudomonas aeruginosa, Aspergillus oryzae* and *Aspergillus niger.*	Dose level ranging from 50 μg/ml to 250 μg/ml	Alcoholic extracts of *Madhuca longifolia* leaves inhibited all the bacterial strains tested. Extracts showed zone of inhibition against all the bacteria, even at a dose of 50 μg/ml of extract exhibited significant zone of inhibition comparable to standard antibiotic (ciprofloxacin) against *B. subtilis.* For *S. aureus* leaf extract exhibited maximum inhibition (20 mm) which was greater than that of standard. For all other bacteria 100 μg/ml concentration of the extract was sufficient to produce effective inhibition.	*Madhuca longifolia* extracts have broad inhibitory activities to pathogenic microorganism and to act as potential antimicrobial agent from natural sources.	Kalaivani M and Jegadeesan M, International Journal of Scientific and Research Publications, 2013; 3(5): 1-3.
10.	Antibacterial activity of the dried inner bark of *Madhuca indica* J.F. GMEL.	*In vitro*	The dried inner bark of *Madhuca indica* was investigated for its possible antibacterial activity against four human pathogenic bacterial strains. The plant extracts were evaluated against some		Among all the extracts the methanolic extracts showed significant antibacterial activity against most of the tested microbes. The most susceptible microorganism was *Staphylococcus aureus* (24 mm zone of inhibition in methanolic extract) followed by *Bacillus subtilis* (20 mm zone of inhibition in methanolic extract) again followed by *Escherichia coli* (15 mm zone of inhibition in methanolic extract) and *Staphylococcus epidermidis* (10 mm zone of inhibition in methanolic extract).	The inner bark of the plant *Madhuca indica* is a potential source for antibacterial activity and provides some idea about phytochemical evaluation on *Madhuca indica.*	Nimbekar T *et al.,* Bulletin of Environment, Pharmacology and Life Sciences. 2012; 1(2): 26-29.

Contd...

TABLE 14.2–*Contd....*

Sl.No.	*Title*	*Type of Study*	*Study Design*	*Form and Dose*	*Summarized Efficacy*	*Conclusion*	*Reference*
			gram positive and gram negative bacterial strains like *Bacillus subtilis, Staphylococcus aureus, Staphylococcus epidermidis* and *Escherichia coli* were carried out by the disk diffusion technique. Minimal inhibitory concentration (MIC) values of extracts and antibiotics were comparatively determined by agar dilution method.				
					METHI (*Trigonella foenum graecum* Linn)		
					Estrogenic Activity		
1.	*In vitro* estrogenic activities of fenugreek *Trigonella foenum graecum* seeds.	*In vitro*	FCE (Fenugreek Chloroform Extract) on cell proliferation of estrogen receptor (ER) positive breast cancer cells, MCF-7. ER binding assay (HAP assay); Transfection and reporter assay (DLR assay), and RT- PCR with an estrogen responsive gene pS2.	Chloroform extracts of fenugreek seeds (FCE) concentration range of 20 to 320 μg/ml.	Hytoestrogenic compounds present in FCE. FCE stimulated the proliferation of MCF-7 cells, showed binding to ER and acted as an agonist for ER mediated transcription via ERE. It also induced the expression of estrogen responsive gene pS2 in MCF-7 cells.	Fenugreek seeds have estrogenic activity.	Sreeja S, Anju VS, Sreeja S. Indian J Med Res. 2010; 131: 814-819.
					Anti-cancer Activity		
2.	*Trigonella foenum graecum* (fenugreek) seed extract as an antineoplastic agent.	*In vivo*	Ehrlich ascites carcinoma (EAC) model in Balb-C mice.	Alcohol extract of fenugreek seed; 100 and 200 mg/kg body wt.	Test item both before and after inoculation of EAC cell in mice produced more than 70 per cent inhibition of tumour cell growth with respect to the control. Treatment with the extract was found to enhance both the peritoneal exudate cell and macrophage cell counts. The extract also produced a significant antiinflammatory effect.	*Trigonella foenum graecum* seed extract have antiinflammatory and antineoplastic effects.	Sur P, *et al.* Phytother Res. 2001; 15(3): 257-259.

Contd...

TABLE 14.2–*Contd....*

Sl.No.	*Title*	*Type of Study*	*Study Design*	*Form and Dose*	*Summarized Efficacy*	*Conclusion*	*Reference*
3.	Fenugreek, a naturally occurring edible spice, kills MCF-7 human breast cancer cells via an apoptotic pathway	*In vitro*	Characterization of the downstream apoptotic genes targeted by FCE in MCF-7 human immortalized breast cells.	Fenugreek Seed Extract 50 μg/mL FCE for 24 hours and 48 hrs	FCE effectively killed MCF-7 cells through induction of apoptosis, confirmed by terminal deoxynucleotidyl transferase-mediated dUTP nick end labeling (TUNEL) and RT-PCR assays. When cells were exposed to fenugreek for 24 hra, 23.2 per cent apoptotic cells resulted, while a 48-hour exposure caused 73.8 per cent apoptosis. The induction of apoptosis by FCE is effected by its ability to increase the expression of pro-apoptotic genes.	Fenugreek may be considered as complementary therapy for breast cancer patients due to its influence in suppressing growth of immortalized breast cells without significant toxicity.	Khoja KK, *et al.*; Asian Pac J Cancer Prev. 2011; 12(12): 3299-3304.
4.	Effect of Diosgenin (Fenugreek) on breast cancer cells	*In vitro*	The effect of diosgenin on breast cancer cell lines was investigated.	Different concentration *e.g.* 25μM, 40μM	Cytotoxic assays showed that diosgenin reduced the viability of both the ER positive MCF-7 cells (20.63 per cent at 25μM concentration) and ER negative MDA 231 cells (20.4 per cent at 25μM concentration). A significant induction of apoptosis by diosgenin was observed in both the breast cancer cell lines-MCF-7 (55 per cent at 20μM and 98 per cent at 40μM) and MDA 231 (52 per cent at 20μM and 98 per cent at 40μM). In MCF-7 cells diosgenin induced p53 protein expression and also down regulated ER activation. However, in MDA 231 cells diosgenin increased caspase activation (6-fold) and also down regulated Bcl-2 protein expression.	Fenugreek (Diosgenin) may have a potential as an anti-cancer agent in breast cancer therapy.	Sowmya-lakshmi S, *et al.*, Proc Amer Assoc Cancer Res. 2006; 65(9): 1382.
					Anti-inflammatory Activity		
5.	Anti-inflammatory and anti-melanogenic steroidal saponin glycosides from Fenugreek (*Trigonella foenum-graecum* L.) seeds.	*In vitro*	Inhibitory activities on the production of inflammatory cytokines.	Methanol (MeOH) extract of fenugreek seed	The MeOH extract inhibited the production inflammatory cytokines such as tumor necrosis factor (TNF)-α in cultured THP-1 cells The isolated steroidal saponins strongly suppressed the production of inflammatory cytokines.	Fenugreek seeds have antiinflammatory activity.	Kawabata T, *et al.*, Planta Med. 2011; 77(7): 705-10.
6.	A study of anti-inflammatory activity of alcoholic extract of seeds of *Trigonella foenum graceum* (Fenugreek) on wister strain rat.	*In vivo*	Species: Albino Wister rat; Model: Carageenan Model N=24 Group: 3 Group I control, GR II, III (100mg/kg and 200mg/kg fenugreek extract), GR IV Indomethacin; Dosage: 100, 200 mg/kg body Duration: single dose Parameter: Paw edema observed.	Alcoholic extract of seeds of Fenugreek 100mg and 200 mg/kg b.w	Fenugreek at 100 mg/kg b.w had shown significant difference with control at 3rd, 4th, 6th hr. At 200 mg/kg b.w and Indomethacin (10 mg/kg b.w) showed significant anti-inflammatory effect up to 4 hour.	Fenugreek seeds has anti inflammatory agent.	Datta D and Shanbagh T, IJPRD, 2010; 2(9):

Contd...

TABLE 14.2–*Contd....*

Sl.No.	*Title*	*Type of Study*	*Study Design*	*Form and Dose*	*Summarized Efficacy*	*Conclusion*	*Reference*
					Antioxidant Activity		
7.	Antioxidative activity of extracts from fenugreek seeds (*Trigonella foenum-graecum*).	*In vitro*	Free radical scavenging (DPPH) method and ferric reducing anti-oxidant power.	Ethanolic extracts of fenugreek seed	The decrease in absorbance of DPPH radical is caused by anti-oxidant through the reaction between antioxidant molecule and radical results in the scavenging of the radical by hydrogen donation. A significant decrease in the concentration of DPPH due to scavenging activity of fenugreek extract was noted.	Fenugreek has anti-oxidant activity due to presence of phenolic acid, volatile oil and flavonoids.	Bukhari SB, *et al.* Pak. J. Anal. Environ. Chem. 2008; 9(2): 78-83.
8.	Antioxidant properties of germinated fenugreek seeds.	*In vitro*	FRAP, radical scavenging by DPPH, ferrylmyoglobin/2,2'-azobis-3-ethylbenzthia-zoline-6-sulfonic acid, pulse radiolysis, oxygen radical absorbance capacity and inhibition of lipid peroxidation in mitochondrial prepara-tions from rat liver.	Different fractions of the germi-nated seeds	Aqueous fraction of fenugreek exhibited the highest antioxidant activity compared with other fractions.	Germinated fenugreek seeds has antioxidant activity.	Dixit P, *et al.* Phytother Res. 2005; 19(11): 977-983.
					Menopausal Depression		
9.	Effect of fenugreek seed on menopausal depression: A rando-mized double-blind clinical trial.	Clinical	Type: double-blind controlled randomized trial Sample Size: 60 Subjects with post menopause and depression Study Group: 2 groups Control, and with diosgenine 25mg Duration of study: 8 weeks Parameters Comparative analysis done with SPSS software test.	6 g Fenugreek seed daily containing 25 mg diosgenin for 8 weeks	The Beck Depression Inventory questionnaire was administered before and after treatment to both the groups receiving fenugreek and sorbitol. Mean depression score significantly decreased compared to baseline.	Fenugreek helps in reduction of depression associated with meno-pause.	Torkestani NA, *et al.* Arak Medical University Journal. 2013; 16(74): 1-7.
					Anti-diabetic Activity		
10.	Effect of fenugreek (*Trigonella foenum-graecum* L.) intake on glycemia: a meta-analysis of clinical trials.	Review of trials	PubMed, SCOPUS, the Cochrane Trials Registry, Web of Science, and BIOSIS were searched up to 29 Nov 2013 for	–	A total of 10 trials were identified. Fenugreek significantly changed fasting blood glucose by - 0.96 mmol/l (95 per cent CI: -1.52, -0.40; I2 =80 per cent; 10 trials), 2 hour post load glucose by -2.19 mmol/l (95 per cent CI: -3.19, -1.19; I2=71 per cent; 7 trials) and HbA1c by -0.85 per cent (95 per cent CI: -1.49 per	In the meta-analysis of 10 clinical trials, intake of fenugreek seeds resulted in a significant reduction in fasting	Neelakantan N, *et al.*, Nutr J. 2014; 13(1): 7.

Contd...

TABLE 14.2–*Contd....*

Sl.No.	*Title*	*Type of Study*	*Study Design*	*Form and Dose*	*Summarized Efficacy*	*Conclusion*	*Reference*
			trials of at least 1 week duration comparing intake of fenugreek seeds with a control intervention; Data on change in fasting blood glucose, 2 hour postload glucose, and HbA1c were pooled using random-effects models.		cent, -0.22 per cent; I2=0 per cent; 3 trials) as compared with control interventions. The considerable heterogeneity in study results was partly explained by diabetes status and dose: significant effects on fasting and 2 hr glucose were only found for studies that administered medium or high doses of fenugreek in persons with diabetes.	blood glucose, 2 hr glucose, and HbA1c. No major harmful side effects of fenugreek were reported in all included studies. To conclude, results from clinical trials support beneficial effects of fenugreek seeds on glycemic control in persons with diabetes.	
11.	Effect of *Trigonella foenum-graecum* (fenugreek) seeds on glycaemic control and insulin resistance in type 2 diabetes mellitus: a double blind placebo controlled study.	Double blind placebo controlled study	HOMA model was used in mild to moderate type 2 diabetes mellitus. Twenty five newly diagnosed patients with type 2 diabetes (fasting glucose < 200 mg/dl) were randomly divided into two groups. Group I (n=12) received 1 gm/day hydroalcoholic extract of fenugreek seeds and Group II (n=13) received usual care (dietary control, exercise) and placebo capsules for two months.	Hydro-alcoholic extract of fenugreek seeds 1 gm/day	In group 1 as compared to group 2 at the end of two months, fasting blood glucose and two hour post glucose blood glucose were not different. But area under curve (AUC) of blood glucose as well as insulin was significantly lower ($p < 0.001$). HOMA model derived insulin resistance showed a decrease in percent beta-cell secretion in group 1 as compared to group 2 and increase in percent insulin sensitivity ($p < 0.05$). Serum triglycerides decreased and HDL cholesterol increased significantly in group 1 as compared to group 2 ($p < 0.05$).	Adjunct use of fenugreek seeds improves glycemic control and decreases insulin resistance in mild type-2 diabetic patients. There is also a favorable effect on hypertriglyceridemia.	Gupta A, *et al.*, J Assoc Physicians India. 2001; 49: 1057-1061.
12.	Effect of fenugreek seeds on blood glucose and lipid profiles in type 2 diabetic patients.	Clinical	In the study, 24 type 2 diabetic patients were placed on 10 grams/day powdered fenugreek seeds mixed with yoghurt or soaked in hot water for 8 weeks. Weight, FBS, HbA(1)C, total cholesterol, LDL, HDL and food record were measured before and after the study. The differences observed in food records, BMI and serum variables were analyzed.	Fenugreek seed powder mixed with yogurt dose 10 gm/day for 8 weeks	After exclusion of 6 cases for changing in medication or personal problems, the results of 18 patients (11 consumed fenugreek in hot water and 7 in yoghurt) were studied. Findings showed that FBS, TG and VLDL-C decreased significantly (25 per cent, 30 per cent and 30.6 per cent respectively) after taking fenugreek seed soaked in hot water whereas there were no significantly changes in lab parameters in cases consumed it mixed with yoghurt. BMI, Energy, Carbohydrate, Protein and fat intake remained unchanged during study.	Fenugreek seeds can be used as an adjuvant in the control of type 2 diabetes mellitus in the form of soaked in hot water.	Kassaian N, *et al.*, Int J Vitam Nutr Res. 2009; 79(1): 34-39.

Contd...

TABLE 14.2–*Contd....*

Sl.No.	*Title*	*Type of Study*	*Study Design*	*Form and Dose*	*Summarized Efficacy*	*Conclusion*	*Reference*
13.	The postprandial hypo-glycemic activity of fenugreek seed and seed's extract in Type 2 diabetics: A pilot study.	Clinical study	One hundred sixty-six patients from both genders (females= 91, males =75) aged 40-70 years old suffering from T2D were recruited from different towns in northern Jordan. All patients must be under medical treatment, showing either a Poor (HbA1C >8.5) or Fair control (HbA1C ≤8.5). Patients affected by diabetic complications or with clinical and biochemical data of nephropathy were not enrolled. Also excluded were pregnant worsen, patients with diagnosis of gestational diabetes, insulin-dependent or type 1 diabetics, people who needed to travel frequently, and hospitalized patients. Postprandial plasma glucose level was measured before and 2-hours after the admini-stration of the treatment.	Three groups: FGO (control group: placebo drink), FG2.5 (2.5g of fenugreek), and FG5 (5g of fenugreek). Participants were instructed to drink the extract and chew the seeds.	Accounting for gender, age, education, physical activity, body mass index, glycemic control, and medication, patients in FG5 group showed the greatest decrease in postprandial glucose with a pretest-posttest difference (D) of - 41 ± 6.1 mg/dl. Two-hour plasma glucose dropped for patients in FG2.5, however, the drop was not statistically different from that noticed in the placebo group (D = - 24.8 ± 4.9 mg/dl vs. - 9.8 ± 2.2 mg/dl respectively). There was a dose dependent reduction in the post-prandial blood glucose level such, that the higher the fenugreek dose the lower blood glucose level.	Fenugreek seeds have potential hypoglycemic activity in T2D patients.	Hiba A Bawadi, *et al.*, Pharmacog Magazine: 2009; 5(18): 134-138.
14.	Effect of fenugreek (*Trigonella foenum-graecum* L) seeds on serum parameters in normal and strepto-zotocin-induced diabetic rats.	*In vivo*	Normal and streptozo-tocin-induced diabetic rats. A comparison was made between the action of fenugreek extract and glibenclamide (600 µg/kg), the known antidiabetic drug.	Oral admini-stration of fenugreek extract (0.1, 0.25, and 0.5 g/kg body weight) for 14 days	Treatment of fenugreek extract alleviated body weight loss in diabetic rats. Administrations of the extract significantly decreased serum glucose, total cholesterol, triacylglycerol, urea, uric acid, creatinine, AST, and ALT levels, whereas it increased serum insulin in diabetic rats but not in normal rats. The anti-diabetic effect of the extract was similar to that observed for glibenclamide.	Fenugreek possesses potential anti- diabetic property.	Akram Eidi, Maryam Eidi, Mousa Sokhteh; Nutrition Research Volume 27, Issue 11, Pages 728–733, Nov 2007.

Contd...

TABLE 14.2–*Contd....*

Sl.No.	*Title*	*Type of Study*	*Study Design*	*Form and Dose*	*Summarized Efficacy*	*Conclusion*	*Reference*
15.	Soluble dietary fibre fraction of *Trigonella foenum-graecum* (fenugreek) seed improves glucose homeostasis in animal models of type 1 and type 2 diabetes by delaying carbohydrate digestion and absorption, and enhancing insulin action.	*In vivo*	Normal, type 1 or type 2 diabetic rats	0.5g/kg body weight dose twice a day for 28 days; soluble dietary fibre (SDF) fraction of *T. foenum-graecum.*	It decreased serum glucose, increased liver glycogen content and enhanced total antioxidant status. Intestinal disaccharidase activity and glucose absorption were decreased and gastro-intestinal motility increased by the SDF fraction. Glucose transport in 3T3-L1 adipocytes and insulin action were increased by *T. foenum-graecum.*	SDF fraction of *Trigonella foenum-graecum* significantly improved glucose homeostasis in type 1 and type 2 diabetes by delaying carbohydrate digestion and absorption, and enhancing or mimicking insulin action it is a potential antidiabetic agent.	Hannan JM, Ali L, Rokeya B, Khaleque J *et al.,* Br J Nutr. 2007 Mar; 97(3): 514-21.
16.	Antidiabetic activity of *Trigonella foenum-graecum* L. seeds extract (IND01) in neonatal streptozotocin-induced (n-stz) rats.	*In vivo*	Neonatal streptozotocin-induced rat (n-STZ) model.	Fenugreek Seed Extract 100 mg/kg, oral for 28 days	Fenugreek treatment showed significant reversal of n-STZ-induced changes (rise in SG, decline in body weight and rise in HBA1c). Histology sections of pancreas showed increase in number and size of pancreatic islet β-cells. Fenugreek showed a potential to ameliorate symptoms of DM during progressive deterioration and improvedglycemic functions.	Fenugreek possesses anti-diabetic potential.	Chetan P. Kulkarni, Subhash L. Bodhankar *et al.;* Diabetologia Croatica 41-1, 2012
17.	Insulin sensitizing actions of fenugreek seed polyphenols, quercetin and metformin in a rat model.	*In vivo*	Insulin resistance (IR) model in wistar rats.	Fenugreek seed polyphenolic extract (FPEt (200 mg/kg bw), quercetin (50mg/kg bw) or metformin (50 mg/kg bw) for 45 days	Fructose caused increased levels of glucose, insulin, TG and FFA, alterations in insulin sensitivity indices, enzyme activities and reduced glycogen content. Higher protein tyrosine phosphatase (PTP) activity and lower protein tyrosine kinase (PTK) activity suggest reduced tyrosine phosphorylation status. Administration of FPEt or quercetin improved insulin sensitivity and tyrosine phosphorylation in fructose-fed animals and the effect was comparable with that of metformin.	FPEt and quercetin improved insulin signaling and sensitivity and thereby promoted the cellular actions of insulin thereby can be used as anti-diabetic agent.	Kannappan S, Anuradha CV.; Indian J Med Res. 2009 Apr; 129(4): 401-8.
18.	Evaluation of the anti-diabetic effect of *Trigonella foenum-graecum* seed powder on alloxaninduced diabetic albino rats.	*In vivo*	Alloxan-induced diabetic rats.	Ethanolic extract of Fenugreek seed powder 50mg/100g bodyweight for 48 days	Administrations of the herbal extract decreased blood glucose, serum cholesterol, SGOT and SGPT levels.	*T. foenum-graecum* seed has potential anti-diabetic effect.	Renuka C, Ramesh N and Saravanan K.; International Journal of PharmTech Research; Vol.1, No.4, pp 1580-1584, Oct-Dec 2009.

Contd...

TABLE 14.2–*Contd....*

Sl.No.	*Title*	*Type of Study*	*Study Design*	*Form and Dose*	*Summarized Efficacy*	*Conclusion*	*Reference*
					Hyperglycemia and obesity		
19.	Diosgenin present in fenugreek improves glucose metabolism by promoting adipocyte differentiation and inhibiting inflammation in adipose tissues;	*In vivo*	Diabetic obese KK-Ay mice.	Fenugreek supplements were administered as powdered fenugreek seeds, debitterized powdered fenugreek seeds, or hydro-alcoholic seed extract either in form of capsules or as an ingredient of unleavened bread. Dose range from 1 gm to 100 gm provided in equal doses 2 to 3 times per day.	Treatment with fenugreek ameliorated diabetes. Moreover, fenugreek miniaturized the adipocytes and increased the mRNA expression levels of differentiation related genes in adipose tissues. Fenugreek also inhibited macrophage infiltration into adipose tissues and decreased the mRNA expression levels of inflammatory genes. In addition, diosgenin was identified, a major aglycone of saponins in fenugreek to promote adipocyte differentiation and to inhibit expressions of several molecular candidates associated with inflammation in 3T3-L1 cells. Fenugreek ameliorated diabetes by promoting adipocyte differentiation and inhibiting inflammation in adipose tissues, and its effects are mediated by diosgenin.	Fenugreek containing diosgenin may be useful for ameliorating the glucose metabolic disorder associated with obesity.	Taku Uemura, Shizuka Hirai, Noriko Mizoguch *et al.*; Molecular Nutrition and Food Research, Vol 54, Issue 11, pp. 1596–1608, Nov 2010.
20.	Hypoglycemic and gypolipidemic effect of fenugreek in different forms on experimental rat.	*In vivo*	Streptozotocin induced diabetic rats.	Fenugreek powder (5 per cent in diet), aqua extract, methanolic extract (3mg/kg body weight) and oil (5 per cent in diet). The experiment period was 60 days	Compared with control (+ve) group, the four treated groups showed significant decrease in glucose, glucosalated heamoglobin (HbAIC per cent) and fructoseamine (FA) values, but significant increased in insulin. Also, they produced a significant fall in various serum lipids like total cholesterol, triglycerides, low and very low density lipoprotein cholesterol (LDLc and VLDLc) and increased high density lipoprotein cholesterol (HDLc) in the diabetic rats. On the other side, they showed a significant decrease in liver cholesterol and total lipids and a significant increase in liver triglyceride and glycogen compared to control (+ve) group. Fenugreek powder, aqua extract, methanolic extract and oil groups showed improvement in nutritional status and liver and renal function of diabetic rats.	Fenugreek supplement in any form either powder or extract or oil have potential to control blood glucose and for prevention of diabetes.	Abor M.M. Abd El Rahman; World Applied Sciences Journal 29 (7): 835-841, 2014.

Contd...

TABLE 14.2–*Contd....*

Sl.No.	Title	Type of Study	Study Design	Form and Dose	Summarized Efficacy	Conclusion	Reference
					Hypolipidemic effect		
21.	Hypolipidemic effect of fenugreek: A clinical study.	Clinical	The study was conducted at cardiology Department, Gandhi Hospital, Hyderabad, 18 patients (age between 35-55 years) having lipid related problems with high serum cholesterol reports were selected. The patients were divided into 3 groups of 6 each as follows: Group I received placebo 50 gm (rice powder and Bengal gram powder in equal measures); Group II - placebo 25 gm + Fenugreek 25 gm and Group III –Fenugreek 50 gm. Patients were directed to take each 50 gm pack orally before lunch and dinner every day for 20 days. Blood samples were collected after overnight fasting on 0, 10th and 20th days during test period and estimated for lipid profile.	Fenugreek seed powder 25 to 50 gm powder twice a day.	There were no significant changes in lipid profile of group I patients receiving placebo. In groups II and III (Placebo + Fenugreek 25g and 50 g respectively) serum cholesterol, triglycerides and VLDL levels were significantly decreased when compared to group I.	Fenugreek is having hypolipidemic effect in hypercholesterolaemic patients.	Prasanna M. Indian J Pharmacol 2000; 32: 34-36.
					Hypoglycaemic effect and herb-drug combination		
22.	Clinical observation on *Trigonella foenum-graecum* L. total saponins in combination with sulfonylureas in the treatment of type 2 diabetes mellitus.	Clinical	Sixty-nine T2DM patients whose blood glucose levels were not well controlled by oral sulfonylureas hypoglycemic drug were randomly assigned to the treated group (46 cases) and the control group (23 cases), and were	TFGs three times per day, 6 pills each time for 12 weeks	The efficacy on TCM symptoms was obviously better in the treated group than that in the control group ($P<0.01$), and there were statistically remarkable decreases in aspect of FBG, 2h PBG, HbA1c and CSQS in the treated group as compared to those in the control group ($P<0.05$ or $P<0.01$), while no significant difference was found in BMI, hepatic and renal functions between the two groups ($P>0.05$).	The combined therapy of TFGs with sulfonylureas hypoglycemic drug could lower the blood glucose level and ameliorate clinical symptoms in the treatment of T2DM, and the therapy was relatively safe.	Lu FR, *et al.*, Chin J Integr Med. 2008; 14(1): 56-60.

Contd...

TABLE 14.2–*Contd....*

Sl.No.	*Title*	*Type of Study*	*Study Design*	*Form and Dose*	*Summarized Efficacy*	*Conclusion*	*Reference*
			given TFGs or placebo three times per day, 6 pills each time for 12 weeks, respectively. Meanwhile, the patients continued taking their original hypoglycemic drugs. The following indexes, including effects on traditional Chinese medicine (TCM) symptoms, fast blood glucose (FBG), 2-h post-prandial blood glucose (2h PBG), glycosylated hemoglobin (HbA1c), clinical symptomatic quantitative scores (CSQS), body mass index (BMI), as well as hepatic and renal functions, were observed and compared before and after treatment.				
23.	Effect of fenugreek (*Trigonella foenum-graecum* L.) intake on glycemia: A meta-analysis of clinical trials;	Review of trials	PubMed, SCOPUS, the Cochrane Trials Registry, Web of Science, and BIOSIS were searched up to 29 Nov 2013 for trials of at least 1 week duration comparing intake of fenugreek seeds with a control intervention; Data on change in fasting blood glucose, 2 hour postload glucose, and HbA1c were pooled using random-effects models.	–	A total of 10 trials were identified. Fenugreek significantly changed fasting blood glucose by - 0.96 mmol/l (95 per cent CI: -1.52, -0.40; I2=80 per cent; 10 trials), 2 hour post load glucose by -2.19 mmol/l (95 per cent CI: -3.19, -1.19; I2=71 per cent; 7 trials) and HbA1c by -0.85 per cent (95 per cent CI: -1.49 per cent, -0.22 per cent; I2=0 per cent; 3 trials) as compared with control interventions. The considerable heterogeneity in study results was partly explained by diabetes status and dose: significant effects on fasting and 2 hr glucose were only found for studies that administered medium or high doses of fenugreek in persons with diabetes.	In the meta-analysis of 10 clinical trials, intake of fenugreek seeds resulted in a significant reduction in fasting blood glucose, 2 hr glucose, and HbA1c. No major harmful side effects of fenugreek were reported in all included studies. To conclude, results from clinical trials support beneficial effects of fenugreek seeds on glycemic control in persons with diabetes.	Neelakantan N, *et al.*, Nutr J. 2014; 13(1): 7.

Contd...

TABLE 14.2–*Contd....*

Sl.No.	Title	Type of Study	Study Design	Form and Dose	Summarized Efficacy	Conclusion	Reference
24.	Effect of *Trigonella foenum-graecum* (fenugreek) seeds on glycaemic control and insulin resistance in type 2 diabetes mellitus: A double blind placebo controlled study.	Double blind placebo cont-rolled study	HOMA model was used in mild to moderate type 2 diabetes mellitus. Twenty five newly diag-nosed patients with type 2 diabetes (fasting glucose < 200 mg/dl) were randomly divided into two groups. Group I (n=12) received 1 gm/day hydroalcoholic extract of fenugreek seeds and Group II (n=13) received usual care (dietary control, exercise) and placebo capsules for two months.	Hydro-alcoholic extract of fenugreek seeds 1 gm/day	In group 1 as compared to group 2 at the end of two months, fasting blood glucose and two hour post glucose blood glucose were not different. But area under curve (AUC) of blood glucose as well as insulin was significantly lower ($p < 0.001$). HOMA model derived insulin resistance showed a decrease in percent beta-cell secretion in group 1 as compared to group 2 and increase in percent insulin sensitivity ($p < 0.05$). Serum trigly-cerides decreased and HDL cholesterol increased significantly in group 1 as compared to group 2 ($p < 0.05$).	Adjunct use of fenugreek seeds improves glycemic control and decreases insulin resistance in mild type-2 diabetic patients. There is also a favorable effect on hypertriglyceri-demia.	Gupta A, *et al.*, J Assoc Physicians India. 2001; 49: 1057-1061.
25.	Effect of fenugreek seeds on blood glucose and lipid profiles in type 2 diabetic patients.	Clinical	In the study, 24 type 2 diabetic patients were placed on 10 grams/day powdered fenugreek seeds mixed with yoghurt or soaked in hot water for 8 weeks. Weight, FBS, HbA(1)C, total choles-terol, LDL, HDL and food record were measured before and after the study. The differences observed in food records, BMI and serum variables were analyzed.	Fenugreek seed powder mixed with yogurt dose 10 gm/day for 8 weeks	After exclusion of 6 cases for changing in medication or personal problems, the results of 18 patients (11 consumed fenugreek in hot water and 7 in yoghurt) were studied. Findings showed that FBS, TG and VLDL-C decreased significantly (25 per cent, 30 per cent and 30.6 per cent respectively) after taking fenugreek seed soaked in hot water whereas there were no significantly changes in lab parameters in cases consumed it mixed with yoghurt. BMI, Energy, Carbohydrate, Protein and fat intake remained unchanged during study.	Fenugreek seeds can be used as an adjuvant in the control of type 2 diabetes mellitus in the form of soaked in hot water.	Kassaian N, *et al.*, Int J Vitam Nutr Res. 2009; 79(1): 34-39.
26.	The postprandial hypoglycemic activity of fenugreek seed and seed's extract in type 2 diabetics: A pilot study.	Clinical	One Hundred Sixty-six patients from both genders (females= 91, males =75) aged 40-70 years old suffering from T2D were recruited from different towns in northern Jordan. All patients must be under medical treatment,	Three groups: FGO (control group: placebo drink), FG2.5 (2.5g of fenugreek), and FG5	Accounting for gender, age, education, physical activity, body mass index, glycemic control, and medication, patients in FG5 group showed the greatest decrease in postprandial glucose with a pretest-posttest difference (D) of $- 41 \pm 6.1$ mg/dl. Two-hour plasma glucose dropped for patients in FG2.5, however, the drop was not statistically different from that noticed in the placebo group (D = $- 24.8 \pm 4.9$ mg/dl vs. $- 9.8 \pm 2.2$ mg/dl respectively). There was a dose dependent reduction in the postprandial blood glucose level such, that the higher the fenugreek dose the lower blood glucose level.	Fenugreek seeds have potential hypoglycemic activity in T2D patients.	Hiba A Bawadi, *et al.*, Pharmacog Magazine. 2009; 5(18): 134-138.

Contd...

TABLE 14.2–*Contd....*

Sl.No.	*Title*	*Type of Study*	*Study Design*	*Form and Dose*	*Summarized Efficacy*	*Conclusion*	*Reference*
			showing either a poor (HbA1C >8.5) or fair control (HbA1C ≤8.5). Patients affected by diabetic complications or with clinical and biochemical data of nephropathy were not enrolled. Also excluded were pregnant worsen, patients with diagnosis of gestational diabetes, insulin-dependent or type 1 diabetics, people who needed to travel frequently, and hospitalized patients. Postprandial plasma glucose level was measured before and 2-hours after the administration of the treatment.	(5g of fenugreek). Participants were instructed to drink the extract and chew the seeds.			
					Protective and potential hypoglycaemic activity, herb-drug interaction		
27.	Evaluation of protective action of fenugreek, insulin and glimepiride and their combination in diabetic Sprague Dawley rats.	*In vivo*	Streptozotocin induced diabetic rats.	Fenugreek seed powder 1 g/kg orally once daily for 8 weeks	There was an increase in the concentration of TBARS and protein carbonyls, and decrease in the concentration of GSH and glycogen, and the activity of GST, G6PD, Na/K ATPase and Mg + ATPase in diabetic livers, while treatment groups (insulin, fenugreek and glimepiride) showed significant increase in the above parameters. The histology of liver revealed marked changes in diabetic rats and mild changes in combination treatment groups. The treatment with fenugreek, insulin and glimepiride reduced the oxidative stress, improved the liver parameters in diabetic rats and their combination showed a beneficial effect on liver. Addition of fenugreek seed powder to insulin and glimepiride had positive interaction in improving the liver parameters.	Addition of fenugreek seed powder to insulin and glimepiride had positive interaction in improving the liver parameters in diabetic conditions.	C. Haritha, A. Gopala Reddy, Y. Ramana Reddy *et al.*, J Nat Sci Biol Med. 2013 Jan-Jun; 4(1): 207–212.
					Diabetes related hepatic dyslipidemias		
28.	Diosgenin, the main aglycon of fenugreek, inhibits LXRa activity in HepG2 cells and decreases plasma and	*In vivo*	Male diabetic obese KK-Ay mice.	0.5 per cent and 0.2 per cent fenugreek administered for 4 weeks	Diosgenin (5 and 10 mmol/L) inhibited the accumulation of TG and the expression of lipogenic genes in HepG2 cells. Moreover, diosgenin inhibited the transactivation of liver-X-receptora. These findings suggest that fenugreek ameliorates dyslipidemia by decreasing the hepatic lipid content in diabetic mice and that its	Fenugreek directly affects the liver and improves dysfunctional lipid metabolism in the liver of obese diabetic mice and	Taku Uemura, Tsuyoshi Goto, Min-Sook Kang, *et al.*; The Journal of

Contd...

TABLE 14.2–*Contd....*

Sl.No.	*Title*	*Type of Study*	*Study Design*	*Form and Dose*	*Summarized Efficacy*	*Conclusion*	*Reference*
	hepatic triglycerides in obese diabetic mice.				effect is mediated by diosgenin. Fenugreek, which contains diosgenin, may be useful for the management of diabetes related hepatic dyslipidemias.	that the inhibitory effect of diosgenin on LXR (Liver X recepter) a activation plays an important role in the therapeutic effect of fenugreek on lipid metabolism disorders.	Nutrition, Biochemical, Molecular, and Genetic Mechanisms, 2010 141: 17–23.
					Reno-Protective		
29.	Fenugreek prevents the development of STZ-induced diabetic nephropathy in a rat model of diabetes.	*In vivo*	Streptozotocin induced diabetic rats.	9 g Fenugreek seed powder/kg daily for 12 weeks	Fenugreek reduced blood glucose levels and improved renal functions. Diabetic nephropathy rats demonstrated a significant renal dysfunction, extra–cellular matrix accumulation, pathological alteration, and oxidative stress, while the symptoms were evidently reduced by fenugreek treatment. Furthermore, the up-regulation of TGF-β1 and CTGF at a transcriptional and translational level in DN rats was distinctly inhibited by fenugreek.	Fenugreek reduces blood glucose level and prevents diabetic nephropathy development.	Yingli Jin, Yan Shi *et al.*; Hindawi Publishing Corporation Evidence-Based Complementary and Alternative Medicine; Vol 2014, Article ID 259368, 11 pages.
30.	*Trigonella foenum-graecum* seed extract protects kidney function and morphology in diabetic rats via its antioxidant activity.	*In vivo*	Streptozotocin induced Diabetic nephropathy.	Low (440 mg/kg), medium (870 mg/kg), or high (1740 mg/kg) dose of *T. foenum-graecum* seed aqueous extract (TE); oral intra-gastric intubation for 6 weeks.	In Fenugreek treated DN rats, blood glucose, kidney/body weight ratio, serum creatinine, blood urea nitrogen, 24-hour content of urinary protein, and creatinine clearance were significantly decreased compared with non-treated DN rats. Diabetic rats showed decreased activities of superoxide dismutase and catalase, increased concentrations of malondialdehyde in the serum and kidney, and increased levels of 8-hydroxy-2'-deoxyguanosine in urine and renal cortex DNA. Treatment with TE restored the altered parameters in a dose-dependent manner. All of the ultra-morphologic abnormalities in the kidney of diabetic rats, including the uneven thickening of the glomerular base membrane, were markedly ameliorated by TE treatment.	Fenugreek confers protection against functional and morphologic injuries in the kidneys of diabetic rats by increasing activities of antioxidants and inhibiting accumulation of oxidized DNA in the kidney, suggesting a potential drug for the prevention and therapy of DN.	Nutr Res. 2011 Jul; 31(7): 555-62; Xue W, Lei J, Li X, Zhang R.

Contd...

TABLE 14.2–*Contd....*

Sl.No.	*Title*	*Type of Study*	*Study Design*	*Form and Dose*	*Summarized Efficacy*	*Conclusion*	*Reference*
					Neuroprotective		
31.	Protection of trigonelline on experimental diabetic peripheral neuropathy; Evid Based Complement.	*In vivo*	Streptozotocin- and high-carbohydrate/ high-fat diet-induced diabetic rats.	Trigonelline (40mg/kg) and sitagliptin (4mg/kg) were mixed daily with a vehicle consisting of the standard diet to diabetic animals for 48 weeks.	Serum glucose, serum insulin, insulin sensitivity index, lipid parameters, body weight, sciatic nerve conduction velocity, nociception, glucagon-like peptide-1 receptor mRNA and protein, total and phosphorylated p38 mitogen-activated protein kinases protein expression, malonaldehyde content, and superoxide dismutase activity were altered in diabetic rats, and were near control levels treated with trigonelline. Slight micropathological changes existed in sciatic nerve of trigonelline-treated diabetic rats. These suggest that trigonelline has beneficial effects for diabetic peripheral neuropathy. Trigonelline significantly improved motor and sensory nerve conduction velocity and nociception; it has protection on the neuron tissue (sciatic nerve) by improving micropathological changes of sciatic nerve; reduced oxidative stress in the sciatic nerve; and GLP-1R and p38 MAPK activity were regulated by trigonelline and are likely relevant to the development of DPN.	Treatment with trigonelline reduced both biomarkers significantly, control animals. Because trigonelline corrected both hyper-corrected both hyperglycemia and reduced oxidative stress it was concluded that trigonelline promoted antioxidant activity in Diabetic Peripheral Neuropathy.	Zhou JY1, Zhou SW; Alternat Med. 2012; 164-219.
32.	Antidiabetic and neuro-protective effects of *Trigonella Foenum-graecum* seed powder in diabetic rat brain.	*In vivo*	Alloxan induced diabetic rat.	5 per cent powdered Trigonella seeds in powered rat feed (5 g of dry TSP was mixed with 95 g of powdered rat feed) for 21 days	A significant increase in lipid peroxidation was observed in diabetic brain. The increased lipid peroxidation following chronic hyperglycemia was accompanied with a significant increase in the neurolipofuscin deposition and Ca2+ levels with decreased activities of membrane linked ATPases and antioxidant enzymes in diabetic brain. A decrease in synaptosomal membrane fluidity may influence the activity of membrane linked enzymes in diabetes. Fenugreek treatment reversed the hyperglycemia induced changes to normal levels in diabetic rat brain.	Fenugreek administration amended effect of hyperglycemia on alterations in lipid peroxidation, restoring membrane fluidity, activities of membrane bound and antioxidant enzymes, thereby ameliorating the diabetic complications.	Kumar P., Kale R.K.; Prague Medical Report/ Vol. 113 (2012) No. 1, pp. 33–43 33.
				MUSALI (*Chlorophytum borivilianum* Linn.)			
					Hypolipidemic and Anti-hyperglycemic Activity		
1.	Hypolipidemic effects of *Chlorophytum borivilianum* tubers.	*In vivo*	Hypolipidemic effects of *Chlorophytum borivilianum* tubers (CBT) on rats fed with a high-cholesterol diet.	Ethanolic extract 100, 200 and 500 mg/kg body weight	Serum cholesterol, triglycerides and HDL were measured on days 0, 10 and 20. The changes on the lipid levels of each group were statistically analyzed using Student t' test. Treatment demonstrated dose dependent hypolipidemic activity at different dose levels by decreasing serum lipid. CBT exhibited potential hypolipidemic activity when compared witha standard dose of lovastatin.	There was a significant decrease in serum cholesterol in both the lovastatin and the CBT treated groups.	Deore SL, Khadabadi SS, Indian J Pharmacol.; 40 (S2).

Contd...

TABLE 14.2–*Contd....*

Sl.No.	Title	Type of Study	Study Design	Form and Dose	Summarized Efficacy	Conclusion	Reference
2.	*Chlorophytum borivilianum* root extract maintains near normal blood glucose, insulin and lipid profile levels and prevents oxidative stress in the pancreas of streptozotocin-induced adult male diabetic rats.	*In vivo*	Body weight, blood glucose, HbA1c, insulin, lipid profile levels and glucose homeostasis indices were determined in streptozotocin (STZ)-induced male diabetic rats. Histopathological changes and oxidative stress parameters *i.e.* lipid peroxidation (LPO) and antioxidant enzymes activity levels of the pancreas were investigated.	*C. borivilianum* root aqueous extract (250 and 500 mg/kg/day) was administered to streptozotocin (STZ)-induced male diabetic rats for 28 days	*C. borivilianum* root extract treatment to diabetic rats maintained near normal body weight, blood glucose, HbA1c, lipid profile and insulin levels with higher HOMA-β cell functioning index, number of Islets/pancreas, number of β-cells/Islets however with lower HOMA-insulin resistance (IR) index as compared to non-treated diabetic rats. Negative correlations between serum insulin and blood glucose, HbA1c, triglyceride (TG) and total cholesterol (TC) levels were observed. *C. borivilianum* root extract administration prevented the increase in lipid peroxidation and the decrease in activity levels of superoxide dismutase (SOD), catalase (CAT) and glutathione peroxidase (GPx) with mild histopathological changes in the pancreas of diabetic rats.	*C. borivilianum* root maintains near normal levels of these metabolites and prevented oxidative stress-induced damage to the pancreas in diabetes.	Giribabu N, *et al.*, Int J Med Sci. 2014; 11(11): 1172–1184.
3.	Antidiabetic activity of the aqueous extract of *Chlorophytum borivilianum* L. in streptozotocin induced-hyperglycemic rats - a preliminary study.	*In vivo*	Streptozotocin induced-hyperglycemic rat model was used. The blood glucose levels were measured at 0, 2 h, 4 h and 6 h after oral treatment.	Dose of 250 mg/kg and 500 mg/kg body weight was administered orally	The aqueous extract reduced the blood glucose in STZ-induced diabetic rats from 285.56 to 206.82 mg/dl, 6 h after oral administration of extract (P<0.01). The antidiabetic activity of aqueous extract of *C. borivilianum* was compared with glibenclamide, an oral hypoglycaemic agent (3 mg/kg).	*C. borivilianum* possesses antidiabetic activity.	Mujeeb M, *et al.*, Journal of Pharmacy Research. 2009; 2(1): 51-53.
					Anti-stress Activity		
4.	Screening of antistress properties of *Chlorophytum borivilianum* tuber.	*In vivo*	This property is assessed by swim endurance stress, anorexic test in rats and despair swim test. Cold stress induced Gastric ulceration model was also selected to evaluate antiulcer activity.	Aqueous and alcoholic tuber extracts 30, 100 and 300 mg/kg body weight, orally.	Alcoholic extract significantly increases swimming time and reduces the ulcer index compared to that of control group.	A significant effect (p< 0.001) from 200 mg/kg dose for both the extracts was observed in all four models.	Deore and Khadabadi *et al.* Pharmacologyonline. 2009; 1: 320-328.
5.	Anti-stress and anti-oxidant effects of roots of *Chlorophytum borivilianum* (Santa Pau and Fernandes).	*In vivo* and *in vitro*	Anti-stress activity was evaluated using chronic cold restrain stress model, and *in vitro* DPPH free radical scavenging activity.	Dose of 125 mg/kg and 250 mg/kg body weight was administered orally for 7 days	The aqueous extract of *C. borivilianum* (250 mg/kg) significantly reverted the elevated levels of plasma glucose, triglycerides, and cholesterol and serum corticosterone and also reduced the ulcer index, adrenal gland weight more as effectively as the standard drug (diazepam) in rats. At 125 mg/kg po, it showed a mild anti-stress activity. Under *in vitro* DPPH* free radical scavenging assay and lipid peroxidation assay the extract considerably inhibited, in a dose-dependent manner, the levels of DPPH* free radicals and thiobarbituric acid reactive substances, respectively thus showing significant antioxidant property.	The results suggested that aqueous extract of *C. borivilianum* could be used for the treatment of oxidative stress-induced disorders.	Kenjale RD *et al.*, Indian J Exp Biol. 2007; 45(11): 974-9.

Contd...

TABLE 14.2–*Contd....*

Sl.No.	*Title*	*Type of Study*	*Study Design*	*Form and Dose*	*Summarized Efficacy*	*Conclusion*	*Reference*
					Immunomodulatory Activity		
6.	Immunomodulatory activity of *Chlorophytum borivilianum* Sant. F.	*In vivo*	Ethanolic extract of the roots and its sapogenin were evaluated for their immunomodulatory activity. Effect of azathioprine-induced myelo-suppresion and administration of extracts on hematological and serological parameters was determined.	200 mg/kg bw ethanolic root extract. and 100 mg/kg bw sapogenin extract	Administration of extracts greatly improved survival against *Candida albicans* infection. An increase in delayed-type hyper-sensitivity response (DTH), per cent neutrophil adhesion and *in vivo* phagocytosis by carbon clearance method was observed after treatment with extracts. Immunostimulant activity of ethanolic extract was more pronounced as compared to sapogenins.	Apart from confirming the immunostimulant activity of *C. borivilianum* also, presents evidence for the presence of the substance other than sapogenins which induce stimulation of immune response in treated animals.	Thakur M *et al.*, Evid Based Complement Alternat Med. 2007; 4(4): 419–423.
					Anti-inflammatory Activity		
7.	Antiinflammatory activity of methanolic extract of *Chlorophytum borivilianum.*	*In vivo*	To evaluate the anti-inflammatory activity of methanolic extracts of leaves and roots of *Chlorophytum borivilianum* in rats using carrageenan induced paw edema model.	Methanolic extract of leaf and roots were administered orally in the concentration of 200 mg/kg	Both the methanolic extracts (leaves and roots) produced statistically significant inhibition of edema induced by carrageenan at all doses when compared to the control groups. The percentage inhibition was also noted for both the methanolic extract.	Both the extracts of *C. borivilianum* possesses anti-inflammatory activity.	Chakraborthy GS, *et al.*, Journal of Pharmacy Research. 2008; 1(1): 58-60.
8.	Effect of *Chlorophytum borivilianum* on adjuvant induced arthritis in rats.	*In vivo*	Anti-arthritic effect of aqueous and alcoholic extracts of *Chlorophytum borivilianum* tubers on complete Freund's adjuvant (CFA) induced arthritis has been studied in wistar albino rats.	Aqueous and alcohol extracts were administered at dose of 30, 100, 300 mg/kg body weight p.o.	During arthritic condition, aqueous and alcoholic extracts of *Chlorophytum borivilianum* significantly reduced the paw volume; inhibited body weight loss compared to vehicle treated control rats.	Extracts of *Chlorophytum borivilianum* shows antiarthritic activity.	Deore S.L and Khadabadi S.S. Annals of Biological Research, 2010; 1(1): 36-40.
9.	Phytochemical and anti-microbial studies of *Chlorophytum borivilianum.*	*In vitro*	Extracts of leaves and stems of *Chlorophytum borivilianum* San and Fern (aerial parts) were subjected to preliminary phytochemical screening for the presence of plant.		The leaf extract of *Chlorophytum borivilianum* displayed over-whelming concentration dependent antimicrobial properties, inhibiting the growth of *Staphylococcus aureus* and *Bacillus cereus*, far above that of ampicillin used in the study at a concentration of 1.0 g/ml. The extract was less sensitive to the 2 Gram negative bacteria in the assay. In the antifungal assay, the growth of *Aspergillus niger* and *Candida albicans*, used were inhibited in the	The extract exhibited a weak activity against *C. albicans* as well as *A. niger*. Both plant parts seem to justify their ethnomedical uses.	Chakraborthy GS, *et al.*, Pharmacophore. 2014; 5(2): 258-261.

Contd...

TABLE 14.2–*Contd....*

Sl.No.	*Title*	*Type of Study*	*Study Design*	*Form and Dose*	*Summarized Efficacy*	*Conclusion*	*Reference*
			secondary metabolites and *in vitro* antibacterial and antifungal studies using agar disc diffusion technique.		same manner comparable to voriconazole the reference drug included in the study. The methanol extract of stem also displayed a concentration related antibacterial activity, inhibiting the growth of S. aureus comparable to ampicillin at 1.0 g/ml. The extract was least active against Escherichia coli with a mild activity at 1.0 g/ml.		
10.	Antibacterial activities of crude extracts of *Chlorophytum borivilia-num* to bacterial pathogens.	*In vitro*	Antibacterial properties of different extracts of *Chlorophytum borivilia-num* were studied. Ethanol, ethyl acetate, acetic acid and water were used to prepare the extract. The antibacterial activity of different extracts were carried out against four bacteria, *Staphylococcus aureus, Escherichia coli, Pseudo-monas aeruginosa* and *Bacillus subtilis*, by agar cup diffusion method. Zone of inhibition produced by different extracts were measured.		Acetic acid extract of *C. borivilianum* showed antibacterial activity against all the tested bacteria in the order of sensitivity as *Staphylococcus aureus>Pseudomonas aeruginosa>Escherichia coli> Bacillus subtilis*. The antibacterial activity of *Staphylococcus aureus* was sensitive with 6, 24, 12 and 8 mm zone of inhibition at 10 mg mL^{-1} of water, acetic acid, ethanol and acetone extract respectively. For, *Pseudomonas aeruginosa* zone of inhibition is 8, 20, 12 and 10 mm for water, acetic acid, ethanol and acetone. *Escherichia coli* revealed no zone of inhibition for water extract whereas it possess 18, 10, 2 mm zones of inhibition at 10 mg mL^{-1} for acetic acid, ethanol and acetone respectively. *Bacillus subtilis* showed 3, 20, 9 and 4 mm zone of inhibition at 10 mg mL^{-1} for different extracts.	These results showed that the extract has a wide range of anti-bacterial property than the other extracts.	Sundaram S *et al.*, Research Journal of Medicinal Plants. 2011; 5(3): 343-347.
					MUSTA (*Cyperus rotundus* Linn)		
					Hypolipidemic Activity		
1.	Lipid lowring activity of alcoholic extraxt of *Cyperus rotundus*	*In vivo*	Hypolipidemic effects of alcoholic extract of rhizome of *Chlorophytum borivilianum* on high fat diet induced hyperlipid-aemic rat of Wistar strain.	Ethanolic extract 70 mg, 140 mg and 280/kg bw/ day for 15 days.	Treatment with standared and different doses, extract exerted statistically significant ($p< 0.05$) reduction in serum total cholesterol. LDL TG and HDL level at end of 15 days of intervention.	Result demonstrated statically significant reduction in serum lipid profile.	R S Chandatre *et al.* IJRPC. 2011; 1(4).

Contd...

TABLE 14.2–*Contd....*

Sl.No.	*Title*	*Type of Study*	*Study Design*	*Form and Dose*	*Summarized Efficacy*	*Conclusion*	*Reference*
2.	Medicinal use and pharmacological activities of *Cyperus rotundus* Linn.	*In vivo*	High fat diet induced hyperlipidaemia in Wistar rats.	Aqueous extract at dose level of 100 mg/kg/day, 200 mg/kg/day, 400 mg/kg/day respectively, orally for 15 days.	Serum was separated from blood by centrifugation for ten minutes at three thousand rpm, subsequently analyzed for total cholesterol. There was significant reduction in serum total cholesterol.	A significant effect of extracts was observed.	S R Shivapalan *et al.*, International Journal of Scientific and Research 2013; 3(5).
3.	Preventive role of *Cyperus rotundus* rhizomes extract on Age associated changes in glucose and lipids.	*In vivo*	Young (3-4 months, 120-150 g) and aged (22-24 months, 380-410 g) Wistar male albino rats were used for the experiments. Rats fed with a high-cholesterol diet (pellet rat feed).	500mg kg^{-1} b.wt/day for 30 days	Serum of aged rats showed a significant increase in total cholesterol, triglycerides, LDL cholesterol and VLDL cholesterol and a decrease in HDL cholesterol as compare to young control rats.	Present study suggests the possible utility of *C. rotundus* rhizomes as therapeutic agent in reducing risk factor for age associated degenerative diseases diabetes mellitus and cardio vascular diseases.	Nagulendran *et al.*, Pharmacologyonline. 2007; 2: 318-325.
					Antidepressant Activity		
4.	New iridoid glycosides with antidepressant activity isolated from *Cyperus rotundus.*	*In vivo*	Evaluate the antidepressant activity of two new iridoid glycosides, named Rotunduside G and rotunduside H by using two despair mice models (Forced swim test and tail suspension test).	Test samples were administered at dose 50 mg/kg intragastrically	In the despair mice models, Rotunduside G and Rotunduside H displayed significant antidepressant activity at the dosage of 50 mg/kg intragastrically (i.g.), which was close to the positive control fluoxetine. Treatment with Rotunduside G and Rotunduside H significantly reduced the immobility periods in mice.	Rotunduside G and Rotunduside H having significant antidepressant activity.	Zhou ZL *et al.*, Chem Pharm Bull (Tokyo) 2016; 64(1): 73-7.
5.	New cycloartane glycosides from the rhizomes of *Cyperus rotundus* and their antidepressant activity.	*In vivo*	Evaluate the antidepressant activity of two new iridoid glycosides, named Cyprotusides A and Cyprotusides B by using two despair mice models (Forced swim test and tail suspension test).	Test samples were administered at dose 50 mg/kg intragastrically	In the despair mice models, Cyprotusides A and Cyprotusides B displayed significant antidepressant activity at the dosage of 50 mg/kg intragastrically (i.g.), which was close to the positive control fluoxetine. Treatment with Cyprotusides A and Cyprotusides B significantly reduced the immobility periods in mice.	Cyprotusides A and Cyprotusides B having significant antidepressant activity.	Zhou ZL *et al.*, J Asian Nat Prod Res. 2016; 18(7): 662-8.

Contd...

TABLE 14.2–*Contd....*

Sl.No.	*Title*	*Type of Study*	*Study Design*	*Form and Dose*	*Summarized Efficacy*	*Conclusion*	*Reference*
6.	Phenolic glycosides from the rhizomes of *Cyperus rotundus* and their antidepressant activity.	*In vivo*	The ethanol extract and its fractions of *Cyperus rotundus* were evaluated for antidepressant activities in mice using Forced swimming test, Tail suspension test and Open-field test.	Test compounds were given i.g. at dose of 50 mg/kg. Fluoxetine at dose of 20 mg/kg was given i.p.	In the despair mice models, rotunduside F displayed significant antidepressant activity at the dosage of 50 mg/kg (i.g.), Treatment with rotunduside F at 50 mg/kg which significantly reduced duration of immobility time in FST and TST produced no significant difference in number of crossing activity of mice in OFT.	Administration of rotunduside F is able to produce an antidepressant like effect in FST and TST.	San-qing Lin *et al.*, J Korean Soc Appl Biol Chem. 2015; 58(5): 685–691.
					Antidiabetic Activity		
7.	Antidiabetic activity of ethanolic extract of *Cyperus rotundus* rhizomes in streptozotocin-induced diabetic mice.	*In vivo*	Ethanolic extract of *Cyperus rotundus* (EECR) rhizomes was evaluated for antidiabetic activity in streptozotocin (STZ)-induced diabetic swiss mice.	Ethanolic extract of *C. rotundus* at doses of 250 and 500 mg/kg body weight p.o. by oral gavages for a period of 21 days	The ethanolic extract at dose levels of 250 and 500 mg/kg body weight revealed significant antidiabetic activity improvement in body weight, and reduction in elevated biochemical parameters such as SGPT, SGOT, cholesterol, and triglyceride levels.	EECR is capable of exhibiting significant anti-hyperglycemic activity in STZ-induced diabetic mice. The extracts also showed improvement in body weight; biochemical parameters such as SGOT, SGPT, and lipid profile and so might be valuable in diabetes treatment.	Singh P, *et al.*, J Pharm Bioallied Sci. 2015; 7(4): 289–292.
8.	Antidiabetic activity of hydro-ethanolic extract of *Cyperus rotundus* in alloxan induced diabetes in rats.	*In vivo*	Investigations were carried out to evaluate the effect of hydro-ethanolic extract of *Cyperus rotundus* on alloxan induced hyperglycemia in rats.	Oral daily administration of 500 mg/kg of the extract (once a day for seven consecutive days)	Oral daily administration of hydro-ethanolic extract of *Cyperus rotundus* revealed significant antidiabetic activity improvement in body weight, and reduction in elevated biochemical parameters such as SGPT, SGOT, cholesterol, and triglyceride levels. This antihyperglycemic activity can be attributed to its antioxidant activity as it showed the strong DPPH radical scavenging action *in vitro*.	The antihyperglycemic activity of *C. rotundus* may be due toits free radical scavenging activity against alloxan induced free radicals.	Raut NA and Gaikwad NJ. Fitoterapia. 2006; 77(7-8): 585-8.
					Antiulcer Activity		
9.	Experimental evaluation of antiulcer activity of *Cyperus rotundus.*	*In vivo*	Antiulcer activity of *C. rotundus* was studied in guinea pigs where gastric ulcers were induced by adminis-	*C. rotundus* was administered in the doses of 1.25 gm/kg	*C. rotundus* at the doses of 1.25 gm/kg and 2.5 gm/kg p.o. showed significant reduction in the ulcer Index when compared to control group and the results were comparable to ranitidine (20 mg/kg). Ranitidine also significantly reduced the ulcer index in both the models. The *C. rotundus* tuber powder showed	The antiulcer activity of *C. rotundus* can be attributed to its significant antioxidant, antihistaminic, mast cell	Arshad M *et al.*, Asian Journal of Biochemical and Pharmaceutical

Contd...

TABLE 14.2–*Contd....*

Sl.No.	*Title*	*Type of Study*	*Study Design*	*Form and Dose*	*Summarized Efficacy*	*Conclusion*	*Reference*
			tration of histamine (50 mg base i.p.) and in rats by administration of aspirin (500 mg/kg orally).	and 2.5 gm/kg orally 45 min prior to histamine and one hour prior to aspirin.	significant antiulcer activity in both histamine and aspirin induced gastric ulcer models in guinea pigs and rats and these results were comparable with those of ranitidine.	stabilizing and cytoprotective activity.	Research. 2012; 2(2): 261-268.
					Anti-inflammatory, Anti arthritic Activity		
10.	Antiinflammatory, anti-arthritic, analgesic and anticonvulsant activity of *Cyperus* essential oils.	*In vivo*	Evaluate the effects of oils in antiinflammatory (carrageenan induced), antiarthritic (formal dehyde induced), analgesic (formalin induced writhing) and anticonvulsant (MES produced convulsion)	Animals received dose of oils (250 mg/kg, 500 mg/kg of Cr and 200 mg/kg, 400 mg/kg of Ce p.o.).	The results showed dose dependentactivity, indicated by reduction in pawedema in antiinflammatory and antiarthritic activity, and significant reduction (p<0.01)in the MES induced convulsion in comparisonto control.	Essential oil possesses a good anti-inflammatory, antiarthritic, analgesic and anticonvulsant activities.	Biradar S, *et al.*, International Journal of Pharmacy and Pharmaceutical Sciences. 2010; 2(4): 112-115.
					NEEM (*Azadirahta indica* A.Juss)		
					Anti-microbial and anti oxidative activity		
1.	Analysis of anti-bacterial and anti oxidative activity of *Azadirachta indica* bark using various solvents extracts.	*In vivo*	Anti-bacterial activity was performed by agar well diffusion method and the results were expressed as the average diameter of zone of inhibition of bacterial growth around the well against pathogenic *Salmonella paratyphi* and *Salmonella typhi* using various solvent extracts.	*Azadirachta indica* stem bark ethanol and methanol extracts	The ethanol and methanol extracts showed better anti-bacterial activity with zone of inhibition (20–25 mm) when compared with other tested extracts and standard antibiotic Erythromycin (15 mcg) with zone of inhibition (13–14 mm). Extracts of *A. indica* stem bark also exhibited significant antioxidant activity, thus establishing the extracts as an antioxidant.	The results obtained in this study give some scientific support to the *A. indica* stem bark for further investigation of compounds and in future could be used as drug.	R.A. Akeel *et al.*, Saudi Journal of Biological Sciences (2017) 24, 11–14.
2.	Effect of neem (*Azadirachta indica* A.Juss) leaf extract on resistant *Staphylococcus aureus* biofilm formation and *Schistosoma mansoni* worms.	*In vitro*	*In vitro* effects of neem leaf ethanolic extract (NeemEE) on Methicillin-resistant *Staphylococcus aureus* (MRSA) biofilm and planktonic aggregation formation, and against *S. mansoni* worms.	*Azadirachta indica* leaf extract	Testing Neem EE sub-inhibitory concentrations, a significant biofilm adherence inhibition from 62.5 μg/mL for a sensitive *S. aureus* and 125 μg/mL for two MRSA strains was observed. AFM images revealed that as the Neem EE concentration increases (from 250 to 1000 μg/mL) decreased ability of a chosen MRSA strain to form large aggregates. In relation of anti-schistosoma assay, the extract caused 100 per cent mortality of female worms at a concentration of 50 μg/mL at 72 h of	Neem leaf ethanolic extract presented inhibitory effect on MRSA biofilm and planktonic aggregation formation, and anthelmintic activity against *S. mansoni* worms	P.V. Quelemes *et al.*, Journal of Ethnopharmacology, (2015) 175: 287–294.

Contd...

TABLE 14.2–*Contd....*

Sl.No.	Title	Type of Study	Study Design	Form and Dose	Summarized Efficacy	Conclusion	Reference
					incubation, while 300 µg/mL at 24h of incubation was required to achieve 100 per cent mortality of male worms. The extract also caused significant motor activity reduction in *S. mansoni*. For instance, at 96 h of incubation with 100 µg/mL, 80 per cent of the worms presented significant motor activity reduction. By the confocal microscopy analysis, the dorsal surface of the tegument of worms exposed to 300 µg/mL (male) and 100 µg/mL (female) of the extract showed severe morphological changes after 24h of treatment.		
3.	Antibacterial efficacy of Neem (*Azadirachta indica*) extract against *Enterococcus faecalis*: An *in vitro* study.	*In vitro*	Neem leaf extract, 2 per cent chlorhexidine, 3 per cent sodium hypochlorite were used to assess the antimicrobial efficiency. Agar well diffusion test was used to study the antimicrobial efficacy with saline as control. The zone of inhibition was recorded, tabulated, and analyzed statistically.	*Azadirachta indica* leaf extract	All the three medicaments showed well-defined and comparable zones of inhibition around their respective wells. All values were significantly higher than the control group. Analysis of variance showed significant difference between zone diameters of chlorhexidine, neem leaf extract, and 3 per cent sodium hypochlorite against *E. faecalis* ($p < 0.05$).	From the present study, it can be concluded that neem leaf extract shows comparable zones of inhibition with that of chlorhexidine and sodium hypochlorite. Neem leaf extract has significant antimicrobial activity against *E. faecalis* and thus opens the perspectives for the use of neem extract as an intracanal medication.	M. Mustafa, J Contemp Dent Pract. 2016 Oct 1; 17(10): 791-794.
					Cardiorenal dysfunction protection		
4.	Preconditioning with *Azadirachta indica* ameliorates cardiorenal dysfunction through reduction in oxidative stress and extracellular signal regulated protein kinase signalling.	*In vivo*	Intestinal ischaemia-reperfusioninjury-induced cardiorenal dysfunction. Materials and methods: Sixty rats were divided into 6 groups; each containing 10. Corn oil was orallyadministered to group A (control) rats for 7 days without intestinal ischaemia-reperfusion injury. Group B underwent intestinal	Methanol extract of *A. indica* leaves 100 and 200 mg/kg	The cardiac and renal hydrogen peroxide increased significantly whereas serum xanthine oxidaseand myeloperoxidase levels were significantly elevated ($p < 0.05$) in IIRI only when compared to he control. The cardiac and renal reduced glutathione, glutathione peroxidase, protein thiol, non-protein thiol and serum nitric oxide (NO) decreased ($p < 0.05$) significantly following IIRI. Immunohistochemicalevaluation of cardiac and renal tissues showed reduced expressions of the extracellular signal regulated kinase (ERK1/2) in rats with IIRI only. However, pre-treatment with *A. indica* and vitamin C significantly reduced markers of oxidative stress and inflammation together with improvement in antioxidant status. Also, reduced serum NO level was normalised in rats pre-treated with A. indica andvitamin C with concomitant higher expressions of cardiac and renal ERK1/2.	Together, *A. indica* and vitamin C prevented IRI-induced cardiorenal dysfunction via reduction in oxidative stress, improvement in antioxidant defence system and increase in the ERK1/2expressions. Therefore, *A. indica* can be a useful chemopreventive agent in the prevention and treatment of conditions	T.O. Omóbòwálé *et al.*, Journal of Ayurveda and Integrative Medicine (2016) 7: 209-217.

Contd...

TABLE 14.2–*Contd....*

Sl.No.	Title	Type of Study	Study Design	Form and Dose	Summarized Efficacy	Conclusion	Reference
			ischaemia-reperfusion injury (IIRI) without any pre-treatment. Groups C, D, E andF were pre-treated orally for 7 days with 100 mg/kg AI (100 and (200 mg/kg) vitamin C (100 and 200 mg/kg) respectively and thereafter underwent IIRI on the 8th day.			associated with intestinal ischaemia-reperfusion injury.	
					Antifilarial activity		
5.	*In vitro* antifilarial activity of *Azadirachta indica* aqueous extract through reactive oxygen species enhancement.	*In vitro*	*In vitro* efficacy of AEA was evaluated against *S. cervi* through estimation of relative motility value, dye exclusion test and MTT assay. Visible morphological alterations were monitored using conventional microscopic techniques in microfilariae and haematoxylin-eosin stained sections of AEA-treated adults.	Aqueous preparation from the *Azadirachta indica* leaves	Enhancement of reactive oxygen species in *S. cervi* treated with AEA was established through alteration in the activity of glutathione S-transferase, superoxide dismutase, catalase, peroxidase and level of superoxide anion and reduced glutathione.	*In vitro* filaricidal activity of AEA is possibly through disturbing redox homeostasis by down-regulating and altering the level of some key antioxidants and regulatory enzymes like reduced glutathione, glutathione S-transferase, superoxide dismutase, catalase and glutathione peroxidase of *S. cervi*.	N. Mukherjee *et al.*, Asian Pac J Trop Med. 2014 Nov; 7(11): 841-8.
					Anti-diabetic and renoprotective activity		
6.	Beneficial effect of *Azadirachta indica* on advanced glycation end-product in streptozotocin-diabetic rat.	*In vivo*	Extract was subjected to *in vitro* bioassays to evaluate advanced glycation end-products formation. Bovine serum albumin (BSA)-glucose, BSA-methylglyoxal, Amadori-rich protein, glycated hemoglobin, oxidation, and glycation of LDL were determined. Doses of extract was given at streptozotocin-	Chloroform extract from leaves of *Azadirachta indica* 200mg/kg/d by oral gavage were administered once daily for 30d	AI exhibits protective action in BSA against glycation formation, GHb, protein levels, and LDL against glycation and oxidation. The renal glucose level decreases a 3.9mg/g wet tissue. TBA-reactive substance showed a significant decrease to 1.82mmol/mg protein. In addition, AI showed inhibitory activity against AGEs formation, methylglyoxal, and glycolaldehyde levels in kidney. Treatment with AI in rat tail tendon produced a reduction in cross-linking of collagen proteins. The antiglycation activities of *A. indica* were attributed in part to their antioxidant activity. AI alleviated oxidative stress under diabetic conditions through the inhibition of lipid peroxidation prevents the onset renal damage.	*A. indica* is an inhibitor AGE formation, and oxidative stress with a renoprotective effect, which are considered to play important roles in diabetic kidney disease.	R.M. P. Gutierrez and M.O.D.J. Martinez., Pharm Biol. 2014 Nov; 52(11): 1435-44.

Contd...

TABLE 14.2–*Contd....*

Sl.No.	*Title*	*Type of Study*	*Study Design*	*Form and Dose*	*Summarized Efficacy*	*Conclusion*	*Reference*
			induced diabetic rats. After this period, renal damage (TBARS), glucose, methylglyoxal, glycolal-dehyde, and tail tendon collagen were investigated.				
					Wound healing potential		
7.	Effects of *Azardirachta indica* on vascular endothelial growth factor and cytokines in diabetic deep wound.	*In vitro*	Deep surgical wounds in streptozotocin-induced mild diabetic rats.	50 per cent ethanol *A. indica* leaves extract (500mg/kg) administered orally, once daily for ten days	*A. indica* leaves extract reversed the increased serum glucose, cholesterol, and triglycerides, food and water intake, and tissue free radicals, myeloperoxidase and, cytokines, but increased body weight, tissue antioxidants, total collagen, and vascular endothelial growth factor contents.	The result indicated an improvement in wound healing by *A. indica* leaves extract in diabetic rats through enhanced angiogenesis mediated through the inhibition of hyperglycemia, oxidative stress, and down- and upregulation of inflammatory mediators and growth factor expression.	M.K. Gautam *et al.*, Planta Med. 2015 Jun; 81(9): 713-21.
					Antitumor activity		
8.	Neem tree (*Azadirachta indica*) extract specifically suppresses the growth of tumors in H22-bearing Kunming mice.	*In vivo*	H22 cells-bearing Kumming mice were generated by injecting H22 cells subcutaneously into the right forelimb armpit of the mice. NTE, carboxymethyl cellulose (CMC, 1 per cent) as blank control and cyclo-phosphamide (CTX, 20 mg/kg) as positive control.	Neem tree (*Azadirachta indica*) extract (NTE) daily for 27 days (150, 300, and 600 mg/kg body weight) by intragastric administration	The antitumor effect of NTE was evaluated by assessment of survival rate, body weight, tumor volume and weight, tumor histology, thymus and spleen indexes, and liver histology. The tumor weight and volume in groups of NTE and CTX were significantly lower than those in the CMC group. The survival rate in the NTE group receiving the high dose (600 mg/kg) was significantly higher than that in the CTX and CMC groups. Compared with CTX, NTE was observed to have a tumor-specific cytotoxicity without impairing the normal liver tissue. Additionally, the higher indexes of thymus and spleen indicated that NTE could facilitate the growth of immune organs.	The results indicate that NTE is a promising candidate for the anti-tumor treatment with high efficacy and safety.	Z. He. *et al.*, Z Naturforsch C. 2016; 71(7-8): 201-8.

Contd...

TABLE 14.2–*Contd....*

Sl.No.	*Title*	*Type of Study*	*Study Design*	*Form and Dose*	*Summarized Efficacy*	*Conclusion*	*Reference*
					Anti-colitis activity		
9.	*Azadirachta indica* attenuates colonic mucosal damage in experimental colitis induced by trinitrobenzene sulfonic acid.	*In vivo*	Trinitrobenzene sulfonic acid-induced colitis in rats.	50 per cent ethanol extract of dried leaves of *Azadirachta indica* extract (500 mg/kg) was administered orally, once daily for 14 days	Intracolonic trinitrobenzene sulfonic acid increased colonic mucosal damage and inflammation, diarrhea, but decreased body weight which were reversed by Azadirachta indica extract and sulfasalazine (positive control) treatments. *Azadirachta indica* extract showed antibacterial activity. *Azadirachta indica* extract and sulfasalazine enhanced the antioxidants but decreased free radicals and myeloperoxidase activities affected in trinitrobenzene sulfonic acid-induced colitis.	*Azadirachta indica* extract, thus seemed to be effective in healing trinitrobenzene sulfonic acid-induced colitis in rats.	M.K. Gautam *et al.*, Indian J Pharm Sci. 2013 Sep; 75(5): 602-6.
					Anti-inflammatory and antinociceptive activities		
10.	Anti-inflammatory and antinociceptive activities of azadirachtin in mice.	*In vivo*	Carrageenan-induced paw edema and fibrovascular tissue growth induced by subcutaneous cotton pellet implantation were used to investigate the anti-inflammatory activity of azadirachtin in mice. Zymosan-induced writhing and hot plate tests were employed to evaluate the antinociceptive activity. To explore putative mechanisms of action, the level of tumor necrosis factor-α in inflammatory tissue was measured and the effect induced by opioidergic and serotonergic antagonists was evaluated.	Azadirachtin (120 mg/kg)	Previous per os administration of azadirachtin significantly reduced the acute paw edema induced by carrageenan. However, the concomitant increase of the paw concentration of tumor necrosis factor-α induced by this inflammatory stimulus was not reduced by azadirachtin. Azadirachtin (6, 60, and 120 mg/kg) inhibited the proliferative phase of the inflammatory response, as demonstrated by the reduced formation of fibrovascular tissue growth. It also inhibited the nociceptive response in models of nociceptive (hot plate) and inflammatory (writhing induced by zymosan) pain. The activity of azadirachtin (120 mg/kg) in the model of nociceptive pain was attenuated by a nonselective opioid antagonist, naltrexone (10 mg/kg, i.p.), but not by a non-selective serotonergic antagonist, cyproheptadine.	This study demonstrates the activity of azadirachtin in experimental models of nociceptive and inflammatory pain, and also in models of acute and chronic inflammation. Finally, multiple mechanisms, including the inhibition of the production of inflammatory mediators and activation of endogenous opioid pathways, may mediate azadirachtin activities in experimental models of inflammation and pain.	D.G. Soares *et al.*, Planta Med. 2014 Jun; 80(8-9): 630-6.

Contd...

TABLE 14.2–*Contd....*

Sl.No.	Title	Type of Study	Study Design	Form and Dose	Summarized Efficacy	Conclusion	Reference
					PIPPALI (*Piper longum* Linn)		
					Immunomodulatory Activity		
1.	Immunomodulatory and antitumor activity of *Piper longum* Linn. and piperine.	*In vivo*	To assess the immunomodulatory and antitumor activity *Piper longum* Linn. and piperine.	*Piper longum* extract and piperine	Administration of *Piper longum* extract and piperine increased the total WBC count to 142.8 and 138.9 per cent, respectively, in Balb/c mice. The number of plaque forming cells also enhanced significantly by the administration of the extract (100.3 per cent) and piperine (71.4 per cent) on 5th day after immunization. Bone marrow cellularity and α-esterase positive cells were also increased by the administration of Piper longum extract and piperine.	Alcoholic extract of the fruits of the plant *Piper longum* and its component piperine was studied for their immunomodulatory and antitumor activity.	E.S Sunila, G Kuttan *et al.*, Journal of Ethnopharmacology Volume 90, Issues 2–3, February 2004, Pages 339–346.
					Antidepressant Activity		
2.	Piperine from the fruits of *Piper longum* with inhibitory effect on monoamine oxidase and antidepressant-like activity.	*In vivo*	The activity guided isolation and inhibitory effect of piperine on MAO activity in mouse brain. also antidepressant-like activity of piperine in the *in vivo* tail suspension test.	Piperine	Piperine showed an inhibitory effect against MAO-A (IC50 value: 20.9mM) and MAO-B (IC50 value: 7.0 mM). The inhibition by piperine was found to be reversible by dialysis of the incubation mixture. In addition, the immobility times in the tail suspension test were significantly reduced by piperine, similar to that of the reference antidepressant fluoxetine, without accompanying changes in ambulation when assessed in an open-field.	Piperine, an active piperidine alklaoid from *P. longum*, reversibly inhibited mouse brain MAO-A and MAO-B activities. In addition, piperine has *in vivo* antidepressant-like activity against the tail suspension test.	Seon A LEE *et al.*, Chem. Pharm. Bull. (2005) 53(7): 832—835.
					Hepatoprotective activity		
3.	Hepatoprotective activity of *Piper longum* traditional milk extract on carbon tetrachloride induced liver toxicity in Wistar rats.	*In vivo*	Study was designed to investigate the hepatoprotective activity of *Piper longum* milk extract.	(200 mg/day p.o. for 21 days)	Significant hepatoprotective effect was observed in CCl4 induced hepatic damage as evident from decreased level of serum enzymes, total bilirubin and direct bilirubin.	Study demonstrated a significant hepatoprotective and antioxidant activity of *Piper longum* milk extract.	Patel and Shah *et al.*, 2009 Boletín Latinoamericano y del Caribe de Plantas Medicinales y Aromáticas, 8 (2), 121 - 129.

Contd...

TABLE 14.2–*Contd....*

Sl.No.	*Title*	*Type of Study*	*Study Design*	*Form and Dose*	*Summarized Efficacy*	*Conclusion*	*Reference*
					Antiinflammatory Activity		
4.	Antiinflammatory activity of *Piper longum* fruit oil.	*In vivo*	Anti-inflammatory activity of the *Piper longum* dried fruit's oil was studied in rats using the carrageenan-induced right hind paw edema method.	Essential oil of *Piper longum* at different doses 0.5 ml/kg and 1 ml/kg orally	*Piper longum* oil had significant antiinflammatory activity in rats. This antiinflammatory activity was dose dependant and found to be statistically significant at the higher concentration, 1 ml/kg.	The antiinflammatory activity of *P. longum* oil is more potent at 1 ml/kg than at 0.5 ml/kg. The most remarkable point of this study was that the oil of *P. longum* at 1 ml/kg produced more inhibition of edema than the antiinflammatory drug, ibuprofen.	Kumar A. *et al.*, Indian J Pharm Sci. 2009; 71(4): 454–456.
5.	Anti-inflammatory activity of two varieties of Pippali (*Piper longum* Linn.).	*In vivo*	Anti-inflammatory activity of the *Piper longum* dried fruit's was studied in rats using the Carrageenan-induced paw edema, Formal-dehyde-induced paw edema.	The test drugs were admini-stered orally at a dose of 200 mg/kg	Among the two different test samples studied, it was found that Chhoti variety of Pippali suppressed inflammation of both acute and sub-acute phase, while Badi variety of Pippali only of acute phase.	*Chhoti* variety of *Pippali* suppressed inflammation of both acute and sub-acute phase while Badi variety of Pippali only of acute phase.	Kumari M, *et al.*, Ayu. 2012; 33(2): 307–310.
					Adaptogenic Activity		
6.	Preventive potentials of piperlongumine and a *Piper longum* extract against stress responses and pain.	*In vivo*	Foot shock stress induced hyperthermia (FSIH) test, Tail suspension test, Potentiation of pento-barbitone induced hypnosis, Hot plate test, Acetic acid induced writhing test and Plasma glucose, insulin, and cortisol level and organs weights.	Graded oral daily doses (1, 4, 16, 64 and 256 mg/kg) of piper-longumine or *Piper longum* extract (PLE) suspended in 0.3 per cent CMC	After their single oral doses no effects of piperlongumine or PLE or doxycycline were observed in the footshock stress induced hyperthermia test or in hot plate test. However, significant effects of piperlongumine and PLE in both the tests were observed after their 5 or more daily doses. Both of them also dose dependently suppressed daily handling and repetitive testing triggered alterations in body weights and core temperatures. Their doxy-cycline like antidepressant activity in tail suspension test and aspirin like analgesic effects in acetic acid writhing test were observed after their 11 daily 5 mg/kg oral dose.	Piperlongumine is another bioactive secondary metabolite of *P. longum* and other plants of piper species with stress response suppressing, analgesic, and anti-inflammatory activities. Its bactericidal activities can also contribute to its thera-peutically interesting bio-activity profile.	Yadav V *et al.*, J Tradit Complement Med. 2016; 6(4): 413–423.

Contd...

TABLE 14.2–*Contd....*

Sl.No.	*Title*	*Type of Study*	*Study Design*	*Form and Dose*	*Summarized Efficacy*	*Conclusion*	*Reference*
7.	Long lasting preventive effects of piperlongumine and a *Piper longum* extract against stress triggered pathologies in mice.	*In vivo*	Mice in treated groups were subjected to a stress induced hyperthermia on the 1st, 5th, 7th, and 10thday. Treated mice were then subjected to tail suspension test on the 11thday. Alteration in body weights, core temperatures, and gastric ulcers triggered by occasional exposures to foot shocks were determined.	Oral treatments with 5 mg/kg piperlongumine or *Piper longum* extract or 50 mg/kg doxycycline (DOX) for 10 consecutive days	DOX like long-lasting protective effects of PL and PLE against gradual alterations in body weights, basal temperatures and transient hyperthermic responses triggered by foot shocks during the post-treatment days were observed. Altered responses of stressed mice in tail suspension test observed 1 day after the last foot-shock exposures and gastric ulcers and other pathologies quantified 1 day after the test were also suppressed in PL or PLE or DOX pretreated groups.	PL and crude PLE are DOX like long-acting desensitizers of stress triggered co-morbidities. Reported observations add further experimental evidences justifying traditionally known medicinal uses of *P. longum* and other plants of the Piperaceae family, and reveal that PL is also another very long acting and orally active inducer of stress resistance. Efforts to confirm stress preventive potentials of low dose plant-derived products enriched in PL or piperine like amide alkaloids in volunteers and patients can be warranted.	Yadav V *et al.*, J Intercult Ethnopharmacol. 2015; 4(4): 277–283.
8.	Anti-stress and nootropic activity of aqueous extract of *Piper longum* fruit, estimated by noninvasive biomarkers and Y-maze test in rodents.	*In vivo*	To evaluate anti-stress activity in rats subjected to forced swim stress one hour after daily treatment of *P. longum* extract. Anti-stress and nootropic activity activities of aqueous extract of *P. longum* fruit extract were estimated as locomotor and working memory in rats in a Y-maze apparatus.	*P. longum* at doses of 100, 200 and 300 mg/kg one hour prior to induction of stress	Daily administration of aqueous extract of *P. longum* prior to induction of stress increased the stress-induced urinary biomarker levels in a dose-dependent manner. *P. longum* treatment showed significant dose-dependent variation in non-invasive biomarker levels in urine samples of rats taken after 24 h. Cognition, determined by working memory and locomotor activity results, were shown to be dose-dependent.	The results of this study suggest anti-stress and nootropic activity effect of *P. longum* in rodents.	Kilari EK *et al.*, Environmental and Experimental Biology 2015; 13: 25–31.

Contd...

TABLE 14.2–*Contd....*

Sl.No.	Title	Type of Study	Study Design	Form and Dose	Summarized Efficacy	Conclusion	Reference
					Antidepressant Activity		
9.	Piperine from the fruits of *Piper longum* with inhibitory effect on monoamine oxidase and antidepressant-like activity.	*In vitro* and *In vivo*	Tail suspension test and estimation of monoamine oxidase A and B.	Test animals were treated orally with 1, 3, 9 mg/kg of piperine	Piperine showed an inhibitory effect against MAO-A (IC50 value: 20.9 microM) and MAO-B (IC50 value: 7.0 microM). Kinetic analyses by a Lineweaver-Burk plot clearly indicated that piperine competitively inhibited MAO-A and MAO-B with Ki values of 19.0+/-0.9 microM and 3.19+/-0.5 microM, respectively. The inhibition by piperine was found to be reversible by dialysis of the incubation mixture. In addition, the immobility times in the tail suspension test were significantly reduced by piperine, similar to that of the reference antidepressant fluoxetine, without accompanying changes in ambulation when assessed in an open-field.	Piperine possesses potent antidepressant-like properties that are mediated in part through the inhibition of MAO activity, and therefore represent a promising pharmacotherapeutic candidate as an anti-depressant agent.	Lee SA *et al.*, Chem Pharm Bull (Tokyo). 2005; 53(7): 832-5.
10.	Antidepressant activity of *Piper nigrum* fruit extract and comparison with Imipramine in mice models.	*In vivo*	Forced swim test and Tail suspension test were used as animal models of depression.	Test animals were treated orally with 250 and 500 mg/kg of *P. nigrum* fruit extrac	Immobility time in Forced swim test and Tail suspension tests was significantly decreased for piper nigrum fruit extract and Imipramine treated groups compared to control group. High dose extract activity was comparable to standard drug Imipramine, but the low dose was less effective. Gross motor activity test demonstrated that treatment does not modified the locomotor activity of mice, which indicates that the plant extract exerts antidepressant effects without modifying significantly locomotor activity.	The aqueous extract of *Piper nigrum* fruit having antidepressant activity.	G. Srinivas Rao *et al.*, Journal of Pharmacy Research 2012; 5(7): 3910-3912.
					PUNARNAVA (*Boerhaavia diffusa* L. nom. cons.)		
					Antidiabetic and Hypolipidemic activities		
1.	Antidiabetic and anti-hyperlipidemic activity of roots of *Boerhaavia diffusa* on streptozotocin induced diabetic rats.	*In vivo*	to evaluate the anti-diabetic activity potential of *Boerhaavia diffusa* roots against streptozotocin (STZ) induced experimental rats.	The animals were treated orally with ethanolic extracts of *Boerhaavia diffusa* at 100, 200 and 400 mg/kg, body weight doses for 14 days.	Ethanolic extract of *B. diffusa* was found to reduce blood sugar in streptozotocin induced diabetic rats. Reduction in blood sugar could be seen from 7th day after continuous administration of the extract. The effect of extracts of *B. diffusa* on serum lipid profile like Total cholesterol, triglycerides, low density, very low density and high density lipoprotein were also measured in the diabetic and non-diabetic rats. There was significant reduction in Total cholesterol, LDL cholesterol, VLDL cholesterol and improvement in HDL cholesterol in diabetic rats.	Ethanolic extract of *B. diffusa* had significantly decreased Total Cholesterol, Triglycerides, VLDL, and LDL with increase in HDL which is having a protective function for the heart compared with diabetic control group.	Murti *et al.*, Pharmacology-online 2011; 1: 15-21.
2.	Evaluation of weight reduction and anti-cholesterol activity of Punarnava root extract	*In vitro* and *In vivo*	To evaluate the weight reduction and hyper-cholesteremic potential extract induced with	100, 200, 400 mg/ kg bw	Significant reduction in body weight, visceral fat pad weight as well as lipidprofiles, liver and kidney marker enzyme levels exhibited as anti-obese activity and hypolipidemic activity of BDRE in experimental animals supplemented with high fat diets.	Study scientifically supported its traditional uses of as antiobesity activity by normalizes	Mohammad Khalid, Asian Pacific Journal of

Contd...

TABLE 14.2–*Contd....*

Sl.No.	Title	Type of Study	Study Design	Form and Dose	Summarized Efficacy	Conclusion	Reference
	against high fat diets induced obesity in experimental rodent.		high fat diets in experimental rodents.			the elevated body weight and organ fat pad weight as well as antilipidemic property by lowering the altered levels of lipid profile in female Sprague-Dawley.	Tropical Biomedicine (2012) 1-6.
3.	Antidiabetic activity of *Boerhaavia diffusa* L.: effect on hepatic key enzymes in experimental diabetes.	*In vivo*	To investigate the effects aqueous solution of *Boerhaavia diffusa* L. leaf extract on blood glucose concentration and hepatic enzymes in normal and alloxan induced diabetic rats.	Daily oral administration of aqueous solution of *Boerhaavia diffusa* L. leaf extract (BLEt) (200 mg/kg) for 4 weeks	A significant decrease in blood glucose and significant increase in plasma insulin levels were observed in normal and diabetic rats treated with BLEt. Treatment with BLEt resulted in a significant reduction of glycosylated haemoglobin and an increase in total haemoglobin level. The activities of the hepatic enzymes such as hexokinase was significantly increased and glucose-6-phosphatase, fructose-1,6-bisphosphatase were significantly decreased by the administration of BLEt in normal and diabetic rats. An oral glucose tolerance test (OGTT) was also performed in the same groups, in which there was a significant improvement in glucose tolerance in rats treated with BLEt.	The aqueous extract of *Boerhaavia diffusa* leaves were found to exhibit a significant hypoglycemic and antihyperglycemic activity in normal and alloxan-diabetic rats.	Pari L *et al.*, J Ethnopharmacol. 2004; 91(1): 109-13.
					Immunomodulatory Activity		
4.	Immunomodulatory activities of Punarnavine, an alkaloid from *Boerhaavia diffusa.*	*In vitro* and *In vivo*	The effect of Punarnavine on the immune system was studied using Balb/c mice.	Intraperitoneal administration of Punarnavine (40 mg/kg body weight)	It was found to enhance the total WBC count on 6th day. Bone marrow cellularity and number of α-esterase positive cells were also increased by the administration ofPunarnavine. Punarnavine also showed enhanced proliferation of splenocytes, thymocytes and bone marrow cells both in the presence and absence of specific mitogens *in vitro* and *in vivo*. More over administration of Punarnavine significantly reduced the LPS induced elevated levels of proinflammatory cytokines such as TNF-α, IL-1β, and IL-6 in mice.	These results indicate the immunomodulatory activity of Punarnavine.	Kanjoormana Aryan Manu, Immunopharmacology and Immunotoxicology, 2009; 31(3): 377–387.
					Anticonvulsant Activity		
5.	Anti-convulsant activity of *Boerhaavia diffusa*: Plausible role of calcium channel antagonism.	*In vitro*	Study was designed to investigate the methanolic root extract of *B. diffusa* and its different fractions including liriodendrin-rich fraction for exploring the possible role of liriodendrin in its anti-convulsant activity.	Methanolic extract (1000, 1500 and 2000 mg kg^{-1} intraperitoneally	The crude methanolic extractof B. diffusa and only its liriodendrin-rich fraction showed a dose-dependent protection against PTZ-induced convulsions. Theliriodendrin-rich fraction also showed significant protection against seizures induced by BAY k-8644. These findings reiterated the anti-convulsant activity of methanolic extract of B. diffusa roots.	Concluded that the observed anticonvulsant activity was due to its calcium channel antagonistic action as this activity was retained only in the liodendrin-rich fraction.	Mandeep Kaur and Rajesh Kumar Goel, Evidence-Based Complementary and Alternative Medicine Volume 2011, Article ID 310420, 7 pages.

Contd...

TABLE 14.2–*Contd....*

Sl.No.	*Title*	*Type of Study*	*Study Design*	*Form and Dose*	*Summarized Efficacy*	*Conclusion*	*Reference*
					Anti-inflammatory Activity		
6.	Evaluation of anti-inflammatory, analgesic, antipyretic and antiulcer activity of Punarnavasava: An ayurvedic formulation of *Boerhavia diffusa.*	*In vivo*	Evaluation of Punarnavasava, an Ayurvedic liquid dosage form for its anti-inflammatory, analgesic, antipyretic, and antiulcer activity by using carrageenan-induced rat paw edema, cotton pellet-induced granuloma, formalin-induced paw licking, yeast induced hyperpyrexia and pyloric ligated ulcers in rats respectively.	Punarnavasava A, B, or C in doses of 2.5 ml/kg and 5.0 ml/kg orally	Punarnavasava, a formulation mainly containing *B. diffusa* inhibited carrageenan-induced paw edema, cotton pellet induced granuloma, formalin-induced pain, and pyloric ligation-induced ulceration in rats.	Punarnavasava possesses promising analgesic and anti-nociceptive properties. Both central and peripheral inhibitory mechanisms are involved in relieving pain and inflammation. The formulation also possessed antacid activity. These effects of Punarnavasava are possibly due to inhibition of proinflammatory cytokines by flavonoids.	Gharate M. and Kasture V, Orient Pharm Exp Med 2013; 13: 121–126.
7.	Anti-inflammatory effect of ethanolic extract of *Boerhavia diffusa* leaves in Wistar rats.	*In vitro* and *In vivo*	Anti-inflammatory activity of ethanol extract of *Boerhavia diffusa* leaves (EBDL) was evaluated by using Nitric Oxide free radical scavenging assay and protein denaturation inhibition assay and Carrageenan induced hind paw edema and Cotton pellet induced granuloma in Wistar rats.	The extract at the concentration of 200 mg/kg body weight (BW) and 400 mg/kg. BW were administered orally to the treatment groups and indomethacin used as standard drug.	Preliminary phytochemical evaluation revealed the presence of alkaloids, flavonoids, phenols, tannins, carbohydrates, saponins, glycosides, and proteins. *In vitro* anti-inflammatory effect of ethanol extract at various concentrations was confirmed by the results obtained in nitric oxide free scavenging activity, protein denaturation and proteinase inhibition assays. The reduced thickness of the paw volume measured at different intervals in treated group exhibited the *in vivo* anti-inflammatory effect of EBDL in both carrageenan induced inflammation and cotton pellet induced granuloma.	*B. diffusa* extract displayed significant potency in anti-inflammatory action in dose dependent manner.	Muthu S *et al.,* Malaya Journal of Biosciences 2014; 1(2): 76–85.
8.	Evaluation of anti-inflammatory effect of aqueous extract of *Boerhaavia diffusa* leaves in rats.	*In vivo*	The anti-inflammatory activity was evaluated by carrageenan induced rat paw edema to determine the activity on acute inflammation and cotton pellet induced granuloma to determine activity on sub-acute inflammation.	*Boerhaavia diffusa* extract at doses of 200 mg/kg and 400 mg/kg B.W. for seven days	*Boerhaavia diffusa* aqueous leaves extract treated groups with 200mg/kg and 400mg/kg bodyweight exhibited dose dependent and significant anti-inflammatory activity in acute (carrageenan-induced rat paw edema) and sub-acute inflammation (cotton pellet induced granuloma).	*Boerhaavia diffuse* extract showed significant anti-inflammatory action in acute and sub-acute experimental models and the activity was dose dependent.	Sudhamadhuri A and Kalasker V, International Journal of Research in Health Sciences. 2014; 2(2): 517-521.

Contd...

TABLE 14.2–*Contd....*

Sl.No.	*Title*	*Type of Study*	*Study Design*	*Form and Dose*	*Summarized Efficacy*	*Conclusion*	*Reference*
					Diuretic Activity		
9.	Evaluation of diuretic activity of an alcoholic extracts of *Boerhaavia diffusa* and *Anisochilus carnosus* in rats.	*In vivo*	To evaluate diuretic effect of alcoholic extracts of stem and leaves of *Boerhaavia diffusa* (AEBD) and leaves of *Anisochilus carnosus* (AEAC) in normal rats. The diuretic effect was evaluated by measuring urine volume, sodium and potassium content in urine.	The extracts were administered to experimental rats orally at doses of 150 and 300 mg/kg of AEBD and 200 and 400mg/kg of AEAC. Furosemide was used as a standard drug at a dose of 20 mg/kg	Urine volume was significantly increased by the doses of AEBD and AEAC in comparison to control group. While the excretion of sodium also increased by the test drugs. The diuretic effect of the extracts was comparable to that of standard drug.	*Boerhaavia Diffusa* and *Anisochilus Carnosus* showed significant diuretic activity. The experimental evidence obtained in the laboratory model could provide a rationale for the traditional use of this plants as diuretic.	Venkatesh P *et al.*, International Journal of Drug Development and Research, 2012; 4(4): 239-242.
					Hepatoprotective activity		
10.	*Boerhaavia diffusa*: A study of its hepatoprotective activity.	*In vivo*	To evaluate hepatoprotective activity alcoholic extract of whole plant *Boerhaavia diffusa* against experimentally induced carbon tetrachloride hepatotoxicity in rats and mice.	Aqueous suspensions of BD (500 mg/kg) in 2 per cent gum acacia were prepared and administered to animals through gastric intubation.	Administration of BD significantly shortened the Ccl4-produced increase in hexobarbitone "sleeping time" and significantly lowered the increase in SGPT, SGOT and bilirubin levels. The extract also produced an increase in normal bile flow in rats suggesting a strong choleretic activity.	The alcoholic extract of *B. diffusa* whole plant has been shown to be a potent and safe anti-hepatotoxic drug.	Chandan BK *et al.*, Journal of Ethnopharmacology, 1991; 31(3): 299-307.
11.	Hepatoprotective activity of *Boerhavia diffusa* Linn. (Nyctaginaceae) against ibuprofen induced hepatotoxicity in Wistar albino rats.	*In vivo*	To evaluate the hepatoprotective activity of different parts of *Boerhavia diffusa* Linn. (Nyctaginaceae) such as root and aerial parts	Methanolic extract of different parts of *Boerhavia diffusa* such as root and	The activities of natural antioxidant enzymes like superoxide dismutase (SOD), catalase (CAT), glutathione peroxidase (GPx), and Glutathione-S-transferase (GST) were decreased significantly. The methanol extract (85 per cent) of the root and aerial part of *Boerhavia diffusa* L. (500 mg/kg. b. wt.) produced remarkable changes in affected hepatic cell architecture and	Root of *Boerhavia diffusa* L. possesses more hepatoprotective efficacy than the aerial part of the same plant. The results suggest that the	Jayavelu A *et al.*, International Journal of Pharma Research and Review,

Contd...

TABLE 14.2–*Contd....*

Sl.No.	*Title*	*Type of Study*	*Study Design*	*Form and Dose*	*Summarized Efficacy*	*Conclusion*	*Reference*
			against ibuprofen (IB) induced hepatotoxicity in Wistar albino rats. Parameters of liver marker enzymes such as ALT, AST, ALP, and bilirubin was estimated.	aerial parts was administered at dose 500 mg/kg, p.o. for 30 days	restored nearly normal structure and functions of hepatic cells. Similarly the different parts of the *Boerhavia diffusa* L. (500 mg/kg. b. wt.) restored the altered biochemical parameters of liver marker enzymes close to normal control levels.	hydro alcoholic (15 : 85 per cent) extract of *Boerhavia diffusa* L. possesses significant potential effect as a hepatoprotective agent.	2013; 2(4): 1-8.
				RASANA (*Pluchea lanceolate* (DC.) Oliv. and Hiern)			
					Anti- inflammatory and Anti-arthritic Activity		
1.	Comparative analysis of *in vitro* anti-inflammatory and *in vivo* and *in vitro* anti-arthritic activity in methanolic extract of *Pluchea lanceolata* oliver and hiern.	*In vitro* and *In vivo*	To study the comparative analysis of *in vitro* anti-arthritic and anti-inflammatory activities in methanolic extracts of *in vivo* (leaf, stem and root) and *in vitro* (callus) plant parts of *Pluchea lanceolata.*	0.05 ml of test solution of each plant extracts (leaf, stem, root, callus) of various concentrations (100, 250, 500, 1000 μg/ml).	The methanolic extracts of all plant parts exhibited notable anti-inflammatory activity and remarkable anti-arthritic action. The membrane stabilization was found to be maximum in leaves (86.8 per cent at a dose of 1000μg/ml) and that of protein denaturation was also found to be maximum in leaves (70.85 per cent at a dose of 1000μg/ml) as compare to other *in vivo* (stem and root) and *in vitro* (callus) plant parts.	Isolation and the use of active constituents from *in vivo* and *in vitro* plant parts of *Pluchea lanceolata* in treating inflammations and rheumatism.	Deepika Arya and Vidya Patni International Journal of Biological and Pharmaceutical Research. 2013; 4(9): 676-680.
2.	Anti-inflammatory activity of *Pluchea lanceolata*: Isolation of an active principle.	*In vivo*	Carrageenan induced Paw inflamation.	Each of Ethanol, Hexane and Aqueous extracts of *Pluchea lanceolate* aerial parts was administered orally at dose of 200 mg/kg in mice and 100 mg/kg in rats, except for the hexane fraction, which was also evaluated at lower doses		The results of the present study indicate that the ethanol extract of *P. lunceolata* has anti-inflammatory activity in mice as well as rats.	Srivastava V *et al.*, Int. J. Crude Drug Res., 1990; 28(2): 135-137.

Contd...

TABLE 14.2–*Contd....*

Sl.No.	Title	Type of Study	Study Design	Form and Dose	Summarized Efficacy	Conclusion	Reference
3.	Phenylethanoids in the herb of *Plantago lanceolata* and inhibitory effect on arachidonic acid-induced mouse ear edema.	*In vivo*	Arachidonic acid-induced mouse ear edema.	The five phenylethanoids, acteoside (1), cistanoside F(2), lavandulifolioside (3), plantamajoside (4) and isoacteoside (5) were isolated from the herb of *Plantago lanceolata*	Compounds 1, the major phenylethanoid in the herb of *P. lanceolata* L., and 4, the major phenylethanoid in the herb of *P. asiatica* L., showed inhibitory effects on arachidonic acid-induced mouse ear edema.	*Plantago lanceolata* has potential anti-inflammatory activity.	Murai M *et al.*, Planta Med. 1995 Oct; 61(5): 479-80.
					Analgesic activity		
4.	Pharmacological evaluation of medicinal plants for their analgesic activity in mice.	*In vivo*	Analgesic activity was assessed on intact mice by tail flick latency via the tail immersion method.	Extracts were given orally in doses of 300, 500 and 1000 mg/kg	The selected parts of four medicinal plants, *Achillea millefolium, Hibiscus rosasinensis, Linum usitatissimum* and *Pluchea lanceolata* were extracted in absolute methanol to determine their analgesic activity. The analgesic activities of these plant extracts were compared with acetylsalicylic acid (300 mg/kg) which was used as the standard drug. 0.9 per cent saline was administered to the control group of animals.	*Linum usitatissimum* and *P. luchea* lanceolata possessed highly significant analgesic activity, while *Achillea millefolium* and *Hibiscus rosa sinensis* did not show any significant effects.	Fayyaz Ahmad *et al.*, Medical Journal of the Islamic Republic of Iran, 10 (2), August 1996.
					Immunosuppressive Activity		
5.	Immunosuppressive properties of *Pluchea lanceolata* leaves.	*In vitro* and *In vivo*	Preliminary screening of the *Pluchea lanceolata* 50 per cent ethanolic extract (PL) was carried out with basic models of immunomodulation, such as, the humoral antibody response (hemagglutination antibody titers), cell-mediated immune response (delayed-type hypersensitivity), skin	Freshly prepared as a homogenized suspension of *Pluchea lanceolata* extract (PL) in doses of 50, 100, 200, 400, 600, and 800 mg/kg and *Pluchea*	Oral administration of PL at doses of 50 to 800 mg/kg in mice, with sheep red blood cells (SRBC) as an antigen, inhibited both humoral and cell-mediated immune responses, as evidenced by the production of the circulating antibody titer and delayed-type hypersensitiviy reaction results, respectively, and the immune suppression was statistically significant in Balb/C mice. PL also decreased the process of phagocytosis both *in vitro* (31.23 per cent) and ex vivo (32.81 per cent) and delayed the graft rejection time (30.76 per cent). To study the T-cell-specific activities, chloroform, n-butanol, and water fractions from *P. lanceolata* were tested for T-cell specific immunosuppressive evaluation, wherein only the chloroform fraction (PLC) showed significant suppression of CD8+/CD4+ T-cell surface markers and intra-	*P. lanceolata* inhibited the humoral antibody response and cell-mediated immune responses. Flow cytometric studies also revealed the down regulation of pro-inflammatory cytokines and this is suggestive of its possible therapeutic usefulness in the treatment of the inflam-	Bhagwat DP, Indian J Pharmacol. 2010; 42(1): 21–26.

Contd...

TABLE 14.2–*Contd....*

Sl.No.	Title	Type of Study	Study Design	Form and Dose	Summarized Efficacy	Conclusion	Reference
			allograft rejection test, *in vitro* (*C. albicans* method), and *in vivo* phagocytosis (carbon clearance test). The extract was then fractionated with chloroform, n-butanol, and water to receive the respective fractions by partitioning. These fractions were employed for flow cytometry to study the T-cell specific immuno-suppressive potential of these fractions.	*lanceolata* chloroform fraction (PLC) in doses of 25, 50, 100, and 200 mg/kg in 1 per cent w/v gum acacia and administered orally, once daily, for the duration of the experiment on Balb/C mice.	cellular Th1 (IL-2 and IFN-Y) cytokines at 25 – 200 mg/kg p.o. doses. PLC, however, did not show significant suppression of the Th2 (IL-4) cytokine.	matory states of the body and autoimmune disorders like arthritis.	
					Antioxidant Activity		
6.	*In vitro* antioxidant activity of root extracts of *Pluchea lanceolata.*	*In vitro*	*In vitro* antioxidant activity of methanolic and aqueous root extracts of *Pluchea lanceolata* was determined by DPPH free radical scavenging assay and hydrogen peroxide scavenging activity.		The antioxidant activity of extracts towards hydrogen peroxide was also reported. A 88.43 per cent and 79.44 per cent of inhibition of hydrogen peroxide was observed with PME and PAE respectively, when compared with control, at a concentration of 0.1 mg/ml using ascorbic acid as standard and positive control on analysis with UV-Visible Spectrophotometer.	The results conclude that the extracts are a potential source of anti-oxidants of natural origin and may be a candidate for treating pathologies related to free radical oxidation due to its overall anti-oxidant effect in scaven-ging free radicals and active oxygen species.	Sharma Kr Surendra and N Goyal/ JPBMS, 2011, 10 (18).
7.	*In vivo* and *in vitro* determination of total phenols, ascorbic acid content and antioxidant activity in *Pluchea lanceolata* Oliver and Hiern.	*In vitro*	Determination of *in vitro* antioxidant activity as well as the phenols and ascorbic acid content in *in vivo* (leaf, stem, and root) and *in vitro* (un-organised static callus cultures) plant parts of *Pluchea lanceolata.* Anti-		Antioxidant potential was found to be maximum (93.55 per cent) in leaf parts and minimum in stem (54.74 per cent). Lower IC50 value indicates high antioxidant activity which was maximum in stem (665.90±0.23) and minimum in leaf (123.37± 0.16). Total phenols and ascorbic acid were also higher in *in vivo* tissues (leaf) as compared to *in vitro* tissues. In the present study excellent DPPH radical scavenging activity (RSA) was found in all extracts of plant. Antioxidant activity has been associated with development of reducing power.	These primary findings showed that *P. lanceolata* possesses higher levels of phenolic and ascorbic acid constituents that are responsible for anti-oxidant activity.	Arya D, *et al.*, International Journal of Pharmaceutical Sciences and Research. 2015; 875-879.

Contd...

TABLE 14.2–*Contd....*

Sl.No.	Title	Type of Study	Study Design	Form and Dose	Summarized Efficacy	Conclusion	Reference
			oxidant activity was analysed in terms of DPPH radical scavenging assay, total phenolic contents were estimated by folin ciocalteau phenol reagent method and ascorbic acid (vitamin c) was analysed by 2, 4-dichlorophenol indo-phenol dye method.				
8.	Antimalarial and safety evaluation of *Pluchea lanceolata* (DC.) Oliv. and Hiern: *In vitro* and *in vivo* study.	*In vitro* and *In vivo*	To evaluate antimalarial and safety profile of *Pluchea lanceolata*: An *in vitro*, *in vivo* for its ethnopharmacological validation		Hexane extract and TxAc showed promising antimalarial activity *in vitro* and *in vivo* condition. TxAc attributed in inhibition of the pro-inflammatory cytokines as well as afford to significant increase in the blood glucose and haemoglobin level when compared with vehicle treated infected mice.	Present study validates the ancient Indian traditional knowledge and use of *Pluchea lanceolata* as an anti-malarial agent.	Mohanty S *et al.*, J Ethno-pharmacol. 2013; 149(3): 797-802.
					Nephroprotective Activity		
9.	*Pluchea lanceolata* protects against Benzo(a) pyrene induced renal toxicity and loss of DNA integrity.	*In vitro* and *in vivo*	Benzo(a)pyrene (B(a)P) administration leads to depletion of renal glutathione and its metabolizing enzymes. Post-mitochondrial supernatant and micro-some preparation. Catalase activity, assay for glutathione-S-transferase activity, glutathione peroxidase activity, quinone reductase activity, Estimation of reduced glutathione, Glucose-6-phosphate dehydrogenase activity, Gel electrophoresis and DNA fragmentation	Total extract of PL was administered orally at doses 100 and 200 mg/kg b.wt for seven days	Pretreatment with PL (100 and 200 mg/kg b.wt) restored renal glutathione content and its dependent enzymes significantly with simultaneous increase in catalase (CAT), quinone reductase (QR) in mouse kidney. Prophylactic administration of PL prior to B (a) P administration significantly decreased the malondial-dehyde(MDA), H2O2 and xanthineoxidase (XO) levels at a signi-ficance of $p<0.001$, at both the doses. PL extract pretreated groups showed marked inhibition in B(a)P induced micronuclei formation in mouse bone marrow cells with simultaneous restoration of DNA integrity, *viz.* alkaline unwinding assay and DNA damage shown by gel-electrophoresis. HPTLC confirms the presence of quercetin in plant extract which could be responsible for PL protecting efficacy.	PL has the efficacy of a potent antioxidant against B(a)P induced renal oxidative stress, loss of DNA integrity and micronuclei induction. The overall antioxidant and anti-clastogenic efficacy of PL are probably due to the presence of flavanols like quercetin and isorhamnetin. Induction of antioxidant armory to suppress oxidative stress may be a possible mechanism of PL in modulating B(a)P toxicity.	Jahangir T *et al.*. Interdiscip Toxicol. 2013 Mar; 6(1): 47–54.

Contd...

TABLE 14.2–*Contd....*

Sl.No.	Title	Type of Study	Study Design	Form and Dose	Summarized Efficacy	Conclusion	Reference
					Anti-asthmatic potential		
10.	Evaluation of anti-asthmatic potential of Ethyl acetate fraction of *Pluchea lanceolata.*	*In vitro*	The ant-asthmatic potential of ethyl acetate fraction was evaluated by *in vitro* animal model in isolated guinea pig tracheal chain preparation.	Dose of 100 µg/ml of ethyl acetate fraction	Ethyl acetate fraction that showed significant relaxant action against histamine induced contraction. The ethyl acetate fraction showed significant anti-asthmatic activity of 57.81±1.22 at the dose of 100µg and can be used for its anti-asthmatic properties.	The ethyl acetate fraction of *Pluchea lanceolata* showed maximum anti-asthmatic potential. It showed anti-asthmatic activity in a dose dependent manner and can be used as anti-asthmatic agent for human welfare.	Arora *et al.*, Pharmacology-online 2011; 2: 1126-1133.
					SALACIA (*Salacia oblonga* Wall.)		
					Anti-diabetic activity		
1.	Extract of *Salacia oblonga* lowers acute glycemia in patients with type 2 diabetes.	Clinical	Randomized, double-blinded crossover study; Sample: Sixty-six patients with diabetes with mean age 61 yrs; Effect on postprandial glycemia and insulinemia after ingestion of a high-carbohydrate meal was studied; Samples were collected at baseline (before meal consumption) and 30, 45, 60, 90, 120, 150, and 180 min after the start of meal consumption.	240 mg and 480 mg of *Salacia oblonga* extract	Both doses of the Salacia extract significantly lowered the post-prandial positive area under the glucose curve (14 per cent for the 240 mg extract and 22 per cent for the 480 mg extract) and the adjusted peak glucose response (19 per cent for the lower dose and 27 per cent for the higher dose of extract) to the control meal. In addition, both doses of the herbal extract significantly decreased the postprandial insulin response, lowering both the positive area under the insulin curve and the adjusted peak insulin response (14 per cent and 9 per cent, respectively, for the 240 mg extract; 19 per cent and 12 per cent, respectively, for the 480 mg extract) in comparison with the control meal.	*Salacia oblonga* lowers acute glycemia and insulinemia in persons with type 2 diabetes after a high-carbohydrate meal.	Am J Clin Nutr. (2007) 86(1): 124-130.; Williams JA, Choe YS, Noss MJ *et al.*
2.	Effect of NR-Salacia on post-prandial hyper-glycemia: A randomized double blind, placebo-controlled, crossover study in healthy volunteers.	Clinical	Randomized, double-blind, placebo-controlled, cross-over study; Salacia was administered before carbohydrate-rich diet. A 6-point plasma glucose profile was performed at different time intervals up to 180 min. Sample: 30 healthy volunteers aged between 18-45 years.	Hydro alcoholic extract of roots and stems of *Salacia chinensis* - 1000 mg, Single dose	Results showed that NR-Salacia treatment significantly lowered plasma glucose level at 90 min, and the percentage reduction in glucose concentration was found to be 13.32 as compared to placebo group. A 33.85 per cent decrease in the plasma glucose positive incremental area under curve (AUC) (0 to 180 min) was observed in comparison to placebo.	Salacia lowered the post-prandial plasma glucose levels after a carbo-hydrate-rich meal and can be used as an oral hypoglycemic agent.	Pharmacogn Mag. 2013 Oct; 9(36): 344-9.; Koteshwar P Raveendra KR, *et al.*

Contd...

TABLE 14.2–*Contd....*

Sl.No.	Title	Type of Study	Study Design	Form and Dose	Summarized Efficacy	Conclusion	Reference
3.	A double blind randomised placebo controlled cross over study of a herbal preparation containing Salacia *reticulate* in the treatment of type 2 diabetes.	Clinical	Randomised single centre double blind cross over clinical trial.	Standard preparation of salacia for 3 months followed by placebo in similar tea bags for a further 3 months ($n = 28$) or in reverse order ($n = 23$). All patients received detailed advice on diet, exercise and lifestyle modification	There were no significant differences between the two groups in age, body mass index, male/female ratio, glycaemic control and baseline laboratory tests. All patients completed both arms of the trial. The HbA1C at the end of drug treatment was significantly lower than after treatment with placebo (6.29±S.D. 1.02 versus 6.65±S.D. 1.04; $P = 0.008$). A statistically significant fall in HBA1c was seen with the active drug compared to a rise in HbA1C with the placebo group (0. 54±S.D. 0.93) versus –0.3±S.D. 1.05; $P < 0.001$. The daily mean dose of Glibenclamide fell by 1.89 (S.D. 6.2) mg in the drug treated group but rose by 2.25 mg in the placebo treated group ($P = 0.07$). The differences in the metformin dose were not significantly significant in the two groups.	Salacia is an effective and safe treatment for type 2 diabetes.	M.H.S. Jayawardenaa, N.M.W. de Alwisa *et al.*, Journal of Ethnopharmacology (2005) 97: 215–218.
4.	Antidiabetic and hypolipidemic effect of *Salacia oblonga* in streptozotocin induced diabetic rats.	*In vivo*	Streptozotocin (STZ) induced diabetic rats; RBG, fasting serum insulin, plasma HbA1c and the lipid profile was estimated.	Hydroalcoholic root extract of Salacia 50 and 100 mg/kg body wt/day, for a period of 16 weeks.	Results showed 45 per cent decrease in the RBG after the treatment with the higher dose of Salacia extract, whereas a 44 per cent decrease was observed with the lower dose as compared to the diabetic control. Serum insulin was significantly increased in all the treated groups as compared to the diabetic control. Plasma HbA1c was significantly decreased. The serum Triacyl Glycerol (TG) levels were signifi¬cantly decreased in the treated rats as compared to the diabetic control. A significant increase in HDL-cholesterol in the diabetic rats as a result of the 100mg/kg SOE treatment was seen.	Salacia improves the glycaemic parameters after a prolonged treatment. The serum TG levels were normalized on treatment.	J Clin Diagn Res. 2012 Dec; 6(10): 1685-7; Bhat BM, CVR, *et al.*
5.	Anti-peroxidative and hypoglycaemic activity of *Salacia oblonga* extract in diabetic rats.	*In vivo*	Hypoglycaemic activity and anti-lipid peroxidative activity on Streptozotocin (STZ) induced hyperglycaemic rats	Petroleum ether extract of the root bark of *Salacia oblonga*	The extract showed significant hypoglycaemia which was supported by an insulin assay. A detailed biochemical study (thiobarbituric acid reactive substances, hydroperoxides, conjugated dienes, glutathione, superoxide dismutase, catalase, glutathione peroxidase, and glutathione reductase) in the renal tissue of diabetic animals treated with SOB demonstrated promising anti-lipid peroxidative activity.	*S. oblonga* root bark possesses anti-diabetic and anti-peroxidative principles, and may be of value in the treatment of diabetes and associated renal complications.	Pharm Biol. 2000; 38(2): 101-5; Krishnakumar K, Augusti KT, *et al.*

Contd...

TABLE 14.2–*Contd....*

Sl.No.	*Title*	*Type of Study*	*Study Design*	*Form and Dose*	*Summarized Efficacy*	*Conclusion*	*Reference*
6.	Beneficial effects of mangiferin isolated from *Salacia chinensis* on biochemical and hematological parameters in rats with streptozotocin-induced diabetes.	*In vivo*	Streptozotocin induced diabetic rats.	Mangiferin given orally in the dose of 40 mg/kg body weight/ day for 30 days	Mangiferin treated diabetic rats significantly ($p<0.05$) lowered the level of blood glucose, it also altered the levels of biochemical parameters including urea, uric acid, and creatinine. Toxicological parameters including AST, ALT and ALP were also significantly reduced after treatment with mangiferin in diabetic rats. Similarly, the levels of red blood, white blood cells and their functional indices were significantly improved through the administration of mangiferin.	Mangiferin present in *S. chinensis* possesses antidiabetic properties.	Pak J Pharm Sci. 2014 Jan; 27(1): 161-7; Sellamuthu PS, Arulselvan P *et al.*
7.	*Salacia oblonga* extract increases glucose transporter 4-mediated glucose uptake in L6 rat myotubes: role of mangiferin.	*In vitro*	2-deoxy-D-glucose uptake were assayed in muscle L6-myotubes and 3T3-adipocytes. In L6-myotubes, the amount and translocation of glucose transporters were assayed. phosphorylation status of key components of signaling pathways that are involved in the molecular mechanisms regulating glucose uptake was analysed.	*S. oblonga* extract	*S. oblonga* extract increased 2-deoxy-D-glucose uptake by 50 per cent in L6-myotubes and 3T3-adipocytes. In L6-myotubes, the extract increased up to a 100 per cent the GLUT4 content, activating GLUT4 promoter transcription and its translocation to the plasma membrane. Mangiferin effects were concomitant with the phosphorylation of 5'-AMP-activated protein kinase without the activation of PKB/Akt. The effect of mangiferin on 2-deoxy-D-glucose uptake was blocked by GW9662, an irreversible PPAR-gamma antagonist.	*S. oblonga* extract and mangiferin exert their antidiabetic effect by increasing GLUT4 expression and translocation in muscle cells. These effects are probably mediated through two independent pathways that are related to 5'-AMP-activated protein kinase and PPAR-gamma.	Clin Nutr. 2009 Oct; 28(5): 565-74; Girón MD, Sevillano N *et al.*
8.	Anti-diabetic activity of a leaf extract prepared from *Salacia reticulata* in mice.	*In vivo*	Absorption of sugars in normal and type 1 diabetic mice.	Oral administration of water extract at a dose of 1.0 mg/ mouse	The simultaneous oral administration of the extract with maltose or sucrose inhibited the postprandial elevation of the plasma glucose and insulin levels and intestinal alpha-glucosidase activities. In addition, the supply of a 0.01 per cent solution of the extract as drinking water prevented the elevation of the plasma glucose level and intestinal alpha-glucosidase activities. Treatment with Salacia also prevented the elevation of the plasma, pancreatic, and kidney lipid peroxide levels, lowering of the plasma insulinevel, and elevation of the kidney aldose reductase activities in diabetic mice.	*S. reticulata* could be a beneficial food material for the prevention of diabetes and obesity because of its multiple effects.	Biosci Biotechnol Biochem. 2009 May; 73(5): 1096-104; Yoshino K, Miyauchi Y *et al.*
9.	Mechanisms of blood glucose-lowering effect of aqueous extract from stems of Kothala himbutu	*In vivo*	Gene expression profiles were assessed by DNA microarray and RT-PCR analyses of RNA from	Aqueous extracts of Kothala himbutu	DNA microarray and RT-PCR analyses revealed that gluconeogenic fructose-1,6-bisphosphatase (FBP) was decreased compared with the control in KTE-treated KK-Ay mice. RT-PCR analysis using cultured liver cells treated with KTED and/or actinomycin	Salacia has anti-diabetic potential with possible mechanism by gluconeogenic gene regulation	J Ethnopharmacol. 2009 Jan 21; 121(2): 234-40;

Contd...

TABLE 14.2–*Contd....*

Sl.No.	Title	Type of Study	Study Design	Form and Dose	Summarized Efficacy	Conclusion	Reference
	(*Salacia reticulata*) in the mouse.		the liver of KK-Ay diabetic mice.	(*Salacia reticulate*) (KTE) stems for 4 weeks	D or cycloheximide, revealed that KTED directly decreased FBP mRNA levels via destabilization of the mRNA. One compound in KTE, mangiferin, was demonstrated to dose-dependently down-regulate FBP mRNA. This study suggest that the mangiferin in KTE acts directly on liver cells and down-regulates the glucone-ogenic pathway through regulation of FBP expression, thereby decreasing fasting blood glucose levels in mice.		FNA Im R, Mano H *et al.*
					Hypolipedemic activity		
10.	*Salacia reticulata* improves serum lipid profiles and glycemic control in patients with prediabetes and mild to moderate hyperlipidemia: A double-blind, placebo-controlled, randomized trial.	Clinical	Randomized, double-blind, placebo-controlled study; Sample: 29 patients with prediabetes and mild to moderate hyper-lipidemia; Efficacy was evaluated in terms of change in lipid profile and glycemic levels.	Salacia leave and root bark extracts 500mg/day for 6 weeks	A statistical significant reduction was observed in low-density lipoprotein cholesterol and fasting blood sugar (FBS) levels at week 3 and 6 when treated with root bark extract. The leaves extract-treated group showed statistically significant reduction in FBS levels at week 6 only.	Salacia is beneficial in the management of prediabetes and mild to moderate hyperlipi-demia.	J Med Food. 2013 Jun; 16(6): 564-8; Shivaprasad HN, Bhanumathy M, *et al.*
11.	Complementary treat-ment of obesity and overweight with *Salacia reticulata* and vitamin D.	Clinical	Randomized open-label study conducted on 40 healthy participants aged 30 - 60 years, physically active, with a body mass index (BMI) of 25 - 45; Both groups (A and B) received a guideline for lifestyle and fitness training for 4 weeks. Group B additionally took *Salacia capsule.*	200 mg of *Salacia reticulata* and 1.6 μg (*i.e.* 64 IU) Vitamin D3 three times/ day with the meals.	Significant weight and body-fat reduction within 4 weeks. Group A lost 1.8 kg or 2.1 per cent, group B lost 5.3 kg or 6.1 per cent ($p = 0.03$), therefore BMI reduction was achieved. While Group A lost 1.4 per cent of body fat, group B reduced it by 4.5 per cent ($p = 0.01$).	Combination of Salacia and Vitamin D might be highly valuable and potent to treat over-weight and obesity, in addition to a modifying lifestyle program.	Int J Vitam Nutr Res. 2013; 83(4): 216-23; Ofner M, Tomaschitz A *et al.*
12.	*Salacia reticulata* has therapeutic effects on obesity.	*In vivo* and *in vitro*	12-week-old TSOD mice with obesity and meta-bolic disorders and in mature 3T3-L1 adipocytes.		*S. reticulata* therapy produced a reduction in body weight and mesenteric fat accumulation, an improvement in abnormal glucose metabolism, and an increase in adiponectin level in plasma. It produced suppression of intracellular triacylglycerol accumulation and enhancement of glycerol release. The mRNA expressions of lipogenesis factor were down-regulated, while the expressions of lipolysis factor and adiponectin were up-regulated. Enhancement of the expression of total AMP-activated protein kinase α and phosphorylated AMPKα in mature adipocytes was seen.	*S. reticulata* has thera-peutic effects on obesity and metabolic disorders.	J Nat Med. 2014 Oct; 68(4): 668-76.; Shimada T, Nakayama Y *et al.*

Contd...

TABLE 14.2–*Contd....*

Sl.No.	*Title*	*Type of Study*	*Study Design*	*Form and Dose*	*Summarized Efficacy*	*Conclusion*	*Reference*
13.	*Salacia oblonga* ameliorates hypertriglyceridemia and excessive ectopic fat accumulation in laying hens.	*In vivo*	Laying hens and pre-adolescent pullets, a unique animal model with a very high rate of triglyceride synthesis in the liver.	0 per cent, 0.5 per cent, or 1 per cent of salacia root water extract for 4 weeks.	SOR extract treatment inhibited body weight increase without affecting food intake. Importantly, this treatment substantially attenuated hypertriglyceridemia and inhibited increases in triglyceride contents in the non-adipose tissues. Importantly, this treatment substantially attenuated hypertriglyceridemia and inhibited increases in triglyceride contents in the non-adipose tissues.	SOR ameliorates hyper-triglyceridemia and excessive ectopic fat accumulation in laying hens. Salacia has triglyceride-lowering property possibly via hepatic mechanisms.	J Ethno-pharmacol. 2012 Jun 26; 142(1): 221-7; Wang J, Rong X *et al.*
14.	Antihyperlipidemic activity of *Salacia chinensis* root extracts in triton-induced and atherogenic diet-induced hyperlipidemic rats.	*In vivo*	Triton induced and atherogenic diet-induced hyperlipidemic rats.	Oral administration of 500 mg/kg of the chloroform extract and alcoholic extract of *S. chinensis* root	A significant reduction in serum lipid parameters like total cholesterol, triglycerides, low density lipoprotein (LDL), very low density lipopreotein (VLDL) and increase in high density lipoprotein (HDL) in hyperlipidemic rats of both models as compared to hyperlipidemic control statistically.	Salacia has antihyper-lipidemic potential.	Indian J Pharmacology 2012, Jan 44(1): 88-92; Sikarwar MS, Patil MB.
					Nephroprotective Activity		
15.	Nephroprotective role of *Salacia chinensis* in diabetic CKD patients: A pilot study.	Clinical	30 diabetic CKD patients; Measures of renal function: Serum creatinine and creatinine clearance; markers of endothelial dysfunction: Interleukin-6 and serum Homocysteine, and lipid profile were measured.	*Salacia chinensis* 1000 mg twice-daily for 6 months	There was stabilization of renal function as measured by serum creatinine and creatinine clearance in patients who received *Salacia chinensis* compared to placebo (P value < 0.05), suggesting that *Salacia chinensis* may retard the progression of chronic kidney disease. Similarly, there was significant decline in both serum homocysteine and IL-6 levels. [P value < 0.05 for both].	*Salacia chinensis* may be advocated as an adjunctive drug in the diabetic patients as it controls post-prandial hyperglycemia, treats lipid abnormalities, and modulates cardiovascular risk factors. It has reno-protective activity too.	Indian J Med Sci. 2010 Aug; 64(8): 378-84; Singh RG, Rathore S.S.
16.	Renoprotective role of sapta chakra (*Salacia chinesis*) in diabetic nephropathy with special reference to endothelial dysfunction.	Clinical	Sample size: 35 patients of Diabetic Nephropathy on standard treatment including antihypertensive, phosphate binder, vitamin and mineral supplementation, hypoglycemic drugs and diabetic measures out of which 18 were on the drug and 17 on the placebo which was randomly given.	Salacia 1 gm twice daily for 6 months.	Males predominated in the study with common symptoms of weakness, anorexia, vomiting and oedema. The renal function was quite static in the drug group. The drug group had significant reduction in post parandial blood sugar but no effect on HDL, LDL and apolipoprotein B. The hemocysteine and IL-6 was significantly reduced in the Salacia group. Similarly, C. reactive protein, Endothelin and TNF-alpha was significantly reduced in the drug group. The alpha amino glycosidase activity and amylase activity were statistically significant in the drug group.	*Salacia Chinesis* seems to have protecting role in patients of diabetic nephropathy.	RG Singh, Ranjeet, Usha, A Agrawal *et al.*, Indian Journal of Nephrology; Jul-Sep 2007, Vol. 17 Issue 3, p. 103.

Contd...

TABLE 14.2–*Contd....*

Sl.No.	Title	Type of Study	Study Design	Form and Dose	Summarized Efficacy	Conclusion	Reference
17.	The ayurvedic medicine *Salacia oblonga* attenuates diabetic renal fibrosis in Rats: Suppression of angiotensin II/AT1 signaling.	*In vivo*	Zucker diabetic fatty (ZDF) rats.	Aqueous extract from Salacia (100 mg/kg, p.o., 6 weeks)	It diminished renal glomerulosclerosis and interstitial fibrosis. SO also reduced renal salt-soluble,acid-soluble and salt-insoluble collagen contents. These changes were accompanied by normalization of hypoalbuminemia and BUN. Gene profiling revealed that the increase in transcripts encoding the glomerulosclerotic mediators was suppressed by SO. In rat-derived mesangial cells, similar to the effect of the AT1 antagonist telmisartan, SO and its major component mangiferin suppressed the stimulatory effect of angiotensin II on proliferation and increased mRNA expression and/or activities of mediators.	SO attenuates diabetic renal fibrosis, at least in part by suppressing anigiotensin II/AT1 signaling showing renoprotective effect.	Hindawi Publishing Corporation, Evidence-Based Complementary and Alternative Medicine Volume 2011, 12 pages; Lan He, Yanfei Qi *et al.*
18.	Nephroprotective and antioxidant activities of *Salacia oblonga* on acetaminophen-induced toxicity in rats.	*In vivo*	Acetaminophen induced toxicity in rats.	Ethanol extract of *S. oblonga* 250 and 500 mg/kg	The results showed that APAP significantly increases the levels of serum urea, creatinine, and reduces levels of uric acid concentration. The salacia extract reduces these by increasing anti-oxidative responses as assessed by biochemical and histopathological parameters.	Salacia possesses nephro-protective and antioxidant effects.	Nat Prod Res 2011 Nov; 25(19): 1876-80; Palani S, Raja S *et al.*
					Cardioprotective Activity		
19.	*Salacia oblonga* root decreases cardiac hypertrophy in Zucker diabetic fatty rats: Inhibition of cardiac expression of angiotensin II type 1 receptor.	*In vivo* and *in vitro*	Male Zucker diabetic fatty (ZDF) rats; angiotensin II-stimulated embryonic rat heart-derived H9c2 cells and neonatal rat cardiac fibroblasts.	Water extract of *Salacia oblonga* 100 mg/kg orally for 7 weeks	SOE-treated ZDF rats showed less cardiac hypertrophy (decrease in weights of the hearts and left ventricles and reduced cardio-myocyte cross-sectional areas). SOE treatment suppressed cardiac overexpression of ANP, brain natriuretic peptide (BNP) and AT(1) mRNAs and AT(1) protein in ZDF rats. SOE (50-100 microg/ml) and MA (25 micromol) suppressed angiotensin II-induced ANP mRNA overexpression and protein synthesis in H9c2 cells. They also inhibited angiotensin II-stimulated [(3)H] thymidine incorporation by cardiac fibroblasts.	SOE decreases cardiac hypertrophy, at least in part by inhibiting cardiac AT(1) over-expression showing potential cardioprotective role in obesity and diabetes-associated cardiac hypertrophy.	Diabetes Obes Metab. 2008 Jul; 10(7): 574-85; Huang TH, He L *et al.*
20.	Pharmaceutical investigation on *Salacia macrosperma.*	*In vivo*	Fasted Rabbits (each N=5); alloxan induced hyperglycemic rats (each N=6); and normal and hypodynamic isolated frog heart (N=8).	Chloroform, ethanol (95 per cent) and aqueous extracts of roots of *Salacia macrosperma* 200mg/kg weight	Ethanolic extract has significant hypoglycemic activity in fasted rabbits, showed a mean blood sugar level reduction of 89.2 2 mg/ 100 ml which was significant when compared with mean variation in blood sugar levels of control group. Alcoholic extract showed considerable positive ionotropic activity and increased cardiac output without affecting heart rate both in normal and hypodynamic isolated frog heart.	Salacia is having hypo-glycemic effect and have positive effect on heart by improving contr-actions.	Ancient Science Life, Vol. IX. No. 4, April 1990, pages 215-219; Venkateswarlu V, Kokate CK.

Contd...

TABLE 14.2–*Contd....*

Sl.No.	*Title*	*Type of Study*	*Study Design*	*Form and Dose*	*Summarized Efficacy*	*Conclusion*	*Reference*
					SARPAGANDHA (*Rauwolfia serpentina* (L.) Benth. ex Kurz)		
					Sleep quality/Anti-hypertensive Activity		
1.	A comparative study of Sarpagandha vati and Vaachadi Yoga in the management of essential hypertension.	Clinical	**Study design** - Open, Comparative study. **Study duration** - One month.	Sarpagandha Vati (400 mg) twice a day with luke warm water	Sarpagandha showed the statically significant improvement in Blood pressure, and other symptoms such as Santapa and Krodh-aprachurata (Feeling of tension), Bhrama (Vertigo), Shirahashula (Headache), Klama (Fatigue and general tiredness) Anidra (Sleep-lessness) Tandra (Drowsiness) Sleep duration of the patients was increased in sarpagandha group None of the patients complained of depression or bradycardia None of the drug shows toxic or adverse reaction for the period of one month or during follow up.	Sarpagandha Vati is helpful in management of essential hypertension.	Anil NS *et al.*, OMICS, open access scientific reports, 2012.
					Anti-hypertensive Activity		
2.	A clinical trial of *Rauwolfia serpentina* in essential hypertension.	Clinical	**Study design** - Open trial **Sample Size** - Fifty cases **Study duration** - Four weeks.	One tab, thrice daily	Within a week of *R. serpentina* therapy, 77 per cent of cases showed a drop of systolic blood pressure ranging from 2 to 38 mm., with an average drop of 13 mm. A drop of 10 mm. or over was noted in 40 per cent of cases.In the case of the diastolic blood pressure, 73 per cent of cases displayed a drop ranging from 2 to 18 mm., with an average drop of 6 mm.; a diastolic response of 5 mm. or over was noted in 35 per cent. In 73 per cent of cases, there was a drop of both systolic and diastolic blood pressure after one week of therapy.	*R. serpentina* is effective and safe in management of essential hypertension.	Rustom Jal Vakil *et al.*, Heart J 1949; 11: 4 350-355.
3.	The use of *Rauwolfia serpentina* in hypertensive patients.	Clinical	**Study design** - Open trial **Sample Size** - 100. **Study duration** - One month to Year.	*R. serpe-ntina* Tab. 1 to 3 tablets a day	✫ *Rauwolfia serpentina* decrease the blood pressure in moderate hypertensive patients ✫ It causes **sedation**, and usually **improves sleep**, although occasionally it causes nightmares. ✫ Its chronic effects may not be fully apparent in less than six weeks. ✫ It produces no serious side effects.	*Rauwolfia serpentina* is effective in reducing blood pressure for moderate hypertensive patients.	Wilkins, R.W. *et al.* N Engl J Med. 1953; 248(2): 48-53.
					Antidiarrheal Activity		
4.	Antidiarrhoeal activity of leaf methanolic extract of *Rauwolfia serpentina.*	*In vivo*	*R. serpentina* leaf methanol extracts were administered to castor oil induced diarrhoea mice to determine its antidiarrhoeal activity.	Doses of 100, 200 and 400 mg/kg *R. serpentina* leaf methanol extracts were administered orally	All doses of the extract and the reference drug atropine sulphate (3 mg/kg, i.p.) produced a dose-dependent reduction in intestinal weight and fluid volume. The extracts also significantly reduced the intestinal transit in charcoal meal test when compared to diphenoxylate HCl (5 mg/kg, p.o.).	The extract of *R. serpentina* leaves has a significant antidiarrhoeal activity and supports its traditional uses in herbal medicine.	II Ezeigbo *et al.*, Asian Pac J Trop Biomed. 2012 Jun; 2(6): 430–432.

Contd...

TABLE 14.2–*Contd....*

Sl.No.	Title	Type of Study	Study Design	Form and Dose	Summarized Efficacy	Conclusion	Reference
					Antidiabetic Activity		
5.	Methanolic root extract of *Rauwolfia serpentina* Benth improves the glycemic, antiatherogenic, and cardioprotective Indices in alloxan-induced diabetic mice.	*In vivo*	Evaluate the effect of methanolic root extract (MREt) of *Rauwolfia serpentina* on alloxan-induced diabetic Wister male mice.	Doses of 10, 30 and 60 mg/kg *R. serpentina* methanolic root extract were administered orally for 14 days	MREt found effective in improving the body weights, glucose and insulin levels, insulin/glucose ratio, glycosylated and total hemoglobin in test groups as compared to diabetic control. Similarly, significantly decreased levels of total cholesterol, triglycerides, low-density lipoprotein (LDL-c), and very low-density lipoprotein (VLDL-c) cholesterols were found in test groups. Significant lipolysis with improved glycogenesis was also found in liver tissues of all test groups. ALT levels were found normal in all groups.	*R. serpentina* is an effective antidiabetic agent as it improves glycemic, antiatherogenic, and cardioprotective indices in alloxan-induced diabetic mice.	Azmi MB and Qureshi SA, Adv Pharmacol Sci. 2012; 376-429.
6.	*Rauwolfia serpentina* ameliorates hyperglycemic, haematinic and antioxidant status in alloxan- induced diabetic mice.	*In vivo*	To investigate the effect of methanolic root extract (MREt) of *Rauwolfia serpentina* on hyperglycemic, haematinic and antioxidative dysfunction associated with alloxan-induced diabetes.	Doses of 10, 30 and 60 mg/kg *R. serpentina* methanolic root extract were administered orally for 14 days	MREt significantly reduced blood glucose level by improving the body weights, glycosylated hemoglobin (HbA1c) to total hemoglobin (Hb) ratio, red blood cell (RBC) and white blood cell (WBC) counts, packed cell volume (PCV), mean corpuscular volume (MCV), mean corpuscular hemoglobin (MCH) and mean corpuscular hemoglobin concentration (MCHC) in test groups. Beside this, extract decreases the percent inhibition of catalase (CAT) and superoxide dismutase (SOD) enzymes and restores the liver function by recovering the total protein concentration and normalizing the levels of alanine transaminase (ALT), aspartate transaminase (AST) and alkaline phosphatase (ALP) in test mice.	Methanolic root extract (MREt) of *Rauwolfia serpentina* ameliorates hyperglycemic, haematinic and antioxidant status in alloxan-induced diabetic mice.	Azmi MB and Qureshi SA, Journal of Applied Pharmaceutical Science, 2013; 3(07): 136-141.
7.	Hypoglycaemic and hypolipidemic activities of *Rauwolfia serpentina* in alloxan-induced diabetic rats.	*In vivo*	Investigate the effect of methanolic root extract of *Rauwolfia serpentina* on glucose, total cholesterol (TC), triglycerides (TG) and alanine aminotransferase (ALT) using alloxan induced diabetic rat model.	Dose of 30 mg/kg *R. serpentina* methanolic root extract were administered orally	Treatment with *R. serpentina* methanolic root extract significantly decreased blood glucose level from 0 to 4 h (94-106 mg dL^{-1}) in test rats as compared to diabetic control. Similarly, TG, TC and ALT were also significantly decrease in test group.	The methanolic root extract of *R. serpentina* was found hypoglycaemic hypolipidemic and hepato-protective in alloxan-induced diabetic rats.	Qureshi SA *et al.*, International Journal of Pharmacology, 2009; 5: 323-326.
					Antioxidant and anti-inflammatory Activity		
8.	Isolation and extraction of flavonoid from the leaves of *Rauwolfia serpentina* and evaluation of DPPH-scavenging antioxidant potential.	*In vitro*	In this study antioxidant potential of the leaves of the leaves of *Rauwolfia serpentina* was determined using DPPH method. The extract was analyzed phytochemically for the presence of flavonoids.		Phytochemical studies revealed the presence of flavonoidal structure, by using chromatographic and spectroscopic techniques, 3,5,7,4'- tetrahydroxy flavone *i.e.*, Kaempherol is identified. Antioxidant potential was determined by DPPH method.	Presences of flavonoid Kaempherol in the leaves of *Rauwolfia serpentina* which have free radical scavenging activity due to which leaves of this plant possess antioxidant properties.	Gupta J and Gupta A, Orient J Chem 2015; 31(S1): 231-235.

Contd...

TABLE 14.2–*Contd....*

Sl.No.	*Title*	*Type of Study*	*Study Design*	*Form and Dose*	*Summarized Efficacy*	*Conclusion*	*Reference*
9.	Studies on methanolic extract of Rauvolfia species from Southern Western Ghats of India – *In vitro* antioxidant properties, characterization of nutrients and phytochemicals.	*In vitro*	Antioxidant potentials using various *in vitro* models such as total antioxidant capacity, DPPH radical scavenging activity, reducing power and superoxide anion scavenging activity.		*Rauvolfia serpentina* exhibited the highest total phenolic content while *Rauvolfia tetraphylla* had highest flavonoid content among the five species. *R. serpentina* showed the highest DPPH radical scavenging activity and also highest pigment composition and vitamin E content, while *Rauvolfia densiflora* showed highest level of vitamin C content and metal chelating activity among the five species. *R. tetraphylla* revealed the highest concentration of carotene. Lycopene was found in very low amounts while comparing with other nutrient compositions and the maximum amount was in *R. tetraphylla* and least amount was in *Rauvolfia beddomei*.	The studied medicinal plants revealed interesting antioxidant properties, nutrients and phyto-chemicals such as phenolics, flavonoids, vitamins and carotenoids that could provide scientific evidence for some folk uses in the treatment of diseases related to the production of ROS and oxidative stress.	Nair VD *et al.*, Industrial Crops and Products. 2012; 39: 17–25.
10.	Chemical characterization and anti-inflammatory effect of rauvolfian, a pectic polysaccharide of Rauvolfia callus.	*In vitro* and *In vivo*	Crude rauvolfian RS was purified using membrane ultrafiltration to yield the purified rauvolfian RSP in addition to glucan as admixture from the callus. Anti-inflammatory activity was determined by using Acetic acid induced colitis model in Male A/HeJ mice, Determination of colonic myeloperoxidase (MPO) activity and Evaluation of adherent colonic mucus	Animals were singly treated orally with the polysaccharide samples [rauvolfian (RS and RSP) (0.2 ml)] dissolved in water using flexible rubber catheter two days before induction of colitis.	A peroral pretreatment of mice with the crude and purified samples of rauvolfian (RS and RSP) was found to decrease colonic macroscopic scores, the total area of damage, and tissue myelo-peroxidase activity in colons as compared with a colitis group. RS and RSP were shown to stimulate production of mucus by colons of the colitis mice.	The pectin rauvolfian inhibiting colonic inflammation induced by rectal infusion of acetic acid in mice was isolated from the Rauvolfia callus. RSP appeared to be an active constituent of the parent RS.	Popov SV *et al.*, Biochemistry (Mosc). 2007; 72(7): 778-84.

Contd...

TABLE 14.2–*Contd....*

Sl.No.	Title	Type of Study	Study Design	Form and Dose	Summarized Efficacy	Conclusion	Reference
					SATAVARI (*Asparagus racemosus* Willd.)		
					Galactogogue activity		
1.	A comparative study on galactogogue property of milk and aqueous decoction of *Asparagus racemosus* in rats.	*In vivo*	Eighteen lactating dams were used for this experiment. Females were divided into three experimental groups: Group I - Normal control, Group II - Milk decoction, Group III -Aqu. decoction. It is given orally with syringe. Milk production was measured from day 3 to day 15 of lactation.	100 mg/kg/ body weight; Milk decoction Aqueous decoction	Female rats that received oral doses of milk decoction of AR during their first lactation produced about 27 per cent more milk than controls. Pup weight gain was also significantly higher than that in the control group. Aqueous decoction of AR in same dose also produced more milk than control.	Shatavari (*Asparagus racemosus*) is having a potent galactogogue property with milk extract being the best than any other form of dosage.	Rajesh Garg and V.B. Gupta; International Journal of Pharmacognosy and Phytochemical Research 2010; 2(2); 36-39.
2.	Effect of Shatavari (*Asparagus racemosus*) on milk production and immune-modulation in Karan Fries crossbred cows.	*In vivo*	Pregnant cows were divided based on most divided based on most probable milk production ability, body weight and parity into a control non-supplemented, NS group (n=5) and an experimental Shatavari supplemented, ARS group (n=5). ARS group cows were fed Shatavari root power, while NS group cows served as control.	Shatavari root power 100 mg/kg from 60 days till parturition once in the morning.	The milk yield was significantly more ($P<0.01$) in ARS group than the NS group. Colostrum protein, total solids, SNF ($P<0.05$) and total immunoglobulin level was higher ($P<0.01$) in ARS group in comparison to NS group. Cows of ARS group took less time to expel placental membranes ($P<0.05$) and had less service period and service/conception ($P<0.05$) than the NS group.	Prepartum supplementation of Shatavari significantly increased milk yield, colostrum's total immunoglobulin and reduced total milk cholesterol, service period and service/ conception in ensuing lactation.	Santosh Kumar, *et al.*, Indian Journal of Traditional Knowledge 2014; 13(2): 404-408.
3.	Shatavari: Potentials for galactogogue in dairy cows.	*In vivo*	Total 20 animals (10 buffaloes and 10 cross breed cows) were selected and were fed powder of shatavari roots.	50 g in concentrates once in a day for a period of 60 days	The overall milk production were increased 1.06±0.17 kg (11.47 per cent) daily and average milk production in buffaloes and cows were increased 0.8±0.34 kg (9.0 per cent),1.32±0.15 kg (12.72 per cent) respectively as compared to their previous production.	Shatavari is effective for increase in milk production and beneficial for economic milk production.	Behera PC, *et al.*, Indian Journal of Traditional Knowledge 2013; 12(1): 9-17

Contd...

TABLE 14.2–*Contd....*

Sl.No.	*Title*	*Type of Study*	*Study Design*	*Form and Dose*	*Summarized Efficacy*	*Conclusion*	*Reference*
4.	Effect of *Asparagus racemosus* rhizome (Shatavari) on mammary gland and genital organs of pregnant rat.	*In vivo*	The extract was administered orally to adult pregnant female albino rats.	Alcoholic extract of Asparagus rhizome; 300mg/kg for 15 days (days 1-15 of gestation)	The macroscopic findings revealed a prominence of the mammary glands, a dilated vaginal opening and a transversely situated uterine horn in the treated group of animals. The weight of the uterine horns of the treated group was found to be significantly higher ($p < 0.001$). Hyperplasia of the glandular and muscular tissue and hypertrophy of the glandular cells were observed in the genital organs. The parenchyma of the genital organs showed abundant glycogen granules with dilated blood vessels and thickening of the epithelial lining. The oviduct in the treated group showed hypertrophied muscular wall, whereas the ovary revealed no effect of the drug.	Shatavari has oestrogenic effect on the female mammary gland and genital organs.	Pandey SK, *et al.*, Phytother Res. 2005; 19(8): 721-724.
5.	A double-blind randomized clinical trial for evaluation of galactogogue activity of *Asparagus racemosus* Willd.	Clinical	Double blind study; Sample Size: 60 subjects, Groups: Research group feed with shatavari powder and another control group given same amount of rice powder	60mg/kg of *Asparagus racemosus* dose filled in capsules and given divided into 3 for 30 days.	Prolactin level was found to increase in research group three times than control group which shows galactogogue action of Shatavari. Secondary parameter tested are increase weight in infant, mother, satisfaction of mother while lactation, overall wellbeing and happiness of baby.	Show positive results in reseach group than control gr shows improvement in lactation or galactogogue action in satavari.	Mradu Gupta and Badri Shaw; Iranian Journal of Pharmaceutical Research 2011; 10 (1): 167-172.
					Antiulcer Activity		
6.	Antisecretory and antiulcer activity of *Asparagus racemosus* Willd. against indomethacin plus phyloric ligation-induced gastric ulcer in rats.	*In vivo*	Evaluate the antiulcer activity of *Asparagus racemosus* Willd. (methanolic extract) and its action against indomethacin (a non-steroidal anti-inflammatory drug) plus pyloric ligation (PL)-induced gastric ulcers in rats.	*Asparagus racemosus* (Shatavari) crude extract (100 mg/kg/day orally) for fifteen days	Treatment with *Asparagus racemosus* significantly reduced ulcer index when compared with control group. The reduction in gastric lesions was comparable to a standard antiulcer drug Ranitidine (30 mg/kg/day orally). Crude extract also significantly reduced volume of gastric secretion, free acidity and total acidity. A significant increase in total carbohydrate (TC) and TC/total protein (TP) ratio of gastric juice was also observed.	*Asparagus racemosus* was found to be an effective antiulcerogenic agent, whose activity can well be compared with that of ranitidine hydrochloride. *Asparagus racemosus* causes an inhibitory effect on release of gastric hydrochloric acid and protects gastric mucosal damage.	Bhatnagar M and Sisodia SS, J Herb Pharmacother. 2006; 6(1): 13-20.
7.	Antiulcer and antioxidant activity of *Asparagus racemosus* Willd and *Withania somnifera* Dunal in rats.	*In vivo*	In rats gastric ulcer was induced by the indomethacin (NSAID) and swim (restraint) stress treatment.	*A. racemosus* as well as *W. somnifera* methanolic extract (100 mg/kg BW/day p.o.) given orally for 15 days	Treatment with *Asparagus racemosus* and *Withania somnifera* significantly reduced the ulcer index, volume of gastric secretion, free acidity, and total acidity. A significant increase in the total carbohydrate and total carbohydrate/protein ratio was also observed. Study also indicated an increase in antioxidant defense, that is, enzymes superoxide dismutase, catalase, and ascorbic acid, increased significantly, whereas a significant decrease in lipid peroxidation was observed.	*A. racemosus* was more effective in reducing gastric ulcer in indomethacin-treated gastric ulcerative rats, whereas *W. somnifera* was effective in stress-induced gastric ulcer.	Bhatnagar M *et al.*, Ann N Y Acad Sci. 2005; 1056: 261-78.

Contd...

TABLE 14.2–*Contd....*

Sl.No.	Title	Type of Study	Study Design	Form and Dose	Summarized Efficacy	Conclusion	Reference
					Anti-diarrheal Activity		
8.	Anti-diarrhoeal potential of *Asparagus racemosus* wild root extracts in laboratory animals.	*In vivo*	Ethanolic and aqueous extracts of *Asparagus racemosus* was evaluated for its antidiarrhoeal potential against several experimental models like castor oil-induced diarrhoea model in rats, astrointestinal tract motility after charcoal meal administration and PGE2 induced intestinal fluid accumulation (enteropooling).	The ethanol and aqueous extracts of *Asparagus racemosus* (150, 200 and 250 mg/kg, p.o.) were administered orally	The plant extracts showed significant inhibitor activity against castor oil induced diarrhoea and PGE2 induced enteropooling in rats when tested at 200 mg/kg. Both extracts also showed significant reduction in gastrointestinal motility in charcoal meal test in rats.	The results point out the possible anti-diarrhoeal effect of the plant extracts and substantiate the use of this herbal remedy as a non-specific treatment for diarrhoea in folk medicine.	Venkatesan N *et al.*, J Pharm Sci. 2005 Feb 25; 8(1): 39-46.
9.	Cytotoxicity, analgesic and antidiarrhoeal activities of *Asparagus racemosus.*	*In vivo*	The test for analgesic activity of the crude ethanol extract was performed using acetic acid induced writhing model in mice. On the other hand, antidiarr-hoeal test of the EEAR was done according to the model of castor oil induced diarrhoea in mice and brine shrimp lethality bioassay was used to determine the cytotoxic activity of ethanol extract of the plant.	The ethanol *Asparagus racemosus* (250 and 500 mg/kg, p.o.) were administered orally	In acetic acid induced writhing in mice, the ethanol extract exhibited significant inhibition of writhing reflex 67.47 per cent at the dose of 500 mg/kg body weight. The plant extract showed antidiarrhoeal activity in castor oil induced diarrhoea in mice. It increased mean latent period and decreased the frequency of defecation with number of stool count at the dose of 250 and 500 mg/kg body weight, respectively comparable to the standard drug Loperamide at the dose of 50 mg/kg body weight. In addition to these, the brine shrimp lethality test showed the significant cytotoxic activity of the plant extract (LC50: 10 μg/ml and LC90: 47.86 μg/ml).	The results point out the possible anti-diarrhoeal, and analgesic effect of the plant extracts and substantiate the use of this herbal remedy as a non-specific treatment for diarrhoea in folk medicine.	Karmakar UK, *et al.*, Journal of Applied Sciences, 12: 581-586.
10.	Algesic activity of aqueous aD alcohol root extracts of *Asparagus racemosus* Willd.	*In vivo*	The present study was undertaken to evaluate the analgesic activity of the aqueous and alcohol root extracts of *Asparagus racemosus* using eddy's hot plate and heat conduction method.	All the animals were treated orally with aqueous and ethanolic extracts of 150 and 250 mg/kg body weight respectively	In eddy's hot plate method the aqueous extract showed significant analgesic activity at the doses of 150 mg/kg and 250 mg/kg and alcohol extract showed significant analgesic activity at the doses of 150 and 250 mg/kg. In heat conduction method both extracts showed significant analgesic activity at the doses of 150 and 250 mg/kg as compared to control group.	The aqueous and alcohol root extracts of *Asparagus racemosus* Willd. have significant analgesic activity.	Fasalu Rahiman OM *et al.*, Pharmacology-online 2011; 2: 558-562.

Contd...

TABLE 14.2–*Contd....*

Sl.No.	*Title*	*Type of Study*	*Study Design*	*Form and Dose*	*Summarized Efficacy*	*Conclusion*	*Reference*
					TAGAR (*Valeriana wallichii*)		
					Sleep Enhancer and antidepressant Activity		
1.	*Valeriana wallichii* root extract improves sleep quality and modulates brain monoamine level in rat.	*In vivo*	Species : Sprague-Dawley rats N= 24 Group: 4 Group I Saline control, GR II/III/IV 100, 200, 300 mg VW. Diazepam as +Ve control Dosage: 100, 200 and 300 mg/kg Duration: 2 months. Parameter: sleep-wake profile and level of brain monoamines was investigated.	*Valeriana wallichii* (VW) aqueous root extract 100, 200 and 300 mg/kg body weight	A significant decrease of sleep latency and duration of wakefulness were observed with VW at doses of 200 and 300 mg/kg. Duration of NREM sleep as well as duration of total sleep was increased significantly after treatment with VW at the doses of 200 and 300 mg/kg. VW also increased EEG slow wave activity during NREM sleep at the doses of 200 and 300 mg/kg.	VW water extract has a sleep quality improving effect which may be dependent upon levels of monoamines in cortex and brainstem.	Surjit Sahu, *et al.*, Phytomedicine. 2012; 19(10): 924-929.
2.	Adaptogenic activity of *Valeriana wallichii* using cold, hypoxia and restraint multiple stress animal model.	*In vivo*	Species: Sprague-Dawley rats. Model; Cold Hypoxia Restrained (CHR) stress model N=54. Group: 2 (Gr I N=24, Gr II N=30 again subdivided into 5 Groups. One of is control and rest 4 with 50/100/200/500mg/kg of VW extract) Duration: Single dose Parameter: Total phenols, antioxidants and flavonoids was investigated.	Different doses 50, 100, 200 and 500 mg/kg single dose	The maximal effective adaptogenic dose was observed to be 200 mg/kg body weight. The animals supplemented with the extract were found to adapt much faster as indicated by the improved malondialdehyde (MDA) and lactate dehydrogenase (LDH) levels as well as reduced superoxide dismutase (SOD) and catalase (CAT) levels in comparison to control.	*V. wallichii* had adaptogenic activity.	Priyanka Sharma, *et al.* Biomedicine and Aging Pathology, 2012: 2(4): 198–205.
3.	Initial exploratory observational pharmacology of *Valeriana wallichii* on stress management: A clinical report.	Clinical	Clinical Study in hospital based clinical set-up; Sample: Thirty-three subjects (20 male and 13 female; average age 34.2 years; Hamilton's Brief Psychiatric Rating Scale (BPRS); Visits baseline (day 0), mid-term (day 30) and final (day 60).	500 mg Plant extract/ capsule, twice daily, p.o. after meal	*V. wallichii* not only significantly attenuated stress and anxiety, but also significantly improved depression and also enhanced the willingness to adjustment. It did not alter memory, concentration or attention of the volunteers.	*V. wallichii* may be useful in the treatment of stress related disorders in human.	Dipankar Bhattacharya, *et al.* Nepal Medical College Journal 2007; 9(1):

Contd...

TABLE 14.2–*Contd....*

Sl.No.	Title	Type of Study	Study Design	Form and Dose	Summarized Efficacy	Conclusion	Reference
4.	Antidepressant effect of *Valeriana wallichii* patchouli alcohol chemotype in mice: Behavioural and biochemical evidence.	*In vivo*	Forced swim test; and neurotransmitter levels in mouse forebrain.	Dichloro-methane extract of *Valeriana wallichii*; 10, 20 and 40 mg/kg; orally for two weeks	Single administration of extract (40 mg/kg) significantly inhibited the immobility period in mice. Similarly, chronic administration of extract (20 and 40 mg/kg) significantly reduced the immobility period and significantly increased the levels of norepinephrine and dopamine in mouse forebrain.	Valeriana have anti-depressant effect and it significantly increased the norepinephrine and dopamine levels in forebrain.	Saha SP, *et al.* J Ethno-pharmacol., 2011; 135(1): 197–200.
5.	Antidepressant effect of *Valeriana wallichii* patchouli alcohol chemotype in mice: Behavioural and biochemical evidence.	*In vivo*	Forced swim test; and neurotransmitter levels in mouse forebrain.	Dichloro-methane extract of *Valeriana wallichii*; 10, 20 and 40 mg/kg; orally for two weeks	Single administration of extract (40 mg/kg) significantly inhibited the immobility period in mice. Similarly, chronic administration of extract (20 and 40 mg/kg) significantly reduced the immobility period and significantly increased the levels of norepinephrine and dopamine in mouse forebrain.	Valeriana have anti-depressant effect and it significantly increased the norepinephrine and dopamine levels in forebrain.	Sangeeta Pilkhwal Saha, *et al.* J Ethno-pharmacol. 2011; 135: 197–200.
					Neuroprotective Effect		
6.	GABAergic effect of valeric acid from *Valeriana wallichii* in amelioration of ICV STZ induced dementia in rats.	*In vivo*	To investigate the neuro-protective effect of rhizomes of *V. wallichii* containing valeric acid and its possible mechanism of action in amelioration of intra-cerebro ventricular streptozotocin induced neurodegeneration in Wistar rats.	*V. wallichii* extract 100 and 200 mg/kg, p.o. (suspended in 1 per cent CMC solution) and valeric acid 20 and 40 mg/kg, i.p (suspended in 1 per cent Tween 80 solution) were given to animals treated with ICV-STZ from day 3rd to 21st days once daily.	Treatment with *V. wallichii* extract 100 and 200 mg/kg and valeric acid 20 and 40 mg/kg significantly decreased the escape latency and retention transfer latency, as compared to intracerebro ventricular streptozotocin group. Plant extract and valeric acid also decreased the level of lipid peroxidation and restored gluta-thione level in rat brains. Administration of picrotoxin signi-ficantly reversed the effects produced by plant extract and valeric acid in intracerebro ventricular streptozotocin treated rats.	Valeric acid present in *V. wallichii* has significant GABAergic effect in amelioration of experimental dementia.	Vishwakarma S *et al.*, Rev bras farmacogn. 2016; 26 (4): 484-489. Doi: 10.1016/j. bjp.2016.02.008.

Contd...

TABLE 14.2–*Contd....*

Sl.No.	*Title*	*Type of Study*	*Study Design*	*Form and Dose*	*Summarized Efficacy*	*Conclusion*	*Reference*
				Picrotoxin (2 mg/kg, i.p., suspended in 1 per cent Tween 80 solution) was given as GABA-A antagonist 30 min before valeric acid administration daily.			
7.	Neuroprotective effect of *Valeriana wallichii* rhizome extract against the neurotoxin MPTP in C57BL/6 mice.	*In vivo*	*V. wallichii* rhizome extract was tested for anti-parkinsonian activity in MPTP induced PD mice. evaluate the behavioural scores of PD mice, evaluate the glutathione and antioxidant system in PD mice. The levels of striatal dopamine, mid brain tyrosine hydroxylase positive (TH+) cell count, TH protein expression, reactive oxygen species (ROS), lipid peroxidation (LPO), antioxidants and inflammatory cytokines were analysed. Mid brain glial fibrillary acidic protein (GFAP) expression was assessed by immunohistochemistry and western blotting. Also mid brain histopathological analysis was performed.	PD induced mice were treated orally with three different doses (50, 100 and 200 mg/kg body weight of VWE for 14 days and their behavioural changes were studied on days 0, 8, 13 and 21.	VWE treatment significantly recuperated the altered behavioural test scores, striatal dopamine levels, mid brain TH+ cell count and TH protein levels, increased GFAP expression and the histopathological changes observed in PD mice. Similarly, diminished levels of antioxidants, elevated levels of ROS, LPO and inflammatory cytokines were also significantly ameliorated following VWE treatment. The effective dose of VWE was found to be 200 mg/kg BW.	*V. wallichii* rhizome extract has the potential to mitigate oxidative stress and inflammatory damage in PD.	Sridharan S *et al.*, Neurotoxicology. 2015; 51: 172-83.

Contd...

TABLE 14.2–*Contd....*

Sl.No.	*Title*	*Type of Study*	*Study Design*	*Form and Dose*	*Summarized Efficacy*	*Conclusion*	*Reference*
					Anti-spasmodic and anti-hypertensive effect		
8.	Antispasmodic and blood pressure lowering effects of *Valeriana wallichii* are mediated through K^+ channel activation.	*Ex vivo*	Rabbit jejunum preparations; normo tensive anaesthetized rats; rabbit aortic preparations.	Crude extract of *Valeriana wallichii* rhizome 0.1-3.0 mg/mL	Valeriana caused relaxation of spontaneous contractions. When tested against high K induced contractions it produced weak inhibitory effect, while caused complete relaxation of the contractions induced by low K. In guinea pig ileum, the plant extract produced similar results as in rabbit jejunum. Intravenous administration produced fall in arterial blood pressure in normotensive anaesthetized rats and this effect was partially blocked by glibenclamide. In rabbit aortic preparations, plant extract also caused a selective and glibenclamide-sensitive relaxation of low K-induced contractions.	*Valeriana wallichii* have antispasmodic and hypotensive effects possibly through K(ATP) channel activation.	Gilani AH, *et al.*, J Ethno-pharmacol. 2005; 100(3): 347-52.
					Antidiarrhoeal Activity		
9.	Antidiarrhoeal and bronchodilatory potential of *Valeriana wallichii.*	*In vivo*	The crude extract of *V. wallichii* was investigated for anti-diarrhoeal activity using castor oil-induced diarrhoea in mice, and bronchodilatory activity using guinea-pig trachea.	Crude extract of *Valeriana wallichii* was given at 300 and 600mg/kg p.o.	Pre-treatment of animals with Vw.Cr produced 20 per cent protection against diarrhoea at 300 mg/kg and 60 per cent protection at 600 mg/kg. In guinea-pig trachea, Vw.Cr concen-tration dependently (0.03-3.0mgmL{[1]) relaxed the low K^+ (25mM)-induced contractions, with a mild effect on the contr-actions induced by high K^+ (80mM). In the presence of gliben-clamide, the relaxation of low K^+-induced contractions was prevented.	These results indicate that *V. wallichii* exhibits antidiarrhoeal and bronchodilatory activities, possibly through K+ channel activation, and thus reveal its medicinal usefulness in hyperactive gut and airway disorders such as diarrhoea and asthma.	Khan AU and Gilani AH, Nat Prod Res. 2012; 26(11): 1045-9.
					Adaptogenic Activity		
10.	Adaptogenic activity of *Valeriana wallichii* using cold, hypoxia and restraint multiple stress animal model.	*In vivo*	The present investigation was undertaken to evaluate the adaptogenic potential of aqueous lyophilized root extract of *V. wallichii* of Indian Himalayan region using Cold-Hypoxia-Restraint (C-H-R) animal model.	Different doses (50, 100, 200 and 500 mg/kg) of *V. wallichii* extract were administered as single overnight dose	The maximal effective adaptogenic dose was observed to be 200 mg/kg body weight which was used for cumulative study and biochemical analysis on attaining rectal temperature (Trec) 23 °C. The animals supplemented with the extract were found to adapt much faster as indicated by the improved malondialdehyde (MDA) and lactate dehydrogenase (LDH) levels as well as reduced superoxide dismutase (SOD) and catalase (CAT) levels in comparison to control.	The supplementation of aqueous root extract of *V. wallichii* had adapto-genic activity as assessed by C-H-R animal model.	Sharma P, *et al.*, Biomedicine and Aging Pathology. 2012; 2(4): 198–205.

Contd...

TABLE 14.2–*Contd....*

Sl.No.	*Title*	*Type of Study*	*Study Design*	*Form and Dose*	*Summarized Efficacy*	*Conclusion*	*Reference*
					TURMERIC (*Curcuma longa* Linn)		
					Anti-Arthritic Activity		
1.	Turmeric extracts containing curcuminoids prevent experimental rheumatoid arthritis.	*In vivo*	To determine the *in vivo* efficacy of well-characterized curcuminoid-containing turmeric extracts in the prevention or treatment of arthritis using streptococcal cell wall (SCW)-induced arthritis.	23 mg total curcuminoids/kg/d	A clinical measure of joint swelling, was used as the primary endpoint for assessing the effect of extracts on joint inflammation. An essential oil-depleted turmeric fraction containing 41 per cent of the three major curcuminoids was efficacious in preventing joint inflammation when treatment was started before, but not after, the onset of joint inflammation. A commercial sample containing 94 per cent of the three major curcuminoids was more potent in preventing arthritis than the essential oil-depleted turmeric fraction when compared by total curcuminoid dose per body weight.	(1) Document the *in vivo* antiarthritic efficacy of an essential oil-depleted turmeric fraction and (2) suggest that the three major curcuminoids are responsible for this antiarthritic effect, while the remaining compounds in the crude turmeric extract may inhibit this protective effect.	Funk JL *et al.*, J Nat Prod. 2006; 69(3): 351-5.
2.	Anti-inflammatory and anti-hyalurondase activity of the volatile oil of *Curcuma longa* Linn. (Haldi).	*In vivo*	Anti-inflammatory activity of the volatile oil of *Curcuma longa* has been compared with cortisone through a battery of tests.	Volatile oil	The volatile oil inhibited carrageenin and formaldehyde induced hind paw oedema and arthritis in rats. Further, the oil was found to inhibit both the exudative as well as the proliferative phases of the inflammatory reaction, as evidenced by significant reduction of exudates and the weight of the inflammatory pouch in granuloma pouch test.	The anti-exudative effect of the volatile oil of *C. longa* has been attributed partly to its anti-hyaluronidase activity as evidenced by inhibition of the diffusion capability of the hyaluronidase enzyme by the oil. The experimental data substantiates the claim for its anti-rheumatic effect.	N. Mishra and S.S. Gupta, Journal of Research in Ayurveda and Siddha. 1997; 18(1-2): 56-62.
3.	A comparative study of curcumin and soluble curcumin as antioxidant, anti-inflammatory and platelet aggregation inhibitors.	*In vitro*	The aim of this study is to compare the anti-oxidant, anti-inflammatory and platelet aggregation inhibition activity of CUR and S-CUR.	Curcumin and soluble curcumin	The *in vitro* models used are: (1) sodium nitrite-induced haemoglobin oxidation; (2) anti-inflammatory activity using macrophage activation with lipopolysaccharide and tumour necrosis factor a (TNF-a); (3) platelet aggregation inhibition using standard aggregating agents such as arachidonic acid and collagen. Results indicate that S-CUR shows antioxidant and anti-inflammatory activity at lower concentrations than CUR. The antioxidant activity of S-CUR is obtained in the range of 5–20 μM, whereas CUR is active in the range of 20–60 μM. The	Results indicate that S-CUR shows antioxidant and anti-inflammatory activity at lower concentrations than CUR. The antioxidant activity of S-CUR is obtained in the range of 5–20 μM, where as CUR is active in the	P N Mertia, A D B Vaidya, 4th World Ayurveda Congress and Arogya Expo proceedings Pp. 63, 9-13 December 2010,

Contd...

TABLE 14.2–*Contd....*

Sl.No.	*Title*	*Type of Study*	*Study Design*	*Form and Dose*	*Summarized Efficacy*	*Conclusion*	*Reference*
					anti-inflammatory activity studied in macrophage model showed a complete inhibition of TNF-a release by S-CUR at 30 μM.	range of 20–60 μM. The anti-inflammatory activity studied in macrophage model showed a complete inhibition of TNF-a release by S-CUR at 30 μM.	Bengaluru, Karnataka, India
					Immunomodulatory Activity		
4.	Immunomodulatory activity of curcumin: Suppression of lymphocyte proliferation, development of cell-mediated cytotoxicity, and cytokine production *in vitro.*	*In vitro*	Investigated the effect of curcumin on mitogen/ antigen induced proliferation of splenic lymphocytes, induction of cytotoxic T lymphocytes (CTLs), lymphokine activated killer (LAK) cells, and the production of cytokines by T lymphocytes and macrophages.	Curcumin	Mitogen, interleukin-2 (IL-2) or alloantigen induced proliferation of splenic lymphocytes, and development of cytotoxic T lymphocytes is significantly suppressed at 12.5-30 micromol/L curcumin. The generation of LAK cells at similar concentrations was less sensitive to the suppressive effect of curcumin compared to the generation of antigen specific CTLs. Curcumin irreversibly impaired the production of these immune functions, since lymphoid cells failed to respond to the activation signals following 8h pretreatment with curcumin. Curcumin also inhibited the expression/production of IL-2 and interferon-gamma (IFN-gamma) by splenic T lymphocytes and IL-12 and tumor necrosis factor-alpha (TNF-alpha) by peritoneal macrophages irreversibly. Curcumin inhibited the activation of the transcription factor nuclear factor kappaB (NF-kappaB) without affecting the levels of constitutively expressed NF-kappaB.	Curcumin most likely inhibits cell proliferation, cell-mediated cytotoxicity (CMC), and cytokine production by inhibiting NF-kappaB target genes involved in induction of these immune responses.	Gao X *et al.*, Biochem Pharmacol. 2004; 68(1): 51-61.
5.	Hepatoprotective and immunomodulatory properties of aqueous extract of *Curcuma longa* in carbon tetra chloride intoxicated Swiss albino mice.	*In vivo*	To evaluate the hepatoprotective and immunotherapeutic effects of aqueous extract of turmeric rhizome in CCl_4 intoxicated Swiss albino mice. To study the liver function, the transaminase enzymes (SGOT and SGPT) and bilirubin level were measured in the serum of respective groups. For assaying the immunotherapeutic.	Turmeric at a dose of 50 mg/kg bw for 15 days.	The result of present study suggested that CCl_4 administration increased the level of SGOT and SGPT and bilirubin level in serum. However, the aqueous extract of turmeric reduced the level of SGOT, SGPT and bilirubin in CCl_4 intoxicated mice. Apart from damaging the liver system, CCl_4 also reduced non-specific host response parameters like morphological alteration, phagocytosis, nitric oxide release, myeloperoxidase release and intracellular killing capacity of peritoneal macrophages. Administration of aqueous extract of *C. longa* offered significant protection from these damaging actions of CCl_4 on the non-specific host response in the peritoneal macrophages of CCl_4 intoxicated mice.	*C. longa* has immunotherapeutic properties along with its ability to ameliorate hepatotoxicity.	Sengupta M *et al.*, Asian Pac J Trop Biomed. 2011 Jun; 1(3): 193–199.

Contd...

TABLE 14.2–*Contd....*

Sl.No.	Title	Type of Study	Study Design	Form and Dose	Summarized Efficacy	Conclusion	Reference
			action of *Curcuma longa* (*C. longa*), nonspecific host response parameters like morphological alteration, phagocytosis, nitric oxide release, myeloperoxidase release and intracellular killing capacity of peritoneal macrophages were studied from the respective groups.				
6.	Immunomodulatory activity of curcumin.	*In vivo*	Curcumin, an active ingredient present in *Curcuma longa*, was analysed for the immuno-modulatory activity in Balb/c mice.	Curcumin	Curcumin administration was found to increase the total WRC count (15,290) significantly on the 12th day Group of animals treated with vehicle alone showed results similar to that of normal animal (10,130 on 12th day). Curcumin increased the circulating antibody titre (512) against SRBC. Curcumin administration increased the plaque forming cells (PFC) in the spleen and the maximum number of PFC was observed on the 6th day (1,130 PFC/10(6) spleen cells) after immunization with SRBC. Rone marrow cellularity (16.9x10(6) cells/femur) and alpha-esterase positive cells (1,622/4000 cells) were also enhanced by Curcumin administration. A significant increase in macrophage phagocytic activity was also observed in Curcumin treated animals ($P<0.001$).	These results indicate the immunostimulatory activity of Curcumin.	Antony S *et al.*, Immunol Invest. 1999; 28(5-6): 291-303.
					Antiulcer Activity		
7.	Evaluation of turmeric (*Curcuma longa*) for gastric and duodenal antiulcer activity in rats.	*In vivo*	An ethanol extract of turmeric was studied in rats for its ability to inhibit gastric secretion and to protect gastro-duodenal mucosa against the injuries caused by pyloric ligation, hypo-thermic-restraint stress, indomethacin, reserpine and cysteamine adminis-tration and cystodestruc-tive agents including 80 per cent ethanol, 0.6 M HCl, 0.2 M NaOH and 25 per cent NaCl.	500 mg/kg of the extract was adminis-tered orally	Treatment with *Curcuma longa* extract produced significant anti-ulcerogenic activity in rats subjected to hypothermic-restraint stress, pyloruic ligation and indomethacin and reserpine administration. The extract had a highly significant protective effect against cystodestructive agents. The reduction in the intensity of ulceration of cysteamine-induced duodenal ulcers was not found to be statistically significant.	Turmeric extract not only increased the gastric wall mucus significantly but also restored the non-protein sulfhydryl (NP-SH) content in the glandular stomachs of the rats.	Rafatullah S *et al.*, J Ethno-pharmacol. 1990; 29(1): 25-34.

Contd...

TABLE 14.2–*Contd....*

Sl.No.	*Title*	*Type of Study*	*Study Design*	*Form and Dose*	*Summarized Efficacy*	*Conclusion*	*Reference*
8	Anti-ulcer activity of curcumin on experimental gastric ulcer in rats and its effect on oxidative stress/antioxidant, IL-6 and enzyme activities.	*In vivo*	The rats were divided into four groups and fasted for 2 days with free access to water. On the third day, the animals were fasted for a further 24 h with no access to water followed by surgery. Nineteen hours after ulcer induction, the rats were killed by decapitation. Stomach was opened along the greater curvature and ulcerative lesions were counted. Total juice acidity, neutrophils activity, mitochondrial activity, total antioxidants, paraoxonase (PON 1)/arylesterase and total peroxides were evaluated. DNA fragmentation (per cent) and pro-inflammatory cytokine IL-6 level were measured. The level of different gastro-cytoprotective effectors including total antioxidants and paraoxonase (PON 1)/ arylesterase activities was measured.	Rats received different doses of curcumin (20, 40, and 80 mg/kg) or vehicle by oral gavage	The anti-ulcer activity of curcumin was displayed by attenuating the different ulcerative effectors including gastric acid hypersecretion, total peroxides, myeloperoxiase (MPO) activity, IL-6 and apoptotic incidence.	Curcumin appears to have a propitious protective effect against gastric ulcer development.	Biomed Environ Sci. 2009 Dec; 22(6): 488-95.
9.	Gastroprotective activity of essential oils from turmeric and ginger.	*In vivo*	Turmeric (*Curcuma longa*) and ginger (*Zingiber officianale*) are widely used in Asian countries as traditional medicine and food ingredients. In the present study, we have	100, 500 and 1000 mg/kg by oral gavage	TEO and GEO inhibited ulcer by 84.7 per cent and 85.1 per cent, respectively, as seen from the ulcer index. Reduced antioxidant enzymes such as GPx, SOD, catalase and GSH produced by alcohol administration were significantly increased by simultaneous administration of TEO and GEO. Histopathological examination showed that ethanol-induced lesions such as necrosis, erosion and hemorrhage of the stomach wall were significantly reduced after oral administration of essential oils.	TEO and GEO could reduce the gastric ulcer in rat stomach as seen from the ulcer index and histopathology of the stomach. Moreover, oxidative stress produced by ethanol was found to	Liju VB, *et al.*, J Basic Clin Physiol Pharmacol. 2015; 26(1): 95-103.

Contd...

TABLE 14.2–*Contd....*

Sl.No.	*Title*	*Type of Study*	*Study Design*	*Form and Dose*	*Summarized Efficacy*	*Conclusion*	*Reference*
			evaluated the gastro-protective activity of turmeric essential oil (TEO) and ginger essential oil (GEO) in rats. Turmeric and ginger were evaluated for their antiulcer activity against ethanol-induced ulcers in male Wistar rats. Ethanol was used to induce gastric ulcer in Wistar rats. Parameters such as ulcer index, histopathology and levels of antioxidant enzymes such as glutathione peroxidase (GPx), super-oxide dismutase (SOD), catalase and glutathione (GSH) levels were measured to assess the degree of protection produced by the essential oils.			be significantly reduced by TEO and GEO.	
					Antidiabetic Activity		
10.	Hypoglycemic effects of turmeric (*Curcuma longa* L. rhizomes) on genetically diabetic KK-Ay mice.	*In vivo*	The turmeric (*Curcuma longa* L. rhizomes) EtOH extract significantly suppressed an increase in blood glucose level in type 2 diabetic KK-A(y) mice.	Treated rats were fed a diet containing the EtOH extract (0.2 or 1.0 g/100 g diet) for 4 weeks	In an *in vitro* evaluation, the extract stimulated human adipocyte differentiation in a dose-dependent manner and showed human peroxisome proliferator-activated receptor (PPAR)-gamma ligand-binding activity in a GAL4-PPAR-gamma chimera assay. The main constituents of the extract were identified as curcumin, demethoxycurcumin, bisdemethoxycurcumin, and ar-turmerone, which had also PPAR-gamma ligand-binding activity.	Turmeric is a promising ingredient of functional food for the prevention and/or amelioration of type 2 diabetes and that curcumin, demethoxy-curcumin, bisdemeth-oxycurcumin, and ar-turmerone mainly contribute to the effects via PPAR-gamma activation.	Kuroda M *et al.*, Biol Pharm Bull. 2005; 28(5): 937-9.

Contd...

TABLE 14.2–*Contd....*

Sl.No.	*Title*	*Type of Study*	*Study Design*	*Form and Dose*	*Summarized Efficacy*	*Conclusion*	*Reference*
11.	Hypoglycemic, hypolipidemic and antioxidant properties of combination of Curcumin from *Curcuma longa* Linn. and partially purified product from *Abroma augusta* Linn. in streptozotocin induced diabetes.	*In vivo*	The effect of the aqueous extract mixture on blood glucose, lipid peroxidation (LPO) and the antioxidant defense system in rat tissues like liver, lung, kidney and brain was studied for 8 weeks in streptozotocin induced diabetic rats.	Oral administration (300 mg/kg) of the combination of aqueous extract of turmeric whose active ingredient is Curcumin and Abromine mixed with diet for 8 weeks.	The administration of an aqueous extract of turmeric and abromine powder resulted in a significant reduction in blood glucose and an increase in total haemoglobin. The aqueous extract also resulted in decreased free radical formation in the tissues studied. The decrease in thiobarbituric acid reactive substances (TBARS) and increase in reduced glutathione (GSH), superoxide dismutase (SOD) and catalase (CAT) clearly showed the antioxidant property of the mixture.	The mixture of the two plants have shown anti-diabetic activity and also reduced oxidative stress in diabetes. A combination of *Abroma augusta* and *Curcuma longa* also restored the other general parameters in diabetic animals.	Ali Hussain HEM, Indian J Clin Biochem. 2002; 17(2): 33–43.
					TVAKA (*Cinnamomum zeylanicum* Breyne)		
					Anti-inflammatory Activity		
1.	Anti-inflammatory and anti-arthritic activity of type-A procyanidine-polyphenols from bark of *Cinnamomum zeylanicum* in rats.	*In vivo*	To evaluate potential of TAPP extracted from Cinnamon (*Cinnamomum zeylanicum*) bark in animal models of inflammation and rheumatoid arthritis in rats.	4, 8 and 25 mg/kg, p.o. for inflammation and 8 mg/kg, p.o. for anti-arthritic activity.	Analgesic activity was evaluated in Randall–Selitto assay in AIA rats. TAPP showed significant anti-inflammatory effect. TAPP treatment in established arthritic rats showed significant reversal of changes induced in AIA with respect to body weight drop (cachexia), ankle diameter, arthritic score, serum C-reactive protein (CRP) levels.	TAPP, type-A procyanidine polyphenols isolated from the bark of *C. zeylanicum* showed anti-inflammatory and anti-arthritic effects in animal models without ulcerogenicity potential.	S. Vetal *et al.*, Food Science and Human Wellness (2013) 2: 59–67.
2.	Antioxidant activity of cinnamon (*Cinnamomum zeylanicum,* Breyne) extracts.		The objective of this study was to evaluate the antioxidant activity of cinnamon extracts.		The etheric (0.69 mg), methanolic (0.88 mg) and aqueous (0.44 mg) cinnamon extracts, inhibited the oxidative process in 68 per cent; 95.5 per cent and 87.5 per cent respectively. The BHT control inhibited 80 per cent oxidation. The spray reagents (1) beta-carotene/linoleic acid and (2) Fe Cl_3/K_3 Fe $(CN)_4$ 1 per cent sol, showed spots in T.L.C. with antioxidant activity (1) and blue color (2), indicating the presence of phenolic compounds with Rf values of 0.50. Five fractions were obtained by column partition with antioxidant activity and the presence of phenolic compounds.	These results suggest that the cinnamon extracts can be used as food antioxidant together with the improvement of food palatability.	Mancini-Filho J *et al.,* Boll Chim Farm. 1998 Dec; 137(11): 443-7.

Contd...

TABLE 14.2–*Contd....*

Sl.No.	*Title*	*Type of Study*	*Study Design*	*Form and Dose*	*Summarized Efficacy*	*Conclusion*	*Reference*
3.	*Cinnamomum zeylanicum* extract inhibits pro-inflammatory cytokine TNFμ: *in vitro* and *in vivo* studies.	*In vivo* and *In vitro*	This investigation was carried out to evaluate the anti-inflammatory potential of hydro-alcoholic bark extract of *C. zeylanicum* (OA4-50) and its effect on Tumor necrosis factor-a (TNF-a) secretion and gene expression.		Ethanol extract of *C. zeylanicum* showed suppression of intra-cellular release of TNF- in murine neutrophils as well as leukocytes in pleural fluid. The extract was found to inhibit TNF- gene expression in LPS-stimulated human PBMCs at 20 μg/ml concentration.	A potent anti-inflammatory activity of cinnamon extract is suggestive of its anti-arthritic activity, which could be confirmed in various models of arthritis.	Kalpana Joshi *et al.* Research in Pharmaceutical Biotechnology Vol. 2(2), pp.14-21, April 2010.
					Antidiabetic Activity		
4.	Effects of the polyphenol content on the anti-diabetic activity of *Cinnamomum zeylanicum* extracts.	*In vivo* and Clinical	This investigation was carried out in streptozotocin-induced diabetic rats and in 15 human volunteers or diabetic patients who had elevated fasting blood glucose levels; but not involved in any medication.	Polyphenol enhanced extracts of *Cinnamomum zeylanicum* barks was administered into rats at dose 200 mg/kg b.w. for 30 days and for Humans 125 mg, 2 times per day for 30 days	Oral administration of the extract for acute study a significant decrease in fasting blood sugar level. Treatment with polyphenol-rich cinnamon extracts has also reduced the elevated creatinine and urea levels. The treatment with cinnamon polyphenol extracts significantly reduced the diabetes induced hypercholesterolemia and hypertriglyceridemia with maximum efficacy.	*C. zeylanicum* extracts containing 45 per cent polyphenol (procyn Z-45) and 75 per cent polyphenol content showing maximum antidiabetic potential and the beneficial effects of polyphenol-rich cinnamon extracts over standard aqueous extracts by placebo controlled randomized double blind clinical trials on respective populations is also warranted.	IM K *et al.*, Food Funct. 2014; 5(9): 2208-20.
5.	Antidiabetic activity of alcoholic extract of *Cinnamomum zeylanicum* leaves in alloxon induced diabetic rats.	*In vivo*	Investigate the anti-diabetic potential of ethanolic extract of *Cinnamomum zeylanicum* leaves in Alloxon induced diabetic rats.	Ethanolic extract of *C. zeylanicum* leaves (100, 150 and 200mg/kg/day p.o.) for 7 days	On single oral administration of the extract for acute study a significant decrease in fasting blood sugar level was observed at dose 150 and 200 mg/kg. The maximum reduction in blood glucose was observed after 5 hr at dose 200 mg/kg. In sub-acute treatment, day, the extract at dose of 150 and 200 mg/kg of body weight showed significant reduction in blood glucose level as compared to that of diabetic control group.	The present study suggests that ethanolic extract of *C. zeylanicum* leaves possess a potent antidiabetic property as it significantly reduced the fasting blood sugar level in alloxon induced diabetic rats.	Tailang M *et al.*, People's Journal of Scientific Research 2014; 1: 9-11.

Contd...

TABLE 14.2–*Contd....*

Sl.No.	Title	Type of Study	Study Design	Form and Dose	Summarized Efficacy	Conclusion	Reference
6	Antidiabetic activity of alcoholic extract of *Cinnamomum zeylanicum* leaves in alloxon induced diabetic rats.	*In vivo*	Evaluate antidiabetic potential of ethanolic extract of *Cinnamomum zeylanicum* leaves in alloxon induced diabetic rats.	Ethanolic extract of *C. zeylanicum* leaves (100, 150 and 200 mg/kg body weight/day respectively) orally for seven days	Treatment with ethanolic extract of *Cinnamomum zeylanicum* leaves significantly decreased the fasting blood sugar level. The maximum reductionin blood glucose was observed at dose 200 mg/kg body weight.	Ethanolic extract of *C. zeylanicum* leaves having antidiabetic potential and used as an alternative for the management of diabetes.	Vyas N *et al.*, Journal of Pharmacy Research 2009; 2(12): 1867-1868.
					Antiulcer Activity		
7.	Anti-secretagogue and antiulcer effects of Cinnamon Cinnamomum zeylanicum in rats.	*In vivo*	To evaluate the gastric antisecretory and antiulcer activities of 'Cinnamon' using pylorus ligation (Shay) rat model, necrotizing agents and indomethacin induced ulceration in rats. Histopathological assessment was done on gastric tissue of rats. Gastric wall mucus and nonprotein-sulfhydryl contents were also estimated.	The aqueous suspension of cinnamon (250 and 500 mg/kg) was administered through oral route	Cinnamon suspension pretreatment decreased the basal gastric acid secretion volume and rumenal ulceration in pylorus ligated rats. The suspension effectively inhibited gastric hemorrhagic lesions induced by 80 per cent ethanol, 0.2 M sodium hydroxide, and 25 per cent sodium chloride. The cinnamon suspension also showed antiulcer activity against indomethacin. Pretreatment with cinnamon suspension offered a dose-dependent protection against various histological indices. Treatment of rats with cinnamon replenished the ethanol-induced decreased levels of gastric wall mucus and nonprotein-sulfhydryl concentrations.	The gastroprotection of cinnamon observed in this study is attributed to its effect through inhibition of basal gastric secretion (attenuation of aggressive factors) and stimulation mucus secretion (potentiation of defensive factors); and increase in non protein-sulfhydryl concentration probably due to prostaglandin-inducing abilities mediated through its antioxidant property.	Alqasoumi S, Journal of Pharmacognosy and Phytotherapy 2012; 4(4): 53-61.
8.	Antiulcer effect of *Cinnamomum zeylanicum* bark in rats.	*In vivo*	To evaluate antiulcer activities of 'Cinnamon' using Aspirin induced ulcer, Acetic acid induced ulcer, pylorus ligation induced ulcer rat model, Ethanol induced ulcer, Cold restraint stress induced ulcer, Indomethacin induced ulcer, Cysteamine induced duodenal ulcer rat models.	The aqueous suspension of cinnamon bark (10, 50 and 100 mg/kg) was given through oral route	Cinnamon suspension pretreatment decreased the basal gastric acid secretion volume and rumenal ulceration in pylorus ligated rats. Pretreatment with Cinnamon bark suspension inhibits gastric damage caused by ulcerogenic agents.	*Cinnamomum zeylanicum* bark produced an increase in healing of gastric ulcers and also prevented the development of duodenal ulcers in rats indicating that Cinnamon possess both gastric cytoprotective and antisecretory effect.	Asad M, Res J Biol Sci. 2014; 9(5): 182-187.

Contd...

TABLE 14.2–*Contd....*

Sl.No.	Title	Type of Study	Study Design	Form and Dose	Summarized Efficacy	Conclusion	Reference
9.	Anti-ulcer effect of cinnamon and chamomile aqueous extracts in rat models.	*In vivo*	Ethanol-induced gastric ulcer in rats followed by Gastric ulcer index, Determination of gastric juice volume and acidity, and Histopathological examination.	Cinnamon and chamomile aqueous extracts at doses of 100, 200, 300, 400mg/kg of body weight for seven days.	Cinnamon and chamomile aqueous extracts at the different tested doses had gastroprotective effects on acute experimental gastric ulcer in rats. Antiulcer effect of cinnamon and chamomile aqueous extracts was higher than that of antiulcer drug. Aqueous extract chamomile had much more favorable antiulcer effect, compared to aqueous extract of cinnamon.	Water extracts of cinnamon and chamomile had potential antiulcer effect, which was superior to the respective effect observed with Zantac. Chamomile extracts were more superior to cinnamon in its protection of the stomach.	Rezq AA and Elmallh MM, J Am Sci. 2010; 6(2): 209-16.
					Immunomodulatory Activity		
10.	Immunomodulatory activity of *Cinnamomum zeylanicum* bark.	*In vivo*	The immunomodulatory effect of *Cinnamomum zeylanicum* was studied using different experimental models such as carbon clearance test, cyclophosphamide induced neutropenia, neutrophil adhesion test, effect on serum immunoglobulins, mice lethality test and indirect hemagglutination test	The bark extracts were administered orally at doses of 10 and 100 mg/kg. Levamisole (2.5 mg/kg p.o.) was used as standard drug.	The low dose of cinnamon bark (10 mg/kg p.o.) produced only an increase in serum immunoglobulins levels while the high dose of cinnamon bark (100mg/kg p.o.) decreased *Pasteurella multocida*-induced mortality by 17 per cent, increased the phagocytic index in carbon clearance test, increased neutrophil adhesion, increased serum immunoglobulin levels and antibody titer values.	The results of the study substantiate the belief that cinnamon is an immune systembooster.	Niphade SR, Pharmaceutical Biology, 2009; 47(12): 1168–1173.
					VACHA (*Acorus calamus* Linn.)		
					Hypolipidemic Activity		
1.	Hypolipidemic activity of *Acorus calamus* L. in rats.	*In vivo*	The efficacy of the *A. calamus* extract in lowering serum cholesterol and triglycerides effects.	Ethanolic extract 100 and 200 mg/kg)	Administrations of the 50 per cent ethanolic extract (100 and 200 mg/kg) as well as saponins (10 mg/kg) isolated from the extract demonstrated significant hypolipidemic activity. On the contrary, the aqueous extract showed hypolipidemic activity only at a dose of 200 mg/kg.	The alcoholic extract of *A. calamus* contains saponins which plays a role in hyperlipidemia.	Parab RS *et al.* Oterapia. 2002; 73(6): 451-455.
					Antidiabetic Activity		
2.	Antidiabetic activity of methanol extract of *Acorus calamus* in STZ induced diabetic rats.	*In vivo*	Oral glucose tolerance test and streptozotocin (STZ) induced diabetic rat model.	200 mg/kg of AC extract was administered orally	Oral administration of AC methanol extract showed significant restoration of the levels of blood glucose level. After 21 days of treatment, blood glucose, lipid profile (total cholesterol, LDL and HDL-cholesterol), glucose 6-phosphatase, fructose 1,6 bis	AC methanol extract possess potent antihyperglycemic activity in normal and STZ induced	Prisilla DH *et al.*, Asian Pacific Journal of

Contd...

TABLE 14.2–*Contd....*

Sl.No.	Title	Type of Study	Study Design	Form and Dose	Summarized Efficacy	Conclusion	Reference
				to diabetic rats for 21 days.	phosphatase levels and hepatic markers enzymes (aspartate aminotransferase, alanine aminotransferase, alkaline phosphatase) were decreased when compared with diabetic control. Plasma insulin, tissue glycogen, glucose-6-phosphate dehydrogenase levels were increased significantly compared to diabetic control. Concurrent histopathological studies of the pancreas showed comparable regeneration by extract which were earlier necrosed by STZ.	diabetic rats and so might be of useful in the management of diabetes.	Tropical Biomedicine. 2012; S941-S946.
					Immunomodulatory Activity		
3.	*In vitro* evaluation for immunomodulatory activity of *Acorus calamus* on human neutrophils.	*In vitro*	Petroleum ether, chloro-form, benzene, alcoholic and aqueous extracts of *Acorus calamus* leaves were tested for various phytoconstituents. The immunomodulatory activity was screened by *in vitro* immune-stimulant activity by phagocytic stimulation (slide method) and nitro-blue test.		As per preliminary phytochemical investigation alcoholic extract showed the presence of flavonoids, tannins, triterpenoids, and proteins. From phagocytic stimulation test and nitro-blue tetrazolium test petroleum ether, alcoholic extract and volatile oil showed highly significant at 5-15 mg/ml concentration.	The present study showed that *Acorus calamus* stimulate cell mediated immune system by modulating the neutrophil function.	Ravichandiran V and Patil VS, Int. Res. J. Pharm. 2015; 6 (7): 450-452.
					Cardioprotective Activity		
4.	Control clinical trial of the lekhaniya drug vaca (*Acorus calamus*) in case of ischemic heart disease.	Clinical			In the clinical trial on 45 patients of ischemic heart disease at the OPD of S.S Hospital BHU, the efficacy of the drug *Acorus calamus* was tested. The patient was divided randomly in the three groups. To the first group the trial drug in a dose of 1.5 3 g/day in divided dose for three month was given. The second d group was given purified 'guggulu' while the third group which was the control group was given a capsule containing lactose powder. There was an encouraging improvement in the first and second groups.	The drug was found to be effective in the improvement of chest pain, dyspnoea on effort, reduction of body weight index, improving in ECG decreasing serum cholesterol, decreasing SLDL (serum low density lipoproteins) and increasing SHDL (serum high density lipoproteins).	Mamgain P, Singh RHJ. Res Ayur Siddha. 1994; 15: 35-51.

Contd...

TABLE 14.2–*Contd....*

Sl.No.	*Title*	*Type of Study*	*Study Design*	*Form and Dose*	*Summarized Efficacy*	*Conclusion*	*Reference*
					Antidepressant Activity		
5.	Antidepressant-like effects of *Acorus calamus* in forced swimming and tail suspension test in mice.	*In vivo*	Tail suspension test (TST) and forced swimming test (FST) in mice were used to evaluate the antidepressant activity of methanolic extract of rhizomes of *A. calamus.*	Methanolic extracts (50 and 100 mg/kg i.p.) were administered daily for 7 days. Imipramine 5 mg/kg was used as standard antidepressant drug.	Test extracts of *A. calamus* decreased immobility periods significantly in a dose dependent manner in both TST and FST. The observed results were also comparable with known standard drug *i.e.* imipramine. The flavonoid apigenin, which selectively binds with high affinity to the central benzodiazepines receptor, possesses important anxiolytic and antidepressant activities. The review of literature reveals that the *A. calamus* contains saponin, glycosides, tannin and flavonoid.	Methanolic extract of *A. calamus* rhizomes shows antidepressant activity probably through interaction with adrenergic, dopaminergic serotonergic and γ-aminobutyric acid (GABA) nergic system. Both the models have been proved to be equally valuable for demonstration of substances with a potential antidepressant activity.	Pawar VS *et al.,* Asian Pacific Journal of Tropical Biomedicine. 2012; S17-S19.
6.	Experimental evaluation of antidepressant effect of Vacha (*Acorus calamus*) in animal models of depression.	*In vivo*	The experimental study was done in rats to evaluate their Open Field Behavior (OFB), High Plus Maze (HPM) activity and 5-HT receptor syndrome, before and after feeding Vacha.	The dosage of Vacha extract was 18 mg/kg. Vacha extract was dissolved in distilled water and the suspension was prepared with the addition of 2 per cent gum acacia powder. In the case of Vacha, 720 mg extract was dissolved in 100 ml distilled water.	Concurrent Vacha administration in the depression model prevented the development of behavioral deficit in ambulation and rearing due to stress. Similarly, in High Plus Maze Test (HPMT), exploratory activity of rat was restored with Vacha administration. In adopted model of depression, when the animal was subjected to Vacha administration, the behavioural deficit was prevented very well as compared to stressed group. While eliciting the 5-HT syndrome, only two components out of five were influenced by Vacha, indicating that Vacha does not sensitize postsynaptic 5-HT1A receptors, which explains the behavioral deficit prevention in stressed rat group.	Vacha possesses marked antidepressant effect but the anxiolytic effect of Vacha is not marked, and Vacha did not produce any significant change in 5-HT1A receptor sensitivity.	Tripathi A.K. and Singh R.H., Ayu. 2010; 31(2): 153–158.

Contd...

TABLE 14.2–*Contd....*

Sl.No.	*Title*	*Type of Study*	*Study Design*	*Form and Dose*	*Summarized Efficacy*	*Conclusion*	*Reference*
7.	*Acorus calamus* Linn. rhizomes extract for antidepressant activity in mice model.	*In vivo*	To evaluate the anti-depressant activity of aqueous-ethanolic extract of *Acorus calamus*, by forced swimming test (FST) and tail suspension test (TST).	*Acorus calamus* rhizomes extract (ACE) in doses 75mg/kg and 150 mg/ kg was administered once daily for 15 days	Pretreatment with ACE (75mg/kg and 150 mg/kg orally, for 15 days) significantly reduced the immobility time in both FST and TST, indicating antidepressant activity. The extract did not show significant effect on locomotor activity in mice. Doxazosin significantly attenuated the extract induced antidepressant effect. The doses of ACE significantly decreased corticosteroid levels, a measure for evaluating antidepressant activity as compared to control group.	The antidepressant effect of ACE normalizing the over-activity of hypothalamic pituitary-arenal (HPA) axis system and through interaction with adrenergic and dopaminergic systems.	De A and Singh M, ARPB, 2013; 3(4): 520-525.
8.	Effect of rhizome extract of *Acorus calamus* on depressive condition induced by forced swimming in mice.	*In vivo*	Evaluated the anti-depressant properties of *A. calamus* rhizome in a forced swimming test (FST) of mice model. The levels of cortisol monoamine oxidase and neurotransmitters were analyzed using standard methods.	Three doses of methanol extract of rhizome (200, 400 and 600 mg extract/kg b.wt) and imipramine (15 mg/kg b.wt), a positive control, were orally administered once a day for the consecutive period of 14 days in Balb/c mice.	The anti-depressant effect was observed maximum at the dose of 200 mg/kg. b.wt that caused 23.82 per cent reduction in immobility period. The extract also significantly attenuated the FST-induced elevation of plasma cortisol, monoamine oxidase activity and returned the altered levels of neurotransmitters near to the normal levels in brain.	The extract of *A. calamus* rhizome has anti-depressant-like activity which is mediated by modulating the central neurochemical as well as HPA (hypothalamic-pituitary-adrenal) axis in response to stress induced by FST.	Ilaiyaraja N, International Journal of phytomedicine 2012; 4: 319-325.
9.	Anticonvulsant activity of raw and classically processed Vacha (*Acorus calamus* Linn.) rhizomes.	*In vivo*	Anticonvulsant activity of raw and Shodhita (classically processed) Vacha rhizomes were screened against Maximal Electro Shock (MES) seizure model to assess the effect of classical purificatory procedure on pharmacological action of Vacha.	The dose of Vacha as per Ayurvedic Pharmacopoiea of India is 120 mg per day	Pretreatment with both raw and classically processed Vacha samples exhibited significant anticonvulsant activity by decreasing the duration of tonic extensor phase. Further classically processed Vacha statistically decreased the duration of convulsion and stupor phases of MES-induced seizures.	The anticonvulsant activity of raw Vacha and subjecting to classical Shodhana procedure did not alter the efficacy of Vacha rhizomes instead it enhanced the activity profile of the Vacha.	Bhat SD *et al.*, Ayu. 2012; 33(1): 119-22.

Contd...

TABLE 14.2–*Contd....*

Sl.No.	Title	Type of Study	Study Design	Form and Dose	Summarized Efficacy	Conclusion	Reference
					Antiulcer Activity		
10.	Anti-secretagogue, anti-ulcer and cytoprotective properties of *Acorus calamus* in rats.	*In vivo*	The ethanolic extract of *A. calamus* was studied in rats for its ability to inhibit gastric secretion and to protect gastroduodenal mucosa against the injuries caused by pyloric ligation, indomethacin, reserpine and cysteamine administration and cytodestructive agents including 80 per cent ethanol, 0.6 M HCl, 0.2 M NaOH and 25 per cent NaCl.	500 mg/kg of the extract was orally administered	An oral dose of 500 mg/kg of the extract showed significant anti-secretory and anti-ulcerogenic activity in rats subjected to pyloric ligation, indomethacin, and reserpine and cysteamine administration. The extract had a highly significant protective effect against cytodestructive agents.	These findings support the use of calamus for the treatment of gastropathy in traditional medicine.	Rafatullah S *et al.*, Fitoterapia 1994; 65(1): 19-23.
					VASA (*Adhatoda vasika* Nees)		
					Antitussive effect		
1.	Antitussive effect of *Adhatoda vasica* extract on mechanical or chemical stimulation-induced coughing in animals.	*In vivo*	The antitussive activity of *Adhatoda vasica* (AV) extract in anaesthetized guinea pigs and rabbits and in unanaesthetized guinea pigs.	Orall admistration	AV was shown to have a good antitussive activity. Intravenously, it was 1/20–1/40 as active as codeine on mechanically and electrically induced coughing in rabbits and guinea-pigs.	Oral administration to the guinea-pig the antitussive activity of AV was similar to codeine against coughing induced by irritant aerosols.	Dhuley JN *et al.*, J Ethnopharmacol. 1999 Nov 30; 67(3): 361-5.
					Anti-inflammatory and Analgesic Activity		
2.	Evaluation of anti-inflammatory and analgesic activities of ethanolic extract of roots *Adhatod avasica* Linn.	*In vivo*	The study designed to evaluate the anti-inflammatory and analgesic activities of ethanolic extract of roots of *Adhatoda vasica* Linn. (Acanthaceae).	200 and 400 mg/kg, oral administration	Significantly ($P < 0.05–0.01$)inhibited both carrageenan- and formalin-induced inflammation. Also, the acute treatment of extract produced asignificant ($P < 0.05–0.01$) antinociceptive effect in the acetic acid-induced writhing, formalin-induced pain licking and hot-plate-induced pain.	Ethanolic extract of roots of *Adhatoda vasica* Linn. possess remarkable anti-inflammatory and analgesic activities.	Wahid A Mulla *et al.* Int.J. Pharm. Tech. Res. 2010, 2(2).

Contd...

TABLE 14.2–*Contd....*

Sl.No.	Title	Type of Study	Study Design	Form and Dose	Summarized Efficacy	Conclusion	Reference
3.	A study on antioxidant and anti-inammatory activity of vasicine against lung damage in rats.	*In vivo*	To investigate antioxidant and anti-inflammatory potential of vasicine isolated from leaves of *Adhatoda vasica* in murine model of asthma. Lung damage was induced by the subcutaneous injection of 1 ml of saline containing 1 mg of ova-albumin and 20 mg aluminum hydroxidqut the same time l ml of *Bordetella pertussis* vaccine containing 6 x 10^9 heat killed organism was given intraperitoheal as adjuvant for 21 days, twice per day.	Treatment group induced with asthma was simultaneously followed by treatment with vasicine (0.2 mg/kg body weight) from second day after sensitization.	The treatment with vasicine in experimental rats resulted in the marked recovery of the pathology of the disease. After treatment with vasicine significant decrease in lipid peroxidation and similarly significant increase in antioxidants superoxide dismutase, catalase, glutathione peroxidase and reduced glutathione was recorded.	Lipid peroxidation and oxidative stress elicited by ovaalbumin and aluminium hydroxide was considerably reduced with vasicine treatment.	Srinivasarao D, *et al.* Indian J Allergy Asthma Immunol 2006; 20(1) : 1-7.
4.	Anti-oxidant and anti-inflammatory activity of *Adhatoda vasica* flowers.	*In vitro*	The flowers of *Adhatoda vasica* is extracted with 90 per cent ethanol and evaluated for antioxidant activities by 2,2-Diphenyl 1-picryl hydrazyl solution (DPPH), 2,2'-azino-bis (3-ethylbenzthiazoline-6-sul- phonic acid (ABTS) assay and anti-inflammatory activities by human blood cell (HRBC) membrane stabilization method and Inhibition of albumin denaturation method.		The ethyl acetate fractions of *Adhatoda vasica* flowers can be considered as good sources of antioxidants, anti-inflammatory and can be incorporated into the drug formulations.	The anti-oxidants, anti-inflammatory activity of the compound isolated from ethyl acetate fractions of *Adhatoda vasica* flowers.	Muruganantham N, *et al.*, IAJPR. 2015; 5(11): 3444-3450.

Contd...

TABLE 14.2–*Contd....*

Sl.No.	Title	Type of Study	Study Design	Form and Dose	Summarized Efficacy	Conclusion	Reference
					Hepatoprotective activity		
5.	Hepatoprotective activity of ethyl acetate extract of *Adhatoda vasica* in swiss albino rats.	*In vivo*	The hepatoprotective activity of Ethyl acetate extract of *Adhatoda vasica*, investigated against CCl_4 induced liver damage in Swiss albino rats.	100mg/kg and 200mg/kg	Pre-treatment of rats with the ethyl acetate Extract of *Adhatoda vasica* (100 mg/kg and 200mg/kg) prior to the CCl_4 dose at 1 ml/kg statistically lowered the three serum level enzymes and also Bilirubin. Histopathlogical observations also coincided with the above results, however 200mg/kg dose was found to be more active.	Ethyl acetate extract of *Adhatoda vasica* has potent hepatoprotective effect against CCl_4 - induced liver damage.	Ahmad R, Raja V, Sharma M., IJCRR. 2013; 5(6): 16-21.
					Antioxidant and Antimicrobial Activity		
6.	Antioxidant and anti-microbial activity of leaf extract of *Adhatoda vasica* against the bacteria isolated from the Sputum samples of asthmatic patients.	*In vitro*	In the present study antioxidant and anti-microbial activity of aqueous and methanolic extracts of *Adhatoda vasica* were evaluated against the bacteria isolated from the sputum samples of asthmatic patients.		Anti-oxidant activity was observed to maximum in methanolic extract as compared to aqueous extract of *Adhatoda vasica*. *Adhatoda vasica* showed a broad spectrum of antibacterial activities against Gram positive (*Staphylococcus aureus* and *Streptococcus pneumoniae*) bacterial species in comparison to the Gram-negative (*E. coli* and *Klebsiella pneumoniae*) bacterial species.	The aqueous and metha-nolic extract of *Adhatoda vasica* has significant amounts of antioxidant and antimicrobial agents.	Inderjit Kaur *et al.*, International Journal of Drug Research and Technology 2012, Vol. 2 (3), 273-278.
7.	Anti-ulcer activity of *Adhatoda vasica* Nees.	*In vivo*	To study the anti-ulcer activity of *Adhatoda vasica* leaves using two ulcer models (1) Ethanol-induced, and (2) Pylorus ligation plus aspirin-induced models.	Animals were dosed with the leaf powder (500 mg/kg in 0.2 per cent agar)	*Adhatoda vasica* leaf powder showeda considerable degree of anti-ulcer activity in experimental rats when compared with a control. The highest degree of activity (80 per cent) was observed in the ethanol-induced ulceration model.	The plant also has immense potential as an anti-ulcer agent of great therapeutic relevance.	Shrivastava N *et al.*, J Herb Pharmacother. 2006; 6(2): 43-9.
8.	Antimicrobial, anti-oxidant, and cytotoxic properties of vasicine acetate synthesized from vasicine isolated from *Adhatoda vasica* L.	*In vitro*	The alkaloid Vasicine was isolated from ethanolic extract of the leaves of *A. vasica* using column chromatography. Vasicine acetate was obtained by acetylation of Vasicine.		Vasicine acetate exhibited good zone of inhibition against bacteria: 10 mm against *E. aerogenes*, 10 mm against *S. epidermidis*, and 10mm against P. aeruginosa. Vasicine acetate showed minimum inhibitory concentration values against bacteria: *M. luteus* (125μg/mL), *E. aerogenes* (125μg/mL), *S. epidermidis* (125μg/mL), and *P. aeruginosa* (125μg/mL). The radical scavenging activity of Vasicine acetate was the maximum at 1000 μg/mL (66.15 per cent). The compound showed prominent cytotoxic activity *in vitro* against A549 lung adenocarcinoma cancer cell line. Quantification of Vasicine and Vasicine acetate by HPLC-DAD analysis showed their contents to be 0.2293 per cent and 0.0156 per cent, respectively, on dry weight basis of the leaves.	Vasicine acetate has immense potential as an antimicrobial agent and could be probed further in drug discovery programme.	Duraipandiyan V *et al.*, BioMed Research International. 2015 Article ID 727304.

Contd...

TABLE 14.2–*Contd....*

Sl.No.	*Title*	*Type of Study*	*Study Design*	*Form and Dose*	*Summarized Efficacy*	*Conclusion*	*Reference*
					Immunomodulatory Activity		
9.	Immunomodulatory activity of various extracts of *Adhatoda vasica* Linn. in experimental rats.	*In vivo*	Methanolic, chloroform and diethyl ether extracts of leaves of Indian medicinal plant *Adhatoda vasica* Linn. were pharmacologically validated for its immunomodulatory properties in experimental rats using Neutrophil adhesion test in rats and Delayed type hypersensitivity (DTH) methods.	The rats were treated orally with *C. deodara* wood oil and Adhatoda extracts at the doses of 100 and 400 mg/kg/day for 8 day.	Oral administration of extracts at a dose of 400 mg/kg in adult male Wister rats significantly increased the percentage neutrophil adhesion to nylon fibers. It extracts were also found to induce Delayed Type Hypersensitivity reaction by sheep erythrocytes.	The extracts of *A. vasica* Linn positively modulate the immunity of the host.	Vinothapooshan G and Sundar K. African Journal of Pharmacy and Pharmacology. 2011; 5(3): 306-310.
10.	Immunomodulatory potential of *Adhatoda vasica.*		To explore the immunomodulatory potential of AV in male swiss albino mice. Effect of alcoholic extract of AV leaf on haematological profile, splenic lymphocytes and peritoneal macrophages was studied 5, 10, 15 and 20 days after treatment. Its effect on macrophage phagocytic index, *E.coli* induced abdominal peritonitis and SRBC induced delayed type hypersensitivity was also evaluated.	*Adhatoda vasica* leaves extract was administered at dose 500 mg/kg, p.o. for 20 days	AV showed significant increase in total WBC, blood lymphocytes, splenic lymphocytes and peritoneal macrophages. It also showed significant protection against *E.coli* induced abdominal peritonitis.	AV leaf extract may have significant immunopotentiating and immunoprophylactic activity.	Thaakur SR, Asian Journal of Microbiology, Biotechnology and Environmental Sciences, 2007; 9(3): 553-557.

15

Ayurvedic Pharmacopoeia of India: A Ready Reckoner

Arun Gupta, Chandra Kant Katiyar, J.L.N. Sastry and Satyajyoti Kanjilal

Ayurvedic Pharmacopoeia of India (API) is an authentic and legal source for providing the standards of quality of raw materials as well as finished products.

Volume I of API covers the raw materials spread over 8 volumes while API part II covers finished products in two volumes. Referring these volumes sometimes becomes very cumbersome because one plant may be mentioned in one volume and second plant may be mentioned in some other volume.

The authors felt the necessity of developing a concise version of API for quick ready reference to offer convenience This compilation includes:

1. API page number
2. Common Ayurvedic names in alphabetical order
3. Botanical name
4. Family/chemical formulae
5. Part used
6. Identity, purity and strength/chemical properties
7. Constituents
8. Properties and actions
9. Important formulations
10. Therapeutic uses and
11. Dose

Since this is a concise version some of the aspects like description have not been included in this chapter and if need be original API volumes may be referred for details.

Hope this compilation would be useful to the users.

-----

* Authors have taken liberty to change spelling of certain letters to depict spoken Sanskrit language. For examples 'c' has been replaced with 'ch'.

Sl. No.	*Part and Volume*	*Page No.*	*Common Ayurvedic Name*	*Botanical Name/ Chemical Name*	*Family/ Chemical Formula*	*Part Used*	*Identity, Purity and Strength/Chemical Properties*	*Constituents*	*Properties and Action*	*Important Formulations*	*Therapeutic Uses*	*Dose*
1	Part-I; Vol. I	1	AJAGANDHA	*Cleome gynandra* Linn. Syn. *Gynandropsis gynandra* (Linn.) Briquet	Capparidaceae	Seed	**Foreign matter**- NMT 2 per cent; **Total ash**- NMT 7 per cent; **Acid-insoluble ash**- NMT 0.4 per cent; **Alcohol-soluble extractive**-NLT 16 per cent; **Water-soluble extractive**- NLT 7 per cent	Fixed oil, Essential oil and Oleoresin	*Rasa* - Katu; *Guna*- Laghu, Ruksa; *Virya*- Sita; *Vipaka*- Katu; *Karma*- Hrdya, Dipana, Vatahara, Pittala, Sulaghni	Narayana churna	Gulma, Asthila, Krmiroga, Kandu, Karnaroga	1-3 g of the drug in powder form
2	Part-I; Vol. I	2	AJAMODA	*Apium leptophyllum* (Pers.) F.V.M. Ex Benth.	Umbelliferae	Dried Aromatic Fruits	**Foreign matter (including stalk)** - NMT 5 per cent; **Total ash**- NMT 14 per cent; **Acid-insoluble ash**- NMT 4 per cent; **Alcohol-soluble extractive**-NLT 14 per cent; **Water-soluble extractive**- NLT 3 per cent **Volatile oil** - NLT 2 per cent	Essential oil and Fixed oil.	*Rasa* - Katu, Tikta; *Guna*- Laghu, Ruksa; *Virya*- Usna; *Vipaka*- Katu; *Karma*- Vidahi, Kaphavatajit, Dipana, Rucikrt, Krmijit, Sulaghna	Ajamodarka; Ajamodadi Churna	Aruci, Adhmana, Gulma, Hikka, Chardi, Krmi roga, Sula	1-3 g of the drug in powder form
3	Part-I; Vol. I	4	AMALAKI (Fresh)	*Emblica officinalis* Gaertn.	Euphorbiaceae	Fresh Fruit Pulp	**Foreign matter**- NMT 2 per cent; **Total ash**- NMT 7 per cent; **Acid-insoluble ash**- NMT 2 per cent; **Alcohol-soluble extractive (on dried basis)** -NLT 40 per cent; **Water-soluble extractive**- NLT 50 per cent **Moisture content** - NLT 80 per cent	Ascorbic acid and Tannins	*Rasa* - Amla, Kasaya, Madhura, Tikta, Katu; *Guna*- Ruksa, Laghu; *Virya*- Sita; *Vipaka*- Madhura; *Karma*-Tridosajit, Vrsya, Rasayana, Caksusya	Chyawanaprasa	Raktapitta, Amlapitta, Prameha, Daha	10-20g of the drug; 5-10 ml of fresh juice
4	Part-I; Vol. I	5	AMALAKI (Dried)	*Emblica officinalis* Gaertn. Syn. *Phyllanthus emblica* Linn.	Euphorbiaceae	Pericarp of Dried Mature Fruits	**Foreign matter (including seed and sed coat)**- NMT 3 per cent; **Total ash**- NMT 7 per cent; **Acid-insoluble ash**- NMT 2 per cent; **Alcohol-soluble extractive**-NLT 40 per cent; **Water-soluble extractive**- NLT 50 per cent	Ascorbic acid and Gallotannins	*Rasa* - Amla, Kasaya, Madhura, Tikta, Katu; *Guna*- Ruksa, Laghu; *Virya*- Sita; *Vipaka*- Madhura; *Karma*- Tridosajit, Vrsya, Rasayana, Caksusya	Chyawanaprasa; Dhatri lauha; Dhatryadi ghrta; Triphala churna	Raktapitta, Amlapitta, Prameha, Daha	3-6 g of the drug in powder form

Sl. No.	*Part and Volume*	*Page No.*	*Common Ayurvedic Name*	*Botanical Name/ Chemical Name*	*Family/ Chemical Formula*	*Part Used*	*Identity, Purity and Strength/Chemical Properties*	*Constituents*	*Properties and Action*	*Important Formulations*	*Therapeutic Uses*	*Dose*
5	Part-I; Vol. I	6	ARAGVADHA	*Cassia fistula* Linn.	Leguminosae	Pulp obtained from Fruits (devoid of Seeds, Septa and peces of pericarp)	**Foreign matter**- NMT 2 per cent; **Total ash**- NMT 6 per cent; **Acid-insoluble ash**- NMT 1 per cent; **Alcohol-soluble extractive**-NLT 15 per cent; **Water-soluble extractive**- NLT 46 per cent	Sugar, Mucilage, Pectin and Anthraquinone	*Rasa* - Madhura, Tikta; *Guna*- Guru; *Virya*- Usna; *Vipaka*- Madhura; *Karma*- Recana	Aragvadhadi kvatha churna	Vibandha, Udavartta, Gulma, Sula, Udararoga, Hrdroga, Prameha	5-10g of the drug in powder form
6	Part-I; Vol. I	8	ARKA (Root)	*Calotropis procera* (Ait.) R. Br.	Asclepiadaceae	Dried Root	**Foreign matter**- NMT 2 per cent; **Total ash**- NMT 4 per cent; **Acid-insoluble ash**- NMT 1 per cent; **Alcohol-soluble extractive**-NLT 2 per cent; **Water-soluble extractive**- NLT 8 per cent	Glycoside (Calotropin)	*Rasa* - Katu, Tikta; *Guna*- Laghu; *Virya*- Usna; *Vipaka*- Katu; *Karma*- Kaphavatahrt, Dipana, Bhedana, Krmighna, Vranahara, Visaghna, Kusthaghna	Mahavisa-garbha taila; Dhanvantara ghrta	Kandu, Kushta, Krmiroga, Gulam, Udararoga, Vrana, Svasa	1-3 of the drug for decoction
7	Part-I; Vol. I	10	ARKA (Leaf)	*Calotropis procera* (Ait.) R. Br.	Asclepiadaceae	Dried leaves	**Foreign matter**- NMT 2 per cent; **Total ash**- NMT 21 per cent; **Acid-insoluble ash**- NMT 5 per cent; **Alcohol-soluble extractive**-NLT 5 per cent; **Water-soluble extractive**- NLT 24 per cent	Glycoside (Calotropin)	*Rasa* - Katu, Tikta; *Guna*- Laghu, Sara, Snigdha; *Virya*- Usna; *Vipaka*- Katu; *Karma*- Vatahrt, Dipana, Krmighna, Sopha, vranahara, Visaghna, Bhedana, Svasahara	Arkalavana	Sotha, Kandu, Kustha, Vrana, Krmiroga, Gulma, Slesmodara roga, Pliharoga, Arsa, Svasa	250-750 mg of the drug in powder form

Sl. No.	*Part and Volume*	*Page No.*	*Common Ayurvedic Name*	*Botanical Name/ Chemical Name*	*Family/ Chemical Formula*	*Part Used*	*Identity, Purity and Strength/Chemical Properties*	*Constituents*	*Properties and Action*	*Important Formulations*	*Therapeutic Uses*	*Dose*
8	Part-I; Vol. I	12	ASANA	*Pterocarpus marsupium* Roxb.	Leguminosae	Heart-Wood	**Foreign matter**- NMT 2 per cent; **Total ash**- NMT 2 per cent; **Acid-insoluble ash**- NMT 0.5 per cent; **Alcohol-soluble extractive**-NLT 7 per cent; **Water-soluble extractive**-NLT 5 per cent	Alkaloids and Resin	*Rasa* - Kasaya, Katu, tikta; *Guna*- Laghu, Ruksa; *Virya*- Usna; *Vipaka*- Katu; *Karma*- Kaphapittasamaka, Galadosaghna, Kesya, Tvaccya, Stambhana, Kusthaghna, Rasayana, Raktasodhana	Nyagrodhadi churna; Asanabilvadi taila	Pandu, Prameha, Medodosa, Kustha, Krmiroga	50-100g of the drug for decoction
9	Part-I; Vol. I	14	ASOKA	*Saraca asoca* (Rosc.) Dc. Willd. Syn. *Saraca indica* Linn.	Leguminosae	Dried Stem Bark	**Foreign matter**- NMT 2 per cent; **Total ash**- NMT 11 per cent; **Acid-insoluble ash**- NMT 1 per cent; **Alcohol (90 per cent)-soluble extractive**-NLT 15 per cent; **Water-soluble extractive**- NLT 11 per cent	Tannins and a Crystalline glycoside	*Rasa* - Kasaya, Tikta; *Guna*- Laghu, Ruksa; *Virya*- Sita; *Vipaka*- Katu; *Karma*- Grahi, Varnya, Hrdya, Sothahara, Visaghna	Asokarista; Asokaghrta	Asrgdara, Apaci, Daha, Raktadosa, Sotha	20-30 g of the drug for decoction
10	Part-I; Vol. I	15	ASVA-GANDHA	*Withania somnifera* Dunal.	Solanaceae	Dried Mature Roots	**Foreign matter**- NMT 2 per cent; **Total ash**- NMT 7 per cent; **Acid-insoluble ash**- NMT 1 per cent; **Alcohol (25 per cent) soluble extractive**-NLT 15 per cent;	Alkaloids and Withanolides	*Rasa* - Tikta, Kasaya; *Guna*- Laghu; *Virya*- Usna; *Vipaka*- Madhura; *Karma*- Vatakaphapaha, Balya, Rasayana, Vajikarana	Asvagandha-dyarista; Asvagandhadi leha; Balasvagandha laksadi taila	Ksaya, Daurbalya, Vataroga, Sotha, Klaibya	3-6 g of the drug in powder form
11	Part-I; Vol. I	17	ASVATTHA	*Ficus religiosa* Linn.	Moraceae	Dried Bark	**Foreign matter**- NMT 2 per cent; **Total ash**- NMT 7 per cent; **Acid-insoluble ash**- NMT 0.3 per cent; **Alcohol-soluble extractive**-NLT 8 per cent; **Water-soluble extractive**-NLT 9 per cent	Tannins	*Rasa* - Kasaya; *Guna*- Guru, Ruksa; *Virya*- Sita; *Vipaka*- Katu; *Karma* - Kaphapittavinasi, Varnya, Samgrahi, Bhagna-sandhanakara, Mutrasam-grahaniya	Nyagrodhadi kvatha churna; Nyagrodhadi churna	Vatarakta, Raktapitta, Vrana, Yonidosa, Prameha	20-30 g of the drug for decoction

Sl. No.	*Part and Volume*	*Page No.*	*Common Ayurvedic Name*	*Botanical Name/ Chemical Name*	*Family/ Chemical Formula*	*Part Used*	*Identity, Purity and Strength/Chemical Properties*	*Constituents*	*Properties and Action*	*Important Formulations*	*Therapeutic Uses*	*Dose*
12	Part-I; Vol. I	19	ATASI	*Linum usitatissimum* Linn.	Linaceae	Dried, Ripe Seeds	**Foreign matter**- NMT 1 per cent; **Total ash**- NMT 5 per cent; **Acid-insoluble ash**- NMT 2 per cent; **Alcohol-soluble extractive**-NLT 30 per cent; **Water-soluble extractive**- NLT 15 per cent; **Fixed oil** - NLT 25 per cent	Fixed oil, Mucilage and Protein	*Rasa* - Madhura, Tikta; *Guna*- Snigdha, Guru; *Virya*- Usna; *Vipaka*- Katu; *Karma*- Vataghna, Acaksusya	Sarsapadi pralepa	Siroroga, Krmiroga, Kustha, Prameha	3-6 g of the drug in powder form
13	Part-I; Vol. I	20	ATIBALA	*Abutilon indicum* (Linn.) Sweet	Malvaceae	Root	**Foreign matter**- NMT 2 per cent; **Total ash**- NMT 8 per cent; **Acid-insoluble ash**- NMT 3 per cent; **Alcohol-soluble extractive**-NLT 3 per cent; **Water-soluble extractive**-NLT 9 per cent	Asparagin	*Rasa* - Madhura; *Guna*- Snigdha; *Virya*- Sita; *Vipaka*- Madhura; *Karma*- Grahi, Vatahara, Balya, Vrsya	Bala taila; Narayana taila; Maha narayana taila	Meha, Vatarakta, Raktapitta	3-6 g of the drug in powder form
14	Part-I; Vol. I	22	ATIVISA	*Aconitum heterophyllum* Wall. ex. Royle	Ranunculaceae	Dried Tuberous Roots	**Foreign matter**- NMT 2 per cent; **Total ash**- NMT 4 per cent; **Acid-insoluble ash**- NMT 1 per cent; **Alcohol-soluble extractive**-NLT 6 per cent; **Water-soluble extractive**-NLT 24 per cent	Alkaloids (Atisine, Dihydroatisine, Hetisined and Heteratisine)	*Rasa* - Tikta, Katu; *Guna*- Laghu, Ruksa; *Virya*- Usna; *Vipaka*- Katu; *Karma*- Dipana, Pacana, Samgrahika, Kaphapittahara	Rodhrasava; Siva gutika; Lakasmi-narayana rasa; Mahavisa-garbha taila; Rasnairan-dadi kvatha churna; Sudarsana churna; Pancatikta guggulu ghrta; Bala Chatur-bhadrika churna	Jvara, Kasa, Chardi, Amatisara, Krmiroga	0.6-2.0 g of the drug in powder form.
15	Part-I; Vol. I	24	BABBULA	*Acacia nilotica* (Linn.) Willd. ex. Del. sp. *Indica* (Benth.) Brenan, Syn. *Acacia arabica* Willd.	Leguminosae	Dried Mature Stem Bark	**Foreign matter**- NMT 2 per cent; **Total ash**- NMT 15 per cent; **Acid-insoluble ash**- NMT 2 per cent; **Alcohol-soluble extractive**-NLT 6 per cent; **Water-soluble extractive**-NLT 4 per cent	Tannins and Gum	*Rasa* - Kasaya; *Guna*- Guru, Ruksa, Visada; *Virya*- Sita; *Vipaka*- Katu; *Karma*- Grahi, Kaphahara, Visaghna	Mrtasanjivani sura; Babbularista	Kustha, Krmiroga, Atisara, Kasa	20-30g of the drug for decoction.

Sl. No.	*Part and Volume*	*Page No.*	*Common Ayurvedic Name*	*Botanical Name/ Chemical Name*	*Family/ Chemical Formula*	*Part Used*	*Identity, Purity and Strength/Chemical Properties*	*Constituents*	*Properties and Action*	*Important Formulations*	*Therapeutic Uses*	*Dose*
16	Part-I; Vol. I	25	BAKUCI	*Psoralea corylifolia* Linn.	Leguminosae	Dry Ripe Fruits	**Foreign matter**- NMT 2 per cent; **Total ash**- NMT 8 per cent; **Acid-insoluble ash**- NMT 2 per cent; **Alcohol-soluble extractive**-NLT 13 per cent; **Water-soluble extractive**- NLT 11 per cent	Essential oil, Fixed Oil, Psoralen, Psoralidin, Isopsoralen and bakuchiol.	*Rasa* - Tikta, Katu; *Guna*- Ruksa; *Virya*- Sita; *Vipaka*- Katu; *Karma*- Slesmasrapittanut, Grahi, Vranapaha, Hrdya	Somaraji taila; Avalgujadi lepa	Svitra, Kustha, Krmiroga, Jvara, Meha	3-6 g of the drug in powder form
17	Part-I; Vol. I	26	BIBHITAKA	*Terminalia belerica* Roxb.	Combretaceae	Pericarp of Dried Ripe Fruits.	**Foreign matter**- NMT 2 per cent; **Total ash**- NMT 7 per cent; **Acid-insoluble ash**- NMT 1 per cent; **Alcohol-soluble extractive**-NLT 8 per cent; **Water-soluble extractive**- NLT 35 per cent	Gallic acid, Tannic acid and Glycosides	*Rasa* - Kasaya; *Guna*- Ruksa, Laghu; *Virya*- Usna; *Vipaka*- Madhura; *Karma*- Kaphapittajit, Bhedaka, Krminasana, Caksusya, Kesya, Kasahara	Triphala churna; Triphaladi taila; Lavangadi vati	Svarabheda, Netraroga, Kasa, Chardi, Krmiroga, Vibandha	3-6 g of the drug in powder form
18	Part-I; Vol. I	27	BILVA	*Aegle marmelos* Corr.	Rutaceae	Pulp of Entire, Unripe or Half Ripe Fruit	**Total ash**- NMT 4 per cent; **Acid-insoluble ash**- NMT 1 per cent **Alcohol-soluble extractive**-NLT 6 per cent; **Water-soluble extractive**- NLT 50 per cent	Marmalosin, Tannins, Mucilage, Fatty oil and Sugar	*Rasa* - Katu, Tikta, kasaya; *Guna*- Laghu, Ruksa; *Virya*- Usna; *Vipaka*- Katu; *Karma*- Dipana, Pacana, Grahi, Pittakrt, Vatakaphahara, Balya	Bilvadi leha; Brhatgangadhara churna	Pravahika, Agnimandya, Grahaniroga	3-6g of the drug in powder form
19	Part-I; Vol. I	28	CANDR-ASURA	*Lepidium sativum* Linn.	Cruciferae	Dried Seeds	**Foreign matter**- NMT 2 per cent; **Total ash**- NMT 8 per cent; **Acid-insoluble ash**- NMT 0.5 per cent; **Water-soluble extractive**- NLT 13 per cent	Alkaloids, Essential oil, Fixed oil and Mucilage	*Rasa* - Katu, Tikta; *Guna*- Laghu, Ruksa, Tiksna; *Virya*- Usna; *Vipaka*- Katu; *Karma*- Balapustivivar-dhana, Vataslesmahrt	Kasturyadi (vayuu) gutika	Hikka, Atisara, Vatarakta	3-6 g of the drug in powder form

Sl. No.	*Part and Volume*	*Page No.*	*Common Ayurvedic Name*	*Botanical Name/ Chemical Name*	*Family/ Chemical Formula*	*Part Used*	*Identity, Purity and Strength/Chemical Properties*	*Constituents*	*Properties and Action*	*Important Formulations*	*Therapeutic Uses*	*Dose*
20	Part-I; Vol. I	29	CITRAKA	*Plumbago zeylanica* Linn.	Plumbaginaceae	Dried Mature Roots	**Foreign matter**- NMT 3 per cent; **Total ash**- NMT 3 per cent; **Acid-insoluble ash**- NMT 1 per cent; **Alcohol-soluble extractive**-NLT 12 per cent; **Water-soluble extractive**- NLT 12 per cent	Plumbagin	*Rasa* - Katu *Guna*- Laghu, Ruksa, Tiksna; *Virya*- Usna; *Vipaka*- Katu; *Karma*- Dipana, Pacana, Grahi, Kaphavatahara, Arsohara, Sulahara, Sothahara	Citrakadi vati; Citrakaharitaki; Citrakadi churna	Agnimandya, Grahani roga, Arsa, Udara sula, Gudasotha	1-2 g of the drug in powder form
21	Part-I; Vol. I	30	DHANYAKA	*Coriandrum sativum* Linn.	Umbelliferae	Dried Ripe Fruits	**Foreign matter**- NMT 2 per cent; **Total ash**- NMT 6 per cent; **Acid-insoluble ash**- NMT 1.5 per cent; **Alcohol-soluble extractive**-NLT 10 per cent; **Water-soluble extractive**- NLT 19 per cent; **Volatile oil** - NLT 0.3 per cent v/w	Essential oil (Coriandrol)	*Rasa* - Katu, Madhura, Tikta, Kasaya; *Guna*- Laghu, Snigdha; *Virya*- Usna; *Vipaka*- Madhura; *Karma*- Dipana, Pacana, Grahi, Tridosanut, Mutrala, Caksusya, Hrdya	Dhanyapancakvatha churna	Jvara, Trsna, Chardi, Daha, Ajirna, Atisara	1-3 g of the drug in powder form
22	Part-I; Vol. I	32	DHATAKI	*Woodfordia fruticosa* (Linn.) Kurz.	Lythraceae	Flowers	**Foreign matter**- NMT 2 per cent; **Total ash**- NMT 10 per cent; **Acid-insoluble ash**- NMT 1 per cent; **Alcohol-soluble extractive**-NLT 7 per cent; **Water-soluble extractive**-NLT 28 per cent	Tannin and Glucoside	*Rasa* - Kasaya, Katu; *Guna*- Laghu; *Virya*- Sita; *Vipaka*- Katu; *Karma*- Grahi, Visaghna, Garbhasthapana, Krminut, Sandhaniya	Brhat gangadhara churna	Atisara, Trsna, Visarpa, Vrana, Raktapitta,	3-6 g of the drug in powder form
23	Part-I; Vol. I	34	ERANDA (Root)	*Ricinus communis* Linn.	Euphorbiaceae	Dried, Mature Roots	**Foreign matter**- NMT 2 per cent; **Total ash**- NMT 8 per cent; **Acid-insoluble ash**- NMT 1 per cent; **Alcohol-soluble extractive**-NLT 3 per cent; **Water-soluble extractive**-NLT 9 per cent	Alkaloid (Ricinine)	*Rasa* - Madhura; *Guna*- Guru, Snigdha; *Virya*- Usna; *Vipaka*- Madhura; *Karma*- Vrsya, Vatahara, Amapacana	Gandharva-hastadi kvatha churna; Vatari guggulu; Gandhar-vahasta taila	Amavata, Sotha, Vastisula, Katisula, Udararoga, Jvara	20-30 g of the drug for decoction

Sl. No.	Part and Volume	Page No.	Common Ayurvedic Name	Botanical Name/ Chemical Name	Family/ Chemical Formula	Part Used	Identity, Purity and Strength/Chemical Properties	Constituents	Properties and Action	Important Formulations	Therapeutic Uses	Dose
24	Part-I; Vol. I	36	GAMBHARI	*Gmelina arborea* Roxb.	Verbenaceae	Dried Mature Root and Root Bark	**Foreign matter**- NMT 2 per cent; **Total ash**- NMT 5 per cent; **Acid-insoluble ash**- NMT 0.3 per cent; **Alcohol-soluble extractive**-NLT 7 per cent; **Water-soluble extractive**-NLT 20 per cent	Alkaloids and Lignans (Arboreal, Isoarboreal and Related lignans).	*Rasa* - Tikta, Kasaya; *Guna*-Guru; *Virya*- Usna; *Vipaka*- Katu; *Karma*- Dipana, Pacana, Bhedana, Medhya, Tridosajit, Sothahara, Visaghna, Jvarahara	Dasamularista; Dasamula-haritaki; Dasamula ghrta; Dasamula satpalaka ghrta	Jvara, Trsna, Daha, Arsa, Sotha	20-30g of the drug for decoction.
25	Part-I; Vol. I	38	GOKSURA (Root)	*Tribulus terrestris* Linn.	Zygophyllaceae	Root	**Foreign matter**- NMT 2 per cent; **Total ash**- NMT 13 per cent; **Acid-insoluble ash**- NMT 3 per cent; **Alcohol-soluble extractive**-NLT 4 per cent; **Water-soluble extractive**-NLT 10 per cent	Alkaloids and Saponins	*Rasa* - Madhura; *Guna*- Guru, Snigdha; *Virya*-Sita; *Vipaka*-Madhura; *Karma*-Vatanut, Vrsya, Brmhana, Mutrala	Sahacaradi taila; Dasamula kvatha churna; Dasamula-katutraya kvatha churna; Dasamula-pancakoladi kvatha churna	Kasa, Svasa, Sularoga, Hrdroga, Vataroga, Mutrakrcchra, Asmari	20-30 g of the drug for decoction
26	Part-I; Vol. I	40	GOKSURA (Fruit)	*Tribulus terrestris* Linn.	Zygophyllaceae	Dried, Ripe, Entire Fruit	**Foreign matter**- NMT 2 per cent; **Total ash**- NMT 15 per cent; **Acid-insoluble ash**- NMT 2 per cent; **Alcohol-soluble extractive**-NLT 6 per cent; **Water-soluble extractive**-NLT 10 per cent	Potassium nitrate, Sterols, Sapogenin with Pyroketone ring (diosgenin), Gitogenin and Hecogenins	*Rasa* - Madhura; *Guna*- Guru, Snigdha; *Virya*-Sita; *Vipaka*-Madhura; *Karma*-Vatanut, Vrsya, Brmhana, Asmarihara, Vastisodhana	Goksuradi guggulu; Traikantaka ghrta; Draksadi cruna	Kasa, Svasa, Asmari, Mutrakrcchra, Prameha, Arsa, Sularoga, Hrdroga, Daurbalya	3-6g the drug in powder form, 20-30g of the drug for decoction
27	Part-I; Vol. I	41	GUDUCI	*Tinospora cordifolia* (Willd.) Miers.	Menisper-maceae	Dried, Matured Pieces of Stem	**For dried drug: - Foreign matter**- NMT 2 per cent; **Total ash**- NMT 16 per cent; **Acid-insoluble ash**-NMT 3 per cent; **Alcohol-soluble extractive**-NLT 3 per cent; **Water-soluble extractive**- NLT 11 per cent; **For fresh drug: - Foregin matter** - Nil; **Moisture content**- 75 per cent	Terpenoids and Alkaloids	*Rasa* - Tikta, Kasaya; *Guna*-Laghu; *Virya*-Usna; *Vipaka*-Madhura; *Karma*-Tridosasamaka, Samgrahi, Balya, Dipana, Rasayana, Raktasodhaka, Jvaraghna	Amrtarista; Amrtottara kvatha churna; Guduci taila; Guducyadi churna; Guduci sattva; Chinnodbhavadi kvatha churna	Kustha, Vatarakta, Jvara, Kamala, Pandu, Prameha	3-6g of the drug in powder form, 20-30g of the drug for decoction

Sl. No.	*Part and Volume*	*Page No.*	*Common Ayurvedic Name*	*Botanical Name/ Chemical Name*	*Family/ Chemical Formula*	*Part Used*	*Identity, Purity and Strength/Chemical Properties*	*Constituents*	*Properties and Action*	*Important Formulations*	*Therapeutic Uses*	*Dose*
28	Part-I; Vol. I	43	GUGGULU	*Commiphora wightii* (Arn.) Bhand; Syn. *Balsamo-dendron mukul* Hook. ex Stocks (*Commiphora mukul* Engl.) (Fam. Burseraceae)	Burseraceae	Exudate	**Foreign matter**- NMT 4 per cent; **Total ash**- NMT 5 per cent; **Acid-insoluble ash**- NMT 1 per cent; **Alcohol-soluble extractive**-NLT 27 per cent; **Water-soluble extractive**- NLT 53 per cent; **Volatile oil** - NLT 1 per cent, v/w	Essential oil, Gum, Resin, Steroids	*Rasa* - Tikta, Katu, Kasaya; *Guna*- Laghu, Sara, Visada; *Virya*- Usna; *Vipaka*- Katu; *Karma*- Vatabalasajit, Rasayana, Varnya, Balya, Bhagnasan-dhanakrt, Medohara	Yogaraja gugulu; Vatari guggulu; Simhanada guggulu; Kaisora guggulu; Mahayogaraja guggulu; Candraprabha vati	Vatavyadhi, Amavata, Granthi, sopha, Gandamala, Medoroga, Prameha, Kustha	2-4g of the drug
29	Part-I; Vol. I	44	GUNJA (Seed)	*Abrus precatorius* Linn.	Leguminosae	Seeds	**Foreign matter**- NMT 2 per cent; **Total ash**- NMT 3 per cent; **Acid-insoluble ash**- NMT 0.5 per cent; **Alcohol-soluble extractive**-NLT 3 per cent; **Water-soluble extractive**-NLT 15 per cent	An Albuminous substance (Abrine and Abralin)	*Rasa* - Tikta, Kasaya; *Guna*- Ruksa, Laghu, Tiksna; *Virya*- Usna; *Vipaka*- Katu; *Karma*- Vatapittajvarapaha, Kesya, Kandughna, Vranapaha, Garbhanirodhaka	Mrtasanjivani gutika; Gunjabhadra rasa	Kustha, Vrana, Vatavyadhi, Indralupta	60-180 mg of the drug in powder form
30	Part-I; Vol. I	45	HARIDRA	*Curcuma longa* Linn.	Zingiberaceae	Dried and Cured Rhizomes	**Foreign matter**- NMT 2 per cent; **Total ash**- NMT 9 per cent; **Acid-insoluble ash**- NMT 1 per cent; **Alcohol-soluble extractive**-NLT 8 per cent; **Water-soluble extractive**-NLT 12 per cent; **Volatile oil** - NLT 4 per cent v/w	Essential oil and a Colouring matter (Curcumin).	*Rasa* - Tikta, Katu; *Guna*- Ruksa; *Virya*- Usna; *Vipaka*- Katu; *Karma*- Kaphapittanut, Visaghna, Varnya, Kusthaghna, Krmighna, Pramehanasaka	Haridra Khanda	Visavikara, Kustha, Vrana, Tvagroga, Prameha, Pandu, Sitapitta, Pinasa	1-3g of the drug in powder form

Sl. No.	*Part and Volume*	*Page No.*	*Common Ayurvedic Name*	*Botanical Name/ Chemical Name*	*Family/ Chemical Formula*	*Part Used*	*Identity, Purity and Strength/Chemical Properties*	*Constituents*	*Properties and Action*	*Important Formulations*	*Therapeutic Uses*	*Dose*
31	Part-I; Vol. I	47	HARITAKI	*Terminalia chebula* Retz.	Combretaceae	Pericarp of Mature Fruits	**Foreign matter**- NMT 1 per cent; **Total ash**- NMT 5 per cent; **Acid-insoluble ash**- NMT 5 per cent; **Alcohol-soluble extractive**-NLT 40 per cent; **Water-soluble extractive**- NLT 60 per cent	Tannins, Anthraquinones and Polyphenolic compounds	*Rasa* - Kasaya, Katu, Tikta, Amla, Madhura; *Guna*- Laghu, Ruksa; *Virya*- Usna; *Vipaka*- Madhura; *Karma*- Sarvadosa-prasamana, Rasayana, Caksusya, Dipana, Anulomana, Hrdya, Medhya	Abhayarista; Agastya haritaki; Rasayana; Citraka haritaki; Danti haritaki; Dasamula haritaki; Brahma rasayana; triphala churna; Triphaladi taila; Abhaya lavana; Pathyadi lepa	Vibandha, Aruci, Udavarta, Gulma, Udararoga, Arsa, Pandu, Sotha, Jirnajvara, Visamajvara, Prameha, Siroroga, Kasa, Tamaka Svasa, Hrdroga	3-6 g of the drug in powder form
32	Part-I; Vol. I	49	HINGU	*Ferula foetida* Regel., *Ferula narthex* Boiss,	Umbelliferae	Oleo-gum-resin Obtained from Rhizomes and Roots	**Foreign matter**- NMT 2 per cent; **Total ash**- NMT 15 per cent; **Acid-insoluble ash**- NMT 3 per cent; **Alcohol(90 per cent) soluble extractive**-NLT 50 per cent; **Water-soluble extractive**- NLT 50 per cent	Essential oil, Gum and Resin	*Rasa* - Katu; *Guna*-Tiksna; *Virya*- Usna; *Vipaka*- Katu; *Karma*- Rucya, Dipana, Pacana, Anulomana, Krmighna, Vatakapha-prasamana, Hrdya	Hingvastaka churna; Hingvadi churna; Hinguvacadi churna	Agnimandya, Adhmana, Anaha, Gulma, Sularoga, Udararoga, Hrdroga, Krmiroga	125-500 mg of the drug
33	Part-I; Vol. I	51	JATAMAMSI	*Nardostachys jatamansi* DC.	Valerianaceae	Rhizomes	**Foreign matter**- NMT 5 per cent; **Total ash**- NMT 9 per cent; **Acid-insoluble ash**- NMT 5 per cent; **Alcohol-soluble extractive**-NLT 2 per cent; **Water-soluble extractive**-NLT 5 per cent; **Volatile oil** - NLT 0.1 per cent, v/w	Essential oil and Resinous matter	*Rasa* - Tikta, Kasaya; *Guna*-Laghu; *Virya*- Sita; *Vipaka*- Katu; *Karma*-Tridosanut, Medhya, Varnya, Nidrajanana, Kusthaghna	Jatamamsyarka	Kustha, Daha, Visarpa, Manasaroga, Anidra	2-3 g of the drug in powder form, 5-10 g of the drug for decoction

Sl. No.	Part and Volume	Page No.	Common Ayurvedic Name	Botanical Name/ Chemical Name	Family/ Chemical Formula	Part Used	Identity, Purity and Strength/Chemical Properties	Constituents	Properties and Action	Important Formulations	Therapeutic Uses	Dose
34	Part-I; Vol. I	53	JATIPHALA	*Myristica fragrans* Houtt.	Myristicaceae	Endosperm of Dried Seeds (kernels of fruits)	**Foreign matter**- NMT 1 per cent; **Total ash**- NMT 3 per cent; **Acid-insoluble ash**- NMT 0.5 per cent; **Alcohol-soluble extractive**-NLT 11 per cent; **Water-soluble extractive**- NLT 7 per cent; **Ether-soluble extractive** - NLT 25 per cent v/w; **Volatile oil**- NLT 5 per cent v/w	Essential oil and Fixed oil.	*Rasa* - Tikta, Katu; *Guna*- Laghu, Tiksna; *Virya*- Usna; *Vipaka*- Katu; *Karma*- Dipana, Grahi, Mukhakledanasaka, Mukhadaurgandhyanasaka, Kaphavatapaha, Vrsya	Jatiphaladi churna	Atisara, Grahani, Chardi, Mukharoga, Pinasa, Kasa, Svasa, Sukrameha	0.5-1.0g of the drug in powder form
35	Part-I; Vol. I	55	KAMPILLAK	*Mallotus philippinensis* Muell. Arg.	Euphorbiaceae	Hairs of Fruit	**Foreign matter**- NMT 2 per cent; **Total ash**- NMT 6 per cent; **Acid-insoluble ash**- NMT 4 per cent; **Alcohol-soluble extractive**-NLT 50 per cent; **Water-soluble extractive**- NLT 1.0 per cent	Resinous colouring matter (Rottlerin).	*Rasa* - Katu; *Guna*- Laghu, Ruksa, Tiksna; *Virya*- Usna; *Vipaka*- Katu; *Karma*- Virecana, Vranapaha, Krmighna	Dhanvantara ghrta; Misraka sneha	Vibandha, Krmiroga, Adhmana, Gulma, Vrana	0.5-1.0 g of the drug in powder form
36	Part-I; Vol. I	56	KANCANARA	*Bauhinia variegata* Blume	Leguminosae	Dried Stem Bark	**Foreign matter**- NMT 2 per cent; **Total ash**- NMT 11 per cent; **Acid-insoluble ash**- NMT 0.2 per cent; **Alcohol-soluble extractive**-NLT 2 per cent; **Water-soluble extractive**- NLT 6 per cent	Tannins	*Rasa* - Kasaya; *Guna*- Laghu, Ruksa; *Virya*- Sita; *Vipaka*- Katu; *Karma* -Tridosahara, Grahi, Dipana, Gandavrddhihara	Kancanara-guggulu	Krmiroga, Gandamala, Apaci, Gudabhramsa, Vrana	20-30 g of the drug for decoction
37	Part-I; Vol. I	58	KANKOLA	*Piper cubeba* Linn. f.	Piperaceae	Mature Dried Fruits	**Foreign matter**- NMT 2 per cent; **Total ash**- NMT 8 per cent; **Acid-insoluble ash**- NMT 1 per cent; **Alcohol-soluble extractive**-NLT 14 per cent; **Water-soluble extractive**- NLT 11 per cent	Essential oil (cubebin)	*Rasa* - Katu, Tikta; *Guna*- Laghu, Tiksna; *Virya*- Usna; *Vipaka*- Katu; *Karma*- Dipana, Pacana, Rucya, Kaphavatahara, Mukhadaurgandhyahara, Vastisodhana	Dasamularista; Kumaryasava	Aruci, Mukharoga, Mutrakrcchra, Sula	1-2 g of the drug in powder form

Sl. No.	*Part and Volume*	*Page No.*	*Common Ayurvedic Name*	*Botanical Name/ Chemical Name*	*Family/ Chemical Formula*	*Part Used*	*Identity, Purity and Strength/Chemical Properties*	*Constituents*	*Properties and Action*	*Important Formulations*	*Therapeutic Uses*	*Dose*
38	Part-I; Vol. I	59	KANTAKARI	*Solanum surattense* Burm. f., Syn. *Solanum xanthocarpum* Schrad. and Wendl.	Solanaceae	Mature Dried Whole Plant	**Foreign matter- NMT 2 per cent; Total ash- NMT 9 per cent; Acid-insoluble ash- NMT 3 per cent; Alcohol-soluble extractive-NLT 6 per cent; Water-soluble extractive-NLT 16 per cent**	Glucoalkaloids and Sterols.	*Rasa* - Katu, Tikta; *Guna*- Laghu, Ruksa; *Virya*- Usna; *Vipaka*- Katu; *Karma*- Dipana, Pacana, Amadosanasaka, Kanthya, Sothahara	Kantakar-yavaleha; Pancatiktaka ghrta; Vyaghriharitaki	Svasa, Kasa, Jvara, Aruci, Pinasa, Parsvasula, Svarabheda	20-30 g of the drug for decoction
39	Part-I; Vol. I	62	KANYASARA	*Aloe barbadensis* Mill. Syn. *Aloe vera* Tourn. ex. Linn; *Aloe indica* Royle.	Liliaceae	Dried Juice of Leaves	**Foreign matter**- NMT 2 per cent; **Total ash**- NMT 5 per cent; **Acid-insoluble ash**- NMT 2 per cent; **Alcohol-soluble extractive**-NLT 80 per cent; **Water-soluble extractive**- NLT 60 per cent; **Moisture content** - NMT 10 per cent of its weight when dried to constant weight at 105°,	Anthraquinone, Glycoside	*Rasa* -Tikta; *Guna*- Ruksa; *Virya*- Usna; *Vipaka*- Katu; *Karma*- Bhedi, Pittanirharana, Rajahpravartaka, Jvaranut	Rajahpravartini vati; Cukkum-tippalyadi gutika	Udararoga, Kastartava, Jvara, Yakrdvikara	125-500mg of the drug in powder form
40	Part-I; Vol. I	63	KARANJA	*Pongamia pinnata* (Linn.) Merr, Syn. *Pongamia glabra* Vent.	Leguminosae	Seeds	**Foreign matter**- NMT 1 per cent; **Total ash**- NMT 3 per cent; **Acid-insoluble ash**- NMT 0.1 per cent; **Alcohol-soluble extractive**-NLT 23 per cent; **Water-soluble extractive**- NLT 13 per cent	Fixed oil, Flavones and Traces of Essential oil.	*Rasa* - Katu, Tikta; *Guna*- Tiksna; *Virya*- Usna; *Vipaka*- Katu; *Karma*- Kaphavataghna, Krmijit, Kusthaghna, Vranasodhana	Aragvadhadi Kvath churna; Pathyadilepa	Vrana, Krmi, Kustha	0.25-g of the drug in powder form; 5-10 g of the drug for decoction

Sl. No.	*Part and Volume*	*Page No.*	*Common Ayurvedic Name*	*Botanical Name/ Chemical Name*	*Family/ Chemical Formula*	*Part Used*	*Identity, Purity and Strength/Chemical Properties*	*Constituents*	*Properties and Action*	*Important Formulations*	*Therapeutic Uses*	*Dose*
41	Part-I; Vol. I	64	KARVIRA	*Nerium indicum* Mill. Syn. *Nerium odorum* Soland	Apocynaceae	Dried Leaves	**Foreign matter**- NMT 2 per cent; **Total ash**- NMT 9 per cent; **Acid-insoluble ash**- NMT 1 per cent; **Alcohol-soluble extractive**-NLT 20 per cent; **Water-soluble extractive**- NLT 20 per cent	Cardiac Glucoside (Oleandrin)	*Rasa* - Katu, Tikta, Kasaya; *Guna*- Tiksna, Laghu, Ruksa; *Virya*- Usna; *Vipaka*- Katu; Prabhava- Hrdya; *Karma*- Jvarapaha, Caksusys, Kusthaghna, Kandughana, Krmighna, Vranapaha, Svasahara	Kasisadi taila	Jvara, Vrana, Kustha, Kandu, Krmiroga, Netraroga, Tamakasvasa, Hrdroga,	30-125 mg of the drug in powder form
42	Part-I; Vol. I	66	KARKATA-SRNGI	*Pistacia chinensia* Burgo, *Pistacia integerrima* Stew. ex Brandis, *Rhus succedanea* Linn.	Anacardiaceas	Gall-like excres-cences formed by insects on the Leaves, Petioles and Branches	**Foreign matter**- NMT 2 per cent; **Total ash**- NMT 7 per cent; **Acid-insoluble ash**- NMT 0.2 per cent; **Alcohol-soluble extractive**-NLT 30 per cent; **Water-soluble extractive**- NLT 30 per cent	Essential oil, Tannins and Resinous matters.	*Rasa* - Kasaya, Tikta; *Guna*- Guru; *Virya*- Usna; *Vipaka*- Katu; *Karma*- Kaphavatahara, Kasahara, Urdhvavatajit, Hikkanigrahana	Balacatur-bhadrika churna	Jvara, Svasa, Kasa, Hikka, Ksaya, Aruci, Chardi	3-6 g of the drug in powder form
43	Part-I; Vol. I	67	KARPASA	*Gossypium herbaceum* Linn.	Malvaceae	Seeds	**Foreign matter**- NMT 2 per cent; **Total ash**- NMT 5 per cent; **Acid-insoluble ash**- NMT 0.1 per cent; **Alcohol-soluble extractive**-NLT 14 per cent; **Water-soluble extractive**- NLT 8 per cent	Fixed oil, Resin and Sterols	*Rasa* - Madhura; *Guna*- Snigdha, Guru; *Virya*- Sita; *Vipaka*- Madhura; *Karma*- Stanyajanana, Vrsya, Kaphakara, Hrdya	Karpasasthvadi taila	Daha, Srama, Bhranti, Murccha, Stanyaksya	3-6g of the drug in powder form

Sl. No.	*Part and Volume*	*Page No.*	*Common Ayurvedic Name*	*Botanical Name/ Chemical Name*	*Family/ Chemical Formula*	*Part Used*	*Identity, Purity and Strength/Chemical Properties*	*Constituents*	*Properties and Action*	*Important Formulations*	*Therapeutic Uses*	*Dose*
44	Part-I; Vol. I	68	KASERU	*Scirpus kysoor* Roxb.	Cyperaceae	Rhizome	**Foreign matter**- NMT 2 per cent; **Total ash**- NMT 8 per cent; **Acid-insoluble ash**- NMT 3 per cent; **Alcohol-soluble extractive**-NLT 4 per cent; **Water-soluble extractive**-NLT 9 per cent	Starch, Saponins, Sugars and Progesterone	*Rasa* - Madhura, Kasaya; *Guna*- Guru; *Virya*- Sita; *Vipaka*- Madhura; *Karma*- Pittaghna, Dahaghna, Sukrakara, Stanyakara, Caksusya, Grahi, Rucikara	Saubhagyasunthi	Daha, Netraroga, Aruci, Atisara, Sukraksya, Stanyaksaya, Daurbalya	5-10 g of the drug in powder form
45	Part-I; Vol. I	69	KETAKI	*Pandanus tectorius* Soland. ex-Parkinson	Pandanaceae	Dried Roots	**Foreign matter**- NMT 2 per cent; **Total ash**- NMT 11 per cent; **Acid-insoluble ash**- NMT 2 per cent; **Alcohol-soluble extractive**-NLT 9 per cent; **Water-soluble extractive**-NLT 16 per cent	Essential Oil	*Rasa* - Tikta, Madhura, Katu; *Guna*- Laghu; *Virya*- Usna; *Vipaka*- Katu; *Karma*- Varnya, Kesya, Daurgandh-yanasana, Balya, Rasayana, Dardhyakara, Saukhyakara, Kaphapaha, Caksusya	Triphaladi taila	Gulma, Kapharoga, Netraroga	20-30 g of the drug for decoction
46	Part-I; Vol. I	70	KHADIRA	*Acacia catechu* (Linn.f.) Willd.	Mimosaceae	Dried Pieces of Heart-Wood	**Foreign matter**- NMT 2 per cent; **Total ash**- NMT 2 per cent; **Acid-insoluble ash**- NMT 0.2 per cent; **Alcohol-soluble extractive**-NLT 1 per cent; **Water-soluble extractive**-NLT 3 per cent	Catechin, Catechu-Tannic acid and Tannin	*Rasa* - Tikta, Kasaya; *Guna*- Laghu, Ruksa; *Virya*- Sita; *Vipaka*- Katu; *Karma*- Kaphapittahara, Raktasodhaka, Kusthaghna, Medohara, Krmighna, Dantya	Khadirarista; Arimedadi taila; Khadiradi gutika	Kustha, Vrana, Sotha, Prameha	20-30g of the drug for decoction

Sl. No.	*Part and Volume*	*Page No.*	*Common Ayurvedic Name*	*Botanical Name/ Chemical Name*	*Family/ Chemical Formula*	*Part Used*	*Identity, Purity and Strength/Chemical Properties*	*Constituents*	*Properties and Action*	*Important Formulations*	*Therapeutic Uses*	*Dose*
47	Part-I; Vol. I	71	KIRATATIKTA	*Swertia chirata* Buch. Ham.	Gentianaceae	Whole Plant	**Foreign matter**- NMT 2 per cent; **Total ash**- NMT 6 per cent; **Acid-insoluble ash**- NMT 1 per cent; **Alcohol (60 per cent) soluble extractive**- NLT 10 per cent; **Water-soluble extractive**- NLT 10 per cent; **Absence of tannin** - on additon of Ferric Chloride to aqueous or alcoholic extract no lue black colour develops **Assay**- Refer API	Xanthones, Xanthone Glycoside and Mangiferine (flavonoid).	*Rasa* - Tikta; *Guna*- Laghu, Ruksa; *Virya*- Sita; *Vipaka*- Katu; *Karma*- Jvaraghna, Vranasodhana, Saraka, Trsnapaha, Raktasodhaka, Kaphapittahara	Sudarsana churna; Chinnodbhavadi kvatha churna	Jvara, Trsna, Daha, Sotha, Kustha, Vrana, Krmiroga, Kandu, Meha	1-3g of the drug in powder form; 20-30 g of the drug for decoction
48	Part-I; Vol. I	73	KRSNAJIRAKA	*Carum carvi* Linn.	Umbelliferae	Dried Ripe Fruits	**Foreign matter**- NMT 2 per cent; **Total ash**- NMT 9 per cent; **Acid-insoluble ash**- NMT 1.5 per cent; **Alcohol-soluble extractive**-NLT 2 per cent; **Water-soluble extractive**- NLT 12 per cent; **Volatile oil** - NLT 3.5 per cent, v/w,	Essential oils (Carvone and Carvacrol)	*Rasa* - Katu; *Guna*- Laghu; *Virya*- Usna; *Vipaka*- Katu; *Karma*- Pacana, Dipana, Samgrahi, Jvaraghna, Rucya, Caksusya, Sothahara	Jirakadyarista; Jirakadi Modaka	Agnimandya, Adhmana, Jirnajvara, Grahaniroga, Krmiroga	1-3 g of the drug in powder form
49	Part-I; Vol. I	75	KULATTHA	*Vigna unquiculata* (Linn.) Walp. Syn. *Dolichos biflorus* Linn.	Leguminosae	Dry Seeds	**Foreign matter**- Nil **Total ash**- NMT 5 per cent; **Acid-insoluble ash**- NMT 1 per cent; **Alcohol-soluble extractive**-NLT 3 per cent; **Water-soluble extractive**- NLT 12 per cent	An Enzyme (Urease) and Oil	*Rasa* - Kasaya; *Guna*- Laghu, Sara; *Virya*- Usna; *Vipaka*- Katu; *Karma*- Vidahi, Svedasamgrahaka, Krmihara, Kaphavatahara	Saptasara kvatha churna; Dhanvantara taila	Asmari, Nastartava	12 g of the drug in powder form for decoction
50	Part-I; Vol. I	76	KUSTHA	*Saussurea lappa* C.B. Clarke	Compositae	Dried Roots	**Foreign matter**- NMT 2 per cent; **Total ash**- NMT 4 per cent; **Acid-insoluble ash**- NMT 1 per cent; **Alcohol-soluble extractive**-NLT 12 per cent; **Water-soluble extractive**- NLT 20 per cent	Essential oil, Alkaloid (Saussurine) and Bitter Resin	*Rasa* - Katu, Tikta; *Guna*- Laghu; *Virya*- Usna; *Vipaka*- Katu; *Karma*- Kaphavatajit, Sukrala, Raktasodhaka, Varnya	Kottamacukkadi taila	Vatarakta, Visarpa, Kushta, Kasa, Svasa	0.2-1.0 g of the drug in powder form

Sl. No.	*Part and Volume*	*Page No.*	*Common Ayurvedic Name*	*Botanical Name/ Chemical Name*	*Family/ Chemical Formula*	*Part Used*	*Identity, Purity and Strength/Chemical Properties*	*Constituents*	*Properties and Action*	*Important Formulations*	*Therapeutic Uses*	*Dose*
51	Part-I; Vol. I	78	KUTAJA	*Holarrhena antidysenterica* (Roth) A. DC.	Apocynaceae	Dried Stem Bark	**Foreign matter**- NMT 2 per cent; **Total ash**- NMT 7 per cent; **Acid-insoluble ash**- NMT 1 per cent; **Alcohol (60 per cent) soluble extractive**-NLT 18 per cent; **Water-soluble extractive**- NLT 10 per cent	Conessine and Related Alkaloids	*Rasa* - Tikta, Kasaya; *Guna*- Laghu, Ruksa; *Virya*- Sita; *Vipaka*- Katu; *Karma*- Dipana, Sangrahi, Kaphapittasamaka	Kutajarista; Kutajavaleha; Kutajaghana vati	Pravahika, Atisara, Jvaratisara, Arsa, Kustha, Trsna	20- 30 g of the drug for decoction
52	Part-I; Vol. I	80	LAVANGA	*Syzygium aromaticum* (Linn.) Merr. and L.M. Perry Syn. *Eugenia aromatica* Kuntze, *Eugenia caryophyllata* Thunb.	Myrtaceae	Dried Flower Bud	**Foreign matter**- NMT 2 per cent; **Total ash**- NMT 7 per cent; **Acid-insoluble ash**- NMT 1 per cent; **Alcohol-soluble extractive**-NLT 3 per cent; **Water-soluble extractive**- NLT 9 per cent; **Volatile oil** - NLT 15 per cent	Essential Oils (Eugenalacetate and Caryophyllene)	*Rasa* - Tikta, Katu; *Guna*- Laghu, Tiksna; *Virya*- Sita; *Vipaka*- Katu; *Karma*- Dipana, Pacana, Rucya, Kaphapittasamaka, Sulahara, Kasahara	Lavangadi vati; Lavangadi churna	Kasa, Svasa, Hikka, Ksaya, Adhmana, Trsna, Chardi, Amlapitta	0.5 - 2.0 g of the drug in powder form
53	Part-I; Vol. I	82	LODHRA	*Symplocos racemosa* Roxb.	Symplocaceae	Dried Stem Bark	**Foreign matter**- Nil **Total ash**- NMT 12 per cent; **Acid-insoluble ash**- NMT 1 per cent; **Alcohol-soluble extractive**-NLT 9 per cent; **Water-soluble extractive**- NLT 15 per cent	Alkaloids (Loturine and Colloturine) and Red colouring matter	*Rasa* - Kasaya; *Guna*- Laghu; *Virya*- Sita; *Vipaka*- Katu; *Karma*- Kaphapittanut, Grahi, Caksusya	Rodhrasava (Lodhrasava); Pusyanuga churna; Brhat gangadhara-churna	Raktapitta, Atisara, Sotha, Pradara, Netraroga	3-5 g of the drug in powder form; 20-30 g of the drug for decoction
54	Part-I; Vol. I	84	MADANA	*Xeromphis spinosa* (Thunb.) Keay, Syn. *Randia dumetorum* Lam.	Rubiaceae	Dried Fruit	**Foreign matter**- NMT 2 per cent; **Total ash**- NMT 6 per cent; **Acid-insoluble ash**- NMT 0.25 per cent; **Alcohol-soluble extractive**-NLT 19 per cent; **Water-soluble extractive**- NLT 16 per cent	Essential Oil, Saponin, Tannin and Resin.	*Rasa* - Madhura, Tikta; *Guna*- Laghu, Ruksa; *Virya*- Usna; *Vipaka*- Katu; *Karma*- Vamana, Lekhana	Pippalyadi taila	Gulma, Vidradhi, Kustha, Slesma jvara, Pratisyaya	0.5 - 1.0 g of the drug in powder form for decoction; 3 - 6 g of the drug for induction of vomitting.

Sl. No.	*Part and Volume*	*Page No.*	*Common Ayurvedic Name*	*Botanical Name/ Chemical Name*	*Family/ Chemical Formula*	*Part Used*	*Identity, Purity and Strength/Chemical Properties*	*Constituents*	*Properties and Action*	*Important Formulations*	*Therapeutic Uses*	*Dose*
55	Part-I; Vol. I	86	MISREYA	*Foeniculum vulgare* Mill.	Umbelliferae	Dried Ripe Fruits	**Foreign matter**- NMT 2 per cent; **Total ash**- NMT 12 per cent; **Acid-insoluble ash**- NMT 15 per cent; **Alcohol-soluble extractive**-NLT 4 per cent; **Water-soluble extractive**-NLT 1 per cent; **volatile oil** - NLT 1.4 per cent v/w	Essential oil and Fixed oil	*Rasa* -Madhura, Katu, Tikta; *Guna*- Laghu, Ruksa; *Virya*- Sita; *Vipaka*- Madhura; *Karma*- Dipana, Vatapittahara, Balya, Anulomana, Amadosahara	Misreyarka; Pancasakara churna	Agnimandya, Sula, Kasa, Raktadosa, Pravahika, Arsa	3 - 6 g of the drug in powder
56	Part-I; Vol. I	88	NYAGRODHA	*Ficus bengalensis* Linn.	Moraceae	Dried Mature Stem Bark	**Foreign matter**- NMT 2 per cent; **Total ash**- NMT 8 per cent; **Acid-insoluble ash**- NMT 3 per cent; **Alcohol-soluble extractive**-NLT 6 per cent; **Water-soluble extractive**-NLT 8 per cent	Tannins, Glycosides and Flavonoids	*Rasa* - Kasaya; *Guna*- Guru, Ruksa; *Virya*- Sita; *Vipaka*- Katu; *Karma*- Kaphapittajit, Vranapaha, Varnya, Stambhana, Mutrasan-grahaniya, Dahaghna Yonidosahrt	Nyagrodhadi kvatha churna; Nyagrodhadi churna	Daha, Trsna, Raktapitta, Vrana, Visarpa, Yonidosa, Prameha	3 - 6g of the drug in powder form
57	Part-I; Vol. I	90	PASANA-BHEDA	*Bergenia ciliata* (Haw.) Sternb., Syn. *Bergenia ligulata* (Wall.) Engl.	Saxifragaceae	Rhizomes	**Foreign matter**- NMT 2 per cent; **Total ash**- NMT 13 per cent; **Acid-insoluble ash**- NMT 0.5 per cent; **Alcohol-soluble extractive**-NLT 9 per cent; **Water-soluble extractive**-NLT 15 per cent	Tannic acid, Gallic acid and Glucose	*Rasa* - Tikta, Kasaya; *Guna*- Laghu; *Virya*- Sita; *Vipaka*- Katu; *Karma*- Asmarighna, Bhedana, Vastisodhana, Mutravirecaniya	Asmarihara kasaya churna; Mutravirecaniya Kasaya churna	Meha, Mutrakicchra, Asmari	3 - 6 g of the drug in powder form; 20-30 g of the drug for decoction

Sl. No.	*Part and Volume*	*Page No.*	*Common Ayurvedic Name*	*Botanical Name/ Chemical Name*	*Family/ Chemical Formula*	*Part Used*	*Identity, Purity and Strength/Chemical Properties*	*Constituents*	*Properties and Action*	*Important Formulations*	*Therapeutic Uses*	*Dose*
58	Part-I; Vol. I	92	PATHA	*Cissampelos pareira* Linn.	Menis-permaceae	Roots	**Foreign matter**- NMT 2 per cent; **Total ash**- NMT 7 per cent; **Acid-insoluble ash**- NMT 1 per cent; **Alcohol-soluble extractive**-NLT 11 per cent; Water-soluble extractive- NLT 13 per cent	Alkaloids, Saponin and Quartenary ammonium bases, Flavonol and Sterol.	*Rasa* - Tikta, Katu; *Guna*- Laghu, Tiksna; *Virya*- Usna; *Vipaka*- Katu; *Karma*- Tridosasamana, Raktasodhaka, Visaghna, Bhagnasan-dhanakrt, Grahi, Stanyasodhana	Pusyanuga churna; Pradarantaka lauha; Sarasvata ghrta; Brhat gangadhara churna; Stanyasodhana kasaya churna	Sularoga, Atisara, Kustha, Kandu, Jvara, Chardi, Stanyadusti	3 - 6 g of the drug in powder form;
59	Part-I; Vol. I	94	PUGA	*Areca catechu* Linn.	Palmae	Dried Ripe Seed	**Foreign matter**- NMT 1 per cent; **Total ash**- NMT 3 per cent; **Acid-insoluble ash**- NMT 0.4 per cent; **Alcohol-soluble extractive**-NLT 19 per cent; **Water-soluble extractive**- NLT 10 per cent	Alkaloid (Arecoline) Tannins and Fats.	*Rasa* - Kasaya; *Guna*- Ruksa, Guru; *Virya*- Sita; *Vipaka*- Katu; Prabhava-Mohakrt *Karma*- Dipana, Kaphapittajit, Kledanasana, Malabhedi, Mukhasodhana, Vikasi	Pugakhanda	Mukhavikara, Aruci, Yonisaithilya, Svetapradara	1-2 g of the drug in powder form
60	Part-I; Vol. I	95	PUNARNAVA (RAKTA)	*Boerhaavia diffusa* Linn.	Nyctaginaceae	Dried Mature Whole Plant	**Foreign matter**- NMT 2 per cent; **Total ash**- NMT 15 per cent; **Acid-insoluble ash**- NMT 6 per cent; **Alcohol-soluble extractive**-NLT 1 per cent; **Water-soluble extractive**-NLT 4 per cent	Alkaloid (Punarnavine).	*Rasa* - Madhura, Tikta, Kasaya; *Guna*- Ruksa; *Virya*- Usna; *Vipaka*- Madhura; *Karma* - Vataslesmahara, Mutrala, Sothahara, Anulomana	Punarnavastaka kvatha churna; Punarnavasava; Punarnavadi mandura; Sukumara ghrta; Sothaghna lepa.	Pandu, Sotha	20-30 g of the drug for decoction

Sl. No.	*Part and Volume*	*Page No.*	*Common Ayurvedic Name*	*Botanical Name/ Chemical Name*	*Family/ Chemical Formula*	*Part Used*	*Identity, Purity and Strength/Chemical Properties*	*Constituents*	*Properties and Action*	*Important Formulations*	*Therapeutic Uses*	*Dose*
61	Part-I; Vol. I	97	SAPTAPARNA	*Alstonia scholaris* (Linn.) R. Br.	Apocynaceae	Stem Bark	**Foreign matter**- NMT 2 per cent; **Total ash**- NMT 11 per cent; **Acid-insoluble ash**- NMT 3 per cent; **Alcohol-soluble extractive**-NLT 4 per cent; **Water-soluble extractive**-NLT 12 per cent	Alkaloids (Echitamine, Ditamine and Echitamidine).	*Rasa* - Tikta, Kasaya *Guna*- Sara, Snigdha *Virya*- Usna *Vipaka*- Katu; *Karma*- Tridosaghna, Dipana, Anulomana, Raktasodhaka, Kusthaghna, Jvaraghna	Aragvadhadi kvatha churna; Amrtarista; Vajraka taila	Sula, Gulma, Krmiroga, Kustha, Jvara, Sandrameha	20-30g of the drug for decoction
62	Part-I; Vol. I	99	SATI	*Hedychium spicatum* Ham. ex smith	Zingiberaceae	Sliced Dried Rhizomes	**Foreign matter**- NMT 1 per cent; **Total ash**- NMT 8 per cent; **Acid-insoluble ash**- NMT 2 per cent; **Alcohol-soluble extractive**-NLT 4 per cent; **Water-soluble extractive**-NLT 8 per cent	Essential oil	*Rasa* - Katu, Tikta, Kasaya; *Guna*- Laghu, Tiksna; *Virya*- Usna; *Vipaka*- Katu; *Karma*- Kaphavataghna, Mukhasodhana, Grahi, Sulahara	Agastyaharitaki rasayana; Satyadi churna	Kasa, Svasa, Mukharoga, Sula, Chardi, Kandu	1-3 g of the drug in powder form
63	Part-I; Vol. I	100	SNUHI	*Euphorbia neriifolia* Linn.	Euphorbiaceae	Stem	**Foreign matter**- NMT 2 per cent; **Total ash**- NMT 8 per cent; **Acid-insoluble ash**- NMT 1 per cent; **Alcohol-soluble extractive**-NLT 5 per cent; **Water-soluble extractive**-NLT 15 per cent	Resin, Gum and Triterpenes	*Rasa* - Katu, Tikta; *Guna*- Guru, Tiksna; *Virya*- Usna; *Vipaka*- Katu; *Karma*- Tiksnavirecana, Bhedana, Amakaphavatahara	Citrakadi taila; Abhaya lavana; Avittoladi bhasma; Vajraksara	Gulma, Udararoga, Meha, Kustha, Sotha	125-250 mg of the drug in powder form
64	Part-I; Vol. I	101	SUKSMAILA	*Elettaria cardamomum* (Linn.)	Zingiberaceae	Seed of Dried Fruits	**Foreign matter**- Nil **Total ash**- NMT 6 per cent; **Acid-insoluble ash**-NMT 4 per cent; **Alcohol-soluble extractive**-NLT 2 per cent; **Water-soluble extractive**- NLT 10 per cent; **Volatile oil** - NLT 4 per cent, v/w	Essential oil	*Rasa* - Katu, Madhura; *Guna*- Laghu; *Virya*- Sita; *Vipaka*- Madhura; *Karma*- Rocana, Dipana, Anulomana, Hrdya, Mutrala	Eladi modaka; Eladi churna; Sitopaladi churna	Kasa, Svasa, Aruci, Chardi, Mutrakrcchra	250-500mg of the drug in powder form

Sl. No.	*Part and Volume*	*Page No.*	*Common Ayurvedic Name*	*Botanical Name/ Chemical Name*	*Family/ Chemical Formula*	*Part Used*	*Identity, Purity and Strength/Chemical Properties*	*Constituents*	*Properties and Action*	*Important Formulations*	*Therapeutic Uses*	*Dose*
65	Part-I; Vol. I	103	SUNTHI	*Zingiber officinale* Roxb.	Zingiberaceae	Dried Rhizome	**Foreign matter**- NMT 1 per cent; **Total ash**- NMT 6 per cent; **Water-soluble ash**- NMT 1.5 per cent; **Alcohol (90 per cent) soluble extractive**-NLT 3 per cent; **Water-soluble extractive**- NLT 10 per cent	Essential oil, Pungent constituents (Gingerol and Shogaol), Resinous matter and Starch.	*Rasa* - Katu; *Guna*- Laghu, Snigdha; *Virya*- Usna; *Vipaka*- Madhura; *Karma*- Dipana, Pacana, Anulomana, Amadosahara, Vatakaphapaha, Hrdya	Saubhag-yasunthi; Trikatu Churna; Saubhagya vati; Vaisvanara churna.	Agnimandya, Adhmana, Pandu, Svasa, Udararoga, Amavata	1-2 g of the drug in powder form
66	Part-I; Vol. I	105	SVARNAPATRI	*Cassia angustifolia* Vahl.	Leguminosae	Dried Leaves	**Foreign matter**- NMT 1 per cent; **Total ash**- NMT 14 per cent; **Acid-insoluble ash**- NMT 2 per cent; **Alcohol-soluble extractive**-NLT 3 per cent; **Water-soluble extractive**- NLT 25 per cent	Anthraquinone, Glucoside, Flavonoids, Steroids, and Resin	*Rasa* - Katu, Tikta, Kasaya; *Guna*- Laghu, Ruksa, Tiksna; *Virya*- Usna; *Vipaka*- Katu; *Karma*- Recana	Pancasakara churna; Sarivadyasava	Vibandha, Udararoga	0.5-2g of the drug in powder form
67	Part-I; Vol. I	106	SVETAJIRAKA	*Cuminum cyminum*, Linn.	Umbelliferae	Ripe Fruits	**Foreign matter**- NMT 2 per cent; **Total ash**- NMT 8 per cent; **Acid-insoluble ash**- NMT 1 per cent; **Alcohol-soluble extractive**-NLT 7 per cent; **Water-soluble extractive**- NLT 15 per cent	Essential oil	*Rasa* - Katu; *Guna*- Laghu, Ruksa, Tiksna; *Virya*- Usna; *Vipaka*- Katu; *Karma*- Rucya, Dipana, Pacana, Grahi, Krmighna, Kaphavatahara	Jirakadyarista; Jirakadimodaka; Hingvadi churna; Hinguvacadi churna	Agnimandya, Atisara, Krmiroga	1-3g of the drug in powder form
68	Part-I; Vol. I	107	SVETA SARIVA	*Hemidesmus indicus* (Linn.) R. Br.	Asclepiadaceae	Root	**Foreign matter**- NMT 2 per cent; **Total ash**- NMT 4 per cent; **Acid-insoluble ash**- NMT 0.5 per cent; **Alcohol-soluble extractive**-NLT 15 per cent; **Water-soluble extractive**- NLT 13 per cent	Essential oil, Saponin, Resin, Tannins, Sterols and Glucosides	*Rasa* - Madhura; *Guna*- Snigdha, Guru; *Virya*- Sita; *Vipaka*- Madhura; *Karma*- Tridosanasana, Dipana, Raktasodhaka, Amanasana, Visaghna, Jvarahara	Sarivadyasava	Aruci, Agnimandya, Atisara, Kasa, Svasa, Kandu, Kustha, Jvara, Raktavikara	20-30 g of the drug for decoction

Sl. No.	*Part and Volume*	*Page No.*	*Common Ayurvedic Name*	*Botanical Name/ Chemical Name*	*Family/ Chemical Formula*	*Part Used*	*Identity, Purity and Strength/Chemical Properties*	*Constituents*	*Properties and Action*	*Important Formulations*	*Therapeutic Uses*	*Dose*
69	Part-I; Vol. I	109	TAGARA	*Valeriana wallichii* Dc.	Valerianaceae	Dried Rhizome	**Foreign matter**- NMT 2 per cent; **Total ash**- NMT 12 per cent; **Acid-insoluble ash**- NMT 10 per cent; **Alcohol (60 per cent) soluble extractive**- NMT30 per cent; **Water-soluble extractive**- NLT 19 per cent	Essential oil	*Rasa* - Tikta, Katu, Kasaya; *Guna*- Laghu, Snigdha; *Virya*- Usna; *Vipaka*- Katu; *Karma*- Tridosahara, Visaghna, Raktadosahara, Manasadosahara	Dhanvantara tila; Mahanarayana taila; Devadarva-dyarista; Jatiphaladi churna	Apasmara, Unmada, Siroroga, Netraroga	1-3g of the drug in powder form
70	Part-I; Vol. I	111	TAMALAKI	*Phyllanthus fraternus* Webst. Syn. *Phyllanthus niruri* Hook. f. non Linn.	Euphorbiaceae	Root, Stem and Leaf	**Foreign matter**- NMT 2 per cent; **Total ash**- NMT 16 per cent; **Acid-insoluble ash**- NMT 7 per cent; **Alcohol-soluble extractive**-NLT 3 per cent; **Water-soluble extractive**- NLT 13 per cent	Phyllanthin	*Rasa* - Kasaya, Tikta, Madhura; *Guna*- Laghu, Ruksa; *Virya*- Sita; *Vipaka*- Madhura; *Karma*- Rocana, Dahanasani, Pittasamaka, Mutrala	Citraka haritaki; Madhuyastyddi taila; Pippalyddi ghrta; Chyawanaprasa; Satavariguda	Trsa, Kasa, Amlapitta, Pandu, Ksaya, Ksata, Kustha, Prameha, Mutraroga	10-20ml. of the drug in juice form; 3-6 of the drug in powder form
71	Part-I; Vol. I	113	TVAK	*Cinnamomum zeylanicum* Blume.	Lauraceae	Dried Inner Bark	**Foreign matter**- NMT 2 per cent; **Total ash**- NMT 3 per cent; **Acid-insoluble ash**- NMT 2 per cent; **Alcohol-soluble extractive**-NLT 2 per cent; **Water-soluble extractive**- NLT 3 per cent; **Volatile oil** - NLT 1 per cent, v/w	Essential oil, Tannin and Mucilage	*Rasa* - Katu, Tikta, Madhura; *Guna*- Ruksa, Laghu, Tiksna; *Virya*- Usna; *Vipaka*- Katu; *Karma*- Kaphavatahara, Visaghna, Kanthasuddhikara, Rucya	Sitopaladi churna; Caturjata churna	Mukhasosa, Trsa, Kantha-mukharoga, Pinasa, Krmiroga, Vastiroga, Arsa, Hrdroga	1-3 g of the drug in powder form
72	Part-I; Vol. I	115	TVAKPATRA	*Cinnamomum tamala* (Buch. Ham.) Ness and Eberm.	Lauraceae	Dried Mature Leaves	**Foreign matter**- NMT 2 per cent; **Total ash**- NMT 5 per cent; **Acid-insoluble ash**- NMT 1 per cent; **Alcohol-soluble extractive**-NLT 6 per cent; **Water-soluble extractive**- NLT 9 per cent; **Volatile oil** - NLT 1 per cent v/w	Essential oils (d-α phellandrene and eugenol)	*Rasa* - Katu, Madhura; *Guna*- Laghu, Picchila, Tiksna; *Virya*- Usna; *Vipaka*- Katu; *Karma*- Rucya, Kaphavatahara, Arsoghna	Citrakadi taila, Kasisadi taila, Vajraka taila	Aruci, Hrllasa, Arsa, Pinasa	1-3g of the drug in powder form

Sl. No.	*Part and Volume*	*Page No.*	*Common Ayurvedic Name*	*Botanical Name/ Chemical Name*	*Family/ Chemical Formula*	*Part Used*	*Identity, Purity and Strength/Chemical Properties*	*Constituents*	*Properties and Action*	*Important Formulations*	*Therapeutic Uses*	*Dose*
73	Part-I; Vol. I	117	UDUMBARA	*Ficus racemosa* Linn. Syn. *Ficus glomerata* Roxb.	Moraceae	Dried Bark	**Foreign matter**- NMT 2 per cent; **Total ash**- NMT 14 per cent; **Acid-insoluble ash**- NMT 1 per cent; **Alcohol-soluble extractive**-NLT 7 per cent; **Water-soluble extractive**-NLT 9 per cent	Tannins	Rasa- Kasaya; *Guna*- Ruksa, Guru; *Virya*- Sita; *Vipaka*- Katu; *Karma*- Mutrasa-mgrahaniya, Vranasodhaka, Vranaropaka, Medohara, Kaphapittasamaka, Raktastambhana	Nyagrodhadi Kvatha churna, Mutrasam-grahaniya kasaya churna	Raktapitta, Daha, Medoroga, Yonidosa	3-6 g of the drug in powder form; 20-30 g of the drug for decoction
74	Part-I; Vol. I	119	UPAKUNCIKA	*Nigella sativa* Linn.	Ranunculaceae	Seeds	**Foreign matter**- NMT 2 per cent; **Total ash**- NMT 6 per cent; **Acid-insoluble ash**- NMT 0.2 per cent; **Alcohol-soluble extractive**-NLT 20 per cent; **Water-soluble extractive**- NLT 15 per cent	Essential oil, Fixed oil, Resin, Saponin and Tannin	*Rasa* -Katu, Tikta; *Guna*- Laghu, Ruksa; *Virya*- Usna; *Vipaka*- Katu; *Karma*- Rucya, Samgrahi, Caksusya, Garbhasaya-visodhana, Pittala, Dipana, Pacana, Medhya, Hrdya, Vatakaphapaha, Krmighna	Narayana churna, Kankayana gutika	Gulma, Adhmana, Atisara, Krmiroga	1-3 g of the drug in powder form
75	Part-I; Vol. I	120	VARUNA	*Crataeva nurvala* Buch-Ham.	Capparidaceae	Dried Stem Bark	**Foreign matter**- NMT 2 per cent; **Total ash**- NMT 13 per cent; **Acid-insoluble ash**- NMT 1 per cent; **Alcohol-soluble extractive**-NLT 1 per cent; **Water-soluble extractive**-NLT 8 per cent	Saponin and Tannin	Rasa- Tikta, Kasaya; *Guna*- Laghu, Ruksa; *Virya*- Usna; *Vipaka*- Katu; *Karma*- Dipana, Bhedi, Vataslesmahara	Varunadi kvatha churna	Asmari, Mutrakrcchra, Gulma, Vidradhi	20-30 g of the drug for decoction
76	Part-I; Vol. I	122	VASA	*Adhatoda vasica* Ness	Acanthaceae	Fresh Dried Mature Leaves	**Foreign matter**- NMT 2 per cent; **Total ash**- NMT 21 per cent; **Acid-insoluble ash**- NMT 1 per cent; **Alcohol-soluble extractive**-NLT 3 per cent; **Water-soluble extractive**-NLT 22 per cent	Alkaloids and Essential oil	*Rasa* - Tikta, Kasaya; *Guna*- Laghu; *Virya*- Sita; *Vipaka*- Katu; *Karma*- Kaphapittahara, Raktasamgrahika, Kasaghna, Hrdya	Vasakasava, Vasavaleha	Kasa, Svasa, Ksaya, Raktapitta, Prameha, Kamala, Kustha	10-20 ml of juice of fresh leaves; 10-20 g of the dried drug for decoction

Sl. No.	*Part and Volume*	*Page No.*	*Common Ayurvedic Name*	*Botanical Name/ Chemical Name*	*Family/ Chemical Formula*	*Part Used*	*Identity, Purity and Strength/Chemical Properties*	*Constituents*	*Properties and Action*	*Important Formulations*	*Therapeutic Uses*	*Dose*
77	Part-I; Vol. I	123	VIDANGA	*Embelia ribes* Burm. f.	Myrsinaceae	Dried Mature Fruits	**Foreign matter**- NMT 2 per cent; **Total ash**- NMT 6 per cent; **Acid-insoluble ash**- NMT 1.5 per cent; **Alcohol-soluble extractive**-NLT 10 per cent; **Water-soluble extractive**- NLT 9 per cent	Benzoquinones, Alkaloid (Christembine), Tannin and Essential oil.	*Rasa* - Katu, Tikta; *Guna*- Ruksa, Laghu, Tiksna; *Virya*- Usna; *Vipaka*- Katu; *Karma*- Krminasana, Dipana, Anulomana, Vatakaphapaha	Vidangarista, Vidanga lauha, Vidangadi lauha	Krmiroga, Adhmana, Sula, Udararoga	5-10 g of the drug in powder form
78	Part-I; Vol. I	125	VIJAYA	*Cannabis sativa* Linn.	Cannabinaceae	Dried Leaves	**Foreign matter**- NMT 2 per cent; **Total ash**- NMT 15 per cent; **Acid-insoluble ash**- NMT 5 per cent; **Alcohol (90 per cent)soluble extractive**- NLT 10 per cent; **Water-soluble extractive**- NLT 13 per cent	Resin (Cannabinols, Particularly tetrahy-drocannabinol)	*Rasa* - Tikta; *Guna*- Laghu, Tiksna; *Virya*- Usna; *Vipaka*- Katu; *Karma*- Dipana, Pacana, Grahi, Kaphahara, Vajikara, Vakvardhana, Nidrajanana, Vyavayi	Jatiphaladi churna, Madanananda modaka	Agnimandya, Atisara, Grahaniroga, Klaibya, Anidra	125-250mg of the drug in powder form
79	Part-I; Vol. I	127	YASTI	*Glycyrrhiza glabra* Linn.	Leguminosae	Dried, Unpeeled, Stolon and Root	**Foreign matter**- NMT 10 per cent; **Acid-insoluble ash**- NMT 2.5 per cent; **Alcohol-soluble extractive**-NLT 10 per cent; **Water-soluble extractive**- NLT 20 per cent	Glycyrrhizin, Glycyrrhizic acid, Glycyrrhetinic acid, Asparagine, Surgars, Resin and Starch.	*Rasa* - Madhura; *Guna*- Guru, Snigdha; *Virya*- Sita; *Vipaka*- Madhura; *Karma*- Vatapittajit, Raktaprasadana, Balya, Varnya, Vrsya, Caksusya	Eladi gutika, Yastimadhuka taila, Madhuyastyadi taila	Kasa, Svarabheda, Ksaya, Vrana, Vatarakta	2-4 g of the drug in powder form
80	Part-I; Vol. I	129	YAVANI	*Trachys-permum ammi* (Linn.) Sprague ex Turril Syn. *Carum copticum* Benth and Hook. f. *Ptychotis ajwan* DC.	Umbellifelrae	Dried Fruit	**Foreign matter**- NMT 5 per cent; **Total ash**- NMT 9 per cent; **Acid-insoluble ash**- NMT 0.2 per cent; **Alcohol-soluble extractive**-NLT 2 per cent; **Water-soluble extractive**- NLT 13 per cent; **Volatile Oil** - NLT 2.5 per cent	Essential oil and Fixed oil	*Rasa* - Katu, Tikta; *Guna*- Ruksa, laghu, Tiksna; *Virya*- Usna; *Vipaka*- Katu; *Karma*- Dipana, Pacana, Rucya, Anulomana, Sulahara, Krmighna	Yavani Sadava	Adhmana, Anaha, Udararoga, Gulma, Krmiroga, Sula,	3-6 g of the drug in powder form

Sl. No.	Part and Volume	Page No.	Common Ayurvedic Name	Botanical Name/ Chemical Name	Family/ Chemical Formula	Part Used	Identity, Purity and Strength/Chemical Properties	Constituents	Properties and Action	Important Formulations	Therapeutic Uses	Dose
81	Part-I; Vol. II	1-2	AKARA-KARABHA (Root)	*Anacyclus pyrethrum* DC	Asteraceae	Dried Roots	**Foreign matter**- NMT2 per cent; **Total ash**- NMT 10 per cent; **Acid-insoluble ash**- NMT 2 per cent; **Alcohol-soluble extractive**-NLT 8 per cent; **Water-soluble extractive**- NLT 22 per cent;	Volatile oil and Alkaloid (Pyrethrin).	*Rasa* - Katu; *Guna*- Ruksa, Tiksna; *Virya*- Usna; *Vipaka*- Katu; *Karma*- Vatahara, Pittahara, Kaphahara, Sukrala, Vajikara, Svedakara, Dipana, Buddhivardhaka, Balakaraka	Kumaryasava, Kasturyadi (Vayu) Gutika, Nagavallabha Rasa	Pratisyaya, Sotha, Ajirna, Kasa, Svasa, Grdhrasi, Paksaghata, Udararoga, Nastartava, Sularoga, Dantasula	0.5-1 g of the drug in powder form
82	Part-I; Vol. II	3-4	AKSODA (Cotyledon)	*Juglans regia* Linn.	Juglandaceae	Dried Cotyle-ndon	**Foreign matter**- NMT 5 per cent; **Total ash**- NMT 2 per cent; **Acid-insoluble ash**- NMT 0.5 per cent; **Alcohol-soluble extractive**-NLT 10.0 per cent; **Water-soluble extractive**- NLT 7.0 per cent	Walnut oil and Tannin	*Rasa* - Madhura; *Guna*- Guru, Snigdha, Sara; *Virya*- Usna; *Vipaka*- Madhura; *Karma*- Vatahara, Kaphakara, Brnhana, Sukral, Balya, Vrsya, Vistambhi, Hrdya	Amrtaprasa Ghrta	Ksata, Ksaya, Vataroga	10-25 g
83	Part-I; Vol. II	5-6	AMRATAK (Stem Bark)	*Spondias pinnata* Linn. f. Kurz. Syn. *S. mangifera* Willd.; *S. acuminata* Roxb. Non Gamble	Anacardiaceae	Dried Stem Bark	**Foreign matter**- NMT 1 per cent; **Total ash**- NMT 13 per cent; **Acid-insoluble ash**- NMT 0.5 per cent; **Alcohol-soluble extractive**-NLT 3 per cent; **Water-soluble extractive**- NLT 7 per cent; **T.L.C.** - Refer API	Tannin and Starch	*Rasa* - Kasaya, Amla; *Guna*- Guru, Sara; *Virya*- Usna; *Vipaka*- ...; *Karma*- Vatahara, Pittakara, Kaphakara, Rucikrt, Kanthya, Amadosahara Hrdya	Dadhika Ghrta	Raktapitta, Ksaya, Data, Ksata	5-10 g of the drug in powder form for decoction

Sl. No.	Part and Volume	Page No.	Common Ayurvedic Name	Botanical Name/ Chemical Name	Family/ Chemical Formula	Part Used	Identity, Purity and Strength/Chemical Properties	Constituents	Properties and Action	Important Formulations	Therapeutic Uses	Dose
84	Part-I; Vol. II	7-9	APAMARGA (Whole Plant)	*Achyranthes aspera* Linn.	Amaranthaceae	Dried Whole Plant	**Foreign matter**- NMT 2 per cent; **Total ash**- NMT 17 per cent; **Acid-insoluble ash**- NMT 5 per cent; **Alcohol-soluble extractive**-NLT 2 per cent; **Water-soluble extractive**-NLT 12 per cent	Saponins	*Rasa* - Katu, Tikta; *Guna*- Tiksna, Sara; *Virya*- Usna; *Vipaka*- Katu; *Karma*- Kaphahara, Vatahara, Medohara, Chedana, Dipana, Pacana, Vamaka, Sirovirecana	Apamargaksara, Apamargaksara Taila, Abhya Lavana, Gudapippali, Jyotismati Taila	Sula, Udara roga, Apaci, Arsa, Kandu, Medoroga	20-50 g of the drug for decoction
85	Part-I; Vol. II	10-11	APARAJITA (Root)	*Clitoria ternatea* Linn.	Fabaceae	Dried Root	**Foreign matter**- NMT 2 per cent; **Total ash**- NMT 5 per cent; **Acid-insoluble ash**- NMT 2 per cent; **Alcohol-soluble extractive**-NLT 5 per cent; **Water-soluble extractive**-NLT8 per cent; **T.L.C.** - Refer API	Tannin, Starch, Resin, Taraxerol and Taraxerone	*Rasa* - Tikta, Kasaya, Katu; *Guna*-. *Virya*- Sita *Vipaka*- Katu; *Karma*- Vatahara, Pittahara, Kaphahara, Kanthya, Medhya, Caksusya, Visahara, Buddhiprada	Misraka Sneha, Vataraktantaka Rasa	Mutraroga, Kustha, Sotha, Vrana, Sula	1-3 g of the drug in powder form
86	Part-I; Vol. II	12-14	ARDRAKA (Rhizome)	*Zingiber officinale* Rose.	Zingiberaceae	Fresh Rhizome	**Foreign matter**- NMT 0.5 per cent; **Total ash**- NMT 8 per cent; **Acid-insoluble ash**- NMT 1 per cent; **Alcohol-soluble extractive**-NLT 5 per cent; **Water-soluble extractive**-NLT 2 per cent; **Moisture content** - NMT 90 per cent; **T.L.C.** - Refer API	Volatile oil containing Cineole zingiberol, and sesquiterpene like Zingiberene, Bisobolene and Sesqui phellandrene, Gingerosl in the Oleo-Resin	*Rasa* - Katu; *Guna*- Tiksna, Ruksa, Guru; *Virya*- Usna; *Vipaka*- Madhura; *Karma*- Vatahara, Kaphahara, Rocana, Dipana, Bhedana, Svarya, Hrdya, Vrsya,	Ardraka Khandavaleha, Saraswatarista	Vibandha, Anaha, Sula, Sopha, Kantharoga	2-3 ml. of the drug in juice form with honey

Sl. No.	Part and Volume	Page No.	Common Ayurvedic Name	Botanical Name/ Chemical Name	Family/ Chemical Formula	Part Used	Identity, Purity and Strength/Chemical Properties	Constituents	Properties and Action	Important Formulations	Therapeutic Uses	Dose
87	Part-I; Vol. II	15-16	ARIMEDA (Stem Bark)	*Acacia leucophloea* Willd.	Fabaceae	Dried Stem Bark	**Foreign matter**- NMT 2 per cent; **Total ash**- NMT 11 per cent; **Acid-insoluble ash**- NMT 1 per cent; **Alcohol-soluble extractive**-NLT 14 per cent; **Water-soluble extractive**- NLT 13 per cent; **T.L.C.** - Refer API	n-hexacosanol, β-amyrin, β-sitosterol and Tannin.	*Rasa* - Kasaya, Tikta; *Guna*- Usna; *Virya*- Usna; *Vipaka*- Katu; *Karma*- Kaphasosaka, Medasosaka, Visanasana	Khadiradi Gutika (Mukharoga), Arimedadi Taila (for external use *i.e.* Kavalagraha and Nasya)	Kustha, Meha, Mukharoga, Kandu, Visajavrana, Sopha, Atisara, Visarpa, Pandu, Dantaroga, Kasa, Krmi, Udardapra samana	40 g for decoction; 3-5 g in powder form
88	Part-I; Vol. II	17-18	ARJUNA (Stem Bark)	*Terminalia arjuna* W. and A.	Combretaceae	Stem Bark	**Foreign matter**- NMT 2 per cent; **Total ash**- NMT 25 per cent; **Acid-insoluble ash**- NMT 1 per cent; **Alcohol-soluble extractive**-NLT 20 per cent; **Water-soluble extractive**- NLT 20 per cent	Tannins	*Rasa* - Kasaya; *Guna*- Ruksa; *Virya*- Sita; *Vipaka*- Katu; *Karma*- Kaphahara, Pittahara, Hrdya, Vrananasana, Bhagnasan-dhanakara, Vyanga hara	Parthadyarista, Nagarjunabhra Rasa, Arjuna Ghrta	Hrdroga, Ksataksaya, Medoroga, Prameha, Vrana, Trsa, Vyanga.	3-6 g of the drug in powder form
89	Part-I; Vol. II	19-20	BHALLATAKA (Fruit)	*Semecarpus anacardium* Linn.	Anacardiaceae	Mature Fruit	**Foreign matter**- NMT 1 per cent; **Total ash**- NMT 4 per cent; **Acid-insoluble ash**- NMT 0.5 per cent; **Alcohol-soluble extractive**-NLT 11 per cent; **Water-soluble extractive**- NLT 5 per cent	A Tarry Oil containing Anacardic Acid, Non-Volatile Alcohol (Cardol).	*Rasa* - Madhura, Katu, Tikta, Kasaya; *Guna*- Laghu, Tiksna, Snigdha; *Virya*- Usna; *Vipaka*- Madhura; *Karma*- Vatahara, Kaphahara, Dipana, Pacana, Chedi, Bhedi, Medhya	Bhallataka Rasayana, Bhallatakadi Modaka, Amrta Bhallataka Leha, Sanjivani Vati	Anaha, Grahani, Gulma, Arsa, Krimi, Kustha	1.2 g of the drug in ksirapaka form

Sl. No.	*Part and Volume*	*Page No.*	*Common Ayurvedic Name*	*Botanical Name/ Chemical Name*	*Family/ Chemical Formula*	*Part Used*	*Identity, Purity and Strength/Chemical Properties*	*Constituents*	*Properties and Action*	*Important Formulations*	*Therapeutic Uses*	*Dose*
90	Part-I; Vol. II	21-24	BHRNGARAJA (Whole Plant)	*Eclipta alba* Hassk.	Asteraceae	Whole Plant	**Foreign matter**- NMT 2 per cent; **Total ash**- NMT 22 per cent; **Acid-insoluble ash**- NMT 11 per cent; **Alcohol-soluble extractive**-NLT 5 per cent; **Water-soluble extractive**-NLT 15 per cent	Alkaloids, Ecliptine and Nicotine	*Rasa* - Katu, Tikta; *Guna*- Ruksa, Tiksna; *Virya*- Usna; *Vipaka*- Katu; *Karma*-Vatahara, Kaphahara, Amahara, Balya, Rasayana, Kesya, Tvacya, Dantaya, Caksusya, Visahara	Bhrngamalakadi Taila, Bhrngaraja Taila, Nili Bhrngadi Taila, Bhrngarajasava, Tekaraja marica.	Yakrdroga, Krmiroga, Sotha, Pandu, Svasa, Kasa, Sirah sula, Hrdroga	3-6 ml of the drug in juice form; 12-36 g of the drug in powder form for decoction
91	Part-I; Vol. II	25-26	BRAHMI (Whole Plant)	*Bacopa monnieri* (Linn.) Wettst., Syn. *Herpestis monnieria* (Linn.) H.B. and K.	Scrophu-lariaceae	Dried Whole Plant	**Foreign matter**- NMT 2 per cent; **Total ash**- NMT 18 per cent; **Acid-insoluble ash**- NMT 6 per cent; **Alcohol-soluble extractive**-NLT 6 per cent; **Water-soluble extractive**-NLT 15 per cent	Alkaloids	*Rasa* - Tikta, Kasaya, Madhura; *Guna*- Laghu, Sara; *Virya*- Sita; *Vipaka*- Madhura; *Karma*; *Karma*- Vatahara, Kaphahara, Rasayana, Ayusya, Medhya, Matiprada, Swarya, Prajasthapana, Visahara, Mohahara	Saraswatarista, Brahmi Ghrta, Ratnagiri Rasa, Brahmi Vati, Saraswata Churna, Smrtisagara Rasa.	Kustha, Jwara, Sopha, Pandu, Prameha, Manasavikara	1-3 g in powder form
92	Part-I; Vol. II	27-28	BRHATI (Root)	*Solanum indicum* Linn.	Solanaceae	Dried Root	**Foreign matter**- NMT 2 per cent; **Total ash**- NMT 6.5 per cent; **Acid-insoluble ash**- NMT 1 per cent; **Alcohol-soluble extractive**-NLT 3 per cent; **Water-soluble extractive**-NLT 4 per cent	Steroidal Alkaloids and Steroids.	*Rasa* - Katu, Tikta; *Guna*- Laghu; *Virya*- Usna; *Vipaka*- Katu; *Karma*- Vatahara, Kaphahara, Dipana, Pacana, Hrdya, Grahi	Dasamula Ghrta, Dasamularista	Hrdroga, Jvara, Svasa, Sula, Agnimandya	10-20 g of the drug for decoction

Sl. No.	Part and Volume	Page No.	Common Ayurvedic Name	Botanical Name/ Chemical Name	Family/ Chemical Formula	Part Used	Identity, Purity and Strength/Chemical Properties	Constituents	Properties and Action	Important Formulations	Therapeutic Uses	Dose
93	Part-I; Vol. II	29-30	CAVYA (Stem)	*Piper retrofractum* Vahl. Syn. *P. chaba* Hunter non blume., *P. officinarum* DC.	Piperaceae	Dried Stem	**Foreign matter**- NMT 2 per cent; **Total ash**- NMT 10 per cent; **Acid-insoluble ash**- NMT 1.5 per cent; **Alcohol-soluble extractive**-NLT 3 per cent; **Water-soluble extractive**-NLT 6 per cent	Alkaloids, Glycosides and Steroids.	*Rasa* - Katu; *Guna*- Laghu, Ruksa, Tiksna; *Virya*- Usna; *Vipaka*- Katu; *Karma*- Vatahara, Kaphahara, Dipana, Pacana, Recana, Bhedana	Pranada Gutika, Candramrta Rasa	Arsa, Krimi, Pliha Roga, Gulma, Anaha, Udara Roga, Sula	1-2 g of the drug in powder form
94	Part-I; Vol. II	31-32	DADIMA (Seed)	*Punica granatum* Linn.	Punicaceae	Dried Seed	**Foreign matter**- NMT 2 per cent; **Total ash**- NMT 4 per cent; **Acid-insoluble ash**- NMT 0.5 per cent; **Alcohol-soluble extractive**-NLT 20 per cent; **Water-soluble extractive**- NLT 35 per cent; **T.L.C.** - Refer API	Madhura Rasa - Madhura -madhura (Kasaya-nurasa); Guna- Madhura - Laghu, Snigdha; Madhura Amla- Laghu; Virya- Usna; Vipaka- Madhura; Karma- Vatahara, Pittahara, Kaphahara, Tarpana, Sukrala, hrdya, Kanthya, Mukhagan-dhahara, Grahi, medhya, Balya Madhura Amla Rasa- Madhura, Amla; Guna- Laghu; Virya- -; Vipaka - -; Karma- Vatahara, Pittahara, Kaphahara,	Dadimastaka Churna, Dadima Ghrta, Dadhika Ghrta, Bhaskara Lavana, Sukra Matrka Vati	Trsna, Daha, Jwara	5-10 g of the drug in powder form	Dadi-mastaka Churna, Dadima Ghrta, Dadhika Ghrta, Bhaskara Lavana, Sukra Matrka Vati

Sl. No.	*Part and Volume*	*Page No.*	*Common Ayurvedic Name*	*Botanical Name/ Chemical Name*	*Family/ Chemical Formula*	*Part Used*	*Identity, Purity and Strength/Chemical Properties*	*Constituents*	*Properties and Action*	*Important Formulations*	*Therapeutic Uses*	*Dose*
								Tarpana, Sukrala, hrdya, Kanthya, Mukhagan-dhahara, Grahi, medhya, Balya				
95	Part-I; Vol. II	33-34	DARU-HARIDRA (Stem)	*Berberis aristata* DC.	Berberidaceae	Dried Stem	**Foreign matter**- NMT 2 per cent; **Total ash**- NMT 14 per cent; **Acid-insoluble ash**- NMT 5 per cent; **Alcohol-soluble extractive**-NLT 6 per cent; **Water-soluble extractive**-NLT 8 per cent	Alkaloids	*Rasa* - Tikta; *Guna*- Ruksa; *Virya*- Usna; *Vipaka*- ....; *Karma*- Stanya Sodhana, Stanya Dosahara, Dosa Pacana.	Asvagandha-dyarista, Bhrngaraja Taila, Khadiradi Gutika, Khadirarista, Jatyadi Taila, Triphala Ghrta	Amatisara, Medoroga, Urustambha, Kapharoga, Karnaroga, Mukharoga, Netraroga, Kandu, Vrana, Meha	5-10 ml of the drug in kvatha form
96	Part-I; Vol. II	35-37	DRONAPUSPI (Whole Plant)	*Leucas cephalotes* Spreng.	Lamiaceae	Dried Whole Plant	**Foreign matter**- NMT 2 per cent; **Total ash**- NMT 17 per cent; **Acid-insoluble ash**- NMT 6 per cent; **Alcohol-soluble extractive**-NLT 5 per cent; **Water-soluble extractive**-NLT 14 per cent	Alkaloid, Glycoside, β-sitosterol and Flavonoid.	*Rasa* - Madhura, Lavana, Katu; *Guna*- Guru, Ruksa, Tiksna; *Virya*- Usna; *Vipaka*- Madhura; *Karma*- Vatakara, Pittakara, Kaphahara, Bhedani, Rucya	Plihari Vatika, Gorocanadi Vati	Kamala, Sotha, Tamaka svasa, Kasa, Agnimandya, Visamajvara	1-3 g of the drug in powder form; 5-10 ml. of the drug in juice form
97	Part-I; Vol. II	38-39	ERVARU (Seed)	*Cucumis melo* var. *utilissimus* duthie and Fuller Syn. *C. utilissimus* Roxb.	Cucurbitaceae	Seeds	**Foreign matter**- NMT 1 per cent; **Total ash**- NMT 4 per cent; **Acid-insoluble ash**- NMT 0.5 per cent; **Alcohol-soluble extractive**-NLT 10 per cent; **Water-soluble extractive**- NLT 5 per cent; **T.L.C.** - Refer API	Oil and Sugars	*Rasa* - Madhura, Tikta; *Guna*- Guru, Ruksa; *Virya*- Sita; *Vipaka*- Madhura; *Karma*- Vatakara, Kaphakara, Pittahara, Rucya, Dipana, Bhedi, Raktadosakara, Grahi	Dadhika Ghrta	Asmari, Mutrakrcchra, Gulma, Raktapitta, Trsna, Daha, Jvara	3-6 g of seeds

Sl. No.	*Part and Volume*	*Page No.*	*Common Ayurvedic Name*	*Botanical Name/ Chemical Name*	*Family/ Chemical Formula*	*Part Used*	*Identity, Purity and Strength/Chemical Properties*	*Constituents*	*Properties and Action*	*Important Formulations*	*Therapeutic Uses*	*Dose*
98	Part-I; Vol. II	40-41	GAJAPIPPALI (Fruit)	*Scindapsus officinalis* Schoott.	Araceae	Dried Trans-versely cut Pices of Mature Female Spadix	**Foreign matter**- NMT 2 per cent; **Total ash**- NMT 14 per cent; **Acid-insoluble ash**- NMT 1.5 per cent; **Alcohol-soluble extractive**-NLT 3 per cent; **Water-soluble extractive**-NLT 11 per cent; **T.L.C.** - Refer API	Glucosides *viz.* Scindapsin A and Scindapsin B, Sugars and Fixed oil.	*Rasa* - Katu; *Guna*- Ruksa; *Virya*- Usna; *Vipaka*- Katu; *Karma*- Vatahara, Kaphahara, Agnivardhaka, Kanthya, Dipana, Malavisosana, Stanya, Varnya	Punarnavasava, Sivagutika, Mahayogaraja Guggulu, Prasarini taila, Candra-prabhavati	Svasa, Krmiroga, Atisara, Kantah Roga	2-3g in extract (Phant) form
99	Part-I; Vol. II	42-43	GAMBHARI (Fruit)	*Gmelina arborea* Roxb.	Verbenaceae	Dried Fruit	**Foreign matter**- NMT 1 per cent; **Total ash**- NMT 6 per cent; **Acid-insoluble ash**- NMT 0.4 per cent; **Alcohol-soluble extractive**-NLT 8 per cent; **Water-soluble extractive**-NLT 25 per cent; **T.L.C.** - Refer API	Butyric acid, Tartaric acid, Alkaloid, Resin and Saccharine	*Rasa* - Madhura, Amla, Kasaya; *Guna*- Guru, Snigdha, Sara; *Virya*- Sita; *Vipaka*- Madhura; *Karma*- Vatahara, Pittahara, Rasayana, Brmhana, Kesya, Medhya, Sukrala, Hrdya	Arvindasava, Draksadi kvatha Churna	Rakta pitta, Daha, Trsna, Ksata, Ksaya, Mutrakrcchra, Hrdroga	1-3 g of the drug in powder form
100	Part-I; Vol. II	44-45	GANGERU (Stem Bark)	*Grewia tenax* (Forsk.) Aschers and Schwf., Syn. *Grewia populifolia* Vahl,	Tiliaceae	Dried Stem Bark	**Foreign matter**- NMT 1 per cent; **Total ash**- NMT 9 per cent; **Acid-insoluble ash**- NMT 1 per cent; **Alcohol-soluble extractive**-NLT 6 per cent; **Water-soluble extractive**-NLT 8 per cent; **T.L.C.** - Refer API	Sugar, Tannin and Sterols (Triacontan-1-ol, α-amyrin, β-amyrin *etc.*).	*Rasa* - Madhura, Amla, Katu, Tikta, Kasaya; *Guna*- Guru; *Virya*- Usna; *Vipaka*- Katu; *Karma*- Tridosahara, Sangrahaka	Jirakadi Modaka	Vrana, Pittavikara	2-3 g of the drug in powder form
101	Part-I; Vol. II	46-47	GUNJA (Root)	*Abrus precatorius* Linn.	Fabaceae	Dried Root	**Foreign matter**- NMT 2 per cent; **Total ash**- NMT 9 per cent; **Acid-insoluble ash**- NMT 2.5 per cent; **Alcohol-soluble extractive**-NLT 4 per cent; **Water-soluble extractive**-NLT 10 per cent	Glucoside (Glycyrrhizin)	*Rasa* - Madhura, Tikta; *Guna*- Ruksa, Sita; *Virya*- Sita; *Vipaka*- Madhura; *Karma*- Vatahara, Pittahara, Kesya	Nili Bhrngadi Taila	Indralupta, Mukhasosa, Sula	1-3g of the drug in powder form

Sl. No.	Part and Volume	Page No.	Common Ayurvedic Name	Botanical Name/ Chemical Name	Family/ Chemical Formula	Part Used	Identity, Purity and Strength/Chemical Properties	Constituents	Properties and Action	Important Formulations	Therapeutic Uses	Dose
102	Part-I; Vol. II	48-49	IKSU (Stem)	*Saccharum officinarum* Linn.	Poaceae	Dried Stem	**Foreign matter**- NMT 2 per cent; **Total ash**- NMT 6 per cent; **Acid-insoluble ash**- NMT 2.5 per cent; **Alcohol-soluble extractive**-NLT 15 per cent; **Water-soluble extractive**- NLT 17 per cent	Sucrose	*Rasa* - Madhura; *Guna*- Sara, Snigdha, Guru; *Virya*- Sita; *Vipaka*- Madhura; *Karma*- Vatahara, Pittahara, Kaphahara, Mutrala, Balya, Vrsya, Brmhana	Bala Taila, Navaratnaraja-mrganka Rasa	Raktapitta, Mutra Ksaya	200-400 ml. in the juice form
103	Part-I; Vol. II	50-51	INDRA-VARUNI (Root)	*Citrullus colocynthis* Schrad.	Cucurbitaceae	Dried Root	**Foreign matter**- NMT 2 per cent; **Total ash**- NMT 8 per cent; **Acid-insoluble ash**- NMT 2 per cent; **Alcohol-soluble extractive**-NLT 6.5 per cent; **Water-soluble extractive**- NLT 20 per cent; **T.L.C.** - Refer API	Saponin and traces of Alkaloid	*Rasa* - Tikta, Katu; *Guna*- Laghu, Sara; *Virya*- Usna; *Vipaka*- Katu; *Karma*- Pittahara, Kaphahara, Recana	Abhayarista, Rodhrasava, Mrtasanjivani Sura, Brhatmanji-sthadi Kvatha Churna, Narayana Churna, Misraka Sheha, Triphaladi Taila, Mahavisagarbha Taila	Kamala, Pliharoga, Svasa, Kasa, Kustha, Gulma, Krmiroga, Prameha, Visavikara, Vrana, Apaci, Gandamala	1-3 g of the drug in powder form
104	Part-I; Vol. II	52-53	INDRA-VARUNI (Leaf)	*Citrullus colocynthis* Schrad.	Cucurbitaceae	Dried Leaves	**Foreign matter**- NMT 2 per cent; **Total ash**- NMT 18 per cent; **Acid-insoluble ash**- NMT 6 per cent; **Alcohol-soluble extractive**-NLT 7 per cent; **Water-soluble extractive**-NLT 18 per cent; **T.L.C.** - Refer API	Colocynthin, Traces of an Alkaloid and Flavonoids	*Rasa* -Katu, Tikta; *Guna*- Laghu, Sara; *Virya*- Usna; *Vipaka*- Katu; *Karma*- Pittahara, Kaphahara, Recana	Nilibhrngadi Taila	Kesapata, Palita, Kustahroga	For external use only
105	Part-I; Vol. II	54-55	JAMBU (Seed)	*Syzygium cuminii* (Linn.) Skeels Syn. *Eugenia Jambolana* Lam.; *E. Cuminii* Druce.	Myrtaceae	Dried Seeds	**Foreign matter**- NMT 1 per cent; **Total ash**- NMT 5 per cent; **Acid-insoluble ash**- NMT 1 per cent; **Alcohol-soluble extractive**-NLT 6 per cent; **Water-soluble extractive**-NLT 15 per cent; **T.L.C.** - Refer API	Glycoside (Jamboline), Tannin, Ellagic acid and Gallic acid.	*Rasa* - Madhura, Amla, Kasaya; *Guna*- Guru, Ruksa; *Virya*- Sita; *Vipaka*- Katu; *Karma*- Vatala, Pittahara, Kaphahara, Vistambhi, Grahi	Pusyanuga Churna.	Madhumeha, Udakameha	3-6 g of the drug in powder form

Sl. No.	*Part and Volume*	*Page No.*	*Common Ayurvedic Name*	*Botanical Name/ Chemical Name*	*Family/ Chemical Formula*	*Part Used*	*Identity, Purity and Strength/Chemical Properties*	*Constituents*	*Properties and Action*	*Important Formulations*	*Therapeutic Uses*	*Dose*
106	Part-I; Vol. II	56-57	JAMBU (Stem Bark)	*Syzygium cuminii* (Linn.) Skeels Syn. *Eugenia jambolana* Lam.; *E. cuminii* Druce.	Myrtaceae	Dried Stem Bark	**Foreign matter**- NMT 2 per cent; **Total ash**- NMT 11 per cent; **Acid-insoluble ash**- NMT 1 per cent; **Alcohol-soluble extractive**-NLT 9 per cent; **Water-soluble extractive**-NLT 11 per cent	Tannins	*Rasa* - Kasaya; *Guna*- Ruksa; *Virya*- Sita; *Vipaka*- Katu; *Karma*- Kaphahara, Pittahara, Vatala, Grahi, Stambhaka, Krmidosaghna	Usirasava	Atisara, Raktapitta	10-20 g of the drug for decoction
107	Part-I; Vol. II	58-59	JAYAPALA (Seed)	*Croton tiglium* Linn.	Euphorbiaceae	Dried Seed	**Foreign matter**- NMT 2 per cent; **Total ash**- NMT 3 per cent; **Acid-insoluble ash**- NMT 0.5 per cent; **Alcohol-soluble extractive**-NLT 15 per cent; **Water-soluble extractive**- NLT 7 per cent; **T.L.C.** - Refer API	Fixed oil, Resins and Phorbol esters.	*Rasa* - Madhura; *Guna*- Guru, Snigdha; *Virya*- Sita; *Vipaka*- Madhura; *Karma*- Pittahara, Kaphahara, Recana	Icchabhedi Rasa, Asvakancuki Rasa	Udararoga, Vibandha, Jvara	6-12 mg of the drug in powder form
108	Part-I; Vol. II	60-61	JAYANTI (Leaf)	*Sesbania sesban* (Linn.) Merr., Syn. *S. aegyptiaca* Pers.	Fabaceae	Fresh and Dried Leaf	**Foreign matter**- NMT 2 per cent; **Total ash**- NMT 11 per cent; **Acid-insoluble ash**- NMT 2 per cent; **Alcohol-soluble extractive**-NLT 7 per cent; **Water-soluble extractive**-NLT 25 per cent; **T.L.C.** - Refer API	Protein, Calcium and Phosphorus	*Rasa* - Katu, Tikta; *Guna*- Laghu; *Virya*- Usna; *Vipaka*- Katu; *Karma*- Vatahara, Pittahara, Kaphahara, Kanthasodhana, Rasayana	Ratnagiri Rasa, Vajrakapata Rasa	Galaganda, Mutrakrcchra, Visaroga	3-6 g in powder form
109	Part-I; Vol. II	62-63	JYOTISMATI (Seed)	*Celastrus paniculatus* Willd.	Celastraceae	Dried, Brownish-Orange, Ripe Seeds, Deviod of Capsule Wall	**Foreign matter**- NMT 2 per cent; **Total ash**- NMT 6 per cent; **Acid-insoluble ash**- NMT 1.5 per cent; **Alcohol-soluble extractive**-NLT 20 per cent; **Water-soluble extractive**- NLT 9 per cent; **Oil contents** - NLT 45 per cent; **T.L.C.** - Refer API	Alkaloids, Oil and Tannins	*Rasa* - Katu, Tikta; *Guna*- Sara, Usna, Tiksna; *Virya*- Usna; *Vipaka*- Katu; *Karma*- Vatahara, Kaphahara, Vamaka, Virecaka, Sirovirecanopaga, Dipana Prabhava-Meddhya	Smrtisagara Rasa, Jyotismati Taila	Vatavyadhi, Smrtidaurbalya, Switra	Seed : 1-2 g; Oil: 5-15 drops

Sl. No.	*Part and Volume*	*Page No.*	*Common Ayurvedic Name*	*Botanical Name/ Chemical Name*	*Family/ Chemical Formula*	*Part Used*	*Identity, Purity and Strength/Chemical Properties*	*Constituents*	*Properties and Action*	*Important Formulations*	*Therapeutic Uses*	*Dose*
110	Part-I; Vol. II	64-65	KADAMBA (Stem Bark)	*Anthocephalus cadamba* Miq., Syn. *A. indicus* A. Rich.	Rubiaceae	Dried Stem Bark	**Foreign matter**- NMT 2 per cent; **Total ash**- NMT 9 per cent; **Acid-insoluble ash**- NMT 15 per cent; **Alcohol-soluble extractive**-NLT 3 per cent; **Water-soluble extractive**-NLT 5 per cent; **T.L.C.** - Refer API	Alkaloids, Steroids, Fats and Reducing Sugars	*Rasa* - Kasaya, Madhura, Lavana; *Guna*- Ruksa; *Virya*- Sita; *Vipaka*- Katu; *Karma*- Vatahara, Pittahara, Vranaropana, Vedanasthapana	Nygrodhadi Kvatha Churna, Grahanimihira Taila	Daha, Yonidosa, Vrana, Raktapitta, Visavrana (Dansaja Vrana)	0.5-1.5g of the drug in powder form
111	Part-I; Vol. II	66-68	KAKAMACI (Whole Plant)	*Solanum nigrum* Linn.	Solanaceae	Dried Whole Plant	**Foreign matter**- NMT 2 per cent; **Total ash**- NMT 16 per cent; **Acid-insoluble ash**- NMT 7 per cent; **Alcohol-soluble extractive**-NLT 4 per cent; **Water-soluble extractive**-NLT 15 per cent; **T.L.C.** - Refer API	Alkaloids and Saponins	*Rasa* - Tikta, Katu; *Guna*- Sara, Snigdha, Laghu; *Virya*- Usna; *Vipaka*- Katu; *Karma*- Vatahara, Pittahara, Kaphahara, Bhedana, Rasayana, Vrsya, Svarya, Hrdya	Hrdayarnava Rasa, Maha visagarbha Taila, Rasaraja Rasa	Kustha, Kandu, Arsa, Prameha, Sotha, Hrdroga, Jvara, Hikka, Chardi, Netraroga	5-10 ml of the drug in juice form
112	Part-I; Vol. II	69-70	KAMALA (Flower)	*Nelumbo nucifera* Gaertn. Syn. *Nelumbium speciosum* Willd.	Nymphaeaceae	Dried Flowers (devoid of stalk)	**Foreign matter**- NMT 2 per cent; **Total ash**- NMT 12 per cent; **Acid-insoluble ash**- NMT 3 per cent; **Alcohol-soluble extractive**-NLT 6 per cent; **Water-soluble extractive**-NLT 14 per cent	Alkaloid (Nelumbine)	Rasa- Madhura, Tikta, Kasaya; *Guna*- Sita, Laghua; *Virya*- Sita; *Vipaka*- Madhura; *Karma*- Kaphahara, Pittahara, Santapahara, Varnya, Mutra Virajaniya	Aravindasava, Catura Kaval Ghrta.	Trsna Daha, Raktapitta, Visarpa, Visavikara	12-24 g of the drug for decoction

Sl. No.	*Part and Volume*	*Page No.*	*Common Ayurvedic Name*	*Botanical Name/ Chemical Name*	*Family/ Chemical Formula*	*Part Used*	*Identity, Purity and Strength/Chemical Properties*	*Constituents*	*Properties and Action*	*Important Formulations*	*Therapeutic Uses*	*Dose*
113	Part-I; Vol. II	71-72	KAPITTHA (Fruit Pulp)	*Feronia limonia* (Linn.) Swingle Syn. *F. elephantum* Correa	Rutaceae	Dried Pulp of Mature Fruit	**Foreign matter**- NMT 2 per cent; **Total ash**- NMT 6 per cent; **Acid-insoluble ash**- NMT 1 per cent; **Alcohol-soluble extractive**-NLT 12 per cent; **Water-soluble extractive**- NLT 25 per cent; T.L.C. - Refer API	Citric acid and Mucilage	Ripe Pulp: *Rasa* - Ripe Pulp - Madhura, Amla, kasaya; *Guna*- Ripe Pulp - Laghu; *Virya*- Ripe Pulp -Sita; *Vipaka*- Ripe Pulp - Madhura; *Karma* -Ripe Pulp- Vrsya, Pittavatahara, Sangrahi, Vrananasaka; Unripe Pulp: *Rasa* - Amla, Kasaya; *Guna*- Guru; *Virya*- Usna; *Vipaka*- Amla; *Karma* -Kaphaghna, Grahi Vatala, Lekhana	Kapitthastaka Churna, Yavanyadi Churna	Ripe - Trsa, Hikka, Svasa, Vami, Unripe - Grahani Roga, Agnimandya	1-3 g of the drug in powder form
114	Part-I; Vol. II	73-74	KARAMARDA (Stem Bark)	*Carissa carandas* Linn.	Apocynaceae	Dried Stem Bark	**Foreign matter**- NMT 2 per cent; **Total ash**- NMT 12 per cent; **Acid-insoluble ash**- NMT 3 per cent; **Alcohol-soluble extractive**-NLT 4 per cent; **Water-soluble extractive**-NLT 8 per cent; T.L.C. - Refer API	Glycosides and β-sitosterol	*Rasa* - Amla; *Guna*- Guru, Sara; *Virya*- Usna; *Vipaka*- Katu; *Karma*- Vatahara, Pittakara, Kaphahara	Marma Gutika	Kusthahara	48 g of the drug for decoction
115	Part-I; Vol. II	75-76	KARANJA (Root Bark)	*Pongamia pinnata* (Linn.) Merr., Syn. *P. glabra* Vent.	Fabaceae	Dried Root Bark	**Foreign matter**- NMT 1 per cent; **Total ash**- NMT 11 per cent; **Acid-insoluble ash**- NMT 2 per cent; **Alcohol-soluble extractive**-NLT 3.5 per cent; **Water-soluble extractive**- NLT 17 per cent; T.L.C. - Refer API	Flavones Kanugin, Demethoxy-kanugin	*Rasa* - Katu, Tikta, Kasaya; *Guna*- Tiksna; *Virya*- Usna; *Vipaka*- Katu; *Karma*- Kaphahara, Vatahara, Pittahara, Kandughna, Visaghna, Vranasodhana	Prabhanjana Vimardana Taila	Kustha, Kandu, Dustavrana, Prameha, Yonirog, Krmiroga, Antravidradhi	1-3 g of the drug for decoction

Sl. No.	*Part and Volume*	*Page No.*	*Common Ayurvedic Name*	*Botanical Name/ Chemical Name*	*Family/ Chemical Formula*	*Part Used*	*Identity, Purity and Strength/Chemical Properties*	*Constituents*	*Properties and Action*	*Important Formulations*	*Therapeutic Uses*	*Dose*
116	Part-I; Vol. II	77-78	KARANJA (Root)	*Pongamia pinnata* (Linn.) Merr., Syn. *P. glabra* Vent.	Fabaceae	Dried Root	**Foreign matter**- NMT 2 per cent; **Total ash**- NMT 8 per cent; **Acid-insoluble ash**- NMT 1 per cent; **Alcohol-soluble extractive**-NLT 1 per cent; **Water-soluble extractive**-NLT 7 per cent	Karnajin, Kanugin, Demethoxy-kanugin, Pongachromene and Tera-o-methylfisetin.	*Rasa* - Katu, Tikta, Kasaya; *Guna*- Tiksna; *Virya*- Usna; *Vipaka*- Katu; *Karma*- Kaphahara, Vatahara, Pittahara, Kandughna, Visaghna, Vranasodhana	Dhanvantara Ghrta	Kustha, Kandu, Dustavrana, Prameha, Yoniroga, Krmiroga, Antravidradhi, Vidradhi	1-2 g of the drug in powder form
117	Part-I; Vol. II	79-80	KARANJA (Stem Bark)	*Pongamia pinnata* (Linn.) Merr., Syn. *P. glabra* Vent.	Fabaceae	Dried Stem Bark	**Foreign matter**- NMT 1 per cent; **Total ash**- NMT 13 per cent; **Acid-insoluble ash**- NMT 1.5 per cent; **Alcohol-soluble extractive**-NLT 3 per cent; **Water-soluble extractive**-NLT 18 per cent	Flavones and Furanoflavones like Karanjin, Pongapin, Demethoxy-kanugin, Kanugin, Pinnatin, Tetra-o-methylfisetin, Gamatin, 5-methoxy-furano (2", 3",7 : 8), flavone and 5-Methoxy-3'4' Methylene dioxyfurano (2", 3",7: 8) flavone and two new Furano compounds Glabra-I and Glabra-II. It also contains alkaloids and Triterpenoid saponin	*Rasa* - Katu, Tikta, Kasaya; *Guna*- Tiksa; *Virya*- Usna; *Vipaka*- Katu; *Karma*- Kaphahara, Vatahara, Pittahara, Kandughna, Visaghna, Vranasodhana	Brhanmanjis-thadi Kvatha Churna, Mustakaranjadi Kvatha Churna	Kustha, Kandu, Dustavrana, Prameha, Yoniroga, Krmiroga, Antravidradhi, Vidradhi	1-2 g of the drug in powder form

Sl. No.	*Part and Volume*	*Page No.*	*Common Ayurvedic Name*	*Botanical Name/ Chemical Name*	*Family/ Chemical Formula*	*Part Used*	*Identity, Purity and Strength/Chemical Properties*	*Constituents*	*Properties and Action*	*Important Formulations*	*Therapeutic Uses*	*Dose*
118	Part-I; Vol. II	81-82	KARANJA (Leaf)	*Pongamia pinnata* (Linn.) Merr., Syn. *P. glabra* Vent.	Fabaceae	Dried Leaf	**Foreign matter**- NMT 2 per cent; **Total ash**- NMT 11 per cent; **Acid-insoluble ash**- NMT 3.5 per cent; **Alcohol-soluble extractive**-NLT 10 per cent; **Water-soluble extractive**- NLT 16 per cent	A new Furano-flavone-3'-methoxy pongapin in addition to Karanjin, Kanjone and its two isomers 7-methoxy-furano-(4", 5", - 6,5) - flavone and 8-methoxy-furano-(4",5",-6-5)-Flavone and 8-methoxy-furano-(4", 5"-6,7)-flavone.	*Rasa* - Katu, Tikta, Kasaya; *Guna*- Tiksna; *Virya*- Usna; *Vipaka*- Katu; *Karma*- Vatahara, Kaphahara, Pittavardhaka, Bhedana, Kandughana, Krimihara, Sothahara	Jatyadi Taila	Kustha, Krmiroga, Vrana, Kandu	For external use only
119	Part-I; Vol. II	83-84	KARAVA-LLAKA (Fresh Fruit)	*Momordica charantia* Linn.	Cucurbitaceae	Fresh Fruit	**Foreign matter**- Nil; **Total ash**- NMT 8.5 per cent; **Acid-insoluble ash**- NMT 0.6 per cent; **Alcohol-soluble extractive**-NLT 6 per cent; **Water-soluble extractive**- NLT 28 per cent; **T.L.C.** - Refer API	Alkaloid (Momoridicine) and Glycosides.	*Rasa* - Tikta, Katu; *Guna*- Laghu; *Virya*- Usna; *Vipaka*- Katu; *Karma*- Vatahara, Kaphahara, Raktadosahara, Dipana, Hrdya, Bhedi	Mahavisagarbha Taila	Kustha, Prameha, Kamala, Pandu, Krmiroga, Raktavikara, Jvara, Svasa, Kasa, Aruci	10-15 ml juice of fresh drug
120	Part-I; Vol. II	85-87	KATUKA (Rhizome)	*Picrorhiza kurroa* Royle ex Benth.	Scrophularia-ceae	Dried Rhizome with Root	**Foreign matter**- NMT 2 per cent; **Total ash**- NMT 7 per cent; **Acid-insoluble ash**- NMT 1 per cent; **Alcohol-soluble extractive**-NLT 10 per cent; **Water-soluble extractive**- NLT 20 per cent; **T.L.C.** - Refer API	Glucoside (Picrorhizin).	*Rasa* - Tikta, Katu; *Guna*- Laghu; *Virya*- Usna; *Vipaka*- Katu; *Karma*- Pittahara, Dipani, Bhedini, Hrdya, Jvarahara	Arogyavardhini Gutika, Tiktaka Ghrta, Sarvajvarahara Lauha, Mahatikataka Ghrta	Kamala, Svasa, Daha, Jvara, Kustha, Visamajvara, Arocaka	1-3 g of the drug in powder form

Sl. No.	*Part and Volume*	*Page No.*	*Common Ayurvedic Name*	*Botanical Name/ Chemical Name*	*Family/ Chemical Formula*	*Part Used*	*Identity, Purity and Strength/Chemical Properties*	*Constituents*	*Properties and Action*	*Important Formulations*	*Therapeutic Uses*	*Dose*
121	Part-I; Vol. II	88-90	KOKILAKSA (Whole Plant)	*Asteracantha longifolia* Nees. Syn. *Hygrophila spinosa*. T. Anders	Acanthaceae	Dried Whole Plant	**Foreign matter**- NMT 2 per cent; **Total ash**- NMT 9 per cent; **Acid-insoluble ash**- NMT 1 per cent; **Alcohol-soluble extractive**-NLT 4 per cent; **Water-soluble extractive**-NLT 20 per cent	Alkaloids	*Rasa* - Madhura, Amala, Tikta; *Guna*- Picchila, Snigdha; *Virya*- Sita; *Vipaka*- Madhura; *Karma*- Balya, Vrsya, Mutrala, Vajikara, Santarpana, Rucya	Panaviraladi Bhasma (Ksara)	Amavata Sotha, Trsna, Vatarakta	3-6 g of the drug in powder form
122	Part-I; Vol. II	91-92	KOKILAKSA (Root)	*Asteracantha longifolia* Ness. Syn. *Hygrophila spinosa*. T. Anders	Acanthaceae	Dried Root	**Foreign matter**- NMT 2 per cent; **Total ash**- NMT 12 per cent; **Acid-insoluble ash**- NMT 1 per cent; **Alcohol-soluble extractive**-NLT 4 per cent; **Water-soluble extractive**-NLT 8 per cent	Essential Oil	*Rasa* - Madhura, Amla, Tikta; *Guna*- Picchila, Snigdha; *Virya*- Sita; *Vipaka*- Madhura; *Karma*- Vatahara, Kaphahara, Mutrala, Vrsya	Rasnairandadi Kvatha Churna, Vastyamayan-taka Ghrta	Amavata Sotha, Asmari, Vatarakta, Pittatisara	3-6 g of the drug for decoction
123	Part-I; Vol. II	93-94	KOKILAKSA (Seed)	*Asteracantha longifolia* Ness. Syn. *Hygrophila spinosa*. T. Anders	Acanthaceae	Dried Seed	**Foreign matter**- NMT 2 per cent; **Total ash**- NMT 15 per cent; **Acid-insoluble ash**- NMT 8 per cent; **Alcohol-soluble extractive**-NLT 10 per cent; **T.L.C.** - Refer API	An yellow semi-drying oil, enzymes like Diastase, Lipase, Protease and an Alkaloid.	*Rasa* - Madhura; *Guna*- Snigdha, Picchila; *Virya*- Sita; *Vipaka*- Madhura; *Karma*- Kaphahara, Vrsya, Balya, Ruchya, Santarpana	Yakrt Sulavinasini Vatika, Vastyamayan-taka Ghrta.	Vatarakta, Sotha, Pittasmari	3-6 g of the drug in powder form
124	Part-I; Vol. II	95-97	KOZUPPA (Whole Plant)	*Portulaca oleracea* Linn.	Portulacaceae	Dried Whole Plant	**Foreign matter**- NMT 5 per cent; **Total ash**- NMT 30 per cent; **Acid-insoluble ash**- NMT 5 per cent; **Alcohol-soluble extractive**-NLT 3 per cent; **Water-soluble extractive**-NLT 19 per cent; **T.L.C.** - Refer API	Protein, Carbohydrates, Vitamin C and Mucilage	Rasa- Amla; *Guna*- Sara, Guru, Ruksa; *Virya*- Usna; *Vipaka*- Amla; *Karma*- Vatahara, Pittakara, Kaphahara, Caksusya, Vanidosahara	Marma Gutika	Vrana, Gulma, Prameha, Sotha, Arsa, Agnimandya	3-6 g of the drug in powder form

Sl. No.	*Part and Volume*	*Page No.*	*Common Ayurvedic Name*	*Botanical Name/ Chemical Name*	*Family/ Chemical Formula*	*Part Used*	*Identity, Purity and Strength/Chemical Properties*	*Constituents*	*Properties and Action*	*Important Formulations*	*Therapeutic Uses*	*Dose*
125	Part-I; Vol. II	98-101	LAJJALU (Whole Plant)	*Mimosa pudica* Linn.	Fabaceae	Dried Whole Plant	**Foreign matter**- NMT 2 per cent; **Total ash**- NMT 10 per cent; **Acid-insoluble ash**- NMT 5 per cent; **Alcohol-soluble extractive**-NLT 9 per cent; **Water-soluble extractive**-NLT 9 per cent; **T.L.C.** - Refer API	Alkaloid	*Rasa* - Kasaya, Tikta; *Guna*- Laghu, Ruksa; *Virya*- Sita; *Vipaka*- Katu; *Karma*- Kaphahara, Pittahara, Grahi	Samangadi Churna, Kutajavaleha, Pusyanuga Churna, Brhat Gangadhara Churna	Raktapita, Atisara, Yoniroga, Sopha, Daha, Svasa, Vrana, Kustha	10-20 g of the drug for decoction
126	Part-I; Vol. II	102-103	MADHUKA (Flower)	*Madhuca indica* J.F. Gmel. Syn. *M. latifolia* (Roxb.) Macbride, *Bassia latifolia* Roxb.	Sapotaceae	Flower Usually Without Stalk or Calyx	**Foreign matter**- NMT 2 per cent; **Total ash**- NMT 5 per cent; **Acid-insoluble ash**- NMT 0.5 per cent; **Alcohol-soluble extractive**-NLT 25 per cent; **Water-soluble extractive**- NLT 70 per cent; **Moisture content** - NMT 10 per cent	Sugars	*Rasa* - Madhura; *Guna*- Guru; *Virya*- Sita; *Vipaka*- Madhura; *Karma*- Vatahara, Pittakara, Sukrala, Sramahara, Balya, Ahrdya	Madhukasava, Draksadi Kvatha Churna, Eladi Modaka	Trsna, Daha, Srama, Svasa, Ksata, Ksaya	10-15 g of the drug
127	Part-I; Vol. II	104-106	MATSYAKSI (Whole Plant)	*Alternanthera sessilis* (Linn.) R. Br., Syn. Atriandra Lam., *A. triandra* Lam., *A. denticulata* R. Br., *A. nodiflora* R. Br., *A. repens* Gmel., non Link.	Amaranthaceae	Dried Whole Plant	**Foreign matter**- NMT 2 per cent; **Total ash**- NMT 10 per cent; **Acid-insoluble ash**- NMT 4.5 per cent; **Alcohol-soluble extractive**-NLT 3 per cent; **Water-soluble extractive**-NLT 19 per cent; **T.L.C.** - Refer API	Sugar, Saponins and Sterols	Rasa- Tikta, Kasaya, Madhura; *Guna*- Laghu; *Virya*- Situ; *Vipaka*- Katu; *Karma*- Vatahara, Pitthara, Kaphahara, Grahi	Traikantaka Ghrta	Kustha, Raktavikara, Pittavikara	2-3 g of the drug in powder form
128	Part-I; Vol. II	107-108	METHI (Seed)	*Trigonella foenum-graecum* Linn.	Fabaceae	Seeds	**Foreign matter**- NMT 2 per cent; **Total ash**- NMT 4 per cent; **Acid-insoluble ash**- NMT 0.5 per cent; **Alcohol-soluble extractive**-NLT 5 per cent;	Alkaloid, Sapogenins and Mucilage	Rasa- Tikta; *Guna*- Snigdha; *Virya*- Usna; *Vipaka*- Katu; *Karma*- Vatahara, Kaphahara, Dipana, Rucya	Mustakarista, Mrtasanjivani Sura	Aruci, Jvara, Grahani, Prameha	3-6 g of the drug in powder form

Sl. No.	*Part and Volume*	*Page No.*	*Common Ayurvedic Name*	*Botanical Name/ Chemical Name*	*Family/ Chemical Formula*	*Part Used*	*Identity, Purity and Strength/Chemical Properties*	*Constituents*	*Properties and Action*	*Important Formulations*	*Therapeutic Uses*	*Dose*
129	Part-I; Vol. II	109-11	MULAKA (Whole Plant)	*Raphanus sativus* Linn.	Brassicaceae	Whole Plant	**Foreign matter**- NMT 2 per cent; **Total ash**- NMT 18 per cent; **Acid-insoluble ash**- NMT 1 per cent; **Alcohol-soluble extractive**-NLT 30 per cent; **Water-soluble extractive**- NLT 22 per cent	Glucoside, Volatile Oil (containging butyl crotonyl isothiocyanate sulphide) with a typical radish odour.	Rasa- Katu, Tikta; *Guna*- Laghu, Tiksna; *Virya*- Usna; *Vipaka*- Katu; *Karma*- Vatahara, Pittahara, Kaphahara, Dipana, Pacana, Rucya, Svarya, Hrdya	Mulakaksara, Gandhaka Vati, Hajarulayahuda Bhasma.	Gulma, Arsa, Agnimandya, Pinasa, Udavarta	20-40 ml of the drug in juice form
130	Part-I; Vol. II	112-113	MULAKA (Root)	*Raphanus sativus* Linn.	Brassicaceae	Fresh Root	**Foreign matter**- NMT 2 per cent; **Total ash**- NMT 24 per cent; **Acid-insoluble ash**- NMT 2 per cent; **Alcohol-soluble extractive**-NLT 36 per cent; **Water-soluble extractive**- NLT 33 per cent; **T.L.C.** - Refer API	Glucoside, Methylmer-captan and Volatile Oil	Rasa- Katu, Tikta; *Guna*- Laghu, Tiksna; *Virya*- Usna; *Vipaka*- Katu; *Karma*- Vatahara, Pittahara, Kaphahara, Dipana, pacana, Rucya, Svarya, Hrdya	Candana-balalaksadi Taila, Mulaka ksara	Jvara, Svasa, Kasa, Pinasa, Galaroga, Vrana, Dadru, Netraroga, Gulma, Arsa, Agnimandya, Udavarta	15-30 ml of the drug in the juice form
131	Part-I; Vol. II	114-115	MURA (Root)	*Selinum candollei* DC. Syn. *S. tenuifolium* Wall. ex DC.	Apiaceae	Dried Root	**Foreign matter**- NMT 2 per cent; **Total ash**- NMT 9 per cent; **Acid-insoluble ash**- NMT 3.5 per cent; **Alcohol-soluble extractive**-NLT 9 per cent; **Water-soluble extractive**-NLT 17 per cent	Dihydropyrano-coumarines (identified as Isopteryxin and Anomalin), Sucrose and Mannitol.	Rasa- Katu, Tikta, Kasaya, Madhura; *Guna*- Laghu; *Virya*- Sita; *Vipaka*- Madhura; *Karma*- Vatahara, Pittahara	Arvindasava, Karpuradyarista	Jvara, Daha, Bhrama, Murchha, Svasa, Trsa	1-3 g of the drug in powder form
132	Part-I; Vol. II	116-117	MURVA (Root)	*Marsdenia tenacissima* Wight. and Arn.	Asclepiadaceae	Dried Root	**Foreign matter**- NMT 2 per cent; **Total ash**- NMT 5 per cent; **Acid-insoluble ash**- NMT 0.5 per cent; **Alcohol-soluble extractive**-NLT 7 per cent; **Water-soluble extractive**-NLT 14 per cent	Resin	*Rasa* - Madhura, Tikta; *Guna*- Guru, Sara; *Virya*- Usna; *Vipaka*- Madhura; *Karma*- Vatahara, Pittahara, Kaphahara, Visaghna	Aragvadhadi Kvatha Churna, Patoladi Kvatha Churna, Prameha Mihira Taila, Sudarsana Churna.	Jvara, Medoroga, Meha, Mukha Sosa, Krmiroga, Hrdroga, kandu, Arsa, Raktapitta, Trsna	2-6 g of the drug in powder form; 10-20 g of the drug for decoction

Sl. No.	*Part and Volume*	*Page No.*	*Common Ayurvedic Name*	*Botanical Name/ Chemical Name*	*Family/ Chemical Formula*	*Part Used*	*Identity, Purity and Strength/Chemical Properties*	*Constituents*	*Properties and Action*	*Important Formulations*	*Therapeutic Uses*	*Dose*
133	Part-I; Vol. II	118-119	NAGAKESARA (Stamen)	*Mesua ferrea* Linn.	Guttiferae	Dried Stamens	**Foreign matter**- NMT 2 per cent; **Total ash**- NMT 6 per cent; **Acid-insoluble ash**- NMT 3 per cent; **Alcohol-soluble extractive**-NLT 15 per cent; **Water-soluble extractive**- NLT 12 per cent	Essential oil and Oleo-resin	*Rasa* - Tikta, Katu, Kasaya; *Guna*- Laghu, Ruksa; *Virya*- Usna; *Vipaka*- Katu; *Karma*-Kaphahara, Varnya, Vastivata-mayaghna, Urdhajatrugata-rogahara	Candana-balalaksadi Taila, Kumaryasava, Nagakesaradi Churna	Vatarakta, Sopharoga, Vastiroga, Raktapitta	1-3 g of the drug in powder form
134	Part-I; Vol. II	120-121	NILI (Leaf)	*Indigofera tinctoria* Linn.	Fabaceae	Dried Leaf	**Foreign matter**- NMT 2 per cent; **Total ash**- NMT 10 per cent; **Acid-insoluble ash**- NMT 2 per cent; **Alcohol-soluble extractive**-NLT 7.5 per cent; **Water-soluble extractive**- NLT 25 per cent	Glycoside (Indican).	*Rasa* - Tikta, Katu; *Guna*- Sara; *Virya*- Usna; *Vipaka*- Katu; *Karma*- Vatahara, Kaphahara, Recani, Kesya	Nilibhrngadi Taila (for external use only), Mahapancagavya Ghrta.	Amavata, Vatarakta, Udararoga, Udavarta, Pliharoga, Gulam, Jvara, Kasa, Visavikara, Krmiroga	50-100 g of decoction
135	Part-I; Vol. II	122-123	NILI (Root)	*Indigofera tinctoria* Linn.	Fabaceae	Dried Root	**Foreign matter**- NMT 2 per cent; **Total ash**- NMT 6 per cent; **Acid-insoluble ash**- NMT 0.7 per cent; **Alcohol-soluble extractive**-NLT 3 per cent; **Water-soluble extractive**- NLT 4 per cent; **T.L.C.** - Refer API	Glycoside (Indican).	*Rasa* - Tikta, Katu; *Guna*- Sara; *Virya*- Usna; *Vipaka*- Katu; *Karma*- Vatahara, Kaphahara, Recani, Kesya, Bhrama Mohahara	Arvindasava, Triphaladi Taila.	Vatarakta, Amavata, Udavarta, Udararoga, Pliharoga, Visavikara, Kasa, Gulma, Krimiroga	48 g of drug for decoction
136	Part-I; Vol. II	124-125	NIMBA (Leaf)	*Azadirachta indica* A. Juss Syn. *Melia azadirachta* Linn.	Meliaceae	Dried Leaf	**Foreign matter**- NMT 2 per cent; **Total ash**- NMT 10 per cent; **Acid-insoluble ash**- NMT 1 per cent; **Alcohol-soluble extractive**-NLT 13 per cent; **Water-soluble extractive**- NLT 19 per cent	Triterpenoids and Sterols	*Rasa* - Tikta, Ruksa; *Guna*- Laghu; *Virya*- Sita; *Vipaka*- Katu; *Karma*- Vatala, Pittanasaka, Grahi	Kasisadi Ghrta, Jatyadi Ghrta, Arogyavardhini Gutika, Nimbapatra-diupanaha, Pancaguna Taila.	Jvara, Amasotha, Vrana, Kustha, Prameha, Netraroga, Krmiroga, Visaroga	1-3 g of the drug in powder form.; 10-20 g of the drug for decoction

Sl. No.	*Part and Volume*	*Page No.*	*Common Ayurvedic Name*	*Botanical Name/ Chemical Name*	*Family/ Chemical Formula*	*Part Used*	*Identity, Purity and Strength/Chemical Properties*	*Constituents*	*Properties and Action*	*Important Formulations*	*Therapeutic Uses*	*Dose*
137	Part-I; Vol. II	126-127	NIMBA (Stem Bark)	*Azadirachta indica* A. Juss. Syn. *Melia azadirachta* Linn.	Meliaceae	Stem Bark	**Foreign matter**- NMT 2 per cent; **Total ash**- NMT 7 per cent; **Acid-insoluble ash**- NMT 1.5 per cent; **Alcohol-soluble extractive**-NLT 6 per cent; **Water-soluble extractive**-NLT 5 per cent; **T.L.C.** - Refer API	Bitter principles Nimbin and Nimbiol	*Rasa* - Tikta; *Guna*- Laghu, Ruksa; *Virya*-Sita; *Vipaka*-Katu; *Karma*-Kaphahara, Pittahara, Visaghna, Kandughna, Vranaso-dhanakara, Hrdaya-vidahasantikara	Nimbadi Kvatha Churna, Nimbadi Churna, Pancanimba Churna, Pancatikta Guggulu Ghrta, Pathyadi Kwatha (Sadanga) Churna, Sudersana Churna.	Vrana, Kustha, Prameha, Kandu, Krmiroga, Jvara, Daha, Rakta pitta	2-4 g of the drug in powder form, Decoction should be used externally
138	Part-I; Vol. II	128-130	PALASA (Stem Bark)	*Butea monosperma* (Lam.) Kuntze	Fabaceae	Dried Stem Bark	**Foreign matter**- NMT 2 per cent; **Total ash**- NMT 12 per cent; **Acid-insoluble ash**- NMT 1.5 per cent; **Alcohol-soluble extractive**-NLT 10 per cent; **Water-soluble extractive**- NLT 14 per cent; **T.L.C.** - Refer API	Kinotannic acid and Gallic acid	*Rasa* - Kasaya, Katu, Tikta; *Guna*- Sara, Snigdha; *Virya*-Usna; *Vipaka*-Katu; *Karma*-Kaphavatasamaka, Agnidipaka, Saraka, Vrsya	Palasa Ksara, Nyagrodhadi Kwatha Churna, Mahanarayana Taila	Grahani, Gulma, Arsa, Vrana, Krmiroga	5-10 g of the drug in powder form for decoction
139	Part-I; Vol. II	131-132	PARIBHADRA (Stem Bark)	*Erythrina indica* Lam.	Fabaceae	Dried Stem Bark	**Foreign matter**- NMT 2 per cent; **Total ash**- NMT 13 per cent; **Acid-insoluble ash**- NMT 1 per cent; **Alcohol-soluble extractive**-NLT 2.5 per cent; **Water-soluble extractive**- NLT 7 per cent	Alkaloids and Resins	*Rasa* - Tikta, Katu; *Guna*- Sara; *Virya*- Usna; *Vipaka*- Katu; *Karma*- Vatahara, Kaphahara, Medohara, Krmighna	Nyagrodhadi Churna, Abhaya Lavana, Narayana Taila	Krmiroga, Sotha, Karnaroga	6-12 g of the drug in powder form.; 12-24 g of the drug for decoction
140	Part-I; Vol. II	133-134	PIPPALIMULA (Stem)	*Piper longum* Linn.	Piperaceae	Dried, cut, Stem Stem	**Foreign matter**- NMT 2 per cent; **Total ash**- NMT 5.5 per cent; **Acid-insoluble ash**- NMT 0.2 per cent; **Alcohol-soluble extractive**-NLT 4.0 per cent; **Water-soluble extractive**- NLT 12 per cent; **T.L.C.** - Refer API	Alkaloids (Piperine, Piper-longumine, Piper-longuminine *etc.*) Essential oils)	*Rasa* - Katu; *Guna*-Laghu, Ruksa; *Virya*- Usna; *Vipaka*- Katu; *Karma*- Vatahara, Kaphahara, Dipana, Pacana, Vatanulomana, Vulaprasamana, Rucya	Pancakola Churna, Dasamula Taila, Dasamula-pancakoladi Kvatha Churna, Dasamula-satpalaka Ghrta	Udararoga, Anaha, Gulma, Krmiroga, Vataroga	0.5-1 g of the drug in powder form

Sl. No.	*Part and Volume*	*Page No.*	*Common Ayurvedic Name*	*Botanical Name/ Chemical Name*	*Family/ Chemical Formula*	*Part Used*	*Identity, Purity and Strength/Chemical Properties*	*Constituents*	*Properties and Action*	*Important Formulations*	*Therapeutic Uses*	*Dose*
141	Part-I; Vol. II	135-136	PLAKSA (Stem Bark)	*Ficus lacor* Buch.- Ham. = *F. lucescens* Blume., Syn. *F. infectoria* Roxb.	Moraceae	Dried Stem Bark	**Foreign matter**- NMT 1 per cent; **Total ash**- NMT 10 per cent; **Acid-insoluble ash**- NMT 1.5 per cent; **Alcohol-soluble extractive**-NLT 5 per cent; **Water-soluble extractive**-NLT 6 per cent	Sterols, Sugar, Tannin, Alkaloid and Saponin	*Rasa* - Katu, Kasaya; *Guna*- Ruksa; *Virya*- Sita; *Vipaka*- Sita; *Karma*- Pittahara, Kaphahara, Medohara, Stambhana, Dahahara, Sramahara, Samgrahi, Bhagnasadhaka, Yonidosahara	Nyagrodhadi Kvatha Churna, Nalpamaradi Taila, Marma Gutika	Raktapitta, Murcha, Vrana, Yoniroga, Sotha, Visarpa, Atisara	50 g of the drug in powder form for decoction
142	Part-I; Vol. II	137-140	PRASARINI (Whole Plant)	*Paederia foetida* Linn.	Rubiaceae	Whole Plant	**Foreign matter**- NMT 2 per cent; **Total ash**- NMT 21 per cent; **Acid-insoluble ash**- NMT 6 per cent; **Alcohol-soluble extractive**-NLT 2 per cent; **Water-soluble extractive**-NLT 9 per cent	Alkaloids, Volatile Oil	*Rasa* - Tikta; *Guna*- Guru, Sara; *Virya*- Usna; *Vipaka*- Katu; *Karma*- Vatahara, Vrsya, Balakrt, Sandhankrt	Prasarini Taila, Dasamularista	Vataroga, Vatarakta	2-4 g of the drug in powder form
143	Part-I; Vol. II	141-142	PRIYALA (Seed)	*Buchanania lanzan* Spreng. Syn. *B. latifolia* Roxb.	Anacardiaceae	Seed	**Foreign matter**- NMT 2 per cent; **Total ash**- NMT 4 per cent; **Acid-insoluble ash**- NMT 0.5 per cent; **Alcohol-soluble extractive**-NLT 10 per cent; **Water-soluble extractive**- NLT 7 per cent; **T.L.C.** - Refer API	Albuminoids, Oil and Starch	*Rasa* - Madhura; *Guna*- Guru, Snigdha, Sara; *Virya*- Sita; *Vipaka*- Madhura; *Karma*- Vatahara, Pittahara, Kaphakara, Sukrakara, Bhagnasadhaka, Sramahara, Brmhana, Vrsya, Balya, Hrdya, Amavardhaka	Pugakhanda, Priyala Taila	Raktapitta, Daha, Ksata, Ksaya	10-20 g of the drug in powder form

Sl. No.	*Part and Volume*	*Page No.*	*Common Ayurvedic Name*	*Botanical Name/ Chemical Name*	*Family/ Chemical Formula*	*Part Used*	*Identity, Purity and Strength/Chemical Properties*	*Constituents*	*Properties and Action*	*Important Formulations*	*Therapeutic Uses*	*Dose*
144	Part-I; Vol. II	143-144	PRIYANGU (Inflorescence)	*Callicarpa macrophylla* Vahl.	Verbenaceae	Dried Inflore-scence	**Foreign matter**- NMT 2 per cent; **Total ash**- NMT 8 per cent; **Acid-insoluble ash**- NMT 2 per cent; **Alcohol-soluble extractive**-NLT 10 per cent; **Water-soluble extractive**- NLT 14 per cent	Glycosides, Terpenes, Phenolic compound, Resin and Saponin.	*Rasa* - Tikta, Kasaya; *Guna*- Ruksa; *Virya*- Sita; *Vipaka*- Katu; *Karma*- Vatahara, Pittahara, Rakta Prasadana, Daurgandhyahara, Purisasam-grahaniya, Mutravirajaniya, Sandhaniya, Vranaropana	Khadiradi Gutika (Mukharoga), Eladi Churna, Kanaka Taila, Kumkumadi Taila, Nilikadya Taila	Daha, Jvara, Rakta-pitta, Pakvatisara, Svedadhikya	1-3 g of the drug in powder form
145	Part-I; Vol. II	145-146	SALI (Root)	*Oryza sativa* Linn.	Poaceae	Dried Root	**Foreign matter**- NMT 5 per cent; **Total ash**- NMT 21 per cent; **Acid-insoluble ash**- NMT16 per cent; **Water-soluble extractive**- NLT 3 per cent	Sugars	*Rasa* - Madhura, Kasaya; *Guna*- Snigdha, Guru, Laghu; *Virya*- Sita; *Vipaka*- Madhura; *Karma*- Vatahara, Pittahara, Kaphahara, Sukrala, Baddhalpavarcasa, Brmhana, Mutrala, Balya, Varnakrt, Svarya, Rucya, Caksusya, Hrdya, Stanyajanana	Brahma Rasayana, Stanyajanana Kasaya Churna.	Stanyaksaya, Mutrakrcchra	50 g of the drug for decoction
146	Part-I; Vol. II	147-149	SANKHA-PUSPI (Whole Plant)	*Convolvulus pluricaulis* Choisy	Convolvulaceae	Whole Plant	**Foreign matter**- NMT 2 per cent; **Total ash**- NMT 17 per cent; **Acid-insoluble ash**- NMT 8 per cent; **Alcohol-soluble extractive**-NLT 6 per cent; **Water-soluble extractive**- NLT 10 per cent	Alkaloid	*Rasa* - Tikta, Katu, Kasaya; *Guna*- Sara; *Virya*- Sita; *Vipaka*- Katu; *Karma*- Pittahara, Kaphahara, Rasayana, Medhya, Balya, Mohanasaka, Ayusya	Agastyaharitaki, Rasayana, Brahma, Rasayana, Brahmi Ghrta, Manasmitra Vataka, Gorocanadi Vati, Brahmi Vati	Manasaroga, Apasmara	3-8 g of the drug in powder form

Sl. No.	Part and Volume	Page No.	Common Ayurvedic Name	Botanical Name/ Chemical Name	Family/ Chemical Formula	Part Used	Identity, Purity and Strength/Chemical Properties	Constituents	Properties and Action	Important Formulations	Therapeutic Uses	Dose
147	Part-I; Vol. II	150-152	SAPTALA (Whole Plant)	*Euphorbia dracunculoides* Lam.	Euphorbiaceae	Dried Whole Plant	**Foreign matter**- NMT 2 per cent; **Total ash**- NMT 11 per cent; **Acid-insoluble ash**- NMT 1 per cent; **Alcohol-soluble extractive**-NLT 5 per cent; **Water-soluble extractive**-NLT 10 per cent; **T.L.C.** - Refer API	Glyco-alkaloid (Euphorbine)	Rasa- Tikta, Kasaya; *Guna*- Laghu, Ruksa, Tiksna, Vikasi *Virya*- Sita; *Vipaka*- Katu; *Karma*- Vatala, Pittahara, Kaphahara, Raktadosahara, Vidbhedini	Brahmi Ghrta, Misraka Sneha, Narayana Churna	Gulma, Udavartta, Anaha, Udararoga, Vibandha, Visarpa	50 g of the drug for decoction
148	Part-I; Vol. II	153-154	SATAHVA (Fruit)	*Anethum sowa* Roxb. ex Flem. Syn. *A. graveolens* Linn. Var. *sowa* Roxb., *A. graveolens* DC., *Peucedanum sowa* Roxb., *P. graveolens* Benth.	Apiaceae	Dried Ripe Fruit	**Foreign matter**- NMT 5 per cent; **Total ash**- NMT 14 per cent; **Acid-insoluble ash**- NMT 1.5 per cent; **Alcohol-soluble extractive**-NLT 4 per cent; **Water-soluble extractive**-NLT 15 per cent; **Volatile oil** - NLT 3 per cent; **T.L.C.** - Refer API	Essential Oil	Rasa- Katu, Tikta; *Guna*- Snigdha; *Virya*- Usna; *Vipaka*- Katu; *Karma*- Vatahara, Kaphahara, Dipana, Sulaprasamana	Brhat phala Ghrta, Gorocanadi Vati, Narayana Churna, Sadbindu Taila	Jvara, Netra roga, Vrana, Sula, Atisara	3-6 g of the drug in powder form
149	Part-I; Vol. II	155-157	SIGRU (Leaf)	*Moringa oleifera* Lam. Syn. *Moringa pterygosperma* Gaertn.	Moringaceae	Dried Leaf	**Foreign matter**- NMT 2 per cent; **Total ash**- NMT 16 per cent; **Acid-insoluble ash**- NMT 4 per cent; **Alcohol-soluble extractive**-NLT 8 per cent; **Water-soluble extractive**-NLT 22 per cent; **T.L.C.** - Refer API	Carbohydrate, Protein, Carotene and Ascorbic acid	*Rasa* - Madhura; *Guna*- Guru, Ruksa, Tiksna; *Virya*- Sita; *Vipaka*- Madhura; *Karma*- Vatahara, Pittahara, Medohara, Sukra nasaka, Krmihara, Brmhana, Caksusya, Sirovirecaka	Visatinduka Taila, Ekangavira Rasa, Ratnagiri Rasa	Sopha, Krmiroga, Medoroga, Pliharoga, Vidradhi, Gulma, Galaganda	10-20 ml of the fresh drug in juice form

Sl. No.	*Part and Volume*	*Page No.*	*Common Ayurvedic Name*	*Botanical Name/ Chemical Name*	*Family/ Chemical Formula*	*Part Used*	*Identity, Purity and Strength/Chemical Properties*	*Constituents*	*Properties and Action*	*Important Formulations*	*Therapeutic Uses*	*Dose*
150	Part-I; Vol. II	158-159	STHULAELA (Seed)	*Amomum subulatum* Roxb.	Zingiberaceae	Dried Seed	**Foreign matter**- NMT 1 per cent; **Total ash**- NMT 4 per cent; **Acid-insoluble ash**- NMT 1.5 per cent; **Alcohol-soluble extractive**-NLT 5 per cent; **Water-soluble extractive**-NLT 14 per cent; **Volatile oil** - NLT 1 per cent (v/v)	Volatile Oil (rich in cineole)	*Rasa* - Katu, Tikta; *Guna*- Laghu, Ruksa, Tisksna; *Virya*- Usna; *Vipaka*- Katu; *Karma*- Vatahara, Kaphahara, Rocaka, Dipani, Mukhasodhaka, Angamar-daprasamana	Sarivadyasava, Karpuradyarka, Kalyanaka Ghrta, Vastyama-yantaka Ghrta, Manasamitra Vataka	Svasa, Kasa, Trsna, Chardi, Mukharoga, Hrllasa, Kandu	0.5-1 g of the drug in powder form
151	Part-I; Vol. II	160-161	TEJOVATI (Stem Bark)	*Zanthoxylum armatum* DC. Syn. *Z. alatum* roxb.	Rutaceae	Dried Stem Bark	**Foreign matter**- NMT 2 per cent; **Total ash**- NMT 12 per cent; **Acid-insoluble ash**- NMT 1.5 per cent; **Alcohol-soluble extractive**-NLT 8.5 per cent; **Water-soluble extractive**- NLT 13 per cent	A bitter crystalline principle identical with Berberine, a Volatile Oil and Resin.	Rasa- Katu, Tikta; *Guna*- Ruksa; *Virya*- Usna; *Vipaka*- Katu; *Karma*- Vatahara, Kaphahara, Dipana, Pacana, Rucya, Medhya	Pancatikta Guggulu Ghrta, Kalaka Churna (Lepa).	Svasa, Kasa, Mukharoga, Amavata, Aruci, Hikka	10-20 g of the drug for decoction
152	Part-I; Vol. II	162-164	TULASI (Whole Plant)	*Ocimum sanctum* Linn.	Lamiaceae	Whole Plant	**Foreign matter**- NMT 2 per cent; **Total ash**- NMT 10 per cent; **Acid-insoluble ash**- NMT 1.5 per cent; **Alcohol-soluble extractive**-NLT 4 per cent; **Water-soluble extractive**-NLT 8 per cent; **T.L.C.** - Refer API	Essential Oil	*Rasa* - Katu, Tikta, Kasaya; *Guna*- Tiksna, Ruksa, Laghu; *Virya*- Usna; *Vipaka*- Katu; *Karma*- Pittavardhini, Vatahara, Kaphahara, Hrdya, Dipana, Rucya, Durgandhihara	Tribhuvanakirti Rasa, Mukta-pancamrta Rasa, Muktadi Mahanjana, Manasamitra Vataka	Svasa, Kasa, Hikka, Chardi, Krmiroga, Parsva Sula, Kustha, Asmari, Netraroga	1-3 ml of the drug in juice form; 1-2 g of the drug in powder form (seed)
153	Part-I; Vol. II	165-167	TULASI (Leaf)	*Ocimum sanctum* Linn.	Lamiaceae	Dried Leaf	**Foreign matter**- NMT 2 per cent; **Total ash**- NMT 19 per cent; **Acid-insoluble ash**- NMT 3 per cent; **Alcohol-soluble extractive**-NLT 6 per cent; **Water-soluble extractive**-NLT 13 per cent; **T.L.C.** - Refer API	Essential Oil (Carvacrol, Caryophyllence, Nerol and Camphene *etc.*)	*Rasa* - Katu, Tikta, Kasaya; *Guna*- Laghu, Ruksa, Tiksna; *Virya*- Usna; *Vipaka*- Katu; *Karma*- Vatahara, Kaphahara, Pittahara, Dipani, Hrdya, Krmighna	Manasamitra Vataka, Tribhuvana Kirti Rasa, Mukta Pancamrt Rasa, Mahajvarankusa Rasa	Svasa, Kasa, Pratisyaya, Parsvasula, Aruci, Hikka, Krmiroga, Kustha	2-3 g of the drug in powder form

Sl. No.	Part and Volume	Page No.	Common Ayurvedic Name	Botanical Name/ Chemical Name	Family/ Chemical Formula	Part Used	Identity, Purity and Strength/Chemical Properties	Constituents	Properties and Action	Important Formulations	Therapeutic Uses	Dose
154	Part-I; Vol. II	168-170	VACA (Rhizome)	*Acorus calamus* Linn.	Araceae	Dried Rhizome	**Foreign matter**- NMT 1 per cent; **Total ash**- NMT 7 per cent; **Acid-insoluble ash**- NMT 1 per cent; **Alcohol-soluble extractive**-NLT 9 per cent; **Water-soluble extractive**-NLT 16 per cent; **Volatile oil** - NLT 2 per cent; **T.L.C.** - Refer API	Volatile Oil (principal constituents of the Volatile oil are Asamyl alcohol, Eugenol and Asarone), also contains a bitter principle Acorin (Glucoside), Starch and Tannin.	*Rasa* - Katu, Tikta; *Guna*- Laghu, Tiksna; *Virya*- Usna; *Vipaka*- Katu; *Karma*- Vatahara, Kaphahara, Mala Mutravisodhani, Dipani, Kanthya, Krmihara, Vamaka, Medhya	Vacadi Taila, Vaca Lasunadi Taila, Saraswata, Churna, Saraswata rista, Manasmitra Vataka, Candra Prabha Vati, Khadiradi Vati, Hinguvacadi Churna	Apasmara, Unmada, Vibandha, Adhamana, Sula, Krana srava, Kasa, Svasa, Smrti daurbalya.	60-120 mgs of the drug in powder form.; 1-2 g of the drug in powder form for inducing vomiting
155	Part-I; Vol. II	171-172	VATSANABHA (Root)	*Aconitum chasmanthum* Stapf. ex Holmes	Ranunculaceae	Dried Roots	**Foreign matter**- NMT 2 per cent; **Total ash**- NMT 5.5 per cent; **Acid-insoluble ash**- NMT 2 per cent; **Alcohol-soluble extractive**-NLT 8 per cent; **Water-soluble extractive**-NLT 24 per cent; **T.L.C.** - Refer API	Alkaloids	*Rasa* - Madhura; *Guna*- Usna, Ruksa, Tiksna, Laghu; Vikasi, Vyavayi, Yogavahi *Virya*- Usna; *Vipaka*- Madhura; *Karma*- Tridosahara, Rasayana, Svedala, Pittasantapakaraka	Tribhuwanakirti Rasa, Sutasekhara Rasa, Anandabhairava Rasa, Vatavidh-wansana Rasa, Mahavisagarbha Taila.	Sannipata, Vatakaphajvara, Vataroga, Jvaratisara, Kantharoga.	15-30 mg of the drug in powder form (Note: It is dangerous to exceed the normal dose)
156	Part-I; Vol. II	173-174	VIDARI (Tuberous Root)	*Pueraria tuberosa* DC.	Fabaceae	Sliced and Dried Pieces of Tuberous Root	**Foreign matter**- NMT 2 per cent; **Total ash**- NMT 17 per cent; **Acid-insoluble ash**- NMT 4.5 per cent; **Alcohol-soluble extractive**-NLT 4 per cent; **Water-soluble extractive**-NLT 24 per cent	Gluconic and Malic acids	*Rasa* - Madhura; *Guna*- Snigdha, Guru; *Virya*- Sita; *Vipaka*- Madhura; *Karma*- Vatahara, Pittahara, Stanyada, Sukrala, Mutrala, Jivaniya, Rasayana, Brmhaniya, Svarya, Varnya, Balya	Vidaryadikvatha Churna, Vidaryadi Ghrta, Marma Gutika, Manmathabhra Rasa, Pugakhanda (Aparah)	Daha, Raktapitta, Agnmarda, Daurbalya, Sosa	3-6 g of the drug in powder form

Sl. No.	*Part and Volume*	*Page No.*	*Common Ayurvedic Name*	*Botanical Name/ Chemical Name*	*Family/ Chemical Formula*	*Part Used*	*Identity, Purity and Strength/Chemical Properties*	*Constituents*	*Properties and Action*	*Important Formulations*	*Therapeutic Uses*	*Dose*
157	Part-I; Vol. II	175-176	YAVA (Fruit)	*Hordeum vulgare* Linn. *H. sativum* Pers.	Poaceae	Dried Fruit	**Foreign matter**- NMT 2 per cent; **Total ash**- NMT 4 per cent; **Acid-insoluble ash**- NMT 1.5 per cent; **Water-soluble ash** - NMT 4 per cent; **Alcohol-soluble extractive**-NLT 2.5 per cent; **Water-soluble extractive**- NLT 5.5 per cent; **T.L.C.** - Refer API	Starch, Sugars, Fats, Proteins (Albumin, Globulin, Prolamin and Glutilin) also contain Flavone Glycosides viz, Orientoside, Orientin, Vitexin *etc.*	*Rasa* - Kasaya, Madhura; *Guna*- Ruksa, Guru, Picchila, Mrdu; *Virya*- Sita; *Vipaka*- Katu; *Karma*- Vatakrt, Pittahara, Kaphahara, Medahara, Balya, Vrsya, Svarya, Varnya, Sthairyakara, Purisakrt, Mutrahara, Lekhana	Agastyaharitaki Rasayana, Eladya Modaka, Dahika Ghrta, Dhanvantara Ghrta, Gandharvahasta Taila, Dhanvantra Taila, Brhatmasa Taila, Sarsapadi Pralepa, Kayasthadya Vartti.	Medoroga, Prameha, Trsna, Urustambha, Kantharoga, Svasa, Kasa, Pinasa, Tvagroga	100-200 g of the drug
158	Part-I; Vol. II	177-179	YAVASAKA (Whole Plant)	*Alhagi pseudalhagi* (Bieb). Desv.	Fabaceae	Whole Plant	**Foreign matter**- NMT 2 per cent; **Total ash**- NMT 13.5 per cent; **Acid-insoluble ash**- NMT 2.5 per cent; **Alcohol-soluble extractive**-NLT 2 per cent; **Water-soluble extractive**-NLT 10 per cent	Sugars (Melizitose, Sucrose, Invert Sugars)	*Rasa* - Madhura, Tikta, Kasaya; *Guna*- Laghu, Sara; *Virya*- Sita; *Vipaka*- Madhura; *Karma*- Kaphahara, Pittahara, Dipana, Balakrt	Chinnodbhavadi Kvatha Churna, Gandhar-vahastadi Kvatha Churna, Bharangyadi Kvatha Churna, Arimedadi Taila	Trsna, Chardi, Kasa, Jwara, Vatarakta, Raktapitta, Visarpa	20-50 g of the drug in powder form for decoction
159	Part-I; Vol. III	1-2	ADHAKI (Root)	*Cajanus cajan* (Linn.) Millsp.	Fabaceae	Dried Root	**Foreign matter**- NMT 2 per cent; **Total ash**- NMT 3.5 per cent; **Acid-insoluble ash**- NMT 0.7 per cent; **Alcohol-soluble extractive**-NLT 2 per cent; **Water-soluble extractive**-NLT 4 per cent; **T.L.C.** - Refer API	Saponins and Reducing Sugars	*Rasa* - Kasaya, Madhura; *Guna*- Ruksa, Laghu; *Virya*- Sita; *Vipaka*- Katu; *Karma*- Vatakara, Pittahara, Kaphahara, Grahi, Varnya, Rucikara, Visaghna	Mahapancagavya Ghrta, kankayana Gutika	Raktavikara	2-6 g of the drugn in powder form

Sl. No.	*Part and Volume*	*Page No.*	*Common Ayurvedic Name*	*Botanical Name/ Chemical Name*	*Family/ Chemical Formula*	*Part Used*	*Identity, Purity and Strength/Chemical Properties*	*Constituents*	*Properties and Action*	*Important Formulations*	*Therapeutic Uses*	*Dose*
160	Part-I; Vol. III	3-4	AGNI-MANTHA (Root)	*Clerodendrum phlomidis* Linn.	Verbenaceae	Dried Mature Roots	**Foreign matter**- NMT 2 per cent; **Total ash**- NMT 6 per cent; **Acid-insoluble ash**- NMT 1 per cent; **Alcohol-soluble extractive**-NLT 2 per cent; **Water-soluble extractive**-NLT 5 per cent; **T.L.C.** - Refer API	Sterols	*Rasa* - Katu, Tikta, Kasaya; *Guna*- Laghu, Ruksha; *Virya*- Usna; *Vipaka*- Katu; *Karma*- Vatahara, Kaphahara, Svayathuhara	Dasamularista, Dasamula Kvatha Churna, Indukanta Ghrta, Dhanvantara Ghrta, Gorocanadi Vati, Narayana Taila	Sotha, Pandu, Arsa, Vatavikara, Vibandha, Agnimandya, Adhmana, Gulma, Mutrakrcchra, Mutraghata	12-24 g of the drug inpowder form for decoction
161	Part-I; Vol. III	5-6	AMBASTHAKI (Root)	*Hibiscus sabdariffa* Linn.	Malvaceae	Dried Roots	**Foreign matter**- NMT 2 per cent; **Total ash**- NMT 11 per cent; **Acid-insoluble ash**- NMT 3 per cent; **Alcohol-soluble extractive**-NLT 2 per cent; **Water-soluble extractive**-NLT 5 per cent; **T.L.C.** - Refer API	Sterols and Polysaccharides	*Rasa* - Madhura, Amla, Tikta, Kasaya *Guna*- Laghu *Virya*-. *Vipaka*- Amla; *Karma*- Pittahara, Kaphahara, Asthisandhanaka, Vranaropana, Rucikara, Dipana, Kanthasodhana	Pusyanuga Churna	Pakvatisara, Kapharoga, Galaroga, Vataroga, Asthibhagna, Vrana	5-10 g
162	Part-I; Vol. III	7-8	AMRA (Seed)	*Mangifera indica* Linn.	Anacardiaceae	Dried Seed	**Foreign matter**- NMT 1 per cent; **Total ash**- NMT 3 per cent; **Acid-insoluble ash**- NMT 0.5 per cent; **Alcohol-soluble extractive**-NLT 10 per cent; **Water-soluble extractive**- NLT 10 per cent; **T.L.C.** - Refer API	Tannins-Pyrogallo-tannins	*Rasa* - Kasaya, Madhura; *Guna*- Ruksa; *Virya*- Sita; *Vipaka*- Katu; *Karma*- Samgrahi, Vatakara, krmighna	Pusyanuga churna, Brhat Gangadhara Churna	Atisara, Pravahika, Chardi, Daha, Tvagroga	1-2 g of the drug in powder form
163	Part-I; Vol. III	9-10	AMRA (Stem Bark)	*Mangifera indica* Linn.	Anacardiaceae	Dried Stem Bark	**Foreign matter**- NMT 2 per cent; **Total ash**- NMT 9 per cent; **Acid-insoluble ash**- NMT 2 per cent; **Alcohol-soluble extractive**-NLT 20 per cent; **Water-soluble extractive**- NLT 14 per cent; **T.L.C.** - Refer API	Tannins--Protocatechuic Acid, Catechin, Mangiferin, Alanine, Glycine, α-aminobutyric acid, Kinic and Shikimic Acids	*Rasa* - Kasaya; *Guna*- Laghu, Ruksa; *Virya*- Sita; *Vipaka*- Katu; *Karma*- Grahi, Kaphapittasamaka, Vranaropana, Rucya	Nyagrodhadi Churna, Nyagrodhadi Kwatha Churna, Candanasava, Grahanimihira Taila, Mutra sangrahaniya Kasaya Churna	Atisara, Vrana, Agnimandya, Grahani, Prameha, Yoni roga	3-6 g of powder.; 25-50 g for decoction

Sl. No.	*Part and Volume*	*Page No.*	*Common Ayurvedic Name*	*Botanical Name/ Chemical Name*	*Family/ Chemical Formula*	*Part Used*	*Identity, Purity and Strength/Chemical Properties*	*Constituents*	*Properties and Action*	*Important Formulations*	*Therapeutic Uses*	*Dose*
164	Part-I; Vol. III	11-12	AMRATAK (Stem)	*Spondias pinnata* (Linn. F.) Kurz Syn. *S. mangifera* Willd., *S. acuminata* Roxb. non Gamble	Anacardiaceae	Dried Stem	**Foreign matter**- NMT 1 per cent; **Total ash**- NMT 6 per cent; **Acid-insoluble ash**- NMT 0.5 per cent; **Alcohol-soluble extractive**-NLT 2 per cent; **Water-soluble extractive**-NLT 5 per cent; **T.L.C.** - Refer API	Tannins	*Rasa* - Kasaya, Amla; *Guna*-Guru; *Virya*- Usna; *Vipaka*- Katu; *Karma*- Vataghna, Saraka	Dadhika Ghrta	Daha, Ksaya, Rakta Vikara, Atisara	1-3 g of powder
165	Part-I; Vol. III	13-14	APAMARGA (Root)	*Achyranthes aspera* Linn.	Amaranthaceae	Dried Root	**Foreign matter**- NMT 1 per cent; **Total ash**- NMT 9 per cent; **Acid-insoluble ash**- NMT 1 per cent; **Alcohol-soluble extractive**-NLT 2 per cent; **Water-soluble extractive**-NLT 10 per cent; **T.L.C.** - Refer API	Saponins	*Rasa* - Tikta, Katu; *Guna*- Laghu, Ruksa, Tiksna, Sara; *Virya*- Usna; *Vipaka*- Katu; *Karma*- Dipana, Pacana, Rucya, Vatahara, Kaphanasaka, Medohara, Mutrala, Vantihara	Agastya Haritaki Rasayana, Maha Pancagavya Ghrta, Vastyama-yantaka Ghrta, Maha Visagarbha Taila, Apamarga Ksara, Ksara Taila, Panaviraladi Ksara	Chardi, Adhmana, Kandu, Sula, Apaci, Granthi, Bhagandara, Hrdaroga, Jwara, Switra, Vadhirya, Udara roga, Yakrt roga, Danta roga, Rakta vikara	5-10 g
166	Part-I; Vol. III	15-16	ARALU (Stem Bark)	*Ailanthus excelsa* Roxb.	Simarubaceae	Dried Stem Bark	**Foreign matter**- NMT 2 per cent; **Total ash**- NMT 8.5 per cent; **Acid-insoluble ash**- NMT 0.5 per cent; **Alcohol-soluble extractive**-NLT 1.5 per cent; **Water-soluble extractive**- NLT 5.5 per cent; **T.L.C.** - Refer API	β-sitosterol, Quassinoids, Ailantic Acid, 2-6 dimethoxy-Benzoquinone and Melanthin.	*Rasa* - Tikta, Kasaya; *Guna*-Ruksa; *Virya*-Katu; *Vipaka*-Sita; *Karma*-Kaphapitta, Samaka, Dipana, Pacana, Grahi, Vranasodhana	Pusyanuga Churna, Brhat Gangadhara Churna, Aralu Putapaka	Atisara, Krmi, Arsa, Sannipata Jwara, Bhrama, Tvakroga, Chardi, Kustha, Pravahika, Grahani, Prameha, Swasa, Gulma, Musaka Visaja Roga	1-3 g

Sl. No.	Part and Volume	Page No.	Common Ayurvedic Name	Botanical Name/ Chemical Name	Family/ Chemical Formula	Part Used	Identity, Purity and Strength/Chemical Properties	Constituents	Properties and Action	Important Formulations	Therapeutic Uses	Dose
167	Part-I; Vol. III	17-18	ARKA (Stem Bark)	*Calotropis procera* (Ait.) R. Br.	Asclepiadaceae	Dried Stem Bark	**Foreign matter**- NMT 2 per cent; **Total ash**- NMT 12 per cent; **Acid-insoluble ash**- NMT 1.5 per cent; **Alcohol-soluble extractive**-NLT 7 per cent; **Water-soluble extractive**-NLT 15 per cent; **T.L.C.** - Refer API	α-and-β-calotropeols, β-amyrin, Giganteol, a Colourless wax small amount of Tetracyclic Terpenes and Traces of Sterols.	*Rasa* - Katu, Tikta; *Guna*- Laghu, Ruksa, Tiksna, Sara; *Virya*- Usna; *Vipaka*- Katu; *Karma*- Sodhana, Virecan, Vatahara, Dipana, Lekhan, Ropana	Abhaya Lavana, Arka Lavana	Udararoga, Kustha, Kandu, Vrana, Pliharoga, Gulma, Arsa, Krmiroga	0.5 - 1 g in powder form
168	Part-I; Vol. III	19-20	ASANA (Stem Bark)	*Pterocarpus marsupium* Roxb.	Fabaceae	Dried Stem Bark	**Foreign matter**- NMT 2 per cent; **Total ash**- NMT 18 per cent; **Acid-insoluble ash**- NMT 1.5 per cent; **Alcohol-soluble extractive**-NLT 7.5 per cent; **Water-soluble extractive**- NLT 11.5 per cent; **T.L.C.** - Refer API	Tannins and Gum Kino (which contains Kino-Tannic Acid, 1-epicatechin and a reddish brown colouring matter)	*Rasa* - Ksaya, Katu, Tikta; *Guna*- Laghu, Ruksa; *Virya*- Usna; *Vipaka*- Katu; *Karma*- Saraka, Vatartidosanut, Galadosaghna, Kesya, Tvacya, Raktaman-dalnasini, Slesmahara, Pittahara	Narasingha Ghrta Rasayana	Pandu, Prameha, Medodosa, Kustha, Krmiroga, Switra, Madhumeha, Sthoulya	32-50 g of the drug for decoction
169	Part-I; Vol. III	21-22	ASTHI-SAMHRTA (Stem)	*Cissus quadran-gularis* Linn.	Vitaceae	Dried Stem	**Foreign matter**- NMT 2 per cent; **Total ash**- NMT 22 per cent; **Acid-insoluble ash**- NMT 1.5 per cent; **Alcohol-soluble extractive**-NLT 4 per cent; **Water-soluble extractive**-NLT 20 per cent; **T.L.C.** - Refer API	Calcium Oxalate, Carotene and Ascorbic Acid	*Rasa* - Katu, Madhura; *Guna*- Laghu, Ruksa, Sara; *Virya*- Usna; *Vipaka*- Madhura; *Karma*- Dipana, Vataslesmahara, Asthisan-dhanakara, Caksusya, Vrsya	Laksadi Guggulu	Krmi, Arsa, Asthibhagna, Sandhi Cyuta	10-20 ml. (Svarasa); 3-6 g (Powder)

Sl. No.	*Part and Volume*	*Page No.*	*Common Ayurvedic Name*	*Botanical Name/ Chemical Name*	*Family/ Chemical Formula*	*Part Used*	*Identity, Purity and Strength/Chemical Properties*	*Constituents*	*Properties and Action*	*Important Formulations*	*Therapeutic Uses*	*Dose*
170	Part-I; Vol. III	23-24	ATMAGUPTA (Seed)	*Mucuna prurita* Hook., Syn. *M. pruriens* Baker.	Fabaceae	Dried Mature Seed	**Foreign matter**- NMT 1 per cent; **Total ash**- NMT 5 per cent; **Acid-insoluble ash**- NMT 1 per cent; **Alcohol-soluble extractive**-NLT 3 per cent; **Water-soluble extractive**-NLT23 per cent; **Fixed oil** - NLT 3 per cent; **T.L.C.** - Refer API	Fixed Oil, Alkaloid and 3,4-dihydroxy-phenylalanine	*Rasa* - Madhura, Tikta; *Guna*- Guru, Snigdha; *Virya*- Sita; *Vipaka*- Madhura; *Karma*- Vatasamana, Vrsya, Kaphanasaka, Pittanasaka, Raktadosanasaka, Brhana, Balya	Brhat Masa Taila	Vatavyadhi, Kampavata, Klaivya, Raktapitta, Dustavrana, Daurbalya	3-6 g
171	Part-I; Vol. III	25-26	BHARANGI (Root)	*Clerodendrum serratum* (Linn.) Moon	Verbenaceae	Dried Roots	**Foreign matter**- NMT 2 per cent; **Total ash**- NMT 11 per cent; **Acid-insoluble ash**- NMT 1 per cent; **Alcohol-soluble extractive**-NLT 6 per cent; **Water-soluble extractive**-NLT 12 per cent; **T.L.C.** - Refer API	Saponins	*Rasa* - Katu, Tikta, Kasaya; *Guna*- Laghu, Ruksa; *Virya*- Usna; *Vipaka*- Katu; *Karma*- Vatahara, Kaphahara, Dipana, Pacana, Swasahara, Rucya	Ayaskrti, Kanakasawa, Dasamularista, Rasnadi Kwatha Churna, Dhanwantara Ghrta, Maha Vatagajankusa Rasa	Gulma, Jwara, Swasa, Kasa, Yaksma, Pinasa, Sotha, Hikka, Raktadosa	3-6 g of powder.; 10-20 g of kwatha churna
172	Part-I; Vol. III	27-28	BIJAPURA (Fresh Fruit)	*Citrus medica* Linn.	Rutaceae	Fresh Fruit	**Foreign matter**- Nil **Total ash**- NMT 5 per cent; **Acid-insoluble ash**- NMT 0.2 per cent; **Alcohol-soluble extractive**-NLT 20 per cent; **Water-soluble extractive**- NLT 45 per cent; **T.L.C.** - Refer API	Volatile oil	*Rasa* - Amla, Madhura; *Guna*- Laghu, Snigdha; *Virya*- Usna; *Vipaka*- Amla; *Karma*- Vatahara, Pittahara, Kaphahara, Dipana, Hrdya, Kantha Sodhaka, Jihvasodhaka, Varnanasaka, Medhya, Chardigrahana	Ksara Taila, Hinguwadi Churna, Kankayana Gutika, Tarunarka Rasa, Sankha Dravaka, Madiphala Rasayana	Rakatapitta, Swasa, Kasa, Aruci, Trsna, Udara roga, Vibandha, Madatyaya, Hikka, Agnimandya	10-20 ml. of juice

Sl. No.	*Part and Volume*	*Page No.*	*Common Ayurvedic Name*	*Botanical Name/ Chemical Name*	*Family/ Chemical Formula*	*Part Used*	*Identity, Purity and Strength/Chemical Properties*	*Constituents*	*Properties and Action*	*Important Formulations*	*Therapeutic Uses*	*Dose*
173	Part-I; Vol. III	29-31	BILVA (Root)	*Aegle marmelos* Corr.	Rutaceae	Dried Root	**Foreign matter**- NMT 1 per cent; **Total ash**- NMT 6 per cent; **Acid-insoluble ash**- NMT 1 per cent; **Alcohol-soluble extractive**-NLT 7 per cent; **Water-soluble extractive**-NLT 7 per cent; **T.L.C.** - Refer API	Auraptene, Coumarins, Glycosides	*Rasa* - Madhura; *Guna*- Laghu; *Virya*- Sita; *Vipaka*- Madhura; *Karma*- Mutrala, Tridosaghna	Manasa Mitra Vataka, Amrtarista, Dantyadyarista, Agastya Haritaki Rasayana, Dasamularista, Dasamula Kwatha Churna, Bilvadi Leha	Vatavyadhi, Sotha, Sula, Agnimandya, Chardi, Mutrakrcchra, Amavata	2-6 g of the drug in powder form
174	Part-I; Vol. III	32-35	BIMBI (Whole Plant)	*Coccinia indica* W. and A. = *C. cordifolia* Cogn. Syn. *Cephalandra indica* Naud.	Cucurbitaceae	Dried Whole Plant	**Foreign matter**- NMT 2 per cent; **Total ash**- NMT 21 per cent; **Acid-insoluble ash**- NMT 2 per cent; **Alcohol-soluble extractive**-NLT 3 per cent; **Water-soluble extractive**-NLT 14 per cent; **T.L.C.** **- Refer API**	Saponins and Fixed Oil in seeds.	*Rasa* - Tikta, Madhura; *Guna*- Guru, Ruksa; *Virya*- Sita; *Vipaka*- Katu; *Karma*- Vatakara, Pittahara, Atirucya, Lekhhana, Stambhana, Vibandhadh-manakara, Chardikara	Vastya-mayantaka Ghrta	Kasa, Swasa, Jwara, Raktavikara, Daha, Sopha, Pandu	3-6 g of the drug in powder form; 5-10 ml. (Svarasa)
175	Part-I; Vol. III	36-38	CANGERI (Whole Plant)	*Oxalis corniculata* Linn.	Oxalidaceae	Dried Whole Plant	**Foreign matter**- NMT 2 per cent; **Total ash**- NMT 20 per cent; **Acid-insoluble ash**- NMT 10 per cent; **Alcohol-soluble extractive**-NLT 5 per cent; **Water-soluble extractive**-NLT 13 per cent; **T.L.C.** - Refer API	Vitamin C, Carotene, Tartaric Acid, Citric Acid and Malic Acid	*Rasa* - Amla, Kasaya; *Guna*- Laghu, Ruksa; *Virya*- Usna; *Vipaka*- Amla; *Karma*-Grahi, Pittakara, Dipana, Agnivardhaka, Rucikara, Vatahara, Kaphahara	Cangeri Ghrta	Grahani, Arsa, Kustha, Atisara	5-10 ml. (Svarasa). It is also used externally

Sl. No.	Part and Volume	Page No.	Common Ayurvedic Name	Botanical Name/ Chemical Name	Family/ Chemical Formula	Part Used	Identity, Purity and Strength/Chemical Properties	Constituents	Properties and Action	Important Formulations	Therapeutic Uses	Dose
176	Part-I; Vol. III	39-40	CIRABILVA (Fruit)	*Holoptelea integrifolia* Planch.	Ulmaceae	Dried Fruit	**Foreign matter**- NMT 1 per cent; **Total ash**- NMT 9 per cent; **Acid-insoluble ash**- NMT 1 per cent; **Alcohol-soluble extractive**-NLT 10 per cent; **Water-soluble extractive**- NLT 13 per cent; **T.L.C.** - Refer API	Fixed Oil	*Rasa* - Tikta, Kasaya; *Guna*- Laghu, Ruksa; *Virya*- Usna; *Vipaka*- Katu; *Karma*- Pittahara, Stambhaka	Piyusavalli Rasa, Gandhar-vahastadi Kwatha Churna	Chardi, Arsa, Krmi, Kustha, Prameha	1-3 g
177	Part-I; Vol. III	41-42	DANTI (Root)	*Baliospermum montanum* Muell.-Arg.	Euphorbiaceae	Dried Root	**Foreign matter**- NMT 2 per cent; **Total ash**- NMT 10 per cent; **Acid-insoluble ash**- NMT 3 per cent; **Alcohol-soluble extractive**-NLT 1.5 per cent; **Water-soluble extractive**- NLT 3 per cent; **T.L.C.** - Refer API	β-sitosterol and Triterpenoids, Resinous Glycosides, Phorbol Esters.	*Rasa* - Katu; *Guna*- Tiksna, Sara, Laghu; *Virya*- Usna; *Vipaka*- Katu; *Karma*- Kaphahara, Raktadosahara, Vidahara, Dipana, Rocaka, Sodhaka, Vikasi Vrana	Dantyadyarista, Punarnava Mandura, Abhayarista, Kankayana Gutika, Dantiharitaki, Kalyanaka Ksara, Kaisora Guggulu	Tvakadosa, Daha, Sotha, Udararoga, Sularoga, Krimi, Arsa, Asmari, Kandu, Kustha, Vrana, Pliha, Vrddhi, Gulma, Kamala	1-3 g of the drug in powder form
178	Part-I; Vol. III	43-44	DHATTURA (Seed)	*Datura metel* Linn.; Syn. *D. fastuosa* L., *D. alba* Ramph; *D. cornucopaea* Hort.	Solanaceae	Dried Seeds	**Foreign matter**- NMT 2 per cent; **Total ash**- NMT 6 per cent; **Acid-insoluble ash**- NMT 1 per cent; **Alcohol-soluble extractive**-NLT 5 per cent; **Water-soluble extractive**- NLT 7 per cent; **T.L.C.** - Refer API	Alkaloids- Tropane Alkaloids- Hyoscyamine *etc.* and Fixed Oil.	*Rasa* - Madhura, Katu, Kasaya, Tikta; *Guna*- Tiksna, Ruksa, Guru; *Virya*- Usna; *Vipaka*- Katu; *Karma*- Madakari, Kaphahara, Visahara, Krmihara, Vranahara, Kanduhara, Bhramahara, Varnya, Vamaka	Kanakasava, Suta Sekhara rasa, Jwarankusa Rasa, Laksmi vilasa Rasa (Naradiya), Kanakasundara Rasa, Dugdha Vati, Piyusavalli Rasa	Krmi, Yuka, Liksa	30-60 mg

Sl. No.	*Part and Volume*	*Page No.*	*Common Ayurvedic Name*	*Botanical Name/ Chemical Name*	*Family/ Chemical Formula*	*Part Used*	*Identity, Purity and Strength/Chemical Properties*	*Constituents*	*Properties and Action*	*Important Formulations*	*Therapeutic Uses*	*Dose*
179	Part-I; Vol. III	45-46	DRAKSA (Fruit)	*Vitis vinifera* Linn.	Vitaceae	Dried Mature Fruits	**Foreign matter**- NMT 2 per cent; **Total ash**- NMT 3 per cent; **Acid-insoluble ash**- NMT 0.2 per cent; **Alcohol-soluble extractive**-NLT 25 per cent; **Water-soluble extractive**- NLT 70 per cent; **Loss on drying** - NMT 15 per cent **T.L.C.** - Refer API	Malic, Tartaric and Oxalic Acids, Carbohydrates and Tannins.	*Rasa* - Madhura, Kasaya; *Guna*- Guru, Sara, Snigdha; *Virya*- Sita; *Vipaka*- Madhura; *Karma*- Brmhana, Caksusya, Vrsya, Vatapittahara, Swarya	Draksasava, Draksarista, Draksavaleha, Draksadi Kwatha Churna, Draksadi Churna, Eladi Gutika	Trsna, Jwara, Kasa, Swasa, Daha, Sosa, Kamala, Raktapitta, Ksata Ksina, Vibandha, Arsa, Agnimandya, Madatyaya, Pandu, Udavarta, Asya Sosa, Vatarakta	5 -10 g of the drug
180	Part-I; Vol. III	47-48	DURVA (Root)	*Cynodon dactylon* (Linn.) Pers.	Poaceae	Dried Fibrous Root	**Foreign matter**- NMT 2 per cent; **Total ash**- NMT 7 per cent; **Acid-insoluble ash**- NMT 3 per cent; **Alcohol-soluble extractive**-NLT 1 per cent; **Water-soluble extractive**-NLT 5 **per cent; T.L.C.** - Refer API	Phenolic Phytotoxins and Flavonoids.	*Rasa* - Kasaya, Madhura, Tikta; *Guna*- Laghu; *Virya*- Sita; *Vipaka*- Madhura; *Karma*- Kaphapittasamaka, Raktapittanasaka, Dahaghna, Atisaraghna, Sramahara	Balasvagandha Laksadi Taila, Madhuyastyadi Taila, Marma Gutika, Manasa Mitra Vataka, Candrakala Rasa	Raktapitta, Trsnaroga, Daharoga, Visarpa, Tvakaroga, Arocaka, Duhswapna, Bhutaroga, Raktapitta, Chardi, Murccha, Raktapradara, Mutra Daha	5-10 ml. (Svarasa)
181	Part-I; Vol. III	49-50	ERANDA (Fresh Leaf)	*Ricinus communis* Linn.	Euphorbiaceae	Fresh Leaf	-	-	*Rasa* - Madhura, Katu, Kasaya; *Guna*- Snigdha, Tiksna, Suksma; *Virya*- Usna; *Vipaka*- Madhura; *Karma*- Kaphavatasamaka, Vrsya, Krmighna, Pittaprakopaka, Raktaprakopaka, Yakrtutejaka	Caturbhuja Rasa, Caturmukha Rasa, Cintamani Caturmukha Rasa	Krmi, Mutrakrcchra, Gulma, Vatavyadha, Vasti Sula, Arocaka, Vidradhi	10-20 ml (Svarasa) 2-3 g (Powder)

Sl. No.	*Part and Volume*	*Page No.*	*Common Ayurvedic Name*	*Botanical Name/ Chemical Name*	*Family/ Chemical Formula*	*Part Used*	*Identity, Purity and Strength/Chemical Properties*	*Constituents*	*Properties and Action*	*Important Formulations*	*Therapeutic Uses*	*Dose*
182	Part-I; Vol. III	51-52	ERANDA (Seed)	*Ricinus communis* Linn.	Euphorbiaceae	Dried Seed	**Foreign matter**- NMT 2 per cent; **Total ash**- NMT 4 per cent; **Acid-insoluble ash**- NMT 1 per cent; **Alcohol-soluble extractive**-NLT 36 per cent; **Water-soluble extractive**- NLT 6 per cent; **Fixed oil** - NLT 37 per cent; **T.L.C.** - Refer API	Fixed Oil	*Rasa* - Madhura, Katu, Kasaya; *Guna*- Snigdha, Tiksna, Suksma; *Virya*- Usna; *Vipaka*- Madhura; *Karma*- Dipana, Amapacana, Vidbhedana, Anulomana, Srotosodhana, Vayasthapana, Medohara	Brhat Saindhavadi Taila, Gandhar-vahastadi Taila, Simhanada Gaggulu, Misraka Sneha	Amavata, Vibandha, Yakrt Roga, Plihodara, Arsa, Kati sula, Grdhrasi	1/2-3 g (Powder)
183	Part-I; Vol. III	53-54	GAMBHARI (Stem)	*Gmelina arborea* Roxb.	Verbenaceae	Dried Stem	**Foreign matter**- NMT 2 per cent; **Total ash**- NMT 3 per cent; **Acid-insoluble ash**- NMT 0.3 per cent; **Alcohol-soluble extractive**-NLT 1 per cent; **Water-soluble extractive**- NLT 4 per cent; **T.L.C.** - Refer API	Lignans	*Rasa* - Madhura, Tikta, Kasaya, Katu; *Guna*- Guru; *Virya*- Usna; *Vipaka*- Madhura; *Karma*- Vatahara, Pittahara, Kaphahara, Dipana, Pacana, Bhedani, Medhya, Virecanopaga, Visahara, Sramahara	Karpuradi Kuzambu (Laghu), Candanasava, Dantadyarista Usirasava	Sopha, Jwara, Daha, Trsna, Raktadosa, Visavikara, Arsa, Sula, Raktapitta, Bhrama, Sosa, Ama Sula	5-10 g of the drug for decoction
184	Part-I; Vol. III	55-57	GOJIHVA	*Onosma bracteatum* Wall.	Boraginaceae	Dried Leaf	**Foreign matter**- NMT 2 per cent; **Total ash**- NMT 26 per cent; **Acid-insoluble ash**- NMT 4 per cent; **Alcohol-soluble extractive**-NLT 1 per cent; **T.L.C.** - Refer API	Tannin and Sugars	*Rasa* - Kasaya, Tikta, Madhura; *Guna*- Laghu; *Virya*- Sita; *Vipaka*- Madhura; *Karma*- Vatala, Pittahara, Kaphahara, Hrdya, Grahi	Manasa Mitra Vataka, Gojihvadi Kwatha	Raktapitta, Kustha, Jwara, Swasa, Kasa, Aruci, Prameha, Raktavikara, Vrana, Danta roga	3-6 g of the drug in powder form

Sl. No.	Part and Volume	Page No.	Common Ayurvedic Name	Botanical Name/ Chemical Name	Family/ Chemical Formula	Part Used	Identity, Purity and Strength/Chemical Properties	Constituents	Properties and Action	Important Formulations	Therapeutic Uses	Dose
185	Part-I; Vol. III	58-59	GRANTHI-PARNI (Root)	*Leonotis nepetaefolia* R. Br.	Lamiaceae	Root	**Foreign matter**- NMT 2 per cent; **Total ash**- NMT 5 per cent; **Acid-insoluble ash**- NMT 1 per cent; **Alcohol-soluble extractive**-NLT 2 per cent; **Water-soluble extractive**-NLT 4 per cent; **T.L.C.** - Refer API	Sterols	*Rasa* - Tikta; *Guna*- Laghu, Tiksna; *Virya*- Usna; *Vipaka*- Katu; *Karma*- Dipana, Kaphavatahara, Daurgandh-yanasana	Brhat Guduci Taila, Mrtasanjivani Sura	Swasa, Kandu, Visa	5-10 g of the drug in powder form
186	Part-I; Vol. III	60-62	HAMSAPADI (Whole Plant)	*Adiantum lunulatum* Burm.	Polypodiaceae	Dried Whole Plant	**Foreign matter**- NMT 2 per cent; **Total ash**- NMT 16 per cent; **Acid-insoluble ash**- NMT 11 per cent; **Alcohol-soluble extractive**-NLT 3 per cent; **Water-soluble extractive**-NLT 5 per cent; **T.L.C.** - Refer API	–	*Rasa* - Kasaya, Tikta; *Guna*- Guru; *Virya*- Sita; *Vipaka*- Katu; *Karma*- Raktavikarahrta, Visaghna	Madhuyastyadi Taila, Manasa Mitra Vataka, Mukta Pancamrta Rasa, Swarnabhupati Rasa, Kalakuta Rasa	Visarpa, Vrana, Daha, Atisara, Luta Visa, Bhuta Graha, Kaksa Sphota, Rakta Vikara	1-3 g
187	Part-I; Vol. III	63-64	HAPUSA (Fruit)	*Juniperus communis* Linn.	Cupressaceae	Dried Fruit	**Foreign matter**- NMT 1 per cent; **Total ash**- NMT 5 per cent; **Acid-insoluble ash**- NMT 0.5 per cent; **Alcohol-soluble extractive**-NLT 12 per cent; **Water-soluble extractive**- NLT 9 per cent; **T.L.C.** - Refer API	Essential Oil and Flavonoids	*Rasa* - Tikta, Katu, Kasaya; *Guna*- Guru, Mrdu; *Virya*- Usna; *Vipaka*- Katu; *Karma*- Agnidipaka, Vatanasaka, Kaphanasaka, Visaghna	Kumaryasava, Saptavinsitika Guggulu, Dadhika Ghrta, Narayana Churna, Trayo-dasanga Guggulu, Pradarantaka Lauha, Nitya-nanda Rasa	Pittodara, Arsa, Grahani, Gulma, Sula, Krmi, Vatodara, Pliharoga	2-6 g in powder form
188	Part-I; Vol. III	65-66	INDRA-VARUNI (Fruit)	*Citrullus colocynthis* Schrad.	Cucurbitaceae	Dried/ Peeled cut Pieces of the Fruit	**Foreign matter**- NMT 2 per cent; **Total ash**- NMT 14 per cent; **Acid-insoluble ash**- NMT 7 per cent; **Light Petroleum soluble -matter:** - On continuous extraction with light petroleum (b.p 40° to 60°) and drying at 100°c, not more than 3.0 per cent; **T.L.C.** - Refer API	Resins-Resinous Glycosides (Colocynthin and Colocyn-thitin), A Phytosterol Citrullol, Pectin and Albuminoids, Cucurbitacins-Cucurbitacin E and I.	*Rasa* - Tikta; *Guna*- Laghu, Ruksa, Tiksna; *Virya*- Usna; *Vipaka*- Katu; *Karma*- Vamaka, Recana, Krmighna, Slesmahara, visahara	Jawaraghni Gutika (II)	Krmiroga, Kamala, Swasa, Kasa, Kustha, Gulma, Udararoga	0.125-0.5 g of powder.; 0.25-0.5 g of powder

Sl. No.	*Part and Volume*	*Page No.*	*Common Ayurvedic Name*	*Botanical Name/ Chemical Name*	*Family/ Chemical Formula*	*Part Used*	*Identity, Purity and Strength/Chemical Properties*	*Constituents*	*Properties and Action*	*Important Formulations*	*Therapeutic Uses*	*Dose*
189	Part-I; Vol. III	67-68	INDRAYAVA (Seed)	*Holarrhena antidysenterica* Wall.	Apocynaceae	Dried Seeds	**Foreign matter**- NMT 2 per cent; **Total ash**- NMT 8 per cent; **Acid-insoluble ash**- NMT 3 per cent; **Alcohol-Soluble** extractive-NLT 12 per cent; **T.L.C.** - Refer API	Alkaloids- Steroidal Alkaloid, Conessine *etc.*, Fats, Tannin and Resin.	*Rasa* - Katu, Tikta; *Guna*- Laghu, Ruksa; *Virya*- Sita; *Vipaka*- Katu; *Karma*- Dipana, Tridosasamaka, Sangrahi	Panca Nimba Churna, Palasa Bijadi Churna, Laughu Gangadhara Churna, Krmi Kuthara Rasa, Piyusavalli Rasa, Jwaraghni Gutika, Siddha Praneswara Rasa, Ahiphenasava	Atisara, Kustha, Jwaratisara, Krmi, Visarpa, Grahani, Raktatisara, Sula, Chardi, Twakroga, Daha	3-6 g (Churna).; 20-30 g (Decoction)
190	Part-I; Vol. III	69-70	ISVARI (Root)	*Aristolochia indica* Linn.	Aristolochiaceae	Dried Root	**Foreign matter**- NMT 2 per cent; **Total ash**- NMT 4 per cent; **Acid-insoluble ash**- NMT 1 per cent; **Alcohol-soluble extractive**-NLT 2 per cent; **Water-soluble extractive**- NLT 3 per cent; **T.L.C.** - Refer API	Alkaloids, Essential Oils, Bitter Principles and Fixed Oil.	*Rasa* - Tikta, Katu, Kasaya; *Guna*- Laghu, Ruksa; *Virya*- Usna; *Vipaka*- Katu; *Karma*- Kaphavatasamaka, Sothahara, Raksoghna, Grahabadhaghna	Mahavisagarbha Taila, Gorocanadi Gutika	Sarpavisa, Luta Visa, Jalagardabha, Vrscikavisa, Jwara, Krmi, Vrana	1-2 g (For external use also)
191	Part-I; Vol. III	71-72	JATI (Leaf)	*Jasminum officinale* Linn.	Oleaceae	Dried Leaves	**Foreign matter**- NMT 2 per cent; **Total ash**- NMT 6 per cent; **Acid-insoluble ash**- NMT 0.5 per cent; **Alcohol-soluble extractive**-NLT 18 per cent; **Water-soluble extractive**- NLT 25 per cent; **T.L.C.** - Refer API	Resin, Salicylic Acid, Alkaloid (jasminine) and Essential Oil	*Rasa* - Tikta, Kasaya; *Guna*- Laghu, Snigdha, Mrdu; *Virya*- Usna; *Vipaka*- Katu; *Karma*- Sirovirecana, Caksusya	Jatyadi Taila, Jatyadi Ghrta, Vasanta Kusumakara Rasa	Siroroga, Aksiroga, Visaroga, Kustha, Vrana, Arsa, Mukhapaka, Putikarna, Stana Sotha, Raktavikara	10-20 g of powder for decoction
192	Part-I; Vol. III	73-74	KADALI (Rhizome)	*Musa paradisiaca* Linn.	Musaceae	Fresh Rhizome	**T.L.C.** - Refer API	Fixed Oil and 4 α- Methyl Sterol Ketone	*Rasa* - Madhura, Kasaya; *Guna*- Sita, Guru, Ruksa; *Virya*- Sita; *Vipaka*- Madhura; *Karma*- Balya, Kaphahara, Pittahara, Dipana, Rucya, Kesya	Abhraka Bhasma (Sataputi), Ksara Taila	Krmi, Kustha, Karna Sula, Somaroga, Amlapitta, Daha, Raktavikara, Rajodosa, Mutrakrcchra	10-20 g in powder form; 10-20 ml in juice form

Sl. No.	*Part and Volume*	*Page No.*	*Common Ayurvedic Name*	*Botanical Name/ Chemical Name*	*Family/ Chemical Formula*	*Part Used*	*Identity, Purity and Strength/Chemical Properties*	*Constituents*	*Properties and Action*	*Important Formulations*	*Therapeutic Uses*	*Dose*
193	Part-I; Vol. III	75-76	KAKAJANGHA (Root)	*Peristrophe bicalyculata* Ness	Acanthaceae	Dried Root	**Foreign matter**- NMT 2 per cent; **Total ash**- NMT 9 per cent; **Acid-insoluble ash**- NMT 2 per cent; **Alcohol-soluble extractive**-NLT3 per cent; **Water-soluble extractive**-NLT 7 per cent; **T.L.C.** - Refer API	Volatile Oil	*Rasa* - Tikta, Kasaya; *Guna*- Sara, Picchila; *Virya*- Sita; *Vipaka*- Katu; *Karma*- Pittahara, Kaphahara, Varnya	Aragvadhadi Kvatha Churna	Vrana, Jwara, Raktapitta, Kandu, Krmi, Kustha, Raktavikara, Visa Vikara, Siddhma, Slipada, Balagraha, Aikahikjvara, Badhirya, Anidra, Rajayjaksma, Pradara, Dantkrimi, Sarpvisa	1-3 g in powser form
194	Part-I; Vol. III	77-78	KAKANASIKA (Seed)	*Martynia annua* Linn. Syn. *M. diandra* Glox.	Martyniaceae	Dried Seed	**Foreign matter**- NMT 2 per cent; **Total ash**- NMT 3 per cent; **Acid-insoluble ash**- NMT 1 per cent; **Alcohol-soluble extractive**-NLT 3 per cent; **Water-soluble extractive**-NLT 3 per cent; **T.L.C.** - Refer API	Fixed Oil- (Semidrying type)	*Rasa* - Madhura; *Guna*- Sita; *Virya*- Sita; *Vipaka*- Madhura; *Karma*- Pittaghna, Dardhyakara, Rasayana	Cyavanprasa, Avaleha, Tryusanadi Ghrta	Palita	2-5 g
195	Part-I; Vol. III	79-80	KAKOLI (Tuberous Root)	*Lilium polyphyllum* D. Don	Liliaceae	Dried Tuberous Root	**Foreign matter**- NMT 1 per cent; **Total ash**- NMT 7 per cent; **Acid-insoluble ash**- NMT 1 per cent; **Alcohol-soluble extractive**-NLT 5 per cent; **Water-soluble extractive**-NLT 7 per cent; **T.L.C.** - Refer API	Sugars	*Rasa* - Madhura; *Guna*- Guru, Sita; *Virya*- Sita; *Vipaka*- Madhura; *Karma*- Vatahara, Pittahara, Sukrala, Brmhana	Brhat Aswagandha Ghrta, Brhat Chagaladya Ghrta, Dasamularista, Siva Gutika, Amrtaprasa Ghrta	Raktapitta, Sosa, Jwara, Swasa, Kasa, Ksaya, Daha	3-6 g

Sl. No.	*Part and Volume*	*Page No.*	*Common Ayurvedic Name*	*Botanical Name/ Chemical Name*	*Family/ Chemical Formula*	*Part Used*	*Identity, Purity and Strength/Chemical Properties*	*Constituents*	*Properties and Action*	*Important Formulations*	*Therapeutic Uses*	*Dose*
196	Part-I; Vol. III	81-83	KAMALA (Rhizome)	*Nelumbo nucifera* Gaertn. Syn. *Nelumbium nelumbo* Druce, *N. Speciosum* Willd.	Nymphaeaceae	Dried Rhizome	**Foreign matter**- NMT 2 per cent; **Total ash**- NMT 14 per cent; **Acid-insoluble ash**- NMT 3.5 per cent; **Alcohol-soluble extractive**-NLT 1.5 per cent; **Water-soluble extractive**- NLT 6.5 per cent; **T.L.C.** - Refer API	Starch and Reducing Sugars	*Rasa* - Tikta, Madhura, kasaya, Katu, Lavana; *Guna*- Guru, Ruksa; *Virya*- Sita; *Vipaka*- Madhura; *Karma*- Pittahara, Kaphahara, Rucya, Vistambhakara, Vrsya, Caksusya, Varnya, Krmighna, Dahasamaka, Raktadustihara, Durjara, Stanyajanana, Sangrahi, Mutravirecaniya, Visaghana, Vatakara	Guducyadi Modaka	Daha, Trsna, Chardi, Raktapitta, Murcha, Kasa, Vatagulma, Visarpa, Visphota, Mutrakrchra, Dansodbhava, Jwara, Bhrama, Sosa, Hrdroga	10-20 ml. of the drug in juice form; 5-10 g of the drug in powder form.
197	Part-I; Vol. III	84-85	KARAVIRA (Root)	*Nerium indicum* Mill, Syn. *N. odorum* Soland	Apocynaceae	Dried Root	**Foreign matter**- NMT 1 per cent; **Total ash**- NMT 7.5 per cent; **Acid-insoluble ash**- NMT 3.5 per cent; **Alcohol-soluble extractive**-NLT 8 per cent; **Water-soluble extractive**-NLT 8 per cent; **T.L.C.** - Refer API	Glycosides-Cardiac Glycosides and Resinous Matter	*Rasa* - Katu, Tikta, Kasaya; *Guna*- Laghu, Ruksa, Tiksna; *Virya*- Usna; *Vipaka*- Katu; *Karma*- Sothaghna, Krmighna, Kandughna, Kusthhara, Sirovirecana, Caksusya	Brhanmaricadya Taila, Karaviradya Taila	Vrana, Upadansa, Kustha, Jalodara, Kandu	30-125 mg of the drug in powder form
198	Part-I; Vol. III	86-87	KARAMARDA (Root)	*Carissa carandas* Linn.	Apocynaceae	Dried Root	**Foreign matter**- NMT 2 per cent; **Total ash**- NMT 5 per cent; **Acid-insoluble ash**- NMT 1 per cent; **Alcohol-soluble extractive**-NLT 4 per cent; **Water-soluble extractive**-NLT 7 per cent; **T.L.C.** - Refer API	Glycosides-Cardiac Glycosides.	*Rasa* - Katu, Tikta *Guna*- Laghu, Ruksa *Virya*- Sita *Vipaka*- Katu; *Karma*- Vamaka, Mutrala	Marma Gutika	Mutra Roga, Visphota, Vidradhi, Vrana	1-3 g of the drug in powder form

Sl. No.	*Part and Volume*	*Page No.*	*Common Ayurvedic Name*	*Botanical Name/ Chemical Name*	*Family/ Chemical Formula*	*Part Used*	*Identity, Purity and Strength/Chemical Properties*	*Constituents*	*Properties and Action*	*Important Formulations*	*Therapeutic Uses*	*Dose*
199	Part-I; Vol. III	88-89	KASA (Root Stock)	*Saccharum spontaneum* Linn.	Poaceae	Dried Root Stock	**Foreign matter**- NMT 2 per cent; **Total ash**- NMT 7 per cent; **Acid-insoluble ash**- NMT 4 per cent; **Alcohol-soluble extractive**-NLT 3 per cent; **Water-soluble extractive**- NLT 4 per cent; **T.L.C.** - Refer API	–	*Rasa* - Madhura, Tikta; *Guna*- Sara; *Virya*- Sita; *Vipaka*- Madhura; *Karma*- Pittahara, Balakrt, Vrsya, Srmahara, Rucikrt	Karpuradyarka, Brahma Rasayana, Sukumara Ghrta, Traikantaka Ghrta, Trnapancamula Kvatha Churna, Mutravirecaniya Kasaya Churna, Stanyajanana kasaya Churna, Asmarihara Kasaya Churna	Raktapitta, Mutarakrcchra, Asmari, Daha, Raktadosa, Sosa, Ksaya	3-6 g of the drug in powder form
200	Part-I; Vol. III	90-91	KATPHALA (Fruit)	*Myrica esculenta* Buch.-Ham. ex D. Don Syn. *M. nagi* Hook. f.	Myricaceae	Dried Fruit	**Foreign matter**- NMT 1 per cent; **Total ash**- NMT 5 per cent; **Acid-insoluble ash**- NMT 2.5 per cent; **Alcohol-soluble extractive-NLT 15 per cent; Water-soluble extractive**- NLT 17 per cent; **T.L.C.** - Refer API	Waxy Material	*Rasa* - Katu, Tikta, Kasaya; *Guna*- Laghu, Tiksna; *Virya*- Usna; *Vipaka*- Katu; *Karma*- Kaphavatahara, Dahahara, Mukharogasamaka, Dhatuvikarajit, Rucya	Brhatphala Ghrta, Pusyanuga Churna, Arimedadi Taila, Bala Taila, Mahavisagarbha Taila, Khadiradi Gutika (Mukha Roga), Khadiradi Gutika (Kasa), Maha Vatagajan Kusa Rasa	Gulma, Meha, Jwara, Arsa, Grahani, Pandu Roga, Hrllasa, Mukha Roga, Kasa, Swasa	3-5 g
201	Part-I; Vol. III	92-93	KATPHALA (Stem Bark)	*Myrica esculenta* Buch.-Ham. ex D. Don, Syn. *M. nagi* Hook. f.	Myricaceae	Dried Stem Bark	**Foreign matter**- NMT 2 per cent; **Total ash**- NMT 4 per cent; **Acid-insoluble ash**- NMT 1 per cent; **Alcohol-soluble extractive**-NLT 13 per cent; **Water-soluble extractive**- NLT 12 per cent; **T.L.C.** - Refer API	Tannin and Glycosides	*Rasa* - Katu, Tikta, Kasaya; *Guna*- Laghu, Tiksna; *Virya*- Usna; *Vipaka*- Katu; *Karma*- Kaphavatahara, Dahahara, Mukharogasamaka, Dhatuvikarajit, Kathaphaladi Nasya	Brhatphala Ghrta, Pusyanuga Churna, Arimedadi Taila, Bala Taila, Mahavisagarbha Taila, Khadiradi Gutika (Mukha Roga), Maha Vatagajankusa Rasa	Gulma, Meha, Jwara, Arsa, Grahani, Pandu Roga, Hrallasa, Mukha Roga, Kasa, Swasa, Agnimandhya, Aruchi, Kantharoga	3-5 g

Sl. No.	*Part and Volume*	*Page No.*	*Common Ayurvedic Name*	*Botanical Name/ Chemical Name*	*Family/ Chemical Formula*	*Part Used*	*Identity, Purity and Strength/Chemical Properties*	*Constituents*	*Properties and Action*	*Important Formulations*	*Therapeutic Uses*	*Dose*
202	Part-I; Vol. III	94-95	KOLA (Fruit Pulp)	*Zizyphus mauritiana* Lam. Syn. *Z. jujuba* Lam.	Rhamnaceae	Dried Fruit Pulp	**Foreign matter**- NMT 1 per cent; **Total ash**- NMT 4.5 per cent; **Acid-insoluble ash**- NMT 0.2 per cent; **Alcohol-soluble extractive**-NLT 25 per cent; **Water-soluble extractive**- NLT 45 per cent; **T.L.C.** - Refer API	Vitamin C, Sugar and Minerals	*Rasa* - Madhura, Amla, Kasaya *Guna*- Guru, Snigdha *Virya*- Usna *Vipaka*- Madhura; *Karma*- Grahi, Vatahara, Rucya, Dipana, Pacana	Dhanvantara Taila, Yavani Sadhava	Daha, Raktavikara, Trsna, Aruci	3-6 g (Dried Pulp)
203	Part-I; Vol. III	96-97	KOLA (Stem Bark)	*Zizyphus mauritiana* lam. Syn. *Z. jujuba* Lam.	Rhamnaceae	Dried Stem Bark	**Foreign matter**- NMT 2 per cent; **Total ash**- NMT 13 per cent; **Acid-insoluble ash**- NMT 15 per cent; **Alcohol-soluble extractive**-NLT 6 per cent; **Water-soluble extractive- NLT 6 per cent; T.L.C.** - Refer API	Tannins and Alkaloids	*Rasa* - Kasaya *Guna*- Laghu, Ruksa *Virya*- Sita *Vipaka*- Katu; *Karma*- Visphotasamani, Stambhana, Vranasodhana	Nyagrodhadi Kwatha Churna	Tvaka, Raktatisara, Vrana	3-5 g (powder); 10-20 g (Decoction)
204	Part-I; Vol. III	98-101	KOSATAKI (Whole Plant)	*Luffa acutangula* (Linn.) Roxb.	Cucurbitaceae	Dried Whole Plant	**Foreign matter**- NMT 2 per cent; **Total ash**- NMT 16 per cent; **Acid-insoluble ash**- NMT 4 per cent; **Alcohol-soluble extractive**-NLT 6 per cent; **Water-soluble extractive**- NLT 13 per cent; **T.L.C.** - Refer API	Bitter Principles, Saponins, Sapogenins and Fixed Oil.	*Rasa* - Tikta, Katu, Alpa Kasaya *Guna*- Tiksna, Laghu *Virya*- Sita *Vipaka*- Katu; *Karma*- Kaphapittaghna, Malavisodhani, Vamanopaga, Tridosahara	Abhaya Lavana	Kustha, Pandu, Pliharoga, Sopha, Gulma, Adhmana, Garavisa, Arsa, Kamala, Gandamala	5-10 g
205	Part-I; Vol. III	102-103	KUMUDA (Flower)	*Nymphaea alba* Linn.	Phaeaceae	Dried Flowers	**Foreign matter**- NMT 2 per cent; **Total ash**- NMT 18 per cent; **Acid-insoluble ash**- NMT 9 per cent; **Alcohol-soluble extractive**-NLT 3 per cent; **Water-soluble extractive**- NLT 20 per cent; **T.L.C.** - Refer API	Alkaloids and Glycosides	*Rasa* - Madhura, Kasaya, Tikta *Guna*- Laghu, Snigdha, Picchila *Virya*- Sita *Vipaka*- Madhura; *Karma*- Vatahara, Pittahara, Stambhana, Hrdya, Garbha Sthapana, Balya, Sramahara	Triphaladi Taila, Bala Asvagandha Laksadi Taila	Raktadosa, Daha, Hrdroga, Raktapitta	3-6 g

Sl. No.	*Part and Volume*	*Page No.*	*Common Ayurvedic Name*	*Botanical Name/ Chemical Name*	*Family/ Chemical Formula*	*Part Used*	*Identity, Purity and Strength/Chemical Properties*	*Constituents*	*Properties and Action*	*Important Formulations*	*Therapeutic Uses*	*Dose*
206	Part-I; Vol. III	104-105	KUSA (Root Stock)	*Desmostachya bipinnata* Stapf.	Poaceae	Dried Root Stock	**Foreign matter**- NMT 2 per cent; **Total ash**- NMT 9 per cent; **Acid-insoluble ash**- NMT 7 per cent; **Alcohol-soluble extractive**-NLT 3 per cent; **Water-soluble extractive**-NLT 5 per cent; **T.L.C.** - Refer API	Terpenes	*Rasa* - Madhura, Kasaya; *Guna*- Laghu; *Virya*- Sita; *Vipaka*- Madhura; *Karma*- Kaphapittahara, Mutrala	Karpuradyarka, Sukumara Ghrta, Asmarihara Kasaya Churna, Trnapancamula Kwatha Churna, Mutravirecaniya Kasaya Churna, Stanyajanana Kasaya Churna	Mutrakrcchra, Visarpa, Daha, Asmari, Trsna, Bastiroga, Pradararoga, Raktapitta	50-100 g of powder for decoction
207	Part-I; Vol. III	106-107	LANGALI (Tuberous Root)	*Gloriosa superba* Linn.	Liliaceae	Dried Tuberous Root	**Foreign matter**- NMT 2 per cent; **Total ash**- NMT 6 per cent; **Acid-insoluble ash**- NMT 1 per cent; **Alcohol-soluble extractive**-NLT 5 per cent; **Water-soluble extractive**-NLT 15 per cent; **T.L.C.** - Refer API	Alkaloids and Resins	*Rasa* - Tikta, Kasaya, Katu; *Guna*- Sara, Tiksna; *Virya*- Usna; *Vipaka*- Katu; *Karma*- Vatahara, Pittahara, Kaphahara, Garbhapatana	Nirgundi Taila, Kasisadi Taila, Mahavisagarbha Taila	Kustha, Sopha, Arsa, Vrana, Sula, Krmi, Bastisula, Garbha, Salya, Vatavyadhi	125-250 mg of purified drug
208	Part-I; Vol. III	108-109	LASUNA (Bulb)	*Allium sativum* Linn.	Liliaceae	Bulb	**Foreign matter**- NMT 2 per cent; **Total ash**- NMT 4 per cent; **Acid-insoluble ash**- NMT 1 per cent; **Alcohol-soluble extractive**-NLT 2.5 per cent; **Loss on drying**- NLT 60 per cent; **Volatile oil** - NLT 0.1 per cent; **T.L.C.** - Refer API	Volatile Oil containging Allyl Disulphide and Diallyl Disulphide. It also contains Allin, Allcin, Mucilage and Albumin.	*Rasa* - Katu, Madhura *Guna*- Guru, Snigdha, Tiksna, Sara, Picchila *Virya*- Usna *Vipaka*- Katu; *Karma*- Vatahara, Kaphahara, Pitta dusanakara, Raktadosahara, Bhagnasandha-nakara, Dipana, Rasayena, Balya, Hrdya, Vrsya, Varnya, Medhya, Jantughna, Kanthya, Asthi Mamsa Sandhankar, Caksusya	Lasunadi Vati, Lasunadi Ghrta and Vaca Lasunadi Taila	Jirna, Jwara, Krmiroga, Gulma, Kustha, Arsa, Kasa, Swasa, Pinasa, Sula, Karnasula Vatavyadhi, Hikka, Medoroga, Yoni vyapata, Visucika, Pliha, Vrddhi, Ksaya, Visama Jwara, Apasmara, Unniada, Sasa, Sopha, Hrdroga, Vatsula, Trikasula, Vrana Krmi	3 g of the drug

Sl. No.	*Part and Volume*	*Page No.*	*Common Ayurvedic Name*	*Botanical Name/ Chemical Name*	*Family/ Chemical Formula*	*Part Used*	*Identity, Purity and Strength/Chemical Properties*	*Constituents*	*Properties and Action*	*Important Formulations*	*Therapeutic Uses*	*Dose*
209	Part-I; Vol. III	110-111	MAHABALA (Root)	*Sida rhombifolia* Linn.	Malvaceae	Dried Roots	**Foreign matter**- NMT 2 per cent; **Total ash**- NMT 8 per cent; **Acid-insoluble ash**- NMT 3 per cent; **Alcohol-soluble extractive**-NLT 1 per cent; **Water-soluble extractive**-NLT 4 per cent; **T.L.C.** - Refer API	Alkaloids (Vasicinone and Vasicine)	*Rasa* - Madhura *Guna*- Guru, Snigdha, Picchila *Virya*- Sita *Vipaka*- Madhura; *Karma*- Vataghna, Pittaghna, Grahi, Sukravrddhikara, Ojovardhaka, Kantivardhaka, Balya	Mahavisagarbha Taila, Navratna Rajamrganka Rasa	Sukraksaya, Ksata, Ksaya, Visamajwara, Daurbalya, Vatavyadhi, Vatarakta, Raktapitta, Sopha	3-6 g of the drug in powder form
210	Part-I; Vol. III	112-114	MANJISTHA (Stem)	*Rubia cordifolia* Linn.	Rubiaceae	Dried Stem	**Foreign matter**- NMT 2 per cent; **Total ash**- NMT 12 per cent; **Acid-insoluble ash**- NMT 0.5 per cent; **Alcohol-soluble extractive**-NLT 3 per cent; **Water-soluble extractive**-NLT 17 per cent; **T.L.C.** - Refer API	Glycosides	*Rasa* - Kasaya, Tikta, Madhura *Guna*- Guru *Virya*- Usna *Vipaka*- Katu; *Karma*- Kaphapittasamaka, Varnya, Swarya, Visa, Sothaghna, Kusthaghna, Pramehaghna, Vrsya, Krmighna, Stambhan, Artavajanana, Rasanyana, Sonitasthapana	Arvindasava, Aswagan-dharista, Usirasava, Candanasava, Brhanman-jisthadi Kwatha, Manjisthadi Taila, Khadiradi Gutika (Mukha)	Yoni Roga, Aksi Roga, Slesmaja Sotha, Karan Roga, Manjistha Meha, Raktatisara, Kustha, Visarpa, Prameha, Sarpavisa, Bhagna, Arsa, Vyanga	2-4 g of the drug
211	Part-I; Vol. III	115-117	MARICA (Fruit)	*Piper nigrum* Linn.	Piperaceae	Dried Fruit	**Foreign matter**- NMT 2 per cent; **Total ash**- NMT 5 per cent; **Acid-insoluble ash**- NMT 0.5 per cent; **Alcohol-soluble extractive**-NLT 6 per cent; **Water-soluble extractive**-NLT 6 per cent; **T.L.C.** - Refer API	Alkaloids (Piperine, Chavicine, Piperidine, Piperetine) and Essential Oil	*Rasa* - Katu, Tikta; *Guna*- Laghu, Ruksa, Tiksna; *Virya*- Usna; *Vipaka*- Katu; *Karma*- Slesmahara, Pittakara, Kaphavatajit Vatahara, Chedana, Dipana, Rucya, Jantunasana, Medohara, Chedi, Hrdroga, Vataroga	Maricadi Gutika, Maricadi Taila, Trikatu Churna	Swasa, Sula, Krmiroga, Tvagroga	250 mg/g of the drug in powder form

Sl. No.	*Part and Volume*	*Page No.*	*Common Ayurvedic Name*	*Botanical Name/ Chemical Name*	*Family/ Chemical Formula*	*Part Used*	*Identity, Purity and Strength/Chemical Properties*	*Constituents*	*Properties and Action*	*Important Formulations*	*Therapeutic Uses*	*Dose*
212	Part-I; Vol. III	118-120	MASAPARNI (Whole Plant)	*Teramnus labialis* Spreng.	Fabaceae	Dried Whole Plant	**Foreign matter**- NMT 2 per cent; **Total ash**- NMT 7 per cent; **Acid-insoluble ash**- NMT 0.5 per cent; **Alcohol-soluble extractive**-NLT 3 per cent; **Water-soluble extractive**-NLT 7 per cent; **T.L.C.** - Refer API	Glycosides	*Rasa* - Tikta, Madhura; *Guna*- Laghu, Ruksa; *Virya*- Sita; *Vipaka*- Madhura; *Karma*- Vatapittasamaka, Kaphavardhaka, Grahi, Balya, Vrsya, Sukrala	Amrtaprasa Ghrta, Asoka Ghrta, Vidaryadi Ghrta, Dhanwantara Ghrta, Narayana Taila, Brhat Masa Taila, Bala Taila, Mahanarayana Taila	Atisara, Pravahika, Vatapitta Jwara, Sukralpata, Raktapitta, Raktavikara, Daha, Sotha, Sirahsula	5-10 g of the powder
213	Part-I; Vol. III	121-122	MASURA (Seed)	*Lens culinaris* Medic.	Fabaceae	Dried Seed	**Foreign matter**- NMT 1 per cent; **Total ash**- NMT 3 per cent; **Acid-insoluble ash**- NMT 0.5 per cent; **Alcohol-soluble extractive**-NLT 6 per cent; **Water-soluble extractive**-NLT 10 per cent; **T.L.C.** - Refer API	Flavonoids and Vitamins	*Rasa* - Madhura, Kasaya *Guna*- Laghu, Ruksa *Virya*- Sita *Vipaka*- Madhura; *Karma*- Sangrahi, Kaphapittasamaka, Vatamayakara, Varnya, Balya	–	Atisara, Muttrakrcchra, Jwara, Raktapitta	10-20 g
214	Part-I; Vol. III	123-124	MUDGA (Seed)	*Phaseolus radiatus* Linn.	Fabaceae	Dried Seeds	**Foreign matter**- NMT 1 per cent; **Total ash**- NMT 4 per cent; **Acid-insoluble ash**- NMT 1 per cent; **Alcohol-soluble extractive**-NLT 1.5 per cent; **Water-soluble extractive**- NLT 10 per cent; **T.L.C.** - Refer API	Saponin, Starch, Albuminoids and Oil	*Rasa* - Madhura, Kasayas *Guna*- Laghu, Ruksa *Virya*- Sita *Vipaka*- Madhura; *Karma*- Pittahara, Kaphahara, Grahi, Balaprada, Varnya, Netrya	Balahathadi Taila, Marma Gutika, Kayasthyadi Vati	Jwara, Netra Roga, Amlalpitta	50 -100 g for yusa
215	Part-I; Vol. III	125-126	MULAKA (Seed)	*Raphanus sativus* Linn.	Brassicaceae	Dried Seed	**Foreign matter**- NMT 2 per cent; **Total ash**- NMT 5.5 per cent; **Acid-insoluble ash**- NMT 1.5 per cent; **Alcohol-soluble extractive**-NLT 4.5 per cent; **Water-soluble extractive**- NLT 11 per cent; **T.L.C.** - Refer API	Fixed Oil and Volatile Oil	*Rasa* - Katu, Tikta, Kasaya; *Guna*- Laghu, Tiksna; *Virya*- Usna; *Vipaka*- Katu; *Karma*- Visahara, Vataslesmahara, Hrdya, Vahnidipana, Kanthya, Grahi,	Sarsapadi Lepa	Gulma, Hrdroga, kantha roga, Sidhmakustha Jwara, Swasa, Nasika roga, Aksi roga, Anartava	1-3 g of the drug in powder form

Sl. No.	*Part and Volume*	*Page No.*	*Common Ayurvedic Name*	*Botanical Name/ Chemical Name*	*Family/ Chemical Formula*	*Part Used*	*Identity, Purity and Strength/Chemical Properties*	*Constituents*	*Properties and Action*	*Important Formulations*	*Therapeutic Uses*	*Dose*
									Kaphavatahara, Grabthasaya-samikocaka, Kaphanissaraka, Mural, Pacaka, Vitanulomana, Mrdurecaka			
216	Part-I; Vol. III	127-128	MUNDITIKA (Leaf)	*Sphaeranthus indicus* Linn.	Asteraceae	Dried Leaf	**Foreign matter**- NMT 2 per cent; **Total ash**-NMT 28 per cent; **Acid-insoluble ash**- NMT 7 per cent; **Alcohol-soluble extractive**-NLT 3 per cent; **Water-soluble extractive**-NLT 12 per cent; **T.L.C.** - Refer API	Essential Oil	*Rasa* - Katu, Madhura, Tikta, Kasaya; *Guna*-Laghu; *Virya*-Usna; *Vipaka*-Katu; *Karma*-Vatakaphahara, Medhya, Arsadosa, Vinasaka, Visaghna	Navaratnaraja Mrganka Rasa, Arka Mundi	Gandamala, Apaci, Kustha, Krmi, Pandu, Slipada, Medaroga, Apasmara, Kasa, Mutrakrcchra, Twak Roga, Stana Saithalya, Yoniroga, Amatisara, Amaroga, Vataroga, Gudaroga, Pliharoga, Chardi, Amavata, Gatra-durgandhya, Suryavarta, Ardhava-bhedaka	3-6 g of the drug
217	Part-I; Vol. III	129-130	MUSTA (Rhizome)	*Cyperus rotundus* Linn.	Cyperaceae	Dried Rhizome	**Foreign matter**- NMT 2 per cent; **Total ash**-NMT 8 per cent; **Acid-insoluble ash**- NMT 4 per cent; **Alcohol-soluble extractive**-NLT 5 per cent; **Water-soluble extractive**-NLT 11 per cent; **Volatile oil** - NLT 1 per cent; **T.L.C.** - Refer API	Volatile Oil	*Rasa* - Tikta, Katu, Kasaya; *Guna*- Laghu, Ruksa *Virya*-Sita; *Vipaka*-Katu; *Karma*-Pittakaphahara, Sthoulyahara, Sothahara, Dipana, Pacana Grahi,	Musakarista, Mustakadi Kwatha, Asokarista, Mustakadi Churna, Mustakadi, Mustakadi Lehya, Dhamya Pancaka	Agnimandya, Ajerna, Trsna, Jwara, Sangrahani, Swasa, Kasa, Mutrakrcchra, Vamana, Stanyavikara, Sutikaroga, Atisara,	3-6 g (Powder); 20-30 ml (Kwatha)

Sl. No.	*Part and Volume*	*Page No.*	*Common Ayurvedic Name*	*Botanical Name/ Chemical Name*	*Family/ Chemical Formula*	*Part Used*	*Identity, Purity and Strength/Chemical Properties*	*Constituents*	*Properties and Action*	*Important Formulations*	*Therapeutic Uses*	*Dose*
									Trsnanigrahana, Krmighna, Tvakadosahara, Jwaraghna, Visaghna	Kvatha Churna, Piyusavalli rasa, Gulmakatanala Rasa, Mahalaksadi Taila, Shadanga-paneeya	Amavata, Krmiroga	
218	Part-I; Vol. III	131-133	NAGAVALLI (Leaf)	*Piper betle* Linn.	Piperaceae	Leaf	**Foreign matter**- NMT 2 per cent; **Total ash**- NMT 17 per cent; **Acid-insoluble ash**- NMT 3 per cent; **Alcohol-soluble extractive**-NLT 10 per cent; **Water-soluble extractive**- NLT 20 per cent; **T.L.C.** - Refer API	Essential Oil, Amino Acids, Vatamins and Enzymes	*Rasa* - Kasaya, Tikta, Katu; *Guna*- Tiksna, Sara, Laghu, Visada; *Virya*- Usna; *Vipaka*- Katu; *Karma*- Rucya, Balya, Slesmahara, Mukhadour-gandhyahara, Mukhamalahara, Vata Hara, Sramahara, Raktapittakarni, Svaryam, Vrsya	Lokanatha Rasa, Puspadhanva Rasa, Brhat Sarwaj-warahara Lauha, Laghu Sutasekhara Rasa, Brhat Visamaj-warantaka Rasa	Kandu, Hrllasa, Agnimandya, Jwara, Hrdroga, Swarabheda	10-20 ml of Swarasa
219	Part-I; Vol. III	134-135	NARIKELA (Endosperm)	*Cocos nucifera* Linn.	Arecaceae	Dried Ensdo-sperm	**Foreign matter**- Nil; **Total ash**- NMT 2.5 per cent; **Acid-insoluble ash**- NMT 0.5 per cent; **Alcohol-soluble extractive**-NLT 13 per cent; **Water-soluble extractive**- NLT 10 per cent; **Fixed oil** - NLT 59 per cent; **T.L.C.** - Refer API	Fixed Oil	*Rasa* - Madhura; *Guna*- Guru, Snigdha; *Virya*- Sita; *Vipaka*- Madhura; *Karma*- vatahara Pittahara, Kaphakara, Balya, Vrsya, Brmhana, Hrdya, Bastisodhaka, Vistambhi	Narikela Khanda, Narikela Lavan	Daha, Ksata, Ksaya, Raktapitta, Trsna, Sosa, Sula	10-20 g of the drug in powder form

Sl. No.	Part and Volume	Page No.	Common Ayurvedic Name	Botanical Name/ Chemical Name	Family/ Chemical Formula	Part Used	Identity, Purity and Strength/Chemical Properties	Constituents	Properties and Action	Important Formulations	Therapeutic Uses	Dose
220	Part-I; Vol. III	136-137	NICULA (Fruit)	*Barringtonia acutangula* (Linn.)	Lecythidaceae	Dried Fruit	**Foreign matter**- NMT 1 per cent; **Total ash**- NMT 7 per cent; **Acid-insoluble ash**- NMT 1 per cent; **Alcohol-soluble extractive**-NLT 5 per cent; **Water-soluble extractive**-NLT 9 per cent; **T.L.C.** - Refer API	Saponins and Sapogenins	*Rasa* - Tikta, Kasaya, Katu; *Guna*- Ruksa, Laghu; *Virya*- Usna; *Vipaka*- Katu; *Karma*- Samgrahi, Vranasodhana, Kaphahara, Recaka, Raksoghna, Visaghna, Vamaka, Vatahara,	Mahapancagavya Ghrta, Laksmi Vilasa Rasa (Nardiya), Nyagrodhadi Gana Kwatha	Raktapitta, Amatisara, Caksusrava, Galganda, Bhutabadha, Grahabadha, Prameha	1-3 g
221	Part-I; Vol. III	138-141	NILI (Whole Plant)	*Indigofera tinctoria* Linn.	Fabaceae	Dried Whole Plant	**Foreign matter**- NMT 2 per cent; **Total ash**- NMT 5.2 per cent; **Acid-insoluble ash**- NMT 1.0 per cent; **Alcohol-soluble extractive**-NLT 2.5 per cent; **Water-soluble extractive**- NLT 7.5 per cent; **T.L.C.** - Refer API	Glycoside (Indican).	*Rasa* - Tikta, Katu; *Guna*- Sara; *Virya*- Usna; *Vipaka*- Katu; *Karma*- Vatahara, Kaphahara, Recani, Kesya, Visaghana, Jantughna	Nilikadya Taila, Gorocanadi Vati	Vata Rakta, Udararoga, Pliharoga, Krmiroga, Moha, Bhrama, Udavartta, kativata, Kasa, Amaroga, Visodara, Jwara, Ksaya, Krmidanta	10-20 g of the drug for decoction
222	Part-I; Vol. III	142-144	NIRGUNDI (Leaf)	*Vitex negundo* Linn.	Verbenaceae	Dried Leaf	**Foreign matter**- NMT 2 per cent; **Total ash**- NMT 8 per cent; **Acid-insoluble ash**- NMT 1 per cent; **Alcohol-soluble extractive**-NLT 10 per cent; **Water-soluble extractive**- NLT 20 per cent; **T.L.C.** - Refer API	Alkaloids and Essential Oil	*Rasa* - Tikta, Katu, Kasaya; *Guna*- Laghu; *Virya*- Usna; *Vipaka*- Katu; *Karma*- Kaphasamaka, Vatasamaka, Sophahara, Kesya, Caksusyam Visaghna, Smtriprada, Anulomna	Vatagajankusa Rasa, Mahavata Vidhvansana Rasa, Ykrtptihara Lauha, Dasamula Taila, Trivikrama Rasa, Nirgundi Taila, Tribhuvan Kirti Rasa, Visa Tinduka Taila	Sula, Sopha, Vatavyadhi, Amavata, Kustha, Kandu, Kasa, Pradara, Adhmana, Piha roga, Gulma, Aruci, Krmi, Vrana, Nadi Vrana, Karnasula, Sutika, Jwara	10-20 ml (Swarasa)

Sl. No.	*Part and Volume*	*Page No.*	*Common Ayurvedic Name*	*Botanical Name/ Chemical Name*	*Family/ Chemical Formula*	*Part Used*	*Identity, Purity and Strength/Chemical Properties*	*Constituents*	*Properties and Action*	*Important Formulations*	*Therapeutic Uses*	*Dose*
223	Part-I; Vol. III	145-146	PADMAKA (Heart Wood)	*Prunus cerasoides* D.Don	Rosaceae	Heart Wood	**Foreign matter**- NMT 1 per cent; **Total ash**- NMT 1 per cent; **Acid-insoluble ash**- NMT 0.5 per cent; **Alcohol-soluble extractive**-NLT 3 per cent; **Water-soluble extractive**-NLT 1 per cent; **T.L.C.** - Refer API	Flavonoids.	*Rasa* - Kasaya, Tikta; *Guna*- Laghu; *Virya*- Sita; *Vipaka*- Katu; *Karma*- Garbhasthapana, Rucya, Vatala	Khadiradi Gutika, Guducyadi Kwatha Churna, Brhacchagatadya Ghrta, Satavaryadi Ghrta, Guducyadi Taila, Usirasava, Candanasava, Dasamularista, Mrtasanjivani Sura, Karpuradhyarka	Visphota, Daha, Kustha, Raktapitta, Vami, Trsa, Bhrama, Visarapa	1-3 g (Churna)
224	Part-I; Vol. III	147-148	PATALA (Root)	*Stereos-permum suaveolens* DC.	Bignoniaceae	Dried Root	**Foreign matter**- NMT 2 per cent; **Total ash**- NMT 8 per cent; **Acid-insoluble ash**- NMT 6 per cent; **Alcohol-soluble extractive**-NLT 10 per cent; **Water-soluble extractive**- NLT 20 per cent; **T.L.C.** - Refer API	Bitter Substances, Sterols, Glycosides and Glyco-Alkaloids.	*Rasa* - Kasaya, Tikta; *Guna*- Laghu, Ruksa; *Virya*- Anusna; *Vipaka*- Katu; *Karma*- Tridosahara, Rucya	Amrtarista, Dasamularista, Bharangi Guda, Indu Kanta Ghrta, Dhanwantari Taila, Dasamula Kwatha Churna	Swasa, Sotha, Arsa, Chardi, Hikka, Trsa, Amlapitta, Rakta Vikara, Mutravikara, Agnidagdha, Vrana Ruja, Visphota, Medoroga	5-10 g (Powder); 25-50 ml (Deco-ction)
225	Part-I; Vol. III	149-150	PHALGU (Fruit)	*Ficus hispida* Linn. f.	Moraceae	Dried Fruits	**Foreign matter**- NMT 1 per cent; **Total ash**- NMT 13 per cent; **Acid-insoluble ash**- NMT 1.5 per cent; **Alcohol-soluble extractive**-NLT 3 per cent; **Water-soluble extractive**-NLT 12 per cent; **T.L.C.** - Refer API	Tannins and Saponins.	*Rasa* - Madhura, Amla, Katu, Tikta, Kasaya *Guna*- Snigdha, Guru *Virya*- Sita *Vipaka*- Madhura; *Karma*- Vatahara Pittahara, Kaphahara, Mansakara, Sukrakara, Mala Stambhana, Trptikaraka, Grahi, Brmihana, Vistambhi	Citrakadi Taila	Vrana, Sveta Kusta, Pandu, Arsa, Kamala, Atisara, Daha, Ksata, Visaroga, Tvakaroga, Raktavikara, Kandu, Kustha, Sopha, Raktapitta, Vatapittajaroga	10-20 g

Sl. No.	*Part and Volume*	*Page No.*	*Common Ayurvedic Name*	*Botanical Name/ Chemical Name*	*Family/ Chemical Formula*	*Part Used*	*Identity, Purity and Strength/Chemical Properties*	*Constituents*	*Properties and Action*	*Important Formulations*	*Therapeutic Uses*	*Dose*
226	Part-I; Vol. III	151-152	PHALGU (Root)	*Ficus hispida* Linn. f.	Moraceae	Dried Root	**Foreign matter**- NMT 1 per cent; **Total ash**- NMT 7 per cent; **Acid-insoluble ash**- NMT 1.5 per cent; **Alcohol-soluble extractive**-NLT 6 per cent; **Water-soluble extractive**-NLT 6 per cent; **T.L.C.** - Refer API	Alkaloids	*Rasa* - Tikta, Kasaya *Guna*- Guru, Sita *Virya*- Sita *Vipaka*- Madhura; *Karma*- Pittahara, Kaphahara, Malastambhaka	Mahapancagavya Ghrta	Svitra, Kandu, Kustha, Vrana, Raktapitta, Sopha, Pandu, Raktavikara, Kamala, Arsa	1-3 g of the drug in powder form
227	Part-I; Vol. III	153-154	PRAPUN-NADA (Seed)	*Cassia tora* Linn.	Fabaceae	Dried Seed	**Foreign matter**- NMT 2 per cent; **Total ash**- NMT 5 per cent; **Acid-insoluble ash**- NMT 0.2 per cent; **Alcohol-soluble extractive**-NLT 7 per cent; **Water-soluble extractive**-NLT 14 per cent; **T.L.C.** - Refer API	Anthra-quinones, Fixed Oil.	*Rasa* - Katu; *Guna*- Laghu, Ruksa; *Virya*- Usna; *Vipaka*- Katu; *Karma*- Kaphavata-samaka Krmighna, Recana, Lekhana, Kusthaghna, Visaghana Tvaka, Varna-prasadakaram, Twacya	Nimbadi Churna, Kasisadi Ghrta, Maha Visagarbha Taila, Brhanmariayadi Taila	Kaphavatajanya vikara, Kustha, Vrana Vikara, Dadru, Paksaghata, Vibandha, Gulma, Krmi, Pama, Kandu Swasa, Kasa	1-3 g of powder
228	Part-I; Vol. III	155-156	RAKTA-CANDANA (Heart Wood)	*Pterocarpus santalinus* Linn. f.	Fabaceae	Heart Wood	**Foreign matter**- NMT 2 per cent; **Total ash**- NMT 2 per cent; **Acid-insoluble ash**- NMT 0.3 per cent; **Alcohol-soluble extractive**-NLT 3 per cent; **Water-soluble extractive**-NLT 1 per cent; **T.L.C.** - Refer API	Glycosides, Colouring Matter	*Rasa* - Tikta, Madhura; *Guna*- Guru, Ruksa; *Virya*- Sita; *Vipaka*- Katu; *Karma*- Pittahara, Netraroga, Visaghna, Vrsya	Candana Bala Laksadi Taila, Candanadi Lauha	Chardi, Trsna, Raktadosahara, Twara, Vrana	3-6 g of the drug (powder)

Sl. No.	*Part and Volume*	*Page No.*	*Common Ayurvedic Name*	*Botanical Name/ Chemical Name*	*Family/ Chemical Formula*	*Part Used*	*Identity, Purity and Strength/Chemical Properties*	*Constituents*	*Properties and Action*	*Important Formulations*	*Therapeutic Uses*	*Dose*
229	Part-I; Vol. III	157-158	RAKTA-PUNARNAVA (Root)	*Boerhaavia diffusa* Linn.	Nyctaginaceae	Dried Root	**Foreign matter**- NMT 2 per cent; **Total ash**- NMT 10 per cent; **Acid-insoluble ash**- NMT 0.8 per cent; **Alcohol-soluble extractive**-NLT 4 per cent; **Water-soluble extractive**-NLT 10 per cent; **T.L.C.** - Refer API	Alkaloid, Hentriacontane, β-sitosterol, Ursolic Acid.	*Rasa* - Tikta, Kasaya, Katu, Madhura; *Guna*- Laghu, Ruksa, Sita, Sara; *Virya*- Usna; *Vipaka*- Katu; *Karma*- Sophaharra, Kaphaghna, Dipana, Vatakara, Pittahara	Kumaryasava, Dahika Ghrta, Dhanvantara Ghrta, Punarna-vadyarista	Sopha, Pandu, Hrdroga, Kasa, Arsa, Vrana, Urahksatasula, Sotha	1-3 g of powder; 10-20 ml (Fresh Juice)
230	Part-I; Vol. III	159-161	RAMA-SITALIKA (Whole Plant)	*Amaranthus tricolor* Linn.; Syn. *A. gangeticus* Linn.; *A. melancholicus* Linn. *A. polygamus* Linn. Hook. f., *A. tristis* Linn.;	Amaranthaceae	Dried Whole Plant	**Foreign matter**- NMT 2 per cent; **Total ash**- NMT 17 per cent; **Acid-insoluble ash**- NMT 2.6 per cent; **Alcohol-soluble extractive**-NLT 3 per cent; **Water-soluble extractive**-NLT 17 per cent; **T.L.C.** - Refer API	Fatty Oils, Sitosterol, Calcium and Magnesium	*Rasa* - Madhura, Tikta; *Guna*- Ruksa, Kincit Guru, Sara; *Virya*- Sita; *Vipaka*- Katu; *Karma*- Pittahara	Candrakali Rasa	Daha, Sosa, Visphota, Vrana	10-20 ml of the drug in juice form
231	Part-I; Vol. III	162-164	RASNA (Leaf)	*Pluchea lanceolata* Oliver and Hiern.	Asteraceae	Dried Leaf	**Foreign matter**- NMT 2 per cent; **Total ash**- NMT 22 per cent; **Acid-insoluble ash**- NMT 7 per cent; **Alcohol-soluble extractive**-NLT 8 per cent; **Water-soluble extractive**-NLT 23 per cent; **T.L.C.** - Refer API	Flavonoids-Quercetin and Isorhamnetin	*Rasa* - Tikta; *Guna*- Guru; *Virya*- Usna; *Vipaka*- Katu; *Karma*- Kaphavatahara, Amapacana	Dasamularista. Devadarvarista, Karpasasthyadi Taila, Rasnadi Kwatha Churna, Rasnaairandadi Kwatha Churna	Sotha, Vatavyadhi, Swasa, Kasa, Jwara, Udararoga, Sidhma, Adhyavata, Amavata, Vatarakta	25 -50 g (Decoction)
232	Part-I; Vol. III	165-168	SAHACARA (Whole Plant)	*Barleria prionitis* Linn.	Acanthaceae	Dried Whole Plant	**Foreign matter**- NMT 2 per cent; **Total ash**- NMT 7 per cent; **Acid-insoluble ash**- NMT 1 per cent; **Alcohol-soluble extractive**-NLT 4 per cent; **Water-soluble extractive**-NLT 10 per cent; **T.L.C.** - Refer API	Alkaloids, β-sitosterol, Potassium.	*Rasa* - Madhura, Tikta (Amla) *Guna*- Snigdha *Virya*- Katu *Vipaka*- Usna; *Karma*- Kaphahara, Kesya, Kasa, Rajana, Visahara	Sahacaradi Taila, Nilikadya Taila, Astavarga Kvatha Churna, Rasnarandadi Kvatha Churna	Kustha, Kandu, Vatarakta, Palit	50-100 g of the drug for decoction

Sl. No.	Part and Volume	Page No.	Common Ayurvedic Name	Botanical Name/ Chemical Name	Family/ Chemical Formula	Part Used	Identity, Purity and Strength/Chemical Properties	Constituents	Properties and Action	Important Formulations	Therapeutic Uses	Dose
233	Part-I; Vol. III	169-171	SAHADEVI (Whole Plant)	*Vernonia cinerea* Lees.	Asteraceae	Dried Whole Plant	**Foreign matter**- NMT 2 per cent; **Total ash**- NMT 14 per cent; **Acid-insoluble ash**- NMT 2 per cent; **Alcohol-soluble extractive**-NLT 4 per cent; **Water-soluble extractive**-NLT 15 per cent; **T.L.C.** - Refer API	Saponins, Sapogenins, Flavonoids.	*Rasa* -Tikta, Katu *Guna*- Laghu, Ruksa *Virya*-Usna *Vipaka*-Katu; *Karma*-Kaphavatasamaka, Sothahara, Swaraghna, Nidrakara	Candrakala Rasa, Alamottadi Kashayam (S.Y.)	Jwara, Visamajwara, Sidhma, Visphota, Bhutabadha, Grahabadha, Sphotaka, Pradara, Slipada	10-20 ml Swarasa; 5-10 g (Powder for external use only)
234	Part-I; Vol. III	172-173	SAILEYA (Lichen)	*Parmelia perlata* (Huds.) Ach	Parmeliaceae	Whole Thallus	**Foreign matter**- NMT 2 per cent; **Total ash**- NMT 9 per cent; **Acid-insoluble ash**- NMT 3 per cent; **Alcohol-soluble extractive**-NLT 4 per cent; **Water-soluble extractive**-NLT 5 per cent; **T.L.C.** - Refer API	Lichen acids-Atranorin and Lecanoric acid	*Rasa* - Tikta, Kasaya; *Guna*-Laghu, Snigdha; *Virya*- Sita; *Vipaka*- Katu; *Karma*- Hrdya, Kaphapittahara, Rucya, Stambhaka, Pittahara	Vasacandanadi Taila, Jirakadi Modaka, Saubhagya Sunthi, Candanadi Taila, Dhanvantara Taila, Narayana Taila, Mahanarayana Taila, Tarksya Guda, Agarvadya Taila, Saileyadi Taila, Mrtasanjwani Sura, Dnjana Vati	Kandu, Kustha, Asmari, Daha, Visa, Hrllasa, Trsna, Vrana, Hrdayaroga, Rakta Vikara, Swasa, Jwara, Mutrakrchra, Mutraghata, Sriah Sula	1-3 g
235	Part-I; Vol. III	174-175	SAKA (Heart Wood)	*Tectona grandis* Linn. f.	Verbenaceae	Dried Heart Wood	**Foreign matter**- NMT 1 per cent; **Total ash**- NMT 2 per cent; **Acid-insoluble ash**- NMT 5 per cent; **Alcohol-soluble extractive**-NLT 3 per cent; **Water-soluble extractive**-NLT 1.5 per cent; **T.L.C.** - Refer API	Resin, Essential Oil, Fatty Oil and Tectoquinone	*Rasa* - Kasaya; *Guna*- Laghu, Ruksa; *Virya*- Sita; *Vipaka*- Katu; *Karma*- Pittahara, Kaphahara, Raktaprasadana, Garbhasth-airyakara	Ayaskrti	Kustha, Raktapitta, Mutraroga, Pandu, Prameha, Medoroga, Daha, Srama, Trsna, Krmiroga, Garbhasrava, Garbhapatana	3-6 g of the drug in powder form; 30-60 g of the drug for decoction

Sl. No.	*Part and Volume*	*Page No.*	*Common Ayurvedic Name*	*Botanical Name/ Chemical Name*	*Family/ Chemical Formula*	*Part Used*	*Identity, Purity and Strength/Chemical Properties*	*Constituents*	*Properties and Action*	*Important Formulations*	*Therapeutic Uses*	*Dose*
236	Part-I; Vol. III	176-177	SAKHOTAKA (Stem Bark)	*Streblus asper* Lour.	Moraceae	Stem Bark	**Foreign matter**- NMT 2 per cent; **Total ash**- NMT 15 per cent; **Acid-insoluble ash**- NMT 2 per cent; **Alcohol-soluble extractive**-NLT 3 per cent; **Water-soluble extractive**-NLT 12 per cent; **T.L.C.** - Refer API	Glycosides, Saponins and Sapogenins.	*Rasa* - Tikta, Kasaya; *Guna*- Ruksa, Laghu; *Virya*- Usna; *Vipaka*- Katu; *Karma*- Vataslesmahara, Medohara, Sothahara	Brhanman-jisthadi Kwatha Churna	Raktapitta, Arsa, Slipada, Apaci, Prameha, Kustha, Gandamala	1-3 g (Powder); 10-20 g (for decoction)
237	Part-I; Vol. III	178-180	SALAPARNI (Root)	*Desmodium gangeticum* DC.	Fabaceae	Dried Root	**Foreign matter**- NMT 2 per cent; **Total ash**- NMT 6 per cent; **Acid-insoluble ash**- NMT 2 per cent; **Alcohol-soluble extractive**-NLT 1 per cent; **Water-soluble extractive**-NLT 6 per cent; **T.L.C.** - Refer API	Alkaloids	*Rasa* - Tikta, Madhura; *Guna*- Guru; *Virya*- Usna; *Vipaka*- Madhura; *Karma*- Tridosahara, Balya, Angamarda-prasamana, Vrsya, Sukha-prasawakara Sarvadosahara, Vatadosajit, Rasayani, Bhramhara, Visahara,	Dasamularista, Indukanta Ghrta, Amrtaprasa Ghrta, Dasamulasa-tapalaka Ghrta, Dhanwantara Taila, Narayana Taila, Mahavisagarbha Taila, Mahanarayana Taila	Jwara, Meha, Arsa, Chardi, Sopha, Swasa, Kasahara, Krmi, Rajayaksma, Netra roga, Hrdaya roga, Rakta Gata Vata, Vata Ardhvabhedaka, Mudha Garbha	5-10 g of the drug in powder form; 10-20 g for decoction
238	Part-I; Vol. III	181-182	SALI (Fruit)	*Oryza sativa* Linn.	Poaceae	Dried Fruit	**Foreign matter**- NMT 2 per cent; **Total ash**- NMT 6 per cent; **Acid-insoluble ash**- NMT 5 per cent; **Alcohol-soluble extractive**-NLT 1 per cent; **Water-soluble extractive**-NLT 1 per cent; **T.L.C.** - Refer API	Carbohydrate-Starch.	*Rasa* - Madhura, Anuras, Kasaya; *Guna*- Snigdha, Laghu; *Virya*- Sita; *Vipaka*- Madhura; *Karma*- Swalpa Vatakara, Swalpa Kapha Kara, Pittahara, Hrdya, Rucikara, Vrsya, Mutral, Brhamma, Visaghna, Baddhavarcasaka, Swarya	Lasunadi Ghrta, Dadhika Ghrta, Tamdulodanam	Jwara, Trsna, Vrana, Atisara, Balatisara, Pradara	100 ml Tandu-lodaka

Sl. No.	*Part and Volume*	*Page No.*	*Common Ayurvedic Name*	*Botanical Name/ Chemical Name*	*Family/ Chemical Formula*	*Part Used*	*Identity, Purity and Strength/Chemical Properties*	*Constituents*	*Properties and Action*	*Important Formulations*	*Therapeutic Uses*	*Dose*
239	Part-I; Vol. III	183-184	SALMALI (Stem Bark)	*Bombax ceiba* Linn. Syn. *B. malabaricum* DC., *Salmalia malabarica* Schott. and Endl.	Bombacaceae	Mature Stem Bark	**Foreign matter**- NMT 1 per cent; **Total ash**- NMT 13 per cent; **Acid-insoluble ash**- NMT 2 per cent; **Alcohol-soluble extractive**-NLT 2 per cent; **Water-soluble extractive**-NLT 7 per cent; **T.L.C.** - Refer API	Saponins, Tannins and Gums	*Rasa* - Madhura, Kasaya; *Guna*- Laghu, Singdha, Picchila; *Virya*- Sita; *Vipaka*- Madhura; *Karma*: Sothahara, Dahaprasamana, Pittahara, Vatahara, Kaphavardhaka	–	Raktapitta, Vrana, Daha, Yuvanapidika	5-10 g (Powder)
240	Part-I; Vol. III	185-186	SANA (Seed)	*Crotolaria juncea* Linn.	Fabaceae	Dried Seed	**Foreign matter**- NMT 2 per cent; **Total ash**- NMT 5 per cent; **Acid-insoluble ash**- NMT 0.5 per cent; **Alcohol-soluble extractive**-NLT 5.5 **per cent; Water-soluble extractive**- NLT 16 per cent; **T.L.C.** - Refer API	A bitter principle 'Corchorin'	*Rasa* - Katu, Tikta, Amla, Kasaya; *Guna*- Ruksa, Tiksna; *Virya*- Usna; *Vipaka*- Katu; *Karma*-Vatahara, Kaphahara, Pittahara, Garbha Anulomaka, Vantikrt, Rakta Pravartaka	Sarsapadi, Pralepa, Dasamuladya Ghrta, Muktadya Churna, Kulatthadya Ghrta	Agnimandya, Jwara, Hrdroga, Mukharoga, Raktadosa, Carma Roga, Timra, Angamarda, Garbhasra-bakara	1-3 g of the drug in powder form
241	Part-I; Vol. III	187-188	SARA (Root)	*Saccharum bengalense* Retz. Syn. *S. sara* Roxb., *S. munja* Roxb.	Poaceae	Dried Roots	**Foreign matter**- NMT 2 per cent; **Total ash**- NMT 6 per cent; **Acid-insoluble ash**- NMT 4 per cent; **Alcohol-soluble extractive**-NLT 3 per cent; **Water-soluble extractive**-NLT 3.5 per cent; **T.L.C.** - Refer API	Sugars	*Rasa* - Madhura, Tikta, Kasaya; *Guna*- Laghu; *Virya*- Anusna; *Vipaka*- Madhura; *Karma*- Kaphahara, Trtdosahara, Balya, Vrsya, Caksusya, Dahahara, Trsnahara	Trnapancamula Kvatha Churna, Brahma Rasayana, Sukumara Ghrta	Daha, Aksiroga, Trsna, Visarpa, Mutrakrcchra, Bastisula, Murcha, Bhrama	20-50 g of Kvatha Churna fro decoction; 6-10 g (Powder)

Sl. No.	*Part and Volume*	*Page No.*	*Common Ayurvedic Name*	*Botanical Name/ Chemical Name*	*Family/ Chemical Formula*	*Part Used*	*Identity, Purity and Strength/Chemical Properties*	*Constituents*	*Properties and Action*	*Important Formulations*	*Therapeutic Uses*	*Dose*
242	Part-I; Vol. III	189-190	SARALA (Heart Wood)	*Pinus roxburghii* Sargent	Pinaceae	Dried Heart Wood	**Foreign matter**- NMT 1 per cent; **Total ash**- NMT 1 per cent; **Acid-insoluble ash**- NMT 0.3 per cent; **Alcohol-soluble extractive**-NLT 5 per cent; **Water-soluble extractive**-NLT 1 per cent; **T.L.C.** - Refer API	Oleo-resin and Flavonoids.	*Rasa* - Madhura, Tikta, Katu; *Guna*- Laghu, Snigdha, Tiksna; *Virya*- Usna; *Vipaka*- Katu; *Karma*- Kaphavatasamaka, Vranasodhaka, Swedahara	Karpuradyarka, Rajanyadi Churna, Sudarsana Churna	Karnaroga, Kantha roga, Aksiroga, Daha, Murccha, Vrana, Kasa, Swarabhramsa, Yuka	1-3 g in powder form
243	Part-I; Vol. III	191-192	SARALA (Root)	*Pinus roxburghii* Sargent.	Pinaceae	Dried Root	**Foreign matter**- NMT 2 per cent; **Total ash**- NMT 1 per cent; **Acid-insoluble ash**- NMT 0.3 per cent; **Alcohol-soluble extractive**-NLT 8 per cent; **Water-soluble extractive**-NLT 3 per cent; **T.L.C.** - Refer API	Resin - Oleo-resin	*Rasa* - Madhura, Tikta, Katu; *Guna*- Laghu, Snigdha, Tiksna; *Virya*- Usna; *Vipaka*- Katu; *Karma*- Kaphavatasamaka, Vranasodhaka, Swedahara	Karpuradyarka, Rajanyadi Churna, Sudarsana Churna	Karna roga, Kantha roga, Aksi roga, Daha, Vrana, Kasa, Swarabhramsa	1-3 g in powder form
244	Part-I; Vol. III	193-194	SARSAPA (Seed)	*Brassica campestris* Linn.	Brassicaceae	Dried Seed	**Foreign matter**- NMT 2 per cent; **Total ash**- NMT 5 per cent; **Acid-insoluble ash**- NMT 0.5 per cent; **Alcohol-soluble extractive**-NLT 8 per cent; **Water-soluble extractive**-NLT 16 per cent; **Fixed oil** - NLT 35 per cent; **T.L.C.** - Refer API	Fixed Oil	*Rasa* - Katu, Tikta; *Guna*- Tiksna, Snigdha; *Virya*- Usna; *Vipaka*- Katu; *Karma*- Kaphahara, Vatahara, Pittakara, Dipana, Vidaha, Hrdya	Maha Yogaraja Guggulu, Karpasasthyadi Taila, Kumakumadi Taila, Prabhanjana Vimardana Taila, Vajraka Taila	Kandu, Kustha, Kosthakrmi, Grahabadha	0.5-1 g in paste form
245	Part-I; Vol. III	195-196	SAPATRIKA (Flower)	*Rosa centifolia* Linn.	Rosaceae	Dried Flower	**Foreign matter**- NMT 2 per cent; **Total ash**- NMT 7.5 per cent; **Acid-insoluble ash**- NMT 1 per cent; **Alcohol-soluble extractive**-NLT 15 per cent; **Water-soluble extractive**- NLT 24 per cent; **T.L.C.** - Refer API	Essential Oil	*Rasa* - Tikta, Kasaya; *Guna*- Laghu; *Virya*- Sita; *Vipaka*- Katu; *Karma*- Vatahara, Pittahara, Kaphahara, Sukrakara, Netrya, Dipana, Hrdya, Varnya	Vasanta Kusumakara Rasa, Tarunarka (Gulabajala), Pravala Pisti, Mukta Pisti, Zahara Mohara Pisti, Trnakanta Mansi Pisti	Kustha, Daha, Mukhasphota, Raktapitta, Raktavikara	3-6 g of the drug inpowder form

Sl. No.	*Part and Volume*	*Page No.*	*Common Ayurvedic Name*	*Botanical Name/ Chemical Name*	*Family/ Chemical Formula*	*Part Used*	*Identity, Purity and Strength/Chemical Properties*	*Constituents*	*Properties and Action*	*Important Formulations*	*Therapeutic Uses*	*Dose*
246	Part-I; Vol. III	197-198	SIMSAPA (Heart Wood)	*Dalbergia sissoo* Roxb.	Fabaceae	Dried Heart Wood	**Foreign matter**- NMT 1 per cent; **Total ash**- NMT 2 per cent; **Acid-insoluble ash**- NMT 0.1 per cent; **Alcohol-soluble extractive**-NLT 1 per cent; **Water-soluble extractive**-NLT 7 per cent; **T.L.C.** - Refer API	Fixed Oil, Essential Oil, Tannins and Flavonoids	*Rasa* - Katu, Tikta, Kasaya; *Guna*- Guru, Picchila; *Virya*- Usna; *Vipaka*- Katu; *Karma*- Vatahara, Pittahara, Kaphahara, Medohara, Kaphavisosana, Medovisosana, Sukradosahara, Varnya, Saiya, Rucikara Gabrhapatini Sosahai Pipana	Ayaskri, Narasimiha Ghtra, Mahakhadira Ghrta	Kustha, Krmi, Daha, Svitra, Vrana, Mutrasarkara, Basti roga, Hikka, Prameha, Arsa, Jwara, Gulma, Asmari, Atisara, Rakta Vikara, Sosa, Sopha, Pandu, Chardi, Pinasa, Dusta Vrana, Vasameha, Sarvajwara	1.5 -10 g of the drug in powder form; 10-20 g for decoction
247	Part-I; Vol. III	199-200	SIMSAPA (Stem Bark)	*Dalbergia sissoo* Roxb.	Fabaceae	Dried Stem Bark	**Foreign matter**- NMT 2 per cent; **Total ash**- NMT 14 per cent; **Acid-insoluble ash**- NMT 2 per cent; **Alcohol-soluble extractive**-NLT 5 per cent; **Water-soluble extractive**-NLT 7 per cent; **T.L.C.** - Refer API	Flavonoids	*Rasa* - Kasaya, Katu, Tikta; *Guna*- Laghu, Ruksa; *Virya*- Usna; *Vipaka*- Katu; *Karma*- Tridosahara, Vransodhana, Garbhapatkar, Balya, Rucikara, Medoara, Vamaka	Narasimhaghrta Rasayana	Kustha, Svitra, Krmi, Bastiroga, Dusta, Vrana, Daha, Kandu, Hikka, Sopha, Visarpa, Pinasa	3-6 g of the drug in powder form; 50-100 ml. of the drug for decoction
248	Part-I; Vol. III	201-202	SIRISA (Stem Bark)	*Albizzia lebbeck* Benth.	Fabaceae	Stem Bark	**Foreign matter**- NMT 1 per cent; **Total ash**- NMT 8 per cent; **Acid-insoluble ash**- NMT 1 per cent; **Alcohol-soluble extractive**-NLT 12 per cent; **Water-soluble extractive**- NLT 6 per cent; **T.L.C.** - Refer API	Saponins and Tannins	*Rasa* - Tikta, Kasaya, Madhura, Katu; *Guna*- Laghu; *Virya*- Anusna; *Vipaka*- Katu; *Karma*- Visaghna, Tvagdosa, Tridosahara, Sothahara, Varnya	Vajraka Taila, Dasanga lepa, Ayaskrti, Devadarvarista, Brhanmaricyadi Taila	Pama, Kustha, Kandu, Visarpa, Kasa, Vrana, Sotha, Swasa, Musaka Visa, Sita Pitta, Visamajwara, Pratisyaya, Visadusti, Suryavarta, Ardha-vabhedaka, Krmi roga, Netrabhiasanda	25-50 g (Kwatha); 3-6 g (Churna)

Sl. No.	*Part and Volume*	*Page No.*	*Common Ayurvedic Name*	*Botanical Name/ Chemical Name*	*Family/ Chemical Formula*	*Part Used*	*Identity, Purity and Strength/Chemical Properties*	*Constituents*	*Properties and Action*	*Important Formulations*	*Therapeutic Uses*	*Dose*
249	Part-I; Vol. III	203-204	STHAU NEYAK (Leaf)	*Taxus baccata* Linn.	Taxaceae	Dried Leaf	**Foreign matter**- NMT 2 per cent; **Total ash**- NMT 6 per cent; **Acid-insoluble ash**- NMT 1.5 per cent; **Alcohol-soluble extractive**-NLT 10 per cent; **Water-soluble extractive**- NLT 16 per cent; **T.L.C.** - Refer API	Alkaloids- Taxine, Ephedrine, Glycoside, Tannins, Resin, Reducing Sugars and Formic Acid	*Rasa* - Katu, Tikta, Madhura; *Guna*- Snigdha, Guru; *Virya*- Sita; *Vipaka*- Madhura; *Karma*- Medhya, Sukravardhaka, Kaphahara, Vatahara, Pittasamaka, Jantughna, Varna Prasadana, Lomasanjanana	Mahanarayana Taila, Bala Taila	Rakta Vikara, Trsna, Tila Kalaka, Daha, Kustha, Krmi Roga, Pidika, Arbuda (Karkata)	1-3 g of the drug in powder form
250	Part-I; Vol. III	205-206	SURANA (Corm)	*Amorphophallus campanulatus* (Roxb.) Blume.	Araceae	Dried Corm	**Foreign matter**- NMT 1 per cent; **Total ash**- NMT 8 per cent; **Acid-insoluble ash**- NMT 2 per cent; **Alcohol-soluble extractive**-NLT 3 per cent; **Water-soluble extractive**- NLT 9 per cent; **T.L.C.** - Refer API	Betulinic Acid, β-sitosterol, Stigmasterol, Lupeol, Triacontane, Glucose, Galactose, Rhamnose and Xylose.	*Rasa* - Katu, Kasaya; *Guna*- Laghu, Ruksa, Visada; *Virya*- Usna; *Vipaka*- Katu; *Karma*- Vatakara Pittakara, Kaphahara, Dipana, Vistambhi, Rucya, Gudakilahrt, Raktapittakara, Dadrukara, Kusthakara	Suranavaloha, Surnavataka, Samudradya Churna	Arsa, Plihagulma, Swasa, Kasa, Asthila	2-10 g of the drug in powder form
251	Part-I; Vol. III	207-208	SVETA-CANDANA (Heart Wood)	*Santalum album* Linn.	Santalaceae	Dried Heart Wood	**Foreign matter**- NMT 1 per cent; **Total ash**- NMT 1 per cent; **Acid-insoluble ash**- NMT 0.2 per cent; **Alcohol-soluble extractive**-NLT 8 per cent; **Water-soluble extractive**- NLT 1 per cent; **Volatile oil** - NLT 1.5 per cent; **T.L.C.** - Refer API	Volatile oil (α-and β-santalol)	*Rasa* - Tikta, Madhura *Guna*- Laghu, Ruksa *Virya*- Sita *Vipaka*- Katu; *Karma*- Pittahara, Kaphahara, Durgandhahara, Dahaprasamana, Varnya, Hrdya, Trsnahar, Vrsya, Krmighna, Visaghna	Ayaskrti, Asvagandhadyarista, Sarivadyasava, Arimedadi Taila, Baladhatryadi Taila, Marma Gutika, Candanasava, Candanadi Churna Candanadi Taila	Sosa, Daha, Raktapitta, Raktarsa, Hikka, Vamana, Raktatisara, Pradara, Sukrameha, Netra Roga, Mutraghata, Bhrama, Raktavikara, Krmi Roga	3-6 g of the drug in powder form

Sl. No.	*Part and Volume*	*Page No.*	*Common Ayurvedic Name*	*Botanical Name/ Chemical Name*	*Family/ Chemical Formula*	*Part Used*	*Identity, Purity and Strength/Chemical Properties*	*Constituents*	*Properties and Action*	*Important Formulations*	*Therapeutic Uses*	*Dose*
252	Part-I; Vol. III	209-210	SYONAKA (Root)	*Oroxylum indicum* Vent.	Bignoniaceae	Dried Root	**Foreign matter**- NMT 1 per cent; **Total ash**- NMT 5 per cent; **Acid-insoluble ash**- NMT 1 per cent; **Alcohol-soluble extractive**-NLT 20 per cent; **Water-soluble extractive**- NLT 42 per cent; **T.L.C.** - Refer API	Flavonoids and Tannins.	*Rasa* - Kasaya, Tikta *Guna*- Laghu, Ruksa *Virya*- Sita *Vipaka*- Katu; *Karma*- Kaphapittasamaka, Dipana, Grahi	Amrtarista, Dantyadyarista, Dasamularista, Narayana Taila, Dhanawantara Ghrta, Brahma Rasayana, Dasamula Kwatha Churna, Chyawanaprasa, Awaleha	Vatatisara, Kasa, Aruci, Basti roga, Amavata, Udara roga, Urustambha, Vatavyadhi, Karna roga, Sotha	5-10 g in powder form; 25-50 g in decoction
253	Part-I; Vol. III	211-212	TALA (Inflorescene)	*Borassus flabellifer* Linn.	Araceae	Dried male Inflorescence	**Foreign matter**- NMT 1 per cent; **Total ash**- NMT 7.5 per cent; **Acid-insoluble ash**- NMT 1.5 per cent; **Alcohol-soluble extractive**-NLT 4 per cent; **Water-soluble extractive**-NLT 8 per cent; **T.L.C.** - Refer API	Kernels contian Galacto-mannan (Poly-saccharide).	*Rasa* - Madhura; *Guna*- Sita, Guru, Snigdha; *Virya*- Sita; *Vipaka*- Madhura; *Karma*- Sukrala, Brmhana, Vrsya, Tarpaka, Sirovirecaka, Vastisuddhikara, Medakara, Vatahara, Pittahara, Vrannasaka, Krmighna	Avlttoladi Bhasma (Ksara), Panviraladi Bhasma, (Tala Puspodbhava Ksara) Guda Pippali	Raktapitta, Urahksata, Swasa, Daha, Krmi, Mutrakrcchra, Sophaghna, Vandhyakara	1-3 g
254	Part-I; Vol. III	123-214	TRIVRT (Root)	*Operculina turpethum* (Linn.) Silva manso Syn. *Ipomoea turpethum* R. Br.	Convolvulaceae	Dried Root	**Foreign matter**- NMT 2 per cent; **Total ash**- **NMT 10 per cent; Acid-insoluble ash**- NMT 1.5 per cent; **Alcohol-soluble extractive**-NLT 10 per cent; **Water-soluble extractive**- NLT 8 per cent; **T.L.C.** - Refer API	Resinous Glycosides	*Rasa* - Madhura, Katu, Tikta, Kasaya; *Guna*- Ruksa, Laghu, Tiksna; *Virya*- Usna; *Vipaka*- Katu; *Karma*- Vatala, Virecana, Kaphapittahara, Sukhavirecanaka, Pittahara, Jwarahara	Hrdyavirecana Leha, Aswagan-dharista, Avipattikara Churna, Manibhadra Guda	Malabandha, Gulma, Udara Roga, Jwara, Sopha Pandu, Pliha, Vrana, Krmi, Kustha, Kandu	1-3 g of the drug in powder form

Sl. No.	*Part and Volume*	*Page No.*	*Common Ayurvedic Name*	*Botanical Name/ Chemical Name*	*Family/ Chemical Formula*	*Part Used*	*Identity, Purity and Strength/Chemical Properties*	*Constituents*	*Properties and Action*	*Important Formulations*	*Therapeutic Uses*	*Dose*
255	Part-I; Vol. III	215-216	TUMBINI (Fresh Fruit)	*Lagenaria siceraria* (Mol.) Standl. Syn. *L. leucantha* Rusby., *L. vulgaris* ser.	Cucurbitaceae	Fresh Fruit	**Foreign matter**- Nil; **Total ash**- NMT 12 per cent; **Acid-insoluble ash**- NMT 0.6 per cent; **Alcohol-soluble extractive**-NLT 10 per cent; **Water-soluble extractive**- NLT 25 per cent; **T.L.C.** - Refer API	Saponin and Fatty Oil	*Rasa* - Madhura; *Guna*- Snigadha; *Virya*- Sita; *Vipaka*- Madhura; *Karma*- Pittahara, Kaphahara, Bhedaka, Rucikara, Hrdya, Vrsya	Mahavisagarbha Taila	Jwara, Kasa, Svasa, Visa roga, Sopha, Vrana, Sula	10-20 ml of fresh drug in juice form
256	Part-I; Vol. III	217-218	UDUMBARA (Fruit)	*Ficus glomerata* Roxb. Syn. *F. racemosa* Linn.	Moraceae	Dried Fruit	**Foreign matter**- NMT 1 per cent; **Total ash**- NMT 9 per cent; **Acid-insoluble ash**- NMT 0.5 per cent; **Alcohol-soluble extractive**-NLT 3 per cent; **Water-soluble extractive**-NLT 15 per cent; **T.L.C.** - Refer API	β-sitosterol, Lupeol Acetate and Carbohydrates	*Rasa* - Madhura, Kasaya; *Guna*- Ruksa, Guru; *Virya*- Sita; *Vipaka*- Madhura; *Karma*- Pittahara, Kaphahara, Varnya, Vrana Ropana, Vrana Sodhana, Bhagna Sandhanaka, Raktadosahara	Raktapitta, Murccha, Daha, Trsna, Pradara, Granthi Roga	Raktapitta, Murccha, Daha, Trsna, Pradara, Granthi Roga	10-15 g of the drug in powder form
257	Part-I; Vol. III	219-220	USIRA (Root)	*Vetiveria zizanioides* (Linn.) Nash	Poaceae	Root	**Foreign matter**- NMT 2 per cent; **Total ash**- NMT 9 per cent; **Acid-insoluble ash**- NMT 6 per cent; **Alcohol-soluble extractive**-NLT 4 per cent; **Water-soluble extractive**-NLT 5 per cent; **Volatile oil** - NLT 1 per cent; **T.L.C.** - Refer API	Essential Oil	*Rasa* - Tikta, Madhura; *Guna*- Laghu, Snigdha; *Virya*- Sita; *Vipaka*- Madhura; *Karma*- Vataghna, Dabaklantihara, Pittaghna, Pacana, Stambhana, Kaphapttahrt	Usirasava, Yogarajaguggulu, Sadanga Kwatha Churna	Jwara, Trsna, Mutrakrcchra, Vrana	3-6 g of the drug in powder form for infusion
258	Part-I; Vol. III	221-223	UTPALA (Flower)	*Nymphaea stellata* Willd.	Nymphaceae	Dried Flower	**Foreign matter**- NMT 2 per cent; **Total ash**- NMT 8 per cent; **Acid-insoluble ash**- NMT 0.5 per cent; **Alcohol-soluble extractive**-NLT 5 per cent; **Water-soluble extractive**-NLT 22 per cent; **T.L.C.** - Refer API	Tannins	*Rasa* - Madhura, Kasaya; *Guna*- Pichila, Snigdha; *Virya*- Sita; *Vipaka*- Madhura; *Karma*- Rucya, Rasayana, Kesya, Dahapaustikara, Medhya, Daha, Dradhykara,	Asokarista, Arvindasava, Usirasava, Candanasava, Kalyanaka Ghrta, Samangadi Churna, Kanaka Taila, Jatyadi Taila,	Pipasa Daha, Raktapitta, Chardi, Murccha, Hrdraoga, Mutra Kecchra, Jwaratisara	3-6 g of the drug

Sl. No.	*Part and Volume*	*Page No.*	*Common Ayurvedic Name*	*Botanical Name/ Chemical Name*	*Family/ Chemical Formula*	*Part Used*	*Identity, Purity and Strength/Chemical Properties*	*Constituents*	*Properties and Action*	*Important Formulations*	*Therapeutic Uses*	*Dose*
									Pittanasaka, Raktaprasadak	Tungadrumadi Taila, Manjesthadi Taila, Candanadi Lauha, Triphala Ghrta		
259	Part-I; Vol. IV	1-3	ADHAKI (Seed)	*Cajanus cajan* Linn.	Fabaceae	Dried Seed	**Foreign matter**- NMT 2 per cent; **Total ash**- NMT4 per cent; **Acid-insoluble ash**- NMT 0.5 per cent; **Protein content** - NLT 20 per cent; **T.L.C.** - Refer API	–	*Rasa* - Kasya, Madhura; *Guna*- Ruksa, Laghu; *Virya*- Sita; *Vipaka*- Katu; *Karma*- Vatakara, Kaphahara, Pittakara, Medohara, Sangrahi, Varnya, Visapaha, Stanyavrddhi	Kankayana Gutika	Atisthaulya, Raktavikara, Raktapitta, Visaroga, Sthaulya, Medoroga, Arsa	As directed by the physician
260	Part-I; Vol. IV	4-5	AGARU (Heart Wood)	*Aquilaria agallocha* Roxb.	Thymelacaceae	Dried Heart wood	**Foreign matter**- NMT 1 per cent; **Total ash**- NMT 13 per cent; **Acid-insoluble ash**- NMT 0.5 per cent; **Alcohol-soluble extractive**-NLT 1 per cent; **Water-soluble extractive**- NLT 2 per cent; **T.L.C.** - Refer API	Essential Oil	*Rasa* - Katu, Tikta; *Guna*- Snigdha, Tiksna, Laghu; *Virya*- Usna; *Vipaka*- Katu; *Karma*- Tvacya, Pittalam, Vatahara, Kaphahara, Sirovirecana	Madhukasava, Mrdvikasava, Karpuradyarka, Chyawanaprasa Avahela Anu Taila, Candanadi Taila, Khadiradi Gutika, Svasahara Kasaya Churna, Guducyadi Taila	Kustha, Karna Roga, Aksi roga, Visa, Swasa	1-3 g
261	Part-I; Vol. IV	6-7	AKLARI (Endosperm)	*Lodoicea maldivica* Pers. Syn. *L. seychellarum* Labill.	Arecaceae	Dried Endo-sperm	**Foreign matter**- NMT 2 per cent; **Total ash**- NMT 2 per cent; **Acid-insoluble ash**- NMT 0.4 per cent; **Alcohol-soluble extractive**-NLT 0.3 per cent; **Water-soluble extractive**- NLT 4 per cent; **T.L.C.** - Refer API	Sugars and Sterols.	*Rasa* - Madhura, Katu; *Guna*- Laghu; *Virya*-Usna; *Vipaka*- Katu; *Karma*- Vatahara, Kaphahara, Hrdya, Visaghna, Trsnanigrahana, Sitaprasamana, Agnidiptikara	Gorocanadi Vati, Mrtasanjivani Gutiaka, Javahara Mohara	Visucika, Hrdroga, Sita Jvara	5-10 g of the drugs in powder form

Sl. No.	*Part and Volume*	*Page No.*	*Common Ayurvedic Name*	*Botanical Name/ Chemical Name*	*Family/ Chemical Formula*	*Part Used*	*Identity, Purity and Strength/Chemical Properties*	*Constituents*	*Properties and Action*	*Important Formulations*	*Therapeutic Uses*	*Dose*
262	Part-I; Vol. IV	8-9	APARAJITA (Leaf)	*Clitoria ternatea* Linn.	Fabaceae	Dried Leaf	**Foreign matter**- NMT 2 per cent; **Total ash**- NMT 15 per cent; **Acid-insoluble ash**- NMT 4 per cent; **Alcohol-soluble extractive**-NLT 7 per cent; **Water-soluble extractive**-NLT 15 per cent; **T.L.C.** - Refer API	Glycosides- Flavonal glycosides and Resin glycosides	*Rasa* - Tikta, Katu, Kasaya *Guna*- Laghu *Virya*- Sita *Vipaka*- Katu; *Karma*- Medhya, Kanthya, Caksusya, Pittopadravanasini, Tridosa Samaka, Visapaha, Grahaghni	Vata Raktantaka Rasa	Kustha, Mutradosa, Sotha, Vrana, Visa, Unmada, Ardhava Bhedaka, Sula, Graha badha, Amadosa, Raktatisara, Bhrama, Swasa, Kasa, Jwara, Daha, Vamana	Root Powder 1-3 g; Seed Powder 1-3 g; Leaf Powder 2-5 g
263	Part-I; Vol. IV	10-11	ATMAGUPTA (Root)	*Mucuna prurita* Hook. Syn. M. *pruriens* (L.) DC.	Fabaceae	Dried Root	**Foreign matter**- NMT 1 per cent; **Total ash**- NMT 6 per cent; **Acid-insoluble ash**- NMT 1 per cent; **Alcohol-soluble extractive**-NLT 4 per cent; **Water-soluble extractive**-NLT 5 per cent; **T.L.C.** - Refer API	Choline	*Rasa* - Tikta, Kasaya; *Guna*- Guru, Srigdha; *Virya*- Sita; *Vipaka*- Katu; *Karma*- Pittahara, Kaphahara, Vrsya, Brhana, Balya, Yonisamkirnikara, Vajikarana	–	Dusta Vrana, Pakwatisara, Raktapitta, Kustha, krsata, Sitapitta, Vatavyadhi, Yoni Sithilata	3-6 g of the drug in the powder form for decoction
264	Part-I; Vol. IV	12-13	BILVA (Stem Bark)	*Aegle marmelos* Corr.	Rutaceae	Dried Stem Bark	**Foreign matter**- NMT 1 per cent; **Total ash**- NMT 10 per cent; **Acid-insoluble ash**- NMT 1 per cent; **Alcohol-soluble extractive**-NLT 4 per cent; **Water-soluble extractive**-NLT 9 per cent; **T.L.C.** - Refer API	Coumarins and Sterols	*Rasa* - Kasaya, Tikta, Madhura; *Guna*- Tiksna, Ruksa, Laghu; *Virya*- Usna; *Vipaka*- Katu; *Karma*- Dipaniya, Kaphahara, Vatahara, Samgrahi, Pittakara, Visaghna	Pusyanuga Churna, Grahani Mihira Taila, Sudarsana Churna, Candanadi Taila, Anu Taila	Chardi, Vatavyadhi, Sula, Sotha, Atisara, Raktatisara, Kuksisula Amasula, Arsa, Medoroga, Grahaniroga, Madhumeha, Pravahika	15-30 ml

Sl. No.	*Part and Volume*	*Page No.*	*Common Ayurvedic Name*	*Botanical Name/ Chemical Name*	*Family/ Chemical Formula*	*Part Used*	*Identity, Purity and Strength/Chemical Properties*	*Constituents*	*Properties and Action*	*Important Formulations*	*Therapeutic Uses*	*Dose*
265	Part-I; Vol. IV	14-15	CAMPAKA (Flower)	*Michelia champaca* Linn.	Magnoliaceae	Dried Buds and Flowers including Calyx	**Foreign matter**- NMT 2 per cent; **Total ash**- NMT 11 per cent; **Acid-insoluble ash**- NMT 1.5 per cent; **Alcohol-soluble extractive**-NLT 9 per cent; **Water-soluble extractive**-NLT 12 per cent; **T.L.C.** - Refer API	Volatile Oil	*Rasa* - Katu, Tikta, Kasaya, Madhura *Guna*- Laghu, Ruksa *Virya*- Sita *Vipaka*- Katu; *Karma*- Pittagit, Kaphapittasara nasaka, Visaghna, Hrdya	Candana-balalaksadi Taila, Baladhatryadi Taila	Krmi, Mutrakrcchra, Vatarakta, Kustha, Kandu, Vrana	Puspa Churna 1-3 g
266	Part-I; Vol. IV	16-17	CINCA (Fruit Pulp)	*Tamarindus indica* Linn.	Fabaceae	Fruit Pulp	**Foreign matter**- NMT 1 per cent; **Total ash**- NMT 4 per cent; **Acid-insoluble ash**- NMT 0.5 per cent; **Alcohol-soluble extractive**-NLT 46 per cent; **Water-soluble extractive**- NLT 59 per cent; **T.L.C.** - Refer API	Inorganic acids, Sugars, Saponin and Bitter principle - Tamarindinca	*Rasa* - Amla, Madhura, Kasaya; *Guna*- Guru, Ruksa, Sara; *Virya*- Usna, *Vipaka*- Amla; *Karma*- Kaphavatanut, Dipana, Bastisuddhikara, Bhedi, Vistambhi, Dipana, Hrdya	Sankha Dravak, Sankhavati	Udararoga, Agnimandya, Arocaka, Paktisula, Trsna, Klama, Srama, Bhranti, Krmi, Karnasula, Nadivrana	4-10 g of the drug
267	Part-I; Vol. IV	18	DADIMA (Fresh Fruit)	*Punica granatum* Linn.	Punicaceae	Fresh Fruit	–	–	*Rasa* - Amala, Madhura, Kasaya; *Guna*- Laghu, Singdha; *Virya*- Usna; *Vipaka*- Madhura; *Karma*- Vatahara, Pittahara, Kaphahara, Dipana, Pacana, Rucya, Grathi, Mukhagandhahara, Hrdya, Medhya, Sramahara, Sukrala, Tarpaka, Varcovibadhaniya, Balya, Medhya	Dadhika Ghrta, Dadimastaka Churna, Bhaskaralavana Churna, Brhacchagaladya Ghrta	Daha, Jvara, Trsna, Kasa, Amavata, Atisara, Raktapitta, Arocaka	15-30 ml

Sl. No.	*Part and Volume*	*Page No.*	*Common Ayurvedic Name*	*Botanical Name/ Chemical Name*	*Family/ Chemical Formula*	*Part Used*	*Identity, Purity and Strength/Chemical Properties*	*Constituents*	*Properties and Action*	*Important Formulations*	*Therapeutic Uses*	*Dose*
268	Part-I; Vol. IV	19-20	DADIMA (Fruit Rind)	*Punica granatum* Linn.	Punicaceae	Dried Fruit Rind	**Foreign matter**- NMT 2 per cent; **Total ash**- NMT 4 per cent; **Acid-insoluble ash**- NMT 0.4 per cent; **Alcohol-soluble extractive**-NLT 9 per cent; **Water-soluble extractive**-NLT 20 per cent; **T.L.C.** - Refer API	Tannic acid, Sugar and Gum	*Rasa* - Kasaya, Amla; *Guna*-Laghu, Singdha; *Virya*- Anusna; *Vipaka*- Katu; *Karma*-Vata kaphahara, Vranaropaka, Grahi	Khadiradi Gutika, Mrtsanjivani Sura, Kalyanaka Ghrta, Maricadi Gutika, Nilikadya Taila	Daha, Jvara, Kantharoga, Mukha-daurgandha, Aruci, Amlapitta, Atisara, Pravahika, Raktapitta, Raktavikara, Kasa	Powder 3-6 g
269	Part-I; Vol. IV	21-22	DADIMA (Leaf)	*Punica granatum* Linn.	Punicaceae	Dried Leaf	**Foreign matter**- NMT 2 per cent; **Total ash**- NMT 10.5 per cent; **Acid-insoluble ash**- NMT 2 per cent; **Alcohol-soluble extractive**-NLT 12 per cent; **Water-soluble extractive**- NLT 25 per cent; **T.L.C.** - Refer API	Tannins and β-sitosterol	*Rasa* - Kasaya, Tikta; *Guna*-Laghu; *Virya*- Sita; *Vipaka*- Kasaya; *Karma*-Kaphahara, Dipana, Rucya	–	Aruci, Agnimandya, Atisara, Pravahika, Krmi, Raktapitta, Kasa, Jvara, Mukhapaka	Patra Svarasa 5-10 ml; Patra Kalka 5-10 g
270	Part-I; Vol. IV	23-24	DEVADARU (Heart Wood)	*Cedrus deodara* (*Roxb.*) Loud.	Pinaceae	Dried Heart Wood	**Foreign matter**- NMT 1 per cent; **Total ash**- NMT 2 per cent; **Acid-insoluble ash**- NMT 1 per cent; **Alcohol-soluble extractive**-NLT 7 per cent; **Water-soluble extractive**-NLT 1.5 per cent; **T.L.C.** - Refer API	Terpenoids, Flavonoids and Glycosides	*Rasa* - Tikta; *Guna*- Laghu, Snigdha; *Virya*-Usna *Vipaka*- Katu; *Karma*- Vatahara, Kaphahara, Dustavrana Sodhaka	Khadirarista, Dasmularista, Devadarvarista, Mrtasanjivani Sura, Karpuradyarka, Pramehamihira Taila, Candanadi Churna, Sudarsana Churna, Narayana Taila, Pradarantaka Lauha, Vataraktanaka Lauha, Mahavisagarbha Taila	Vibandha, Adhamana, Sotha, Tandra, Kikka, Jvara, Prameha, Pinasa, Kasa, Kandu, Krmi, Kustha, Amavata, Raktavikara, Sutikaroga	3-6 g of the drug in the powder form

Sl. No.	*Part and Volume*	*Page No.*	*Common Ayurvedic Name*	*Botanical Name/ Chemical Name*	*Family/ Chemical Formula*	*Part Used*	*Identity, Purity and Strength/Chemical Properties*	*Constituents*	*Properties and Action*	*Important Formulations*	*Therapeutic Uses*	*Dose*
271	Part-I; Vol. IV	25-27	DHATTURA (Whole Plant)	*Datura metel* Linn. Syn. D. *fastuosa* L.;	Solanaceae	Dried Whole Plant	**Foreign matter**- NMT 2 per cent; **Total ash**- NMT 16 per cent; **Acid-insoluble ash**- NMT 4 per cent; **Alcohol-soluble extractive**-NLT 4 per cent; **Water-soluble extractive**-NLT 15 per cent; **T.L.C.** - Refer API	Alkaloids (Hyoscine) and two withanolide Glucosides (Dhatura-metelin A and B	*Rasa* - Katu, Kasaya, Madhura, Tikta; *Guna*-Tiksna, Guru; *Virya*- Usna; *Vipaka*- Katu; *Karma*- Madakari, Kaphahara, Agni, Vrddhikara, Varnya, Jangama Vishahara	Kanakasawa, Ekangaviara Rasa, Puspadhanwa Rasa, Tribhuvana Kirti Rasa, Sri Jayamangala Rasa, Laghu Visagarbha Taila, Visatinduka Taila, Dhattura Taila	kasa, Swasa, Jwara, Kustha, Vrana, Mutrakrecchra, Twak dosa, Yika liksa, krmi, Alarka, Visa, Karma, Nadi, Kandu, Indralupta, Padadaha, Stanuthita pida, Unmada	100-200 mg
272	Part-I; Vol. IV	28-30	DURVA (Whole Plant)	*Cynodon dactylon* (Linn.) Pers.	Poaceae	Dried Whole Plant	**Foreign matter**- NMT 2 per cent; **Total ash**- NMT 9 per cent; **Acid-insoluble ash**- NMT 4.5 per cent; **Alcohol-soluble extractive**-NLT 3 per cent; **Water-soluble extractive**-NLT 9.5 per cent; **T.L.C.** - Refer API	Phenolic Phytotoxins (Ferulic, Syringic, P-coumaric, Vanillic, P-hydroxy-benzoic and O-hydroxy-phenil acetic acid	*Rasa* - Kasaya, Madhura, Tikta; *Guna*- Laghu; *Virya*- Sita; *Vipaka*- Madhura; *Karma*- Pittahara, Kaphahara, Sramahara, Rucya	–	Raktapitta, Trsna, Chardi, Daha, Murccha, Visarpa, Raktavikara, Tvak Roga, Atisara, Kaphaja Jvara, Vataja Jvara, Jvara, Nasagata Raktapitta	Svarasa: 10-20 ml.p
273	Part-I; Vol. IV	31-32	GAMBHARI (Stem Bark)	*Gmelina arborea* Linn.	Verbenaceae	Dried Stem Bark	**Foreign matter**- NMT 1 per cent; **Total ash**- NMT 11 per cent; **Acid-insoluble ash**- NMT 0.3 per cent; **Alcohol-soluble extractive**-NLT 8 per cent; **Water-soluble extractive**-NLT 23 per cent; **T.L.C.** - Refer API	Alkaloids, in traces.	*Rasa* - Tikta, Katu, Madhura; *Guna*-Guru; *Virya*- Usna; *Vipaka*- Katu; *Karma*-Kaphahara, Sothahara, Dipana, Pacana, Medhya, Bhedana, Visahara, Daha Prasamana	Candanasava	Sula, Arsa, Jvara, Raktapitta, Trsna, Bhrama, Sotha	3-5 g

Sl. No.	*Part and Volume*	*Page No.*	*Common Ayurvedic Name*	*Botanical Name/ Chemical Name*	*Family/ Chemical Formula*	*Part Used*	*Identity, Purity and Strength/Chemical Properties*	*Constituents*	*Properties and Action*	*Important Formulations*	*Therapeutic Uses*	*Dose*
274	Part-I; Vol. IV	33-34	IKSU (Root Stock)	*Saccharum officinarum* Linn.	Poaceae	Root Stock	**Foreign matter**- NMT 2 per cent; **Total ash**- NMT 8 per cent; **Acid-insoluble ash**- NMT 5 per cent; **Alcohol-soluble extractive**-NLT 4 per cent; **Water-soluble extractive**-NLT 7 per cent; **T.L.C.** - Refer API	–	*Rasa* - Madhura; *Guna*- Sara, Guru, Snigdha; *Virya*- Sita; *Vipaka*- Madhura; *Karma*- Bramhana, Vrsya, Vatasamaka, Kaphakara, Pittahara, Mutrala, Balya	Trnapancamula Kvatha, Sukumara Ghrta, Brahma Rasayana	Raktapitta, Mutrakrcchra, Ojoksaya, Nasa rakta srava, Grahani, Pandu, Ksataja Kasa, Visarpa	15-30 g in decoction form
275	Part-I; Vol. IV	35-36	KADALI (Flower)	*Musa paradisiaca* Linn.	Musaceae	Dried Flower	**Foreign matter**- NMT 1 per cent; **Total ash**- NMT 15 per cent; **Acid-insoluble ash**- NMT 1 per cent; **Alcohol-soluble extractive**-NLT 3 per cent; **Water-soluble extractive**-NLT 18 per cent; **T.L.C.** - Refer API	Saponins, Tannins, reducing and non-reducing Sugars, Sterols and Triterpenes	*Rasa* - Kasaya, Madhura, Tikta; *Guna*- Mrdu, Grahi, Dipana; *Virya*- Usna; *Vipaka*- Madhura; *Karma*- Pittanasaka, Ruca, Kaphaghna, Balya, Vrsya, Stambhaka	Hemanatha Rasa	Krmi, Swasa, Roga, Raktapitta, Pradara	10-20 g
276	Part-I; Vol. IV	37-38	KARCURA (Rhizome)	*Curcuma zedoaria* Rosc.	Zingiberaceae	Dried Pieces of Rhizome	**Foreign matter**- NMT 2 per cent; **Total ash**- NMT 7 per cent; **Acid-insoluble ash**- NMT 2 per cent; **Alcohol-soluble extractive**-NLT 4 per cent; **Water-soluble extractive**-NLT 10 per cent; **Volatile oil** - NLT 2 per cent; **T.L.C.** - Refer API	Essential Oil and Resin.	*Rasa* - Katu, Tikta; *Guna*- Laghu, Tiksna; *Virya*- Usna; *Vipaka*- Katu; *Karma*- Vatahara, Kaphara, Rucya, Dipana Mukhavisadyakara	Karcuradi Churna (Karcuradi Lepa), Karpuradyarka, Sutasekhara Rasa	Hikka, Swasa, Kasa, Kustha, Arsa, Gulma, Jvara, Vrana, Pliha, Galganda, Krmi	1-3 g of the drug in powder form
277	Part-I; Vol. IV	39-40	KASTURI-LATIKA (Seed)	*Hibiscus abelmoschus* Linn. Syn. *Abelmoschus moschatus* Medik	Malvaceae	Seed	**Foreign matter**- NMT 2 per cent; **Total ash**- NMT 5 per cent; **Acid-insoluble ash**- NMT 0.3 per cent; **Alcohol-soluble extractive**-NLT 10 per cent; **Water-soluble extractive**- NLT 9 per cent; **Fixed oil** - NLT 10 per cent; **T.L.C.** - Refer API	Fixed Oil and Volatile Oils	*Rasa* - Katu, Tikta, Madhura; *Guna*- Laghu; *Virya*- Sita; *Vipaka*- Madhura; *Karma*- Caksusya, Chedini, Vrsya, Kaphahara, Mukhadaurgan-dhyanasaka, Vasti Visodhani	Karpuradyarka	Trsna, Vasti Roga, Mukha Roga.	2-4 g of the drug in powder form

Sl. No.	*Part and Volume*	*Page No.*	*Common Ayurvedic Name*	*Botanical Name/ Chemical Name*	*Family/ Chemical Formula*	*Part Used*	*Identity, Purity and Strength/Chemical Properties*	*Constituents*	*Properties and Action*	*Important Formulations*	*Therapeutic Uses*	*Dose*
278	Part-I; Vol. IV	41-42	KATAKA (seed)	*Strychnos potatorum* Linn. f.	Loganiaceae	Dried Seed	**Foreign matter**- NMT 2 per cent; **Total ash**- NMT 2 per cent; **Acid-insoluble ash**- NMT 0.5 per cent; **Alcohol-soluble extractive**-NLT 1 per cent; **Water-soluble extractive**-NLT 5 per cent; **T.L.C.** - Refer API	Alkaloids.	*Rasa* - Madhura, Tikta, Kasaya; *Guna*- Guru, Sita; *Virya*- Usna; *Vipaka*- Katu; *Karma*- Caksusya, Vatahara, Slesmahara, Visaghna, Pittala, Asu-drstiprasadakrt,	Dasmularista, Niruryari Gutika	Mutrakrcchra, Mutrasmari, Krmi, Aruci, Trsna, Sula, Netra Roga, Sarkara meha, Rakta abhisyanda, Prameha, Vrscika Visa, Apasmara	3-6 g
279	Part-I; Vol. IV	43-44	KHARJURA (Dried Fruit)	*Phoenix dactylifera* Linn.	Araceae	Dried Fruit with Seeds Removed	**Foreign matter**- NMT 2 per cent; **Total ash**- NMT 3 per cent; **Acid-insoluble ash**- NMT 1 per cent; **Alcohol-soluble extractive**-NLT 20 per cent; **Water-soluble extractive**- NLT 74 per cent; **T.L.C.** - Refer API	Sugars, Tannins and Vitamins.	*Rasa* - Madhura, Kasaya; *Guna*- Guru, Snigdha; *Virya*- Sita; *Vipaka*- Madhura; *Karma*- Vatahara, Pittahara, Kaphahara, Hrdya, Tarpana, Balya, Brmhana, Vrsya	Draksadi Churna, Eladya Modaka, Eladi gutika, Siva Gutika (Laghu)	Kasaya, Ksata Ksaya, Daha, Raktapitta, Murccha, Trsna, Madatyaya, Abhighata, Kasa, Svasa, Srama, Gulma, Jvara, Mukha, Vairasya, Hikka, Prameha, Pittasula	10-15 g
280	Part-I; Vol. IV	45-46	KHARJURA (Fresh Fruit)	*Phoenix dactylifera* Linn.	Araceae	Ripe and Mature Fruit with Seed Removed	**Foreign matter**- NMT 1 per cent; **Total ash**- NMT 3 per cent; **Acid-insoluble ash**- NMT 0.5 per cent; **Alcohol-soluble extractive**-NLT 20 per cent; **Water-soluble extractive**- NLT 65 per cent; **T.L.C.** - Refer API	Sugars, Protein and Vitamins.	*Rasa* - Madhura, Kasaya; *Guna*- Guru, Snigdha; *Virya*- Sita; *Vipaka*- Madhura; *Karma*- Vatahara, Pittahara, Kaphahara, Mansavardhaka, Sukrakarna, Rucikara, Hrdya, Balya, Tarpaka, Kosthagata vayunasaka, Vamaka, Ksudha Sramahara	Draksadi Churna, Eladya Modaka, Eladi Gutika, Siva Gutika (Laghu)	Ksata Ksaya, Raktapitta, Jvaratisara, Trsna, Kasa, Swasa, Murccha, Madatya, Daha, Abhighat	10-50 g

Sl. No.	*Part and Volume*	*Page No.*	*Common Ayurvedic Name*	*Botanical Name/ Chemical Name*	*Family/ Chemical Formula*	*Part Used*	*Identity, Purity and Strength/Chemical Properties*	*Constituents*	*Properties and Action*	*Important Formulations*	*Therapeutic Uses*	*Dose*
281	Part-I; Vol. IV	47-49	KRSNASARIVA (Root)	*Cryptolepis buchanani* Roem. and Schult.	Asclepiadaceae	Dried Root	**Foreign matter**- NMT 2 per cent; **Total ash**- NMT 6 per cent; **Acid-insoluble ash**- NMT 1.5 per cent; **Alcohol-soluble extractive**-NLT 8 per cent; **Water-soluble extractive**-NLT 7 per cent; **T.L.C.** - Refer API	Alkaloids	*Rasa* - Madhura, Tikta; *Guna*- Guru, Snigdha; *Virya*- Sita; *Vipaka*- Madhura; *Karma*- Sukrakara, Kaphanasaka, Visaghna Rucya, Sangrahi, Rakta Vikara Nasaka, Ama Visaghna, Tridosahara, Trsnahara	Satavari Guda, Kalyanaka Ghrta, Triphala Ghrta, Brhat Phala Ghrta, Maha Kalyanaka Ghrta, Maha Tiktaka Ghrta, Maha Pancagavya Ghrta, Vatasyama-yantaka Ghrta, Candanadi Taila, Brhaccha-galadya Ghrta	Agnimandya, Aruci, Svasa, Kasa, Jvara, Prameha, Mukha Daurgandhya, Atisara, Kustha, Kandu, Pradara, Vata Rakta, Dehadur-gandha, Raktapitta	5-10 g
282	Part-I; Vol. IV	50-51	KUNDURU (Exudate)	*Boswellia serrata* Roxb.	Burseraceae	Exudate	**Foreign matter**- NMT 5 per cent; **Total ash**- NMT 10 per cent; **Acid-insoluble ash**- NMT 8 per cent; **Alcohol-soluble extractive**-NLT 45 per cent; **Water-soluble extractive**- NLT 28 per cent; **T.L.C.** - Refer API	Oleo-gum-resins	*Rasa* - Madhura, Katu, Tikta *Guna*- Guru, Tiksna, Snigdha *Virya*- Usna *Vipaka*- Madhura; *Karma*- Kaphapittahara, Kaphahara, Vatahara, Rakta Stambhara, Balya, Swedahara	Karpuradyarka, Jirakadi Modaka, Bala Tila, Bala Guducyadi Taila	Swasa, Pittabhisyanda, Pradara, Jwara, Sarkarameha, Vrsana Sula, Mukha roga, Uka	1-3 g
283	Part-I; Vol. IV	52-54	KUNKUMA (Style and Stigma)	*Crocus sativus* Linn.	Iridaceae	Dried Style and Stigma	**Foreign Organic matter**- NMT 2 per cent; **Styles** - NMT 10 per cent; **Loss on drying** - NMT 14 per cent of its weight, when dried at 100°C; **Ash**- NMT 7.5 per cent; **Acid-insoluble ash**- NMT 1 per cent	Essential Oils, bitter Glycoside, Picrocrocin and Crocin	*Rasa* - Katu, Tikta *Guna*- Snigdha *Virya*- Usna *Vipaka*- Katu; *Karma*- Varnya, Slesmahara, Vatahara, Rasayana, Visaghna, Jantuhara	Karpuradyarka, Balarka Rasa, Yakuti, Kunkumandya Taila, Mahanarayana Taila, Pusyanuga Churna	Vyanga, vrana, Siroroga, Drasti Roga, Chardi, Kasa, Kantha Roga, Sidhma, Mutrasotha, Udavartta, Mutraghata, Suryavartta, Ardhava bhedaka	25-50 mg

Sl. No.	*Part and Volume*	*Page No.*	*Common Ayurvedic Name*	*Botanical Name/ Chemical Name*	*Family/ Chemical Formula*	*Part Used*	*Identity, Purity and Strength/Chemical Properties*	*Constituents*	*Properties and Action*	*Important Formulations*	*Therapeutic Uses*	*Dose*
284	Part-I; Vol. IV	55-56	KUSMANDA (Fruit)	*Benincasa hispida* (Thunb.) Cogn.	Cucurbitaceae	Dried Piece of Fruits	**Foreign matter**- NMT 1 per cent; **Total ash**- NMT 12 per cent; **Acid-insoluble ash**- NMT 1 per cent; **Alcohol-soluble extractive**-NLT 10 per cent; **Water-soluble extractive**- NLT 24 per cent; **T.L.C.** - Refer API	Fatty Oil.	*Rasa* - Madhura, Amla; *Guna*- Laghu; *Virya*- Sita; *Vipaka*- Madhura; *Karma*- Dipana, Hrdya, Bastisodhaka, Vrsya, Balya, Mhana, Tridosahara, Jirnanga Pusti Pradama, Bastisodhaka, Sramsana, Arocakahara, Vatapittajit	Kusmandaka Rasayana, Dhatryadi Ghrta, Vastya-mayantaka Ghrta	Mutraghata, Prameha, Mutrakrcchra, Asmari, Trsa, Manasa Vikara, Malabandh	5-10 g
285	Part-I; Vol. IV	57-58	MADAYANTI (Leaf)	*Lawsonia inermis* Linn.	Lythraceae	Dried Leaves	**Foreign matter**- NMT 2 per cent; **Total ash**- NMT 11 per cent; **Acid-insoluble ash**- NMT 3 per cent; **Alcohol-soluble extractive**-NLT 18 per cent; **Water-soluble extractive**- NLT 25 per cent; **T.L.C.** - Refer API	Glycosides, Colouring matter (Lawsone), Hennotannic acid, Essential Oil containing β-Ionone.	*Rasa* - Tikta, Kasaya *Guna*- Laghu, Ruksa *Virya*- Sita *Vipaka*- Katu; *Karma*- Kaphasamaka, Pittasamaka	Madayantyadi Churna	Jwara, Kandu, Raktapitta, Kamala, Raktapittahara, Kustha, Mutrakrcchra, Bhrama, Vrana	5-10 ml
286	Part-I; Vol. IV	59-60	MAHANIMBA (Stem Bark)	*Melia azedarach* Linn.	Meliaceae	Dried Stem Bark	**Foreign matter**- NMT 1 per cent; **Total ash**- NMT 11 per cent; **Acid-insoluble ash**- NMT 1 per cent; **Alcohol-soluble extractive**-NLT 6 per cent; **Water-soluble extractive**- NLT 7 per cent; **T.L.C.** - Refer API	Tannins and Alkaloids.	*Rasa* - Tikta, Kasaya, Katu *Guna*- Ruksa, *Virya*- Sita *Vipaka*- Katu, *Karma*- Grahi, Kaphajit, Pittajjit, Rakta Vikarajit, Dahanasaka, Pittakaphahara, Raktadahahara	Brhanmanjstha-dikvatha Churna, Maha visagarbha Taila	Prameha, Kustha, Hrllasa, Svasa, Gulma, Arsa, Musika visa, Visuci, Bhrama, Chardi, Visama jvara	5-10 g

Sl. No.	*Part and Volume*	*Page No.*	*Common Ayurvedic Name*	*Botanical Name/ Chemical Name*	*Family/ Chemical Formula*	*Part Used*	*Identity, Purity and Strength/Chemical Properties*	*Constituents*	*Properties and Action*	*Important Formulations*	*Therapeutic Uses*	*Dose*
287	Part-I; Vol. IV	61-63	MANDU-KAPARNI (Whole Plant)	*Centella asiatica* (Linn.) Urban. *Syn. Hydrocotyle asiatica* Linn.	Apiaceae	Dried Whole Plant	**Foreign matter**- NMT 2 per cent; **Total ash**- NMT 17 per cent; **Acid-insoluble ash**- NMT 5 per cent; **Alcohol-soluble extractive**-NLT 9 per cent; **Water-soluble extractive**-NLT 20 per cent; **T.L.C.** - Refer API	Glycosides -Saponnin Glycosides	*Rasa* - Tikta, Kasaya, Madhura, Katu; *Guna*- Laghu, Sara; *Virya*- Sita; *Vipaka*- Madhura; *Karma*- Kaphapittahara, Hrdya, Medhya, Swarya, Rasayana, Dipana, Varnya, Visaghna, Ayusya, Balya, Smratiprada	Brahama Rasayana	Raktapitta, Kustha, Meha, Jwara, Swasa, Kasa, Aruci, Pandu, Sotha, Kandu, Raktadosa	3-6 g
288	Part-I; Vol. IV	64-66	MAYAKKU (Gall)	*Quercus infectoria* Oliv.	Fagaceae	Dried Gall	**Foreign matter**- Nil **Total ash**- NMT 2 per cent; **Acid-insoluble ash**- NMT 0.5 per cent; **Alcohol-soluble extractive**-NLT 60 per cent; **Water-soluble extractive**- NLT 55 per cent; **Total Tannin Content** - NLT 50 per cent; **T.L.C.** - Refer API	Tannic Acid, Starch and Sugars	*Rasa* - Kasaya; *Guna*- Laghu, Ruksa; *Virya*- Sita; *Vipaka*- Katu; *Karma*- Pittahara, Kaphahara, Dipana, Grahi	Gorocanadi Vati, Asthisan-dhanaka Lepa	Atisara, Grahani, Pravahika, Sveta Pradara, Arsa, Danta Roga, Mukha Roga, Yoni Kanda	1-3 g of the drug in powder form
289	Part-I; Vol. IV	67-69	MUDGA-PARNI (Whole Plant)	*Vigna trilobata* (L.) Verde. Syn. *Phaseolus trilobus* Ait.	Fabaceae	Dried Whole Plant	**Foreign matter**- NMT 2 per cent; **Total ash**- NMT 11.5 per cent; **Acid-insoluble ash**- NMT 1.5 per cent; **Alcohol-soluble extractive**-NLT 3 per cent; **Water-soluble extractive**-NLT 11 per cent; **T.L.C.** - Refer API	Sterols	*Rasa* - Tikta, Madhura; *Guna*- Ruksa, Laghu; *Virya*- Sita; *Vipaka*- Madhura; *Karma*- Sukradosahara, Kaphahara, Pittahara, Caksusya, Sukrala, Visaghna, Rasayana, Garbhasthapana	Amrtaprasa Ghrta, Asoka Ghrta, Vidaryadi Ghrta, Dhanvantra Taila, Brahma Rasayana, Bala Taila, Mahanarayana Taila, Ratnagiri Rasa	Daha, Jwara, Vatarakta, Pitta daha, Kasa, Musika visa, Ksaya, Krmi, Ksat Sotha, Kustha, Pradara, Madya Trsna	3-5 g

Sl. No.	*Part and Volume*	*Page No.*	*Common Ayurvedic Name*	*Botanical Name/ Chemical Name*	*Family/ Chemical Formula*	*Part Used*	*Identity, Purity and Strength/Chemical Properties*	*Constituents*	*Properties and Action*	*Important Formulations*	*Therapeutic Uses*	*Dose*
290	Part-I; Vol. IV	70-72	MUNDITIKA (Whole Plant)	*Sphaeranthus indicus* Linn.	Asteraceae	Dried Whole Plant	**Foreign matter**- NMT 2 per cent; **Total ash**- NMT 23 per cent; **Acid-insoluble ash**- NMT 9 per cent; **Alcohol-soluble extractive**-NLT 2 per cent; **Water-soluble extractive**-NLT 6 per cent; **T.L.C.** - Refer API	Essential Oil, Sterols and Alakaloids.	*Rasa* - Katu, Madhura, Tikta, Kasaya *Guna*- Laghu *Virya*- Usna *Vipaka*- Katu; *Karma*- Vatahara, Medhya, Kaphapittanut, Rucya, Swarya, Rasayana, Visaghna	Mundi Arka, Vata gajankusa Rasa, Ratnagiri Rasa, Nava ratnaraya Mrganka Rasa	Apau, Mutrakrcichra, Krmi roga, Vatarakta, Pandu, Yoni roga, Amatisara, Kasa, Slipada, Apasmara, Pliharoga, Medoroga, Guda roga, Prameha, Chardi	10-20 ml Swarasa
291	Part-I; Vol. IV	73-74	NYAGRODHA JATA (Aerial Root)	*Ficus bengalensis* Linn	Moraceae	Dried Aerial Root	**Foreign matter**- NMT 2 per cent; **Total ash**- NMT 7 per cent; **Acid-insoluble ash**- NMT 1 per cent; **Alcohol-soluble extractive**-NLT 3 per cent; **Water-soluble extractive**-NLT 4 per cent; **T.L.C.** - Refer API	Tannins.	*Rasa* - Kasaya, Madhura, *Guna*- Ruksa, Guru *Virya*- Sita *Vipaka*- Madhura; *Karma*- Pittahara, Kaphahara, Grahi, Stambhaka, Varna, Bhagnasandha-nakara, Sodhana, Ropana, Kesyam	Kumkumadi Taila, Rasa Sindura, Abhraka Bhasma (marana), Svarna Sindura, Naga Bhasma/ Vanga Bhasma (Jaranartha), Taila Moorchana	Raktapitta, Trsna, Daha, Yoniroga, Bhagandara, Visarpa	2-5 g of the drug in powder form
292	Part-I; Vol. IV	75	NIMBU (Fresh Fruit)	*Citrus limon* (Linn.) Burm. f. Syn. *C. medica* var. *limonum*	Rutaceae	Fresh Fruit	–	–	*Rasa* - Amla; *Guna*- Laghu; *Virya*- Usna; *Vipaka*- Amla; *Karma*- Vatahara, Pittakara, Kaphahara, Dipana, Pacana	Varisosana Rasa, Vasanta Malati Rasa, Vamga Bhasma, Kasisa Bhasma, Gandhaka Vati, Samkha Vati, Ajirnkantaka Rasa, Kalakuta Rasa, Mahasamkha Vati, Nasika Churna	Trsna, Vatika sula, Chardi, Vibandha, Krmi, Aruci, Agnimandya, Udara roga, Visucika	6-12 g of the drug in juice form

Sl. No.	*Part and Volume*	*Page No.*	*Common Ayurvedic Name*	*Botanical Name/ Chemical Name*	*Family/ Chemical Formula*	*Part Used*	*Identity, Purity and Strength/Chemical Properties*	*Constituents*	*Properties and Action*	*Important Formulations*	*Therapeutic Uses*	*Dose*
293	Part-I; Vol. IV	76-77	NIRGUNDI (Root)	*Vitex negundo* Linn.	Verbenaceae	Dried Root	**Foreign matter**- NMT 1 per cent; **Total ash**- NMT 3 per cent; **Acid-insoluble ash**- NMT 0.2 per cent; **Alcohol-soluble extractive**-NLT 5 per cent; **Water-soluble extractive**- NLT 9 per cent; **T.L.C.** - Refer API	Alkaloid (Nishindine).	*Rasa* - Tikta, Kasaya, Katu; *Guna*- Laghu, Ruksa; *Virya*- Sita/Usna (Nila); *Vipaka*- Katu; *Karma*- Pittavinasana, Kesya, Netrya, Slesmaha, Vatahara, Pidahara	Maha Visagrabha Taila, Manasa mitra Vataka	Sula Roga, Kasa, Kustha, Kandu, Pradara, Adhmana, Krmi Roga, Slesmaja Jvara	10-20 ml
294	Part-I; Vol. IV	78-79	PALASA (Flower)	*Butea monosperma* (Lam.) Kuntze, Syn. *B. frondosa* Koeign ex Roxb.	Fabaceae	Dried Flower	**Foreign matter**- NMT 2 per cent; **Total ash**- NMT 7 per cent; **Acid-insoluble ash**- NMT 1 per cent; **Alcohol-soluble extractive**-NLT 7 per cent; **Water-soluble extractive**- NLT 20 per cent; **T.L.C.** - Refer API	Glycosides and Flavonoids.	*Rasa* - Katu, Tikta, Kasaya, Madhura *Guna*- Laghu, Ruksa, Sara *Virya*- Sita *Vipaka*- Madhura; *Karma*- Pittahara, Kaphahata, Dipana, Vatahara, Trsnasamaka, Rakta Stambhana, Mutrala, Kusthaghna, Sandhaniya, Dahaprasamana Grahi	Kunkumadi Taila, Vanga Bhasma (Jarana (b))	Raktavikara, Mutrakrcchra, Grahi, Krmi, Meha, Daha, Vata Rakta, Kustha, Trsna, Raktapitta, Pliharoga, Gulma, Grahani, Netrasula, Krmi, Kandu, Arsa	3-6 g of the drug in powder form
295	Part-I; Vol. IV	80-81	PALASA (Gum)	*Butea monosperma* (Lam.) Kuntze Syn. *B. frondosa* Koen. ex Roxb.	Fabaceae	Dried Gum	**Foreign matter**- NMT 2 per cent; **Total ash**- NMT 3 per cent; **Acid-insoluble ash**- NMT 1 per cent; **Alcohol-soluble extractive**-NLT 69 per cent; **Water-soluble extractive**- NLT 63 per cent; **T.L.C.** - Refer API	Anthocyanins and Tannins.	*Rasa* - Kasaya, Katu, Tikta *Guna*- Sara, Snigdha *Virya*- Usna *Vipaka*- Katu; *Karma*- Dipana, Vrsya, Bhagnasan-dhanakrt, Vatahara, Slesmahara	Bala Taila	Grahani, Gulma, Arsa, Krmi Roga, Gudaroga, Asthibhaga, Vrana, Pliha Roga	0.5 to 1.5 g

Sl. No.	*Part and Volume*	*Page No.*	*Common Ayurvedic Name*	*Botanical Name/ Chemical Name*	*Family/ Chemical Formula*	*Part Used*	*Identity, Purity and Strength/Chemical Properties*	*Constituents*	*Properties and Action*	*Important Formulations*	*Therapeutic Uses*	*Dose*
296	Part-I; Vol. IV	82-83	PALASA (Seed)	*Butea monosperma* (Lam.) Kuntze, Syn. B. *frondosa* Koen. ex Roxb.	Fabaceae	Dried Seed	**Foreign matter**- NMT 1 per cent; **Total ash**- NMT 7 per cent; **Acid-insoluble ash**- NMT 0.5 per cent; **Alcohol-soluble extractive**-NLT 9 per cent; **Water-soluble extractive**-NLT 25 per cent; **Hexane-soluble extractive (by soxhlet extraction)** - NLT 15 per cent; **T.L.C.** - Refer API	Fixed Oil, Enzymes and small quantities of Resins and Alkaloids	*Rasa* - Katu, Tikta, kasaya; *Guna*- Sara, Snigdha; *Virya*- Usna; *Vipaka*- Katu; *Karma*- Vatahara, Pittahara, Kaphahara, Dipana, Vrsya, Asthisandhanaka, Sangrahi	Ayaskrti, Krmimudgara Rasa, Krmikuthara Rasa, Palasa bijadi Churna, Palasa Arka	Vrana, Gulma, Grahani, Arsa, Krmi Roga, Basti Roga, Pliha Roga, Dadru, Kandu, Tvaka Roga, Prameha, Timira Roga, Netrabhisyanda, Garbhadhana-nivaranartha	3 g of the drug in powder form
297	Part-I; Vol. IV	84-86	PARPATA (Whole Plant)	*Fumaria parviflora* Lam.	Fumaraceae	Dried Whole Plant	**Foreign matter**- NMT 2 per cent; **Total ash**- NMT 30 per cent; **Acid-insoluble ash**- NMT 10 per cent; **Alcohol-soluble extractive**-NLT 7 per cent; **Water-soluble extractive**-NLT 29 per cent; **T.L.C.** - Refer API	Alkaloids, Tannins, Sugars and Salt of Potassium.	*Rasa* - Tikta; *Guna*- Laghu; *Virya*- Sita; *Vipaka*- Katu; *Karma*- Samgrahi, Pittahara, Kaphahara, Raktadosahara, Rocaka	Pacanamrt Kwatha Churna, Tiktaka Ghrta, Mahatiktaka Ghrta, Nalpamaradi Taila, Brhat Manjisthadi Kwatha Churna, Patoladi Ghrta, Parpatadi Kwatha, Sadangapaniya, Brhat Garbha, Cintamani Rasa	Chardi, Raktapitta, Mada, Bhrama, Jvara, Trsna, Daha, Raktavikara, Glani	1-3 g
298	Part-I; Vol. IV	87-88	PATALAI (Stem Bark)	*Stereos-permum chelonoides* (L. f.) DC.	Bignoniaceae	Dried Stem Bark	**Foreign matter**- NMT 2 per cent; **Total ash**- NMT 8 per cent; **Acid-insoluble ash**- NMT 1 per cent; **Alcohol-soluble extractive**-NLT 12.5 per cent; **Water-soluble extractive**- NLT 25 per cent; **T.L.C.** - Refer API	Gum and a bitter substance.	*Rasa* - Tikta, Katu, Kasaya, Madhura *Guna*- Guru, Visada *Virya*- Usna *Vipaka*- Katu; *Karma*- Tridosahara, Dipana, Raktadosahara, Visaghna, Trsaghna, Hrdya, Rasayana, Adhodhaga-dosahara	Amrtarista, Dantyadhyarista, Dasamularista, Indukanta Ghrta	Svayathu, Sannipata, Hikka, Vami, Arocaka, Svasa, Adhman, Dagdhavrana, Vrana, Mutraghata, Sotha	3-6 g in powder form, 10-30 g for decoction in dividing dose

Sl. No.	Part and Volume	Page No.	Common Ayurvedic Name	Botanical Name/ Chemical Name	Family/ Chemical Formula	Part Used	Identity, Purity and Strength/Chemical Properties	Constituents	Properties and Action	Important Formulations	Therapeutic Uses	Dose
299	Part-I; Vol. IV	89-90	PATTANGA	*Caesalpinia sappan* Linn.	Caesalpiniaceae	Dried Heart Wood	**Foreign matter**- NMT 1 per cent; **Total ash**- NMT 1 per cent; **Acid-insoluble ash**- NMT 0.2 per cent; **Alcohol-soluble extractive**-NLT 2 per cent; **Water-soluble extractive**- NLT 8 per cent; **T.L.C.** - Refer API	Brasilin, Essential oils, Saponin Glycoside, Amino Acids, and Sugars.	*Rasa* -Modhura, Tikta; *Guna*- Ruksa; *Virya*- Sita; *Vipaka*- Katu; *Karma*- Varnya, Pittahara, Dosahara	Arimedadi Taila, Karpuradyarka, Kunkumadi Taila	Vrana, Daha, Rakta dosa, Pradara, Mukharoga	5-10 g
300	Part-I; Vol. IV	91-92	PIPPALI (Fruit)	*Piper longum* Linn.	Piperaceae	Dried, Immature, Catkin-like Fruits with Bracts	**Foreign matter**- NMT 2 per cent; **Total ash**- NMT 7 per cent; **Acid-insoluble ash**- NMT 0.5 per cent; **Alcohol-soluble extractive**-NLT 5 per cent; **Water-soluble extractive**- NLT 7 per cent; **T.L.C.** - Refer API	Essential Oil and Alkaloids	*Rasa* - Katu, Tikta, Madhura *Guna*- Snighdha and Laghu *Virya*- Anusna *Vipaka*- Madhura; *Karma*- Vatahara, Kaphahara, Dipana, Rucya, Rasayana, Hrdya, Vrsya, Tridosahara, Recana	Gudapippali, Amrtarista, Ayasakrti, Asvagandha-dyarista, Kumaryasava, Candanasava, Cayavanaprasa avaleha, Siva Gutika, Kaisora Guggulu	Svasa, Kasa, Pliha Roga, Gulma, Jvara, Prameha, Arsa, Ksaya, Udara Roga, Hikka, Trsna, Krmi, Kustha, Sula, Ama Vata, Amadosa	1-3 g
301	Part-I; Vol. IV	93-94	PLAKSA (Fruit)	*Ficus lacor* Buch.-Ham. Syn. *F. lucescens* Blume., *F. infectoria* Roxb.	Moraceae	Dried Fruit	**Foreign matter**- NMT 2 per cent; **Total ash**- NMT 9 per cent; **Acid-insoluble ash**- NMT 1 per cent; **Alcohol-soluble extractive**-NLT 5 per cent; **Water-soluble extractive**- NLT 15 per cent; **T.L.C.** - Refer API	Amino Acids	*Rasa* - Kasaya, Madhura *Guna*- Sita *Virya*- Sita *Vipaka*- Katu; *Karma*- Pittahara, Kaphahara	–	Daha, Raktapitta, Murccha, Srama, Pralapa, Bhrama, Sotha	5-10 g
302	Part-I; Vol. IV	95-96	PRIYALA (Stem Bark)	*Buchanania lanzan* spreng. Syn. *B. latifolia* roxb.	Anacardiaceae	Dried Stem Bark	**Foreign matter**- NMT 2 per cent; **Total ash**- NMT 18 per cent; **Acid-insoluble ash**- NMT 1 per cent; **Alcohol-soluble extractive**-NLT 14 per cent; **Water-soluble extractive**- NLT 15 per cent; **T.L.C.** - Refer API	Alkaloids, Tannins, Saponins, Reducing Sugars, Triterpenoids and Flavonoids	*Rasa* - Madhura; *Guna*- Guru, Snigdha, Sara; *Virya*- Sita; *Vipaka*- Madhura; *Karma*-Vaitahara, Pittahara, Dahahara, Raktaprasadana, Hrdya, Vrsya, Virecanopaga	Nyagrodhadi Kvatha Churna, Asoka Ghrta	Jvara, Trsa, Raktatisara, Raktapitta	5-10 g

Sl. No.	Part and Volume	Page No.	Common Ayurvedic Name	Botanical Name/ Chemical Name	Family/ Chemical Formula	Part Used	Identity, Purity and Strength/Chemical Properties	Constituents	Properties and Action	Important Formulations	Therapeutic Uses	Dose
303	Part-I; Vol. IV	97-98	PRIYANGU (Fruit)	*Callicarpa macrophylla* Vahl.	Verbenaceae	Dried Fruit	**Foreign matter**- NMT 2 per cent; **Total ash**- NMT 6.5 per cent; **Acid-insoluble ash**- NMT 1 per cent; **Alcohol-soluble extractive**-NLT 3 per cent; **Water-soluble extractive**-NLT 10 per cent; **T.L.C.** - Refer API	Fixed Oil	*Rasa* - Madhura, Tikta, kasaya; *Guna*- Ruksa, Sitala, Guru; *Virya*- Sita; *Vipaka*- Katu; *Karma*- Pittahara, Kaphahara, Sangrahi, Balakrta, Udrikta raktaprasadana	Jirakadi Modaka, Brhatphala Ghrta, Brhacchagaladya Ghrta, Vyaghri Taila	Jvara, Daha, Chardi, Raktadosa, Bhrama, Vataroga, Vaktrajadya	1-2 g of the drug in powder form
304	Part-I; Vol. IV	99-101	PRSNIPARNI (Whole Plant)	*Uraria picta* Desv.	Fabaceae	Dried Whole Plant	**Foreign matter**- NMT 2 per cent; **Total ash**- NMT 11 per cent; **Acid-insoluble ash**- NMT 4 per cent; **Alcohol-soluble extractive**-NLT 7 per cent; **Water-soluble extractive**-NLT 8 per cent; **T.L.C.** - Refer API	Alkaloids, Reducing sugars and Sterols.	*Rasa* - Madhura, Katu, Amala, Tikta; *Guna*- Laghu, Sara; *Virya*- Usna; *Vipaka*- Madhura; *Karma*- Tridosahara, Vrsya, Dipana, Sangrahi Vatahara, Sothahara, Angamardapra Samana, Sandhaniya, Jivanu-nasaka, Balavardhaka	Angamarda-prasamana Kasaya Churna, Amrtarista, Dasamula Taila, Vyaghritaila, Madhyama Narayana Taila, Sirah Suladi Vajra Rasa, Dasamularista	Daha, Jvara, Svasa, Raktavikara, Vataroga, Unmada, Chardi, Kasa, Raktatisara, Atisara, Vrana, Raktarsa, Kaphaja-madatyayaja, Trsna, Nataprabala, Vatarakta, Ekahika Jwara, Pilla (Netra Roga), Asthibhagna	20-50 g powder for decoction
305	Part-I; Vol. IV	102-103	PUSKARA (Root)	*Inula racemosa* Hook. f.	Asteraceae	Dried Root	**Foreign matter**- NMT 2 per cent; **Total ash**- NMT 5 per cent; **Acid-insoluble ash**- NMT 0.6 per cent; **Alcohol-soluble extractive**-NLT 10 per cent; **Water-soluble extractive**- NLT 20 per cent; **T.L.C.** - Refer API	Essential Oil	*Rasa* - Tikta, Katu; *Guna*- Laghu; *Virya*- Usna; *Vipaka*- Katu; *Karma*- Kaphavatjit	Mahanarayana Taila, Kankayana Gutika, Manasa-mitravataka, Dasmularista, Kumaryasava, Lodhrasava, Rasnadi Kvatha Churna	Hikka, Kasa, Svasa, Parsvasula, Sopha, Ardita, Pandu, Aruci, Jvara, Adhmana	1-3 g of the drug in powder form

Sl. No.	*Part and Volume*	*Page No.*	*Common Ayurvedic Name*	*Botanical Name/ Chemical Name*	*Family/ Chemical Formula*	*Part Used*	*Identity, Purity and Strength/Chemical Properties*	*Constituents*	*Properties and Action*	*Important Formulations*	*Therapeutic Uses*	*Dose*
306	Part-I; Vol. IV	104-105	RUDRAKSA (Seed)	*Elaeocarpus sphaericus* Gaertn. K. Schum	Elaeo-carpaceae	Seeds	**Foreign matter**- NMT Nil per cent; **Total ash**- NMT 1.2 per cent; **Acid-insoluble ash**- NMT 0.4 per cent; **Alcohol-soluble extractive**-NLT 2 per cent; **Water-soluble extractive**-NLT 1 per cent; **T.L.C.** - Refer API	Fixed Oil and Fatty Acids	*Rasa* - Madhura *Guna*- Snigdha, Sthula *Virya*- Usna *Karma*- Raksoghna, Arogyaprda, Medya, Hrdyam (Somnasya Karah),	Gorocanadi Vat, Cukkumti-ppalyadi Gutika, Dhanwantara Gutika, Svarnamukladi Gutiaka, Mrtasanjivani Gutika	Matisudhikar, Uccaraktacapa, Prgyaparadha, Hrdyaroga, Romantika, Manasroga, Anidra	1-2 g internally
307	Part-I; Vol. IV	106-107	SARJA (Exudate)	*Vateria indica* Linn.	Diptero-carpaceae	Resinous Exudate	**Foreign matter**- Nil per cent; **Total ash**- NMT 0.1 per cent; **Acid-insoluble ash**- Negligible; **Alcohol-soluble extractive**-NLT 60 per cent; **T.L.C.** - Refer API	Resins	*Rasa* - Katu, Tikta, Kasaya; *Guna*- Snigdha; *Virya*- Usna; *Vipaka*- Katu; *Karma*- Varnya, Vatahara, Kaphaghna, Krmighna, Visaghna, Svedahara	Kaccuradi Churna Lepa, Pinda Taila, Lavangadi Churna	Pandu, Karna Roga, Prameha, Kustha, Badhirya, Vrana, Atisara, Visphota, Medoroga, Grahani, Vata Rakta, Ksudraroga, Lippa, Manasa Roga, Musika Visa, Vidradhi, Dagdhaka, Yoni Roga, Rakta Dosa, Krmi Roga	1-2 g internal, external
308	Part-I; Vol. IV	108-109	SATAVARI (Root)	*Asparagus recemosus* Willd.	Liliaceae	Tuberous Root	**Foreign matter**- NMT 1 per cent; **Total ash**- NMT 5 per cent; **Acid-insoluble ash**- NMT 0.5 per cent; **Alcohol-soluble extractive**-NLT 10 per cent; **Water-soluble extractive**- NLT 45 per cent; **T.L.C.** - Refer API	Sugar, Glycosides, Saponin and Sitosterol	*Rasa* - Madhura, Tikta; *Guna*- Snigdha, Guru; *Virya*- Sita; *Vipaka*- Madhura; *Karma*- Vrsya, Sukraja, Balya, Medhya, Rasayana, Kaphavataghna, Pittahara, Vatahara, Stanyakara, Hrdya, Netrya, Sukrala, Agnipustikara	Satavari Guda, Brahma Rasayana, Puga Khanda, Saubha-gyasunthi, Mahanarayana Taila, Brhaecha-gaeadya Ghrta, Satavari Ghrta, Satavari Kalpa, Asvagandha-rista, Narasimiha Churna	Sotha, Ksaya, Parinama Sula, Gulma, Atisara, Raktatisara, Raktavikara, Mutrarakta, Amlapitta, Arsa, Vatajvara, Svarabheda, Naktandhya, Vatarakta, Raktapitta, Visarpa, Sutika Roga, Stanya Dosa, Stanya Ksaya	3-6 g of the drug

Sl. No.	*Part and Volume*	*Page No.*	*Common Ayurvedic Name*	*Botanical Name/ Chemical Name*	*Family/ Chemical Formula*	*Part Used*	*Identity, Purity and Strength/Chemical Properties*	*Constituents*	*Properties and Action*	*Important Formulations*	*Therapeutic Uses*	*Dose*
309	Part-I; Vol. IV	110-111	SIGRU (Root Bark)	*Moringa oleifera* Lam. Syn. *Moringa pterygosperma* Gaertn.	Moringaceae	Dried Root Bark	**Foreign matter**- NMT 2 per cent; **Total ash**- NMT 18 per cent; **Acid-insoluble ash**- NMT 10 per cent; **Alcohol-soluble extractive**-NLT 3 per cent; **Water-soluble extractive**-NLT 11 per cent; **T.L.C.** - Refer API	Alkaloids and Essential Oil	*Rasa* - Katu, Tikta, Madhura *Guna*- Laghu, Ruksa, Tiksna, Sara *Virya*- Usna *Vipaka*- Katu; *Karma*- Vatahara, Kaphara, Pittakara, Medohara, Sukral, Dipana, Pacana, Hrdya, Sophaghna, Caksusya, Samgrahi, Hrdya, Rocana, Visaghna	Prabhanjana Vimardana Taila, Sarasvata Ghrta, Vastyamaya-ntaka Ghrta, Kasara Taila Manikya Rasa	Sopha, Krmiroga, Medoroga, Pliha Roga, Vidradhi, Galaganda, Mukhajadya, Grathi, Visarpa, Asmari Vrana Vikara, Mutra Sarkara, Kustha, Ksata, Karnasula, Antarvidradhi	25-50 g of the drug in powder form
310	Part-I; Vol. IV	112-113	SIGRU (Seed)	*Moringa oleifera* Lam. Syn. *M. Pterygosperma* Gaertn.	Moringaceae	Dried Seed	**Foreign matter**- NMT 2 per cent; **Total ash**- NMT 5 per cent; **Acid-insoluble ash**- NMT 0.8 per cent; **Alcohol-soluble extractive**-NLT 12 per cent; **Water-soluble extractive**- NLT 24 per cent; **T.L.C.** - Refer API	Fixed Oil	*Rasa* - Katu, Tikta *Guna*- Laghu, Ruksa, Tiksna, *Virya*- Usna *Vipaka*- Katu; *Karma*- Vatahara, Kaphahara, Hrdya, Caksusya, Sangrahi, Dipana	Sudarsana Churna, Sothaghna Lepa, Sarsapadi Pralepa, Sarvajvarahara Lauha	Krmiroga, Netraroga, Sotha, Vidradhi, Medoroga, Gulma, Pliharoga, Galaganda, Vrana, Mukhajadya, Siroroga, Vataroga, Atinidra	5-10 g of the drug in powder form
311	Part-I; Vol. IV	114-115	SIGRU (Stem Bark)	*Moringa oleifera* Lam. Syn. *M. ptery-gosperma* Gaertn.	Moringaceae	Dried Stem Bark	**Foreign matter**- NMT 2 per cent; **Total ash**- NMT 11 per cent; **Acid-insoluble ash**- NMT 1 per cent; **Alcohol-soluble extractive**-NLT 1 per cent; **Water-soluble extractive**-NLT 5 per cent; **T.L.C.** - Refer API	Sterols and Terpenes	*Rasa* - Katu, Tikta, Madhura; *Guna*- Laghu, Tiksna, Ruksa, Picchila, Sara; *Virya*- Usna; *Vipaka*- Katu; *Karma*- Dipana, Hrdya, Vidahkrt, Visaghna, Sukrala, Rocana, Caksusya, Kaphaghna, Vataghna, Sophaghna, Sirovirecanopaga, Pittotklesaka	Karpasasthyadi Taila, Ksara Taila, Visatimduka Taila, Kanda Lawana Sarasvata Ghrta, Sarsapadi Pralepa Vastya-mayantaka Ghrta, Sveta Karvira Pallavadya Taila	Krmi, Vidradhi, Pliha Roga, Gulma, Hrdaya Roga, Aksi Roga, Medoroga, Apaci, Galaganda, Vrana Sotha, Arsa, Bhagandara, Drsti Roga, Sarvapida Nivarani	Stem Bark juice 10-20 ml, Stem bark Powder 2-5 g

Sl. No.	*Part and Volume*	*Page No.*	*Common Ayurvedic Name*	*Botanical Name/ Chemical Name*	*Family/ Chemical Formula*	*Part Used*	*Identity, Purity and Strength/Chemical Properties*	*Constituents*	*Properties and Action*	*Important Formulations*	*Therapeutic Uses*	*Dose*
312	Part-I; Vol. IV	116-117	SRNGATAKA (Dried Seed)	*Trapa natans* Linn. var. *bispinosa* (Roxb.) Makino. Syn. *T. bispinosa* Roxb. *T. quadrispinosa* Wall.	Trapaceae	Dried Seeds	**Foreign matter**- NMT 2 per cent; **Total ash**- NMT 3 per cent; **Acid-insoluble ash**- NMT 0.3 per cent; **Water-soluble extractive**- NLT 8 per cent; **T.L.C.** - Refer API	Starch and Protein	*Rasa* - Madhura, Kasaya; *Guna*- Guru; *Virya*- Sita; *Vipaka*- Madhura; *Karma*- Pittahara, Vrsya, Sramahara, Sukrakara, Grahi, Stanyajanan, Rakta Stambhaka, Garbhasdhapana	Saubhagya Sunthi, Amrtaprasa Ghrta, Pugakhanda	Raktapitta, Daha, Garbha Srava, Sopha (external), Mutrakrchra, Asthibhagna Vatavyadhi, Prameha, Visarpa, Trsna	5-10 mg of the dry in powder form
313	Part-I; Vol. IV	118-119	SRUVAVRKSA (Leaf)	*Flacourtia indica* Merr. Syn. F. *ramontchi* L. Herit.	Flacourtiaceae	Dried Leaf	**Foreign matter**- NMT 2 per cent; **Total ash**- NMT 9 per cent; **Acid-insoluble ash**- NMT 0.6 per cent; **Alcohol-soluble extractive**-NLT 4 per cent; **Water-soluble extractive**- NLT 12 per cent; **T.L.C.** - Refer API	Tannin and Sugar	*Rasa* - Madhura, Amla, Tikta; *Guna*- Laghu; *Virya*- Sita; *Vipaka*- Madhura; *Karma*- Pittahara, Kaphahara, Dipana, Pacana	Aragvadhadi Kvatha Churna	Raktavikara, Sopha, Kamala	50-100 g for decoction
314	Part-I; Vol. IV	120-121	SRUVAVRKSA (Stem Bark)	*Flacourtia indica* Merr. Syn. *F. ramontchi* L. Herit.	Flacourtiaceae	Dried Stem Bark	**Foreign matter**- NMT 1 per cent; **Total ash**- NMT 16 per cent; **Acid-insoluble ash**- NMT 0.6 per cent; **Alcohol-soluble extractive**-NLT 6 per cent; **Water-soluble extractive**- NLT 11 per cent; **T.L.C.** - Refer API	Tannin and Flacourtin, a phenolic glucoside ester	*Rasa* - Tikta; *Guna*- Laghu, Tiksna *Virya*- Sita *Vipaka*- Katu; *Karma*- Pittahara, Kaphahara, Dipana	Aragvadhadi Kvatha Churna	Raktavikara, Sopha (Sotha), Dusta Vrana	50-100 g of the drug for decoction
315	Part-I; Vol. IV	122-123	TALAMULI (Rhizome)	*Curculigo orchioides* Gaertn.	Amaryllidaceae	Dried Rhizome	**Foreign matter**- NMT 2 per cent; **Total ash**- NMT 9 per cent; **Acid-insoluble ash**- NMT 2 per cent; **Alcohol-soluble extractive**-NLT 3 per cent; **Water-soluble extractive**- NLT 17 per cent; **T.L.C.** - Refer API	Tannin, Resin, Sapogenin and Alkaloid	*Rasa* - Madhura, Tikta *Guna*- Guru, Picchila *Virya*- Usna *Vipaka*- Madhura; *Karma*- Vrsya, Bramhana, Rasayana, Pustiprada, Balaprada, Sramahara, Pitta hara, Daha hara	Gandharvahastadi Kvatha Churna, Candanadi Churna	Arsa, Vataroga, Karsya, Kstaksina	3-6 g of the drug in powder form

Sl. No.	*Part and Volume*	*Page No.*	*Common Ayurvedic Name*	*Botanical Name/ Chemical Name*	*Family/ Chemical Formula*	*Part Used*	*Identity, Purity and Strength/Chemical Properties*	*Constituents*	*Properties and Action*	*Important Formulations*	*Therapeutic Uses*	*Dose*
316	Part-I; Vol. IV	124-125	TALISA (Leaves)	*Abies webbiana* Lindl.	Pinaceae	Dried Needle Like Leaves	**Foreign matter**- NMT 2 per cent; **Total ash**- NMT 6 per cent; **Acid-insoluble ash**- NMT 0.5 per cent; **Alcohol-soluble extractive**-NLT 14 per cent; **Water-soluble extractive**- NLT 15 per cent; **T.L.C.** - Refer API	Essential Oil and Alkaloid	*Rasa* - Tikta, Katu, Madhura *Guna*- Laghu, Tiksna *Virya*- Usna *Vipaka*- Katu; *Karma*- Vatakaphapaham, Slesmapittajit, Dipana, Hrdya	Talisadi Churna, Bhaskara Lavana, Pranada Gutika, Jatiphaladi Churna, Puga Khanda, Draksadi Churna, Talisadi Modaka	Swasa, Kasa, Gulma, Agnimandya, Amadosa, Ksaya, Hikka, Chardi, Krmi, Mukharoga, Aruci	2-3 g of the drug in powder form
317	Part-I; Vol. IV	126-127	TILA (Seed)	*Sesamum indicum* Linn.	Pedaliaceae	Dried Seeds	**Foreign matter**- NMT 2 per cent; **Total ash**- NMT 9 per cent; **Acid-insoluble ash**- NMT 1.5 per cent; **Alcohol-soluble extractive**-NLT 20 per cent; **Water-soluble extractive**- NLT 4 per cent; **Fixed oil** - NLT 35 per cent; **T.L.C.** - Refer API	Fixed Oil	*Rasa* - Madhura, Tikta, Kasaya, Katu *Guna*- Guru, Snigdha, Suksma, Vyavai *Virya*- Usna *Vipaka*- Madhura; *Karma*- Snehana, Svraka, Saedopaga, Balya, Vataghna, Kusthakara, Pittala, Vitbardhaka, Mutrabandhaka, Medhavardhala Agnivardhaka, Sangrahi, Kesya, Avasadakar, Kesa Krsnakara, Kasa Vardhaka, Karnpalivaidhaka, Kaphakopaka, Mrdurecaka, Vrana Samsodhaka, Vrana pacaka, Vrana dahanasaka, Bhagna prasadhaka, Rasayana, Visaghna, Vajikara, Varnya, Agnibala Vardhaka	Jatiphaladya Churna, Narsimha Churna, Samangadi Churna, Haridradi Lepa, Vrsya Pupalika Yoga, Nagaradi Yoga, Tiladi Upanaha, Tiladi Yoga, Priyaladi Yoga, Mustadi Upanaha, Sunthyadi Churna, Pathyadi Gutika, Hingavadi Yoga, Paniya Ksara, Bhallatakadi Modaka	Udavarta, Yonisula, Gulma, Udara Anaha, Sirah Sula, Parsva Sula, Amasula, Raktarsa, Guda bhramsa, Kasa, Svasa, Pravahika, Visarpa, Hikka, Pinasa, Vatarakta, Pradara, Asmarai, Nadi vrana, Kustha, Svitra, Granthi, Upadamsa, Vidaraka, Alasa, Khalitya, Palitya, Aksi Roga, Pratisyaya, Sankhaka, Sakuni, Graha, Kumara, Pitrmesagraha, Atisara, Raktatisara, Ksaya, Krmi, Mutraghata,	Powder 5-10 g/day

Sl. No.	Part and Volume	Page No.	Common Ayurvedic Name	Botanical Name/ Chemical Name	Family/ Chemical Formula	Part Used	Identity, Purity and Strength/Chemical Properties	Constituents	Properties and Action	Important Formulations	Therapeutic Uses	Dose
											Dantaroga, Dantaharasa, Vatika Mukharoga, Atidagdha, Trsna, Pliharoga, Galganda, Musika Dansa, Karnapali Sora	
318	Part-I; Vol. IV	128-129	TULASI (Seed)	*Ocimum sanctum* Linn.	Lamiaceae	Seeds	**Foreign matter**- NMT 2 per cent; **Total ash**- NMT 8 per cent; **Acid-insoluble ash**- NMT 2 per cent; **Alcohol-soluble extractive**-NLT 4 per cent; ; **T.L.C.** - Refer API	Fixed Oil and Mucilage	*Rasa* - Katu, Tikta, Kasaya *Guna*- Laghu, Ruksa, Tiksna *Virya*- Usna *Vipaka*- Katu; *Karma*- Vatahara, Kaphahara, Pittahara, Rucikrt, Dipana, Dahakrt, Krmighna, Visahara, Vranasodhaka, Hrdya	Muktadi mahanjana	Swasa, Kasa, Hikka, Parsvasula, Kustha, Mutarakrchra, Pratisyaya, Aruci, Puthigandha, Gara Visa, Sopha, Krmi, Rakta Vikara, Jantuvisa, Bhuta Roga	1-2 g of the seed in powder form
319	Part-I; Vol. IV	130-131	TUMBURU (Fruit)	*Zanthoxylum armatum* DC. Syn. *Z. alatum* Roxb.	Rutaceae	Dried Fruit	**Foreign matter**- NMT 2 per cent; **Total ash**- NMT 8.5 per cent; **Acid-insoluble ash**- NMT 1 per cent; **Alcohol-soluble extractive**-NLT 8 per cent; **Water-soluble extractive**-NLT 10 per cent; **T.L.C.** - Refer API	Essential Oil	*Rasa* - Katu, Tikta; *Guna*- Laghu, Ruksa, Tiksna; *Virya*- Usna; *Vipaka*- Katu; *Karma*- Rucya, Dipana, Pacana, Vatahara, Kaphahara, Lalapraseka, Cimicimayanam, Rasana Samvedaka	Saptavimsati Guggulu, Dadhika Ghrta, Maha Visagarbha Taila, Hingavadi Taila	Swasa, Kasa, Ardita, Kaphaja Roga, Hrdroga, Kantha Roga, Arsa, Hikka, Agnimandya, Asya Roga, Danta Roga	2-4 g

Sl. No.	Part and Volume	Page No.	Common Ayurvedic Name	Botanical Name/ Chemical Name	Family/ Chemical Formula	Part Used	Identity, Purity and Strength/Chemical Properties	Constituents	Properties and Action	Important Formulations	Therapeutic Uses	Dose
320	Part-I; Vol. IV	132-133	UTINGANA (Seed)	*Blepharis persica* (Burm. F.) O. *Kuntze*. Syn. *B. edulis* Pers.	Acanthaceae	Dried Mature Seed	**Foreign matter**- NMT 2 per cent; **Total ash**- NMT 7 per cent; **Acid-insoluble ash**- NMT 1.5 per cent; **Alcohol-soluble extractive**-NLT 16 per cent; **Water-soluble extractive**- NLT 23 per cent; **T.L.C.** - Refer API	Glycosides and Tannin	*Rasa* - Madhura, Tikta; *Guna*- Guru, Snigdha, Picchila; *Virya*- Usna; *Vipaka*- Madhura; *Karma*- Vrsya, Mutrala	Kumaryasava	Mutrakrcchra, Klaibya	3-6 g of the drug in powder form
321	Part-I; Vol. IV	134-135	VARAHI (Rhizome)	*Dioscorea bulbifera* Linn.	Dioscoreaceae	Dried cut Pices of Rhizome	**Foreign matter**- NMT 2 per cent; **Total ash**- NMT 6 per cent; **Acid-insoluble ash**- NMT 1 per cent; **Alcohol-soluble extractive**-NLT 3 per cent; **Water-soluble extractive**-NLT 9 per cent; **T.L.C.** - Refer API	Saponins-Steroidal Saponins.	*Rasa* - Madhura, Tikta, Katu; *Guna*- Laghu; *Virya*- Usna; *Vipaka*- Katu; *Karma*- Rasayana, Slesmaghna, Balya, Vrsya, Svarya, Varnya, Ayvardhana, Agnivrdhi Kara, Pittakara	Vastyamayantaka Ghrta, Narasimha Churna, Pancanimba Churna	Kustha, Kandu, Prameha, Krmi	3-6 g
322	Part-I; Vol. IV	136-137	VARSABHU (Root)	*Trianthema portula-castrum* Linn. Syn. *T. monogyna* Linn., *T. obcordata* Roxb.	Aizoaceae	Dried Root	**Foreign matter**- NMT 2 per cent; **Total ash**- NMT 11 per cent; **Acid-insoluble ash**- NMT 2 per cent; **Alcohol-soluble extractive**-NLT 2 per cent; **Water-soluble extractive**-NLT 11 per cent; **T.L.C.** - Refer API	Glycoside	*Rasa* - Tikta, Kasaya, Katu, Madhura; *Guna*- Ruksa, Laghu; *Virya*- Usna; *Vipaka*- Katu; *Karma*- Vatahara, Kaphahara, Dipana, Mutrala, Bhedana, Rucya	Suskamulaka Taila, Kumaryasava, Dhanvantara Ghrta, Sukumara, Ghrta, Punarnavadyarista	Sopha, Pandu, Arsa, Udara Roga, Gulma Jvara, Garvisa, Vasti Sula, Hrdroga, Urahksatad, Agnimandya, Ykrt avam Pliha Roga	2-5 g of the drug in powder form
323	Part-I; Vol. IV	138-139	VASA (Root)	*Adhatoda zeylanica* Medic. Syn. *A. vasica* Nees	Acanthaceae	Dried Root	**Foreign matter**- NMT 1 per cent; **Total ash**- NMT 5 per cent; **Acid-insoluble ash**- NMT 1 per cent; **Alcohol-soluble extractive**-NLT 4 per cent; **Water-soluble extractive**-NLT 10 per cent; **T.L.C.** - Refer API	Alkaloids (Vasicine and Vasicinol) and Oil	*Rasa* - Tikta, Kasaya; *Guna*- Laghu, Snigdha; *Virya*- Sita; *Vipaka*- Katu; *Karma*- Raktasodhaka, Pittahara, Kaphahara, Svara, Vivardhaka, Vatakrt, Hrdya	Brhat Manjisthadi Kvatha Churna, Panca tikta Ghrta, Chyawanaprasa Avaleha, Kanakasava	Kustha, Vata Roga, Krmi, Svasa, Kasa, Jvara, Chardi, Meha, Ksaya, Raktapitta, Trsna	3-6 g

Sl. No.	*Part and Volume*	*Page No.*	*Common Ayurvedic Name*	*Botanical Name/ Chemical Name*	*Family/ Chemical Formula*	*Part Used*	*Identity, Purity and Strength/Chemical Properties*	*Constituents*	*Properties and Action*	*Important Formulations*	*Therapeutic Uses*	*Dose*
324	Part-I; Vol. IV	140-142	VISAMUSTI (Seed)	*Strychnos nux-vomica* Linn.	Fabaceae	Dried Seed	**Foreign matter**- NMT 1 per cent; **Total ash**- NMT 2 per cent; **Acid-insoluble ash**- NMT 0.2 per cent; **Alcohol-soluble extractive**-NLT 4 per cent; **Water-soluble extractive**- NLT 12 per cent; **Assay** - NLT 1.2 per cent of strychnine; **T.L.C.** - Refer API	Alkaloids, Indole Alkaloids, Strychnine and Burcine, Monoterpenoid Glycoside (Loganin), α, β-colubrine, Vomicine.	*Rasa* - Tikta, Katu; *Guna*- Ruksa, Laghu, Tiksna; *Virya*- Usna; *Vipaka*- Katu; *Karma*- Grahi, Madakaraka, Vatalam, Kaphanasaka, Pittanasaka, Raktadosa Nasaka, Vranasodhana, Parama Vedanahara, Agniret, Rujahara, Jantunasana	Visatinduka Taila, Mahavisagarbha Taila, Agnitundi Vati, Ekangavira Rasa, Visatinduka Vati, Krmimudgara Rasa, Navajivana Rasa	Agnimandya, Ardita, Paksaghata, Visucika, Nadi Daurbahy, Kustha, Arsa, Klaibya, Grdhrasi, Kandu, Vrana	60-125 mg powder of the sodhita drug
325	Part-I; Vol. IV	143-145	VRSCIKALI (Whole Plant)	*Tragia involucrata* Linn.	Euphorbiaceae	Dried Whole Plant	**Foreign matter**- NMT 2 per cent; **Total ash**- NMT 14 per cent; **Acid-insoluble ash**- NMT 3 per cent; **Alcohol-soluble extractive**-NLT 4 per cent; **Water-soluble extractive**- NLT 11 per cent; **T.L.C.** - Refer API	–	*Rasa* - Katu; *Guna*- Usna; *Virya*- Usna; *Vipaka*- Katu; *Karma*- Vatakara, Suddhikrt, Balya, Hrtsuddhikrt	Vidaryadi Kvatha Churna, Vidaryadi Ghrta	Raktapitta, Vibandha, Arocaka	3-6 g
326	Part-I; Vol. IV	146-148	YAVA (Whole Plant)	*Hordeum vulgare* Linn. Syn. *H. sativum* Pers.	Poaceae	Dried Whole Plant	**Foreign matter**- NMT 1 per cent; **Total ash**- NMT 8.5 per cent; **Acid-insoluble ash**- NMT 4 per cent; **Alcohol-soluble extractive**-NLT 7 per cent; **Water-soluble extractive**- NLT 8 per cent; **T.L.C.** - Refer API	Proteins, Carbohydrate, free Amino-acids, Vitamins, Tannins and Flavonoid glycosides-Luteolin and Orientin.	*Rasa* - Madhura; *Guna*- Ruksa, Aguru, Mrdu; *Virya*- Sita; *Vipaka*- Katu, ; *Karma*- Kaphapittahara, Medhavardhaka, Swara vardhaka varna vardhaka, Lekhana, Medohara, Vatahara Vrsaya	–	Pinasa, Swasa, Kasa, Urustambha	10-20 g

Sl. No.	Part and Volume	Page No.	Common Ayurvedic Name	Botanical Name/ Chemical Name	Family/ Chemical Formula	Part Used	Identity, Purity and Strength/Chemical Properties	Constituents	Properties and Action	Important Formulations	Therapeutic Uses	Dose
327	Part-I; Vol. V	1	AMRA HARIDRA (Rhizome)	*Curcuma amada* Roxb.	Zingiberaceae	Rhizome	**Foreign matter**- NMT 1 per cent; **Total ash**- NMT 12 per cent; **Acid-insoluble ash**- NMT 2 per cent; **Alcohol-soluble extractive**-NLT 9 per cent; **Water-soluble extractive**-NLT 14 per cent; **Starch** -NLT 16 per cent; **Essential oil** - NLT 1 per cent; **T.L.C.** - Refer API	Volatile oil (α-pines, δ-camphor). α-curcumene, 1-β curcumene, Phytosterol	*Rasa* - Madhura, Tikta *Guna*- Laghu, Sara *Virya*- Sita *Vipaka*- Katu; *Karma*- Pittahara, Kaphahara, Vrsya, Ruciprada, Dipani	Asthisan-dhanaka Lepa	Kandu, Vrana, Kasa, Svasa, Hikka, Jvara, Abhighataja, Sopha, Karnasula, Sannipata	2-4 g
328	Part-I; Vol. V	3	ANISUNA (Fruit)	*Pimpinella anisum* Linn.	Apiaceae	Dried Fruit	**Foreign matter**- NMT 2 per cent; **Total ash**- NMT 8 per cent; **Acid-insoluble ash**- NMT 1 per cent; **Alcohol-soluble extractive**-NLT 15 per cent; **Water-soluble extractive**- NLT 30 per cent; **T.L.C.** - Refer API	Volatile oil, Fixed oils and Protein	*Rasa* - Tikta, Katu *Guna*- Tiksna, Laghu *Virya*- Usna *Vipaka*- Katu; *Karma*- Vatanulomaka, Raksoghna, Kaphahara, Artavajanana	Brahmi Vati	Sula, Adhmana, Kaphavikara, Mutraghata, Balagraha	1-3 g; Q. S. for dhupa-nartha [fumi-gation]
329	Part-I; Vol. V	5	ANKOLAH (Leaf)	*Alangium salviifolum* (Linn.f.) Wang. Syn. *A. lamarckii* Thw.;	Alangiaceae	Dried Leaf	**Foreign matter**- NMT 2 per cent; **Total ash**- NMT 10 per cent; **Acid-insoluble ash**- NMT 1 per cent; **Alcohol-soluble extractive**-NLT 5 per cent; **Water-soluble extractive**-NLT 15 per cent; **T.L.C.** - Refer API; **Assay**- Refer API	Alkaloids (Alangi-marckine, Deoxy-tubulosine, Ankorine), Campesterol, Episterol, Stigmast-5,22,25-trien-3β- ol, Alangidiol and Isoalangi-diol.	*Rasa* - Tikta, Katu, Kasaya *Guna*- Laghu, Snigdha, Tiksna, Sara *Virya*- Usna *Vipaka*- Katu; *Karma*- Vatahara, Kaphahara, Vamaka, Recaka, Vranasodhaka, Mutrala, Parada Sodhra, Jvarghna	(No formulations)	Matsyavisa, Amavata, Jvara, Kantharoga, Sotha, Sopha, Sula, Krmi, Visarpa, Graha badha, Raktavikara, Musakavisa, Jantuvisa, Lutavisa, Kukkuravisa, Visarikara	2-10 g

Sl. No.	*Part and Volume*	*Page No.*	*Common Ayurvedic Name*	*Botanical Name/ Chemical Name*	*Family/ Chemical Formula*	*Part Used*	*Identity, Purity and Strength/Chemical Properties*	*Constituents*	*Properties and Action*	*Important Formulations*	*Therapeutic Uses*	*Dose*
330	Part-I; Vol. V	8	ARAGVADHA (Stem Bark)	*Cassia fistula* Linn.	Fabaceae	Stem Bark	**Foreign matter**- NMT 2 per cent; **Total ash**- NMT 13 per cent; **Acid-insoluble ash**- NMT 1 per cent; **Alcohol-soluble extractive**-NLT 25 per cent; **Water-soluble extractive**- NLT 18 per cent; **T.L.C.** - Refer API	Anthra-quinones, Tannins, Sterols.	*Rasa* - Tikta *Guna*- Guru *Virya*- Sita *Vipaka*- Katu; *Karma*- Vatahara, Pittahara, Kosthasuddhikara	Avittoladi Bhasma Ksara, Manasamitra Vataka	Gandamala, Upadamsa, Kustha, Aruci, Vibandha, Sula, Kamala, Hrdroga, Raktapitta, Vatarakta, Sotha, Mutrakrcchra, Daha, Jvara, Udaravikara, Krmi, Prameha, Gulma, Vrana, Kandu, Grahani, Asmari	50-100 ml kvatha
331	Part-I; Vol. V	10	ASPHOTA (Root)	*Vallaris solanacea* Kuntze Syn. *V. heynei* Spreng.	Apocynaceae	Dried Root	**Foreign matter**- NMT 2 per cent; **Total ash**- NMT 8 per cent; **Acid-insoluble ash**- NMT 0.7 per cent; **Alcohol-soluble extractive**-NLT 6 per cent; **Water-soluble extractive**- NLT 11 per cent; **T.L.C.** - Refer API	–	*Rasa* - Tikta, Kasaya *Guna*- Laghu, Ruksa *Virya*- Usna *Vipaka*- Katu; *Karma*- Vatahara, Vranasodhaka	Vajraka Taila, Abhaya Lavana	Asmari, Sula, Mutrakrcchra, Putana-grahavista (Balaroga), Kustha, Grahani, Svasa, Musaka Visavikara, Arsa, Vrana	3-6 g
332	Part-I; Vol. V	12	BASTANTRI (Root)	*Argyreia nervosa* (Burm.f.) Boj. Syn. *A. Speciosa* Sweet.	Convolvulaceae	Dried Root	**Foreign matter**- NMT 1 per cent; **Total ash**- NMT 11 per cent; **Acid-insoluble ash**- NMT 0.8 per cent; **Alcohol-soluble extractive**-NLT 4 per cent; **Water-soluble extractive**- NLT 8 per cent; **T.L.C.** - Refer API	–	*Rasa* - Katu, Tikta, Kasaya *Guna*- Sara, Laghu *Virya*- Usna *Vipaka*- Katu; *Karma*- Kaphavatahara, Adhobhagahara, Vrsya, Rasayana, Ayurvrdhikara, Balya, Medhya, Rucya, Svarya, Kanthya, Asthisandhana	Misraka Sneha	Gulma, Mutrakrchra, Aruci, Hrdruja, Anaha, Udavarta, Arsa, Udara, Grahabadha, Sula, Vataruja, Raktapitta, Vatarakta, Amavata, Sopha, Meha, Vatarsa,	3-5 g

Sl. No.	*Part and Volume*	*Page No.*	*Common Ayurvedic Name*	*Botanical Name/ Chemical Name*	*Family/ Chemical Formula*	*Part Used*	*Identity, Purity and Strength/Chemical Properties*	*Constituents*	*Properties and Action*	*Important Formulations*	*Therapeutic Uses*	*Dose*
									Kari, Agnikara, Kntikara, Visaghna		Svayathu, Krmi, Pandu, Ksaya, Kasa, Unmada, Apasmara, Pratitum, Slipada	
333	Part-I; Vol. V	14	BHURJAH (Stem Bark)	*Betula utilis* D. Don syn. *B. bhojpattra* Wall.	Betulaceae	Stem Bark	**Foreign matter**- NMT 2 per cent; **Total ash**- NMT 2.1 per cent; **Acid-insoluble ash**- NMT 1.1 per cent; **Alcohol-soluble extractive**-NLT 19 per cent; **Water-soluble extractive**- NLT 0.8 per cent; **T.L.C.** - Refer API	Betulin, Lupeol and 3β-aetoxy-12-oleanen-28-oic acid.	*Rasa* - Katu, Kasaya *Guna*- Laghu *Virya*- Usna *Vipaka*- Katu; *Karma*- Tridosasamana, Bhutaraksakara, Visaghna, Balya, Slesmahara, Medohara	Ayaskrti	Karnaroga, Raktapitta, Kustharoga, Raksoghna-dhupana, Vrana, Aparapatana, Garbhasanga, Granthivisarpa, Balagraha	1-3 g
334	Part-I; Vol. V	16	CANDA (Root)	*Angelica archangelica* Linn.	Apiaceae	Dried Root	**Foreign matter**- NMT 2.0 per cent; **Total ash**- NMT 7 per cent; **Acid-insoluble ash**- NMT 1.2 per cent; **Alcohol-soluble extractive**-NLT 10 per cent; **Water-soluble extractive**- NLT 12 per cent; **Volatile oil** - NLT 0.3 per cent; **T.L.C.** - Refer API	Essential Oil: Containing limonene, α-phellandrene, Pinene, p-cymene, Terpinolene, Myrcene, Fenchone, Linalool, α-terpineol, Cadinene, Borneol, β-caryo-phyllene, Bisabolol, Angelica lactone, and other mono and sesqui-terpenes. Other constituents include selimone, Archangelin	*Rasa* - Katu *Guna*- Laghu, Tiksna *Virya*- Usna *Vipaka*- Katu; *Karma*- Vatahara, kaphahara, Svasahara, Mutrala, Varnaprasadaka, Svedaghna, Kandughna, Visaghna, Daurgandhahara	Manjisthadi Taila	Sotha, Svasa, Apasmara, Hikka, Arsa, Kandu, Pidaka, Kotha	1-3 g

Sl. No.	Part and Volume	Page No.	Common Ayurvedic Name	Botanical Name/ Chemical Name	Family/ Chemical Formula	Part Used	Identity, Purity and Strength/Chemical Properties	Constituents	Properties and Action	Important Formulations	Therapeutic Uses	Dose
335	Part-I; Vol. V	18	CORAKAH (Root and Root Stock)	*Angelica glauca* Edgw.	Apiaceae	Dried Mature Root and Root Stock	**Foreign matter**- NMT 1 per cent; **Total ash**- NMT 6.5 per cent; **Acid-insoluble ash**- NMT 2 per cent; **Alcohol-soluble extractive**-NLT 14 per cent; **Water-soluble extractive**- NLT 30 per cent; **Volatile oil** - NLT 0.4 per cent; **T.L.C.** - Refer API	Oxypeucedanin, 3-Butylidene phthalide, 3-butylidene dihydro-phthalide [(E-and(Z)-ligustilide] and dimers of Butyl phthalides [Angiolide, Angelicolide].	*Rasa* - Madhura, Tikta, Katu *Guna*- Laghu, Tiksna, Ruksa *Virya*- Usna *Vipaka*- Katu; *Karma*-Vatahara, Kaphahara, Medohara, Swedahara, Hrdya, Sajnasthapana, Dipana, Pacana, Vranaprasadana, Vamaka	Guducyadi Modaka, Balasvagan-dhalaksadi Taila, Mahanarayana Taila	Kandu, Pitika, Kotha, Kustha, Jvara, Visaroga, Vrana, Raktadosa, Agnimandya, Sirah Sula, Unmada, Apasmara, Hikka, Svasa, Pratisyaya, Sitajvara, Balaroga	3-6 g
336	Part-I; Vol. V	21	DARBHA (Root)	*Imperata cylindrica* (Linn.)	Poaceae	Root	**Foreign matter**- NMT 2 per cent; **Total ash**- NMT 4 per cent; **Acid-insoluble ash**- NMT 3 per cent; **Alcohol-soluble extractive**-NLT 2 per cent; **Water-soluble extractive**-NLT 4 per cent; **T.L.C.** - Refer API	Contains five Triterpenoids *viz.* Cylindrin, Arundoin, Fernenon, Isoburneol, and Simiarenol	*Rasa* - Madhura, Kasaya *Guna*- Laghu, Snigdha *Virya*- Sita *Vipaka*- Madhura; *Karma*- Tridosahara, Rasayana, Mutravirecaniya, Stanyajanana, Pipasahara, Kusthaghna, Dahaprasamana, Vamaka	Karpuradyarka, Brahmarasayana, Traikantaka Ghrta, Sukumara Ghrta	Mutrakrcchra, Asmari, Mutraghata, Bastisula, Trsa, Daha, Raktapradara, Raktarsa, Pradara, Raktapitta, Jvara, Visarpa, Pittabhisyanda	10-20 g for decoction
337	Part-I; Vol. V	23	DHANVA-YASAH (Whole Plant)	*Fagonia cretica* Linn. Syn. *F. arabica* Linn., *F. bruguieri* DC.	Zygophyllaceae	Dried Whole Plant	**Foreign matter**- NMT 2 per cent; **Total ash**- NMT 10 per cent; **Acid-insoluble ash**- NMT 0.4 per cent; **Alcohol-soluble extractive**-NLT 5 per cent; **Water-soluble extractive**-NLT 10 per cent; **T.L.C.** - Refer API	Alkaloids (Harmine), Amino acids(Alanine, Glycine, Leucine, Arginine isoleucine, Lysine, Phenylalanine, Proline, Tyrosine and Valline); Terpenoids of Oleanane group.	*Rasa* - Madhura, Tikta, Kasaya, Katu *Guna*- Laghu, Sara *Virya*- Sita *Vipaka*- Madhura; *Karma*- Kaphahara, Vatahara, Pittahara, Medohara	Duralabhadi Kvatha, Duralabhadi Kasaya, Rasnadi Kvatha Churna (Maha), Tiktaka Ghrta, Usirasava, Kanta-karyavaleha, Maha-pancagavya Gyhrta, Dasamularista, Punarnavasava	Atisara, Grahani, Daha, Jvara, Visamajvara, Trsna, Moha, Murccha, Madaroga, Raktapitta, Raktavikara, Kustha, Visarpa, Vatarakta, Bhrama, Gulma, Chardi, Kasa, Mutraghata	5-10 g powder; 40-80 ml phanta

Sl. No.	Part and Volume	Page No.	Common Ayurvedic Name	Botanical Name/ Chemical Name	Family/ Chemical Formula	Part Used	Identity, Purity and Strength/Chemical Properties	Constituents	Properties and Action	Important Formulations	Therapeutic Uses	Dose
338	Part-I; Vol. V	26	DRAVANTI (Seed)	*Jatropha glandulifera* Roxb.	Euphorbiaceae	Dried Seed	**Foreign matter**- NMT 2 per cent; **Total ash**- NMT 6 per cent; **Acid-insoluble ash**- NMT 0.3 per cent; **Alcohol-soluble extractive**-NLT 9 per cent; **Water-soluble extractive**-NLT 7 per cent; **Fatty oil** - NLT 9 per cent; **T.L.C.** - Refer API	Jatrophin, Jatropholone A, Fraxetin, Coumarino-lignan (I).	*Rasa* - Katu *Guna*- Laghu, Tiksna, Snigdha *Virya*- Usna *Vipaka*- Katu; *Karma*- Pittahara, Kaphahara, Recaka, Vidabhedana, Dipana, Visaghna	Misraka sneha	Raktavikara, Kandu, Kustha, Sotha, Pandu, Gulma, Udara, Anaha, Udavarta, Ajirna, Sula, Hrdroga, Grahaniroga, Trsna, Jvara, Garavisa, Prameha, Bhagandara, Amavata, Paksaghata, Urustambha, Granthi, Parsvasula, Pliharoga, Dustavrana, Dustaapaci	250-500 mg after purifi-cation
339	Part-I; Vol. V	28	DUGDHIKA (Whole Plant)	*Euphorbia prostrata* W. Ait.	Euphorbiaceae	Whole Plant	**Foreign matter**- NMT 1 per cent; **Total ash**- NMT 11 per cent; **Acid-insoluble ash**- NMT 0.2 per cent; **Alcohol-soluble extractive**-NLT 11 per cent; **Water-soluble extractive**- NLT 27 per cent; **T.L.C.** - Refer API	Glucoside, Galactoside, β-sistosterol, Compesterol, Stigmasterol, Cholesterol	*Rasa* - Katu, Tikta, Madhura, Lavana *Guna*- Guru, Ruksa, Tiksna *Virya*- Usna *Vipaka*- Katu; *Karma*- Kaphahara, Garbhakaraka, Mutrala, Vistambhini, Grahi, Malastambhaka, Dhatuvrddhikara, Vrsya, Hrdya	Gaganasundara Rasa	Kustha, Krmi, Svasa, Pravahika, Raktapitta, Prameha, Raktarsa, Palita, Danta-ghuna, Dadru, Sphota	5-10 g

Sl. No.	*Part and Volume*	*Page No.*	*Common Ayurvedic Name*	*Botanical Name/ Chemical Name*	*Family/ Chemical Formula*	*Part Used*	*Identity, Purity and Strength/Chemical Properties*	*Constituents*	*Properties and Action*	*Important Formulations*	*Therapeutic Uses*	*Dose*
340	Part-I; Vol. V	31	ELAVALUKAM (Seed)	*Prunus avium* Linn.f.	Rosaceae	Dried Mature Seed	**Foreign matter**- NMT 2 per cent; **Total ash**- NMT 3 per cent; **Acid-insoluble ash**- NMT 0.1 per cent; **Alcohol-soluble extractive**-NLT 14 per cent; **Water-soluble extractive**- NLT 16 per cent; **T.L.C.** - Refer API	Prunasin (D-mandelo-nitrile-β-glucoside), Quercetin-3-O-rutinosyl-7,3-O-biglucoside, Kaempferol-3-O-rutinosyl-4'-di-O-glucoside and 6-ethoxy-kaempferol.	*Rasa* - Kasaya *Guna*- Laghu, Ruksa *Virya*- Sita *Vipaka*- Katu; *Karma*- Kaphahara, Yonidosahara, Varnya, Stambhana, Sukrasodhaka, Vedanasthapana, Visaghna	Asvagandha Taila	Kandu, Vrana, Chardi, Aruci, Kasa, Hrdroga, Raktapitta, Kustha, Krmiroga, Mukharoga, Medoroga, Trsna, Arsa, Pandu, Unmada, Jvara, Daha	3-6 g
341	Part-I; Vol. V	33	GANDIRA (Root)	*Coleus forskohlii* Briq. Syn. *C. barbatus* Benth.	Lamiaceae	Dried Mature Root	**Foreign matter**- NMT 2 per cent; **Total ash**- NMT 9 per cent; **Acid-insoluble ash**- NMT 1.5 per cent; **Alcohol-soluble extractive**-NLT 16 per cent; **Water-soluble extractive**- NLT 23 per cent; **Essential oil** - NLT 0.1 per cent; **Cp; epmp**; - NLT 0.15 per cent; **T.L.C.** - Refer API	Diterpene, Coleonol, Coleosol, Deoxy-coleonol, Forskohlin, Naphthopyrone, Coleoforsine.	*Rasa* - Katu, Kasaya, Tikta *Guna*-Ruksa, Tiksna, Sara *Virya*- Usna *Vipaka*- Katu; *Karma*- Vatahara, Kaphahara, Tridosahara, Vranasodhana, Vidahi	Krmighna Kasaya Churna	Sotha, Arsa, Kasa, Krmi, Kustha, Dusta Vrana, Hutavisa, Gulma, Udara, Pliiharoga, Sula, Mandagni, Mutrabandha, Malabandha	3-5 g
342	Part-I; Vol. V	35	GAVEDHUKA (Root)	*Coix lachryma-jobi* Linn. Syn. *C. lachryma* Linn.	Gramineae	Dried Root	**Foreign matter**- NMT 2 per cent; **Total ash**- NMT 4 per cent; **Acid-insoluble ash**- NMT 1 per cent; **Alcohol-soluble extractive**-NLT 10 per cent; **Water-soluble extractive**- NLT 10 per cent; **T.L.C.** - Refer API	Benzox-azolinones, Amino acids (Leucine, Tyrosine, Histadin, Arginine and Coicin)	*Rasa* - Katu, Madhura *Guna*- Laghu, Ruksa *Virya*- Sita *Vipaka*- Katu; *Karma*- Kaphahara, Pittahara, Mutrala, Karsniya	Visnu Taila	Mutrakrcchra, Netra-Masurika, Pittaja Chardi, Sthaulya	3-6 g

Sl. No.	*Part and Volume*	*Page No.*	*Common Ayurvedic Name*	*Botanical Name/ Chemical Name*	*Family/ Chemical Formula*	*Part Used*	*Identity, Purity and Strength/Chemical Properties*	*Constituents*	*Properties and Action*	*Important Formulations*	*Therapeutic Uses*	*Dose*
343	Part-I; Vol. V	37	GHONTA (Fruit)	*Ziziphus xylopyrus* Willd.	Rhamnaceae	Fruit	**Foreign matter**- NMT 1 per cent; **Total ash**- NMT 12 per cent; **Acid-insoluble ash**- NMT 1 per cent; **Alcohol-soluble extractive**-NLT 3 per cent; **Water-soluble extractive**-NLT 2 per cent; **T.L.C.** - Refer API	The pulp of the fruit contains reducing Sugars, Sucrose, Citric acid, Carotene, Vitamin C and Tannins.	*Rasa* - Kasaya, Katu, Madhura *Guna*- Laghu *Virya*- Usna *Vipaka*- Katu; *Karma*- Vatakaphahara, Visaghna	Aragvadhadi Kvatha Churna	Vrana, Kandu, Kustha, Raktavikara, Svayathu, Prameha, Nadivarana, Dustavrana, Vamana, Jvara	3-6 g
344	Part-I; Vol. V	39	GUNDRAH (Rhizome and Root)	*Typha australis* Schum. and Thonn. Syn. *T. angustata* Bory and Chaub.,	Typhaceae	Rhizome with Root	**Foreign matter**- NMT 2 per cent; **Total ash**- NMT 10 per cent; **Acid-insoluble ash**- NMT 4 per cent; **Alcohol-soluble extractive**-NLT 6 per cent; **Water-soluble extractive**-NLT 8 per cent; **T.L.C.** - Refer API	Flavonoids (Quercetin, Isorhamnetin-3-O-rutinoside); Sterols (β-sitosterol, Lanosterol, Cholesterol)	*Rasa* - Kasaya, Madhura *Guna*- Guru *Virya*- Sita *Vipaka*- Madhura; *Karma*- Pittasamasamana, Vatahara, Stanyasodhaka, Stanyajanana, Sukrasodhaka, Rajosodhaka, Mutravirecaniya, Mutrasodhaka	Mutravirecaniya Kasaya Churna, Stanyajanana Kasaya Churna	Raktapitta, Asmari, Sarkara, Mutraghata, Mutrakrcchra, Stanya Ksaya	3-6 g
345	Part-I; Vol. V	41	HIMSRA (Root)	*Capparis spinosa* Linn.	Capparidaceae	Root	**Foreign matter**- NMT 1 per cent; **Total ash**- NMT 13 per cent; **Acid-insoluble ash**- NMT 5 per cent; **Alcohol-soluble extractive-NLT 1 per cent; Water-soluble extractive**-NLT 2 per cent; **T.L.C.** - Refer API	The Roots contian Alkaloid sachydrine. Gluco-brassicin, Neogluco-brassicin and 4-methoxy-glucobrassicin have also been identified in the roots.	*Rasa* -Katu, Tikta *Guna*- Laghu, Ruksa *Virya*- Usna *Vipaka*- Katu; *Karma*- Vatahara, Kaphahara, Dipani, Rucya	Amratadi Taila, Kutikhadi Vatika, Himsradya Ghrta	Vatavikara, Kasa, Svasa, Galagandha, Gulma, Arsa, Amavata, Grdhrasi, Vatarakta, Raktagranthi, Vatikayoniroga, Vatasopha, Vrana, Granthi	1-3 g

Sl. No.	*Part and Volume*	*Page No.*	*Common Ayurvedic Name*	*Botanical Name/ Chemical Name*	*Family/ Chemical Formula*	*Part Used*	*Identity, Purity and Strength/Chemical Properties*	*Constituents*	*Properties and Action*	*Important Formulations*	*Therapeutic Uses*	*Dose*
346	Part-I; Vol. V	43	HINGUPATRI (Leaf)	*Ferula jaeschkeana* Vatke	Apiaceae	Dried Leaf	**Foreign matter**- NMT 2 per cent; **Total ash**- NMT 13.0 per cent; **Acid-insoluble ash**- NMT 1.10 per cent; **Alcohol-soluble extractive**-NLT 10 per cent; **Water-soluble extractive**- NLT 30 per cent; **T.L.C.** - Refer API	–	*Rasa* - Katu, Tikta *Guna*- Tiksna *Virya*- Usna *Vipaka*- Katu; *Karma*- Pacana, Hrdya, Vatakaphahara, Rucikara	Kumaryasava	Hrdroga, Bastisula, Vibandha, Garbhani, Arsa, Gulmaroga, Krmi, Pliharoga, Apasmara, Unmada	3-6 g
347	Part-I; Vol. V	45	ITKATA (Root)	*Sesbania bispinosa* W. F. Wight	Fabaceae	Dried Root	**Foreign matter**- NMT 1 per cent; **Total ash**- NMT 5 per cent; **Acid-insoluble ash**- NMT 1 per cent; **Alcohol-soluble extractive**-NLT 2 per cent; **Water-soluble extractive**- NLT 6 per cent; **T.L.C.** - Refer API	Amino acids such as Lysine, Arginine, Histidine.	*Rasa* - Madhura *Guna*- Snigdha, Guru *Virya*- Sita *Vipaka*- Madhura; *Karma*- Pittahara, Vatahara, Mutravirecaniya, Stanyajanana	Mutravirecaniya Churna, Stanyajanana Kasaya Churna.	Kasa, Pratisyaya, Jvara, Netraroga, Asmari, Pittas-mari, Sarkara, Mutrakrcchra, Mutraghata, Mutraruja	3-6g
348	Part-I; Vol. V	47	ITKATA (Stem)	*Sesbania bispinosa* W. F. Wight	Fabaceae	Dried Stem	**Foreign matter**- NMT 1 per cent; **Total ash**- NMT 5 per cent; **Acid-insoluble ash**- NMT 1 per cent; **Alcohol-soluble extractive**-NLT 2 per cent; **Water-soluble extractive**- NLT 8 per cent; **T.L.C.** - Refer API	Amino acids such as Lysine, Arginine, Histidine.	*Rasa* - Madhura *Guna*- Snigdha, Guru *Virya*- Sita *Vipaka*- Madhura; *Karma*- Vatahara, Pittahara, Slesmaprakopaka, Stanyajanana Mutravirecaniya	Candanadi Taila (Caraka)	Kasa, Pratisyaya, Jvara, Netraroga, Asmari, Pittas-mari, Sarkara, Mutrakrcchra, Mutraghata, Mutraruja	3-6 g
349	Part-I; Vol. V	49	JALAPIPPALI (Whole Plant)	*Phyla nodiflora* Greene syn. *Lippia nodiflora* Mich.	Verbenaceae	Dried Whole Plant	**Foreign matter**- NMT 2 per cent; **Total ash**- NMT 27 per cent; **Acid-insoluble ash**- NMT 5 per cent; **Alcohol-soluble extractive**-NLT 4 per cent; **Water-soluble extractive**- NLT 12 per cent; **T.L.C.** - Refer API	Flavonoids namely Nodiflorin A and Nodiflorin B, Nodifloretin, Lippiflorins A and B.	*Rasa* - Katu, Tikta, Kasaya *Guna*- Ruksa, Tiksna, *Virya*- Sita *Vipaka*- Katu; *Karma*- Pittahara, Kaphahara, Mutral, Jvaraghna, Sukarala, Mukhasodhani, Dipani, Hrdya, Caksusya, Sangrahi, Rucya, Visaghna	Akika, Pisti, Akika Bhasma	Raktaroga, Daha, Vrana, Svasa, Bhrama, Murchha, Trsa, Raktadosa, Krmi, Jvara, Pittatisara, Visarpa	2 to 3g powder, 1/2 to 2 ml juice

Sl. No.	*Part and Volume*	*Page No.*	*Common Ayurvedic Name*	*Botanical Name/ Chemical Name*	*Family/ Chemical Formula*	*Part Used*	*Identity, Purity and Strength/Chemical Properties*	*Constituents*	*Properties and Action*	*Important Formulations*	*Therapeutic Uses*	*Dose*
350	Part-I; Vol. V	52	JIVAKAH (Pseudo-bulb)	*Malaxis acuminata* D. Don syn. *Microstylis wallichii* Lindl.	Orchidaceae	Dried and Fresh Pseudo-bulb	**Foreign matter-** NMT 2 per cent; **Total ash-** NMT 3 per cent; **Acid-insoluble ash-** NMT 0.5 per cent; **Alcohol-soluble extractive-**NLT 4 per cent; **Water-soluble extractive-** NLT 12 per cent; **Starch** - NLT 19 per cent; **T.L.C.** - Refer API	Alcohol (Ceryl alcohol), Glucose, Rhamnose and Diterpenes.	*Rasa* - Madhura *Guna*- Snigdha, Picchila *Virya*- Sita *Vipaka*- Madhura; *Karma*-Vatahara Pittahara, Dhatuvardhaka, Sukrala, Brmhana, Balya, Snehopaga, Jivaniya, Rasayana	Dasamularista, Chyawanaprasa, Brahma Rasayana, Sivagutika, Amrtaprasa Ghrta, Asoka Ghrta, Dhanvantra Taila, Bala Taila, Manasamitra Vataka, Guducyadi Taila, Brhat Asvagandha Ghrta	Raktapitta, Daha, Ksaya, Raktavikara, Karsya, Svasa, Kasa, Sosa	5-10 g
351	Part-I; Vol. V	54	KADARAH (Heart Wood)	*Acacia suma* Buch.-Ham.	Mimosaceae	Heart wood	**Foreign matter-** NMT 2 per cent; **Total ash-** NMT 4 per cent; **Acid-insoluble ash-** NMT 2 per cent; **Alcohol-soluble extractive-**NLT 2 per cent; **Water-soluble extractive-** NLT 8 per cent; **T.L.C.** - Refer API	An Alkaloid diaboline, β-sitosterol, Stigmasterol, Oleanolic acid and its 3β-acetate, a Saponin containing Oleanolic acid, Galactose, Mannose.	*Rasa* - Tikta; *Guna*- Visada; *Virya*- Sita; *Vipaka*- Katu; *Karma*- Kaphahara, Varnya, Pittahara, Raktasodhaka	Ayaskrti	Madhumeha, Mukharoga, Udarda, Kandu, Medodosa, Vrana, Pandu, Kustha, Svitra, Raktadosa	2-6 g
352	Part-I; Vol. V	56	KAKAJANGHA (Seed)	*Peristrophe bicalyculata* (Retz.) Nees	Acanthaceae	Dried Mature Seed	**Foreign matter-** NMT 2 per cent; **Total ash-** NMT 6 per cent; **Acid-insoluble ash-** NMT 0.1 per cent; **Alcohol-soluble extractive-**NLT 10 per cent; **Water-soluble extractive-** NLT 20 per cent; **T.L.C.** - Refer API	–	*Rasa* - Tikta, Kasaya; *Guna*- Sara, Picchila; *Virya*- Usna; *Vipaka*- Katu; *Karma*- Kaphapittanut, Krmighna, Varnya, Vranahara, Visaghna	Mahavisagarbha Taila	Visamajvara, Badhirya, Raktapitta, Pandu, Pradara, Jvara, Kandu, Sosa, Ksata Ksina, Jantakrmi, Grahani, Dustavrana, Slipada, Sidhma, Sarpavisa,	1-3 g

Sl. No.	Part and Volume	Page No.	Common Ayurvedic Name	Botanical Name/ Chemical Name	Family/ Chemical Formula	Part Used	Identity, Purity and Strength/Chemical Properties	Constituents	Properties and Action	Important Formulations	Therapeutic Uses	Dose
											Sastraksata, Galaganda, Apaci, Balagraha, Pratisyaya	
353	Part-I; Vol. V	58	KAKANAJA (Fruit)	*Physalis alkekengi* Linn.	Solanaceae	Dried Mature Fruit	**Foreign matter**- NMT 2 per cent; **Total ash**- NMT 6 per cent; **Acid-insoluble ash**- NMT 1 per cent; **Alcohol-soluble extractive**-NLT 10 per cent; **Water-soluble extractive**- NLT 22 per cent; **T.L.C.** - Refer API	Auroxanthin, Mutatoxanthin, Phydalein, Zeaxanthin, β-cryptoxanthin from the Calyx of the Fruit; Glycoalkaloids detected in the seeds but Alkaloids were absent in fruit	*Rasa* - Madhura, Tikta *Guna*- Ruksa *Virya*- Sita *Vipaka*- Katu; *Karma*- Vatahara, Dahasamaka, Balya, Mutrala, Virecana, Sulanasini, Raktavidravani	Lauha Rasayana	Puyameha, Tamakasvasa, Vrana, Visarpa, Kandu, Sopha, Kasa, Svasa, Jvara	5-10 g in the powder form
354	Part-I; Vol. V	60	KALIYAKA (Root and Stem)	*Coscinium fenestratum* (Gaertn.) Colebr.	Menisper-maceae	Dried Root and Stem	**Kaliyaka Root Foreign matter**- NMT 1 per cent; **Total ash**- NMT 2 per cent; **Acid-insoluble ash**- NMT 0.4 per cent; **Alcohol-soluble extractive**-NLT 11 per cent; **Water-soluble extractive**- NLT 10 per cent; **Total alkaloid as berberine chloride**- NLT 2 per cent; **T.L.C.** - Refer API **Kaliyaka Stem-** **Foreign matter**- NMT 1 per cent; **Moisture content** -NMT 6 per cent; **Total ash**- NMT 3 per cent; **Acid-insoluble ash**- NMT 2 per cent; **Alcohol-soluble extractive**-NLT 3 per cent; **Water-soluble extractive**- NLT 8 per cent; **Total alkaloid as berberine chloride** - NLT 1 per cent; **T.L.C.** - Refer API	Alkaloids- berberine, Palmitine, Jatorrhizine, Proto-berberine, N, N-di-lindacarpine, Thalifendine and Columbamine.	Kaliyaka (Root) *Rasa* - Kasaya; *Guna*- Laghu, Ruksa; *Virya*- Sita; *Vipaka*- Katu; *Karma*- Slesmasama-samana, Pittahara, Dipana, Pacana, Anulomaka, Raktasodhaka Kaliyaka (Stem) *Rasa* - Tikta; *Guna*- Laghu, Ruksa; *Virya*- Sita, *Vipaka*- Katu; *Karma*- Slesmasama-samana, Pittahara, kaphamedohara, Dipana, Pacana	NA	Kaliyaka (Root): Tikta-Usna, Raktapitta, Jirna Jvara, Prameha, Krmi, Ajirna, Adhmana, Kamala, Agnimandya, Vrana, Vyonga Kliyaka (Stem) : Kustha; Prameha; Panduroga; Jvara; Ajirna; Agnimandya; Adhamana Yakrt Vikana; Krmi; Daha; Asmari; Upadamsa; Vrana; Yuvanapidaka; Vyanga	Kaliyaka (Root)-1-3 g; Kaliyaka (Stem)- 2-6g

Sl. No.	*Part and Volume*	*Page No.*	*Common Ayurvedic Name*	*Botanical Name/ Chemical Name*	*Family/ Chemical Formula*	*Part Used*	*Identity, Purity and Strength/Chemical Properties*	*Constituents*	*Properties and Action*	*Important Formulations*	*Therapeutic Uses*	*Dose*
355	Part-I; Vol. V	63	KAPITANA (Stem Bark)	*Thespesia populnea* (L.) Soland. ex Correa *syn. Hibiscus populneus* Linn.	Malvaceae	Stem Bark	**Foreign matter**- NMT 2 per cent; **Total ash**- NMT 13 per cent; **Acid-insoluble ash**- NMT 2 per cent; **Alcohol-soluble extractive**-NLT 3 per cent; **Water-soluble extractive**- NLT 2 per cent; **T.L.C.** - Refer API	Flavonoids, Steroids and Sesqui-terpenoidal quinines	*Rasa* - Kasaya; *Guna*- Laghu, Ruksa; *Virya*- Sita; *Vipaka*- Katu; *Karma*- Vatahara, Pittahara, Kaphahara, Mutrasam-grahaniya, Stambhana, Medohara, Sandhaniya, Sukrala, Samgrahi, Bhagna-sandhanakrta, Pumsavanam	Nyagrodhadi Kvatha Churna	Raktapitta, Prameha, Raktavikara, Yoniroga, Daha, Trsa, Medoroga, Vrana, Sotha, Tvakroga, Balavisarpa, Pama, Kandu, Dadru	50-100 ml kvatha
356	Part-I; Vol. V	65	KARKASA (Root)	*Momordica dioica* Roxb. ex Willd.	Cucurbitaceae	Root	**Foreign matter**- NMT 1 per cent; **Total ash**- NMT 8 per cent; **Acid-insoluble ash**- NMT 2 per cent; **Alcohol-soluble extractive**-NLT 3 per cent; **Water-soluble extractive**- NLT 31 per cent; **T.L.C.** - Refer API	α-eleostearic acid, 2-acetyl-5-chloropyrrole.	*Rasa* - Tikta; *Guna*- Laghu, Tiksna; *Virya*- Sita; *Vipaka*- Katu; *Karma*- Kaphahara, Pittahara, Vranasodhaka, Rucikara, Rasayana	Hiraka rasayana, Visanasaka yoga (Ayurved Prakash), Kakadani taila, Kalagnirudra rasa, Sannipata vidhvanisa rasa, Candrarudra rasa	Visarpa, Sarpavisavikara, Mutrakrcchra, Sarpavisa, Jvara, Kasa, Svasa, Hikka, Arsa, Ksaya, Raktarsa, Madhumeha, Netraroga, Siroroga, Kamala, Asmari	3-6 g
357	Part-I; Vol. V	67	KARNA-SPHOTA (Seed)	*Cardio-spermum halicacabum* Linn.	Sapindaceae	Seed	**Foreign matter**- NMT 2 per cent; **Total ash**- NMT 5 per cent; **Acid-insoluble ash**- NMT 0.5 per cent; **Alcohol-soluble extractive**-NLT 21 per cent; **Water-soluble extractive**- NLT 5 per cent; **Fixed oil** - NLT 20 per cent; **T.L.C.** - Refer API	Fixed oil.	*Rasa* - Tikta, Katu; *Guna*- Laghu, Ruksa, Tiksna; *Virya*- Sita; *Vipaka*- Katu; *Karma*- Vatahara, Mutrala, Kesya, Medhya, Visaghna	Amatisaranasaka Yoga, Vasadilepa, Nagaradi Taila, Lausunadi Kasaya	Jvara, Sopha, Pandu, Sula Vrddhi, Sanhi-Vata, Graha-Badha, Bhutabadha, Visabadha	1-2 g

Sl. No.	*Part and Volume*	*Page No.*	*Common Ayurvedic Name*	*Botanical Name/ Chemical Name*	*Family/ Chemical Formula*	*Part Used*	*Identity, Purity and Strength/Chemical Properties*	*Constituents*	*Properties and Action*	*Important Formulations*	*Therapeutic Uses*	*Dose*
358	Part-I; Vol. V	69	KARNA-SPHOTA (Root)	*Cardio-spermum halicacabum* Linn.	Sapindaceae	Root	**Foreign matter**- NMT 2 per cent; **Total ash**- NMT 7 per cent; **Acid-insoluble ash**- NMT 1 per cent; **Alcohol-soluble extractive**-NLT 9 per cent; **Water-soluble extractive**-NLT 15 per cent; **T.L.C.** - Refer API	–	*Rasa* - Tikta, Katu; *Guna*- Tiksna, Laghu, Ruksa; *Virya*- Sita; *Vipaka*- Katu; *Karma*- Vatahara, Kaphasamaka, Rasayana, Kesya, Medhya, Vamaka, Mutrala, Virecaka, Visaghna	Aragvadhadi Kvatha Churna	Jvara, Pandu, Kamala, Sula, Vrddhi, Smrti Ksaya, Sandhi-Vata, Kustha, Sarpavisa, Musikavisa, Jvarayukta-Kasa, Indralupta, Sannipatodara, Asmari, Sopha, Bhuta-badha, Grahabadha	1-3 g
359	Part-I; Vol. V	71	KATTRNA (Whole Plant)	*Cymbopogon citratus* (DC.) Stapf syn: *Andropogon Citratus* DC.	Poaceae	Whole Plant	**Foreign matter**- NMT 2 per cent; **Total ash**- NMT 11 per cent; **Acid-insoluble ash**- NMT 6 per cent; **Alcohol-soluble extractive**-NLT 5 per cent; **Water-soluble extractive**-NLT 12 per cent; **T.L.C.** - Refer API	Essential oil containing Citral as major component besides Geraniol and other Terpenes.	*Rasa* - Katu, Tikta; *Guna*- Tiksna, Laghu, Ruksa; *Virya*- Usna; *Vipaka*- Katu; *Karma*- Vatahara, Kaphahara, Sitaprasamana, Stanyajanana, Dipana, Recana, Visaghna, Mukhasodhana, Avrsya, Caksusya, Rucikaraka, Vamihara	Masabaladi Kvatha Churna	Kustha, Krmi, Arocaka, Santapa, Daha, Vami, Kasa, Svasa, Dadru, Udara, Bhutabadha, Grahabadha, Udarda	3-6 g
360	Part-I; Vol. V	74	KEBUKA (Rhizome)	*Costus speciosus* (Koerning ex Retz.) Smith.	Zingiberaceae	Dried Rhizome	**Foreign matter**- NMT 2 per cent; **Total ash**- NMT 20 per cent; **Acid-insoluble ash**- NMT 5 per cent; **Alcohol-soluble extractive**-NLT 3 per cent; **Water-soluble extractive**-NLT 12 per cent; **T.L.C.** - Refer API	Steroidal Saponins such as (Tigogenin and Diosgenin).	*Rasa* - Tikta; *Guna*- Laghu, Ruksa; *Virya*- Sita; *Vipaka*- Katu; *Karma*- Pittahara, Kaphahara, Dipana, Pacana, Grahi, Krmighna, Hrdya, Raktasodhaka, Garbhasaya, Sankocaka	Krmighna Kvatha Churna	Kaphapittaja vikara, Agnimandya, Grahani, Krmiroga, Raktavikara, Slipada, Prameha, Svitra, Kustha, Jvara, Kasa, Kamala, Arsa, Kaphaja, Mutrakrcchra	3-6 g (after puri-fication)

Sl. No.	*Part and Volume*	*Page No.*	*Common Ayurvedic Name*	*Botanical Name/ Chemical Name*	*Family/ Chemical Formula*	*Part Used*	*Identity, Purity and Strength/Chemical Properties*	*Constituents*	*Properties and Action*	*Important Formulations*	*Therapeutic Uses*	*Dose*
361	Part-I; Vol. V	76	KHAKHASA (Seed)	*Papaver somniferum* Linn.	Papaveraceae	Seed	**Foreign matter**- NMT 1 per cent; **Total ash**- NMT 8 per cent; **Acid-insoluble ash**- NMT 1.5 per cent; **Alcohol-soluble extractive**-NLT 7 per cent; **Water-soluble extractive**-NLT 13 per cent; **Fixed oil** - NLT 19 per cent; **T.L.C.** - Refer API	Fixed oil containing Esters of Linoleic, Palmitic, Oleic acids.	*Rasa* - Madhura; *Guna*- Guru; *Virya*- Sita; *Vipaka*- Madhura; *Karma*- Vatahara, Rucya, Stambhana, Vedanasthapana, Vrsya, Balya, Varnya	Abhyadi Gutika, Abhrakadi Vati, Asvani Kumar Rasa	Kasa, Atisara	5-10 g
362	Part-I; Vol. V	78	KHATMI (Root)	*Althaea officinalis* Linn.	Malvaceae	Root	**Foreign matter**- NMT 2 per cent; **Moisture content** - NMT 8 per cent; **Total ash**- NMT 7 per cent; **Acid-insoluble ash**- NMT 1.5 per cent; **Alcohol-soluble extractive**-NLT 8 per cent; **Water-soluble extractive**- NLT 21 per cent; **T.L.C.** - Refer API	Galacturonic acid, Galactose, Glucose, Xylose and Rhamnose, Polysaccharide althaea mucilage-O, Asparaginene, Betaine, Lecithin and Phytosterol, Polysaccharides.	*Rasa* - Madhura *Guna*- Snigdha, Picchila, Guru *Virya*- Sita *Vipaka*- Madhura; *Karma*- Vatahara, Pittahara, Slesmasaraka, Mutrala, Vedanasthapana, Kaphaghna	Gojihvadi Kvatha Churna	Kasa, Pratisyaya, Mutradaha, Mutrasayasotha, Kantharoga, Mutrakrcchra, Antrasotha, Daha, Raktapitta,	3-6 g
363	Part-I; Vol. V	80	KHATMI (Seed)	*Althaea officinalis* Linn.	Malvaceae	Dried Seed	**Foreign matter**- NMT 2 per cent; **Total ash**- NMT 8 per cent; **Acid-insoluble ash**- NMT 1.5 per cent; **Alcohol-soluble extractive**-NLT 10 per cent; **Water-soluble extractive**- NLT 18 per cent; **T.L.C.** - Refer API	Glucose, Sucrose, Galactose and Mannose, Linoleic acid, Isobutylalcohol, Limonene, Phellandrene, γ-toluerldehyde, Citral, Terpeneol, β-sitosterol	*Rasa* - Madhura, *Guna*- Snigdha, Picchila, Guru *Virya*- Sita *Vipaka*- Madhura; *Karma*- Vatahara, Pittahara, Slesma saraka, Mutrala, Vedanasthapana, Slesma kala Snehakara	Gojihvadi Kvatha Churna	Pratisyaya, Kasa, Mutrakrcchra, Mutradaha, Kantharoga	3-6 g

Sl. No.	*Part and Volume*	*Page No.*	*Common Ayurvedic Name*	*Botanical Name/ Chemical Name*	*Family/ Chemical Formula*	*Part Used*	*Identity, Purity and Strength/Chemical Properties*	*Constituents*	*Properties and Action*	*Important Formulations*	*Therapeutic Uses*	*Dose*
364	Part-I; Vol. V	82	KHUBKALAN (Seed)	*Sisymbrium irio* Linn.	Brassicaceae	Seed	**Foreign matter**- NMT 2 per cent; **Total ash**- NMT 5 per cent; **Acid-insoluble ash**- NMT 1 per cent; **Alcohol-soluble extractive**-NLT 22 per cent; **Water-soluble extractive**- NLT 14 per cent; **Fixed oil - NLT 20 per cent; T.L.C.** - Refer API	Fixed oil and Isorhamnetin	*Rasa* - Katu; *Guna*- Snigdha, Guru, Picchila; *Virya*- Usna; *Vipaka*- Katu; *Karma*- Vatahara, Kaphahara, Balya, Svedakara, Sothahara	Gojihvadi Kvatha Churna	Jvara, Kasa, Vatajanya Vikara, Svasa, Svarabheda, Daurbalya, Kaphavikara	3-6 g
365	Part-I; Vol. V	84	KODRAVAH (Grain)	*Paspalum scrobiculatum* Linn.	Poaceae	Dehusked and well-matured Caryopsis	**Foreign matter- NMT 2 per cent; Total ash- NMT 6 per cent; Acid-insoluble ash**- NMT 4 per cent; **Alcohol-soluble extractive**-NLT 3 per cent; **Water-soluble extractive**-NLT 2 per cent; **T.L.C.** - Refer API	Hydrocarcons Hentria-contanol, Hentria-contanone; Sterols such as β-β-Sitosterol, Campestrol.	*Rasa* - Kasaya, Madhura *Guna*- Ruksa, Laghu *Virya*- Sita *Vipaka*- Katu; *Karma*- Pittahara, Kaphahara, Grahi, Lekhana, Visaghna	Nadivranahara aturyadi lepa, Nadivranahara aturyadi taila	Raktapitta, Vrana, Atisthaulya, Annadravasula, Prameha, Medovrddhi, Nadivrana, Jalodara	50-100 g
366	Part-I; Vol. V	86	KSIRAKAKOLI (Bulb)	*Fritillaria roylei* Hook.	Liliaceae	Dried Whole Bulb	**Foreign matter**- NMT 2 per cent; **Total ash**- NMT 3 per cent; **Acid-insoluble ash**- NMT 0.5 per cent; **Alcohol-soluble extractive**-NLT 4 per cent; **Water-soluble extractive**-NLT 14 per cent; **T.L.C.** - Refer API	Alkaloids Kashimirine (Imperialine), Peimine, Peimisine, Propeimine, Peimiphine and Peimitidine	*Rasa* - Madhura *Guna*- Guru, Snigdha *Virya*- Sita *Vipaka*- Madhura; *Karma*- Vatahara, Pittahara, Rasayana, Brmhana, Sukravardhaka, Vrsya, Stanyajanana, Kaphakara, Trsahara, Basti visodhani, Visaghna	Dasamularista, Sivagutika, Brhataphala Ghrta, Brhat-guduci Taila, Brhatmasa Taila, Manasamitra Vataka, Rasaraja Rasa	Raktapitta, Daha, Sosa, Jvara, Ksaya, Raktadosa, Raktaroga, Hrdroga, Svasa, Kasa, Vaatarakta, Yoni Vyapad, Vatavyadhi, Vatapittaruja, Ksaya, Hrdroga	3-5 g in the powder form

Sl. No.	*Part and Volume*	*Page No.*	*Common Ayurvedic Name*	*Botanical Name/ Chemical Name*	*Family/ Chemical Formula*	*Part Used*	*Identity, Purity and Strength/Chemical Properties*	*Constituents*	*Properties and Action*	*Important Formulations*	*Therapeutic Uses*	*Dose*
367	Part-I; Vol. V	88	KSHIRA-VIDARI (Root)	*Ipomoea digitata* Linn. Syn. *Ipomoea paniculata* (Linn.) R. Br.	Convolvulaceae	Dried Root	**Foreign matter**- NMT 2 per cent; **Total ash**- NMT 6 per cent; **Acid-insoluble ash**- NMT 1 per cent; **Alcohol-soluble extractive-NLT 20 per cent; Water-soluble extractive**- NLT 8 per cent; **T.L.C.** - Refer API	Glycosides, Steroids, Tannins and Fixed oil	*Rasa* - Madhura, Kasaya, Tikta *Guna*- Snigdha, Guru *Virya*- Sita *Vipaka*- Madhura; *Karma*- Vatahara, Vrsya, Brmhana, Atimutrala, Balya, Svarya, Varnya, Stanyajanana, Rasayana, Jivaniya	Sivagutika	Stanyavikara, Pittaja sula, Raktavikara, Maha-vatavyadhi, Mutraroga, Varana, Bhagna	5-10 g
368	Part-I; Vol. V	90	KULANJANA (Rhizome)	*Alpinia galanga* Willd.	Zingiberaceae	Dried Rhizome	**Foreign matter**- NMT 2 per cent; **Total ash**- NMT 5 per cent; **Acid-insoluble ash**- NMT 2 per cent; **Alcohol-soluble extractive-NLT 6 per cent; Water-soluble extractive**-NLT 13 per cent; **Starch** - NLT 22 per cent; **Essential oil**- NLT 0.4 per cent; **T.L.C.** - Refer API	Essential oil, Containing α- pinene, β-pinene, Limonene, Cineol, Terpine-4-ol and α-terpineol.	*Rasa* -Katu, Tikta *Guna*- Guru *Virya*- Usna *Vipaka*- Katu; *Karma*- Vatahara, Kaphahara, Pacani, Rucya, Svarya, Hrdya, Kanthya, Mukha Sodhaka, Visaghna	Brahmi Vati, Rasnadikasaya, Rasnadarvadi Kasaya, Rasnapancakam, Rasna saptakam, Rasnasunthyadi Kasaya, Rasnairandadi Kasaya	Pratisyaya, Svasa, Hikka, Sopha, Vataja Sula, Udararoga, Kampa, Vismajvara, Kaphajakasa, Asiti, Vatavyadhi, Mahakustha	1-3 g powder
369	Part-I; Vol. V	93	KUMBHIKAH (Seed)	*Careya arborea* Roxb.	Lecythidaceae	Dried Seed	**Foreign matter**- NMT 2 per cent; **Total ash**- NMT 4 per cent; **Acid-insoluble ash**- NMT 1 per cent; **Alcohol-soluble extractive**-NLT 7 per cent; **Water-soluble extractive**-NLT 15 per cent; **T.L.C.** - Refer API	Saponins (Five Sapogenols-careyagenol A, B, C, D and E); Sterols, α-spinosterol and α-spinosterone.	*Rasa* - Katu, Kasaya; *Guna*- Ruksa; *Virya*- Usna; *Vipaka*- Katu; *Karma*- Kaphahara, Vatahara, Grahi, Vrana Ropana	Marma Gutika	Vatika Kasa, Kustha, Prameha, Krmi, Visaroga, Pakvatisara, Vrana, Nadivrana	2-6 g powder
370	Part-I; Vol. V	95	LATA-KARANJA (Seed)	*Caesalpinia bonduc* (Linn.) Roxb.	Caesalpiniaceae	Seed	**Foreign matter- NMT 1 per cent; Total ash- NMT 5 per cent; Acid-insoluble ash**- NMT 1 per cent; **Alcohol-soluble extractive**-NLT 26 per cent; **Water-soluble extractive**- NLT 4.0 per cent; **T.L.C.** - Refer API	Seeds contain bitter substance Phytosterenin, Bonducin, Saponin, Phytosterol, Fixed oil, Starch and Sucrose. Seed	*Rasa* - Tikta, Kasaya *Guna*- Laghu, Ruksa *Virya*- Usna *Vipaka*- Katu, *Karma* - Vatahara, Pittahara, Kaphahara, Dipana,	Aragvadhadi Kvatha Churna, Kuberaksadi Vati	Visamajvara, Sutikajvara, Sula, Gulma, Kasa, Meha, Vatavikara, Tvakroga, Sotha, Vrana, Udarasula, Svasa,	1-3 g

Sl. No.	*Part and Volume*	*Page No.*	*Common Ayurvedic Name*	*Botanical Name/ Chemical Name*	*Family/ Chemical Formula*	*Part Used*	*Identity, Purity and Strength/Chemical Properties*	*Constituents*	*Properties and Action*	*Important Formulations*	*Therapeutic Uses*	*Dose*
								also contain α,β,γ,δ and ζ caesalpins.	Vedanasthapaka, Artavajanana, Vranaropana		Raktatisara, Kustha, Amavata, Sandhivata, Agnimandya, Pravahika, Arsa, Yakrtpliharoga, Chardi, Krmi	
371	Part-I; Vol. V	98	LAVALIPHALA (Fruit)	*Phyllanthus acidus* (Linn.) Skeels *syn. Cicca acida* Linn. Merrill	Euphorbiaceae	Dried Fruit	**Foreign matter**- NMT 2 per cent; **Total ash**- NMT 6 per cent; **Acid-insoluble ash**- NMT 0.5 per cent; **Alcohol-soluble extractive**-NLT 7 per cent; **Water-soluble extractive**-NLT 15 per cent; **T.L.C.** - Refer API	Triterpenoids (β-amyrin, Phyllanthol) and Gallic acid	*Rasa* - Madhura, Amla, Kasaya; *Guna*- Ruksa, Guru, Visada; *Virya*- Sita; *Vipaka*- Madhura; *Karma*- Pittahara, Kaphahara, Vatakara, Grahi, Rakta Stambhana, Hrdya, Rucikara	Draksasava	Asmari, Arsa, Aruci	10-20 g
372	Part-I; Vol. V	100	MADHULIKA (Root)	*Eleusine corocana* (L.) Gaertn.	Poaceae	Dried Root	**Foreign matter**- NMT 2.5 per cent; **Total ash**- NMT 5.5 per cent; **Acid-insoluble ash**- NMT 1.3 per cent; **Alcohol-soluble extractive**-NLT 3 per cent; **Water-soluble extractive**-NLT 8 per cent; **T.L.C.** - Refer API	Flavonoids, Orientin, Isoorientin, Vitexin, Isovitexin, Violanthin, Lucenin-1, Tricin, Keto acids, Polysaccharide and the free Sugars, β-sitosterol glucoside.	*Rasa* - Madhura, Kasaya, Tikta; *Guna*- Laghu; *Virya*- Sita; *Vipaka*- Madhura; *Karma*- Pittahara, Tridosasamaka, Raktadosahara, Vrsya, Rasayana	Amlapittantaka modaka, Amrta guggulu, Asvagandhadi leha, Kusthadi Kvatha, Katutumbyadi taila	Trsna, Karapada daha, Vrkkasmari, Svasa, Kasa, Jvaropdrava	5-10 g

Sl. No.	*Part and Volume*	*Page No.*	*Common Ayurvedic Name*	*Botanical Name/ Chemical Name*	*Family/ Chemical Formula*	*Part Used*	*Identity, Purity and Strength/Chemical Properties*	*Constituents*	*Properties and Action*	*Important Formulations*	*Therapeutic Uses*	*Dose*
373	Part-I; Vol. V	102	MAHAMEDA (Rhizome and Root)	*Polygonatum cirrhifolium* Royle	Liliaceae	Dried Rhizome and Root	**Foreign matter**- NMT 3 per cent; **Total ash**- NMT 3.5 per cent; **Acid-insoluble ash**- NMT 1 per cent; **Alcohol-soluble extractive**-NLT 4.5 per cent; **Water-soluble extractive**- NLT 70 per cent; **T.L.C.** - Refer API	Glucose, Sucrose.	*Rasa* - Madhura; *Guna*- Guru, Snigdha; *Virya*- Sita; *Vipaka*- Madhura; *Karma*- Kaphavardhaka, Vatahara, Pitahara, Vrsya, Sukravardhaka, Stanyajanna, Brmhana, Jivaniya, Rucya	Dasamularista, Sivagutika, Amrtaprasa Ghrta, Asoka Ghrta, Dhanvantara Taila, Brhatmasa Taila, Mahanarayana Taila, Vasacandanadi Taila	Jvara, Raktavikara, Ksaya, Daha, Raktapitta, Balaroga, Kamala, Ksata, Ksina	3-6 g
374	Part-I; Vol. V	104	MADHU-SNUHI (Tuberous Root)	*Smilax china* Linn.	Liliaceae	Tuberous Root	**Foreign matter**- NMT 2 per cent; **Total ash**- NMT 0.6 per cent; **Acid-insoluble ash**- NMT 0.06 per cent; **Alcohol-soluble extractive**-NLT 0.8 per cent; **Water-soluble extractive**- NLT 5 per cent; **T.L.C.** - Refer API	Saponins, Sarsaponin and Parallin, which yield Isomeric sapogenins, Sarsapogenin and Smilogenin. It also contains Sitosterol and Stigmasterol in the free form and as Glucosides.	*Rasa* - Tikta; *Guna*- Laghu, Ruksa; *Virya*- Usna *Vipaka*- Katu; *Karma*- Tridosahara, Rasayana, Sothahara, Vedanasthapana, Nadibalya, Dipana, Anulomana, Raktasodhaka, Vrsya, Sukrasodhaka, Mutrala, Svedajanana	Madhusnuhi Rasayana, Copacinyadi Churna	Vibandha, Adhmana, Sula, Krmi, Kustha, Puyameha, Sukravikara, Vatavyadhi, Phiranga, Unmada, Apasmara, Sandhivata, Kampavata, Gandamala	3-6 g powder
375	Part-I; Vol. V	106	MEDASAKAH (Stem Bark)	*Litsea chinensis* Lam. Syn. *L. glutinosa* (Lour.) C.B. Robins. *L. sebifera* Pers.	Lauraceae	Stem Bark	**Foreign matter**- NMT 2 per cent; **Total ash**- NMT 8 per cent; **Acid-insoluble ash**- NMT 1 per cent; **Alcohol-soluble extractive**-NLT 5 per cent; **T.L.C.** - Refer API	Alkaloids (Laurotetaline, Actinodaphine, Boldine, Norboldine, Sebiferine and Litseferine).	*Rasa* - Katu, Tikta, Kasaya *Guna*- Laghu, Snigdha *Virya*- Usna *Vipaka*- Katu: *Karma*- Vatahara, Kaphahara, Dipana, Stambhana, Bhagnaprasadhaka	Asthisandha-naka Lepa	Sotha, Sula, Vatavikara, Agnimandya, Atisara, Raktasrva, Asthibhanga	5-10 g powder

Sl. No.	Part and Volume	Page No.	Common Ayurvedic Name	Botanical Name/ Chemical Name	Family/ Chemical Formula	Part Used	Identity, Purity and Strength/Chemical Properties	Constituents	Properties and Action	Important Formulations	Therapeutic Uses	Dose
376	Part-I; Vol. V	108	MEDASAKAH (Wood)	*Litsea chinensis* Lam. Syn. *L. glutinosa* (Lour) C.B. Robins, *L. sebifera* Pers.	Lauraceae	wood	**Foreign matter**- NMT 2 per cent; **Total ash**- NMT 3 per cent; **Acid-insoluble ash**- NMT 1 per cent; **Alcohol-soluble extractive**-NLT 1.5 per cent; **Water-soluble extractive**- NLT 2 per cent; **T.L.C.** - Refer API	Alkaloids (Laurotetanine, Actinodaphine, Boldine, Norboldine).	*Rasa* - Katu, Tikta, Kasaya; *Guna*- Laghu, Snigdha; *Virya*- Usna; *Vipaka*- Katu; *Karma*- Vatahara, Kaphahara, Dipana, Stambhana	Aileyaka Taila (Citrakadi Taila), Vataghna Lepa (Cintamani Rasa)	Sotha, Sula, Vatavikara, Agnimandya, Atisara, Raktasrava	1-3 g powder
377	Part-I; Vol. V	110	MESASRNGI (Leaf)	*Gymnema sylvestre* R. Br.	Asclepiadaceae	Dried Leaf	**Foreign matter**- NMT 2 per cent; **Total ash**- NMT 12 per cent; **Acid-insoluble ash**- NMT 2 per cent; **Alcohol-soluble extractive**-NLT 7 per cent; **Water-soluble extractive**- NLT 28 per cent; **T.L.C.** - Refer API	Triterpenoid saponins of Gymnemic acid A, B, Ca and D with Sugar-residues such as glucuronic acid, Galacturonic acid, Ferulic and Angelic acids attrached as Carboxylic acids. Several Isopropylene derivatives of Gymne-magenin, a Hexahydro-terpene, Gymne-magenin, Gymnemic acid. The leaves also contain Betaine, Choline, Gymnamine alkaloids, Inositol, d-quercitol.	*Rasa* - Tikta, Kasaya; *Guna*- Ruksa, Laghu; *Virya*- Usna; *Vipaka*- Katu; *Karma*- Vatahara, Kaphahara, Visaghna, Dipana, Caksusya, Sramasana	Ayaskrti, Nyagrodhadi Churna, Mahavisagarbha Taila, Mrtasanjivani Sura	Svasa, Kasa, Sula, Kustha, Prameha, Krmi, Vrana, Sopha, Hrdroga, Dantakrmi, Netraroga	3-6 g

Sl. No.	Part and Volume	Page No.	Common Ayurvedic Name	Botanical Name/ Chemical Name	Family/ Chemical Formula	Part Used	Identity, Purity and Strength/Chemical Properties	Constituents	Properties and Action	Important Formulations	Therapeutic Uses	Dose
								Hydrocarbons such as Nonacosane, Hentri-acontane, Tritriacontane, Pentatri-acontane, Phytin, Resin, Tartaric acid, formic acid, Butyric acid, Amino acids such as Leucine, Isoleucine, Valine, Alanine, γ-Butyric acid				
378	Part-I; Vol. V	113	MESASRNGI (Root)	*Gymnema* sylvestre R. Br.	Asclepiadaceae	Root	**Foreign matter**- NMT 2 per cent; **Total ash**- NMT 6 per cent; **Acid-insoluble ash**- NMT 1 per cent; **Alcohol-soluble extractive**-NLT 5 per cent; **Water-soluble extractive**-NLT 14 per cent; **T.L.C.** - Refer API	–	*Rasa* - Kasaya, Tikta; *Guna*- Laghu, Ruksa; *Virya*- Usna; *Vipaka*- Katu; *Karma*- Vatahara, Kaphahara, Mutrala, Dipana, Sirovirecaka, Sramsana	Maha Visagarbha Taila, Nyagrodhadi Churna, Mrtasanji-vanisura	Kustha, Prameha, Kasa, Krmiroga, Vrana, Visavikara, Mutrakrcchra, Svasa, Hrdroga, Raktavikara, Daha, Aksisula, Vidradhi, Vatahara	50-100 ml decoction; 1-2 g powder
379	Part-I; Vol. V	115	NANDI (Root)	*Ficus arnottiana* Miq.	Moraceae	Dried Root	**Foreign matter**- NMT 2 per cent; **Total ash**- NMT 5 per cent; **Acid-insoluble ash**- NMT 0.5 per cent; **Alcohol-soluble extractive**-NLT 4 per cent; **Water-soluble extractive**-NLT 8 per cent; **T.L.C.** - Refer API	–	*Rasa* - Madhura, Tikta, Kasaya; *Guna*- Laghu; *Virya*- Usna (alpa); *Vipaka*- Katu; *Karma*- Pittahara, Kaphahara, Grahi, Medohara, Bhagnasandhana	Nyagrodhadi Kvatha Churna	Raktapitta, Raktavikara, Visavikara, Daha, Kaphavikara, Vrana, Bhagna, Yonidosa	10-20 g powder; 30-50 g decoction

Sl. No.	*Part and Volume*	*Page No.*	*Common Ayurvedic Name*	*Botanical Name/ Chemical Name*	*Family/ Chemical Formula*	*Part Used*	*Identity, Purity and Strength/Chemical Properties*	*Constituents*	*Properties and Action*	*Important Formulations*	*Therapeutic Uses*	*Dose*
380	Part-I; Vol. V	117	NILAJHINTI (Root)	*Barleria strigosa* Willd.	Acanthaceae	Root	**Foreign matter**- NMT 1 per cent; **Total ash**- NMT 6 per cent; **Acid-insoluble ash**- NMT 1 per cent; **Alcohol-soluble extractive**-NLT 6 per cent; **Water-soluble extractive**-NLT 1 per cent; **T.L.C.** - Refer API	–	*Rasa* - Tikta, Madhura; *Guna*-Snigdha; *Virya*-Usna; *Vipaka*-Katu; *Karma*-Vatakaphahara, Kesaranjana, Visaghna, Mutrala, Kesya, Garbhavrddhi Kara	Manikya Rasa	Kustha, Vatarakta, Kandu, Mutrakrcchra, Raktavikara, Vatajanyaksaya, Musikavisa, Siragranthi, Dantaroga, Kasa, Sotha	10-20 ml swarasa; 50-100 ml kvatha
381	Part-I; Vol. V	119	NIMBA (Root Bark)	*Azadirachta indica* A. Juss. Syn. *Melia azadirachta* Linn.	Meliaceae	Dried Root Bark	**Foreign matter**- NMT 2 per cent; **Total ash**- NMT 15 per cent; **Acid-insoluble ash**- NMT 3 per cent; **Alcohol-soluble extractive**-NLT 6 per cent; **Water-soluble extractive**-NLT 7 per cent; **T.L.C.** - Refer API	Tetranortri-terpenoids, Margocin, Nimbidiol, Nimbolicin, Azadirinin	*Rasa* - Tikta *Guna*-Laghu *Virya*- Sita *Vipaka*- Katu, *Karma*- Pittahara, Kaphahara, Sitagrahi, Rucya Dipana, Visaghna, Kandughna, Ahrdya, Vranasodhana	Amrtastaka, Astangadasanga lanha	Chardi, Kustha, Raktapitta, Prameha, Hrllasa, Dusta Vrana, Trsa, Jvara, Daha, Kasa, Svasa, Sotha, Kaphavikara, Krmiroga, Aruci, Grahani, Yakrtvikara, Hrdayavidaha, Vamana	3-6 g
382	Part-I; Vol. V	121	NIMBA (Flower)	*Azadirachta indica* A. Juss. Syn. *Melia azadirachta* Linn.	Meliaceae	Dried Flower	**Foreign matter**- NMT 2 per cent; **Total ash**- NMT 14 per cent; **Acid-insoluble ash**- NMT 5 per cent; **Alcohol-soluble extractive**-NLT 5 per cent; **Water-soluble extractive**-NLT 12 per cent; **T.L.C.** - Refer API	15-acetoxy-7-deacetoxy-dihydro-azadirone (neeflone), Nonacosane (saturated hydrocarbon)	*Rasa* - Tikta; *Guna*- Laghu; *Virya*- Sita; *Vipaka*- Katu; *Karma*- Pittahara, Kaphahara, Vatakara, Kusthaghna, Krmighna, Caksusya, Visaghna, Grahi	Kusthakalamla rasa, Kustha sailendra rasa, Krmivinasana rasa	Kustha, Aruci, Prameha, Krmi, Kaphapittaja vikara, Daha, Jvara, Visamajvara, Netraroga, Raktavikara, Phiranga, Sotha, Srama, Trsna, Kasa, Vrana, Chardi, Kandu Vrana, Hrllasa, Hrdayavidaha	2-4 g puspa churna; 10-20 ml puspa svarasa

Sl. No.	*Part and Volume*	*Page No.*	*Common Ayurvedic Name*	*Botanical Name/ Chemical Name*	*Family/ Chemical Formula*	*Part Used*	*Identity, Purity and Strength/Chemical Properties*	*Constituents*	*Properties and Action*	*Important Formulations*	*Therapeutic Uses*	*Dose*
383	Part-I; Vol. V	123	NIMBA (Fruit)	*Azadirachta indica* A. Juss. Syn. *Melia azadirachta* Linn.	Meliaceae	Fruit	**Foreign matter**- NMT 2 per cent; **Total ash**- NMT 8 per cent; **Acid-insoluble ash**- NMT 2 per cent; **Alcohol-soluble extractive**-NLT 16 per cent; **Water-soluble extractive**- NLT 19 per cent; **T.L.C.** - Refer API	Fixed oil containing Diterpenoids and Triterpenoids (Limonoids); Nimbin, Gedunin, Azadirachtin; Nimbidinin, Salanin.	*Rasa* - Tikta *Guna*- Tiksna, Laghu, Snigdha *Virya*- Usna *Vipaka*- Katu; *Karma*- Vatahara, Kaphahara, Bhedaniya, Hrdayadahahara, Visaghna, Rasayana, Pacana	Arsoghnivati (seed), Palasabijadi Churna (seed)	Krmi, Kustha, Prameha, Gulma, Arsa, Palitya, Netraruja, Raktapitta, Ksata Ksaya, Siroroga, Jvara, Aruci, Daha, Chardi, Hrllasa, Vrana, Sotha, Visavikara, Vibandha, Khalitya, Gandamala	1-2 g churna; 5-10 drops of oil
384	Part-I; Vol. V	125	PALASAH (Seed)	*Butea monosperma* (Lam.) Kuntze, syn. *B. frondosa* Roxb.	Fabaceae	Seed	**Foreign matter**- NMT 1 per cent; **Total ash**- NMT 8 per cent; **Acid-insoluble ash**- NMT 0.5 per cent; **Alcohol-soluble extractive**-NLT 20 per cent; **Water-soluble extractive**- NLT 25 per cent; **Protein** - NLT 18 per cent; **Fatty oil** - NLT 6 per cent; **T.L.C.** - Refer API	Fatty oil, Amono acids.	*Rasa* - Kasaya, Tikta, Katu *Guna*- Laghu, Snigdha, Sara *Virya*- Usna *Vipaka*- Katu; *Karma*- Tridosahara, Dipana, Vrsya, Bhedana, Bhagnasan-dhanakara, Garbhanirodhaka, Rasayana	Krmimudgara Rasa, Ayaskrti	Krmi, Vrana, Gulma, Gudajaroga, Arsa, Raktavikara, Vatarakta, Udararoga, Kasa, Kandu, Tvakroga, Prameha, Yonidosa, Sukradosa, Mutrakrcchra, Kustha, Pama, Dadru, Daha, Pliharoga, Atisara, Netrasukra, Sula, Medoroga, Pandu, Asmari, Vrscikavisa	0.5-1 g

Sl. No.	*Part and Volume*	*Page No.*	*Common Ayurvedic Name*	*Botanical Name/ Chemical Name*	*Family/ Chemical Formula*	*Part Used*	*Identity, Purity and Strength/Chemical Properties*	*Constituents*	*Properties and Action*	*Important Formulations*	*Therapeutic Uses*	*Dose*
385	Part-I; Vol. V	127	PALASAH (Flower)	*Butea monosperma* (Lam.) Kuntze syn. *B. frondosa* Roxb.	Fabaceae	Dried Flower	**Foreign matter**- NMT 1 per cent; **Total ash**- NMT 10 per cent; **Acid-insoluble ash**- NMT 1 per cent; **Alcohol-soluble extractive**-NLT 15 per cent; **Water-soluble extractive**- NLT 32 per cent; **T.L.C.** - Refer API	Coumarins and Glycosides, Cumaranone glycosides, Butrin, Isobutrin, Mono-spermoside, Isomono-spermoside, Carbo-methoxy-3, 6-dioxo-5-hydro-1, 2, 4-triazine, Coreopsin, Isocoreopsin.	*Rasa* - Katu, Tikta, Kasaya, Madhura; *Guna*- Laghu, Ruksa, Sara; *Virya*- Sita; *Vipaka*- Madhura; *Karma*- Pittahara, Kaphahara, Dipana, Trsnasamaka, Rakta Stambhana, Mutrala, Kusthaghna, Sandhaniya, Dahaprasamana, Grahi	Kunkumadi taila, Vanga Bhasma (Jarana (b)	Raktavikara, Mut rakrcchra, Daha, Vatarakta, Kustha, Trsna, Raktapitta, Pliharoga, Gulma, Grahani, Krmi, Kandu, Arsa, Pittabhisyanda, Netrasukara	3-6 g
386	Part-I; Vol. V	130	PARASI-KAYAVANI (Seed)	*Hyoscyamus niger* Linn.	Solanaceae	Seed	**Foreign matter**- NMT 2 per cent; **Total ash**- NMT 4 per cent; **Acid-insoluble ash**- NMT 1 per cent; **Alcohol-soluble extractive**-NLT 16 per cent; **Water-soluble extractive**- NLT 10 per cent; **T.L.C.** - Refer API	Tropane Alkaloids Hyoscyamine, (its Racemic mixture and Atropine and Hyoscine.	*Rasa* - Tikta, Katu; *Guna*- Ruksa, Guru; *Virya*- Usna; *Vipaka*- Katu; *Karma*- Vatahara, Kaphahara, Pittakara, Madaka, Vedanasthapana, Pacaka, Grahi, Dipana, Nidrakara	Sarpagandha-ghana Vati	Rajahkrcchra, Sighrapatana, Svpanadosa, Udarasula, Anaha, Gulma, Krmi, Asmari, Kasa, Svasa, Anidra, Unmada, Sula, Sandhisual	125-500 mg
387	Part-I; Vol. V	132	PATTURA (Whole Plant)	*Aerva lanata* (Linn.) juss.	Amaranthaceae	Whole plant	**Foreign matter**- NMT 2 per cent; **Total ash**- NMT 17 per cent; **Acid-insoluble ash**- NMT 2 per cent; **Alcohol-soluble extractive**-NLT 2 per cent; **Water-soluble extractive**- NLT 11 per cent; **T.L.C.** - Refer API	α-amyrin, β-sitosterol, β-Sitosterol palmitate, Compesterol, Chrysin, Flavonoid glycosides and Tannins.	*Rasa* - Tikta, Kasaya; *Guna*- Laghu, Tiksna; *Virya*- Usna; *Vipaka*- Katu; *Karma*- Vatahara, Kaphahara, Mutravirecana, Krmighna	Satavaryadi Ghrta	Asmari, Mutrakrcchra	50-100 ml in the from of decoction

Sl. No.	*Part and Volume*	*Page No.*	*Common Ayurvedic Name*	*Botanical Name/ Chemical Name*	*Family/ Chemical Formula*	*Part Used*	*Identity, Purity and Strength/Chemical Properties*	*Constituents*	*Properties and Action*	*Important Formulations*	*Therapeutic Uses*	*Dose*
388	Part-I; Vol. V	135	PILUH (Fruit)	*Salvadora persica* Linn. var. wightiana (Planch. ex Thw.) Verdc, syn. *S. persica* Linn.	Salvadoraceae	Fruit	**Foreign matter**- NMT 2 per cent; **Total ash**- NMT 15 per cent; **Acid-insoluble ash**- NMT 4 per cent; **Alcohol-soluble extractive**-NLT 12 per cent; **Water-soluble extractive**- NLT 40 per cent; **T.L.C.** - Refer API	β-sitosterol, sterol glycoside, Benzyle isothioagnate, Traces of alkaloid, Fixed oil, Sugar and Fat, Non-Saponifiable portion of oil consists of Dibenzylurea and Dibenz-lethiourea	*Rasa* - Madhura, Tikta, Katu; *Guna*- Laghu, Snigdha, Tiksna; *Virya*- Usna; *Vipaka*- Katu; *Karma*- Vatahara, Kaphahara, Bhedana, Virecana, Sothahara, Vednasthapana, Sirovirccaka, Dipana, Vidahi, Rasayana	Misrakasneha	Gulma, Asmari, Mutrakrcchra, Jvara, Sarpavisa, Arsa, Bastivikara, Udararoga, Visavikara, Anaha	3-6 g
389	Part-I; Vol. V	137	PILUH (Leaf)	*Salvadora persica* Linn. Var. *wightiana* (Planch.ex Thw.) Verdc, syn. S. *persica Linn.*	Salvadoraceae	Leaf	**Foreign matter**- NMT 2 per cent; **Total ash**- NMT 27 per cent; **Acid-insoluble ash**- NMT 1 per cent; **Alcohol-soluble extractive**-NLT 5 per cent; **Water-soluble extractive**- NLT 40 per cent; **T.L.C.** - Refer API	β-Sitosterol, Gluco-tropaeolin, Terpenes and Flavonoids.	*Rasa* - Katu, Tikta *Guna*- Laghu, Snigdha, Tiksna, Sara *Virya*- Usna *Vipaka*- Katu; *Karma*- Vatahara, Kaphahara, Bhedana, Virecana, Sothahara, Vedanasthapana, Sirovirecaka, Dipana, Vidahi, Rasayana	Pilu Taila	Gulma, Asmari, Mutrakrcchra, Jvara, Sarpavisa, Arsa, Bastivikara, Anaha, Udararoga, Udavarta, Vatarakta, Yonivyapat, Krmi, Nadivrana, Dustavrana, Vrana, Vransotha, Mukhapaka, Madyaja Trsna, Pliharoga, Sarva Kustha, Bhagandara, Apaci	3-6 g

Sl. No.	*Part and Volume*	*Page No.*	*Common Ayurvedic Name*	*Botanical Name/ Chemical Name*	*Family/ Chemical Formula*	*Part Used*	*Identity, Purity and Strength/Chemical Properties*	*Constituents*	*Properties and Action*	*Important Formulations*	*Therapeutic Uses*	*Dose*
390	Part-I; Vol. V	140	PILUH (Root Bark)	*Salvadora persica* Linn. var. *wightiana* (Planch.ex Thw.) Verdc, syn. *S. persica* Linn.	Salvadoraceae	Root Bark	**Foreign matter**- NMT 2 per cent; **Total ash**- NMT 15 per cent; **Acid-insoluble ash**- NMT 6 per cent; **Alcohol-soluble extractive**-NLT 2 per cent; **Water-soluble extractive**- NLT 25 per cent; **T.L.C.** - Refer API	β-sitosterol and Elementral γ-monoclinic sulphur (S-8) and Glucotropaeolin isolated from root.	*Rasa* - Katu, Tikta, Madhura; *Guna*- Laghu, Snigdha, Tiksna, Sara; *Virya*- Usna; *Vipaka*- Katu; *Karma*- Vatahara, Kaphahara, Bhedana, Virecana, Sothahara, Vedanasthapana, Sirovirecaka, Dipana, Vidahi, Rasayana	Arsakuthara Rasa, Vaidurya Rasayana, Chitrakhadiya Taila, Triphaladi Gutika, Naracaka Churna, Vilvakhadhi Lepa, Pippalyadi Gutika	Gulma, Asmari, Mutrakrcchra, Jvara, Sarpavisa, Arsa, Bastivikara, Anaha, Udararoga, Udavarta, Vatarakta, Yonivyapat, Krmi, Nadivrana, Dustavrana, Vrana, Vranasotha, Mukhapaka, Madyaja Trsna, Pliharoga, Sarva Kustha, Bhagandara, Apaci	10-20 g for decoction
391	Part-I; Vol. V	142	POTAGALA (Root)	*Typha elephantina* Roxb.	Typhaceae	Dried Root	**Foreign matter**- NMT 2 per cent; **Total ash**- NMT 5 per cent; **Acid-insoluble ash**- NMT 2 per cent; **Alcohol-soluble extractive**-NLT 7 per cent; **Water-soluble extractive**- NLT 20 per cent; **T.L.C.** - Refer API	β-sitosterol, Cholestrol, Quercetin and Ianosterol	*Rasa* - Madhura, Kasaya, Tikta *Guna*- Laghu, Snigdha, *Virya*- Sita *Vipaka*- Madhura; *Karma*- Pittahara, Kaphahara, Vrsya, Caksusya, Mutrala, Grahi, Vranaropana	Sukumara Ghrta	Daha, Raktavikara, Vatarakta, Visarpa, Raktapitta, Bastisotha, Mutrakrcchra, Asmari, Sopha, Sukradaurbalya, Vrana	10-20 g for decoction

Sl. No.	Part and Volume	Page No.	Common Ayurvedic Name	Botanical Name/ Chemical Name	Family/ Chemical Formula	Part Used	Identity, Purity and Strength/Chemical Properties	Constituents	Properties and Action	Important Formulations	Therapeutic Uses	Dose
392	Part-I; Vol. V	144	PUDINAH (Aerial Part)	*Mentha viridis* Linn. *Syn. M. spicata* var. *viridis* Linn.	Lamiaceae	Aerial Part	**Foreign matter**- NMT 2 per cent; **Total ash**- NMT 14 per cent; **Acid-insoluble ash**- NMT 4 per cent; **Alcohol-soluble extractive**-NLT 2 per cent; **Water-soluble extractive**-NLT 7 per cent; **Essential oil** - NLT 0.2 per cent; **T.L.C.** - Refer API	Essential oil (0.2 to 0.8 percent) containing Terpene such as Carvone (60 per cent) and Limonene (10 per cent) as major constituents.	*Rasa* - Katu; *Guna*- Laghu, Ruksa, Tiksna; *Virya*- Usna; *Vipaka*- Katu; *Karma*- Vatahara, Kaphahara, Dipana, Mutrala, Rocana, Balya	Pudinarka	Adhmana, Sula, Chardi, Krmi, Jvara, Jirna Jvara, Mutrakrcchra, Kastartava, Prasutijvara, Aruci, Kasa, Hikka, Svasa, Mada, Agnimandya, Visucika, Atisara, Grahani, Ajirna, Vaktrajadya	5-10 ml patra svarasa; 20-40 ml phanta; 1-3 drops taila
393	Part-I; Vol. V	146	PULLANI (Leaf)	*Calycopteris floribunda* Lam.	Combretaceae	Leaf	**Foreign matter**- NMT 2 per cent; **Total ash**- NMT 6 per cent; **Acid-insoluble ash**- NMT 1 per cent; **Alcohol-soluble extractive**-NLT 7 per cent; **Water-soluble extractive**-NLT 8 per cent; **T.L.C.** - Refer API	Octacesanol, Sitosterol, Calycopterin, 3'O-methyl-calycopterin, 4-0 methyl-calycopterin, ellagic acid Quercetin and Proantho-cyanidin.	*Rasa* - Tikta *Guna*- Laghu, Ruksa *Virya*- Usna *Vipaka*- Katu; *Karma*- Pittahara, Kaphahara, Bhedini, Vibandhahara	Marma Gutika	Krmi, Pandu, Kustha, Jvara	3-6 g
394	Part-I; Vol. V	148	PULLANI (Root)	*Calycopteris floribunda* Lam.	Combretaceae	Root	**Foreign matter**- NMT 2 per cent; **Total ash**- NMT 2.5 per cent; **Acid-insoluble ash**- NMT 0.5 per cent; **Alcohol-soluble extractive**-NLT 4 per cent; **Water-soluble extractive**-NLT 3 per cent; **T.L.C.** - Refer API	Octacesanol, Sitosterol, Calycopterin, 3'O-methyl-calycopterin, 4-O methyl-calycopterin, Ellagic acid, Gossoypol and Quercetin	*Rasa* - Tikta; *Guna*- Laghu, Ruksa; *Virya*- Usna; *Vipaka*- Katu; *Karma*- Pittahara, Kaphahara, Bhedini, Vibandhahara	Marma Gutika	Krmi, Pandu, Kustha, Jvara	3-6 g

Sl. No.	*Part and Volume*	*Page No.*	*Common Ayurvedic Name*	*Botanical Name/ Chemical Name*	*Family/ Chemical Formula*	*Part Used*	*Identity, Purity and Strength/Chemical Properties*	*Constituents*	*Properties and Action*	*Important Formulations*	*Therapeutic Uses*	*Dose*
395	Part-I; Vol. V	150	PULLANI (Stem)	*Calycopteris floribunda* Lam.	Combretaceae	Stem	**Foreign matter**- NMT 2 per cent; **Total ash**- NMT 5 per cent; **Acid-insoluble ash**- NMT 1 per cent; **Alcohol-soluble extractive**-NLT 2 per cent; **Water-soluble extractive**-NLT 2.5 per cent; **T.L.C.** - Refer API	Octacesanol, Sitosterol, Calycopterin, 3'O-methyl-calycopterin, 4-O methyl-lcalycopterin, Ellagic acid	*Rasa* - Tikta; *Guna*- Laghu, Ruksa; *Virya*- Usna; *Vipaka*- Katu; *Karma*- Pittahara, Kaphahara, Bhedini, Vibandhahara	Marma Gutika	Krmi, Pandu, Kustha, Jvara	3-6 g
396	Part-I; Vol. V	152	PUTI-KARANJA (Stem Bark)	*Caesalpinia crista* Linn.	Caesalpiniaceae	Dried Stem Bark	**Foreign matter**- NMT 2 per cent; **Total ash**- NMT 6 per cent; **Acid-insoluble ash**- NMT 1 per cent; **Alcohol-soluble extractive**-NLT 7 per cent; **Water-soluble extractive**-NLT 10 per cent; **T.L.C.** - Refer API	Flavonoid, Saponins and Alkaloids	*Rasa* - Tikta, Kasaya, Katu *Guna*- Laghu, Ruksa, *Virya*- Usna *Vipaka*- Katu; *Karma*- Slesmasamsamana, Sothahara, Dipana, Anulomana, Lekhaniya, Bhedaniya, Krmighna, Visaghna, Aparapatana	Indukanta Ghrta, Visnu Taila, Pramehamihira Taila	Kustha, Prameha, Arsa, Kandu, Pakva-Sopha, Vrana, Tvakroga, Slipada, Vataja Sula, Udara, Gulma, Sula, Masurika, Amlapitta, Svitra, Sarira-durgandha	50-100 ml in the form of decoction
397	Part-I; Vol. V	154	RENUKA (Fruit)	*Vitex negundo* Linn.	Verbenaceae	Dried Fruit	**Foreign matter**- NMT 1 per cent; **Total ash**- NMT 5 per cent; **Acid-insoluble ash**- NMT 1 per cent; **Alcohol-soluble extractive**-NLT 3 per cent; **Water-soluble extractive**-NLT 2 per cent; **T.L.C.** - Refer API	Seed contain Hydrocarbons such as *n*-tritriacontane, n-hentri-acontane, *n*-pentatri-acontane and Nonacosane. Other constituents of the seeds include β-sitosterol, *p*-hydroxy-benzoic acid and 5 oxyiso-phthalic acid.	*Rasa* - Tikta, Katu; *Guna*- Laghu; *Virya*- Sita; *Vipaka*- Katu; *Karma*- Pittakara, Vatahara, Kaphahara, Dipani, Medhya, Pacani, Garbhapatini, Mukhavai-malyakara, Visaghna	Candanadi Taila, Pramehamihira Taila, Dasamularista, Sarsvatarista, Mahayogaraja Guggulu, Anutaila, Balasvagandha laksadi Taila, Vasacandanadi Taila	Trsna, Kandu, Daha, Kasa, Netraroga, Daurbalya, Dadru, Klaibya, Gulma	1-3 g

Sl. No.	*Part and Volume*	*Page No.*	*Common Ayurvedic Name*	*Botanical Name/ Chemical Name*	*Family/ Chemical Formula*	*Part Used*	*Identity, Purity and Strength/Chemical Properties*	*Constituents*	*Properties and Action*	*Important Formulations*	*Therapeutic Uses*	*Dose*
398	Part-I; Vol. V	157	RIDDHI (Tuber)	*Habenaria intermedia* D.Don	Orchidaceae	Dried Tuber	**Foreign matter**- NMT 2 per cent; **Total ash**- NMT 5 per cent; **Acid-insoluble ash**- NMT 1 per cent; **Alcohol-soluble extractive**-NLT 14 per cent; **Water-soluble extractive**- NLT 22 per cent; **T.L.C.** - Refer API	–	*Rasa* - Madhura; *Guna*- Guru, Snigdha, Picchila; *Virya*- Sita; *Vipaka*- Madhura; *Karma*- Vatahara, Pittahara, Rasayana, Sukrajanana, Vrsya, Ojovardhaka, Tridosasamaka	Amrtaprasa Ghrta, Asoka Ghrta, Chagaladya Ghrta, Dasamularista	Ksaya, Raktavikara, Jvara, Murccha	3-6 g
399	Part-I; Vol. V	159	ROHISA (Whole Plant)	*Cymbopogon martinii* (Roxb.) Wats.	Poaceae	Dried Leaf, Stem and Root	**Foreign matter**- NMT 2 per cent; **Total ash**- NMT 14 per cent; **Acid-insoluble ash**- NMT 7 per cent; **Alcohol-soluble extractive**-NLT 5 per cent; **Water-soluble extractive**-NLT 7 per cent; **Essential oil** - NLT 0.2 per cent; **T.L.C.** - Refer API	Essential oil (0.5 percent) containing terpenes such as Geraniol, Geranyl acetate, Citronellol, Linalool, Geranyl butyrate, Myrcene, α-and β-pinene	*Rasa* - Katu, Tikta; *Guna*- Laghu, Ruksa, Tiksna; *Virya*- Usna; *Vipaka*- Katu; *Karma*- Pittahara, Kaphavatasamaka, Balagrahahara, Pumstvaghna	Bala Taila, Masabaladi Kvatha Churna	Kasa, Hrdroga, Sula, Raktapitta, Apasmara, Pinasa, Kaphajvara, Kantha roga, Jvara, Aruci, Kustha, Katisula, Prameha, Vrscika-Visa	10-20 g
400	Part-I; Vol. V	162	RUMI-MASTAGI (Resin)	*Pistacia lentiscus* Linn.	Anacardiaceae	Resin	**Foreign matter**- NMT 2 per cent; **Total ash**- NMT 2.6 per cent; **Acid-insoluble ash**- NMT 0.34 per cent; **Alcohol-soluble extractive**-NLT 94.0 per cent; **Water-soluble extractive**- NLT 0.5 per cent; **T.L.C.** - Refer API	Resin, Volatile oil, a Bicyclic terpenoid and Fatty acids.	*Rasa* - Madhura; *Guna*- Laghu, Ruksa; *Virya*- Usna; *Vipaka*- Madhura; *Karma*- Kaphahara, Mutrala, Vrsya, Vajikarana, Rakta Samgrahika, Dipana, Varnya, Mukhadurgan-dhanasaka, Dasans-thiratakara	Eladi, Kameda, Sukrama Vati	Mutrakrchra, Kasa, Svasa, Adhmana, Agnimandya, Grahani, Raktasrava, Vatapittaja Vikara, Sotha	1-2 g

Sl. No.	*Part and Volume*	*Page No.*	*Common Ayurvedic Name*	*Botanical Name/ Chemical Name*	*Family/ Chemical Formula*	*Part Used*	*Identity, Purity and Strength/Chemical Properties*	*Constituents*	*Properties and Action*	*Important Formulations*	*Therapeutic Uses*	*Dose*
401	Part-I; Vol. V	164	SARALA (Exudate)	*Pinus roxburghii* Sargent syn. *P. longifolia* Roxb.	Pinaceae	Exudate Obtained by Tapping the Wood	**Foreign matter**- NMT 1 per cent; **Total ash**- NMT 0.6 per cent; **Acid-insoluble ash**- NMT 0.40 per cent; **Alcohol-soluble extractive**-NLT 74 per cent; **Water-soluble extractive**- NLT 0.15 per cent; **Volatile oil** - NLT 18 per cent; **T.L.C.** - Refer API	1-α-pinene, 1-β-pinene, Care-3-ene, Longifolene and other Mono and Sesquiterpenes.	*Rasa* - Katu, Tikta, Kasaya; *Guna*- Laghu, Tiksna, Snigdha; *Virya*- Usna; *Vipaka*- Katu; *Karma*- Vatahara, Kaphahara, Dipana, Durgandhahara, Dustavrana-sodhaka, Visaghna, Varna-prasadana, Raksoghana	Amrtaprasa Churna, Kustadi Taila	Jatrurdha-varoga, Sved-daurgandhya, Vatavyadhi, Agnimandya, Adhmana, Krmiroga, Murccha, Kustha, Tvakroga, Karnasula, Kantharoga, Sotha, Nadivrana, Kandu, Kotha, Pidaka, Urustambha, Yukaroga, Grahabadha, Yonidosa	1-3 g
402	Part-I; Vol. V	166	SARPA-GANDHA (Root)	*Rauwolfia serpentina* (Linn.) Benth ex Kurz	Apocynaceae	Air Dried Root	**Foreign matter**- NMT 2 per cent; **Total ash**- NMT 8 per cent; **Acid-insoluble ash**- NMT 1 per cent; **Alcohol-soluble extractive**-NLT 4 per cent; **Water-soluble extractive**-NLT 10 per cent; **T.L.C.** - Refer API	Rauwolfia contains Indole alkaloids, such as Reserpinine, Serpentinine and Ajmalicine.	*Rasa* - Tikta, Katu; *Guna*-Ruksa, Laghu; *Virya*- Usna; *Vipaka*- Katu; *Karma*- Vatahara, Kaphahara, Mutral, Dipani, Rucya, Pacani, Nidraprada, Visaghna, Kamavasadaka, Hrdavasadaka	Sarpagandhadi Churna, Sarpagan-dhayoga, Sarpagandha Vati, Sarpagandha Ghana Vati	Madaroga, Yonisula, Jvara, Sula, Krmiroga, Anidra, Unmada, Apasmara, Bhrama, Raktavata, Bhutabadha, Manasaroga, Visucika, Vrana	1-2 g

Sl. No.	Part and Volume	Page No.	Common Ayurvedic Name	Botanical Name/ Chemical Name	Family/ Chemical Formula	Part Used	Identity, Purity and Strength/Chemical Properties	Constituents	Properties and Action	Important Formulations	Therapeutic Uses	Dose
403	Part-I; Vol. V	168	SVETA-PUNARNAVA (Root)	*Borhaaria verticillata* Poir	Nyctaginaceae	Root	**Foreign matter**- NMT 1 per cent; **Total ash**- NMT 16 per cent; **Acid-insoluble ash**- NMT 4 per cent; **Alcohol-soluble extractive**-NLT 7 per cent; **Water-soluble extractive**- NLT 2 per cent; **T.L.C.** - Refer API	–	*Rasa* - Tikta, Madhura; *Guna*- Ruksa, Laghu; *Virya*- Usna; *Vipaka*- Madhura; *Karma*- Vatahara, Kaphahara, Pittasamaka, Agnidipaka, Visaghna, Jvarahara	Kumaryasava (A), Punarnavadyarista, Dhanvantara Ghrta, Dadhika Ghrta	Pandu, Visavikara, Sotha, Sopha, Udararoga, Hrdroga, Kasa, Urahksata, Sula, Rakta Vikara, Paittika Jvara, Caturthikajvara, Pliharoga, Vatakantaka, Vidradhi, Alarkavisa, Vrscikavisa, Sarpavisa, Musakavisa	5-15 g
404	Part-I; Vol. V	170	TAILA-PARNAH (Leaf)	*Eucalyptus globulus* Labill.	Myrtaceae	Leaf	**Foreign matter**- NMT 1 per cent; **Total ash**- NMT 9 per cent; **Acid-insoluble ash**- NMT 1 per cent; **Alcohol-soluble extractive**-NLT14 per cent; **Water-soluble extractive**- NLT 21 per cent; **Essential oil** - NLT 2 per cent; **T.L.C.** - Refer API	Essential oil containing Terpenes such as 1,8-cineole, Camphene, Sabinene, Myrcene, P-menthone, α-and γ-terpinene, Fenchone, α-β-thujone, Citral, Verbenone	*Rasa* - Katu, Tikta, Kasaya; *Guna*- Laghu, Snigdha; *Virya*- Usna; *Vipaka*- Katu; *Karma*- Vatahara, Kaphahara, Dipana, Pacana, Hrdya, Mutrala, Durgandhinasaka, Agnimandya, Balaprada	Ekadasasatikaprasanini Failam, Mahasugandhika Taila, Pancavaktra Rasa, Pancaguna Taila, Martandabhairava Rasa, Jvaramari Rasa	Krmi, Jirnakasa, Pratisyaya, Svarabheda, Visamajvara, jvara, Sula, Puyameha, Ksaya, Svasa, Bastiroga, Pravahika, Pliharoga, Hrdroga, Agnimandya	1-2 g
405	Part-I; Vol. V	172	TINISAH (Wood)	*Ougeinia oojeinensis* (Roxb.) Hochr. Syn. *O. dalbergioides* Benth.	Fabaceae	Wood	**Foreign matter**- NMT 1 per cent; **Total ash**- NMT 7 per cent; **Acid-insoluble ash**- NMT 1.5 per cent; **Alcohol-soluble extractive**-NLT 5 per cent; **Water-soluble extractive**- NLT 2 per cent; **T.L.C.** - Refer API	Flavonoids mainly Homoferreirin and Ougeinin.	*Rasa* - Kasaya *Guna*- Laghu, Ruksa, *Virya*- Sita *Vipaka* -Katu; *Karma*- Rasayana, Pittahara, Kaphasosana, Medohara, Kusthaghna, Visaghna, Vranaropana, Sonitasthapana	Ayaskrti	Sotha, Kustha, Atisara, Raktatisara, Pravahika, Raktavikara, Raktapitta, Prameha, Svitra, Vrana, Krmi, Panduroga, Medoroga, Daha	50-100 ml kvatha

Sl. No.	*Part and Volume*	*Page No.*	*Common Ayurvedic Name*	*Botanical Name/ Chemical Name*	*Family/ Chemical Formula*	*Part Used*	*Identity, Purity and Strength/Chemical Properties*	*Constituents*	*Properties and Action*	*Important Formulations*	*Therapeutic Uses*	*Dose*
406	Part-I; Vol. V	174	TINTIDIKAH (Aerial Part)	*Rhus parviflora* Roxb.	Anacardiaceae	Mature dried Aerial Part	**Foreign matter**- NMT 2 per cent; **Total ash**- NMT 5 per cent; **Acid-insoluble ash**- NMT 0.7 per cent; **Alcohol-soluble extractive**-NLT 10 per cent; **Water-soluble extractive**- NLT 12 per cent; **T.L.C.** - Refer API	Tannins (Gallic acid); Flavones (Myricetin, Quercetin, Myricitrin, Quercitrin, Kampferol), Glycosides (Isorhmnetin-3-α-L-arabinoside).	*Rasa* - Amla; *Guna*- Laghu, Ruksa; *Virya*- Usna; *Vipaka*- Amla; *Karma*- Vatahara, Kaphavatahara, Pittakara, Rocana, Dipana, Grahi, Jvaraghna	Yavani sadava, Hinguvacadi Churna, Sri Ramabana Rasa	Vatavikara, Atisara, Agnimandya, Aruci, Trsna, Pravahika	3-6 g
407	Part-I; Vol. V	177	TRAPUSAM (Seed)	*Cucumis sativus* Linn.	Cucurbitaceae	Dried Seed	**Foreign matter**- NMT 2 per cent; **Total ash**- NMT 6 per cent; **Acid-insoluble ash**- NMT 1 per cent; **Alcohol-soluble extractive**-NLT 5 per cent; **Water-soluble extractive**- NLT 7 per cent; **T.L.C.** - Refer API	Fixed Oil and Sugars	*Rasa* - Tikta, Madhura; *Guna*- Singdha, Guru; *Virya*- Sita; *Vipaka*- Mahura; *Karma*- Vatapittahara, Kaphakara, Mutrala, Balya, Abhisyandi, Mutrabasti-visodhaka, Agnisadana	Dadhika Ghrta	Mutraghata, Mutrakrcchra, Raktapitta, Daurbalya, Daha, Raktavikara, Anidra, Sirahsula, Chardi, Sitajvara	3-6 g powder
408	Part-I; Vol. V	179	TUNI (Stem Bark)	*Cedrela toona* Roxb.	Meliaceae	Stem Bark	**Foreign matter**- NMT 2 per cent; **Total ash**- NMT 14 per cent; **Acid-insoluble ash**- NMT 1 per cent; **Alcohol-soluble extractive**-NLT 12 per cent; **Water-soluble extractive**- NLT 9 per cent; **T.L.C.** - Refer API	Triterpenoids	*Rasa* - Tikta, Kasaya, Madhura; *Guna*- Laghu; *Virya*- Sita; *Vipaka*- Katu; *Karma*- Pittahara, Kaphahara, Grahi, Bhagna-sandhanaka, Medohara	Nyagrodhadi Kvatha Churna	Bala Pravahika, Vrana, Daha, Yoniroga, Kandu, Kustha, Gandamala, Raktavikara, Raktapitta, Svetakustha, Prameha, Visavikara, Medovikara	3-6 g kvatha - 10-20 ml

Sl. No.	*Part and Volume*	*Page No.*	*Common Ayurvedic Name*	*Botanical Name/ Chemical Name*	*Family/ Chemical Formula*	*Part Used*	*Identity, Purity and Strength/Chemical Properties*	*Constituents*	*Properties and Action*	*Important Formulations*	*Therapeutic Uses*	*Dose*
409	Part-I; Vol. V	181	VANDA (Leaf)	*Dendrophthoe falcata* (Linn. f.) Ettingsh. syn. *Loranthus falcatus* Linn. f.	Loranthaceae	Dreid Leaf	**Foreign matter**- NMT 1 per cent; **Total ash**- NMT 14 per cent; **Acid-insoluble ash**- NMT 4 per cent; **Alcohol-soluble extractive**-NLT 3 per cent; **Water-soluble extractive**-NLT 3 per cent; **T.L.C.** - Refer API	Leaves contain Flavonoids such as Quercetin, Quercetrin; Tannins comprising of Gallic and Chebulinic acid	*Rasa* - Kasaya, Tikta, Madhura; *Guna*- Laghu, Ruksa; *Virya*- Sita; *Vipaka*- Katu; *Karma*- Pittahara, Kaphahara, Vatahara, Mutravirecaniya, Sukrajanana, Vrsya, Rasayana, Grahi, Vranaropana, Raksoghna, Sramahara, Netrya, Grahanasana, Mangalakara, Garbhasthapana,	No formulation	Raktapitta, Vrana, Visaroga, Vandhyatva, Hikka, Visamajvara, Bhagandara, Vata-smari, Mutraroga	10-20 ml juice
410	Part-I; Vol. V	183	VANDA (Stem)	*Dendrophthoe falcata* (Linn. f.) Ettingsh. syn. *Loranthus falcatus* Linn. f.	Loranthaceae	Dried Stem	**Foreign matter**- NMT 1 per cent; **Total ash**- NMT 5 per cent; **Acid-insoluble ash**- NMT 1 per cent; **Alcohol-soluble extractive**-NLT 3 per cent; **Water-soluble extractive**-NLT 3 per cent; **T.L.C.** - Refer API	Young shoots contain nearly 10 per cent tannins and the stem contains β-amyrin-O-acetate, Oelonolic acid its Methyl ester acetate, β-sitosterol and Stigmasterol	*Rasa* - Kasaya, Tikta, Madhura; *Guna*- Laghu, Ruksa; *Virya*- Sita; *Vipaka*- Katu; *Karma*- Pittahara, Kaphahara, Vatahara, Mutravirecaniya, Sukrajanana, Vrsya, Rasayana, Grahi, Vranaropana, Raksoghna, Sramahara, Netrya, Grahanasana, Mangalakara, Garbhasthapana	No formulation	Raktapitta, Vrana, Visaroga, Vandhyatva, Hikka, Visamajvara, Bhagandara, Vata-smari, Mutraroga	10-20 ml juice

Sl. No.	*Part and Volume*	*Page No.*	*Common Ayurvedic Name*	*Botanical Name/ Chemical Name*	*Family/ Chemical Formula*	*Part Used*	*Identity, Purity and Strength/Chemical Properties*	*Constituents*	*Properties and Action*	*Important Formulations*	*Therapeutic Uses*	*Dose*
411	Part-I; Vol. V	185	VANDA (Aerial Root)	*Dendrophthoe falcata* (Linn.f.) Ettingsh. syn. *Loranthus falcatus* Linn. f.	Loranthaceae	Dried Aerial Root	**Foreign matter**- NMT 1 per cent; **Total ash**- NMT 6 per cent; **Acid-insoluble ash**- NMT 1 per cent; **Alcohol-soluble extractive**-NLT 12 per cent; **Water-soluble extractive**- NLT 1 per cent; **T.L.C.** - Refer API	Catechin and Leucocynidin in the bark	*Rasa* - Kasaya, Tikta, Madhura *Guna*- Laghu, Ruksa *Virya*- Sita *Vipaka*- Katu; *Karma*- Pittahara, Kaphahara, Vatahara, Mutravirecaniya, Sukrajanana, Vrsya, Rasayana, Grahi, Vranaropana, Sramahara, Netrya, Grahanasana, Mangalakara, Garbhasthapana	Mutravirecaniya Kasaya Churna	Raktapitta, Vrana, Visaroga, Vandhyatva, Hikka, Visamajvara, Bhagandara, Vata-smari, Mutraroga	10-20 ml juice
412	Part-I; Vol. V	187	VANDA (Flower)	*Dendrophthoe falcata* (Linn.f.) Ettingsh. syn. *Loranthus falcatus* Linn. f.	Loranthaceae	Flower	**Foreign matter**- NMT 1 per cent; **Total ash**- NMT 8 per cent; **Acid-insoluble ash**- NMT 1 per cent; **Alcohol-soluble extractive**-NLT 20 per cent; **Water-soluble extractive**- NLT 4 per cent; **T.L.C.** - Refer API	–	*Rasa* - Kasaya, Tikta, Madhura *Guna*- Laghu, Ruksa *Virya*- Sita *Vipaka*- Katu; *Karma*- Pittahara, Kaphahara, Vatahara, Mutravirecaniya, Sukrajanana, Vrsya, Rasayana, Grahi, Vranaropana, Raksoghna, Sramahara, Netrya, Grahanasana, Garbhasthapana	No formulation	Raktapitta, Vrana, Visaroga, Vandhyatva, Hikka, Visamajvara, Bhagandara, Vata-Smari, Mutraroga	10-20 ml juice

Sl. No.	Part and Volume	Page No.	Common Ayurvedic Name	Botanical Name/ Chemical Name	Family/ Chemical Formula	Part Used	Identity, Purity and Strength/Chemical Properties	Constituents	Properties and Action	Important Formulations	Therapeutic Uses	Dose
413	Part-I; Vol. V	189	VANDA (Fruit)	*Dendrophthoe falcata* (Linn.f.) Ettingsh. syn. *Loranthus falcatus* Linn. f.	Loranthaceae	Fruit	**Foreign matter**- NMT 1 per cent; **Total ash**- NMT 8 per cent; **Acid-insoluble ash**- NMT 1 per cent; **Alcohol-soluble extractive**-NLT 17 per cent; **Water-soluble extractive**- NLT 5 per cent; **T.L.C.** - Refer API	–	*Rasa* - Kasaya, Tikta, Madhura *Guna*- Laghu, Ruksa *Virya*- Sita *Vipaka*- Katu; *Karma*- Pittahara, Kaphahara, Vatahara, Visaghna, Vrsya, Rasayana, Grahi, Vranaropana, Raksoghna, Sramahara, Grahanasana,	No formulation	Raktapitta, Vrana, Arsa, Vatavikara, Asmari, Mutrasarkara, Mutrakrcchra, Mutraghata, Mutraruja, Garbhasrava, Kantharoga, Vatarakta, Sopharoga, Amatisara, Netraroga, Visamjvara, Slipada	10-20 ml
414	Part-I; Vol. V	191	VANYAJIRAKA (Fruit)	*Centratherum anthel-minticum* (L.) Kuntze	Asteraceae	Dried Fruit	**Foreign matter**- NMT 2.0 per cent; **Total ash**- NMT 7.5 per cent; **Acid-insoluble ash**- NMT 4.5 per cent; **Alcohol-soluble extractive**-NLT 20 per cent; **Water-soluble extractive**- NLT 14 per cent; **T.L.C.** - Refer API	Sterols, Avenasterol and Vernosterol, a Bitter principle, Essential oil, Resins and Fixed oil consisting of Myristic, Palmitic, Stearic, Oleic, Linoleic and Vernolic acids.	*Rasa* - Tikta, Katu, Kasaya *Guna*- Laghu, Tiksana *Virya*- Usna *Vipaka*- Katu; *Karma*- Vatahara, Kaphahara, Jantunasaka, Mutrala, Dipana, Stambhana, Netrya	Madhusnuhi Rasayana,	Svasa, Kasa, Hikka, Jvara, Kustha, Vrana, Kandu, Svitrakustha, Krmi, Sopha, Sula, Gulma, Mutraghata, Raktavikara	1-3 g
415	Part-I; Vol. V	193	VIDARI-KANDA (Tuber)	*Pueraria tuberosa* DC.	Fabaceae	Dried Tuber	**Foreign matter**- NMT 2 per cent; **Moisture content** - NMT 10 per cent; **Total ash**- NMT 11 per cent; **Acid-insoluble ash**- NMT 1 per cent; **Alcohol-soluble extractive**-NLT 13 per cent; **Water-soluble extractive**- NLT 22 per cent; **Starch** - NLT 14 per cent; **T.L.C.** - Refer API	Pterocarpan-tuberosin, Ptero-carpanone-hydroxy-tuberosone, two pterocarpenes-anhydro-tuberosin and 3-O-methylan-hydro-.	*Rasa* - Madhura; *Guna*- Guru, Snigdha; *Virya*- Sita; *Vipaka*- Madhura; *Karma*- Vatahara, Pittahara, Hrdya, Brhana, Vrsya, Mutral, Balya, Stanyadu, Svarya, Vajikarana,	Marmagutika, Nityananda Rasa, Sarasvatarista, Satavaryadi Ghrta, Asvagandha-dyarista, Mahavisagarbha Taila	Raktapitta, Sukraksaya, Raktadosa, Daha, Ksaya, Kasa, Sula, Mutrakrcchra, Visarpa, Visamajvara	3-6 g

Sl. No.	*Part and Volume*	*Page No.*	*Common Ayurvedic Name*	*Botanical Name/ Chemical Name*	*Family/ Chemical Formula*	*Part Used*	*Identity, Purity and Strength/Chemical Properties*	*Constituents*	*Properties and Action*	*Important Formulations*	*Therapeutic Uses*	*Dose*
								tuberosin, and a coumestan tuberostan. An isoflavone-puerarone and a coumestan-puerarostan	Varnya, Jivaniya, Rasayani			
416	Part-I; Vol. V	195	VIRALA (Stem Bark)	*Diospyros exsculpta* Buch.-Ham. *syn. D. tomentosa* Roxb.	Ebenaceae	Dried Stem Bark	**Foreign matter**- NMT 2 per cent; **Total ash**- NMT 15 per cent; **Acid-insoluble ash**- NMT 5 per cent; **Alcohol-soluble extractive**-NLT 1.5 per cent; **Water-soluble extractive**- NLT 2 per cent; **T.L.C.** - Refer API	Triterpenoids (Lupeol, Betulin, Betulinic acid, Oleanolic acid) and Sterol.	*Rasa* - Madhura, Kasaya, Tikta; *Guna*- Guru, Snigdha; *Virya*- Usna; *Vipaka*- Madhura; *Karma*-Pittahara, Kaphahara, Grahi, Jihvajadyakara, Vranaropana, Savarnakara	Nayagrodhadi Kvatha Churna	Udarda, Prameha, Raktapitta, Aruci, Atisara, Vibandha, Pittaroga, Karanasrava, Vrana, Agnidagdha Vrana, Atidagdha Vrana, Bhagna, Trsa, Daha, Yoniroga, Medoroga	5-10 g
417	Part-I; Vol. V	197	VISALA (Root)	*Trichosanthes bracteata* (Lam.) Voigt	Cucurbitaceae	Root	**Foreign matter**- NMT 2 per cent; **Total ash**- NMT 14 per cent; **Acid-insoluble ash**- NMT 3 per cent; **Alcohol-soluble extractive**-NLT 1 per cent; **Water-soluble extractive**- NLT 4 per cent; **T.L.C.** - Refer API	Saponins, Trichosanthin	*Rasa* - Katu, Tikta *Guna*- Laghu, Ruksa *Virya*- Usna *Vipaka*- Katu; *Karma*- Pittahara, Kaphahara, Prasutikrta, Vamaka, Visaghna	Paniya Kalyanaka Ghrta, Visaladi Churna	Jvara, Amadosa, Prameha, Antarvrddhi, Kustha, Stanapida, Kamala, Slipada, Vrddhi, Plihodara, Svasa, Kasa, Gulma, Gandamaya, Granthi, Vrana, Mudhagarbha	1-3 g

Sl. No.	Part and Volume	Page No.	Common Ayurvedic Name	Botanical Name/ Chemical Name	Family/ Chemical Formula	Part Used	Identity, Purity and Strength/Chemical Properties	Constituents	Properties and Action	Important Formulations	Therapeutic Uses	Dose
418	Part-I; Vol. V	199	VYAGHRA-NAKHA (Fruit)	*Capparis sepiaria* Linn. Syn. *C. zeylanica* Linn. f.	Capparidaceae	Fruit	**Foreign matter**- NMT 2 per cent; **Total ash**- NMT 8 per cent; **Acid-insoluble ash**- NMT 1 per cent; **Alcohol-soluble extractive**-NLT 30 per cent; **Water-soluble extractive**- NLT 26 per cent; **T.L.C.** - Refer API	Thioglucoside glucocapparin, n-triacontane, α-amyrin and Fixed oil.	*Rasa* - Katu, Tikta, Kasaya, Madhura *Guna*- Ruksa Laghu *Virya*- Usna *Vipaka*- Katu; *Karma*- Vatahara, Kaphahara, Varnya, Visaghna, Kandughna	Bala Taila	Visavikara, Sarpavisa, Kandu, Pidaka, Kotha, Bhrama, Pravahika, Raktapradara, Kustha, Vrana, Jvara, Graharoga, Vatavikara, Mukha-durgandha	2-6 g
419	Part-I; Vol. VI	1	ADARI (Leaf)	*Acacia pennata* (L.) Willd. Syn. *Mimosa pennata* L. (Dam. Mimosaceae)	Mimosaceae	Dried Tender Leaves	**Foreign matter**- NMT 2 per cent; **Total ash**- NMT 7 per cent; **Sulphated ash** - 11 per cent; **Acid-insoluble ash**- NMT 1 per cent; **Alcohol-soluble extractive**-NLT 8 per cent; **Water-soluble extractive**-NLT 18 per cent; **T.L.C.** - Refer API	Octa-decadienoic, Octa-decanoic, Palmitic and Penta-decanoic acids,; Lupeol, α-spinasterol, β-sitosterol and Tannins	*Rasa* - Kasaya, Katu, Tikta *Guna*- Laghu, Ruksa *Virya*- Sita *Vipaka*- Katu; *Karma*- Kasahara, Pittasamaka	Used as single drug	Jvara (Fever). Raktadosa (Disorder of blood), Agnimandya (Digestive impairment)	Churna (Powder): 3-6 g
420	Part-I; Vol. VI	3	AMRA-GANDHI-GUGGULU (Leaf)	*Balsamodendron caudata* Mauch. Syn. *Commiphora caudata* Engl. *Protium caudatum* W. and A.	Burseraceae	Leaves	**Foreign matter**- NMT 1 per cent; **Total ash**- NMT 9 per cent; **Acid-insoluble ash**- NMT 3 per cent; **Alcohol-soluble extractive**-NLT 6 per cent; **Water-soluble extractive**-NLT 13 per cent; **Fixed oil** - NLT 2 per cent; **T.L.C.** - Refer API	Guggulsterones	*Rasa* - Tikta, Katu *Guna*- Laghu, Snigdha, Visada, Suksma, Sara, Sugandhi *Virya*- Usna *Vipaka*- Katu; *Karma*- Hrdya, Pratidusaka, Kapha-vatahara, Vranaropana, Vranasodhana	Used as single drug	Amavata (Rheumatism), Angamarda (Body ache), Gandamala (Cervical lymphadenitis), Kustha (Leprosy/ diseases of skin). Padadari (Chaffed/ Cracked soles/ Rhagades), Prameha (Metabolic disorder), Sandhisotha (Arthritis),	Svarasa (Juice): 5-10 ml

Sl. No.	*Part and Volume*	*Page No.*	*Common Ayurvedic Name*	*Botanical Name/ Chemical Name*	*Family/ Chemical Formula*	*Part Used*	*Identity, Purity and Strength/Chemical Properties*	*Constituents*	*Properties and Action*	*Important Formulations*	*Therapeutic Uses*	*Dose*
											Sotha (Inflammation), Vatarakta (Gout), Vataroga (Disease due to Vata dosa), Visarpa (Erysepales), Vrana (Ulcer)	
421	Part-I; Vol. VI	5	ARANYA-SURANA (Tuber)	*Synantherias sylvatica* Schott Gen. Aocja Syn. *Amorphophallus sylvaticus* (Roxb.) Kunth.	Araceae	Dried Tuber	**Foreign matter**- NMT 1 per cent; **Total ash**-NMT 8 per cent; **Acid-insoluble ash**- NMT 2 per cent; **Alcohol-soluble extractive**-NLT 4 per cent; **Water-soluble extractive**-NLT 14 per cent; **T.L.C.** - Refer API	–	*Rasa* - Katu, Kasaya *Guna*- Ruksa, Tiksna *Virya*- Usna *Vipaka*- Katu; *Karma*- Krmighna, Arsoghna, Rucya, Vedanahara	Used as single drug	Granthisotha (Lymphadenitis), Arbuda (Tumor), Vicarcika (Eczema), Udararoga (Diseases of abdomen), Slipada (Filariasis), Arsa (Piles)	Churna (Powder): 5-10 g after sodhana
422	Part-I; Vol. VI	7	ARAROTA (Rhizome)	*Maranta arundinacea* L.	Marantaceae	Dried Rhizome	**Foreign matter**- NMT 2 per cent; **Total ash**- NMT 5 per cent; **Sulphated ash** -NMT 7 per cent; **Acid-insoluble ash**- NMT 1 per cent; **Alcohol-soluble extractive**-NLT 1 per cent; **Water-soluble extractive**-NLT 12 per cent; **T.L.C.** - **Refer API**	Starch (25-30 per cent), Dextrin and Sugars.	*Rasa* - Madhura *Guna*- Guru, Snigdha *Virya*- Sita *Vipaka*- Madhura; *Karma*- Pittahara, Balya, Vrsya	Used as single drug	Kasa (Cough), Svasa (Asthma) Daha (Burning sensation), Trsna (thirst), Ksaya (Pthisis), Agnimandya (Digestive impairment), Raktadosa (Disorders of blood)	Churna (Powder): 5-10 g

Sl. No.	*Part and Volume*	*Page No.*	*Common Ayurvedic Name*	*Botanical Name/ Chemical Name*	*Family/ Chemical Formula*	*Part Used*	*Identity, Purity and Strength/Chemical Properties*	*Constituents*	*Properties and Action*	*Important Formulations*	*Therapeutic Uses*	*Dose*
423	Part-I; Vol. VI	9	ASTHISRN-KHALA (Aerial Part)	*Cissus quadr-angularis* L.	Vitaceae	Dried Aerial Part	**Foreign matter**- NMT 1 per cent; **Total ash**- NMT 20 per cent; **Acid-insoluble ash**- NMT 3 per cent; **Alcohol-soluble extractive**-NLT 7 per cent; **Water-soluble extractive**-NLT 2 per cent; **T.L.C.** - Refer API	Triterpenoids: 7-oxo-onocer-8-ene-3β, 21α-diol; Friedelan-3-one; Taraxerol,; Isopent-acosanoic acid; β-sitosterol.	*Rasa* - Katu, Kasaya, Madhura *Guna*- Laghu, Sara, Snigdha, Picchila *Virya*- Usna *Vipaka*- Madhura; *Karma*- Balya, Kaphahara, Krmighna, Pacana, Sandhaniya, Stambhana, Vatahara, Vrsya	Asthisanghatika Yoga, Asthisamhara Vatika, Asthisamhara tailam	Arsa (Piles), Asthibhagna (Bone fracture), Krmi (Worm infestation), Netraroga (Disease of the Eye), Svasa (Asthma), Urustambha (Stiffness in thigh muscles), Vrana (Ulcer)	Svarasa (Juice): 10 to 20 ml, Ardra kalka (paste): 10 to 20 g
424	Part-I; Vol. VI	12	BHUTAKESI (Fruit)	*Selinum vaginatum* C.B Clarke	Apiaceae	Dried Fruits	**Foreign matter**- NMT 2 per cent; **Total ash**- NMT 8 per cent; **Acid-insoluble ash**- NMT 2 per cent; **Alcohol-soluble extractive**-NLT 7 per cent; **Water-soluble extractive**-NLT 17 per cent; **T.L.C.** - Refer API	Essential oil and Coumarins.	*Rasa* - Tikta, Katu, Kasaya *Guna*- Laghu, Ruksa *Virya*- Sita *Vipaka*-Katu; *Karma*-Tridosaghna, Vedanahara, Raksoghna, Kesya, Kantiprada	Candanadi Taila	Apasmara (Epilepsy), Bhrama (Vertigo), Jvara (Fever), Ksaya (Pthisis), Svasa (Asthma), Murccha (Syncope), Raktagata vata (Hypertension), Raktapitta (Bleeding disorder), Trsa (Thirst), Vatavyadhi (Disease due to Vata dosa)	Churna (powder): 1 to 3 g

Sl. No.	*Part and Volume*	*Page No.*	*Common Ayurvedic Name*	*Botanical Name/ Chemical Name*	*Family/ Chemical Formula*	*Part Used*	*Identity, Purity and Strength/Chemical Properties*	*Constituents*	*Properties and Action*	*Important Formulations*	*Therapeutic Uses*	*Dose*
425	Part-I; Vol. VI	14	BHUTAKESI (Rhizome)	*Selinum vaginatum* C.B Clarke	Apiaceae	Dried Rhizomes	**Foreign matter**- NMT 2 per cent; **Total ash**- NMT 8 per cent; **Acid-insoluble ash**- NMT 4 per cent; **Alcohol-soluble extractive**-NLT 23 per cent; **Water-soluble extractive**- NLT 7 per cent; **T.L.C.** - Refer API	Coumarins: Vaginatin, Selinidin, Vaginol, Vaginidin and Archangelone.	*Rasa* - Kasaya, Tikta *Guna*- Sugandhi, Ruksa *Virya*- Usna *Vipaka*- Katu; *Karma*- Tridosahara, Vedanahara, Raksoghna, Kesya, Vranasodhana	Used as single drug	Apasmara (Epilepsy), Jvara (Fever), Kasa (Cough), Krmi (Helminthiasis), Pratisyaya (Coryza), Ucca Raktacapa (Hypertension), Unmada (Mania/ Psychosis), Vatavyadhi (Diseases due to Vata dosa)	Churna (powder): 3 to 6 g
426	Part-I; Vol. VI	16	BIJAPATRA (Whole Plant)	*Adiantum capillus-veneris* L.	Adiantaceae (Polypodiaceae)	Dried Whole Plant	**Foreign matter**- NMT 2 per cent; **Total ash**- NMT 15 per cent; **Acid-insoluble ash**- NMT 10 per cent; **Water-soluble extractive**- NLT 4 per cent; **Alcohol-soluble extractive**- NLT 10 per cent **T.L.C.** - Refer API	Adiantone; Adiantoxide; Astragalin; Nicotiflorin; Isoquercitrin; Rutin; Kaempferol-3-O-rutinoside; 1-Caffeylglucose and Sulphate esters of 1-coumaryl-glucose and 1-coumaryl-galactose; Kaempferol-3-glucuronide; Quercetin; β-sitosterol; Stigmasterol; Campesterol	*Rasa* - Kasaya, Katu *Guna*- Guru *Virya*- Sita *Vipaka*- Katu; *Karma*- Kanthya, Kaphahara, Kaphapittasamaka, Mutrajanana, Rasayana, Stambhana, Visaghna, Vranaropana	Used as single drug	Agnirohini (Acute stage of diphtheria), Angamarda (Body ache) Apasmara (Epilepsy), Atisara (Diarrhoea), Bhrama (Vertigo), Daha (Burning sensation), Gulma (Abdominal lump), Jvara (Fever), Kasa (cough), Lutavisa (Spider bite), Mutrakrcchra (Dysuria), Raktapitta (Bleeding disorders),	Churna (powder): 1 to 3 g, Svarasa (juice): 10 to 20 g

Sl. No.	*Part and Volume*	*Page No.*	*Common Ayurvedic Name*	*Botanical Name/ Chemical Name*	*Family/ Chemical Formula*	*Part Used*	*Identity, Purity and Strength/Chemical Properties*	*Constituents*	*Properties and Action*	*Important Formulations*	*Therapeutic Uses*	*Dose*
											Raktavikara (Disorders of blood), Sosa (Emaciation), Sotha (Oedema), Svasa (Asthma), Svarbheda (hoarseness of voice), Visarpa (Erysepales), Vrana (Ulcer)	
427	Part-I; Vol. VI	19	BIMBI ((Leaf)	*Coccinia grandis* (L.) Vogit Syn. *C.Cordifolia* Cogn, *C.indica* W and A, *Cephalandra indica* Naud.	Cucurbitaceae	Dried Leaves	**Foreign matter**- NMT 2 per cent; **Total ash**- NMT 6 per cent; **Acid-insoluble ash**- NMT 2 per cent; **Alcohol-soluble extractive**-NLT 15 per cent; **Water-soluble extractive**- NLT 38 per cent; **T.L.C.** - Refer API	Alkaloids such as Cephalandrine A, Cephalandrine B, Cephalandrine, β-sitosterol and Triacontane	*Rasa* - Madhura, Kasaya, Tikta *Guna*- Laghu *Virya*- Sita *Vipaka*- Katu; *Karma*- Grahi, Kaphapittahara, Vatakara	Bimbaghrta, Tundighrta	Kamala (Jaundice), Madhumeha (Diabetes mellitus), Puyameha (Urinary infection)	Svarasa (juice): 10 to 20 ml, Churna (powder): 3 to 6 g
428	Part-I; Vol. VI	21	BIMBI (Stem)	*Coccinia grandis* (L.) Vogit Syn. *C.Cordifolia* Cogn, *C.indica* W and A, *Cephalandra indica* Naud.	Cucurbitaceae	Dried Stem	**Foreign matter**- NMT 2 per cent; **Total ash**- NMT 10 per cent; **Acid-insoluble ash**- NMT 1 per cent; **Alcohol-soluble extractive**-NLT 33 per cent; **Water-soluble extractive**- NLT 40 per cent; **T.L.C.** - Refer API	Alkaloids such as Cephal-andrine-A, Cephal-andrine-B and β-sitosterol, Triacontane	*Rasa* - Kasaya, Tikta, Madhura *Guna*- Laghu *Virya*- Sita *Vipaka*- Katu; *Karma*- Grahi, Kaphapittahara, Vatakara	Used as single drug	Aruci (Tastelessness), Prameha (Metabolic disorder), Pravahika (Dysentery), Raktapitta (Bleeding disorder)	Churna (powder): 3 to 6 g

Sl. No.	*Part and Volume*	*Page No.*	*Common Ayurvedic Name*	*Botanical Name/ Chemical Name*	*Family/ Chemical Formula*	*Part Used*	*Identity, Purity and Strength/Chemical Properties*	*Constituents*	*Properties and Action*	*Important Formulations*	*Therapeutic Uses*	*Dose*
429	Part-I; Vol. VI	23	BRHAT DUGDHIKA (Whole Plant)	*Euphorbia hirta* L. Syn. *E. pilulifera* Auct. non L.	Euphorbiaceae	Dried Whole Plant	**Foreign matter**- NMT 2 per cent; **Total ash**- NMT 12 per cent; **Acid-insoluble ash**- NMT 7 per cent; **Alcohol-soluble extractive**-NLT 3 per cent; **Water-soluble extractive**-NLT 10 per cent; **T.L.C.** - Refer API	Flavonoids, Ellagotannins and Triterpenoids.	*Rasa* - Katu, Tikta, Madhura *Guna*- Ruksa, Guru, Tiksna *Virya*- Usna *Vipaka*- Katu; *Karma*- Garbhakaraka, Kaphahara, Mutrala, Slesmanissaraka, Stanya, Vrsya, Vistambhi	Used as single drug	Dadru (Taeniasis), Krmi (Worm infestation), Kasa (Cough), Kustha (Leprosy/ diseases of skin), Mutrakrcchra (Dysuria), Puyameha (Urinary infection), Sula (Pain/colic), Tamakasvasa (Bronchial asthma)	Churna (powder): 1 to 3 g, Svarasa (juice): 10 to 20 drops
430	Part-I; Vol. VI	26	BRHATI (Whole Plant)	*Solanum anguivi* Lam. Syn. *S. indicum* L.	Solanaceae	Dried Whole Plant	**Foreign matter**- NMT 2 per cent; **Total ash**- NMT 12 per cent; **Acid-insoluble ash**- NMT 5 per cent; **Alcohol-soluble extractive**-NLT 4 per cent; **Water-soluble extractive**-NLT 6 per cent; **T.L.C.** - Refer API	Steroidal Saponins: Protodiscin saponin C, Indioside A, B, C, D and E; Solafuranone.	*Rasa* - Tikta, Katu *Guna*- Laghu, Ruksa, Tiksna *Virya*- Usna *Vipaka*- Katu; *Karma*- Dipana, Grahi, Hrdya, Kaphahara, Kesya, Pacana, Vatahara, Vedanasthapana	Dasamularista, Dasamulakvatha	Amadosa (Products of impaired digestion and metabolism), Agnimandya (Digestive impairment), Aruci (Tastelessness), Chardi (Emesis), Hrdroga (Heart disease), Hikka (Hiccup), Jvara (Fever), Krmi (Worm infestation/ Helminthiasis), Kasa (Cough), Kustha (Leprosy/ diseass of skin),	Churna (powder): 3 to 6 g, Kvatha (decoction): 40 to 80 ml

Sl. No.	*Part and Volume*	*Page No.*	*Common Ayurvedic Name*	*Botanical Name/ Chemical Name*	*Family/ Chemical Formula*	*Part Used*	*Identity, Purity and Strength/Chemical Properties*	*Constituents*	*Properties and Action*	*Important Formulations*	*Therapeutic Uses*	*Dose*
											Netraroga (Diseases of the eye), Pratisyaya (Rhinitis), Svarabheda (Hoarseness), Svasa(Asthma), Sula (Pain)	
431	Part-I; Vol. VI	29	CANAKA (Whole Plant)	*Cicer arietinum* L.	Fabaceae	Whole Plant	**Foreign matter**- NMT 2 per cent; **Total ash**- NMT 12 per cent; **Acid-insoluble ash**- NMT 3 per cent; **Alcohol-soluble extractive**-NLT 9 per cent; **Water-soluble extractive**-NLT 14 per cent; **T.L.C.** - Refer API	Flavonoids such as, quercetin, Isoquercetin, Kaempferol-3-glucoside, Astragalin, Populnin, Biochenin-A-7-glucoside, Isorhamnetin, Protensein, Garbanzol and Cyanogenic glycosides.	*Rasa* - Kasaya, Lavana, Amla *Guna*- Ruksa, Laghu *Virya*- Sita *Vipaka*- Katu; *Karma*- Vatakara, Pittahara, Kaphahara, Vistambhi, Balya, Rucikara, Adhmanakaraka	Kravyada Rasa, Canakamla, Canakadi Lepa	Annadravasula (Gastric ulcer), Chardi (Emesis), Daha (Burning sensation), Jvara (Fever), Kasa (Cough), Pinasa (Chronic rhinitis/ sinusitis), Prameha (Metabolic disorder), Sosa (Emaciation), Svasa (Asthma), Trsna (Thirst), Udara (Disease of abdomen)	Churna (powder): 5 to 20 g

Sl. No.	*Part and Volume*	*Page No.*	*Common Ayurvedic Name*	*Botanical Name/ Chemical Name*	*Family/ Chemical Formula*	*Part Used*	*Identity, Purity and Strength/Chemical Properties*	*Constituents*	*Properties and Action*	*Important Formulations*	*Therapeutic Uses*	*Dose*
432	Part-I; Vol. VI	32	DARU-HARIDRA (Fruit)	*Berberis aristata* DC.	Berberidaceae	Dried Fruit	**Foreign matter**- NMT 2 per cent; **Total ash**- NMT 7 per cent; **Acid-insoluble ash**- NMT 1 per cent; **Alcohol-soluble extractive**-NLT 13 per cent; **Water-soluble extractive**- NLT 14 per cent; **T.L.C.** - Refer API	Alkaloids, Berberine, Oxyberberine, Berbamine, Palmatine, Jatrorhizine, Tetrahydro-palmitine *etc.*	*Rasa* - Madhura, Amla *Guna*- Laghu, Ruksa *Virya*- Sita *Vipaka*- Katu; *Karma*- Rucya, Pittasamana, Vistambhi	Used as single drug	Amatisara (Diarrhoea due to indigestion), Aruci (Tastelessness), Hrllasa (Nausea), Jvara (Fever), Pittaja-atisara (Diarrhoea due to Pitta dosa), Raktavikara (Disorders of blood), Trsna (Thirst), Vamana (Emesis), Visavikara (Disorders due to poison), Yakrtodara (Enlargement of Liver/ Hepatomegaly)	Churna (powder): 3 to 5 g
433	Part-I; Vol. VI	34	DHAVA (Fruit)	*Anogeissus latifolia* Wall.	Combretaceae	Dried Fruits	**Foreign matter**- NMT 2 per cent; **Total ash**- NMT 4 per cent; **Acid-insoluble ash**- NMT 0.1 per cent; **Alcohol-soluble extractive**-NLT 0.4 per cent; **Water-soluble extractive**- NLT 8 per cent; **T.L.C.** - Refer API	Tannins, Gallic acid, Saponins, and Flavonols like Quercetin and Myricetin.	*Rasa* - Madhura, Kasaya *Guna*- Ruksa, Guru *Virya* - Sita *Vipaka*- Katu; *Karma*- Pittahara, Kaphahara, Rucya, Dipana, Vatakara	Used as single drug	Asmari (Calculus), Arsa (Piles), Mutrakrcchra (Dysuria), Medoroga (Obesity), Pandu (Anemia), Prameha (Metabolic disor), RaKtavikara (Disorders of blood), Upadamsa (Soft chancre)	Churna (powder): 5 tc 10 g

Sl. No.	*Part and Volume*	*Page No.*	*Common Ayurvedic Name*	*Botanical Name/ Chemical Name*	*Family/ Chemical Formula*	*Part Used*	*Identity, Purity and Strength/Chemical Properties*	*Constituents*	*Properties and Action*	*Important Formulations*	*Therapeutic Uses*	*Dose*
434	Part-I; Vol. VI	36	DHAVA (Stem Bark)	*Anogeissus latifolia* Wall.	Combretaceae	Dried Stem Bark	**Foreign matter**- NMT 2 per cent; **Total ash**- NMT 11 per cent; **Acid-insoluble ash**- NMT 1 per cent; **Alcohol-soluble extractive**-NLT 11 per cent; **Water-soluble extractive**- NLT 20 per cent; **T.L.C.** - Refer API	Phenolic compounds such as Ellagic acid, Flavellagic acid, and Flavonols like Quercetin, Myricetin and Procyandin along with Gallotannins, Shikimic acid, Quinic acid, Amino acids, Alanine and Phenylanine.	*Rasa* - Madhura, Kasaya *Guna*- Ruksa, Guru *Virya*- Sita *Vipaka*- Katu; *Karma*- Pittahara, Kaphahara, Rasayana, Dipana, Medoghna	Ayaskrti, Nyagrodhadi Churna,	Asmari (calculus), Arsa (piles), Karnasrava (Otorrhoea), Kustha (Leprosy/ diseases of skin), Mutrakrcchra (Dysuria), Medoroga (Obesity), Pandu (Anaemia), Prameha (Metabolic disorder), Raktavikara (Disorders of blood), Upadamsa (Soft chancre), Visarpa (Erysepales)	Kvatha (decoction): 30 to 50 ml
435	Part-I; Vol. VI	38	DVIPANTARA DAMANAKA (Whole Plant)	*Artemisia absinthium* L.	Asteraceae	Dried Whole Plant	**Foreign matter**- NMT 2 per cent; **Total ash**- NMT 14 per cent; **Acid-insoluble ash**- NMT 7 per cent; **Alcohol-soluble extractive**-NLT 5 per cent; **Water-soluble extractive**-NLT 11 per cent; **Volatile oil** - 0.1 per cent; **T.L.C.** - Refer API	Volatile oil (which contain α-pinene, β-pinene, β-phellandrene, Thujone, Azulene, Sabinyl acetate, *etc.*) and bitter principles absinthin and iso-absinthin.	*Rasa* - Tikta *Guna*- Laghu, Ruksa, Tiksna *Virya*- Usna *Vipaka*- Katu; *Karma*- Artavajanana, Dipana, Kaphahara, Krmighna, Mutrala, Sothahara, Sugandhi, Vatahara, Vedanasthapana	Used as single drug	Agnimandya (Digestive impairment), Apasmara (Epilepsy), Jirnajvara (Chronic fever), Jalodara (Ascites), Krmi (Worm infestation), Kastartava (Dysmenorrhoea), Karnasula (Otalgia),	Churna (powder): 1 to 2 g

Sl. No.	Part and Volume	Page No.	Common Ayurvedic Name	Botanical Name/ Chemical Name	Family/ Chemical Formula	Part Used	Identity, Purity and Strength/Chemical Properties	Constituents	Properties and Action	Important Formulations	Therapeutic Uses	Dose
											Mutrakrcchra (Dysuria), Paksaghata (Paralysis/ Hemiplegia), Pliharoga (Splenic disease), Sandhisotha (Arthritis), Sotha (Inflammation), Udararoga (Diseases of abdomen), Vataroga (Disease due to Vata dosa) Yakrt roga (Liver disorder)	
436	Part-I; Vol. VI	41	DVIPANTARA SATAVARI (Root)	*Asparagus Officinalis* L.	Liliaceae	Dried Roots	**Foreign matter**- NMT 1 per cent; **Total ash**- NMT 10 per cent; **Acid-insoluble ash**- NMT 2.5 per cent; **Alcohol-soluble extractive**-NLT 9 per cent; **Water-soluble extractive**-NLT 24 per cent; **T.L.C.** - Refer API	Saponin glycosides, β-sitosterol, Saccharopine, 2-aminoadipic acid, Asparagusic acid, Dihydro-asparagusic acid, *S-A*cetyl dihydro-asparagusic acid, Spirostanol glucoside, Sarsasapogenin glycoside, Asparasaponin I and Asparasaponin	*Rasa* - Madhura *Guna*- Snigdha, Guru *Virya*- Sita *Vipaka*- Madhura; *Karma*- Hrdya, Mutrala, Pittahara, Vrsya, Vajikarana	Used as single drug	Asmari (Calculus), Kamala (Jaundice), Mutrakrcchra (Dysuria), Sotha (Inflammation), Vatararakta (Gout)	Churna (powder): 3 to 6 g

Sl. No.	*Part and Volume*	*Page No.*	*Common Ayurvedic Name*	*Botanical Name/ Chemical Name*	*Family/ Chemical Formula*	*Part Used*	*Identity, Purity and Strength/Chemical Properties*	*Constituents*	*Properties and Action*	*Important Formulations*	*Therapeutic Uses*	*Dose*
								II and Nine steroid glucosides named as Asparagosides A, B, C, D, E, F, G, H and I.				
437	Part-I; Vol. VI	43	ELAVALUKAM (Root)	*Prunus avium* L.	Rosaceae	Roots	**Foreign matter**- NMT 2 per cent; **Total ash**- NMT 8 per cent; **Acid-insoluble ash**- NMT 1 per cent; **Alcohol-soluble extractive**-NLT 7 per cent; **Water-soluble extractive**-NLT 5 per cent; **T.L.C.** - Refer API	Cyanogenic glycoside like D-Mandel-onitril-β-glucoside (Prunasin).	*Rasa* - Kasaya, Tikta *Guna*- Laghu *Virya*- Sita *Vipaka*-Katu; *Karma*-Kaphahara, Pittahara, Sukrasodhana, Vedanasthapana, Vamana	Used as single drug	Arsa (Piles), Aruci (Tastelessness), Krmi roga (Worm infestation), Kandu (Itching), Kustha (Leprosy/ diseases of skin), Vrana (Ulcer), Mutraroga (Urinary diseases), Raktapitta (Bleeding disorder)	Churna (powder): 1 to 3 g
438	Part-I; Vol. VI	45	ELAVA-LUKAM (Stem Bark)	*Prunus avium* L.	Rosaceae	Stem Bark	**Foreign matter**- NMT 1 per cent; **Total ash**- NMT 1 per cent; **Acid-insoluble ash**- NMT 7 per cent; **Alcohol-soluble extractive**-NLT 0.5 per cent; **Water-soluble extractive**- NLT 11 per cent; **T.L.C.** - Refer API	Cyanogenic glycoside like D-mandelo-nitril-β-glucoside (Prunasin), D-mandelo-nitrile-β-gentio-bioside dehydro-wogonin 7-glucoside and Chrysin 7-glucoside are main	*Rasa* - Kasaya, Tikta *Guna*- Laghu *Virya*- Sita *Vipaka*-Katu; *Karma*-Kaphahara, Pittahara Sukrasodhana, Vamana, Vedanasthapana	Used as single drug	Arsa (Piles), Aruci (Tastelessness), Hrdroga (Heart disease), Kandu (Itching), Krmi (Worm infestation), Kustha (Leprosy/ disease of skin), Mutraroga (Urinary diseases), Raktapitta	Churna (powder): 1 to 3 g

Sl. No.	Part and Volume	Page No.	Common Ayurvedic Name	Botanical Name/ Chemical Name	Family/ Chemical Formula	Part Used	Identity, Purity and Strength/Chemical Properties	Constituents	Properties and Action	Important Formulations	Therapeutic Uses	Dose
								components. Tectochrysin, Apigenin 5-glucoside, Genkwanin 5-glucoside and Neosakuranine are the minor components			(Bleeding disorder), Vrana (Ulcer)	
439	Part-I; Vol. VI	47	ERANDA-KARKATI (Fruit)	*Carica papaya* L.	Caricaceae	Dried Pericarp of Mature and Unripe Fruits	**Foreign matter**- NMT 2 per cent; **Total ash**- NMT 14 per cent; **Acid-insoluble ash**- NMT 0.5 per cent; **Alcohol-soluble extractive**-NLT 2 per cent; **Water-soluble extractive**-NLT 25 per cent; **T.L.C.** - Refer API	β-carotene, Papain, Carpaine.	*Rasa* - Tikta, Madhura *Guna*- Laghu *Virya*- Usna *Vipaka*- Madhura; *Karma*- Pittahara, Kaphahara, Dipana, Vatakara, Stanya, Hrdya, Brmhana	Apakva-phalaniryasa Lepa	Krmi (Worm infestation), Kasa (Cough), Raktavikara (Disorder of blood), Svasa (Asthma), Vatararakta (Gout)	Churna (powder): 10 to 20 g
440	Part-I; Vol. VI	49	ERANDA-KARKATI (Root)	*Carica papaya* L.	Caricaceae	Dried Roots	**Foreign matter**- NMT 2 per cent; **Total ash**- **NMT 18 per cent; Acid-insoluble ash**- NMT 1.5 per cent; **Alcohol-soluble extractive**-NLT 3.0 per cent; **Water-soluble extractive**- NLT 15.0 per cent; **T.L.C.** - Refer API	Carpesanine, Carpaine	*Rasa* - Katu, Tikta *Guna*- Laghu, Ruksa, Tiksna *Virya*- Usna *Vipaka*- Katu; *Karma*- Kaphahara, Mutrala	Asmarihara-kasaya Churna	Asmari (Calculus), Arsa (Piles), Aruci (Tastelessness), Krmi roga (Worm infestation), Mutraroga (Urinary diseases), Raktapitta, Raktapradara (Menorrhagia or Metrorr-hagia or both), Tvakroga (Skin diseases), Udarasula (Pain in the abdomen), Vataraktа (Gout), Vrana (Ulcer)	Churna (powder): 2 to 6 g

Sl. No.	*Part and Volume*	*Page No.*	*Common Ayurvedic Name*	*Botanical Name/ Chemical Name*	*Family/ Chemical Formula*	*Part Used*	*Identity, Purity and Strength/Chemical Properties*	*Constituents*	*Properties and Action*	*Important Formulations*	*Therapeutic Uses*	*Dose*
441	Part-I; Vol. VI	51	GANDHA-SIPHA (Whole Plant)	*Pavonia odorata* Willd.	Malvaceae	Whole Plant	**Foreign matter**- NMT 2 per cent; **Total ash**- NMT 9 per cent; **Acid-insoluble ash**- NMT 2 per cent; **Alcohol-soluble extractive**-NLT 4 per cent; **Water-soluble extractive**-NLT 9 per cent; **Fixed oil** - NLT 4 per cent; **T.L.C.** - Refer API	β-sitosterol; Palmitic, Stearic, Oleic, Linoleic, Isovaleric and *n*-caproic acids, α-pinene and Methyl eptenone, Isovalaral-dehyde, Aromadendrin, Azulene, Pavonene, Pavonenol	*Rasa* - Tikta *Guna*- Ruksa, Laghu, Sugandhi *Virya*- Sita *Vipaka*- Katu; *Karma*- Balya, Dipana, Jvaraghna, Kaphahara, Kesya, Mutrala, Pacana, Pittahara	Used as single drug	Aruci (Tastelessness), Atisara (Diarrhoea), Chardi (Emesis), Daha (Burning sensation), Hrdroga (Heart disease), Hrllasa (Nausea), Jvara (Fever), Kustha (Leprosy/ diseases of skin), Raktapitta (Bleeding disorder), Svitra (Leucoderma/ Vitiligo), Trsna (Thirst), Visarpa (Erysepales), Vrana (Ulcer)	Churna (powder): 3 to 6 g
442	Part-I; Vol. VI	54	GRISMA-CHATRAKA (Whole Plant)	*Mollugo cerviana* Seringe	Aizoaceae	Dried Whole Plant	**Foreign matter**- NMT 2 per cent; **Total ash-NMT 9.5 per cent; Acid-insoluble ash**- NMT 4 per cent; **Alcohol-soluble extractive**-NLT 10 per cent; **Water-soluble extractive**- NLT 14 per cent; **T.L.C.** - Refer API	Flavonoid: Orientin, Vitexin and their 2'-*O*-glucosides.	*Rasa* - Tikta *Guna*- Laghu, Ruksa *Virya*- Sita *Vipaka*- Katu; *Karma*- Dipana, Jvaraghna, Trsnahara, Virecana	Used as single drug	Agnimandya (Digestive impairment), Jvara (Fever), Daha (Burning sensation), Kamala (Jaundice), prameha (Metabolic disorder)	Churna (powder): 3 to 6 g

Sl. No.	Part and Volume	Page No.	Common Ayurvedic Name	Botanical Name/ Chemical Name	Family/ Chemical Formula	Part Used	Identity, Purity and Strength/Chemical Properties	Constituents	Properties and Action	Important Formulations	Therapeutic Uses	Dose
443	Part-I; Vol. VI	56	GOKSURA (Whole Plant)	*Tribulus terrestris* L.	Zygophyllaceae	Dried Whole Plant	**Foreign matter**- NMT 2 per cent; **Total ash**- NMT 17 per cent; **Acid-insoluble ash**- NMT 4 per cent; **Alcohol-soluble extractive**-NLT 2 per cent; **Water-soluble extractive**-NLT 12 per cent; **T.L.C.** - Refer API	Alkaloids: Terrestriamide, Tribulusamide A, B; Steroidal saponins: Terrestrosin C, D, E, F, G, H, I, J, and K, Terrestroneoside A and F, Terreside A and B, Terrestroside F; Tribulosaponin A and B, Tribulosin, Protodioscin saponin C, Prototribestin, Terrestrosin J, Isoterrestrosin B, Flavonoid glycosides: Isorhamnetin-3-gentiotrioside, Quercetin-3-gentiobioside-7-glucoside,; amide: moupinamide.	*Rasa* - Madhura, Tikta *Guna*- Guru, Snigdha *Virya*- Usna *Vipaka*- Madhura; *Karma*- Balya, Brmhana, Dipana, Kaphahara, Kesya, Mutrala, Pittahara, Sothahara, Vrsya, Vatahara, Vedanasthapana	Chyawanaprasa Avaleha, Dasmula Kvatha, Rasnadi Kvatha, Dasamula Satpalaka Ghrta	Amavata (Rheumatism), Amlapitta (Hyperacidity), Antravrddhi (Hernia), Asmari (Calculus), Ardita (Facial palsy), Arsa (Piles), Hrdroga (heart disease), Indralupta (Alopecia), Jvara (Fever), Kasa (Cough), Mutraghata (Urinary obstruction), Mutrakrcchra (Dysuria), Paksaghata (Paralysis/ Hemiplegia), Pradara (excessive vaginal discharge), Prameha (Metabolic disorder), Raktapitta, Sula (Pain/ Colic), Sotha (Oedema), Svasa, Sutikaroga (Puerperal disorders), Sitapitta (Urticaria), Vatarakta (Gout)	Churna (powder) 3: 6 g; Kvatha (decoction): 50 to 100 ml

Sl. No.	Part and Volume	Page No.	Common Ayurvedic Name	Botanical Name/ Chemical Name	Family/ Chemical Formula	Part Used	Identity, Purity and Strength/Chemical Properties	Constituents	Properties and Action	Important Formulations	Therapeutic Uses	Dose
444	Part-I; Vol. VI	59	GRANTHI-MULA (Rhizome)	*Alpinia calcarata* Rose.	Zingiberaceae	Rhizome	**Foreign matter**- NMT 2 per cent; **Total ash**- NMT 7 per cent; **Acid-insoluble ash**- NMT 3 per cent; **Alcohol-soluble extractive**-NLT 5 per cent; **Water-soluble extractive**-NLT 6 per cent; **T.L.C.** - Refer API	Volatile oil rich in Methyl cinnamate, Cineol, Camphor.	*Rasa* - Katu, Tikta *Guna*- Laghu, Ruksa, Tiksna *Virya*- Usna *Vipaka*- Katu; *Karma*- Kaphaghna, Svarya, Sothahara, Sulaghna	Used as single drug	Amavata (Rheumatism), Hikka (Hiccup), Kasa (Cough), Prameha (Metabolic disorder), Svasa (Asthma), Sandhisula (Joint pain), Sula (Pain/ Colic)	Churna (powder): 1 to 3 g
445	Part-I; Vol. VI	61	GULADAUDI (Leaf)	*Chrysanthemum indicum* L.	Asteraceae	Dried Leaves	**Foreign matter**- NMT 2 per cent; **Total ash**- NMT 21 per cent; **Acid-insoluble ash**- NMT 4 per cent; **Alcohol-soluble extractive**-NLT 10 per cent; **Water-soluble extractive**- NLT 22 per cent; **T.L.C.** - Refer API	Sesquiterpene lactones- Angeloyl-cumambrin B, Arteglasin A and Angleoy-lajadin. Essential oil from aerial parts contain di-and Sesquiter-penoids α-copaene, β-elumene, β-carophyllene, β- farnesene, β-humulene, Germacrene-D, α-silenene, Curcumene, Calamenene, γ-cadinene and T-murolol, and Monoterpenoids myrcene, 1,8-cineol and Bornyl acetate.	*Rasa* - Tikta, Kasaya *Guna*- Laghu, Ruksa *Virya*- Sita *Vipaka*- Katu; *Karma*- Pittahara, Ropana, Sulaprasamana, Hrdya	Used as single drug	Ardhava-bhedaka (Hemicrania/ Migraine), Mukhasphota (Ulcer in the mouth), Sirahsula (Headache), Tvakroga (Skin diseases), Vrana (Ulcer), Yuvanapidika (Pimples/Acne vulgaris)	Churna (powder): 3 to 6 g

Sl. No.	*Part and Volume*	*Page No.*	*Common Ayurvedic Name*	*Botanical Name/ Chemical Name*	*Family/ Chemical Formula*	*Part Used*	*Identity, Purity and Strength/Chemical Properties*	*Constituents*	*Properties and Action*	*Important Formulations*	*Therapeutic Uses*	*Dose*
								Chrysanthe-none and Chrysanthe-nin glucoside. Aerial parts also contain lignans sesamin and fargesin, and flavonoid penduletin.				
446	Part-I; Vol. VI	63	HARITA-MANJARI (Whole Plant)	*Acalypha indica* L.	Euphorbiaceae	Dried Whole Plant	**Foreign matter**- NMT 2 per cent; **Total ash**- NMT 14 per cent; **Acid-insoluble ash**- NMT 1 per cent; **Alcohol-soluble extractive**-NLT 3 per cent; **Water-soluble extractive**-NLT 10 per cent; **T.L.C.** - Refer API	Alkaloids: Acalyphine, Quinine, Amides such as Acalyphamide, Sterols, a Flavonol kaempferol and Cyanogenic glycoside.	*Rasa* - Tikta, Katu *Guna*- Laghu, Ruksa *Virya*- Usna *Vipaka*- Katu; *Karma*- Kaphaghna, Vamaka, Sransana, Krmighna, Mutrala, Tvakdosahara, Amadosahara	Used as single drug	Agnimandya (Digestive impairment), Dantasula (Toothache), Karnasula (Otalgia), Kasa (Cough), Sandhisotha (Arthritis), Svasa (Asthma), Vibandha (Constipation)	Churna (powder): 3 to 5 g; Svarasa (juice): 5 to 10 ml, 1 to 3 drops in karnasula
447	Part-I; Vol. VI	66	HASTISUNDI (Aerial Part)	*Heliotropium indicum* L.	Boraginaceae	Dried Aerial Part	**Foreign matter**- NMT 2 per cent; **Total ash**- NMT 12 per cent; **Acid-insoluble ash**- NMT 2 per cent; **Alcohol-soluble extractive**-NLT 4 per cent; **Water-soluble extractive**-NLT 12 per cent; **T.L.C.** - Refer API	Pyrrolizidine Alkaloids (Heliotrine, Indicine N-oxide), Tannins.	*Rasa* - Katu, Tikta *Guna*- Tiksna, Laghu *Virya*- Usna *Vipaka*- Katu; *Karma*- Jvaraghna, Vedanahara	Used as single drug	Sannipatajvara (High fever due to vitiation of all dosas), Sula (Pain/Colic)	Churna (powder): 3 to 6 g

Sl. No.	*Part and Volume*	*Page No.*	*Common Ayurvedic Name*	*Botanical Name/ Chemical Name*	*Family/ Chemical Formula*	*Part Used*	*Identity, Purity and Strength/Chemical Properties*	*Constituents*	*Properties and Action*	*Important Formulations*	*Therapeutic Uses*	*Dose*
448	Part-I; Vol. VI	68	INDIVARA (Rhizome)	*Monochoria vaginalis* Presl. Syn. *Pontederia vaginalis* Burm. f	Pontederiaceae	Rhizome	**Foreign matter**- NMT 1 per cent; **Total ash**- NMT 15 per cent; **Acid-insoluble ash**- NMT 5 per cent; **Alcohol-soluble extractive**-NLT 7 per cent; **Water-soluble extractive**-NLT 10 per cent; **Fixed oil** - NLT 1 per cent; **T.L.C.** - Refer API	Stigmasterol 3-O-beta-D-glucopyranoside	*Rasa* - Madhura *Guna*- Guru, Snigdha *Virya*- Sita *Vipaka*- Madhura; *Karma*- Brmhana, Balya, Dahaprasamana, Pittasamaka, Vrsya, Vata-Kaphavardhaka	Used as single drug	Daha (Burning sensation), Daurbalya (Weakness), Dhatuksya (Tissue wasting), Raktapitta (Bleeding disorder), Yakrtvikara (Disorder of liver)	Churna (powder) 3 to 6 g
449	Part-I; Vol. VI	70	JALAKUMBHI (Whole Plant)	*Pistia stratiotes* L.	Araceae	Dried Whole Plant	**Foreign matter**- NMT 6 per cent; **Total ash**- NMT 52 per cent; **Acid-insoluble ash**- NMT 35 per cent; **Alcohol-soluble extractive**-NLT 5 per cent; **Water-soluble extractive**-NLT 2 per cent; **T.L.C.** - Refer API	Flavonoids like Vicenin, Lucenin and Cyanidina-3-glucoside.	*Rasa* - Madhura, Tikta, Katu *Guna*- Laghu, Ruksa, Sara *Virya*- Sita *Vipaka*- Madhura; *Karma*- Balya, Mutrajanana, Sothahara, Tridosahara	Jalakumbhi-bhasma-prayogah	Arsa (Piles), Daha (Burning sensation), Galaganda (Goitre), Jvara (Fever), Kustha (Leprosy/ diseases of skin), Mutrakrcchra (Dysuria), Sosa (Emaciation), Raktapitta (Bleeding disorder)	Churna (powder): 3 to 5 g; Svarasa (juice): 10 to 20 ml
450	Part-I; Vol. VI	73	JIVANTI (Root)	*Leptadenia reticulata* W. and A.	Asclepiadaceae	Dried Roots	**Foreign matter**- NMT 2 per cent; **Total ash**- NMT 14 per cent; **Acid-insoluble ash**- NMT 1.5 per cent; **Alcohol-soluble extractive**-NLT 5 per cent; **Water-soluble extractive**-NLT 3 per cent; **T.L.C.** - Refer API	Hentria-contanol, α-and β-amyrin, Stigmasterol, β-sitosterol and Flavonoids-diosmetin and Luteolin.	*Rasa* - Madhura, Kasaya *Guna*- Laghu, Snigdha *Virya*- Sita *Vipaka*- Madhura; *Karma*- Rasayana, Balya, Caksusya, Grahi, Vrsya, Brmhana, Stanyajanana, Visaghna, tridosahara	Chyawanaprasa, Brahmarasayana, Amrtaprasa ghrta, Asokaghrta, Brhatmasataila, Marmagutika, Manasa-mitravataka, Svasahara Kasayachurna, Guducyaditaila	Atisara (Diarrhoea), Daha (Burnign sensation), Jvara (Fever), Ksaya (Pthisis), Kasa (Cough), sosa (Emaciation), Mukharoga (disease of mouth),	Churna (powder): 3 to 6 g

Sl. No.	*Part and Volume*	*Page No.*	*Common Ayurvedic Name*	*Botanical Name/ Chemical Name*	*Family/ Chemical Formula*	*Part Used*	*Identity, Purity and Strength/Chemical Properties*	*Constituents*	*Properties and Action*	*Important Formulations*	*Therapeutic Uses*	*Dose*
											Naktandhya (Night blindness), Netraroga (Diseases of the eye), Raktapitta (Bleeding disorder), Trsna (Thirst), Urahksata (Pulmonary caviation), Vrana (Ulcer)	
451	Part-I; Vol. VI	75	KANTA-KIGULMA (Aerial Part)	*Lycium barbarum* L. Syn. *L. europeaum*	Solanaceae	Aerial Parts	**Foreign matter**- NMT 2 per cent; **Total ash**- NMT 15 per cent; **Acid-insoluble ash**- NMT 2 per cent; **Alcohol-soluble extractive**-NLT 4.5 per cent; **Water-soluble extractive**- NLT 20 per cent; **T.L.C.** - Refer API	Tropane alkaloid like Atropine, Streoidal Sapogenin like Diosgenin and Flavonoids like Quercetin and Rutin.	*Rasa* - Tikta *Guna*- Laghu, Ruksa *Virya*- Sita *Vipaka*- Katu; *Karma*- Caksusya, Dipaniya, Mutrala	Used as single drug	Agnimandya (Digestive impairment), Dantasula (Toothache), Jalodara (Ascites), Kandu (Itching), Raktarsa (Bleeding piles)	Churna (powder): 2 to 5 g
452	Part-I; Vol. VI	78	KARAPHSA (Root)	*Apium graveolens* L.	Apiaceae	Dried Roots	**Foreign matter**- NMT 5 per cent; **Total ash**- NMT 10 per cent; **Acid-insoluble ash**- NMT 2 per cent; **Alcohol-soluble extractive**-NLT 9 per cent; **Water-soluble extractive**-NLT 10 per cent; **Volatile oil** - NLT 0.05 per cent; **T.L.C.** - Refer API	α-pinene, β-pinene, Limonene, Pentylbenzene, β-selinen, 3-*n*-butyl phthalide.	*Rasa* - Katu, Kasaya *Guna*- Laghu, Ruksa *Virya*- Usna *Vipaka*- Katu; *Karma*- Dipana, Kaphahara, Mutrala, Svedajanana, Vatahara	Used as single drug	Asmari (Calculus), Bastiroga (Diseases of urinary system), Grdhrasi (Sciatica), Hikka (Hiccup), Jalodara (Ascites), Kaphaja Siroroga (Catarrhal Siroroga/ Sinusitis),	Churna (powder): 5 to 7 g

Sl. No.	*Part and Volume*	*Page No.*	*Common Ayurvedic Name*	*Botanical Name/ Chemical Name*	*Family/ Chemical Formula*	*Part Used*	*Identity, Purity and Strength/Chemical Properties*	*Constituents*	*Properties and Action*	*Important Formulations*	*Therapeutic Uses*	*Dose*
											Kaphajvara (Fever due to Kapha dosa), Mutraghata (Urinary obstruction/ retention of urine), Mastiska-daurbalya (Neurosthenia), Prsthasula (Lumbago), Parsvasula (Intercostal neuralgia and pleurodynia), Sarvanga sopha (anasarca), Sula (Pain), Udarasula (Pain in the abdomen), Udararoga (diseases of abdomen), Vatarakta (Gotu), Yakrtpliha Vikara (Diseases of liver and spleen)	

Sl. No.	Part and Volume	Page No.	Common Ayurvedic Name	Botanical Name/ Chemical Name	Family/ Chemical Formula	Part Used	Identity, Purity and Strength/Chemical Properties	Constituents	Properties and Action	Important Formulations	Therapeutic Uses	Dose
453	Part-I; Vol. VI	80	KATUGULMA (Whole Plant)	*Toddalia asiatica* (L.) Lam. Syn. *Toddalia aculeata* Pers.	Rutaceae	Whole Plant	**Foreign matter**- NMT 2 per cent; **Total ash**- NMT 6 per cent; **Acid-insoluble ash**- NMT 0.4 per cent; **Alcohol-soluble extractive**-NLT 5 per cent; **Water-soluble extractive**- NLT 3 per cent; **T.L.C.** - Refer API	Alkaloids; Toddaline, Toddalinine, Skimmianine and Berberine. Other constituents include Citric acid, an Oil, Resin, Pectin and Starch	*Rasa* - Katu, Tikta, *Guna*- Laghu, Ruksa *Virya*- Usna *Vipaka*- Katu; *Karma*- Pacana, Dipana, Sitaprasamana, Sothaghna, Svedana	Used as single drug	Agnimandya (Digestive impairment), Kapha-vatavyadhi (Disorders due to Kapha and Vata dosa), Angamarda (Bodyache), Atisara (Diarrhoea), Jvara (Fever), Krmi (Worm infestation), Kustha (Leprosy/ diseases of skin), Visamajavara (Intermittent fever)	Churna (powder): 0.5 to 2 g
454	Part-I; Vol. VI	83	KESARAJA (Whole Plant)	*Wedelia chinensis* Merril Syn. *Wedelia calendulacea* Less	Asteraceae	Dried Whole Plant	**Foreign matter**- NMT 2 per cent; **Total ash**- NMT 9.5 per cent; **Acid-insoluble ash**- NMT 1 per cent; **Alcohol-soluble extractive**-NLT 17 per cent; **Water-soluble extractive**- NLT 31 per cent; **T.L.C.** - Refer API	Coumestan (Mixture of wedelolactone and Demethyl-wedelolactone); Norwedelic acid, Norwe-delolactone, tri-*o*-methylwe-delolactone and β-amyrin.	*Rasa* - Katu, Tikta, Kasaya *Guna*- Tiksna *Virya*- Usna *Vipaka*- Katu; *Karma*- Vatahara, Kaphahara, Mutrala, Hrdya, Vrsya, Svedakara, Kesya, Balya	Grahanimihira taila, Asokaghrta, Brhta Visamaj-varantaka lauha	Arsa (Piles), Atisara (Diarrhoea), Daurbalya (Weakness), Hrdroga (Heart disease), Indralupta (Alopecia), Jvara (Fever), Krmi (Helminthiasis), Kamala (Jaundice), Kasa (Cough), Pandu (Anaemia), Plihavrddhi (Splenomegaly),	Churna (powder): 3 to 6 g

Sl. No.	*Part and Volume*	*Page No.*	*Common Ayurvedic Name*	*Botanical Name/ Chemical Name*	*Family/ Chemical Formula*	*Part Used*	*Identity, Purity and Strength/Chemical Properties*	*Constituents*	*Properties and Action*	*Important Formulations*	*Therapeutic Uses*	*Dose*
											Sirahsula (Headache), Slipada (Fliariasis), Stiroga (Gynaecological disorders), Sula (Pain/ Colic), Svasa (Asthma), Vrana (Ulcer)	
455	Part-I; Vol. VI	86	KETAKI (Stilt Root)	*Pandanus odoratissimus* Roxb. Syn. *P. fascicularis* Lamk. *P. tectorius* Soland. ex Parkinson	Pandanaceae	Stilt Root	**Foreign matter**- NMT 2 per cent; **Total ash**- NMT 4 per cent; **Acid-insoluble ash**- NMT 0.1 per cent; **Alcohol-soluble extractive**-NLT 4 per cent; **Water-soluble extractive**-NLT 8 per cent; **T.L.C.** - Refer API	Physcion; *p*-hydroxy-benzoic acid, Cirsilineol, *n*-triacontanol, β-sitosterol, Stigmasterol, Campesterol, Daucosterol, Stigmast-4-en-3, 6-dione, Andamarine, Piperidine	*Rasa* - Tikta, Kasaya, Madhura *Guna*- Laghu, Snigdha *Virya*- Sita *Vipaka*- Katu; *Karma*- Balya, Dehadardhyakara, Hrdya, Pittasamaka, Rasayana, Stambhana	Balaketakyadi Kasaya	Gulma (Abdominal lump), Jvara (Fever), Mutrakrcchra (Dysuria), Pradara (Excessive vaginal discharge), Raktapitta (Bleeding disorder), Tvakroga (Skin diseases)	Churna (powder): 1 to 2 g; Kvatha (deco-ction): 30 to 50 ml
456	Part-I; Vol. VI	88	KITAMARI (Leaf)	*Aristolochia bracteolata* Lam. Syn. *A. bracteata* Retz.	Aristolochiaceae	Leaves	**Foreign matter**- NMT 2 per cent; **Total ash**- NMT 10 per cent; **Acid-insoluble ash**- NMT 1.3 per cent; **Alcohol-soluble extractive**-NLT 12.8 per cent; **Water-soluble extractive**- NLT 25.5 per cent; **Fixed oil** - NLT 5.3 per cent; **T.L.C.** - Refer API	Aristolochic acid, Magnoflorine; *N*-acetylnor-nuciferine; Aristolactam; β-sitosterol and Ceryl alcohol	*Rasa* - Tikta *Guna*- Laghu, Ruksa, Tiksna *Virya*- Usna *Vipaka*- Katu; *Karma*- Dipana, Garbhasayottejaka, Kapahara, Kasahara, Krmighna, Kusthaghna, Rucya, Vatahara, Virecana, Visaghna, Vranasodhana	Used as single drug	Krmi (Worm infetation), Kastartava (Dysmeno-rrhoea), Sandhisula (Joint pain), Sitapitta (Urticaria), Sotha (Oedema), Tvakroga (Leprosy/Skin disorders),	Churna (powder): 1 to 3 g

Sl. No.	Part and Volume	Page No.	Common Ayurvedic Name	Botanical Name/ Chemical Name	Family/ Chemical Formula	Part Used	Identity, Purity and Strength/Chemical Properties	Constituents	Properties and Action	Important Formulations	Therapeutic Uses	Dose
											Visamajvara (Intermittent fever), Vicarcika (dry and weeping eczema), Vrana (Ulcer)	
457	Part-I; Vol. VI	90	KUMARI-VETRA (Rhizome)	*Calamus thwaitesii* Becc.	Arecaceae	Rhizomes	**Foreign matter**- NMT 2 per cent; **Total ash- NMT 6 per cent; Acid-insoluble ash**- NMT 3 per cent; **Alcohol-soluble extractive**-NLT 8 per cent; **Water-soluble extractive**-NLT 7 per cent; **Fixed oil** - NLT 0.98 per cent; **T.L.C.** - Refer API	No report on the chemical constituents of the Rhizome is available.	*Rasa* - Kasaya, Tikta *Guna*- Laghu, Ruksa *Virya*- Sita *Vipaka*- Katu; *Karma*- Dahaprasamana, Grahi, Jvaraghna, Kusthaghna, Pittahara, Vranya	Used as single drug	Atisara (Diarrhoea), Jvara (Fever), Kustha (Leprosy/ diseases of skin), Prameha (Metabolic disorder), Raktapitta (bleeding disorder), Visarpa (Erysepales), Vrana (Ulcer)	Churna (powder): 3 to 6 g
458	Part-I; Vol. VI	92	KUSUMBHA (Fruit)	*Carthamus tinctorius* L.	Asteraceae	Dried Fruit	**Foreign matter**- NMT 2 per cent; **Total ash**- NMT 4.5 per cent; **Acid-insoluble ash**- NMT 1 per cent; **Alcohol-soluble extractive**-NLT 7 per cent; **Water-soluble extractive**-NLT 8 per cent; **T.L.C.** - Refer API	Lignan glucoside (Matairesinol, mono-glucoside), Glucose, Maltose, Raffinose, Luteolin-7-O-glucoside, N-(P-coumaroyl) tryptamine, Campesterol, Cholestrol, β-sitosterol and its Glucoside, Δ^7-stigmasterol,	*Rasa* - Madhura, Kasaya, Tikta, Katu *Guna*- Snigdha, Guru *Virya*- Usna *Vipaka*- Katu; *Karma*- Mutrala, Sarvado-saprakopaka, Svedajanana, Vidahi, Virecana	Used as single drug	Amavata (Rheumatism), Asmari (Calculus), Daurbalya (Weakness), Kamala (Jaundice), Kastartava (Dysmen-orrhoea), Mutrakrcchra (Dysuria), Pratisyaya (Coryza), Raktapitta (Bleeding disorder)	Churna (powder): 2 to 4 g

Sl. No.	Part and Volume	Page No.	Common Ayurvedic Name	Botanical Name/ Chemical Name	Family/ Chemical Formula	Part Used	Identity, Purity and Strength/Chemical Properties	Constituents	Properties and Action	Important Formulations	Therapeutic Uses	Dose
								Myristo-oleo-linolein, Myristo-dilinolein, Palmito-oleolinolein, Palmito-dilinolein, Stearo-oleolinolein, Stearo-dilinolein, Dioleolinolein, Oleo-dilinolein, Trilinolein.				
459	Part-I; Vol. VI	94	KUSUMBHA (Leaf)	*Carthamus tinctorius* L.	Asteraceae	Dried Leaves	**Foreign matter**- NMT 2 per cent; **Total ash**- NMT 19 per cent; **Acid-insoluble ash**- NMT 2 per cent; **Alcohol-soluble extractive**-NLT 20 per cent; **Water-soluble extractive**- NLT 23 per cent; **T.L.C.** - Refer API	Hinesol-β-D-fucopyranoside, 1-Pentadecene.	*Rasa* - Madhura, Kasaya *Guna*- Ruksa, Laghu *Virya*- Usna *Vipaka*- Katu; *Karma*- Vatakara, Pittakara, Kaphahara, Dipana, Madanasaka, Balya	Used as single drug	Asmari (Calculus), Badhirya (deafness), Daurbalya (Weakness), Mutrakrcchra (Dysuria), Mutravikara (Urinary diseases), Netraroga (Diseases of the eye), Pralapa (Delirium), Prameha (Metabolic disorder), Raktavikara (Disorders of blood), Yoniroga (Disease of female genital tract), Pradara (Excessive discharge)	Churna (powder): 2 to 4 g

Sl. No.	*Part and Volume*	*Page No.*	*Common Ayurvedic Name*	*Botanical Name/ Chemical Name*	*Family/ Chemical Formula*	*Part Used*	*Identity, Purity and Strength/Chemical Properties*	*Constituents*	*Properties and Action*	*Important Formulations*	*Therapeutic Uses*	*Dose*
460	Part-I; Vol. VI	96	KUSUMBHA (Flower Head)	*Carthamus tinctorius* L.	Asteraceae	Dried Flower head	**Foreign matter**- NMT 2 per cent; **Total ash**- NMT 7 per cent; **Acid-insoluble ash**- NMT 1 per cent; **Alcohol-soluble extractive**-NLT 6 per cent; **Water-soluble extractive**-NLT 14 per cent; **T.L.C.** - Refer API	Contains a dye of Flavonoid, Carthamin.	*Rasa* - Madhura, Kasaya *Guna*- Ruksa, Laghu *Virya*- Katu *Vipaka*- Usna; *Karma*- Kaphahara, Svedajanana, Dipana, Kesaranjana, Visaghna	Used as single drug	Kastartava (Dysmeno-rrhoea), Kasa (Cough), Mutrakrcchra (Dysuria), Pratisyaya (Coryza), Raktapitta (Bleeding disorder), Romantika (Measles), Svasa (Asthma), Visphotaka (Blisterous eruption), Yoniroga (Disease of female genital tract)	Churna (powder): 2 to 4 g
461	Part-I; Vol. VI	99	LAGHU HARITA-MANJARI (Root)	*Acalypha fruticosa* Forsk.	Euphorbiaceae	Roots	**Foreign matter**- NMT 2 per cent; **Total ash**- NMT 4 per cent; **Acid-insoluble ash**- NMT 0.5 per cent; **Alcohol-soluble extractive**-NLT 2 per cent; **Water-soluble extractive**-NLT 5 per cent; **Fixed oil** - NLT 1 per cent; **T.L.C.** - Refer API	Arjunolic acid	*Rasa* - Tikta, Katu *Guna*- Laghu, Snigdha *Virya*- Usna *Vipaka*- Katu; *Karma*- Dipana, Kaphahara, Pacana, Sramsana, Vamana, Vrana ropana	Used as single drug	Agnimandya (Digestive impairment), Vrana (Ulcer)	Churna (powder): 3 to 6 g

Sl. No.	*Part and Volume*	*Page No.*	*Common Ayurvedic Name*	*Botanical Name/ Chemical Name*	*Family/ Chemical Formula*	*Part Used*	*Identity, Purity and Strength/Chemical Properties*	*Constituents*	*Properties and Action*	*Important Formulations*	*Therapeutic Uses*	*Dose*
462	Part-I; Vol. VI	101	LAGHUPATRA VARSABHU (Whole Plant)	*Trianthema decandra* L.	Ficoidaceae (Aizoaceae)	Whole Plant	**Foreign matter**- NMT 2 per cent; **Total ash**- NMT 22 per cent; **Acid-insoluble ash**- NMT 8 per cent; **Alcohol-soluble extractive**-NLT 12 per cent; **Water-soluble extractive**- NLT 27 per cent; **T.L.C.** - Refer API	Saponins and Alkaloid Punarnavine	*Rasa* - Tikta *Guna*- Ruksa *Virya*- Usna *Vipaka*- Katu; *Karma*- Kaphahara, Mutrala, Sramsana, Sulaghna	Used as single drug	Amavata (Rheumatism), Apasmara (Epilepsy), Ardhava-bhedaka (Migrain/ Hemicrania), Hrdayaroga (Heart disease), Kamala (Jaundice), Kasa (Cough), Pandu (Anaemia), Sotha (Oedema), Svasa (Asthma), Urahksata (Chest wound), Vrana (Ulcer)	Churna (powder): 3 to 6 g
463	Part-I; Vol. VI	104	LOHITA-NIRYASA (Exudate)	*Dracaena cinnabari* Balf.f.	Agavaceae	Exudate of Stem	**Foreign matter**- NMT 2 per cent; **Total ash**- NMT 2 per cent; **Acid-insoluble ash**- NMT 8 per cent; **Alcohol-soluble extractive**-NLT 95 per cent; **Water-soluble extractive**- NLT 2 per cent; **T.L.C.** - Refer API	2-Hydroxy-chalcone, 7-Hydroxy-3-(3-hydroxy-4-methoxy-benzyl) Chroman, S)-7, 3'-Dihydroxy-4'-Methoxy-flavan and 4-Hydroxy-2-methoxy-dihydro-chalcone	*Rasa* - Kasaya *Guna*- Laghu, Ruksa *Virya*- Sita *Vipaka*- Katu; *Karma*- Raktastambhana, Sangrahi, Vranaropana	Used as single drug	Atisar (Diarrhoea), Pravahika (Dysentery), Raktarsa (Bleeding piles), Raktapitta (Bleeding disorder), Rakta-Pradara (menorrhagia or metrorrhagia or both), Raktasrava (Bleeding disorder), Vrana (Ulcer)	Churna (powder): 1 to 2 g

Sl. No.	*Part and Volume*	*Page No.*	*Common Ayurvedic Name*	*Botanical Name/ Chemical Name*	*Family/ Chemical Formula*	*Part Used*	*Identity, Purity and Strength/Chemical Properties*	*Constituents*	*Properties and Action*	*Important Formulations*	*Therapeutic Uses*	*Dose*
464	Part-I; Vol. VI	106	MADHAVI (Flower)	*Hiptage benghalensis* L.	Malpighiaceae	Dried Flowers	**Foreign matter**- NMT 2 per cent; **Total ash**- NMT 10 per cent; **Acid-insoluble ash**- NMT 0.5 per cent; **Alcohol-soluble extractive**-NLT 10 per cent; **Water-soluble extractive**- NLT 30 per cent; **T.L.C.** - Refer API	No report on the chemical consitituents of the flower is available.	*Rasa* - Madhura, Katu, Tikta *Guna*- Laghu *Virya*- Sita *Vipaka*- Madhura; *Karma*- Tridosaghna, Kusthaghna	Candrakala rasa,	Agnimandya (Digestive impairment), Krmi roga (Worm infestation), Kandu (Itching), Pama (Eczema), Raktapitta (Bleeding disorder), Sthaulya (Obesity), Tvakroga (Skin diseases)	Churna (powder): 3 to 6 g
465	Part-I; Vol. VI	108	MATSYA-PATRIKA (Whole Plant)	*Merremia tridentata* (L.) Hall. f. Syn. *Ipomoea tridentata* (L.) Roth.	Convolvulaceae	Whole Plant	**Foreign matter**- NMT 2 per cent; **Total ash**- NMT 10 per cent; **Acid-insoluble ash**- NMT 1 per cent; **Alcohol-soluble extractive**-NLT 5 per cent; **Water-soluble extractive**- NLT 14 per cent; **T.L.C.** - Refer API	Flavonoids like Diosmetin, Luteolin, Diosmetin-7-*O*-β-glucoside and Luteolin-7-O-β glucoside.	*Rasa* - Tikta, Kasaya *Guna*- Guru, Sara *Virya*- Usna *Vipaka*- Katu; *Karma*- Vatahara, Arsoghna, Bhedana, Sandhaniya, Sara, Vrsya	Prasaranida-taila (keraliya)	Arsa (Piles), Dhatuksya (Tissue wasting), Paksaghata (Paralysis/ Hemiplegia), Sandhisotha (Arthritis), Sotha (Inflammation), Vibandha (Constipation), Vrrana (Ulcer)	Churna (Powder): 3 to 6 g; Svarasa (juice): 5 to 10 ml
466	Part-I; Vol. VI	111	MEDA (Rhizome)	*Polygonatum cirrhifolium* Royle	Liliaceae	Dried Rhizome	**Foreign matter**- NMT 2 per cent; **Total ash**- NMT 5 per cent; **Acid-insoluble ash**- NMT 1 per cent; **Alcohol-soluble extractive**-NLT 25 per cent; **Water-soluble extractive**- NLT 62 per cent; **T.L.C.** - Refer API	Steroidal saponins (Diosgenin), Proteins and Resins.	*Rasa* - Madhura *Guna*- Snigdha, Picchila, Guru *Virya*- Sita *Vipaka*- Madhura; *Karma*- Balya, Brmhana, Garbhada, Jivaniya, Kaphavardhaka, Paustika, Pittahara,	Dasamularista, Asoka Ghrta	Balaroga (Disease of children), Bhagandara (Fistula-in-ano), Gulma (Abdominal Lump), Kamala (Jaundice), Karsya	Churna (powder): 3 to 6 g

Sl. No.	*Part and Volume*	*Page No.*	*Common Ayurvedic Name*	*Botanical Name/ Chemical Name*	*Family/ Chemical Formula*	*Part Used*	*Identity, Purity and Strength/Chemical Properties*	*Constituents*	*Properties and Action*	*Important Formulations*	*Therapeutic Uses*	*Dose*
									Stanyajanana, Vrsya		(Emaciation), Kasa (Cough), Ksaya (Pthisis), Naktandhya (Night blindness), Netrasrava (Chronic dacrocystitis or Epiphora), Rajayaksma (Tuberculosis), Raktapitta (Bleeding disorder), Sosa (Emaciation), Svasa (Asthma), Timira (Cataract), Visarpa (Erysepales)	
467	Part-I; Vol. VI	113	NADIHINGU (Exudate)	*Gardenia gummifera* L. f. Syn. *G. arborea* Roxb.	Rubiaceae	Dried Resinous Exudate from the Shoot Tip	**Foreign organic matter** - NMT 2 per cent. **Solubility:** Insoluble in water and slightly soluble in most of the organic solvents; dissolves in strong acids, turning brown to reddish brown, as it gets charred. **Identification test:** Refer API **TLC:** Refer API	Gardenin, 3',4',5' apigenin, Demethoxy-sudachitin and 3',5'-dihydoxy-4'-Methoxy-wogonin	*Rasa* - Kantu; *Guna*- Tiksna; *Virya*- Usna; *Vipaka*- Katu; *Karma*- Vatahara, Kaphahara, Dipana, Vatanulomaka, Pacana	Used as single drug	Adhmana (Flatulence with gurgling sound), Agnimandya (Digestive impairment), Ajirna (Indigestion), Amadosa (Products of impaired digestion and metabolism), Aruci (Tastelessness), Gulma (Abdominal lump),	Churna (powder): 1 to 3 g

Sl. No.	Part and Volume	Page No.	Common Ayurvedic Name	Botanical Name/ Chemical Name	Family/ Chemical Formula	Part Used	Identity, Purity and Strength/Chemical Properties	Constituents	Properties and Action	Important Formulations	Therapeutic Uses	Dose
											Hikka (Hiccup), Krmi (Helminthiasis), Medoroga (Obesity), Udarasula (Pain in the abdomen)	
468	Part-I; Vol. VI	115	NAHI (Whole Plant)	*Enicostemma axillare* (Lam.) A. Raynal. Syn. *E. littorale* Blume, *E. hysoppifolium* (Willd.) Verd.	Gentianaceae	Whole Plant	**Foreign matter**- NMT 1 per cent; **Total ash**- NMT 9 per cent; **Acid-insoluble ash**- NMT 3 per cent; **Alcohol-soluble extractive**-NLT 16 per cent; **Water-soluble extractive**- NLT 28 per cent; **Fixed oil** - NLT 5 per cent; **T.L.C.** - Refer API	Flavonoids like Genkwanin, Apigenin, Isovitexin, Swertisin, Saponarin, Swertiamarin, Betulin, Enicoflavin, Gentiocrucine, Gentianine, Erythro-centaurine, Ephelic acid Glycoside, Sylswerti-sioside, Isoswertisin-5-*O*-glucoside, Sylswertisin-5-O-glucoside.	*Rasa* - Tikta *Guna*- Laghu, Ruksa *Virya*- Usna *Vipaka*- Katu; *Karma*- Vatanulomaka, Pittahara, Kaphahara, Dipana, Pacana, Visaghna	Vayucchaya Surendra Taila	Krmi (Worm infestation), Sotha (Oedema), Madhumeha (Diabetes mellitus), Medoroga (Obesity), Prameha (Metabolic disorder), Raktavikara (Disorders of blood), Tvakroga (Skin diseases), Visamajvara (Intermittent fever), Vibandha (Constipation), Yakrtdaurbalya (Poor function of liver)	Churna (powder): 1 to 3 g

Sl. No.	*Part and Volume*	*Page No.*	*Common Ayurvedic Name*	*Botanical Name/ Chemical Name*	*Family/ Chemical Formula*	*Part Used*	*Identity, Purity and Strength/Chemical Properties*	*Constituents*	*Properties and Action*	*Important Formulations*	*Therapeutic Uses*	*Dose*
469	Part-I; Vol. VI	118	NIKOCAKA (Kernel)	*Pinus gerardiana* Wall.	Coniferae	kernels	**Foreign matter**- NMT 2 per cent; **Total ash**- NMT 3 per cent; **Acid-insoluble ash**- NMT 0.2 per cent; **Alcohol-soluble extractive**-NLT 28 per cent; **Water-soluble extractive**- NLT 18 per cent; **Fixed oil** - NLT 43 per cent; **T.L.C.** - Refer API	Palmitic, Stearic, Oleic and Linoleic acids; Palmito-dilinolein, Stearodilinolein, Palmito-oleolinolein, Stearo-oleolinolein trilinolein, Oleodilinolein, Dioleolinolein and Triolein.	*Rasa* - Madhura *Guna*- Snigdha, Guru *Virya*- Usna *Vipaka*- Madhura; *Karma*- Slesma-nihsaraka, Brmhana, Balya, Dhatuvardhana, Kaphakara, Pittakara, Raktaprasadaka, Uttejaka, Vrsya, Vatahara	Used as single drug	Amavata (Rheumatism), Apasmara (Epilepsy), Ardita (Facial plasy), Hikka (Hiccup), Kasa (Cough), Ksata (Wound), Ksaya (Pthisis), Katisula (Lower backache), Pandu (Anaemia), Parsvasula (Intercostal neuralgia and pleurodynia), Paksavadha (Paralysis/ Hemiplegia), Sandhivata (Arthritis due to vata dosa), Svasa (Asthma), Vatarakta (Gotu)	Churna (powder): 10 to 20 g
470	Part-I; Vol. VI	120	PANASA (Root Bark)	*Artocarpus heterophyllus* Lamk. Syn. *A integrifolia* L.f	Moraceae	Dried Root Bark	**Foreign matter**- NMT 2 per cent; **Total ash**- NMT 19 per cent; **Acid-insoluble ash**- NMT 10 per cent; **Alcohol-soluble extractive**-NLT 7 per cent; **Water-soluble extractive**- NLT 3 per cent; **T.L.C.** - Refer API	β-Sitosterol, Cycloartenone, Cycloartenol, Tannins.	*Rasa* - Kasaya, Tikta *Guna*- Laghu, Ruksa *Virya*- Sita *Vipaka*- Katu; *Karma*- Grahi, Pittahara, Stambhana, Tvakdosahara, Vatavardhaka, Vistambhakaraka	Used as single drug	Atisara (Diarrhoea), Daha (Burning sensation), Raktapitta (Bleeding disorder), Sotha (Inflammation), Tvakroga (Skin diseases)	Churna (powder): 3 to 6 g

Sl. No.	*Part and Volume*	*Page No.*	*Common Ayurvedic Name*	*Botanical Name/ Chemical Name*	*Family/ Chemical Formula*	*Part Used*	*Identity, Purity and Strength/Chemical Properties*	*Constituents*	*Properties and Action*	*Important Formulations*	*Therapeutic Uses*	*Dose*
471	Part-I; Vol. VI	122	PAPATAH (Root)	*Pavetta indica var. tomentosa* Hook. Syn. *P. tomentosa* Roxb.	Rubiaceae	Root Pieces	**Foreign matter**- NMT 2 per cent; **Total ash**- NMT 3 per cent; **Acid-insoluble ash**- NMT 1 per cent; **Alcohol-soluble extractive**-NLT 5 per cent; **Water-soluble extractive**- NLT 9 per cent; **T.L.C.** - Refer API	Fixed oil	*Rasa* - Tikta *Guna*- Laghu, Ruksa *Virya*- Sita *Vipaka*- Katu; *Karma*- Balya, Kaphaghna, Mutrala, Varnya, Virecana	Used as single drug	Kamala (Jaundice), Kandu (Itching), Mutraroga (Urinary diseases), Sotha (Inflammation), Udararoga (Diseases of abdomen), Vibandha (Constipation), Visphota (Blister)	Churna (powder): 3 to 6 g
472	Part-I; Vol. VI	124	PARNAYAVANI (Leaf)	*Coleus amboinicus* Lour. Syn. *C. aromaticus* Benth.	Lamiaceae	Leaves	**Foreign matter**- NMT 2 per cent; **Total ash**- NMT 16 per cent; **Acid-insoluble ash**- NMT 2 per cent; **Alcohol-soluble extractive**-NLT 7 per cent; **Water-soluble extractive**- NLT 23 per cent; **Fixed Oil** - NLT 2.8 per cent; **T.L.C.** - Refer API	Oleanolic acid; Crategolic acid; Pomolic acid; Euscaphic acid; Tormentic acid; ursolic acid and 2α,3α,19α,23-oxalacetic acid; Cirsimaritin; Sitosterol glucoside; Salvingenin; Quercetin; 6-methoxy-genkwanin; Chrysoeriol; Ethyl salicylate; γ-terpinene; β-salinene; Luteolin; Apigenin; Eriodyctol; Þ-cymene; α and β-pinene; Taxifolin; Thymol;	*Rasa* - Katu, Tikta *Guna*- Laghu, Ruksa, Tiksna *Virya*- Usna *Vipaka*- Katu; *Karma*- Dipana, Kapahahara, Malasangrahini, Pacana, Rucya, Vatahara, Vedanasthapana, Visaghna	Used as single drug	Adhmana (Flatulence with gurgling sound), Agnimandya (Digestive impairment), Ajirna (Indigestion), Aruci (Tastelessness), Atisara (Diarrhoea), Grahani roga (Colitis/ Ulcerative colitis), Gulma (Abdominal lump), Hikka (Hiccup), Hrdyadaurbalya (Weakness of the heart), Jirnasvasa (Chronic asthma),	Svarasa (Juice): 5 to 10 ml

Sl. No.	*Part and Volume*	*Page No.*	*Common Ayurvedic Name*	*Botanical Name/ Chemical Name*	*Family/ Chemical Formula*	*Part Used*	*Identity, Purity and Strength/Chemical Properties*	*Constituents*	*Properties and Action*	*Important Formulations*	*Therapeutic Uses*	*Dose*
								Carvacrol; Myrcene, 1,8-cineol; Eugenol; β-caryo-phyllene.			Kasa (Cough), Krmi (Worm infestation), Mutrakrcchra (Dysuria), Mutraroga (Urinary diseases), Mutrasmari (Urinary calculus), Svasa (Asthma), Udararoga (Diseases of abdomen), Unmada (Mania/ Psychosis), Visucika (Gastro-enteritis with piercing pain)	
473	Part-I; Vol. VI	127	PATRASNUHI (Latex)	*Euphorbia nivulia* Buch.-Ham.	Euphorbiaceae	Fresh or Dried Latex	**Foreign matter**- NMT 2 per cent; **Total ash**- NMT 2 per cent; **Acid-insoluble ash**- NMT 0.12 per cent; **Alcohol-soluble extractive**-NLT 29 per cent; **Water-soluble extractive**- NLT 7 per cent; **Fixed oil** - NLT 21 per cent; **T.L.C.** - Refer API	Cyclonivuliaol, Cycloartenol, Cycloeucalenol, Cycloart-25-en-3- β-24-diol	*Rasa* - Katu *Guna*- Lagu, Tiksna, Snigdha *Virya*- Usna *Vipaka*- Katu; *Karma*- Bhedana, Dahakara, Lekhana, Virecana	Used as single drug	Arsa (Piles), Bhagandara (Fistula-in-ano), Kustha (Leprosy/ Diseases of skin), Svasa (Asthma), Udararoga (Diseases of Abdomen)	Ksira (latex): 125 to 250 mg

Sl. No.	Part and Volume	Page No.	Common Ayurvedic Name	Botanical Name/ Chemical Name	Family/ Chemical Formula	Part Used	Identity, Purity and Strength/Chemical Properties	Constituents	Properties and Action	Important Formulations	Therapeutic Uses	Dose
474	Part-I; Vol. VI	129	PINDATA-GARA (Rhizome)	*Asarum europaeum* L.	Aristolochiaceae	Dried Rhizomes	**Foreign matter**- NMT 2 per cent; **Total ash**- NMT 6 per cent; **Acid-insoluble ash**- NMT 2 per cent; **Alcohol-soluble extractive**-NLT 20 per cent; **Water-soluble extractive**- NLT 25 per cent; **T.L.C.** - Refer API	α-agrofuran, Chalcone, Diglycoside, α-asarone, Diasarone-1, Diasarone-2, *trans and cis*-isoasarones. Fixed oil and Volatile oil	*Rasa* - Katu, Amla, Kasaya *Guna*- Laghu *Virya*- Usna *Vipaka*- Katu *Karma*- Kaphaghna, Nadibalya, Sirovirecana, Svayathuvilayana, Svedajanana, Tiksnavirecana, Vamaka, Visaghna	Used as single drug	Amavata (Rheumatism), Anartava (Amenorrhoea), Apasmara (Epilepsy), Ardita (Facial palsy), Avarodhajanya Kamala (Obstructive Jaundice), Grdhrasi (Sciatica), Jalodara (Ascites), Mutravarodha (Uninary obstraction), Netraroga (Diseases of the eye), Paksavadha (Paralysis/ Hemiplegia), Parsvasula (Intercostal Neuralgia and Pleurodynia), Pliha (Splenic disease), Sula (Pain/Colic), Yakrtasotha (Hepatitis)	Churna (powder): 1 to 3 g

Sl. No.	*Part and Volume*	*Page No.*	*Common Ayurvedic Name*	*Botanical Name/ Chemical Name*	*Family/ Chemical Formula*	*Part Used*	*Identity, Purity and Strength/Chemical Properties*	*Constituents*	*Properties and Action*	*Important Formulations*	*Therapeutic Uses*	*Dose*
475	Part-I; Vol. VI	131	PITA-KANCANARA (Bud)	*Bauhinia racemosa* Lamk	Caesalpiniaceae	Dried, Mature Flower Bud	**Foreign matter**- NMT 2 per cent; **Total ash**- NMT 6 per cent; **Acid-insoluble ash**- NMT 1 per cent; **Alcohol-soluble extractive**-NLT 16 per cent; **Water-soluble extractive**- NLT 28 per cent; **T.L.C.** - Refer API	Flavonoids like Quercetin, Isoquercetin.	*Rasa* - Madhura, Kasaya *Guna*- Snigdha, Guru *Virya*- Sita *Vipaka*- Madhura; *Karma*- Pittakaphasamaka, Sangrahi, Kaphavatahara, Pittahara	Used as single drug	Bhutavikara (Psychotic syndrome), Daha (Burning sensation), Galaganda (Goitre), Gandamala (Cervical lymphadenitis), Prameha (Metabolic disorder), Raktavikara (Disorders of blood), Trsna (Thirst), Vidaha (Burning sensation), Visamjvara (Intermittent fever)	Churna (powder): 1 to 3 g
476	Part-I; Vol. VI	133	RAKTA CITRAKA (Root)	*Plumbago indica* L. Syn. *P. rosea* L.	Plumbaginaceae	Dried Roots	**Foreign matter**- NMT 2 per cent; **Total ash**- NMT 12 per cent; **Acid-insoluble ash**- NMT 1 per cent; **Alcohol-soluble extractive**-NLT 5 per cent; **Water-soluble extractive**- NLT 10 per cent; **T.L.C.** - Refer API	Quinones and Naphthaquinones such as Isoshinanolone, Plumbagic acid Vanillic acid and Zeylanone.	*Rasa* - Katu, Tikta *Guna*- Laghu, Ruksa, Tiksna *Virya*- Usna *Vipaka*- Katu, Tikta; *Karma*- Dipana, Grahi, Pacana, Rasayana, rucya	Used as single drug	Arsa (Piles), Grahani (Malabsorption syndrome), Kasa (Cough), Krmi (Helminthiasis), Kustha (Leprosy/ diseases of skin), Pandu (Anaemia), Sikatameha (Lithuria), Sotha (Oedema), Sula (Pian)	Churna (powder): 0.5 to 2 g

Sl. No.	Part and Volume	Page No.	Common Ayurvedic Name	Botanical Name/ Chemical Name	Family/ Chemical Formula	Part Used	Identity, Purity and Strength/Chemical Properties	Constituents	Properties and Action	Important Formulations	Therapeutic Uses	Dose
477	Part-I; Vol. VI	135	ROHITAKA (Stem Bark)	*Tecomella undulata* (Sm.) Seem.	Bignoniaceae	Dried Stem Bark	**Foreign matter**- NMT 2 per cent; **Total ash**- NMT 12 per cent; **Acid-insoluble ash**- NMT 1 per cent; **Alcohol-soluble extractive**-NLT 10 per cent; **Water-soluble extractive**- NLT 15 per cent; **T.L.C.** - Refer API	Tecomin (Veratroyl β-D-glucoside), n-triacontane, *n*-heptacosane, *n*-nonacosane, *n*-triacontanol, *n*-octacosanol, β-sitosterol.	*Rasa* - Katu, Kasaya, Tikta *Guna*- Laghu, Ruksa, Sara *Virya*- Sita *Vipaka*- Katu; *Karma*- Vatahara, Kaphahara, Rucya, Raktaprasadana, Medohara, Stanya, Visaghna	Rohitakarista, Rohitaka Lauha, Yakrtsula vinasini Vatika	Gulma (Abdominal lump), Krmi (Helminthiasis), Kamala (Jaundice), Karnaroga (Disease of ear), Kustha (Leprosy/ diseases of skin), Medoroga (Obesity), Netraroga (Diseases of eye), Plihodara (Splenomegaly), Prameha (Metabolic disorder), Raktavikara (Disorders of blood), Sula (Pain/Colic), Svetapradara (Leucorrhoea), Vibandha (Constipation), Vrana (Ulcer), Yakrtroga (Liver disorders)	Churna (powder): 3 to 6 g, Kvatha (decoction): 50 to 100

Sl. No.	*Part and Volume*	*Page No.*	*Common Ayurvedic Name*	*Botanical Name/ Chemical Name*	*Family/ Chemical Formula*	*Part Used*	*Identity, Purity and Strength/Chemical Properties*	*Constituents*	*Properties and Action*	*Important Formulations*	*Therapeutic Uses*	*Dose*
478	Part-I; Vol. VI	137	SALA (Heart Wood)	*Shorea robusta* Gaertn.	Diptero-carpaceae	Dried Heart Wood	**Foreign matter**- NMT 2 per cent; **Total ash**- NMT 2 per cent; **Acid-insoluble ash**- NMT 0.7 per cent; **Alcohol-soluble extractive**-NLT 6 per cent; **Water-soluble extractive**-NLT 1.5 per cent; **T.L.C.** - Refer API	Bergenin, Shoreaphenol, Chalcone, 4'-Hydroxy-chalcone-4-*O*-β-D-gluco-pyranoside, 12α-Hydroxy-3-oxo-olenano-28, 13-Lactone.	*Rasa* - Kasaya *Guna*- Ruksa *Virya*- Usna *Vipaka*- Katu; *Karma*- Kaphahara, Medohara, Vranasodhana, Grahi, Visaghna, Vedanasthapana, Stambhana, Krmighna	Ayaskrti, Eladi ghrta	Agnidaha (Burns), Kandu (Itching)., Krmi (Helminthiasis), Kustha (Leprosy/ diseases of skin), Pandu (Anaemia), Prameha (Metabolic disorder), Raktavikara (Disorders of blood), Sotha (Oedema), Upadamsa (Syphilis/ Soft chancre), Vatavyadhi (Disease due to vata dosa), Visavikara (Disorders due to poison), Vidradhi (Abscess), Vrana (Ulcer), Yoniroga (Disease of female genital tract), Karnaroga (Disease of ear), Badhirya (Deafness), Ashibhanga (Bone fracture)	Churna (powder): 3 to 6 g; Kvatha (deco-ction): 50 to 100 ml

Sl. No.	Part and Volume	Page No.	Common Ayurvedic Name	Botanical Name/ Chemical Name	Family/ Chemical Formula	Part Used	Identity, Purity and Strength/Chemical Properties	Constituents	Properties and Action	Important Formulations	Therapeutic Uses	Dose
479	Part-I; Vol. VI	139	SALAPARNI (Whole Plant)	*Desmodium gangeticum* DC	Fabaceae	Dried Whole Plant	**Foreign matter**- NMT 2 per cent; **Total ash**- NMT 8 per cent; **Acid-insoluble ash**- NMT 2.5 per cent; **Alcohol-soluble extractive**-NLT 6 per cent; **Water-soluble extractive**-NLT 10 per cent; **T.L.C.** - Refer API	Alkaloids; Flavonoids, Desmocarpan, Desmocarpin, Pterocarpan, Desmodin, Gangetin, Gangetinin,; Others: 2-(NN-dimethylamino) acetophenone	*Rasa* - Tikta, Madhura *Guna*- Guru, Snigdha *Virya*- Usna *Vipaka*- Madhura; *Karma*- Balya, Brmhana, Mutrala, Rasayna, Tridosahara, Vatahara, Vrsya	Dasamularista, Dasamulakvatha	Arsa (Piles), Atisara (Diarrhoea), Chardi (Emesis), Jvara (Fever), Kasa (Cough), Krmi (Worm infestation), Ksata (Wound), Mutrakrcchra (Dysuria), Prameha (Metabolic disorder), Santapa (Emotional stress), Sosa (Cachexia), Sotha (Inflammation), Sukradaurbalya (Seminal stress), Svasa (Asthma), Vataroga (Disease due to vata dosa), Visamjvara (Intermittent fever), Visavikara (Disorders due to poision)	Churna (powder): 6 to 12 g; Kvatha (deco-ction): 50 to 100 ml

Sl. No.	*Part and Volume*	*Page No.*	*Common Ayurvedic Name*	*Botanical Name/ Chemical Name*	*Family/ Chemical Formula*	*Part Used*	*Identity, Purity and Strength/Chemical Properties*	*Constituents*	*Properties and Action*	*Important Formulations*	*Therapeutic Uses*	*Dose*
480	Part-I; Vol. VI	142	SAMI (Leaf)	*Prosopis cineraria* Druce Syn. *P. spicigera* L.	Leguminosae -Mimosaceae	Leaves	**Foreign matter**- NMT 1 per cent; **Total ash**- NMT 7 per cent; **Acid-insoluble ash**- NMT 1 per cent; **Alcohol-soluble extractive**-NLT 14 per cent; **Water-soluble extractive**- NLT 21 per cent; **Fixed oil** - NMT 4 per cent; **T.L.C.** - Refer API	Rich in Tannin, Volatile fatty acid	*Rasa* - Tikta, Katu, Kasaya *Guna*- Laghu, Ruksa *Virya*- Sita *Vipaka*- Katu; *Karma*- Arsoghna, Krmighna, Kaphapittahara, Kusthaghna, Recaka, Sangrahaka, Vatakara	Used as single drug	Arsa (Piles), Atisara (Diarrhoea), Balagraha (Psychotic syndrome of children), Bhrama (Vertigo), Krmi (Worm infestation), Kasa (Cough), Kustha (Leprosy/ Deisease of Skin), Netraroga (Disease of the Eye), Raktapitta (Bleeding disorder), Svasa (Asthma), Visavikara (Disorders due to Poison)	Churna (powder): 3 to 5 g
481	Part-I; Vol. VI	145	SAURA-BHANIMBA (Leaf)	*Murraya koenigii* (L.) Spreng Syn. *M. koenigii.* Spreng	Rutaceae	Dried Leaves	**Foreign matter**- NMT 2 per cent; **Total ash**- NMT 12 per cent; **Acid-insoluble ash**- NMT 2 per cent; **Alcohol-soluble extractive**-NLT 20 per cent; **Water-soluble extractive**- NLT 34 per cent; **T.L.C.** - Refer API	Alkaloids like Koenidine, Koenigine, Koenimbine, Mahanimbine, Muconine murrayacine and Volatile oils.	*Rasa* - Kasaya, Tikta, Madhura *Guna*- Laghu, Snigdha *Virya*- Sita *Vipaka*- Katu; *Karma*- Kaphapittahara, Rucya, Dipana, Pacana, Visaghna, Varnya	Used as single drug	Arsa (Piles), Atisara (Diarrhoea), Chardi (Emesis), Daha (Burning Sensation), Dusta Vrana (Non-Healing Ulcer), Jvara (Fever), Kandu (Itching), Krmi (Helminthiasis), Kustha (Leprosy/	Churna (powder): 3 to 6 g; Svarasa (juice): 10 to 20 ml

Sl. No.	Part and Volume	Page No.	Common Ayurvedic Name	Botanical Name/ Chemical Name	Family/ Chemical Formula	Part Used	Identity, Purity and Strength/Chemical Properties	Constituents	Properties and Action	Important Formulations	Therapeutic Uses	Dose
											Diseases of Skin), Prameha (Metabolic disorder), Pravahika (Dysentery), Sula (Pain/ Colic), Sosa (Emaciation), Sopha (Oedema), Svitra (Leucoderma/ Vitligo)	
482	Part-I; Vol. VI	148	SITIVARAKA (Seed)	*Celosia argentea* L.	Amaranthaceae	Seeds	**Foreign matter**- NMT 2 per cent; **Total ash**- NMT 5 per cent; **Acid-insoluble ash**- NMT 1 per cent; **Alcohol-soluble extractive**-NLT 3 per cent; **Water-soluble extractive**-NLT 7 per cent; **T.L.C.** - Refer API	Nonpeptide, Celogenamide, Celosian, and Acidic Polysaccharide.	*Rasa* - Kasaya, Madhura, Katu *Guna*- Ruksa, Guru, Sara *Virya*- Sita *Vipaka*- Madhura; *Karma*- Tridosahara, Bastisodhaka, Samgrahi, Mutrala, Vrsya, Snehana, Medhya, Rsayana	Used as single drug	Asmari (Calculus), Arsa (Piles), Atisara (Diarrhoea), Gulma (Abdominal lump), Hrdroga (Heart disease), Jvara (Fever), Mutraghata (Urinary obstruction), Mutrakrcchra (Dysuria), Pliharoga (Spleenic disease), Raktavikara (Disorders of Blood), Sopha (Oedema)	Churna (powder): 3 to 6 g

Sl. No.	*Part and Volume*	*Page No.*	*Common Ayurvedic Name*	*Botanical Name/ Chemical Name*	*Family/ Chemical Formula*	*Part Used*	*Identity, Purity and Strength/Chemical Properties*	*Constituents*	*Properties and Action*	*Important Formulations*	*Therapeutic Uses*	*Dose*
483	Part-I; Vol. VI	150	SIVA-NILI (Root and Stem)	*Indigofera aspalathoides* Vahl ex DC.	Fabaceae	Dried Roots and Stem	**Foreign matter**- NMT 2 per cent; **Total ash**- NMT 8 per cent; **Acid-insoluble ash**- NMT 3 per cent; **Alcohol-soluble extractive**-NLT 8 per cent; **Water-soluble extractive**-NLT 13 per cent; **T.L.C.** - Refer API	Fixed oil	*Rasa* - Tikta, Kasaya *Guna*- Laghu, Ruksa *Virya*- Usna *Vipaka*- Katu; *Karma*- Kapha Vatahara, Kesya, Kusthaghna	Used as single drug	Amavata (Rheumatism), Arumsika (Dandruff), Dantasula (Tooth ache), Gulma (Abdominal lump), Kustha (Leprosy/ Diseases of Skin), Pliharoga (Spleenic disease), Udararoga (Diseases of Abdomen), Vatarakta (Gotu), Vidradhi (Abscess), Visarpa (Erysepales)	Churna (powder): 3 to 6 g
484	Part-I; Vol. VI	152	SLESMATAKA (Fruit)	*Cordia dichotoma* Forst. F. Syn.*C. obliqua* Willd., C. *myxa* Roxb.	Boraginaceae	Dried, Ripe Fruits	**Foreign matter**- NMT 2 per cent; **Total ash**- NMT 9 per cent; **Acid-insoluble ash**- NMT 1 per cent; **Alcohol-soluble extractive**-NLT 7 per cent; **Water soluble extractive** -NLT 30 per cent; **T.L.C.** - Refer API	β-sitosterol, Palmitic, Stearic and Oleic acids.	Ama Phala *Rasa* - Madhura, Tikta, Kasaya; *Guna*- Laghu, Ruksa; *Virya*- Sita; *Vipaka*-Katu; *Karma*- Pittahara, Kaphahar, Grahi; Pakva-phala *Rasa* - Madhura *Guna*- Snigdha, Guru *Virya*- Sita *Vipaka*-Madhura; *Karma*- Pittahara, Brmhana, Vrsya, Rucya, Caksusya, Kesa-Krsnikarna	Gojihvadi Kvatha Churna	Jvara (Fever), Kasa (Cough), Krmi (Worm infestation), Pratisyaya (Coryza), Raktadosa (Disorders of blood), Raktapitta (Bleeding disorder), Sukradaurbalya (Seminal stress), Svasa (Asthma), Trsna (Thirst),	Pakva phala panaka (syrup of ripened fruit): 10 to 20 ml

Sl. No.	*Part and Volume*	*Page No.*	*Common Ayurvedic Name*	*Botanical Name/ Chemical Name*	*Family/ Chemical Formula*	*Part Used*	*Identity, Purity and Strength/Chemical Properties*	*Constituents*	*Properties and Action*	*Important Formulations*	*Therapeutic Uses*	*Dose*
											Upadamsa (Syphilis/ Soft chancre), Vatapittajanya Vikara (Disorders due to vata and Pitta dosa)	
485	Part-I; Vol. VI	154	SLESMATAKA (Stem Bark)	*Cordia dichotoma* Forst. f. Syn. *C. obliqua* Willd., *C. myxa* Roxb.	Boraginaceae	Dried Stem Bark	**Foreign matter**- NMT 2 per cent; **Total ash**- NMT 17 per cent; **Acid-insoluble ash**- NMT 0.6 per cent; **Alcohol-soluble extractive**-NLT 9 per cent; **Water-soluble extractive**-NLT 4 per cent; **T.L.C.** - Refer API	Gallic acid and β-sitosterol	*Rasa* - Madhura, Tikta, Kasaya, Katu *Guna*- Ruksa, Picchila *Virya*- Sita *Vipaka*- Katu; *Karma*- Pittahara, Kaphahara, Kesya, Vistambhi, Grahi, Krmighna, Pacana, Visaghna	Used as single drug	Amadosa (Semi-disgested food metabolites), Bahuvrana (Multiple injuies/Ulcers), Drkjata-masurika (Occular mainfestation of small pox), Krmi-sula (Colic due to worm Infestation), Kustha (Leprosy/ Diseases of Skin), Lutavisa (Spider bite), Masurika (Small pox), Raktadosa (Disorders of Blood), Tvak-roga (Skin diseases), Visarpa (Erysepelas), Visphota (Blister), Vrana (Ulcer).	Kvatha (decoction): 50 to 100 ml

Sl. No.	Part and Volume	Page No.	Common Ayurvedic Name	Botanical Name/ Chemical Name	Family/ Chemical Formula	Part Used	Identity, Purity and Strength/Chemical Properties	Constituents	Properties and Action	Important Formulations	Therapeutic Uses	Dose
486	Part-I; Vol. VI	156	SLIPADARI-KANDA (Tuber)	*Typhonium trilobatum* Schoot.	Araceae	Fresh or Dry Tuber	**Foreign matter**- NMT 1 per cent; **Total ash**- NMT 3 per cent; **Acid-insoluble ash**- NMT 1 per cent; **Alcohol-soluble extractive**-NLT 9 per cent; **Water-soluble extractive**-NLT 21 per cent; **T.L.C.** - Refer API	β-sitosterol and unidentified Sterols.	*Rasa* - Katu, Kasaya *Guna*- Tiksna, Ruksa *Virya*- Usna *Vipaka*- Katu; *Karma*- Arsoghna, Sothahara, Lekhana, Visaghna, Dipana, Pacana, Sulaprasamana	Used as single drug	Agnimandya (Digestive Impairment), Arbuda (Tumor), Arsa (Piles), Raktarsa (Bleeding Piles), Sotha (Odema), Sarpadamsa (Snake bite), Slipada (Fliariasis), Udararoga (Diseases of abdomen)	Churna (powder): 5 to 10 g daily dose after Sodhana
487	Part-I; Vol. VI	158	SPHITAKITARI (Rhizome)	*Dryopteris filix-mas* (L.) Schott. Syn. *Aspidium filix-mas* L.	Dryopteridaceae	Dried Rhizome	**Foreign matter**- NMT 2 per cent; **Total ash**- NMT 5 per cent; **Acid-insoluble ash**- NMT 0.1 per cent; **Alcohol-soluble extractive**-NLT 12 per cent; **Water-soluble extractive**- NLT 13 per cent; **Fixed oil** - NLT 3 per cent; **T.L.C.** - Refer API	Filicin; α-flavaspidic acid; Albaspidin; Filixic acid; Hexadeca aspidinol; Dropterin; Filmarone; β-Aspidin; 9-aliphatic alcohols and 3 Sterols.	*Rasa* - Katu *Guna*- Laghu, Ruksa *Virya*- Usna *Vipaka*- Katu; *Karma*- Lekhana, Virecana	Used as single drug	Jvara (Fever), Sphita Krmi (Tape worm), Vatarakta (Gotu)	Churna (powder): 1 to 3 g
488	Part-I; Vol. VI	160	SPRKKA (Whole Plant)	*Anisomeles malabarica* (L.) R. Br. Ex Sims	Lamiaceae	Dried Entire Plant	**Foreign matter**- NMT 2 per cent; **Total ash**- NMT 7 per cent; **Acid-insoluble ash**- NMT 2 per cent; **Alcohol-soluble extractive**-NLT 6 per cent; **Water-soluble extractive**-NLT 11 per cent; **T.L.C.** - Refer API	Triterpenic acid, Betulinic acid, two Diterpenoids *viz.*, Ovatodiolide and Anisomelic acid, Aerial parts contain five 14 membered	*Rasa* - Tikta, Katu, Kasaya *Guna*- Ruksa, Laghu, Tiksna *Virya*- Usna *Vipaka*- Katu; *Karma*- Vatahara, Kaphahara, Varnaprasadana, Anulomana, Lekhana, Visaghni	Sahacaradi Taila, Bala Taila, Baladhatryadi Taila	Asmari (Calculus), Kandu (Itching), Kaphavikara (Disorders due to vitiation of kapha dosa), Kasa (Cough), Kotha (Ringworm/	Churna (powder): 3 to 5 g

Sl. No.	Part and Volume	Page No.	Common Ayurvedic Name	Botanical Name/ Chemical Name	Family/ Chemical Formula	Part Used	Identity, Purity and Strength/Chemical Properties	Constituents	Properties and Action	Important Formulations	Therapeutic Uses	Dose
								Macrocylic diterpenes namely Anisomelode, β-sitosterol, Malabaric acid, 2-Acetoxy-malabaric acid, Anisomelyl acetate and Anisoelol, a Terpenoid, Anisomelin and a Flavone 4, 5-dihydroxy-3,6, 7-trimethoxy-flavone.			Impetigo/ Erythema), Mutrakrcchra (Dysuria), Pidaka (Carbuncle), Prameha (Metabolic disorder), Svasa (Asthma), Vrana (Ulcer)	
489	Part-I; Vol. VI	163	SRUVAVRKSA (Fruits)	*Flacourtia indica* (Burm.f.) Merr. Syn. F. *ramontchii Herit.*	Flacourtiaceae	Dried Fruit	**Foreign matter**- NMT 1 per cent; **Total ash**- NMT 5 per cent; **Acid-insoluble ash**- NMT 2 per cent; **Alcohol-soluble extractive**-NLT 20 per cent; **Water-soluble extractive**- NLT 21 per cent; **T.L.C.** - Refer API	Flacourside, and on Methyl 6-*O*-(*E*)-*P*-coumaroyl gluco-pyranoside, and 6-*O*-(*E*)-*P*-coumaroyl gluco-pyranose	*Rasa* -Madhura, Tikta *Guna*-Tiksna, Laghu, ruksa *Virya*-Usna *Vipaka*-Katu; *Karma*-Kaphahara, Mutral, Pacana, Pittahara, Rucya, Visaghna	Used as single drug	Agnimandya (Digestive impairment), Kamala (Jaundice), Plihavrddhi (Splenomegaly), Prameha (Metabolic disorder), Raktavikara (Disorders of blood), Sotha (Inflammation), Yakrdroga (Diseases of liver)	Churna (powder): 5 to 10 g

Sl. No.	*Part and Volume*	*Page No.*	*Common Ayurvedic Name*	*Botanical Name/ Chemical Name*	*Family/ Chemical Formula*	*Part Used*	*Identity, Purity and Strength/Chemical Properties*	*Constituents*	*Properties and Action*	*Important Formulations*	*Therapeutic Uses*	*Dose*
490	Part-I; Vol. VI	165	STHULAELA (Fruit)	*Amomum subulatum* Roxb.	Zingiberaceae	Dried Fruits	**Foreign matter**- NMT 2 per cent; **Total ash**- NMT 7 per cent; **Acid-insoluble ash**- NMT 2 per cent; **Alcohol-soluble extractive**-NLT 6 per cent; **Water-soluble extractive**-NLT 18 per cent; **Volatile oil** - NLT 1 per cent; **T.L.C.** - Refer API	Volatile oil predominantly containing Cineol with other Constituents such as α-pinene, β-pinene, Sabinene, Myrcene, α-terpinene, β-terpinene, Limonene, *P*-cymene, Terpineol, α-terpineol, δ-terpineol and Nerolidol.	*Rasa* - Tikta, Katu *Guna*- Laghu, Ruksa *Virya*- Usna *Vipaka*- Katu; *Karma*- Anulomana, Dipana, Hrdya, Kaphahara, Mutrala, Pittasaraka, Sirahsodhaka, Vatahara	Used as single drug	Aruci (Tastelessness), Bastivikara (Bladder disorder), Chardi (Emesis), Dantaroga (Disease of tooth), Hrllasa (Nausea), Kandu (Itching), Kantharoga (Disease of throat), Kasa (Cough), Mukharoga (Disease of mouth), Raktapitta (Bleeding disorder), Raktavikara (Disorders of blood), Siroroga (Disease of head). Sula (Pain/ Colic), Svasa (Asthma), Trsa (Thirst), Tvakroga (Skin diseases), Visavikara (Disorders due to poison), Vrana (Ulcer).	Churna (powder): 1 to 3 g

Sl. No.	Part and Volume	Page No.	Common Ayurvedic Name	Botanical Name/ Chemical Name	Family/ Chemical Formula	Part Used	Identity, Purity and Strength/Chemical Properties	Constituents	Properties and Action	Important Formulations	Therapeutic Uses	Dose
491	Part-I; Vol. VI	167	SUKANASA (Rhizome)	*Corallocarpus epigaeus* Benth. ex Hook. f. Syn. *Bryonia epigaea* Rotter; *Rhyncocarpa epigaea* Naud and *Aechmandra epigaea* Arn.	Cucurbitaceae	Rhizomes	**Foreign matter**- NMT 2 per cent; **Total ash**- NMT 4.5 per cent; **Acid-insoluble ash**- NMT 1.0 per cent; **Alcohol-soluble extractive**-NLT 5 per cent; **Water-soluble extractive**- NLT 14 per cent; **Fixed il** - NLT 1 per cent; **T.L.C.** - Refer API	Bryonin; Epigaeusyl ester; Corallo-carpuscala-rolide; Corallo-carpenoyl ester; Dotriacont-22,25-dio-10-one.	*Rasa* - Katu, Tikta *Guna*- Laghu, Ruksa, Tiksna *Virya*- Usna *Vipaka*- Katu; *Karma*- Sothahara, Vamana, Virecana, Visaghna	Kasmaryadi ghrta	Ama Vata (Rheumatism), Aruci (Tastelessness), Atisara (Diarrhoea), Daha (Burning sensation), Hikka (Hiccup), Jirna Antrasotha (Chronic intestinal pain), Jirnajvara (Chronic fever) Jvara (Fever), Kasa (Cough), Krmi roga (Worm infestation), Pravahika (Dysentery), Sarpa visa (Snake poison), Sotha (Inflammation), Svasa (Asthma) Vatakapha Jvara (Fever due to vata and kapha dosa), Visphotaka (Blisterous eruption), Vrana (Ulcer), Yoni roga (Disease of female genital tract)	Churna (powder): 3 to 5 g

Sl. No.	Part and Volume	Page No.	Common Ayurvedic Name	Botanical Name/ Chemical Name	Family/ Chemical Formula	Part Used	Identity, Purity and Strength/Chemical Properties	Constituents	Properties and Action	Important Formulations	Therapeutic Uses	Dose
492	Part-I; Vol. VI	169	SVETA VETASA (Leaf)	*Salix alba* L.	Salicaceae	Dried Leaves	**Foreign matter**- NMT 2 per cent; **Total ash**- NMT 10.5 per cent; **Acid-insoluble ash**- NMT 0.6 per cent; **Alcohol-soluble extractive**-NLT 11 per cent; **Water-soluble extractive**- NLT 66 per cent; **T.L.C.** - Refer API	Amentoflavone, Apigenin, (+)-Catechin, (+)-Gallo-catechin, Iso-quercetrin, Rutin, Narcissin, Iso-rhamnetin-3-*O*-β-*D*-glucoside, Salicin, Fragilin, Salicortin.	*Rasa* - Tikta, Kasaya *Guna*- Laghu, Ruksa *Virya*- Sita *Vipaka*- Katu; *Karma*- Grahi, Jvaraghna, Kaphahara, Mutrala, Raksoghna, Vedanasthapana, Vrana Sodhana	Used as single drug	Amavata (Rheumatism), Svitra (Leucoderma/ Vitiligo), Atisara (Diarrhoea), Kamala (Jaundice), Karnaroga (Disease of ear), Pravahika (Dysentery). Raktasthivana (Haemoptysis), Raktapitta (Bleeding disorder), Vatarakta (Gout)	Churna (powder): 3 to 6 g; Kasaya (decoction): 50 to 100 ml
493	Part-I; Vol. VI	171	TAKKOLA (Fruit)	*Illicium verum* Hook. f.	Magnoliaceae	Fruits	**Foreign matter**- NMT 2 per cent; **Total ash**- NMT 4 per cent; **Acid-insoluble ash**- NMT 2 per cent; **Alcohol-soluble extractive**-NLT 13 per cent; **Water-soluble extractive**- NLT 21 per cent; **Volatile oil** - NLT 3 per cent; **T.L.C.** - Refer API	Essential oils, Flavonol glycosides, and Veranisatins A, B, and C.	*Rasa* - Madhura, Katu *Guna*- Laghu, Snigdha, Tiksna *Virya*- Usna *Vipaka*- Madhura; *Karma*- Kaphahara, Dipana, Pacana, Vatanulomana, Mutrala, Vataghna, Kosthavata-Samana, Vedanahara	Karpuradi Churna	Adhmana (Flatulence with gurgling sound), Aruci (Tastelessness), Gulma (Abdominal Lump), Mukha-durgandha (Halitosis), Sandhivata (Arthritis), Sula (Pain/Colic)	Churna (powder): 250 to 625 mg

Sl. No.	*Part and Volume*	*Page No.*	*Common Ayurvedic Name*	*Botanical Name/ Chemical Name*	*Family/ Chemical Formula*	*Part Used*	*Identity, Purity and Strength/Chemical Properties*	*Constituents*	*Properties and Action*	*Important Formulations*	*Therapeutic Uses*	*Dose*
494	Part-I; Vol. VI	173	TINDUKA (Fruit)	*Diospyros peregrina* Gurke Syn. *Diospyros embryopteris* L.	Ebenaceae	Unripe and Ripe Fruits	**Foreign matter**- NMT 2 per cent; **Total ash**- NMT 6 per cent; **Acid-insoluble ash**- NMT 2 per cent; **Alcohol-soluble extractive**-NLT 10 per cent; **Water-soluble extractive**- NLT 16 per cent; **T.L.C.** - Refer API	Alkanes and triterpenoids. Seed contains Hexacosane and β-sitosterol, β-sitosterol glucoside, Gallic acid and Betulinic acid. Fatty oil (32 per cent), unsaponified matter and β-amyrin.	Pakva phala *Rasa* - Madhura, *Guna*-uru, Snigdha, *Virya*- Sita *Vipaka*- Madhura, *Karma*- Pittahara, Kaphahara, Durjana, Pustikara, Apakva-phala *Rasa* - Kasaya *Guna*- Laghu, Ruksa *Virya*- Sita *Vipaka*- Katu; *Karma*-- Vataprakopaka, Grahi, Lekhana	Used as single drug	Pakva Phala = Asmari (Calculus), Aruci (Tastelessness), Kapharoga (Disease due to kapha dosa), Prameha (Metabolic disorder), Raktadosa (Disorders of blood) Apakva Phala = Atisara (Diarrhoea), Bhagna (Fracture), Daha (Burning sensation), Kustha (Leprosy/ diseases of skin), Sotha (Oedema), Medoroga (Obesity), Pravahika (Dysentery), Raktapitta (Bleeding disorder), Udarda (Urticaria), Vrana (Ulcer)	Pakva phala (ripe)- Churna (powder): 5 to 10 g; Apakva phala (unripe)- Churna (powder): 4 to 8 g

Sl. No.	Part and Volume	Page No.	Common Ayurvedic Name	Botanical Name/ Chemical Name	Family/ Chemical Formula	Part Used	Identity, Purity and Strength/Chemical Properties	Constituents	Properties and Action	Important Formulations	Therapeutic Uses	Dose
495	Part-I; Vol. VI	175	TRAYAMANA (Rhizome)	*Gentiana kurroo* Royle	Gentianaceae	Rhizome	**Foreign matter**- NMT 2 per cent; **Total ash**- NMT 7 per cent; **Acid-insoluble ash**- NMT 2 per cent; **Alcohol-soluble extractive**-NLT 28 per cent; **Water-soluble extractive**- NLT 13 per cent; **T.L.C.** - Refer API	Bitter Crystalline Glycoside-Picrorhizin (3 to 4 per cent) cathartic acid. Secoiridoids like Picroside A and Kutuoside.	*Rasa* - Tikta, Kasaya *Guna*- Sara *Virya*- Usna *Vipaka*- Katu; *Karma*- Pittahara, Kaphahara, Visaghna	Trayamana ghrta, Trayamana kvatha, Maha paisacika ghrta	Atisara (Diarrhoea), Bhrama (Vertigo), Gulma (Abdominal lump), Hrdroga (Heart disease), Jvara (Fever), Raktapitta (Bleeding disorder, Raktavikara (Disorders of Blood), Sula (Pain/Colic), Sutikasula (Postpartum abdominal Pain), Trsna (Thirst), Visarpa (Erysepelas)	Churna (powder): 1 to 3 g
496	Part-I; Vol. VI	177	TRIPAKSI (Whole Plant)	*Coldenia procumbens* L.	Boraginaceae	Whole Plant	**Foreign matter**- NMT 1 per cent; **Total ash**- NMT 13 per cent; **Acid-insoluble ash**- NMT 2 per cent; **Alcohol-soluble extractive**-NLT 10 per cent; **Water-soluble extractive**- NLT 17 per cent; **fixed oil** - NLT 3 per cent; **T.L.C.** - Refer API	Steroid Glycosides	*Rasa* - Tikta, Kasaya *Guna*- Laghu, Ruksa *Virya*- Usna *Vipaka*- Katu; *Karma*- Kaphaghna, Pacana, Sothaghna, Vatahara	Used as single drug	Amavata (Rheumatism), Vidradhi (Abscess)	Churna (powder): 3 to 6 g

Sl. No.	*Part and Volume*	*Page No.*	*Common Ayurvedic Name*	*Botanical Name/ Chemical Name*	*Family/ Chemical Formula*	*Part Used*	*Identity, Purity and Strength/Chemical Properties*	*Constituents*	*Properties and Action*	*Important Formulations*	*Therapeutic Uses*	*Dose*
497	Part-I; Vol. VI	180	TUVARAKA (Seed)	*Hydnocarpus pentandra* (Buch.-Ham.) Oken Syn. *H. laurifolia* (Dennst.) Sleummer., *H. wightiana* Blume	Flacourtiaceae	Dried Seeds	**Foreign matter**- NMT 2 per cent; **Total ash**- NMT 4 per cent; **Acid-insoluble ash**- NMT 1 per cent; **Alcohol-soluble extractive**-NLT 35 per cent; **Water-soluble extractive**- NLT 12 per cent; **T.L.C.** - Refer API	Apigenin, Hydnocarpin, Isohydno-carpine methoxyhy-dnocarpin and Fixed oils.	*Rasa* - Tikta, madhura, Kasaya *Guna*- Snighdha, Tiksna *Virya*- Usna *Vipaka*- Katu; *Karma*- Vatahara, Kaphahara, Rasayana, Ubhayato-bhagahara	Tuvaraka Taila	Anaha (Distension of abdomen due to obstruction to passage of urine and stools), Arsa (Piles), Grdhrasi (Sciatica), Gandamala (Cervical lymphadenitis), Gulma (Abdominal lump), Jvara (Fever), Kandu (Itching), Kaphavataja roga (Disorders due to kapha and vata dosa), Krmi (Helminthiasis), Kustha (diseases of Skin), Sotha (Oedema), Prameha (Metabkolic disorder), Raktavikara (Disorders of blood), Tvakroga (Skin diseases), Udara (Urticaria), Udavarta (Partial obstruction), Vrana (Ulcer)	Churna (powder): 1 to 3 g

Sl. No.	*Part and Volume*	*Page No.*	*Common Ayurvedic Name*	*Botanical Name/ Chemical Name*	*Family/ Chemical Formula*	*Part Used*	*Identity, Purity and Strength/Chemical Properties*	*Constituents*	*Properties and Action*	*Important Formulations*	*Therapeutic Uses*	*Dose*
498	Part-I; Vol. VI	182	USANDI (Whole Plant)	*Glinus lotoides* L. Syn. *Mollugo hirta* Thub., *M. lotoides* Kuntz.	Aizoaceae	Whole Plant	**Foreign matter**- NMT 2 per cent; **Total ash**- NMT 12 per cent; **Acid-insoluble ash**- NMT 1 per cent; **Alcohol-soluble extractive**-NLT 8 per cent; **Water-soluble extractive**-NLT 23 per cent; **Fixed oil - NLT 3 per cent; T.L.C.** - Refer API	Mollugogenol A, B, C, D, E, F and G; Mollugocin A and B; β-and γ-sitosterol glucosides; Oleanolic acid; Flavonoids like Apigenin-8-C-glucoside; Apigenin-7-rhamno-glucoside; Pelargonidin-3-sophorsido-7-glucoside; Sulfuretin; Vicenin 2; Vitexin.	*Rasa* - Kasaya, Tikta *Guna*- Laghu, Ruksa *Virya*- Sita *Vipaka*- Katu; *Karma*- Jvaraghna, Kapha-pittahara, Paustika, Sothahara, Stambhana, Udardaprasamana	Used as single drug	Atisara (diarrhoea), Raktapitta (bleeding disorder), Udararoga (Diseases of abdomen), Vidradhi (Abscess), Vrana (Ulcer)	Churna (powder): 3 to 6 g
499	Part-I; Vol. VI	185	VAJRANNA (Leaf Base)	*Pennisetum typhoides* (Burm.) Stapf and C.E. Hubb, syn. *P. Typhoideum* Rich., *P. spicatum* Roem and Schult	Poaceae (Graminae)	Dried Sheathy Leaf Bases	**Foreign matter**- NMT 2 per cent; **Total ash**- NMT 15 per cent; **Acid-insoluble ash**- NMT 12 per cent; **Alcohol-soluble extractive**-NLT 6 per cent; **Water-soluble extractive**-NLT 15 per cent; **T.L.C.** - Refer API	Flavonoid, Alkaloids, Tannins, Phenols and Saponin	*Rasa* - Madhura, Kasaya *Guna*- Ruksa, Guru *Virya*- Usna *Vipaka*- Amla; *Karma*- Balya, Durjara, Hrdya, kaphavatahara, Pittahara, Pumstvahara, Vatakara	Used as single drug	Prameha (Metabolic disorder), Saitya (Coldness), Santar-panajanya roga (Disorders due to Obesity), Sthaulya (Obesity)	Svarasa (Juice): 10 to 200 ml
500	Part-I; Vol. VI	187	VALUKA-SAKA (Leaf)	*Gisekia pharnaceoides* L. Syn. *G. molluginoides* Wt.	Aizoaceae	Dried Leaves	**Foreign matter**- NMT 2 per cent; **Total ash**- NMT 12 per cent; **Sulphated ash** - NMT 20 per cent; **Acid-insoluble ash**- NMT 1 per cent; **Alcohol-soluble extractive**-NLT 7 per cent; **Water-soluble extractive**-NLT 30 per cent; **T.L.C.** - Refer API	Oxalic, Tartaric, Citric and Succinic acids besides Triacontane, Myristone, Tetracosanol and Dotriacontane.	*Rasa* - Tikta, Kasaya *Guna*- Laghu, Ruksa *Virya*- Sita *Vipaka*- Katu; *Karma*- Anulomana, Krmighna, Kusthaghna, Durgandhanasana	Lavangadya churna	Kandu (Itching), Krmi (Helminthiasis), Kustha (Leprosy/ Diseases of skin), Raktapitta (Bleeding disorder)	Churna (powder): 3 to 6 g

Sl. No.	*Part and Volume*	*Page No.*	*Common Ayurvedic Name*	*Botanical Name/ Chemical Name*	*Family/ Chemical Formula*	*Part Used*	*Identity, Purity and Strength/Chemical Properties*	*Constituents*	*Properties and Action*	*Important Formulations*	*Therapeutic Uses*	*Dose*
501	Part-I; Vol. VI	189	VANYA-ASVAGOLA (Fresh Leaf)	*Plantago lanceolata* L.	Plantaginaceae	Fresh Leaves	**Foreign matter**- NMT 2 per cent; **Total ash**- NMT 24.5 per cent; **Acid-insoluble ash**- NMT 1.7 per cent; **Alcohol-soluble extractive**-NLT 12 per cent; **Water-soluble extractive**- NLT 35 per cent; **T.L.C.** - Refer API	Chlorogenic acid, Chrysophanic acid, Emodin, Luteolin, Plantaginin, Scutellarin, Aesculetin	*Rasa* - Kasaya, madhura *Guna*- Snigdha, Guru *Virya*- Sita *Vipaka*- Madhura; *Karma*- Mutrala, Rakta-stambhana, Rasayana, Sothahara, Srmsana, Vedanasamaka	Used as single drug	Arsa (Piles), Karanasula (Otalgia), Asragdara (Menorrhagia or Metrorrhagia or both), Dantasula (Toothache), Kasa (Cough), Raktasrava (Hemmorhage), Sotha (Oedema), Svasa (Asthma), Vrana (Ulcer).	Patra Svarasa (Leaf juice): 5 to 10 ml
502	Part-I; Vol. VI	191	VETRA (Rhizome)	*Calamus rotang* L.	Arecaceae	Rhizomes	**Foreign matter**- NMT 2 per cent; **Total ash**- NMT 3 per cent; **Acid-insoluble ash**- NMT 1 per cent; **Alcohol-soluble extractive**-NLT 10 per cent; **Water-soluble extractive**- NLT 9 per cent; **T.L.C.** - Refer API	Saponins, Alkaloids and Flavonoids.	*Rasa* - Katu, Tikta *Guna*- Laghu *Virya*- Sita *Vipaka*- Katu; *Karma*- Chedana, Dipana, Kaphahara, Mutrala, Pittahara, Visaghna	Used as single drug	Arsa (Piles), Aruci (Tastelessness), Asmari (Calculus), Daha (Burning sensation), Jvara (Fever), Kasa (Cough), Kustha (Leprosy/ diseases of skin), Mutrakrcchra (Dysuria), Prameha (Metabolic disorder), Pravahika (Dysentery), Raktapitta (Bleeding disorder),	Kvatha (decoction): 50 to 100 ml; Churna (powder): 5 to 10 g

Sl. No.	*Part and Volume*	*Page No.*	*Common Ayurvedic Name*	*Botanical Name/ Chemical Name*	*Family/ Chemical Formula*	*Part Used*	*Identity, Purity and Strength/Chemical Properties*	*Constituents*	*Properties and Action*	*Important Formulations*	*Therapeutic Uses*	*Dose*
											Sotha (Inflammation), Trsna (Thirst), Tvakroga (Skin diseases), Visarpa (Erysepelas), Yoniroga (Disease of femal genital tract)	
503	Part-I; Vol. VI	193	VISANIKA (Whole Plant)	*Pergularia daemia* (Forsk) Chiov. Syn. *Daemia extensa* (Jacq.) R.Br.	Asclepiadaceae	Whole Plant	**Foreign matter**- NMT 2 per cent; **Total ash**- NMT 11 per cent; **Acid-insoluble ash**- NMT 1 per cent; **Alcohol-soluble extractive**-NLT 6 per cent; **Water-soluble extractive**- NLT 14 per cent; **T.L.C.** - Refer API	Several cardenolides such as calotropin, calactin, calotropagenin, uzarigenin, coroglaucigenin and triterpenoids, β-amyrin and lupeol	*Rasa* - Katu, Kasaya *Guna*- Laghu, Ruksa, Visada *Virya*- Anusna *Vipaka*- Katu; *Karma*- Kaphanihsaraka, Dipana, Virecana, Kusthaghna	Used as single drug	Mahakustha (Group of major skin diseases), Agnimandya (Digestive impairment), Vibandha (Constipation), Yonidosa (Disorder of female genital tract), Svasa (Asthma), Sotha (Inflammation), Mutrakrcchra (Dysuria)	Churna (powder): 1 to 3g
504	Part-I; Vol. VI	196	VRANTA-MLAPHALA (Fruit Rind)	*Garcinia pedunculata* Roxb.	Guttifeare	Fruit Rind	**Foreign matter**- NMT 2 per cent; **Total ash**- NMT 3 per cent; **Acid-insoluble ash**- NMT 2 per cent; **Alcohol-soluble extractive**-NLT 39 per cent; **Water-soluble extractive**- NLT 42 per cent; **Fixed oil** - NLT 1 per cent; **T.L.C.** - Refer API	Pedunculol; Garcinol; Cambogin.	*Rasa* - Amala, Kasaya *Guna*- Ruksa, Tiksna, Snigdha, Laghu *Virya*- Usna *Vipaka*- Amla; *Karma*- Anulomaka, Bhedana, Dipana, Kaphahara, Mutrala, Pacana, Vatahara	Used as single drug	Anaha (Distension of abdomen due to intestinal obstruction), Ajirna (Indigestion), Asmari (Calculus), Arsa (Piles), Aruci (Tastelessness),	Svarasa (juice): 5 to 10 ml

Sl. No.	*Part and Volume*	*Page No.*	*Common Ayurvedic Name*	*Botanical Name/ Chemical Name*	*Family/ Chemical Formula*	*Part Used*	*Identity, Purity and Strength/Chemical Properties*	*Constituents*	*Properties and Action*	*Important Formulations*	*Therapeutic Uses*	*Dose*
											Gulma (Abdominal lump), Hrdroga (Heart disease), Hikka (Hiccup), Krmi (Worm infestation), Kasa (Cough), Pliharoga (Splenic disease), Sula (Pain/ Colic), Svasa (Asthma), Udavarta (upward movement of gases), Vibandha (Constipation)	
505	Part-I; Vol. VI	198	VRSCIKA-KANDA (Rhizome)	*Doronicum hookeri* C.B. Clarke	Asteraceae	Dried Rhizomes	**Foreign matter**- NMT 2 per cent; **Total ash**- NMT 4 per cent; **Acid-insoluble ash**- NMT 0.7 per cent; **Alcohol-soluble extractive**-NLT 6.6 per cent; **Water-soluble extractive**- NLT 20 per cent; **T.L.C.** - Refer API	Essential oil	*Rasa* - Tikta *Guna*- Ruksa, Laghu, Sugandhi *Virya*- Usna *Vipaka*- Katu; *Karma*- Anulomana, Kaphahara, Visaghna, Hrdbalya, Jvaraghna	Used as single drug	Anaha (Distension of abdomen due to intestinal obstruction), Ardita (Facial palsy), Damsavisa (Poisoning due to bites), Garbhasayasula (Uterine pain), Hrdroga (Heart disease), Paksavadha (Paralysis/ Hemiplegia), Udarasula (Pian in the Abdomen),	Churna (powder): 1 to 3g

Sl. No.	*Part and Volume*	*Page No.*	*Common Ayurvedic Name*	*Botanical Name/ Chemical Name*	*Family/ Chemical Formula*	*Part Used*	*Identity, Purity and Strength/Chemical Properties*	*Constituents*	*Properties and Action*	*Important Formulations*	*Therapeutic Uses*	*Dose*
											Vrscika Damsa (Scorpian bites), Vataroga (Disease due to Vata dosa), Vatika Unmada (Mania/ Psychosis), Granthikajvara (Bubonic plague)	
506	Part-I; Vol. VI	200	DARUSITA TAILA (Cinnamomum Oil)	*Cinnamomum zeylanicum* Blume	Lauraceae	Distilled from the Dried Inner Bark of the Shoots of Coppiced Tree	**Optical rotation** - 0° to - 2° **Refractive index** - 1.573 to 1.600 **Weight per ml** - 1.000 to 1.040 g **Assay** - NLT 55.0 per cent, w/w and NMT70.0 per cent, w/w of cinnamaldehyde, C_9H_8O; **Microbial limits** - Complies with API; **Pesticide residue** - Complies with API;	–	*Rasa* - Madhura, Tikta, Katu *Guna*- Laghu, Ruksa, Tiksna *Virya*- Usna *Vipaka*- Katu; *Karma*- Artavapravartaka, Balya, Dantya, Dipana, Kanthya, Mukhadur-gandhanasana, Pacana, Pittahara, Pratidusaka, Sugandhi, Sukrajanana, Uttejaka, Vatahara, Vatanulomaka, Vranasodhaka, Vranaropaka	Used as single drug	Adhmana (Fatulence with gurgling sound), Amadosa (Products of impaired digestion and metabolism), Amasaya sula (Peptic ulcer), Antrika Pratidusaka (Enteritis), Arsa (Piles), Chardi (Emesis), Dantasula (Toothache), Dhvajabhanga (Failur of penile erection), Krmi (Helminthiasis/ worm infestation), Ksayaj vrana (Tubercular wound), Mukhasosa (Dryness of mouth),	1 to 3 drops

Sl. No.	*Part and Volume*	*Page No.*	*Common Ayurvedic Name*	*Botanical Name/ Chemical Name*	*Family/ Chemical Formula*	*Part Used*	*Identity, Purity and Strength/Chemical Properties*	*Constituents*	*Properties and Action*	*Important Formulations*	*Therapeutic Uses*	*Dose*
											Nadisula (Acute pain of nervine origin), Pinasa (Chronic rhinitis/ sinusitis), Pratisyaya (Coryza), Rajayaksma (Tuberculosis), Raktavikara (Disorders of blood), Sula (Pain) Trsna (Thrist), Vrscika dansa (Scorpion bile)	
507	Part-I; Vol. VI	202	GANDHA-PURA PATRA TAILA	*Gaultheria fragrantissima* Wall.	Ericaceae	Steeping and Fermentation of Fresh Leaves	**Identification** - Taike 2 ml of oil, add a drop of *ferric chloride solution*, a violet colour is produced.; **Specific gravity** - At 15.5°, 1.180 to 1.187; **Optical rotation** - At 25°, 0° to -1° **Refractive index** - At 20°, 1.537 to 1.539; **Assay** -Determination of esters (methyl salicylate $C_8H_8O_3$) - NLT 98 per cent	–	*Rasa* - Madhura, Tikta, Katu *Guna*- Tiksna, Snigdha *Virya*- Usna *Vipaka*- Katu; *Karma*- Putihara, Sangrahi, Svedala, Uttejaka, Vatahara, Vatanulomaka, Vedanasthapana	Used as single drug	Amavata (Rheumatism), Ankusa krmi (Hookworm infestation), Atisara (Diarrhoea), Dantasula (Toothache), Grdhrasi (Sciatica), Jvara (Fever), Nadisula (Acute pain of nervine origin), Udarakrmi (Intestinal worms), Vatarakta (Gotu)	0.1 to 0.5 ml

Sl. No.	*Part and Volume*	*Page No.*	*Common Ayurvedic Name*	*Botanical Name/ Chemical Name*	*Family/ Chemical Formula*	*Part Used*	*Identity, Purity and Strength/Chemical Properties*	*Constituents*	*Properties and Action*	*Important Formulations*	*Therapeutic Uses*	*Dose*
508	Part-I; Vol. VI	204	GOGHRTA (Clarified Cow's Butter)	-	–	Carlified cow's Butter Derived from Cow's Milk	**Specific gravity** - At 25°, 1.01995; **Reichert Meissel Value** - 24-28, **Moisture** - NMT 0.5 per cent; **Saponification Value** - NMT 225 **Iodin Value** - NMT 35 **Unsaponifiable matter** - NMT 1.5 **Carotene** - NLT 2000 IU **Microbial limits** - Complies with API; **Heavy Metals** - Complies with API	–	*Rasa* - Madhura *Guna*- Guru, Snigdha, Mrdu *Virya*- Sita *Vipaka*- Madhura; *Karma*- Agnidipana, Anabhisyandi, Ayusya, Balya, Caksusya, Dipana, Hrdya, Kantiprada, Medhya, Ojovardhaka, Rasayana, Rucya, Slesmavardhana, Snehana, Sukravardhaka, Tejobalakara, Tvacya, Vatapitta-prasamana, Vayahsthapna, Visahara, Vrsya	Brahmi ghrta, Triphala ghrta, Asoka ghrta, Eladi ghrta, Cangeri ghrta, Amrta ghrta	Agnidagdha (Fire burns), Amlapitta (Hyperacidity), Apasmara (Epilepsy), Aruci (Tastelessness), Grahani (Malabsorption syndrome), Jirnajvara (Chronic fever), Karnasula (Otalgia), Ksataksina (Debility due to chest injury), Mada (Intoxication), Murccha (Syncope), Sirahsula (Headache), Smrtinasa (Loss of memory), Sosa (Cachexia), Unmada (Mania/ Psychosis), Visamajvara (Intermittent fever), Visarpa (Erysepales), Visavikara (Disorders due to poison), Yonisula (Pain in female genital tract)	5 to 20 ml

Sl. No.	*Part and Volume*	*Page No.*	*Common Ayurvedic Name*	*Botanical Name/ Chemical Name*	*Family/ Chemical Formula*	*Part Used*	*Identity, Purity and Strength/Chemical Properties*	*Constituents*	*Properties and Action*	*Important Formulations*	*Therapeutic Uses*	*Dose*
509	Part-I; Vol. VI	206	GUDA (Jaggery)	*Saccharum officinarum* L.	Poaceae	Concen-trating Juice expressed from the Stems	**Loss on Drying** - NMT 10 per cent; **(Other thanthat of the liquid or semi-liquid variety) Total ash** - NMT 6 per cent **Acid-insoluble ash** - NMT 0.5 per cent **Water-insoluble matter** - NMT 2 per cent **Total sugars** - NLT 90 per cent; **Sucrose** - NLT 60 per cent; **Sulphur dioxide concentration** - NMT 70 ppm; **Heavy metals** - Complies with API; **Microbial limits** - Complies with API; **Pesticide residue** - Complies with API;	-	*Rasa* - Madhura *Guna*- Snigdha, Isatksariya *Virya*- Natisita *Vipaka*- Madhura; *Karma*- Svadukara, Rakta sodhaka, Natipittajit, Kaphavrddhikara, Vataghna, Krmivrddhikara, Balya, Vrsya, Medovrddhikara	Sarivadyasava, Kumaryasava, Madhukasava	Vataroga (Disease due to vata dosa), Daurbalya (Weakness), Dhatuksaya (Tissue wasting)	5 to 30 g
510	Part-I; Vol. VI	208	JALA (Potable Water)	-	–	–	**Colour (Hazen Units)** - **NMT 5; Odour** - **None; Taste** - Agreeable and refreshing; **Turbidity** (NTU) - NMT 5; **pH** - 6.5-8.5 **Alkalinity (mg/l)** - NMT 200 **Total hardness(as $CaCO_3$) (mg/l)** - NMT 300; **Iron (as Fe) (mg/l)** -NMT 0.3; **Chlorides (as CL) (mg/l)** - NMT 250; **Residual, free Chlorine (mg/l)** - NMT 0.2; **Dissolved Solids (mg/l)**- NMT 500; **Calcium (as Ca) (mg/l)**- NMT 75; **Copper (as Cu) (mg/l)** - 0.05; **Manganese (as Mn) (mg/l)** - NMT 0.1; **Sulphate (as SO_4) (mg/l)** - NMT 200; **Nitrate (as NO_3) (mg/l)**- NMT 45; **Fluoride (as F) (mg/l)** -	-	*Rasa* - Madhura *Guna*- Laghu *Virya*- Sita *Vipaka*- Madhura; *Karma*- Ahaladana, Alasyahara, Balya, Buddhiprada, Dipana, Hrdya, Hrtbalakara, Kaphahara, Klamahara, Medohara, Nidrahara, Pacana, Pathya, Pittasamaka, Rucya, Santarpana, Saumya, Sramahara, Tarpana, Vatahara, Visahara, Vrsya	Kvatha, Hima, Phanta, Asava, Arista	Ajirna (Dyspepsia), Bhranti (Mental confusion), Chardi (Emesis), Daha (Burning sensation), Krodha (Anger), Moha (Delusion), Mukhasosa (Dryness of mouth), Murccha (Syncope), Sosa (Cachexia). Tandra (Drowsiness), Trsna (Thirst), Vibandha (Constipation), Visavikara	Q.S

Sl. No.	Part and Volume	Page No.	Common Ayurvedic Name	Botanical Name/ Chemical Name	Family/ Chemical Formula	Part Used	Identity, Purity and Strength/Chemical Properties	Constituents	Properties and Action	Important Formulations	Therapeutic Uses	Dose
							NMT 1; **Phenolic compounds (as C_6H_5OH) (mg/l)** - NMT 0.001; **Heavy Metals** - Complies with API; **Arsenic** - Complies with API; **Microbial Limits: Colifrom organisms** - Absent; **E.coli** - Absent; **Pesticides (mg/l)** - Absent;				(Disorders due to poison)	
511	Part-I; Vol. VI	210	KARPURA (Natural camphor)	*Cinnamomum camphora* (L.) Nees and Eberm. and *Ocimum kilimand-scharicum* Guerke	Lamiaceae	leaves, chipped wood and root and whole plant	**Identification** - Volatilises at ordinary temperature and readily burns with a smoky flame; **Melting Range**- 174° to 179°; **Specific Optical Rotation**- +41° + 43° **(Synthetic Camphor is the optically inactive, racemic form)**; **Non-volatile Matter** - NMT 0.05 per cent; **Pesticide residue** - Complies with API; **Assay** - Refer API	-	*Rasa* - Tikta, Katu, Madhura *Guna*- Laghu, Tiksna, Snigdha *Virya*- Sita *Vipaka*- Katu; *Karma*- Caksusya, Durgandhanasaka, Hrdya, Lekhana, Madakaraka, Medya, Pacana, Tridosahara, Vedanasthapana, Vrsya	Karpura rasa, Karpurasava, Arka Kapura, Khadiradivati, Mrdvikarista	Adhmana (Flatulence with gurgling sound), Agnimandya (Digestive impairment), Amavata (Rheumatism), Aruci (Tastelessness) Atisara (Diarrhoea), Daha (Burning sensation), Dantapuya (Pyorrhoea), Dantasula (Toothache), Jirnapratisyaya (Chronic sinusitis), Kandu (Itching), Kantharoga (Disease of throat), Kasa (Cough), Klaibya (Male impotence),	125 to 375 mg

Sl. No.	*Part and Volume*	*Page No.*	*Common Ayurvedic Name*	*Botanical Name/ Chemical Name*	*Family/ Chemical Formula*	*Part Used*	*Identity, Purity and Strength/Chemical Properties*	*Constituents*	*Properties and Action*	*Important Formulations*	*Therapeutic Uses*	*Dose*
											Krmi (Helminthiasis/ Worm infestation), Kustha (Disease of Skin), Medoroga (Obesity), Parsvasula (Intercoastal neuralgia and Pleurodynia), Sandhisuala (Joint pain), Svasa (Asthma), Trsna (Thirst), Tvakroga (Skin diseases), Vicarcika (Eczema), Visavikara (Disorders due to poison), Visucika (Gastro- enteritis with piercing pain), Vrkkaroga (Renal disorder)	
512	Part-I; Vol. VI	112	LAVANGA TAILA (Clove Oil)	*Syzygium aromaticum* Merril and Perry Syn. *Eugenia caryophyllus* (Spreng) Sprague	Myrtaceae	Volatile oil obtained by Expression or steam distillation from dried, unopened flower buds	**Specific gravity** - 1.047 - 1.060; **Optical rotation** - 0° to - 1.5°; **Refractive index** - 1.528 to 1.537; **Weight per ml** - 1.041 to 1.054 g; **Microbial limits** - Complies with API; **Pesticide residue** - Complies with API	–	*Rasa* - Katu, Tikta *Guna*- Snigdha, Laghu *Virya*- Sita *Vipaka*- Katu; *Karma*- Agnikrt, Kaphaghna, Mukhasodhaka, Durgandhanasana, Vaktrakledanasana	Used as single drug	Trsna (Thirst), Garbhinichardi (Morning sickness), Dantavestaroga (Gingivitis), Kaphajanya pida (Pain due to kapha dosa)	2 to 6 drops

Sl. No.	*Part and Volume*	*Page No.*	*Common Ayurvedic Name*	*Botanical Name/ Chemical Name*	*Family/ Chemical Formula*	*Part Used*	*Identity, Purity and Strength/Chemical Properties*	*Constituents*	*Properties and Action*	*Important Formulations*	*Therapeutic Uses*	*Dose*
513	Part-I; Vol. VI	214	MADHU (Honey)	-	–	Sweet Fluid Produced by the Honeybees	**Wt. per ml at 25°C** - NLT 1.35; **Moisture content (LOD)** - NMT 25 per cent by wt; **Reducing sugars** - NMT 65 per cent by wt; **Sucrose** - NMT 5.0 per cent by wt; **Fructose-Glucose ratio** - NMT 1 per cent by wt; **Ash** - NMT 0.50 per cent by wt; **Acidity (expressed as Formic acid)**- NMT 0.2 per cent by wt; **Fiehe's Test**- Negative; **Aniline Chloride Test** - Negative; **Heavy metals** - Complies with API; **Microbial limits** - Complies with API; **Pesticide residue** - Complies with API	–	*Rasa* - Madhura, Kasaya *Guna*- Laghu (Susruta), Guru (Caraka), Ruksa, Picchila, Yogavahi *Virya*- Sita *Vipaka*- Katu; *Karma*- Agnidipana, Caksusya, Pittaprasamana, Prasadana, Ropana, Sandhana, Slesmaprasamana, Sodhana, Tridosaprasamana, Vatapittaghna, Visaghna	Madhukasava, Chyawanaprasa, Kutajavaleha	Arsa (Piles), Atisara (Diarrhoea), Chardi (Emesis), Daha (Burning sensation), Hikka (Hiccup), Kasa (Cough), Krmi (Helminthiasis/ Worm infestatin), Ksata (Wound), Ksaya (Pthisis), Kustha (Diseases of skin), Medoroga (Obesity), Prameha (Incresed frequency and turbidity of urine), Raktapitta (Bleeding, disorder), Raktavikara (Disorders of blood), Svasa (Asthma), Trsna (Thirst), Visavikara (Disorders due to poison)	1 to 10 ml

Sl. No.	Part and Volume	Page No.	Common Ayurvedic Name	Botanical Name/ Chemical Name	Family/ Chemical Formula	Part Used	Identity, Purity and Strength/Chemical Properties	Constituents	Properties and Action	Important Formulations	Therapeutic Uses	Dose
514	Part-I; Vol. VI	216	PEPPERMINT-SATVA (Menthol)	Various species of Mentha	Lamiaceae	–	**Acidity or Alkalinity** - A Solution in alcohol is neutral to litmus. **Non-volatile matter** - NMT 0.05 per cent; **Melting range** - Between 42° and 44°; **Specific optical rotation** - Between -49° and -50°; **Congealing range** - Between 27° and 28°; on prolonged stirring the temperature rises between 30° and 32°	–	*Rasa* - Tikta, Katu *Guna*- Tiksna, Snigdha, Laghu, Visada *Virya*- Usna *Vipaka*- Katu; *Karma*- Dipana, Kaphahara, Mukha-sodhana, Pacana, Putihara, Sulaprasamana, Uttejaka, Vatahara, Vedanasthapana	Used as single drug	Ajirna (Dyspepsia), Dantasula (Toothache), Jirna jvara(Chronic fever), Kaphaja vikara (Disorders due to kapha dosa), Mukha-Roga (Diseases of mouth), Udarasula (Pain in the abdomen), Sula (Pain/Colic), Vrana (Ulcer)	10 to 30 mg
515	Part-I; Vol. VI	218	SARKARA (Sugar)	-	–	–	**Moisture content** - NMT 1.5 per cent by wt.; **Acid-Insoluble Ash** - NMT 0.7 per cent wt.; **Sucrose** - NMT 93 per cent by wt; **Sulphur dioxide** - Absent; **Calcium Oxide** - NMT 100 (mg/100g); **Heavy Metal** - Complies with API; **Microbial Limit** - Complies with API; **Pesticide Residue** - Complies with API; **Storage** - Should be stored in air tight container	–	*Rasa* - Madhura *Guna*- Snigdha *Virya*- Sita *Vipaka*- Madhura; *Karma*- Caksusya, Dhatuvardhaka, Hrdya, Pittahara, Vatahara, Vrsya	Chyawanaprasa, Vasavaleha, Kanta-karyavaleha	Arsa (Piles), Aruci (Tastelessness), Bhrama (Vertigo), Chardi (Emesis), Daha (Burning sensation), Daurbalya (Weakness), Jvara (Fever), Krmi (Helminthiasis/ Worm Infestation), Ksata (Wound), Madatyaya (Alcoholism), Moha (Delusion), Murccha (Syncope),	5 to 30 g

Sl. No.	Part and Volume	Page No.	Common Ayurvedic Name	Botanical Name/ Chemical Name	Family/ Chemical Formula	Part Used	Identity, Purity and Strength/Chemical Properties	Constituents	Properties and Action	Important Formulations	Therapeutic Uses	Dose
											Raktapitta (Bleeding disorder), Raktasruti (Haemorrage), Raktavikara (Disorders of blood), Srama (Fatigue/ lethargy) Trsna (Thirst), Vatarakta (Gout), Visavikara (Disorders due to poison)	
516	Part-I; Vol. VI	220	SARSAPA TAILA (Mustard Oil)	*Brassica campestris* L.	Brassicaceae	Fixed Oil expressed from clean and Healthy Seeds	**Specific gravity at 15°** - 0.9140-0.9206; **Refractive Index at 40°** - 1.4630 -1.4670; **Essential Oil content** - NLT 0.4 per cent; **Acid value** - NMT 6.0; **Iodine value** - Between 115 and 125; **Saponification value** - Between 190 and 198; **Unsaponifiable matter** - NMT 1.5 per cent by wt. **Test for Sulphur** - Positive; **Test for Argemone oil** - Negative; **Heavy Metals**- Complies with API; **Microbial Limits** - Complies With API; **Pesticide residue** - Complies with API	-	*Rasa* - Tikta, Katu *Guna*- Snigdha, Tiksna, Laghu *Virya*- Usna *Vipaka*- Katu; *Karma*- Dipana, Garbhasayottejaka, Kaphara, Krmighna, Lekhana, Mutrajanana, Snehana, Tvacya, Vatahara, Vedanasthapana, Vidahi	Unmatta Taila, Pancanana Taila, Sinduradya Taila, Jirakadya Taila, Arkamanhsila Taila	Angamarda (Bodyache), Arsa (Piles), Dantapuya (Pyorrhoea), Dustakrmi (Worm infestation), Karnaroga (Disease of ear), Kandu (Itching), Kotha (Urticaria), Krmi (Helminthiasis/ Worm infestation), Kustha (Leprosy/ diseases of skin), Netraroga (Diseases of eyes), Pliha roga	5 to 10 ml

Sl. No.	Part and Volume	Page No.	Common Ayurvedic Name	Botanical Name/ Chemical Name	Family/ Chemical Formula	Part Used	Identity, Purity and Strength/Chemical Properties	Constituents	Properties and Action	Important Formulations	Therapeutic Uses	Dose
											(Splenic disease), Siroroga (Disease of Head), Slipada (Filariasis), Svetakustha (Leucoderma), Tvakroga (Skin disease), Vata vikara (Disorder due to Vata dosa), Vrana (Ulcer)	
517	Part-I; Vol. VI	222	TAILAPARNA TAILA (Eucalyptus Oil)	*Eucalyptus globulus* Labill.	Myrtaceae	Steam Distil-lation of the Fresh Leaves	**Wt, per ml at 25°** - 0.901 to 0.920 g; **Optical rotation** - +5° to + 10°; **Refractive Index (at 25°)** - 1.457 to 1.469; **Assay** - Not less than 60.0 per cent, w/w of Cineole $C_{10}H_{18}O$	–	*Rasa* - Katu, Tikta, Kasaya *Guna*- Laghu, Snigdha *Virya*- Usna *Vipaka*- Katu; *Karma*- Anulomana, Dipana, Durgandhanasaka, Kaphanihsaraka, Krmighna, Mutrala, Pacana, Pratidusaka, Putihara, Sulaghna, Svedajanana, Uttejaka, Vatahara, Vedanasthapaka, Visamajvara-pratibandhaka	Pancaguna Taila	Agnimandya (Digestive impairment), Amvata (Rheumatism), Bala pratisyaya (Sinusitis in children), Bastisotha (Cystitis), Dusta vrana (Non-healing Ulcer), Jirnapuyameha (Chronic pyaemia), Jvara (Fever), Kasa (Cough), Krmi (Helminthiasis/ Worm infestation), Pinasa (Chronic rhinitis/ Sinusitis), Pratisyaya (Coryza),	1 to 5 drops

Sl. No.	*Part and Volume*	*Page No.*	*Common Ayurvedic Name*	*Botanical Name/ Chemical Name*	*Family/ Chemical Formula*	*Part Used*	*Identity, Purity and Strength/Chemical Properties*	*Constituents*	*Properties and Action*	*Important Formulations*	*Therapeutic Uses*	*Dose*
											Sandhivata (Osteoarthritis), Sirahsula (Headache), Sutika Jvara (Puerperal fever), Svasa (Asthma), Tvakroga (Skin disease), Yaksma (Tuberculosis)	
518	Part-I; Vol. VI	224	TILA TAILA (Sesamum Oil)	*Sesamum indicum* L.	Pedaliaceae	Clean and Healthy Seeds	**Specific gravity** - 0.9160 - 0.9190; **Refractive index (at 40°)** - 1.4650 to 1.4665; **Wt. per ml (at 25°)** - 0.916 to 0.921 g; **Acid value** - NMT 2.0; **Iodine value** - Between 103 and 116; **Saponification value** - Between 188 and 195; **Unsaponifiable matter** - NMT 1.5 per cent; **Cottonseed oil** - Absent; **Microbial limits** - Complies with API; **Pesticide residue** - Complies with API;	–	*Rasa* - Madhura Anurasa- Tikta, Kasaya *Guna*- Snigdha, Guru, Suksma, Vyavayi, Visada, Sara, Vikasi *Virya*- Usna *Vipaka*- Madhura; *Karma*- Balya, Caksusya, Dipana, Garbhasaya Sodhana, Kesya, Medhya, Sandhaniya, Snehana, Stanyajanana, Tvak prasadana, Vatahara, Vranaropana, Vranasodhana, Vrsya	Narayana Taila, Mahalaksadi Taila, Bala Taila	Agnidagdha (Fire burns), Ardita (Ficial palsy), Bhagna (Fracture), Dantasula (Toothache), Kandu (Itching), Karnasula (Otalgia), Krmi (Helminthiasis/ Worm infestation), Daurbalya (Weakness), Paksaghata (Paralysis/ Hemiplegia), Puyameha (Gonorrhoea), Sirahsula (Headache), Sula (Pain), Vatavikara (Disorders due to vata dosa), Vrana (Ulcer)	5 to 20 ml

Sl. No.	Part and Volume	Page No.	Common Ayurvedic Name	Botanical Name/ Chemical Name	Family/ Chemical Formula	Part Used	Identity, Purity and Strength/Chemical Properties	Constituents	Properties and Action	Important Formulations	Therapeutic Uses	Dose
519	Part-I; Vol. VI	226	YAVANISATVA (Thymol)	*Thymus vulgaris* L. and *Trachyspermum ammi* (L.)	Lamiaceae	Crystalline phenolic component, obtained from the volatile oil	**Melting range** - Between 48° and 51°C; **Non-volatile matter** - NMT 0.05 per cent; **Acidity or Alkalinity** - 4.0 w/v solution in Alcohol (50 per cent) in neutral to litmus solution; **Assay** - Thymol contains NLT 99 per cent of $C_{10}H_{14}O$,	–	*Rasa* - Katu, Tikta *Guna*- Laghu, Ruksa, Tiksna *Virya*- Usna *Vipaka*- Katu; *Karma*- Dipana, Lekhana, Pacana, Partidusaka, Slesmaghna, Sulaghna, Uttejaka, Vatanulomana, Vedanasamaka, Visaghna	Used as single drug.	Ajirna (Dyspepsia), Amavata (Rheumatism), Anaha (Distension of abdomen due to intestinal obstruction), Ankusa krmi (Hookworm infestation), Aruci (Tastelessness), Balatisara (Infantile diarrhoea), Chardi (Emesis), Dantasula (Toothache), Gulma (Abdominal lump), Krmi (Helminthiasis/ Worm infestation), Mutrakrcchra, Plihodara, Sandhisula, Sirahsula, Tvakroga (Skin disease), Udara (Diseases of abdomen), Udarasula (Pain abdomen), Vatarsa (Dry Piles), Visucika (Gastro-entritis with piercing Pain)	25 to 125 mg

Sl. No.	*Part and Volume*	*Page No.*	*Common Ayurvedic Name*	*Botanical Name/ Chemical Name*	*Family/ Chemical Formula*	*Part Used*	*Identity, Purity and Strength/Chemical Properties*	*Constituents*	*Properties and Action*	*Important Formulations*	*Therapeutic Uses*	*Dose*
520	Part-I; Vol. VII	1	ABHRAKA (Biotite Mica)	Ferro-magnesium silicate	K (Mg Fe)$_3$ X(Si$_3$AlO$_{11}$) X (OH)$_2$}	–	**Assay**- NLT 50 per cent Silica by gravimetric method; **Heavy metals and Arsenic:** Should not contain more than the stated limits for the following: - **Lead**- NMT 45ppm, **Arseninc**- NMT 3ppm, **Cadmium**- NMT 2ppm; **Other Elements: May contain the following within ±20 per cent of the stated limits Iron**- 6 per cent **Almuinium**- 5 per cent **Magnesium**- 9 per cent **Potassium**-5 per cent	–	–	–	Refer monograph of bhasma	–
521	Part-I; Vol. VII	3	AKIKA (Agate)	Silica mineral	SiO_2	–	**Assay**- NLT 95 per cent silica (SiO_2) when analysed by by gravimetric method; **Heavy metals and Arsenic:** Should not contain more than the stated limits for the following: **Arsenic**- NMT 190 ppm, **Cadmium** - NMT 1.6 ppm	–	–	–	Refer monograph of pisti or bhasma	–
522	Part-I; Vol. VII	5	GAIRIKA (Red Ochre)	Oxide of Iron	Fe_2O_3	–	**Assay** - Gairika should not contain less than 16 per cent Fe, or NLT 21 per cent Fe_2O_3 by gravimetric method; **Heavy metals and Arsenic**: Gairika should not contain more than the stated limits for the following: **Lead**- NMT 6ppm, **Arsenic**- NMT 2ppm, **Cadmium**- NMT 2 ppm, **Other Element: Magnesium** -1 per cent **Titanium** - 1 per cent	–	Rasa- Madhura, Kasaya; *Guna*- Snigdha, Visada; *Virya*- Sita; *Vipaka* - Madhura; *Karma*- Pitta-Nasaka, Balya, Vrana, Ropana, Netrya, Kaphajit	Kunkumadi taila, Bhrngaraja taila, Tutthadi lepa, Maha Jvarankusa rasa, Laghu Sutasekhara rasa, Kamadudha rasa (Muktika yukta)	Netra roga (Diseases of eyes); Raktapitta (Bleeding disorder); Hikka (Hiccup); Vamana (Vomiting); Visa vikara (Disorders due to poison); Rakta pradara	250-500mg of suddha *Gairika*

Sl. No.	Part and Volume	Page No.	Common Ayurvedic Name	Botanical Name/ Chemical Name	Family/ Chemical Formula	Part Used	Identity, Purity and Strength/Chemical Properties	Constituents	Properties and Action	Important Formulations	Therapeutic Uses	Dose
											(Menorrhagia or Metrorrhagia or both); Kandu (Itching); Jvara (Fever); Daha (Burning sensation); Udara roga (Diseases of abdomen)	
523	Part-I; Vol. VII	8	GANDHAKA (Sulphar)	Sulphur	S	–	**Solubility:** Insoluble in water as well as any acid, but soluble in carbon-di-sulphide. **Assay-** Should contain not less than 90 per cent Sulphur by gravimetric method; **Heavy Metals and Arsenic:** Should not contain more than the stated limits for the following: **Arsenic** = 1ppm, **Cadmium** = 2ppm	–	Rasa- Madhura, Katu, Tikta, Kasaya, *Guna-* Usna, Sara, Snigdha; *Virya-* Usna; *Vipaka-* Katu; *Karma-* Rasayana, Dipana, Pacana, Visahara, Kaphahara, Balya, Medya, Pittala, Caksusya, Krmihara, Sutajit, Kusthahara, Amasosahara, Sutendra; *Virya-* Prada, Vatahara	Mahagandhaka Vati, Pancamrta parpati, Candrakala rasa, Tarunarka rasa, Rasa parpati, Gandhaka rasayana	Kandu (Itching); Kustha (Diseases of the Skin), Visarpa (Erysepales), Dadru (Taeniasis), Amavata (Rheumatism); Kapha roga (Disease due to Kapha dosa); Garavisa (Slow/ accumulated Poison), Pliha roga (Splenic disease), Ksaya (Pthisis), Kasa (Cough), Svasa (Asthma), Netra roga (Diseases of eyes), Vata roga (Diseases due to Vata dosa)	125mg-1g of suddha Gandhaka

Sl. No.	Part and Volume	Page No.	Common Ayurvedic Name	Botanical Name/ Chemical Name	Family/ Chemical Formula	Part Used	Identity, Purity and Strength/Chemical Properties	Constituents	Properties and Action	Important Formulations	Therapeutic Uses	Dose
524	Part-I; Vol. VII	11	GODANTI (Selenite)	Selenite	$CaSO_4.2H_2O$	–	**Assay** - Calcium oxide (CaO) NLT 20 per cent or 30 per cent by gravimetric method; **Heavy metals: Lead**- NMT 6ppm, **Arsenic** - NMT 1ppm; **Cadmium** - NMT 4ppm;	–	–	–	Refer monograph of bhasma	–
525	Part-I; Vol. VII	13	GOMEDA (Garnet)	Almandite	$Fe_3Al_2 (SiO_4)_3$	–	**Assay**- Should contain NLT 35% SiO_2, NLT 10% Alumina (Al_2O_3), NLT 5% Iron (Fe) by gravimetric method; **Heavy metals and Arsenic: Mercury**- NMT 7 ng/g, **Arsenic**- NMT 2 ppm, **Cadmium** = traces; **Other Elements: Calcium** = 0.10% ± 20%, **Magnesium** = 0.10% ± 20%, **Manganese** = 11% ± 20%	–	–	–	Refer monograph of pisti or bhasma	–
526	Part-I; Vol. VII	15	JAHARA-MOHARA (Serpentine)	Hydrous silicate of Magnesium	$Mg_6(Si_4O_{10})X(OH)_8$	–	**Assay**- Should contain NLT 30% Magnesium oxide by gravimetric method; NLT 30% Silica (SiO_2), NLT 5% Ferric Oxide (Fe_2O_3), NLT 5% calcium oxide (CaO) by gravimetric method; **Heavy metals and Arsenic:** *Jaharamohara* should not contain more than the stated limits for the following: - **Arsenic** - NMT 2ppm, **Cadmium** - NMT 3ppm, **Other Elements:** May contain the follwing within ±20% of the stated limits: 0.15% Nickel when analysed by Atomic Absorption Spectrophotometer	–	–	–	JAHARA-MOHARA is used as a pisti, the details pf which are given in Monograph of pisti	–

Sl. No.	*Part and Volume*	*Page No.*	*Common Ayurvedic Name*	*Botanical Name/ Chemical Name*	*Family/ Chemical Formula*	*Part Used*	*Identity, Purity and Strength/Chemical Properties*	*Constituents*	*Properties and Action*	*Important Formulations*	*Therapeutic Uses*	*Dose*
527	Part-I; Vol. VII	17	KANTA LAUHA (Iron Ore)	Ferric oxide	Fe_3O_4	–	**Assay**- *Kanta Lauha* is ore from should contain NLT 60 per cent Iron (Fe) when analysed by gravimetric method; **Heavy metals and Arsenic**: *Kanta Lauha* is ore from should not contain more than the stated limits for the following: Arsenic- NMT 2 ppm, Cadmium- NMT 7 ppm; **Other Elements**: *Kanta Lauha* is ore from may contain the following within ± 20 per cent of the stated limits: - Zinc - 95 ppm, Manganese - 500 ppm, Silver - 5 ppm	–	–	–	*Kanta Lauha* is used in form of *bhasma*, the details of which are given in the Monograph of *bhasma*	–
528	Part-I; Vol. VII	19	KASISA (Ferrous Sulphate)	Ferrous Sulphate	$FeSO_4.7H_2O$	–	**Assay**- *Kasisa* should contain NLT 25 per cent Iron, NLT 15 per cent Sulphur; NLT 45 per cent $SO_{4;}$ **Heavy metals and Arsenic** : Kasisa should not contain more than the stated limits for the follwiing: - **Arsenic** = 2 ppm, **Cadmium** = 2 ppm **Other Elements:** May contain the follwing within ± 20 per cent of the stated limits: **Copper (cu)** - 188 ppm When analysed by Atomic Absrption Spectrophotometer method	–	Rasa- Amla, Tikta, Kasaya, *Guna*- Usna, *Virya*- Usna, *Vipaka*- Katu, *Karma*- Vata-Kaphahara, Kesya, Netrya, Rajah, Pravartaka, Krsnikarana, Sankocaka (Astringent), Balya, Kesya Ranjana, Rakta vardhaka	Rajahpravartini vati, Kasisadi taila, Kasisadi ghrta, Sankha dravaka, Plihari vatika	Kandu (Itching), Visa roga (Disease due to poison), Mutrakrcchra (Dysuria), Asmari (Calculus), Svitra (Leucoderma/ Vitiligo); Pitta Apasmara (Epilepsy due to Pitta dosa); Pandu (Anaemia), Plihavrddhi (Splenomegaly); Krmi (Helminthiasis/ worm infestation); Gudabhramsa	60-250mg

Sl. No.	*Part and Volume*	*Page No.*	*Common Ayurvedic Name*	*Botanical Name/ Chemical Name*	*Family/ Chemical Formula*	*Part Used*	*Identity, Purity and Strength/Chemical Properties*	*Constituents*	*Properties and Action*	*Important Formulations*	*Therapeutic Uses*	*Dose*
											(Prolapse of rectum); Visarapa (Erysepales), Netra roga (Disease of eyes), Slesma roga (Disease due to kapha dosa)	
529	Part-I; Vol. VII	22	KHATIKA (Kaolinite)	Kaolinite	$(Al_2(Si_2O_5)(OH)_4)$	–	**Assay** - *Khatika* should contain NMT 50 per cent Silica (SiO2), and NLT 30 per cent Alumina (Al_2O_3) by gravimetric method; **Heavy metals and Arsenic:** *Khatika* should not contain more than the stated limits for the following : - **Lead** = 15 ppm, **Arsenic** = 2 ppm, **Cadmium** = 7 ppm; **Other Elements:** May contain the folowing within ±20 per cent of the stated limits: - **Calcium**- 0.32 per cent, **Magnesium**- 0.78 per cent **Manganese** - 20 ppm **Chromium**- 250 ppm	–	Rasa- Tikta, Madhura, *Guna*- Sita, *Virya*- Sita, *Vipaka*- Madhura, *Karma*- Pitta samaka, Vrana ropana, Kapha-daha-rakta dosaghni, Svedadisravahara	Dasana samskara churna (for external)	Sotha (Inflammation), Netra roga (Diseases of eyes); Atisara (Diarrhoea)	1/2-1 g of suddha Khatika for both internal and external use

Sl. No.	*Part and Volume*	*Page No.*	*Common Ayurvedic Name*	*Botanical Name/ Chemical Name*	*Family/ Chemical Formula*	*Part Used*	*Identity, Purity and Strength/Chemical Properties*	*Constituents*	*Properties and Action*	*Important Formulations*	*Therapeutic Uses*	*Dose*
530	Part-I; Vol. VII	25	MANDURA (Iron Slag)	Oxide-cum-silicate of iron	Fe_2SiO_4	–	**Assay**- *Mandura* should contain NLT 30 per cent Iron (Fe) by gravimetric method; *Mandura* Should contain NLT 30 per cent Silica by gravimetric method, Mandura Should show NLT 80 per cent **fayalite (**Fe_2SiO_4**)** when studies through XRD method; **Heavy metals and Arsenic:** *Mandura* should not contain morte than the stated limits for the following- **Arsenic** - NMT 6ppm, **Cadmium**- NMT 8 ppm, **Other Elements:** May contain the following within ±20 per cent of the stated limits: - **Copper**- 0.45 per cent **Zinc**- 50 ppm **Silver**- 7 ppm	–	–	–	*Mandura*is used in form of bhasma, the details of which are given in the Refer monograph of bhasma	NA
531	Part-I; Vol. VII	27	RAJATA (Silver metal)	Silver- white metal	Ag	–	**Assay**- *Rajata* should contain NLT 98.5 per cent Silver (Ag) when analysed by A.A.S. **Heavy metals and Arsenic:** In *Rajata* Mercury, Lead, Arsenic and Cadmium should be absent. **Other Elements:** May contain the following within ±20 per cent of the stated limits: - **Copper** = 1.40 per cent **Sulphur** = traces **Gold** = 0.001 per cent	–	–	–	*Rajata* is used in form of bhasma, the details of which are given in the monograph of bhasma	NA

Sl. No.	*Part and Volume*	*Page No.*	*Common Ayurvedic Name*	*Botanical Name/ Chemical Name*	*Family/ Chemical Formula*	*Part Used*	*Identity, Purity and Strength/Chemical Properties*	*Constituents*	*Properties and Action*	*Important Formulations*	*Therapeutic Uses*	*Dose*
532	Part-I; Vol. VII	29	SAMUDRA LAVANA (Sea Salt)	Sea salt	NaCl	–	**Assay**- *Samudra lavana* should contain NLT 35 per cent Sodium (Na) when analysed by flame photometry; *Samudra Lavana* should contain NLT 58 per cent Chlorine (Cl); **Heavy metals and Arsenic:** *Samudra Lavana* Should not contain more than the stated limits for the following: **Lead** = 12 ppm, **Arsenic** = 4 ppm, **Cadmium** = 4ppm	–	Rasa- Lavana; *Guna*- Snigdha, Laghu, Usna; *Virya*- Nati Usna/Nati Sitala; *Vipaka*- Madhura; *Karma*- Vata-hara, Hrdya, Bhedi, Rucikara, Dipana, Kaphahara, Sulaghna, Avidahi, Isat Pittala, Snehana, Pacana, Kledana, Balya	Lvana Bhaskara churna, Samudradya churna, Narayana churna, Mahasankha vati, Kalyanaka guda	Ajirna (Dyspepsia), Sosa (Cachexia), Jirna carma roga (Chronic skin diseases), Galaganda (Goiter), Pandu (Anaemia), Pratisyaya (Coryza)	According to formu-lation
533	Part-I; Vol. VII	31	SAUVI-RANJANA (Lead Ore)	Lead Ore	PbS	–	**Assay**- **S***auviranjana* in ore form shoud contain NLT 50 per cent Lead(pb) when analysed by A.A.S, *Sauviranjana in ore form shoud contain*NLT 10 per cent Sulphur when analysed by gravimetric method, *Sauviranjana* in ore form should contain NLT 500 ppm Silver (Ag) when analysed by by A.A.S.; **Heavy metals and Arsenic:** *Sauviranjana* is ore form should not contain more than the stated limits for the follwing: - **Arsenic** = 2 ppm, **Cadmium** = 22 ppm, **Other Elements:** *Sauviranjana* in ore form may contain the following within ±20 per cent of the stated limits: **Copper** = 70 ppm **Gold** = 0.10 ppm **Zinc** = 20 ppm	–	Rasa- Tikta, Kasaya, Katu; *Guna*- Snigdha; *Virya*- Sita; *Vipaka*- Madhura; *Karma*- Grahi, Vrana Sodhana, Ropana, Rajorodhaka	Irimedadi taila (for external use), Nayanamr-tanjana (for external use)	Suddha Sauviranjana (External): Netra roga (Diseases of eye); Bhasma of Sauviranjana: Raktapitta (Bleeding disorder); Visa dosa(Disorders due to poison), Hikka (Hiccup); Rajorodha (Obstruction of menstrual flow), Raktapradara* (menorrhagia or metrorrhagia or both)	60-125 mg of the bhasma; *Pre-caution: it should not be used for more than three days in Rakta-pradara.

Sl. No.	*Part and Volume*	*Page No.*	*Common Ayurvedic Name*	*Botanical Name/ Chemical Name*	*Family/ Chemical Formula*	*Part Used*	*Identity, Purity and Strength/Chemical Properties*	*Constituents*	*Properties and Action*	*Important Formulations*	*Therapeutic Uses*	*Dose*
534	Part-I; Vol. VII	34	SVARNA (Gold metal)	Gold metal	Au	–	**Assay**- *Svarna* Should contain NLT 99.99 per cent Gold (Au) when by Atomic Absorption Spectorometer; **Heavy metals and Arsenic:** In *Svarna* Mercury, Lead, Arsenic and Cadmium should be absent	–	–	–	*Svarna* is used in the form of *bhasma*, the details of which are given in the monograph of *bhasma*	NA
535	Part-I; Vol. VII	36	SVARNA-MAKSIKA (Copper Ore)	Chalcopyrite	$CuFeS_2$	–	**Assay**- *Svarnamaksika* in ore form should contain NLT 5 per cent Copper (Cu), *Svarnamaksika* in ore from should contain NLT 20 per cent Iron and 12 per cent Sulphur analysed by gravimetric method; **Heavy metals and Arsenic** : *Svarnamaksika* in ore form should contain not more than the stated limits for the following: - **Lead** = 70 ppm, **Arsenic** = 1 ppm, **Cadmium** = 3 ppm; **Other Elements:** *Svarnamaksika* in ore form may contain the following within ±20 per cent of the stated limits: - **Gold** = 0.70 ppm **Silver** = 48 ppm **Zinc** = 800 ppm	–	–	–	*Svarnamaksika* is used in the form of *bhasma*, the details of which are given monograph of *bhasma*	NA

Sl. No.	Part and Volume	Page No.	Common Ayurvedic Name	Botanical Name/ Chemical Name	Family/ Chemical Formula	Part Used	Identity, Purity and Strength/Chemical Properties	Constituents	Properties and Action	Important Formulations	Therapeutic Uses	Dose
536	Part-I; Vol. VII	38	SVARNA-MAKSIKA-SANDRITA (Copper Concentrate)	Chalcopyrite	$CuFeS_2$	–	**Assay** - *Sandrita Svarnamaksika* should contain NLT 12 per cent Copper (Cu), *Sandrita Svarnamaksika* should contain NLT 25 per cent Iron and 28 per cent Sulphur when analysed by by gravimetric method; **Heavy metals and Arsenic:** *Sandrita Svarnamaksika* should contain not more than the stated limits for the following: **Arsenic** = 100 ppm, **Cadmium** = 5 ppm, **Other Elements:** *Sandrita Svarnamaksika* may contain the following within ± 20 per cent of the stated limits: **Gold** = 0.40 ppm **Silver** = 50 ppm **Zinc** = 0.20 ppm	–	–	–	*Svarnamaksika Sandrita* is used in the form of *bhasma*, the details of which are given monograph of *bhasma*	–
537	Part-I; Vol. VII	40	TAMRA (Copper Metal)	Copper Metal	NA	–	**Assay**- *Tamra* should contain NLT 99.5 per cent Copper (Cu); **Heavy metals and Arsenic:** *Tamra* should not contain more than the stated limits for the folowing: - Lead= 5 ppm, **Arsenic** = 1 ppm, **Cadmium** = NMT 5 ppm, **Other Elements:** May contain the following within ± 20 per cent of the stated limits: - **Zinc** = 25ppm **Silver** = 10 ppm **Gold** = 135 ppb	–	–	–	*Tamra*is used in the form of *bhasma*, the details of which are given monograph of *bhasma*	–

Sl. No.	*Part and Volume*	*Page No.*	*Common Ayurvedic Name*	*Botanical Name/ Chemical Name*	*Family/ Chemical Formula*	*Part Used*	*Identity, Purity and Strength/Chemical Properties*	*Constituents*	*Properties and Action*	*Important Formulations*	*Therapeutic Uses*	*Dose*
538	Part-I; Vol. VII	42	TANKANA (Borax)	Borax	$Na_2B_4O_7.10H_2O$	–	**Assay** -*Tankana* should contain NLT 35 per cent B_2O_3 (Boron trioxide), *Tankana* should contain NLT 15 per cent Sodium (Na),; **Heavy metals and Arsenic:** *Tankana* should not contain more than the stated limits for the following: - **Arsenic** = 5 ppm, **Cadmium** = 4 ppm	–	Rasa- Katu; *Guna*- Ruksa, Usna, Tiksna, Saraka; *Virya*- Usna; *Vipaka*- Katu; *Karma*- Hrdya, Balya, Saraka, Kapha nissaraka, Dipana, Stri puspajanana, Mudhagarbha-pravartaka	Anandabhairava rasa, Candramrta rasa, Icchabhedi rasa, Saubhagya vati, Tribhuvanakirti rasa	Kasa (Cough), Svasa (Asthma), Vata roga (Diseases due to Vata dosa), Sthavara Visa (Poisoning by plant or mineral), Adhmana (Flatulence with gurgling sound), Vrana (wound/ulcer)	125-250 mg
539	Part-I; Vol. VII	45	TUTTHA (Copper Sulphate)	Copper Sulphate	$Cu SO_4.5H_2O$	–	**Assay** - *Tuttha* should contain NLT 20 per cent Copper, *Tuttha* should contain NLT 15 per cent Sulphur and NLT 50 per cent SO_4;, **Heavy metals and Arsenic:** Tuttha should not contain more than the stated limits for the following : - **Lead** = 226 ppm, **Arsenic** = NMT 4 ppm, **Cadmium** = NMT 97 ppm, **Other Elements:** May contain the following within ±20 per cent of the stated limits: **Iron** 4 per cent when analysed by gravimetric method	–	Rasa- Katu, Kasaya, Madhura; *Guna*- Laghu, Sara; *Virya*- Usna, Sita; *Vipaka*- Katu; *Karma*- Kaphapittahara, Lekhana, Bhedana, Balya, Tridosaghna, Rasayana, Rucikara, Vamaka, Varnya, Garavisahara, Sulaghna, Caksusya, Asmarihara, Kandughna, Ksara*Karma*kara, Arsoghna, Krmighna	Jatyadi taila, Nityanda rasa, Jatyadi ghrta, Maha visagarbha taila, Kasisadi ghrta	Krmi (Helminthiasis/ worm infestation), Prameha (Increased frequency and turbidity of urine); Medoroga, Sula (Pain/ colic), Kustha (Diseases of the skin), Svasa (Asthma), Amlapitta, Tvak roga (Skin disease), Svitra (Leucoderma/ Vitiligo), Arsa (Piles), Vrana (Ulcer/wound), Nadi Vrana (Sinus), Netra roga (Diseases of eyes), Dusta Vrana (Non-healing ulcer)	15-30 mg

Sl. No.	Part and Volume	Page No.	Common Ayurvedic Name	Botanical Name/ Chemical Name	Family/ Chemical Formula	Part Used	Identity, Purity and Strength/Chemical Properties	Constituents	Properties and Action	Important Formulations	Therapeutic Uses	Dose
540	Part-I; Vol. VII	48	VAIKRANTA (Tourmaline)	Sodium aluminium borosilicate	(Ca, Na)X(Mg, Al_6) X{B_3 A_{12} Si_6 $X(O,OH)_{30}$}	–	**Assay-** *Vaikranta* should contain NLT 6 per cent B_2O_3 When analysed by ICPA method, *Vaikranta* should contain NLT 12 per cent ferric oxide (Fe_2O_3), NLT 30 per cent Alumina (Al_2O_3), NLT 30 per cent Silica (SiO_2) When analysed by gravimetric method,; **Heavy metals and Arsenic:** *Vaikranta* should not contain more than the stated limits for teh following : - **Lead** = 11 ppm, **Arsenic** = 4 ppm, **Cadmium** = 2 ppm	–	–	–	*Vaikranta* is used in the form of *bhasma*, the details of which are givenRefer monograph of *bhasma*	–
541	Part-I; Vol. VIII	1	AMALAKI	*Phyllanthus emblica* L. syn. *Emblica officinalis* Gaertn.	Euphorbiaceae	Dried pericarp of matue fruits	**Quantitative parameters: Foreign matter**- NMT 3.0 per cent (Including seed and seed coat; **Loss on drying**- NMT 12 per cent; **Total ash**- NMT 7.0 per cent; **Acid-insoluble ash**- NMT 2.0 per cent; **Alcohol-soluble extractive**-NLT 40.0 per cent; **Water-soluble extractive**- NLT 50.0 per cent; **T.L.C.** - Refer API **Assay** - Refer API	Tannins, Gallic acid, Ellagic acid, Phyllemblic acid, Emblicol, Alkaloids, Phyllantidine and Phyllantine, Pectins, Minerals	Rasa- Amla, Kasaya, Madhura, Tikta, Katu; *Guna*- Ruksa, Laghu; *Virya*- Sita; *Vipaka*- Madhura; *Karma*- Tridosajit, Vrsya, Rasayana, Caksusya	Cyavanprasa, Dhatri lauha, Dhatryadi ghrta, Triphala churna	Raktapitta (Bleeding disorders); Amlapitta (Hyperacidity); Prameha (Increased frequency and turbidity of urine); Daha (Burning sensatin)	3 to 6 g of the drug in powder form
542	Part-I; Vol. VIII	6	AMALAKI (POWDER)	*Phyllanthus emblica* L. syn. *Emblica officinalis* Gaertn.	Euphorbiaceae	Powder	**Identity, Purity and Strength** - Complies with the tests for identity, Purity, Strength and Thin-layer chromatography as stated under Amalaki; **Assay:** complies with the limits for Assay as per method stated under Amalaki	–	–	–	–	–

Sl. No.	*Part and Volume*	*Page No.*	*Common Ayurvedic Name*	*Botanical Name/ Chemical Name*	*Family/ Chemical Formula*	*Part Used*	*Identity, Purity and Strength/Chemical Properties*	*Constituents*	*Properties and Action*	*Important Formulations*	*Therapeutic Uses*	*Dose*
543	Part-I; Vol. VIII	7	AMALAKI HYDRO-ALCOHOLIC EXTRACT	*Phyllanthus emblica* L. syn. *Emblica officinalis* Gaertn.	Euphorbiaceae	Hydro-alcoholic Extract	**Quantitative parameters: Loss on drying** - NMT 5.0 per cent, **Total ash**- NMT 10.0 per cent, **Acid-insoluble ash**- NMT 1.5 per cent, **pH** - 3.0-4.5, **Total soluble solids** - NLT 90.0 per cent; T.L.C. - Refer API Assay - Refer API	–	–	–	–	–
544	Part-I; Vol. VIII	10	AMALAKI WATER EXTRACT	*Phyllanthus emblica* L. syn. *Emblica officinalis* Gaertn.	Euphorbiaceae	Dried and powdered extract	**Quantitative parameters: Loss on drying** - NMT 7.0 per cent, **Total ash**- NMT 10.0 per cent, **Acid-insoluble ash**- NMT 1.25 per cent, **pH** - Less than 3.5, **Total soluble solids** - NLT 90.0 per cent; **T.L.C.** - Refer API Assay - Refer API	–	–	–	–	–
545	Part-I; Vol. VIII	13	ARJUNA	*Terminalia arjuna* W. and A.	Combretaceae	Stem bark	**Foreign Matter**- NMT 2 per cent; **Loss on drying**- NMT 12.0 per cent; Total ash- NMT 25.0 per cent **Acid-insoluble ash**- NMT 1 per cent; **Alcohol-soluble extractive**- NLT 20 per cent; **Water-soluble extractive** - NLT 20.0 per cent **T.L.C.** - Refer API **Assay** - Refer API	Tannins, Arjunin, Arjunic acid, Arjunolic acid, Arjunetin, Arjunolitin, Friedelin, Terminoic acid, Arjungenin and Arjunglucosides	Rasa- Kasaya; *Guna*- Ruksa; *Virya*- Sita; *Vipaka*- Katu; *Karma*- Kaphahara, Pittahara, Hrdya, Vranasana, Bhagna-sandhanakara, Vyangahara	Prathadyarista, Nagarjunabhra rasa, Arjuna ghrta	Hrdroga (Heart disease); Ksataksaya (Emaciation due to injury); Medoroga (Obesity); prameha (Increased frequency and turbidity of urine), Vrana (Ulcer), Trsa (Thirst), Vyanga (Dark shade on face due to stress and excessive exercise/ localized hyper pigmentation of skin)	3 to 6 g of the drug in powder form

Sl. No.	Part and Volume	Page No.	Common Ayurvedic Name	Botanical Name/ Chemical Name	Family/ Chemical Formula	Part Used	Identity, Purity and Strength/Chemical Properties	Constituents	Properties and Action	Important Formulations	Therapeutic Uses	Dose
546	Part-I; Vol. VIII	19	ARJUNA (Powder)	*Terminalia arjuna* W. and A.	Combretaceae	Powder	**Identity, Purity and Strength:** Complies with the test for Identity, Purity, Strength and Thin-layer chromatography as stated under Arjuna, **Assay:** Complies with the limits for Assay as per method stated under Arjuna	–	–	–	–	–
547	Part-I; Vol. VIII	20	ARJUNA HYDRO-ALCOHOLIC EXTRACT	*Terminalia arjuna* W. and A.	Combretaceae	Dried and Powder extract	**Quantitative parameters: Loss on drying** - NMT 7.0 per cent, **Total ash**- NMT 5.0 per cent, **Acid-insoluble ash**- NMT 0.5 per cent, **pH** - 4.5-5.5, **Total soluble solids** - NLT 90.0 per cent; **T.L.C.** - Refer API **Assay** - Refer API	–	–	–	–	–
548	Part-I; Vol. VIII	24	ARJUNA WATER EXTRACT	*Terminalia arjuna* W. and A.	Combretaceae	Dried and Powder extract	**Quantitative parameters: Loss on drying** - NMT 7.0 per cent, **Total ash**- NMT 7.0 per cent, **Acid-insoluble ash**- NMT 0.5 per cent, **pH** - 4.5-5.5, **Total soluble solids** - NLT 90.0 per cent; **T.L.C.** - Refer API **Assay** - Refer API	–	–	–	–	–
549	Part-I; Vol. VIII	28	ASVA-GANDHA	*Withania somnifera* (L.) Dunal.	Solanaceae	Dried Mature Roots	**Quantitative parameters: Foreign matter**-NMT 2.0 per cent; **Loss on drying**-NMT 12.0 per cent; **Total ash**- NMT 7.0 per cent; **Acid-insoluble ash**- NMT 1.0 per cent; **Alcohol-soluble extractive**- NLT 15.0 per cent; **Water-soluble extractive**- NLT 7.0 per cent, **T.L.C.** - Refer API **Assay** - Refer API	Alkaloids, Withanone, Withaferin A, Withanolides and Withanosides	Rasa- Tikta, Kasaya; *Guna*- Laghu; *Virya*- Usna; *Vipaka*- Madhura; *Karma*- Vatakaphapaha, Balya, Rasayana, Vajikarana	Asvangan-dhadyarista, Asva-gandhadi leha, Balasvan-gandha Laksadi taila	Ksaya (Pthisis); Daurbalya (Weakness); Vataroga (Disease due to vat dosa), Sotha (Inflammation); Klaibya (Male impotence)	3 to 6 g of the drug in powder form

Sl. No.	*Part and Volume*	*Page No.*	*Common Ayurvedic Name*	*Botanical Name/ Chemical Name*	*Family/ Chemical Formula*	*Part Used*	*Identity, Purity and Strength/Chemical Properties*	*Constituents*	*Properties and Action*	*Important Formulations*	*Therapeutic Uses*	*Dose*
550	Part-I; Vol. VIII	33	Asvagandha (Powder)	*Withania somnifera* (L.) Dunal.	Solanaceae	Powder	**Identity, Purity and Strength:** Complies with the test for Identity, Purity, Strength and Thin-layer chromatography as stated under Asvagandha. **Assay**- Complies with the limits for Assay as per method stated under Asvagandha	–	–	–	–	–
551	Part-I; Vol. VIII	34	Asvagandha Hydro-Alcoholic Extract	*Withania somnifera* (L.) Dunal.	Solanaceae	Hydro-alcoholic Extract	**Quantitative parameters: Loss on drying** - NMT 5.0 per cent, **Total ash**- NMT 16.0 per cent, **Acid-insoluble ash**- NMT 2.0 per cent, **pH** - 4.5-5.5, **Total soluble solids** - NLT 90.0 per cent; **T.L.C.** - Refer API **Assay** - **Refer API**	–	–	–	–	–
552	Part-I; Vol. VIII	37	Asvagandha Water Extract	*Withania somnifera* (L.) Dunal.	Solanaceae	Dried and powdered extract	**Quantitative parameters: Loss on drying** - NMT 7.0 per cent, **Total ash**- NMT 12.5 per cent, **Acid-insoluble ash**- NMT 2.0 per cent, **pH** - 4.5-5.5, **Total soluble solids** - NLT 90.0 per cent; **T.L.C.** - Refer API **Assay** - Refer API	–	–	–	–	–
f553	Part-I; Vol. VIII	40	BIBHITAKA	*Terminalia belerica* Roxb.	Combretaceae	Pericarp of Ripe fruit	**Quantitative parameters: Foreign matter**- NMT 2.0 per cent; **Loss on drying**- NMT 12.0 per cent; **Total ash**- NMT 7.0 per cent; **Acid-insoluble ash**- NMT 1.0 per cent; **Alcohol-soluble extractive** - NLT 8.0 per cent; **Water-soluble extractive**- NLT 35.0 per cent, **T.L.C.** - Refer API **Assay** - Refer API	Bellericagenin A and B, Bellericaside A and B, Termilignan, Gallic acid, Ellagic acid, Ethyl gallate, Chebulagic acid, Corilagin, 1,3,6-Trigalloyl glucose,	Rasa- Kasaya, *Guna*- Ruksa, Laghu, *Virya*- Usna; *Vipaka*- Madhura, *Karma*- Kaphapittajit, Bhedaka, Krminasana, Caksusya, Kesya, Kasahara	Triphala churna, Triphaladi taila, Lavangadi vati	Svarabheda (Hoarseness of voice); Netraroga (Disease of eye), Kasa (Cough); Chardi (Emesis); Krmiroga (Worm infestation),	3 to 6 g of the drug in powder form

Sl. No.	*Part and Volume*	*Page No.*	*Common Ayurvedic Name*	*Botanical Name/ Chemical Name*	*Family/ Chemical Formula*	*Part Used*	*Identity, Purity and Strength/Chemical Properties*	*Constituents*	*Properties and Action*	*Important Formulations*	*Therapeutic Uses*	*Dose*
								Bellericanin, Phyllembin and Thannilignam			Vibandha (Constipation)	
554	Part-I; Vol. VIII	45	BIBHITAKA (POWDER)	*Terminalia belerica* Roxb.	Combretaceae	Powder	**Identity, Purity and Strength:** Complies with the test for Identity, Purity, Strength and Thin-layer chromatography as started under Bibhitaka, **Assay**- Complies with the limits for Assay as per method stated under Bibhitaka	–	–	–	–	–
555	Part-I; Vol. VIII	46	BIBHITAKA HYDRO-ALCOHOLIC EXTRACT	*Terminalia belerica* Roxb.	Combretaceae	Dried and Powder extract	**Quantitative parameters: Loss on drying** - NMT 7.0 per cent, **Total ash**- NMT 10.0 per cent, **Acid-insoluble ash**- NMT 1.25 per cent, **pH** - 3.5-5.5, **Total soluble solids** - NLT 90.0 per cent; **T.L.C.** - Refer API **Assay** - Refer API	–	–	–	–	–
556	Part-I; Vol. VIII	49	BIBHITAKA WATER EXTRACT	*Terminalia belerica* Roxb.	Combretaceae	Dried and Powder extract	**Quantitative parameters: Loss on drying** - NMT 5.0 per cent, **Total ash**- NMT 10.0 per cent, **Acid-insoluble ash**- NMT 1.5 per cent, **pH** - 3.5-5.5, **Total soluble solids** - NLT 90.0 per cent; **T.L.C.** - Refer API **Assay** - Refer API	–	–	–	–	–

Sl. No.	*Part and Volume*	*Page No.*	*Common Ayurvedic Name*	*Botanical Name/ Chemical Name*	*Family/ Chemical Formula*	*Part Used*	*Identity, Purity and Strength/Chemical Properties*	*Constituents*	*Properties and Action*	*Important Formulations*	*Therapeutic Uses*	*Dose*
557	Part-I; Vol. VIII	52	BHRNGARAJA	*Eclipta alba* Hassk.	Asteraceae	Whole Plant	**Quantitative parameters: Foreign matter**- NMT 2.0 per cent; **Loss on drying**- NMT 12.0 per cent; **Total ash**- NMT 22.0 per cent; **Acid-insoluble ash**- NMT 11.0 per cent; **Alcohol-soluble extractive**- NLT 5.0 per cent; **Water-soluble extractive**-NLT 15.0 per cent **T.L.C.** - Refer API **Assay** - Refer API	Alkaloids, Ecliptine, Nicotine and Wedelolactone	Rasa- Katu, Tikta, Guan- Ruksa, Tiksna, *Virya*- Usna, *Vipaka*- Katu, *Karma*- Vatahara, Kaphahara, Amahara, Balya, Rasayana, Kesya, Tvacya, Dantya, Caksusya, Visahara	Bhrngamalakadi taila, Bhrngaraja taila, Nilibhrngadi taila, Bhrngarajasava	Yakrdroga (Disease of liver); Krmiroga (Worm infestation), Sotha (Inflammation), Pandu (Anaemia), Svasa (Asthma), Kasaya (Cough), Sirahsula (Headache), Hrdroga (Heart disease)	3 to 6 ml of the drug in the juice form.; 12 to 36 g of the drug in the powder form for decoction
558	Part-I; Vol. VIII	58	BHRNGARAJA (POWDER)	*Eclipta alba* Hassk.	Asteraceae	Powder	**Identity, Purity and Strength:** Complies with the tests for Identity, Purity, Strength and Thin-layer chromatography as stated under Bhrngaraja, **Assay**- Complies with the limits for Assay as per method stated under Bhrngaraja	–	–	–	–	–
559	Part-I; Vol. VIII	59	BHRNGARAJA HYDRO-ALCOHOLIC EXTRACT	*Eclipta alba* Hassk.	Asteraceae	Dried and Powder extract	**Quantitative parameters: Loss on drying** - NMT 5.0 per cent, **Total ash**- NMT 10.0 per cent, **Acid-insoluble ash**- NMT 1.5 per cent, **pH** - 5.0-7.5, **Total soluble solids** - NLT 90.0 per cent; **T.L.C.** - Refer API **Assay** - Refer API	–	–	–	–	–

Sl. No.	*Part and Volume*	*Page No.*	*Common Ayurvedic Name*	*Botanical Name/ Chemical Name*	*Family/ Chemical Formula*	*Part Used*	*Identity, Purity and Strength/Chemical Properties*	*Constituents*	*Properties and Action*	*Important Formulations*	*Therapeutic Uses*	*Dose*
560	Part-I; Vol. VIII	62	BHRNGARAJA WATER EXTRACT	*Eclipta alba* Hassk.	Asteraceae	Dried and Powder extract	**Quantitative parameters: Loss on drying** - NMT 6.0 per cent, **Total ash**- NMT 10.0 per cent, **Acid-insoluble ash**- NMT 1.5 per cent, **pH** - 5.0-7.5, **Total soluble solids** - NLT 90.0 per cent; **T.L.C.** - Refer API **Assay** - Refer API	–	–	–	–	–
561	Part-I; Vol. VIII	65	BRAHMI	*Bacopa monnieri* (L.) Wettst. Syn. *Herpestis monnieria* (L.) H. B. and K.	Scro-phulariaceae	Dried Whole Plant	**Quantitative parameters: Foreign matter:** NMT 2.0 per cent; **Loss on drying:** NMT 12.0 per cent : **Total ash:** NMT 18.0 per cent; **Acid-insoluble ash:** NMT 6.0 per cent; **Alcohol-soluble extractive:** NLT 6.0 per cent; **Water-soluble extractive**- NLT 15.0 per cent, **T.L.C.** - Refer API **Assay** - Refer API	Brahmine, Herpestine, α-Mannitol, Hersaponin, Monnierin, luteolin, Apigenin, Bacopasaponins and Bacosides	Rasa- Tikta, Kasaya, Madhura; *Guna*- Laghu, Sara; *Virya*-Sita; *Vipaka*- Madhura; *Karma*- Vatahara, Kaphahara, Rasayana, Ayusya, Medhya, Matiprada, Svarya, Prajasthapana, Visahara, Mohahara	Sarasvatarista, Brahmi ghrta, Ratnagiri rasa, Brahmi vati, Sarasvata churna, Smrtisagara rasa	Kustha (Diseases of skin); Jvara (Fever); Sopha (Oedema); Pandu (Anaemia); Prameha (Increased frequency and turbidity of urine); Manasavikara (Mental disorders)	1-3 g in powder form
562	Part-I; Vol. VIII	70	BRAHMI (Powder)	*Bacopa monnieri* (L.) Wettst. Syn. *Herpestis monnieria* (L.) H. B. and K.	Scro-phulariaceae	Powder	**Identity, Purity and Strength:** Complies with the tests for Identity, urity, Strenght and Thin-layer chromatography as stated under Brahmi, **Assay** - Complies with the limits for Assay as per method stated under Brahmi	–	–	–	–	–

Sl. No.	*Part and Volume*	*Page No.*	*Common Ayurvedic Name*	*Botanical Name/ Chemical Name*	*Family/ Chemical Formula*	*Part Used*	*Identity, Purity and Strength/Chemical Properties*	*Constituents*	*Properties and Action*	*Important Formulations*	*Therapeutic Uses*	*Dose*
563	Part-I; Vol. VIII	71	BRAHMI HYDRO-ALCOHOLIC EXTRACT	*Bacopa monnieri* (L.) Wettst. Syn. *Herpestis monnieria* (L.) H. B. and K.	Scro-phulariaceae	Dried and Powder extract	**Quantitative parameters: Loss on drying** - NMT 7.0 per cent, **Total ash**- NMT 35.0 per cent, **Acid-insoluble ash**- NMT 5.0 per cent, **pH** - 5.0-6.5, **Total soluble solids** - NLT 90.0 per cent; **T.L.C.** - Refer API **Assay** - Refer API	–	–	–	–	–
564	Part-I; Vol. VIII	74	BRAHMI WATER EXTRACT	*Bacopa monnieri* (L.) Wettst. Syn. *Herpestis monnieria* (L.) H. B. and K.	Scro-phulariaceae	Dried and Powder extract	**Quantitative parameters: Loss on drying** - NMT 7.0 per cent, **Total ash**- NMT 35.0 per cent, **Acid-insoluble ash**- NMT 5.0 per cent, **pH** - 4.5-6.5, **Total soluble solids** - NLT 90.0 per cent; **T.L.C.** - Refer API **Assay** - Refer API	–	–	–	–	–
565	Part-I; Vol. VIII	77	HARIDRA	*Curcuma longa* L.	Zingiberaceae	Dried and cured rhizomes	**Quantitative parameters: Foreign matter**- NMT 2.0 per cent; **Loss on drying**- NMT 12.0 per cent; **Total ash**- NMT 9.0 per cent; **Acid-insoluble ash**- NMT 1.0 per cent; **Alcohol-soluble extractive**- NLT 8.0 per cent; **Water-soluble extractive**- NLT 12.0 per cent; **Volatile oil**- NLT 4.0 per cent **T.L.C.** - Refer API **Assay** - Refer API	Curcuminoids, Curcumin, Essential oil, ar-termerone, α-and β termerone, Curcumol	Rasa-Tikta, Katu, *Guna*-Ruksa; *Virya*-Usna; *Vipaka*-Katu; *Karma*- Kaphapittanut, Visaghna, Varnya, Kusthaghna, Krmighna, Pramehanasaka	Haridra khanda	Visavikara (Disorders due to poison); Kustha (Disease of skin); Vrana (Ulcer); Tvagroga (Skin disease); Prameha (Increased frequency and turbidity of urine), Pandu (Anaemia), Sitapitta (Urticaria), Pinasa (Chronic rhinitis/ sinusitis)	1 to 3 g of the drug in powder form

Sl. No.	*Part and Volume*	*Page No.*	*Common Ayurvedic Name*	*Botanical Name/ Chemical Name*	*Family/ Chemical Formula*	*Part Used*	*Identity, Purity and Strength/Chemical Properties*	*Constituents*	*Properties and Action*	*Important Formulations*	*Therapeutic Uses*	*Dose*
566	Part-I; Vol. VIII	82	HARIDRA (POWDER)	*Curcuma longa* L.	Zingiberaceae	Powder	**Identity, Purity and Strength:** Complies with the tests for Identity, Purity, Strength and Thin-layer chromatography as stated under Haridra, **Assay:** Complies with the limits for Assay as per method stated under Haridra	–	–	–	–	–
567	Part-I; Vol. VIII	83	HARIDRA HYDRO-ALCOHOLIC EXTRACT	*Curcuma longa* L.	Zingiberaceae	Dried and Powder extract	**Quantitative parameters: Loss on drying** - NMT 7.0 per cent, **Total ash**- NMT 30.0 per cent, **Acid-insoluble ash**- NMT 1.0 per cent, **pH** - 5.0-7.0, **Total soluble solids** - NLT 90.0 per cent; **T.L.C.** - Refer API **Assay** - Refer API	–	–	–	–	–
568	Part-I; Vol. VIII	86	HARIDRA WATER EXTRACT	*Curcuma longa* L.	Zingiberaceae	Dried and Powder extract	**Quantitative parameters: Loss on drying** - NMT 7.0 per cent; **Total ash**- NMT 30.0 per cent; **Acid-insoluble ash**- NMT 1.0 per cent; **pH - 5.0-7.0; Total soluble solids** - NLT 90.0 per cent; **T.L.C.** - Refer API; **Assay** - Refer API	–	–	–	–	–

Sl. No.	Part and Volume	Page No.	Common Ayurvedic Name	Botanical Name/ Chemical Name	Family/ Chemical Formula	Part Used	Identity, Purity and Strength/Chemical Properties	Constituents	Properties and Action	Important Formulations	Therapeutic Uses	Dose
569	Part-I; Vol. VIII	89	HARITAKI	*Terminalia chebula* Retz.	Combretaceae	Pericarp of Mature Fruits	**Quantitative parameters: Foreign matter**- NMT 1.0 per cent; **Loss on drying**- NMT 12.0 per cent; **Total Ash**- NMT 6.0 per cent; **Acid-insoluble ash**- NMT 3.0 per cent; **Acid-insoluble ash**- NMT 3.0 per cent; **Alcohol-soluble extractive**- NLT 40.0 per cent; **Water-soluble extractive** - NLT 60.0 per cent; **T.L.C.** - Refer API; **Assay** - Refer API	Tannins, Chebulagic acid, Chebulic acid, Chebulinic acid, Ellagic acid, Gallic acid, Terchebin, Ellagi tannin, Terchebulin and Syryngic acid	Rasa- Kasaya, Katu, Tikta, Amla, Madhura; *Guna*- Laghu, Ruksa, *Virya*-Usna, *Vipaka*-Madhura, *Karma*- Sarvado-saprasamana, Rasayana, Caksusya, Dipana, Anulomana, Hrdya, Medhya	Abhayarista, Agastya haritaki rasayana, Citraka haritaki, Danti haritaki, Dasamula haritaki, Brahma rasayana, Triphala churna, Triphaladi taila, Abhaya lavana, Pathyadi lepa	Vibandha (Constipation), Aruci (Tastelessness), Udavarta (Upward movement of gases), Gulma (Abdominal lump), Udararoga (Diseases of abdomen), Arsa (Piles), Pandu (Anaemia), Sotha (Inflammation), Jirnajvara (Chronic fever), Visamajvara (Intermittent fever), Prameha (Increased frequency and turbidity of urine), Siroroga (Diseases of head), Kasa (cough), Tamakasvasa (Bronchial asthma), Hrdroga (Heart disease)	3 to 6 g of the drug in powder form

Sl. No.	Part and Volume	Page No.	Common Ayurvedic Name	Botanical Name/ Chemical Name	Family/ Chemical Formula	Part Used	Identity, Purity and Strength/Chemical Properties	Constituents	Properties and Action	Important Formulations	Therapeutic Uses	Dose
570	Part-I; Vol. VIII	94	HARITAKI (POWDER)	*Terminalia chebula* Retz.	Combretaceae	Powder	**Identity, Purity and Strength:** Complies with the test for Identity, Purity, Strength and Thin-layer chromatography as stated under Haritaki, **Assay:** Complies with the limits for Assay as per method stated under Haritaki	–	–	–	–	–
571	Part-I; Vol. VIII	95	HARITAKI HYDRO-ALCOHOLIC EXTRACT	*Terminalia chebula* Retz.	Combretaceae	Dried and Powder extract	**Quantitative parameters: Loss on drying** - NMT 7.0 per cent, **Total ash-** NMT 10.0 per cent, **Acid-insoluble ash-** NMT 1.5 per cent, **pH** - 3.0-4.5, **Total soluble solids** - NLT 90.0 per cent; **T.L.C.** - Refer API **Assay** - Refer API	–	–	–	–	–
572	Part-I; Vol. VIII	98	HARITAKI WATER EXTRACT	*Terminalia chebula* Retz.	Combretaceae	Dried and Powder extract	**Quantitative parameters: Loss on drying** - NMT 7.0 per cent, **Total ash-** NMT 10.0 per cent, **Acid-insoluble ash-** NMT 1.5 per cent, **pH - 3.0-4.5, Total soluble solids** - NLT 90.0 per cent; **T.L.C.** - Refer API Assay - Refer API	–	–	–	–	–
573	Part-I; Vol. VIII	101	KALAMEGHA	*Andrographis paniculata* (Burm.f.) wall.ex Ness	Acanthaceae	Aerial Part	**Quantitative parameters: Foreign matter-** NMT 2.0 per cent; **Loss on drying** - NMT 12.0 per cent; **Total ash-** NMT 15 per cent; **Acid-insoluble ash-** NMT 3.0 per cent; **Alcohol-soluble extractive** - NLT 12.0 per cent; **Water-soluble extractive** - NLT 19.0 per cent; T.L.C. - Refer API Assay - Refer API	Andro-grapholide, Neoandro-grapholide, Andro-grapanin, 14-deoxy-11, 12-didehydro-androgra-pholide	Rasa- Tikta, *Guna*- Laghu; Ruksa *Virya*- Sita; *Vipaka*- Katu; *Karma*- Dipana, Jvaraghna, Krmighna, Kasaghna, Pacana, Raktamoksana	Kalameghasava	Ajirna (Dyspepsia), Arsa (Piles), Atisara (Diarrhoea), Jvara (Fever), Kandu (Itching), Kamala (Jaundice), Kustha	1 to 3 g of the drung in powder form

Sl. No.	*Part and Volume*	*Page No.*	*Common Ayurvedic Name*	*Botanical Name/ Chemical Name*	*Family/ Chemical Formula*	*Part Used*	*Identity, Purity and Strength/Chemical Properties*	*Constituents*	*Properties and Action*	*Important Formulations*	*Therapeutic Uses*	*Dose*
											(Disease of skin), Prameha (Increased frequency and turbidity of urine); Pravahika (Dysentery), Tvakvikara (Skin disorders); Vrana (Wound), Yakrtvikara (Disorders of liver)	
574	Part-I; Vol. VIII	107	KALAMEGHA (POWDER)	*Andrographis paniculata* (Burm.f.) wall.ex Ness	Acanthaceae	Powder	**Identity, Purity and Strength:** Complies with the tests for Identity,Purity, Strength and Thin-layer chromatography as stated under kalamegha, **Assay**- Complies with the limits for Assay as per method stated under Kalamegha	–	–	–	–	–
575	Part-I; Vol. VIII	108	KALAMEGHA HYRO-ALCOHOLIC EXTRACT	*Andrographis paniculata* (Burm.f.) wall.ex Ness	Acanthaceae	Dried and powdered extract	**Quantitative parameters: Loss on drying** - NMT 6.0 per cent, **Total ash**- NMT 28.0 per cent, **Acid-insoluble ash**- NMT 5 per cent, **pH** - 6.0-8.0, **Total soluble solids** - NLT 90.0 per cent; **T.L.C.** - Refer API **Assay** - Refer API	–	–	–	–	–

Sl. No.	*Part and Volume*	*Page No.*	*Common Ayurvedic Name*	*Botanical Name/ Chemical Name*	*Family/ Chemical Formula*	*Part Used*	*Identity, Purity and Strength/Chemical Properties*	*Constituents*	*Properties and Action*	*Important Formulations*	*Therapeutic Uses*	*Dose*
576	Part-I; Vol. VIII	111	KALAMEGHA WATER EXTRACT	*Andrographis paniculata* (Burm.f.) wall.ex Ness	Acanthaceae	Dried and powdered extract	**Quantitative parameters: Loss on drying** - NMT 6.0 per cent, **Total ash**- NMT 32.0 per cent, **Acid-insoluble ash**- NMT 7.0 per cent, **pH** - 5.5-7.5, **Total soluble solids** - NLT 90.0 per cent; **T.L.C.** - Refer API **Assay** - Refer API	–	–	–	–	–
577	Part-I; Vol. VIII	114	KANTAKARI	*Solanum surattense* Burm. f. syn. *Solanum xanthocarpum* Schrad. and Wendl.	Solanaceae	Dried Whole Plant	**Foreign matter**- NMT 2.0 per cent; **Loss on drying** - NMT 12.0 per cent; **Total ash**- NMT 9.0 per cent; **Acid-incoluble ash**- NMT 3.0 per cent; **Alcohol-soluble extractive** - NLT 6.0 per cent; **Water-soluble extractive**- NLT 16.0 per cent; **T.L.C.** - Refer API **Assay** - Refer API	Solamargine, α-Solamargine, Solasonine, Solasodine, Sterols, Cycloartenol, Morcarpesterol	Rasa- Katu, Tikta, Guan- Ruksa, Laghu, *Virya*- Usna, *Vipaka*- Katu, *Karma*- Dipana, Pacana, Amadosanasaka, Kanthya, Sothahara	Kanta-karyavaleha, Pancatik-taka ghrta, Vyaghri-harataki	Svasa (Asthma), Kasa (Cough), Jvara (Fever), Aruci (Tastelessness), Pinasa (Chronic rhinits/ sinusitis) Parsvasula (Intercostal neuralgia and pleurodynia), Svarabheda (Hoarseness of voice)	20 to 30 g of the drug for decoction
578	Part-I; Vol. VIII	120	KANTAKARI (POWDER)	*Solanum surattense* Burm. f. syn. *Solanum xanthocarpum* Schrad. and Wendl.	Solanaceae	Powder	**Identity, Purity and Strength:** Complies with the tests for Identity, Purity, Strength and Thin-layer chromatography as stated under Kantakari, **Assay**- Complies with the limits for Assay as per method stated under Kantakari	–	–	–	–	–

Sl. No.	*Part and Volume*	*Page No.*	*Common Ayurvedic Name*	*Botanical Name/ Chemical Name*	*Family/ Chemical Formula*	*Part Used*	*Identity, Purity and Strength/Chemical Properties*	*Constituents*	*Properties and Action*	*Important Formulations*	*Therapeutic Uses*	*Dose*
579	Part-I; Vol. VIII	121	KANTAKARI HYDRO-ALCOHOLIC EXTRACT	*Solanum surattense* Burm. f. syn. *Solanum xanthocarpum* Schrad. and Wendl.	Solanaceae	Dried and powdered extract	**Quantitative parameters: Loss on drying** - NMT 7.0 per cent, **Total ash**- NMT 10.0 per cent, **Acid-insoluble ash**- NMT 1.5 per cent, **pH - 5.0-7.0, Total soluble solids** - NLT 90.0 per cent; **T.L.C.** - Refer API **Assay** - Refer API	–	–	–	–	–
580	Part-I; Vol. VIII	124	KANTAKARI WATER EXTRACT	*Solanum surattense* Burm. f. syn. *Solanum xanthocarpum* Schrad. and Wendl.	Solanaceae	Dried and Powder extract	**Quantitative parameters: Loss on drying** - NMT 6.0 per cent, **Total ash**- NMT 10.0 per cent, **Acid-insoluble ash**- NMT 1.5 per cent, **pH** - 5.0-7.0, **Total soluble solids** - NLT 90.0 per cent; **T.L.C.** - Refer API **Assay** - Refer API	–	–	–	–	–
581	Part-I; Vol. VIII	127	MANDU-KAPARNI	*Centella asiatica* (L.) Urban. Syn. *Hydrocotyle asiatica* L.	Apiaceae	Dried Whole plant	**Foreign matter**- NMT 2.0 per cent; **Loss of drying**- NMT 10.0 per cent; **Total ash**- NMT 17.0 per cent **Acid-insoluble ash**- NMT 5.0 per cent; **Alcohol-soluble extractive**- NLT 9.0 per cent; **Water-soluble extractive**- NLT 20.0 per cent; T.L.C. - Refer API Assay - Refer API	Madecassoside, Asiaticoside, Madecassic acid, Terminolic acid and Asiatic acid	Rasa- Katu, Tikta, Kasaya, Madhura Guan- Sara, Laghu, *Virya*- Sita, *Vipaka*- Madhura, *Karma*- Kaphapittahara, Hrdya, Medhya, Svarya, Rasayana, Dipana, Varnya, Visaghna, Ayusya, Balya, Smrtiprada	Brahma rasayana	Raktapitta (Bleeding disorders); Kustha (Diseases of skin); Meha (Excessive flow of urine); Jvara (Fever), Svasa (Asthma), kasa (Cough), Aruci (Tastelessness), Pandu (Anaemia), Sotha (Inflammation), Kandu (Itching), Raktadosa (Disorders of blood)	3 to 6 g

Sl. No.	*Part and Volume*	*Page No.*	*Common Ayurvedic Name*	*Botanical Name/ Chemical Name*	*Family/ Chemical Formula*	*Part Used*	*Identity, Purity and Strength/Chemical Properties*	*Constituents*	*Properties and Action*	*Important Formulations*	*Therapeutic Uses*	*Dose*
582	Part-I; Vol. VIII	133	MANDU-KAPARNI (POWDER)	*Centella asiatica* (L.) Urban. Syn. *Hydrocotyle asiatica* L.	Apiaceae	Powder	**Complies with the tests of Identity,** Purity, Strength and Thin-layer chromatography as stated under Mandukaparni, **Assay:** Complies with the limits for Assay as per method stated under Mandukaparni	–	–	–	–	–
583	Part-I; Vol. VIII	134	MANDU-KAPARNI HYDRO-ALCOHOLIC EXTRACT	*Centella asiatica* (L.) Urban. Syn. *Hydrocotyle asiatica* L.	Apiaceae	Dried and Powder extract	**Quantitative parameters: Loss on drying** - NMT 7.0 per cent, **Total ash-** NMT 22.0 per cent, **Acid-insoluble ash-** NMT 5.0 per cent, **pH** - 4.5-6.0, **Total soluble solids** - NLT 90.0 per cent; **T.L.C.** - Refer API **Assay** - Refer API	–	–	–	–	–
584	Part-I; Vol. VIII	137	MANDU-KAPARNI WATER EXTRACT	*Centella asiatica* (L.) Urban. Syn. *Hydrocotyle asiatica* L.	Apiaceae	Dried and powdered extract	**Quantitative parameters: Loss on drying** - NMT 7.0 per cent, **Total ash-** NMT 22.0 per cent, **Acid-insoluble ash-** NMT 5.0 per cent, **pH** - 5.0-6.5, **Total soluble solids** - NLT 90.0 per cent; **T.L.C.** - Refer API **Assay** - Refer API	–	–	–	–	–

585	Part-I; Vol. VIII	140	SATAVARI	*Asparagus racemosus* Willd.	Liliaceae	Tuberous roots	**Foreign matter**- NMT 1.0 per cent; **Loss of drying**- NMT 10.0 per cent; **Total ash**- NMT 5.0 per cent **Acid-insoluble ash**- NMT 5.0 per cent; **Alcohol-soluble extractive**- NLT 10.0 per cent **Water-soluble extractive**- NLT 45.0 per cent, T.L.C. - Refer API Assay - Refer API	Sugar, Glycosides, Saponin, Sitosterol and Shatavarins	Rasa- Tikta, Madhura Guan- Snigdha, Guru *Virya*- Sita, *Vipaka*- Madhura, *Karma*- Vrsya, Balya, Medhya, Rasayana, Kaphavataghna, Pittahara, Vatahara, Stanyakara, Hrdya, Netrya, Sukrala, Agnipustikara	Satavari guda, Brahma rasayana, Puga khanda, Saubhagyasunthi, Maha-narayana taila, Brhacchagaladya ghrta, Satavari ghrta, Satavari Kalpa, Asvangandharista, Narasimha churna	Sotha (Inflammation), Ksaya (Pthisis), Parinamasula (Duodenal ulcer), Gulma (Abdominal lump), Atisara (Diarrhoea), Raktatisara (Diarrhoea with blood), Raktavikara (Disorders of blood), Mutrarakta (Haematuria), Amlapitta (Hyperacidity), Arsa (Piles), Vatajvara (Fever due to vata dosa), Svarabheda (Hoarseness of voice), Naktandhya (Night blindness), Vatarakta (Gout), Raktapitta (Bleeding disorders), Visarpa (Erysepales), Sutika roga (Puerperal diseases), Stanya Dosa (Disorders of breast milk) Stanya Ksaya (Decrease in breast milk)	3 to 6 g of the drug

Sl. No.	Part and Volume	Page No.	Common Ayurvedic Name	Botanical Name/ Chemical Name	Family/ Chemical Formula	Part Used	Identity, Purity and Strength/Chemical Properties	Constituents	Properties and Action	Important Formulations	Therapeutic Uses	Dose
586	Part-I; Vol. VIII	144	SATAVARI (POWDER)	*Asparagus racemosus* Willd.	Liliaceae	Powder	**Identity, Purity and Strength:** Complies with the tests for Identity, Purity, Strength and Thin-layer chromatography as stated under Satavari, **Assay-** Complies with the limits for Assay as per method stated under Satavari	–	–	–	–	–
587	Part-I; Vol. VIII	145	SATAVARI HYDRO-ALCOHOLIC EXTRACT	*Asparagus racemosus* Willd.	Liliaceae	Dried and powdered extract	**Quantitative parameters: Loss on drying** - NMT 5.0 per cent, **Total ash**- NMT 5.0 per cent, **Acid-insoluble ash**- NMT 1.0 per cent, **pH** - 4.5-6.5, **Total soluble solids** - NLT 90.0 per cent; **T.L.C.** - Refer API **Assay** - Refer API	–	–	–	–	–
588	Part-I; Vol. VIII	147	SATAVARI WATER EXTRACT	*Asparagus racemosus* Willd.	Liliaceae	Dried and powdered extract	**Quantitative parameters: Loss on drying** - NMT 5.0 per cent, **Total ash**- NMT 5.0 per cent, **Acid-insoluble ash**- NMT 1.0 per cent, **pH** - 4.5-6.5, **Total soluble solids** - NLT 90.0 per cent; **T.L.C.** - Refer API **Assay** - Refer API	–	–	–	–	–

Sl. No.	Part and Volume	Page No.	Common Ayurvedic Name	Botanical Name/ Chemical Name	Family/ Chemical Formula	Part Used	Identity, Purity and Strength/Chemical Properties	Constituents	Properties and Action	Important Formulations	Therapeutic Uses	Dose
589	Part-I; Vol. VIII	149	TAMALAKI	*Phyllanthus fraternus* Webst. Syn. *Phyllanthus niruri* Hook. f, non L.	Euphorbiaceae	Roots, Stems and Leaves	**Foreign matter**- NMT 2.0 per cent; **Loss of drying**- NMT 12.0 per cent; **Total ash**- NMT 16.0 per cent **Acid-insoluble ash**- NMT 7.0 per cent; **Alcohol-soluble extractive**- NLT 3.0 per cent **Water-soluble extractive**- NLT 13.0 per cent, **T.L.C.** - Refer API **Assay** - Refer API	Phyllanthin, Hypo-phyllanthin	Rasa- Kasaya, Tikta, Madhura; *Guna*-Laghu, Ruksa; *Virya*-Sita; *Vipaka*-Madhura; *Karma*-Rocana, Dahanasani, Pittasamaka, Mutrala	Citraka haritaki, Madhuyastyadi taila, Pippalyadi ghrta, Chyawanaprasa, Satavariguda	Trsa (Thirst), Kasa (Cough), Amlapitta (Hyperacidity), Pandu (Anaemia), Ksaya (Pthisis), Ksata (Wound), Kustha (Diseases of sikn), Prameha (Increased frequency and turbidity of urine), Mutraroga (Urinary diseases)	10 to 20 ml of the drug in juice form; 3 to 6 g of the drug in powder form
590	Part-I; Vol. VIII	154	TAMALAKI (POWDER)	*Phyllanthus fraternus* Webst. Syn. *Phyllanthus niruri* Hook. f, non L.	Euphorbiaceae	powder	**Identity, Purity and Strength:** Complies with the tests for Identity, Purity, Strength and Thin-layer chromatography as started under Tamalaki, **Assay**- Complies with the limits for Assay as per method stated under Tamalaki	–	–	–	–	–
591	Part-I; Vol. VIII	155	TAMALAKI HYDRO-ALCOHOLIC EXTRACT	*Phyllanthus fraternus* Webst. Syn. *Phyllanthus niruri* Hook. f, non L.	Euphorbiaceae	Dried and powdered extract	**Quantitative parameters: Loss on drying** - NMT 6.0 per cent, **Total ash**- NMT 16.5 per cent, **Acid-insoluble ash**- NMT 3.5 per cent, **pH** - 4.5-5.5, **Total soluble solids** - NLT 90.0 per cent; **T.L.C.** - Refer API **Assay** - Refer API	–	–	–	–	–

Sl. No.	*Part and Volume*	*Page No.*	*Common Ayurvedic Name*	*Botanical Name/ Chemical Name*	*Family/ Chemical Formula*	*Part Used*	*Identity, Purity and Strength/Chemical Properties*	*Constituents*	*Properties and Action*	*Important Formulations*	*Therapeutic Uses*	*Dose*
592	Part-I; Vol. VIII	158	TAMALAKI WATER EXTRACT	*Phyllanthus fraternus* Webst. Syn. *Phyllanthus niruri* Hook. f, non L.	Euphorbiaceae	Dried and powdered extract	**Quantitative parameters: Loss on drying** - NMT 7.0 per cent, **Total ash**- NMT 25.0 per cent, **Acid-insoluble ash**- NMT 7.5 per cent, **pH - 4.5-5.5, Total soluble solids** - NLT 90.0 per cent; **T.L.C.** - Refer API **Assay** - Refer API	–	–	–	–	–
593	Part-I; Vol. VIII	161	VASA	*Justicia adhatoda* L. syn. *Adhatoda zeylanica* medicus	Acanthaceae	Fresh Dried Mature leaves	**Quantitative parameters: Foreign matter**- NMT 2.0 per cent; **Loss of drying**- NMT 12.0 per cent; **Total ash**- NMT 21.0 per cent **Acid-insoluble ash**- NMT 1.0 per cent; **Alcohol-soluble extractive**- NLT 3.0 per cent **Water-soluble extractive**- NLT 22.0 per cent, **T.L.C.** - Refer API **Assay** - Refer API	Vasicine, Vasicinone, Deoxyvasicine, Adhatonine, Vasicinol	Rasa- Kasaya, Tikta, *Guna*-Laghu, *Virya*-Sita; *Vipaka*-Katu; *Karma*-Kaphapittahara, Raktasangrahika, Kasaghna, Hrdya	Vasakasava, Vasavaleha	Kasa (Cough), Svasa (Asthma), Ksaya (Pthisis), Raktapitta (Bleeding disorders), Prameha (Increased frequency and turbidity of urine), Kamala (Jaundice), Kustha (Diseases of skin)	10 to 20 ml of juice of fresh leaves, 10 to 20 g of the dried drug for decoction
594	Part-I; Vol. VIII	166	VASA (POWDER)	*Justicia adhatoda* L. syn. *Adhatoda zeylanica* medicus	Acanthaceae	Powder	**Identity, Purity and Strength:** Complies with the tests for Identity, Purity, Strength and Thin-layer chromatography as stated under Vasa, **Assay**-Complies with the limits for Assay as per method stated under Vasa	–	–	–	–	–

Sl. No.	*Part and Volume*	*Page No.*	*Common Ayurvedic Name*	*Botanical Name/ Chemical Name*	*Family/ Chemical Formula*	*Part Used*	*Identity, Purity and Strength/Chemical Properties*	*Constituents*	*Properties and Action*	*Important Formulations*	*Therapeutic Uses*	*Dose*
595	Part-I; Vol. VIII	167	VASA HYDRO-ALCOHOLIC EXTRACT	*Justicia adhatoda* L. syn. *Adhatoda zeylanica* medicus	Acanthaceae	Dried and powdered extract	**Quantitative parameters: Loss on drying** - NMT 6.0 per cent, **Total ash**- NMT 25.0 per cent, **Acid-insoluble ash**- NMT 1.0 per cent, **pH** - 6.0 - 8.0, **Total soluble solids** - NLT 90.0 per cent; **T.L.C.** - Refer API **Assay** - Refer API	–	–	–	–	–
596	Part-I; Vol. VIII	170	VASA WATER EXTRACT	*Justicia adhatoda* L. syn. *Adhatoda zeylanica* medicus	Acanthaceae	Dried and powdered extract	**Quantitative parameters: Loss on drying** - NMT 6.0 per cent, **Total ash**- NMT 32.0 per cent, **Acid-insoluble ash**- NMT 1.0 per cent, **pH** - 6.0 - 8.0, **Total soluble solids** - NLT 90.0 per cent; **T.L.C.** - Refer API **Assay** - Refer API	–	–	–	–	–
597	Part-I; Vol. VIII	173	YASTI	*Glycyrrhiza glabra* L.	Fabaceae	Dried, Unpeeled, Stolon and Roots	**Quantitative parameters: Foreign matter**- NMT 2.0 per cent; **Loss of drying**- NMT 12.0 per cent; **Total ash**- NMT 10.0 per cent **Acid-insoluble ash**- NMT 2.5 per cent; **Alcohol-soluble extractive**- NLT 10.0 per cent **Water-soluble extractive**- NLT 20.0 per cent **T.L.C.** - Refer API **Assay** - Refer API	Glycyrrhizin, Glycyrrhetinic acid, Glycyrrhetol, Glabrolide, Isoglabrolide, Asparagine, Sugars, Resin and Starch	Rasa- Madhura *Guna*-Guru, Snigdha *Virya*-Sita; *Vipaka*-Madhura *Karma*-Vatapittajit, Raktaprasadana, Balya, Varnya, Vrsya, Caksusya	Eladi gutika, Yastimadhuka taila, Madhuyastyadi taila	Kasa (Cough), Svarabheda (Hoarseness of voice), Ksaya (Pthisis), Vrana (Ulcer), Vatarakta (Gout)	2 to 4 g of the drug in powder form
598	Part-I; Vol. VIII	178	YASTI (POWDER)	*Glycyrrhiza glabra* L.	Fabaceae	Powder	**Identity, Purity and Strength:** Complies with the test for Identity, Purity, Strength and Thin-layer chromatography as stated under yasti, **Assay**- Complies with the limits for Assay as per method stated under Yasti	–	–	–	–	–

Sl. No.	*Part and Volume*	*Page No.*	*Common Ayurvedic Name*	*Botanical Name/ Chemical Name*	*Family/ Chemical Formula*	*Part Used*	*Identity, Purity and Strength/Chemical Properties*	*Constituents*	*Properties and Action*	*Important Formulations*	*Therapeutic Uses*	*Dose*
599	Part-I; Vol. VIII	179	YASTI HYDRO-ALCOHOLIC EXTRACT	*Glycyrrhiza glabra* L.	Fabaceae	Dried and powdered extract	**Quantitative parameters: Loss on drying** - NMT 5.0 per cent, **Total ash**- NMT 7.0 per cent, **Acid-insoluble ash**- NMT 0.5 per cent, **pH** - 5.0 - 6.5, **Total soluble solids** - NLT 90.0 per cent; **T.L.C.** - Refer API **Assay** - Refer API	–	–	–	–	–
600	Part-I; Vol. VIII	182	YASTI WATER EXTRACT	*Glycyrrhiza glabra* L.	Fabaceae	Dried and powdered extract	**Quantitative parameters: Loss on drying** - NMT 5.0 per cent, **Total ash- NMT 10.0 per cent, Acid-insoluble ash**- NMT 0.5 per cent, **pH** - 5.5 - 6.5, **Total soluble solids** - NLT 90.0 per cent; **T.L.C.** - Refer API **Assay** - Refer API	–	–	–	–	–

16

Challenges in Developing Pharmaceutical Dosage Forms of Ayurvedic Medicines

Chandra Kant Katiyar and Avinash Narwaria

Development of Pharmaceutics in Ayurveda is an interesting exploration of evolution of newer and newer dosage forms by the ancient Ayurvedic experts.

Evolution of dosage forms in Ayurveda is generally presumed to be empirical, however, it has solid scientific thinking behind the same when we analyse the same in the current context. For example *Churna*, crude powder of the plant contains cellular and fibrous material which is important for several Pharmacological actions like purgation. *Triphala* and *Isabgol* are used in crude form basis this rationale. *Avaleha*, semi solid dosage form is another contribution of Ayurveda which was used mostly to provide therapeutic benefits along with the nutrition. These products were manufactured with sugar or jaggery which contributed not only to organoleptic characteristics but also provided energy and nutrition. There is evidence of concept of new drug delivery system (NDDS) in the *Bhaishyaja Kalpana*.

Currently dosage forms of modern medicine are limited to tablet, capsule, powder, injection, nasal drops, patches *etc.* while in the ancient times Ayurvedic texts have mentioned more than 25 types of dosage forms and if we include food formats the number is close to 150.

Evolution of certain dosage forms also provides proof of understanding of site of Pharmacological action. For example if the drug has to cross Blood Brain Barier (BBB) it has to have lipids as carrier and *Brahmi Ghrit* which is based on fat appears to be a perfect recipe to cross BBB to provide therapeutic effect of enhancing the memory.

Ayurvedic Pharmaceutics/*Bhaisajya Kalpana* started with five basic dosage forms which were *Swarasa* (Fresh Juice), *Churna* (Powder), *Him* (Cold water infusion), *Phant* (Hot water infusion) and *Kwath* (Decoction) but all of them were having short shelf-life and were required in large dose to exhibit therapeutic actions. Later on keeping in view the need of innovation to make these dosage forms more palatable and patient friendly, dosage form like *Vati, Gutika*, and *Avaleha* were evolved. *Kwath Kalpana* was very effective but had a very short shelf-life. Ancient sages had great expertise in fermentation to make *Sura* in parallel. Taking lead fro this Ayurvedic experts using their innovative acumen also developed *Asav-Arishtas* which were self fermented preparations of *Kwath* or *Him*. This Dosage form of *Asav Arishta* is a unique contribution of Ayurveda. Benefits of this dosage forms are not only long shelf-life but continued extraction of the active compounds of the plants due to presence of alcohol which also acts as carrier of molecules, makes the products more potent and faster acting. The rate of fermentation of *Asav* also reveals a uniqueness of this dosage form. Slow rate of fermentation for prolonged time may be converting certain chemical compounds in the component herbs from prodrugs to drugs by chemical transformation through biological route to make such products more effective.

Later on in 11th Century when the *Acharyas* like *Nagarjuna* came to know that Mercury is used to convert Copper into Gold, our ancient *Rishis* thought to apply this technology also on the human being and it gave birth to a new discipline in Ayurveda called *Rasa Shastra* which was based upon journey from *Dhatuvada* to *Dehavada*. All the *Rasa* (mercurial) preparations were having benefits of requiring very small dose (125mg – 250mg only), unlimited shelf life and most importantly even they were "*Prakiti neutral*". It created a revolution in Ayurveda similar to advent of Antibiotics in Modern Medicine.

Earlier there was no concept of commercial production of Ayurvedic medicines. The *Vaidyas* used to prepare ayurvedic products for their own patients. Later on commercial production of Ayurvedic medicines started in 19th Century leading to new challenges like availability of raw materials in bulk, need for manufacturing with high volume in minimum time requiring application of modern pharmaceutical technology and machineries, *etc.* Since commercial production required batch to batch consistency there arose the need of quality standards of not only finished products but also of the raw materials besides process standardization. This acted as trigger and the Govt. of India responded by forming Ayurvedic Pharmacopoeia Committee to develop and provide quality standards for ayurvedic raw materials and products.

Certain *Vaidyas* receive application of modern manufacturing technology to ancient ayurvedic dosage form as compromise with the authentic Ayurveda. However, the other more pragmatic *Vaidyas* took it in a positive manner since they were not able to do manufacturing of products on their own due to various reasons.

With the application of modern technology and mechanization in manufacturing of Ayurvedic products several other challenges cropped up *e.g.* textbooks have provided ratio of ingredients in the formulations in parts and therefore, mention of batch size has been missed.

With the advent of extraction technology, companies started to use extracts to make tablets and there was no guidance and guiding documents for use of suitable excipients.

One important but least discussed issue is basis of recommended dose and schedule of Ayurvedic products. In case of Allopathic Medicines the frequency of dose is decided basis Pharmacokinetic profile of the molecule. In case of Ayurvedic products, as per *Sharangdhar Samhita* all Ayurvedic products should be taken only in the morning unless otherwise directed. It might be providing two benefits. Since herbs contain hundreds of compounds, it is not possible to decide frequency basis Pharmacokinetics as all compounds may be having different half-life. Therefore, ancient *Rishis* may have thought safe to recommend the intake of drugs only in the morning and once in a day. Second benefit is that if the drug is working for 12 hours and most of the compounds have to pass through liver, liver gets time to take rest unlike Allopathic medicine, where the dosage frequency is decided basis Pharmacokinetics and liver is over burdened by the chemical onslaught round the clock which may lead to damage of liver.

With the advent of more efficient methods of extraction of plant material Ayurvedic preparations are now being formulated with more potent extracts in better look and feel and more convenient dosage forms. Techniques like Super Critical extraction, Solvent extraction and Vacuum evaporation methods have facilitated the commercial feasibility of better quality Ayurvedic formulations keeping its holistic nature intact yet providing standardized, more stable and efficacious products.

Application of modern processing methods, advanced machinery and packaging technologies have given the opportunities of making consumer friendly dosage forms like capsules, film coated tablets, emulgels, pillets, patches, aerosols, rubs and balms, granules, syrups *etc.*

Challenges in Pharmaceutics of Ayurveda include process standardisation, development of non-destructive method of analysis of *bhasmas*, application of ligand technology to understand and develop standards of *bhasmas*, development of faster fermentation techniques to provide similar efficacious *Asav Arishta* in shorter cycle time, decontamination techniques for *Churnas* to prevent microbial contamination, determining optimum particle size for *churna* and *kwath*, whether *gugglus* should have short disintegration time or long, determination of uniform formula for developing unit dosage forms of various formats.

With the enhancing awareness and interest of consumers worldwide Ayurveda cannot afford to remain limited in its own boundary. People have already started looking at its products critically and have started demanding application of higher standards of not only product quality but also related to their safety and efficacy.

Controversy of presence of heavy metals in certain Ayurvedic products has diverted the attention from efficacy to safety. Industry, Academia, Researchers and Regulators have to interact regularly to draw strategies to counter these challenges. Otherwise, one paper published in JAMA would continue forcing knee jerk reaction from Govt. of India and one report of Lord Waltan committee questioning research based evidence of efficacy of Ayurvedic products would continue to haunt the fraternity (Katiyar CK, 2015).

Design of Commonly Used Dosage Forms

A medicine should be properly formulated for administration to patients keeping in mind the need of consumers and property of ingredients used to deliver desirable efficacy and safety. Ayurveda mentions a wide range of dosage forms which would have been developed basis patient's need and availability of raw materials. Kwatha, Kalka, Swarasa, Hima and Phant are preparations for instant use. Dosage forms that can be stored for certain period of time are:

i. Solid dosage forms: Pill, Vati Gutika
ii. Semi-solid dosage forms: Avaleha, Pak, Ghrita, Lepa
iii. Liquid dosage forms: Asava, Arishta, Arka, Taila, Dravaka, Panaka
iv. Powder dosage forms: Churna, Bhasma, Satva, Mandura, Pisti, Parpati, Lavana, Kshara

Keeping pace with changing times and recent technological advancements Ayurvedic medicines are also being developed as per modern practices and regulations. There are proprietary Ayurvedic products available in the form of capsules, film coated tablets, gel, cream, ointment, lotions *etc.* which were not mentioned in traditional Ayurvedic literature but have been developed using new technologies to provide better efficacy, convenience and improved shelf-life.

Developing an Ayurvedic dosage form provides some additional challenges to the formulator due to complex nature of herbal active materials. Ayurvedic preparations are either in single herb form or in multiple ingredient form generally having admixtures of herbs and extracts. This challenge further intensifies due to natural variations in plant material itself, lack of standardization and insufficient research work on physical, chemical, and biological characteristics of active ingredients to be used in preparing an Ayurvedic medicine. In addition, stability study, appearance, palatability, preservation, impurities like microbial load and heavy metals further complicate the issue.

The proper design and formulation of a dosage form requires consideration of physico-chemical and therapeutic properties of all of the drug substances and excipients to be used in fabricating the product. The active ingredients and excipients used must be compatible and produce a product that is stable, efficacious, palatable, easy to administer, and safe for consumption.

To minimize the issues certain points need to be considered at the time of conceptualization of dosage form and product development. Instead of blindly following the pharma drug development steps the formulator should also evaluate the physical characteristics and compatibility of active ingredients through well-designed pre-formulation. Pre-formulation work include evaluating the ingredients as well as prototype formulations for physical properties such as color, appearance, taste and palatability, particle size, solubility, partition coefficient, disintegration, stability, *etc.*

Products with Extracts Vs Products with Powders

One of the most common issues is whether to use herbs in powder form or its extract. In most cases delivery of therapeutic dose per unit dosage form in powder form is not possible. Use of extracts based on activity guided fraction is the remedy but nature of extract and its standardization based on efficacy is very important. However, plant ingredients when used in concentrated or dried extract form give other challenges like high disintegration time of tablets or capsules, hygroscopic behaviour in the form of powder or granules *etc.* Careful selection of adsorbents, diluents and disintegrating agents through pre-formulation studies can resolve this issue. Dark unpleasant color and bitterness is another issue while formulating a liquid oral dosage form. Selection of correct fraction after removal of chlorophyll and undesired part using discoloration techniques may help improving color of extract. Bitterness can be reduced using various bitter masking agents.

When plant material is to be used in powder form importance should be given to balance the powder fineness and retention of active ingredients specially the volatile material from plant material due to heat generated during grinding process. Hence use of moderately coarse powder in case of volatile oil containing plant drugs is suggested. Hard plant material like heart wood, bark and seeds can be finely grinded using a pulverizer. All herbo-mineral or mineral formulations should be finely grinded.

Selection of Excipients

In any dosage form role of active ingredients is to deliver the intended efficacy of the product. Similarly, excipients used in a dosage form play an equally important role. Excipients used in any product largely affect the physical properties and release of medicament thus indirectly affecting product's efficacy and stability. Use of excipients also plays a major role as processing aids for smooth running of products in manufacturing facilities. In other words excipients used in a dosage form has multi-faceted role like appearance, palatability, physical characteristics, stability, efficacy delivery *etc.* Hence selection of excipients should be done very carefully considering the desired benefits from product.

For Ayurvedic medicines choices of right kind of excipients could be made from ingredients mentioned in Indian Pharmacopoeia or Food Safety Standards of India.

Standardization Related Issues

Although lot of publications are available on phytochemistry of plants they are still insufficient to establish an ingredient Vs activity relationship in most of Ayurvedic medicines. It is necessary to further co-relate the bioactivity of plant material with pharmacological studies. Limited knowledge on standardization, bio-availability, pharmacokinetics and mode of action of Ayurvedic drugs are major challenges in developing a safe and effective dosage form.

Lack of standardization carries further challenges for Ayurvedic dosage forms like maintaining batch to batch consistency, taste, appearance, difficulty in validation of test procedures and manufacturing procedures.

Emerging Trends in Designing Dosage Forms

Development of proprietary Ayurvedic dosage forms in-line with allopathic formulations is now being practiced by industrial research laboratories. Latest techniques and tools for pre-formulation, formulation development, testing quality parameters, stability studies, monitoring quality at various stages of manufacturing and proof of efficacy studies is gradually being implemented for Ayurvedic products also.

Developing Ayurvedic dosage form involves various steps, starting from authentication of raw materials, chemical quality standardization, safety and preclinical studies, clinical trials. Systematic standardization of active botanical ingredients needs broader considerations and seriously required, although one cannot readily apply the typical modern pharmaceutical Pharmacopoeial standards. The concept of active markers in the process of standardization needs a flexible approach in favour of the complex nature of herbal materials. For Ayurvedic medicines, newer and updated guidelines of standardization are required.

Conducting Stability and Testing Related Issues

Due to lack of standardization efficacy based stability testing is not possible in all the cases. A formulator is left with the option of evaluating the dosage form for its physico-chemical parameters and behaviour of marker compounds during the shelf-life study. This may or may not be a confirmatory study on assuring efficacy of the product during its assigned shelf-life. A more systematic approach for conducting shelf-life by using active compound based study is needed. In many cases it is not possible to quantify the marker compounds in finish formulation due to very small quantity which is often below detectable limits of highly sophisticated instruments like HPLC, HPTLC and GCMS.

Stability Testing of Ayurvedic Medicines

Stability study of a pharmaceutical dosage form is an important part of its development process to arrive at shelf life by collecting data on physio-chemical parameters and overall acceptability. Selection of suitable packing material, storage and shipping conditions are also decided basis results of stability study.

The shelf life can be determined by the guidelines issued in Ayurvedic Pharmacopoeia of India, Part I, Volume VIII to assign a shelf life to new and existing Ayurvedic Medicine.

General Information on Stability

Shelf life (expiry date) is mandatory requirement for all licensed Ayurvedic medicines. The stability depends on various factors like the nature of the product, the ingredients of the products, the packaging material *etc.* Stability studies are carried out to demonstrate that the medicine will remain suitable for consumption during storage period when kept under the recommended condition(s) mentioned on the packaging. On the product label, if there is no mention about any specific storage condition, then it is assumed that the product can be stored at room temperature (below 30°C).

Various approaches can be followed to monitor the stability of the product. The first approach is to store the sample of same batch material at standard storage and accelerated storage conditions and test them periodically. Based on the evaluation of the results, the expiry date or shelf life may be determined. The second approach, also known as the "cross sectional approach", is to select samples from batches manufactured over a period of last five years spanning six months and evaluate them simultaneously. Based on the result obtained the expiry date or shelf life may be determined.

For other details like selection of batches, container closure system, one should refer API, Volume VIII or suitable regulatory guidelines (WHO/ICH/OECD *etc.*) based upon country of registration.

Specification and Stability Indicating Parameters

Specification is a list of tests, reference to analytical procedures and proposed acceptance criteria.

Stability study should include testing of those attributes of the drug that are susceptible to change during storage and are likely to influence quality, safety, and/or efficacy. The testing should cover as appropriate, the physical, chemical, biological, and microbiological attributes. Validated stability-indicating analytical procedures should be applied. Whether and to what extent replication should be performed will depend on the results from validation studies.

The physical parameters included in the specification need not be limited to colour, odour, appearance, shape and taste only. The chemical parameters should include colour reaction, *p*H value, weight variation, disintegration, bulk density, extractive values, estimation of active or marker or category compound by suitable methods and chromatographic profiling. A suitable bioassay may be employed wherever possible.

The limits of acceptance for the products should be those specified in pharmacopoeia. If limits are not available these should be derived from release specification. Shelf life acceptance criteria should be derived from consideration of all available stability information. It may be appropriate to have justifiable differences between the shelf life and release acceptance criteria based on the stability evaluation and the changes observed on storage. Any differences between the release and shelf life acceptance criteria for antimicrobial preservative content should be supported by a validated correlation of chemical content and preservative effectiveness demonstrated during development of the product in its final formulation (except for preservative concentration) intended for marketing.

Evaluation

The purpose of stability is to establish, based on testing a minimum of at least three batches of the drug a retest period applicable to all future batches for the drug substance, or a shelf life and label storage instructions applicable for all future batches of the drug product manufactured and packed under similar circumstances.

An ASU drug can be considered to be stable if "no significant change" occurs during at any time of testing at accelerated storage condition or at real time storage condition.

"Significant change" for a drug is defined as

1. A + or – 20 per cent change from the initial assay value (If the drug is analysed for its marker). A + or -15 per cent change from the initial assay value (If the drug analysed for its active compound).
2. Appearance of new spots in Identification by TLC (when compared with the sample stored in less than 10%) or complete disappearance of existing spot.
3. The physico-chemical parameters (moisture, ash, particle size) shall not vary beyond 25 per cent of the initial value.
4. Failure to meet acceptance criteria for appearance (Physical attributes, and functionality tests, *e,g.*, colour, phase separation, caking, hardness).

Testing Frequency

For long-term studies frequency of testing should be sufficient to establish the stability profile of the drug. For drug with proposed shelf life of more than 12 months, the frequency of testing at long-term storage condition should normally be every 6 months over first year, and the second year and annually thereafter through the proposed re-test period or shelf life.

At the accelerated storage condition, a minimum of three time points including the initial and final time points (*e.g.* 0, 3 and 6 months) from a 6 months study is recommended.

Reduced designs *i.e.* matrixing or bracketing, where the resting frequency is reduced or certain factor combinations are not tested at all, can be applied if justified.

However, testing frequency can be increased basis product nature and understanding susceptibility of product degradation profile. Container-closure system would also to be considered for time-point establishment. Bulk containers are also suggested to undergo in-use stability testing as per their dosage recommendation.

Storage Condition

The choice of test conditions defined in this guidance document is based on ICH climatic zone IV where India belongs. Recommended storage conditions are:

Sl.No.	Study	Storage conditions	Minimum time
1	Accelerated	40°± 2°/75 per cent RH ± 5 per cent	6 months
2	Long-term	30° ± 2°/65 per cent RH ± 5 per cent	12 months

Other storage conditions are allowable if suitably justified and/or recommended by countries where the drug product is to be registered. For products which are temperature-sensitive, lower temperature can be designated as the recommended storage condition and long-term storage temperature. The accelerated testing should be then carried out at least 10°C more than the long-term storage condition along with appropriate relative humidity condition for that temperature.

The reference (control) samples for the above study should be stored in a temperature less than 10°C.

Stability Protocols

Stability Study Protocol for Tablet (Guti/Vatika)

A. Product Details

Product Code/Name	STABILITY STUDY PROTOCOL	Protocol No. Initiation Date: Study Duration:

B. Stability Study Plan

Time Points (In Months)	*Date of Withdrawal*	*Sampling Plan*		
		40°C/75 per cent RH	*30°C/65 per cent RH*	*2 - 8° C*
0 M (Initial)	DD/MM/YYYY	*	–	XX packs of each batch to be kept in 2° - 8° C as control sample. To be withdrawn as and when required.
3 M	DD/MM/YYYY	#	*	
6 M	DD/MM/YYYY	#	#	
9 M	DD/MM/YYYY	–	*	
12 M	DD/MM/YYYY	–	#	
18 M	DD/MM/YYYY	–	*	
24 M	DD/MM/YYYY	–	#	
36 M	DD/MM/YYYY		#	
48 M	DD/MM/YYYY		#	
60 M	DD/MM/YYYY		#	

* All quality parameters to be done except Microbiological testing.
# All quality parameters to be done.
– No sample is being kept for analysis.

Packing Material Composition

(Example)

1. ---μ/---GSM (Grams per Square Meter) PVC (Polyvinyl Chloride) / PVDC (Polyvinylidene Chloride) & XXμ Aluminium Foil.
2. HDPE (High Density Polyethylene) Bottle with PP (Polypropylene) Cap

C. Tentative Specification for Analysis

Sl.No.	*Test Parameters*	*Tentative Specification*	*Acceptance Limits*
1.	**Description**	XXXXXXXX	XXXXXXXX
2.	**Average Weight (mg)**	XXX-XXX	XXX-XXX
3.	**LOD at 105°C**	XXX-XXX	XXX-XXX
4.	**Disintegration Time**	NMT XX minutes	NMT XX minutes
5.	**Length (mm)**	XXX-XXX	XXX-XXX
6.	**Width (mm)**	XXX-XXX	XXX-XXX
7.	**Thickness (mm)**	XXX-XXX	XXX-XXX
8.	**Content (per cent w/w) of Total Group of compound (*e.g.* Polyphenolics, Saponins)**	XXX-XXX	XXX-XXX (± 15 per cent of initial values)
9.	**Assay of marker compound (Example-Curcumin)**	XXX-XXX	XXX-XXX (± 20 per cent of initial values)
10.	**TLC finger printing (Identification)**	Should comply wrt control sample	Should comply wrt control sample
11.	**Microbiology**		
	A. Total Plate Count (cfu/g)	NMT 10^4	NMT 10^5
	B. Yeast and Mould (cfu/g)	NMT 10^2	NMT 10^3
	C. Pathogens		
	a. E.coli	Absent	Absent
	b. S.aureus	Absent	Absent
	c. Salmonella sp.	Absent	Absent
	d. P. aeruginosa	Absent	Absent

Stability Study Protocol for Churna

A. Product Details

Product Code/Name	**STABILITY STUDY PROTOCOL**	Protocol No.
		Initiation Date:
		Study Duration:

B. Stability Study Plan

Time Points (In Months)	*Date of Withdrawal*	*Sampling Plan*		
		40°C/75 per cent RH	*30°C/65 per cent RH*	*2 - 8° C*
0 M (Initial)	DD/MM/YYYY	*	–	XX packs of each batch to be kept in 2° - 8° C as control sample. To be withdrawn as and when required.
3 M	DD/MM/YYYY	#	*	
6 M	DD/MM/YYYY	#	#	
9 M	DD/MM/YYYY	–	*	
12 M	DD/MM/YYYY	–	#	
18 M	DD/MM/YYYY	–	*	
24 M	DD/MM/YYYY	–	#	
36 M	DD/MM/YYYY		#	
48 M	DD/MM/YYYY		#	
60 M	DD/MM/YYYY		#	

* All quality parameters to be done except Microbiological testing.
\# All quality parameters to be done.
– No sample is being kept for analysis.

Packing Material Composition (Example)

XX ml HDPE (High Density Polyethylene) Bottle with PP (Poly Propylene) Cap

C. Tentative Specification for Analysis

Sl.No.	*Test Parameters*	*Tentative Specification*	*Acceptance Limits*
1.	**Description**	XXXXXXXX	XXXXXXXX
2.	**LOD at 105⁰C**	XXX-XXX	XXX-XXX
3.	**Water Soluble Extractive (per cent w/w)**	XXX-XXX	XXX-XXX (± 15 per cent of initial values)
4.	**Alcohol Soluble Extractive (per cent w/w)**	XXX-XXX	XXX-XXX (± 15 per cent of initial values)
5.	**Content (per cent w/w) of Total Group of compound (*e.g.* Polyphenolics, Saponins)**	XXX-XXX	XXX-XXX (± 15 per cent of initial values)
6.	**Assay of marker compound (Example-Curcumin)**	XXX-XXX	XXX-XXX (± 20 per cent of initial values)
7.	**TLC finger printing (Identification)**	Should comply wrt control sample	Should comply wrt control sample
8.	**Microbiology**		
	A. Total Plate Count (cfu/g)	NMT 10^4	NMT 10^5
	B. Yeast and Mould (cfu/g)	NMT 10^2	NMT 10^3
	C. Pathogens		
	a. E.coli	Absent	Absent
	b. S.aureus	Absent	Absent
	***c. Salmonella* sp.**	Absent	Absent
	d. P. aeruginosa	Absent	Absent

Stability Study Protocol for Asava-Arishta

A. Product Details

Product Code/Name	**STABILITY STUDY PROTOCOL**	Protocol No.
		Initiation Date:
		Study Duration:

B. Stability Study Plan

Time Points (In Months)	*Date of Withdrawal*	*Sampling Plan*		
		40⁰C/75 per cent RH	*30⁰C/65 per cent RH*	*2 - 8⁰C*
0 M (Initial)	DD/MM/YYYY	*	–	XX bottles to be kept in 2° - 8° C as control sample. To be withdrawn as and when required.
3 M	DD/MM/YYYY	#	*	
6 M	DD/MM/YYYY	#	#	
9 M	DD/MM/YYYY	–	*	
12 M	DD/MM/YYYY	–	#	
18 M	DD/MM/YYYY	–	*	
24 M	DD/MM/YYYY	–	#	
36 M	DD/MM/YYYY		#	
48 M	DD/MM/YYYY		#	
60 M	DD/MM/YYYY		#	

* All quality parameters to be done except Microbiological testing.
# All quality parameters to be done.
– No sample is being kept for analysis.

Packing Material Composition

(Example)

1. XX g glass/PET (Polyethylene terephthalate) bottle
2. XX mm Aluminium Cap

C. Tentative Specification for Analysis

Sl.No.	*Test Parameters*	*Tentative Specification*	*Acceptance Limits*
1.	**Description**	XXXXXXXX	XXXXXXXX
2.	**Specific gravity (Weight/ml) at 25°C (g/ml)**	XXX-XXX	XXX-XXX
3.	**pH**	XXX-XXX	XXX-XXX
4.	**Colour Index**	XXX-XXX	XXX-XXX
5.	**Alcohol content (per cent v/v)**	XXX-XXX	XXX-XXX
6.	**Absence of Methanol**	Should Comply	Should Comply
7.	**Total reducing sugar (per cent w/v)**	Not less than XX	Not less than XX
8.	**Total non-reducing sugar (per cent w/v)**	Not more than XX	Not more than XX
9.	**Total Solids (per cent w/v)**	Not less than XX	Not less than XX
10.	**Acidity (ml)**	XXX-XXX	XXX-XXX
11.	**Identification (By Thin Layer Chromatography)**	Should comply wrt control sample	Should comply wrt control sample
12.	**°Brix**	XXX-XXX	XXX-XXX
13.	**Estimation of active (*e.g.* Total Phenolics)**	XXX-XXX	XXX-XXX (± 15 per cent of initial values)

Sl.No.	*Test Parameters*	*Tentative Specification*	*Acceptance Limits*
14.	**Assay (per cent w/w) of marker compound**	XXX-XXX	XXX-XXX (± 20 per cent of initial values)
15.	**Taste Panel Evaluation**	Should comply wrt control sample	Should comply wrt control sample
16.	**Microbiology**		
	A. Total Plate Count (cfu/g)	NMT 10^4	NMT 10^5
	B. Yeast and Mould (cfu/g)	NMT 10^2	NMT 10^3
	C. Pathogens		
	a. E.coli	Absent	Absent
	b. S.aureus	Absent	Absent
	***c. Salmonella* sp.**	Absent	Absent
	d. P. aeruginosa	Absent	Absent

Stability Study Protocol for Avaleha

A. Product Details

Product Code/Name	**STABILITY STUDY PROTOCOL**	Protocol No.
		Initiation Date:
		Study Duration:

B. Stability Study Plan

Time Points (In Months)	*Date of Withdrawal*	*Sampling Plan*		
		40⁰C/75 per cent RH	*30⁰C/65 per cent RH*	*2 - 8⁰ C*
0 M (initial)	DD/MM/YYYY	*	–	X Containers kept as control sample. To be withdrawn as and when required.
3 M	DD/MM/YYYY	#	*	
6 M	DD/MM/YYYY	#	*	
9 M	DD/MM/YYYY	–	*	
12 M	DD/MM/YYYY	–	#	
18 M	DD/MM/YYYY	–	*	
24 M	DD/MM/YYYY	–	#	
36 M	DD/MM/YYYY	–	#	

* Microbiological test not required.
# Microbiological test required.
– No sample is being kept for analysis.

Packing Material Details
(Example)

HDPE (High Density Polyethylene) Container with PP(Poly Propylene) Cap

C. Tentative Specification for Analysis

Sl.No.	*Test Parameters*	*Tentative Specification*	*Acceptance Limits*
1.	**Description**	XXXXXXXXX	XXXXXXXXX
2.	**Loss On Drying at 105⁰C**	XX – XX	XX – XX (± 25 per cent of initial values)
3.	**pH**	XX – XX	XX – XX (± 25 per cent of initial values)
4.	**Thin Layer Chromatography**	Should comply wrt control sample	Should comply wrt control sample
5.	**Content of total group of compound (per cent w/w)(Example-Polyphenols, Saponins, Glycosides)**	XXX-XXX	XXX-XXX (± 15 per cent of initial values)
6.	**Total Reducing Sugar (per cent w/w)**	XX – XX	XX – XX (± 25 per cent of initial values)
7.	**Total Non-reducing Sugar (per cent w/w)**	XX – XX	XX – XX(± 25 per cent of initial values)
8.	**Water Soluble Extractive (per cent w/w)**	XX – XX	XX – XX(± 25 per cent of initial values)
9.	**Alcohol Soluble Extractive (per cent w/w)**	XX – XX	XX – XX(± 25 per cent of initial values)
10.	**Taste Panel Evaluation**	Should comply wrt control sample	Should comply wrt control sample
11.	**Microbiology**		
	A. Total Plate Count (cfu/g)	NMT 10^4	NMT 10^5
	B. Yeast and Mould (cfu/g)	NMT 10^2	NMT 10^3
	C. Pathogens		
	a. E. coli	Absent	Absent
	b. S. aureus	Absent	Absent
	***c. Salmonella* sp.**	Absent	Absent
	d. P. aeruginosa	Absent	Absent

Stability Study Protocol for Capsules

A. Product Details

Product Code/Name	STABILITY STUDY PROTOCOL	Protocol No. Initiation Date: Study Duration:

B. Stability Study Plan

Time Points (In Months)	*Date of Withdrawal*	*Sampling Plan*		
		40°C/75 per cent RH	*30°C/65 per cent RH*	*2 - 8° C*
0 M (Initial)	DD/MM/YYYY	*	–	XX packs to be kept in 2° - 8° C as control sample. To be withdrawn as and when required.
3 M	DD/MM/YYYY	#	*	
6 M	DD/MM/YYYY	#	#	
9 M	DD/MM/YYYY	–	*	
12 M	DD/MM/YYYY	–	#	
18 M	DD/MM/YYYY	–	*	
24 M	DD/MM/YYYY	–	#	
36 M	DD/MM/YYYY		#	
48 M	DD/MM/YYYY		#	
60 M	DD/MM/YYYY		#	

* All quality parameters to be done except Microbiological testing.
# All quality parameters to be done.
– No sample is being kept for analysis.

Pack Material Composition
(Example)

1. ---µ/--- GSM (Gram per Square Meter) PVC (Polyvinyl Chloride)/PVDC (Polyvinylidene Chloride) & ---µ Aluminium Foil.
2. HDPE (High Density Polyethylene) Bottle with PP (Poly Propylene) Cap

C. Tentative Specification for Analysis

Sl.No.	*Test Parameters*	*Tentative Specification*	*Acceptance Limits*
1.	**Description**	XXXXXXXX	XXXXXXXX
2.	**Average Fill Weight (mg)**	XXX-XXX	XXX-XXX
3.	**Average Capsule Weight (mg)**	XXX-XXX	XXX-XXX
4.	**LOD at 105°C**	XXX-XXX	XXX-XXX
5.	**Disintegration Time**	**NMT 30 minutes**	**NMT 30 minutes**
6.	**Locked Length (mm)**	XXX-XXX	XXX-XXX
7.	Width (mm)	XXX-XXX	XXX-XXX
8.	**Content (per cent w/w) of Total Group of compound (*e.g.* Polyphenolics, Saponins)**	XXX-XXX	XXX-XXX (± 15 per cent of initial values)
9.	**Assay of marker compound (Example-Curcumin)**	XXX-XXX	XXX-XXX (± 20 per cent of initial values)
10.	**TLC finger printing (Identification)**	Should comply wrt control sample	Should comply wrt control sample
11.	**Microbiology**		
	A. Total Plate Count (cfu/g)	NMT 10^4	NMT 10^5
	B. Yeast and Mould (cfu/g)	NMT 10^2	NMT 10^3
	C. Pathogens		
	a. E. coli	Absent	Absent
	b. S. aureus	Absent	Absent
	***c. Salmonella* sp.**	Absent	Absent
	d. P. aeruginosa	Absent	Absent

Stability Study Protocol for Panak (Syrup)

A. Product Details

Product Code/Name	STABILITY STUDY PROTOCOL	Protocol No.
		Initiation Date:
		Study Duration:

B. Stability Study Plan

Time Points (In Months)	*Date of Withdrawal*	*Sampling Plan*		
		40⁰C/75 per cent RH	*30⁰C/65 per cent RH*	*2 - 8⁰ C*
0 M (initial)	DD/MM/YYYY	*	–	X bottles kept as control sample. To be withdrawn as and when required.
3 M	DD/MM/YYYY	#	*	
6 M	DD/MM/YYYY	#	#	
9 M	DD/MM/YYYY	–	*	
12 M	DD/MM/YYYY	–	#	
18 M	DD/MM/YYYY	–	*	
24 M	DD/MM/YYYY	–	#	
36 M	DD/MM/YYYY		#	

*All quality parameters to be done except Microbiological testing.
All quality parameters to be done.
– No sample is being kept for analysis.

Pack Details

(Example)

----- coloured, Glass bottle with ROPP (Roll on Pilfer Proof) Cap

C. Tentative Specification for Analysis

Sl.No.	*Test Parameters*	*Tentative Specification*	*Acceptance Limits*
1.	**Description**	XXXXXXXXX	XXXXXXXXX
2.	**Weight/ml at 25⁰C**	XX - XX	XX - XX
3.	**pH**	XX - XX	XX - XX
4.	**Colour Index**	XX - XX	XX - XX
5.	**Acidity (ml)**	XX - XX	XX - XX
6.	**Viscosity (cps)**	XX - XX	XX - XX
7.	**Brix**	XX - XX	XX - XX
8.	**Total Sugar (per cent w/w)**	XX - XX	XX - XX
9.	**Thin Layer Chromatography**	Should comply wrt control sample	Should comply wrt control sample
10.	**Content of total group of compound (per cent w/w)(Example-Polyphenolics, Saponins, Glycosides)**	XXX-XXX	XXX-XXX (± 15 per cent of initial values)
11.	**Microbiology**		
	A. Total Plate Count (cfu/g)	NMT 10^4	NMT 10^5
	B. Yeast and Mould (cfu/g)	NMT 10^2	NMT 10^3
	C. Pathogens		
	a. E. coli	Absent	Absent
	b. S. aureus	Absent	Absent
	c. *Salmonella* sp.	Absent	Absent
	d. P. aeruginosa	Absent	Absent

Stability Study Protocol for Taila/Ghrita

A. Product Details

Product Code/Name	STABILITY STUDY PROTOCOL	Protocol No. Initiation Date: Study Duration:

B. Stability Study Plan

Time Points (In Months)	*Date of Withdrawal*	*Sampling Plan*		
		40°C/75 per cent RH	*30°C/65 per cent RH*	*2 - 8° C*
0 M (initial)	DD/MM/YYYY	*	–	X bottles kept as control sample. To be withdrawn as and when required.
3 M	DD/MM/YYYY	#	*	
6 M	DD/MM/YYYY	#	#	
9 M	DD/MM/YYYY	–	*	
12 M	DD/MM/YYYY	–	#	
18 M	DD/MM/YYYY	–	*	
24 M	DD/MM/YYYY	–	#	
36 M	DD/MM/YYYY		#	

*All quality parameters to be done except Microbiological testing.
All quality parameters to be done.
– No sample is being kept for analysis.

Pack Details

(Example)

Glass bottle with ROPP (Roll on Pilfer Proof) Cap

C. Tentative Specification for Analysis

Sl.No.	*Test Parameters*	*Tentative Specification*	*Acceptance Limits*
1.	**Description**	XXXXXXXXX	XXXXXXXXX
2.	**Refractive Index**	XX - XX	XX - XX
3.	**Weight/ml (g/ml)**	XX - XX	XX - XX
4.	**Acid Value**	XX - XX	XX - XX
5.	**Iodine Value**	XX - XX	XX - XX
6.	**Peroxide Value**	XX - XX	XX - XX
7.	**Congealing Point**	XX - XX	XX - XX
8.	**Saponification Value**	XX - XX	XX - XX
9.	**Thin Layer Chromatography**	Should comply wrt control sample	Should comply wrt control sample
10.	**Content of total group of compound (per cent w/w)(Example-Polyphenolics, Saponins, Glycosides)**	XXX-XXX	XXX-XXX (± 15 per cent of initial values)
11.	**Microbiology**		
	A. Total Plate Count (cfu/g)	NMT 10^4	NMT 10^5
	B. Yeast and Mould (cfu/g)	NMT 10^2	NMT 10^3
	C. Pathogens		
	a. E. coli	Absent	Absent
	b. S. aureus	Absent	Absent
	c. *Salmonella* sp.	Absent	Absent
	d. P. aeruginosa	Absent	Absent

Stability Study Protocol for Cream/Gel/Ointment/Lotion

A. Product Details

Product Code/Name	STABILITY STUDY PROTOCOL	Protocol No. Initiation Date: Study Duration:

B. Stability Study Plan

Time Points (In Months)	Date of Withdrawal	Sampling Plan		
		40ºC/75 per cent RH	30ºC/65 per cent RH	2 - 8º C
0 M (Initial)	DD/MM/YYYY	*	–	XX number of pack to be kept in 2° - 8° C as control sample. To be withdrawn as and when required.
3 M	DD/MM/YYYY	#	*	
6 M	DD/MM/YYYY	#	#	
9 M	DD/MM/YYYY	–	*	
12 M	DD/MM/YYYY	–	#	
18 M	DD/MM/YYYY	–	*	
24 M	DD/MM/YYYY	–	#	
36 M	DD/MM/YYYY		#	

*All quality parameters to be done except Microbiological testing.
All quality parameters to be done.
– No sample is being kept for analysis.

Packing Material Composition
(Example)

1. HDPE (High Density Polyethylene) Bottle with PP (Poly Propylene) Cap
2. Laminated tubes

C. Tentative Specification for Analysis

Sl.No.	Test Parameters	Tentative Specification	Acceptance Limits
1.	**Description**	XXXXXXXX	XXXXXXXX
2.	**Weight/ml (g/ml)**	XX - XX	XX - XX
3.	**Viscosity**	XXX-XXX	XXX-XXX
4.	**pH**	XX - XX	XX - XX
5.	**Iodine Value**	XX - XX	XX - XX
6.	**Peroxide Value**	XX - XX	XX - XX
7.	**Acid Value**	XX - XX	XX - XX
8.	**Saponification Value**	XX - XX	XX - XX
9.	**Content (per cent w/w) of Total Group of compound (*e.g.* Polyphenolics, Saponins)**	XXX-XXX	XXX-XXX (± 15 per cent of initial values)
10.	**Assay of marker compound (Example-Curcumin)**	XXX-XXX	XXX-XXX (± 20 per cent of initial values)
11.	**TLC/GC finger printing (Identification)**	Should comply wrt control sample	Should comply wrt control sample
12.	**Microbiology**		
	A. Total Plate Count (cfu/g)	NMT 10^4	NMT 10^5
	B. Yeast and Mould (cfu/g)	NMT 10^2	NMT 10^3
	C. Pathogens		
	a. E. coli	Absent	Absent
	b. S. aureus	Absent	Absent
	***c. Salmonella* sp.**	Absent	Absent
	d. P. aeruginosa	Absent	Absent

Reference

Katiyar CK (2015): Challenges of Pharmaceutics in Ayurveda. Annals of Ayurvedic Medicine Vol 4, Issue 3-4, July-Dec

17

Herb-Drug Interaction Studies on Ayurvedic Medicines

Subrata Pandit and Satyajyoti Kanjilal

Introduction

Ayurvedic medicine is one of the world's oldest medical systems, originated in India more than 3,000 years ago. Ayurvedic medicine endorses the use of herbal components, special diets, for curing of disease and maintain of healthy life (NCCIH, 2016). There is a widespread misconception that "natural" always "safe" and a common belief that remedies from natural origin are safe and without any risk (Mukherjee, 2003). Patients frequently take herbal medicinal products in combination with other conventional medicines for additive effects. The medical and scientific literature is replete with *in vitro* and *in vivo* reports suggesting that the concomitant oral administration of Ayurvedic medicinal products and modern allopathic drugs or over-the-counter products may affects on human drug metabolizing enzymes (DMEs) and significantly increase the risk for clinically significant adverse reactions.

With limited regulatory oversight and strong advertising in media and web, many herbal products and food supplements made their way into patients' self-prescribed therapy. Moving away from the 'synthetic' world towards 'natural' world, patients are increasingly seeking herbal remedies to maintain their overall health and well-being. Most patients receive their information about herbs and supplements from sources other than their healthcare provider (Staines SS., 2011).

The increased usage of herbs as dietary supplements, over-the-counter products and prescription medicine, initiated the need for the development of clinical and scientific data for their quality and safety evaluation. Awareness rises in recent years on metabolic clearance and disposition of conventional pharmaceuticals by herbal remedies and other phytoconstituents (Strandell *et al.*, 2004). People frequently consume herbs as food in combination with prescribed modern medicine which may cause both pharmacokinetic and pharmacodynamic herb/food–drug interaction. Scientific reports demonstrated that Ayurvedic herbs are often administered in combination with therapeutic drugs, raising possibilities for herb–drug interactions (Mukherjee *et al.*, 2012). This leads to encourage studies on DMEs, especially cytochrome P450 (CYP450) enzymes (Takanaga *et al.*, 2000). Using *in vitro, in vivo* assays and clinical trial with CYP450 a large number of Ayurvedic herbal extracts as well as isolated compounds can be conveniently estimated for adverse drug reaction prediction (Pandit *et al.*, 2016).

1. Adverse effects and Interactions with Ayurvedic Medicine

All medicinal agents have potentially unexpected effects including toxicity, herbs are not exception of that. The risk of adverse effects may be influenced by age, gender, genetics, nutrition status, and concurrent disease states and treatments. In clinical practice recognizing adverse effects of Ayurvedic medicine is not routine and their reporting is less frequent (Baxter *et al.*, 2009). It is important to be aware of any substances that have the potential to cause toxicities or any adverse reaction when interact with synthetic allopathic medications.

The interaction potential should be based on evidence bases depending on in vitro in vivo data and clinical trial. An interaction may involving the herb component to cause an increase/decrease in the amount of drug in the blood stream *i.e.* interfere in metabolic clearance and distribution. Drug-herb interactions are based on the same pharmacokinetic and pharmacodynamic principles as drug-drug interactions (Kuhn MA, 2002; Scott and Elmer, 2002; Williamson EM 2003;).

2. Herb-Drug Interactions

Over the past few decade, there has been an increased global interest in develop scientifically validated safe and effective traditional systems of medicine. A drug interaction is defined as any modification caused by another exogenous chemical (drug, herb or food) in the diagnostic, therapeutic or other action of a drug in or on the body (Eisenberg *et al.*, 1998). When herbs take part such interaction coined as herb-drug interaction. Patients often consume Ayurvedic medicines concomitantly with conventional medicine for duel benefit (Pandit *et al.*, 2016). The popularity of herbal medicinal products makes it important to understand potential interactions between herbs and prescribed drugs. Reports indicated that the practice of poly pharmacy or multi drug therapy especially with patients with multiple complications has resulted in some serious clinical herb-drug interaction (Hemaiswarya and Double, 2006 and Nadler, *et al.*, 2003). All Herbal medicinal products (even single-herb products) contain mixtures of pharmacologically active constituents so like hood herb-drug interaction study is important (Fugh-Berman and Ernst, 2001). Understanding of proper mechanism of herb-drug interaction leads to right path for the physicians for successful treatment.

2.1. Mechanism of Herb-Drug Interaction

Perceived effectiveness, fewer side effects and relatively low cost, of Ayurvedic medicine are being used for the management of numerous medical conditions. Contrary to popular belief that "herbal medicines are safe," they may cause significant toxic effects and altered pharmaceutical outcomes when coadministered with conventional medicines in multidrug combination therapy (Oga *et al.*, 2016). These interactions are predominantly of clinically relevance when DMEs and xenobiotic transporters, which are responsible for the fate of many drugs, are induced or inhibited, sometimes resulting in unexpected outcomes.

The possibilities of herb-drug interaction are limitless, because numerous over-the-counter products, hundreds of herbs, vitamins and minerals supplements

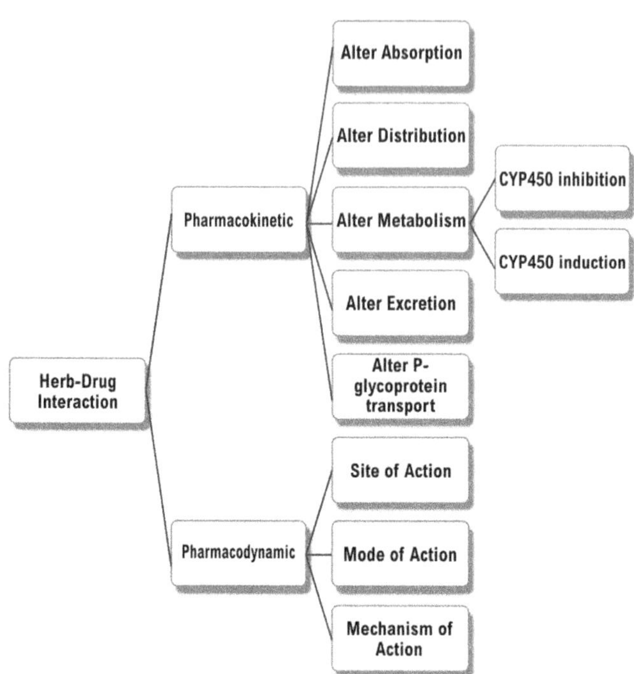

FIGURE 17.1: Herb-Drug Interaction Pathway.

and unique chemical substances of allopathic medicine are available in market. Herb-drug interaction considered playing a significant role in altering pharmacokinetic and/or pharmacodynamic mechanisms of drugs. Although the underlying mechanisms for the adverse drug effects by concomitant herbal medicines need to be explore. Clinically important herb-drug interaction depends on many factors associated with the particular herb, drug and patient (Hu *et al.*, 2005). Interactions between herbs and drugs may increase or decrease the pharmacological or toxicological effects of either component. Health care providers need to be aware of herb and supplement use by their patients due to possible herb–drug interactions (Eisenberg *et al.*, 1998). The mechanisms for drug interaction can be divided into two general categories such as pharmacokinetic interaction includes absorption, distribution, metabolism, and excretion of a drug and pharmacodynamic interactions which includes combined pharmacological effects of a drug. The pathway of herb-drug interaction has been depicted in Figure 17.1.

2.1.1. Pharmacokinetic Interactions

Since herbal medicines contain more than one pharmacologically active ingredient and are commonly used with many prescribed drugs, there potential herb-drug interactions are inevitable. Herb-drug interactions must not be ignored; nor can they always be taken at face value. Variety of reported herb-drug interactions are of pharmacokinetic origin, arising from the effects of herbal medicines on metabolic enzymes and/or drug transporters. Alteration in drug metabolism or transport can result in changes in absorption, distribution, metabolism, and excretion profile of drug. The resultant alterations caused by the herbs may or may not alter the dose-response relationship despite the change in the plasma levels and/or drug disposition profile of the drugs. In human body such kind of interaction occurs in several places. In Liver many drugs metabolized either in order to become therapeutically active or to alter polarity to be removed from the bloodstream. Some drugs are eliminated from the bloodstream through the kidney. Herbs that affect the functioning of liver and kidney can change the level of drug in the blood.

Primary mechanisms of herbal drug interaction involve either induction or inhibition of intestinal drug efflux pumps (e. g. efflux proteins such as P-glycoprotein (P-gp) and multiple resistance proteins (MRPs) and intestinal and hepatic metabolism by CYP450 (Ioannides, 2002; Evans, 2000; Wilkinson, 1997). Clinically significant pharmacokinetic herb-drug interactions parameters such as the area under the plasma concentration-time curve (AUC), the maximum plasma concentration (Cmax) or the elimination half-life (t1/2) of the concomitant drug alter (Meng and Liu, 2014).

2.1.1.1. Alter Absorption

Herbs and drugs can effect of absorption of each other into the bloodstream. Herbs may affect the pathway of drug absorbed, leading to changes in the amount of drug that enters the bloodstream. Further herbs can change the physical environment of the stomach, such as alteration of pH level. Affecting intestinal motility also take part in alteration of absorption. Herbs may chemically bind to drugs, causing them to remain in the stomach instead of entering the bloodstream (Staines SS; 2011). For example, herbs such as aloe leaf, guar gum and senna, which are common ingredients in herbal weight-loss products, exert a laxative effect that may decrease intestinal transit time and reduce drug absorption. St. John's Wort induces intestinal P-glycoprotein, which may decrease the absorption of common P-glycoprotein substrates, such as digoxin (Williamson EM, 2003; Zeping *et al.*, 2005). Such kind of interaction may be reduced by consuming of drugs 1 hour before or 2 hours after the herb (Kuhn MA, 2002). Absorption of drugs can be impaired when herbs like psyllium, rhubarb, flaxseed and aloe gel containing hydrocolloidal fibers, gums, and mucilage are taken together. These herbals may bind to drug molecules preventing their absorption and, subsequently, reduce systemic availability of the compounds (Obiageri, 2012).

2.1.1.2. Alter Distribution

A drug with high plasma protein binding affinity that has a small volume of distribution may be displaced by herb competing for the same binding sites. For example pain-reducing salicylates from meadowsweet and black willow may increases adverse effects of warfarin and carbamazepine by interfering with their binding site. Drug displacement from protein-bound forms, by concurrent herb administration, causes an increase in serum drug levels and which may lead to an increase in toxic effect (Scott GN, Elmer; 2002; Staines SS; 2011). These products should not be taken concurrently. Although displacement of protein-bound drugs has been described as a mechanism for potential adverse drug interactions, more research need to carry out to get clear-cut idea on specific interference on protein-binding sites (Obiageri O O, 2012).

2.1.1.3. Altered Metabolism

Alter metabolism is the most important and widely studied mechanism of herb-drug interactions. Two important processes involved in metabolic alteration model. Herbs interfere with the function of DMEs related to oxidative metabolism by CYP450 and/or the efflux drug transporter P-glycoprotein, with fewer evidence of the involvement of other enzymes such as glutathione S-transferases and uridine diphosphoglucuronyl transfereases (UGTs) (Boullata J. 2005).

2.1.1.3.1. Cytochrome P450 and Drug Metabolism

CYP450 is the most important Phase I DMEs. This haemoproteins is responsible for the metabolism of a variety of xenobiotics including therapeutic drugs and some important endogenous substances. Relative abundance of different CYP450 superfamily in human hepatic smooth endoplasmic reticulum have been reported as 30 per cent CYP3A4, 13 per cent CYP1A2, 7 per cent CYP2E1, 4 per cent CYP2A6, 2 per cent CYP2D6, 20 per cent CYP2C, and 1 per cent CYP2B6 (Figure 17.2) (Zhou *et al.*, 2003; Pelkone *et al.*, 2008).

Till date more than 1000 CYP450s have been identified and about 50s are active in human (Nelson, Dr, 2009). CYP450 is responsible for nearly 90 per cent of the drugs metabolism in hepatic and intestinal level in human. Among the identified CYP450 isoforms, 5 human CYP isoforms such as CYP1A2, 2C9, 2D6, 2E1 and 3A4 are important as they are responsible for 80 per cent of the CYP450 mediated drug metabolism (Daly, 2004; Ingelman-Sundberg, M., 2004). Among these CYP3A4 isozyme is most important, nearly 50

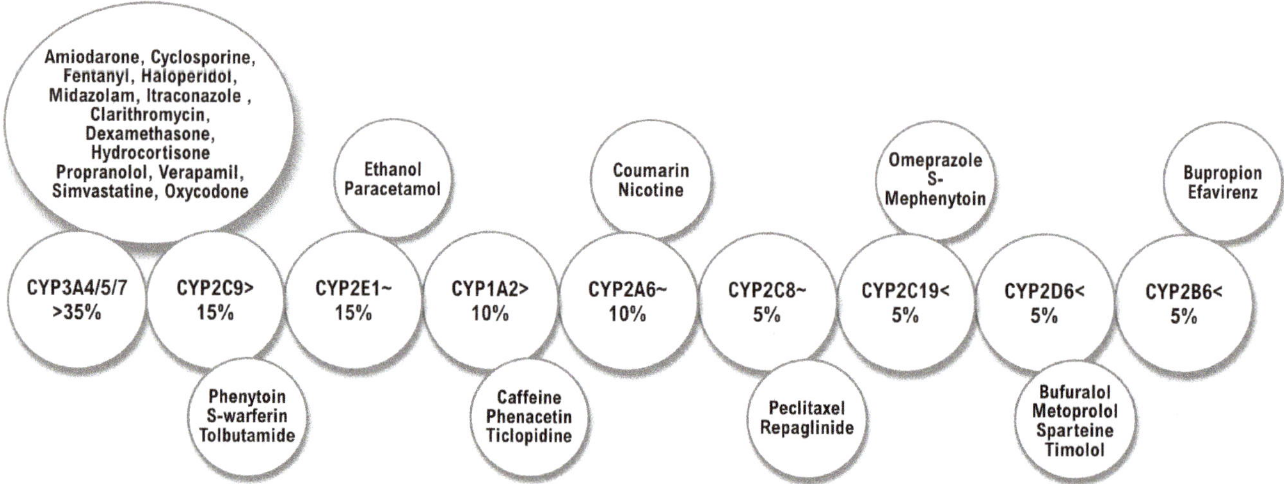

FIGURE 17.2 : Relative Abundance of CYP450 in Human Liver and some Substrate (Modified from Pelkone *et al.*, 2008).

per cent of marketed drugs are metabolized by CYP3A4 only. Drug interaction generally involves CYP enzyme inhibition or induction; interfere in metabolic clearance of drug. Herbs that can cause CYP450 metabolism mediated drug interactions are referred as either inhibitors or inducers of CYPs (Mukherjee *et al.*, 2012). Inhibitors or inducers interfere with the specific substrate metabolized by different CYP450. Relative abundance and some common substrate of major CYP450 have been represented in Figure 17.2. Any inhibitory effect of herbal extracts on CYP450 may result in enhanced plasma and tissue concentrations of drug substrate and leading to toxicity, while any inductive effect may cause reduced drug concentrations leading to decreased drug efficacy and treatment failure (Figure 17.3).

2.1.1.3.2. CYP450 Inhibition

The most common mechanism underlying herb-drug interaction is the inhibition potential of CYP450 enzymes. Inhibition of metabolism of drug by herbs may cause considerable elevations in the exposure of drug or both herbs and drugs which puts the patient at risk to serious adverse effects. The onset of CYP450 inhibition usually occurs relatively faster than induction effect. Modulation of intestinal and hepatic CYP450 is the key to alter systemic drug concentration. Consumption of herbs that are capable of modulating CYP450 may cause clinically relevant herb–drug interactions and alter drug bioavailability (Izzo and Ernst, 2001).

CYP450 inhibition divided into reversible, quasi-irreversible and irreversible interactions (Lin and Lu 1998). Irreversible inhibition or mechanism-based inhibition is relatively rare in nature. Irreversible inhibition is long lasting because new catalytically active enzymes must be *de novo* synthesized. Clinically important drugs that can inhibit CYP450 irreversibly include the antibacterial agent clarithromycin (CYP3A4), the sex steroid, gestodene (CYP3A4) and the anti-depressant, paroxetine (CYP2D6) (Bertelsen *et al.*, 2003; Zhou *et al.*, 2005). Quasi-irreversible inhibition is reversible *in vitro*, but inhibitor-CYP complex is stable enough to be irreversibly inhibited *in vivo* (Lin and Lu, 1998). In reversible inhibition the inhibitor binds to the CYP450 enzyme by weak bonds and do not permanently inhibit the enzyme's activity or enzyme activity can be restoring in the absence of inhibitor (Lin and Lu 1998; Hollenberg 2002; Pelkonen *et al.*, 2008). Reversible inhibition can be further divided into competitive,

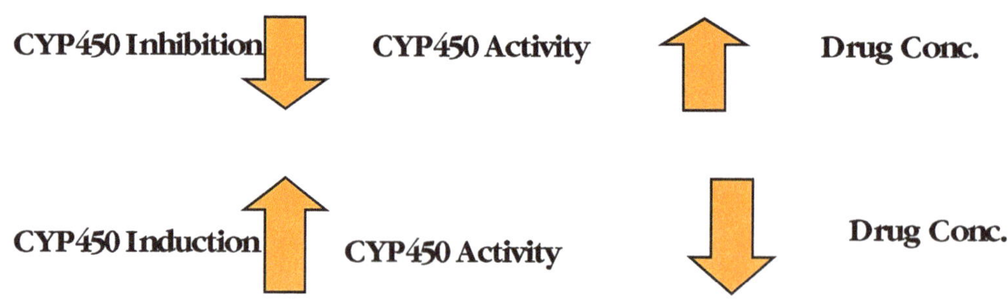

FIGURE 17.3: Fate of Cytochrome P450 Inhibition and Induction.

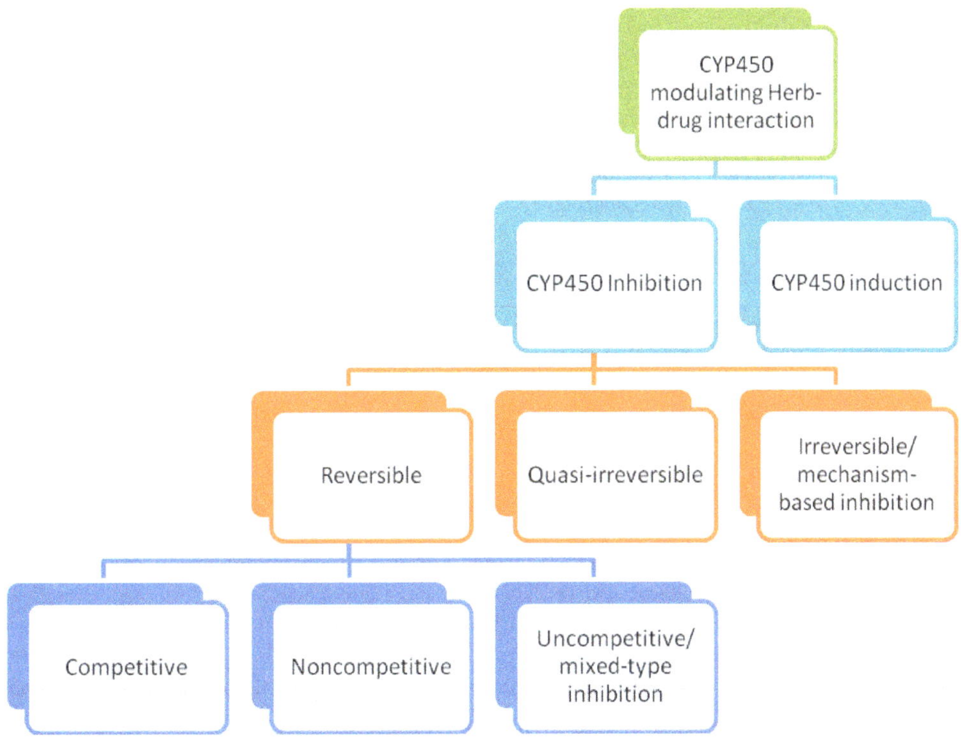

FIGURE 17.4: CYP450 Modulating Herb-Drug Interaction.

noncompetitive and uncompetitive or mixed-type inhibition (Pelkonen *et al.,* 2008). Reports are available on CYP450 inhibitory activity on Ayurvedic herbs. CYP450 inhibition mediated adverse drug reaction on Ayurvedic plants and phytoconstituents have been listed in Tables 17.1 and Table 17.2 respectively. CYP 450 mediated herb-drug interaction flow diagrams have been described in Figure 17.4.

2.1.1.3.3. CYP Induction

Human body has the ability to reduce environmental xenobiotic pressure by accelerating the function of specific DMEs. Increased CYP450 enzyme synthesis is mediated by ligand-activated transcription factors and receptors such as the aryl hydrocarbon receptor (AhR), pregnane X receptor (PXR) and constitutive androstane receptor (Pelkonen *et al.,* 2008). In addition to enhanced transcription of CYP450 gene, the concentrations of intracellular CYPs may be elevated by a decreased rate of protein degradation (Hollenberg 2002). In some cases, non-receptor-mediated induction processes may be involved. Many receptors that involve CYP450 enzyme transcription are presently unknown (Zhu 2010). Common inducible CYP450 enzymes include CYP1A1, 1A2, 2A6, 2C9, 2C19, 2E1 and 3A4. In humans, the antimicrobial agent, rifampicin, is a potent enzyme inducers (Kanebratt *et al.,* 2008).

Drug interactions mediated by P450 induction are significantly less common than those mediated by CYP450 inhibition. The interaction is less likely to result in safety issues but may impact efficacy of one or more medications. Consumption of herbs can induce CYP450. CYP450 induction studies on Ayurvedic plants have been summarize in Table 17.1. CYP450 induction leads to increases in the rate of metabolite production and hepatic biotransformation, decreases in serum half-life and drug response (Kalra, 2007).

2.1.1.3.4. Altered Renal Excretion

Change in renal clearance of a drug is another potential mechanism of herbal-drug interactions. Herbals that can inhibit tubular uptake or in other words that can interfere with the renal clearance of drugs considered as potential pharmacokinetic herb-drug interactions. Herbal diuretics are quite weak and unlikely to cause large problems in excretion. However, chronic ingestion of licorice may result in hypokalemia and water retention and accordingly may interfere with various medications including antihypertensive and antiarrhythmic agents (Staines SS, 2011). Numerous drugs are able to either enhance or inhibit the excretion of other drugs. In case of Ayurvedic medicine there is limited alter renal excretion mechanism need to exercise more.

TABLE 17.1: CYP450 Induction or Inhibition Mediated Herb-Drug Interaction of Ayurvedic Plants

Sl. No.	*Local Name*	*Scientific Name*	*Isozyme Tested*	*Positive Control*	*Inhibition/ Induction*	*Test Procedure/ Parameters*	*Test Vehicle*	*Results*	*References*
1.		*Andrographis paniculata*	CYP1A2, CYP2C9, CYP2D6, CYP3A4	α-napthoflavone, Sulfaphenazole, Quinidine, Ketoconazole	Inhibition	*In vitro*, IC50 value, fluorimetric high throughput screening	Recombinant human liver microsome	Less inhibition compared to positive inhibitors, less likely to produce significant drug interactions.	Kar *et al.*, 2016
2.		*Bacopa monnieri*	CYP1A2, CYP2C9, CYP2D6, CYP3A4	α-napthoflavone, Sulfaphenazole, Quinidine, Ketoconazole	Inhibition	*In vitro*, IC50 value, fluorimetric high throughput screening	Recombinant human liver microsome	Less inhibition compared to positive inhibitors, less likely to produce significant drug interactions	Kar *et al.*, 2016
3.		*Centella asiatica*	CYP1A2, CYP2C9, CYP2D6, CYP3A4	α-napthoflavone, Sulfaphenazole, Quinidine, Ketoconazole	Inhibition	*In vitro*, IC50 value, fluorimetric high throughput screening	Recombinant human liver microsome	Less inhibition compared to positive inhibitors, less likely to produce significant drug interactions	Kar *et al.*, 2016
4.	Chirata	*Swertia chirata* Buch-Ham (Gentianaceae)	CYP2D6, CYP3A4	Quinidine, Ketoconazole	Inhibition	*In vitro*, IC50 value, fluorescence microplate assay	Recombinant human liver microsome	Less inhibition compared to positive inhibitors, less likely to produce significant drug interactions	Ahmed *et al.*, 2016
5.	Trikatu	*Rasayana*	CYP2D6, CYP3A4	Quinidine, Ketoconazole	Inhibition	*In vitro*, IC50 value, fluorescence microplate assay	cDNA expressed recombinant human microsome	Very less inhibition potential	Harwansh *et al.*, 2014
6.	Long pepper	*Piper longum* L. (Piperaceae)	CYP2D6, CYP3A4	Quinidine, Ketoconazole	Inhibition	*In vitro*, IC50 value, fluorescence microplate assay	cDNA expressed recombinant human microsome	Very less inhibition potential	Harwansh *et al.*, 2014
7.	Black pepper	*Piper nigrum* L. (Piperaceae)	CYP2D6, CYP3A4	Quinidine, Ketoconazole	Inhibition	*In vitro*, IC50 value, fluorescence microplate assay	cDNA expressed recombinant human microsome	Very less inhibition potential	Harwansh *et al.*, 2014
8.	Ginger	*Zingiber officinale* Rosc. (Zingibera-ceae)	CYP2D6, CYP3A4	Quinidine, Ketoconazole	Inhibition	*In vitro*, IC50 value, fluorescence microplate assay	cDNA expressed recombinant human microsome	Very less inhibition potential	Harwansh *et al.*, 2014
9.		*Moras alba L.* (Moraecae)	CYP1A2, CYP2C9, CYP2D6, CYP3A4	α-napthoflavone, Sulfaphenazole, Quinidine, Ketoconazole	Inhibition	*In vitro*, IC50 value, Fluorogenic assay	cDNA expressed recombinant human microsome	Possibilities of herb-drug interaction are very less	Kar *et al.*, 2015
10.		*Moringa oleifera* Lam. (Family: Moringaceae)	CYP2D6, CYP3A4	Quinidine, Ketoconazole	Inhibition	*In vitro*, IC50 value, fluorescence microplate assay	Recombinant human liver microsome	less potential to inhibit the CYP isozymes, so traditional use of this plant may be safe.	Ahmed *et al.*, 2015

Contd...

Table 17.1-*Contd...*

Sl. No.	*Local Name*	*Scientific Name*	*Isozyme Tested*	*Positive Control*	*Inhibition/ Induction*	*Test Procedure/ Parameters*	*Test Vehicle*	*Results*	*References*
11.		*Aegle marmelos* (L.) Correa (Rutaceae)	CYP1A2, CYP2C9, CYP2D6, CYP3A4	α-napthoflavone, Sulfaphenazole, Quinidine, Ketoconazole	Inhibition	*In vitro*, IC50 value, Fluorogenic assay	cDNA expressed recombinant human microsome	Significantly lesser (p<0.001, 0.01) interaction potential than positive control, Unlikely to causes clinically relevant interaction	Bahadur *et al.*, 2015
12.		*Triphala*	CYP2D6, CYP3A4	Quinidine, Ketoconazole	Inhibition	*In vitro*, IC50 value, Fluorogenic assay	cDNA expressed recombinant human microsome	Triphala did not involves in drug interaction i.e. traditional use are safe	Ponnusankar *et al.*, 2010
13.		*Emblica officinalis* Gaertn. (Euphorbiaceae)	CYP2D6, CYP3A4	Quinidine, Ketoconazole	Inhibition	*In vitro*, IC50 value, Fluorogenic assay	cDNA expressed recombinant human microsome	Significantly less inhibitory activity (p < 0.001) on CYP isoforms compared to positive control.	Ponnusankar *et al.*, 2010
14.		*Terminalia belerica* Linn. (Combretaceae)	CYP2D6, CYP3A4	Quinidine, Ketoconazole	Inhibition	*In vitro*, IC50 value, Fluorogenic assay	cDNA expressed recombinant human microsome	Significantly less inhibitory activity (p < 0.001) on CYP isoforms compared to positive control.	Ponnusankar *et al.*, 2010
15.		*Terminalia chebula* Retz (Combretaceae)	CYP2D6, CYP3A4	Quinidine, Ketoconazole	Inhibition	*In vitro*, IC50 value, Fluorogenic assay	cDNA expressed recombinant human microsome	Significantly less inhibitory activity (p < 0.001) on CYP isoforms compared to positive control.	Ponnusankar *et al.*, 2010
16.		*Terminalia chebula* Retz (Combretaceae)	-	Ketoconazole	Inhibition	Cytochrome P450-carbon monoxide complex assay	Rat liver microsome	Mild inhibition effect on drug metabolizing enzymes	Ponnusankar *et al.*, 2010a
17.	Sweet flag	*Acorus calamus* Linn (AC) (Family: Acoraceae)	CYP2D6, CYP3A4	Quinidine, Ketoconazole	Inhibition	*In vitro*, IC50 value, Fluorogenic assay	cDNA expressed recombinant human microsome	Extract showed higher IC50 value than positive inhibitors	Pandit *et al.*, 2010
18.	Licorice / Yasti	*Glycyrrhiza glabra* L. (Family: Fabaceae)	CYP2D6, CYP3A4	Quinidine, Ketoconazole	Inhibition	*In vitro*, IC50 value, Fluorometric assay	cDNA expressed recombinant human CYP2D6 and CYP3A4	Weak interaction potential with drug metabolizing enzymes	Pandit *et al.*, 2011
19.		*Capsicum annuum* (Family: Solanaceae)	CYP1A2, CYP2C9, CYP2D6, CYP3A4	α-napthoflavone, Sulfaphenazole, Quinidine, Ketoconazole	Inhibition	*In vitro*, IC50 value, Fluorogenic assay	Baculosome	Likely to inhibit drug metabolizing enzymes, but less likely to produce significant drug interactions.	Pandit *et al.*, 2012
20.		*Murraya koenigii* (Family: Rutaceae)	CYP1A2, CYP2C9, CYP2D6, CYP3A4	α-napthoflavone, Sulfaphenazole, Quinidine, Ketoconazole	Inhibition	*In vitro*, IC50 value, Fluorogenic assay	Baculosome	likely to inhibit drug metabolizing enzymes, but less likely to produce	
21.		*Zingiber officinale*	CYP1A2, CYP2C9, CYP2D6, CYP3A4	α-napthoflavone, Sulfaphenazole, Quinidine, Ketoconazole	Inhibition	*In vitro*, IC50 value, Fluorogenic assay	Baculosome	likely to inhibit drug metabolizing enzymes, but less likely to produce	

Contd...

Table 17.1–*Contd...*

Sl. No.	*Local Name*	*Scientific Name*	*Isozyme Tested*	*Positive Control*	*Inhibition/ Induction*	*Test Procedure/ Parameters*	*Test Vehicle*	*Results*	*References*
22.		*Coleus forskohlii* Briq. (Family: Lamiaceae) extracts and fractions	CYP2B, CYP2C, CYP3A	Dexamethasone, rifampin	Induction	*In vitro*, fold m-RNA expression, polymerase chain reaction, followed by agarose gel electrophoresis	mRNA expression in hepatocyte	Not involved in CYP450 induction based drug interaction	Pandit *et al.*, 2016
23.		*Ridayarishta* (Ingredients of Arjuna-rishta & Ashwagandha-rishta)	CYP1A2, 2C19, 2D6 and 3A4	Furafylline, Tranylcypromine, Quinidine and Ketoconazole	Inhibition	*In vitro*, IC50 value, fluorimetric high throughput screening	cDNA expressed recombinant human microsome	Formulation alone and cocktail with amlodipine besilate, atenolol, atorvastatin, metformin, glipizide, glimepiride had negligible or insignificant effect on CYP450 inhibition	Pandit *et al.*, 2016

TABLE 17.2: CYP Inhibition Mediated Herb-Drug Interaction of Phyto-constituents from Ayurvedic Plants.

Sl. No.	*Phyto-constituents*	*Isozyme Tested*	*Positive Control*	*Inhibition/ Induction*	*Test Procedure/ Parameters*	*Test Vehicle*	*Results*	*References*
1.	Andrographolide	CYP1A2, CYP2C9, CYP2D6, CYP3A4	α-napthoflavone, Sulfaphenazole, Quinidine, Ketoconazole	Inhibition	*In vitro*, IC50 value, fluorimetric high throughput screening	Recombinant human liver microsome	Less inhibition compared to positive inhibitors, less likely to produce significant drug interactions.	Kar *et al.*, 2016
2.	Bacoside A	CYP1A2, CYP2C9, CYP2D6, CYP3A4	α-napthoflavone, Sulfaphenazole, Quinidine, Ketoconazole	Inhibition	*In vitro*, IC50 value, fluorimetric high throughput screening	Recombinant human liver microsome	Less inhibition compared to positive inhibitors, less likely to produce significant drug interactions.	Kar *et al.*, 2016
3.	Asiaticoside	CYP1A2, CYP2C9, CYP2D6, CYP3A4	α-napthoflavone, Sulfaphenazole, Quinidine, Ketoconazole	Inhibition	*In vitro*, IC50 value, fluorimetric high throughput screening	Recombinant human liver microsome	Less inhibition compared to positive inhibitors, less likely to produce significant drug interactions.	Kar *et al.*, 2016
4.	Ursolic acid	CYP2D6, CYP3A4	Quinidine, Ketoconazole	Inhibition	*In vitro*, IC50 value, fluorescence microplate assay	Recombinant human liver microsome	Less inhibition compared to positive inhibitors, less likely to produce significant drug interactions.	Ahmed *et al.*, 2016
5.	6-gingerol	CYP2D6, CYP3A4	Quinidine, Ketoconazole	Inhibition	*In vitro*, IC50 value, fluorescence microplate assay	cDNA expressed recombinant human microsome	Very less inhibition potential	Harwansh *et al.*, 2014
6.	Piperine	CYP2D6, CYP3A4	Quinidine, Ketoconazole	Inhibition	*In vitro*, IC50 value, fluorescence microplate assay	cDNA expressed recombinant human microsome	Very less inhibition potential	Harwansh *et al.*, 2014
7.	Chlorogenic acid	CYP1A2, CYP2C9, CYP2D6, CYP3A4	α-napthoflavone, Sulfaphenazole, Quinidine, Ketoconazole	Inhibition	*In vitro*, IC50 value, Fluorogenic assay	cDNA expressed recombinant human microsome	Possibilities of herb-drug interaction are very less	Kar *et al.*, 2015; Ahmed *et al.*, 2015

Contd...

Table 17.2–*Contd...*

Sl. No.	*Phyto-constituents*	*Isozyme Tested*	*Positive Control*	*Inhibition/ Induction*	*Test Procedure/ Parameters*	*Test Vehicle*	*Results*	*References*
8.	Imperatorin	CYP1A2, CYP2C9, CYP2D6, CYP3A4	α-napthoflavone, Sulfaphenazole, Quinidine, Ketoconazole	Inhibition	*In vitro*, IC50 value, Fluorogenic assay	cDNA expressed recombinant human microsome	Significantly lesser ($p<0.001, 0.01$) interaction potential than positive control, Unlikely to causes clinically relevant interaction	Bahadur *et al.*, 2015
9.	Gallic acid	CYP2D6, CYP3A4	Quinidine, Ketoconazole	Inhibition	*In vitro*, IC50 value, Fluorogenic assay	cDNA expressed recombinant human microsome	Significantly less inhibitory activity ($p < 0.001$) on CYP isoforms compared to positive control.	Ponnu-sankar *et al.*, 2010
10.	α-asarone	CYP2D6, CYP3A4	Quinidine, Ketoconazole	Inhibition	*In vitro*, IC50 value, Fluorogenic assay	cDNA expressed recombinant human microsome	Extract showed higher IC50 value than positive inhibitors	Pandit *et al.*, 2010
11.	Glycyrrhizin	CYP2D6, CYP3A4	Quinidine, Ketoconazole	Inhibition	*In vitro*, IC50 value, Fluorogenic assay	cDNA expressed recombinant human microsome	Weak interaction potential with drug metabolizing enzymes	Pandit *et al.*, 2010
12.	Capsaicin	CYP1A2, CYP2C9, CYP2D6, CYP3A4	α-napthoflavone, Sulfaphenazole, Quinidine, Ketoconazole	Inhibition	*In vitro*, IC50 value, Fluorogenic assay	Baculosome	Likely to inhibit drug metabolizing enzymes, but less likely to produce significant drug interactions.	Pandit *et al.*, 2012
13.	Mahanine	CYP1A2, CYP2C9, CYP2D6, CYP3A4	α-napthoflavone, Sulfaphenazole, Quinidine, Ketoconazole	Inhibition	*In vitro*, IC50 value, Fluorogenic assay	Baculosome	Likely to inhibit drug metabolizing enzymes, but less likely to produce significant drug interactions.	
14.	Mahanimbine	CYP1A2, CYP2C9, CYP2D6, CYP3A4	α-napthoflavone, Sulfaphenazole, Quinidine, Ketoconazole	Inhibition	*In vitro*, IC50 value, Fluorogenic assay	Baculosome	Likely to inhibit drug metabolizing enzymes, but less likely to produce significant drug interactions.	
15.	6-gingerol	CYP1A2, CYP2C9, CYP2D6, CYP3A4	α-napthoflavone, Sulfaphenazole, Quinidine, Ketoconazole	Inhibition	*In vitro*, IC50 value, Fluorogenic assay	Baculosome	Likely to inhibit drug metabolizing enzymes, but less likely to produce significant drug interactions.	
16.	Forskolin	CYP2B, CYP2C, CYP3A	Dexamethasone, rifampin	Induction	*In vitro*, fold m-RNA expression, polymerase chain reaction, followed by agarose gel electrophoresis	mRNA expression in hepatocyte	Not involved in CYP450 induction based drug interaction	Pandit *et al.*, 2016
17.	1-deoxy-forskolin	CYP2B, CYP2C, CYP3A	Dexamethasone, rifampin	Induction	*In vitro*, fold m-RNA expression, polymerase chain reaction, followed by agarose gel electrophoresis	mRNA expression in hepatocyte	Not involved in CYP450 induction based drug interaction	Pandit *et al.*, 2016
18.	1,9-dideoxy-forskolin	CYP2B, CYP2C, CYP3A	Dexamethasone, rifampin	Induction	*In vitro*, fold m-RNA expression, polymerase chain reaction, followed by agarose gel electrophoresis	mRNA expression in hepatocyte	Not involved in CYP450 induction based drug interaction	Pandit *et al.*, 2016

2.2. Interactions that Involve Drug Transporter or P-glycoprotein

P-gp is an ATP-driven efflux pump capable of transporting a wide variety of structurally diverse compounds from the cell interior into the extracellular space. P-gp was initially assumed to play a role in modulating cellular permeability ('P' stands for permeability), it was later demonstrated to be an ATP-dependent ef?ux pump (Shirasaka *et al.,* 2006). P-gp constitutively expressed in intestine, kidney, liver, brain microvascular endothelia, placenta, adrenal cortex, testis, uterus, lymphocytes and hematopoietic cells and plays important roles in drug absorption, elimination and distribution [5-7, MATSUNAGA *et al.,* 2006]. P-gp expression prevents absorption and aid in the elimination of xenobiotics, or prevents exposure of sensitive tissues (*i.e.* the brain or fetus) to xenobiotic agents. P-gp expression in the liver and kidneys is thought to function in aiding the elimination of drugs from the blood [15, 16].

P-gp has great impact on the pharmacokinetic pro?le of a variety of drugs (SHIRASAKA *et al.,* 2006). Intestinal expression of P-gp may affect the oral bioavailability of drug molecules that are substrates for this transporter. A variety of *in vitro* assays have been used to classify compounds as substrates, inhibitors or inducers of P-gp (8—14). P-gp inhibitory activity by ?avonoids and other polyphenols has been reported. Several reports are available for pharmokinetic profile through P-gp transport models on western medicinal plants and phytoconstituents. It has been reported that St John's wort induced intestinal P-glycoprotein (P-gp) in humans (3.3mg/kg/d) (KOBAYASHI *et al.,* 2004). Few Ayurvedic plants have been evaluated for pharmacokinetic profile through P-gp study model. Pharmacokinetic properties of glabridin study suggested that glabridin is a substrate for P-gp and both P-gp/MDR1-mediated efflux and first-pass metabolism contribute to the low oral bioavailability of glabridin (Cao *et al.,* 2007).

2.3. Pharmacodynamic Interaction

Pharmacodynamics studies provide information about how drugs bring therapeutic effects and how they cause side effects. Pharmacodynamics considers the sites, modes, and mechanisms of actions of drugs. Physicians and other healthcare practitioner are able to provide effective and safe therapeutic care to their patients from pharmacodynamics information (Ebadi M. 2007). Pharmacodynamic interactions occur when the drug's effect is being altered by interfering of other drug or herb. Herbs and drugs may work together (additive/synergistically) or in opposition (antagonistically) when consumed together in collaborative therapy. These interactions are not due to an alteration in the plasma concentration of drug/herbs but rather because of the net therapeutic or toxic effect. For example, individually, herb and drug may have the similar toxic effects, when consume together, they cause increased side effects. These effects can refer to alteration in therapeutic effect, or change in the toxicity levels and adverse side effects as well.

In case of allopathic drugs pharmacodymanic interaction study is common. In Ayurevedic medicine pharmacodynamics interaction studies are less. Study was conducted to investigate pharmacokinetics and pharmacodynamics profile ginger and warfarin on blood clotting status. Study confirmed that ginger did not significantly affect pharmacokinetics or pharmacodynamics of warfarin in healthy subjects (Jiang *et al.,* 2005). Pharmacodynamic and pharmacokinetic study of a well-known Ayurvedic formulary Trikatu was evaluated by Lala *et al.,* 2004. Study concluded that Trikatu pre-treatment might decrease the bioavailability of certain drugs probably through a drug–herb interaction. Components of Trikatu were interacting with the drug, which might result in modifying both, the pharmacokinetic and the pharmacodynamic aspects of the drug and the herbal formulation (Lala *et al.,* 2004).

3. *In vitro* and *In vivo* Approaches for Evaluation of Herbal-Drug Interactions

In vitro, in vivo and clinical trial assay are used to evaluate herbal drug interactions. Recently, there has been an increase in the use of *in silico* approaches to drug metabolism, drug transport and drug interaction studies. Most research on herb-drug interactions has focused on the *in vitro* evaluation of herbal constituents in liver microsomal systems, supersomes, cytosols, expressed enzymes or cell culture systems such as transfected cell lines, primary cultures of human hepatocytes and tumor derived cells (MacGregor *et al.,* 2001). These studies are valuable for evaluating multiple products and multiple components, provide mechanistic information about any potential interaction and are simple to conduct. It has certain limitations a typical of higher concentrations than clinical relevance; it does not account for the poor bioavilability or the binding of the same *in vivo* to plasma proteins (Venkataramanan *et al.,* 2006). Studies have been carried out *in vivo* in animals (normal, transgenic, humanized) and in humans (primarily healthy individuals). Most of the studies reported have used the commercially available products or a crude extracts of the herbal product or isolated purified individual components. These studies so far have paid particular attention to the effect of herbal components on CYP450 enzymes. Only a small number of studies have examined the effects of herbal products on phase II metabolism or drug transport.

4. Discussion

Since the use of herbs in daily life has become quite prevalent, issues of the safety of co-administration of such products together with Western medicines should be brought into attention. Although the pharmacokinetics and pharmacodynamics of Western medicines are well-known, the activities of any co-administered herbal products have not been well studied due to their complex components and variability. Most reports on drug-drug or herb-drug interactions focus more on pharmacokinetics than on the pharmacodynamics. However, both effects cannot be ignored in practice, especially for interactions that may occur between a single component Western medicine and a multicomponent herbal product. Herb-drug interactions are essential considerations that need to be addressed by undertaking high quality scientific research and conducting thorough systematic literature reviews (Zuo *et al.*, 2015).

Ayurvedic medicines are complex mixture of multiple ingredients. Chemical constituents responsible for pharmacological activity are many and complex and the majority of them have not been identified. For these reasons, there is a major difficulty in determining the clinical pharmacokinetic and pharmacodynamic effects when Ayurvedic are implicated. In addition to the chemical complexity of Ayurvedic medicine, many patients take these "natural" products concomitantly with prescription drugs in multidrug combination therapy for fast relief or better efficacy. Herbal medicine can interact adversely with prescription medicines, with danger of injury and even death without a shadow of doubt (WHO 2006).

Physicians prescribing drugs wish to see scienti?c documentation as pharmacology, therapeutic efficacy and safety of Ayurvedic medicine. Though Ayurvedic medicine has been proven their safety through traditional practice, physicians yearn to ?nd out how Ayurvedic drugs, proven ef?cacious for thousands of years. A relevant safety concern associated with the use of herbal medicines is the risk of interactions with prescription medications (Izzo, 2005; Izzo, 2004; Brazier and Levine, 2003; Izzo and Ernst, 2001; Fugh-Berman, 2001; Markowitz and DeVane, 2001; Williamson, 2003). Herbal drug interactions can results in unexpected concentration of therapeutic drug and lead to the undesired effects. Thus, contrary to the popular belief that "natural are safe" (Kaufman *et al.*, 2002); herbal medicines can cause significant toxic effects, drug interactions and even morbidity or mortality (Parmar, 2005). Pharmacokinatic and pharmacodynamics study need to carry out for Ayurvedic products to establish evidence-based safety profile to promote in international commerce.

5. Conclusion

In conclusion, interactions between Ayurvedic medicines and prescribed drugs can occur and may lead to serious clinical consequences. Both pharmacokinetic and/or pharmacodynamic mechanisms have been observed to play a role in these interactions. The clinical importance of herb-drug interactions depends on many factors associated with the particular herb, drug and the patient. Herb-drug interaction study needs to be carried out to consume safe Ayurvedic medication without any adverse drug reaction.

References

Baxter K, Driver S, Williamson E. *Stockley's herbal medicines interactions*. London, Pharmaceutical Press.2009.

Bertelsen, K. M., K. Venkatakrishnan, L. L. Von Moltke, R. S. Obach and D. J. Greenblatt, 2003. Apparent mechanism-based inhibition of human CYP2D6 in vitro by paroxetine: comparison with fluoxetine and quinidine. Drug metabolism and disposition: the biological fate of chemicals 31(3): 289-293.

Boullata J. Natural Health Product Interactions with Medication. *Nutr Clin Pract*. 2005; 20(1): 33- 51.

Daly, A. K., 2004. Pharmacogenetics of the cytochromes P450. *Curr Top Med Chem*, 1733-44.

David R Nelson. 2009. The Cytochrome P450 Homepage. *Hum Genomics*. 2009; 4(1): 59–65.

Dietary Supplements—A Framework for Evaluating Safety. Institute of Medicine. National Academic Press, 2005. pp. 235-246.

Ebadi M. 2007. *Pharmacodynamic the basic of herbal medicine*. 2nd edition. CRC Press. London. 25-29.

Eisenberg DM, Davis RB, Ettner SL. Trends in alternative medicine use in the United States, 1990–1997: results of a follow-up national survey. *JAMA*. 1998; 280: 1569–1575.

Fugh-Berman A, Ernst E. Herb-drug interactions: review and assessment of report reliability. *Br J Clin Pharmacol*. 2001 Nov;52(5): 587-95.

HemaIswarya, S.,Doble, M., 2006. Potential synergism of natural products in the treatment of cancer. *Phytother Res*. 20, 239-49.

Hollenberg, P. F., 2002. Characteristics and common properties of inhibitors, inducers, and activators of CYP enzymes. *Drug metabolism reviews* 34(1-2): 17-35.

Hu Z, Yang X, Ho PC, Chan SY, Heng PW, Chan E, Duan W, Koh HL, Zhou S. 2005. Herb-drug interactions: a literature review. *Drugs*, 65(9): 1239-82.

Hu Z, Yang X, Ho PC, Chan SY, Heng PW, Chan E, Duan W, Koh HL, Zhou S. 2005. Herb-drug interactions: a literature review. *Drugs*, 65(9): 1239-82.

Ingelman-Sundberg, M., 2004. Pharmacogenetics of cytochrome P450 and its applications in drug therapy: the past, present and future. *Trends Pharmacol Sci.* 25, 193-200.

Izzo AA, Ernst E. 2001. Interactions between herbal medicines and prescribed drugs: a systematic review. Drugs.;61(15): 2163-75.

Jiang X, Williams KM, Liauw WS, Ammit AJ, Roufogalis BD, Duke CC, Day RO, McLachlan AJ. 2005. Effect of ginkgo and ginger on the pharmacokinetics and pharmacodynamics of warfarin in healthy subjects. *Br J Clin Pharmacol.* 2005: 59(4): 425-32.

Jie Cao, Xiao Chen, Jun Liang, Xue-Qing Yu, An-Long Xu, Eli Chan, Wei Duan, Min Huang, Jing-Yuan Wen, Xi-Yong Yu, Xiao-Tian Li, Fwu-Shan Sheu, and Shu-Feng Zhou. 2007. Role of P-glycoprotein in the Intestinal Absorption of Glabridin, an Active Flavonoid from the Root of *Glycyrrhiza glabra*. DMD 35: 539–553.

Jon C Tilburt, Ted J Kaptchuk. 2016. Herbal medicine research and global health: an ethical analysis. Bulletin of the World Health Organization.

Kanebratt, K. P., U. Diczfalusy, T. Backstrom, E. Sparve, E. Bredberg, Y. Bottiger, T. B. Andersson and L. Bertilsson, 2008. Cytochrome P450 induction by rifampicin in healthy subjects: determination using the Karolinska cocktail and the endogenous CYP3A4 marker 4betahydroxycholesterol. *Clinical pharmacology and therapeutics* 84(5): 589-594.

Kuhm M. 1998. Pharmaceutical, Pharmakokinetic, and Pharmacodynamic phases of drug action. In Kuhn, M. (ed). Pharmacotherapeutics: A Nursing Process Approach, 4th edn. Philadelphia, PA: F. A. Davis Co., pp. 36-51.

Kuhn MA. Herbal Remedies: Drug-Herb Interactions. *Crit Care Nurse.* 2002; 22(2): 22-32.

L.G. Lala, P.M. D'Mello, S.R. Naik. 2004. Pharmacokinetic and pharmacodynamic studies on interaction of "Trikatu" with diclofenac. *Journal of Ethnopharmacology* 91 (2004) 277–280.

Lin, J. H. and A. Y. Lu, 1998. Inhibition and induction of cytochrome P450 and the clinical implications. *Clinical pharmacokinetics* 35(5): 361-390.

Meng Q, Liu K. Pharmacokinetic interactions between herbal medicines and prescribed drugs: focus on drug metabolic enzymes and transporters. *Curr Drug Metab.* 2014; 15(8): 791-807.

Michiya KOBAYASHI, Hiroshi SAITOH, Shujiro SEO, Veronika BUTTERWECK, and Sansei NISHIBE. Apocynum venetum Extract Does Not Induce CYP3A and P-Glycoprotein in Rats. *Biol. Pharm. Bull.* 27(10) 1649–1652.

Nadler, E. P. *et al.*, 2003. Monotherapy versus multi-drug therapy for the treatment of perforated appendicitis in children. *Surg Infect* (Larchmt). 4, 327-33.

Obiageri O. Obodozie (2012). Pharmacokinetics and Drug Interactions of Herbal Medicines: A Missing Critical Step in the Phytomedicine/Drug Development Process, Readings in Advanced Pharmacokinetics - Theory, Methods and Applications, Dr Ayman Noreddin (Ed.), ISBN: 978-953-51-0533-6.

Oga EF, Sekine S, Shitara Y, Horie T. Pharmacokinetic Herb-Drug Interactions: Insight into Mechanisms and Consequences. *Eur J Drug Metab Pharmacokinet.* 2016 Apr;41(2): 93-108. doi: 10.1007/s13318-015-0296-z.

Pelkonen, O., M. Turpeinen, J. Hakkola, P. Honkakoski, J. Hukkanen and H. Raunio, 2008. Inhibition and induction of human cytochrome P450 enzymes: current status. Archives of toxicology 82(10): 667-715.

Scott GN, Elmer GW. Update on Natural Product-Drug Interactions. *Am J Health-Syst Pharm.* 2002; 59(4): 339-347.

Staines, S.S., 2011. Herbal medicines: adverse effects and drug-herb interactions. *Journal of the Malta College of Pharmacy Practice*, 17, 38-42

Tamihide MATSUNAGA, Eiji KOSE, Sachiyo YASUDA, Hirohiko ISE, Uichi IKEDA, and Shigeru OHMORI. 2006. Determination of P-Glycoprotein ATPase Activity Using Luciferase. *Biol. Pharm. Bull.* 29(3) 560-564 (2006).

Watkins PB (1997) The barrier function of CYP3A4 and P-glycoprotein in the small bowel. *Adv Drug Deliv Rev* 27: 161–170

Williamson EM. Drug interactions between herbal medicines and prescription medicines. *Drug Safety.* 2003; 26(15): 1075-1092.

Yoshiyuki SHIRASAKA, Yuko ONISHI, Aki SAKURAI, Hiroshi NAKAGAWA, Toshihisa ISHIKAWA, and Shinji YAMASHITA. 2006. Evaluation of Human P-Glycoprotein (MDR1/ABCB1) ATPase Activity Assay Method by Comparing with in Vitro Transport Measurements: Michaelis–Menten Kinetic Analysis to Estimate the Af?nity of P-Glycoprotein to Drugs P-glycoprotein (P-gp)-mediated efflux and CYP3A4-mediated me- tabolism play important roles in influencing the oral bioavailability of their substrates (Watkins, 1997).

Zeping H *et al.*, Herb-Drug Interactions: a Literature Review. *Drugs* 2005; 65(9): 1239-1282.

Zhang W, and Lim L-Y. 2008. Effects of Spice Constituents on P-Glycoprotein-Mediated Transport and CYP3A4-Mediated Metabolism *in vitro*. *DMD* 36: 1283–1290.

Zhong Zuo, Min Huang, Isadore Kanfer, Moses S. S. Chow, and William C. S. Cho. 2015. Herb-Drug Interactions: Systematic Review, Mechanisms, and Therapies. Evidence-Based Complementary and Alternative Medicine. Article ID 239150, 1. http: //dx.doi.org/10.1155/2015/239150

Zhou, S., S. Yung Chan, B. Cher Goh, E. Chan, W. Duan, M. Huang and H. L. McLeod, 2005. Mechanism-based inhibition of cytochrome P450 3A4 by therapeutic drugs. *Clinical pharmacokinetics* 44(3): 279-304.

Zhu, B. T., 2010. On the general mechanism of selective induction of cytochrome P450 enzymes by chemicals: some theoretical considerations. Expert opinion on drug metabolism and toxicology 6(4): 483-494.

Index

A

D

E

F

G

H

Q

R

S

T

U

V

W

X

Y

Z

www.ingramcontent.com/pod-product-compliance
Lightning Source LLC
Chambersburg PA
CBHW061328190326
41458CB00011B/3928
9781787150058